# ECOLOGICAL STEWARDSHIP

# Ecological Stewardship

## A Common Reference for Ecosystem Management

### Volume II

- **Biological and Ecological Dimensions**
- **Humans as Agents of Ecological Change**

*Editors*
R.C. Szaro, N.C. Johnson, W.T. Sexton & A.J. Malk

*A practical reference for scientists and resource managers*

ELSEVIER
SCIENCE LTD

ELSEVIER SCIENCE Ltd
The Boulevard, Langford Lane
Kidlington, Oxford OX5 1GB, UK



First edition 1999

*Library of Congress Cataloging in Publication Data*
A catalog record from the Library of Congress has been applied for.

*British Library Cataloguing in Publication Data*
A catalogue record from the British Library has been applied for.

ISBN: 0-08-043206-9 (Set: Volumes I–III)

∞ The paper used in this publication meets the requirements of ANSI/NISO Z39.48-1992 (Permanence of Paper).

Printed in The Netherlands.

# Cooperating Organisations

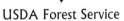

USDA Forest Service

USDI National Oceanic and Atmospheric Administration

USDI Bureau of Land Management

USDI Fish & Wildlife Service

USDI Geological Survey
USDI National Biological Service

USDI National Park Service

WRI
World Resources Institute

# Cooperating Foundations

American Fisheries Society
American Forests
Boise Cascade Corporation
Bullit Foundation
Consultative Group on Biological Diversity
Hispanic Association of Colleges and Universities
Liz Claiborne & Art Ortenberg Foundation
Moriah Fund
National Fish and Wildlife Foundation
National Forest Foundation
National Parks and Conservation Association
Pacific Rivers Council
Pinchot Institute for Conservation
Society of American Archeology
The Henry P. Kendall Foundation
The Nature Conservancy
The Pew Charitable Trusts
Tides Foundation
University of Arizona
W. Alton Jones Foundation

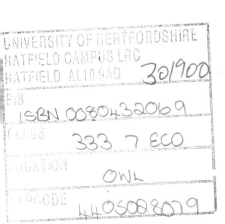

# Foreword

by

Jack Ward Thomas, *Boone and Crocket Professor of Wildlife Conservation, School of Forestry, University of Montana, USA; Chief Emeritus U.S. Forest Service*

This book is a milestone on the road to a more holistic management approach to renewable natural resources. When Forest Service Chief F. Dale Robertson declared in 1992 that the agency would henceforth follow a path of "ecosystem management", I thought it was a bold move, boldly taken and at the right time. Clearly, the old multiple-use management paradigm and the associated land-use planning did not serve well when constantly impacted by the listing of one threatened or endangered species after another.

Many have been and are frightened by this new concept and considered it a revolution in land management. This was no revolution. In fact, the move to ecosystem management was simply another step — albeit a large step — in an evolutionary process. The only differences from the management approach already in place were in terms of simultaneously considering, in the course of planning and management, several new or modified factors. The areas under consideration were usually larger and more logical in terms of defining factors (ecological, economic, social, political), longer time frames are involved, additional variables are considered, and it is clearly recognized that people and their needs and desires are critical factors.

Some have argued that the term is nebulous, that ecosystems can be anything and everything, and this is but one more "buzz word". That could be true enough if the process stopped there. Ecosystem management is a concept that must always be placed in context. That context includes a clearly identified area, a prescribed time period, definition of variables that will be considered, and the detailing of the needs and desires of people and the extent to which they can be met within economic, ecological, and legal constraints. In the light of a clearly defined context, most of such charges dissipate.

Ecosystem management, in my opinion at least, is simply a concept of land management whose time has come. The concepts of the interrelationships of humans and nature are present in the religious texts underlying many religions. Some philosophers have pondered similar questions. The "subversive science" of ecology has laid the technical foundation. The now well established concern with the retention of biodiversity in the process of land management — particularly the management of the public's lands — has produced a technical management requirement to meet a political objective. And, the combination of computers and remote sensing capability have combined with the above mentioned factors to make ecosystem management a feasible means to meet expanded land management demands.

I do not believe there is any rational retreat from this continuous and rapidly evolving concept. In a hearing before a Senate Committee on Energy and Natural Resources, I was asked where in the law it said the Forest Service could use "ecosystem management". I responded that I saw no prohibition against the approach, noted that the purpose of the Endangered Species Act was the preservation of the ecosystems upon which threatened and endangered species depended, and mentioned the judicial statement that there was no alternative that would meet the myriad requirements of law.

The Senator said he didn't understand it, didn't like it, and thought he might proceed to make "ecosystem management" against the law. I suggested that, before he did that, he might go out to the beach and practice by ordering the tide to roll back. Neither of us laughed.

It became clear to me in the course of such hassles that the concept of ecosystem management required a firm and detailed intellectual foundation to support the structure that was being put in place. This foundation would require structural stones representing many aspects of science — including natural sciences, social sciences (including economics and political sciences), and law. I felt that once this foundation was in place we could implement the concepts of ecosystem management in the next round of planning for National Forests.

I felt this process needed to be as inclusive of various technical fields and experts as possible, jointly sponsored by government and the private sector, jointly planned, and jointly executed. That was accomplished — but not without much time spent in allaying suspicion and building trust among and between participants.

The first step was the organization of this effort — including the securing of funding and commitment of the participants. The second step was an intense two-week workshop in Arizona where the various teams that put together the chapters in this book did their initial work. Each team received feedback from the overall group. When the workshop ended, the teams continued to work to produce their assigned chapters. What you have in your hands is the results of this critical step in the evolution of the concept of ecosystem management. I hope and believe that this is a foundation upon which land managers can build and rely. It is a strong foundation that will bear the weight that is placed upon it. It is what I hoped for. I trust that the authors of the chapters, the organizers of this effort, the sponsors, and the participants in the workshop will look back in a decade and say with pride, "I was there."

I, and all who contributed to this effort, owe two of my Forest Service colleagues a word of appreciation and recognition for a superb effort — Drs William Sexton and Robert Szaro. I asked much, perhaps too much, of them and they delivered even more than I asked.

# Contents

# Preface

*by*

William T. Sexton, *Deputy Director, Ecosystem Management, U.S. Forest Service*
Robert C. Szaro, *Special Assistant, Ecosystem Management Research, U.S. Forest Service*
Nels C. Johnson, *Deputy Director, Biological Resources Program, World Resources Institute*
Andrew J. Malk, *Associate, Biological Resources Program, World Resources Institute*

*"The difficulty lies, not so much in developing new ideas, as in escaping from the old ones"*
— John Maynard Keynes

For two weeks in December of 1995, over 350 natural resource managers, ecologists, economists, sociologists, and administrators gathered in Tucson, Arizona, to begin developing a knowledge base for the stewardship of lands and waters in the United States. As Jack Ward Thomas, then Chief of the U.S. Forest Service, noted at the opening of the workshop, better access to scientific findings and management experiences is a key to promoting wider use of ecosystem approaches in natural resource management. The teams of resource managers and researchers assembled in Tucson were invited on the basis of their knowledge and experience on issues central to implementing ecosystem approaches to natural resources management. Over the next two years, these teams labored to synthesize both current scientific understanding and resource management experiences on ecological, social, economic, and resource management topics. This reference is the final product of their efforts. It is intended to be useful to resource managers in the field, planners and decision makers, and researchers seeking to better understand ecosystems and the people who interact with them.

The contributors, the management examples, and the audience of this reference are largely located in the United States. Nevertheless, as organizations around the world increasingly turn to ecosystem management, bioregional management, and other integrated approaches to natural resource management, we hope they will find the process and information generated by the Ecological Stewardship Project useful in their own context.

## DISTINCTIVE FEATURES OF THE ECOLOGICAL STEWARDSHIP REFERENCE

The process used to develop the *Ecological Stewardship* reference was distinctive in several respects. First, recognizing that no one organization could marshall the understanding and experience to comprehensively address ecosystem approaches, partnerships became an essential feature of the Ecological Stewardship project. Second, recognizing that implementation depends on good science and lessons from the field, the project strived to link both research and management perspectives on the topics it addressed. Third, recognizing that accuracy is a key feature of a reference work, the project used a peer-reviewed forum to encourage discussion and documentation of competing views, philosophies, and facts. Fourth, recognizing that the reference needs to be useful in virtually any natural resource management setting, the project sought diverse examples and avoided prescriptive recommendations and site-specific goals. Fifth, recognizing the need to make the reference widely accessible, the project has used the Internet both to facilitate comments on draft chapters and to display the final document. Finally, recognizing the large volume of information contained in the reference and the limited time of resource managers and other users, the project has developed a concise key findings volume and a CD-ROM version of the full text.

In particular, users of this reference should understand the key role partnerships played in developing this document, and the importance of the review process.

### Partnerships

This reference exists only because a large number of government agencies, private foundations, interest groups, and individuals committed themselves to share their knowledge, experiences, time, and financial resources. Two types of partnerships were especially important. First, a partnership between public agencies and private organizations helped to ensure the broad usefulness of the reference. Although managers of federal lands and waters are the primary audience, we are confident the information it contains will also be useful to managers of state and private lands, and to researchers and analysts in universities and non-governmental organizations. Second, a partnership between researchers and resource managers helps to ensure that the reference is both scientifically sound and addresses the issues most often encountered by natural resource management practitioners.

#### *A Public–Private Partnership*

The need to develop partnerships between public and private organizations was driven by several factors. First, neither the U.S. Forest Service nor any other federal agency had all of the experience or experts needed to synthesize state-of-the-art knowledge about ecosystem approaches. Second, to be most useful, the reference needed to be relevant to a range of organizations involved in natural resources management. A partnership between public agencies and between public and private organizations has helped to ensure broad ownership of the document. Third, the significant cost in human and financial resources to develop the reference would be difficult for any one organization to bear. Financial commitments by private foundations and commitments of human resources by various federal agencies were vital to making the Ecological Stewardship Project work.

The use of partnerships characterized the project from the beginning (Box 1). Dozens of individuals from public agencies and private organizations

**Box 1. Chronology of Major Steps in the Ecological Stewardship Project**

| | |
|---|---|
| *April–September 1994:* | Project concept developed by Jack Ward Thomas and the eight foundations of the Consultative Group on Biodiversity. |
| *October 1994–August 1995:* | Public meetings to identify topics and workshop participants. |
| *December 1995:* | Ecological Stewardship Workshop, Tucson, Arizona. |
| *May 1996:* | Cooperative Agreement signed between U.S. Forest Service and World Resources Institute to develop Ecological Stewardship reference. |
| *June 1996* | Rough draft chapters reviewed by editorial team. |
| *January–April 1997:* | Expert and public review of draft chapters. |
| *April–May 1997:* | Review panel meetings in Portland, Oregon. |
| *December 1997:* | Final chapter drafts for Volumes II and III submitted to World Resources Institute. |
| *May 1998:* | Publication agreement developed with Elsevier Science Ltd. |
| *April–August 1998:* | Preparation of Key Findings volume. |
| *September 1999:* | Publication of Ecological Stewardship Reference by Elsevier Science Ltd. |

met regularly in the first months of the effort to identify topics to be addressed, and to identify individuals who could contribute as authors and participants in the Tucson workshop. Eight private foundations and several companies made generous financial contributions early on. The workshop itself involved participants from federal and state agencies, universities, private resource management companies, and non-governmental organizations. Among the authors are staff from six federal agencies, several state agencies, over a dozen universities, several forest products companies, and several non-governmental organizations. To ensure the credibility of the effort, the U.S. Forest Service engaged the World Resources Institute to develop an independent review process and oversee development of the reference materials. Finally, to enhance marketing and accessibility, Elsevier Science Ltd was engaged to publish the reference.

The broad partnership used to develop this reference is representative of the cooperation essential to any sound ecosystem approach to natural resources management. Not surprisingly, examples of collaborative partnerships are found throughout the document. Many chapters stress both technical and practical reasons for using partnerships in ecosystem approaches to natural resource management. Developing a knowledge base, like other elements of an ecological approach, requires close collaboration between many resource management interests and organizations. During its four years of development, the Ecological Stewardship Project engaged literally hundreds of individuals and dozens of organizations.

### *A Partnership of Researchers and Managers*

Communication between scientists and resource managers is a fundamental requirement of ecosystem approaches to natural resources management. A sound scientific foundation is needed to develop credible management options that both work and can withstand legal challenges. Managers, like scientists, are concerned with the validity of their information. A common thread for decision making, therefore, is that all stakeholders have access to the most scientifically sound information available. At the same time, some of the most important challenges facing resource managers are related to social, economic, political and institutional factors. Research can shed important light on these factors, but implementation experiences are needed to provide insights into what works and what does not.

From the start, the Ecological Stewardship Project has sought to foster a partnership between researchers and managers. Teams of researchers and managers were assembled to examine both the science and the management experiences behind the topics addressed in this reference. At the Tucson workshop, scientists and managers critiqued each others' outlines before writing began. When review drafts were completed, steps were taken to ensure that each chapter was reviewed by both researchers and managers. Staff from the World Resources Institute visited with natural resource managers in Montana, Maryland, and West Virginia to seek their views on how the information could be made most useful to them. The format and content of the Key Findings volume (Volume I) are a direct result of those conversations. Ecosystem approaches afford new opportunities to strengthen partnerships between researchers and managers in ways that will benefit both science and management. It is our hope that this reference will contribute to that process.

## A Rigorous and Public Review Process

Organizers of the Ecological Stewardship Project viewed an independent and rigorous review process as essential to the project's credibility. To achieve this, the U.S. Forest Service and the World Resources Institute (WRI) entered into a cooperative agreement in which WRI would design and manage an independent peer review process and oversee the development of the manuscript.

In consultation with the U.S. Forest Service and a steering group composed of representatives of federal and state agencies, non-governmental organizations, and industry, WRI designed a three-stage peer review process. The first review stage, completed in August 1996, provided authors with general comments from the editorial team. The main goal was to strengthen the links between research and management chapters on the same topic, and identify revisions needed before chapters were ready for a more formal review.

A formal review process for the revised chapters took place between January and April of 1997. Each chapter was reviewed by at least three, and typically more, recognized experts including both researchers and managers. At the same time, WRI created a World Wide Web site to facilitate public access to the draft papers. Notices were placed in a number of professional journals (e.g., *Conservation Biology, Society and Natural Resources, Journal of Forestry, Journal of Range Management,* and *Journal of Ecological Economics*) and on computer list-serves to invite broader participation in the review process.

To facilitate the review, WRI divided the chapters into six thematic groups, each with a coordinator to summarize reviewer comments and help authors to identify the most important revisions. The groups and their coordinators included: *Public Values and Expectations* (Dean Bibles, Department of Interior); *Social and Cultural Dimensions* (Margaret Shannon, University of Syracuse); *Humans as Agents of Ecological Change* (Nels Johnson, World Resources Institute); *Biological and Ecological Dimensions* (Gary Meffe, University of Florida);

*Economic Dimensions* (Robert Mendelsohn, Yale University), and *Data and Information Management Tools* (Francisco Dallmeier, Smithsonian Institution).

After the reviews had been received and summarized by the coordinators, a "review panel" was convened for each thematic group. Chaired by the thematic group coordinator, these panels brought together lead authors from each chapter in the group and outside experts. The panel meetings accomplished two objectives. First, they enabled authors to meet with their coordinator to agree on how to revise their final drafts. Second, they enabled the lead authors and outside experts assigned to their topic to summarize the key findings from their chapters. These summaries provided the "raw" material used by the editorial team to draft the Key Findings volume.

## ORGANIZATION OF THIS REFERENCE

The complete *Ecological Stewardship* work consists of a Key Findings volume (Volume I) and two volumes containing 54 chapters that address 31 separate topics (Volumes II & III). The original plan was to have a total of 30 topics, each addressed by two chapters — one on the scientific foundations of the topic, the other on management perspectives and experiences. This symmetry did not survive for several reasons. For several topics, the science and the management author teams decided to integrate their chapters. In several other cases, the

**Box 2. Topics Addressed by the "Ecological Stewardship" Reference**

The 31 topics addressed in the reference are grouped in six thematic sections

Volume II

BIOLOGICAL AND ECOLOGICAL DIMENSIONS
- Genetic and species diversity
- Population viability
- Ecosystem/landscape diversity
- Ecological functions and processes
- Role of disturbance and temporal dynamics
- Scale phenomena
- Ecological classification

HUMANS AS AGENTS OF ECOLOGICAL CHANGE
- Human role in evolution of North American ecosystems
- Cultural heritage management
- Ecosystem sustainability and condition
- Ecological restoration
- Producing and using natural resources

Volume III

PUBLIC EXPECTATIONS, VALUES, AND LAW
- Public expectations and shifting values
- Processes for achieving consensus
- Regional cooperation
- Evolving public agency beliefs and behaviors
- Legal perspectives

SOCIAL AND CULTURAL DIMENSIONS
- Cultural/social diversity and resource use
- Cultural heritage management
- Social/cultural classification
- Social system functions and processes

ECONOMIC DIMENSIONS
- Shifting demands for natural resources
- Economic interactions at local/regional/national scales
- Ecological economics
- Uncertainty and risk assessment
- Economic tools and incentives for ecological stewardship

INFORMATION AND DATA MANAGEMENT
- Adaptive management
- Monitoring and evaluation
- Data collection, management, and inventory
- Assessment methods
- Decision support systems/models and analysis

author team was either unable to finish their chapter or was unable to make satisfactory revisions following the review process. For topics where one chapter did not make it to completion, efforts were made to integrate at least some of the missing scientific or management perspective into the remaining chapter. Finally, one topic was added to address a large gap identified by the editorial team.

The topics addressed by the *Ecological Stewardship* are listed in Box 2. For a complete listing of all chapters, please turn to the complete table of contents listed in the back of each volume.

## Key Findings Volume (Volume I)

The Key Findings volume is designed to highlight both scientific findings and management insights for all 31 topics addressed by the *Ecological Stewardship* reference. The topic summaries are not executive summaries of individual chapters nor are they referenced. Rather, they highlight key findings and practical considerations that come from the one or two chapters that address a particular topic. Rough drafts for most of these summaries were prepared by teams assigned to each topic at the panel review meetings in Portland, Oregon. Final drafts were prepared by the editorial team once the chapters had been completed. Wherever possible, the editorial team sought to use language directly from the individual chapters. Otherwise, key findings and practical considerations were distilled from the full content of the relevant chapter(s). To ensure that the topic summaries are faithful to the contents of the chapters on which they are based, the final drafts were sent to each of the lead authors for their review.

We think the Key Findings volume is an extremely useful tool that enables readers to quickly learn about the full range of topics addressed by the *Ecological Stewardship* reference. To learn more about a given topic, we urge readers to turn to the relevant chapters in Volumes II and III.

## Volumes II and III

Volumes II and III contain 54 chapters prepared for the *Ecological Stewardship* reference. Volume II begins with three chapters that provide broad cross-cutting analysis and perspectives on ecosystem approaches to natural resources management. The rest of the volume contains the thematic sections on Biological and Ecological Dimensions (13 chapters) and Humans as Agents of Ecological Change (9 chapters). Volume III contains thematic sections on Public Expectations, Values and Law (7 chapters), Social and Cultural Dimensions (5 chapters), Economic Dimensions (8 chapters) and Information and Data Management (9 chapters). Each chapter is fully referenced, and cross references have been made by the editorial team to other chapters that address issues in greater detail. Each thematic section begins with an 8–10 page overview by the coordinator for that group of papers. The purpose of the thematic overview is to give readers a sense of the linkages between the chapters in the section. We urge users to read the overviews since they place individual chapters in a larger context.

## CD-ROM

Finally, the *Ecological Stewardship* reference is accompanied by a CD-ROM containing the complete text of all three volumes. This CD-ROM is intended to serve as a tool to help users locate specific information, and is searchable by table of contents, keywords and index.

# Acknowledgments

Partnerships have been fundamental to the Ecological Stewardship Project from the initial concept meeting that launched the project in 1994 to the publication of this reference work nearly five years later. Several hundred people have been involved in some facet of the project during that time as planners, authors, reviewers, and workshop organizers. Each of those individuals contributed support, energy, and ideas that helped make this project a success. A number of individuals and organizations, however, merit special recognition for their efforts.

Jack Ward Thomas, Chief of the Forest Service from 1993 to 1996, defined the need for a reference work that addressed both the scientific fundamentals of ecosystem management and lessons and perspectives from resource managers in the field. His vision of the project included a broad-based partnership of public agencies, private foundations, and researchers and managers from state and federal government, universities, non-governmental organizations, and the private sector. Without his energy and commitment, this reference would not exist.

Some of the earliest and most important partners in the Ecological Stewardship Project have been several private foundations including the Bullit Foundation, Henry P. Kendall Foundation, Liz Claiborne & Art Ortenberg Foundation, Moriah Fund, National Fish and Wildlife Foundation, Pew Charitable Trusts, Tides Foundation, and the W. Alton Jones Foundation. In addition to providing vital financial support, they also supplied ideas and guidance during the early conceptualization of the project. Peter Stangel of the National Fish and Wildlife Foundation was instrumental in coordinating support from the foundations and as an advisor and contributor throughout the project's evolution. Financial contributions were also provided by the Hancock Timber Resources Group and the Seneca Jones Timber Company.

Several federal agencies co-sponsored the project including the Bureau of Land Management, U.S. Fish and Wildlife Service, U.S. Geological Survey, U.S. National Park Service, U.S. Forest Service, and the National Oceanic and Atmospheric Administration. Beyond co-sponsorship, these agencies enabled many of their talented staff to actively contribute to the project as workshop participants, authors, and reviewers. In addition, a number of professional societies, conservation groups, and corporations co-sponsored the Tucson Workshop that brought together over 300 participants to plan and outline the contents of this reference. This included the American Fisheries Society, American Forests, Boise Cascade Corporation, Consultative Group on Biodiversity, Hispanic Association of Colleges and Universities, National

Forest Foundation, National Parks and Conservation Association, Pacific Rivers Council, Pinchot Institute for Conservation, Society of American Archeology, The Nature Conservancy, and the University of Arizona. Individuals from these and other organizations participated in a five month series of public meetings held in Washington to plan and shape the project in the summer and fall of 1995.

The December 1995 workshop in Tucson, Arizona brought together all of the authors and a wide range of scientists and resource managers to develop a framework for the topics and papers that make up this reference. Special thanks go to those on the workshop organizing team, particularly Pam Kelty, to the facilitators from various agencies who kept things on track for two intensive weeks, and to the personnel who provided indispensable logistical and administrative support during the workshop. Larry Hamilton, BLM State Director in Montana was exceptional as Master of Ceremonies and guiding light for the workshop; his efforts are greatly appreciated. A small group who helped identify key cross-cutting themes and follow-up issues at the Tucson Workshop included Rai Behnert, Gary Benson, Gordon Brown, Jim Caplan, Bill Civish, Jerry Clark, Joan Comanor, Bob Doppelt, Chris Jauhola, Dennis LeMaster, John Mosesso, Steve Ragone, Chris Risbrudt, Bob Robinson, Peter Stangel, Brad Smith and Dick Smythe. Finally, the University of Arizona served as host and co-sponsor for the workshop. Pat Reid and Carol Wakely at the University were critical to making the Tucson Workshop a success.

A steering group helped the World Resources Institute define the review process and explore publication options. Members of that group included Greg Aplet, Douglas Burger, Robert Dumke, Judy Guse-Noritake, John Haufler, Meg Jensen, John Mosesso, Jeff Olson, Chris Risbrudt, Walt Reid, Larry Ross, Craig Shafer, Margaret Shannon, Steve Thompson, and Henry Whittemore. Their suggestions helped to shape the reference work in several important ways, most notably establishing the importance of having a relatively brief Key Findings volume. Six "thematic group coordinators" — Dean Bibles, Francisco Dallmeier (assisted by Deanna Kloepffer), Nels Johnson, Gary Meffe, Rob Mendelsohn, and Margaret Shannon — were indispensable to the review process and helping authors revise their drafts. We are also grateful to Clint Carlson, Butch Farabee, Steve Kelly, Kniffy Hamilton, Larry Hamilton, Bertie Pearson, and Hal Salwasser in Montana, and to Julie Concannon in West Virginia and to Mark Koenig and Carl Zimmerman in Maryland for their help in arranging interviews with field resource managers. These interviews helped us to design this reference, particularly Volume I.

The reviewers (see list below) are of particular note. They not only reviewed lengthy manuscripts with no compensation, but many of them actually reviewed two manuscripts (one on the research side and one on the management side of the same topic). Their reviews provided the grist for a series of review panel meetings in Portland, Oregon during the spring of 1997. Panel participants (see list below) helped authors focus on key conclusions and findings for their chapters and helped draft the raw materials for the Key Findings volume. Thanks go in particular to Trish White, Oretta Tarkhani, Claudia Tejani, and Teri Aledo who helped organize the intensive three week series of review panel meetings like clockwork.

Robert Hamre and David Johnston edited final draft papers. Their skills added immeasurably to the readability and consistency of the individual chapters in this reference. Carolyne Hutter edited the Key Findings volume and the introductory overviews for the thematic sections. Their skill under tight deadlines, as well as their patience and good humor, are deserving of special recognition here. Trish White stepped in with much needed background research and assistance to put the Key Findings volume together at the very end.

Finally, we are grateful for Mary Malin's enthusiasm and guidance at Elsevier Science. She and others at Elsevier, including Pam Birtles, transformed an enormous amount of manuscript into an attractive and finished product.

## PROJECT SPONSORS

Many private foundations, corporations, organizations and federal agencies played a critical role in making this vision a reality by accelerating the collaborative process and setting the stage for the innovative public–private partnership necessary to develop this reference document. Public–private partnerships are at the core of an ecological approach and this project "lived the vision" as a major element in making this reference possible.

### Private Financial Supporters

Bullitt Foundation
Hancock Timber Resources Group
Liz Claiborne & Art Ortenberg Foundation
Moriah Fund
National Fish and Wildlife Foundation
Seneca Jones Timber Company
The Henry P. Kendall Foundation
The Pew Charitable Trusts
Tides Foundation
W. Alton Jones Foundation

### Agencies and Organizational Supporters

American Fisheries Society
American Forests
Boise Cascade Corporation
Consultative Group on Biological Diversity
Hispanic Association of Colleges and Universities
National Forest Foundation
National Parks and Conservation Association
Pacific Rivers Council
Pinchot Institute for Conservation
Society of American Archeology
The Nature Conservancy
University of Arizona
USDA Forest Service
USDI National Oceanic & Atmospheric Admin
USDI Bureau of Land Management
USDI Fish and Wildlife Service
USDI Geologic Survey
USDI National Biological Service
USDI National Park Service
World Resources Institute

## CHAPTER REVIEWERS AND PANEL PARTICIPANTS

The following individuals generously gave their time to review chapters and/or participate in review panel meetings that were held in Portland, Oregon in April 1997.

James Agee, Paul Alaback, Michael Alvard, Steve Apflebaum, Greg Aplet, Steve Arno, Richard Artley, Jerry Asher, Gordon Bakersville, David Bayles, Jill Belsky, Dean Beyer, Dean Bibles, William Birch, Jr., Myron Blank, John Bliss, Bernard Bormann, Dale Bosworth, Bill Bradley, Martha Brookes, Deborah Brosnan, Gordon Brown, Michael Brown, Perry Brown, Donald Callaway, Henry Campa, David Capen, David Caraher, Clint Carlsen, Connie Carpenter, C. Ronald Carroll, Joyce Casey, Alfonso Peter Castro, Jiquan Chin, Steve Cinnamon, Roger Clark, Beverly Collins, Julie Concannon, Allen Cooperrider, Hanna Cortner, W. Wallace Covington, Paul Cunningham, Francisco Dallmeier, Terry Daniel, Steve Daniels, Roy Darwin, Frank Davis, Robert Dumke, David Edelson, Robert Ewing, Stephen Farber, Scott Farrow, Scott Ferson, Suzanne Fish, Cornelia Flora, Jo Ellen Force, Tim Foresman, Sandra Jo Forney, Lee Frelich, Oz Garten, Paul Geissler, Susan Giannentino, James Gosz, Russell Graham, Dennis Grossman, Elaine Hallmark, Johathan Harrod, Johathan Haufler, Richard Haynes, Robert Hedricks, Gerald Helton, Thomas Hemmerly, Al Hendricks, Karen Holl, Tom Holmes, Anne Hoover, Amy Horne, Mark Hummell, Michael Huston, Donald Imm, Patrice Janiga, David Jaynes, Robert Jenkins, Mark Jensen, Meg Jensen, Kathleen Johnson, David Johnston, Ron Kaufman, Robert Keiter, Jim Kenna, Winnie Kessler, Kate Kitchell, Michael Klise, Deanne Kloepfer, Dennis Knight, Richard Krannick, Linda Langner, Bill Lauenroth, Daniel Leavell, Susan Lees, Gene Lessard, Bernard Lewis, Henry Lewis, Joseph Lint, Bruce Lippke, Doug Loh, Mark Lorenzo, B. John Losensky, Ariel Lugo, Douglas Mac Cleery, David Maddox, Jerry Magee, Marie Magleby, Robert Mandelsohn, Jon Martin, Clay Mathers, Mary Mc Burney, Maureen Mc Donough, Joe McGlincy, Gary Meffe, Gene Namkoong, Judy Nelson, Barry Noon, James O' Connell, Joseph O' Leary, Carlton Owen, J. Kathy Parker, Steve Paustain, Brian Payne, John Peine, Richard Periman, David Perry, Chris Peterson, Dave Peterson, George Peterson, Steward Pickett, Tom Quigley, Carol Raish, Connie Reid , Carl Reidel, Fritz Rennebaum, Keith Reynolds, Thomas Riley, Jeffrey Romm, William Romme, Len Ruggiero, Michael Ruggiero, Linda Rundell, Fred Samson, Don Schwandt, Jim Sedell, Roger Sedjo, Jan Sendzimir, William Sexton, Mark Shaffer, Margaret Shannon, William Shaw, Thomas Sisk, William Snape, Joel Snodgrass, Rhey Soloman, Jim Space, Pat Spoerl, Cleve Steward, Victoria Sturtevant, Alan Sullivan, Robert Szaro, Joseph Tainter, Paul Templet, Timothy Tolle, David Wear, Patricia White, Henry Whittemore, David Williams, Sue Willits,

# Introduction

## The Ecological Stewardship Project: A Vision for Sustainable Resource Management

*by*

William T. Sexton, *Deputy Director, Ecosystem Management, U.S. Forest Service* (Co-executive Secretary of the Ecological Stewardship Project)
Robert C. Szaro, *Special Assistant, Ecosystem Management Research, U.S. Forest Service* (Co-executive Secretary of the Ecological Stewardship Project)
Nels C. Johnson, *Deputy Director, Biological Resources Program, World Resources Institute* (WRI Ecological Stewardship Project Manager)

Just as sustained-yield and multiple-use concepts emerged as frameworks to manage natural resources earlier in this century, the end of the 20th century has become an era of change in the use of knowledge, tools, and strategies to manage lands and waters in the United States. The shift now is to ecological stewardship approaches that enable public and private resource managers to use science plus social guidance to help sustain productivity while maintaining biodiversity and other ecosystem services. This conceptual shift is evolving rapidly, as measured by the emergence of new concepts of resource management, an expanding body of scientific information relevant to management practices, a growing number of ecosystem management efforts around the country, and the development of new policies (Cortner and Moote 1998).

Yet, the concept and practice of integrating ecological knowledge into natural resources management — often referred to as "ecosystem management" — is not altogether new in the United States. For example, Aldo Leopold (Leopold 1949) was advocating the use of ecological knowledge in natural resources management as early as the 1940s and by the 1970s ecological concepts were finding their way into a growing number of publications on wildlife and natural resources management (e.g., Thomas 1979). More recently, the concept of "sustainable development" — often defined as managing natural resources to meet present human needs without compromising the ability of future generations to meet their needs — has led to greater consideration of the long term impacts of natural resource management decisions. Meanwhile, the cumulative impacts of numerous local management decisions have led many scientists and resource managers to conclude that biodiversity, water quality, and other natural resources can only be conserved through cooperative efforts across large landscapes — landscapes that often cross ownership boundaries (e.g., Aplet et al. 1993; PCSD 1996; Szaro and Sexton 1996). Researchers and resource managers in other parts of the world have also recognized the value of integrated regional approaches to conservation and resource management, although terms such as "bioregional management" and "integrated conservation and development projects" may be used in place of "ecosystem management" (Miller 1996).

What distinguishes ecosystem management approaches from earlier strategies and approaches to natural resources management is the integration of environmental, social, and economic knowledge to make sustainable resource management decisions at multiple geographic scales (Sexton et al. 1996). Shifting from single resource or single species management to managing an ecosystem for a variety of resources, including its biodiversity, requires using the best available scientific, social, and economic information (PCSD 1996). Scientific information is needed to identify ecosystem processes essential to the productivity of a wide variety of natural resources. Social and economic information can be used to determine which strategies will best meet public demands and landowner objectives. Voluntary and cooperative efforts at managing natural resources across ownership boundaries characterize successful ecosystem management efforts (Keystone Center 1996).

Ecosystem management, as an approach to understanding and managing the environment, provides resource managers with a broad range of tools to establish partnerships, collect information, analyze management options, and monitor the results of management and policy decisions. It includes understanding resource management options through interpreting the range of social, cultural, and spiritual values, perceptions and goals held by people, as well as the history, status and possible future trends of the physical and biological elements of the landscapes of particular interest. Using an ecosystem approach is not a simple, static prescription for a quick and perfect solution. An ecosystem approach, however, does provide a framework and a much larger tool kit for incorporating the best available science and management experience in a dynamic process of framing and addressing questions about the environment and the human presence and activity that affects it (Sexton 1998). It implies a heuristic process to address how humans think about, understand and analyze options for managing this relationship in sustainable ways. It requires that natural resource managers collect, organize and plan, and make decisions with an extensive array of scientific information and management experience.

Still, the use of ecosystem approaches faces a number of challenges. Most efforts are relatively recent, making it difficult to assess the results of ecosystem management approaches in a range of field situations (Yaffee et al. 1996). For example, nearly all the examples surveyed by the Keystone Dialogue (Keystone Center 1996) started after 1990. Another problem is that the concepts, science, and policies of ecosystem management have outpaced the rate at which managers can develop and implement practical applications as a part of their routine management practices. Much of the rapidly growing scientific literature on the subject is presented in idealized form, only limited quantitative assessments of on-the-ground experiences have been conducted, and ecosystem management concepts are continuing to evolve as field experience provides a means to understand those concepts in a context of real-world implementation. This leaves natural resource managers with limited guidance on how to apply ecosystem management concepts in specific field situations.

The use of ecosystem approaches in day-to-day management, research, and policy development lags behind, in part for want of practical assessment and a summary of the growing scientific, social, and management experience that already exists. Natural resource managers across the country are faced with the daily task of finding effective ways to manage resources sustainably in the context of a wide variety of desired conditions. Ecosystem approaches require more information about larger areas. How then do managers, often stationed in remote field locations, learn about, find, collect, organize and use information across an expanded range of disciplines?

Within the last several years, thousands of publications have addressed issues important to ecosystem approaches to natural resources management such as: the evolution of ecological concepts, perspectives on the results of past resource management practices, biodiversity inventory and conservation, a renewed emphasis on the concept of sustainability and an on-going debate about what is the appropriate interaction of humans with their environment. The result is an enormous pool of ideas, information, and critiques regarding resource management. For any individual, accessing and synthesizing this information and knowledge is difficult, if not impossible. This reference seeks to fill that gap by synthesizing both existing scientific understanding and management experience on a wide range of topics relevant to ecosystem approaches in natural resources management. It provides information, advice, concepts, and experience from many individuals in similar operating situations to aid field managers in their demanding daily routines and to help researchers define future scientific investigations.

## KEY THREADS THAT LINK ECOSYSTEM MANAGEMENT APPROACHES

There is no one universal formula for defining and implementing an ecosystem approach in natural resources management. Discussions regarding the definition of what constitutes an ecosystem approach to natural resource management invariably focus on certain common key themes and linkages (e.g. Interagency Task Force 1995; Keystone Center 1996). During the Tucson workshop, a subset of the participants were asked to step back from the detailed discussions on individual topics and identify the behaviors, attitudes, and values that support a successful ecosystem approach to natural resources management.* The following points are a synthesis of the issues raised by that group.

**Rai Bennett, Gary Benson, Gordon Brown, Jim Caplan, Bill Civish, Jerry Clark, Joan Comanor, Bob Doppelt, Chris Jauhola, Dennis LeMaster, John Mosesso, Steve Ragone, Chris Risbrudt, Rob Robinson, Peter Stangel, Brad Smith, and Dick Smythe.*

### Keeping Everyone "In the Loop"

It is critically important to involve key parties early in the ecosystem management planning process, and keep them productively involved throughout. This seems trite, but in case after case, successful planning efforts involved key parties, and unsuccessful efforts failed to do so. Agency decision-makers should consider involving the public in non-traditional ways, such as in data collection and monitoring. Managers also need to do more than just provide a forum for input; if necessary, they have to draw others into the process. Often, long-term residents have in-depth knowledge of how ecosystems have responded to past influences, and can provide valuable guidance not available elsewhere. Decision-makers must be receptive to such input from the public, and keep the decision-making process open. For some public interest groups, the *process* used in inviting participation is as important as the outcome.

### Sharing Information for Success

To keep key parties actively engaged in the ecological stewardship process, agency personnel must share information about the ecosystem, appropriate laws, agency management goals, the decision-making process, different stakeholder interests, and a host of other factors. It may also be necessary for key players to educate agency personnel about local cultures and historical traditions. This give-and-take of information contributes to each partner's education, and builds the bonds necessary for effective ecological stewardship. A shared literacy and awareness about ecosystems, and about how humans rely on and affect them, can help build receptivity and support for ecological stewardship approaches.

### Fostering a Culture of Ecological Stewardship

Social commitment exists for environmental protection, but social support for ecological stewardship can be improved. Society must be convinced that people are inextricably wedded to ecosystems and, therefore, humans have to take into account how they "draw" upon ecosystems for their personal well-being. An imperfect but useful analogy is how people routinely review and balance their checkbooks; they draw upon their resources to support their lifestyle choices but cannot withdraw beyond a minimum balance. We must also increase acceptance of inherent ecosystem value — the belief that ecosystems have intrinsic value as functioning systems as well as for how their components can be used by people. In addition, this belief asserts that system components and their relationships constitute a valued and valuable asset for all species, especially humans. Ecological stewardship is viewed as a way to bring communities together and improve the quality of natural resources. It is about looking for collaborative approaches among all landowners who desire health and productivity for the lands, waters, and resources they manage.

### Motivated by a Love of Place

Successful approaches to ecosystem management are increasingly community-based — initiated by local people, and motivated by a "love of place." In such cases, ecosystem management is in the local interest, and is a means to achieve the beneficial use of its "natural capital." There is growing realization that a long-term approach to land use and management is generally better ecologically and economically for developing harmonious and sustainable relationships between people and the land. Ecosystem approaches are also a means to build trusting relationships, often among former antagonists, to gain political power in furthering common envisioned ends.

### Interagency Collaboration

Interagency collaboration is essential, and can be improved, if people focus on each agency's relevant legislative authorities as *enabling* cooperation, rather than hindering it. Positive relationships between employees in the various agencies are important to success, and these should be recognized and encouraged through administrative channels. Having a shared vision for ecological stewardship is also critical for interagency collaboration. The history of agency missions has led to a somewhat disjointed federal conservation mandate, but ground-breaking efforts such as the Ecological Stewardship project bring executive leadership teams in direct contact to discuss shared visions and objectives. Many saw the networking at staff and executive levels as a productive means to continue development of a shared vision for ecological stewardship. Effective collaboration could also help overcome budgetary restrictions. Congress and other funding entities should respond more favorably to clearly demonstrated collaborative approaches to land management.

## Private Industry Leadership and Partnership

Partnerships that cross-cut public/private boundaries are also essential for ecological stewardship. Awareness of private property rights was a high priority, and the notion of federal or state dominion over private lands was flatly rejected. In fact, many participants noted that private industry was providing effective models that can strengthen federal land management, and industry was invited to assume the leadership role on some resource issues.

Effective collaboration between the public and private sectors will require a shared vision and strategy for implementation, mutually acceptable and binding rules for collaboration, and willingness to trust and act separately and together to attain this vision. Shared leadership will be critical to collaboration. It recognizes "co-dominance" among partners based on strengths, talents, and expertise. It recognizes the potential for "co-evolution" in personal and institutional growth and agreements. Thus, shared leadership is really "co-leadership." Co-leadership means conscious avoidance of "I win; you lose" approaches. Co-leadership emerges when partners find common vision, establish agreements for how to treat one another, and contribute the resources for joint activity. In the case of federal agencies, laws (e.g., National Environmental Policy Act, Federal Land Policy and Management Act, National Forest Management Act, and Administrative Procedures Act) compel certain minimum actions, and these should be used in a way that maximizes tangible and effective participation by other co-leaders.

## Hierarchical Approaches

Hierarchies are useful in ecological, economic, social, temporal, and political arenas to match the question or issue to the correct scale for analysis or decision-making. For example, in designing projects, knowledge of the characteristics and probable responses of the ecosystem are key to achieving objectives. It is useful to look one level above, to gain knowledge of context, and one level below, to gain understanding of content and processes. Theoretically, a hierarchy in the decision-making process for ecosystem management could include the individual, family, community, county, state, region, national, and global levels.

Agencies also have administrative hierarchies. For example, the Forest Service has ranger districts, forests, regions, the Washington Office, and the Department of Agriculture. Policies and decisions must be understood in the context of the hierarchy in which they are made.

Cultural hierarchies are sometimes overlooked, but are also critically important. Local groups may differ more in their willingness to take risks than will state or national groups. For example, a local development project may pose risks to an endangered species. Local groups may support the project, accepting the risk in favor of economic or other returns. National groups charged with the range-wide stewardship of the endangered species may not be willing to accept that same risk, given the potential impact to the species as a whole. In situations such as this, hierarchical analysis can more clearly portray the effect of decisions at multiple scales, and thereby assist with decision-making.

### Shifting Cultural Values

Values shift over time — groups should not be seen as representing the same point of view on resource issues forever. Also, the public trust is constantly being re-evaluated, and reliance on older stereotypes and role models may greatly delay the implementation of a more holistic approach to resource management. Agencies should work to improve and share data on human demographic trends so that local managers can better track and predict changes that are likely to occur as a result of shifting cultural values. And, agencies must meet rapid cultural changes with rapid and effective personnel changes and training.

### The Role of Science in Ecosystem Management

Everyone understands that good science is critical to good decision-making, but much research is not relevant to the actual decisions that managers must make. In particular, there is increasing need for research that looks at structure, composition, and function simultaneously, and at multiple scales. Because this kind of research is the most expensive to undertake, it is also becoming critical to build constituency support for scientific research. Finally, and perhaps most importantly, much controversy surrounding resource management questions is not due to the lack of science, or disagreements about the state of nature, but instead involve basic disputes about human values. Scientists and decision-makers need to make sure these issues are clearly separated in both the research and decision-making process.

### Laws — The Good, the Bad, and the Ugly

Laws were identified both as facilitators and inhibitors of ecologically-based stewardship. Some saw the Endangered Species Act as a motivator for change to avoid the need for difficult and costly efforts to save individual species. Others encouraged creative use of the administrative flexibility in this and other acts of Congress to demonstrate more quickly the benefits of broad authority. Narrow mandates were seen generally to hinder problem resolution, because they raised the likelihood of litigation. Many reported that when stakeholders resorted to lawsuits, the chances for successful ecologically based stewardship diminished sharply as stakeholders withdrew in anticipation of the courts' command and control. Others pointed out that some laws simply were not cast to solve the current problems of cross-boundary and joint ownership activities on larger landscapes. Thus, jurisdictional issues were seen as needing clarification before new cooperative activities could be undertaken wholeheartedly.

### I Need the Information... Yesterday!

Management decisions, and public education, are often hampered by the lack of current information from the scientific community. This is an important issue, especially given the dynamic nature of ecological stewardship issues. Participants reiterated the need to accelerate information dissemination through informal publications (that do not entail delays of months or years for publication), computer networks, and workshops that bring together the scientific and management communities. Scientists restated the need to ensure that their work was properly peer-reviewed, and implemented correctly. Continued thoughtful and up-front interaction between scientists and managers is the best approach to this challenge.

### Broadening Horizons

Biological and biophysical scientists have a great deal to learn from interactions with social scientists, physicists, economists, and other fields not traditionally associated with "core" natural resource sciences. In addition, ecosystem scientists have to seek greater exposure to philosophers (for example, theologians and ethicists) to help describe appropriate directions for ecosystem science. And finally, ecosystem scientists need to gain insight from creative people working in fields such as literature, drama, movies and television, or photography. Such creative fields often reveal cultural preferences and concerns, and, in their own way, enable scientific ventures.

### Can We Adapt?

Adaptive management, like ecosystem management, is one of those things that many people are already doing without realizing it. Participants felt, however, that "feedback loops" to provide managers with monitoring data and evaluations of their management actions were lacking in many agencies. Without this evaluation, managers have no way to assess the effectiveness of their actions and make necessary modifications. Many agencies have a history of collecting lots of data, but not of effectively using it for adaptive purposes. This was also recognized to be symptomatic of shifting agency goals. With the proper feedback loops, managers can be held accountable for their actions, and are, therefore, more likely to be responsive to local resource needs.

### The Only Place it Matters Is on the Ground

No one at the Tucson workshop wanted to invest time in meetings, publications, or other activities that did not have a direct payoff for on-the-ground ecological stewardship. This was the underlying theme to every presentation and discussion. Clearly, the interactions between science and management authors led to a better understanding of what managers need and what scientists can do. This sort of interaction needs to become the norm, rather than the exception. Furthermore, the on-the-ground needs of the managers should strongly influence the research system.

## CONCLUSION

Perhaps the most challenging aspect of ecosystem approaches to natural resources management is the need for interdisciplinary research and management. Context — environmental, social, and economic — is a backdrop for virtually all decisions in an ecosystem approach. Science is crucial, but reductionism is dangerous. To be effective, specialists often have to know something about issues they have little training or experience in. Economists working on management options for an area with endangered species may not need to be specialists in population viability analysis, but they should understand the basics of extinction processes in order to frame their analysis. Likewise, a restoration ecologist need not be expert with cost–benefit analysis to define a technical restoration plan, but a basic understanding of how economists evaluate costs and benefits will likely improve the feasibility of his or her proposal. And, resource managers often have little choice but to know something about many different issues — from building collaborative partnerships to knowing when to call for expertise in highly technical areas such as restoration ecology or legal advice. Ecosystem approaches simply will not work if researchers and managers isolate themselves from the context in which they work.

The topic summaries in Volume I are a good place to begin learning more about topics that you would like (or need) to know more about. The summaries in Volume I, however, are only previews that highlight findings from the full chapters in Volumes II and III. To make the most of this resource, we urge you to read those chapters for the full context of the topic. We are confident that the *Ecological Stewardship Reference* will help users to resolve specific problems and more fully understand the context for ecosystem approaches at the dawn of the 21st Century.

## REFERENCES

Aplet, G.H., N.C. Johnson, J.T. Olson, and V.A. Sample. 1993. *Defining Sustainable Forestry*. Island Press, Washington, DC.

Cortner, H.J. and M.A. Moote. 1998. *The Politics of Ecosystem Management*. Island Press, Washington, DC.

Keystone Center. 1996. *The Keystone National Policy Dialogue on Ecosystem Management: Final Report*. The Keystone Center, Keystone, CO.

Leopold, A. 1949. *A Sand County Almanac*. Alfred Knopf Publishers, New York, NY.

Miller, K.R. 1996. *Balancing the Scales: Guidelines for Increasing Biodiversity's Chances Through Bioregional Management*. World Resources Institute, Washington, DC.

PCSD. 1996. *Sustainable America: a New Consensus for Prosperity, Opportunity, and a Healthy Environment*. The President's Council on Sustainable Development Final Report. President's Council on Sustainable Development, Washington, DC.

Sexton, W.T. 1998. Ecosystem Management: Expanding the Resource Management "Tool Kit". Special Issue on Ecosystem Management. *Landscape and Urban Planning*, 40: 103–112.

Sexton, W.T., N.C. Johnson, and R.C. Szaro. 1997. The Ecological Stewardship Project: A public–private partnership to develop a common reference for ecosystem management in the United States. In: *Proceedings of the XI World Forestry Congress: Forests, Biological Diversity and the Maintenance of Natural Heritage — Protective and Environmental Functions of Forests. Volume* 2. (Antalya, Turkey), pp. 75–82. United Nations Food and Agriculture Organization, Rome.

Szaro, R.C. and W.T. Sexton. 1996. Biodiversity conservation in the United States. In: A. Breymeyer et al. (eds.), *Biodiversity Conservation in Transboundary Protected Areas*, pp. 57–69. Proceedings of an International Workshop, Bieszczady-Tatry, Poland. National Academy Press, Washington, DC.

Thomas, J.W. 1979. *Wildlife Habitats in Managed Forests of the Blue Mountains of Oregon and Washington*. USDA Forest Service Handbook No. 533. USDA Forest Service, Portland, OR.

Yaffee, S.L., A.F. Phillips, I.C. Frentz, P.W. Hardy, S.M. Maleki, and B.E. Thorpe. 1996. *Ecosystem Management in the United States: An Assessment of Current Experience*. Island Press, Washington, DC.

# ♦ Introduction

# Ecosystem Management: Evolving Model for Stewardship of the Nation's Natural Resources

Hanna J. Cortner, John C. Gordon, Paul G. Risser,
Dennis E. Teeguarden, and Jack Ward Thomas

**Key questions addressed in this chapter**

- *What is ecosystem management?*
- *What historical, social, economic, and scientific forces led to ecosystem management?*
- *What forces will shape ecosystem management in the future?*
- *What steps can be taken to build upon progress and address the challenges facing ecosystem management?*

**Keywords: Social values, Federal Advisory Committee Act, ecosystem management, political process**

## 1 INTRODUCTION

The beginning of the 20th century brought a marked change in American values and approaches to management of the nation's renewable natural resources: unbridled exploitation of resources was giving way to the notion of conserving resources for future generations. Reflecting upon the choice of the word "conservation" as a means of symbolizing the adoption of new goals and values for managing the nation's forests, rangelands, and waters, Gifford Pinchot later wrote: "Having just been born, the new arrival was still without a name. There had to be a name to call it by before we could even attempt to make it known, much less give it a permanent place in the public mind" (Pinchot 1947, p. 326). Although there have been variations in operational philosophies and some periodic deviations toward exploitation, conservation of natural resources has been a continuing theme throughout this century.

Today, as the 21st century approaches, another transition driven by changing social values, new scientific information, and professional experience is rapidly developing. And again, a term is needed to symbolize the changing perspectives, planning approaches, and management practices that are occurring throughout the nation. Several have been suggested — e.g., *ecosystem approach* (Interagency Ecosystem Management Task Force 1995), *ecosystem conservation* (Czeck 1995), *ecological stewardship, watershed management, integrated resource management*, and *landscape ecology* (Risser 1987). At present, *ecosystem management* has become the most widely used term to represent a revised approach for resource management, both publicly and across resource management disciplines. Consequently, it is used throughout this chapter.

We begin this chapter with a brief characterization of the basic features of ecosystem management, including how the concept differs from the dominant multiple-use/sustained yield approach of the last century. Next, we review the historical, social, economic, and scientific forces that have led to adoption of ecosystem management as a new and different approach to resource management for policy-makers, scientists, managers, and the public. We believe that these forces are now so pervasive and persuasive that there is no turning back to the management philosophies, assumptions, and practices that guided us in the past. Our discussion then turns to the future. We propose six evolving realities that will continue to shape ecosystem management. The final section recommends several near-term steps that can be taken to build on the progress that has been made to date, while allowing us to address some of the scientific uncertainties, political conflicts, and managerial challenges that will arise as policy-makers and managers shift from a management style based on one set of premises to management based on a different set.

## 2 ECOSYSTEM MANAGEMENT: WHAT IS IT?

The ecosystem management approach to planning and decision-making has been defined in various ways, which is not surprising given the relatively recent adoption of the concept. For our purposes here, the federal Interagency Ecosystem Management Task Force (1995) definition includes the most commonly identified elements:

- An ecosystem is an interconnected community of living things, including humans, and the physical environment in which they interact.
- The ecosystem approach is a method for sustaining or restoring natural systems and their functions and values. It is goal driven, and it is based on a collaboratively developed vision of desired future conditions that integrates ecological, economic, and social factors. It is applied within a geographic framework defined primarily by ecological boundaries.

Although no standard definition of ecosystem management exists, there is reasonably consistent agreement on the essential elements. It is agreed, for example, that fundamental to ecosystem management is the concept of resource sustainability (see Bormann et al., Vol. III of this book, Franklin 1993b, Christensen et al. 1996). Ecosystem management stresses the sustainability of social welfare and of ecosystem structures and processes. Implicit or explicitly included in the concept is the goal of maintaining biological diversity within the system being managed (Franklin 1993b, Grumbine 1994).

Ecosystem management is a conceptual approach to resource management; it is not a rigid set of management prescriptions or a cookbook with "how to do it" recipes. However, compared to previous policies and practices it incorporates these key features (Interagency Ecosystem Management Task Force 1995):

- More partnerships, greater collaboration
- Broader program perspective
- Broader resource perspective, oriented toward interacting systems
- Broader geographical and longer temporal perspectives
- More dynamic and flexible planning processes
- Stronger reliance on the best science
- More proactive

Put another way, ecosystem management enjoins managers to: manage where you are, manage with people in mind, manage across boundaries, manage based on mechanisms rather than "rules of thumb," and manage without externalities (Gordon 1993b).

Ecosystem management differs from multiple-use/sustained yield, as historically implemented by federal, state, and private agencies, in several important respects (see MacCleery and Le Master, this volume, Kennedy and Dombeck, Vol. III of this book). Generally, the multiple-use/sustained yield approach on general purpose lands, as contrasted to formally classified wilderness areas or wildlife refuges, has focused on sustaining the annual or periodic production of commodity resources such as commercially valuable timber. Moreover, planning tends to be confined to the boundaries of the property or administrative unit at issue. In contrast, ecosystem management does not focus on a single resource output, or even some fixed combination of outputs, but rather on restoring and sustaining ecological structures and functions in perpetuity. This planning typically focuses on a relatively large geographic area defined by ecological and/or physical relationships and commonly cuts across administrative or legal boundaries. Thus, planning and management place sustainability of the system as the central goal and this goal drives the ways in which the system is considered (Franklin 1993b, Christensen et al. 1996).

Ecosystem management does not preclude resource utilization. It is clearly understood that humans are part of the ecosystem, and it is recognized that resource management must be responsive to societal needs, including commodity production. The ecosystem management concept calls for commodity and amenity outputs that are compatible with the goal of maintaining long-term sustainability of ecological relationships (Kessler et al. 1992, Slocombe 1993, Society of American Foresters 1993, Christensen et al. 1996).

Although a number of troublesome legal and institutional issues persist (e.g., private property rights, collaborative processes subject to the Federal Advisory Committee Act, and interjurisdictional coordination) (see Keiter et al., Vol. III of this book, Keiter 1994, Cortner et al. 1996, Meidinger 1997), ecosystem management is consistent with existing federal laws, including, for example: the National Environmental Policy Act (NEPA), the National Forest Management Act (NFMA), the Federal Land Policy Management Act (FLPMA), the Endangered Species Act (ESA), and the Multiple-Use Sustained Yield Act (MUSY) (see Keiter et al., Vol. III of this book). Evolving case law may even be making ecosystem management mandatory. A federal court decision in the case of the northern spotted owl concluded, for example, that the Forest Service and the Bureau of Land Management could only meet their statutory mandates on forest lands in the Pacific Northwest through an ecosystem management approach (see Keiter et al., Vol. III of this book). Previous efforts to plan on a less-than-ecosystem basis had led to numerous court injunctions.

Implementing ecosystem management requires adjusting to new management challenges and accommodating a growing body of scientific information. Thus, central to ecosystem management are the principles of adaptive management and flexibility. Analysis of alternatives will include assessments of a broad range of benefits, costs, and risks from multiple perspectives and in various spatial and temporal scales. All parties will have to operate in an environment characterized by risk and uncertainty, even with the best scientific information available. Joint planning among different agencies sharing responsibility for providing goods and services for multiple generations and for bringing all stakeholders into the problem-definition and decision-making processes is requisite.

At a minimum, the ultimate effectiveness of ecosystem management in planning, decision-making, and on-the-ground management will be measured by whether ecosystem management is:

- scientifically credible
- multi-generationally supportive
- politically and legally acceptable
- publicly embraced
- technologically sound
- economically feasible

and by whether:

- ecosystem and resource sustainability are achieved, and
- species and habitat diversity are maintained

Collectively, these criteria represent a significantly different approach to the policies and practices employed throughout much of the past century.

## 3 THE HISTORICAL TRAJECTORY OF ECOSYSTEM MANAGEMENT

By 1994 more than 18 federal agencies had adopted ecosystem management as a guiding policy (Morrissey et al. 1994). However, ecosystem management is not just a recent product of the 1990s. Its origins transcend multiple presidential administrations and are based on changing social values, evolving science in many disciplines, and professional experience derived from implementation of legislative mandates (Cortner and Moote 1999).

## 3.1 Evolving Social Values

To a large extent, ecosystem management shares its roots with environmentalism and the evolving socio-political values captured by that movement. Some observers consider Rachel Carson to be the mother of the environmental movement, for her 1962 book, *Silent Spring*, was among the first to raise significantly public consciousness about the damages that indiscriminate use of human technologies could have on natural environments. Heightened public concerns about pesticides and pollution of air and water were made graphic by images of the burning Cuyahoga River in Ohio, oil soaked birds on the beaches of Santa Barbara, CA, and smog-filled air in Birmingham, AL. Writers such as Barry Commoner, Wallace Stegner, Paul Ehrlich, and Edward Abbey published environmental perspectives on pollution and land-use issues, and environmental interest groups formed to advance environmental policy objectives. Events such as Earth Day in April of 1970 galvanized public support. With the organizational and political impetus in place for change, President Richard Nixon created the Environmental Protection Agency (EPA) in 1970, and Congress, with bi-partisan support, passed environmental statutes such as the Federal Environmental Pesticide Control Act (1972), the Clean Water Act (1972), and the Clean Air Act (1970).

In addition to expressing concern over the quality of air and water resources, the public also became increasingly concerned about the visual and environmental impacts of forest management practices such as clearcutting, the protection of unique places, oceans and coastlines, and the extirpation of plant and animal species. These concerns were addressed in legislation such as the Wilderness Act (1964), the Federal Land Policy Management Act (1976), the Forest and Rangeland Renewable Resources Planning Act (1974), the National Forest Management Act (1976), the Coastal Zone Management Act (1972), the Wild and Scenic Rivers Act (1968), and the Endangered Species Act (1973). Certain lands were put under protective status; elaborate and publicly-open planning and environmental assessment processes were established; and legal and administrative frameworks were designed to make it more difficult for local development interests to influence public lands management.

Whereas the flurry of federal environmental legislation in the late 1960s and throughout the 1970s institutionalized the public's embrace of environmental values, it also tended to centralize decision-making in Washington. Parties near the resource no longer maintained a situational advantage in making decisions. Washington-based agency leadership and Washington-based interest groups held center stage. This separated decisions with local impacts from local interest, history, and culture.

Public support for environmental programs and values remained high during the conservative administrations of Presidents Ronald Reagan and George Bush (Dunlap and Mertig 1992, Hendee and Pitstick 1992). A number of policy initiatives during these administrations foreshadowed the emergence of ecosystem management. For example, the missions of construction-oriented agencies such as the Bureau of Reclamation and the Army Corps of Engineers became more openly environmentally friendly, shifting from an emphasis on construction of large-scale water projects to water conservation and remediation of the adverse impacts of previously built projects. Major revisions expanded the scope of the Clean Water Act and Clean Air Act; non-point source pollution, acid rain, ozone depleting chemicals, and toxic releases all came under tougher regulatory purview (Rosenbaum 1995). In the area of public lands management, several major additions were made to the National Wilderness System. And largely in response to the spotted owl controversy, the Forest Service made a major policy shift, reducing clearcutting as a standard timber harvest practice and committing itself to an ecological approach to management of national forests and grasslands (Robertson 1992).

Some of the policy initiatives of the 1980s, such as improved accounting procedures to weed out below-cost timber sales and cost-sharing requirements for federally financed water projects, utilized concerns about the size of the federal budget to reduce environmental impacts and achieve environmental goals. Emphasis on market approaches thus married resource conservation with the need for financial economy and political decentralization (O'Toole 1988, Anderson and Leal 1991). It became increasingly recognized that to conserve financial resources the nation would have to manage its land and water resources with extensive cooperation among public agencies at all levels and, as possible, with private individuals and organizations.

The Reagan–Bush administrations' critique of big Washington government and the need to decentralize decision-making to those communities closest to the resource also began to resonate broadly: with local economies and communities hit hard by federal laws and decisions that favored resource protection over resource extraction; with scientists who saw the need to tailor better human wants and needs to an understanding of the dynamics of ecosystem processes; and with agency leaders who realized that the problems they experienced in fulfilling their mandated public involvement requirements could never be corrected unless they dispersed and shared power.

Reliance on decentralized decision-making and collaborative partnerships was woven into the philosophy of ecosystem management. Such collaborations and partnerships took many forms. Local watershed groups formed to manage jointly communities of place (Natural Resources Law Center 1996). The Park Service began to experiment with the concept of partnership parks that included privately owned parcels within park boundaries (Commission on Research and Resource Management Policy 1989, U.S. Department of the Interior, National Park Service 1992). An amendment to the Endangered Species Act in 1982 facilitated the development of cooperatively-based private-public habitat conservation plans that take a landscape approach (Thornton 1991). The Forest Service's New Perspective program, EPA's National Estuary Program, and the Fish and Wildlife Service's North American Waterfowl Management Program also set powerful examples of collaborative management (see Yaffee, and Johnson et al., Vol. III of this book).

Actions at state, local, and international levels, and in the private sphere, demonstrated that acceptance of ecosystem management principles was not limited to the federal sector (Samson and Knopf 1996). State and local resource agencies began to implement ecosystem management projects and more landscape level studies (Carey and Elliott 1994, Wodraska 1995, Brown and Marshall 1996). The Western governors, for example, began to examine new state roles in water resources and public lands management, initiatives which stressed sustainability and more effective public participation (Western Governors' Association 1992, 1996). Many other ecosystem management projects originated in response to neither federal or state initiatives, but because of locally defined issues and needs (Yaffee et al. 1996).

The international community addressed global issues such as biodiversity, global warming, deforestation, and ozone depletion, and reaffirmed its commitment to the ideals of sustainable development spelled out in the Brundtland report (World Commission on Environment and Development 1987). The United Nations Conference on Environment and Development held in Rio de Janeiro in 1992 —attended by 178 countries and over 3,000 non-governmental organizations — resulted in binding agreements on climate change and biological diversity, and statements of principles on sustainable development and forestry.

Many ecosystem management-related initiatives also emerged from the private sector. Private organizations like the Keystone Center and American Forests' Forest Policy Center organized dialogues around the theme of implementing ecosystem management (Sample et al. 1995, Keystone Center 1996). Industry began to embrace "green production," which uses a total systems concept to minimize emissions, effluent, and the use of virgin materials, as a business strategy to attract investors and increase competitiveness (Hart 1994, Bhat 1996). And a major forest products association, the American Forest and Paper Association, established a set of sustainable forestry principles and implementation guidelines to which members are expected to adhere (American Forest and Paper Association 1994, Wallinger 1995).

While the foundation for ecosystem management had been building for many years, the Clinton Administration made it explicit policy. In 1993, that administration established both a White House Task Force on Ecosystem Management and the President's Council on Sustainable Development.

Today, policies and programs embracing ecosystem management as a new approach to management are evident across all levels of government and in the private as well as public sphere. They are the result of fundamental changes in social values toward human-nature relationships and an evolving public appreciation of the complexities of ecological relationships. Synergistically, these shifts in public values have been both a response to advances in scientific knowledge about ecosystems as well as a driver prompting science to rethink traditional epistomologies, assumptions, and methods.

## 3.2 Evolving Scientific Knowledge

Ecological science, with its emphasis on holistic understanding of the interconnections among key components of the system, is an important element of ecosystem management (Holling 1992). While the term *ecosystem* is generally attributed to Sir Arthur Tansley in 1935, as pointed out by Hagen (1992), Golley (1993), and Christensen et al. (1996), scientific writers discussed underlying concepts much earlier.

In the United States, conservation leaders from John Muir to Bob Marshall and Aldo Leopold were early proponents of many of the same themes that characterize ecosystem management today. Aldo Leopold's (1949) "Golden Rule of Ecology" and his land ethic championed ecological integrity and a changed human relationship to nature and land as early as 1933 (Meine 1988). During the next 60 plus years, several scholars published books and articles that attempted to bring ecological concepts into natural resource management. In 1970 Lynton Caldwell, who is also known as the principal intellectual architect of NEPA's environmental impact statement requirement, specifically advocated the use of the ecosystem as a criterion for public land policy (Caldwell 1970). Plentiful historical examples of attempts to develop and apply ecological approaches in the natural resource management disciplines (Salwasser 1994) as well as in a number of social sciences and humanities disciplines (Slocombe 1993, Francis 1993) can therefore be cited. Nonetheless, only recently did such approaches become the *dominant* mode by which scientists and managers looked at problems, framed their research programs, or designed management projects.

As social values and law began to demand greater emphasis on the environmental consequences of various actions, ecological approaches began to become more salient. Scientific research began to pay more attention to the impacts of environmental change and to the societal and ecological benefits of non-commodity and amenity resources. A more diverse cadre of scientific disciplines, e.g., landscape architects and wildlife biologists, was integrated into the staffs of resource management agencies. As this occurred, the scientific base underlying management began to change significantly.

Beginning in the late 1980s various professional groups addressed the need to adopt new approaches to professional practice and research (National Research Council 1990, Lubchenco et al. 1991, Society of American Foresters 1993). Two new disciplines, conservation biology (Soule 1985, 1986, Primack 1993, Meffe and Carroll 1994) and ecological economics (see Farber and Bradley, Vol. III of this book, Costanza 1991), emerged out of critiques that the parent disciplines —wildlife biology and economics — were too oriented to the utilitarian uses of resources and failed to deal adequately with global ecological issues. These new disciplines are avowedly normative and committed to the goal of sustainability. Conservation biologists, for example, clearly believe that biological diversity should be preserved and pursued as a policy (Meffe and Viederman 1995).

By the 1980s there was an accumulation of scientific evidence that mandated revisions in thinking about how resources are studied and managed. Long-term studies in places like the Hubbard Brook and H.J. Andrews experimental forests yielded insights about the functions and dynamics of ecosystems (Likens et al. 1977, Bormann and Likens 1979). Because results challenged the traditional view of the biological value of old-growth forests (Maser et al. 1979), scientists began to test less invasive forest harvest practices (Franklin 1989, Gillis 1990). Large-scale disturbances, such as those associated with the eruption of Mount St. Helens and the Yellowstone fires, provided laboratories for study of ecosystem dynamics and pointed to the role that biological legacies played in recovery processes (Maser et al. 1979, Franklin 1990).

Science began to place new emphases on restoration ecology and disturbance ecology (see Engstrom, Covington et al., Kenna et al., this volume). Scientific research also identified the adverse consequences of fragmentation of natural systems into small, relatively isolated units and the important role of late seral stage forest stands in large-scale systems. Studies showing that protected areas alone would not provide adequate habitat for free-roaming animals (Newmark 1985, Agee and Johnson 1988, Grumbine 1990) made it necessary to examine biophysical conditions and human activity outside the protected area's boundaries (Freemuth 1991, Franklin 1993a, Halvorson and Davis 1996, Weeks 1997).

Scientists realized the need to study ecosystem processes holistically at larger spatial scales, and managers tested cooperative ventures in managing resources across a patchwork of jurisdictions and ownership patterns. Science and policy became linked in places such as the Northern Lands (Harper et al. 1990), the Greater Yellowstone Ecosystem (Clark and Minta 1994), and the Great Lakes Ecosystem (Dworsky 1993, Francis and Regier 1995). The more scientists came to understand some of the complexities of ecosystems, the more they realized they did not understand. To cope with the biophysical and social uncertainties, they embraced the design of adaptive management systems (see Bormann et al., Vol. III of this book, Holling 1978, Walters 1986, Lee 1993, Gunderson et al. 1995).

## 3.3 Professional Experience

Professional experience and learning based on efforts to implement resource management programs is a third seam in the evolution toward an ecosystem perspective for planning and decision-making. To be sure, professional perspectives evolve in response to shifts in societal values and scientific knowledge. But because they are often on the cutting edge of testing new technologies or implementing complex and sometimes ambiguous legislation, the resource professions themselves — ecologists, foresters, hydrologists, wildlife biologists, entomologists, pathologists, geologists, landscape architects, and planners, to name but a few — have played a central role in developing planning and decision-making models that incorporate a whole systems approach to resource management.

Managers came to realize that the old models of how to study and manage resources required revision. The basic concept, for example, that trees could be treated as a commercial crop, managed under short rotations, and, at the same time, adequately achieve other social goals, such as recovery of species or protection of high quality watershed systems, was no longer adequate (Gordon 1994). Managers began to manage more explicitly for ecological values, and both managers and professional associations began renewed discussions about land stewardship and ethical obligations to the land (Reeves et al. 1992).

Experience with implementing legal mandates also showed the necessity of taking a more holistic approach to management. For example, recovery plans under the Endangered Species Act that focus on a single-listed endangered species rather than on habitat for a whole set of interacting species raised concern that in some situations undesirable biological results for other species may result. The Lands Unsuitable for Mining Program (LUMP) forced a detailed and comprehensive examination of integrated environmental concerns, and allowed the public to participate in designating lands that could not be mined. Certainly, the estimation and evaluation of cumulative effects of management practices, as required by NEPA and state laws, also figured prominently in the evolving trend toward utilization of broader geographic scales.

An example of a more "holistic" view is the long-standing trend toward larger spatial units for analysis and planning of the national forests. During the 1900s, geographically-defined regulatory units called "working circles" were used for administrative purposes to establish a sustainable level of timber harvest within a national forest and to control other forest operations (Davis 1966). Such units were largely set up to supply timber to a nearby, rural timber manufacturing center with access by railroad to urban markets. The working circles were relatively small by today's planning standards, typically in the range of 100 to 250 thousand acres.

Gradually the working circles and spatial scale became larger, reflecting improvements in transportation, recognition that integrating resource uses required assessments at watershed or regional levels, economies of scale in planning and operations, and synergistic effects in computing long-term sustained yield levels (Teeguarden 1973). In 1976 the National Forest Management Act directed that a single, integrated resource management plan be prepared for each unit, or combination of units, of the national forest system; in effect, the Congress created a legally defined spatial scale for planning and decision-making. Plans had to comply with a complex set of statutory requirements, including sustainability and biodiversity.

The success and failures (both technically and politically) of the national forest planning effort have been the focus of much debate (USDA Forest Service 1990, U.S. Congress, Office of Technology Assessment 1992). Yet significant organizational and professional development occurred as a result of the planning experience. The lessons learned and the planning technologies developed will have considerable value for the next round of planning.

Advances and experience with information technologies, such as remote sensing and GIS, made it possible to gather and portray ecological data at a variety of scales and to integrate ecological data with human land-use patterns, including projected land uses (Sample 1994). These information technologies, as well as the explosion of personal computing capabilities and electronic communication opportunities, have made scientific information more accessible throughout organizations and across the nation and the world. Multiple constituencies can now more effectively participate in analysis and decision-making. Widespread communication processes mean that more parties can know about issues, participate in discussion about them, and collaborate in implementing and monitoring management prescriptions.

Finally, just like managers in the corporate world and other sectors of public service, resource managers began examining their role as leaders in modern society. New leadership concepts addressed the need to replace the traditional, hierarchical, bureaucratic model with organizational forms that could adapt to more complex and collaboration management situations (Meidinger 1997). These concepts advised managers to develop strong personal ethics and communication skills, embrace change by being flexible and open to new ideas, and share responsibility with others (Berry and Gordon 1993, Sirmon et al. 1993). Leadership became increasingly linked with the need and passion for a land ethic, the reason of good science, and a connection to the evolving social values of the larger society (Cornett 1995).

## 4 THE FUTURE

Ecosystem management concepts will continue to evolve, requiring efforts by all participants to sharpen those concepts. As this evolution continues, it will be bounded and shaped by six contextual realities:

- Political processes will have a publicly accepted role in ecosystem management;
- Economic considerations will have a publicly accepted role in ecosystem management;
- Additional scales of interest will become necessary considerations;
- Vastly more potentially useful information will be available, with greater and wider access to it;
- Science will increase its ability to detect and predict ecosystem change; and

- Ecosystem management will persist through periodic changes in political, social, and economic ideologies.

## 4.1 Political Processes Will Have a Publicly Accepted Role in Ecosystem Management

When evaluating any policy or proposed management action, the political consequences obviously must be carefully considered. These political considerations, which range from who benefits and who pays to how election chances may be improved or harmed, cannot be avoided (Cortner et al. 1996). Decisions about ecosystems involve value judgments and hence politics. Political considerations are an important component of ecosystem management because ecosystem management explicitly recognizes that humans are part of ecosystems. The collaborative (intra-agency, inter-agency, intergovernmental, and public/private) and adaptive management features of ecosystem management will require more consensual and more participatory, and hence more overtly political, decision processes.

It must also be recognized that there is rarely one "right" answer, in the sense that if we just have more research or if we just consider the issue thoroughly enough, we will find the elusive "correct" answer. Because there may be no "right" answer, the final policy or management action will necessarily include compromises and political accommodations. Compromise and accommodation are necessary because our political culture is fragmented and power is distributed among many branches and levels of government, and between the private and public sectors, with many conflicting views on what should be the "right" answer.

Furthermore, institutions will need the capacity to respond and adapt to new knowledge, multiple sources of knowledge, changing public attitudes, and lessons learned from on-the-ground management experiences. Instead of primary reliance on technical experts, who traditionally — and erroneously — were presumed to be neutral, ecosystem management requires that authority and power be distributed to a wider array of participants. This will also require a more explicit recognition of the political facets of resource decision-making. Under ecosystem management, the leadership role of the resource manager goes beyond that of authoritative expert, and will include educator, mediator, technician, conflict manager, public relations specialist, scientist, or some combination of these roles (Cortner et al. 1996).

The political scene changes over time, thus decision-making will be influenced by current political conditions. Political considerations will need to be measured with a variety of tools, and this information must be invoked alongside technical recommendations such as erosion rates and habitat fragmentation. Thus, ecosystem management must be developed by processes that are expressly political, where the political considerations are not hidden or subtle, but rather are exposed. As political considerations are explicitly evaluated in the process, it will be important to identify clearly the sources and motivations for each political position.

As ecosystem management is developed with overt political considerations, it will be necessary to ensure that political considerations are explained and evaluated with expertise levels commensurate with the quality of economic and technical considerations. Political considerations should not dominate the process because simple inclusion of politics does not diminish the importance of technical information and analyses. For example, scientists must be involved in (a) formulating the significant issues; (b) conducting the research and analyzing data; and (c) predicting the technical consequences of any decisions that involve political choices.

The ecosystem management process must recognize that as we learn more about ecosystems, management approaches and structures must change in an adaptive manner. Similarly, as political conditions change, there will be strong political reasons to reassess existing management plans and processes. Therefore, when plans/actions are promulgated, it will be useful to contemplate the possible subsequent political factors that call for such reassessment.

The centrality of politics means that ecosystem managers must learn to understand politics and civics better, and that politicians must better understand ecosystems and resource management (Holling 1995). It also means building better bridges between policy, science, and management; such bridges are evolving. For example, science-based assessments are emerging as an analysis tool (see Lessard et al., Graham et al., Vol. III of this book), and suggestions that the practice be institutionalized at universities have been made (Gordon 1993a). Science-based assessments attempt to use the methods and people of science to answer time-bounded questions arising outside science, usually as a result of the political process, formal or informal. One such assessment, the Scientific Panel on Late Successional Ecosystems, (the "Gang of Four" report) (Johnson et al. 1991), formed the basis for the approach and results of the Northwest Forest Plan (the "President's Plan"), which moved the controversy over the northern spotted owl and other old-growth related species out of the courts and into proactive management on the ground (Gordon and Lyons 1997). Some characteristics of successful assessments are emerging:

- They should be conducted at an early stage of development of a key political question;
- They should be structured as "if-then" statements and their results should be presented as a matrix of alternatives subject to political decision;
- They should never attempt to define a single "right" answer, but provide options and projected outcomes;
- They should be based on sound, published scientific results in so far as possible;
- The credibility of the assessment depends on the credibility of those doing it and the process they employ;
- The creation of assessments should not be confused with "public involvement"; and
- Scientists involved must recognize the validity and inevitability of public involvement and political choice.

Scientists alone will not resolve these issues, for public participation in both research and decision-making will continue to expand. How to restructure the scientific research enterprise to accommodate active and constructive citizen participation in research design, the conduct of research, and the evaluation and monitoring of results, creating in effect a "civic science", will become increasingly imperative. Civic science is a catalyst for the adaptive management component of ecosystem management (Shannon and Antypas 1997).

## 4.2 Economic Considerations Will Have a Publicly Accepted Role in Ecosystem Management

The systems perspective of ecosystem management obviously requires consideration of the economic and financial consequences of alternative strategies and policies. The public generally expects that resources, whether natural, social, or financial, are efficiently and equitably allocated within and between competing investment needs. These same expectations will apply to ecosystem management.

Presently, the future appears to offer the prospect of declining real-dollar expenditures for natural resources management and increasing demand for competing investments in such areas as education, health, public safety, infrastructure, and aid to the poorer countries of the world. The scale and intensity of ecosystem management will therefore be influenced by objective, effective documentation of associated costs and benefits. Furthermore, many if not all ecosystem management decisions will require trade-offs involving both environmental values and changes in market-based consumption patterns (see MacCleery and Le Master, this volume). Analysis of these trade-offs will be an important feature of the planning and decision-making process (see Horne et al., Vol. III of this book). The costs of managing the environment based on ecological prescriptions will be large and require significant investments in labor and capital. As traditional sources of capital (such as revenues from commodity production) diminish, new capitalization strategies will need to be developed to fund resource management programs.

To integrate economic consequences in ecosystem management better, analytical tools, including cost–benefit analyses that address the value of unpriced ecosystem services, must be improved to the point where such analyses are regarded as credible by decision-makers and public stakeholders. There will, of course, continue to be points of contention in analysis methodology because of the many externalities and the difficulty of quantifying ecological services and values. Furthermore, some people who hold strong moral convictions about what is right or wrong in the management of natural environments often have difficulty in attaching any economic weight to non-market resources and amenity values.

Economic and monetary incentives, as well as penalties, will also have significant public policy roles in promoting ecosystem management objectives on private land (see Sedjo et al., Vol. III of this book). A number of states, including California, already have such programs in the form of direct regulation, financial assistance, preferential taxation, financial or civil penalties, and encouragement of cooperative planning across multiple ownerships.

Risk and uncertainty is inherent in ecosystem management as in any management approach. Economic analysis, indeed all types of evaluations, must explicitly consider the range of possible outcomes and, if feasible, their probability. Differences in risk, whether expressed quantitatively or qualitatively, should be included in comparison of alternative management programs. Decision-makers must also consider both their immediate, short-term consequences and also long-term impacts. Finally, if alternatives are expected to have different distributional effects, beneficial or negative, across different social groups (racial, ethnic, gender, and income class), equity concerns will also need to be analyzed and considered during the decision-making process.

## 4.3 Additional Scales of Interest Will Become Necessary Considerations

The ecosystem management process has already expanded the geographic and temporal scales of consideration (see Haufler et al. and Caraher et al., this volume). It is recognized that local ecosystems are tied to adjacent ecosystems via hydrology and migration routes, and that public lands should be managed in ways that are reasonably coordinated with management of adjoining private lands. Although intense local views are often expressed about how an ecosystem should be managed, federal lands also have a national constituency. Thus local management decisions will affect, and be affected by, this national constituency.

It is to be expected that the scales of consideration will continue to expand in the future. The reasons for this expansion are many and will arise from multiple sources. First, as science provides better understanding of global biological-chemical-physical processes, more connections will be obvious (e.g., El Nino effect on annual plant productivity in the western United States). Second, as markets and international treaties and agreements become more global, analysis of economic consequences of ecosystem management will also become more global. Third, as more innovative economic and policy tools are used, more constituencies will become involved. Finally, as science becomes better able to detect and predict ecosystem changes (e.g., gene flow and metapopulation dynamics), more spatial and temporal considerations will be regarded as important. Models will include wider and longer-term dynamics of ecosystem properties. More landscape-scale efforts like those designed to sustain large carnivores, such as grizzly bears and wolves, will occur, and many will require not only regional, but international cooperation as well.

Because additional scales of consideration will occur, the ecosystem management process needs to include explicitly the provisions for these scales and a clear evaluation of the relative importance of each (see Haufler et al., this volume). It will be important to include these additional scales, but not allow the increased complexity caused by their inclusion to distract the decision-making process from the most essential ingredients (see Caraher et al., this volume).

The need to understand interactions at larger spatial scales and over longer time frames is also likely to change the conduct of science. The scientific focus will need to include, in many cases, both a comprehensive understanding at the stand or site level and perhaps a less detailed understanding of a broader landscape (Risser 1995). This will also be necessary because ecosystem management will likely be implemented in an economic and political environment in which government agencies will have to operate with fewer dollars and fewer employees.

## 4.4 Vastly More Potentially Useful Information Will Be Available, With Greater and Wider Access To It

Information of all kinds is increasing at an exponential rate, including information resources specifically pertaining to ecosystem management. With the World Wide Web and Internet and the Freedom of Information Act, more information is available at lower cost to people and organizations wishing to participate in ecosystem management. The presentation of information through images and multi-media meets the needs of many current consumers, and it permits the style of presentation to influence the input of the information itself to a greater degree than in the past. This available information is, however, of varying quality, and the need to evaluate, manage, select, and apply this vast information is increasing (see Cooperrider, Vol. III of this book).

Because political, socioeconomic, and technical considerations are integral parts of ecosystem management, the ecosystem management process itself will also need to provide frameworks for legitimate comparative analyses of this burgeoning data set. With the increased information available for multiple topics of concern, setting standards for evaluating information in general, and for ecosystem management practices in particular, will become more important.

## 4.5 Science Will Increase Its Ability To Detect and Predict Ecosystem Change

Ecosystem science has increased in strength by developing theories, conducting experiments, and analyzing results on many processes and structures; these analyses and observations have been made in virtually all ecosystems, although some ecosystem types have been studied more thoroughly than others.

In the last 30 years, the amount of research conducted in ecosystem science has greatly increased. The result of this investment is an expanded capacity to understand and predict the behavior of ecosystems. Two important advances include better instrumentation and methods (e.g., GIS, remote sensing, pollutant detectors) and long-term data sets that allow scientists to determine if individual measurements are unusual, part of a trend, or within a normal variation (see Correll et al. and Reynolds et al., Vol. III of this book).

Notwithstanding the increased capability within the ecosystem science field to detect and predict ecosystem change, surprises will always occur, including, for example, behavior patterns that are different from what is expected, thresholds reached that cause the system to behave in unexpected or vastly different ways, and more complicated connections with other systems that cause surprising behavior (Holling 1986, Kates and Clark 1996). Dealing with such surprises will require greater flexibility and anticipatory management strategies (see Lugo et al., this volume).

Ecosystems are affected by many controls and limiting factors. In some cases, these factors operate in a largely additive manner, but in other cases, they act in non-linear or synergistic ways. Ecosystems have characteristics of resilience, resistance to change, and self-restoration (see Covington et al., this volume). These characteristics are valuable in predicting the consequences of ecosystem management, but the data base is sparse for estimating the rates of change in response to changes in driving variables or controlling factors. Moreover, because of the complex interactions within ecosystems themselves, it is difficult in some cases to isolate and identify the causes for ecosystem behavior. A challenge for future scientific study will be to examine further the cumulative and synergistic changes that affect ecosystems and to apply this understanding to evaluate alternative management strategies.

A major change in the perception of ecosystems is emerging because of the application of systems concepts to classical ecosystem theory (Allen 1993). These are particularly important when applied to forests (Gordon et al. 1993). In the past forests have been viewed as collections of organisms and populations of organisms. With emerging knowledge of mutualism, guilds, and extensive physical connections (e.g., mycorrhizae), the view of ecosystems as collections of organisms is seen to be less valid. To the worm, an aspen clone is an "organism" completely interconnected below ground; to the above-ground view, it is many "organisms" because the stems and crown appear to be individuals. Science is now seeking ways to study ecosystems as units and to compare them across large differences in time and environment (Vogt et al. 1997, and see Bormann et al., Vol. III of this book). Similarly, the term "evosystem" (short for evolutionary system) denotes that the *system* functions as the evolutionary anvil on which biological diversity is hammered out and the matrix in which it is maintained. It is the maintenance of this evolutionary anvil and matrix, and not necessarily the species and populations it creates, that is the principal biological feature of ecosystem management.

### 4.6 Ecosystem Management Will Persist Through Periodic Changes in Political, Social, and Economic Ideologies

In any decision-making process, an ideological framework bounds the decision space that is considered to be the proper course. For example, there is currently debate among those who think the earth's carrying capacity is linked inextricably to natural capital, those who think human ingenuity can overcome the limitations of natural capital, and those who take a more intermediate position and acknowledge the uncertainty and the ability of social and natural systems to evolve over time. Whatever position is chosen, the manager is likely guided to different answers; yet these ideologies are likely to change on the basis of new knowledge or different values defined by society. Other examples that may have an impact on ecosystem management include beliefs as to the appropriate balance among market, regulation, and other policy tools; between centralization and decentralization; among executive, legislative, and judicial branch powers; and between rich and poor.

It is important to recognize prevailing ideologies, put these ideologies in historical context and future perspective, and analyze their impact on ecosystem management decisions. Because ecosystem management involves inclusive public participation, and a broad set of social values, management decisions could be different in the future depending upon ideological influences. Changes in social values need to be tracked and monitored just as are changes in biophysical conditions (see Bliss and Cordell et al., Vol. III of this book, Gerlak et al. 1997).

Ecosystem management itself will change over time, especially with regard to the tools and approaches to be employed. However, the basic concepts and principles of ecosystem management, as outlined here and elsewhere in this volume, will endure because they are responsive to citizen concerns, and because they are widely regarded as representing an enlightened approach to resource management decisions. Likewise, the core issues of sustainability, biodiversity, social welfare, and equality will persist.

## 5 NEXT STEPS

Current interest in ecosystem management is a culmination of historical changes in social values, scientific knowledge, and professional practice. Future imperatives and their implications encourage comprehensive and immediate implementation of ecosystem management as the basis for the policies and practices guiding natural resource management. The social values and scientific body of knowledge that comprise ecosystem management will continue to evolve and mature. Faced with these developments, resource managers need to be proactive rather than reactive. Being proactive will require leadership to build greater scientific and public understanding of current state-of-the art ecosystem management, fill gaps in scientific knowledge, and test and refine applications in daily professional practice. Being reactive, which has occurred far too often in the realm of environmental and natural resources policy (Rosenbaum 1995), means waiting until a crisis of institutional capacity eventually forces action. By this time the available decision space may be quite narrow and the feasible options more expensive and politically contentious.

Proactive implementation of ecosystem management will require changes in law and policies, science and research institutions, land management and organization, educational institutions and students, and citizenry and communities. A number of next steps toward making these changes are recommended below. This list is not inclusive of all the recommendations that have been suggested as means to facilitate implementation of ecosystem management. Many items are the product of considerable thought by previous authors, professional task forces, or stakeholder groups. They are highlighted here because they are responsive to and cut across the future realities discussed above. Furthermore, not only do they involve actions that can be taken in the near term, they also may have multiplier effects that can trigger further improvements in the ecosystem management planning and decision-making process.

## 5.1 Changes in Law and Policies

The President and Congress should establish a joint commission, including representatives of federal agencies to: review existing legislation through the lens of ecosystem management theory and applications to date; identify conflicting legal requirements; identify potential barriers; and recommend to the President and Congress corrective and augmenting legislative action. Congress needs to recognize that ecosystem management involves extensive sharing of information and the interactive involvement of stakeholders in the management of the nation's lands and waters. Therefore, federal rules and policies must be revised to provide structures with incentives to encourage the open exchange of information and the productive collaborative decision-making needed for ecosystem management. Specific attention should be paid to: legislation relevant to federal land management; current missions and jurisdictions of the federal agencies; and the budgetary processes that affect implementation of ecosystem management programs and objectives.

The Federal Advisory Committee Act (FACA) should be reviewed and revised. Enacted to control excessive costs, lack of openness in decision-making, and inappropriate relationships between special interests and federal agencies, the act, as currently interpreted and implemented, is too inflexible. It constrains open discussion with appropriate stakeholders, does not allow for spontaneous changes in direction as agency needs develop, and involves unnecessarily bureaucratic procedures. Although Congress has amended FACA to exempt intergovernmental communication, the law needs further revision so that open and inclusive interactions and exchanges of information among *all* stakeholders can be encouraged, while still guarding against inappropriate behavior.

## 5.2 Changes in Science, Research and Institutions

At the federal level, agencies should initiate joint interdisciplinary research programs on ecosystem management. Such research should be conducted through partnerships among government, university, industry, and non-government organization scientists. An interagency working group should be established to coordinate ecosystem management research across agencies, and to recommend research priorities to the agencies, the Office of Management and Budget, and Congress. In addition, agencies should develop technical assistance teams with skills and resources to link such research to field applications.

Ecosystem management requires new methods of scientific inquiry, including methods that break down traditional functional approaches and encourage systems-thinking. Institutions that fund research should examine opportunities to direct research toward projects that blend qualitative and quantitative methods and the biophysical and the social sciences. Additional opportunities to involve citizens in the design and implementation of research projects should also be more extensively explored. One goal is to harness the energy and concern of a new generation of "citizen ecologists".

## 5.3 Changes in Land Management and Organizations

Agencies should provide the structure and resources for employees to implement ecosystem management. They should develop performance evaluation systems that are designed to reward employees for making measurable progress toward using available science, building partnerships and coalitions, and other critical aspects of implementing ecosystem management "on the ground". Such measures should reward risk-taking and experimentation with innovative arrangements, including experiments with breaking down functional and geographic barriers. Where employees lack critical skills, training programs should be developed to build such capacity.

In no place is the building of partnerships more critical than in integrating management and research. Most important questions related to ecosystem management take place at scales too large, and over time too long, to be completely susceptible of treatment by classic scientific methods involving replication, randomization, and "strong inference" methods (Platt 1966). Rather, they need to be approached by coalitions of managers and scientists linking smaller-scale trials, models, and large-scale trials. The latter will be done in the routine implementation of management decisions (see Bormann et al., Vol. III of this book).

## 5.4 Changes in Educational Institutions

The deans and directors of the natural resources colleges and departments should commission an independent review of the current organizational structures, incentive and reward systems for faculty, and curriculum content and learning experiences for all students. This review should recommend changes needed in research, instruction, and public service to integrate more fully ecosystem management principles and approaches.

This review should consider how best to expand the audience for undergraduate programs not only to train specialists to obtain natural resource jobs targeted by employer groups, but also to educate future citizens. For example, general ecosystem management courses can be offered within the professional schools to attract students from across the campus who may not major in a resource discipline. Such "liberal science" curricula are necessary to build an ecologically literate citizenry, one that can effectively participate in ecosystem management decisions.

Finally, all educators should be involved in creating an ecologically literate citizenry early-on in the education process. Professional societies as well as colleges and universities can begin working with environmental education practitioners and funders of environmental education to develop effective and economical ways in which ecosystem management concepts can be integrated into K-12 educational programs.

## 5.5 Changes in Citizens and Communities

Citizens in a democratic society should be prepared to participate in decision-making that trades one value for another, and to reach agreements with others that may often involve compromise. To do so they have a responsibility to acquire an understanding of all consequences of management alternatives. Governmental agencies, with private partners and other public groups, can develop programs to assist stakeholders in being broadly informed on environmental issues at local, regional, national, and global levels. In developing these programs, however, care must be given that they are truly educational in nature — encouraging dialogue, presenting all sides of an issue, and empowering citizens as partners in the decision-making process — and not simply efforts to convince citizens of the pre-conceived merits of any particular agency solution.

In the case of federal lands, federal land managers must represent the federal interest and not just local wants and needs. Nonetheless, federal land managers must also recognize that local communities matter. They matter because they are often the most directly impacted by management decisions. They also matter because they have critical roles to play in identifying emerging problems and issues, formulating alternative solutions, implementing programs, and monitoring and evaluating outcomes.

Both rural and urban communities should organize to secure the capacities to be full and effective participants in decision-making and to assume greater responsibilities for ecological stewardship. Agencies can assist by working to remove impediments to effective community involvement at all levels of government and by being active and collegial partners. But citizens still have responsibilities to build the organizational capacity to sustain their organizations and to fulfill their responsibilities as informed, civic partners. Although there are outstanding examples of recently developed community-based organizations in resource management, considerable effort still remains to broaden the community base. Citizens must empower citizens to participate actively in ecologically based resource management.

# 6 CONCLUSIONS

A strong and well-developed theoretical basis for ecosystem management exists in the scientific literature, a basis that is reflected in the many papers from the Tucson workshop. This literature ranges across many disciplines. It recognizes that ecosystem management is not just about learning how to move, alter, harvest, or protect physical matter better, but that people are also important parts of ecosystem management. Discussions, such as those by Machlis et al. (this volume), Clark et al. (Vol. III of this book), and Bormann et al. (Vol. III of this book), provide a beginning road map by which managers and scientists can understand and integrate the linkages among natural, cultural, social, and economic resources and the governance structures within which decisions about resources are made. Because ecosystem management fosters continued study of the linkages between the social and physical components of ecosystems, we hope, and believe, that advances in ecosystem theory and practice will eventually drop the artificial distinctions between biophysical and social sciences and/or the hard and the soft sciences, and speak just of "science".

Despite agreements on broad theoretical principles and general types of approaches needed to implement ecosystem management, nonetheless, many questions and uncertainties still persist about the specific methods for achieving these fairly broad goals — uncertainties, for example, about the resources that are needed and available to implement ecosystem management approaches. Although it is evident that many uncertainties in implementing the principles of ecosystem management persist, participants at the workshop expressed the need for more managers and communities to be risk-takers and to learn by doing. Unknowns, data gaps, and areas where there is scientific disagreement cannot be used as excuses for not moving forward. Scientific uncertainty is not unique to ecosystem management. Few non-arguable scientific truths exist; unified science is more ideal than reality.

Moreover, it is unrealistic to expect that ecosystem management tools will apply equally to small parcels, private lands, mixed ownerships, as well as to the arid ecosystems of the West and the humid, industrialized, urbanized ecosystems of eastern North America. Individual agency approaches at all levels of government are also likely to be different because agencies have different mandates, different cultures, and different politics. A country as diverse culturally, socially and physically as ours, with a political system explicitly designed to foster creative tensions and to dissipate power, is unlikely to come to any agreement on a single integrative scientific framework for ecosystem management or on a commonly adopted and applied set of highly prescriptive management protocols. Efforts to derive the ultimate "cookbook" are likely to be futile and contrary to the principles of flexibility and adaptable institutions that are core to ecosystem management. Certainly criteria for evaluating success and standards to ensure public accountability will be mandatory. However, both fairness and a pluralistic political system will militate against a "one for all" cookbook approach across all physical and political landscapes.

Although much remains to be done in ecosystem management in terms of turning theory into practice, there are innovative case examples where managers and scientists — despite uncertainties, data gaps, and nascent performance standards — are using available tools to implement and refine ecosystem management principles. Examples from the workshop included projects initiated by the federal government, state governments, industry, and private organizations. Partnerships among governments, organized interests, and communities are working together to build the institutional capabilities for problem solving.

A powerful confluence of forces has led to the current scientific and policy adoption of ecosystem management as the guiding approach to resource management. These forces are legal, political, economic, sociocultural, scientific, and biophysical. Woven into a complex scientific and institutional fabric that now defines ecosystem management, these forces cannot simply be unraveled. They have created imperatives and opportunities that scientists, ecosystem managers, and the public must respond to today and in our near and long-term future. Everyone has a stake and everyone has a role to play.

## ACKNOWLEDGMENTS

The authors would like to thank Gary Machlis, Professor of Forest Resources and Sociology, College of Forestry, Wildlife and Range Sciences, University of Idaho, and Visiting Chief Social Scientist, National Park Service, who contributed to the team's efforts during the Ecological Stewardship meeting in Tucson in December, 1995, and at a subsequent meeting of the team in Chicago in May, 1996. The team also wishes to thank Margaret Ann Moote, Senior Research Specialist, Udall Center for Studies in Public Policy, University of Arizona, who read and made valuable comments and additions to draft material as well as assisted the team administratively, and Ted Smith of the Kendall Foundation for his careful review and useful suggestions.

## REFERENCES

Agee, J.K., and D.R. Johnson. 1988. *Ecosystem Management for Parks and Wilderness*. University of Washington Press, Seattle, WA.

Allen, T. 1993. Physiology to Ecosystem: Not as Easy as You Think. In: *Proceedings, Twelfth North American Forest Biology*

*Workshop*, August 17–20, 1992, pp. 50–61. Ontario Ministry of Natural Resources, Sault Ste. Marie, Ontario, Canada.

American Forest and Paper Association. 1994. *Sustainable Forestry Principles and Implementation Guidelines*. Washington, D.C.

Anderson, T.L., and D.R. Leal. 1991. *Free Market Environmentalism*. Pacific Research Institute for Public Policy and Westview Press, San Francisco, CA.

Berry, J.K., and J.C. Gordon. 1993. *Environmental Leadership: Developing Effective Skills and Styles*. Island Press, Washington, D.C.

Bhat, V.A. 1996. *The Green Corporation: The Next Competitive Advantage*. Quorum Books, Westport, CT.

Bormann, F.H., and G. Likens. 1979. *Pattern and Processes in a Forested Ecosystem*. Springer-Verlag, New York.

Brown, R.S., and K. Marshall. 1996. Ecosystem management in state governments. *Ecological Applications* 6(3): 721–723.

Caldwell, L.K. 1970. The ecosystem as a criterion for public land policy. *Natural Resources Journal* 10(2): 203–221.

Carey, A.B., and C. Elliott. 1994. Washington Forest Landscape Management Project – Progress Report. Report No. 1. Washington State Department of Natural Resources, Olympia, WA.

Christensen, N.L., et al. 1996. The report of the Ecological Society of America Committee on the Scientific Basis for Ecosystem Management. *Ecological Applications* 6(3): 665–691.

Clark, T.W., and S.C. Minta. 1994. *Greater Yellowstone's Future*. Homestead Publishing, Moose, WY.

Commission on Research and Resource Management Policy in the National Park System. 1989. *National Parks: From Vignettes to a Global View*. National Park Service, Washington, D.C.

Cornett, Z.J. 1995. Birch seeds, leadership, and a relationship with the land. *Journal of Forestry* 93(9): 6–11.

Cortner, H.J., and M.A. Moote. 1999. *The Politics of Ecosystem Management*. Island Press, Washington, DC.

Cortner, H.J., et al. 1996. *Institutional Barriers and Incentives for Ecosystem Management*. PNW-GRT 354. USDA Forest Service Pacific Northwest Research Station, Portland, OR.

Costanza, R. 1991. *Ecological Economics: The Science and Management of Sustainability*. Columbia University Press, New York.

Czech, B. 1995. Ecosystem management is no paradigm shift: Let's try conservation. *Journal of Forestry* 93(12): 17–23.

Davis, K.P. 1966. *Forest Management: Regulation and Valuation*, 2nd ed. McGraw-Hill, New York.

Dunlap, R., and A. Mertig. 1992. *American Environmentalism: The U.S. Environmental Movement, 1970–1990*. Taylor and Francis, Bristol, PA.

Dworsky, L.B. 1993. Ecosystem management: Great Lakes perspective. *Natural Resources Journal* 33(2): 347–362.

Francis, G. 1993. Ecosystem management. *Natural Resources Journal* 33(2): 315–345.

Francis, G.R., and H.A. Regier. 1995. Barriers and Bridges to the Great Lakes Basin Ecosystem. In: L.H. Gunderson et al. (eds.), *Barriers and Bridges to the Renewal of Ecosystems and Institutions*, pp. 239–291. Columbia University Press, New York.

Franklin, J. 1989. Toward a new forestry. *American Forests* 95(11&12): 37–44.

Franklin, J.F. 1990. Biological legacies: A critical management concept from Mount St. Helens. *Trans. 55th North American Wildlife and Natural Resources Conference* 1990: 216–219.

Franklin, J.F. 1993a. Preserving biodiversity: Species, ecosystems or landscapes? *Ecological Applications* 3(2): 202–205.

Franklin, J.F. 1993b. The Fundamentals of Ecosystem Management with Applications in the Pacific Northwest. In: G. Aplet et al. (eds.), *Defining Sustainable Forestry*, pp. 127–144. Island Press, Washington, D.C.

Freemuth, J.C. 1991. *Islands Under Siege: National Parks and the Politics of External Threats*. University of Kansas Press, Lawrence, KS.

Gerlak, A.K., et al. 1997. Responding to Social Values in Natural Resource Decision Making. Unpublished paper prepared for the USDA Forest Service North Central Station.

Gillis, A.M. 1990. The new forestry. *BioScience* 40: 558–662.

Golley, F.B. 1993. *A History of the Ecosystem Concept in Ecology*. Yale University Press, New Haven, CT.

Gordon, J.C. 1993a. Assessments I Have Known: Toward Science-based Forest Policy? In: B. Shelby and S. Arbogast (eds.), *Communications, Natural Resources and Policy: 1993 Starker Lectures*, pp. 28–38. Oregon State University College of Forestry, Corvallis, OR.

Gordon, J.C. 1993b. Ecosystem Management: An Idiosyncratic Overview. In: G.A. Aplet et al. (eds.), *Defining Sustainable Forestry*, pp. 240–244. Island Press, Washington, D.C.

Gordon, J.C. 1994. From vision to policy: A role for foresters. *Journal of Forestry* 92(7): 16–19.

Gordon, J.C., et al. 1993. Physiology and Genetics of Ecosystems: A New Target? Or Forestry Contemplates an Entangled Bank. In *Proceedings, Twelfth North American Forest Biology Workshop*, August 17–20, 1992, pp. 1–14. Ontario Ministry of Natural Resources, Sault Ste. Marie, Ontario, Canada.

Gordon, J.C., and J. Lyons. 1997. The Emerging Role of Science and Scientists in Ecosystem Management. In: K.A. Kohm and J.F. Franklin (eds.), *Creating a Forestry for the 21st Century: The Science of Ecosystem Management*, pp. 447–453. Island Press, Washington, D.C.

Grumbine, R.E. 1990. Viable populations, reserve size, and federal lands management: A critique. *Conservation Biology* 4(2): 127–134.

Grumbine, R.E. 1994. What is ecosystem management? *Conservation Biology* 8: 27–38.

Gunderson, L.H., et al. 1995. *Barriers and Bridges to the Renewal of Ecosystems and Institutions*. Columbia University Press, New York.

Hagen, J.B. 1992. *An Entangled Bank: The Origins of Ecosystem Ecology*. Rutgers University Press, New Brunswick, NJ.

Halvorson, W.L., and G.E. Davis. 1996. *Science and Ecosystem Management in the National Parks*. University of Arizona Press, Tucson, AZ.

Harper, S.C., et al. 1990. *The Northern Forest Lands Study of New England and New York*. USDA Forest Service, Rutland, VT.

Hart, S.L. 1994. How green production might sustain the world. *Illahee* 1: 4–14.

Hendee, J.C., and R.C. Pitstick. 1992. The growth of environmental and conservation-related organizations: 1980–1991. *Renewable Resources Journal* 10(2): 6–19.

Holling, C.S. 1978. *Adaptive Environmental Assessment and Management*. Wiley, New York.

Holling, C.S. 1986. Local Surprise and Global Change in Sustainable Development of the Biosphere. In: W.C. Clark

and R.E. Munn (eds.), *The Resilience of Terrestrial Ecosystems*, pp. 292–317. Cambridge University Press, New York.
Holling, C.S. 1992. Cross-scale morphology, geometry, and dynamics of ecosystems. *Ecological Monographs* 62: 447–502.
Holling, C.S. 1995. What Barriers? What Bridges? In: L.H. Gunderson et al. (eds.), *Barriers and Bridges to the Renewable of Ecosystems and Institutions*, pp. 14–16. Columbia University Press, New York.
Interagency Ecosystem Management Task Force. 1995. *The Ecosystem Approach: Healthy Ecosystems and Sustainable Economies.* Report of the Interagency Ecosystem Management Task Force. Vol I. Overview. National Technical Information Service, Arlington, VA.
Johnson, K.N., et al. 1991. Alternatives for Management of Late-successional Forests of the Pacific Northwest: A Report to the U.S. House of Representatives, Committee on Agriculture, Subcommittee on Forests, Family Farms, and Energy, Committee on Merchant Marines and Fisheries, Subcommittee on Fisheries and Wildlife, Conservation, and the Environment. Oregon State University College of Forestry, Corvallis, OR.
Kates, R.W., and W.C. Clark. 1996. Environmental surprise: Expecting the unexpected. *Environment* 38(2): 6–12, 28–34.
Keiter, R. 1994. Beyond the boundary line: Constructing a law of ecosystem management. *University of Colorado Law Review* 65(2): 293–333.
Kessler, W.B., et al. 1992. New perspectives for sustainable natural resources management. *Ecological Applications* 2(3): 221–225.
Keystone Center. 1996. The Keystone National Policy Dialogue on Ecosystem Management: Final Report. Keystone Center, Keystone, CO.
Lee, K.N. 1993. *Compass and Gyroscope: Integrating Science and Politics for the Environment.* Island Press, Washington, D.C.
Leopold, A. 1949. *A Sand County Almanac.* Oxford University Press, New York.
Likens, G.E., et al. 1977. *Biogeochemistry of a Forested Ecosystem.* Springer-Verlag, New York.
Lubchenco, J., et al. 1991. The sustainable biosphere initiative: An ecological research agenda. *Ecology* 72(2): 371–412.
Maser, C., et al. 1979. Dead and Down Woody Material. In J.W. Thomas (ed.), *Wildlife Habitats in Managed Forests: The Blue Mountains of Oregon and Washington*, pp. 78–95. USDA Agric. Handbook 553. USDA Forest Service, Washington, D.C.
Meffe, G.K., and C.R. Carroll. 1994. *Principles of Conservation Biology.* Sinauer Associates, Sunderland, MA.
Meffe, G.K., and S. Viederman. 1995. Combining science and policy in conservation biology. *Renewable Resources Journal* 13(3): 15–18.
Meidinger, E.E. 1997. Organizational and Legal Challenges for Ecosystem Management. In: K.A. Kohm and J.F. Franklin (eds.) *Creating a Forestry for the 21st Century: The Science of Ecosystem Management*, pp. 361–379. Island Press, Washington, D.C.
Meine, C. 1988. *Aldo Leopold: His Life and Work.* University of Wisconsin Press, Madison, WI.
Morrissey, W. A., et al. 1994. *Ecosystem Management: Federal Agency Activities.* Congressional Research Service, Washington, D.C.
National Research Council. 1990. *Forestry Research: Mandate for Change.* National Academy Press, Washington, D.C.
Natural Resources Law Center. 1996. *The Watershed Source Book: Watershed-Based Solutions to Natural Resource Problems.* University of Colorado Natural Resource Law Center, Boulder, CO.
Newmark, W.D. 1985. Legal and biotic boundaries of Western North American national parks: A problem of congruence. *Biological Conservation* 33: 197–205.
O'Toole, R. 1988. *Reforming the Forest Service.* Island Press, Washington, D.C.
Pinchot, G.P. 1947. *Breaking New Ground.* Island Press, Washington, D.C.
Platt, J.R. 1966. Strong inference. *Science* 146: 347–353.
Primack, R.B. 1993. *Essentials of Conservation Biology.* Sinauer Associates, Sunderland, MA.
Reeves, G.H., et al. 1992. *Ethical Questions for Resource Managers.* PNW-GRT 288. USDA Forest Service Pacific Northwest Research Station, Portland, OR.
Risser, P.G. 1987. Landscape Ecology: State of the Art. In: M.G. Turner (ed.), *Landscape Heterogeneity and Disturbance*, pp. 3–14. Springer-Verlag, New York.
Risser, P.G. 1995. Ecotones: The status of the science. *BioScience* 45(5): 318–325.
Robertson, F. Dale. 1992. Ecosystem Management of the National Forests and Grasslands. Memo to Regional Foresters and Station Directors, June 4. USDA Forest Service, Washington D.C.
Rosenbaum, W.A. 1995. *Environmental Politics and Policy*, 3rd ed. Congressional Quarterly, Washington, D.C.
Salwasser, H. 1994. Ecosystem management: Can it sustain diversity and productivity. *Journal of Forestry* 92(8): 6–10.
Sample, V.A. 1994. *Remote Sensing and GIS in Ecosystem Management.* Island Press, Washington, D.C.
Sample, V.A., et al. 1995. *Building Partnerships for Ecosystem Management on Mixed Ownership Landscapes.* American Forests Forest Policy Center, Washington, D.C.
Samson, F.B., and F.L. Knopf. 1996. *Prairie Conservation.* Island Press, Washington, D.C.
Shannon, M.A., and A.R. Antypas. 1997. Open Institutions: Uncertainty and Ambiguity in 21st-Century Forestry. In: K.A. Kohm and J.F. Franklin (eds.), *Creating a Forestry for the 21st Century: The Science of Ecosystem Management*, pp. 437–445. Island Press, Washington, D.C.
Sirmon, J., et al. 1993. Communities of interests and open decisionmaking. *Journal of Forestry* 91(7): 17–21.
Slocombe, D.S. 1993. Environmental planning, ecosystem science, and ecosystem approaches for integrating environment and development. *Environmental Management* 17(3): 289–303.
Society of American Foresters. 1993. Task Force Report on Sustaining Long-term Forest Health and Productivity. Society of American Foresters, Bethesda, MD.
Soule, M.E. 1985. What is conservation biology? *BioScience* 35(11): 727–734.
Soule, M.E. 1986. *Conservation Biology: The Science of Scarcity and Diversity.* Sinauer Associates, Sunderland, MA.
Teeguarden, D.E. 1973. Forest regulation: The geographic basis. *Journal of Forestry* 72(4): 217–220.
Thornton, R.D. 1991. Searching for consensus and predictability: Habitat conservation planning under the Endangered Species Act of 1973. *Environmental Law* 21: 605–626.
U.S. Congress, Office of Technology Assessment. 1992. *Forest*

*Service Planning: Accommodating Uses, Producing Outputs, and Sustaining Ecosystems*. OTA-F-505. U.S. Government Printing Office, Washington D.C.
U.S. Department of the Interior, National Park Service. 1992. *Issues and Recommendations from the 75th Anniversary Symposium: Vail CO, October 10, 1991*. National Park Service, Washington D.C.
USDA Forest Service. 1990. *Critique of Land Management Planning* (11 vols.). USDA Forest Service Policy Analysis Staff, Washington D.C.
Vogt, K.A. 1997. *Ecosystems: Balancing Science with Management*. Springer-Verlag, New York.
Wallinger, S. 1995. A Commitment to the future: AF&PA's sustainable forestry initiative. *Journal of Forestry* 93(1): 16–19.
Walters, C.J. 1986. *Adaptive Management of Renewable Resources*. Macmillan, New York.
Weeks, W.W. 1997. *Beyond the Ark: Tools for an Ecosystem Approach to Conservation*. Island Press, Washington, D.C.
Western Governors' Association. 1992. *Pioneering New Solutions: Directing Our Destiny*. Western Governors' Association, Denver, CO.
Western Governors' Association. 1996. Future Management of the National Forests and Public Lands. Resolution 96-011, June 23. Western Governors' Association, Denver, CO.
Wodraska, J.R. 1995. The touch of the master's hand. *Hydata* (newsletter of the American Water Resources Association) 14(4): 12–13.
World Commission on Environment and Development. 1987. *Our Common Future*. Oxford University Press, New York.
Yaffee, S.L. et al. 1996. *Ecosystem Management in the United States: An Assessment of Current Experience*. Island Press, Washington, DC.

## THE AUTHORS

**Hanna J. Cortner**
*Professor, School of Renewable Natural Resources, 325 BioSciences East, University of Arizona, Tucson, AZ 85721-0043, USA*

**John C. Gordon**
*Pinchot Professor, Yale School of Forestry and Environmental Studies, Yale University, New Haven, CT 06511, USA*

**Paul G. Risser**
*President, Oregon State University, Corvallis, OR 97331, USA*

**Dennis E. Teeguarden**
*S.J. Hall Professor Emeritus of Forestry Economics, University of California Berkeley, Berkeley, CA 94803, USA*

**Jack Ward Thomas**
*Boone and Crockett Professor, School of Forestry, University of Montana, Missoula, MT 59812, USA*

# The Human Ecosystem as an Organizing Concept in Ecosystem Management

Gary E. Machlis, Jo Ellen Force, and William R. Burch, Jr.

**Key questions addressed in this chapter**

- *How important is the ecosystem concept to ecosystem management?*
- *How can the biological and social sciences be integrated to the benefit of ecosystem management?*
- *What are the key components of a human ecosystem?*
- *How can the human ecosystem as an organizing concept be applied — by scientists, resource managers, decision makers, and the public?*

**Keywords: Human ecology, human ecosystem, social institutions**

*"...the ultimate challenge for Ecology is to integrate and synthesize the ecological information available from all levels of inquiry into an understanding that is meaningful and useful to managers and decision makers."* G.E. Likens (1992)

## 1 INTRODUCTION

In the early decades of the 21st century, a major challenge will likely confront the natural resource professions. Depending upon one's source, somewhere around 2020, the globe will contain 6 to 8 billion humans (Demeny 1990, WRI 1994). There is little evidence that this human population, with its ever-increasing expectations, will experience a voluntary redistribution of resources from the well-to-do regions, classes, and persons to the poorer regions, classes, and persons. Our past hopes have been based upon technologically induced supply increases from a finite resource base that would permit some trickle down effects. In the United States, trends suggest more polarization between rich and poor, and increased struggle over resources.

Consequently, the natural resource professions will need to intensify their search for models of resource systems that include the forces driving infinite human desires, along with the more limited possibilities of satisfying those desires with increased natural resource productivity. Human variables as both the cause and consequence of system change will need to be joined to the traditional biophysical concerns of the forester, agriculturist, range manager, and park superintendent.

Since 1990, "ecosystem management" has carried the most hopes for finding some coherent and comprehensive means for systematically fitting human demand within biophysical and sociopolitical realities (Endnote 1). The organization and description of a comprehensive ecosystem management model is only just underway; the inclusion of *Homo sapiens* is unrealized. Biologists have focused on "impact" measures of humans, a strategy that puts our species outside the ecosystem as, at most, a permanent perturbation. Social scientists have largely focused on idiosyncratic "human dimensions" outside of and immune to biological reality. The traditional academic divisions have played at intellectual balkanization, seeking advances in territory rather than a more inclusive paradigm truly helpful to resource management professionals. Such a paradigm would be a new kind of life science, one that treats the biosocial reality of human beings as a serious part of its approach toward ecosystem management.

In this chapter, we describe one version of what a human ecological perspective might offer ecosystem managers. We propose the *human ecosystem* as an organizing concept for ecosystem management (Machlis et al. 1997). Because our goal is to describe the human ecosystem as a useful structure, we emphasize description of component parts (as a trophic level model might do) rather than critical processes (as a model of succession might do). First, we outline the genealogy of the human ecosystem idea; we draw upon a long tradition of intellectual risk-taking by many biologists and social scientists. Next, we present the critical elements of a human ecosystem model, followed by a detailed description of the individual variables and their relevance to ecosystem management. We conclude with specific suggestions as to how the human ecosystem model can play a useful role in managing the natural resources of the 21st century.

We make no pretense of resolving the larger issues surrounding environment/human studies. Our belief is that the most sustainable joining of biology and social science will come only when both approach the task as equals, with mutual respect for the theory and methods of the other. Our effort is a statement of ecology from a human perspective, with due consideration for biologically centered ecosystem models and social science-derived constructs. Our hope is a fusion that transcends the arcane division of the biophysical and the sociocultural — one that is truly ecological.

## 2 BACKGROUND

Studies of the patterns and processes in ecosystems emphasize the diversity and complexity of the elements affecting the systems. This was the central theme of Lindeman's breakthrough lake studies (1942) and the pioneering Hubbard Brook ecosystem research (Likens et al. 1977, Bormann and Likens 1979). Their emphasis was upon the dynamics of ecosystems in terms of flows, exchanges, and cycles of factors such as materials, nutrients, and energy. From the Hubbard Brook data, Likens et al. note that: "a vast number of variables, including biologic structure and diversity, geologic heterogeneity, climate, and season, control the flux of both water and chemicals through ecosystems." (1977:2)

More recently, Golley has also emphasized ecosystem complexity:

> "...the ecosystem consists of co-evolved suites of organisms... there are keystone species that provide special environments for many other groups. There also are social organisms, such as ants, that form yet another pattern of organization. This means that the actual organization of an ecosystem is much more complex than the network model suggests. Indeed, the organiza-

tion of a large city might be a better model than the systems models of textbooks, the links of which if very complicated look like a bowl of spaghetti." (Golley 1993: 203)

It is partly this complexity that has caused biologists to generally exclude human behavior from their models, and social scientists to remain largely at the level of metaphor (Endnote 2).

As the social sciences emerged as self-conscious disciplines in the 19th century, they struggled with the problem of how much of human behavior should be attributed to our biological nature and how much should be attributed to our social nature. Obviously, humanity shares characteristics with the animal kingdom, particularly the large non-human primates. At the same time, there is a sense of great difference. Causal priority seems to shift from one polarity to another. For some, human behavior is determined by genes or anatomy or chemistry. For others, human behavior is determined by norms, or moral values, or the mind, linguistic constructs, demography, or God's grand design. Human ecology is particularly at risk in such discussions, as it tries to account for environmental variables and biological pre-dispositions, and to merge these with social variables unique to humans such as symbolic language, elaborate normative systems, values, and meanings.

Sorokin (1928), in criticizing the application of organismic analogies to human society, captured the reality of social science attempts to include the biological domain. He noted,

> "...sociology has to be based on biology; that the principles of biology are to be taken into consideration in an interpretation of social phenomena; that human society is not entirely an artificial creation; and that it represents a kind of a living unity different from a mere sum of the isolated individuals. These principles could scarcely be questioned. They are valid." (Sorokin 1928:207)

He went on to critique organicist, bio-social, geographic determinist and demographic approaches to explaining human behavior and the patterns and processes of human society. To Sorokin, all such explanations suffered from too much dependence upon analogy and too strong a desire for single causes. Yet, each of the mainstream theories critiqued by Sorokin and those that have emerged since have had to find some rationale for attributing, incorporating, excluding, or compartmentalizing the priorities of environment, biology and human culture. Each theory must assume that the observed regularities in human social life has an explanation.

Mainstream social theories have tended to cluster around certain biophysical and environmental determinants as key. For example, the structure of a society and its processes of stability and change have long been attributed to "carrying capacity" levels as population presses against resource constraints (Sumner and Keller 1927, Durkheim 1933, Catton 1982). Or the structures and processes may be attributed to spatial differences in resource "meanings" (Park and Burgess 1921, Hawley 1950 and 1986). Or the ecological processes and environmental conditions may be considered as aspects of symbolic systems (Wirth 1928, Firey 1960). Or the variety of human organizational patterns and processes may be seen as shaped by environmental variation (Duncan 1964, Selznick 1966). Or societal patterns and processes are mediated by adaptive technologies, for which the cultural elements exhibit a poor or better ability to accommodate to the technological modifiers (Ogburn 1950, Cottrell 1955). Or the structure of political power may determine (Marx 1972) and in turn be shaped (Weber 1968, West 1982) by characteristics of natural resources.

Our point is twofold. The first is that (contrary to contemporary commentary such as Dunlap [1994]) environmental sociology is neither a recent product of sociologists nor distinct from mainstream social theory. The second is to remind the reader that either explicitly or implicitly, traditional mainstream social theory must make an accommodation to the dilemma of reconciling social biological facts in understanding our species. An ecology that includes humans is like other zoological studies in that it begins with the biological and environmental conditions of the observed species (Udry 1995), rather than a determined assertion as to how little such factors matter in explaining the observed behavioral patterns. Indeed, our goal is movement toward a unified theory of ecology (Allen and Hoekstra 1992) that can ultimately account for the ecologies of all life forms. And a critical starting point is the ecosystem concept.

## 3 THE ROOTS OF HUMAN ECOLOGY AND THE HUMAN ECOSYSTEM

The ecosystem was formally defined by Sir Arthur Tansley in 1935 (Endnote 3), and brought into common application by E.P. Odum's use of the ecosystem as an organizing concept in his 1953 text *Fundamentals of Ecology*. Several contemporary histories of the ecosystem idea have been published, notably Hagen's *An Entangled Bank* (1992) and Golley's *A History of the Ecosystem Concept in Ecology* (1993). Both limit their discussions to the rise of a biological ecology that excludes *Homo sapiens*.

The roots of a *human* ecology lie primarily in general ecology, sociology, and anthropology, as documented by comprehensive literature reviews (Micklin 1977, Field and Burch 1988) and texts (Hawley 1950 and 1986). The application of general ecological principles to human activity was sparked by sociologists at the University of Chicago in the 1920s and 30s. Sociologists Park and Burgess drew analogies between human and non-human communities, describing society's symbiotic and competitive relationships as an organic web (Faris 1967). Simultaneously, anthropologists such as Steward (1955), Bennett (1976), and others began to employ the ecosystem as a tool for organizing fieldwork and research. While the Chicago "school" treated the community (and for them that meant the city) as a key unit of analysis, the limited focus on spatial relationships and urban life eventually led to a search for a more holistic framework.

That search (active in the 1950s and 60s) led to the POET model. This model defined the human ecosystem as the interaction between *population, organization,* and *technology* in response to the *environment* (Duncan 1964, Catton 1982). These were to be human ecology's "master variables"; their interaction the human ecologist's central concern. A derivative model (Ehrlich and Ehrlich 1970, Dietz and Rosa 1994) modified the interactions to estimate environmental impacts, rather than describe human ecosystems.

In the 1980s and early 90s, anthropologists such as Moran (1990), sociologists such as Burch, his colleagues and students (Burch and DeLuca 1984), and ecologists such as H. T. Odum (1983) and E.P. Odum (1993) employed the human ecosystem as a theoretical framework. It was applied to archeological research (Butzer 1990), energy policy (Burch and DeLuca 1984), threats to national parks (Machlis and Tichnell 1985), and anthropogenic impacts upon biodiversity (Machlis 1992).

## 4 THE HUMAN ECOSYSTEM DEFINED AND DESCRIBED

In this chapter, the human ecosystem is defined as *a coherent system of biophysical and social factors capable of adaptation and sustainability over time.* For example, a rural community can be considered a human ecosystem if it exhibits boundaries, resource flows, social structures, and dynamic continuity. Human ecosystems can be described at several spatial scales and/or units of analysis, and these are hierarchically linked. Hence, a family unit, community, county, region, nation, even the planet can fruitfully be treated as a human ecosystem.

While the scale of human ecosystems can vary, there are several essential elements. Figure 1 outlines these elements in a basic model of a human ecosystem. A set of *critical resources* are required, in order to provide the system with necessary supplies. These resources are of three kinds: (1) *natural resources* (such as energy, fauna, wood or water); (2) *socioeconomic resources* (such as labor or capital); and (3) *cultural resources* (such as myths and beliefs). These resources keep the human ecosystem functioning and partially regulate human activity;

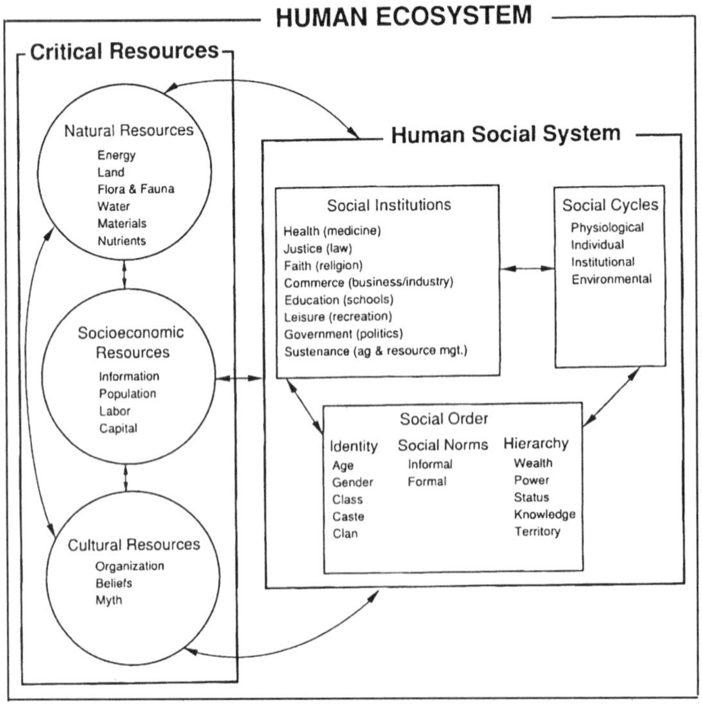


Fig. 1. Working model of the human ecosystem.

their flow and distribution are critical to sustainability. All exhibit great richness in structure, components, and characteristics. Natural resources in particular provide a diverse range of "ecosystem services" all not perceived or understood by contemporary scientific knowledge. Some critical resources may be indigenous to the local area (and used locally or exported), others may be imported from adjacent or far away locales. For example, urban sources of investment capital and national media sources of information are integral parts of rural human ecosystems, as are other distantly produced but critical supplies.

The flow and use of these critical resources is regulated by the *social system*, the set of general social structures that guide much of human behavior. The social system is composed of three subsystems. The first is a set of *social institutions*, defined as collective solutions to universal social challenges or needs. Such solutions may include organizations, rules for behavior, accepted practices and required or expected activities. For example, the collective challenge of maintaining human health leads to medical institutions, which can range from traditional shamans to modern hospital systems, rural health cooperatives, and preventive care. Other social institutions deal with universal challenges such as justice (which leads to law), faith (which leads to religion), and sustenance (which leads to agriculture and resource management).

The second subsystem is a series of *social cycles*, which are the temporal patterns for allocating human activity. Time is both a fixed resource as well as a key organizing tool for human behavior. Some cycles may be physiological (such as diurnal patterns); others institutional (permitted hunting seasons). Still others may be specific to the individual (such as grave-yard shifts) or environment (such as climate change). Social cycles significantly influence the distribution of critical resources. An example is the set of collective rhythms within a community or culture that organize its calendar, festivals, harvests, fishing seasons, business days, and so forth.

The third subsystem is the *social order*, which is a set of cultural patterns for organizing interaction among people and groups. The social order includes three key mechanisms for ordering behavior: personal *identities* (such as age or gender), *norms* (rules for behaving) and *hierarchies* (of wealth or power, for example). Hence, certain predictions about interaction are created when one can identify the age, gender, status and power of individuals or groups, and such expectations allow the social system to function.

The social order provides high predictability in much of human behavior. Taken together, social institutions, social cycles, and the social order constitute the social system. Combined with the flow of critical resources, this creates the human ecosystem. Each of these elements substantially influences the others. For example, changes in the flow of energy (such as an embargo and resultant rationing) may alter hierarchies of power (those with fuel get more) and norms for behavior (such as informal sanctions against wasting fuel).

Adaptation (not necessarily Darwinian) is continuous in human ecosystems (Bennett 1976); social institutions adapt to changes in resource flows and in turn alter such flows. The result is a perpetually dynamic system. For example, political institutions may adapt to the increased demands on forest resources by altering decision-making processes (such as increased public participation), and by altering the resource flow (as when the legal system issues injunctions against timber cutting). Adaptation is used here in a non-valued sense; what is adaptive (or advantageous) for one institution or social group may be maladaptive (or harmful) for another (Bennett 1976, 1993).

Finally, a particular human ecosystem may be hierarchically nested within human ecosystems at different scales. Hence, the rural community as a human ecosystem may be linked to a larger watershed, region, and state, and to smaller human ecosystems such as clans or households. Changes in a human ecosystem at one scale may have effects at larger and smaller scales. For example, a rise in rural unemployment may impact family health conditions, increase demands upon community doctors, and deplete state medical funds.

This human ecosystem model provides an organizing framework for ecosystem management. Each of the key components is discussed below.

## 5 KEY COMPONENTS IN THE HUMAN ECOSYSTEM

In this section, we identify and describe the key components in human ecosystems as shown earlier in Fig. 1. For each component or variable, we: (a) provide a general definition or description, (b) suggest ways the variable can be measured, and (c) give selected (and by no means comprehensive) examples of how the variable may influence other components of the human ecosystem.

### 5.1 Natural Resources

#### Energy

Energy is the ability to do work or create heat; energy flows are transfers of work and heat over space and time. Energy is a critical natural resource, and its influence upon social systems is well-documented (see for example Rosa et al. 1988). As Cottrell (1955) notes, the energy available to humans "limits what we can do, and influences what we will do." Energy flows vary by type of source (hydroelectricity, petrol, natural gas, solar, nuclear, wood, and so forth) as well as quality (high or low entropy) and flow (continuous, cyclical, or interruptible). An important element is the locus and scale of control (external or internal, local or global, multinational or household). Energy can be measured by heat output (kcal), patterns of consumption (kcal/ per household) or economic value ($/kcal). Changes in energy flows can dramatically alter social cycles and the social order (witness the disruptions caused by North American oil shortages in 1973 and 1979), and can force social institutions (such as the recreation industry or agriculture) to make significant adaptations.

## Land

Land includes terrestrial surface, subsoil, and underground features. Land is a critical resource for both its economic and cultural value (Zelinsky 1973). It can be characterized by ownership patterns (public, private or mixed), cover (vegetation or plant community types), use (such as agricultural, forestry, urban, and so forth), and economic value. Changes in land use can often be measured in hectares/land cover-land use type. Such changes often follow restricted and predictable trajectories, as forested land is altered to agricultural and then urban uses (Turner et al. 1990). Land ownership powerfully influences many social institutions (sustenance and commerce are examples), and changes in land use often are reflected in altered hierarchies of wealth, power and/or territory through shifts in land tenure and property rights.

## Water

Water includes surface, subsurface, and marine supplies. Ground water (quickly renewed) and aquifers (a form of capital stock not easily renewable) can both be integrated into human ecosystems. Water provides a wide range of ecosystem services (Postel and Carpenter 1997, Daily 1997). Water resources can be characterized by quality, flow (acre-feet/second), distribution patterns, and cyclical trends (such as wet years or drought periods). The control and distribution of water is a major source of economic, social and political power, particularly in arid regions (Reisner 1986). Changes in water quality can impact social institutions such as health and commerce; water rights are crucial to maintaining social order; access to water influences wealth by increasing property values and land productivity.

## Materials

Materials include basic products derived largely from natural resources. Examples include fertilizers (petrol as a source), dimension lumber (wood), silver and other minerals (ore), plastic (oil), glass, concrete, cocaine, and denim. The variety of materials used by human ecosystems varies by culture, stage of economic development, and consumption patterns. Common measures include economic value/unit and/or the flow of raw product (by ton, pound, ounce, or milligram). Much of the sustenance and commerce institutions are based on the production, distribution, and exchange of materials. When flows are altered, norms for use can be impacted (conservation incentives increase with price), and certain materials may be critical for specific institutions, such as precious gems for industrial use, or coca paste for the illegal drug trade (Morales 1989).

## Nutrients

Nutrients include the full range of food sources used by a human population. The range of tolerance for nutrient gain or loss is relatively small in *Homo sapiens* (Clapham 1981), making food a critical resource on a continuous basis. Such resources may vary by culture (religious proscriptions may make certain foodstuffs inedible) as well as climate, and both the caloric value and nutritional supplies (such as amino acids) are critical. Modern human ecosystems include a wide range of imported foods (witness espresso coffee beans from Brazil being brewed in Montana gas stations), and few are self-reliant even for short, seasonal periods. The need for food resources certainly influences sustenance institutions such as agriculture. Food carries mythic connotations (the spiritual value of salmon to several indigenous tribes in the North American northwest; the turkey as celebratory poultry). Hence, changes in nutrient flows can alter human health, social norms and cultural beliefs.

## Flora and Fauna

Flora and fauna are critical resources beyond their function as nutrient and material sources; a wide range of flora has ecological, sociocultural and economic value. Plants are vital sources of pharmacopoeia (Wilson 1992), myth (the cedars of Lebanon and the redwoods of California are examples) and status (the American lawn; see Bormann et al. (1993)). Fauna, including domesticated livestock, pets, feral animals, and wildlife, have significant economic value through activities as wide-ranging as hunting, birdwatching, pet keeping, and (in some cultures) the production of aphrodisiacs. Flora and fauna can be valued biologically (such as species richness, number of endemic species, population size, genetic diversity), economically (dollar value per bushel, boardfoot, pelt, head, horn, or hoof), and/or culturally (proportion of citizens interested in preserving a species). Changes in flora and fauna, such as the threat of extinction or overpopulation, can lead to changes in nutrient supplies, myth, law, sustenance (particularly wildlife management and farming practices), and social norms toward the natural world.

## 5.2 Socioeconomic Resources

### Information

Information is a necessary supply for any biophysical or social system. Information flow (and its potential for feedback) is central to general systems theory (von Bertalanffy 1968), sociobiology (Wilson 1975, 1978), and human ecology (Hawley 1950, Burch and DeLuca 1984). Information may be coded and transmitted in numerous ways: genes, "body language," oral traditions, electronic (digital data), print (local weeklies, national dailies, news magazines), film, radio, and television. It can be measured by both transmission rates (such as amount of local radio programming) and/or consumption patterns (such as newspaper circulation rates). Information flow can significantly alter numerous components of social systems, such as educational institutions or hierarchies of knowledge. Its impact upon other critical resources is also substantial (for example, the importance of maps in land management).

### Population

Population includes both the number of individuals and the number of social groups and cohorts within a social system. Population as a socioeconomic resource includes the consumption impacts of people, as well as their creative actions (accreting knowledge, engaging in sexual behavior, providing labor, and so forth). Human population growth is a dominant factor influencing much of human ecology (Hawley 1986) and social systems (Durkheim 1933), both historically (Turner et al. 1990) and within contemporary nation-states, regions and cities. Growth can be measured by natural increases (births over deaths/year) as well as migration flows. While population can act as an ecosystem stressor, it is also a supply source for many critical components within human ecosystems, such as labor, information (including genetic code), and social institutions (Geertz 1963).

### Labor

Labor has many definitions; in the human ecosystem model it is defined as the individual's capacity for work (economists sometimes label this as labor power; see Thompson 1983). Applied to raw materials and machinery, labor can create commodities, and is a critical socioeconomic resource. There are many measures: labor time needed to create a unit of economic value (hrs/$100 value), labor value (measured in real wages), labor output (units of production per worker or hour labor), or surplus labor capacity (unemployment rates) are examples. Labor is critical to human ecosystems both for its energy and information content; that is, both relatively unskilled yet physically demanding labor (such as harvesting crops) and specialized, sedentary skills (such as resource planning or stock brokering) have economic and sociocultural importance. Changes in labor (such as increased unemployment) can impact a variety of social institutions and hierarchies from health care to income distribution.

### Capital

Capital can have a range of meanings. A narrow definition treats capital as the "durable physical goods produced in the economic system to be used for the production of other goods and services" (Eckaus 1972). Other definitions include 'human capital,' financial capital and so forth (McConnell 1975). In the human ecosystem model, capital is defined as the economic instruments of production, i.e. financial resources (money or credit supply), technological tools (machinery) and resource values (such as underground oil). Hence, technology, a critical variable in the POET model, is considered a form of capital available for application in the human ecosystem. These instruments of production provide the basic materials for producing (with labor inputs) commodities. Capital is a critical socioeconomic resource; its influence over production, consumption, transformation of natural resources and creation of by-products (such as pollution) is significant. Capital is often measured in dollar values, either for commodities produced or the stock of capital on hand. Changes in capital, either in its mix of sources (a new processing plant or mill) or output (a reduction in profits earned by the plant or mill), can alter social institutions as well as hierarchies of wealth, class identities, and other features of the human social system.

## 5.3 Cultural Resources

### Organization

In the human ecosystem model, organization is defined as the capacity to create social structures. It is treated as a cultural resource, for it provides the structural flexibility needed to create and sustain human social systems. That is, our species' special ability to create numerous and complex organizational forms is a necessary skill in interacting with nature and society (Wilson 1978). It is a *cultural* resource because there is demonstrated wide variation among cultures in how these generic organizing skills are employed. For example, citizens of the United States are willing to create, continually and often, new organizations to deal with collective issues: building a water supply system (irrigation districts), managing education (school boards), caring for the poor (welfare societies) and so forth. Organization can be measured by its diversity (the range of organizational types), intensity (the number of organizations), or saturation (the percent of population that claim membership). Organization is critical to natural resource management; ecosystem management is itself an experiment in new ways of organizing the relations between human and non-human domains.

### Beliefs

Beliefs are statements about reality that are accepted by an individual by thought or conviction as true (Theodorson and Theodorson 1969, Boudon and Bourricaud 1989); citizens may have the belief that forests are being overcut, that water quality is low, that certain salmon stocks may not be endangered. (Beliefs differ from values, which are opinions about the desirability of a condition). Beliefs arise from many sources: personal observation, mass media, tradition, ideologies, testimony of others, faith, logic and science. Beliefs (stated or unstated) are crucial to human ecosystem functioning, for they supply a set of "social facts" (Durkheim 1938) that individuals, social groups and organizations use in interacting with the world. Hence, environmental interest groups and industry associations rely on a public set of beliefs concerning environmental crises (which may or may not be factual) to energize and increase their membership. Beliefs can be measured by their ideological content (liberal or conservative), their intensity (the proportion of a population to feel strongly about a belief), and their public acceptance (the proportion of a population that share a similar belief). As beliefs change, social institutions are often forced to respond. For example, the changing public beliefs concerning the safety of nuclear power has led to a decline in nuclear power production in the U.S. (Dunlap et al. 1993).

### Myths

To the human ecologist, myths are narrative accounts of the sacred and symbolic in a society; they legitimate social arrangements (Malinowski 1948), explain collective experiences (Burch 1971), and contribute to worldview. Hence, myths are an important supply variable because they provide reasons and purposes for human action. Myths are critical to human ecosystems as guides to appropriate and predictable behavior (witness Smokey Bear's admonitions about fire); they give meaning to and rationale for a wide range of social institutions and social ordering mechanisms. For example, the myth of "manifest destiny" provided U.S. citizens at the turn of the century with a rationale for the permanent and private development of the American west; indigenous tribal groups simultaneously called on traditional myths to legitimate their role as temporary stewards of communal land (Worster 1992). Myths operate at various scales: national myths (such as the manifest destiny), community myths (a timber town's story of how and why it was founded), and clan myths (a family's story of its early matriarchs). Myths are difficult but not impossible to measure: festivals, symbols, legends are all indicators of myth supply. A change in myth (such as reduced perception of community self-reliance) can impact social institutions (such as faith) and a variety of social norms, as well as resource use (such as wilderness).

## 5.4 Social Institutions

### Health (medicine)

The health care institution encompasses the full range of organizations and activities that deal with the health needs of a human ecosystem. Health care in modern industrial societies is relatively complex, including primary care (personal and family health maintenance, outpatient activities by general practitioners), secondary care (such as services of specialists) and tertiary care (such as hospital procedures involving surgery [Rodwin 1984]). Informal systems of care provide alternatives: those that cannot afford health insurance may nevertheless access health care through such means (such as midwives). Health care institutions are often measured by capacity (the number of doctors or hospitals per 1000 population) or outcomes (such as infant mortality rates). In rural communities, primary care is often available locally; secondary and tertiary care are often provided on a regional basis. Hence, relatively small changes in the health institution (a doctor's retirement, the closing of a pharmacy) may have direct and indirect effects that ripple through the social system.

### Justice (law)

The collective problem of justice faces all human social systems; its role in human ecosystems is critical. Two forms are central: distributive justice (who should get what, such as property rights [Rawls, 1971]) and corrective justice (how should formal norms be enforced, such as rules for punishment [Runciman 1966]). The legal system can be measured by both its practitioners (such as the number of lawyers or judges/1000 population) and its performance (number of trials or convictions). The contemporary legal system plays an important role in ecosystem management — the courts influencing distributive justice through timber sale appeals and injunctions, and meting punishment for resource crimes (such as poaching). Changes in legal institutions, such as new procedures for appeal or new laws (the revision of the Endangered Species Act is an example) can dramatically and directly impact the use of natural resources, the development of capital, and other components of the human ecosystem.

## Faith (religion)

To the human ecologist, religion as an institution has two components: (1) a system of organizations and rituals that bind people together into social groups (Durkheim 1938), and (2) a coherent system of beliefs and myths (Weber 1930). Both are critical to human ecosystem functioning. Religion, like other social institutions, can be measured by diversity (range of religious practices), capacity (number of churches) or participation (percent population claiming membership). Religion impacts the social system in many ways, altering social cycles (religious holidays), providing identity for both caste and clan, influencing beliefs and myths and influencing (indeed, often prescribing) behavior. A change in faith (such as increased worship after a natural disaster) can have significant bearing on how effectively social systems adapt to new ecological and socioeconomic conditions.

## Commerce (business/industry)

All societies require a system for exchanging goods and services, and the institution of commerce is central to this exchange (Durkheim 1933). Commerce includes not only the exchange medium but the organizations that manage exchange, such as banks, markets, warehouses, retail outlets and so forth. Modern industrialized societies (including their rural regions) rely on a mix of exchange styles; the typical U.S. rural community usually conducts its commerce through a mix of cash, credit, and barter (Machlis and Burch 1983). The informal economy (I'll fix your car if you'll fill my garden) may expand with stresses in the formal economy. Commerce can be measured as capacity (such as the percent of production capacity utilized, the number of banks) and/or as a flow (the number of transactions or the dollar value of a gross regional or local product). Commerce in rural areas, particularly in the west, is largely dependent upon local natural resources (be it water, energy, timber, scenery or other values [West 1982]); a change in commerce can create a cascading set of impacts upon other social institutions (such as sustenance), the social order (shifts in wealth or power), social cycles (as in a recession), and on critical resources (such as land or labor).

## Education (schools)

Individual *Homo sapiens* are born into the world sorely lacking in the knowledge needed to survive, adapt, and interact with others. Hence, education (the transmission of knowledge) is a ubiquitous collective challenge: we must educate our young. While significant learning takes place in the home and on the streets, the educational institution largely functions through the school system, including public and private schools, teachers, school boards, and parent organizations (Bidwell and Friedkin 1988). Education can be measured as a density (teacher/student ratios), input (dollars expended/student), and an output (percent of high school seniors graduating). Changes in the educational system directly impact other components of the social system (such as the timing of leisure activities, the distribution of knowledge, the availability of skilled labor). Dramatic changes in the institution (such as rural school consolidation) can have significant effects on the entire human ecosystem.

## Leisure (recreation)

Leisure (the culturally influenced ways we use our non-work time) is an important institution in all but the harshest human ecosystems (Cheek and Burch 1976). Several studies suggest industrialized societies have *less* leisure time per capita than agricultural or pastoral ones (Burch and DeLuca 1984, Schor 1992). In industrialized societies, the recreation institution includes formally managed leisure opportunities (bowling alleys, wilderness areas, movie-going, hunting and fishing) as well as less formal pursuits (socializing, sex, courtship, resting) and specialized activities (holidays, festivals, and so forth). Leisure can be measured as an amount (hours per day per capita), as a level of participation (percent of adults with hunting permits), or as a range (number of festivals or special events). Changes in leisure can impact human ecosystems in several ways, through direct impacts upon commerce (a boom or bust in the tourist industry), by changing social norms (a decline in festival attendance or a change in gender participation), and by altering physical systems (such as human impacts upon wilderness areas).

### Government (politics)

The political subsystem is at once a central component of human ecosystems and a result of other components (such as organization, myths, legal institutions, and so forth [Shell 1969]). Politics as an institution is a collective solution to the need for decision-making at scales larger than clan or caste. It includes the modes of interaction between political units (such as states and counties), the processes of decision-making within political units (such as elections and legislative action), and the participation of citizens in political action (campaigns, party activity, referendum, and so forth). Government can be measured by its resources (tax receipts, authorized expenditures, and employees are examples) and/or its actions (laws passed, hearings held, and so forth). As governments at several scales control critical natural resources (such as the federal government's forestland), changes in government action or process (such as revision to the Endangered Species Act) can have a significant influence upon human ecosystems.

### Sustenance (agriculture and resource management)

The provision of sustenance (food, potable water, energy, shelter, and other critical resources) is a central and collective challenge facing all social systems (Hawley 1950). The management of that challenge and the production of necessary supplies requires agricultural and resource management institutions of some complexity (Field and Burch 1988). Irrigation districts, farmer's cooperatives, timber companies, tree farm associations, extension offices, federal management agencies and environmentally oriented interest groups are all components of the sustenance institution. Measures include organizational capacity (number of agents/ farm, acres in production), output (measured in dollar values or crop tonnage), and range of sustenance products (number of crops or timber types). As agriculture and resource management are the chief methods for transforming critical resources into necessary social system supplies, their importance to human ecosystem functioning is key. Changes in production, efficiency, or distribution can generate effects throughout human ecosystems.

## 5.5 Social Cycles

### Physiological cycles

*Homo sapiens* has evolved a series of physiological cycles that deeply influence human behavior. For example, diurnal cycles of night and day create peaks of labor and rest; menstrual cycles control reproduction patterns. The life cycle is roughly similar across cultures: birth, childhood, labor, marriage, child-rearing, retirement from active labor, and death. Each stage of the life cycle creates expectations and norms for behavior (including the use of resources [Burch and DeLuca 1984]). Measurement can include the proportion of the population at each stage of the life cycle. These cycles create predictable patterns of activity within the human ecosystem: park-going during daylight hours, increases in energy demands during early morning hours (for showering, cooking, heating and so forth), rituals at each juncture of life cycle stages (such as weddings and funerals). While physiological cycles may rarely change, they substantively impact human ecosystem functioning at several scales.

### Individual cycles

Beyond physiological cycles, individuals may follow time cycles that are personal and idiosyncratic. Examples are graveyard shifts for certain workers (such as bakers or police), part-time or seasonal work (such as agricultural field labor or lumbering), and personal patterns of recreation activity (weekend hiking or camping). These cycles impact social institutions (such as leisure) and the use of natural resources. They can be measured by such indicators as employment patterns (for example, the proportion of part-time to full time workers). Changes in individual cycles can reflect alterations in labor needs, social institutions or hierarchies of wealth. For example, displaced mill workers may have to travel farther from home for employment, changing family time and budgets.

### Institutional cycles

Each of the social institutions described above have (or create) social cycles that control the flow of relevant activities (Burch and DeLuca 1984). The legal institution, for example, creates court seasons and trial days; the recreation and sustenance institutions create hunting and fishing seasons. These institutional cycles are critical to human ecosystem functioning, for they provide guidance and predictability to the ebb and flow of human action. Institutional cycles can be measured in terms of frequency (the number of times persons or groups participate), duration (such as the length of a hunting season), proportion (the percentage of the population involved), or intensity (the depth of the meaning assigned to the cycle, such as the funeral of a national leader). Changes in institutional cycles may directly impact the use of natural resources (for example, a year-round school calendar diversifying park-going patterns), and, importantly, the conduct of commerce (such as fishing seasons, field-burning periods, or fiscal year cycles of funding).

### Environmental Cycles

Not all cycles are socially constructed: environmental cycles are natural patterns (though influenced by anthropogenic factors) that can significantly influence the human ecosystem (Bormann and Likens 1979, Turner et al. 1990). Environmental cycles include seasons, drought periods, El Niño patterns, biogeochemical cycles, short-term successional stages, and long-term climatological change. Drought cycles in the western U.S., for example, impact natural resources such as wildlife and forests, the capital needs for dams, reservoirs, and other storage devices, agricultural institutions, litigation over water rights, and many other components of the human ecosystem. The cycles can be measured by duration (such as length of growing season) or occurrence (the proportion of years in a decade with low precipitation). Changes in environmental cycles, such as the end of a drought or the movement of the seasons, can alter ecosystem and social system responses, often significantly.

## 5.6 Social Order

### Identity

One of the key ways that social systems maintain coherence and the ability to function is through the use of identity. In sociological terms, identity is often ascriptive — it is assigned by society based on birth or circumstances rather than through the individual's actions or achievements. Caste or race, for example, is ascriptive: one is born into a racial category which then follows and constrains the individual throughout the life course. These identities are used (often through stereotyping or other generalizations) to differentiate people and manage interactions: African-Americans claim affinity to one another (by the ascription of race), Chinese make similar claims to each other, both groups identify differences between them, and so forth. Other identities are less ascriptive, such as class: individuals can alter their class through changes in wealth, education, occupation, and so forth.

Several forms of identity are critical to human ecosystems. *Age* is important, for much of human activity is age-dependent (Eisenstadt 1956): certain occupations (such as mining) are mainly for the young; certain recreation activities (such as white-water sports) are likewise often specialized by age. *Gender* (the socially constructed masculine and feminine roles) is important, both for its crucial impact on social norms and for its differential effects upon social institutions — women and men having different access to capital, health care, wealth, power, and other features of the social systems (Weitz 1977). *Class* is important, though its definition is problematic (Abercrombie et al. 1988). Some social scientists define class in purely economic terms (based on occupation or income); others include sociocultural concerns (such as education or social norms). *Caste* (an anthropological term for race/ethnic groupings) is significant for reasons described above. Finally, *clan* (the extended family or tribal group) is crucial, both as a predictor of interaction (most recreation, for example, takes place with family members) and as a source of support. Clans routinely provide health care, financial assistance, even natural resources (such as food or other supplies) to members in need.

These identities can be measured in terms of diversity (the range of ethnic or age groups in a community) and/or distribution (the proportion of non-Caucasians within a population, the ratio of working-age individuals to dependents). Changes in identity usually impact social systems through an alteration in social norms; an influx of young people, Jews, women, and blue-collar workers leads to shifts in what is expected as well as what people do; these shifts further alter the human ecosystem.

### Social Norms

Norms are rules for behavior, what Abercrombie et al. (1988) call the "guidelines for social action." *Informal norms* are administered through community or social group disapproval: deviating from the norm is noticed but sanctions are modest. Speaking too loud in a museum or too soft at a football game are examples (as are norms for behavior in campgrounds, along trails, or on fishing boats). The full range of etiquettes for eating, socializing, courtship, and so forth are also informal norms. *Formal norms* are more serious and institutionalized; formal norms are usually codified in laws that not only prohibit certain actions but also prescribe punishments for breaking such norms (Wrong 1994). Misdemeanor and felony laws are examples. Sometimes a community's informal norms may conflict with its formal (legal) norms. The results are "folk crimes," i.e., activities that are against the law but not considered harmful by the population. Some kinds of wildlife poaching or illegal woodcutting are folk crimes (Scialfa 1992).

Norms can be measured by both their adherence (the proportion of a population following a social convention, such as marriage before childbirth) and/or deviance (the number of felonies per capita). Changes in social norms can impact social institutions (divorce directly impacts health and justice for women), and alter resource use.

### Hierarchy

An important mechanism for social differentiation and for managing the social order, is hierarchy. In almost all social systems, hierarchy is ubiquitous; inequality of access is a consistent fact across communities, regions, nations, and civilizations. Five sociocultural hierarchies seem critical to ecosystem functioning: wealth, power, status, knowledge, and territory.

*Wealth* is access to and control of material resources in the form of natural resources, capital (money), and credit. The distribution of wealth is a central feature of social inequality and has human ecosystem impacts; the rich have more life opportunities than the poor. *Power* is the ability to alter others' behavior, either by coercion or deference (Wrong 1988, Mann 1984). The powerful (often elites with political or economic power) can have access to resources denied the powerless; an example is politicians that make land-use decisions and personally profit from these decisions at the expense of other citizens. *Status* is access to honor and prestige (Lenski 1984, Goode 1978); it is the relative position of an individual (or group) on an informal hierarchy of social worth. Cultures may vary as to whom is granted high status (e.g. teachers are given high status in China, modest status in the U.S.); those with high status can often influence natural resource management through persuasion. Status is distributed unequally, even within small communities, and high-status individuals (such as ministers) may not necessarily have access to wealth or power.

*Knowledge* is access to specialized information (technical, scientific, religious, and so forth); not all individuals within a social system have such access. Knowledge provides advantages in terms of access to critical resources and the services of social institutions. Finally, *territory* is access to property rights (such as land tenure and water rights); individuals or groups with territory control such rights over a given space. Hierarchies of territory are created when some have strong land tenure (large tracts with secure ownership) and others weak tenure or are landless. This can vary by region. For example, in the U.S. arid west, water rights (granted by historical priority) may be especially crucial as access to water limits development (Reisner 1986).

These critical hierarchies can be measured in several ways. Wealth can be measured by indicators such as the range of incomes or the proportion of the population that is below the poverty line. The distribution of power can be indirectly measured by certain decision-making activities, such as elections. It can also be measured by levels of domination and subordination — the disproportion of blacks and latinos in prison or on death row, "glass ceilings" faced by women workers, the persistence of spouse abuse, and the relationship between timber workers and company executives. Status can be measured by public polling techniques that capture public opinion; knowledge can be indicated by educational attainment. Territory can be measured by ownership patterns, the distribution of land by size (i.e. the proportion of landholders with large tracts), or the distribution of water rights (by acre/feet). Changes in hierarchies, by altering who has access to critical resources and social institutions, can dramatically alter the human ecosystem.

## 6 POTENTIAL APPLICATIONS OF THE HUMAN ECOSYSTEM MODEL

This human ecosystem model is, we hope, neither an oversimplification nor caricature of the complexity that undergirds human ecosystems occurring in the world. Parts of the model are orthodox to specific disciplines: there is little new in attributing importance to energy or capital as necessary resources. Other portions of the model are less commonplace to resource managers (though still not original) — myth as a cultural resource, justice as a critical institution, and others. We believe the model is a reasonably coherent whole, and a useful organizing concept for ecosystem management. There are several potential applications.

First, the model could be employed as an organizing framework for social impact assessments (SIAs) associated with ecosystem management plans. Such plans will be broader and more multi-scaled than the traditional development projects that have been subject to SIAs, and the model may guide resource managers and their social science partners in capturing a full range of possible impacts. For example, changes in land use (such as a shift from timbering to recreation) may impact a full range of social institutions in ways that ecosystem managers and citizens need to anticipate.

Second, the model could serve as a guide for the development of social indicators for ecosystem management. Social indicators have a long tradition in the social sciences and in social policy decisions; their use in natural resource management is at present experimental. Yet there is both precedent and potential in constructing a set of social indicators for human ecosystems. Resource managers already employ biophysical indicators of stream quality, tree growth, soil erosion, and so forth. They use these indicators to guide decision-making and to monitor the effects of on-the-ground actions — and when done in a systematic way, define the result as "adaptive management."

In a recent paper (Force and Machlis 1997), we describe how this human ecosystem model was used to select a set of social indicators for monitoring ecosystems management in the Upper Columbia River Basin. County level data were collected on 39 social indicators derived from variables in the model. The data came from several accessible sources (such as the U.S. Census), and were displayed as a map series in *An Atlas of Social Indicators for the Upper Columbia River Basin* (Machlis et al. 1995).

The results can be used to compare across counties within the ecoregion, search for unique or generalized conditions, monitor change over time, and evaluate human ecosystem responses to resources management decisions and actions throughout the Basin. Similar efforts have utility for other critical regions (adjusting the selected indicators as appropriate), from South Florida to Chesapeake Bay to the Mississippi Delta to Puget Sound.

Third, the human ecosystem model can serve as a basis for monitoring other programs directly tied into the activities of natural resource agencies. By collecting and learning from data related to the model's variables, management alternatives that meet local needs for sustenance and long-term requirements for sustainability may be devised. For example, the emphasized role of social institutions in ecosystem order (from health to business to faith) suggests that the inclusion of local leaders beyond the typical political and special interest representatives may have significant benefits for public planning. Human ecosystems with weak or sound institutions may respond very differently to the manager's plan for altered timber harvests, special management zones, wilderness areas, and other forms of ecosystem manipulations; predicting such variation is an important ecosystem management skill.

Fourth, the model could serve as an introduction to the human ecological sciences for current and future ecosystem managers. Current resource managers, often trained in the postwar disciplines of wildlife, recreation or forest management, must struggle to overcome their professional concentrations and "trained incapacities." Future ecosystem managers now in the professional schools are being told that there is a new paradigm for resource management being developed. Yet they are being shown that the traditional faculties and departments are not suitable for mastering and synthesizing the broad range of technical and sociopolitical skills needed to enact this new paradigm. The human ecosystem model could function as a basic teaching tool — its description, analysis, application, and critique providing a bridge between the courses, departments, and faculties involved in ecosystem management education.

Fifth, the human ecosystem concept offers an intellectual crossroads for social scientists working on issues related to ecosystem management. Because it is derived from numerous disciplines and explicitly multi-scaled, there is opportunity for economists, anthropologists, geographers, political scientists, sociologists, and others to link their work and findings and contribute to the model's overall improvement.

Even further, the human ecosystem as an organizing concept is an invitation to cooperation with biophysical scientists, for many of the model's critical variables function in ways being discovered and described by landscape ecologists, botanists, hydrologists, and others. Frank Golley notes:

> "It is not clear to me where ecology ends and the study of the ethics of nature begins, nor is it clear to me where biological ecology ends and human ecology begins. These divisions become less and less useful. Clearly, the ecosystem, for some at least, has provided a basis for moving beyond strictly scientific questions to deeper questions of how humans should live with each other and the environment. In that sense, the ecosystem concept continues to grow and develop as it serves a larger purpose." (Golley 1993: 205)

From our own experience, we suspect such efforts at interdisciplinary, mutual learning to be simultaneously difficult and exhilarating.

## 7 CONCLUSION

The human ecosystem has great potential as an organizing concept for ecosystem management. Our model of the human ecosystem, and our selection of variables and the importance we place on them, are, of course, preliminary. The model must be tested, applied, revised — that is, it must go through the same "adaptive management" cycles required of ecosystem management techniques being applied to the nation's forests, grasslands, parks, and preserves.

But, more broadly, the human ecosystem is a necessary building block in a true life science — one that attempts to grasp the full complexity of the earth's dominant species. Such a life science is difficult to distinguish at this time. It is doubtful that it will spring from the determinist arguments currently joined in academe, from genes to gender. It is also doubtful that it will be discovered in the specialized researches of the wildlife ecologist or the zoologist. More likely, it will evolve, in advances and retreats, in response to the great necessity of our species to come to an accommodation with our powers, desires, weaknesses, and limits. For the fate of human ecosystems is our own.

## ENDNOTES

1. There are numerous definitions of ecosystem management, as well as vigorous debate (see for example the August 1994 issue of the *Journal of Forestry*). Moote et al. (1994), synthesizing the literature, provides a serviceable if generalized working definition:

   "Ecosystem management is a management philosophy which focuses on desired states, rather than system outputs, and which recognizes the need to protect or restore critical ecological components, functions, and structures in order to sustain resources in perpetuity." (Moote et al. 1994: 1)

   Definitions describe five principles central to ecosystem management: (1) socially defined goals and management objectives; (2) integrated holistic science; (3) broad spatial and temporal scales; (4) adaptable institutions; and (5) collaborative decision-making. The actual practice of ecosystem management may be considerably less inclusive.

2. There is some irony in the historical evidence that early ecologists freely borrowed from the social sciences to construct key concepts. H.F. Cowles described "plant societies." A.G. Tansley borrowed from Herbert Spencer to create his "organism-complex"; he in fact left ecology to study psychology with Sigmund Freud. A. Kerner reasoned from human communities to "plant-species communities"; Forbes to "communities of interest" between predator and prey. F.E. Clements was influenced by Spencer and sociologist Lester Ward. Both H.T. and E.P. Odum were influenced by their father, H.W. Odum, whose sociological study of the American South (1936) was prescient in human ecology. For a review, see Golley (1993) and Hagen (1992).

3. Tansley's definition was exceptionally holistic and hierarchical:

   "But the more fundamental conception is, as it seems to me, the whole *system* (in the sense of physics), including not only the organism-complex, but also the whole complex of physical factors forming what we call the environment of the biome — the habitat factors in the widest sense.

   It is the systems so formed which, from the point of view of the ecologist, are the basic units of nature on the face of the earth.

   These *ecosystems*, as we may call them, are of the most various kinds and sizes. They form one category of the multitudinous physical systems of the universe, which range from the universe as a whole down to the atom." (Tansley 1935: 299)

   Odum's definition is similar, though it includes the proscription of human needs:

   "any entity or natural unit that includes living and nonliving parts interacting to produce a stable system in which the exchange of materials between the living and nonliving parts follows circular paths is an ecological system or ecosystem. The ecosystem is the largest functional unit in ecology, since it includes both organismal (biotic communities) and abiotic environment, each influencing the properties of the other and both necessary for maintenance of life as we have it on the earth. A lake is an example of an ecosystem." (1953: 9)

## REFERENCES


Abercrombie, N., S. Hill, and B.S. Turner. 1988. *The Penguin Dictionary of Sociology*. 2nd ed. Penguin Books, New York.

Allen, T.F.H., and T.W. Hoekstra. 1992. *Toward a Unified Ecology*. Columbia University Press, New York.

Bennett, J.W. 1993. *Human Ecology as Human Behavior*. Transaction Publishers, New Brunswick, NJ.

Bennett, J.W. 1976. *The Ecological Transition: Cultural Anthropology and Human Adaptation*. Pergamon Press, New York.

Bidwell, C.E., and N.E. Friedkin. 1988. The sociology of education. In: N.J. Smelser (ed.), *Handbook of Sociology*, pp. 449–471. Sage, Newbury Park, CA.

Bormann, F.H., D. Balmori, and G.T. Geballe. 1993. *Redesigning the American Lawn*. Yale University Press, New Haven, CT.

Bormann, F.H., and G. Likens. 1979. *Pattern and Processes in a Forested Ecosystem*. Springer-Verlag, New York.

Boudon, R., and F. Bourricaud. 1989. *A Critical Dictionary of Sociology*. University of Chicago Press, Chicago.

Burch, W.R., Jr. 1971. *Daydreams and Nightmares: A Sociological Essay on the American Environment*. Harper and Row, New York.

Burch, W. R., Jr., and D.R. DeLuca. 1984. *Measuring the Social Impact of Natural Resource Policies*. University of New Mexico Press, Albuquerque, NM.

Butzer, K.W. 1990. A human ecosystem framework for archaeology. In: E.F. Moran (ed.), *The Ecosystem Approach in Anthropology: From Concept to Practice*, pp. 91–30. The University of Michigan Press, Ann Arbor, MI.

Catton, W.R., Jr. 1982. *Overshoot: The Ecological Basis of Revolutionary Change*. University of Illinois Press, Urbana, IL.
Cheek, N.H., Jr., and W.R. Burch, Jr. 1976. *The Social Organization of Leisure in Human Society*. Harper and Row, New York.
Clapham, W.B., Jr. 1981. *Human Ecosystems*. Macmillan, New York.
Cottrell, W.F. 1955. *Energy and Society: The Relation Between Energy, Social Change, and Economic Development*. Greenwood Press, Westport, CT.
Daily, Gretchen C. 1997. *Nature's Services: Societal Dependence on Natural Ecosystems*. Island Press, Washington, DC.
Demeny, P. 1990. Population. In: B.L. Turner II et al. (eds.), *The earth as transformed by human action*, pp. 41–54. Cambridge University Press, Cambridge, MA.
Dietz, T., and E.A. Rosa. 1994. Rethinking the environmental impacts of population, affluence and technology. *Human Ecology Review* 1: 277–300.
Duncan, O.D. 1964. Social organization and the ecosystem. In: F. Robert (ed.), *Handbook of Modern Sociology*, pp. 36–82. Rand McNally, New York.
Dunlap, R.E., and W.R. Catton, Jr. 1994. Struggling with human exemptionalism: The rise, decline, and revitalization of environmental sociology. *The American Sociologist* 25: 5–30.
Dunlap, R.E., M.E. Kraft, and E.A. Rosa (eds.). 1993. *Public Reactions to Nuclear Waste: Citizen's Views of Repository Siting*. Duke University Press, Durham, NC.
Durkheim, E. 1938. *The Rules of Sociological Method*, 8th ed. The Free Press, New York.
Durkheim, E. 1933. *The Division of Labor in Society*. The Free Press, Glencoe, IL.
Eckaus, R.S. 1972. *Basic Economics*. Little, Brown and Company, Boston.
Ehrlich, P.R., and A.H. Ehrlich. 1970. *Population, Resources, Environment: Issues in Human Ecology*. W.H. Freeman, San Francisco.
Eisenstadt, S.N. 1956. *From Generation to Generation*. The Free Press, Glencoe, IL.
Faris, R.E.L. 1967. *Chicago Sociology 1920–1932*. University of Chicago Press, Chicago.
Field, D.R., and W.R. Burch, Jr. 1988. *Rural Sociology and the Environment*. Social Ecology Press, Middleton, WI.
Firey, W. 1960. *Man, Mind and Land: A Theory of Resource Use*. The Free Press, Glencoe, IL.
Force, J.E., and G.E. Machlis. 1997. The human ecosystem, Part II: Social indicators for ecosystem management. *Society & Natural Resources*, 10: 369–382.
Geertz, C. 1963. *Agricultural Involution*. University of California Press, Berkeley, CA.
Golley, F.B. 1993. *A History of the Ecosystem Concept in Ecology*. Yale University Press, New Haven, CT.
Goode, W.J. 1978. *The Celebration of Heroes: Prestige as a Social Control System*. University of California Press, Berkeley, CA.
Hagen, J. 1992. *An Entangled Bank: The Origins of Ecosystem Ecology*. Rutgers University Press, New Brunswick, NJ.
Hawley, A.H. 1950. *Human Ecology: A Theory of Community Structure*. The Ronald Press, New York.
Hawley, A.H. 1986. *Human Ecology: A Theoretical Essay*. University of Chicago Press, Chicago.
Lenski, G.E. 1984. *Power and Privilege: A Theory of Social Stratification*. University of North Carolina Press, Chapel Hill, NC.
Likens, G.E. 1992. The ecosystem approach: Its use and abuse. In: O. Kinne (ed.), *Excellence in Ecology*. Ecology Institute, Oldendorf/Luhe, Germany.
Likens, G.E., F.H. Bormann, R.S. Pierce, J.S. Eaton, and N.M. Johnson. 1977. *Biogecohemistry of a Forested Ecosystem*. Springer-Verlag, New York.
Lindeman, R.L. 1942. The trophic-dynamic aspect of ecology. *Ecology* 23: 399–418.
Machlis, G.E. 1992. The contribution of sociology to biodiversity research and management. *Biological Conservation* 61: 161–170.
Machlis, G.E., and W.R. Burch, Jr. 1983. Relations between strangers: Cycles of structure and meaning in tourist systems. *Sociological Review* 31: 666–692.
Machlis, G.E., J.E. Forces, and W.R. Burch, Jr. 1997. The Human Ecosystem, part I: The Human Ecosystem as an organizing concept in ecosystem management. *Society and Natural Resources* 10: 347–367.
Machlis, G.E., J.E. Force, and J.E. McKendry. 1995. *An Atlas of Social Indicators for the Upper Columbia River Basin*. Idaho Forest, Wildlife and Range Experiment Station, Moscow, ID. Contribution Number 759, University of Idaho.
Machlis, G.E., and D.L. Tichnell. 1985. *The State of the World's Parks: An International Assessment of Resource Management, Policy, and Research*. Westview Press, Boulder, CO.
Malinowski, B. 1948. *Magic, Science and Religion and Other Essays*. The Free Press, Glencoe, IL.
Mann, M. 1984. *The Sources of Social Power: Volume 1, A History of Power from the Beginning to A.D. 1760*. Cambridge University Press, New York.
Marx, K.G. 1972. *Theories of Surplus Value*. Vol. 3. Lawrence and Wishart, London.
McConnell, C.R. 1975. *Economics*, 6th ed. McGraw-Hill, New York.
Micklin, M. 1977. The ecological perspective in the social sciences: A comparative overview. Paper presented at Conference on Human Ecology, October, Seattle, WA.
Moote, M.A., S. Burke, H.J. Cortner, and M.G. Wallace. 1994. *Principles of Ecosystem Management*. Water Resources Research Center, College of Agriculture, The University of Arizona.
Morales, E. 1989. *Cocaine: White Gold Rush in Peru*. The University of Arizona Press, Tucson, AZ.
Moran, E.F. 1990. Ecosystem ecology in biology and anthropology: A critical assessment. In: E.F. Moran (ed.), *The Ecosystem Approach in anthropology: From concept to practice*, pp. 3–40. The University of Michigan Press, Ann Arbor, MI.
Odum, E.P. 1953. *Fundamentals of Ecology*. Saunders, Philadelphia.
Odum, E.P. 1993. *Ecology and Our Endangered Life-support Systems*. Sinauer, Sunderland, MA.
Odum, H.T. 1983. *Systems Ecology: An Introduction*. John Wiley & Sons, New York.
Odum, H.W. 1936. *Southern Regions of the United States*. University of North Carolina Press, Chapel Hill, NC.
Ogburn, W.F. 1950. *Social Change*. Viking Press, New York.
Park, R.E., and E.W. Burgess, 1921. *Introduction to the Science of Sociology*. University of Chicago Press, Chicago.
Postel, S. and S. Carpenter. 1997. Freshwater Ecosystem Services. In: G.C. Daily (ed.), *Nature's Services: Societal Depend-*

*ence on Natural Ecosystems*, pp. 195–214. Island Press, Washington, DC.

Rawls, J. 1971. *A Theory of Hustice*. The Belknap Press of Harvard University Press, Cambridge, MA.

Reisner, M. 1986. *Cadillac Desert: The American West and Its Disappearing Water*. Penguin Books, New York.

Rodwin, V.G. 1984. *The Health Planning Predicament: France, Quebec, England and the United States*. University of California Press, Berkeley, CA.

Rosa, E.A., G.E. Machlis, and K.M. Keating. 1988. Energy and society. *Annual Review of Sociology* 14: 149–172.

Runciman, W.G. 1966. *Relative Deprivation and Social Justice*. Routledge & Kegan Paul, London.

Schor, J.B. 1992. *The Overworked American: The Unexpected Decline of Leisure*. Basic Books, New York.

Scialfa, M. 1992. An ethnographic analysis of poachers and poaching in northern Idaho and eastern Washington. Master's thesis, University of Idaho, Moscow, ID.

Selznick, P. 1966. *TVA and the Grass Roots*. Harper & Row, New York.

Shell, K.L. 1969. *The Democratic Political Process*. Blaisdell, Waltham, MA.

Sorokin, P.A. 1928. *Contemporary Sociological Theories Through the First Quarter of the Twentieth Century*. Harper & Row, New York.

Steward, J.H. 1955. The concept and method of cultural ecology. In: J. Steward (ed.), *The Theory of Culture Change*, pp. 30–42. University of Illinois Press, Urbana, IL.

Sumner, W.G. and A.G. Keller. 1927. *The Science of Society*. Yale University Press, New Haven, CT.

Tansley, A.G. 1935. The use and abuse of vegetational concepts and terms. *Ecology* 16: 284–307.

Theodorson, G.A., and A.G. Theodorson. 1969. *Modern Dictionary of Sociology*. Thomas Y. Crowell, New York.

Thompson, P. 1983. *The Nature of Work: An Introduction to Debates on the Labour Process*. Macmillan, London.

Turner, B.L., II (ed.), W.C. Clark, R.W. Kates, J.F. Richards, J.T. Mathews, and W.B. Meyer. 1990. *The Earth as Transformed By Human Action: Global and Regional Changes in the Biosphere Over the Past 300 Years*. Cambridge University Press with Clark University, Cambridge.

Udry, J.R. 1995. Sociology and biology: What biology do sociologists need to know? *Social Forces* 73: 1267–1278.

von Bertalanffy, L. 1968. *General Systems Theory*. Braziller, New York.

Weber, M. 1930. *The Protestant Ethic and the Spirit of Capitalism*. Allen & Uniwin, London.

Weber, M. 1968. In: G. Roth and C. Wittich (eds.), trans. G. Roth, E. Fischoff, H. Gerth, A.M. Henderson, F. Kolegar, C.W. Mills, T. Parsons, M. Rheinstein, E. Shils and C. Wittich, *Economy and Society: An Outline of Interpretative Sociology*, Bedminster Press, New York.

Weitz, S. 1977. *Sex Roles: Biological, Psychological and Social Foundations*. Oxford University Press, New York.

West, P.C. 1982. *Natural Resource Bureaucracy and Rural Poverty: A Study in the Political Sociology of Natural Resources*. University of Michigan, Ann Arbor, MI.

Wilson, E.O. 1975. *Sociobiology: The New Synthesis*. The Belknap Press of Harvard University Press, Cambridge, MA.

Wilson, E.O. 1978. *On Human Nature*. Harvard University Press, Cambridge, MA.

Wilson, E.O. 1992. *The Diversity of Life*. W.W. Norton, New York.

Wirth, L. 1928. *The Ghetto*. University of Chicago Press, Chicago.

World Resources Institute. 1994. *World Resources*. Oxford University Press, New York.

Worster, D. 1992. *Under Western Skies: Nature and History in the American West*. Oxford University Press, New York.

Wrong, D.H. 1994. *The Problem of Order: What Unites and Divides Society*. The Free Press, New York.

Wrong, D.H. 1988. *Power: Its Forms, Bases, and Uses*. University of Chicago Press, Chicago.

Zelinsky, W. 1973. *The Cultural Geography of the United States*. Prentice-Hall, Englewood Cliffs, NJ.


## THE AUTHORS


**Gary E. Machlis**
*Department of Forest Resources,*
*College of Forestry, Wildlife and Range Services,*
*University of Idaho,*
*Moscow, ID 83844-1133, USA*

**Jo Ellen Force**
*Department of Forest Resources,*
*University of Idaho,*
*Moscow, ID 83844-1133, USA*

**William R. Burch, Jr.**
*School of Forestry and Environmental Studies,*
*Yale University,*
*New Haven, CT 06511, USA*

# ♦ Biological and Ecological Dimensions

# Biological and Ecological Dimensions — Overview

Gary K. Meffe

Although it should be intuitively obvious that the biological and ecological underpinnings of our world form the ultimate basis for human health, prosperity, and sustainability, in many arenas, humanity has divorced itself from the constraints of the natural world, and placed itself "above" nature. In many ways, we humans act as though natural laws simply do not apply to our species. This is evidenced by a global indifference to exponential population growth, by unlimited western-style consumption of resources and discarding of wastes, by an unceasing march toward destruction of forests, grasslands, coastal zones, and other natural systems, and by political systems that refuse to admit the intractable problems underlying perpetual economic growth. It is also evident in how the global community "packages" its environmental obligations and duties. For example, the *Brundtland Report* (1987) defined global sustainable development as that which "seeks to meet the needs and aspirations of the present without compromising the ability to meet those of the future." Note that ecological sustainability of natural systems plays no role in that definition. The Rio Earth Conference of 1992, and the five-year follow-up at the United Nations only meekly dealt with issues that require deep thought and major action.

Any thoughtful person can only conclude that there can be no long-term human prosperity and happiness without healthy, intact ecosystems that provide the goods and services upon which we all depend (Daily 1996). All of the political posturing, legislation, global conferences, economic forecasts, and aid programs ultimately are in vain without the biological components and ecological processes that give the planet life and make it possible for humanity to exist. This section deals with those biological and ecological aspects, the underlying bases for ecological stewardship and human survival.

Any ecological system may be described in terms of its composition, structure, and function — or, more descriptively, by what is there, how it is distributed, and what it does (Noss 1990). A complete system description should include all three elements; society cannot hope to adequately understand ecological systems with more narrow perspectives that exclude any of these major elements. The papers in this section collectively address composition, structure, and function of ecosystems, although not in a formal, structured way, because composition, structure, and function are intimately related and not easily dissected. All three elements must be simultaneously considered; however, in most of these chapters, one of the elements usually is emphasized over the others.

## COMPOSITION, STRUCTURE, AND FUNCTION

This section begins with several treatments of composition and structure, ranging from genes, through species, to entire ecosystems. Huston et al. discuss genetic and species diversity, the very foundations of biological diversity and function. The store of genetic and species diversity on earth is not only the basis for all life, but also represents the foundation for all potential evolutionary changes and future patterns of life. The destruction of this genetic and species diversity by human activities is an extremely serious threat to ecological sustainability and to the future of humanity. Huston points out that humans rarely interact directly with genetic diversity, although it does surface in loss of crop genetic diversity and recombinant DNA research and applications. Nevertheless, species diversity is humanity's common point of interaction with biological diversity.

Huston et al. develop the nearly counter-intuitive, but centrally important point that plant species diversity typically is highest in unproductive areas not especially favorable for growth, such as dry, infertile soils or nutrient-poor waters. In these areas, no species can become abundant or dominant, which permits many species to coexist, resulting in high overall diversity. The pattern applies as well to some animals, such as insects, that depend directly on plants. Other animals, especially predators, are most diverse in productive conditions where their prey species are abundant. Thus, overall diversity patterns are complex, and not easily amenable to management manipulation. Huston et al., however, offer three general patterns through which management may (intentionally or inadvertently) affect specific components of species diversity:

- overall diversity is augmented by increasing spatial structure and heterogeneity of the environment, but the effects on specific components depend on particular conditions created;
- productivity expansions achieved through water or nutrient addition decrease plant diversity in most cases, but can heighten plant or animal diversity if the essential resource added was previously low; and
- increasing frequency or intensity of disturbance causes greater plant diversity except where productivity is extremely low or disturbance frequency is already high. Animal diversity generally will be decreased by raising their mortality rate, except where a few species are dominant under highly productive conditions.

In spite of the complexity of diversity patterns, the strong effects of environmental conditions on ecosystem processes and species diversity provide a framework for predicting the consequences of any particular management action on species diversity. Huston et al. supply a starting point for understanding why the same management action (such as a harvesting schedule) can have opposite effects on species diversity in different regions. Managing species diversity and evaluating the impacts of any resource management plan on diversity require understanding how geology, topography, climate, and human activities influence the distribution of ecosystem processes and disturbance dynamics across the multiple spatial and temporal scales of landscapes. These issues are discussed in more detail in the subsequent chapters of this section.

Overall diversity is not always the management question of greatest concern. Many times, managers are more interested in maintaining process and function, and a subset — perhaps even a small subset — of species may be the primary focus of a functional approach. From a purely functional perspective, there may be redundancy in species diversity, and a subset of species may perform critical ecosystem functions, while the rest would not be missed if lost. Huston et al. wisely caution that, because managers know little about the functional aspects of many species, managers should never assume redundancy or trivialize a species' importance. Huston et al. go on to discuss the major threats to biodiversity: habitat loss and fragmentation, and invasion and spread of nonnative species. The latter is directly tied to human disturbance, a topic more fully taken up in later papers, but which clearly promotes the invisibility and spread of nonnative species.

Population viability analysis provides an explicit way of estimating risks to the long-term persistence of a species. While the origins of PVA are firmly in the discipline of population biology, including species modeling and island biogeography, it's most often encountered by natural resource managers in the context of identifying species at risk of extinction. Noon et al. review the fundamentals of PVA including population dynamics, genetic structure, and the spatial and temporal dimensions of demographic change. In the United States, the Endangered Species Act (ESA) has fueled much of the research and advances in PVA and made natural resource managers a key audience for PVA studies. To the uninitiated, PVA is a complex and rigorous process. As a result, a virtual small industry of models, software programs, and consultants has arisen to help natural resource agencies, land-owning companies, and researchers meet legal requirements of the ESA. As helpful as these advances have been, Noon et al. strongly caution that PVA should not be viewed as a casual process. There is no substitute for a thorough understanding of population biology and genetics. Furthermore, a lack of information on spatial and temporal relationships between habitat and demographic factors can lead to misleading PVA results. This may actually exacerbate risks to endangered species and management activities alike. PVA is a vital tool, nevertheless, and Noon et al. helpfully point to approaches that researchers and managers can use to more productively use PVA in the face of limited data and considerable uncertainty.

Holthausen et al. explore some of the fascinating lessons learned during the past decade of using PVA in the field for management purposes. Two conclusions are worth emphasizing. First, it is not practical or possible to conduct a PVA for most species in a management setting. Coarse filter approaches, such as the Heritage ranking system developed by The Nature Conservancy and the vulnerability criteria used by the IUCN Species Survival Commission, are a useful way to prioritize species for PVA assessments, and the investments in data, expertise, and time needed to do them right. Second, most examples of published PVA cases are atypical. Since there are data to conduct PVA for less than five percent of species native to the United States (and considerably less in most other parts of the world), managers and researchers should be prepared to use alternative approaches to identify management options that can enhance a species' chance of survival. And, both Noon et al. and Holthausen et al. share the view that substantially more money and time need to be invested in collecting missing data to more fully realize the potential of PVA as a management tool.

Gosz et al. and Concannon et al. address diversity at the ecosystem and landscape levels, where many management decisions take place and good ecological stewardship needs to occur. It is here that the more theoretically based and biologically derived arguments of Huston et al. are put into practice on real landscapes. Gosz et al. recognize that real landscapes are first and foremost ecological in nature, and that human boundaries are generally artificial and ignorant of real ecological boundaries. Thus, management must consider the entire landscape rather than act in a piecemeal fashion driven by political boundaries and budgets. Gosz et al. also recognize the many forces — ranging from large-scale geologic actions to local biotic and abiotic disturbances to anthropogenic activities — that shape landscapes on spatial and temporal scales, thus placing human modifications in the context of other forces. In addition, Gosz et al. establish terminology needed for ecosystem and landscape scales, such as patch, matrix, and landscape boundaries, and standardized measures of these and other terms.

Gosz et al. further emphasize the importance of considering scale, both spatial and temporal, in landscape-level management. Scale is critical, because what may appear homogeneous at one spatial scale can be quite heterogeneous at other scales. If management decisions are made at inappropriate scales with the incorrect level of heterogeneity in mind, then severe consequences can be experienced. Furthermore, Gosz et al. argue that, although landscapes change over time, not all landscape processes occur at the same rate.

## SCALE AND DISTURBANCE

Disturbances on multiple scales, and managing at the appropriate scale are important themes that emerge from the pieces by Gosz et al. and Concannon et al. Complex interactions can occur among disturbance, landscape patterns, patch size and position, natural and artificial boundaries, and other parameters. Disturbance can influence movements of species on the landscape, but species, in turn, can influence disturbance patterns, as, for example, when exotic species respond differently to fire, flood, or insect outbreaks than do native species. Disturbances can facilitate invasion by exotic species, and human actions can enhance that process. Concannon et al. present examples that demonstrate the complex relationships among some of these parameters.

Next, Lugo et al. and Paustian et al. address ecosystem processes and functions. First, Lugo et al. define and distinguish among these terms: process is "a sequence of events or states, one following from and dependent on another, which lead to some outcome," while function is "the role that any given process, species, population, or physical attribute plays in the interrelationships between various ecosystem components or processes." Lugo et al. argue then that processes and functions occur at multiple spatial scales, from microscopic to global, but that the ecosystem is a functional, not a spatial concept, and it functions at many spatial scales. Thus, one does not need to argue about precise ecosystem definitions and spend time defining their boundaries; for management, it is more important to understand that ecosystems function — they accumulate matter and energy, cycle materials, process energy and so forth — and that these actions occur on all scales of management interest. Management alters ecosystem functions, as well as changes composition and structure, though it is the latter two that often gain the most attention. The capacity of ecosystems to absorb anthropogenic changes depends on processes and function, but these are often neglected or poorly understood in management activities.

Two concepts are critical to understanding how ecosystems might respond to human alterations: resilience and resistance. Lugo et al. define resilience as the ability of an ecosystem to "bounce back" from disturbances — i.e. to reattain its former state after being disturbed. In other words, resilience occurs when environmental changes do not exceed the capacities of individual organisms and populations to restore initial conditions. Thus, a forest has high resilience if it continues to be the same type of forest after experiencing a large fire. Resistance is the ability of a system to resist change altogether — i.e. to not be affected by a change. For example, if fishes and other organisms in a flood-prone ecosystem are not significantly affected by floods (i.e., maintain their populations intact), the system is said to have high resistance to flooding. Either mechanism leads to continued ecosystem composition, structure, and function.

Maintaining naturally high ecosystem resilience and resistance, by maintaining usual system composition, structure, and function, allows management greater flexibility to pursue ecosystem products, services, and sustainability. Increasing the intensity or frequency of disturbances, or creating new disturbances foreign to the system, can quickly degrade a system beyond its capacity for resilience. Examples include changing fire regimes, altering runoff patterns into streams, releasing toxic materials, or changing hydroperiods in wetlands. Maintaining ecosystem resilience and resistance requires knowledge of normal, or reference, conditions.

Lugo et al. provide examples of some key indicators to establish frames of reference. Paustian et al. present several management and restoration case studies that focus on and demonstrate a functional approach to ecological stewardship. Both of these papers argue strongly that the most successful and cost-effective resource management approaches will understand ecological processes and work within their constraints rather than work against and fight these processes.

In an expanded discussion of this critical ecosystem process, White et al. and Engstrom et al. further deal with the topic of disturbance and temporal dynamics. White et al. simply define a disturbance as a "discrete event that changes ecosystem structure and resource availability." White et al. indicate the broad range of descriptors of disturbances, such as type, temporal and spatial characteristics, specificity, magnitude, and possible synergisms — parameters that together describe a "disturbance regime." Disturbances may range from minor disruptive events to near or complete removal of normal composition, structure, and function. In the more extreme cases, the "biological legacy" — the amount of living and dead material that remains — will partly determine recovery and may help to maintain ecosystem function. This is why logging practices are now changing in some areas to leave coarse woody debris on the ground rather than collect and burn it; the biological legacy represented by the debris offers cover for many organisms, returns nutrients to the soil, and reduces soil erosion.

Disturbances help to create landscape patterns: this in turn assists with controlling patch processes, which are scale-dependent. For example, the death of a single tree will influence local processes in a forest patch, but will have little effect on the larger stand or watershed. The larger the disturbance, the greater the spatial extent of influence on processes.

Understanding the nature of natural disturbance regimes can greatly aid resource management. Organisms typically adapt to, and even come to depend on, disturbances that occur on a somewhat regular basis over evolutionary time. Thus, serotinous cones that only open and disperse seeds after a fire, germination and rapid growth of longleaf pines after a fire, or rapid colonization of newly formed forest gaps by early successional species, are all adaptations to disturbances, and indicate evolutionary responses. Management that mimics natural disturbances can be more sustainable than that which ignores natural disturbance regimes, but it is difficult to mimic nature precisely. Nevertheless, knowing the historical fire patterns, the flooding regimes of rivers, or the typical size of forest gaps can greatly improve management, bringing it closer to the normal range of variation experienced over evolutionary time, and keeping it within the bounds that define resiliency or resistance.

The concept of scale is woven throughout the topics already discussed, and is integral to all aspects of ecology and natural resource management. This topic is explicitly addressed by Haufler et al. and Caraher et al., who point out that single-scale management has a history of failure, and that no one scale can address all the problems of ecosystem management or lead to good ecological stewardship. The common challenge will be to think about and work at larger spatial and temporal scales than have been traditionally done. This is both a conceptual and technical challenge, for which tools and approaches are continually being developed.

Haufler et al. and Caraher et al. warn that analyses at different scales can lead to different conclusions, as can analyses of different landscapes at the same scale. According to them, ecosystem analysis, and hence management, is scale-dependent. Because of that, scale considerations for good management should include the extent of the spatial landscape, the spatial resolution needed to address the objectives, the timespan for the planning horizon, and the timespan for an historical perspective. The proper spatial extent for good management is problematical (often being constrained by socio-political considerations), but these authors suggest it should be large enough to address population viability, biodiversity, and other ecological objectives, but not too large to be unfeasible, defeat collaborative partnerships, or encompass too much spatial variation. Obviously, there is no ready formula for this — it must be judged on a case-by-case basis. A similar balance in scale must be struck in resolution of mapping and data collection, and planning time spans.

Another management challenge, regardless of the scale(s) at which one is working, is to always look up the temporal and spatial scales to gain perspective and context, and down the scales to gain an understanding of structure and function. It is easy to become focused on a problem at a particular scale and lose sight of how it fits within other ecological, institutional, and socio-economic contexts. An ecosystem approach can be the unifying tool to link different scale levels and hierarchies.

For many people, maps representing the distribution of ecosystems are the most concrete expressions of ecosystem scale and perhaps a starting place for an ecosystem management project. However, few realize all that goes into a useful or scientifically credible ecological classification or that ecological classification figured prominently in the rise of ecology as a science. Grossman et al. review the increasingly sophisticated array of tools and techniques that are being used to classify and map ecosystems for both research and management applications. One basic point stressed is that ecological classification should be designed with a purpose or management objective in mind. That is, there is no one best classification to use for all management or research applications. As with population viability analysis, successful approaches to ecological classification require a thorough understanding of basic ecological principles and the management objectives of the users. Ecological classification is perhaps best viewed as a spatially explicit framework for organizing our knowledge about the distribution of species and features of the physical environment, interactions between species and their environment, and changes in composition and structure over space. Behind every ecological classification map is an enormous set of information.

Stressing the varied roles ecological classification can have in resource management, Carpenter et al. review experiences from the U.S. Forest Service and other organizations where ecological classification has played a key role in ecosystem management planning and implementation. Included are case studies where ecological classification has been used for integrated resource inventories; modeling the cumulative impacts of local management decisions on desired future conditions; regional and landscape assessments for long-term planning; and project monitoring and evaluation. One clear lesson from the case studies is that ecological classification rarely works if resource managers are not engaged with the classification scientists in the early design, field testing, and final revisions of a classification. If the classification produces a system that doesn't jibe with managers' observations and experiences, they are unlikely to use it. And, the act of ecological classification has catalyzed promising management partnerships between state and federal agencies and between the public and private sector that have gone significantly beyond the initial limited goals of developing a shared classification system.

## CONCLUSIONS

The approaches discussed in the chapters in this section provide a guide not only for incorporating biological and ecological dimensions into ecosystem management, but also how to make these dimensions a central focus, given that natural ecosystems clearly must continue to function at some minimal level to support human interests. Good and wise stewardship must, first and foremost, understand and accept the natural limitations inherent in ecological systems. The basic natural laws operating in all ecosystems dictate what humans can and cannot do. It is possible to circumvent some of these over the short-term (such as extracting groundwater at greater than recharge rate, or overgrazing a prairie until all grasses disappear), but eventually these limitations become apparent, often causing great human misery through economic and social disruption. Wishing for or legislating a particular entity or outcome does not make it so; we humans can no more legislate a higher carrying capacity or greater water flow in an ecosystem for the benefit of people than we can legislate reduced gravity for the benefit of a space shuttle launch. It behooves the successful manager (as well as the sustainable society) to look ahead, try and understand these limitations, and work within their constraints, many of which are collectively discussed in these chapters. There are too many examples of not doing this in natural resource management (e.g., water extraction and management in the American southwest; alteration of the entire south Florida water drainage system; fire suppression throughout the American west) to excuse further such behavior.

Despite the base of knowledge provided here, many things are unknown about living systems. The uncertainties and surprises are many, given the nearly infinite complexities of ecosystems. By comparison, "rocket science" — the alleged paragon of brilliance — is simple. In rocketry (or any engineering endeavor), scientists know all the parts, understand their behaviors, and know how they interact; none of these is true for even the simplest ecosystems. Our best engineers would fail dismally if they had to precisely predict specific behaviors of any natural ecosystem for any meaningful period of time; yet much of natural resource management has proceeded with an engineering mentality and a heavy dose of hubris, as though a great deal is known about the systems, which could be freely manipulated without long-term consequences. In the absence of detailed knowledge and abilities, it is advisable to closely adhere to three principles in management of natural systems: (a) the humility principle, which recognizes and accepts the limitations of human knowledge and management abilities; (b) the precautionary principle, which suggests thinking deeply and moving slowly when uncertainty is high; and (c) the reversibility principle, which advises against making irreversible changes. The only sensible approach, then, under such circumstances of great uncertainty, is to manage in an adaptive manner, learn and build upon management knowledge as society proceeds, not accept dogmatic or prescriptive management measures, and carefully listen to what nature says about what it can, and cannot, provide.

## THE AUTHOR

**Gary K. Meffe**
*Department of Wildlife Ecology and Conservation,*
*University of Florida.*
*Gainesville, FL 32611-0420, USA*

# A Functional Approach to Ecosystem Management: Implications for Species Diversity

Michael Huston, Gary McVicker and Jennifer Nielsen

**Key questions addressed in this chapter:**

- *What are the natural levels and patterns of species and genetic diversity that would be expected on different types of landscapes? How might these patterns have been changed by past human uses and management of the landscape? What level of "recovery" or change in diversity can be expected to result from management?*
- *Why is diversity important and what, if anything, does a particular level of diversity tell us about the landscape? How much diversity is "needed" in a particular area?*
- *How stable is species diversity? How is diversity likely to change over time, both in response to specific management actions, and in response to factors beyond the control of management, such as the weather?*
- *What types of management activities have the greatest effect on species or genetic diversity? How can the effects of management be understood and predicted?*
- *How should the need to protect and manage diversity be balanced against other uses of the landscape?*

**Keywords: Species diversity, temporal and population dynamics, exotic species, patterns of diversity, spatial structure, environmental gradients, processes affecting species diversity, measurement analysis of species diversity, comparative advantage, multiple goal management**

# 1 INTRODUCTION

The objective of ecosystem management is to manipulate ecosystems to meet current human needs while maintaining the capability of the ecosystems to meet future needs for their range of goods and services. The premise of this chapter is that it is possible to simplify ecosystem management by focusing on ecosystem functions and processes, rather than ecosystem properties.

In setting objectives for ecosystem management, and evaluating the degree to which those objectives are being achieved, the attainment or maintenance of specific ecosystem functions and process rates (e.g., primary production, nutrient availability, disturbance rate) is likely to prove more definable, stable, and useful than attempting to achieve management goals based on specific ecosystem properties (standing timber or harvest, species diversity, population size). The problem with the use of specific properties as management endpoints is that ecosystems are highly dynamic both in time and space, and specific properties are not only difficult to monitor, but are likely to experience natural variation that makes an accurate evaluation of management success difficult.

Management actions (see Section 7) are typically limited to manipulation of a few fundamental ecosystem processes, all of which are interconnected but usually treated separately. The goal of the following sections is to provide some guidance on: (1) the interconnections of these basic management actions through the ecological processes they affect; (2) the effect of these actions on other objectives of Ecosystem Management, specifically those described as biological diversity; and (3) the spatial distribution of comparative advantage for various ecosystem functions, plus those properties related to biological diversity.

Although the protection of species diversity has only recently become a specific management objective (e.g., National Forest Management Act 1976; 36 CFR 219) it has long been an issue of scientific interest and esthetic appeal. While the Endangered Species Act (1973) was passed to prevent the extinction of species, the protection of endangered species has often resulted in the protection of a diversity of species that are not endangered. Most other environmental protection legislation also has as an explicit or implicit objective: the protection of species diversity.

The new emphasis on "ecosystem management" means that species diversity has become an essential consideration in all resource management decisions. Every resource manager must be aware of the potential effects of management action or inaction on overall species diversity, as well as on particular species of concern.

This chapter provides an overview of the patterns of diversity that are typically found on landscapes, and a review and synthesis of the processes that affect diversity, with a focus on those processes that are most easily manipulated and managed. Even though species diversity is a complex subject that has preoccupied biologists from before Darwin to the present, enough is now known about the regulation of species diversity to provide general guidelines that will enable resource managers to understand the patterns of species diversity on their landscapes sufficiently well to evaluate the potential effects of most management actions. Genetic diversity is less well understood, but may require management consideration under specific circumstances.

## 1.1 Species Diversity

Taxonomists determine whether the differences between populations of organisms are sufficient to classify them as separate species. This determination provides the basis for calculating the number of different species in any group of organisms. Diversity can also be evaluated at the genus or family level. Historically these taxonomic distinctions were made on the basis of morphology (shape, size, color), but they are increasingly being made on the basis of genetic differences. From the perspective of conserving genetic diversity, not all species are necessarily equivalent, nor are all individuals of a single species equivalent. If the goal is to preserve the maximum genetic diversity (and presumably evolutionary potential), then there may be more value in preserving two distantly related species than two closely related species (see Section 1.2.1). Likewise, to preserve the genetic diversity of a species, it will usually be more effective to preserve multiple small populations than a single large population.

In most situations, however, the most important property of species is not their taxonomic status, but rather the functional roles that they play in physical processes and biological interactions. Developing a scientific understanding of species diversity, as well as planning and evaluating management actions in terms of their effects on species diversity, requires a functional approach, in addition to a taxonomic approach, to species diversity.

### 1.1.1 Functional Roles

The total species diversity of a natural ecosystem can be considered at two functional levels. At the highest level is the number of functions performed by identifiable groups of species (the number of "functional types"). The second level is the number of functionally similar (not necessarily identical) species within each

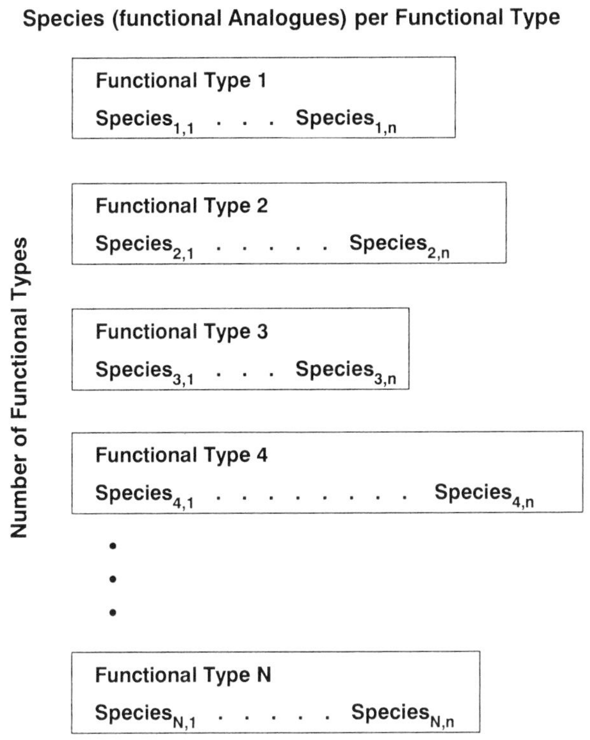


Fig. 1. The two components of total species diversity: the number of different functional types and the number of species within each functional type. Not all functional types will have the same number of species. Because functional types are defined on the basis of resource use, competitive interactions are potentially intense among organisms of the same functional type, but very weak between organisms of different functional types (from Huston 1994)

functional type (Fig. 1). From this perspective, the total number of species in an ecosystem is the number of different functional types multiplied by the number of species in each functional type. Of course, the number of different functional types in an ecosystem depends on how finely functions are subdivided.

The value of the functional approach is that it helps us understand how the different components of biodiversity contribute to ecosystem function, and, in particular, provides some guidance on how different functional groups of organisms are likely to respond to natural or manmade changes in environmental conditions. The broader concept of biodiversity extends to levels both higher and lower than the species. At a level below the species, genetic diversity results from differences between populations of the same species, as well as the genetic differences between separate species. Above the species level, variation in the physical structure of landscapes creates conditions for many different combinations of species, thus increasing both genetic and species diversity.

The "adequate" performance of a particular ecosystem function depends on some combination of sufficient numbers or biomass of organisms in the relevant functional type, and sufficient diversity of species in that functional type. There is little, if any, evidence to demonstrate that a particular number of species is required to perform a function adequately, or that ecosystems function better with more species than with few species (Huston 1997, Hooper and Vitousek 1997). However, if changes in the abundance or diversity of a particular functional group are observed, it is an indication that the ecosystem is changing with regard to that function (and perhaps others). Maintaining adequate ecosystem function through the presence and vitality of all necessary functional types of organisms is a much more practical objective than attempting to maintain a large number of species without regard to their function.

Understanding the functional composition of an ecosystem is essential for predicting how different functional groups will respond to a specific management action. The ability to predict how different functional groups will respond is essential for planning ecosystem management and evaluating its effects, and can also contribute to more effective management of endangered species and species with economic value.

### 1.1.2 Exotic vs. Native Species

The subject of biological diversity raises the issue of whether or not exotic species (i.e., non-native plants and animals) should be included when evaluating biodiversity. Regardless of how this question is answered, exotic species can be a serious problem and are an essential consideration for resource management.

Native species are often valued more highly than exotic species because they represent the unique biological heritage of a region. In many cases, native species may have co-evolved or developed close mutualistic relationships with one another, so that the loss of one may result in the loss of other species dependent on it.

Exotic species are not necessarily bad. Most agricultural crops are exotic species, as are many horticultural or ornamental plants. Exotics are generally considered to be a problem only in specific situations: (1) when they interfere with human activities (as in the case of weeds or aquatic pests); (2) when they change ecosystem function in ways that are perceived to be negative (such as by increasing fire frequency or reducing water availability); and (3) when they cause the extinction or reduce the abundance of native plant or animal species that are valued. When focused on the goal of sustaining native species and their genetic diversity, however, the evidence against exotic species is substantial.

Arguments over inclusion or exclusion of exotic species in measures of biological diversity are not particularly useful. The real question is what effect exotic species are having upon native species and ecosystems. For ecosystem management purposes, dealing with exotics according to their relative threat or value, and striving to maintain or restore the native species of the area, is a more rational approach than trying to achieve a certain number of species or to eliminate all non-native species.

### 1.2 Genetic Diversity

Genetic diversity is the raw material of natural selection and evolution, and thus the foundation for all life on earth. Yet genetic diversity does not exist outside the context of individual organisms, so that conserving the Earth's species diversity will also conserve virtually all of its genetic diversity. Nonetheless, loss of genetic diversity within a single species can be an issue of legitimate concern to natural resource managers. Recently developed methods for evaluating genetic diversity within a species (Hedrick and Miller 1992, Moritz et al. 1995) reveal that species vary greatly in their genetic diversity, and that high genetic diversity is not necessarily "better" than low.

Loss of genetic diversity within a species can be a problem for species survival at several time scales (Meffe 1996). Over relatively short time periods, particularly in very small populations, loss of genetic diversity can result in "inbreeding depression," which results from a number of processes (Mills and Smouse 1994). Over longer time scales, reduced genetic diversity can decrease the probability that a species could successfully adjust to changing environmental (either physical or biological) conditions, and thus avoid extinction, or evolve into one or more additional species (Lynch and Lande 1993).

#### 1.2.1 Phylogenetic Relatedness

In prioritizing either individual species or ecological communities for conservation, the question of genetic relatedness becomes an issue (Humphries et al. 1995, Mills and Smouse 1994, Oldfield 1989). If one endangered species has many close relatives (e.g., other species in the same genus or subgenus) while another is the only representative of a particular genus or family, most conservationists would assign a higher priority to the species with no close relatives, with the goal of preserving maximum genetic diversity. Similarly, if two areas have the same number of species, but one area has species that are distantly related, while the other has closely related species, the first area would have a higher priority for conservation. For this reason, evaluating phylogenetic relatedness has become an important issue in conservation planning (Mayden and Wood 1995, Wayne and Jenks 1991).

### 1.3 Other Biodiversity Levels

The concept of biodiversity also includes the patterns formed on landscapes by different combinations of environmental conditions. These patterns may result from differences in function, structure, color, or other differences in groups of organisms or between different dominant organisms. Landscape patterns of biodiversity occur at scales from a few meters (as around the margin of a temporary pond), to the altitudinal zonation of vegetation on mountainsides, to the latitudinal zonation of ecosystems at the scale of continents.

The patterns of landscape-scale biodiversity range from random (such as treefalls or lightning strikes), to gradual changes over distance along natural gradients of environmental conditions, to highly complex and regular patterns (such as vegetation associated with stream and river drainage systems or with geological structures). These patterns of biodiversity may have functional consequences associated with the movement of water, sediment, or the atmosphere. In general the number and types of species found in a particular habitat are influenced by the numbers and types of species found in surrounding habitats, which are dependent on the number, proximity, and heterogeneity of the surrounding habitats. Thus, these large-scale patterns of biodiversity influence species composition and diversity at the local scale.

## 2. SIGNIFICANCE OF SPECIES DIVERSITY

The issue of biodiversity has dominated the national and international conservation agenda for nearly a decade. Concern has been driven by the perception that extinction rates have recently increased to unprecedented levels, primarily as a result of habitat destruction (Ehrlich and Ehrlich 1981, Lawton and May 1995). Extinction of species has been considered to be a problem for reasons ranging from religious and ethical to industrial and financial. Although most of the philosophical and economic arguments about the value of biodiversity are beyond the scope of this chapter, the relationship of species and genetic diversity to ecosystem properties and processes is of real concern for land management.

## 2.1 Ecosystem Function and Species Diversity

Ecologists have long believed that ecosystem functions (efficiency, productivity, resilience, nutrient cycling) improve with an increasing number of species (Elton 1958, May 1973). Although it is obvious that there is some minimum number of species below which an ecosystem cannot function properly, it is not at all evident what that number is, whether it can be unequivocally defined, and how it may differ among ecosystems. For example, there are numerous cases of unproductive ecosystems with many species, and many highly productive ecosystems have only a few species (see Sections 3.2.3 and 4.4). The position taken here is that the functional properties of the species in an ecosystem are much more important than the total number of species present for ecosystem function and sustainability (rare species may be important for reasons other than ecosystem function, such as medical or agricultural potential, aesthetics, or ethical considerations). However, above some minimum level, there is no evidence that high genetic or species diversity produces a higher level of ecosystem function than does lower diversity (Anderson 1994, Huston 1997, Hooper and Vitousek 1997).

An approach to understanding the relationship between diversity and ecosystem function can be based on a listing of essential ecosystem processes along with a classification of species based on their functional properties, rather than on their taxonomy or phylogeny. Of course, the "required" number of species then depends on how finely ecosystem processes are subdivided, and on how adequately the full set of ecosystem processes is understood. Another problem is that the number of species required to adequately perform a specific function may be impossible to define. There is a temptation to define one or a few species as "essential," and the rest of the species that perform the same (or similar) function as "redundant " (Walker 1992). However, neither ecosystem processes nor the functional properties of all species are understood sufficiently well to make this distinction with any confidence. Nonetheless, enough is known about ecosystems to identify many of the major types of processes, as well as the specific functional components of genetic and species diversity that contribute to those processes.

### 2.1.1 Trophic Structure

Trophic structure refers to the various pathways by which plants, and the organisms that eat plants, process energy. The trophic level essential for life is primary production, the capture and storage of energy in forms that can be used to produce physical structure. In all but a few cases, such as anaerobic soils and the deep ocean, the primary producers are green plants, and the energy source is sunlight. In addition to the carbon compounds formed with solar energy, the plants also take up and store mineral nutrients.

In many ecosystems, the supply of mineral nutrients is limited, and when all available nutrients have been taken up and stored in plant material, plant growth must stop. The only way to release the nutrients from dead plant material is through combustion or decomposition. Decomposers, including fungi, bacteria, and associated invertebrates such as earthworms, perform a critical ecosystem function that is often overlooked. The simplest functioning ecosystem would be plants (perhaps only one species) and decomposers (probably multiple species to allow decomposition to occur rapidly enough to release nutrients needed by plants). Loss of decomposers as a result of fungicides, insecticides, or other causes, can have a major effect on ecosystem processes, generally reducing nutrient cycling and plant growth (McMinn and Crossley 1996, Waring and Schlesinger 1985).

Higher trophic levels, such as herbivores and carnivores, depend completely for their existence on the primary production of plants. Although herbivores and carnivores are not essential for the minimal ecosystem functions of production and decomposition, they have a major effect on the physical structure and appearance of ecosystems, and also on the rates at which primary production and decomposition occur. By consuming plants, herbivores regulate both the amount and form of plant matter in an ecosystem, as well as the actual rate of formation of plant material (Huntly 1991). Herbivores also initiate the decomposition of plant material that they consume, thus speeding the release of mineral nutrients essential for plant growth and producing a fertilization effect.

In addition to consuming plant material, herbivores can affect species diversity by favoring some plant species over others (Bryant et al. 1991). By preventing woody plants from getting big enough to shade out smaller plants, such as grasses, herbivores can maintain a grassland in an environment where the vegetation would otherwise shift to a shrubland or woodland. Often the effect of herbivores is so pervasive that it is not noticed until the herbivores are removed and the ecosystem changes completely to another form. For example, experiments that fence out herbivores in forests or grasslands often create a very different plant community within the exclosure, with different rates of ecosystem processes (McInnes et al. 1992, Norton-Griffiths 1979). When multiple species of plants are present, multiple species of herbivores can produce complex spatial and temporal patterns of vegetation, depending on the abundance and food preferences of the herbivores (Pastor and Cohen 1997).

Wild herbivores can compete directly with humans and domestic livestock for plant production. Substituting domestic herbivores for wild herbivores may not have a major effect on herbivore diversity, but can have a large effect on the plant community because of different feeding preferences. In cases of ecosystems adapted to grazing, the shift from natural grazers to livestock may have little effect on overall productivity and diversity (Milchunas et al. 1988).

Carnivores eat herbivores, decomposers, and other carnivores, and can have major effects on ecosystems. The removal of carnivores can allow herbivores to become too numerous, leading to destruction of trees and other vegetation, damage to streams and riparian areas, and soil degradation and erosion (Fleischner 1994).

With regard to trophic structure, the primary diversity issue is not how many species are present, but rather how the species function with regard to their trophic interactions. Because organisms at different trophic levels, as well as within the same trophic level, interact with one another, any change in trophic structure is likely to have a significant effect on ecosystem properties, with diversity increasing at some trophic levels and decreasing at others.

### 2.1.2 Productivity

Productivity is the net result of the trophic interactions discussed above, but is of such central importance to all human and natural uses of the environment that it deserves separate attention. The two critical aspects of productivity are the rate at which carbon is fixed and the availability of that carbon for consumption or use by humans and wildlife. Both primary and secondary production in natural and managed ecosystems are harvested by humans.

Primary productivity is controlled mainly by the physical properties of climate and soils. Favorable climatic conditions and fertile soils support higher primary production. High primary production supports higher secondary production (higher animal biomass) (Abrams 1993). In most situations, the number of species within any particular trophic level is a response to the level of primary productivity supported by the soils and climate, rather than being the cause of that level of productivity (Huston 1997). The number of species is much less important than the specific properties of the species that are present (Hooper and Vitousek 1997).

The properties of the species present in an ecosystem can alter its level of productivity under a particular set of climatic and soil conditions (Pastor and Post 1986). The properties of plants can influence net primary productivity through the efficiency with which mineral resources and solar energy are used, as well as through the total amount of solar energy captured. For example, shade-intolerant trees can grow rapidly under favorable conditions with abundant light, but generally let a significant amount of light pass through their canopy. If this light is then captured by shade-tolerant species living in the understory, the net primary production of the ecosystem is higher than it would be with only a single type of trees. The types of plants can also affect the palatability, fiber length, and other properties of the plant material. Plants that fix nitrogen or are able to take up scarce nutrients efficiently can also increase the productivity of an ecosystem.

A fundamental property of energy transfer between trophic levels is that typically 90 percent of the energy is lost in the transfer, through digestive inefficiency and respiratory losses. Consequently, the secondary production of herbivores is only approximately 10% of the available primary production by plants, and the secondary production of carnivores is only 10 percent of the energy in the herbivores they consume. As a result, ecosystems with carnivores will have less total animal biomass (specifically, herbivore biomass) than ecosystems without carnivores. This is obviously a major consideration for ranching and other livestock operations. Thus, not only the total amount of animal biomass, but the form of that biomass (e.g., wolf versus moose) is affected by the types of species present.

### 2.1.3 Physical Properties

Physical properties of ecosystems, such as the depth and structure of living and dead roots, the size and leaf mass of plants, and structures made by animals such as dams, burrows, and mounds, have a strong influence on processes related to hydrology and nutrient cycling, human uses of the land for agriculture or other purposes, and the presence and abundance of plants and animals that require specific physical structure (Aber and Melillo 1991).

An increase in the structural complexity of the above and belowground components of an ecosystem will usually increase biodiversity, particularly among organisms that are small in relation to the dominant structural organisms in the system (e.g., soil organisms, insects, birds).

Higher biomass in the above and belowground components of an ecosystem can result in increased rates of a number of ecosystem processes, particularly those related to surface area or volume. For example, uptake of carbon dioxide by leaves, as well as adsorption of toxic air pollutants such as nitric acid vapor or volatile organics, will be higher in situations with high leaf area and high surface area of bark and other adsorptive surfaces. Higher root biomass, particularly if associated with higher root surface area, will generally result in higher net uptake rates of soil nutrients and water. Plant size and canopy structure affect the aerodynamics of the plant atmosphere interface, and can influence such processes as heat exchange, evapotranspiration, deposition of nutrients, and local microclimate.

### 2.1.4 Chemical Properties

In animals, chemical differences among related species are relatively subtle, being limited primarily to differences in color, odor, and specialized secretions such as toxins. However, it is among plants and lower life forms such as bacteria and fungi that chemical differences between species reach their maximum in functional significance. Differences between plant species in the chemistry of leaves and wood have dramatic effects on such ecosystem properties as decomposition rate (Melillo et al. 1982) and resistance to herbivores (Coley et al. 1985). This type of chemical variation can strongly influence soil fertility, cation exchange capacity, and hydrologic storage capacity (Pastor and Post 1986).

Such chemical effects on ecosystem processes are particularly strong where the plant community is dominated by a single species or group of species. Pine-dominated ecosystems tend to decrease in soil fertility over time, due to the acidifying and leaching effects of the humic acids produced by decomposition of the needles (Pastor and Post 1986). These plant chemical properties can potentially influence the quality of water draining through the soils, such as nitrate leaching from high elevation ecosystems (Johnson and Lindberg 1992) or the humic-stained water of "blackwater" rivers and northern lakes. Many of the chemicals that influence decomposition and herbivory also have value for agriculture and medicine. Equally significant chemical differences are found among fungi and bacteria.

### 2.1.5 Temporal Dynamics and Life History Strategies

Plant succession influences both the amount and the form of primary production, as well as the variety and abundance of animals present in an ecosystem. Succession results from a combination of processes that includes growth and competitive exclusion. Successional changes in plant communities can occur with or without shifts in climate, require multiple plant species with different functional properties, and affect the diversity of animals that can survive in the ecosystem.

Although functional differences produce the shifts in dominant species that we perceive as succession, this shift in species composition can help stabilize ecosystem processes such as primary productivity. As physical conditions change, better-adapted species replace species that are less suited to the new conditions than they were to the old conditions. This shift in species results in many ecosystem properties and processes being resistant to change. For example, the leaf area of forest canopies, the standing biomass of the forest, and the rate of primary production, can remain quite stable through shifts in species dominance. The replacement of the American chestnut (*Castanea dentata*) by chestnut oak (*Quercus prinus*) and other species in the eastern deciduous forest was apparently accomplished without major disruptions in many ecosystem processes.

### 2.1.6 Population Dynamics

In addition to the major changes in species composition as a result of successional processes or gradual environmental changes, the abundance of individual species may fluctuate dramatically as well. The sizes of populations can vary for a number of reasons, including variation in physical conditions (e.g., weather), biological conditions (e.g., food, habitat, predators, competitors, parasites, and diseases), and harvesting. Populations of some species undergo cyclic variations, the causes of which may or may not be well understood. Organisms with short lifetimes and potentially rapid population growth rates (such as insects, small rodents, some birds) are more likely to have cycles or fluctuations that can be easily observed and influence short-term management actions than are organisms with long lifespans and slow growth rates (large vertebrates, particularly predators).

Population fluctuations are likely to have a larger influence (and be more likely to be noticed) under two types of conditions. Where there are relatively few species, major fluctuations in population size are much more likely to be noticed than in a situation where there are a great many species (in general, where diversity is high most species will have smaller populations and be less conspicuous). Where productivity is high and the organisms can become very abundant, such as in agricultural fields or forest plantations, pests can quickly achieve high densities when conditions are suitable. For a variety of reasons that will be discussed in detail later, highly productive conditions often have naturally low species diversity, particularly of plants and the organisms (insects and other herbivores) that depend directly on the plants.

## 2.2 Ecosystem Stability

Probably the most important, and most intuitively obvious, consequence of biodiversity is that with a larger number of species, it is more likely that one or more species will survive and grow well enough when conditions change to continue to provide the ecosystem processes that were preformed by different species before the change or disturbance.

Ecosystem resilience relates directly to the issue of the ecosystem function of different species. Having multiple species performing the same function clearly provides a type of security analogous to the redundant backup systems used on spacecraft. Considering a species to be redundant in the performance of a particular function in no way diminishes its importance to the ecosystem, any more than the second or third backup hydraulic system or fuel cell is unimportant for a spacecraft.

Unfortunately, we have virtually no information on the functional roles or redundancy of most species, and few examples of what happens when all of the species that perform a particular function are eliminated. The most striking examples result from the natural, experimental, or accidental addition or elimination of single species. The concept of a "keystone" species (Paine 1966) is based on observations of the dramatic effect of removing a single species of predator or herbivore that had prevented the dominant prey species from eliminating most of the other prey species. In situations such as this, loss of a "keystone" predator results in a reduction of diversity among the prey organisms.

For example, in the last century, overhunting nearly eliminated the sea otter along the Pacific coast of North America. A direct consequence was a dramatic increase in one of its primary food items, sea urchins. An indirect effect of the urchin population explosion was increased herbivory on the giant kelp that provided most of the structure and primary productivity for the complex subtidal community. Only recently, with the reintroduction of the sea otter, have urchin populations been reduced, leading to the recovery of the kelp and the fish communities of the kelp beds (Duggins 1980).

Equally dramatic changes in ecosystems can occur when a new species is introduced into an ecosystem. As discussed in section 2.3 (Exotic Species), there are many examples of introduced species completely altering the structure and properties of an ecosystem. In such cases, it is clear that a particular function is being performed primarily by the added species, and that removal of the introduced species would presumably allow the ecosystem to return to its former state.

These examples of the effects of the addition or removal of single species illustrate the *lack* of ecosystem resilience when the diversity of a particular functional group is low and a single species dominates the function. Beyond the effects of single dominant species, there is virtually no information on how many redundant species are needed to maintain an adequate level of ecosystem resilience, nor on what level of ecosystem resilience is adequate under different circumstances.

Differences in resilience or stability between different ecosystems, or changes in resilience or stability through time, are often influenced much more strongly by variation in environmental conditions (such as drought) than by differences in species composition or diversity. For example, plant communities growing on fertile, well-watered portions of a landscape will recover biomass lost as a result of disturbances much more rapidly than plant communities growing on unproductive areas. Likewise, animal populations with abundant food resources will recover much more rapidly from a population reduction than populations with limited food resources. Rather than the number of species, it is the properties of the dominant species that can lead to different rates of ecosystem processes in regions with the same physical conditions (e.g., Wardle et al. 1997; Grime 1998).

## 2.3 Effects of Exotic Species

Many of the most conspicuous and serious changes in our natural ecosystems have resulted from the invasion of exotic species. These include: (1) the invasion of many southwestern and western springs and water courses by tamarisk from Australia, which can completely dry up wetlands because of its high evapotranspiration rates; (2) the invasion of fields and grasslands throughout the east by Russian thistle; (3) the invasion of annual grasses into western rangelands that has increased the frequency of fires and threatens to eliminate entire native plant communities; and (4) the invasion of the kudzu vine in the southeast, where it covers and kills trees along the edges of forests.

Exotic species brought by settlers have long been established throughout the West, often significantly altering native ecosystems. In what has been called an explosion in slow motion (Asher 1995) a number of these exotic plants are now spreading rapidly and often completely altering or destroying native ecosystems. Spotted knapweed (*Centaurea maculosa*) first reported in Montana in 1920 has since invaded nearly five million acres in that state alone. Cheatgrass (*Bromus tectorum*) is a particular problem in the northern sagebrush steppe regions of Idaho and eastern Oregon. It is highly competitive with native species such as bluebunch wheatgrass (*Agropyron spicatum*) which had already been widely reduced in both abundance and vigor because of persistent livestock grazing. Both cheatgrass and native wheatgrasses germinate in the fall. But unlike the native wheatgrass, cheatgrass continues root growth in the winter giving it an advantage in the spring during competition for soil moisture. Cheatgrass dries quickly following spring growth and leaves abundant dry fuels. This has led to greatly increased frequency of fires. Native species are not adapted to the intensity nor the frequency of the new fire regime. The result is often a pure stand of cheatgrass, which is maintained as new fires sweep through and help enlarge its area of dominance.

A similar phenomenon is evident throughout much of the desert southwest. A variety of exotic grasses is widely established in the Sonoran and Mojave Deserts. They support wildfire regimes in ecosystems thought not to be adapted to fire at all. Long-lived desert plants succumb to wildfires fueled by these invaders. Germination for many of these natives is very infrequent (i.e. during rare climatic events) or is dependent upon microclimates afforded by other natives. Saguaro cactus (*Carnegiea gigantea*), for example, typically germinates under the shade afforded by small trees, such as palo verde (*Cercidium*) or iron wood (*Olneya tesota*), and subsists there for several decades before gaining dominance over the host plant. Such highly specialized relationships among species in desert ecosystems appear to be at great risk because of invasion by exotic grasses.

Huenneke (1995) characterized ecosystem health as the provision of habitat for sustainable populations of native plants and animals; the maintenance of soil, nutrient cycling, and hydrological patterns; the maintenance of disturbance cycles typical of the evolutionary history of the region; and resilience (or natural recovery processes) in response to disturbance. Although some exotic species can occupy native ecosystems with little effect, some obviously have the ability to disrupt some or all components of ecosystem health.

Deleterious effects to native species caused by the introduction of exotic animals have been long known, but the consequences of exotic plant introductions are just beginning to be recognized. Changes to native bird species compositions have been attributed to Eurasian plant invasions (Wilson and Belcher 1989). Severe population reductions have been noted for native kangaroo rats and ground squirrels on areas infested with Russian knapweed (Johnson et al. 1994).

Exotic plants have been shown to alter natural soil properties, energy fixation, and nutrient cycling (Vitousek 1986). Replacement of native grasses with broad leaf exotics changes the root structure from an extensive fibrous root system to one with deeper and more widely spaced tap roots. Such structural changes can profoundly influence micro-climate, water budgets, nitrogen fixation, carbon fixing, nutrient cycles, and soil erosion. Impacts to associated riparian and aquatic systems have also been noted.

If the changes to ecosystem structure and function caused by exotic plants were to expand to landscape scales, as may be underway in parts of the western United States, major ecosystem types might become fragmented, degraded, and perhaps completely lost. Habitat degradation and fragmentation, reduction and isolation of populations, and disrupted inter-species relationships, regardless of their cause, are likely to have serious long-term negative consequences to native species and genetic diversity. Exotic species illustrate how important the properties of individual species can be for the structure and function of ecosystems.

## 2.4 Landscape Structure and Properties

Many environmental problems with air and water quality, as well as their potential solutions through management actions, are directly related to the physical structure of the landscape. Riparian buffers, grassed waterways, and other similar management practices are manipulations of the arrangement of organisms on the landscape in order to improve ecosystem functions related to water quality and flood prevention. Particularly in forestry and agriculture, management control over the arrangement of plants on the landscape can have a significant impact on water quality and quantity, air quality and temperature, and the distribution of both common and rare organisms. The species of plants and animals that occur in a small area are strongly influenced by the conditions in the large area that surrounds them.

## 2.5 Indicators of Environmental Change

Because many organisms are sensitive to environmental changes, the presence and diversity of certain types of plants or animals can indicate that the environment has somehow changed. In some situations, the responses of particular species can provide a great deal of information about the nature of the environmental change itself, as in the case of species sensitive to ozone, ultraviolet radiation, certain pesticides, or low temperatures.

Although a change in the species diversity of a particular functional group of organism is a clear indicator that the environment has changed, diversity is not a simple indicator of whether the environmental change was for the better or worse. In many situations, an "improvement" in environmental conditions can result in a decrease in the species diversity of certain groups of organisms. One of the most striking examples is the decrease in plant diversity that often results from an increase in soil fertility (Huston 1994). Changes in species diversity must be understood before they can be interpreted with regard to the condition of the environment, because any specific change in environmental conditions is likely to increase the diversity of some types of organisms and decrease the diversity of others.

Individual species or type of species can be reliable indicators of certain environmental conditions, and thus indicators of environmental change. Differential sensitivity of aquatic insects and fish to pollution or other types of degradation of water quality can be good indicators of water quality (Karr 1993). Similarly, many plants and animals have specific temperature optima or limits (e.g., freezing) or sensitivity to atmospheric conditions such as ozone concentration. The presence or absence of such species can be a good indicator of recent environmental conditions.

### 2.6 Esthetic Value

One important function of biodiversity is the pleasure that it provides to people through sights, sounds, and odors. These esthetic consequences of biodiversity can be experienced at scales ranging from a backyard garden or a single birdfeeder, to entire mountainsides and vistas of vast grasslands, deserts, or rainforest canopies.

Related to the esthetic values of biodiversity are the various moral concerns surrounding the issue of extinction and human relations with their environment. These issues are too complex to address here, but it is worth noting that a number of philosophers, as well as a growing group of conservative Christians, consider that the moral reasons for protecting biodiversity far outweigh the scientific ones (Randall 1994, Sagoff 1997).

## 3 PATTERNS OF SPECIES DIVERSITY

Strong regularities in the distribution of organisms in relation to their environment make the distribution of many components of biodiversity highly predictable. Understanding those patterns is the first step toward protecting and managing biodiversity. Knowing where to expect high (or low) diversity of certain types of organisms is essential for planning and evaluating management actions.

### 3.1 Abundance Distributions

#### 3.1.1 Rarity and Endemism

In general, species that are widespread tend to be common wherever they occur, whereas most naturally rare species are not particularly widespread. There are several possible combinations of abundance and range size, of which only a few are commonly found in nature. One particularly important type of natural rarity is called endemism.

Endemic species are simply species that are naturally restricted to a small area. Some endemic species are found only on a single hillside or mountain, whereas other species may be defined as endemic because they occur only in a single state, or only on a particular continent. Oceanic islands typically have many endemic species of plant and animals, as is the case with the Hawaiian Islands. Similarly, the island continent of Australia is known for its many unusual endemic species.

Most endangered species or subspecies are endemics (Flather et al. 1994). Because of their limited range, their populations are invariably small, and any type of habitat destruction within their range is likely to affect a large proportion of their total habitat. Consequently, urban, industrial, or agricultural development poses a major threat of extinction to endemic species. Fortunately, endemic species are not uniformly distributed, but tend to be concentrated in specific types of habitats, with much of the earth's surface having a relatively low density of endemic species.

On mainland areas, and even within islands, endemic plant species are typically found in unproductive environments, where low soil nutrients or lack of water prevent more competitive plant species from eliminating them (Huston 1994). The highest levels of plant endemism in the United States are found in the dry, rocky, or sandy parts of Florida, Texas, New Mexico, Arizona, and California. Endemic animal species are often also found in unproductive environments, although spatial isolation, such as mountain ranges or valleys, is usually associated with endemism in both plant and animal species. Because endemic species are found in a specific subset of environments, protecting rare species of this type generally does not conflict with timber management or agriculture on productive lands (Huston 1994).

#### 3.1.2 Invasions and Outbreaks

Although invasions can occur naturally, they more commonly are a result of human activity. Humans increase invasions in two primary ways: by trans-

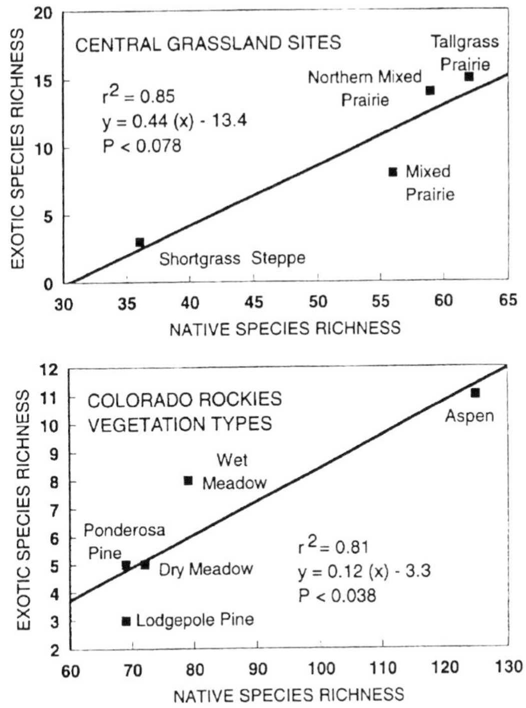


Fig. 2. Patterns of exotic species richness in relation to native species richness in selected herbaceous vegetation types of the Great Plains Grasslands and Colorado Rocky Mountains, based on combined species lists for four 1000 $m^2$ plots in each vegetation type (from Stohlgren et al. in press).

porting species long distances and across barriers that they could not naturally cross, and by disturbing or otherwise altering natural communities in ways that make them more easily invaded and dominated by new species.

Ecologists have long theorized that highly diverse communities are more resistant to invasion: the many species in a high-diversity community would more completely utilize the environmental resources than the fewer species of a low-diversity community, and make it more difficult for new species to invade. Recent experimental evidence apparently supports the invasion resistance of high-diversity plant communities (Tilman 1997), but this pattern is not found in all situations. Extensive vegetation surveys in Great Plains grasslands and meadows in the Rocky Mountains (Stohlgren, in press) demonstrate that the highest rates of invasion of exotic species is found in the areas with the highest diversity of native species (Fig. 2).

These apparently contradictory patterns are clarified when invasion patterns are interpreted in terms of plant biomass or cover. When plant cover is high, as in forests or productive grasslands, invasion rates are low, regardless of species diversity. Likewise, when plant cover is low, as a result of disturbances or low productivity, invading species can easily become established, regardless of the species diversity of the native community. Observations around the world indicate that unproductive habitats, such as infertile or rocky soils, are readily invaded. However, invaders of unproductive areas rarely become dominant unless they are free of natural pests or are able to escape some of the limits to productivity (such as with deep roots to water, or nitrogen fixation). In contrast, productive environments, which are difficult to invade *unless* they have been disturbed, can be dominated by successful invaders or by successful native species (Fig. 3). This pattern of susceptibility to invasion and dominance is strongest for plants, and sufficiently predictable that it can provide some guidance to managers.

Although the patterns of plant invasion are fairly predictable, invasions by organisms at higher trophic levels, such as diseases, pests, and predators, are much less predictable and can have severe and rapid impacts. Epidemics such as chestnut blight, Dutch elm disease, and tracheal mites of bees, as well as the recent invasion of predators in tropical environments, can be

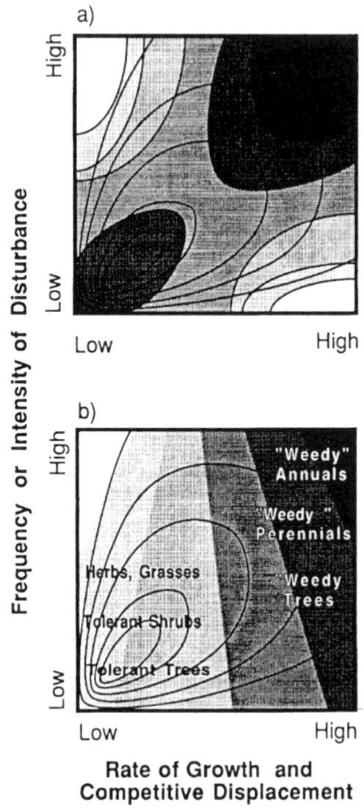


Fig. 3. Predicted community susceptibility to invasions and degree of dominance of invading species in environments classified according to disturbance and productivity (Huston 1979). (a) Predicted susceptibility of communities to invasion. Darker shading indicates higher susceptibility. Note that communities with low diversity are least likely to be invaded successfully. (b) Predicted life histories of successful invaders under various combinations of productivity and disturbance. Shading indicates expected dominance of a community by a successful invader. Note that if invaders alter the disturbance regime by increasing frequency or intensity, the community will shift to lower diversity and higher dominance by the invader (from Huston 1994).

extremely difficult or impossible to control once they are started.

## 3.2 Spatial Structure

A phenomenon as complex as biodiversity cannot be summarized in a few simple numbers. Both the distribution of organisms and the distribution of the environmental conditions that affect the organisms must be classified according to specific spatial scales and specific functional attributes to describe them in a manner that contributes to understanding, rather than to confusion.

The basic statistical descriptor of biodiversity is the number of different types of the units of interest, which may be genes, species, vegetation types, forest types, or stand ages. Species diversity is most simply quantified as the number of species in a given area (often called species richness). Other statistical descriptions of diversity include relative abundance, generally considering situations in which most species have similar abundances to be more diverse than situations in which one species is abundant and most are rare. Various statistics have been developed (e.g., Simpson's Index, the Shannon–Weiner Index) that represent different combinations of the number of species in a sample and their relative abundances (see Magurran 1988 or Huston 1994).

### 3.2.1 Different Scales of Species Diversity

Ecologists have historically (since Whittaker 1960) defined three different spatial scales at which species diversity is affected by three different types of processes.

At the smallest spatial scale is local diversity (called alpha diversity). This is generally considered to be the scale at which the organisms are potentially interacting with each other, through processes such as competition, predation, or mutualism. For grasses and herbs this scale is often considered to be a square meter or smaller, while for trees it may be an area 10 meters in diameter, or even larger for large trees. Studies of local diversity in forests typically use plots from 0.1 to 1.0 hectare in size. At this scale, the increase in species number with increasing sample area is considered to be a statistical consequence of sampling randomly distributed individuals. The physical environment is considered to be uniform at this scale, and plots are usually chosen with homogeneity of soils and other physical properties as some of the criteria.

Homogeneous area are usually not very large, and quite different environments can often be found close together. The second spatial scale for diversity measurement is called between-habitat diversity (called beta diversity). Between-habitat diversity is the result of sampling different environments that have different sets of species. The greater the number of different environments that are sampled, the greater will be the total number of species. The increase in species number with area at this scale results from environmental heterogeneity (as contrasted with the homogeneous environment addressed by local diversity) and the response of organisms to the environment. Between-habitat diversity can be high if there are large differences in the environmental conditions of the different habitats that are sampled, or if the species are very sensitive to environmental conditions and small changes in environmental conditions result in large differences in species composition. In addition to the total number of species found, between-habitat diversity is sometimes described using statistics that quantify how much the species composition changes from one environment to another.

The next larger scale at which a new set of processes is considered to influence diversity is the regional scale (called gamma diversity). At this scale (which may be as large as continents or the entire globe) diversity increases with increasing area because similar habitats in different regions have different species. The full range of habitat differences that contribute to between-habitat diversity is included within this scale, and diversity increases because habitats that are physically similar, but separated by large distances, have different species that are functionally similar but genetically distinct. Diversity at this scale is considered to result primarily from evolutionary processes, with different groups of species evolving (or invading and surviving) in similar habitats that are separated by a sufficient distance that the mixing of species and genes occurs at a very low rate. One of the interesting aspects of regional diversity is that the minimum distance that seems to be required to allow different species to occupy similar environments can vary greatly from one part of the world to another (Huston 1994). At this scale, conservation of biodiversity requires preserving the different distinct groups of species, and cannot be achieved simply by preserving representative environments. On the other hand, although the species may differ from one area to another, the environmental similarities, along with the functional similarities between the groups of species, imply that the same management actions are likely to work in the same way in similar environments in different regions or continents.

### 3.2.2 Environmental Heterogeneity

Spatial variability in environmental conditions is probably the most important single explanation for variations in biodiversity. A combination of wet and dry environments will inevitably have more species that either a wet or a dry environment alone. These differences are usually described in terms of the concept of "niche," with different species using their environment in ways that are sufficiently different to be classified as being different niches. Scientific arguments continue about how niches should be defined, and whether they actually exist, but it is undeniable that, for whatever reason, different species tend to be found in different environments.

Around the world, at all spatial scales, the environmental variability is a major explanation for the diversity found in different areas, particularly at large scales (between-habitat and regional diversity). Diversity tends to be higher in heterogeneous areas than in homogeneous areas, and diversity can often be increased by increasing environmental heterogeneity, through harvest patterns or earth moving.

Within a region or even a relatively small local area, the differences in environmental heterogeneity that are typically manifested in patterns of plant height and species composition are often correlated with variation in the diversity of birds, small mammals, insects, and some groups of plants. Heterogeneity at this scale can be detected in standard satellite imagery, and analyses of the heterogeneity of color patterns in satellite images (French SPOT images with a ground area resolution of 20×20 m) have found significant correlations with the measured species diversity of birds and small mamals (Podolsky 1997). Such patterns in the heterogeneity of vegetation, soils, topography, and other spatial properties are likely to be correlated with many components of species diversity in all regions of the Earth. It is reasonable to expect that most components of biodiversity will increase with increasing spatial (and possibly temporal) heterogeneity in the environment. However, the rate at which diversity increases with heterogeneity is likely to differ from one region to another.

### 3.2.3 Environmental Gradients

Many of the Earth's most dramatic patterns of biodiversity occur along natural environmental gradients, where both environmental conditions and the distribution and diversity of species change gradually over some distance. Examples include changes in species composition and diversity with increasing elevation along a mountainside, with changes in soil fertility along a floodplain, with changes in rainfall from the edge to the interior of a continent, or with decreases in the minimum and average temperatures as one moves from the equator toward either pole. Observing what environmental conditions change along gradients over which species diversity also changes provides important clues about the causes of variation in biodiversity.

Biodiversity changes along virtually all environmental gradients (temperature, precipitation, soil fertility, humidity, light, water chemistry), often for reasons that may seem obvious. However, it is also true that along virtually any gradient, the diversity of some types of organisms will be increasing, whereas that of other types of organisms will be decreasing. Not only does diversity change along gradients, but the performance and properties of individual species may change greatly along the gradient as well. For example, many plants grow to large sizes and may dominate the ecosystem at the favorable end of the gradient of soil fertility and/or moisture, but are present only as small, stunted individuals that may be difficult to find at the

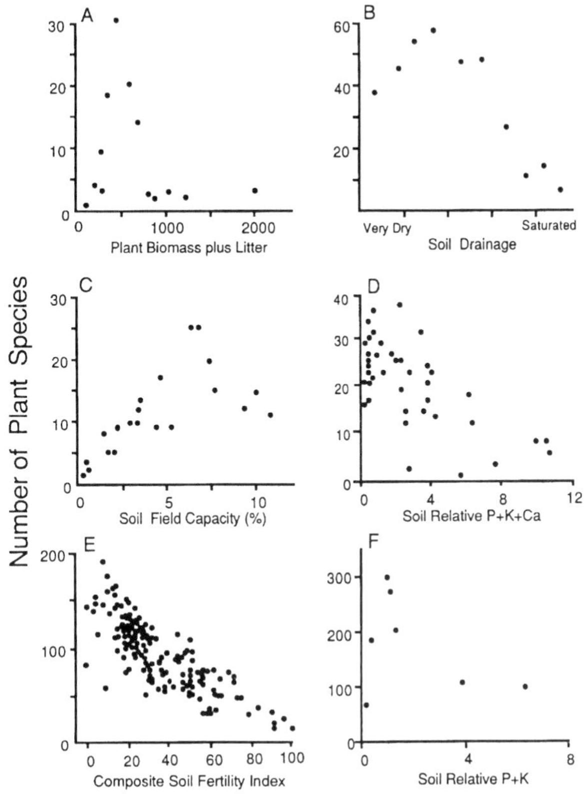


Fig. 4. Patterns of plant species diversity in relation to factors correlated with productivity (e.g., plant biomass, soil fertility) in different vegetation types. (A)–(C) are oldfields and grasslands, values are number of plant species per unit area; (D)–(F) are tropical forests, values are number of tree species per 0.1 ha, number of vascular plants per 0.1 ha, and number of tree species per ha, respectively. Note that in all cases the number of plants species declines at the highest levels of fertility or other factors correlated with productivity, while the maximum number of species is found at intermediate to low levels of the same factors (from Huston 1993, 1994).

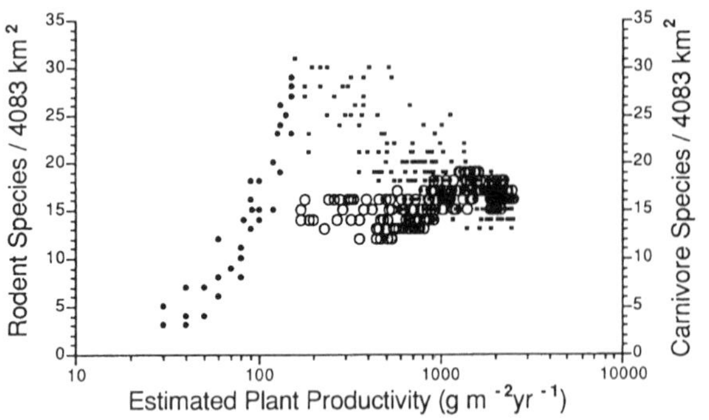


Fig. 5. Animal diversity along a productivity (rainfall) gradient across the southwestern U.S. Small dots indicate number of rodent species in gridded areas of 4083 square kilometers. Open circle indicate number of carnivore species in the same sized area. Note that maximum carnivore diversity occurs at a higher level of plant productivity than maximum rodent diversity (based on Owen 1988).

poor end of the gradient (Smith and Huston 1989). Along such gradients, plant diversity is very low under extremely unfavorable conditions, reaches its maximum under low productivity conditions, and decreases under the most productive conditions where plants are large and a few species dominate (Fig. 4). Similar patterns are found among animals (Fig. 5). In addition to changes in biodiversity, ecosystem processes also change along environmental gradients, with shifts in the rate at which processes occur, as well as in which processes are dominant.

In situations where a management area spans a natural environmental gradient, the effect of any particular management action is likely to change dramatically from one end of the gradient to the other.

### 3.2.4 Isolation and Dispersal

Because species differ greatly in their ability to move and disperse, the distance that organisms must travel to reach a given area has a great effect on what organisms will be found in that area. This effect of isolation is most conspicuous on oceanic islands, such as the Hawaiian and Galapagos Islands. These islands are so far from sources of potential colonizing species on mainland areas, that most species on these islands are endemics which evolved from the few species that reached the islands by chance over the past several million or more years.

Endemic species or subspecies are also found in areas that are isolated by conditions other than water, such as alpine mountaintops in the middle of deserts, or lakes that are far from other lakes.

Over long time scales isolation can lead to speciation. However, over shorter time scales, particularly when isolation results from human activities such as habitat destruction and fragmentation, isolation can lead to extinction by a variety of mechanisms. Isolation can reduce gene flow, leading to problems associated with inbreeding. This problem is likely to be particularly severe in small populations. Isolation can also reduce the probability of dispersal and recolonization of areas where local populations have become extinct. A group of partially isolated local populations that are connected primarily through migration or dispersal is called a "metapopulation" (Hanski and Gilpin 1991). The phenomenon of extinction of local populations followed by recolonization from other local populations can prevent a regional population (the metapopulation) from becoming extinct (as well as prevent genetic problems). Any factors that decrease dispersal and recolonization, including increasing distance (isolation) between local populations, increase the probability that the metapopulation will become extinct.

Although isolation is usually considered to have primarily negative effects on population viability, genetic structure, and species diversity, in many cases some degree of isolation provides some benefits as well. Populations of certain species may escape predators or diseases through isolation, thus allowing them to survive and contribute to regional diversity. For example, certain species of amphibians and insects are found only in ponds or lakes that lack predatory fish (Werner and McPeek 1994). Much work shows that isolated individual plants are less likely to be found and eaten by herbivores (or affected by diseases and pathogens) than are plants in large groups. However, species that escape danger through isolation are likely to be affected severely if or when that danger eventually finds them.

Creation or destruction of isolated habitats are management actions that inevitably affect the survival of specific populations, and thus can affect species diversity. Habitat fragmentation has a variety of negative effects, including the reduction of total habitat area as well as the isolation of remaining fragments. As a general rule, maintaining the isolation of naturally isolated habitats and avoiding the creation of isolated patches of previously widespread habitats are the best management strategies with regard to isolation.

## 3.3 Temporal and Spatiotemporal Patterns

Patterns of biodiversity on landscapes are constantly changing through time as a result of natural processes, with or without human intervention or management actions. Understanding these patterns of spatial and temporal change is essential both for planning management actions, and for evaluating the effects of those actions.

## 3.3.1 Succession

Succession is a temporal pattern of changing abundance of species that results from changes in environmental conditions. The relevant environmental conditions may change as a result of: (1) the actions of the organisms themselves; (2) changes in climatic patterns; or (3) geomorphic processes, such as erosion and deposition, which alter the distribution of resources across a landscape. In general, the rates at which successional changes occur is highest for changes produced by the organisms themselves, and decreases through types 2 and 3 above.

The best-known type of succession is plant succession, which is typically a predictable pattern of increasing plant size through time, as in the sequence of grasses, shrubs, and trees, that results from environmental changes caused by the plants themselves. Changes in ecosystem properties during succession are generally quite predictable, whereas the identity of the species involved can be more variable (although in situations with few species, the changes in species composition are often predictable). Most often, plant succession occurs after the previous plant community has been completely or partially removed by a natural disturbance or by harvesting. This is called secondary succession (Fig. 6). Primary succession occurs on substrates that have not been previously occupied by plants, such as lava flows, sandbars, glacial outwash plains, or landslide scars. Primary succession generally occurs more slowly than secondary succession, because the substrate is often unfavorable for plant growth. Improvement of the substrate, including soil formation, occurs as a result of effects of plants and other organisms, such as lichen, fungi, and nitrogen-fixing bacteria (Fastie 1995). As the fertility of the substrate improves, plant biomass can increase and larger types of plants can survive.

The rates of change during plant succession are related to the growth rates and life spans of the plants involved. Forest succession typically occurs more slowly than succession in prairies or marshes. Succession among algal species in lakes can occur very rapidly, typically completing a similar cycle of seasonal changes in species composition each year.

Successional patterns that result from climatic or geological changes (allogenic succession) occur much more slowly and have a different pattern from successional changes that result primarily from the effects of the organisms themselves (autogenic succession). Autogenic succession generally follows a pattern that is quite predictable in terms of the properties of the organisms that dominate at different times during succession. For example, plant size generally increases during primary and secondary succession, and later successional species tend to be more shade tolerant

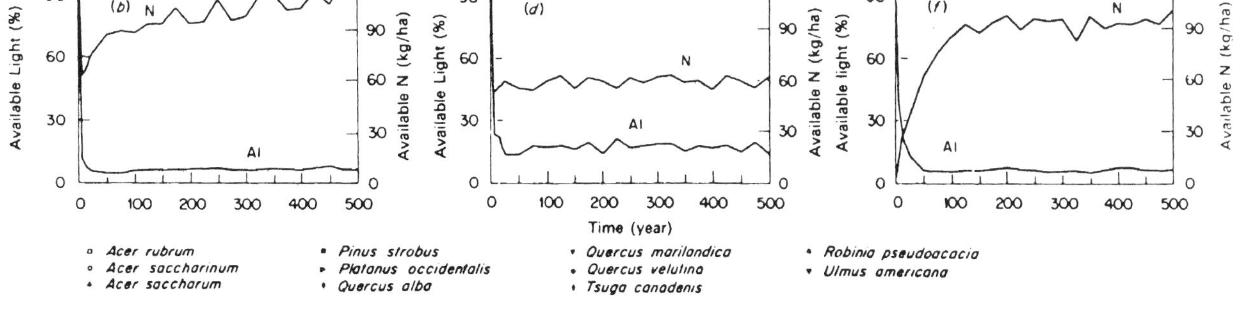

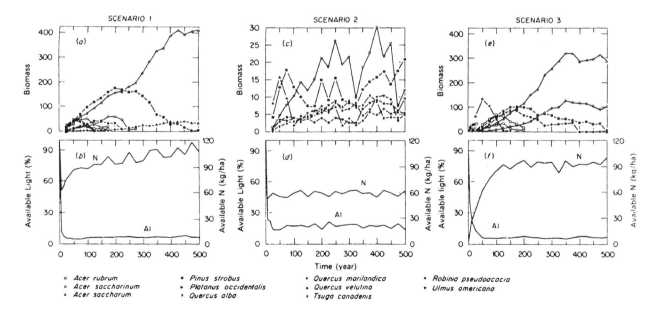

Fig. 6. Effect of environmental conditions on species diversity and composition during secondary succession in eastern deciduous forests. All three successional sequences were produced by an individual-based plant competition model (Pastor and Post, 1986) in which the same set of species were available for establishment in each sequence. The differences between the simulations are in the moisture-holding capacity of the soil (scenario 2 less than scenario 1) and the initial nitrogen content of the soil (scenario 3 higher than scenario 1). The drier soils of scenario 2 result in much slower growth and higher diversity than in scenario 1. The higher nitrogen of scenario 3 results in initial dominance by a nitrogen-fixing legume (*Robinia pseudoacacia*) and a slowed overall rate of successional change. Low panels show changes over succession in the levels of two major resources, available light at ground level (Al) and soil nitrogen (N), both of which are affected by the species composition and the total plant biomass (from Huston and Smith 1987).

than early successional species. Allogenic succession is driven by changes in physical conditions (such as climate) over which the plants have no influence. Consequently, the properties of the species that dominate at any point in the successional sequence can only be predicted if the changes in physical conditions are known in advance.

Manipulation of successional dynamics is one of the most effective means of managing landscape structure and diversity.

### 3.3.2 Functional Types and Guilds

The abundance of different functional types (or guilds) of species inevitably changes over the course of succession. In fact, plant succession can be interpreted as a change in dominance among different functional types of plants (Smith and Huston 1989, Huston 1994). Associated with changes in the structure and species composition of the plant community are changes in the types of animal species that utilize the various resources provided by the plants. For example, bird species diversity generally increases during plant succession, as the size and structural complexity of the plant community, as well as its diversity, increase through time. In addition, certain types of birds (and other animals) are found primarily in early successional vegetation, whereas others are limited to "old growth" vegetation.

The "habitat requirements" of many bird species are well known, and such information is also available for many mammals, reptiles, amphibians, and even insects (Szaro and Balda 1979, Verner et al. 1986, Ralph et al. 1991, DeGraf et al. 1991, Mills et al. 1996). This predictability in the association of animal species with their environment is the basis of habitat assessment methods such as the Habitat Evaluation Procedure (Short and Williamson 1987, Short 1992).

### 3.3.3 Invasion and Replacement

Successional changes in plant species abundance occur even when all species are present in early succession. However, in some situations all potentially available species are not present initially. When such species do arrive, through natural dispersal or other means, they may quickly invade the plant community, changing its structure and altering the pattern of succession that would have occurred in its absence.

Such dispersal and invasion is a natural process when native species are involved. When non-native (exotic) species are involved, however, the process is primarily a human-caused phenomenon, and the effects on the natural community can be undesirable. Susceptibility to invasion by either native or exotic species can be predicted on the basis of the same factors that influence the structure of the natural community (see Section 3.1.2).

### 3.3.4 Landscape Rates of Change

Some types of changes permanently alter the physical structure of the landscape, whereas others affect the plants and animals without causing permanent changes. The most rapid landscape-scale changes occur as a result of disturbances such as floods, landslides, or fires, which may permanently alter the structure of the landscape. Plant succession, generally with a pattern of increasing plant size, occurs continuously on nearly all landscapes. In early succession, species composition and stature of vegetation can change rapidly under productive conditions, and changes can be observed on an annual basis. Under unproductive conditions, or late in succession when plants are large, rates of change are much slower and may only be detectable at timescales of 10 years or longer.

Processes such as soil formation, erosion, and geological uplift or subsidence generally occur at slower rates. The physical processes that affect landscapes and the organisms upon them are discussed in more detail in Gosz et al. and Lugo et al. (this volume).

## 4. PROCESSES AFFECTING SPECIES DIVERSITY

The following sections describe the current state of scientific understanding about the multiple processes influencing genetic and species diversity. This process-level understanding provides the basis for predicting the effects of management actions, and how those effects are likely to vary under different environmental conditions.

### 4.1 Extinction and Area Effects

The process that has the most unequivocal negative effect on species diversity is extinction. Extinction can be defined at different scales. When the last individual of a species dies, diversity is reduced from the global scale to the particular location where the last individual was found. In contrast, local extinction refers to the disappearance of a species from a particular area, while the same species continues to survive in other locations. For example, the cougar is apparently extinct in Great Smoky Mountains National Park, but it still occurs in healthy populations in many parts of North and South America. Fortunately, local extinction is not forever, and can be reversed by natural immigration or intentional re-introduction.

Small populations are more likely to become extinct than are large populations, although some species with initially very large populations, such as the passenger pigeon, have also become extinct (the American bison narrowly escaped the same fate). Species that have small populations because they occur in a relatively small area (i.e., endemic species), or in a rare habitat type, are more likely to have their entire habitat or entire population destroyed by a single catastrophe. The fatal event may be either a natural catastrophe such as a flood or drought, or human-caused habitat destruction such as draining of wetlands or deforestation. Management solutions are to increase the available area of suitable habitat, or transplant individuals to existing areas of suitable habitat where the species does not currently occur because of previous local extinction.

Species with small populations are particularly susceptible to a number of problems: (a) males and females may have a reduced probability of encountering one another for successful reproduction; (b) short-term local increases in mortality rates or decreases in reproductive rates; (c) negative genetic effects of inbreeding; and (d) "demographic stochasticity," the possibility that, by chance, all (or most) offspring will be of a single sex.

Small populations experiencing any of the above problems generally require strong management intervention if they are to survive. Where global extinction is a possibility, a primary remedy is captive breeding programs, ideally coupled with habitat improvement and eventual reintroduction to the "wild." Where local extinction is the concern, inbreeding problems can potentially be solved by introducing individuals from other populations that have a different makeup. This is being attempted for the Florida panther by introducing cougars from Texas.

The majority of known extinctions of plants and animals have occurred among endemic species on islands (Lawton and May 1995), and within the United States, a significant proportion of the threatened and endangered species are endemics (Flather et al. 1994). The small population sizes of most endemic species are a natural phenomenon, and areas with high species diversity often have a high proportion of endemics. This relationship is most conspicuous in tropical rainforests, but is also found in areas of dry, infertile soils, such as in parts of California and similar mediterranean climate regions around the world (Huston 1994).

### 4.1.1 Species-area Curves

The number of species nearly always increases with the size of the area. This is described by the "species-area curve" (Fig. 7), which typically increases steeply from very small areas to areas of intermediate size, and then eventually begins to level off as most of the species are encountered. This pattern is found in many situations, from counts of the total number of species on oceanic islands of different sizes, to increasing the "quadrat size" of vegetation samples in a field. The shape of the species area curve may provide much information about the type of processes influencing species diversity.

Even within a relatively uniform area, it may take a large sample to encounter all the species. Consequently, the shape of the species-area curve is often used to estimate the total number of species that occur in the area. The goal is to determine the number of species at which the curve "levels off," which indicates that no more species are likely to be encountered. A similar pattern occurs with other sampling methods. For example, the number of species of insects that are attracted to lights (often ultraviolet lights are used for sampling insects) usually increases with time, so that

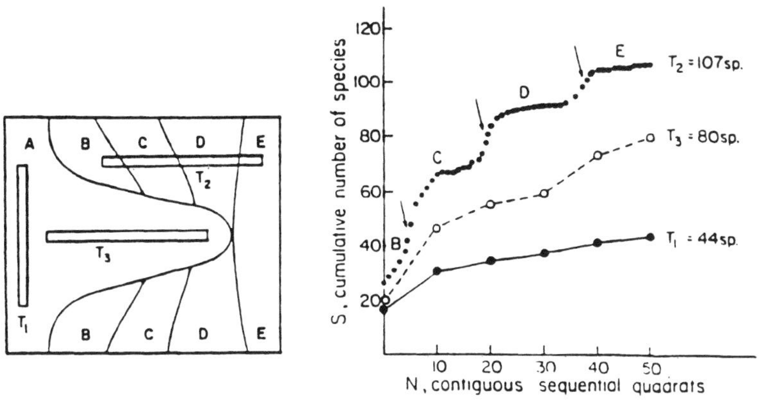


Fig. 7. Species-area curves showing the increase in plant species diversity with increasing sample area in the Judean desert of Israel. Transect T1 is in a homogeneous area and has the lowest species diversity. Transect T2 is in a heterogeneous area with four vegetation zones, and demonstrates a stairstep pattern caused by complete sampling of a single homogeneous habitat, followed by sampling in another habitat. Transect T3 is in a homogeneous area, but shows the influence species from the nearby zoned area (from Shmida and Wilson 1985).

more species are found in a sample collected over 10 hours than in a 1-hour sample in the same area. In cases such as this, "sampling effort" or time may be correlated with the total area that is sampled. A number of methods have been used to estimate the total number of species from these curves (Magurran 1988, Rosenzweig 1995).

If additional habitat types are included in the sample area, a new group of species enters into the sample. This can cause a species-area curve that has leveled off to begin to rise again. until all of the new habitat has been sampled. This component of the total number of species in the sample area is called "between-habitat" diversity (or beta diversity, as described earlier). The rate at which the species-area curve begins to increase is a consequence of the spatial heterogeneity of the area.

## 4.2 Mutualism and Survival

All species depend to some extent on other species for survival. While the individual organism that is the victim of a predator or herbivore generally does not benefit from the action of its consumer, numerous studies have demonstrated a variety of ecologically and evolutionarily beneficial effects of predation and herbivory on populations, species, and ecosystem processes (Owen and Wiegert 1981; Huntley 1991; Pastor et al. 1993).

Other relationships between species are mutually beneficial. In many situations, the benefits of a mutualistic relationship are shared among a group of species, such as the "generalist" pollinators that visit the flowers of many different species of plants. Other mutualistic relationships are much tighter, such as those between a plant that has only a single highly specialized species of pollinator, which depends for its survival on the resources provided by the plant. Loss of pollinators can lead to reproductive failure and extinction. These obligate mutualisms are found most commonly in situations where resources are scarce, and neither species alone could obtain all the resources it needs to survive (Huston 1994). An important example of obligate mutualism is that between many plant species and mycorrhizal fungi, which help the plant obtain scarce soil nutrients in return for sugar that allows the fungus to grow and reproduce. Mycorrhizae are essential for the survival and growth of many tree species, particularly when the trees are growing on poor soils (Perry et al. 1987). Many reforestation projects have failed because they did not assure that planted seedlings had mycorrhizae on their roots (Perry et al. 1987).

Mutualistic relationships are extremely important for diversity in certain situations. In the case of obligate mutualisms between two species, the extinction of either species guarantees that the other species will be lost as well. Some species, such as certain mycorrhizae or plants whose fruit is essential to the survival of many bird species, are particularly critical, because the loss of a single species (or type of species) can lead to the loss of many additional species that depend on it.

## 4.3 Environmental Heterogeneity and Niche Structure

The "niche" is one of the best known concepts in ecology, although not necessarily the best understood. A species' "niche" is considered to be that subset of all components of the environment that it needs for its survival. Typically, a niche is described in terms of the portion of various general environmental resources that a species uses or requires. For example, a bird species might require open grassland and eat insects between 1/2 and 3 centimeters in length. Other niche parameters that could be used to describe a bird species would include nesting requirements, and temperature or elevation range. In the context of the niche, a species' habitat is that portion of the landscape where all of the species' niche requirements are met.

The widely used habitat suitability index (HSI) and related habitat evaluation procedure (HEP) are based on the concept of niche, and represent an attempt to identify and measure those environmental conditions that meet the niche requirements of a species. Habitat suitability indices have been developed for many game and non-game species (Verner et al. 1986; Mills et al. 1996), and provide a good estimate of where a particular environment is suitable for a given species (although not whether that species will actually be found there).

The theoretical basis of the niche concept is that species evolve to use resources that differ from those used by other species to avoid competition for those resources. The "competitive exclusion principle" is based on the assumption that two species that both use the same environmental resources (i.e., have the same niche) cannot coexist, because one will inevitably be a slightly better competitor and will drive the other to extinction (generally, local extinction). Thus, "niche differences" indicate that species have evolved to avoid competing with other similar species. Many data are consistent with this interpretation, including observed differences in food type use, bill length, and body size between coexisting species of animals, and the observation that very similar species do not usually occur together, but rather occupy different areas. These niche differences are particularly conspicuous among

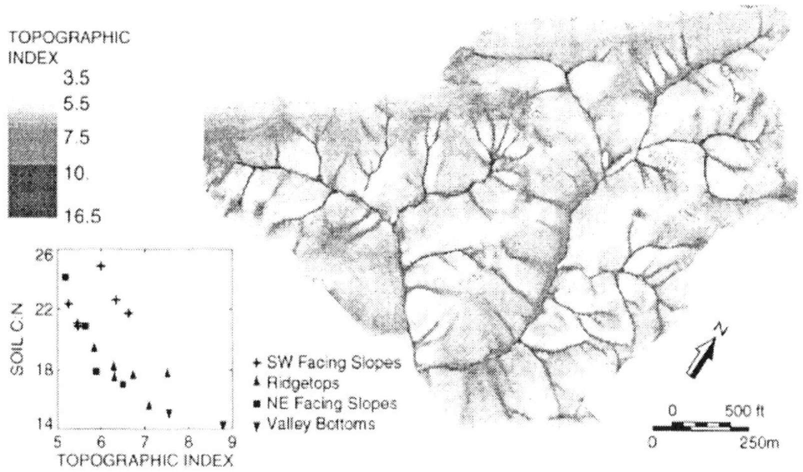


Fig. 8. The relationship between landscape topography and carbon-to-nitrogen ratios in the soils on Walker Branch Watershed, near Oak Ridge, Tennessee. Soil C:N is a good indicator of soil nitrogen dynamics, with lower C:N ratios being associated with higher availability of nitrogen for plants, and potentially higher forest productivity. The map of Walker Branch Watershed was developed from a high resolution digital elevation model (DEM) and is shaded using a hydrologic wetness index based on drainage area and slope (darker shades indicate wetter average conditions, see Beven and Kirby 1979; Beven and Moore 1993). The inset graph shows the strong correlation between soil C:N and the topographic wetness index. The strong effect of soil moisture on forest species composition (e.g., species distributions along a dry to wet gradient) and the significant differences in leaf chemistry between xeric species such as oaks and pines, and mesic species such as maples and tulip poplar, produce the observed pattern of surface soil C:N, which is determined primarily by the C:N chemistry of leaf litterfall (based on Garten et al. 1994).

vertebrates, but are less obvious among plants. Among plants, it is common to find that most species grow best under the same conditions of abundant resources.

Areas that have a wide range of environmental conditions almost always have more species than uniform areas with only a narrow range of conditions. This environmental heterogeneity is the dominant influence on species diversity in most situations. Environmental heterogeneity results from many factors, and different scales of heterogeneity are important for organisms with differing size and mobility. Elevation is a dominant component of environmental heterogeneity at all scales. Even at small scales, such as along a hillside or in a marsh, relatively small differences in elevation can have a major effect on water conditions and soil nutrients (Fig. 8). Areas with a greater range of elevations typically have more species than areas that are relatively level.

Organisms themselves can provide environmental heterogeneity. Plants, in particular, provide much heterogeneity, through food resources for herbivores, and structure provided by trunks, branches, and leaves (McMinn and Crossley 1996). Animals also provide heterogeneity by disturbing or moving soil, killing plants, and concentrating nutrients in dung or carcasses.

## 4.4 Productivity and Survival

Productive areas, with high availability of essential resources, can support high population growth rates which produce enough offspring to disperse to other areas that may be less productive or have higher mortality rates (Fig. 9). Those portions of a landscape have been called "source areas" and "sink areas," respectively (Pulliam 1988). Being able to distinguish source areas from sink areas is obviously critical for conservation planning and resource management, but it is important to recognize that climatic conditions have a strong influence on productivity. An area that is a "source" in a "good" year can be a "sink" in a "bad" year.

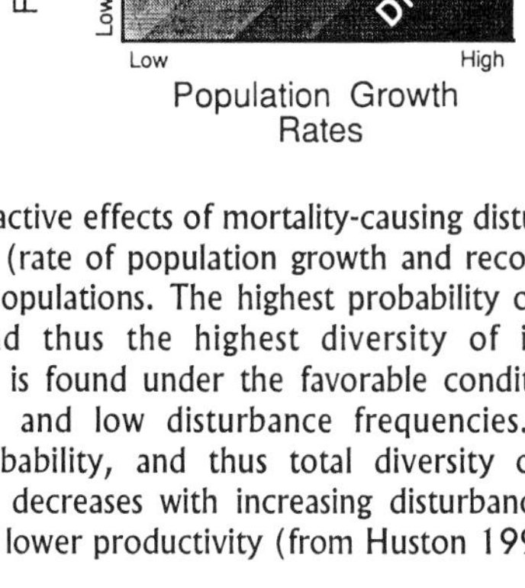

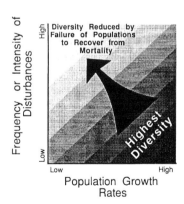

Fig. 9. Interactive effects of mortality-causing disturbances and productivity (rate of population growth and recovery) on the survival of populations. The highest probability of population survival, and thus the highest diversity of independent populations, is found under the favorable conditions of high productivity and low disturbance frequencies. Population survival probability, and thus total diversity of coexisting populations, decreases with increasing disturbance frequency and lower productivity (from Huston 1994).

## 4.5 Productivity and Competition

Along many productivity gradients, both terrestrial and aquatic, the number of some types of species decreases with increasing productivity. This pattern is particularly conspicuous in plants, which suggests that there are fundamental differences in the regulation of animal versus plant diversity. The typical situation on productive soils is that one or a few species of plants grow rapidly, become large, and shade out the smaller species. Under less productive conditions, most individual plants remain small, and species diversity is typically much higher. Plant diversity always increases with increasing productivity beginning at the very lowest levels, but then decreases from intermediate to high productivity, producing what has been called the "hump-backed" diversity-productivity curve (Grime 1973, Al-Mufti et al. 1977) (Fig. 10). This pattern of diversity is found along productivity gradients in all vegetation types, from grasslands to forests. Under unproductive conditions, plants are relatively small, there is much open area, and light is abundant at ground level. Under these conditions, there is little competition for light and species diversity is often high. As productivity increases, plants grow larger, larger species of plants can survive, open area is reduced, and there is little light at ground level. Intense competition for light (and perhaps other resources) reduces species diversity through competitive exclusion.

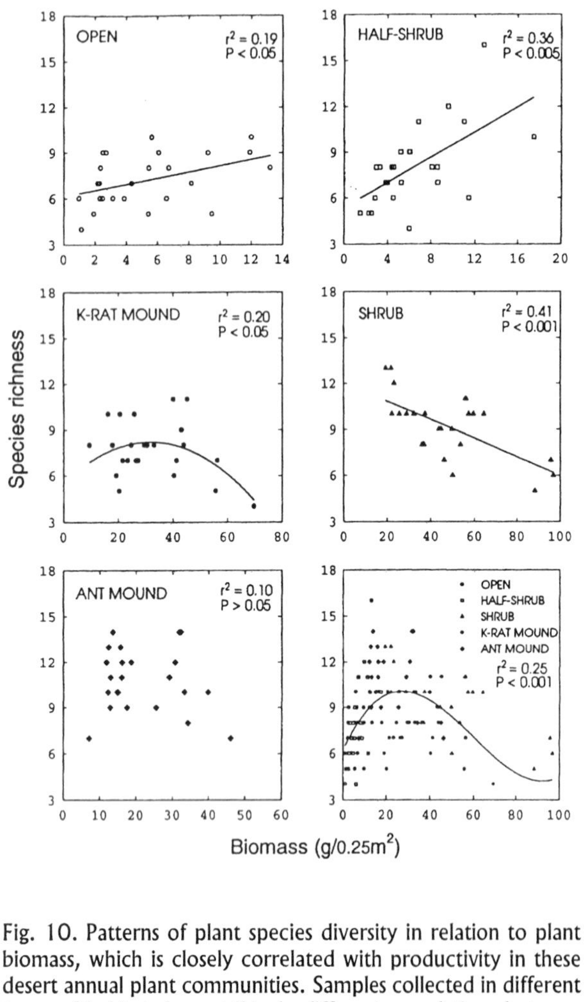


Fig. 10. Patterns of plant species diversity in relation to plant biomass, which is closely correlated with productivity in these desert annual plant communities. Samples collected in different types of habitat show strikingly different correlations between diversity and biomass. However, when the samples from different habitats are assembled into a single biomass gradient, the contradictory patterns are revealed to be different portions of the typical "hump-backed" diversity-productivity curve (from Guo and Berry 1998).

The "hump-backed" pattern of plant diversity is found along productivity gradients (e.g., from the top of a hill with dry, infertile soil, to the bottom of the hill where soils have higher nutrient content and water availability), but has been repeatedly demonstrated with fertilization experiments, beginning with some pasture experiments in England in 1843 (Lawes et al. 1882; see reviews in Huston 1979, Tilman 1993). The same reduction in diversity is occurring on poor soils in heavily industrialized regions, where nitrogen deposition from atmospheric pollution is causing an increase in productivity (Berendse et al. 1993). A similar phenomenon is found in aquatic systems, where addition of nutrients (particularly phosphorus) results in greatly increased algal production and reduced algal diversity (Proulx et al. 1996). This process, called eutrophication, is often associated with reduced water quality and unpleasant odors associated with algal growth and decomposition, as well as reduced oxygen levels and fish kills.

## 4.6 Interactions of Productivity and Disturbance

The importance of mortality-causing disturbances in maintaining species diversity is being increasingly recognized. The role of fire in maintaining the structure and species diversity of plant communities such as prairies, savannas, and woodlands is now well understood, and other disturbances, such as hurricanes and thunderstorm downbursts, are now known to be important factors in forest dynamics in many situations (White et al. this volume, Engstrom et al. this volume). Disturbances can have a similar positive effect on species diversity in some aquatic and marine ecosystems as well (Huston 1994).

Although some level of mortality can sustain higher levels of diversity, disturbances that occur too frequently or at too high a level of mortality can reduce diversity. This "intermediate disturbance hypothesis" suggests that diversity is highest at some intermediate level of disturbance where mortality is sufficient to prevent competitive exclusion, but not so severe that

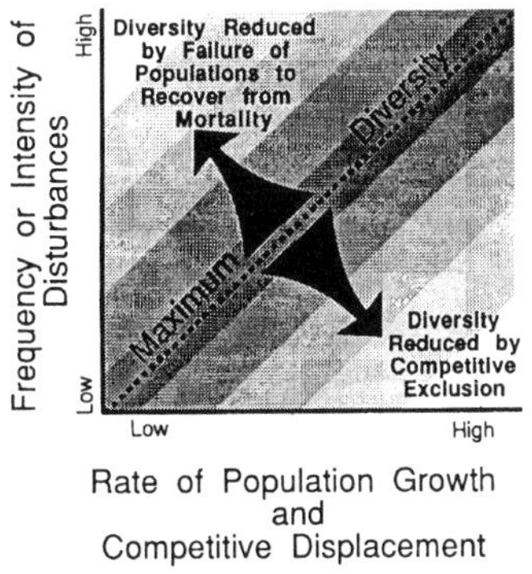


Fig. 11. Interactive effects of mortality-causing disturbances and productivity (rate of population growth and recovery) on the survival of populations of competitors. In addition to the reduction of diversity caused by failure of populations to survive under conditions of high frequency of disturbances and low productivity (as in Fig. 9), diversity can also be reduced by competitive exclusion under high productivity conditions where mortality-causing disturbances are infrequent. Consequently, highest diversity is found under intermediate conditions where both mortality due to failure to recover and mortality due to competitive exclusion are minimized (from Huston 1994, based on Huston 1979).

most species are eliminated (Fox 1979). Both the negative and positive effects of disturbance on species diversity have been documented in a wide range of ecosystems (see Bakker 1989, Huston 1994).

The diversity-enhancing effects of disturbances result from a balance between diversity reduction because of competitive exclusion and diversity reduction because of mortality that exceeds the rate of population recovery. Because both the rate of competitive exclusion and the rate of population recovery are affected by population growth rates and productivity, it is evident that the "intermediate" level of disturbance associated with the highest diversity will change depending on the productivity of the environment and the population growth rates supported by that environment. This interaction between disturbance and productivity was described as the "dynamic equilibrium model of species diversity" (Fig. 11, Huston 1979), which predicts that the same change in disturbance frequency or intensity could have opposite effects on species diversity depending on the level of productivity in the environment (Fig. 12). This interaction means that neither the effects of productivity nor the effects of disturbance can be predicted without knowing the level of the other factor. Under productive conditions an increase in disturbance frequency can cause an *increase* in species diversity (Fig. 12c), whereas under unproductive conditions the same increase in disturbance frequency can cause a *decrease* in species diversity (Fig. 12a). This prediction of a reversal of the effect of distrubance on diversity between productive and unproductive environments has been confirmed by a recent review of grazing studies (Proulx and Mazumder 1998).

Understanding this relationship between productivity and disturbance is essential for making sense out of the complex patterns of diversity found on both natural and managed landscapes. A recent study of species richness in riparian wetlands quantified both the frequency of flooding (a disturbance) and the differences in productivity at many locations in multiple wetland sites in Alaska (Pollock et al. 1998). The highest diversity was found at intermediate levels of both productivity and disturbance frequency, as predicted by the dynamic equilibrium model, and that 78% of the variation in species richness across the sites was explained by this model.

It is important to keep in mind that this complex interaction between productivity and disturbance is found only among species that are potentially competitors.

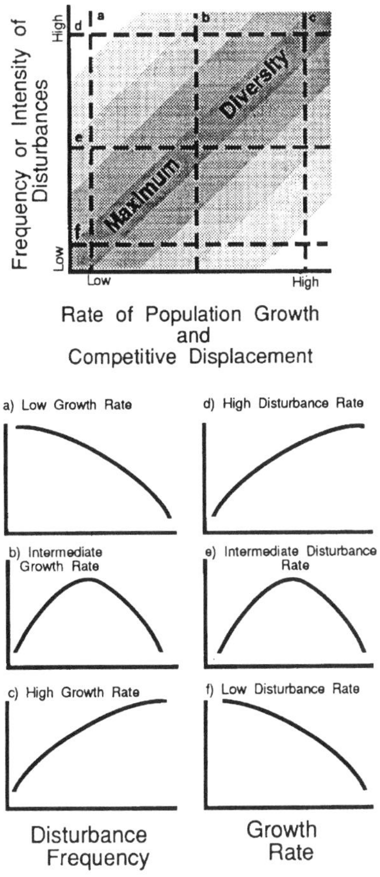


Fig. 12. Variety of diversity responses to changes in disturbance frequency or productivity (such as may result from management activities, or occur naturally along a gradient) that can occur as a result of natural variation in productivity or disturbance regime (as in Fig. 11). Diversity can either increase consistently, decrease consistently, or peak at an intermediate level as either productivity or disturbance frequency are altered. (From Huston 1994).

The interactive effects of productivity and disturbance on plant diversity are likely to influence the diversity of all of the organisms that depend on the resources and heterogeneity provided by plants. This large group of organisms whose diversity is likely to closely track the diversity of plants includes some types of birds, many types of insects, fungi, and other organisms that evolve specialized or host-specific relationships with plants (Fig. 12).

## 4.7 Energy Flow and Trophic Structure

The closest intersection of ecology and physics is in thermodynamics, or the gain, loss, and transfer of energy. Plants obtain their energy from the sun, and all other forms of life ultimately get their energy from plants (a few from specific chemical reactions). To obtain energy, animals must expend energy for hunting, capturing, consuming, and digesting. In addition, since not all the energy content of an animal (or plant) can be converted to a usable form by digestion, consumption at best transfers 10% of the energy contained in the consumed organism to the consumer organism.

As a consequence, predators have less energy available to them than organisms at lower trophic levels. There is generally a much lower biomass (as well as number of individuals) of species at high trophic levels than at lower trophic levels.

Trophic systems can become quite complex at the higher levels, where predators may eat other predators, parasites may parasitize parasites of predators, and decomposing fungi and bacteria, as well as insects, may consume the remains of animals at any trophic level.

Since higher trophic levels have less energy available, and consequently lower productivity and population growth rates, the level of primary (plant) productivity at which diversity is highest is shifted to higher levels for higher trophic levels. This phenomenon, called the "trophic shift" in maximum diversity (Huston 1994), means that a given set of environmental conditions is not likely to result in high diversity at all trophic levels. For example, plant diversity is typically highest under conditions of relatively low primary productivity, but low productivity conditions generally do not support a high biomass or diversity of predators (Fig 13b). It is virtually impossible to find, or create, high diversity of all trophic levels in a single environment.

Resource managers must be aware of the productivity and level of diversity for different types of organisms that their management area is likely to support. High plant diversity may be a reasonable management goal, whereas high predator diversity may be impossible. Maintaining healthy populations of one or a few species of predators may be the most realistic (and important) management goal, rather than a high diversity of predators (or herbivores, or any other trophic level).

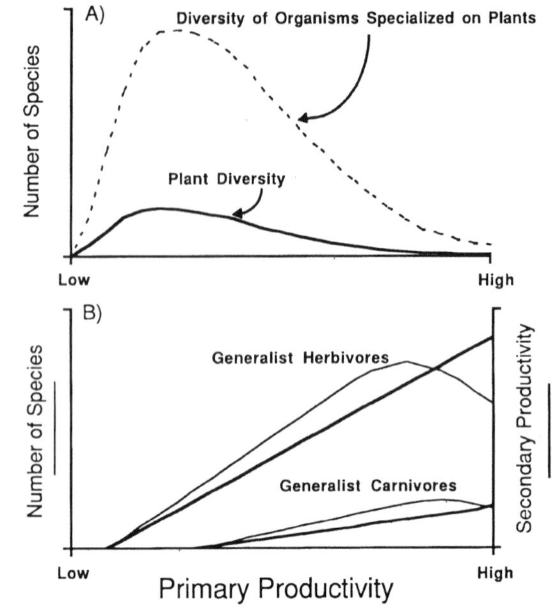


Fig. 13. Expected response of species diversity to variation in primary productivity for (A) species specialized on the structural and chemical heterogeneity of plants; and (B) generalist species (herbivores, fungivores, carnivores) that depend primarily on the total energy content of plants. Specialist species respond primarily to woody plant diversity, which tends to be highest under low productivity conditions. Generalist species respond primarily to total energy availability, independent of the plant diversity (from Huston and Gilbert 1996).

## 5. MEASUREMENT AND ANALYSIS OF SPECIES DIVERSITY

Managing species diversity requires not only information about the number and types of species, but also information about the factors that are likely to influence species diversity, either positively or negatively. Much of this information is also required for other, specific management objectives. Information required for multiple purposes includes data on historical conditions (Periman et al. this volume, Bonnicksen this volume), ecosystem functions (Gosz et al. this volume, Concannon et al. this volume, Lugo et al. this volume, Paustian et al. this volume), population viability (Noon et al. this volume, Holthausen et al. this volume), endangered species (Maddox et al., and Tolle et al. Vol. III of this book), disturbance regimes (White et al. this volume, Engstrom et al. this volume, MacCleery and Le Master this volume, Johnson et al. this volume), climate, soils, and other environmental conditions. Although there is a great deal of general information about the distribution and abundance of species, it is often not sufficiently specific about particular locations to be of direct use to resource managers.

## 5.1 Use of existing data

Most existing information relevant to species diversity and resource management is in the form of range maps. These maps indicate whether a species is potentially present in a general area (e.g., a county), but are generally not of sufficient resolution or detail to indicate whether the species is actually present in a specific local area. An ongoing project, known as GAP Analysis, is attempting to produce detailed habitat maps for all 50 states using satellite images (Scott et al. 1993). As with the range maps, the GAP Analysis habitat maps indicate whether a species is potentially present, not that it is actually present.

Range and habitat maps provide an indication of what species are likely to be present, but do not provide a resource manager with specific indications about the structure and condition of the natural communities for which he is responsible. Site-specific information is essential for resource management. Limited site-specific information is available at a national scale through state Natural Heritage Programs, most of which were started as a cooperative effort between The Nature Conservancy (TNC) and state governments. Most of these programs focus on rare and endangered species, habitats, or ecosystems, and so may provide information relevant to this aspect of management.

One inevitable problem with all general survey data, including the Natural Heritage Programs, is that species can only be known to be present in locations that have been carefully searched. Numerous studies have demonstrated that general maps of species diversity usually indicate the highest diversity in locations where biologists have spent the most time looking for species. Consequently, general maps (or even detailed maps) indicating that a species is not found in a particular location cannot be considered a reliable indication that the species is not present. There is no substitute for site-specific surveys of the distribution and abundance of species in areas that are to be managed.

## 5.2 Sample Design

### 5.2.1 Species Surveys

The strong effects of environmental conditions, specifically disturbance regimes and resource availability, on species diversity require that data on the distribution and abundance of species be collected in conjunction with data on physical environmental conditions. High quality information on certain groups of organisms, such as plants, along with good data on physical conditions (soils, geology, slope, exposure, disturbances), can provide an adequate base of information for managing other species (e.g., insects) that are much more difficult to sample (Colwell and Coddington 1994).

Two different strategies are required for collecting information on species diversity in order to deal with both rare species and common species. While information on rare species is often required by regulations, information on common species is much more useful for general management purposes related to ecosystem function and overall ecosystem condition.

Information on common species is most effectively collected through systematic sampling, such as by using a grid to sample all conditions across a landscape in proportion to their abundance. Grid-based sampling generates much more useful information than "random" sampling in most situations. Random sampling is based on the assumption that sampling occurs within a uniform area, with no underlying spatial pattern. However, most natural landscapes have a great deal of underlying pattern, some components of which are correlated. In such a situation, it is virtually impossible to have sufficient advance knowledge to "stratify" the landscape in such a way that random sampling within uniform areas is feasible. Consequently, sampling on a grid (or surveyed lines) provides the most effective way to relate biological patterns to the underlying physical conditions. Of course, basic information on physical conditions should be collected or obtained for all locations where species are sampled. Distance between grid lines and the area that should be sampled at each "grid point" depends on the size and distribution of the organisms being sampled and the level of spatial resolution that is needed and feasible.

The sample area that is needed to provide an adequate sample at each "grid point" depends on a number of factors, particularly the size of the organisms being sampled and the number of species present. If species diversity is high, a larger spatial area (or larger number of individuals) must be sampled to provide a good estimate of the abundance of particular species as well as the total species diversity. Ideally, preliminary sampling would be used to create a "species-area" curve, which could be used to determine how large an area should be sampled. Issues associated with sampling and estimating diversity are discussed by Magurran (1988) and Stohlgren et al. (1997). For many situations, standard forest mensuration methods are appropriate, although they may need to be modified to include smaller size classes, understory shrubs, herbaceous plants, or other specific components of species diversity (see Maddox et al., Tolle et al., Cooperrider et al., and Correl et al., Vol. III of this book).

A systematic, grid-based sampling plan is unlikely to produce an adequate sample of either rare species or rare community types, however. Both require a sampling plan that focuses on locations where they are most likely to be found. Such a design will produce a map of the known locations of these rare types, as well as give an indication of what proportion of apparently suitable locations are actually occupied.

### 5.2.2 Landscape-scale Components of Diversity

Landscape features, such as hydrologic and geomorphological structures, geology, soil types, and the distribution of successional classes or forest management units, are most effectively quantified using aerial photography and remote imagery from satellites. Geographical information systems are well-suited to evaluating and measuring these features, and for combining this type of information with data from surveys and grid-based samples (Franklin 1995).

## 5.3 Data Analysis

Most data collected on natural or managed landscapes will not be amenable to analysis by the standard statistical methods used for analyzing experimental results, such as analysis of variance (ANOVA). The critical difference between experimental data and sample data from landscapes is that the independent variables on landscapes (e.g., soil fertility or water availability for vegetation productivity, vegetation productivity and/or vegetation structure for wildlife population dynamics) are continuous, varying across a wide range of values, rather than being the discrete levels used in experiments. In addition, samples collected in close proximity to one another are likely to have properties that are highly correlated, which violates one of the fundamental assumptions of most statistical tests.

Effective management requires the ability to predict how a property of interest (e.g., forest productivity) will vary in response to natural variation in environmental conditions or to management practices. The most useful statistical method for this type of prediction is the regression, either simple or multivariate. It is essential to recognize that many important and predictable relationships in nature are non-linear, and consequently may not be detected by linear regressions. Nonlinear regression methods are also available, and visual inspection of data (e.g., scatterplots) is often valuable for determining what statistical method is most appropriate.

Many natural patterns can be most simply interpreted in relation to environmental gradients, either resources such as nitrogen or water, or physical conditions such as temperature. Most ecological properties of management concern will show some pattern along such gradients. Some ecological properties, such as species diversity, may appear to have a linear relationship to environmental conditions when examined along a portion of the environmental gradient, but when the property is evaluated along the entire gradient, the pattern appears as a curve (Fig. 10).

Along most environmental gradients, many factors change in a similar manner. For example, both water and nutrients often increase from the top of a hill downslope toward the valley bottom. When factors are strongly correlated, it is difficult to determine which (if either) is the "controlling" factor. Multivariate methods, such as the many variants of "principal components" analysis, can be useful for evaluating what independent variables are associated with most of the variation in a particular ecological property. A variety of reference books discuss these issues in much more detail (e.g., Gauch 1982). Improved methods for analyzing spatial data, such as from geographical information systems, continue to be developed (Jongman et al. 1995).

## 5.4 Use of Models

The importance of evaluating alternative management plans, as well as the need to foresee future problems, makes the use of models an essential component of ecosystem management. Models that can be useful in management evaluation and planning range in size and complexity from flow charts to simple regressions to detailed computer simulations. Because species diversity and ecosystem dynamics are influenced by multiple processes, models of tree growth and physiology, soil nutrient cycling and hydrology, plant succession and fire dynamics, can all contribute to better ecosystem management.

### 5.4.1 Biophysical Models

The dominant effect of physical conditions on biological diversity and ecosystem processes means that measurements or predictions of spatial and temporal variation in physical conditions are essential for natural resource management (see Gosz et al. this volume, Concannon et al. this volume, Lugo et al. this volume, Paustian et al. this volume). Even though most physical conditions cannot be controlled or managed, their variability sets the bounds within which resource management must take place.

A variety of models are available that address physical environmental conditions relevant to management. These include hydrologic models that predict streamflow and flooding (Beven and Moore 1993), soil moisture models that predict growth conditions for plants (Lin et al. 1994), fire models that predict the intensity and rate of spread of wildfires (Rothermel 1972), and models of nutrient cycling and atmospheric processes (Nielsen et al. 1989, Pastor and Post 1986, Post and Pastor 1990, Johnson and Lindberg 1992, Running and Hunt 1993). Most of these models have been or can be linked to the digital maps of geographical information systems to provide dynamic maps of changing environmental conditions (e.g., Hargrove 1994, Mladenoff and Stearns 1993).

Biophysical models that are linked to GIS can provide invaluable information on how patterns of resource availability, such as soil moisture, are likely to change across a specific landscape in response to seasonal or interannual variation in rainfall, or changes in land use patterns. For example, in the assessment of alternative engineering solutions for restoring the hydrology of the Everglades of South Florida, output from the hydrology models used for water management is linked to satellite-derived vegetation maps to produce fine-scale estimates of how patterns of water depth change seasonally and between wet and dry years. Topographic patterns at the scale of satellite images (with spatial units 30×30 meters) are relevant to the distribution of plants and animals across landscapes. Dynamic models of resources and other properties that vary across landscapes are becoming available for workstations and "high end" personal computers.

### 5.4.2 Population Dynamics Models

Changes in the abundance of a species are of particular concern when the species is threatened or endangered. Models of the population dynamics of a species can be used to evaluate the potential effects of different management plans (e.g., Thomas et al. 1990, Boyce 1997, Franklin 1997, Noon et al. this volume, Holthausen et al. this volume).

Population models were among the earliest types of models developed for ecological issues (e.g., the Lotka–Volterra competition models). These models treat all individuals in the population as identical, and predict changes in population size in response to competition or predation on the basis of a few simple parameters that emphasize differences between populations of different species rather than differences between individuals within a population. Although these models are able to reproduce the general dynamics of such ecological processes as competitive exclusion and predator-prey population cycles, they have been less successful at predicting what would happen to specific populations under real conditions. Part of the problem with these models is the difficulty of accurately measuring the parameters needed in the models. The needed data are difficult and time-consuming to collect, and model results are very sensitive to small differences in the parameters. Much of the current criticism of the relevance of ecological theory to environmental problems (Sarkar 1996) is based on the shortcomings of these models.

Recognition that there were important differences between individuals in a population of a single species led to the development and widespread use of more sophisticated "matrix models." These models represent such factors as the fecundity (reproduction) rate of females of different ages, and the probability of surviving and/or growing from one age or size to the next higher age or size (Caswell 1989). As with Lotka–Volterra models, obtaining accurate estimates of fecundity rates and "transition probabilities" is important, but difficult.

Matrix models address some, but not all, of the variation among individuals within a population. One source of variation that is not easily addressed is differences that result from spatial variation. Plants growing in different positions on a hillside or animals that move from one location to another can experience different levels of food and other resources, as well as other environmental conditions that influence growth, survival, and reproduction. Effects of these differences can accumulate over time, so that individuals of the same age may be very different in size, or individuals of the same size or age may be very different in health.

One approach is to model local populations separately (using age- or size-structured models or more simplified models). These local populations can be mathematically linked in a way that represents migration or dispersal, to form a group of linked populations called a "meta-population." Meta-population models are being increasingly used in situations where habitat fragmentation and isolation separate a previously continuous population into separate local sub-populations that are linked to varying degrees (Hanski and Gilpin 1991). Meta-population models can be linked to specific landscape configurations, such as vegetation maps based on aerial photography or satellite images, to provide the highly specific, localized predictions required for effective resource management.

Another approach to accounting for differences between individuals is to use models in which each individual is represented separately, with parameters that indicate the size, health, age, and reproductive condition of each individual, and change through time. Models of this type generally treat resource acquisition by organisms in great detail, such as the feeding and growth of animals as prey abundance varies through time or from place to place, or the light, water, and nutrient use by plants. These models have been particularly useful in dealing with organisms that vary greatly in size over their lifetime, such as fish and plants. Similar model structures have been used to look at changes in size distributions over time in both plants and fish (Huston and DeAngelis 1987), and models of this type are being increasingly used to investigate stock management issues in fisheries (VanWinkle et al. 1993).

Some of these "individual-based" population models have been used to simulate explicitly the movement of animals across landscape so that the position of simulated animals can be represented on satellite images of real landscape. Models of this type, called "spatially-explicit individual-based models" require a different type of information than traditional population models. The critical parameters in these models are properties of individual organisms, such as food requirements, energy consumption rates, and movement behavior, rather than population-level parameters such as survival rates.

Once the basic parameters for a species have been obtained from field and/or laboratory measurements of individual animals, the model can be applied to many different localities, as long as each locality can be characterized in terms of the properties important to the species. For example, models of this type typically require "GIS" information such as topography, vegetation type, forage or prey availability, temperature, etc., as well as information or estimates of how these environmental conditions vary across the landscape and change through time. Spatially-explicit individual-based models can be linked to biophysical models of landscape processes (see above) to develop the capability to predict how populations might respond to different combinations or patterns of environmental conditions that could occur in the future. Thus, a particular population management plan could be evaluated in terms of how the population might do under various extremes of conditions, such a prolonged droughts or cold winters.

This discussion of population-level models leads directly to models of species diversity, since diversity is simply a consequence of how many populations of different species are present in a given area. Assuming that important interactions such as competition among individuals and predation between populations can be represented, species diversity models can be envisioned simply as a group of population models.

### 5.4.3 Community Dynamics Models

Community dynamics models by definition address multiple species. The dynamics of interest could include changes due to natural succession, to disturbances, to changes in resource levels or environmental conditions, to harvest or other management actions, or to changes in predation rates or the invasion of exotic species. The detail with which each species in a community is represented varies from simple models in which the population of a species is modeled as a single quantity (i.e., the number of individuals of that species, as in the Lotka–Volterra model) to more complex models in which each individual plant or animal is represented separately. This latter type of model is equivalent to the individual-based population models described above.

The ecological processes addressed by community models are most often competition and predation, although other processes such as mutualism (e.g., pollination, nurse plant effects) can also be addressed. Community models often have a spatial component, particularly if the environment is not uniform. The spatial component is important for animals that move across the landscape, and for plants, which may experience very different environmental conditions depending on where they germinate.

Forest dynamics models were among the first individual-based community models developed. These models were originally designed to examine the effects of environmental conditions on forest structure and successional dynamics, but were quickly modified to address harvesting and other sources of mortality such as fire (Shugart et al. 1981). The great advantage of using individual-based models for forests (or any plant community) is that plants vary greatly in size as they grow from seedling to adult, and the amount of light available to any particular plant depends on how tall and how dense its neighbors are. A tall plant intercepts light and shades a small plant whether they are the same species or not. Consequently, the same type of model can be used for both single-species populations and multi-species communities (Huston et al. 1988).

Individual-based plant models can be used to predict successional dynamics under different environmental conditions (Fig. 6) or harvesting regimes (Smith and Huston 1989), and have been developed for a wide range of forest types, and for issues ranging from theoretical ecology (Huston and Smith 1987) to applied (Shugart 1984). Several investigators have developed spatially explicit versions of these models, in which either groups of plants or each individual plant (Pacala et al. 1993) can be assigned a specific spatial location. These models can be used to look at spatial patterns that develop as a result of variation in environmental conditions (e.g., gradients or patches of soil nutrient availability, Huston and Smith 1987, Huston 1994) or processes such as tree mortality or dispersal (Smith and Urban 1988, Pacala and Deutschman 1995).

These models require information on the properties of each species that is modeled (e.g., growth rate, maximum age and size, shade tolerance) as well as of the environmental conditions that affect tree growth (temperature, nutrients, water). Many different versions of individual-based forest succession models have been developed, and there is ongoing improvement both in model formulation and the quality of the species-specific parameters used in the models (Urban et al. 1991, Pacala et al. 1993).

Plant models can also be used to predict animal populations through the effect of habitat structure and food resource on animals. The effects of forest succession on the number and types of birds found in an area are well known, and forest succession models can be used to make specific predictions about how the bird community will change through time in response to changing forest structure (Urban and Smith 1989). Similarly, the influence of the amount and quality of food available on the growth and population dynamics of herbivores can be evaluated using this type of model (Pastor and Cohen 1997). The effects of animals on plants, as a result of herbivory or other forms of damage, can be incorporated in models that predict how forest dynamics differ in response to the presence or absence of herbivores such as deer (Pastor and Cohen 1997).

Complex trophic interactions between plants, herbivores, and carnivores can be modeled using individual-based models or a combination of model types that increase in complexity from lower to higher trophic levels. The complex interactions between predators and their prey generally require information on spatial patterns and the actual locations of the animals. When such features are incorporated into community models, they are often called landscape models.

### 5.4.4 Landscape Models

Landscape models relevant to biodiversity management must combine biophysical models with population or community models. Biophysical models are necessary to describe the distribution of environmental conditions across the topographic, geological, and climatic variation found on all landscapes. These environmental conditions determine the rates of biological responses and ecological processes, including the processes that influence species distributions and diversity, as they are distributed across a landscape (including "waterscapes" and land-water interactions) (Franklin 1995, Concannon et al. this volume).

Although any of the properties discussed in the above sections, including the movement and health of individual organisms, can be addressed in landscape models, landscape models need not include individual-based models. All landscape models, including those with individual-based components, include several different types of models, and have sometimes been called "multi-models." For example, a landscape model of forest productivity must include some representation of climate, including temperature and precipitation. This climate model could be as simple as an input of climate station data or as complex as a climate model that predicted how air temperature and precipitation vary across a mountain range under different seasonal weather patterns. The forest growth component could be as simple as a set of yield curves for different site conditions or as complex as a physiological model of tree growth that simulated photosynthesis, respiration, and evapotranspiration.

Landscape models are typically linked with some sort of GIS, either for printed output or with a dynamic linkage to display the model output as it is running. "Visualizing" the model's output as dynamic or static maps is useful for developing and "debugging" models, as well as for presenting results in a form that is directly useful to managers. In addition, the type of information needed to describe the essential environmental patterns across a landscape includes the maps and satellite images that are the standard data for GIS.

The level of detail needed for the various types of information in a landscape model depends on the questions being asked and on the quality and resolution of data available. For example, rainfall may be relatively uniform over a fairly large area, and thus could be represented as a single number with no spatial variability for every hour, day, or week over which the model is run. However, the variation in topography that channels water away from ridges and into valleys must be modeled in much greater detail if the model is to predict plant growth.

Landscape models are particularly useful for evaluating how a landscape will change over time as a result of natural succession, or of management activities and natural disturbances. The interaction of expanding beaver populations with forest succession and wetland formation has been so modeled (Johnston and Naiman 1990).

In hilly terrain, digital models of topography (DEMs, or digital elevation models) can be used to predict both the distribution of soil moisture resulting from runoff (e.g., Fig. 8), and the pattern of evapotranspiration, which differs from north-facing to south-facing slopes. These landscape patterns of runoff and soil moisture can in turn be used to predict the distribution of plant species, variation in plant growth and productivity, and patterns of soil nitrogen availability, using either regressions derived from previous research, or more detailed models that simulate plant growth, forest succession, and nutrient cycling (Post and Pastor 1990).

Landscape models have been developed to predict the spread of fires in specific regions (Hargrove 1994, Hargrove et al. in press), as well as to predict plant growth (net primary production or NPP) at the broad scale of continents or the entire globe (Running 1994). Models that predict patterns over large areas generally have much simpler representations of most processes, modeling the productivity of large blocks of forest by lumping all species into a single generic plant, rather than treating species separately as in some models.

One of the most detailed landscape models ever developed is the Across Trophic Level System Simulation (ATLSS), which is a group of models of many different types that is being used to help plan and evaluate the restoration of the Florida Everglades (DeAngelis et al. 1998). This "multi-model" uses the output of a detailed, mechanistic hydrologic model (the South Florida Water Management Model, South Florida Water Management District 1997) to provide daily water levels for every 2×2-mile portion of a large area of South Florida. The output of this biophysical model is then processed to provide higher resolution predictions of the pattern of water depth across the various vegetation types of the region. Water depth information is used in models of plant growth and fish populations to predict how deer forage and fish for wading birds varies across the region from week to week and year to year. Individual-based models of deer and wading birds move across the landscape in response to food availability, and grow, reproduce, or starve, depending on how much food they can obtain. A panther model interacts with the deer model, as well as other landscape properties such as water depth and plant cover. This complex landscape model is being used to address a number of issues of concern to resource manager and water use planners. The ATLSS model is being used to make predictions of how wildlife populations may change in the future, in response to known patterns of water management and unknown patterns of rainfall.

Landscape models must be customized for specific locations, using local topography, soils, climate, and other features. Consequently, these models can be valuable tools for resource managers to predict and evaluate the consequences of specific management actions, as well as to foresee the potential consequences of such disasters as fires, droughts, and pest outbreaks. With increasing computer power and better information from satellites and other sources, dynamic landscape models are likely to become an essential tool for resource management and planning at the large scales of national parks and forests.

## 6 COMPARATIVE ADVANTAGE AND MULTIPLE-GOAL MANAGEMENT

The spatial variability in ecosystem functions, as they are influenced by variation in topography, soils, and climate, means that different areas inevitably have a comparative advantage for particular ecosystem functions or services — a basic idea from economic theory. Comparative advantage in ecosystem processes varies at scales ranging from continental and regional to across a single hillside. *Understanding the spatial distribution of comparative advantage for desired ecosystem functions is essential for achieving the goals of ecosystem management.*

### 6.1 Comparative Advantage for Production Functions

Maintaining high and sustainable plant productivity is the key to both successful agriculture and successful forestry. Although "best management practices" can generally increase productivity on a particular site, the greatest differences in productivity between sites are generally the result of physical environmental conditions, summarized in forestry as "site index."

Any production system should be located on the most productive land available, keeping in mind that the most productive land may already be committed to a more profitable crop. A corollary is that for some production systems (e.g., plantation forestry). Sustainable land-use management generally requires planning at the regional or national levels, as well as the local level.

Dedication of the most fertile and productive lands to production forestry inevitably entails some sacrifice of other possible uses of those lands. The loss of plant diversity is likely to be low, because the most productive soils tend to have low plant diversity (Fig. 14), but the diversity of predatory mammals and birds tends to be highest in the most productive areas, so the diversity of these organisms is likely to be reduced by intensive forestry (Fig. 13b). Although the loss of large predators (e.g., grizzly bears, wolves) over most of North America occurred over a century ago, suitable management practices may allow a surprisingly high density and diversity of smaller predators to survive in productive regions. In addition to the conservation

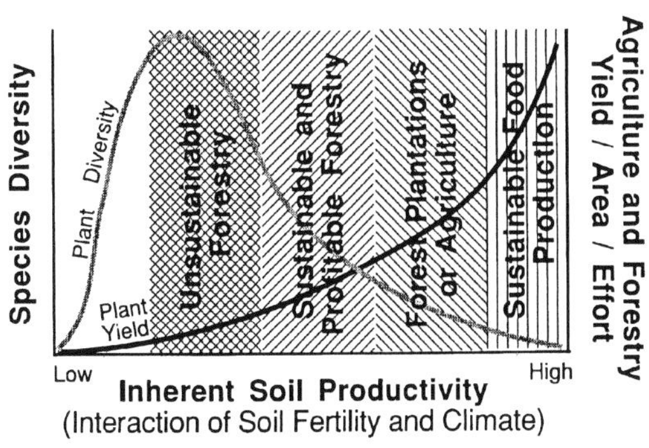


Fig. 14. Pattern of plant species diversity in relation to environmental conditions related to growth and net primary productivity (yield/area). See Fig. 4 for examples of field data. (based on Huston 1994).

benefits, management practices that allow predatory species to survive in agricultural regions benefit agriculture through the reduction of pest species that serve as the prey base for the predators.

Although the loss of plant species due to cultivation of the most productive lands may be low, there may be a significant genetic loss. The loss of genotypes well-suited for growth on productive sites could potentially mean the loss of those traits best suited for high production forestry. Species and genetic conservation should be important management objectives in forest plantation regions, as well as in wild lands.

## 6.2 Comparative Advantage for Biodiversity Components

Because different components of biodiversity reach their highest levels under different environmental conditions, no single location is likely to have high value for all biodiversity components. Fortunately, it is possible to identify those environmental conditions and landscape locations where specific biodiversity components are likely to be high.

One of the key considerations in biodiversity management is the importance of low productivity lands, where plant diversity is often high. Because these lands have little value for harvest-based uses, their comparative advantage for conserving plant diversity is very high (Fig. 14). The diversity of many animals that are closely tied to plant resources, such as most insects and many bird species, is likely to be strongly correlated with plant diversity (Fig. 13a), which further increases the comparative advantage of marginal lands for biodiversity. In addition to high overall diversity, these less productive regions tend to have a high proportion of rare and endangered species, particularly naturally rare endemic species.

Organisms at higher trophic levels, such as predators, require more productive lands to reach their maximum abundance and diversity. The fundamental problem is that the productive lands required to support large herbivores and predators have already been, or are rapidly being, taken over for human uses such as intensive agriculture.

Predator abundance and diversity can be managed in three general patterns: (1) dedicated areas of moderate size with high plant productivity and consequently high densities of prey animals (e.g., coastal lowland swamp forests); (2) large areas of low to moderate plant productivity with scattered and isolated areas of high productivity (e.g., riparian vegetation, wetlands, alluvial soils); and (3) extremely large areas of low productivity that support a low density of predators over a sufficiently large area for adequate population size (e.g., arid western montane and intermontane areas).

Although total diversity may be low on areas with extremely poor soils, these areas may have unique or endemic species. Concentrated areas of high productivity are extremely important in regions of low productivity. In addition to functioning as "islands" that may support rare or endemic species, these "hotspots" often play a critical role in supporting species found throughout the region.

## 6.3 Comparative Advantage for Ecosystem Services

In addition to productivity and biodiversity, other ecosystem processes provide local effects such as shading, humidifying, and temperature moderation, as well as waste water and sewage treatment. Flood reduction, erosion prevention, aquifer recharge, and atmospheric gas balancing and pollutant removal are provided at larger spatial scales. Many of the most valuable ecosystem services are associated with hydrologic processes.

Most of the local-scale ecosystem services can be fit into a land management plan where they are needed, although wetland services such as sewage treatment and nutrient removal will only occur in appropriate geomorphological positions on the landscape.

Regional-scale ecosystem services generally require much larger areas, although the value of the services they provide often justifies the dedication of large areas. One of the most important ecosystem services is the capture, storage, and release of clean water from forested catchments. Much of the water supply for New York City, which is known for its high quality water, is derived from protected forested watersheds. Water storage and erosion prevention by forested catchments are particularly important in mountainous regions, although flood reduction and aquifer recharge are economically important everywhere.

Use of large portions of landscapes for these ecosystem services is compatible with other land uses as well, particularly those related to conservation, recreation, and low-intensity harvesting.

Esthetic and recreational uses are compatible with all but the most intensive harvest-based land uses and the protection of areas that are extremely sensitive because they are easily destroyed or degraded (e.g., small wetlands, montane meadows), harbor sensitive endangered species, or are used for drinking water supplies.

### 6.4 Reconciliation of Economic, Environmental, and Social Needs of Society

The above discussions of the conditions under which different land uses have the greatest comparative advantage illustrate the key to sustainable land use management. The central conclusion is that there is no inherent conflict between the major land-use needs of society. Food production need not be sacrificed for biodiversity conservation or water supply, because these land uses have their highest comparative advantage on different parts of the landscape.

Topographic, climatic, and geological gradients of productivity provide a template over which most of the needs of society (including conservation of species and genetic diversity) can be met without insoluble conflicts. With some flexibility and willingness to compromise, sustainable land use planning can provide "win–win" solutions that simultaneously meet multiple needs of society.

One particularly destructive land use that has no direct economic benefit and inevitably destroys or degrades all ecosystem functions is housing, a problem throughout the world. Because use of land for housing removes that land from agricultural or forest production, as well as destroying or degrading the value of that land for ecosystem services, biodiversity conservation, or recreation, use of land for single-family dwellings is probably the single most environmentally destructive land conversion occurring today. Successfully addressing the issue of housing is one of the major challenges facing sustainable development.

## 7. EFFECTS OF COMMON MANAGEMENT ACTIVITIES ON SPECIES DIVERSITY

### 7.1 Managing Productivity

Productivity is the ecosystem function that supports all life on Earth, and increasing usable productivity is the primary goal of most human manipulations of ecosystems. Virtually all productivity results from the use of solar energy by plants to convert carbon dioxide from the air into sugar, starch, and other carbon-based compounds valued for their energy content or structural properties. Technically, production of these materials by plants is called "primary production," which may be consumed by animals to produce more or larger animals, called "secondary production." *Maintaining the capability of the Earth's ecosystems to continue to produce the current (or higher) levels of primary and secondary productivity is the most basic definition of Ecosystem Sustainability and thus the ultimate goal of Ecosystem Management.*

The rate of primary productivity, along with its spatial and temporal variability, defines the potential and limitations of most human uses of the Earth's surface, along with virtually all of the other ecosystem processes and properties that are the focus of ecosystem management. Within any region, even within small watersheds or a single hillside, there is great variability in potential primary production. Any area can be classified into units based on productivity, and arranged from lowest to highest. These areas will differ in their potential for human uses related to agriculture and forestry, as well as in the rates of ecosystem processes, and such properties as biodiversity.

The "site index" of classical forestry is one representation of productivity used to guide forest planning. Site index directly affects management through determination of the optimal rotation length, which is itself a manipulation of primary productivity. The rotation for maximum return is a function of growth rate, specifically, the size at which the annual rate of wood production begins to decline as a result of respiration by the living wood biomass. The size at which this decline in return begins occurs at younger ages for fast-growing stands than for slowly-growing stands.

Biodiversity is also highly sensitive to net primary productivity, with the diversity of different types of organisms being highest at different levels of productivity. Understanding the productivity conditions under which different organisms have their highest diversity provides important insights into the potential impacts of forest management on biodiversity in areas with different levels of net primary productivity (see Sections 3.2, 4.4, 4.5 and 4.6). Basically, many types of plants have their highest diversity at low levels of primary productivity, levels which may be too low to support tree growth. In forests, the most productive sites tend to have fewer tree species than less productive (but not extremely unproductive) sites. However, the diversity of many types of animals is higher under productive than unproductive conditions. Consequently, at any single location, management is likely to have different impacts on different groups of organisms (that is, different components of biodiversity). Similarly, along a productivity gradient, the effects of management on any single group of organisms is likely to vary.

Enough is known about general productivity responses of biodiversity to estimate the approximate diversity of different types of organisms under different productivity conditions, and to predict how diversity will respond to specific types of management, such as harvest, fire, or fertilization.

### 7.1.1 Effects of Management for Increasing Primary Productivity

*Intentional increases: Fertilization, soil/water management, species and genetic selection, rotation length, other*

Although most management of biological resources is focused on increasing the amount or quality of the harvested productivity, directly increasing potential productivity is usually expensive and used only under limited circumstances. Fertilization and irrigation are the primary methods for increasing the primary productivity of both natural and managed ecosystems, although in greenhouse situations productivity can be increased by increasing the carbon dioxide in the air and manipulating temperature. The currently increasing concentration of carbon dioxide in the Earth's atmosphere, and the potential changes in temperature, are likely to have highly variable effects across the Earth's ecosystems, and are unlikely to produce the strong positive effects on productivity experienced in some commercial greenhouses (Wullschleger et al. 1995).

Fertilization and/or irrigation are used primarily in agriculture and plantation forestry where productivity is already relatively high, and the value of the crop produced is high. In these heavily managed situations, maintaining natural ecosystem processes and properties such as species diversity are generally not issues. Under such conditions, concern is likely to focus on the off-site impacts of fertilization or irrigation on the quality and amount of surface and groundwater.

The effects of irrigation and fertilization on species and genetic diversity will be very different for different types of organisms. Although plants will grow faster and become larger, the total number of species that are able to coexist within a given area will decrease as a result of competitive dominance (see Fig. 11). Thus, plant diversity will decrease, along with that subset of vertebrate, invertebrate, and microbial species closely associated with the plant species that disappear from the area. Virtually every fertilization experiment, except those on extremely poor or sandy soils, has resulted in a decrease in plant species diversity (Huston 1994, 1997).

However, other species, both herbivores and predators, may actually increase in diversity, if they are able to consume some portion of the increased productivity and are not negatively affected by other factors. Both game and non-game wildlife generally increase in both abundance and diversity when productivity is increased in part or all of their environment. In some situations, populations of certain species may increase to the point they become pests, as has happened with deer, waterfowl, blackbirds, rabbits, and various rodents. Species that experience population explosions due increases in productivity may cause decreases in the diversity of similar organisms as a result of competition or other interactions, such as overgrazing. Generally, further increases in the productivity of moderate to highly productive areas will have negative effects on most components of natural species and genetic diversity.

Another method of increasing primary productivity is the introduction of species that may grow faster than the local species. This is often the strategy in plantation forestry. Fast-growing plantation species obviously create an ecosystem that differs dramatically from the local natural systems. Native species of plants, and in some cases animals, can be negatively affected or eliminated under such conditions, although some native or exotic species (including pests) may benefit from the increase in productivity. Productivity can also be increased by managing the amount of biomass in an ecosystem, as by altering rotation length (i.e., plant size at harvest) or grazing intensity. These manipulations increase net primary productivity by reducing the amount of non-producing standing biomass (boles or stems) that converts sugar back into carbon dioxide through respiration. Maintaining areas in an early successional stage is a common management method for increasing the populations of many game species.

Positive diversity responses to increased productivity can be expected under conditions where productivity is extremely low, either due to natural conditions or various forms of environmental degradation. Restoration of vegetative cover and associated wildlife on severely degraded lands, such as reclaimed mining sites, can be greatly enhanced by use of fertilizer (Berger 1990). Even though it will usually be physically and financially impossible, and ultimately undesirable, to increase productivity over large areas of unproductive landscapes, increases in productivity in concentrated areas can have a major positive effect on many components of species and genetic diversity. Under these conditions, areas of high productivity that result from natural or managed concentrations of fertile soils and water (such as springs, riparian areas, and cienegas) often support unique plant and animal communities. Although the local diversity of these plant communities may be low, they contribute an important element to the diversity of the landscape by providing specialized habitat for some species, and a reliable food source for other species that may range widely over large areas.

Concentrated areas of high primary productivity are critical for valued components of diversity in most landscapes, and not only those in arid or infertile regions. Riparian zones and meadows are particularly important, and should be a major focus of restoration efforts in areas that have been degraded by erosion, declining water tables, or other results of poor resource management. This enhancement of productivity in concentrated areas also increases the economic and recreational value of degraded landscapes, moving them toward a condition where higher levels of resource extraction by grazing or hunting can be sustainably maintained.

#### *Unintentional increases: eutrophication of waters, atmospheric deposition and climate change, exotic species*

In addition to intentional manipulations of primary productivity, a number of human activities can result in unintentional increases, often with negative consequences. The most common example is the eutrophication of water as the result of fertilizer, sewage, phosphate detergents, or other sources of nutrients that are usually limiting in natural ecosystems. Eutrophication increases productivity of algae, generally leading to accumulation of high algal biomass, which decomposes and creates problems ranging from toxins to offensive smells to anoxia that kills fish and other aquatic animals. Algal diversity always decreases dramatically as a result of eutrophication, and the diversity of many types of aquatic animals also decreases.

A similar, though less extreme, situation of unintentional eutrophication terrestrial ecosystems involves increased nitrogen input in precipitation or gaseous forms from agricultural activities (releasing ammonia or other nitrogen gases) or internal combustion engines (releasing $NO_x$ forms that are ultimately deposited as nitrate). These problems are particularly severe near major metropolitan centers, such as Los Angeles, where nitrogen deposition may be contributing to the replacement of native coastal sage scrub by exotic grass species (*Bromus* spp., E. Allen, pers. comm.). Although associated pollutants, such as ozone, may have negative effects on net primary productivity, the net result is to change the balance of growth and competitive ability among species, which may cause a total change in species composition and even life form of the vegetation.

Any significant climatic change is likely to affect net primary production positively in some regions and negatively in others. The largest positive changes are likely to result from increased precipitation in certain warm areas, and from warmer temperatures and longer growing seasons in cold areas. Similarly, continued increases in the concentration of atmospheric carbon dioxide will potentially increase NPP if other factors, such as drought, do not counteract the $CO_2$ fertilization effect.

The introduction of exotic species can also increase net primary production. In nearly all cases, the increase in NPP that results from exotic species, atmospheric deposition of nitrogen, or other causes, leads to a reduction of species diversity, particularly of native plants. The effect on native species is particularly severe when the exotic species alter the natural disturbance regime, as is happening throughout the west as *Bromus tectorum* and other exotic grasses invade native shrublands and increase the fire frequency (Billings 1990).

In some cases an increase in productivity can be beneficial to animals. Intentional local increases in food resources can be particularly important for large migratory birds, such as raptors or waterfowl. The population densities of these species can increase significantly, although the total number of species may not be affected.

### 7.1.2 Effects of Management for Decreasing Primary Productivity

#### *Intentional decreases: eutrophication control, rotation length, other*

It is uncommon that a management objective will be to decrease net primary productivity. The primary situations will be where NPP has been increased by some type of nutrient pollution (e.g., sewage, fertilizer runoff) that has negative effects on streams or lakes, typically including a significant decrease in species diversity. Controlling these causes of increased NPP is a prerequisite for restoring the natural diversity. For example, the Everglades Nutrient Removal (ENR) systems being built to reduce the phosphorus levels in water flowing into Everglades National Park are considered to be essential for restoring the natural vegetation and ecology of the region. In these situations, decreasing NPP will generally lead to an increase in plant diversity (algal diversity in lakes or streams), although the population densities and sometimes diversity of higher trophic levels (e.g., predators) may decrease.

Another management action that can reduce NPP is increasing the rotation length for harvesting, and/or maintaining a higher proportion of the landscape in older, higher biomass stands. This has the effect of losing a larger proportion of the total production to respiration, and is likely to also change the quantity and quality of the food available to animals. For example, moose, deer, and elk typically have much higher population densities in landscapes dominated by early successional stands than in old growth areas. This decrease in NPP is likely to have complex effects on biodiversity, increasing the populations and diversity of some groups of organisms, and decreasing others.

*Unintentional decreases: soil erosion, nutrient depletion by harvest, pollution and climate change*

The most serious potential consequence of human misuse of natural resources is reduction of the maximum net primary productivity. Such changes are almost always unintentional consequences of activities related to the harvest of NPP or the extraction of minerals. Decreases in potential productivity may be reversible, but in practice are often irreversible within the timescales of human civilization, as in the case of severe erosion. A decrease in primary productivity may result from management actions to achieve other goals, particularly if management is not based on maintaining sustainable production.

The mechanisms by which the potential productivity of a land area can be decreased include: (1) loss of soil nutrients as a result of erosion or removal during harvest; (2) alteration of soil physical and chemical properties that make the soil less favorable for plant growth (pH change, change in water holding capacity, loss of organic matter); (3) contamination of the soil with chemicals that inhibit plant growth or biogeochemical processes associated with plant production; (4) climate change that makes temperature or soil moisture less favorable for plant growth.

The consequences for human welfare of decreases in the potential net primary production of the land are invariably negative. Throughout human history, a principal reason for the collapse of major civilizations has been the destruction of their land base (Hyams 1952). The continuing destruction of the productive potential of the Earth's ecosystems inevitably means a decrease in the quality of life for future generations.

Although the consequences of decreasing NPP for humans are almost completely negative, the consequences for biodiversity are not necessarily as bad. Depending on the initial level of NPP, a decrease in NPP could potentially increase plant diversity (see Fig. 12). In fact, some of the highest levels of plant diversity are found in areas that have been severely degraded by human use, such as parts of the Mediterranean basion and the Tehuacan Valley in Mexico (J. McAuliffe, pers. comm.). Although such decreases in potential NPP may not have a negative effect on plant diversity, the effect on higher trophic levels is more likely to be negative (see Fig. 13). For example, degradation of rangelands typically has a much greater effect on the carrying capacity of the land for cattle and native herbivores (and their predators) than it does on the species diversity of the plant community.

Decreases in the NPP of land or waters, plus the associated decrease in secondary productivity (including that harvested by humans), alters natural ecosystems to make them less able to meet human needs. Although such reductions in NPP also change the composition and structure of the natural community of plants and animals, they are not likely to result in extinctions, except in unusual situations (such as islands) or for the highest trophic levels (e.g., birds of prey).

The short-term effects of altered NPP on the various components of biodiversity can be predicted on the basis of current scientific understanding, although potential long-term effects (such as on the rate of speciation) are beyond our capability to predict.

## 7.2 Managing Disturbance

In the context of ecosystem dynamics, disturbance is defined as processes, other than natural aging or senescence, that cause mortality of plants or animals. Natural disturbances include fires, floods, droughts, severe freezes, windstorms, ice storms, herbivory, pests, and predation. Many management activities, such as harvesting of plants or animals, are thus disturbances intentionally imposed on ecosystems for human benefit. Other management practices involve the reduction or prevention of natural disturbances such as fire, pests, or predation, with the objective of leaving a larger amount of the resource to be harvested by humans.

Both natural and management-based disturbances have effects on biodiversity that can be understood and predicted using the dynamic equilibrium model (Fig. 12). The critical factor that must be considered is productivity (population growth or recovery rates) of the organisms that experience the disturbance. Depending on the growth rates of the particular organisms, a particular pattern of disturbance can either increase or decrease species diversity. Although the specific details of any particular disturbance are difficult to predict, the general consequences and dynamics of many disturbances are highly predictable.

### 7.2.1 Effects of Management to Control Natural Disturbances

Disturbances that are independent of ecosystem properties, such as lightning, floods, and freezes, are difficult to control or reduce through management. However, ecosystem-dependent disturbances, such as fire and pest outbreaks, tend to be more amenable to management, at least under some conditions. The problem is that the efforts to control these disturbances generally lead to an increase in those ecosystem properties that contribute to the disturbance, with the result that when the disturbance finally occurs, it is much more extensive and more intense than if it had occurred earlier (e.g., the 1988 Yellowstone fires). Thus, the result of fire suppression is the accumulation of sufficient fuel to produce uncontrollable fires, just as the result of controlled burning is the prevention of large fires by using many small fires to prevent fuel accumulation.

Management control of natural disturbances rarely eliminates disturbances entirely, but rather changes the timing, frequency, and/or intensity of the disturbances. Although a decrease in disturbance frequency can either decrease or increase species diversity, depending on productivity and growth rates (see Fig. 12), most of the situations in which ecosystem-dependent disturbances are important are sufficiently productive that a decrease in disturbance frequency leads to a decrease in biodiversity at several scales.

At the stand level, a decrease in disturbance frequency allows the competitive processes of plant succession to proceed further toward the low tree diversity typical of late succession. At the landscape scale, decreased disturbance frequency allows a larger proportion of the landscape to reach a late successional stage, reducing the spatial complexity component of landscape biodiversity. A further consequence of reduced disturbance frequency is that the entire system becomes more likely to support intensive disturbances that affect large areas. These processes are particularly evident when fire is the disturbance, but are also relevant for other disturbance types such as windthrows and pest outbreaks. For example, fire suppression in the sage scrub and chaparral vegetation of southern California has produced a dramatic increase in the size and intensity of burns, in contrast to the fine-scale mosaic of small burns found immediately across the international border in Mexico, where there is no fire suppression (Chou et al. 1993).

In most natural situations, an intentional reduction in disturbance frequency leads to an increase in the biomass or population size of the resource being protected, at least in the short term, but the diversity of that resource will generally decrease. This is the case for plant diversity in forests that are protected from fires, as well as of wildlife and game species diversity when predators are removed. Elimination of either fire or grazers will reduce the species diversity of productive grasslands.

### 7.2.2 Effects of Disturbances Imposed for Management

The majority of management actions involve increasing natural disturbance rates or imposing disturbances (e.g., harvest). In general, increasing disturbance rates in unproductive environments will decrease diversity, while increasing disturbance rates in productive environments will increase diversity.

Most timber management practices focus on fast-growing early successional species in productive habitats, which are conditions in which plant diversity tends to be low naturally. In some situations, shortening the rotation length (increasing the disturbance frequency) may lead to an increase in the diversity of some types of plants (such as sun-loving herbs), but typically the effect will be a decrease in diversity. In many forest management situations, an important component of diversity is the distribution of stand ages (successional stages) across the landscape. Either too long a rotation or too short a rotation will produce a landscape in which most stands are the same age, which reduces an important component of landscape heterogeneity that impacts the species diversity of many types of organisms (Carey et al. 1996).

In many situations, controlled burns reduce the diversity of tree and other woody species, which are relatively slow-growing, while increasing the diversity of the faster-growing herbaceous species (White et al. 1991). In most situations, all types of species cannot be maintained at high levels of diversity in a single area. High diversity can be maintained at larger scales by creating a landscape with patches of different ages, so both herbaceous and woody plant diversity can be maintained. Mowing or bush- hogging can have similar effects on plant diversity as fire. In grasslands or shrublands, more species are generally found in areas that are mowed periodically than in areas that are rarely mowed (Huston 1979, 1994). Pre-harvest thinning can also increase the diversity of certain plant types in managed forests.

Firebreaks and roadbuilding also affect diversity. They produce open habitats that may be important for species that cannot survive in closed forest or shrublands, particularly in productive areas. On the Department of Energy's Oak Ridge Reservation in Tennessee, populations of a threatened plant species, tall larkspur, survive primarily in powerline right-of-ways which are periodically mowed to prevent encroachment of the oak-hickory forest that has eliminated the larkspur from most of the landscape (Mann et al. 1996). However, disturbances such as roads can also allow the invasion of exotic species, which can replace native species and thus reduce a valuable component of natural biodiversity. In unproductive areas where the rate of vegetation recovery from disturbance is low, soil disturbance can greatly reduce plant diversity and produce long-lasting physical scars.

Grazing can have similarly complex effects on plant diversity, generally decreasing plant diversity in unproductive areas, and increasing diversity in productive areas. Because of variations in the palatability and growth rates of different plant species, as well as variation in the preferences and foraging patterns of different grazers, the effects of grazing can be extremely complex. As a result, grazing can be managed to produce a wide range of effects on plant species composition, rangeland productivity, and species diversity.

Excessive rates of disturbance (from natural mortality or harvesting) not only lead to reduced diversity, but can result in the collapse of resource stocks and even the extinction of species. This is occurring with increasing commonness in fisheries around the world, is contributing to the disappearance of species such as rhinoceros and tigers, and even affects some timber species, particularly in the tropics (Uhl et al. 1997).

## 7.3 Managing Ecosystem Structure and Heterogeneity

Multiple-use requirements for land management impose the need to consider landscape properties, such as viewsheds and aesthetic appeal, which are not the primary objective of most management actions. In an increasing number of situations, often involving stream water quality or endangered species, control of the spatial arrangement and physical structure of a landscape may actually become the primary management goal. The implications of the spatial structure and heterogeneity for species and genetic diversity are complex, primarily because different groups of organisms can respond in opposite ways to the same manipulation.

### 7.3.1 Area Requirements and Scale of Management

The number of species found in an area increases with the size of the area as a result of many processes (see Sections 3.1 and 4.1). The rate of increase varies from one environment to another, and differs among groups of organisms. The positive effects of creating "edge" habitat for many game and nongame species have been well known for many years, whereas the negative effects of fragmentation and the associated increased edge for other species have been more recently recognized. Consequently, the size, age, and arrangement of harvested blocks of forest can significantly affect the abundance and even survival of certain plant and animal species.

The area required to support viable populations of a particular species is generally smaller in productive areas and larger in unproductive areas. The low productivity of most of the regions where large carnivores are still present in the United States implies that extremely large areas will be needed to support viable populations.

### 7.3.2 Manipulation of Structure and Heterogeneity

Increasing the structural and spatial heterogeneity of an area almost always has a positive effect on the number of species, and probably on genetic diversity as well. However, this overall increase in species diversity may be accompanied by the decrease or loss of some of the species that may be most highly valued by society. An increase in spatial heterogeneity on a landscape will inevitably fragment and destroy the large blocks of uniform habitat that some species require (Mladenoff et al. 1994). The northern spotted owl is the classic example of this phenomenon (Forsman et al. 1984), but many large carnivores and "forest interior" species also respond similarly. Wherever the historically dominant habitat is being reduced in area and fragmented, it is inevitable that some of the typical species of the region will be threatened.

If the preservation or maintenance of the historical native species of plants and animals is a management objective, achieving this goal will require recreating or preserving the historical spatial arrangement of landscape units (successional stages, forest types, and distribution of processes) over a sufficiently large area that natural disturbances can occur without destroying all of a particular habitat.

Land managers have the option to manipulate spatial structure over a wide range of spatial scales. Structural complexity can be altered within a single habitat type or landscape unit, or complexity can be altered at the larger scale of a landscape or region by manipulating the number, size, and types of patches of different habitat types.

*Within-habitat structural changes*

Managers have a variety of methods that can increase overall diversity or favor the population of a particular species. The primary objective is to alter the structure of the vegetation, either through planting of particular species, standard silvicultural thinnings, thinnings designed to create a specific stand structure, or prescribed burning.

Opening up a stand through thinning or burning will generally increase plant species diversity, particularly in areas where savanna is (or was) a natural vegetation type. Such manipulations can also improve conditions for birds, mammals, and insects that require savannah conditions and vegetation.

"Old growth" forest can also be increased through structural manipulation. Although trees cannot be aged any more rapidly through management, the critical structural feature of many old growth vegetation types is the density and spatial distribution of the largest size classes. Selective thinning can be used to speed the approach of a stand toward a typical old-growth structure (Oliver and Larson 1996), which may be important for such species as spotted owls (Forsman et al. 1984) and red-cockaded woodpeckers.

Manipulation of canopy density, stem size distribution, and understory shrub density allows managers to meet a range of objectives related to species diversity and endangered species. Other structural manipulations include leaving or creating standing dead snags for nesting and shelter; adding artificial nesting boxes or roosts; leaving large woody debris (trunks and large branches), which could favor certain reptiles, amphibians, insects, and small mammals (McMinn and Crossley 1996).

*Manipulation of landscape structure and pattern*

In all managed forests, the size and spatial distribution of cuts and the rotation length have major effects on landscape pattern (Oliver and Larson 1996, Carey et al. 1996). The strong effect of successional stage on wildlife populations provides a powerful predictive tool for land managers.

In addition to manipulating successional stages, managers can also increase the number of different vegetation types on a landscape by creating wetlands or lakes, or enhancing the differences between existing vegetation types (e.g., riparian enhancement, or restoration of degraded vegetation types such as prairies or swamp forests) (Berger 1990). Protection, restoration, or even creation of rare, highly productive habitats such as riparian thickets or forests, or wet meadows, can be critical to the survival of many species in unproductive, arid environments.

The spatial arrangement and shape of vegetation types on a landscape can also be critical. Managers can maximize "interior" habitat area by creating compact units (low perimeter to area ratio), or improve hydrologic performance by managing the vegetation of flowpaths and riparian zones. Corridors or other features or conditions that facilitate movement and dispersal may be important for some species, while isolated patches (or "habitat islands") may be better for other species. Understanding the needs of critical species can provide clear guidance about the most effective landscape configurations.

## 7.4 Species Management

"Single-species management" has been criticized for sacrificing the well-being of many species for the benefit of a single threatened or endangered species. This criticism has been one of the major justifications for the movement toward more holistic, multi-species "ecosystem management." Although there are certainly situations in which management for the benefit of an endangered species has had a negative effect on the natural biodiversity and functioning of an ecosystem, this does not justify a "prohibition" of management that focuses on a single species.

The critical issues in any single- or multi-species management situation are the functional roles of the species under consideration, and the relationships among the individual species, the entire community, and the physical conditions of the local environment. There are many possible situations in which a management focus on a single species can be of great benefit to many other species in the same ecosystem. Examples of such situations are likely to include "keystone species," which are those species whose presence (or absence) potentially causes major shifts in the properties of the ecosystem. In many cases, dominant plants can serve as a "keystone resource," providing food at a critical time of year, or food of a quantity or quality far superior to that of other species. Examples of keystone species that have been eliminated or greatly reduced by pests, pathogens, or direct human activities include the American chestnut, the American elm, the passenger pigeon, the American bison, and the beaver. Near elimination of the large carnivores has allowed deer herds to grow to unmanageable sizes in urban and agricultural areas.

In many cases, management to restore the population of a keystone species is likely to have significant benefits for many additional species that are directly or indirectly. Thus, "single species management" can be an important tool for effective ecosystem management. In other cases, single species management plans must focus on the control or eradication of species that have become threats to the health of the ecosystem, most often invading "exotic" species.

Problem species are not limited to plants and animals imported from other countries. Relocating North American species into ecosystems in which they were not native is also a problem. One of the better known examples is the bullfrog (*Rana catesbeiana*). Originally native only to the eastern United States (generally, east of the Rockies but thought to have been primarily native to the southeastern states), the bullfrog has been widely introduced in the west and southwest where it is recognized as a major factor in the decline of endemic herpetofauna, including other frogs and possibly the Mexican garter snake (*Thamnophis eques*) in southern Arizona (Rosen and Schwalbe 1995). Similar impacts from bullfrogs preying upon native species have been widely observed in the western states by other authors.

The introduction of various fish species for sport is also a major factor in the decline of native aquatic organisms. Non-indigenous species (both native and non-native to North America) are cited as a contributing cause in the extinction of 27 species and 13 subspecies of North American fishes over the past 100 years (Miller et al. 1989).

Management practices can decrease the invasibility of management ecosystems, and also minimize the negative impact of species that do invade. The primary objective is to maintain the combination of environmental conditions that is most favorable to the desired native species, and to minimize disturbances that increase opportunities for invasion (see Sections 3.1.2 and 4.4–4.6).

Major changes in environmental conditions often causes destabilization of the native community, making it more susceptible to invasion by competitive exotics. For example, much of the Platte River system that threads easterly from the Rocky Mountains through the shortgrass prairie was once an open flood plain with widely spaced pockets of trees. Spring flows controlled by dams have since converted much of the Platte to a continuous riparian woodland. Species once isolated to the east now have corridors to move west. Eastern blue jays have become common in cities at the foot of the Rocky Mountains, as have white-tailed deer and other eastern species. Cross-breeding with native western species has already been noted in the case of the white-tail, and is a matter of concern with the blue jay.

Minimizing disturbances, particularly those that remove vegetation and disturb the soil, can reduce the probability of invasion by exotic species, as well as reduce the severity of the problems caused by the invaders.

Once a problem exotic becomes established in an area, its control and eventual elimination can require a substantial effort (see Covington et al. this volume, Kenna et al. this volume).

## 8. CONCLUSION

Scientific understanding of the factors that influence species diversity is sufficient to provide guidance on whether a particular management action (or inaction) is likely to increase or decrease the diversity of a particular group or type of organisms. All landscapes, both managed and unmanaged, are in a constant state of flux, with some species increasing in abundance and others decreasing, in response to both natural processes and human activities. Against this background of change, resource managers can envision their landscapes as a patchwork of areas in different stages of successional development in response to different regimes of productivity and disturbance.

Species diversity changes in complex, but predictable, ways in response to natural variation in environmental conditions, and in response to management actions. The most important pattern of species diversity in response to productivity (or site index) is that the diversity of plants, and many other associated organisms, first increases as productivity increases from very low levels, and then decreases as productivity increases to high levels. This means that the diversity of many types of plants and animals is actually highest under conditions of low to intermediate productivity, rather than under the most productive conditions.

A second pattern that is extremely important to resource management is the effect of disturbances on species diversity. At moderate to high levels of productivity, an increase in disturbance frequency or intensity usually causes an increase in the diversity of plants and associated animals. At low levels of productivity, however, the same increase in disturbance will usually result in a decrease in species diversity. Consequently, the effects of a single type of management activity can be completely different depending on the productivity conditions under which it occurs.

Because both productivity and disturbances vary across landscapes as a result of natural variability in soils and climate, as well as of human activities such as harvesting, hunting, thinning, and fertilizing, patterns of species diversity are highly variable and constantly changing at different rates in different areas. This perspective of landscapes in dynamic equilibrium between disturbance and recovery provides a framework for predicting the consequences of any specific management action, as well as for planning how landscapes can be used to maximize their capability to perform the many services required by society.

## Comment from the Authors

This chapter has been edited and shortened from the original manuscript to meet space requirements. In the process of editing, many figures, comments, and references have been deleted that the authors consider important. Anyone desiring a copy of the original manuscript, with the complete citation list and attributions in the text, should contact M. Huston.

## REFERENCES

Aber, J.D., and J.M. Melillo, 1991. *Terrestrial Ecosystems.* Saunders College Publishing, Philadelphia, Pennsylvania.

Abrams, P.A. 1993. Effects of increased productivity on the abundances of trophic levels. *American Naturalist* 141: 351–371.

Al-Mufti, M.M., C.L. Sydes, S.B. Furness, J.P. Grime, and S.R. Band. 1977. A quantitative analysis of shoot phenology and dominance in herbaceous vegetation. *Journal of Ecology* 65: 759–91.

Anderson, J.M. 1994. Functional attributes of biodiversity in land use systems. In: Greenland D.J., and I. Szabolcs (eds.), *Soil Resilience and Sustainable Land Use*, pp. 267–290. Oxford University Press, Oxford.

Asher, J. 1995. Proliferation of invasive alien plants on Western Federal Lands. *Proceedings of the Alien Plant Invasions Symposium.* Society of Range Management, Annual meeting, Phoenix, AZ, January 17,1995.

Bakker, J.P. 1989. *Nature Management by Grazing and Cutting.* Geobotany 14. Kluwer, Dordrecht, The Netherlands.

Berendse, F., R. Aerts, and R. Bobbink. 1993. Atmospheric nitrogen deposition and its impact on terrestrial ecosystems. In: Vos, C.C., and P. Opdam (eds.), *Landscape Ecology of a Stressed Environment.* pp. 103–121. Chapman and Hall, England.

Berger, J.J. (ed). 1990. *Environmental Restoration.* Island Press, Washington, D.C. 398 pp.

Beven, K.J., and I.D. Moore (eds.). 1993. *Terrain Analysis and Distributed Modeling in Hydrology.* Wiley, New York.

Billings, W.D. 1990. Bromus tectorum, a biotic cause of ecosystem impoverishment in the Great Basin. In: Woodwell, G.M. (ed.), *The Earth in Transition: Patterns and Processes of Biotic Impoverishment.* pp. 301–322. Cambridge University Press, Cambridge.

Bjergo C., C. Boydstun, M. Crosby, S. Kokkanakis, R. Sayers Jr. 1995. Non-native aquatic species in the United States and coastal waters. Our Living Resources, U.S. Department of the Interior, National Biological Service.

Boyce, M.S. 1997. Population viability analysis: adaptive management for threatened and endangered species. In: Boyce, M.S., and A. Haney (eds.), *Ecosystem Management: Applications for Sustainable Forest and Wildlife Resources.* pp. 226–236. Yale University Press, New Haven.

Bryant, J.P., F.D. Provenza, J. Pastor, P.B. Reichardt, T.P. Clausen, and J.T. du Toit. 1991. Interactions between woody plants and browsing mammals mediated by secondary metabolites. *Ann. Rev. Ecol. Systematics* 22: 431–446.

Carey, A.B., C. Elliott, B.R.Lippke, and others. 1996. Washington Forest Landscape Management Project - A Pragmatic, Ecological Approach to Small-landscape Management, Report No. 2, Washington State Department of Natural Resources, Olympia, Washington.

Caswell, H. 1989. *Matrix Population Models.* Sinauer Associates, Sunderland, MA.

Chou, Y.H., R.A. Minnich, and R.J. Dezzani. 1993. Do fire sizes differ between southern California and northern Baja California. *Forest Science* 39: 835–844.

Coley, P.D., J.P. Bryant, and F.S. Chapin III. 1985. Resource availability and plant herbivore defense. *Science* 230: 895–899.

Colwell, R.K., and J.A. Coddington. 1994. Estimating terrestrial biodiversity through extrapolation. *Philosophical Transactions of the Royal Society, London*, B 345: 101–118.

DeAngelis, D.L., L.J. Gross, M.A. Huston, and others. 1998. Landscape modeling for Everglades ecosystem restoration. Ecosystems 1: in press.

DeGraf, R.M., V.W. Scott, R.H. Hamre and others. 1991. Forest and rangeland birds of the United States: natural history and habitat use. Agric. Handb. AH-688. U.S. Department of Agriculture, Forest Service, Washington, D.C. 625 pp.

Duggins, D.O. 1980. Kelp beds and sea otters: an experimental approach. *Ecology* 61: 447–453.

Ehrlich, P.R., and A.H. Ehrlich . 1981. *Extinction: The Causes and Consequences of the Disappearance of Species.* Random House, New York.

Fastie, C.L. 1995. Causes and ecosystem consequences of multiple pathways of primary succession at Glacier Bay, Alaska. *Ecology* 76: 1899–1916.

Flather, C.H., L.A. Joyce, and C.A. Bloomgarden. 1994. Species endangerment patterns in the United States. USDA General Technical Report RM-241.

Fleischner, T.L. 1994. Ecological costs of livestock grazing in western North America. *Conservation Biology* 8: 629–644.

Forsman, E.B., E.C. Meslow, and H.M. Wright. 1984. Distribution and biology of the spotted owl in Oregon. *Wildlife Monographs* 87: 1–64.

Fox, J.F. 1979. Intermediate disturbance hypothesis. *Science* 204: 1344–1345.

Franklin, J. 1995. Predictive vegetation mapping: geographic modelling of biospatial patterns in relation to environmental gradients. *Progress in Physical Geography* 19: 474–499.

Franklin, J.F., and R.T.T. Forman. 1987. Creating landscape patterns by forest cutting: ecological consequences and principles. *Landscape Ecology* 1: 5–18.

Franklin, J.F. 1997. Ecosystem management: an overview. In: Boyce, M.S., and A. Haney (eds.), *Ecosystem Management: Applications for Sustainable Forest and Wildlife Resources*, pp. 21–53. Yale University Press, New Haven.

Gauch, H.G. 1982. *Multivariate Analysis in Community Ecology.* Cambridge University Press, Cambridge.

Grime, J.P. 1973. Competitive exclusion in herbaceous vegetation. *Nature* 242: 344–347

Grime, J.P. 1998. Benefits of plant diversity to ecosystems: immediate, filter, and flexibility effects. *Journal of Ecology* 86: 902–910.

Guo, Q., and W.L. Berry. 1998. Species richness and biomass: dissection of the hump-shaped relationships. *Ecology* 79: 2555–2559.

Hanski, I., and M. Gilpin. 1991. *Metapopulation Dynamics.* Academic Press, London.

Hargrove, W.W. 1994. Using EMBYR, a large-scale probabilistic fire model, to re-create the Yellowstone Forest Lake fire. Environmental Research News, Oak Ridge National Laboratory: http: //www.esd.ornl.gov/ern/embyr/embyr.html.

Hargrove, W.W., R.H. Gardner, M.G. Turner, W.H. Romme, and D.G. Despain. Simulating fire patterns in heterogeneous landscapes. *Ecological Modeling*, in press.

Hedrick, P. W. and P. S. Miller. 1992. Conservation genetics: techniques and fundamentals. *Ecol. Appl.* 2: 30–46.

Hooper, D.U., and P.M. Vitousek. 1997. The effects of plant composition and diversity on ecosystem processes. *Science* 277: 1302–1305.

Huenneke, L.F., 1995. Ecological impacts of plant invasions in rangeland ecosystems. *Proceedings of the Alien Plant Symposium*. Society of Range Management, Annual meeting, Phoenix, AZ, 1/17/95

Humphries, C. J., P. H. Williams, and R. I. Vane-Wright. 1995. Measuring biodiversity value for conservation. *Annu. Rev. Ecol. Syst.* 26: 93–111.

Huntley, N. 1991. Herbivores and the dynamics of communities and ecosystems. *Ann. Rev. Ecology Systematics* 22: 477–504.

Huston, M.A . 1979. A general hypothesis of species diversity. *American Naturalist* 113: 81–101.

Huston, M.A. 1994. *Biological Diversity: The Coexistence of Species on Changing Landscapes.* Cambridge University Press, Cambridge.

Huston, M.A. 1997. Hidden treatments in ecological experiments: re-evaluating the ecosystem function of biodiversity. *Oecologia* 110: 449–460.

Huston, M.A., and T.M. Smith. 1987. Plant succession: life history and competition. *American Naturalist* 130: 168–198.

Huston, M.A., and D.L. DeAngelis. 1987. Size bimodality in monospecific populations: A review of potential mechanisms. *American Naturalist* 129: 678–707.

Huston, M.A, and D.L. DeAngelis. 1994. Competition and coexistence: the effects of resource transport and supply rates. *American Naturalist* 144: 954–977.

Huston, M.A., D.L. DeAngelis, and W.M. Post. 1988. New computer models unify ecological theory. *BioScience* 38: 682–692.

Hyams, E. 1952. *Soil and Civilization.* Harper and Row, New York.

Johnson, D.W., and S.E. Lindberg. 1992. *Atmospheric Deposition and Forest Nutrient Cycling: A Synthesis of the Integrated Forest Study*. Springer-Verlag, New York. 632 pp.

Johnson, K.H., R.A. Olson, T.D. Whitson, R.J. Swearingen, and G.L. Kurz, 1994. Ecological implications of Russian knapweed infestations: small mammal and habitat associations. *Proc. Western Soc. Weed Sci.*, 47: 98–101.

Johnston, C.A., and R.J. Naiman. 1990. Aquatic patch creation in relation to beaver population trends. *Ecology* 71: 1617–1621.

Jongman, R.H., C.J. ter Braak, and P.O. van Tongeren. 1995. *Data Analysis in Community and Landscape Ecology.* Cambridge University Press, Cambridge.

Karr, J.R. 1993. Measuring biological integrity: lessons from streams. In: *Ecological Integrity and the Management of Ecosystems.*, pp. 83–104. St. Lucie Press, Delray, FL.

Keddy, P.A, and P. MacLellan. 1990. Centrifugal organization in forests. Oikos 59: 75–84

Lawes J.B., J.H. Gilbert, and M.T. Masters. 1882. Agricultural, chemical, and botanical results of experiments on the mixed herbage of permanent grasslands, conducted for more than twenty years in succession on the same land. Part II. The botanical results. *Philosophical Transactions of the Royal Society of London*, A&B 173: 1181–1423

Lawton J.H., and R.M. May. 1995. *Extinction Rates.* Oxford University Press, Oxford.

Lin, D.-S., E.F. Wood, P.A. Troch, M. Mancini, and T.J. Jackson. 1994. Comparisons of remotely sensed and model-simulated soil moisture over a heterogeneous watershed. *Remote Sens. Environ.* 48: 159–171.

Lynch, M., and R. Lande. 1993. Evolution and extinction in response to environmental change. In: Karieva, P., R. Huey, and J. Kingsolver (eds.), *Evolutionary, Population, and Community Responses to Global Change*, pp. 243–250. Sinauer Associates, Sunderland, MA.

Magurran, A.E. 1988. *Ecological Diversity and its Measurement.* Princeton University Press, Princeton, NJ.

Mann, L.K., P.D. Parr, L.R. Pounds, and R.L. Graham. 1996. Protection of biota on nonpark public lands: examples from the U.S. Department of Energy Oak Ridge Reservation. *Environmental Management* 20: 207–218.

Mayden, R.L. and R.M. Wood. 1995. Systematics, species concepts, and the evolutionarily significant unit in biodiversity and conservation biology. In: Nielsen, J.L. (ed.), *Evolution and the Aquatic Ecosystem: Defining Unique Units in Population Conservation.*, pp. 58–113. Am. Fish. Soc. Symposium 17, Bethesda, MD.

McInnes, P.F., R.J. Naiman, J. Pastor, and Y. Cohen. 1992. Effects of moose browsing on vegetation and litterfall of the boreal forests of Isle Royale, Michigan, USA. *Ecology* 73: 2059–2075.

McMinn, J.W., and D.A. Crossley (eds). 1996. Biodiversity and Coarse Woody Debris in Southern Forests. USDA Forest Service, General Technical Report SE-94.

Meffe, G.K. 1996. Conserving genetic diversity in natural systems. In: Szaro, R.C., and D.W. Johnston (eds.), *Biodiversity in Managed Landscapes: Theory and Practice*, pp. 41–57. Oxford University Press, New York.

Melillo, J.M., J.D. Aber, and J.F. Muratore. 1982. Nitrogen and lignin control of hardwood leaf litter decomposition dynamics. *Ecology* 63: 621–626.

Milchunas, D.G., Sala, O.E., and W.K. Lauenroth. 1988. A generalized model of the effects of grazing by large herbivores on grassland community structure. *American Naturalist* 132: 87–106.

Miller, R.R., J.D. Williams, and J.E. Williams, 1989. Extinctions of North American fishes during the last century. *Fisheries* 14(6): 22–38.

Mills, L. S. and P. E. Smouse. 1994. Demographic consequences of inbreeding in remnant populations. *American Naturalist* 144: 412–431.

Mills, T.R., M.A. Rumble, and L.D. Flake. 1996. Evaluation of a habitat capability model for nongame birds in the Black Hills, South Dakota. Research Paper RM-RP-323, USDA Forest Service, Fort Collins, Colorado.

Mladenoff, D.J., and F.Stearns. 1993. Eastern hemlock regeneration and browsing in the northern Great Lakes region: a re-examination and model simulation. *Conservation Biology* 7: 889–900.

Mladenoff, D.J., M.A. White, T.R. Crow, and J. Pastor. 1994.

Applying principles of landscape design and management to integrate old-growth forest enhancement and commodity use. *Conservation Biology* 8: 752–762.
Moritz C., S. Lavery, and B. Slade. 1995. Using allele frequency and phylogeny to define units for conservation and management. In: Nielsen, J.L. (ed.), *Evolution and the Aquatic Ecosystem: Defining Unique Units in Population Conservation.*, pp. 249–262. Am. Fish. Soc. Symposium 17, Bethesda, MD.
Oldfield, M. 1989. The value of conserving genetic resources. Sinauer, Sunderland, Massachusetts.
Oliver, C.D., and B.C.Larson. 1996. *Forest Stand Dynamics*, Second Edition. McGraw-Hill, New York.
Owen, D.F., and R.G. Wiegert. 1981. Mutualism between grasses and grazers: an evolutionary hypothesis. *Oikos* 36: 376–378.
Pacala, S., C. Canham, and J. Silander. 1993. Forest models defined by field measurements. I. The design of a northeastern forest simulator. *Canadian Journal of Forest Research* 23: 1980–1988.
Pacala, S., and D. Deutschman.1995. Details that matter: the spatial distribution of individual trees maintains forest ecosystem function. *Oikos* 74: 357–365.
Paine, R.T. 1966. Food web complexity and species diversity. *American Naturalist* 100: 65–75.
Pastor, J.J., J.D. Aber, C.A. McClaugherty, and J.M. Melillo. 1984. Aboveground production and N and P cycling along a nitrogen mineralization gradient on Blackhawk Island, Wisconsin. *Ecology* 65: 256–268.
Pastor, J., and Y. Cohen. 1997. Herbivores, the functional diversity of plant species, and the cycling of nutrients in ecosystems. *Theoretical Population Biology* 51: 165–179.
Pastor, J. and W.M. Post. 1986. Influence of climate, soil moisture, and succession on forest carbon and nitrogen cycles. *Biogeochemistry* 2: 3–17.
Perry, D.A., R. Molina, and M.P. Amaranthus. 1987. Mycorrhizae, mycorrhizospheres, and reforestation: current knowledge and research needs. *Canadian Journal of Forest Research* 17: 929–940.
Podolsky, R. 1997. Software tools for the management and visualization of biodiversity data. The United Nations Development Program. Website: www.undp.org/biod/bio.html.
Pollock, M.M., R.J. Naiman, and T.A. Hanley. 1998. Plant species richness in riparian wetlands — a test of biodiversity theory. *Ecology* 79: 94–105.
Post, W.M., and J. Pastor. 1990. An individual-based forest ecosystem model for projecting forest responses to nutrient cycling and climate change. In: Wensel, L.C., and G.S. Biging (eds.), *Forest Simulation Systems*, pp. 61–73. University of California, Division of Agriculture and Natural Resources, Bulletin 1927.
Proulx, M., and A. Mazumder. 1998. Reversal of grazing impact on plant species richness in nutrient-poor vs. nutrient-rich ecosystems. *Ecology* 79: 2581–2592.
Proulx M., F.R. Pick, A. Mazumder, P.B. Hamilton, and D.R.S. Lean. 1996. Experimental evidence for interactive impacts of human activities on lake algal species richness. *Oikos* 76: 191–195.
Pulliam, H.R. 1988. Sources, sinks, and population regulation. *American Naturalist* 132: 652–661.
Ralph, C.J., P.W.C. Patton, and C.A. Taylor. 1991. Habitat association patterns of breeding birds and small mammals in Douglas-fir/hardwood stands in northwestern California and southwestern Oregon. Gen. Tech. Rep. PNW-285, USDA Forest Service, Pacific Northwest Research Station.
Randall. A. 1994. Thinking about the value of biodiversity. In: Kim, K.C., and R.D. Weaver (eds), *Biodiversity and Landscapes*, pp. 271–286. Cambridge Univ Press, Cambridge.
Rosen P.C., C.R. Schwalbe, 1995. Bullfrogs: Introduced predators in Southwestern wetlands. Our Living Resources, U.S. Department of the Interior, National Biological Service.
Rosenzweig, M. 1995. *Species Diversity in Space and Time.* Cambridge University Press, Cambridge.
Rothermel, R.C. 1972. A mathematical model for predicting fire spread in wildland fuels. Research Paper INT-115. Ogden, Utah. USDA Forest Service.
Running, S.W. 1994. Testing forest-BGC ecosystem process simulations across a climatic gradient in Oregon. *Ecological Applications* 4: 238–247.
Running, S.W., and E.R. Hunt, Jr. 1993. Generalization of a forest ecosystem model for other biomes, BIOME-BGC, and an application for global scale models. In: Ehlringer, J.R., and C.B. Field (eds.), *Scaling Physiological Processes: Leaf to Globe*, pp. 141–158. Academic Press, San Diego, CA.
Sagoff, M. 1997. Do we consume too much? *The Atlantic Monthly* 279: 80–96
Sarkar, S. 1996. Ecological theory and anuran declines. *BioScience* 46: 199–207.
Scott, J.M., F. Davis, B. Csuti, R. Noss, B. Butterfield, G. Groves, H. Anderson, S. Caicco, F. D'Erchia, T.C. Edwards, J. Ulliman, and R.G. Wright. 1993. GAP analysis: a geographic approach to protection of biological diversity. *Wildlife Monographs* 123: 1–41.
Shmida, A., and M.V. Wilson. 1985. Biological determinants of species diversity. *Journal of Biogeography* 12: 1–20.
Short, H.L. 1992. Use of the habitat linear appraisal system to inventory and monitor the structure of habitats. In: McKenzie, D.H., D.E. Hyatt, and V.J. McDonald (eds.), *Ecological Indicators*. pp. 961–974. Elsevier Applied Science, London.
Short, H.L., and S.C. Williamson. 1987. Evaluating the structure of habitat for wildlife. Part I. Development, testing, and application of wildlife-habitat models. U.S. Fish and Wildlife Service, Fort Collins, CO.
Shugart, H.H. 1984. *A Theory of Forest Dynamics.* Springer-Verlag, New York.
Shugart, H.H., M.S. Hopkins, I.P Burgess, and A.T. Mortlock. 1981. The development of a succession model for subtropical rainforest and its application to assess the effect of timber harvest at Wiangarree State Forest, New South Wales. *Journal of Environmental Management* 11: 243–265.
Smith, T.M., and M.A. Huston 1989. A theory of the spatial and temporal dynamics of plant communities. *Vegetatio* 83: 49–69.
South Florida Water Management District. 1997. Documentation for the South Florida Water Management Model. Hydrologic Systems Modeling Division, SFWMD, West Palm Beach, FL.
Stohlgren T.J., G.W. Chong, M.A. Kalkhan, and L.D. Schell. 1997. Rapid assessment of plant diversity patterns: a methodology for landscapes. *Environmental Monitoring and Assessment* 48: 25–43.

Stohlgren T.J., D.A. Binkley, G.W. Chong, M.A. Kalkhan, L.D. Schell, K.A. Bull, Y. Otsuki, G. Newman, M. Bashkin, and Y. Son. Exotic plant species invade hot spots of native plant diversity. *Ecology*, in press.

Szaro, R.C., and R.P. Balda. 1979. Bird community dynamics in a ponderosa pine forest. *Studies in Avian Biology* 3: 1–66.

Thomas, J.W., E.D. Forsman, J.B. Lint, E.C. Meslow, B.R. Noon, and J. Verner. 1990. A conservation strategy for the northern spotted owl. Report of the interagency committee to address the conservation strategy of the northern spotted owl. U.S. Government Printing Office Publication 1990-791-171/20026. Washington, DC.

Tilman, D. 1993. Species richness of experimental productivity gradients: how important is colonization limitation. *Ecology* 74: 2179–2191

Tilman, D. 1997. Community invasibility, recruitment limitation, and grassland biodiversity. *Ecology* 78: 81–92.

Uhl, C., P. Barreto, A. Verissimo, and E. Vidal. 1997. Natural resource management in the Brazilian Amazon: an integrated approach. *BioScience* 47: 160–168.

Urban, D.L., and T.M. Smith. 1989. Microhabitat pattern and the structure of forest bird communities. *American Naturalist* 133: 811–829.

Urban, D.L., G.B. Bonan, T.M. Smith, and H.H. Shugart. 1991. Spatial applications of gap models. *Forest Ecology and Management* 42: 95–110.

VanWinkle, W., K.A. Rose, K.O. Winemiller, D.L. DeAngelis, S.W. Christensen, R.G. Otto, and B.J. Schuter. 1993. Linking life history theory, environmental setting, and individual-based modeling to compare responses of different fish species to environmental changes. *Trans. Am. Fish. Soc.* 122: 459–466.

Verner, J.M., J.R. Lewis, and C.J. Ralph. 1986. *Wildlife 2000*. Univ. Wisconsin Press, Madison, WI. 470 pp.

Vitousek, P.M. 1986.Biological invasions and ecosystem properties: can species make a difference. In: Mooney, H.A., and J.A. Drake (eds), *Ecology of Biological Invasions of North America and Hawaii*, pp.163–176. Springer-Verlag, New York.

Walker, B.H. 1992. Biodiversity and ecological redundancy. *Biological Conservation* 55: 235–254.

Wardle, D.A., O. Zackrisson, G. Hornberg, and C. Gallet. 1997. The influence of island area on ecosystem properties. *Science* 277: 1296–1299.

Waring, R.D., and W.H. Schlesinger. 1985. *Forest Ecosystems: Concepts and Management*. Academic Press, New York.

## THE AUTHORS

**Michael Huston**
*Environmental Sciences Division*
*Oak Ridge National Laboratory*
*Oak Ridge, TN 37831-6335, USA*

**Gary McVicker**
*Bureau of Land Management*
*Colorado State Office*
*2850 Youngfield St.*
*Lakewood, CO 80215, USA*

**Jennifer Nielsen**
*Alaska Biological Science Center*
*USGS/BRD*
*1011 East Tudor Road*
*Anchorage, AK 997503, USA*

Wayne, R.K. and S.M. Jenks. 1991. Mitochondrial DNA analysis implying extensive hybridization of the endangered red wolf *Canis rufus*. *Nature* 351: 565–568.

Werner, E.E., and M.A. McPeek. 1994. Direct and indirect effects of predators on two anuran species along an environmental gradient. *Ecology* 75: 1368–1382.

White, D.L., T.A. Waldrop, and S.M. Jones. 1991. Forty years of prescribed burning on the Santee fire plots: effects on understory vegetation. In: Nodvin, S.C., and D.A. Waldrop (eds.), *Fire and the Environment: Ecological and Cultural Perspectives*, pp. 45–59. Southeastern Forest Experiment Station, Asheville, NC.

Wilson, S.D., J.W. Belcher, 1989. Plant and bird communities of native prairie and introduced Eurasian vegetation in Manitoba, Canada. *Conservation Biol.* 3: 39–44.

Whittaker, R.H. 1960. Vegetation of the Siskiyou Mountains, Oregon and California. *Ecological Monographs* 26: 1–80.

Wullschleger, S.D., W.M. Post, and A.W. King. 1995. On the potential for a $CO_2$ fertilization effect in forests: estimates of the biotic growth factor based on 58 controlled-exposure studies. In Woodwell, G.M., and F.T. Mackenzie (eds.) *Biotic Feedbacks in the Global Climate System*, pp. 85–107. Oxford University Press, New York.

# Population Viability Analysis: A Primer on its Principal Technical Concepts

Barry R. Noon, Roland H. Lamberson, Mark S. Boyce, and Larry L. Irwin

**Key questions addressed in this chapter**

- *What constitutes a reliable population viability analysis*
- *Why population viability analyses are important to ecological stewardship*
- *The basic data requirements for a defensible population viability analysis*
- *Mathematical models that are the theoretical and practical foundations of population viability analyses*
- *Applications of population viability analyses to improved management*

**Keywords: Extinction, risk, stochasticity, genetic structure, populations, dispersal, reproduction, models**

# 1 INTRODUCTION

As natural resource professionals, one of our key responsibilities is to manage ecosystems so as to maintain a full complement of plant and animal species for the use and enjoyment of future human generations. History and the economic principles that govern human commerce, however, clearly indicate that our exploitation of natural resources has the inherent potential to put species at risk of extinction (Clark 1973, Ludwig et al. 1993). To preserve biological diversity, therefore, we must assess those factors that put species at risk of extinction prior to species loss. In general, this is the motivation for population viability analyses (PVAs) — to assess the threats to a species' persistence, and to intervene before declines become irreversible.

Gilpin and Soule (1986) coined the term "population viability analysis" for the comprehensive analysis of all factors that may cause a species to go extinct. "PVA is a structured, systematic, and comprehensive examination of the interacting factors that place a population or species at risk" (Shaffer 1990); and, according to Boyce (1992), PVA entails a formal evaluation of data in the context of models of population dynamics to estimate the likelihood that a population will persist for some arbitrarily chosen time into the future. All these definitions suggest that PVA entails a sophisticated process of data exploration and model development in the context of formal mathematical analyses and computer simulations.

When the available data allows, such refined analyses are preferred. Such data-rich cases provide the best known examples of PVA, and they are often the only examples in the scientific literature. If we were to require this level of data and associated knowledge of the species' life history and ecology, however, we would be able to conduct PVAs for only a handful of species. To be widely useful, we need general principles for the conduct of PVAs that can be used in the more general case of sparse data. The goal of these more streamlined PVAs is still to identify those factors that jeopardize a species' persistence before irreversible declines have occurred. In the absence of extensive data, however, we need to acknowledge the tradeoffs involved. Decisions about species at risk, and what is required to ameliorate these risks, will be made with much less certainty.

## 1.1 Brief History of the Basis for PVAs

The PVA process is closely related to the concept of minimum viable population, an estimate of the minimum number of organisms that constitutes a self-perpetuating population. The term, "minimum viable population" (MVP) was first used in 1949 (Allee et al. 1949) to define a size below which populations will experience great increases in their chances of extinction. Shaffer (1981) defined the minimum viable population arbitrarily as the smallest isolated population having a 99% chance of surviving for 1000 years. Importantly, this definition introduced the concept of *risk* to the PVA process. The MVP concept, in turn, is conceptually linked to early work on species-area relationships (Preston 1948) and the extinction–colonization dynamics of oceanic islands (MacArthur and Wilson 1967). As a general rule, as the area of a sample (or island) decreases, the expected number of species in the sample also decreases. This suggests the existence of a critical threshold area, a minimum area supporting a minimum population size, below which a given species will rapidly become extinct. Those species that persist in small samples are those with small area requirements and/or high colonization rates. The biological interpretation of the species-area phenomenon established a connection between critical population size and minimum habitat area requirements. That is, for a population to persist, it will require a certain area of suitable habitat.

There are many possible factors that put populations at risk of extinction. These include habitat loss, habitat fragmentation, over-harvest, illegal hunting, episodic disease outbreaks, toxic pollution, and the introduction of exotic species that become competitors, predators or radically modify habitat. Of these, changes in habitat amount and distribution represent the most pervasive threats to biodiversity loss in terrestrial ecosystems (Ehrlich 1995). The consequences of habitat destruction and fragmentation include an increased risk of global extinction through failure of individuals to disperse between local populations (recognition of the importance of the spatial structure of populations), edge effects resulting from increased rates of predation and parasitism, inbreeding depression resulting from isolation and reduced gene flow, and the increased significance of random accidents in small, isolated populations.

Given the diversity of factors that can stress local and regional populations, it is useful to distinguish between two broad classes of stressors (Lawton 1995): factors that result in populations being rare and isolated in the first place; and factors that result in eventual extinction once populations have already become small and disjunct. The first set of factors have been called "the ultimate causes of extinction" (Simberloff 1986) or collectively referred to as the "declining population paradigm" (Caughley 1994). The second set of factors represent the "proximate causes of extinction" (Simberloff 1986), or what Caughley (1994) grouped together as consequences of the "small population paradigm." It is also useful to further qualify these two sets of threats by considering ultimate factors as the deterministic events that initially put the population at risk, and proximate factors as the stochastic events leading to eventual extinction.

We believe this broad classification is useful because so much emphasis in PVAs conducted to date has been on the proximate causes of extinction. This concentration is somewhat misdirected because it represents a focus more on the symptoms than the root causes of population decline. For conservation biology to be a more proactive science, that is, to provide the information needed to get out in front of the extinction curve, will eventually require a more focused management response to the ultimate factors that put populations at risk.

## 1.2 Forms of PVA

There is no set format for the conduct of a PVA, nor a single "best" approach. The structure that a given PVA assumes is dictated by data availability, the degree to which the species' ecology and life history are understood, knowledge of the factors that put populations at risk, and the goals and responsibilities of the management agent. Possible response variables in PVA models are diverse and include: time to extinction (zero individuals) or time to quasi-extinction (a given, small number of individuals), probability of population extinction within $x$ years, probability that a population will still persist after $x$ years, and various measures of the occupancy likelihood of suitable habitat patches. As a consequence, the criteria for evaluating viability (e.g., persistence likelihood and time frame) are somewhat arbitrary. Ultimately, data availability, particularly the available measures of population status (i.e., presence/absence, abundance, density, birth rate, etc.) dictate the response variable(s) and the methods of analysis.

It is important to recognize that PVA is a predictive tool only in a probabilistic sense — the output of analyses are generally probability statements about the likelihoods of various population outcomes. PVA is an assessment of the likelihood of future population events based on our current understanding of how populations change in space and time as a consequence of the action of various internal and external factors. External factors generally are the deterministic drivers of change such as habitat reduction or pollution. They affect populations largely independent of current size, sex, and age structure, or dispersal dynamics. Internal factors generally are the stochastic factors of change, which predominate in small, isolated populations. These include the proximate causes of extinction, including demographic and environmental stochasticity, genetic deterioration, and social dysfunction (Lawton 1995). Given the unknown likelihoods of internal and external drivers affecting a population at a given point in time, and our incomplete understanding of population dynamics, the results of PVAs are never precise forecasts of future events. Using PVA as a predictive tool is an unavoidably uncertain process.

## 1.3 Objective of PVA

For the purposes of this paper, we are most interested in the usefulness of PVA as a tool to help solve management problems. Given this focus, we believe that PVA needs to have a clear connection to the management decision making process. Specifically, PVA should be conducted "...to provide insights into how resource management can change parameters influencing the probability of extinction" (Boyce 1992). The results of a PVA should inform managers as to what actions to take to ameliorate threats to a species' persistence, such as the control of invasive species, change in the amount and distribution of suitable habitat, intervention to stop the input of toxic substances, or preemptive control of disease vectors.

In practice this means that, during the PVA process, the measured population variable must be linked to the factor(s) affected by management actions. To specify the relationships and to make them quantitative, the linkage is usually expressed in the form of mathematical equations that relate a response variable (e.g., population size, probability of extinction within $x$ years) to one or more predictor variables (e.g., amount of suitable habitat, mean patch size). The equation(s) establish a cause- effect relationship between a variable that indexes population viability, and the values of the environmental factors that affect viability and are, in turn, affected by management actions.

## 1.4 The Concept of Risk

Essential to a full understanding of the PVA process is the concept of risk. Risk is the probability of a population experiencing an adverse event during a specified time interval. A quantitative characterization of all the possible risks that could impinge upon a population during this interval is termed risk assessment. This is a useful framework for developing a PVA, and the process has been described in detail by Burgman et al. (1993), a reference we highly recommend.

Because of the inescapable uncertainty of environmental events, the likelihood of a population persisting from time $t$ to time $t + \Delta$ is uncertain. During this interval (Δ) the population may be exposed to a number of unpredictable threats to its persistence. Unpredictable does not necessarily mean unknown, but the outcome of these events is only known within the context of some probability distribution. Because all populations are exposed to randomly occurring threats of unpredictable magnitude, the evaluation of the effects of risk events requires a quantitative analysis of a population's dynamics in the context of those events.

One way to envision the assessment of risks to a population in the context of a PVA is as the product of the probabilities of a number of independent events. (In the following the symbol "|" indicates a conditional probability — the likelihood of an event given that the event to the right of the "|" symbol has already occurred). For example, an estimate of the likelihood that a population persists over some time interval Δ could be expressed as: probability (pr) of an adverse event × pr (the event is of a certain magnitude | event) × pr (population is in a given state) × pr (population responds | population state) × pr (of a specific population response | population responds). This simple example makes explicit the uncertain and probabilistic nature of the PVA process.

## 1.5 Sources of Stochasticity

Wild populations are exposed to three major sources of variation and uncertainty. These include demographic and environmental variation, and catastrophic events (Shaffer 1981, Lande 1993). All three of these threats to persistence are lessened with increases in population size, but not necessarily in a linear fashion. Demographic uncertainty arises in small populations as a result of random sampling events — for example, all offspring are of one sex, all adult animals die between breeding periods, or there is a sequence of years without successful reproduction. Because such events arise as a consequence of the product of individual probabilities of death or reproduction, populations quickly escape the risks of demographic uncertainty with increases in size.

Environmental uncertainty arises as a consequence in year-to-year changes in environmental conditions that similarly affect all individuals in the population. One expression of this uncertainty is change from year to year in the expected values of a population's vital rates (birth and death rates). The result can be sequences of poor years, accompanied by low reproduction and/or high mortality, and local extinction events. Populations gain freedom from environmental uncertainty proportional to linear or logarithmic changes in population size (Lande 1993). Regional populations also gain freedom by being spatially structured so that population fluctuations are not synchronous across all local populations (Harrison and Quinn 1989).

Populations are susceptible to catastrophic events to the extent that such events exceed the adaptational limits of the species. Populations gain freedom from these risks by having long-term mean growth rates that are positive (Lande 1993), or by having wide geographic distributions. In the latter case, freedom from catastrophic extinction occurs only because some populations escape the catastrophe. In a recent comparative analysis of the three sources of uncertainty and their effects on extinction risks, Lande (1993) concluded that in "sufficiently large" populations the only general conclusion was that environmental and catastrophic uncertainty pose greater risks of extinction than demographic stochasticity.

The relative significance of environmental or catastrophic risks is difficult to predict; it is a complex function of the mean and variance of a population's growth rates, and the frequency of catastrophes. An important finding (Lande 1993) was that previous estimates of time to extinction (e.g., Goodman 1987, Shaffer 1987) had been too pessimistic, at least for populations with long-term, positive growth rates. The average time to extinction scales as a power of the population's carrying capacity (*K*), a function that scales faster or slower than linearly with *K* depending upon the population's pattern of growth.

The usual case with species at risk of extinction, however, is that their long-term growth rates are near zero, or negative. In this case, average time to extinction increases only as a function of the natural logarithm of carrying capacity, divided by the long-term rate of population decline (Lande 1993). The consequence is that even a large initial population size provides only a small increase in mean time to extinction.

## 1.6 Spatial Structure

Loss and fragmentation of habitat have greatly disrupted otherwise continuous distributions of populations. In fragmented landscapes, populations become distributed across disjunct patches of suitable habitat, separated by intervening habitat of varying degrees of suitability. Knowledge of a regional population's spatial distribution is important — recent advances in modeling the dynamics of plant and animal populations have indicated that the spatial distribution of local populations, and of individuals within local populations, can have profound effects on their dynamics (Lande 1987, Lamberson et al. 1992b, McKelvey et al. 1993; Gilpin 1996). Whether the goal of a PVA is to assess risks to local or regional populations, failure to include spatial information from the total population can lead to erroneous estimates of extinction likelihoods.

The term "metapopulation" has been applied to regional populations that are locally subdivided. Metapopulations are generically referred to as a "population of populations" (Gilpin and Hanski 1991). Population subdivision results in critical differences in biological rates that are relatively unimportant in continuously distributed populations. For example, rates of mating, competition, and other interactions are much higher within than among local populations. The dynamics of metapopulations are complex functions of the number, size, and distribution of local populations, their internal dynamics, and their degree of connectivity.

A population's distribution across the landscape can affect its persistence in both positive and negative ways. Spatial structure can stabilize populations by "risk spreading"; for example, a species whose populations are widely distributed may be temporally decoupled from environmental disturbances and thus have asynchronous population fluctuations. This means that it is unlikely that all local populations will simultaneously be experiencing adverse events. In contrast, over-dispersion of populations can destabilize populations because of threshold effects arising from difficulties in finding suitable habitats and mates (Lande 1987).

Contemporary models explicitly incorporating spatial structure can be generically referred to as metapopulation models. The distributions of real populations, however, are quite variable. Patterns of distribution range from true metapopulations with local populations of similar size connected by dispersal, to core-satellite distributions with a small number of large populations providing colonists to smaller, satellite populations, to mainland–island distributions where a single source population maintains all island populations via colonization (see discussion in Harrison 1994). The key demographic process that often dominates the dynamics of metapopulations is the rate of dispersal among local populations relative to the local rate of extinction. In general, if rates of dispersal among populations are sufficient to stabilize local population fluctuations, prevent local extinction, and allow the colonization of new habitats made available by extinction, regional metapopulations will persist (Gutierrez and Harrison 1996).

## 1.7 Essential Components of a Defensible PVA

All PVAs should have certain elements in common. Where they will differ most is in terms of the details, dictated ultimately by the availability of data. Data-rich PVAs will be more sophisticated, provide more reliable assessments of risks to a species' persistence, and provide clearer guidance to an appropriate management response. In the usual case of sparse data, however, addressing the following issues will still prove useful, even if the final assessment is largely a qualitative one.

- A fundamental understanding of the species' ecology, including what constitutes suitable habitat and some insights into the species' ability to disperse to distant patches of habitat.
- An understanding of what environmental disturbances constitute threats to the species' persistence, and their mechanism of action.
- An understanding of the likely response of the population when exposed to threat(s). That is, to what extent can the species accommodate the perturbation? What are the limits to the species' resilience? Are the population responses to disturbance linear, or nonlinear with steep threshold regions?
- Some knowledge of the likelihood of a threat occurring, and given its occurrence, the expected magnitude of its impact on the population.
- Insights into the deterministic threats to persistence (e.g., timber harvest, water diversions, grazing pressure)
- Insights into the stochastic threats to persistence (e.g., precipitation, temperature, fire, droughts, floods, etc.)
- An assessment of the current state of the population in terms of its vulnerability to disturbance. Where is the population at this point in time (in terms of its resilience) and what is its expected response to disturbance of different types and magnitudes?
- An assessment of both the risks to persistence and possible population responses in a probabilistic context. Estimates of the certainty associated with population forecasts (i.e., the statistical reliability of projections).
- Formal connection to the concepts of decision-making in the context of uncertainty. A clear exposition of the tradeoffs between type I and type II errors and how the error likelihoods were allocated in the decision-making process.
- Assuming that habitat degradation, loss, and fragmentation are the major threats to persistence, PVA must address both the dynamics of the habitat as well as the dynamics of the population. Most PVAs have ignored habitat dynamics, variation in habitat

quality, and the importance of habitat geometry and distribution.

- The results of a PVA are valuable to the extent that they inform the management decision-making process.

### 1.8 Overview of the Remainder of the Chapter

In the following sections we first digress to explain some fundamental principles of population dynamics in the form of simple mathematical models. Our goal is to provide the background necessary to understand the essential population processes that must be addressed in PVA models. We have attempted to illustrate the principles both as simple equations and, whenever possible, in a graphical format. One key point is that changes in a population attribute, for example size, are usually not a simple linear function of change in some landscape attribute, for example habitat amount or its spatial distribution. Understanding the nonlinearities of population dynamics is key to understanding threats to a species' persistence. That is, fixed incremental changes in a factor, such as habitat amount, do not necessary result in constant marginal changes in population size. There exist zones of change in habitat amount, for example, that can result in precipitous changes in population size or persistence likelihoods.

Second, we introduce some basic concepts of population genetics. Our goal here is to emphasize that not all threats to persistence are expressed solely as demographic risks. Researchers are just beginning to understand the relationships between changes in the genetic composition of a population and the expression of those differences through its demographic rates. Therefore, the linkages between genetic changes and population dynamics and persistence are more speculative, but no less important. Finally, it is essential to appreciate that the genetic structure of a population ultimately determines a population's ability to evolve and adapt in the context of changing environmental conditions. A balanced discussion of PVA must include both demographic and genetic models.

### 1.9 A Note on the Use of Mathematics

Estimating birth and death rates or changes in population number or gene frequencies, and understanding what they mean relative to a species' viability, are quantitative exercises that require mathematics. Therefore, our expectation is that individuals engaged in viability assessments possess some minimal degree of expertise in mathematics. For the most part, the level of mathematics used in this chapter does not exceed that gained by undergraduate majors in the sciences after their sophomore year in college (most university curricula in the natural sciences now require the calculus for an undergraduate degree). And, where possible, we have attempted to display equations in graphical form. Therefore, if the equations seem unfamiliar, first study the accompanying graphs. After that, we encourage you to refresh your memory of the relevant mathematics. If you conduct PVAs for your agency, you should be comfortable with all the material in this chapter. If you are unfamiliar with this material, then the viability analyses should be delegated to someone else. Effective management of our natural resources requires numeracy as well as literacy (see Boyce 1991b).

### 1.10 Relation to the Management Chapter

Holthausen et al. (this volume) have written a companion chapter on PVA, emphasizing management issues and implications. The topic overlap between the two chapters is extensive. This chapter is meant to complement Holthausen et al. by providing a mechanistic explanation for the demographic and genetic "rules of thumb" which guide the PVA process. Given this focus, this chapter is more technical and mathematical, requiring the reader to have a deeper background in population dynamics and genetics. As a consequence, this chapter may appeal to a smaller audience. When writing this chapter, however, we identified as our primary reader the practitioner — the individual who actually conducts the viability analyses for his or her agency.

## 2. AN OVERVIEW OF POPULATION DYNAMICS MODELS

A central element of any population viability analysis is a model that describes the dynamics of the population in question. The role this model will play may be nothing more than a structure for organizing the current knowledge of the species (e.g., Salwasser and Marcot 1986), or it may be something as elaborate as a spatially explicit, dynamic map model for the population and the habitat it occupies (e.g., Pulliam et al. 1994). Finally, models may be expanded into multi-species spatial models useful as ecosystem management tools.

### 2.1 Basic Population Models

Fundamentally, population models are simply systems for keeping an accounting of the relationship between recruitment (e.g. births, immigration, etc.) and removal (deaths, migration, etc.). In its simplest form the change in population abundance ($N$) from one time step to the next can be represented as

$$N(t+1) - N(t) = (\text{birth rate})N(t) - (\text{death rate})N(t) = (b - d)N(t)$$

or

$$N(t+1) = [1 + (b - d)]\, N(t)$$

where $1 + (b - d)$ is frequently called the finite rate of increase ($\lambda$) for the population.

In the unusual circumstance where we have a fixed rate of increase, $\lambda$ (i.e., $[b - d]$ is constant), we have that

$$N(t+1) = \lambda\, N(t)$$

which gives us

$$N(t) = \lambda^t N(0)$$

exponential growth (where $N(0)$ is the initial size of the population). If the rate of increase exceeds 1, the population grows; it decreases when $\lambda$ is less than 1; and it remains fixed when $\lambda$ equals 1 (see Fig. 1).

The notion of an exponential growth model with a fixed rate of increase is far too simplistic to be satisfactory for real, unexploited populations. For short time periods, populations may exhibit exponential growth if their size is far from that where growth is limited by environmental carrying capacity. For many commercially harvested populations there is a deliberate attempt by managers to maintain them in this exponential growth phase. For an unexploited population, however, exponential increase implies that the population will grow toward infinite size. To be realistic, either the rate of increase should include some compensatory component (e.g., $\lambda$ decreasing as density increases) or some random variation, or more likely, both.

The rate of increase may itself be a function of many things: availability of nest sites or spawning habitat, year to year variations in the weather, vegetative cover on migration routes, long-term climatic changes, and a multitude of other factors. These issues will be addressed in the sections on density dependence and stochastic models.

The most simple population models assume that all individuals have identical vital rates and occupy a homogeneous space. It is sometimes necessary, however, to subdivide the population into a stage class structure or incorporate some form of spatial structure for the area they occupy (Caswell 1989, Noon and Sauer 1992). Spatial structure becomes necessary when

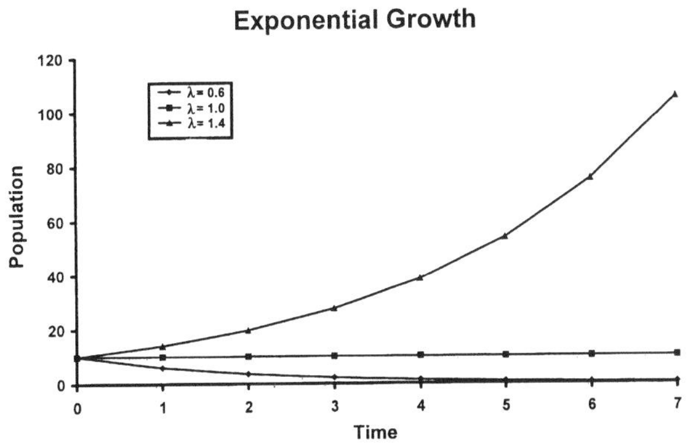


Fig. 1. Changes in population size with time resulting from an exponential growth model using values for the finite rate of increase ($\lambda$: $\lambda > 1.0$; $\lambda = 1.0$; $\lambda < 1.0$).

variation in the suitability of habitat results in the necessity for movement from patch to patch, or results in the isolation of some sub-populations. Models incorporating spatial information, often referred to as patch models, will be considered in Section 2.11. For now, we consider the case of a single population continuously distributed across some space that is assumed homogeneous (that is, uniformly suitable habitat).

Stage classes are necessary when vital rates vary throughout an individual's life span. For example, the stage classes may represent life stages such as juveniles and adults, or the stages may simply be all of the different age classes. For fish populations, stages are often size classes; for trees, stages are often defined by diameter classes. The level of detail represented in a model will be dictated by two factors, the current knowledge of the population being modeled and the specific purpose for that model. There are clear trade-offs between simple and complex models. As models become more complex the number of parameters that must be estimated increases rapidly. The interaction of many poorly known parameters simultaneously increase the probability of propagation of errors and decrease a model's reliability.

## 2.2 Age and Stage Structured Population Models

The basis for structured models is the Leslie matrix projection model (Leslie 1945, 1948) and stage matrix model by Lefkovitch (1965). The Leslie model divides the population into distinct age groups and applies survival and birth rates to each to account for the deaths and births that occur as the population is projected forward into the next time period. The basic accounting of the population in year $t$ is provided by a population vector where the entries correspond to the number of individuals in each age or stage class,

$$\begin{bmatrix} \text{Number of newborns in year } t \\ \text{Number in their second year in year } t \\ \cdot \\ \cdot \\ \cdot \\ \text{Number in their } n\text{th year in year } t \end{bmatrix} = \begin{bmatrix} x_1(t) \\ x_2(t) \\ \cdot \\ \cdot \\ \cdot \\ x_n(t) \end{bmatrix} = X(t)$$

The total population size at time $t$ is obtained by summing the entries in the population vector.

The dynamics of the model are provided by a system of difference equations:

$$x_2(t+1) = s_1 x_1(t)$$
$$x_3(t+1) = s_2 x_2(t)$$
$$\ldots$$
$$x_n(t+1) = s_{n-1} x_{n-1}(t)$$
$$x_1(t+1) = b_2 s_1 x_1(t) + b_3 s_2 x_2(t) + \ldots + b_{n+1} s_n x_n(t)$$

where $s_i$ is the probability that an individual in its $i^{th}$ year of life will survive to its $i{+}1^{th}$ year, and $b_i$ is the number of young born to an individual in age group $i$ during a given year. The first of these equations simply says that the number in the $i{+}1^{th}$ age class next year will be the number of individuals in the $i^{th}$ age class this year multiplied by the probability each survives through the year. The last equation states that the number of newborns next year will equal the total of the expected number of offspring per individual in a given age class times the number of individuals in each age class. To arrive at this form for the equations we are assuming that births occur during a short time period (a birth pulse) just before the census is taken, so the number of potential parents has been reduced by mortality through the year. If, alternatively, we assume that the birth pulse takes place immediately after the census, then the number of parents in each age class is just $x_i(t)$, but we must adjust the number of offspring to account for the number that will have died through the period before the next census. Noon and Sauer (1992) and McDonald and Caswell (1993) discuss the structure of projection matrices for variable time of census relative to the birth pulse.

Assuming a post-birth-pulse census, this system of difference equations can be rewritten in terms of a vector-matrix equation with the age structure vector, $X(t)$, defined above and the matrix, $P$, where

$$P = \begin{bmatrix} s_1 b_2 & s_2 b_3 & s_3 b_4 & \cdot & \cdot & \cdot & 0 \\ s_1 & 0 & 0 & \cdot & \cdot & \cdot & 0 \\ \cdot & & & & & & \\ \cdot & & & & & & \\ \cdot & & & & & & \\ 0 & 0 & 0 & \cdot & \cdot & s_{n-1} & 0 \end{bmatrix}$$

The Leslie matrix model becomes

$$X(t+1) = PX(t) \quad \text{or} \quad X(t) = P^t X(0)$$

Notice that the form of these equations is identical to that of the earlier exponential growth model with $P \propto \lambda$.

Allowing for only $n$ age classes suggests that all individuals that reach the last age class must die during that time interval — the model does not provide any place for them to go. This problem can be resolved by using a stage matrix model. The simplest form of a stage matrix is suitable in the case where we cannot distinguish between individuals within the adult age classes, so it is not possible to obtain age-specific survival rates and fecundities once individuals enter this class. Assuming that we have three stages — juveniles, subadults, and adults — and that the juvenile and subadult stages both last for one time interval, we get the matrix model,

$$P = \begin{bmatrix} s_j b_s & s_s b_a & s_a b_a \\ s_j & 0 & 0 \\ 0 & s_s & s_a \end{bmatrix}$$

This stage matrix model can also be written as a system of difference equations:

$$x_j(t+1) = b_s s_j x_j(t) + b_a s_s x_s(t) + b_a s_a x_a(t)$$

$$x_s(t+1) = s_j x_j(t)$$

$$x_a(t+1) = s_s x_s(t) + s_a s_a(t)$$

Notice, in the particular case shown here, first reproduction occurs as subadults ($1 < \text{age} < 2$), and that the fecundity rate for this group, $b_s$, may differ from that for adults (we are again assuming that the birth pulse occurs immediately before the census is taken).

Other forms of stage matrix models are possible. For example, if a tree grows sufficiently slowly, it may remain in one diameter class for two time periods, or if its growth is sufficiently fast, it may skip over a diameter class in the allotted time interval. An entry in the matrix on the main diagonal represents the probability that an individual in the class represented by that row will remain in that class for another time interval. An entry in the sub-diagonal represents the probability an individual advances from the previous class into the one represented by that row, and an entry in the sub-sub-diagonal represents the probability that an individual skips a stage to enter that stage.

Leslie and stage matrix models with constant parameters are simply highly structured exponential growth models (see Fig. 2). If the proportions of

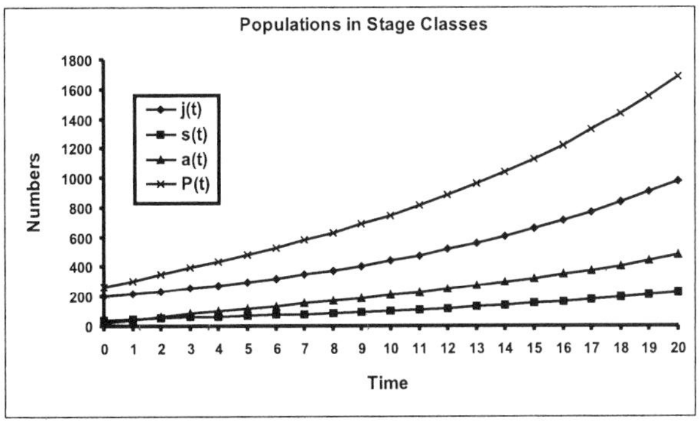


Fig. 2. Growth of the total population ($P(t)$) and numbers in each of the three stage classes ($j(t)$; $s(t)$; and $a(t)$) for a three-stage matrix population model.

individuals in each of the stage classes are stable (this happens quickly if the birth and death rates remain constant) then the rate of increase for the population becomes fixed at a rate which is equal to the dominate *eigenvalue*, $\lambda$, for the matrix (Caswell 1989). The distribution of the stable stage structure is given by the right *eigenvector* corresponding to the dominant eigenvalue. This vector gives the proportion of the total population in each stage (age) class when the rate of population change ($\lambda$) is constant. The left *eigenvector* gives the distribution of reproductive values by stage class. This vector gives the potential future reproductive contribution of the different stages. Once the population structure has stabilized (has reached the same proportions as the above-mentioned eigenvector) these structured matrix (linear) models can be represented by a simple constant growth model

$$X(t+1) = \lambda X(t)$$

where $\lambda$ is the rate of change of the population (dominant eigenvalue). Of course, the one constant truth of real systems is that parameters do not remain fixed for long, so that at any given time interval this model will at best be an approximation of the growth of a real system.

## 2.3 Eigenvalues and Eigenvectors

In the case where the proportions in the various stage classes have stabilized, a structured model takes the form $X(t+1) = AX(t) = \lambda X(t)$ where $\lambda$ is a scalar value (eigenvalue). The matrix acts as a function that performs a linear transformation on the vector $X(t)$, producing some new vector with the same direction (proportions) as the original vector and that has simply been stretched or shrunk by a factor $\lambda$.

The matrix equation $Ax = \lambda x$ may be rewritten as $(A - \lambda I)\,x = \mathbf{0}$ where $I$ is an identity matrix and $\mathbf{0}$ is the zero vector of appropriate size. This equation has nontrivial solutions when the $\det(A - \lambda I) = 0$ (the *characteristic equation*). The eigenvalues are those values of $\lambda$ for which the determinant (det) is zero, and the eigenvectors are the vectors, $x$, that satisfy the equation $(A - \lambda I)\,x = \mathbf{0}$ when $\lambda$ is an eigenvalue.

Any matrix model with the form

$$X(t+1) = PX(t)$$

where $P$ is an $n \times n$ matrix with distinct eigenvalues, has a general solution of the form

$$\mathbf{X}(t) = c_1\mathbf{v}_1\lambda_1^t + c_2\mathbf{v}_2\lambda_2^t + \ldots + c_n\mathbf{v}_n\lambda_n^t$$

where the $c_i$'s are constants to be evaluated from initial conditions, the $\lambda_i$'s are the eigenvalues for $\boldsymbol{P}$, and the $\mathbf{v}_i$'s are the corresponding eigenvectors.

Suppose that one of the eigenvalues is strictly dominant; that is, it is larger than any of the others in absolute value. For the sake of simplicity, let us assume that $\lambda_1$ is the strictly dominant eigenvalue. If $c_1$ is nonzero, then, as $t$ gets large, the first term in this solution will grow much larger than all of the other terms. Thus, for large values of $t$, we have that

$$X(t) \cong c_1\mathbf{v}_1\lambda_1^t$$

It follows then that

$$X(t+1) \cong c_1\mathbf{v}_1\lambda_1^{t+1} = X(t)\lambda_1.$$

This gives us a reliable approximation of the long-term solution to the dynamic system. This approximation is interpreted to mean that the long-term growth rate for the system is given by the dominant eigenvalue $\lambda_1$ and the proportions in the various classes (age or stage structure in a population model) are given by the ratios of the entries in the right eigenvector corresponding to the dominant eigenvalue. To get a reliable approximation of the long-term behavior of such a dynamic system, we need only find the dominant eigenvalue and the corresponding right eigenvector.

On those rare occasions where there is no strictly dominant eigenvalue, the solution to our model may exhibit persistent oscillations. We can recognize when this is a possibility relatively easily, however. For Leslie and stage matrix models, the dominant eigenvalue is always real and positive, and this eigenvalue will be strictly dominant if there are two adjacent birth rates (in the top row of the projection matrix) that are nonzero. Even without this condition, stage matrices with a nonzero entry on the main diagonal will usually have a strictly dominant, positive eigenvalue. As a result, the analysis of the long-term behavior of these models is relatively easy. The long-term growth rate will be equal to the dominant eigenvalue, and stable age structure will have the proportions given by the corresponding right eigenvector.

## 2.4 Sensitivity Analysis

Some species depend more on survival than reproduction for the continued success of the population, while others do the opposite. Sensitivity analysis gives us a quantitative measure of the relative importance of each of the life history parameters to population growth. It can also be an important tool in assessing the effects of errors in estimation of vital rates, ranking the importance of further efforts to estimate particular parameters, evaluating the relative impact of various management options on the growth of a population, and predicting the impact of environmental variations.

The characteristic equation, $\det(A - \lambda I) = 0$, provides a mechanism for relating the parameters in the matrix to the rate of increase for the population. This relationship may be exploited to determine the sensitivity of the growth rate of a population to small changes in life history parameters. Recall that the matrix $A$ is composed of elements $a_{ij}$ occupying position row $i$ and column $j$ in the matrix. The *sensitivity* of $\lambda$ to changes in $a_{ij}$ is given by the partial derivative of $\lambda$ with respect to $a_{ij}$, which can be obtained by implicitly differentiating the characteristic equation (see example in Noon and Biles 1990). It may also be computed from the fact that

$$\frac{\partial\lambda}{\partial a_{ij}} = w_i v_j$$

where $w,v = 1$.

Here $w$ and $v$ are the left and right eigenvectors corresponding to the dominant eigenvalue, and $w_i$ and $v_j$ are their $i^{th}$ and $j^{th}$ entries, respectively.

In many cases the fecundities in a matrix model may be much larger than the survival rates so a given change in each may correspond to very different proportional changes. To remedy this problem, Caswell (1989, p. 132) proposes a measure of the proportional sensitivity of a matrix element, called its elasticity ($e$),

$$e_{ij} = \frac{a_{ij}\partial\lambda}{\lambda\ \partial a_{ij}}$$

Elasticity is the sensitivity of $\lambda$ to the $a_{ij}$ values, scaled to vary between 0 and 1 — the higher the elasticity value, the greater the sensitivity. Since elasticities sum to 1, direct comparisons among the $a_{ij}$'s are possible (deKroon et al. 1986). Elasticities should be used as the measure of sensitivity any time there are large size differences of measurement scale in entries in the matrix model.

Lande (1988a) computed the sensitivities for the vital rates of the northern spotted owl. His model was the same as our three-stage model except that he divided juvenile survival (i.e., the rate of survival from fledging to just before the birth pulse of the next breeding season) into two distinct components: pre-dispersal survival, $s_0$, and dispersal success, $s_d$, resulting in $s_j = s_0\,s_d$. His results were

| Vital rate | $s_0$ | $s_d$ | $s_s$ | $s_a$ | $b_s$ | $b_a$ |
|---|---|---|---|---|---|---|
| Estimated value | 0.60 | 0.19 | 0.71 | 0.942 | 0 | 0.24 |
| Sensitivities | 0.03 | 0.102 | 0.026 | 0.981 | 0.000 | 0.076 |

These results indicated that subadult fecundity ($b_s$) did not play a vital role for this population, so further data collection related to age of first reproduction was not warranted. However, adult survival ($s_a$) was the most sensitive parameter followed by juvenile dispersal success ($s_d$). Any change in management practices that enhanced these rates had the greatest prospects for improving the state of the owl.

Measures of sensitivity, however, should be interpreted cautiously. For the spotted owl, for example, it would be incorrect to infer that those vital rates with low sensitivities (e.g. $s_j$ and $b_a$) should not be considered in management plans. Adult survival is already very high and may not be amenable to further increases. Therefore, increases in $s_j$ or $b_a$ through direct habitat management may be the most likely way to increase the growth rate of spotted owl populations (Noon and Biles 1990). In addition, the relative rank of elasticities can be misleading when the vital rates vary simultaneously and disproportionately (Wisdom and Mills 1997).

## 2.5 Density Dependence

Age and stage structured models provide a more realistic description of population dynamics. In addition, density dependent models provide a more realistic mechanism for portraying the dynamics of population growth. In any population, at some point competition for resources will intervene to limit exponential growth. This may be exhibited by reductions in the quality of the preferred habitat or food supply, or increased competition for nest sites or space in refuges. These will likely first result in reduced birth rates or increased juvenile mortality, followed by overall reduction in survival and lower growth rates.

Density dependence is usually modeled by giving a functional form to the rate of increase, $\lambda$. Many different functional forms have been studied, but the logistic equation below has gained the most attention in spite of the fact it may not be the most appropriate (other forms, sometimes called the discrete logistics model are better behaved mathematically). In the logistic equation we have

$$N(t+1)=\left[1+r\left(1-\frac{N(t)}{K}\right)\right]N(t),\ \text{discrete logistic model}$$

so

$$\lambda(N(t))=1+r\left(1=\frac{N(t)}{K}\right)$$

where $r$ is the intrinsic growth rate, $K$ is the carrying capacity, and $\lambda(N(t))$ = $N_{t+1}/N_t$. The relationship between the finite ($\lambda$) and intrinsic ($r$) growth rates is given by: $\lambda(N(t))$ = $1+r(1-N(t)/k)$. That is, the value of $\lambda(N(t))$ — the value of $\lambda$ when the population size at time $t = N$ — declines linearly with increases in population size (Fig. 3). One difficulty with the discrete logistic model is that it does not necessarily make a direct approach to the carrying capacity, but may overshoot. In fact, if the population is above the carrying capacity at size $N(t)$ = $K(1+r)/r$ or greater, the population will go directly (and unrealistically) to zero in the next time step.

The logistic form of density dependence simply assumes that the rate of increase declines linearly to compensate for increases in population size. Density dependent growth might take several other similar forms:

$$\lambda(N(t))=\frac{a}{b+N(t))}\quad \text{Beverton–Holt model}$$

$$\lambda(N(t))=be^{-aNt}\ \text{Ricker model}$$

These models do not pretend to explicitly model the mechanisms causing the density dependence; instead, they each try to capture the essence of density dependence in two parameters. For each of these examples, the rate of increase is at its maximum when $N(t)$ = 0 (see Fig. 3). This results in a maximum rate of increase for the Beverton–Holt model equal to $a/b$, while the carrying capacity for the Beverton–Holt model is $a - b$. To parameterize any of these three models, it is sufficient to estimate the maximum growth rate and the carrying capacity (see Burgman et al. 1993: 96–98).

For low values of the rate of increase the output of these models are quite similar (Fig. 4); however, as these values increase the model outputs begin to part company. In particular, we see the logistic model begin to overshoot its carrying capacity with the rate of increase equal to three. By the time the rate of increase has been increased to four the logistic model has gone chaotic (Figs. 4, 5 and 6).

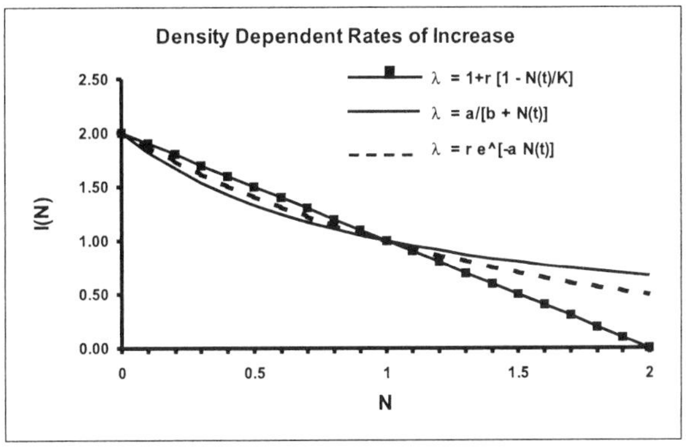


Fig. 3. Changes in the finite rate of growth ($\lambda(N)$) with time as a function of population size ($N$) for three density dependent models.

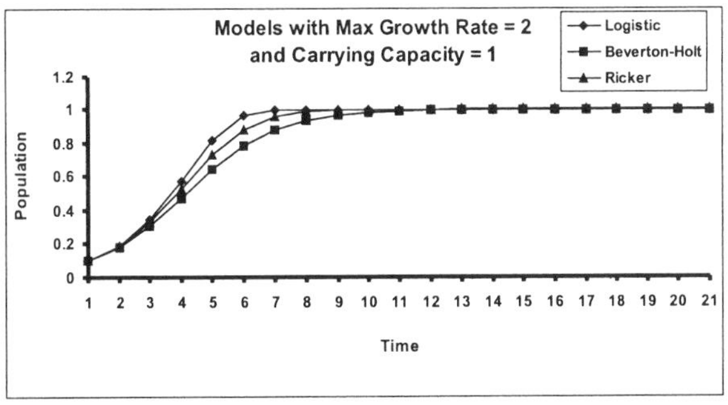


Fig. 4. Changes in population size with time for the logistic, Beverton–Holt, and Ricker growth models when the rate of increase is relatively low.

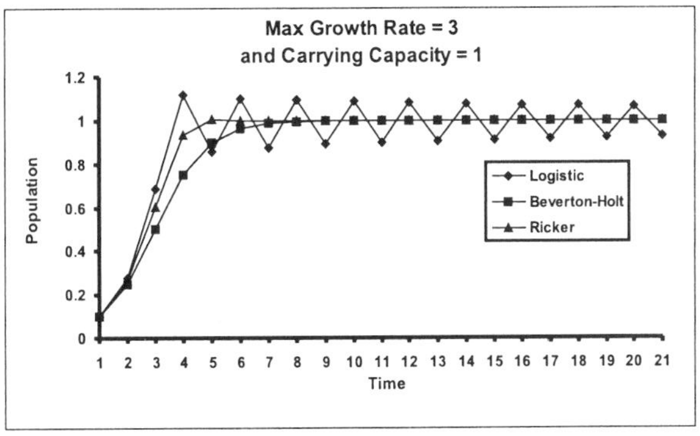


Fig. 5. Changes in population size with time for the logistic, Beverton–Holt, and Ricker models when the rate of increase is set at an intermediate level. The logistic model shows cyclic output.

Models of this form capture the essence of density dependence, and they require little data, so they can be very useful in situations where data are scarce. These are simple models, however, with mostly heuristic value. That is, their results capture the general picture

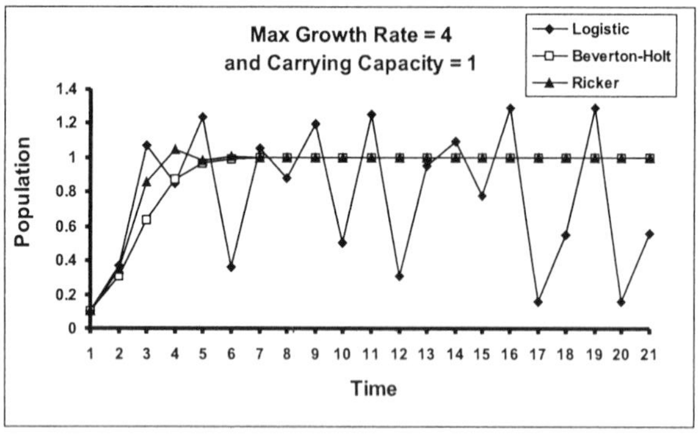


Fig. 6. Changes in population size with time for the logistic, Beverton–Holt, and Ricker models when the rate of increase is high. The logistic model shows chaotic output.

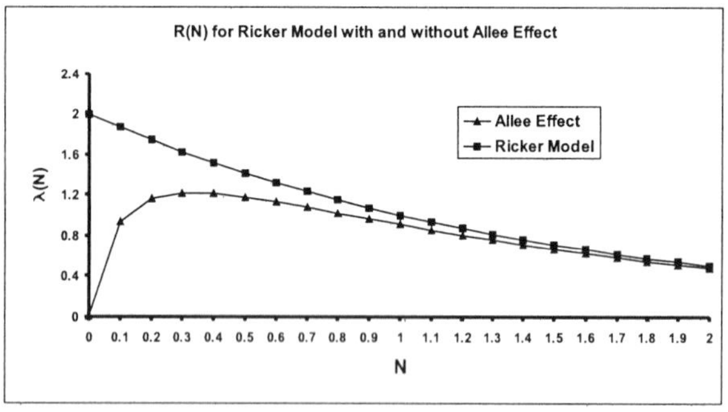


Fig. 7. The finite rate of increase ($\lambda(N)$) for the Ricker model with and without the decompensatory Allee effect.

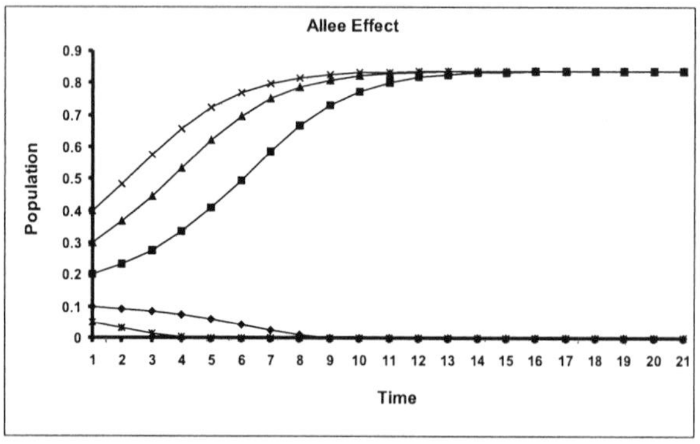


Fig. 8. Changes in population size with time for the Ricker model with the Allee effect when initialized at several different population sizes.

of what is going on with the population, but they can not be expected to provide reliable predictions of future population sizes.

For more reliable insights to population dynamics we must provide a more detailed model. One detail that has been omitted from previous models is the fact that, in some populations, rates of increase begin to drop when the density of the species gets too low. This phenomenon is often referred to as the Allee effect (Allee et al. 1949), or population depensation. Allee effects frequently result from difficulty in securing a mate when population densities are low, but they may also arise from inbreeding depression or even from disruption of social function.

In the cases with an Allee effect, the maximum rate of increase no longer occurs at $N(t) = 0$. Figure 7 provides a comparison of the rate of increase for the Ricker model with and without an Allee effect. The mathematical form chosen here to represent the rate of increase with the Allee effect is

$$\lambda(N(t)) = re^{-aN(t)}\frac{N(t)}{c + N(t)}$$

where $c$ is the population size at which the Allee effect has reduced the rate of increase by half.

The results of the Allee effect incorporated in a Ricker model can be seen graphically in Fig. 8. Here low initial populations tend to extinction while populations initialized with somewhat larger numbers increase to the carrying capacity much as we would expect without the Allee effect. Thus, we expect the Allee effect to be most pronounced at low population densities.

An additional variation on these standard models that needs to be considered is the rate at which density dependent effects enter the dynamics of the population as its numbers increase. Different rates may result from the manner in which competition for resources is expressed. In *contest competition* we have a competition for an allotment of resources sufficient for survival and successful reproduction, and the participants either win or lose. An example of this is the competition for home ranges within territorial species. On the other hand, *scramble competition* results in a sharing of resources, and as density increases it leads to a reduced fitness across the entire population.

The Beverton–Holt model can be modified to account for these various forms and degrees of competition by introducing a new parameter, $q > 0$:

$$N(t+1) = \frac{aN(t)}{b + (N(t))^q} \quad \text{i.e.} \quad \lambda(N(t)) = \frac{a}{b + (N(t))^q}$$

Figure 9 demonstrates the different population levels at which competition begins to impact the rate of increase for different values of the parameter $q$. Contest competition is represented by large values for $q$, while scramble competition results from small values. As $q$ increases, the growth rate at small population size is increased.

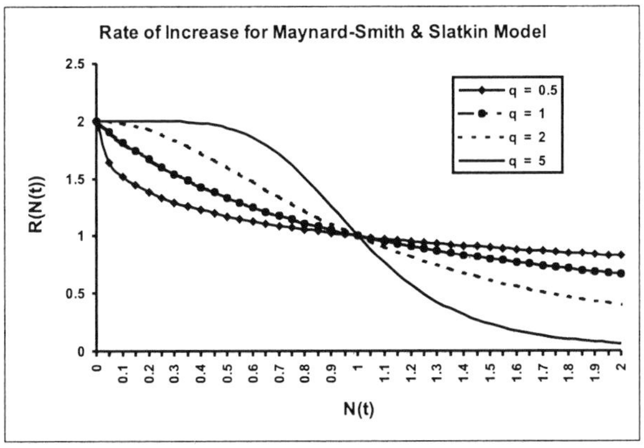


Fig. 9. Rate of increase with time-dependent changes in population size for the Maynard–Smith and Slatkin model for various values of the competition parameter, $q$.

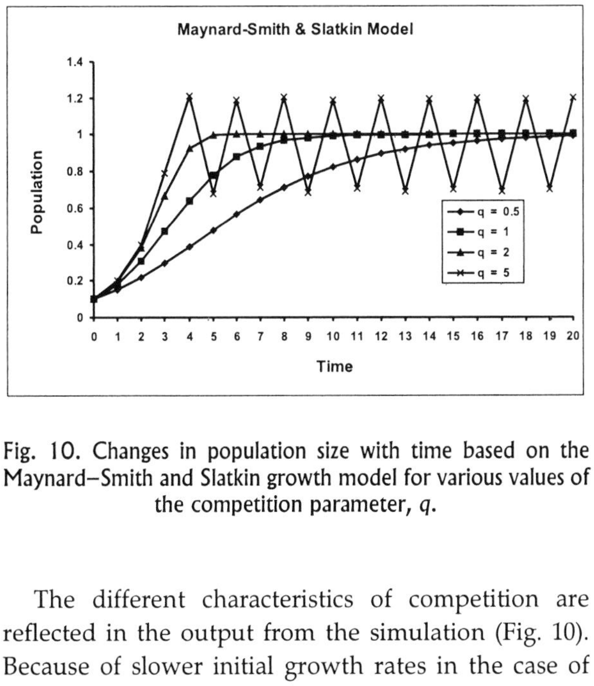


Fig. 10. Changes in population size with time based on the Maynard–Smith and Slatkin growth model for various values of the competition parameter, $q$.

The different characteristics of competition are reflected in the output from the simulation (Fig. 10). Because of slower initial growth rates in the case of scramble competition ($q \leq 1$), population sizes initially lag behind. In contrast, the density-dependent impacts of contest competition only show up at high population sizes as territories begin to fill up. As $q$ gets large this model will oscillate and even exhibit chaotic dynamics for very large values of $q$.

## 2.6 Mechanistic Models for Density Dependence

We have been examining various "off-the-shelf" models that have been widely studied. Each incorporates density dependence in a slightly different way, but the Beverton–Holt and the Ricker models have similar behavior. In addition, we examined some more subtle variations of density dependence, the Allee effect and scramble and contest competition. Using these models for particular applications is simply a problem of estimating the parameters as they pertain to the population in question.

In many cases we may know the source of density dependence and want to model it explicitly. As an example we will take the first model used to aid conservation planning for the northern spotted owl (Thomas et al. 1990, Lamberson et al. 1992b). Here density dependence was thought to be imposed by juvenile dispersal mortality and competition among dispersing birds for suitable territories.

The owl model is similar to the three-stage class model described in the section on age and stage structured models. The important change we introduce is to split juvenile survival into two components: pre-dispersal survival and dispersal success. Pre-dispersal survival was assumed to be a fixed value while we tried to be more explicit in our model for dispersal success. Since very little is known of the actual dynamics of dispersal in spotted owls, we followed Lande (1987) and chose to look at dispersal as a sampling process. We portrayed the landscape as a checkerboard of territories with some suitable and some unsuitable for reproduction. Some proportion of the suitable territories were assumed to be already occupied by owls, thus unavailable. A dispersing owl was allowed to take a sample of $m$ territories in search of one that was both suitable and unoccupied. Dispersal success was given by the probability of finding at least one suitable, unoccupied site in the sample,

$$\text{Prob(success)} = 1 - \left[\frac{\text{numbers of suitable unoccupied sites}}{\text{total number of sites}}\right]^m$$

This formulation also allowed us to simulate ongoing timber harvests by varying the number of suitable sites from one time step to the next.

The search process in this model has two phases. First is the search for a suitable, available site, while the second is the search for a mate. The underlying assumption was that juvenile males dispersed first and attempted to secure territories. This event was followed by the dispersal of females who were searching for single males that had been successful in occupying a site. We used the same general formulation for both search processes, except that in the mate search we replace the "number of suitable unoccupied sites" by the "number of sites occupied by single males". We also assumed that the sample sizes might differ since the hooting of single males would assist females in their search for mates. Given the structure of our simulated landscape, a random array of suitable and unsuitable sites, our response variable was the number or pro-

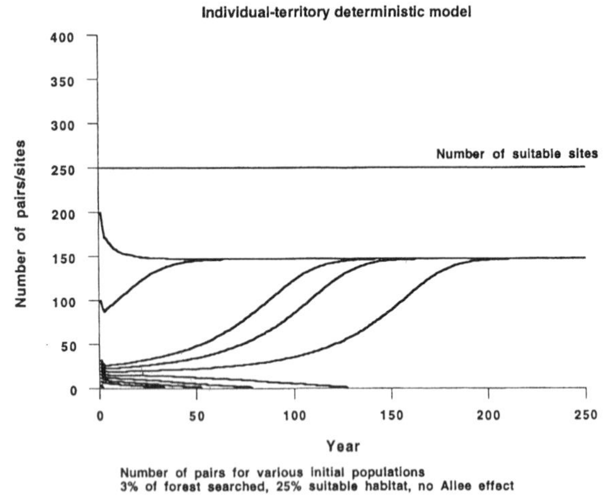


Fig. 11. Trends in number of spotted owl pairs from a 250 year simulation with 25% (250/1000) of sites with suitable habitat, and each owl able to search 30 territories for a suitable unoccupied site. Each simulation was initiated with a different, initial population size.

portion of suitable sites occupied by owl pairs as a function of time.

The first phase of this search process provided a bound on the growth of the population because of the limited number of territories available. The mate search process introduced an Allee effect because of the difficulty in finding mates when the population is at low density. Figure 11 shows results of the model simulations initialized at different population sizes.

## 2.7 Thresholds

There are examples both in models and in nature where what seem to be small changes in habitat characteristics (e.g., amount or distribution) or some vital rate can result in a large change in the dynamics of a species occupying that habitat. One of the most important reasons for building models of threatened or endangered populations is the search for such thresholds. One example is the Allee effect mentioned earlier (see Fig. 11). With the Allee effect, a small reduction in population density can potentially move a population from a size where it is likely to recover and return to near carrying capacity to a density where securing a mate or finding a suitable site becomes problematic. The result is reduced fecundity and ultimately the demise of the population.

Several different models have been proposed for territorial species (Lande 1987, Lamberson et al. 1992b, Carroll and Lamberson 1993, Lamberson and Carroll 1993). Even though these models have quite different structures, they have one common characteristic. They all have a threshold related to the fraction of the landscape that is suitable habitat. If we plot a graph of level of occupancy of suitable habitat vs. percent of habitat that is suitable for any of these models, we see a threshold in percent suitable habitat (Fig. 12). When there is a large amount of suitable habitat available, small reductions in this amount have virtually no effect on the level of occupancy (at 75% of the landscape suitable habitat a reduction of 5% results in about 0.3% drop in occupancy); however for lesser amounts of suitable habitat a small reduction in percent of the landscape that is suitable habitat can result in a sharp drop off in occupancy rate (at 20% suitable, a drop of 5% results in a decline of more than 15% in occupancy; Fig. 12). The solid lines in Fig. 12 show results from an all-female version of the model. The dashed curves on Fig. 13 shows result from a two sex version of the model, showing Allee effects at low population dynamics.

The models shown in Figs. 12 and 13 illustrate an important, and counter-intuitive, biological possibility.

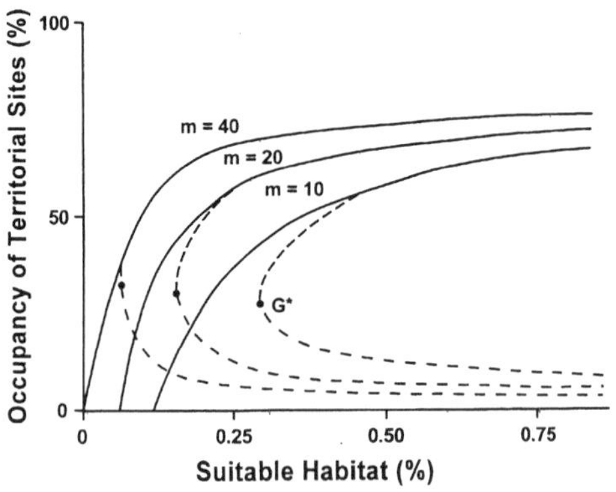


Fig. 12. Percent occupancy of suitable spotted owl territories versus percent of the forest that is suitable habitat, with various numbers of sites searched. The dashed curves result from a two sex version of the model, with an Allee effect (i.e., requires search for mates as well as breeding territories).

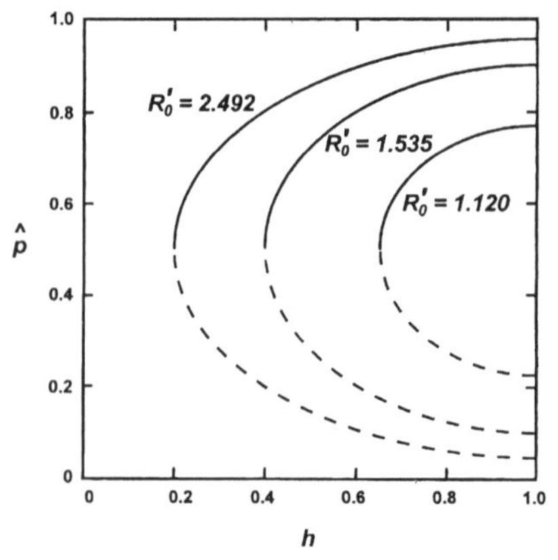


Fig. 13. Percent occupancy of suitable spotted owl territories (p) versus percent of the forest that is suitable habitat (h) with various levels of reproductive potential (Ro). Results are similar to those in Fig. 12 but result from a different model (see Lande 1987).

It is possible for a species to be critically habitat-limited in the presence of suitable, unoccupied habitat. At low densities of suitable habitat, or mates, the uncertainties and survival costs associated with the search process can result in population collapse. The implication for the field biologist is that it would be incorrect to infer that a species is not habitat-limited from the observation of unoccupied habitat.

These results speak more to the analysis of models than their construction. Models need to be thoroughly tested throughout the full range of potential parameterizations to determine any particular sensitivities that we may not have been anticipated. In some cases an unexpected phenomenon may result from peculiarities in the model structure or bugs in the software. In cases like the one above, however, where it occurs in several models with differing structure, the consistent pattern suggests a real effect, and one we might expect to occur in real populations.

## 2.8 Competition Models

The introduction of non-native species is of particular concern in conservation biology, since they may have substantial impact on the populations of native species. Interaction between species can take on three basic forms: *competition,* where the growth rate of each population is reduced; *predator–prey,* where the growth of one (predator) is enhanced and the other (prey) diminished; and *mutualism* or *symbiosis,* where each species benefits from the presence of the other.

In the case where two species compete for the same limited resource (food, breeding sites, territory, etc.) each will inhibit the growth rate of the other. A simple model for this interaction may be developed from the Beverton–Holt model discussed earlier:

$$N(t+1) = \frac{a}{b + N(t)} N(t)$$

$$N_1(t+1) = \frac{a_1}{b_1 + N_1(t) + c_1 N_2(t)} N_1(t)$$

$$N_2(t+1) = \frac{a_2}{b_2 + c_2 N_1(t) + N_2(t)} N_2(t).$$

The term in the denominator of the original B-H model measures the impact of intraspecific competition on the growth of a single population. To account for interspecific competition between two species, we develop a separate equation for each species and extend the competition term by adding a multiple of the other population. Here $c_1$ measures the impact of an individual of species 2 in reducing the growth rate for an individual of species 1; the reverse is true of $c_2$.

This model exhibits either three or four equilibrium points, depending on the relative size of the parameters. Extinction of both species is possible ($N_1 = 0$, $N_2 = 0$), when the growth rates for both species are too small to sustain them. One species may out compete the other and exclude it from the region, with equilibrium points $(a_1–b_1,0)$ or $(0,a_2–b_2)$; this is competitive exclusion, and results in the winning species attaining a population size equal to the single species carrying capacity. For a small range of parameter combinations, there is an equilibrium point that provides for the stable coexistence of both species, with population sizes given by $[(a_1–b_1) - c_1(a_2–b_2)] / [1–c_1 c_2]$, and $[(a_2–b_2) - c_2(a_1–b_1)] / [1–c_1 c_2]$.

Competition models suggest that two species are unlikely to both persist if they strongly compete with each other for some critical limiting resource. Consider the case where an exotic species is introduced to an area containing all the resources necessary to sustain it. The fact that it is an exotic suggests that it may not be faced with some of the factors that would normally limit its growth (natural predators, etc.) so it may be able to establish itself in the new environment and, at the same time, displace some competing native species. Whether a native species can be displaced will depend on two factors, the relative carrying capacities of the two species (or their relative efficiency at using the available resources) and the relative competitive impact of each species on the other. Establishment is not dependent on the size of the initial invading population or its rate of increase (Fig. 14). If the exotic is more efficient than the native in the use of the shared resources and has a strong competitive impact on the natives species, it is likely to succeed in replacing it. If the exotic is not as efficient and does not compete strongly it will not be able to establish itself. For a small

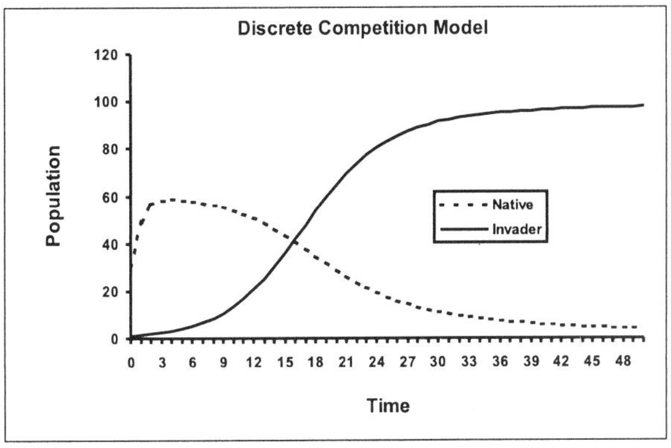


Fig. 14. Changes in population size with time from a discrete competition model for the case where the invading species successfully displaces the native species.

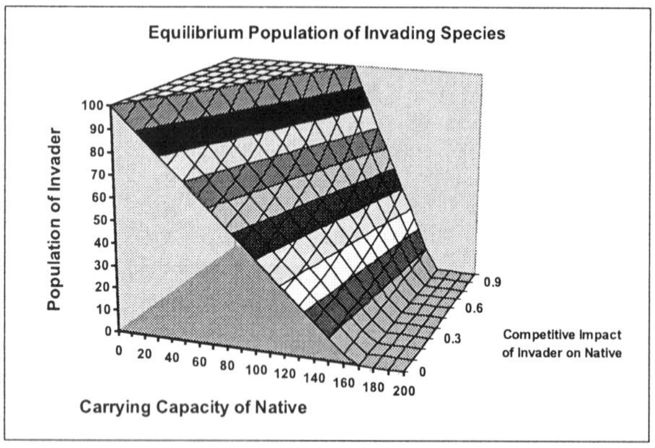


Fig. 15. Changes in equilibrium population size of the invading species as a function of limiting resources available to the native species (as measured by the carrying capacity) and the level of competition for those resources (competitive impact of invader on native species). As the level of competition increases the range where the two species coexist gets much smaller (and may disappear).

range of intermediate parameters, however, both may persist (Fig. 15).

In Fig. 15 the carrying capacity of the invader is fixed at 100 and the competitive impact of the native on the invader is fixed at 0.6. The sloping surface represents the parameter combinations where the two species coexist, while the flat regions are the values where competitive exclusion occurs. Notice that as competitive impact of the invader increases, the range of value where the two coexist decreases. At high levels of invader impact, coexistence is possible only when the carrying capacity of the native species is much higher than the invasive species.

A similar model can be developed for a predator-prey system by changing the sign on $c_1$ and then letting the equation for species 1 represent the predator population. The rate of growth of the predator population then increases with increasing prey availability. To model mutualism, both $c_1$ and $c_2$ should be negative so that each growth rate is enhanced in the presence of the other species.

## 2.9 Stochastic Models

At its foundation, population viability analysis is an assessment of the risks of extinction faced by a particular population as a result of the inherent uncertainty in the processes that characterize our environment. Stochastic models provide a framework for including predictable, chance events in models for the dynamics of a population. Stochasticity may take on many forms, such as variation in litter sizes or sex of offspring, changes in average survival rate from year to year, short term climatic variations, introduction of an exotic disease or other catastrophes, genetic variations, and many others.

To examine stochastic models we return to the basic model from our first section,

$$N(t+1) = \lambda(N(t)$$

The simplest form of a stochastic model assumes that λ is a random variable with a known distribution and some known mean and variance. A λ value is randomly chosen from this distribution at each time step. In Fig. 16 we have plotted 10 realizations of the basic model, where λ was chosen from a uniform distribution having mean 1 and ranging from 0.9 to 1.1.

Even though the growth rate is chosen from a distribution that has a mean of one (implying a stable population, $\lambda = N_{t+1}/N_t$), in two cases we see an increase of more than 40% in population size over 30 years and in another we see a 60% drop in the size of the population. This is an example of *environmental stochasticity*. Environmental stochasticity results from the year to year variations that impact average values of vital rates of all individuals. These variations are attributable to events such as year to year changes in weather patterns and food supply.

*Demographic stochasticity* is a quite different process. It is the variation that we see among individuals within a population. Examples include variation in individual litter sizes, sex of an offspring, and whether a particular individual lives or dies, or reproduces or not. To model demographic stochasticity in annual survival for a population with survival rate equal to $s$, for example, we need to randomly apply this survival rate to each individual in the population separately, and then count the number that survived. The result may not be

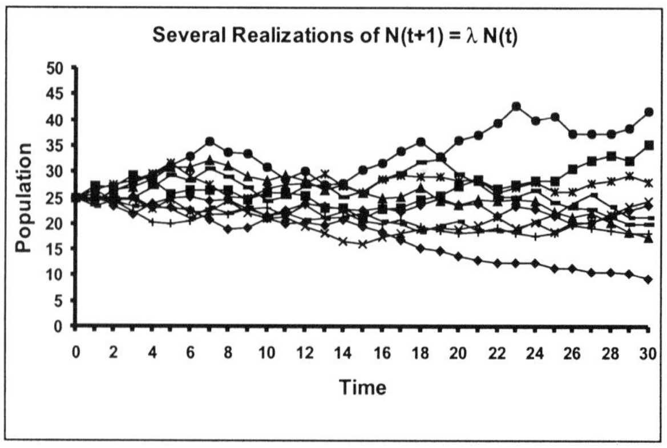


Fig. 16. Changes in population size with time from 10 simulation runs of a stochastic version of the exponential growth model. The finite rate of increase is drawn at each time step from a uniform distribution with mean of one and a range from 0.9 to 1.1.

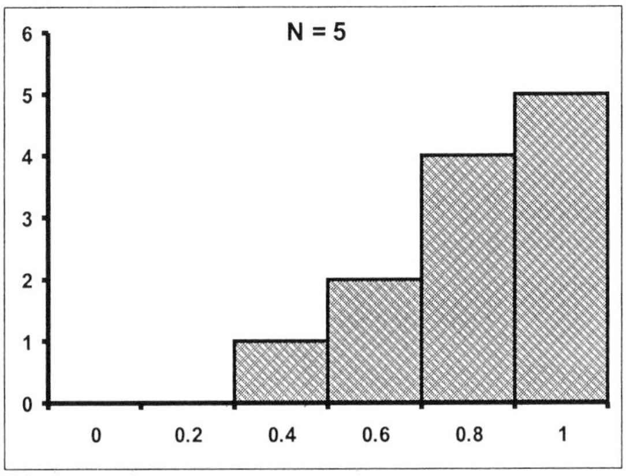


(a)

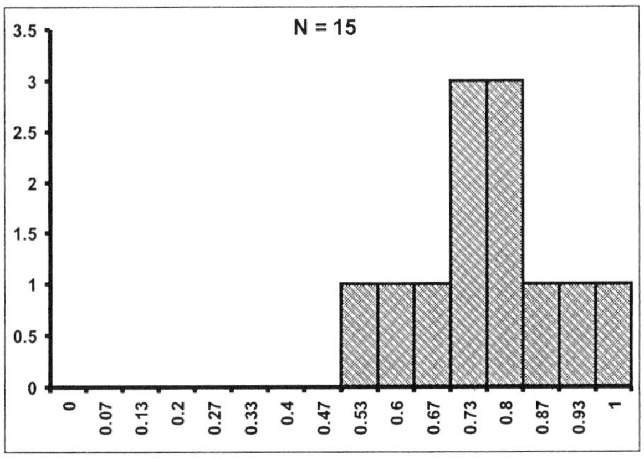


(b)

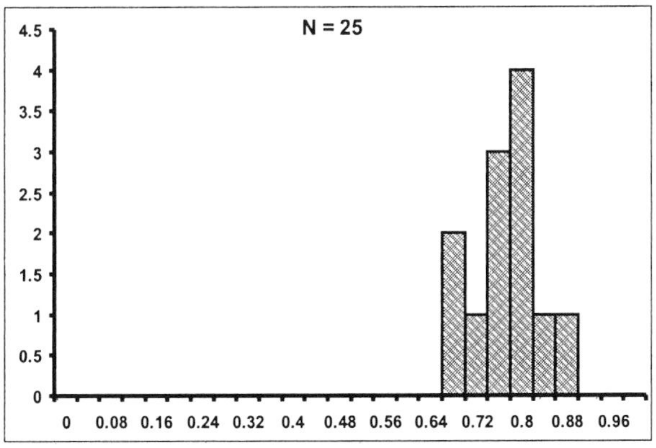


(c)

Fig. 17. Histograms showing the distribution of survival rates based on the results of 12 stochastic simulations where an expected survival rate of 0.8 was applied to each individual at each time step. The simulated population sizes ($N$) varied from (a) 5, (b) 15 to (c) 25. With small populations the mean survival rate that results is often quite far from the expected value (0.8).

an overall survival rate of $s$, and, if the population size is small, it may be quite far from this value (see McCarthy et al. 1994).

Figure 17 is a collection of histograms resulting from 12 replicates of the application of a survival rate of 0.8 to populations ranging in size from 5 to 25. Demographic variation is simulated by sampling from a uniform, random distribution varying from 0 to 1. If the sampled value is $< s$ the individual is assumed to have survived; if $> s$, death is assumed. In the case of a population of 5, the results show 5 of the 12 replicates giving a 100% overall survival rate, while one of them produced a 40% overall survival rate. For a population of 25, we have a much narrower range of results. All of the results are centered around 80% with a low of 68% and a high of 88%.

If the population is very small, it is important to model its dynamics on an individual basis. The reason is that there is a much higher likelihood of sequences of "bad luck" or "good luck" events leading to a result far from expected. Once the population size exceeds about 50, however, the results of individual application of the vital rates approaches the results expected from a collective application of the rates (see also Richter-Dyn and Goel 1972, Gilpin 1992). As a result, it is usually not necessary to account for demographic stochasticity until the population is quite small. Fifty is a frequently used cutoff point for the consideration of demographic stochasticity, but for some life histories demographic stochasticity can remain significant in the range of 50 to 100 individuals (Gilpin 1992).

Stochastic models are basic tools for investigating the possibility of extinction – a population size of zero. Below are several replicates resulting from the basic model, $N(t+1) = \lambda N(t)$, where $\lambda$ is normally distributed with mean of 1.0 and standard deviation of 0.2 (Fig. 18). One of the replicates in this experiment was not included on the graph since it reached a maximum of over 2000 and did not go to extinction until year 296.

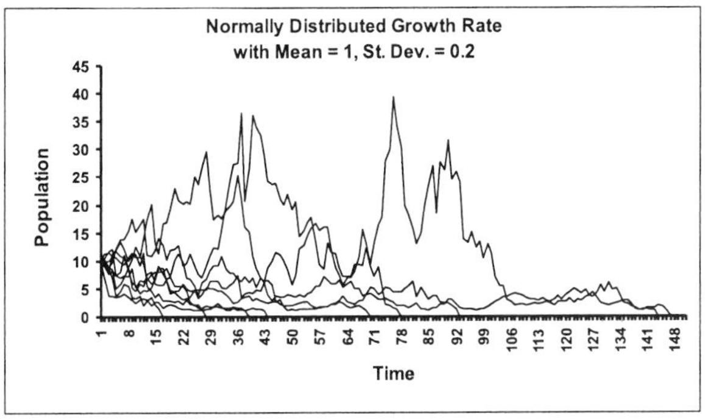


Fig. 18. Changes in population size with time based on several replicates of a stochastic, exponential growth model with mean rate of increase equal to 1.

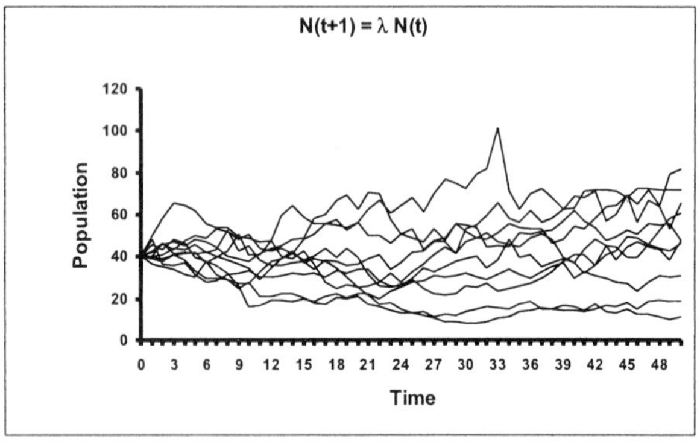


Fig. 19. Changes in population size with time from ten, 50 year simulations of a stochastic, exponential growth model.

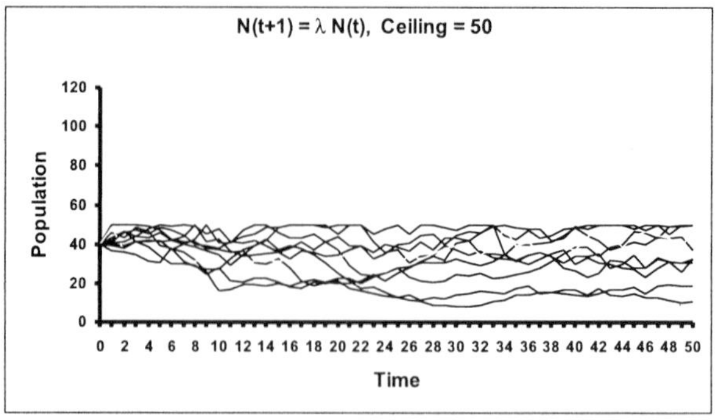


Fig. 20. Stochastic simulations similar to those shown in Fig. 19 except a population size limit of 50 was established.

Notice that all of these trials lead to eventual extinction, even though the mean growth rate was one, indicative of a stable population.

A repeat of this experiment with 10 replicates where each had a mean rate of increase of 1.01 (1% increase per time step) led to a similar result. All led to extinction before 1000 time periods had elapsed, and all but two were extinct by time step 321. The results of this experiment are typical for times to extinction; the mean time to extinction is 254, but only three of the trials reached that point. Seventy percent of the trials reached extinction by step 177. The mean time to extinction is typically skewed to the right of the mode, so it is a poor measure of how long we should expect a population to survive. In this experiment the median time to extinction was 139 time steps, but 40% of the trials did not get past step 71.

A population may be effectively extinct long before it actually reaches zero. In a sexual population, a population size of one is effectively extinct as is a population of any size that is all males or females. Extinction may also occur when the population density gets very low, as seen in our discussion of Allee effects. The size at which a population inescapably enters a trajectory towards extinction has been called its *minimum viable population* (Shaffer 1981). A related, but less formal, concept is the *quasi-extinction* level (size or density) of a population. This value is chosen because it is expected that any population dropping below it will then almost certainly go on to extinction (see Ginzburg et al. 1982). If we use a quasi-extinction level of 2 in the experiment above, 50% of the trials were extinct by year 69, while an extinction level of 1 meant that 50% did not go extinct until step 126.

Another factor that can complicate the dynamics of stochastic models is the fact that it is reasonable to invoke some sort of carrying capacity — an upper limit for the size of the population. We did not provide for a carrying capacity in the stochastic models above; in these models we placed no bounds on their growth. In fact, in the two models immediately above we had trials go from an initial size of 10 to several thousand before they eventually each suffered declines that ended in extinction.

One way to impose a carrying capacity in models of this sort is to place a ceiling on the maximum population size. That is, the population is that given by the model so long as it is less than or equal to the ceiling; however, if the size exceeds the ceiling then the population is assigned the ceiling value.

Figures 19 and 20 both show 10 replicates of the basic model using exactly the same sequence of rates of increase, λ's. The only difference is that for the simulations in Fig. 20 we placed a ceiling of 50 on the population. A casual look suggests that this simply cuts off the highest spikes in the population. However, more than that is going on. Subtle differences that have important consequences which are obscured in these graphs.

In Fig. 21 one of the most successful trials from the first graph is compared with its realization when the ceiling is applied. The result of the ceiling is that we

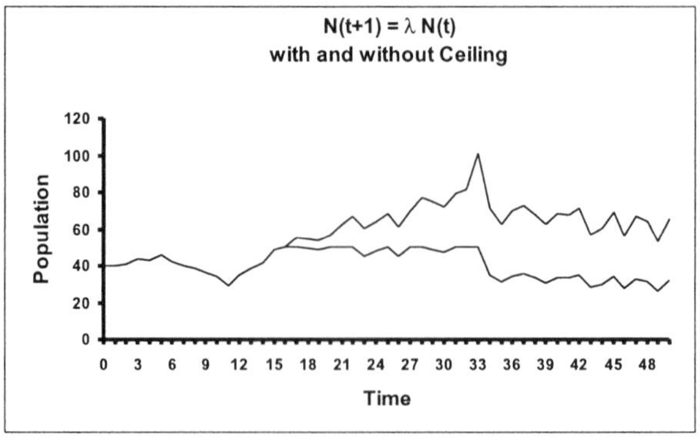


Fig. 21. Specific population trajectories selected from Figs. 19 and 20. The result of the population ceiling is that increases above carrying capacity are not possible, but all of the population decreases remain.

lose those increases when the population is already doing well, but all of the decreases remain. The result is that the population on average is doing less well, and will be much more vulnerable in the case of a series of bad years.

Foley (1994) has approximated the expected time to extinction for this form of model by studying its diffusion approximation. To estimate time to extinction, we first must find the logarithmic transform of the model. That is, we let $n(t) = \ln(N(t))$ and $r(t) = \ln(\lambda(t)) = n(t+1) - n(t)$. Then the expected time to extinction $T_e$ is given by

$$T_e(n_0) = \frac{1}{2sr}\left[e^{2sk}(1 - e^{-2sn_0}) - 2sn_0\right]$$

where $n_0$ is the natural log of the current population estimate, $r$ is the mean growth rate for the transformed populations (average of the $r(t)$'s), $v$ is the variance for the growth rates, $s$ is $r/v$, and $k$ is the natural log of the estimated ceiling population ($k = \ln(K)$).

There can be considerable subtlety in estimating these parameters from real population data because the model assumes that the only applicable growth rates arise when the population is below carrying capacity. Carrying capacities can vary from year to year and populations can occasionally exceed carrying capacities. See Foley (1994) for a discussion of this and several example computations.

Middleton and Nisbet (1997) attack a similar data set with several models. These models include: (1) a closed population with density-dependent vital rates; (2) a regulated, closed population; and (3) a regulated population with immigration and emigration. Their approach can be a very useful analogue to that by Foley (1994) described above. See also Tuljupurkar (1990) who has dealt with the problem of determining the expected time to extinction for models that include age or stage structure.

## 2.10 An Application to Willow Ptarmigan

As an example application of the estimation of expected time to extinction, consider the following estimates of the willow ptarmigan population size on Tanoy Island, Norway (see Fig. 22, Myrberget 1988). The data set seems to have two distinct patterns of population size, 1960 to 1971 and 1972 to 1980. We will apply the procedure of Foley (1994) to estimate expected time to extinction, using the data from each part as well as the entire data set.

Using the entire data set we compute mean $r = -0.08113$ and $v = 0.11606$. Making the assumption that in 1967 the population spiked above the carrying capacity, we assume $K = 160$ or $\ln(K) = k = 5.07517$ and we

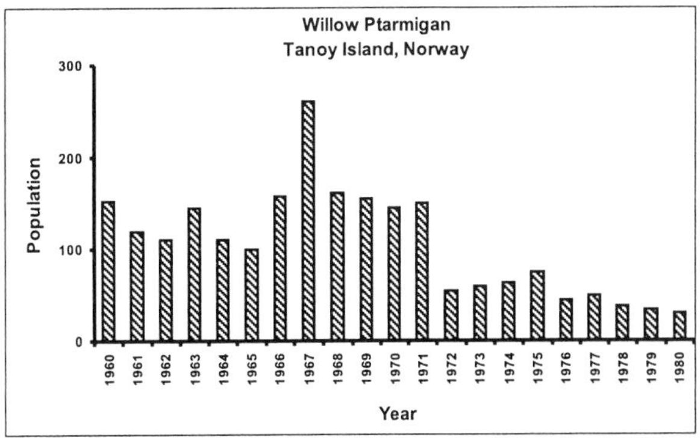


Fig. 22. Changes in the size of the Willow ptarmigan population on Tanoy Island, Norway from 1960 to 1980.

let $N(0) = 30$ or $n_0 = 3.40120$. These values give an expected time to extinction of $T_e = 41$ years. If the 1967 population is eliminated from the computation of $r$ and $v$, we estimate that $T_e = 39.5$ years. The decrease in time to extinction results from the fact that the reduction in the variance of the growth rate (which should increase $T_e$) is balanced by the smaller time series, which results in a slightly greater downward slope (more negative $r$).

Something apparently happened in 1971 to make the system different after that year (Fig. 22), so it is probably more appropriate to do the analysis on just the later years. In fact, in 1971 only about 1/3 of the birds bred; the rest emigrated from the study area. If we take 1971 as our starting point and the largest population value in this series, 75, to estimate $K$, we compute $T_e = 60$ years.

For the sake of comparison, let us do the same computation for the first part of the data set (1960–71). Here mean $r = -0.00120$, very close to zero, so there is virtually no downward trend. The resulting $T_e$ is 269 years, which is a consequence of the near-zero growth rate ($r$) and higher *carrying capacity* ($K$). If the high value in 1967 is removed, $T_e$ is increased to 428 years simply because of the reduction in the variance.

We noted earlier that the expected time to extinction is a poor measure of the viability of a population, since it tends to be skewed by some very long times to extinction that occur with very low probabilities. If we assume that extinction can be represented as a Poisson process with a constant annual rate of extinction equal to $1/T_e$, then we can compute the probability of persistence. (In this case a Poisson process is a random procedure in which the probability of an extinction event within a given time interval is $1/T_e$). The probability of extinction not occurring by time $t$ is equivalent to computing the probability that the population persists to time $t$,

$$P(t) = e^{-t/T_e}$$

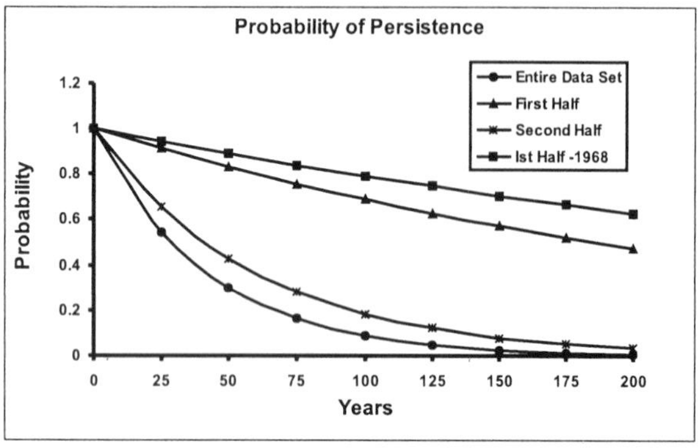


Fig. 23. The probability of the ptarmigan population persisting for times up to 200 years resulting from four different treatments of the data. See text for details of the model used to estimate persistence likelihoods.

Using the 1971–80 population time series results in a *65%* probability of persistence for *25* years and a *19%* probability of persisting for another century. Figure 23 shows persistence probabilities for times up to 200 years for four different cases.

## 2.11 Stochasticity in Structured Models

Returning to the structured models examined earlier, we now address the issue of environmental variation. There are two fundamentally different approaches to incorporating this process into simulation models: element selection method and matrix selection.

In the commonly used element selection approach, the values of the elements in the system matrix (e.g., survival rates in a projection matrix) are selected as random variables. Each element may have its own underlying distribution chosen to represent its known characteristics (e.g., mean and variance). In some cases it may be important to consider the year-to-year co-variation in some parameters. For example, if a "bad year" should occur it may likely result in reduced survival rates in all age classes, though the young and the old are likely to suffer the most.

Usually the two most highly varying parameters are fecundity and the juvenile survival rate, and the variation in these parameters may be highly correlated to some known environmental factors (Armbruster and Lande 1993, Price and Kelly 1994). For example, reproductive rates in many waterfowl species may be strongly tied to spring rainfall on the breeding grounds. In a case like this there may be a known (regression) relationship between spring rainfall and reproductive success that can be exploited. Since long-term rainfall records are usually available, we should be able to extract their basic statistical characteristics. These may then be used to simulate spring rainfall which is coupled with the regression relationship to generate the simulated variation in reproductive success.

The matrix selection method is as the name implies. It involves the selection of a complete matrix from some collection of matrices. The matrix selected determines the behavior of the system for that time step. Since the elements in the matrices are pre-assigned, this approach provides a method of simulating variability when many of the parameters are tightly correlated (see example in Beissinger 1995). This approach can also be used to address behavior over intervals longer than a single time step when there is evidence of temporal correlation. In a structured model with projection matrix $\boldsymbol{P}$, we have

$$N(t+1) = \boldsymbol{P}N(t)$$

However, stochasticity means that $\boldsymbol{P}$ is changing from one time step to the next. As a result the behavior over $t$ time steps can be found by recognizing that

$$N(t) = \boldsymbol{P}_{t-1}\boldsymbol{P}_{t-2}...\boldsymbol{P}_0 N(0)$$

The product of $\boldsymbol{P}_{t-1}\boldsymbol{P}_{t-2}...\boldsymbol{P}_0$ is a matrix describing the results of a projection of $t$ time steps. A random selection from a collection of such products of matrices can used to represent stochasticity in the form of the random variation. This process could represent different sequences of weather cycles, for example.

## 2.12 A Final Word on Stochastic Models

Stochasticity plays an important role in the dynamics of many systems, but it is not always the force that drives the system. It is important during the process of analyzing a model to explore the relationship between the stochastic model and a deterministic version of the same model. If there is no significant difference in their output, a great deal of time and effort may be saved by studying the deterministic version. The model for the dynamics of the northern spotted owl in Lamberson et al. (1992b) is an example where the introduction of stochasticity does not make a significant difference in the results of the analysis (Fig. 24 and 25).

We have not addressed two forms of stochasticity in this section: genetic variation resulting from mating and migration, and precipitous population declines arising from catastrophes. Genetic variation is discussed in Section 3 of this chapter. Catastrophes are usually defined as rare events that impose additional mortality across a population. They are easy to model *if* they are known events with a recognized frequency and severity (e.g., floods, forest fires, etc.). In that case

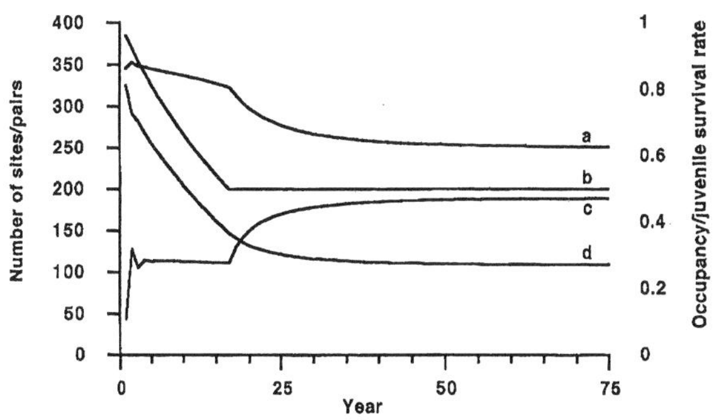


Fig. 24. Changes in the number of suitable sites (a); number of owl pairs (b); number of suitable sites occupied by pairs of owls (c); and juvenile survival rate (d) based on a deterministic version of the individual-territory spotted owl model (Lamberson et al. 1992b).

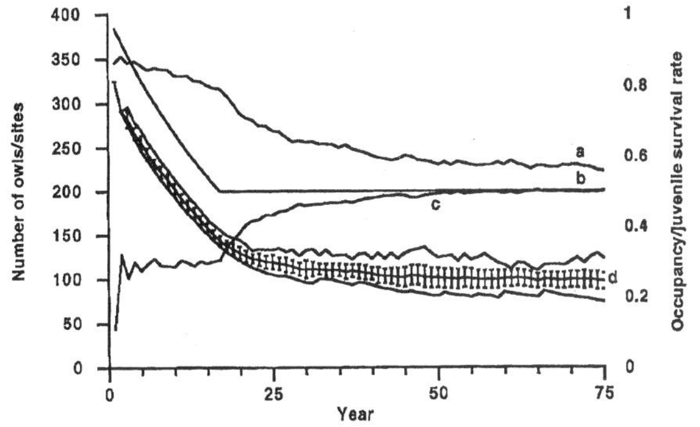


Fig. 25. Changes in the number of suitable sites (a); number of owl pairs (b); number of suitable sites occupied by pairs of owls (c); and juvenile survival rate (d) based on a stochastic version of the individual-territory spotted owl model (Lamberson et al. 1992b). Compare to Fig. 24.

we can simulate them by using the frequency as the probability of occurrence, and then impose a suitable mortality across the population conditioned on a catastrophe occurring. However, many catastrophes arise as a complete surprise. Examples include the accidental introduction of a competing species or predator, the introduction of an exotic disease, or the explosion of a volcano. These may be impossible to model in any adequate fashion.

Shaffer (1987) has studied the relationship between the size of a population and the risk of extinction resulting from three types of stochasticity (see Fig. 26). As discussed in the Introduction, his work has recently been revised by Lande (1993). The results of both Shaffer and Lande show that increased population size quickly mitigates for the risk of extinction due to demographic stochasticity (Fig. 26). For environmental stochasticity, however, the relationship shown in Fig. 26 is only true when the ratio of the growth rate to its variance, $r/\sigma_r^2 = 1.0$; when this quantity is $> 1.0$, time to extinction increases faster than linearly. In the case of catastrophes, many possible relationships between extinction time and initial population size are possible — the determining factors are the long-term growth

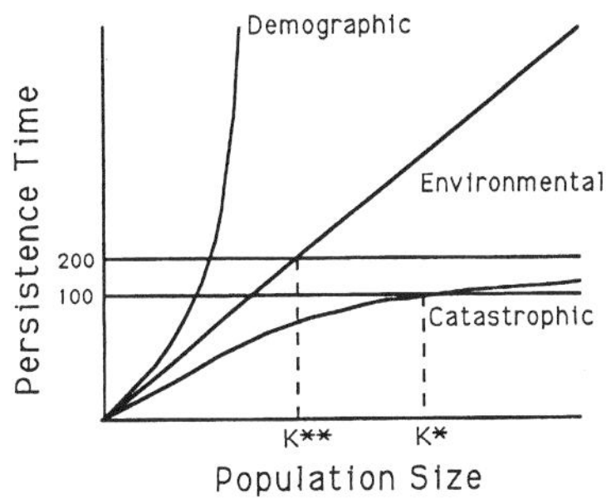


Fig. 26. Expected persistence time as a function of population size when the population is exposed to only one source of stochastic variation: demographic, environmental, or catastrophic.

rate of the population and the frequency of catastrophes. Figure 26 shows just one possible relationship between persistence and population size in the context of environmental variation and catastrophes.

## 2.13 Models Incorporating Spatial Variation

The models examined up to now assume a single population occupying a space that is uniformly suitable for the species we are modeling. Even the most casual observation of our environment suggests that the landscape is heterogeneous, and any habitat type is distributed in patches. In turn, we find the species that have particular habitat preferences are distributed non-uniformly, with their population distribution reflecting the underlying distribution of the preferred habitat (Brown et al. 1995). It is also the case that even within a particular habitat type, some patches are better at promoting survival and reproduction than others. Habitat patches that have expected finite growth rates ($\lambda > 1.0$) and produce an excess of individuals are called "source" habitats; those with $\lambda < 1.0$ and whose populations are sustained by immigration are called "sink" habitats (Pulliam 1988). To capture this level of reality it is necessary to incorporate some level of spatial structure in our models. Adding spatial structure can very quickly add to the complexity of a model, however, so it is important that the structure added be compatible with our current level of knowledge of the system in question, and directly related to the results desired from the model.

The minimal level of spatial structure is a patch model, where the various patches are simply modeled as compartments with the potential for some flow between them. Figures 27 and 28 show a simple example of a two compartment model with no correlation between the growth rates. In the first graph there is no dispersal between patches, so the patches function

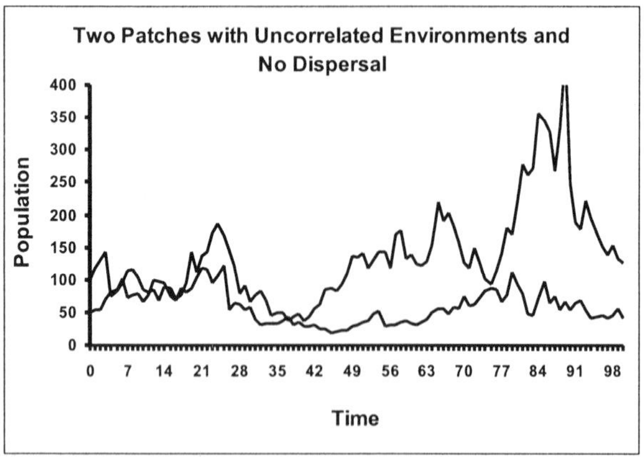


Fig. 27. Changes in population size with time based on a stochastic, two patch model with no correlation between annual rates of increase and no dispersal between patches.

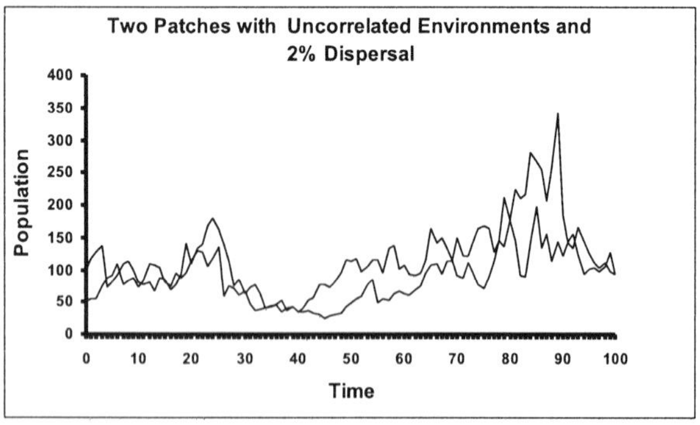


Fig. 28. Changes in population size with time based on the same stochastic, two patch model shown in Fig. 27 (i.e., the same annual growth rate) but allowing a 2% dispersal rates between patches.

completely independently of each other, while in the second case there is 2% dispersal between the two patches in each time interval. As dispersal between patches increases, the dynamics of the two patches become similar. If there is sufficient dispersal between patches, then they effectively become one patch and the need for modeling them as separate entities is reduced.

The most familiar of the two patch models may be the source-sink model (Pulliam 1988); in the simulation, the source patch produces a surplus of individuals, while the sink patch maintains its population by receiving immigrants from the source. In Fig. 29 we show a simulated source-sink system where there is 2% dispersal from the source to the sink population each time interval. The sink population thrives so long as there is excess production in the source area. The sink population approaches zero, however, once dispersal from the source is removed (Fig. 30), graphically demonstrating the importance of even relatively small amounts of dispersal for maintaining regional populations (see Donovan et al. 1995).

The possibility that many regional populations may exists as sets of local source and sink populations is becoming widely recognized (Robinson et al. 1995). Such spatial or temporal variation in population status has made it mandatory to explicitly consider the spatial structure of regional populations. The interpretation of long-term census data from monitoring programs, for example, is particularly difficult if source-sink dynamics are operative (Brown and Robinson 1996).

Extending the idea of a patch model, we can imagine that our patches are laid out in a fashion that approximates an actual landscape. If we provide for the movement of individuals between multiple patches, and the effects of demographic and environmental stochasticity within them, we have generated a *metapopulation model.* Levins (1969, 1970) first coined the term "metapopulation" to describe a collection of equal-sized local, or patch-based, populations connected by dispersal. This concept has subsequently been generalized to describe any structured populations distributed over spatially disjunct patches of habitat

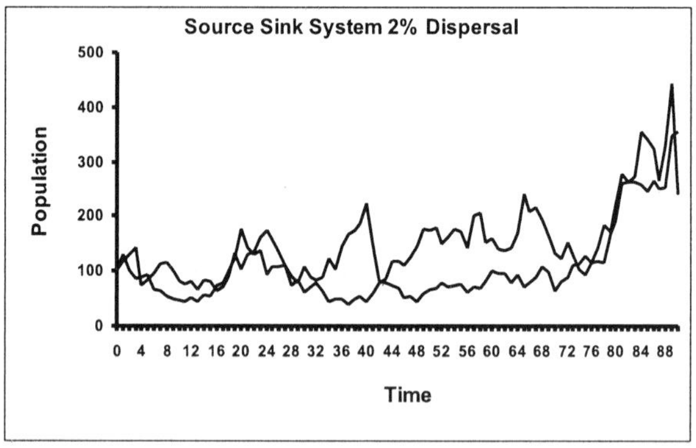


Fig. 29. Changes in population size with time based on a two-patch, source-sink model with a 2% dispersal rate between patches.

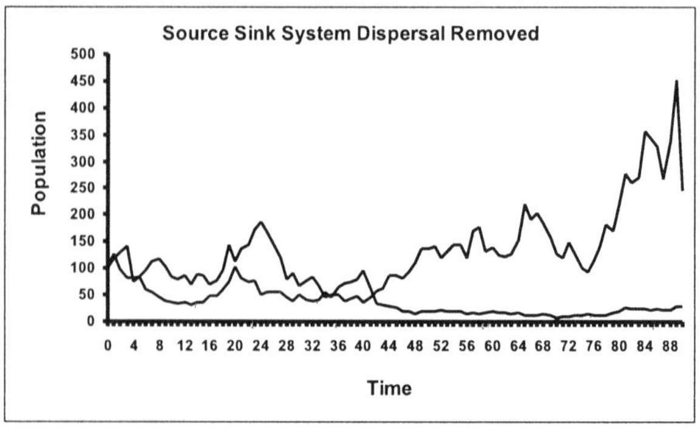


Fig. 30. Output from the same two patch source-sink model shown in Fig. 29 but with no dispersal allowed between patches. Note the collapse of the sink population.

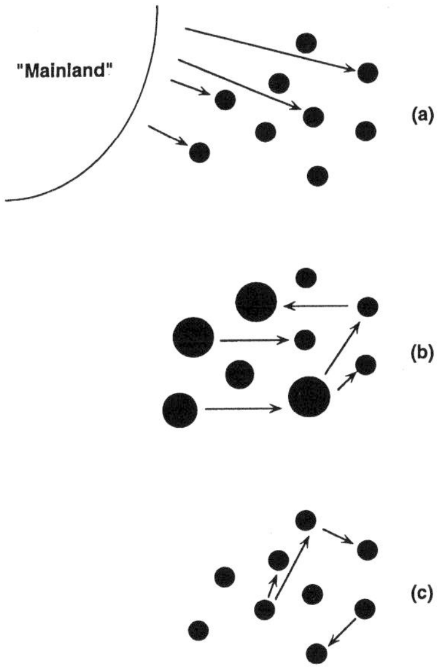


Fig. 31. Three stylized forms of metapopulations: (a) mainland-island model; (b) the core-satellite model; and (c) the classic Levin's metapopulation model.

separated by intervening unsuitable habitat (McCullough 1996). Good examples of the utility of metapopulation models to explore threats to viability are Murphy et al. (1990), Temple (1992), Possingham et al. (1994), and Lindenmayer and Lacy (1995a, 1995b).

In true metapopulations we see occasional extinctions in a few patches, and recolonization of some of the patches where extinctions have occurred earlier. The metapopulation concept is relevant to many threatened and endangered species since their vulnerable status has frequently been the result of habitat loss and fragmentation. Given the diversity of human-induced disturbances, the spatial pattern of a metapopulation can be quite variable, ranging from the classic Levins form, to core satellite distributions (Gotelli 1991), to mainland island distributions (Harrison 1994) (Fig. 31). The key consequence of fragmenting landscapes, regardless of the spatial pattern, is that it forces many species to risk dispersal through unsuitable areas as they move from one patch to another.

In real populations, and model simulations, dispersal may take many forms. Possibly the most common results from juveniles dispersing from their natal territory. Juvenile dispersal is obviously age dependent, and may also be sex specific. This life history suggests the need for a structured population model. Other forms of dispersal include density-dependent dispersal that is motivated by crowding or intraspecific competition in a patch. The amount of emigration from a patch may also depend on patch size. In very large patches, dispersing individuals can move far away from their natal territory and still not leave the patch; this is not possible in smaller patches. Carroll and Lamberson (1993) and others have used the ratio of the area to perimeter of the patch as a measure of the variation in likelihood of dispersal from patches.

Lande (1987) incorporated basic spatial characteristics of the landscape in his extinction model for territorial species (see also Lamberson et al. 1992b, Carroll and Lamberson 1993). In this model he imagined the landscape to be an array of territories, with some suitable and the others not suitable. Among those that were suitable, some were occupied and the remainder were available for occupancy. He then thought of the process of juvenile dispersal as drawing a sample of $n$ territories from the landscape. The individual was successful if the sample included at least one suitable, unoccupied territory. The probability of successful dispersal was described by the following equation,

$$\text{Prob(dispersal success)} = 1 - \left[1 - \frac{\text{number of suitable available territories}}{\text{total number of territories}}\right]^n$$

This sort of model has been extended to a landscape-scale patch model by Carroll and Lamberson (1993) and Lamberson et al. (1994). In these models the patches are assumed to encompass all of the suitable habitat, but individual patches are not necessarily completely suitable — that is, the patches are assumed to be made up of an array of territories with some unsuitable, some suitable and occupied, and the remainder suitable territories available for occupancy. Dispersal success within patches is computed the same as above, but if an individual is unsuccessful in one patch it is possible to search another patch. To search a second patch the dispersing animal must both locate another patch (travel in the correct direction across the unsuitable region between patches) and survive the trip between patches. This probability is estimated with the following equation:

$$\text{Prob(getting to search second patch)} = \text{Prob(locating second patch)}\, e^{-kd}$$

Death between patches is assumed to be a Poisson process, so the probability of successfully moving from one patch to the next declines exponentially with the distance, $d$, between patches at a fixed cost $k$ per unit distance.

This model was specifically built to evaluate the impact of various reserve sizes and geographic arrangements on the long-term viability of the northern spotted owl (Lamberson et al. 1994). Given a fixed land area allocated to a reserve design, the mean occupancy of suitable sites within the reserve was

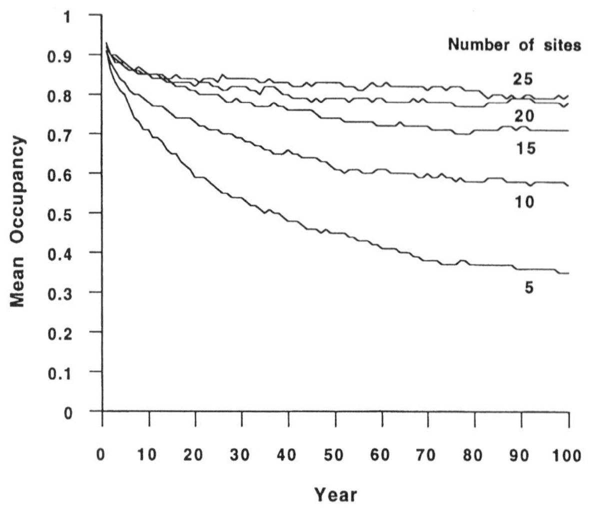


Fig. 32. Changes in occupancy rate with time (proportion of suitable sites occupied by pairs of owls) for various reserve sizes (in terms of number of sites) with the same fraction of the landscape devoted to reserves in each case (see Lamberson et al. 1994).

strongly dependent on the size of patches that composed the reserve (Fig. 32).

Figure 32 clearly demonstrates that the spatial structure of a reserve system can be as important as the amount of land allocated to the reserves. In this case if the habitat patches are numerous but small, the level of occupancy (density of animals in the reserve) will be low so few animals will be protected. As the same reserve area is redistributed in fewer but larger patches, the occupancy level increases so a larger number of animals are protected. As the patches get very large, however, there are diminishing returns in site occupancy. Since large patches will be further apart, dispersal between patches is decreased and local populations become more isolated.

An alternative approach to modeling spatial dynamics is the patch extinction model (Levins 1970) (Fig. 31c). This form of model pictures the patch as the basic unit, and patch extinctions and recolonizations play the role of births and deaths. A patch extinction/recolonization model may be the most appropriate model for a plant such as Furbish's lousewort (*Pedicularus furbishiae)*, a disturbance specialist that colonizes areas along Maine's St. John River disturbed by ice scour or bank slumping. In this case, the colony may be the basic unit in the population model.

Suppose that $p$ represents the fraction of patches occupied by a particular species. Levins (1970) modeled the year to year change in this fraction by simply writing

$$p(t+1) = p(t) + mp(t)[1 - p(t)] - ep(t)$$

where $e$ is the extinction rate, $m$ is the colonization rate, and $1 - p(t)$ is the fraction of patches available to be colonized. This model has a stable equilibrium with $p = 1 - e/m$, which has a positive value only when $m > e$; that is, the colonization rate exceeds the extinction rate. Changing scale from the individual to the local population demonstrates a fundamental equivalence between Lande's (1987) model for the dynamics of individual territorial animals and Levins' (1970) metapopulation model (see Noon and McKelvey 1996).

To adapt the Levins model for the Furbish's lousewort, we would have to incorporate an appropriate dispersal mechanism — one that includes both distance between sites and direction of movement (up or down river). It would also need to include year-to-year variation in colonization and extinction rates, and the rate of creation and loss of suitable sites for germination. However, it is possible to use these basic ideas to build a suitable model (see Menges 1990, Hanski 1994).

## 2.14 Spatially Explicit Models

The models discussed in the previous sections assumed spatial homogeneity or simply distinguished patches that were suitable habitat from those that were unsuitable. Other than these differences, there was no attempt to specify variation in site quality. To model population dynamics in real-world landscapes, which are heterogeneous in the amount, quality, and distribution of resources (e.g., habitat), a spatially explicit modeling approach that includes the arrangement of landscape features is required (Dunning et al. 1995). Spatial models have two distinct scales of resolution: population-based or individual-based. Population-based models assign a population to each patch or cell on the landscape, and every individual within the patch is assumed to have the same birth and death rates. Movement between patches is simulated by some probability of movement per time step, with the probability often dependent upon population density within the patch. In contrast, individual-based models keep track of the location and fate of each individual in each time step. As a result, grid-cell sizes are considerably smaller, requiring habitat maps of much finer resolution. The expected birth and death rates and the likelihood of movement are associated with the individual's location —the habitat quality of the grid cell in which it currently resides. Population-scale inferences from such models are based on the sum of the states and fates of the individual animals.

Several recent models have explicitly considered the spatial variation in habitat quality and its effects on population dynamics. The first step is to "rasterize" the landscape by intersecting it with a grid and explicitly identifying the habitat type or site quality of each cell (Lamberson et al. 1992b, Pulliam et al. 1992, McKelvey et al. 1993, Turner et al. 1995). Then the fitness of an individual is associated with the habitat characteristics of the cell that the individual occupies — or set of cells if variation in quality can be mapped below the scale of a home range or territory. Some individual-based models also simulate the interactions between sympatric species. For example, Turner et al. (1993) constructed a model that simulates the movement, foraging behavior, and mortality of bison (*Bison bison*) and elk (*Cervus elaphus*) wintering in Yellowstone National Park.

In Fig. 33 we show a grid-based habitat map for parts of The Nature Conservancy's Lamphere-Christensen Dunes Reserve (northwestern California), a representation of that area used in a spatially explicit model for a population of Menzies' wallflower *(Erysimum menziesii)* (Lamberson et al. 1992b). This map was formed by intersecting an existing habitat map with a hexagonal grid. Grid cell sizes were scaled to the size of local wallflower populations. Six different habitat types are represented by variations in the shading of the cells. The dot in the center of some cells indicate that cell is also occupied by wallflowers.

The fundamental structure of the model is fairly simple. The simulated landscape is first partitioned into square or hexagonal cells. Then an underlying life history model provides estimates of the birth and death rates of the inhabitants of that cell as a function of its habitat type or site quality. Finally, there is a movement sub-model which characterizes the process of seed dispersal in plant populations or movement of individuals in animal populations, usually as a function of distance between cells.

The life history models are usually stage-structured models of the sort discussed earlier (Section 2.2). However, they are usually applied on an individual basis so that the habitat-specific parameters can reflect the properties of the cell occupied by a particular individual. As an individual moves from one cell to another the expected vital rates for that individual change to reflect the change of cell characteristics. The variation of vital rates among cells can account for factors beyond simple changes in survival and reproduction rates. For example, it can provide for delays in first reproduction in animals in some cell types, and in plants it can provide for failure (or low probability) to germinate.

Perhaps the most important feature of spatially explicit population models is that they incorporate movement as changes in an animal's spatial coordinates, and allow an assessment of how change in location affects population dynamics. The dispersal models

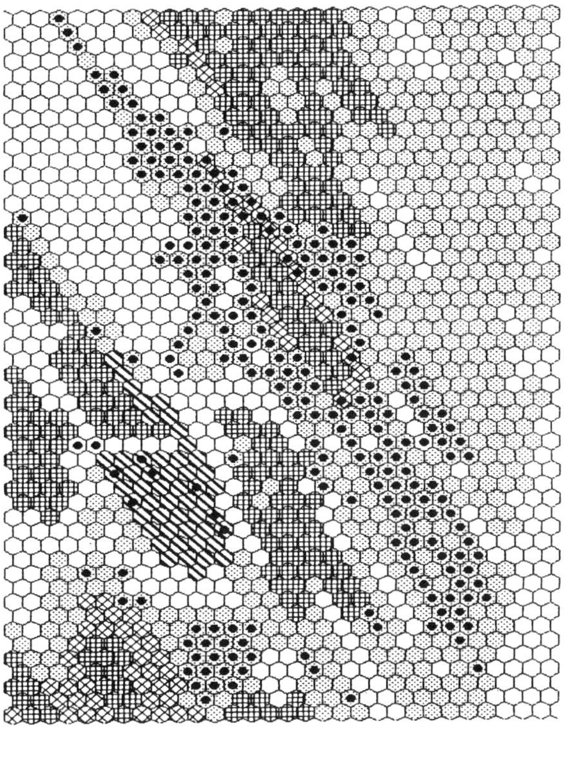

Fig. 33. A hexagonal grid-based map for the distribution of Menzies' wallflower and the surrounding habitat. Different degrees of shading are used to represent six different habitat types. Cells with a black dot were occupied by 1 wallflower individual.

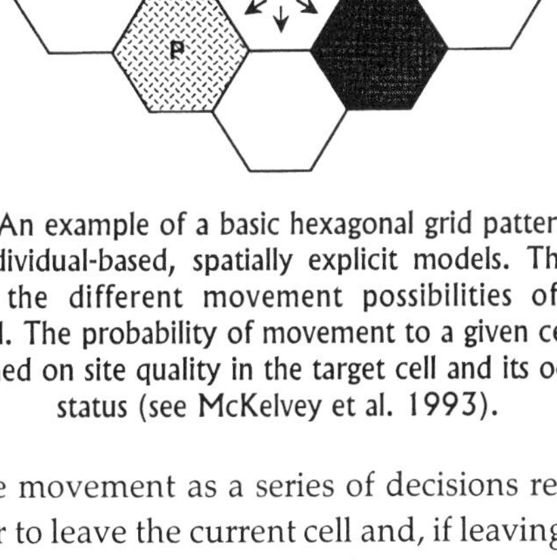

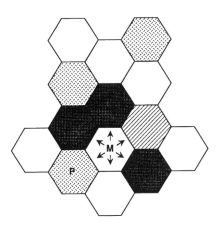

Fig. 34. An example of a basic hexagonal grid pattern used in many individual-based, spatially explicit models. The arrows indicate the different movement possibilities of a male individual. The probability of movement to a given cell can be conditioned on site quality in the target cell and its occupancy status (see McKelvey et al. 1993).

simulate movement as a series of decisions regarding whether to leave the current cell and, if leaving, which of the adjacent cells to enter in the next time step (Fig. 34). Probabilities are assigned for movement from one cell type to another based on the habitat types in each so that moving into preferred habitat is more likely. In addition, some models modify these probabilities by reducing the probability of entering an already occupied cell or giving a preference for moving in the same general direction taken while entering the cell. Each cell (habitat type) has a distinct birth and survival rate associated with it, and this can affect the movement path and the probability of settling in a given cell. Of course, death of the dispersing individual is always a possibility in any given time step.

Spatially explicit models have both advantages and disadvantages over other models. Their chief advantage, demonstrated even by simple models, is that the spatial distribution of populations and their habitats can profoundly affect population dynamics. As a result, spatially explicit models are very good for comparing various options for reserve design for endangered species, and ranking their effectiveness (see Holthausen et al. 1995). Since they are spatially explicit, these results can be directly translated into map-based management decisions. They are also valuable for examining the effects of reserve shape as well as size on local population stability (McKelvey et al. 1993).

Spatially explicit models have many of the limitations of other models we have examined. As a result, we should not expect their output to exactly predict the future dynamics of a population. In addition, spatially explicit models are difficult to parameterize because they require parameters (e.g., survival and fecundity) to be related to variation in habitat type or site quality. Information relating habitat-induced variations in fitness are available for very few species. Finally, these models are wonderfully fun to play with, but they take a very long time to build and analyze.

Uncertainty about the correct structure of the model is always a concern. One way to evaluate robustness is to independently construct two or more models that simulate the dynamics of the same system or population. Faith in the biological insights provided by model output is directly proportional to the degree that inferences drawn from the results of "competing" models are similar. To illustrate this approach, we explored the relationship between occupancy rate and reserve size using the LANDSCAPE model (McKelvey et al. 1993, Holthausen et al. 1995). A comparison of these results with those from same experiment using the patch model discussed earlier (Fig. 32) yielded very similar interpretations.

## 3 INSIGHTS FROM DEMOGRAPHIC MODELS

Factors contributing to variation in population size range from changes in realized birth and death rates arising from sampling variation, to changes in the expected values of these rates from year to year as a consequence of extrinsic environmental factors. When these sources of variance are coupled with those arising from density-dependent effects and spatial variation, the range of possible distributions of population sizes is immense. Given all of this expected variation under "normal" environmental conditions, how do we begin to separate signal from noise? That is, how can we identify populations that are truly at risk and subject to unexpected rates of population decline?

If this task were not challenging enough, there is also the complexity of predicting the response of managed populations to our conservation efforts. Even if an adequate model of a population's dynamics can be developed, the diversity of possible outcomes can make assessment of management practices very complex. For example, a stochastic model of a well-studied species can be used to generate thousands of possible population trajectories. In reality, though, only one of these possible trajectories will actually be realized. Which one is most likely?

### 3.1 The Need for Population Models

Understanding how populations change through time and across space requires a model linking the birth and death processes to changes in population size. Further, an assessment of viability relies on a demographic model of individual populations, the relevant stage structure of the population, and the transition probabilities between stages (e.g., juvenile → adult). Transition probabilities are determined by additions to the population via birth and immigration, and deletions to the population via death and emigration. Estimates of the year-to-year variability in these vital rates must also be included in the model. Demographic models of this form are useful for assessing the risk of population extinction (e.g., Burgman et al. 1993), and for bounding the conservation problem. By bounding we mean limiting the scope of the problem to those factors most likely to put species at risk of extinction, and projecting, through simple models, how the population is most likely to respond to the risk factors.

There are two steps in assigning risks. The first is to identify the deterministic processes that are putting species at risk. Attention should be focused on those factors (i.e., habitat destruction) that have altered the balance between reproduction and survival within stable populations, or the patterns of colonization and extinction that maintain metapopulations (Holsinger 1996). As a first priority, models should be structured according to state variables that reflect the action of these deterministic factors. An example is the deterministic loss and fragmentation of late seral stage forest in the Pacific Northwest and its apparent expression as a declining trend in survival rate of adult female spotted owls (Burnham et al. 1994, 1996).

The second step is to include year-to-year variation in the rates of reproduction, growth, and survival as a result of variation in the physical and biotic environment (Holsinger 1996). Including these sources of environmental stochasticity within the model will provide estimates of the likelihood of population declines simply as a consequence of chance events (successive years of poor reproduction; skewed sex ratios at birth, etc.).

In summary, models are needed to integrate variables that characterize the state of the population with the processes that determine the value of the state variables. These processes can be deterministic or stochastic drivers that act singly or synergistically to put species at risk of extinction. Adding to this complexity, and increasing the need for models, is the reality that a population's response to environmental stress is strongly affected by the spatial distribution of the population and its resources. Finally, models are needed to help choose among competing management alternatives based on differences in their predicted effects on species viability.

## 3.2 Sensitivity Analyses

A first step in developing a management plan for a threatened taxon is to identify those factors that are limiting its distribution and abundance (Holsinger 1996). Demographic models can be used to identify those life history attributes that most influence rates of population growth (see Section 2.4). Identifying critical life history stages is a valuable exercise even if data are too sparse for a complete PVA. Life history sensitivities provide a focus for research, and provide practical insights to managers. As a result, they speed up the process of acquiring the information most useful to conserve populations.

Good examples of the usefulness of formal sensitivity analyses of demographic models include analyses of the California condor (Mertz 1971), the Everglades (snail) kite (Nichols et al. 1980), the green sea turtle (Crouse et al. 1987), the spotted owl (Lande 1988a, Noon and Biles 1990), the Florida scrub jay (McDonald and Caswell 1993), the desert tortoise (Doak et al. 1994), and the prairie chicken (Wisdom and Mills 1997). For each of these threatened species it was possible to identify a deterministic factor, amenable to management intervention, affecting one or more key life history attributes.

## 3.3 Density-Dependence

The significance of density-dependent effects on the dynamics of wildlife populations has been debated for decades (e.g., Hassell 1986; Strong 1986). The primary reason is that density-dependent effects are difficult to measure from census data spanning < 30 years (Gaston and Lawton 1987; Pollard et al. 1987). Despite their elusive character, density-dependent effects are assumed to operate because many populations are regularly observed to narrowly fluctuate around some carrying capacity ($K$). For example, when population size is well below $K$, competition is lessened and survival and reproduction increase. In contrast, when population size if near or above $K$, intraspecific competition is intense, and survival and reproductive rates decline. Collectively, these two density-dependent factors would tend to maintain a population around its carrying capacity.

We have seen (Section 2) that typical matrix expressions of population dynamics have three possible outcomes: (1) exponential growth; (2) exponential decline; or (3) equilibrium. Equilibrium in this case is only possible within a narrow range of parameter values. When density-dependence is added to a Leslie matrix model, the range of values for the birth and survival rates under which equilibrium conditions can occur expands infinitely (Ginzburg et al. 1990). As we have seen previously, the strength of density-dependence and how it is represented in the model can significantly affect simulated population dynamics. Intuitively, since density-dependence moves a population toward K, density-dependent effects usually reduce population fluctuations and lower the risk of extinction. Therefore, models without density-dependence may overestimate extinction likelihoods. Given the difficultly in estimating the form and magnitude of density-dependence, but its common influence on most natural populations, how should density-dependence be treated in PVAs?

This question was explicitly explored by Ginzburg et al. (1990) where extinction risk was assessed for various degrees of density-dependence and for various threshold population sizes (i.e., quasi-extinction thresholds). The results were complex. Based on a Ricker model (Section 2), when there was no density-dependence the risk of extinction was quite high. When very weak density-dependence was introduced to the model, extinction risks were dramatically reduced. However, at higher levels of density-dependence, extinction likelihood increased. The effects of density-dependence were also affected by the structure of the model. If the Beverton–Holt model (Section 2) was used, density-dependence reduced extinction likelihoods at all population sizes.

Given this complexity, are there any general rules when modeling the viability of a given species? Ginzburg et al. (1990) suggest that for species for which there is little life history and demographic data a PVA model without density-dependence should allow a conservative (upper bound) estimation of the probability of extinction. As more data become available to estimate the magnitude of density-dependence, models explicitly incorporating its effects could be developed and upper and lower bounds of extinction likelihood computed for a range of plausible density-dependence levels.

## 3.4 Stochastic Factors

We used simple models to demonstrate the effects of stochastic factors on population dynamics, looking separately at the effects of demographic and environmental variation (Section 2.8). As shown by previous authors (e.g., Shaffer 1987; Lande 1993) it is useful to distinguish demographic, environmental, and catastrophic source of uncertainty because of their very different relationships to population size and growth rate. Catastrophes can be conceived as an extreme form of environmental variation, but it may be useful to view them as a separate source of stochasticity since their expression differs from the "series of bad years" scenario (Holsinger 1996).

One of the most useful ways to summarize the effects of stochasticity on the risk of extinction is by computing the distribution of extinction times under a given set of conditions. This requires specifying a specific stochastic model (see Section 2.9) and performing 100–1000s of simulations. It is important to examine the entire distribution, rather than just the mean time to extinction, because the distribution is strongly skewed to the right — most extinctions occur *before* the mean time. In models with density dependence, the distribution is even more skewed, appearing approximately negative exponential in shape (Gabriel and Burger 1992).

The most important work on the relationship between a population's demography, sources of stochasticity, and the risk of extinction has been done by Lande (1993) and Foley (1994). Lande found that the mean time to extinction follows different scaling laws in response to the different sources of stochasticity. With demographic stochasticity, the average time to extinction varies almost exponentially with carrying capacity ($K$), proportional to $e^{aK}/K$, where $a$ equals twice the mean population growth rate divided by its variance among individuals ($a = 2r/V_r$). With environmental stochasticity, mean extinction time scales as a power of the population's carrying capacity, $K^c$, where $c = 2r/Ve - 1$, and $Ve$ is the environmental variance in $r$. Mean extinction time scales faster or slower than linearly with $K$, depending upon whether the mean growth rate remains large relative to its variance (Lande 1993). Stochasticity due to demographic variation is inversely proportional to population size. In contrast, stochasticity due to environmental variation is often independent of population size. As a result, for sufficiently large populations (e.g., $N > 50$), environmental variation is a greater threat to persistence than demographic variance (Lande 1993).

The relationship between persistence and catastrophes is complex. In general, populations can persist if their growth rate between catastrophic events exceeds the product of the frequency of the catastrophe times its magnitude (Lande 1993). One of the most significant aspects of Lande's research is that he found that previous estimates of mean times to extinction (e.g., Goodman 1987, Shaffer 1987) (Fig. 26) had been too pessimistic, at least for populations with long-term positive growth rates.

## 3.5 Spatial Structure

At any point in time the location of an individual organism, habitat element, or landscape patch is given by its spatial coordinates. Up until the last decade, however, population models largely ignored the spatial context of organisms — models were zero-dimensional (Gilpin 1996), and assumed that organisms were uniformly distributed in a completely homogeneous landscape. This assumption has a long history of acceptance despite the fact that ecologist have long recognized that dispersal, disturbance events, and spatial mosaics profoundly affect the outcome of species interactions (Kareiva 1994).

Ignoring space assumes that animals in the system can move anywhere in the range of the system within one time step, and that all resources are globally available (Gilpin 1996). This assumption ignores the fact that the behavior of an organism is going to be most influenced by its immediate environment, and that it is more likely to interact with neighboring organisms than those far away. In addition to neighborhood effects, organisms disperse through the environment and are affected by habitat types and other organisms encountered during movement. For example, a dispersing animal experiences different resistance (i.e., survival costs) to activity depending on the composition of the landscape encountered during movement. (By themselves, the differential costs to movement and the uncertain nature of survival produce spatial heterogeneity.) Thus, an organisms fate is affected by its immediate neighborhood, its relationships to the distribution of other resources, and the habitats and organisms encountered during movement.

We have seen earlier (Fig. 13) that even the implicit incorporation of space, expressed as a decreasing density of suitable habitat and potential mates, can have pronounced effects on population dynamics. The uncertainty of the dispersal process, coupled with the inherent costs of search and movement, result in strongly non-linear relationships between population size and the density of critical resources, such as habitat. Given the importance of spatial context to population dynamics, and the fact that all management activities are spatially explicit, it is incumbent upon us to incorporate space into most of our viability assessments.

## 4 INSIGHTS FROM GENETIC MODELS

PVA is a tool for evaluating persistence probabilities for populations of animals or plants. Ultimately, of course, the objective is to use PVA to help ensure the perpetuation of genetic diversity on this planet. Seldom have genetic models been used to estimate viability, however, because demographic and ecological processes were thought to present more immediate threats to extinction (Lande 1988b, Boyce 1992, Schemske et al. 1994). Risks associated with environmental and demographic stochasticity at low population sizes, and systematic causes for decline, such as habitat loss (Caughley 1994), appear to be the most urgent reasons for conservation efforts. Caughley notes that not a single species extinction is known to have been caused by genetic mechanisms.

We are concerned, however, that these observations may result in inappropriate approaches to PVA (Nunney and Campbell 1993, Hedrick et al. 1996). Clearly, the loss of genetic variation increases the risk of species extinction, and under some circumstances a PVA must focus on genetic structure of a population to ensure persistence. Lande (1994, 1995) and Lynch (1996) have recently argued that mutation can have consequences to viability at least as large as demographic and environmental stochasticity.

Perhaps the biggest problem with incorporating genetic information into PVAs is the lack of data or understanding of how genetic population processes work for a species (Hamrick and Godt 1996). Because we are increasingly realizing how fundamental such information is to accomplishing conservation objectives, we are refocusing priorities to developing better understanding of conservation genetics (Avise and Hamrick 1996).

### 4.1 Priority for Preserving Genetic Diversity

Because the ultimate objective behind PVA is to develop prescriptions for species survival for the purpose of preserving genetic diversity (Soulé 1987), it seems appropriate that models of genetic variation ought to contribute to the formulation of a PVA. We know that small population size can result in inbreeding depression in some populations, which may increase the risk of extinction for the population (Lande and Barrowclough 1987, Ralls et al. 1986, 1988). We also know that small population size can reduce genetic variation through drift, thereby reducing the raw material for evolutionary change, and that genetic variation can be essential to ensure preadaptation to disease, competition, or predation (Futuyma 1983). But what we do not know is how much and what types of genetic variation should be the target for conservation.

### 4.2 Types of Genetic Variation

Convincing arguments have been made for emphasizing unique evolutionary lineages such as species or subspecies in our conservation efforts (Templeton 1986). But even within a taxonomic group, many forms of genetic variation may respond differentially to particular conservation strategies. Genetic variation is revealed by many techniques, including restriction site analysis of mitochondrial DNA, karyotypy, electrophoresis of allozymes, heritability of quantitative traits (Falconer 1981), and morphological variation (but see James 1983). Although many of these measures can vary almost independently of one another (Lande 1988b, Zink 1991), broad correlations tend to occur among sources of genetic variation (Hartl and Clark 1989).

Genetic variation within populations often is measured by mean heterozygosity, or the proportion of alleles that are heterozygous. Polymorphism (P) is the proportion of loci that are polymorphic in a population, typically tallied when below some arbitrary threshold, say 95%. Heterozygosity, on the other hand, is the proportion of loci at which the average *individual* is heterozygous. At a population level, heterozygosity is often compared with Hardy–Weinberg expectation (2pq) to see if the population is in equilibrium or if the gene frequencies are evolving.

At a larger scale, heterozygosity can be evaluated by distinguishing within from among population or subpopulation variation. The total heterozygosity, $H_t$, is the sum of that occurring within a population, $H_s$, and the variability among populations, $D_{st}$, i.e., $H_t = H_s + D_{st}$ (Nei 1987). If $D_{st}$ is large, we might suspect that conservation priority needs to be placed on maintaining good geographic representation of populations to ensure preservation of the among-population component of variation.

Reasons for maintaining genetic variation merit careful consideration. If preadaptation to future insults from other species (disease, parasites, competitors, predators) is the reason to preserve genetic variation, the focus may better be on preserving rare alleles (Futuyma 1983). For conservation purposes, perhaps the number of alleles per locus is a more relevant measure of genetic variation than mean heterozygosity (Allendorf 1986).

Quantitative traits are most frequently the target of natural selection, therefore Lande and Barrowclough (1987) argue that heritability should be monitored as a measure of genetic variation for conservation. (Heritability is a measure of the degree of covariation between parent and offspring in a given trait that is attributable to the genes.) However, estimates of heritability can be difficult to interpret because the response to selection can be greatly complicated by maternal effects (Atchley and Newman 1989). Relatively low levels of genetic variation may confer substantial heritability to some quantitative traits (Lynch 1985). There is also the difficulty of deciding which quantitative traits should be measured. According to Lande and Barrowclough's (1987) rationale, the most important traits ought to be those that are most frequently the target of natural selection. Yet, these are exactly the traits expected to bear the lowest heritability as a consequence of selection (Falconer 1981, Boyce 1988).

## 4.3 Genetic Structure of Populations

How genetic variation is structured within populations may also bear on PVA (Boecklen 1986). Spatial heterogeneity appears to be one of the most important mechanisms maintaining genetic variation in natural populations (Hedrick 1987). Whether or not this pertains to the importance of inbreeding in natural populations has become a topic of debate (Shields 1982, Ralls et al. 1986), but there is no question that spatial variation in genetic composition of populations can be substantial. We are just beginning to understand the role of population subdivision on genetic structure and heritability (Wade 1991). How significant is local adaptation? How important is coadaptation of gene complexes (Templeton 1986)? Although spatial structuring of genetic variation is complex and interesting, it is not clear that our understanding of genetic structure of most populations is sufficient to use it as a basis for manipulating populations for conservation (Templeton and Georgiadis 1996). Attempts to manage a species by transplanting individuals among subpopulations can be an effective tool to maintain or increase genetic variation within populations (Griffith et al. 1989) but may destroy variance among populations or result in outbreeding depression (Lynch 1996).

The solution to this dilemma requires that we somehow foresee the threats that a species is likely to encounter. If local subpopulations are likely to be threatened by habitat destruction or political unrest, it may be extremely important to maintain geographic variants to ensure that the species can continue to survive in other localities (Templeton 1986). For example, the Canadian Wildlife Service has a program to restore swift foxes (*Vulpes velox*) to their former range in the prairie provinces of Canada by transplanting animals from eastern Wyoming and Colorado. However, occasional swift foxes have been documented in adjacent portions of the United States but they are very rare at the northern extent of their distribution. Potential exists for dispersers from the Canadian transplant program to interbreed with the rare native northern genotypes, with the risk of swamping locally adapted gene complexes (Stromberg and Boyce 1986).

The peril of such genetic exchanges was illustrated in southern Ontario and northern states in the United States where bobwhite (*Colinus virginianus*) populations were supplemented with stocks from further south to increase hunting opportunities. Transplanted quail interbred with native quail, apparently eroding local adaptations that enabled them to survive northern winters, and resulted in widespread declines in northern populations of bobwhites (Clarke 1954, Frankel and Soule 1981). Indeed, the species is currently absent from portions of the northern extent of its natural distribution.

Knowledge of the genetic background of source populations can be of fundamental importance. Conservation biologists recently made a case to the U.S. Fish and Wildlife Service that the small population of bison (*Bos bison*) in Jackson Hole, Wyoming might suffer loss of genetic variability and genetic drift if not supplemented occasionally. The argument was that injecting one individual into the population every generation ought to prevent drift, and because females are more readily accepted into strange herds, cow bison are probably the best for such transplants (Shaw 1993). So the National Elk Refuge arranged to move a single female bison from Custer State Park, South Dakota into Jackson Hole in 1992. Fortunately this individual died before it had opportunity to breed, because in 1993 we learned that about 32% of the bison in Custer State Park carry genes from domestic cattle (*Bos taurus*), probably residual from hybridization trials to create beefalo earlier in this century (Strobeck et al. 1993).

If, on the other hand, future threats due to diseases and parasites are expected, there may be a premium on ensuring the maximum allelic diversity throughout the population. Ideally this diversity should be spatially dispersed to protect species from epidemics that can extirpate local populations, as occurred in black-footed ferrets (*Mustela nigripes*) in the 1980s (May 1986). How such genetic diversity should be structured optimally is a complex problem that has received little research attention.

An approach commonly used in trying to determine a genetic basis for minimum viable populations is to examine effective population size, $N_e$ (Reed et al. 1988, Nunney and Elam 1994). $N_e$ is the number of individuals in an ideal panmictic (i.e., randomly breeding) population that would suffer the same amount of random drift in gene frequencies as in the observed population. Mating systems, unequal sex ratio, and variance in population size are the primary causes for $N_e$ to be substantially less than the actual population size. To account for variance in population size we calculate the harmonic mean among generations:

$$1/N_e = 1/t\,(1/N_1 + 1/N_2 + \ldots + 1/N_t),$$

where subscripted *N*s are population sizes in generations 1, 2, 3, ..., *t*.

The influence of imbalanced sex ratio or variation in mating system is commonly modeled:

$$N_e = 4\,N_f\,N_m/(N_f + N_m)$$

where $N_f$ is the number of breeding females, and $N_m$ is the number of breeding males. Again, deviation from equal numbers of breeding males and females has the effect of reducing $N_e$ below the actual population size. But these approaches to estimating $N_e$ are probably overly simplistic.

As is the case for measures of genetic variation, we have numerous measures of effective population size, depending upon the mechanisms affecting drift. For example, Ewens (1990) reviews the calculation of $N_{ei}$ relative to inbreeding, $N_{ev}$ for the variance in gene frequencies among subpopulations, $N_{ee}$ targeting the rate of loss of genetic variation, and $N_{em}$ for mutation effective population size. Still more measures may be derived. For example, $N_e^{(meta)}$ defines the effective population size in a metapopulation experiencing repeated extinction-recolonization events (Gilpin and Hanski 1991). Each of these basic measures of $N_e$ is then subject to adjustment for unequal sex ratio, age structure (Hill 1972, 1979), demographic structure (Waite and Parker 1996), and variable population size (Harris and Allendorf 1989). There is no sound basis for selecting one of these basic measures of $N_e$ over another, yet as Ewens (1990) shows, they can lead to much different conclusions about MVP. Several PVAs have simply attempted to calculate $N_e$ for populations of 50 and 500 accounting for imbalance in sex ratio, mating system, or population fluctuations. We do not believe that this approach is sufficient to provide meaningful targets for conservation.

## 4.4 Genetic Guidelines for PVA and Population Conservation

In some circumstances, genetic models ought to be the focus of PVA. When reducing genetic erosion is likely to be an important consideration in determining optimal management for a species, genetic models would seem necessary to evaluate the consequences of alternative management scenarios. For example, the African wild dog (*Lycaon pictus*) currently occurs in highly subdivided populations (Ginsberg and Macdonald 1990), and some subpopulations are severely depleted due to epidemics of canine distemper virus. There is concern over proposed programs for transplants, however, because the species shows substantial geographic variation.

Genetic variation is lost in small populations due to random reassortment during sexual reproduction. More generally, this results in a random change in gene frequencies, an effect that accumulates over time when populations are maintained at small numbers. This loss of genetic variation is likely to have negative consequences because of decreases in fitness. Genetic variation affects fitness in populations in three ways (Packer 1979): (1) heterozygous individuals may be more robust, resistant to disease, and have higher survival (Allendorf and Leary 1986), (2) homozygous individuals are more likely to express recessive deleterious alleles that are suppressed in heterozygotes, and (3) at least some genetically variable offspring are likely to be preadapted to unpredictable changes in the environment, including parasites, predators, climate change, or local conditions (the Tangled Bank hypothesis and the Red Queen's hypothesis).

As a consequence of being in a small population, related individuals are likely to breed. Such inbreeding can result in a loss of fitness for reasons noted in the preceding paragraph. A common pattern is that juvenile mortality is elevated among inbred offspring, or fecundity may be reduced (Ralls et al. 1986). Inbreeding depression has been documented repeatedly in natural populations (Jiménez et al. 1994, Keller et al. 1994). Such consequences on demographic parameters have obvious ramifications to population viability, and PVA software has been developed that explicitly incorporates inbreeding depression (Mills and Smouse 1994). One way to accommodate such effects in a PVA is to model certain vital rates as a function of population size. Mills and Smouse (1994) showed that inbreeding depression can appreciably influence the probability of extinction, especially in long-lived species with low potential growth rates.

Although Lande (1988b) first concluded that modeling genetics is not likely to be as important as modeling demographic and environmental processes in the formulation of a PVA, he appears to have changed his mind based on subsequent work on the role of mutation in small populations (Lande 1994, 1995). Mutation can result in inbreeding depression or decreased fitness resulting from the accumulation of mildly deleterious mutations. Empirical studies show that deleterious consequences of mutation are likely to be much greater than the potential benefits from the emergence of adaptive variants. Lande (1995) concludes that mutation effects may be comparable in importance to environmental stochasticity, and recommends managing for MVPs of a few thousand individuals (also see Lynch 1996).

Seldom will we have sufficient data on genetic variation and structure to defend PVAs based on genetic models. Yet, conservation genetics is a dynamic field in which rapid advances are being made (Avise and Hamrick 1996). Surely the preservation of genetic diversity is ultimately the foundation for any PVA, and we expect that some of the greatest opportunities for research advances exist in this field.

## 5 HABITAT-BASED PVA

Although the birth and death rates of a population ultimately determine its viability (i.e., $\lambda \geq 1.0$), the long-term persistence of any population will also depend upon an adequate amount and distribution of habitat. In cases where demographic information is limited and there is reason to believe that habitat conditions limit population growth, habitat-based PVA may be a useful exercise. The key assumption is that changes in amount and distribution of habitat are primary drivers of population dynamics. Even when essential demographic data (i.e., means and variances of the vital rates) are available, a habitat approach is often warranted because the deterministic processes leading to habitat loss and fragmentation may dominate over stochastic demographic effects in influencing population persistence.

A habitat-based PVA should also consider factors that determine the distribution of habitats — landforms, soil types, and climate — and those human-induced factors that lead to habitat change. As a first step, habitat-based PVA should at least include habitat structure and landscape patterns of vegetation components relevant to the species of concern. For example, Holt et al. (1995) suggested that for long-term conservation planning, PVA models should simulate changes in landscapes via vegetation submodels to evaluate risks of extinction associated with proposed land-use practices. Further, to be relevant in ecosystem management on public lands, such models should also relate to existing hierarchical, ecologically based land classification systems (e.g., O'Neill et al. 1986). A reliable habitat-based PVA model would be able to predict future population status by simulating changes in vegetation patterns and biophysical processes at local and regional scales.

In previous sections, we described models with elementary spatial structure (Section 2.11) that represented habitat variation in a coarse fashion. These models considered variation in the density of suitable habitat, and allowed habitat to be arrayed in patches of varying size. In general, habitat occurred in only two categories, suitable for breeding and survival, or unsuitable. This section describes in more detail the basis for developing and applying habitat-based PVA models that more explicitly incorporate habitat geometry and variation in habitat quality.

Even though current knowledge indicates that population–habitat interactions are complex, our initial objective is to build a habitat-based PVA model with minimum data requirements. That is, we encourage PVA model development that confronts the tradeoffs between complex, computationally intensive, and parameter-rich population models (Anderson and Mahoto 1995, Wennergren et al. 1995) and the real world of limited, low-resolution data that underlies most management decisions. Therefore, an ideal habitat-based PVA would be able to balance the tradeoffs between complexity and practicality (e.g., Ackakaya and Atwood 1997).

### 5.1 Habitat As A Surrogate For Populations

The pattern and dynamics of habitat structure and composition can be measured as possible surrogates for direct measurement of population parameters. For now we define habitat as the type and physical arrangement of objects in space (Bell et al. 1991). These objects, tangible and with discrete boundaries, can be either biogenic or geologic in origin. They have defining attributes including type, size, volume, mass, form (geometry), and, if biogenic, characteristic renewal rates. Most relevant to population dynamics are objects that create and modify environments, vary in their effects, and can be affected by human-induced and natural disturbance processes. These objects include dominant geologic features (e.g., mountains, rivers) or biogenic, structure-producing organisms (e.g., coral reefs, trees, kelp forests) that create environments favorable to structure-using organisms. Over time frames relevant to management, the geologic features are assumed to be static. Thus, our focus is primarily on the biogenic components, particularly plants in terrestrial systems, as the dominant objects in the space of landscapes that modify the environment and are in turn affected by human activities.

Focusing on the biogenic components, the dynamics of some subset of the vegetation or aspects of plant community structure and pattern would be directly modeled. Such models could be either non-spatial or spatially explicit. Those aspects of the vegetation that are measured and modeled would be chosen primarily on their ability to provide surrogate information back to the dynamics of the population and provide predictive insights into future responses to habitat change. Preferably, these structural and compositional components would be biogenic elements that are persistent, change or ameliorate the local environment to create conditions suitable to survival and reproduction, and are affected by management actions.

## 5.2 Justification

The justification for using habitat structure and composition as surrogate variables for predicting population response is based on both pragmatic and theoretical arguments. World-wide, most species that have become rare or gone extinct have done so as a consequence of the loss and fragmentation of their habitat (Ehrlich 1995), thus it is logical to directly focus on this stressor. The compelling evidence of increasing extinctions resulting from habitat destruction (Tilman et al. 1994) suggests that the dynamics of many imperiled populations can be described by simple models that track changes in habitat amount and geometry. In such cases, population age-structure and precise spatial pattern may not be necessary to predict the consequences of future land-use (Fahrig 1993).

In addition to directly tracking deterministic factors that put species at risk, habitat-based projection models may anticipate population time lags. For example, Doak (1995) argued that habitat loss could result in eventual precipitous declines in the population of grizzly bears in Yellowstone National Park that would not be detected by population monitoring. A similar conclusion reached by Thomas et al. (1990) for the northern spotted owl lead them to propose direct estimates of birth and survival rates instead of population surveys in order to estimate population trend.

The theoretical argument for a habitat focus is based on the belief that animals respond to habitat in an adaptive fashion (Noon 1986). That is, where an animal selects to live is believed to be an evolved behavioral response stimulated by structural and compositional features of the landscape. The fitness gain accrued by making the "correct" decision is an increase in lifetime reproductive success. According to the conceptual model of Southwood (1977, 1988), by the processes of discrimination and eventual selection, habitat acts as the template that guides the evolution of ecological strategies. Many behavioral studies support the understanding that animals evaluate a habitat's "quality" on the basis of proximal cues received from the environment. Therefore, habitat selection would evolve if different decisions were associated with different probabilities of survival and reproduction.

Casting the theory of animal–habitat relations in an evolutionary framework serves two key purposes. First, it establishes a theoretical justification for attempting to model the relations between animals and habitat structure. Second, it suggests that predictive models relating patterns of a species' distribution to components of habitat structure are likely to be useful. Given such models, management-induced change in specific structural or distributional components of the landscape can be simulated and the likely biotic response estimated. Such tools allow management alternatives to be evaluated in an anticipatory context.

## 5.3 Defining Suitable Habitat

If habitat is defined at a sufficiently fine scale, all species occur over a range of habitat types. This distribution, the occupied (actual or potential) habitat gradient, can be defined in terms of a single variable (e.g., average tree diameter), or in terms of some linear combination of many habitat variables (e.g., a principal components axis). If demographic response can be correlated with habitat variation, then various positions along the gradient can be interpreted in terms of their habitat quality. One possibility is to define habitat quality in terms of the expected population growth rate ($\lambda$), a function of the average birth and death rates (Section 2.1). Suitable habitat is then defined as "habitat of sufficient quality so as to provide for a long-term balance between birth and death rates so as to support a stable ($\lambda = 1.0$) or growing population ($\lambda > 1.0$). Given this definition, habitat quality is equated with the expected value of $\lambda$ and is a continuous function of variation in habitat (Fig. 35).

For a given location along the habitat gradient, habitat quality is the expected value of $\lambda$ at that point. The distinction between suitable and unsuitable habitat is defined as the point along the gradient where the function assumes a value of $\lambda = 1.0$. (This is also the point that distinguishes source from sink habitat). Note that unsuitable habitat may still support survival and birth, but not a rates sufficient to maintain a local population.

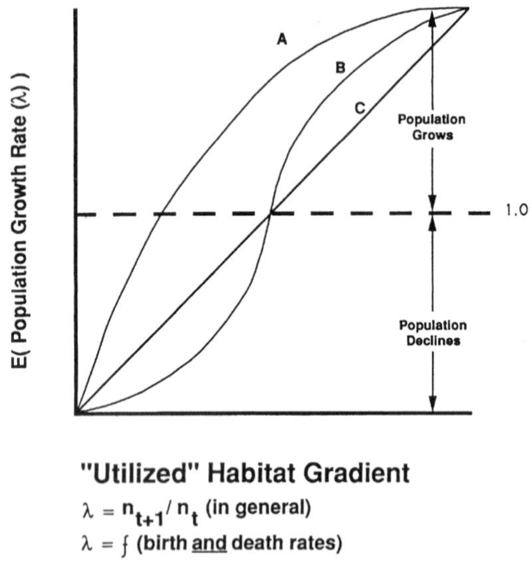


Fig. 35. Changes in expected population growth rate (λ) as a function of location along a habitat gradient. Three hypothetical relationships are shown. The location along the gradient where λ = 1.0 distinguishes sink (λ < 1.0) from source (λ > 1.0) habitats.

Figure 35 is a heuristic model and ignores some critical factors — for example, temporal variation in habitat quality. Nevertheless, the concept is important because it provides a clear link between habitat variation and the evolution of habitat selection. In addition, Fig. 35 shows the connection between habitat quality and the population viability — a viable population is one that has an expected $\lambda \geq 1.0$ with this condition being maintained with probability $p$ for some specified time period $t$.

## 5.4 Spatial Scales of Habitat Evaluation

Habitat selection is a hierarchical process (Johnson 1980) occurring across spatial scales — from the geographic range of a species to coarse habitat categories within the range (e.g., forest, field, marsh), to home ranges within habitat types, to the selection of specific resources within home ranges. Habitat-based PVA must determine interactions between demography and habitat variables on scales of resolution that match those of the animals being modeled (Rosenzweig 1991), because interpretations from studies of habitat selection are a function of the scale at which the research was conducted. Different patterns are likely to emerge when one examines habitat selection at different scales (Wiens et al. 1987), because different ecological mechanisms operate at different scales (Huston 1995).

Despite this understanding, most studies of habitat suitability have been conducted at the home range or within-home range scale. That is, ecologists have focused on the relations between individual patterns of occurrence on the landscape and features of the environment within the immediate neighborhood of these locations. Only recently have studies reflected the fact that the distribution and abundance of most plant and animals are also strongly influenced by the spatial arrangement of suitable habitats at multiple spatial scales (e.g., Flather et al. 1992, Lamberson et al. 1992b, Turner et al. 1993, Short and Turner 1994).

Species respond to environmental patterns at different spatial scales – there is no single correct scale for habitat evaluation sufficient for all species. In addition, the "correct" spatial arrangement of habitat elements is a function of the level of population organization — what may be a fragmented habitat at the scale of the population may not be fragmented at the scale of the individual home range of the same species (Wiens 1996). Thus, habitat quality must be evaluated at a minimum of two spatial scales — at the level of the individual and at the level of the population. A third scale of assessment becomes important when we consider long-term persistence of populations in the context of environmental variation. For this consideration, it is necessary to have many distinct populations, widely distributed across the landscape, whose fluctuations are spatially asynchronous (Den Boer 1981). This scale requires a metapopulation perspective.

A continuing cause of declines in biodiversity and in individual species is the human-induced fragmentation of habitat at local and regional scales (reviewed in Noss and Csuti 1994). As a result, the spatial arrangement of habitats and habitat elements is dynamic and subject to ongoing disturbances. Therefore predictive habitat suitability models will need to explicitly incorporate the dynamic nature of landscape patterns and their life history consequences for individual species. This will require monitoring changes in the status, trend, and geometry of habitat at the level of the individual, the local population, and at the metapopulation scale (see example for spotted owls in Noon and McKelvey 1996). This understanding means that managers and policy makers need to consider the effects of their decisions (i.e., land use) at all three scales — decisions affecting habitat quality and species persistence can no longer be assessed at just the local scale.

## 5.5 Developing Predictive Habitat-Relations Models for Animals

The primary problem in building habitat-based PVAs is to find methods for integrating vegetation dynamics and spatial pattern, geology, and topography with the behavioral process of habitat selection. A mechanistic model should include as predictor variables those features of the landscape that stimulate an animal to settle at a given site, and the attributes of that site that ultimately affect birth and death rates. Adding to the challenge of model-building is the fact that these factors are operative at different scales, and their relative contributions may vary through the range of the species. For example, population segments at the fringes of a species' range are likely to be constrained by their limits of tolerance to the physical environment. A consequence is a greater probability of local extinction for such segments, even though vegetation structure and composition may appear suitable. Predicting population dynamics under such circumstances requires an understanding of the physical processes that operate at larger scales and their interaction with local habitat attributes. A fragmentary look at a species' viability may detect local instability, but fail to detect that it is embedded within a regionally stable system.

Numerous physical and biotic factors might be considered for inclusion in habitat-based PVA, because both natural and man-made variation in habitat conditions can result in complex population dynamics (Dunning et al. 1995). Past approaches have focused almost exclusively on vegetation structure and composition, but other factors may override local habitat effects. For example, some species' densities may increase along a productivity gradient: net primary productivity and potential evapotranspiration correlate strongly with species richness of trees, birds, mammals, amphibians, and reptiles (Currie 1991, Huston 1995). Therefore, topography and geomorphology may need to be addressed because they control the distribution of water and soil nutrients, solar energy, and precipitation, thereby influencing the distribution and continuity of areas of high productivity.

In practice, developing models for plants and animals will be an iterative and initially crude process. Despite its limitations, a logical first step is to explore relations between coarse-resolution habitat data (based on remotely-sensed vegetation patterns) and existing data on patterns of species distribution. Intersecting these data sources will reveal spatial relations between species and habitats. These relations can be quantified by contemporary geostatistical techniques (Rossi et al. 1992), and carried forward in subsequent model-building. These "first-cut" spatial models can be integrated with the many higher resolution wildlife habitat relations models that already exist (e.g., Verner et al. 1986, Morrison et al. 1992). This synthesis will produce preliminary models capable of forecasting changes in animal species distributions based on future changes in vegetation structure and composition. However, these models should be treated as hypotheses to be tested. To the extent they fail to predict, they will need to be revised or discarded.

So far we have discussed simple correlative models relating presence/absence or probability of occurrence to habitat attributes. Predictive PVA models for some well-studied species, however, may be able to relate demographic rates directly to habitat variation (Fig. 35). These models often have a structure that specifies the location of each object of interest (individual, population, or habitat patch) within a landscape, and therefore include information on the spatial relationships between objects (Dunning et al. 1995). Such models , referred to as spatially explicit population models, can be developed at the individual, local population, or metapopulation scales.

A thorough review of wildlife-habitat relationship models is beyond the scope of this paper. A comprehensive discussion can be found in Van Horne and Wiens (1991) and Morrison et al. (1992).

## 5.6 Map-Based Models

Spatial models that explicitly integrate population dynamics with the amount, quality, and distribution of habitat are often called "map-based" models. Models structured in this way allow an assessment of population viability in the context of real-world landscapes by overlaying the distribution of individual animals or populations on habitat (usually vegetation) maps through a GIS interface (e.g., Johnson 1990). Individual animals or populations are then assigned demographic rates based on the quality of the patch or cell in which they reside (see Section 2.14).

By incorporating animal movement, these models can address the effects of fragmentation, isolation, and patch size and shape on population viability. These models allow the manager to determine not only what types of habitats are needed, but also how these habitats need to be arranged across the landscape (Turner et al. 1995). For example, given a fixed land area (equivalent to a fixed amount of money) available for a species' conservation, these models can explore the optimal arrangement of habitats across the landscape to maximize persistence likelihood. Holthausen et al. (1995) used such a map models to solve a spatial optimization problem for the conservation of northern spotted owls on the Olympic Peninsula.

Habitat-based PVAs with this complexity are data-intensive — they require functions relating demography explicitly to landscape conditions via habitat-specific demographic rates (see Figs. 34 and 35). Since these models are sensitive to animal movement, they also assume that a species' dispersal behaviors are well understood. Given these data requirements, models of this detail are only applicable to a small number of well-studied species.

Despite the appeal of map-based models (McKelvey et al. 1993, Dunning et al. 1995, Turner et al. 1995), and the recognition that all management actions are spatially explicit, these models have been criticized. For example, Wennergren et al. (1995) cautioned that we need to question the merits of spatially explicit models for two reasons. First, increasing the complexity of spatial models to include age structure may not enhance their ability to predict population growth in environments that vary over time. Second, modest errors in estimating attributes of dispersal propagate into large errors in predicting the success of individuals in a variety of habitat patch configurations. Further, dispersal predictions used in spatially explicit models are likely to be unreliable until dispersal attributes are known better (Boyce et al. 1994, Wennergren et al. 1995).

## 5.7 Model Validation

Species populations are affected by many factors other than their immediate habitat. For example, disturbance history, availability of source populations, and current weather greatly affect observed population abundance and distribution. Therefore, the correspondence between observed and predicted states based on habitat models is seldom going to be 1:1. The challenge is to estimate the contribution of habitat structure to the state of the population. That is, how much of the variation in population attributes is due to habitat variation. To estimate the habitat contribution requires the influence of habitat structure and composition on organisms (and processes) be teased apart from all other environmental influences; the majority of the structural components of habitats that affect organisms be identified; and habitat structure be examined at a spatial scale relevant to the organism of interest. All these factors must be considered as habitat-based models are constructed, validated, and revised.

Useful PVA models are predictive and lend themselves to some forms of validation. Habitat-based PVAs can simulate anticipated habitat change arising from management or natural disturbance, and predict the population consequences. If these changes occur, the population can be monitored to see if it responds as predicted. Recall that the absence of a 1:1 mapping of habitat to population status requires that other drivers of change be treated as covariates and statistically controlled. If little correspondence is found between the observed and predicted response, then the model is rejected or revised to incorporate new understandings, and validation begins again. Model validation and revision are ongoing and iterative processes.

The reality is that most PVA models are going to be difficult to validate. Those based on habitat as the primary driver of change allow specific predictions of population response to landscape pattern. However, many imperiled species are long-lived vertebrates that will not show short-term responses to habitat change. This inherent time lag between habitat change and population response may lead to the rejection of valid models. The ease of validation will also depend on the resolution of the model output. Models that predict simple presence/absence patterns will be easier to validate than those that predict demographic responses to variations in habitat quality. Models of the former type can be validated with census data; the latter type models will require detailed field investigations. A good discussion of the difficulties of validating wildlife-habitat models is found in Morrison et al. (1992: 255–61).

## 5.8 Statistical Models For Habitat-Based PVA

When a species uses resources disproportionately to their availability, use is said to be selective. Understanding how animals respond to a heterogeneous environment — what resources are used and what are avoided — is essential to understand a species requirements for birth and survival. In addition, differential resource selection is one of the key processes that allow sympatric species to coexist (Rosenzweig 1981). Resource selection functions (RSF) are thus mathematical functions of the attributes of resource units (e.g., habitat characteristics of used patches) such that the value of the unit is proportional to the probability of that unit being used (Manley et al. 1992).

The basic data assumption for habitat selection studies is that the landscape can be partitioned into discrete habitat types. Information on species use of habitat units is usually based on survey data (counts of animals or animal sign). Resource availability may be estimated from a random selection of sites or from an existing habitat map. The subsequent comparison of frequency of use of habitats to their availability provides a measure of selection. For habitat-based PVAs, the habitat types most strongly selected are assumed to be those most essential to population viability.

Boyce et al. (1994) described an application of RSF in developing a habitat-based PVA for northern spotted owls. Their model linked vegetation attributes with an abiotic variable, range of elevation, to predict existing carrying capacity of a landscape in western Oregon. Irwin and Hicks (1995) developed a similar model that incorporated suitable vegetative habitat, variation in elevation, and an index that integrated climax forest association and probability of wildfire.

A typical habitat-based PVA projection is presented in Fig. 36. Boyce et al. (1994) used the trajectory of

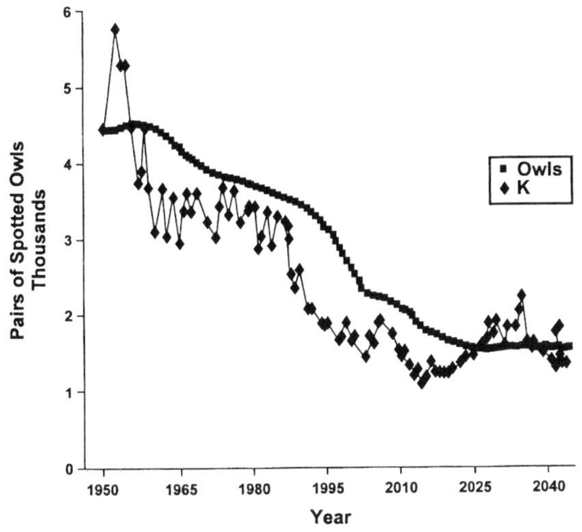


Fig. 36. Changes in the size of owl populations and habitat carrying capacity with time based on a 100-year simulation. Note how the population tracks changes in habitat carrying capacity. Stochasticity enters only through the variable $K(t)$. Modified from Boyce et al. (1994).

northern spotted owl habitat decline as described by USDI (1992) to track changes in carrying capacity ($K(t)$) for 1950 to 1990. From 1990 to 2050 they assumed there was no net change in mean $K(t) = 1{,}780$ owl pairs. Such an assumption is reasonable due to implementation of a regional conservation strategy that provided for forest regrowth and limited timber harvesting (FEMAT 1993). With those assumptions, there were no extinctions over 100 years, even after 100,000 iterations.

Further, with variation imposed only through $K(t)$, Boyce et al. (1994) used the projections in Fig. 36 to calculate 5 years of population growth ($\lambda$) values for the period 1985 to 1990 to compare with estimates of $\lambda$ from detailed demographic data (Burnham et al. 1994). Their calculations suggested a distribution of $\lambda$ values that overlapped substantially with those of Burnham et al. (1994) although Boyce et al. (1994) recognized possible bias in demographic estimates of $\lambda$. This is a weak, but nonetheless useful, test of model validity.

By linking the RSF model with vegetation-growth models and incorporating various landscape management scenarios in a GIS environment, one can compute dynamic estimates of carrying capacity. If it accounts for dispersal, the RSF model then becomes the foundation of a spatially-explicit PVA — it can predict future animal locations and population distributions and sizes via the spatial distribution of habitat. Boyce et al. (1994) thus believed that their predictions carried more power than traditional PVA models because their model required estimating fewer parameters. In future applications of RSF, both deterministic (e.g., habitat loss or restoration) and stochastic (e.g. wildfire, inbreeding potential) factors could also be included. In addition, dynamics involving alternative schedules of proposed vegetation changes over space and through time can be incorporated. Another important feature of RSFs is that outcomes are identifiable over much shorter timeframes (decades) than with typical PVA models, which should facilitate testing of management alternatives. Further, estimates of many of the physical environmental factors that might bear on demography are available from land management agencies (e.g., maps of soil classes, topography, potential evapotranspiration, net primary productivity, precipitation, etc.), or can be derived from other map variables.

In applying RSF, one must heed the admonitions of Rexstad et al. (1988) regarding multivariate analyses. Also, habitat-based PVA constructed from current data and conditions may not perform well under future conditions, with different spatial combinations of vegetative habitat (Conroy et al. 1995). In building a spatial correlation model using data exploratory processes such as RSF, one must recognize that inferences about persistence depend crucially upon model assumptions. For example, RSF models assume that the variables that actually influence population dynamics via habitat selection are correctly identified and/or modeled, are sampled without significant error, and the errors are not propagated through time. Also, habitat-based PVA calls for hierarchical, biophysical habitat classification systems that are appropriately defined and reliable maps that match the scales of resolution of habitat selection by the species being modeled.

## 5.9 Extensions To Traditional Wildlife–Habitat Models

Irwin (1994, in press) suggested that evaluating the influences of the physical environment on animal–vegetation relations might allow development of PVA models that better predict demographic responses to habitat changes. Vertebrates, for example, are more likely to respond to changes in habitat carrying capacity, a function of both abiotic and biotic factors, than to direct manipulation of their numbers (Caughley 1977: 200). Geologic materials and processes, along with weather, provide the dominant abiotic environment, which strongly influences wildlife–habitat interactions, and thereby, a population's viability.

Precipitation, productivity, and soils interact to create enormous differences in rates of nutrient and energy flow through ecosystems, as well as the relative number of functional redundancies within them. And physical factors influence locations, frequencies, and patterns of disturbances, which do not necessarily have to be modeled as stochastic phenomena (Huston

1995). The difference between habitat supporting a source or sink population may simply be availability of an essential nutrient in soils, or subtle differences in rainfall.

Recent research has increasingly demonstrated that animal distribution and abundance are influenced strongly by abiotic factors. In fact, Risser (1994) suggested that ecosystem processes are often controlled primarily by abiotic rather than biotic interactions. For example, Schueck and Marzluff (1995) indicated that variation in weather accounted for more variance in raptor abundance than disturbance by military activities. They noted that failure to account for abiotic factors can result in dubious conclusions, because the same data can produce opposing management ramifications depending on the identity of environmental factors used as covariates. Desert tortoises are associated with specific metamorphic rock types in the Mohave Desert. Red Hills salamanders are found only in areas with particular soils and geologic formations. In combination with structural diversity, elevation and moisture were major factors accounting for habitat selection by birds and butterfly in Glacier National Park, Montana (Debinski and Brussard 1994). And Manley et al. (1992) combined vegetation classes with elevation and aspect as separate layers from a Landsat TM satellite image to construct a seasonal preference map for grizzly bears.

Surprisingly few investigators have included measures of soil or topographic influences in their wildlife–habitat relationship models. Hill (1972) described soil fertility factors that influenced populations of cottontail rabbits (*Sylvilagus floridanus*). Irwin and Cook (1985) found that interactions among vegetation, weather, and topography influenced demography of pronghorn antelope (*Antilocapra americana*) populations. Rice et al. (1993) incorporated precipitation and soil factors as well as attributes of vegetation to predict bobwhite quail (*Colinus virginianus*) population responses to management. Weiss et al. (1988) found that topographic diversity is a prime indicator for habitat quality for the butterfly, *Euphydrayas editha*, and suggested that areas of high local topographic diversity were important for long-term population presistence. And Ormerod et al. (1991) found that clutch size and body mass of the dipper (*Cinclus cinclus*) were inversely related to stream acidity, which was influenced by soil composition.

Johnson (1990) noted that weighting or ordinal models may be a useful means of incorporating physical environmental covariates in GIS overlay operations. For example, GIS layers of soils and topography can be intersected with vegetation cover and telemetry locations, or some index to fitness for each patch. Scores or indices reflecting the overall characteristics of each polygon are derived. By selectively weighting habitat properties and describing spatial variables such as vegetation patch size, shape, and arrangement, both the quality and quantity of the habitat can be estimated. We expect rapid advances in habitat-based PVAs that link GIS tracking of physical and biological environmental variables that interact to control a species' demography and spatial distribution.

## 5.10 Tradeoffs Between Habitat and Population Monitoring

If we measure a surrogate instead of the variable of interest directly (i.e, population growth rate), inference to status and trend of the population becomes indirect. The step back from direct inference inescapably introduces additional uncertainty to estimates of "true" status and trend (i.e., viability) of the population. If the uncertainty associated with surrogate measures is too large, no substitute exists for direct measurement of the population. The results of past attempts to predict population status from habitat attributes suggests that surrogate measures should be viewed with skepticism (e.g., papers in Verner et al. 1986). There are several reasons why inferences from habitat state to population status are uncertain.

The realized habitat distribution of a species is affected by more than just habitat quality. Actual habitat use is affected by the species' density, the density and competitive interactions with other species, the abundance and distribution of biotic resources not directly related to vegetation, history of the population, and abiotic drivers such as climatic variation (Block and Brennan 1993). As a consequence, a species' pattern of habitat use is dynamic in both space and time, which influences the feasibility of using habitat as a surrogate indicator of population status. To some degree, habitat attributes, particularly those components assessed by vegetation measures, will not be sufficient predictors of population status. Some component of population variation (i.e., variation in distribution or abundance) will be unrelated to the current value of habitat attributes.

Despite these caveats, a habitat-based PVA is often the best choice. Obviously, if habitat change is believed to be the key driver of population dynamics, then habitat variation should be the basis of the assessment. Further, it is often habitat attributes — patch size, shape, distribution, and number — that are the basic output of a management plan or conservation strategy. Other factors that influence population dynamics — for example, weather, soils, and topography — are often not amenable to management intervention.

## 6 ADAPTIVE RESOURCE MANAGEMENT

Ecologists do not have a good track record in their attempts to use models to predict the behavior of ecological systems (Eberhardt and Thomas 1991). But when combined with an iterative process of model improvement and validation, a PVA model can provide a progressively more robust understanding of the dynamics of a species and its habitat. Essentially, the PVA model forms an hypothesis of how the system works. This hypothesis is best evaluated with an "experimental" management intervention followed by monitoring to see whether the model predictions were correct. The PVA model should then be modified and the entire process reiterated. Such an iterative process, often called adaptive management, is critical for constructing models that can be reliable tools for management. Adaptive management entails the infusion of the scientific method into natural resource management (Walters 1986).

How should PVA be incorporated into adaptive management protocols? *Active adaptive management* proposes application of different management tactics in time and space, essentially as experiments, to develop a better understanding of the behavior of the system (Walters and Holling 1990). For endangered species applications, it may be possible to implement various management strategies in spatially separated subpopulations, and to then compare the population sizes or demographic rates obtained with those predicted by the PVA model. Active management must be part of such a program, and may encompass a variety of activities such as habitat manipulation, predator or disease control, manipulation of potential competitors, winter provisioning of food, transplanting individuals from other subpopulations to sustain genetic variation, and supplementation of population with releases of captive stock. Monitoring of the genetic and population consequences of such manipulations then provides data to validate and/or refine the PVA model.

Management of grizzly bears (*Ursus arctos*) in the greater Yellowstone ecosystem has proceeded according to an adaptive management protocol. The very first PVA which noted high sensitivity of population growth rate to adult survival, suggested the importance of minimizing adult mortality factors (Shaffer 1978, 1983; Knight and Eberhardt 1985). Aggressive programs to eliminate bear/human conflicts focused on areas where repeated bear mortalities had been documented (Knight et al. 1988). After the recovery program had been implemented and additional data were obtained, Shaffer's original model was updated (Suchy et al. 1985). The program to reduce bear mortality has apparently been so successful (Eberhardt et al. 1994) that federal officials recently have entertained the possibility of delisting grizzly bears and allowing management to revert to respective state and federal agencies (Boyce 1991a).

Another adaptive management program driven by a PVA is being used for endangered populations of *Banksia cumeata* in Western Australia. When their PVA modeling showed seedling survival was a crucial stage in the life history of *Banksia cumeata*, Burgman and Lamont (1992) recommended watering seedlings in several subpopulations to enhance seedling survival. Such programs require careful monitoring, however, because watering or other forms of "enrichment" can have community-level effects that could be counter productive (Rosenzweig 1971). For example, competing species or herbivores might respond more vigorously to watering than the target species, and *B. cumeata* appears sensitive to fungal growth in moist soil (Burgmann, pers. commun.).

For the northern spotted owl, the Interagency Scientific Committee (ISC) explicitly acknowledged the importance of adaptive management approaches for evaluating and updating their conservation strategy, posed as an Appendix in the ISC report (Thomas et al. 1990). Adaptive management would require modeling to predict the consequence of forest management alternatives, followed by implementation of various experimental timber harvest programs and associated landscape manipulations, and finally a monitoring program that would document the consequences for spotted owl populations. Thus far, no such programs have been implemented because litigation has interfered with the ability of management agencies to develop timber harvests, even on an experimental basis.

For several years, the Captive Breeding Specialists Group (CBSG) of the IUCN has been organizing "Population and Habitat Viability Analysis" (PHVA) workshops for various threatened and endangered species. Some of these have been successful at bringing together available data on a species, identifying model structures, and encouraging agency coordination for conservation programs. We caution against reliance on MVP estimates that emerge from such exercises, but if they help to provide structure that will encourage adaptive management approaches, they perform a valuable function.

Federal and state agency restrictions may severely limit our ability to actually perform adaptive management with threatened and endangered species. Any programs that might pose a risk to a threatened or endangered species will meet resistance from agencies charged with protecting the species. Yet, creative manipulations may be allowed if they could be viewed only as enhancing conditions for the species of concern.

Models for PVA probably will face many challenges because of biological uncertainty. Given errors in population parameter estimates and our inability to generate robust population projections, any PVA will be open to question even though the PVA may constitute our best statement of the expected behavior of a population. Such uncertainty was used in court to challenge the proposed adoption of the Interagency Scientific Committee's conservation strategy for the northern spotted owl by the USDA Forest Service. Although biological uncertainty may often frustrate attempts to apply adaptive management (Lee and Lawrence 1986), only through adaptive management can we hope to resolve some of the uncertainty associated with PVA. It is the best we can do, and we know of no better way to gain "reliable knowledge" about managing our natural resources (Romesburg 1981).

## 7 PROMINENT RESEARCH QUESTIONS

Population viability analysis is an evolving analytical process whose effectiveness is limited by many scientific uncertainties. These uncertainties arise from an incomplete understanding of the dynamics of wild populations and the resilience of species to natural and human-induced stresses. There are many ways in which the research community can contribute to improving the PVA process. Below we present a representative (not exhaustive) listing of pressing questions that arise from a quantitative analysis of the factors that put species at risk of extinction.

- How should density-dependence in birth and survival rates be included in viability analyses? It is extremely difficult to measure density dependence in the demographic rates of wild populations. Given the problems of estimation, density-dependence is either omitted from analyses, or its magnitude is based on opinion. There are dangers to each of these approaches. If the population does exhibit density dependent behavior, but this process is not included in the model, then extinction likelihoods are generally overestimated (Ginzburg et al. 1982, 1990). In contrast, invoking density dependence when it is truly absent, or overestimating its effects or failing to incorporate time lags in its expression, will underestimate extinction likelihoods.
- How does one begin to model the multiple threats to persistence faced by many species and their possible synergistic effects? The effects of different threats may not be additive, but multiplicative in their impacts. Further, interactions among population components (e.g., vital rates and population structure), in response to disturbance, may be complex.
- How can we identify threshold (nonlinear) population responses to environmental change before irreversible loss? Processes that generate threshold population behavior include Allee effects in sexual species, uncertainties associated with the search for suitable habitat in fragmented landscapes, and critical population densities required for the social facilitation of breeding behavior. However, these factors are often identified only after a threshold has been passed.
- How do we address the transient behavior of recovering populations? Time lags in the population response of sensitive species arising from slow renewal rates of their habitats, keep them vulnerable to extinction for long periods of time.
- What are the linkages between changes in the genetic structure of populations and their effects on demography ? Given how little we know of this interaction, we have largely defaulted to managing sensitive species on the basis of demographic principles alone.
- How can we move from a single- to a multi-species approach to conservation planning? Little is known about how to conduct multi-species PVAs because we have limited understanding of the dynamics of interspecific interactions (i.e., competition, predator-prey, host-parasite, etc.). A preliminary step towards the analysis of multi-species' sensitivities to habitat loss and fragmentation has been proposed (Noon et al. 1997) but much remains to be done.
- How can we assess the tradeoffs between the amount of habitat and spatial distribution of habitat across the landscape? Often conservation problems are framed in terms of a maximum land area society is willing to devote to species preservation. Given a fixed area, is there an optimal geometry that maximizes the persistence likelihood of species?

## 8 SYNTHESIS AND RECOMMENDATIONS

### 8.1 Identifying Species At Risk

A population viability analysis has at least two distinct components. First, we need a system to assess and rank the degree of extinction risk faced by species within a given management unit. Given this ranking, attention is then directed towards those species at greatest risk. Second, for the species at risk, PVA is an appropriate tool to specifically assess those factors that put them at risk of extinction, and to provide insights into how to mitigate against those factors. So far, we have said little about the first component, but it is a useful adjunct to the PVA process.

Based on the theory of extinction processes for single populations (Section 2.10), and meaningful time frames for conservation action, Mace and Lande (1991) proposed a useful system for categorizing species according to the degree of extinction threat. The categories of threat they identified provide a quantitative assessment of the likelihood of extinction, should current environmental conditions prevail. The three probability-based categories of threat they proposed are:

- *Critical* — 50% probability of extinction within 5 years or 2 generations, whichever is longer.
- *Endangered* — 20% probability of extinction within 20 years or 10 generations, whichever is longer.
- *Vulnerable* — 10% probability of extinction within 100 years.

Information included in the assessment criteria includes estimates of population size, direction of recent trends in population size, number of known subpopulations, exposure to catastrophic events, and trends in habitat. Mace (1993) added two additional criteria to the list: size of the species' geographic range, and a composite statistic based on the number of years with no recruitment, bias in the sex ratio, and population size. In addition to ranking species by risk, the Mace and Lande (1991) system can be used to identify key population data for field workers to collect to better assess extinction risks. We highly recommend use of this system as an initial filter to identify species for which population viability is a concern. (A real world application of this ranking process is found in Mace (1993)). Additional discussion of ways to identify species at risk is found in Mace and Collar (1995) and Holthausen et al. (this volume).

## 8.2 Factors To Consider

Previous sections of this chapter have illustrated that the short-term viability of a species is influenced by many factors, including its population size, sex ratio, age structure, reproductive rate, and geographic distribution. Given the uncertainty with which these factors are known for any population, we have emphasized that a management plan to ensure a species' viability is best viewed as a hypothesis to be tested. The usefulness of the hypothesis is based on the degree to which it can be tested and potentially falsified. Towards this goal, it is useful to list the basic attributes that constitute a management plan for a species in danger of extinction. In almost all cases, a conservation strategy for an imperiled species will take the form of a map with explicit spatial properties (see example in Murphy and Noon 1992). These basic population statistics and map attributes include:

- the number of local populations (habitat patches)
- the size of the local populations (habitat patches)
- the geographic distribution of local populations (habitat patches)
- the degree of spatial correlation in environmental variation between local populations (habitat patches)
- movement rates between local populations (habitat patches)
- the degree of genetic interchange between local populations
- the degree of genetic differentiation among local populations.

Extinction (local or regional) results when environmental change exceeds the capacity of a species to adapt. The adaptation limits of a species can be exceeded either by the magnitude, rate, or duration of environmental change. Thus, the attributes listed above must be considered both in a static and dynamic context. For example, it is not enough to set aside a reserve system that will only provide suitable habitat at some distant time if it is unlikely that the species will survive the transition period.

## 8.3 PVA In the Context of Limited Data

In this concluding section we propose a process for PVA, implicitly based on the principles of demography and population genetics discussed in Sections 2–6, but which lacks the details of an explicit mathematical model. We emphasize that these are recommendations, not absolute rules. This process should prove useful when a detailed PVA is not possible — that is, when the ecology and life history of the species are poorly understood. In this summary we urge you to keep in mind four key factors that affect the expected time to extinction of any population: (1) maximum population size (patch-specific carrying capacity); (2) mean population growth rates; (3) the variance in mean growth rate at each population size; and (4) the relationship between growth rate and population size.

In the absence of definitive life history data, the PVA process must be based on the best available information. This information can often be gained through a series of workshops involving those who have studied the species and are aware of the factors putting it at risk of extinction. For example, a first workshop would include biologists with the goal of coming up with parameter estimates for survival and fecundity, dispersal, and some discussion of possible density-dependent relationships. Even if the parameter values are "guestimates" or "borrowed" from related species, it is important that they be based on the opinions of the biologists. These estimates, plus a general knowledge of life history structure, would allow the development of a simple dynamics model and preliminary analyses. The focus of these analyses would be to identify the sensitivity of model output (e.g., probability of extinction) to assumptions made about population parameters. These sensitivities will be useful to help focus future data summaries and prioritize research efforts.

A second workshop should focus on the factors putting the species at risk of extinction. For example, if habitat decline is the key driver of population change, then the goal of the workshop would be a summary of suspected habitat relationships. Again, the best source of information may be from the field biologists familiar with the species. If habitat is the focus, then it would be important to consider geographic variation in habitat relations, the possibility of describing habitat relations in terms of attributes that can be estimated from remotely sensed data, and the availability of remotely sensed data coverages.

To be most relevant to managers, the following should be organizing themes of the workshops:

*Factors putting species at risk*: The first consideration in a PVA is to identify the deterministic factors that are putting species at risk of extinction. Unless these are addressed and mitigated there is little to be gained from ameliorating more proximate factors of risk. A comprehensive summary of risk factors is found in Marcot and Murphy (1996).

*Population size and stochastic variation*: Perhaps the single most important factor that puts species at risk of extinction is small population size. Because what constitutes a "small population" is a function of the magnitude of environmental variation and a species life history (i.e., mating system, demography and habitat relations), there is no magic number. However, a rough guideline is that a local population should not have an effective population size ($N_e$) < 50. Assuming an $N_e/N$ ratio of 0.2 (Mace and Lande 1991), this corresponds to a census population of 250 individuals.

*Geographic scale*: For most species, three primary spatial scales affect the dynamics of individuals, populations, and genes. These are the territory (or home range), the local population, and the metapopulation. At each of these scales there are both demographic and genetic factors to consider. The most important requirements for viability at these scales are:

- Home range — Sufficient area and habitat for high levels of individual fitness (survival and successful reproduction).
- Local population — Expected birth and survival rates that provide for a mean growth rate ($\lambda$) ≥ 1.0 given background environmental variation; sufficient interchange with other populations to avoid inbreeding depression and deleterious genetic drift.
- Metapopulation — Sufficient spatial connectivity among populations for demographic rescue and gene exchange; sufficient number of local populations to achieve some degree of spatial independence to partially uncoupled local population growth rates and environmental effects (risk spreading).

*Habitat* — Most existing conservation strategies are largely plans for the preservation of habitat. In this context, the three spatial scales discussed above for population processes are also appropriate scales for habitat planning. For example, multiple habitat reserves are preferable to a single large reserve, provided there is some degree of spatial independence and connectivity (Goodman 1987). Thus, conservation strategies are most simply portrayed as maps showing the size, number, shape, and distribution of habitat patches (Murphy and Noon 1992, Noon and McKelvey 1996).

Given that habitat loss and fragmentation is currently the factor most responsible for putting species at risk, this habitat focus is appropriate. Obviously this approach has limits — simply preserving habitats may not be sufficient if populations are threatened by other factors such as predation or competition. Also, even if habitat loss is the key threat, persistence is also unlikely if populations remain small and subdivided while habitat is recovering from disturbance. Whatever the deterministic driver of population decline, two factors must be explicitly considered for a PVA to be reliable: the relationship between the driver of population decline and demographic and genetic variation; and an explicit consideration of these relations at the individual, local population and metapopulation scales. Effective conservation is more likely to result when these factors are evaluated in the context of simple mathematical models.

## REFERENCES

Ackakaya, H.R., and J.L. Atwood. 1997. A habitat-based metapopulation model of the California gnatcatcher. *Conservation Biology* 11: 422–434.

Allee, W.C., A.E. Emerson, O. Park, T. Park, and K.P. Schmidt. 1949. *Principles of Animal Ecology*. Saunders, Philadelphia, PA.

Allendorf, F.W. 1986. Genetic drift and the loss of alleles versus heterozygosity. *Zoo Biology* 5: 181–190.

Allendorf, F.W., and R.F. Leary. 1986. Heterozygosity and fitness in natural populations of animals. In: Soule, M.E. (ed.), *Conservation Biology*, pp. 57–76. Sinauer, Sunderland, MA.

Anderson, M.C., and D. Mahoto. 1995. Demographic models and reserve designs for the California spotted owl. *Ecological Applications* 5: 639–647.

Armbruster, P., and R. Lande. 1993. A population viability analysis for African elephant: how big should reserves be? *Conservation Biology* 7: 602–10.

Atchley, W.R., and S. Newman. 1989. A quantitative genetic perspective on mammalian development. *American Naturalist* 134: 486–512.

Avise, J.C., and J.L. Hamrick. 1996. *Conservation Genetics*. Chapman and Hall, New York, NY.

Beissinger, S.R. 1995. Modeling extinction in periodic environments: Everglades water levels and snail kite population viability. *Ecological Applications* 5: 618–31.

Bell, S.S., E.D. McCoy, and H.R. Mushinsky (eds.). 1991. *Habitat Structure: The Physical Arrangement of Objects in Space*. Chapman and Hall, New York.

Block, W.M,. and L.A. Brennan. 1993. The habitat concept in ornithology: theory and applications. *Current Ornithology* 11: 35–91.

Boeklen, W.J. 1986. Optimal design of nature reserves: consequences of genetic drift. *Biological Conservation* 38: 323–328.

Boyce, M.S. 1988. Evolution of life histories: theory and pattern from mammals. In: Boyce, M.S. (ed.), *Evolution of Life Histories of Mammals*, pp. 3–30. Yale University Press, New Haven, CT.

Boyce, M.S. 1991a. Natural regulation or the control of nature? In: Keiter, R.B. and M.S. Boyce (eds.), *The Greater Yellowstone Ecosystem: Redefining America's Wilderness Heritage*, pp. 183–208. Yale University Press, New Haven, CT.

Boyce, M.S. 1991b. Simulation modelling and mathematics in wildlife management and conservation. In: Maruyama, N. (ed.), *Wildlife Conservation: Present Trends and Perspectives for the 21st Century*, pp. 116–119. Japan Wildlife Research Center, Tokyo.

Boyce, M.S. 1992. Population viability analysis. *Annual Reviews of Ecology and Systematics* 23: 481–506.

Boyce, M.S., J.S. Meyer, and L.L. Irwin. 1994. Habitat-based PVA for the Northern Spotted Owl, pp. 63–85, In: Fletcher, D.J., and B.F.J. Manly (eds.), *Statistics for Ecology and Environmental Monitoring*. Otago Conf. Series 2, University of Otago, Dunedin, New Zealand.

Brawn, J.D., and S.K. Robinson. 1996. Source-sink population dynamics may complicate the interpretation of long-term census data. *Ecology* 77: 3–12.

Brown, J.H., D.W. Mehlman, and G.C. Stevens. 1995. Spatial variation in abundance. *Ecology* 76: 2028–2043.

Burgman, M.A., and B.B. Lamont. 1992. A stochastic model for the viability of *Banksia cumeata* populations: Environmental, demographic and genetic effect. *Journal of Applied Ecology* 29: 719–727.

Burgman, M.A., S. Ferson, and H.R. Akcakaya. 1993. *Risk Assessment in Conservation Biology*. Chapman and Hall, New York.

Burnham, K.P., D.R. Anderson and G.C. White. 1994. Estimation of vital rates of the northern spotted owl. Appendix J. In: *Final Supplemental Environmental Impact Statement on Management of Habitat for Late-successional and Old-Growth Forest Related Species Within the Range of the Northern Spotted Owl, Vol. II*, pp. 1–26. U.S. Government Printing Office, Washington, D.C.

Burnham, K.P., D.R. Anderson, and G.C. White. 1996. Meta-analysis of vital rates of the Northern Spotted Owl. *Studies in Avian Biology* 17: 92–101.

Carroll, J.E., and R.H. Lamberson. 1993. The owl's odyssey: A continuous model for the dispersal of territorial species. *Siam Journal Applied Mathematics* 53: 205–218.

Caswell, H. 1989. *Matrix Population Models: Construction, Analysis, and Interpretation*. Sinauer Associates, Inc., New York.

Caughley, G. 1977. *Analysis of Vertebrate Populations*. John Wiley and Sons, New York.

Caughley, G. 1994. Directions in conservation biology. *Journal of Animal Ecology* 63: 215–244.

Clark, C.W. 1973. The economics of overexploitation. *Science* 181: 630–634.

Clarke, C.H.D. 1954. The bob-white quail in Ontario. Technical Bulletin, Fish & Wildlife Service, No. 2. Ontario Department of Lands and Forests, Ontario Ministry of Natural Resources, Toronto, Ontario, Canada.

Conroy, M.J., Y. Cohen, F.C. James, et al. 1995. Parameter estimation, reliability, and model improvement for spatially explicit models of animal populations. *Ecological Applications* 5: 17–19.

Crouse, D., L. Crowder, and H. Caswell. 1987. A stage-based population model for loggerhead sea turtles and implications for conseration. *Ecology* 68: 1412–1423.

Currie, D.J. 1991. Energy and large-scale patterns of animal- and plant-species richness. *American Naturalist* 137: 27–49.

Debinski, D.M., and P.F. Brussard. 1994. Using biodiversity data to assess species-habitat relationships in Glacier National Park, Montana. *Ecological Applications* 4: 833–843.

deKroon, H., A. Plaisier, J. Van Groenendael, and H. Caswell. 1986. Elasticity: the relative contribution of demographic parameters to population growth rate. *Ecology* 67: 1427–1431.

Den Boer, P.J. 1981. On the survival of populations in a heterogeneous and variable environment. *Oecologia* 50: 39–53.

Doak, D., P. Karieva, and B. Klepetka. 1994. Modeling population viability for the desert tortoise in the western Mohave Desert. *Ecological Applications* 4: 446–60.

Doak, D.F. 1995. Source-sink models and the problem of habitat degradation: general models and applications to the Yellowstone grizzly. *Conservation Biology* 9: 1370–1379.

Donovan, T., R.H. Lamberson, A. Kimber, F.R. Thompson III, and J. Faaborg. 1995. Modeling the effects of habitat fragmentation on source and sink demography of neotropical migrant birds. *Conservation Biology* 9: 1396–1407.

Dunning, J.B., Jr., D.J. Stewart, B.J. Danielson, and others. 1995. Spatially explicit population models: current forms and future uses. *Ecological Applications* 5: 3–11.

Eberhardt, L.L., and J.M. Thomas. 1991. Designing environmental field studies. *Ecological Monographs* 61: 53–73.

Eberhardt, L.L., B.M. Blanchard, and R.R. Knight. 1994. Population trend of the Yellowstone grizzly bear as estimated from reproductive and survival rates. *Canadian Journal Zoology* 72: 360–363.

Ehrlich, P.R. 1995. The scale of human enterprise and biodiversity loss. pp 214–226 In: Lawton, J.H., and R.M.

May (eds.), *Extinction Rates*. Oxford University Press, London, U.K.

Ewens, W.J. 1990. The minimum viable population size as a genetic and a demographic concept. In: Adams, J., D.A. Lam, A.I. Hermalin, and P.E. Smouse (eds.), *Convergent Issues in Genetics and Demography*, pp. 307–316. Oxford University Press, Oxford, U.K.

Fahrig, L. 1992. Relative importance of spatial and temporal scales in a patchy environmnet. *Theoretical Population Biology* 41: 300–314.

Fahrig, L. 1993. When should population models be spatially explicit? *Bulletin of the Ecological Society of America* 74: 230.

Falconer, D.S. 1981. *Quantitative Genetics*. Longman, London, U.K.

Flather, C.H., S.J. Brady, and D.B. Inkley. 1992. Regional habitat appraisals of wildlife communities: A landscape evaluation of a resource planning model using avian distribution data. *Landscape Ecology* 7: 137–147.

Foley, P. 1994. Predicting extinction times from environmental stochasticity and carrying capacity. *Conservation Biology* 8: 124–137.

Forest Ecosystem Management Assessment Team (FEMAT). 1993. Forest ecosystem mangement: An ecological, economic, and social assessment. Portland, OR: USDA, USDI, and others.

Frankel, O.H., and M.E. Soulé. 1981. *Conservation and Evolution*. Cambridge University. Press, Cambridge, U.K. 327 pp.

Futuyma, D. 1983. Interspecific interactions and the maintenance of genetic diversity. In: Schonewald-Cox, C.M., S.M. Chambers, B. MacBryde, and L. Thomas (eds.), *Genetics and Conservation: a Reference for Managing Wild Animal and Plant Populations*, pp. 364–373. Benjamin/ Cummings Publishing Company, Menlo Park, CA.

Gabriel, W., and R. Burger. 1992. Survival of small populations under demographic stochasticity. *Theoretical Population Biology* 41: 44–71.

Gaston, K.J., and J.H. Lawton. 1987. A test of statistical techniques for detecting density dependence in sequential censuses of animal populations. *Oecologia* 74: 404–410.

Gilpin, M.E. 1992. Demographic stochasticity: A Markovian approach. *Journal of Theoretical Biology* 154: 1–8.

Gilpin, M.E. 1996. Metapopulations and wildlife conservation: approaches to modeling spatial structure. In: McCullough, D.R. (ed.), *Metapopulations and Wildlife Conservation*, pp. 11–28. Island Press, California.

Gilpin, M.E., and M.E. Soule. 1986. Minimum viable populations: processes of species extinction. In: Soule, M.E. (ed.), *Conservation Biology: The Science of Scarcity and Diversity*, pp. 19–34. Sinauer.

Gilpin, M.E., and I. Hanski. 1991. *Metapopulation Dynamics*. Academic Press, London, U.K. 336pp.

Ginsberg, J.R., and D.W. Macdonald. 1990. Foxes, wolves, jackals, and dogs: an action plan for the conservation of canids. IUCN/SSC Canid Specialist Group, Morges, Switzerland.

Ginzburg, L.R., L.B. Slobodkin, and K. Johnson. 1982. Quasiextinction probabilities as a measure of impact on population growth. *Risk Analysis* 21: 171–181.

Ginzburg, L.R., S. Ferson, and H.R. Akcakaya. 1990. Reconstructibility of density-dependence and conservative estimates of extinction risks. *Conservation Biology* 4: 63–70.

Goodman, D. 1987. Consideration of stochastic demography in the design and management of biological reserves. *Natural Resources Modeling* 1: 205–234.

Gotelli, N.J. 1991. Metapopulation models: the rescue effect, the propagule rain, and the core satellite hypothesis. *American Naturalist* 138: 768–776.

Griffith, B., J.M. Scott, J. Carpenter, and C. Reed. 1989. Translocations as a species conservation tool: status and strategy. *Science* 245: 477–480.

Gutierrez, R.J., and S. Harrison. 1996. Applying metapopulation theory to spotted owl management: A history and critiques. pp. 167–185. In: McCullough, D.R. *Metapopulations and Wildlife Management*. Island Press, CA.

Haig, S.M., J.R. Beltoff, and D.H. Allen. 1993. Population viability analysis for a small population of red-cockaded woodpeckers and an evaluation of enhancement strategies. *Conservation Biology* 7: 289–301.

Hamrick, J.L., and M.J.W. Godt. 1996. Conservation genetics of endemic plant species. In: Avise, J.C. and J.L. Hamrick (eds.), *Conservation Genetics*, pp. 281–304. Chapman and Hall, New York, NY.

Hanski, I. 1994. A practical model of metapopulation dynamics. *Journal of Animal Ecology* 63: 151–162.

Harris, R.B., and F.W. Allendorf. 1989. Genetically effective population size and large mammals: an assessment of estimators. *Conservation Biology* 3: 181–191.

Harrison, S. 1994. Metapopulations and conservation. In: Edwards, P.J., N.R. Webb, and R.M. May (eds.), *Large-scale Ecology and Conservation Biology*, pp.111–128. Blackwell, Oxford.

Harrison, S., and J.F. Quinn. 1989. Correlated environments and the persistence of metapopulations. *Oikos* 56: 293–298.

Hartl, D.L., and A.G. Clark. 1989. *Principles of Population Genetics*. 2nd ed. Sinauer, Sunderland, MA. 682 pp.

Hassell, M.P. 1986. Detecting density-dependence. *Trends in Ecology and Evolution* 1: 90–93.

Hedrick, P.W. 1987. Genetic polymorphism in heterogeneous environments: a decade later. *Annual Review of Ecology Systematics* 17: 535–566.

Hedrick, P.W., R.C. Lacey, F.W. Allendorf, and M.E. Soulé. 1996. Directions in conservation biology: comments on Caughley. *Conservation Biology* 10: 1312–1320.

Hill, W.G. 1972. Effective size of population with overlapping generations. *Theoretical Population Biology* 3: 278–289.

Hill, W.G. 1979. A note on effective population size with overlapping generations. *Genetics* 92: 317–322.

Holsinger, K.E. 1996. Population biology for policy makers. *Bioscience* 510–520.

Holt, R.D., S.W. Pacala, T.W. Smith, and J. Liu. 1995. Linking contemporary vegetation models with spatially explicit animal population models. *Ecological Applications* 5: 20–27.

Holthausen, R.S., M.G. Raphael, K.S. McKelvey, and others. 1995. The contribution of federal and non-federal habitat to persistence of the northern spotted owl on the Olympic Peninsula, Washington. USDA, Forest Service General Technical Report PNW-GTR-352. Portland, OR. Hughes, S.P. 1995. Two tools for integrating geology into ecosystem studies. *Environmental Geology* 26: 246–251.

Huston, M.A. 1995. *Biological Diversity: The Coexistence of Species on Changing Landscapes*. Cambridge University Press, Cambridge, U.K.

Irwin, L.L. 1994. A process for improving wildlife habitat

models for assessing forest ecosystem health. *Journal of Sustainable Forestry* 2: 293–306.

Irwin, L.L. In press. Abiotic influences on bird habitat-population relationships. In: J. Marzluff and R. Sallabanks (eds.), *Research Needs for Avian Conservation*. Island Press, Covelo, CA.

Irwin, L.L., and J.G. Cooke. 1985. Determining appropriate variables for a habitat suitability model for pronghorns. *Wildlife Society Bulletin* 13: 434–440.

Irwin, L.L. and L.L. Hicks. 1995. Estimating carrying capacity for spotted owls. Chap. 10. In: *Habitat Conservation Plan for the Snoqualmie Pass Aarea*. Plum Creek Timber Co., Seattle, WA.

James, F.C. 1983. Environmental component of morphological differentiation in birds. *Science* 221: 184–186.

Jiménez, J.A., K.A. Hughes, G. Alaks, L. Graham, and R.C. Lacy. 1994. An experimental study of inbreeding depression in natural habitat. *Science* 266: 271–273.

Johnson, D.H. 1980. The comparison of use and availability measures for evaluating resource preference. *Ecology* 61: 65–71.

Johnson, L.B. 1990. Analyzing spatial and temporal phenomena using geographical information systems. *Landscape Ecology* 4: 31–43.

Kareiva, P. 1994. Space: The final frontier for ecological theory. *Ecology* 75: 1.

Keller, L.F., P. Arcese, J.N.M. Smith, W.M. Hochachka, and S.C. Stearns. 1994. Selection against inbred song sparrows during a natural population bottleneck. *Nature* 372: 356–357.

Knight, R.R., and L.L. Eberhardt. 1985. Population dynamics of Yellowstone grizzly bears. *Ecology* 66: 323–334.

Knight, R.R., B.M. Blanchard, and L.L. Eberhardt. 1988. Mortality patterns and population sinks for Yellowstone grizzly bears, 1974–1985. *Wildlife Society Bulletin* 16: 121–125.

Lamberson, R.H., and J. Carroll. 1993. Thresholds for persistence in territorial species. pp 55–62. In: Varbieri, I, E. Grassi, G. Pallotti, et al. (eds.), *Topics In Biomathematics*. World Scientific Publications, Singapore.

Lamberson, R.H., M.C. Liermann, and S.J. McKay. 1992a. A spatially explicit demographic model for the management of the Menzies' Wallflower. Technical Report, Humboldt State University. 43 pp.

Lamberson, R.H., R. McKelvey, B.R. Noon, and C. Voss. 1992b. A dynamic analysis of northern spotted owl viability in a fragmented forest landscape. *Conservation Biology* 6: 505–12.

Lamberson, R.H., B.R. Noon, C. Voss, and K. McKelvey. 1994. Reserve design for territorial species: the effects of patch size and spacing on the viability of the northern spotted owl. *Conservation Biology* 8: 185–195.

Lande, R. 1987. Extinction thresholds in demographic models of territorial populations. *American Naturalist* 130: 624–635.

Lande, R. 1988a. Demographic models of the northern spotted owl. *Oecologia* 75: 601–607.

Lande, R. 1988b. Genetics and demography in biological conservation. *Science* 241: 1455–1460.

Lande, R. 1993. Risks of population extinction from demographic and environmental stochasticity, and random catastrophes. *American Naturalist* 142: 911–927.

Lande, R. 1994. Risk of population extinction from fixation of new deleterious mutations. *Evolution* 48: 1460–1469.

Lande, R. 1995. Mutation and conservation. *Conservation Biology* 9: 782–791.

Lande, R., and G.F. Barrowclough. 1987. Effective population size, genetic variation, and their use in population management. In: Soule, M. (ed.), *Viable Populations for Conservation*, pp. 87–123. Cambridge University Press, Cambridge, U.K.

Lawton, J.H. 1995. Population dynamics principles. In: Lawton, J.H. and R.M. May (eds.), *Extinction Rates*, pp. 147–163. Oxford University Press, Oxford, UK.

Lee, K.N., and J. Lawrence. 1986. Adaptive management: learning from the Columbia River Basin Fish and Wildlife Program. *Environmental Law* 16: 431–460.

Lefkovitch, L.P. 1965. The study of population growth in organisms grouped by stages. *Biometrics* 21: 1–18.

Leslie, P.H. 1945. On the use of matrices in certain population dynamics. *Biometrika* 33: 183–212.

Leslie, P.H. 1948. Some further notes on the use of matrices in population mathematics. *Biometrika* 35: 213–245.

Levins, R. 1969. The effects of random variation on different types of population growth. *Proceedings National Academy of Science (USA)* 62: 1061–1065.

Levins, R. 1970. Extinction. In: Gerstenhaber, M. (ed.), *Some Mathematical Questions in Biology*, pp. 77–107. American Mathematical Society, Providence, RI.

Lindenmayer, D.B., and R.C. Lacy. 1995a. Metapopulation viability of Leadbeater's possum, *Gymnobelideus leadbeateri*, in fragmented old-growth forests. *Ecological Applications* 5: 164–82.

Lindenmayer, D.B., and R.C. Lacy. 1995b. Metapopulation viability of arboreal marsupials in fragmented old-growth forests: comparison among species. *Ecological Applications* 5: 183–99.

Ludwig, D., R. Hilborn, and C. Walters. 1993. Uncertainty, resource exploitation, and conservation: Lesson from history. *Ecological Applications* 3: 547–549.

Lynch, M. 1985. Spontaneous mutations for life-history characters in an obligate parthenogen. *Evolution* 39: 804–818.

Lynch, M. 1996. A quantitative-genetic perspective on conservation issues. In: Avise, J.C. and J.L. Hamrick (eds.), *Conservation Genetics*, pp. 471–501. Chapman and Hall, New York, NY.

MacArthur, R.H., and E.O. Wilson. 1967. *The Theory of Island Biogeography*. Princeton University Press.

Mace, G.M. 1993. An investigation into methods for categorizing the conservation status of species. pp 293–312. In: Edwards, P.J., R.M. May, and N.R. Web (eds.), *Large-scale Ecology and Conservation Biology*. Blackwell Scientific Publications, London.

Mace, G.M., and R. Lande. 1991. Assessing extinction threats: toward a reevaluation of the IUCN threatened species categories. *Conservation Biology* 5: 148–157.

Mace, G.M., and N.J. Collar. 1995. Extinction risk assessment for birds through quantitative criteria. *Ibis* 137: s240–s246.

Manley, T.L., K. Ake, and R.D. Mace. 1992. Mapping grizzly bear habitat using Landsat TM satellite imagery. Remote sensing and natural resource management. *Proceedings Forest Service Remote Sensing Conference* 4: 231–240.

Marcot, B.G., and D.D. Murphy. 1996. On population viability analysis and management. pp. 58–76. In: Szaro, R.C., and D.W. Johnston (eds.), *Biodiversity in Managed Landscapes: Theory and Practice*. Oxford University Press.

May, R.M. 1986. The cautionary tale of the black-footed ferret. *Nature* 320: 13–14.

McCarthy, M.A., D.C. Franklin, and M.A. Burgman. 1994. The importance of demographic uncertainty: an example from the helmeted honeyeater *Lichenostomus melanops cassidix*. *Biological Conservation* 67: 135–42.

McCullough, D.R. (ed.). 1996. *Metapopulations and Wildlife Conservation*. Island Press, Washington, D.C.

McDonald, D.S., and H. Caswell. 1993. Matrix methods for avian demography. *Current Ornithology* 10: 139–185.

McKelvey, K., B.R. Noon, and R.H. Lamberson. 1993. Conservation planning for species occupying fragmented landscapes: The case of the northern spotted owl. pp. 424–450. In: Karieva, P.M., J.G. Kingsolver, and R.B. Huey (eds.), *Biotic Interactions and Global Change*. Sinauer Associates, Sunderland, MA.

Menges, E.S. 1990. Population viability analysis for an endangered plant. *Conservation Biology* 4: 52–62.

Mertz, D.B. 1971. The mathematical demography of the California condor population. *American Naturalist* 105: 437–453.

Middleton, D.A.J., and R.M. Nisbet. 1997. Population persistence times: estimates, models, and mechanics. *Ecological Applications* 7: 107–117.

Mills, L.S., and P.E. Smouse. 1994. Demographic consequences of inbreeding in remnant populations. *American Naturalist* 144: 412–431.

Morrison, M.L., B.G. Marcot, and R.W. Mannan. 1992. *Wildlife-Habitat Relationships: Concepts and Applications*. University of Wisconsin Press, Madison, WI.

Murphy, D.D., and B.R. Noon. 1992. Integrating scientific methods with habitat conservation planning: Reserve design for the Northern Spotted Owl. *Ecological Applications* 2: 3–17.

Murphy, D.D., K.E. Freas, and S.B. Weiss. 1990. An environment-metapopulation approach to population viability analysis for a threatened invertebrate. *Conservation Biology* 4: 41–51.

Myrberget, S. 1988. Demography of an island population of willow ptarmigan in northern Norway. pp 379–422. In: Bergerud, A.T., and M.W. Gratson (eds.), *Adaptive Strategies and Population Ecology of Northern Grouse*. University of Minnesota Press, Minneapolis, MN.

Nei, M. 1987. *Molecular Evolutionary Genetics*. Columbia University Press, New York, NY. 512 pp.

Nichols, J.D., G.L. Hensler, and P.W. Sykes. 1980. Demography of the Everglade kite: implications for population management. *Ecological Modelling* 9: 215–232.

Noon, B.R. 1986. Summary: Biometric approaches to modeling — The researcher's viewpoint. pp. 197–201. In: Verner, J., M.L. Morrison, and C.J. Ralph (eds.), *Wildlife 2000: Modeling Habitat Relationships of Terrestrial Vertebrates*. University of Wisconsin Press, Madison, WI.

Noon, B.R., and C.M. Biles. 1990. Mathematical demography of spotted owls in the Pacific Northwest. *Journal of Wildlife Management* 54: 18–26.

Noon, B.R., and J.R. Sauer. 1992. Population models for Passserine birds: Structure, parameterization, and analysis. pp 441–464. In: McCullough, D.R., and R.H. Barrett (eds.), *Wildlife 2001: Populations*. Elsevier Applied Science, New York.

Noon, B.R., and K.S. McKelvey. 1996. A common framework for conservation planning: Linking individual and metapopulation models. pp. 139–166. In: D.R. McCullough (ed.), *Metapopulations and Wildlife Conservation*. Island Press.

Noon, B.R., K.S. McKelvey, and D.D. Murphy. 1997. Developing an analytical context for multispecies conservation planning. pp. 43–59 In: Pickett, S.T.A., R.S. Osterfield, M. Shachak, and G.E. Likens (eds.), *Ecological Heterogeneity: Implications for Biological Conservation*. Chapman and Hall, New York.

Noss, R.F., and B. Csuti. 1994. Fragmentation. pp. 237–264. In: Meffee, G.K., and C.R. Carroll (eds.), *Principles of Conservation Biology*. Sinauer Associates, MA.

Nunney, L., and K.A. Campbell. 1993. Assessing minimum viable population size: demography meets population genetics. *Trends in Ecology and Evolution* 8: 234–239.

Nunney, L., and D.R. Elam. 1994. Estimating the effective population size of conserved populations. *Conservation Biology* 8: 175–184.

O'Neill, R.V., D.L. DeAngelis, J.B. Waide, and T.F.H. Allen. 1986. A Hierarchical concept of ecosystems. Princeton University Press, Princeton, N.J. 253 pp.

Ormerod, S.J., J.O. O'Halloran, S.D. Gribbon, and S.J. Tyler. 1991. The ecology of dippers *Cinclus cinclus* in relation to stream acidity in upland Wales: breeding performance, calcium physiology, and nestling growth. *Journal of Applied Ecology* 28: 419–433.

Packer, C. 1979. Inter-troop transfer and inbreeding avoidance in *Papio anubis*. *Animal Behaviour* 27: 1–36.

Pollard, E., K.H. Lakhani, and P. Rothery. 1987. The detection of density dependence from a series of annual censuses. *Ecology* 68: 2046–2055.

Possingham, H.P., D.B. Lindenmayer, T.W. Norton, and I. Davies. 1994. Metapopulation viability analysis of the greater glider *Petauroides volans* in a wood production area. *Biological Conservation* 70: 227–36.

Preston, F.W. 1948. The commonness and rarity of species. *Ecology* 29: 254–283.

Price, M.V., and P.A. Kelly. 1994. An age-structured demographic model for the endangered Stephens' kangaroo rat. *Conservation Biology* 8: 810–21.

Pulliam, H.R. 1988. Sources, sinks, and population regulation. *American Naturalist* 132: 652–661.

Pulliam, H.R., J.B. Dunning, Jr., and J. Liu. 1992. Population dynamics in complex landscapes: a case study. *Ecological Applications* 2: 165–177.

Pulliam, H.R., J. Liu, J.B. Dunning, Jr., D.J. Stewart, and T.D. Bishop. 1994. Modeling animal populations in changing landscapes. *Ibis* 137: S120–S126.

Ralls, K., P.H. Harvey, and A.M. Lyles. 1986. Inbreeding in natural populations of birds and mammals. In: Soule, M. (ed.), *Conservation Biology*, pp. 35–56. Sinauer, Sunderland, MA.

Ralls, K., J. Ballou, and A.R. Templeton. 1988. Estimates of lethal equivalents and the cost of inbreeding in mammals. *Conservation Biology* 2: 185–193.

Reed, J.M., P.D. Doerr, and J.R. Walters. 1988. Minimum viable population size of the red-cockaded woodpecker. *Journal of Wildlife Management* 52: 385–91.

Rexstad, E.A., D.D. Miller, C.H. Flather, and others. 1988. Questionable multivariate statistical inference in wildlife habitat and community studies. *Journal of Wildlife Management* 52: 794–798.

Rice, S.M., F.S. Guthery, G.S. Spears, and others. 1993. A precipitation-habitat model for northern bobwhites on semiarid rangeland. *Journal of Wildlife Management* 57: 92–102.

Richter-Dyn, N., and N.S. Goel. 1972. On the extinction of colonizing species. *Theoretical Population Biology* 3: 406–433.

Risser, P.G. 1994. Biodiversity and ecosystem function. *Conservation Biology* 9: 742–746.

Robinson, S.K., F.R. Thompson III, T.R. Donovan, and others. 1995. Regional forest fragmentation and the nestinf success of migratory birds. *Science* 267: 1987–1990.

Romesburg, H.C. 1981. Wildlife science: gaining reliable knowledge. *Journal of Wildlife Management* 45: 293–313.

Rosenzweig, M.L. 1971. Paradox of enrichment: destabilization of exploitation ecosystems in ecological time. *Science* 171: 385–387.

Rosenzweig, M.L. 1981. A theory of habitat selection. *Ecology* 62: 327–335.

Rosenzweig, M.L. 1991. Habitat selection and population interactions: the search for mechanism. *American Naturalist* 137: S5–S28.

Rossi, R.E., D.J. Mulla, A.G. Journel, and E.H. Franz. 1992. Geostatistical tools for modeling and interpreting ecological spatial dependence. *Ecological Monographs* 62: 277–314.

Salwasser, H., and B.G. Marcot. 1986. Viable population planning: a planning framework for viable populations. In: Wilcox, B.A., P.F. Brussard, and B.G. Marcot (eds.), *The Management of Viable Populations: Theory, Applications, and Case Studies.* Center for Conservation Biology, Stanford University.

Schemske, D.W., B.C. Husband, M.H. Ruckelhaus, and others. 1994. Evaluating approaches to the conservation of rare and endangered plants. *Ecology* 75: 584–606.

Schonewald-Cox, C.M., S.M. Chambers, B. MacBryde, and L. Thomas. 1983. *Genetics and Conservation: A Reference for Managing Wild Animal and Plant Populations.* Benjamin/ Cummings Publishing Company, Menlo Park, CA.

Schueck, L.S., and J.M. Marzluff. 1995. Influence of weather on conclusions about effects of human activities on raptors. *Journal of Wildlife Management* 59: 674–682.

Shaffer, M.L. 1978. Determining minimum viable population size: a case study of the grizzly bear (*Ursus arctos* L.). Ph. D. Dissertation, Duke University, Durham, NC. 190 pp.

Shaffer, M.L. 1981. Minimum population sizes for species conservation. *Bioscience* 31: 131–134.

Shaffer, M.L. 1983. Determining minimum viable populations for the grizzly bear. *International Conference on Bear Research and Management* 5: 133–139.

Shaffer, M.L. 1987 Minimum viable populations: coping with uncertainty. In: Soule, M. (ed.), *Viable Populations for Conservation*, pp. 69–86. Cambridge.

Shaffer, M.L. 1990. Population viability analysis. *Conservation Biology* 4: 39–40.

Shaw, J.H. 1993. American bison: a case study in conservation genetics. pp 3–14. In: Walker, R.E. (ed.), *Proceedings North American Public Bison Herds Symposium.* International Bison Conference, LaCrosse, WI.

Shields, W.M. 1982. *Philopatry, Inbreeding, and the Evolution of Sex.* State University New York Press, Albany, NY.

Short, J., and B. turner. 1994. A test of the vegetation mosaic hypothesis: a hypothesis to explain the decline and extinction of Australian mammals.

Simberloff, D. 1986. The proximate causes of extinction. pp. 259–276. In: Raup, D., and D. Jablonski (eds.), *Patterns and Processes in the History of Life.* Springer-Verlag, New York.

Soulé, M. 1987. *Viable Populations for Conservation.* Cambridge University Press, Cambridge, U.K. 189 pp.

Southwood, T.R.E. 1977. Habitat, the templet for ecological strategies. *Journal of Animal Ecology* 46: 337–365.

Southwood, T.R.E. 1988. Tactics, strategies, and templets. *Oikos* 52: 3–18.

Strobeck, C., R.O. Polziehn, and R. Beech. 1993. Genetic relationship between wood and plains bison assayed using mitochondrial DNA sequence. In: Walker, R.E. (ed.), *Proceedings North American Public Bison Herds Symposium*, pp. 209–227. International Bison Conference, LaCrosse, WI.

Stromberg, M.R., and M.S. Boyce. 1986. Systematics and conservation of the swift fox, Vulpes velox, in North America. *Biological Conservation* 35: 97–110.

Strong, D.R. 1986. Density-vague population change. *Trends in Ecology and Systematics* 1: 39–42.

Suchy, W., L.L. McDonald, M.D. Strickland, and S.H. Anderson. 1985. New estimates of minimum viable population size for grizzly bears of the Yellowstone ecosystem. *Wildlife Society Bulletin* 13: 223–228.

Temple, S.A. 1992. Population viability analysis of a sharp-tailed grouse metapopulation in Wisconsin. pp. 750–757. In: McCullough, D.R., and R.H. Barrett (eds.), *Wildlife 2001: Populations.* Elsevier.

Templeton, A.R. 1986. Coadaptation and outbreeding depression. pp. 105–116. In: Soule, M. (ed.), *Conservation Biology.* Sinauer, Sunderland, MA

Templeton, A.R., and N.J. Georgiadis. 1996. A landscape approach to conservation genetics: conserving evolutionary processes in the African Bovidae. pp. 398–430. In: Avise, J.C. and J.L. Hamrick (eds.), *Conservation Genetics.* Chapman and Hall, New York, NY.

Thomas, J.W., E.D. Forsman, J.B. Lint, and others. 1990. *A Conservation Strategy for the Northern Spotted Owl.* U.S. Government Printing Office, Portland, OR., USA.

Tilman, D., R. May, C. Lehman, and M. Nowak. 1994. Habitat destruction and the extinction debt. *Nature* 371: 65–66.

Tuljapurkar, S. 1990. *Population Dynamics in Variable Environments.* Springer-Verlag, New York.

Turner, M.G., Y. Wu, W.H. Romme, and L.L. Wallace. 1993. A landscape simulation modeling of winter foraging by large ungulates. *Ecological Modelling* 69: 163–184.

Turner, M.G., G.J. Arthaud, R.T. Engstrom, and others. 1995. Usefulness of spatially explicit population models in land management. *Ecological Applications* 5: 12–16.

USDI, Fish and Wildlife Service. 1992. Recovery plan for the northern spotted owl — draft. U.S. Government Printing Office, Washington, DC. 662 pp.

Van Horne, B., and J.A. Wiens. 1991. Forest bird habitat suitability models and the development of general habitat models. U.S. Fish and Wildlife Service, Fish and Wildlife Research 8. 31 pp.

Verner, J., M.L. Morrison, and C.J. Ralph (eds.), 1986. *Wildlife 2000: Modeling Habitat Relationships of Terrestrial Vertebrates.* University of Wisconsin Press, Madison, WI.

Wade, M.J. 1991. Genetic variance for rate of population increase in natural populations of flour beetles. *Evolution* 45: 1574–1584.

Waite, T.A., and P.G. Parker. 1996. Dimensionless life histories and effective population size. *Conservation Biology* 10:

1456–1462.
Walters, C.J. 1986. *Adaptive Management of Renewable Resources*. Macmillan Co., New York, NY.
Walters, C.J., and C.S. Holling. 1990. Large-scale management experiments and learning by doing. *Ecology* 71: 2060–2068.
Weiss, S.B., D.D. Murphy, and R.R. White. 1988. Sun, slope, and butterflies: topographic determinants of habitat quality for *Euphdryas editha*. *Ecology* 69: 1486–1496.
Wennergren, U., M. Ruckelshaus, and P. Kareiva. 1995. The promise and limitations of spatial models in conservation biology. *Oikos* 74: 349–356.
Wiens, J.A. 1996. Wildlife in patchy environments: Metapopulations, mosaics, and management. pp. 53–84. In: McCullough, D.R. (ed.), *Metapopulations and Wildlife Conservation*. Island Press, CA.
Wiens, J.A., J.T. Rotenberry, and B. Van Horne. 1987. Habitat occupancy patterns of North American shrubsteppe birds: The effects of spatial scale. *Oikos* 48: 132–147.
Wisdom, M.J., and L.S. Mills. 1997. Sensitivity analysis to guide population recovery: prairie-chickens as an example. *Journal of Wildlife Management* 61: 302–312.
Zink, R.M. 1991. The geography of mitochondrial DNA variation in two sympatric sparrows. *Evolution* 45: 329–339.

## THE AUTHORS

**Barry R. Noon**
*Department of Fishery and Wildlife Biology,*
*Colorado State University,*
*Fort Collins, CO 80523, USA*

**Roland H. Lamberson**
*Department of Mathematics,*
*Humbolt State University,*
*Arcata, CA 95521, USA*

**Mark S. Boyce**
*Department of Fisheries & Wildlife,*
*University of Wisconsin,*
*Stevens Point, WI 54481-3897, USA*

**Larry L. Irwin**
*National Council for Air and Stream*
*Improvement,*
*Darby, MT 59870, USA*

# Population Viability in Ecosystem Management

Richard S. Holthausen, Martin G. Raphael, Fred B. Samson, Dan Ebert, Ronald Hiebert, and Keith Menasco

**Key questions addressed in this chapter**

- ♦ *How are species of concern selected?*
- ♦ *What techniques are available to determine population status?*
- ♦ *How does one identify and quantify risk factors?*
- ♦ *How does one determine the relationship of population dynamics to habitat dynamics?*
- ♦ *How does one define suitable habitat and project trends into the future?*
- ♦ *How does one calculate the probability of a population's persistence for a given unit of time?*
- ♦ *What is the role of risk assessment in PVA?*
- ♦ *What is the appropriate analysis at different scales and for different groups of organisms?*
- ♦ *What are the keys to coordination of PVA across multiple jurisdictions?*
- ♦ *What are the relationships of agency-specific regulations and policies, and regulations that apply to all agencies*

**Keywords: Endangered species, genetics, habitat dynamics, demographics, regulations**

## 1 INTRODUCTION

Regulations implementing the National Forest Management Act (NFMA) of 1976 provided the first formal requirement for land managers to consider the concept of species viability. Significant scientific contributions to the concept of population viability appeared in the early to mid-1980s (Schonewald-Cox et al. 1983, Shaffer 1981, Shaffer 1987, Shaffer and Samson 1985, Soulé 1987, Soulé and Wilcox 1980). Progress on implementation of this requirement was slow with managers struggling to keep pace with the evolving science of population viability (Salwasser et al. 1984, Samson et al. 1985). Early efforts tended to focus on population genetics as the primary extinction threat (Salwasser et al. 1984), but these efforts evolved to consider the full range of demographic, environmental, and genetic extinction factors (Marcot and Holthausen 1987).

Over time, implementation of the viability provision of NFMA has raised many difficult technical and policy questions (Raphael and Marcot 1994). A variety of approaches have been developed that explicitly recognize the limitations imposed by incomplete information and the uncertainty associated with future projections. Policy questions have ranged from the appropriate level of assurance that populations will remain viable, to the role of single species assessments in ecosystem management. In this chapter, we present an overview of the management context for population viability analysis (PVA), a description of alternative approaches that have been used, and a series of case studies. The chapter attempts to answer the following key questions identified by workshop participants and the authors:

1. *How are species of concern selected?* Federal land managers are required by law to make efforts to protect and recover species listed as threatened or endangered. However, it is more difficult to select a full set of species that may be at risk under proposed or possible future management. This would include species that are not listed but may be declining, and may also include species that are indicators of ecosystem integrity. In addition, it may be desirable to apply PVA concepts to species that may pose a serious threat to natural resources.

2. *What techniques are available to determine population status?* Which techniques are most reliable and most cost effective? Are some techniques more appropriate at various scales or for certain types of organisms?

3. *How does one identify and quantify risk factors?* All wild populations are subject to risks due to demographic and environmental uncertainty and catastrophic events. Additional risks, such as chronic loss of habitat, may result from man's intervention in ecosystems. How does one identify those risks that have the most significant effects on the species populations of interest? How does one estimate the likelihood and the level of impact of the risks identified?

4. *How does one determine the relationship of population dynamics to habitat dynamics?* Empirical information that specifically relates population dynamics to habitat dynamics is scarce. What are alternative techniques to approximate this relationship that underlies many population viability projections?

5. *How does one define suitable habitat and project trends into the future?* The primary risk to many species is ongoing loss of habitat, so the ability to define habitat and project trends is critical.

6. *How does one calculate the probability of a population's persistence for a given unit of time?* What can empirical data, modeling, and expert judgment tell us about the probability of persistence of populations?

7. *What is the role of risk assessment in PVA?* How does the formulation of PVA as a risk assessment facilitate the management decision process?

8. *What is the appropriate analysis at different scales and for different groups of organisms?* How should viability be addressed at regional, sub-regional, basin, watershed and local scales? Should the scale of analysis be varied for different classes of species?

9. *What are the keys to coordination of PVA across multiple jurisdictions?* How can data, analysis, and management strategies be shared?

10. *What are the relationships of agency-specific regulations and policies, and regulations that apply to all agencies (e.g., the Endangered Species Act)?* What is the regulatory basis for decisions based on PVA?

## 2 BACKGROUND

How does population viability analysis (PVA) relate to the evolving concepts of ecosystem stewardship? In 1993 Secretary of the Interior Bruce Babbitt announced that the Clinton Administration was shifting federal policy away from a single species approach to one that looks "at entire ecosystems." Jack Ward Thomas, as Chief of the U.S. Forest Service, moved the agency from multiple-use management to ecosystem management with emphasis on the sustainability of systems rather than the production of commodities. The U.S. Fish and Wildlife Service has delineated more ecologically based management units (watersheds) and adopted more holistic management goals for refuges. The U.S. National Park Service is expanding its vision of ecosystem management to include man as part of the system and recognize that parks are not islands and have to be managed in their landscape context. Secretary Babbitt created the National Biological Service (now Biological Resources Division of U.S. Geological Survey) to break down barriers among Interior agencies, encourage information sharing, and promote regional holistic problem solving. Where does single species population analysis fit in this paradigm?

## 2.1 Role of Viability Assessment in Ecosystem Management

Some have argued that single species analysis is the antithesis of ecosystem management (Franklin 1993). However, the ecosystem stewardship paradigm does not exclude investigations of and management for individual species population viability. Rather it calls for a change in why and how PVAs are conducted, and facilitates using PVAs at the appropriate scale. All federal land management agencies continue to have the mandate to protect and restore those species listed as endangered or threatened under the Endangered Species Act (ESA). Ecosystem stewardship does not conflict with this mandate. Rather, ecosystem management reinforces the need to view individual species within the context of the systems that sustain them. This means looking at species populations in relation to other levels of biological organization, scaling management to ecological rather than political boundaries, recognizing the species role in and response to community and landscape processes, and setting goals that are consistent with sustainable systems including the human values associated with those systems (Grumbine 1994).

## 2.2 Scope

A companion chapter in this volume (see Noon et al.) provides information on the scientific background for PVA. The focus of this chapter is on the application and interpretation of PVAs for management decisions. We assume a broad definition of population viability analysis as did Noon et al. A PVA, as interpreted here, is a formal evaluation of data and other information concerning a population to estimate the likelihood that it will persist for some arbitrarily chosen time into the future (Shaffer 1981, Boyce 1992). PVA can range from a simple objective analysis of existing information to the development of complex quantitative models that allow probabilistic predictions of population persistence under numerous possible scenarios. The concepts of PVA can be applied to all plant and animal species in either aquatic or terrestrial systems, and we present examples addressing a variety of taxa. Although PVA has been largely applied to species populations that are rare, or in decline, we believe the same techniques can be useful in decisions about controlling species such as invasive exotics like the zebra mussel or leafy spurge or in analyses of populations of economically important species that are hunted or fished.

## 2.3 Regulatory Environment

Federal regulations implementing NFMA require that habitat be provided to maintain viable populations of all existing native and desired non-native vertebrate species within a planning area (36 CFR 219.19). The regulations define a viable population as:

> "one which has the estimated numbers and distribution of reproductive individuals to insure its continued existence is well distributed in the planning area. In order to insure that viable populations will be maintained, habitat must be provided to support, at least, a minimum number of reproductive individuals and that habitat must be well distributed so that those individuals can interact with others in the planning area."

Several components of this regulation have proven difficult to interpret (Raphael and Marcot 1994). Most people who have worked with the regulation agree that you cannot "insure" the continued existence of a population indefinitely. Consequently, the appropriate level of assurance that a population would be maintained becomes a complex policy, legal, and technical issue. Likewise, the term "well distributed" has proven troublesome. Should the species be maintained throughout its historical distribution, or is it adequate to maintain it within its current distribution? What is the obligation under NFMA to restore a species to its historical range if it has been extirpated from some part of it? Is it adequate to provide habitat that will allow the species to function as a metapopulation, with gaps between local populations, or must habitat for the species be continuously distributed across the landscape? Finally, how should the obligations under NFMA be interpreted when a significant portion of the species range and habitat falls on lands other than National Forests?

Section 7 of the 1973 Endangered Species Act (ESA) as amended provides some of the most powerful tools to conserve species listed under the act, assist with species' recovery, and help protect critical habitat. When a Federal land management agency conducts an activity where threatened or endangered species are involved, by law it must begin section 7 consultation informally or formally. The charge of the federal agencies within section 7 of ESA is to work together to help recover listed species.

The interaction between NFMA and the ESA is not entirely clear. When a species listed under ESA occurs on a National Forest, the forest is obligated to follow both ESA and NFMA. However, the relationship between the viability regulation of NFMA and the goal of recovery under ESA may not be straightforward. In some cases, viability is an apparently stronger standard because of its requirement to maintain well-distributed habitat for a species. Recovery under ESA may allow extirpation of a species from some part of its range. Conversely, in other cases ESA may call for aggressive measures to restore a species to some part of its range, while the obligation for such positive action is less clear under NFMA.

### 2.4 General Summary of Management History and Experiences

Wildlife management was defined by Leopold (1933) as the scientifically based art of manipulating habitat to enhance conditions for selected species or manipulating animal populations to achieve desired ends. In the early 1970s, the U.S. Forest Service subscribed to two general wildlife management outlooks, management for species richness and featured species management (USDA Forest Service 1975).

NFMA formally changed wildlife management for the U.S. Forest Service because it requires provision of "diversity of plant and animal communities and tree species consistent with the overall multiple-use objectives of a planning area." The regulations implementing this provision require the National Forests "to maintain viable populations of all native and desired non native wildlife vertebrate species in the planning area" (36 CFR 219.26). As already noted, a viable population is one that "has the estimated numbers and distribution of reproductive individuals to insure its continued existence is well distributed in the planning area" (36 CFR 219.19).

Implementation of this regulation has proven to be a significant challenge for the Forest Service. As part of recent attempts to rewrite the NFMA regulations (Federal Register, Vol. 60 No. 71, Thursday, April 13, 1995), the agency noted that "viability" had been only a general concept when the NFMA regulations were finalized in 1982. Specific interpretations of viability analysis had not been published in the scientific literature, and there is no indication that the Committee of Scientists who drafted the regulations foresaw the need for extensive viability analyses. The Forest Service argued that the evolution of the paradigm for viability analysis caused the NFMA regulations to become increasingly burdensome. However, the agency also acknowledged (60 FR 18896, April 13, 1995) that the courts had upheld agency contentions that the regulation could be satisfied without complex analyses. Although the Forest Service has wavered in its commitment to the viability provision of the NFMA regulation, the provision has been cited as one of the best ways to maintain biodiversity (Wilcove 1993).

Much has been written about experiences with implementation of the ESA. Problem areas have been similar to those encountered under NFMA, with specific problems including difficulty in defining thresholds of recovery and in dealing with the full diversity of life forms protected by the act (Rohlf 1991). Even so, O'Connell (1992) argued that the act has been largely successful, citing the disappearance of only a handful of species from the wild since ESA was passed in 1973. Wilcove (1993) argued that reauthorization of the act should be a cornerstone in efforts to preserve biodiversity.

## 3 USE OF PVA IN ECOSYSTEM MANAGEMENT

Experience to date suggests that there are three keys to integration of viability analysis with land management decisions. First, to be useful in decision-making, PVAs must be directly linked to factors that are under control of management. When land management decisions are involved, this generally means that PVAs must have a direct link to the availability of species habitat. Since habitat deterioration, fragmentation, or destruction is a primary risk factor for many species, linkage of PVAs to habitat is both logical and necessary.

A second requirement for successfully integrating PVAs into a decision-making framework is linkage of PVAs to predictive models. Land management decisions generally focus on the future status of resources, including populations, under various management scenarios. Providing a link between PVA and systems that are used to make future projections allows full utilization of the PVA in decision-making. This linkage is generally made through habitat and models that are used to project habitat change through time (Raphael et al. 1994).

A third requirement for integration of PVAs in decision-making is selection of an appropriate species or set of species for analysis. The set of species for which PVA is done should include or reasonably represent all species that might be judged to be at risk under feasible management scenarios. Approaches to selecting such a set of species have included the use of management indicator species, keystone species, and functional groups of species. The pros and cons of these concepts are discussed further below, as is the need to consider different taxa and different facets of viability at various scales.

Finally, we believe that PVA represents an area with unique requirements for coordination between management and science. Thus, PVA must involve a science/management team approach from the initial identification and definition of the problem, through the formal analysis of existing data, modeling, selection and execution of management alternatives, and the evaluation of management effectiveness in meeting established goals.

## 3.1 Scales of Analysis

Matching the scale of analysis to the scale of biological processes is key to the success of PVA. Different taxa, and different ecological processes that influence the life histories of those taxa, call for analyses at different scales. Analysis of broad-ranging, broadly-distributed species should generally be done at larger scales than analysis of species that operate within small home ranges and are narrowly-distributed. Unfortunately, the most appropriate scale of analysis often conflicts with the scale of proposed management. For example, biologists are often asked to analyze effects of local projects on the viability of broad-ranging species. Conversely, they may be asked to analyze the effects of regional conservation strategies on local, endemic species. In both cases, mismatch of scale will hinder analysis (Raphael and Marcot 1994, Ruggiero et al. 1994).

In examining questions of scale, it is important to consider both geographic scales and scales of biological organization. Levels of biological organization include individuals, demes, populations, metapopulations, and species. Geographic scales are represented by common units such as stands, watersheds, basins, geographic provinces, and regions. Analysis of each level of biological organization most appropriately occurs at one of the levels of geographic scale. Matching these scales appropriately, and tiering analyses done at different scales, are key to success in using PVA to inform management decisions.

## 3.2 General Process Steps

A variety of options are available for completing an analysis of risk to species over time (Boyce 1992). However, any of these analytical techniques must fit into an overall planning framework to be useful in ecosystem management. We propose six basic steps in a framework for analyzing population viability in conjunction with ecosystem management:

1. Select species of concern
2. Describe population status
3. Describe risk factors
4. Identify suitable habitat amount, distribution, and trends
5. Describe relationship of population dynamics to habitat dynamics
6. Assess likelihood of species persistence

These steps are not a cookbook recipe for analyzing viability. Rather, they simply provide a logical sequence and a set of checkpoints for addressing questions of viability. Considerations pertinent to each step are discussed below.

### 3.2.1 Selecting Species of Concern

Viability analyses have always been directed at species thought to be at risk, but there has not been general agreement about how to identify the species or set of species that ought to be analyzed in support of particular management proposals (Raphael and Marcot 1994). A number of approaches have been tried. The U.S. Forest Service (Sidle and Suring 1986) has attempted to use indicator species in developing management assessments and plans. This approach has been criticized by several authors (Patton 1987, Landres et al. 1988) who pointed out that each species has a unique response to environmental conditions and to changes in those conditions. Similarly, the use of guilds has been proposed as a means to simplify the task of choosing species for management assessments (Hunter 1990). This approach has been criticized on grounds similar to the criticism of indicator species (Morrison et al. 1992). A focus on viability of keystone species has been proposed as a means of making biological protection under the ESA more effective (Rohlf 1991). The keystone species concept, however, has been criticized as being poorly defined and largely undemonstrated in nature (Mills et al. 1993).

Mace and Lande (1991) developed a set of relatively simple criteria that could be used to place species in one of three categories; critical, endangered, and vulnerable. Such categories could be used to determine a set of species on which to focus PVA. Finally, Thomas et al. (1993a, 1993b) decided to complete qualitative viability analyses on a broad range of species rather than narrowing the focus to a selected set (Raphael and Marcot 1994). Thomas et al. (1993a, 1993b) reviewed viability of over 1,000 species of plants and animals in the analysis that supported planning for National Forests and BLM lands in the Pacific Northwest (Meslow et al. 1994).

Currently the best approach to select species for analysis may be simply to screen all species in the planning area for viability concerns (recognizing that there are practical limits to even a simple screening of groups such as arthropods which may contain tens of thousands of species). A coarse screen should consider the following factors:

- General abundance of the species (relative population size)
- Distribution of the species' population, with particular emphasis on population interaction and/or isolation both within the planning area and with populations outside the planning area
- Species life history traits, especially body size, reproductive rate, dispersal ability, and migratory rate
- Habitat specificity
- Suspected trends in abundance and distribution
- Risk factors known for the species

This information could be used in conjunction with a simple rating system (Schonewald-Cox 1983, Rabinowitz et al. 1986, Lehmkuhl and Ruggiero 1991, Mace and Lande 1991, Reed 1992, Mace and Collar 1995) to determine species for which more detailed analysis is appropriate. Several of these rating systems are described later in this chapter (Section 4.2.1).

### 3.2.2 Describe Population Status

Once species have been selected for analysis, their population status in the planning area and in adjacent areas should be described as fully as possible. The objective of this step is to collect information that will contribute to the estimation of species' persistence under proposed management regimes. Information collected could include the following:

- Species' population size, including the estimated number of individuals in the planning area, the number in the interbreeding population of which the planning area is a part, and the number in the species' total population within and beyond the planning area
- Major historical changes in species' range/distribution, preferably displayed in map format and stored in digital format on a geographic information system (GIS)
- Historical decreases or increases in species' population, including information on the reliability of data about the estimated change
- Current population trends, indicating source and reliability of the data
- Species' demographic characteristics, with an indication if possible of those characteristics to which the species' population trend is most sensitive
- Species' genetic characteristics, if known, including insights into the degree of heterozygosity in the population and any genetic differences among subpopulations and/or subspecies

Information on population status is key to assessing viability. Unfortunately, such information is only available for a limited number of species and frequently not available for species for which there is the most concern. Lack of such data does not preclude the use of qualitative forms of PVA, but the level of reliability of an assessment done with little data will necessarily be lower than that of an assessment done with comprehensive data. The consequences of poor or missing data should be recognized within the context of overall risk management for the species. In addition, uncertainty in any of these attributes should be described and incorporated into analyses whenever possible (Taylor 1995).

### 3.2.3 Describe Risk Factors

Both intrinsic and extrinsic factors may cause the extirpation or extinction of species (Soulé 1983, Gilpin and Soulé 1986, Soulé 1987, Mace and Lande 1991). A partial list of risk factors would include:

- Demographic: Random variations in birth and death rates; Population age structure
- Genetic: Inbreeding depression; Genetic drift
- Environmental: Random changes in condition; Loss of habitat (systematic and chronic losses, random catastrophic losses); Biological interactions (predators, competitors, disease, parasites, change in prey populations); Pollutants and toxicants

At some level, every species is likely affected by all or nearly all of these factors. The purpose of looking at risk factors within a PVA is not simply to list all factors that might have some influence on the species, regardless of how minor that influence is. Rather, the review of risk factors should focus on key influences that are most likely to result in species extirpation within the area being analyzed. Determining how the factors might interact is also important. For example, loss of habitat is thought to be the primary cause of declines in northern spotted owl (*Strix occidentalis caurina*) populations in the Pacific Northwest, with the effect of habitat declines occurring mainly through modification of demographic rates (USDI 1992, USDA/USDI 1994). Lande (1993) analyzed interactions among demographic stochasticity, environmental stochasticity, and random catastrophes and provided a basis for determining which one(s) may be of the most importance in a given situation.

### 3.2.4 Identify Suitable Habitat Amount, Distribution, and Trends

When actions of land management agencies are being considered, the direct effect on species is most likely to occur through habitat modification. Therefore, it is critical that a PVA include identification of habitats important to the species and projection of trends in those habitats under proposed management. The process used to identify suitable habitat could include both a review of the literature and consultation with species experts (Thomas et al. 1993a, Marcot et al. 1997). Primary habitats used for reproduction, feeding, cover, and movement should be identified. Important patterns of geographic distribution and spatial patterning of habitat should be described. This could include juxtaposition of habitat (e.g., proximity of nest trees to water for bald eagles), geographic distribution of habitat (e.g., proximity of nest sites to the ocean for marbled murrelets), and landscape composition of habitat (e.g., availability of both nest and foraging sites within a home range area for northern goshawks). Habitat characteristics that are important to the maintenance of population interactions, such as the availability of suitable nesting sites within the juvenile dispersal capability of avian species, should also be described. Key habitat elements other than vegetation (e.g., roads, caves, talus) should also receive attention.

Summarizing this information in a habitat relationships model will aid in making projections of future outcomes for the species under alternative management scenarios. Changes in amount and distribution of species habitat can be projected under each scenario, and those changes can in turn be used as inputs to the habitat relationships model. Habitat projections should include consideration of all habitat influences including management activities, succession, and chronic and catastrophic natural disturbances.

### 3.2.5 Describe Relationship of Population Dynamics to Habitat Dynamics

To tie PVA to the assessment of various land management scenarios, a link must be established at some level between population dynamics and the dynamics of suitable habitat for the species. At a very basic level this could mean distinguishing source habitat from sink habitat or identifying some threshold of habitat amount below which a species is no longer expected to occupy a landscape (e.g., Lande 1987, 1988; With and Crist 1995). If appropriate information is available, this may become a more sophisticated process of predicting changes in demographic rates based on amount and quality of habitat (Raphael et al. 1996). The relationship of demographic rates to habitat could ultimately be used in models that simulate change in habitat and species' response to that change. Such analyses were conducted for northern spotted owls throughout their range (Raphael et al. 1994), on the Coast Ranges of Oregon (Boyce et al. 1994), and on the Olympic Peninsula of Washington (Holthausen et al. 1995). Although the data needed to drive such analyses may not be available for many species, the basic concept of associating changes in population dynamics to changes in habitat should be useful in all PVAs regardless of their sophistication.

### 3.2.6 Assess Likelihood of Species' Persistence

The final step in a PVA is to provide an assessment of the likelihood of species' persistence to some specified point in time (Boyce 1992). Although this assessment is often stated as a simple likelihood (Shaffer 1981), that likelihood generally contains an implicit assumption about the species' distribution. That is, it is not enough to know that a species persists until some point in time. We also need to know the area across which the species persists. The distribution of species persistence should be recognized explicitly in the assessment. This is particularly important when the assessment is conducted in support of the NFMA regulations, which require that species' viability be maintained well-distributed across the planning area. The implied species objective of ecosystem management — to maintain species at a level at which they would continue to carry out ecological functions (Conner 1988) — would also require an assessment of the future distribution of a species in place of a simple projection of its persistence.

The assessment of persistence should utilize all the information gathered on species' status, risk factors, population dynamics, and habitat trends. In addition, the uncertainty associated with the assessment should be stated. All potential risk factors for the species should be addressed, and the cumulative effects of proposed actions and reasonably foreseeable actions should be assessed. The level of uncertainty may be a key element in management decisions that are supported by the PVA and is a key consideration in judging the reliability of any population projection (Taylor

1995). Methods for assessing likely persistence and distribution range from simple qualitative judgements to empirically-validated simulation models. Those methods, and the uncertainties generally associated with each, are discussed below.

## 4 ALTERNATIVE APPROACHES

Managers rarely have all of the information needed to conduct a fully quantitative PVA. There are a number of alternatives that can used in such cases. It should be recognized, however, that the following surrogate approaches will yield less accurate estimates of population viability than a more rigorous PVA. Still, these surrogate approaches will be needed by resource managers in many settings to better understand the potential for species extinction or recovery.

### 4.1 Assessments Based on Habitat Relationships

In the face of missing information, one practical (but risky) alternative is to use inventories of the amount and distribution of suitable habitat as a surrogate for PVA. This method relies on three primary assumptions: (1) that attributes of suitable habitat are known well enough to identify cover types that meet the life requisites of the species; (2) that the amount of suitable habitat is correlated with fitness (Van Horne 1983); and (3) that habitat is limiting so that changes in amount of suitable habitat are correlated with changes in population status. Viability assessments based on habitat inventories are useful to the degree that these assumptions are met, but testing the assumptions may not be possible — after all, if data were available to test fully the assumptions, one could proceed with more sophisticated assessment procedures.

Several approaches have been used to develop relationships between habitat and populations. Expert judgment can be used to develop habitat relationships models. For example, Kangas et al. (1993) described a technique where teams of experts were given a set of land management alternatives and were asked to consider all possible pairs of alternatives and rank which member of each pair was better for the species being evaluated. A matrix of pairwise comparisons was then constructed and analyzed to determine the mean priority of each alternative. Correlations of these priorities with vegetation attributes of the alternatives were used to create a habitat suitability function for the species. Other examples of expert judgment approaches include construction of rule-based expert systems (Marcot 1986) and Bayesian belief networks (Lee, in press). These models have the advantages that they can be built quickly, can be broadly applicable, and are relatively inexpensive.

Semi-quantitative methods based on a combination of literature reviews and professional judgment have been used to build wildlife-habitat relationships models such as Habitat Suitability Index models (Schamberger and Krohn 1982, Cole and Smith 1983), Habitat Capability models (Nelson and Salwasser 1982, Hoover and Wills 1984, Fagen 1988), PATREC models (Williams et al. 1977, Grubb 1988), and wildlife habitat relationships matrix models (Thomas 1979, Brown 1985, Zeiner et al. 1990). These methods have the advantage of being broadly applicable at large geographic scales and they can be adapted to create linkages between GIS maps of cover types and habitat suitability to model responses of species to dynamics of habitat change. Many of these models are available and can be modified (again, based on professional judgement and literature) to fit local conditions. These models suffer from lack of precision and unknown accuracy, especially at smaller geographic scales (watersheds and below). Models have received only limited testing (but see Raphael and Marcot 1986, Thomosma et al. 1991). To assess and improve reliability of these models, they must be tested, revised, and tested again in more than one study area; none, to our knowledge, have received such testing.

Finally, quantitative models can be built from site-specific studies of habitat selection. Each of the semi-quantitative models described above can be developed for a specific planning area through research on empirical relationships between habitat attributes and species' presence, abundance, or vital rates. Newer methods of quantifying habitat selection, such as development of resource selection functions (Manley et al. 1993), are available from which to build potentially useful PVAs (Boyce et al. 1994). The primary disadvantages of this approach are that the cost of doing such research may be prohibitive, new research may take several years to generate useful results and managers may not be able to delay decisions, models can be developed for only a limited number of species, and the geographic scope of such site-specific models may be narrow. Advantages include potentially greater precision and accuracy and the ability to match scales of analysis and model-building with scale of the planning area.

All of these methods share the same shortcoming: population dynamics are not explicitly considered. They may be useful to demonstrate broadly that a species status is likely to decline, improve, or remain unchanged. However, the methods should not be relied upon to make critical determinations in either biologically or socially risky situations. Habitat modeling can be combined with other techniques, such as expert panels or demographic assessments, to provide a more rigorous analysis.

## 4.2 Assessments Based on Current Population Status

Another category of surrogate approaches are those based on information about current population(s) for species of concern.

### 4.2.1 Risk Classifications Based on Population Numbers and Distribution

A variety of simple rules and risk classifications have been developed to allow coarse determinations of species status where little information is available. Some initial rules of thumb for viability were based on the establishment of a threshold for minimum viable population size (Soulé 1980). For example, one frequently cited estimate is that a minimum viable population must have an effective population size >50 for the short term and >500 for the long term (Soulé 1987). These rules were influenced most strongly by genetic considerations, but they were quickly abandoned in favor of consideration of the interactions of genetic, demographic, and environmental effects (Gilpin and Soulé 1986).

More recently, Mace and Lande (1991) proposed a simplified system for assessing threats to species. Their system describes three levels of risk and is flexible in its data requirements and in the types of population units to which it can be applied. The three risk categories, critical, endangered, and vulnerable, are defined as:

- Critical: 50% probability of extinction within 5 years or 2 generations, whichever is longer.
- Endangered: 20% probability of extinction within 20 years or 10 generations, whichever is longer.
- Vulnerable: 10% probability of extinction within 100 years.

Each of these categories is defined using criteria of effective population size, population fragmentation, population decline, population fluctuation, rate of habitat loss, and exploitation or interspecific interactions leading to population decline. The system explicitly recognizes threats because of fluctuating environmental conditions and catastrophic events. Appropriate application of the system is limited to large vertebrates, and the data used can include demographic information, population size information, or habitat information.

Mace and Collar (1995) proposed four criteria for estimating risk to species: (1) an observed, estimated, or projected decline measured from population size, range area, or habitat area; (2) a restricted range that is contracting, fragmented or subject to extreme fluctuation; (3) a small population that is in decline or highly fragmented; (4) a very small population. They propose an optional fifth category (rarely available) of a detailed quantitative analysis of population status that leads to an estimation of extinction risk. Satisfying any one of these criteria qualifies a species for listing in one of the three levels of risk described above.

Reed (1992) proposed a simpler scheme based in part on Rabinowitz' (1981) elements of rarity. In this scheme, a species is classified by two levels of population size (somewhere large, everywhere small). For each population level, the species' distribution is further classified as having broad or restricted habitat associations, and within each of those categories as having a wide or narrow geographic distribution. Each of the eight combinations can be ranked for extinction probability; habitat generalists with broadly distributed large populations are least vulnerable and habitat specialists occurring over a small geographic area with small populations are most vulnerable.

With the exception of the initial rules of thumb for minimum viable populations, these classifications may represent a useful approach to the discussion of population viability when other tools are not available. They may be particularly useful in an initial screening of species to determine those most in need of additional consideration. The classifications are probably less useful in judging how habitat change resulting from land management alternatives would affect risk.

### 4.2.2 Demographic Characteristics

The most powerful information on population status is derived from estimates of vital rates. Such vital rate information is most frequently derived from capture–recapture studies (Pollock et al. 1990) and can include estimates of age-specific survival and fecundity, immigration, and emigration. Data derived from capture–recapture studies can be used to estimate trends over time in these parameters (Lebreton et al. 1992) and overall rates of population increase or decrease (McDonald and Caswell 1993). Demographic data have played a key role in some decisions to list species as threatened or endangered (USDI 1990) and have been proposed for use in delisting species (USDI 1992; USDI 1994).

Although demographic information can be compelling, its limitations must also be recognized. First among these is the expense of collecting the data and the need to collect data over a period of years to allow the analysis of trends. Because of the expense of data collection, it is unlikely that reliable demographic data will ever be collected for many species. Recent work by Wisdom and Mills (1997) used an analysis of elasticity to determine the demographic parameters that have the greatest influence on the finite rate of increase. This technique may allow more cost-effective collection of demographic data in some situations.

The second limitation is the need to restrict inferences about demographic data to both the geographic area and the time period within which the data were collected. Thus, for widely distributed species, demographic data will generally be valid only for the portion of the population actually sampled. The limitation on temporal interpretation of the data is also key. Trends estimated using demographic data are valid for the period of data collection and are not projections of future population trends. The simulation modeling section below discusses ways that demographic data can be tied to habitat information and used in model-based projections of future population trends.

The final limitation on the use of demographic data is the potential for bias in the estimates of survival and reproductive rates and of the overall rate of population increase (Raphael et al. 1996). Knowledge of these potential biases should be used to temper conclusions drawn from demographic analyses.

### 4.2.3 Genetic Considerations

Knowledge of genetic variation ought to contribute to formulation of a PVA. For example, isolation of populations can result in restriction of gene flow and loss of genetic variation with increased risk of inbreeding depression and genetic drift, which may increase risk of extinction (Frankel and Soulé 1981). We do not know, however, how much and what type of genetic variation is most important to preserve (Boyce 1992). Building some method of monitoring and predicting genetic implications of land management decisions on populations at risk is desirable, but suitable methods remain difficult to implement and most models or analyses have not explicitly incorporated genetic information. Until better indicators of genetic condition are proposed, and critical thresholds established, we will not see progress in this area (see Huston et al., this volume, for details on this topic).

## 4.3 Assessments Based on Simulation Models

PVA is probably most often desired by managers to satisfy legal requirements such as ESA and NFMA and to inform decision-makers about the consequences to viability of alternative land management plans or other actions. Therefore, linking population viability to habitat dynamics is crucial. One of the more tractable methods of making this link is through the use of simulation models (Raphael et al. 1994, Akçakaya et al. 1995, Holthausen et al. 1995). The simulation models that are most germane to management questions are those that link population attributes (size, birth, and death rates) to habitat conditions, and thus base future population performance on projected future habitat conditions (e.g., Akçakaya 1992, McKelvey et al. 1993).

The parameters of most interest in an analysis of habitat effects on population performance are fecundity (including the proportion of the population that attempt to breed), recruitment (including juvenile survival), dispersal, and adult survival. Persistence of individuals on home ranges or territories may be used to provide insight into survival if more direct estimates are unavailable (Bart 1995). The relationships between demography and habitat variation at the individual scale can most readily be estimated for fecundity and persistence on home ranges or territories (e.g., Bart 1995). The habitat parameter used by Bart in his work on the northern spotted owl was the percent suitable habitat within single or multiple territory-sized areas. Habitat was used either as a continuous variable, allowing correlation (Bart 1995), or as a categorical variable, allowing investigation for significant differences among categories (Bart and Forsman 1992). Determining the relationship between habitat and survival is the most difficult. Survival values estimated in demographic studies are the result of the capture-recapture histories of large samples of individuals. At small scales, habitat variables can be attached to individual or groups of capture histories using several modeling approaches (Lebreton et al. 1992, Skalski et al. 1993, Conroy 1993).

Simulation models using the relationship of demographic performance to habitat can yield a number of different measures of risk, defined as the likelihood of a population extinction by some specified time under various management scenarios. Such measures include quasi-extinction probabilities (chance of a population decline below a specified level), time to extinction, and likelihood of extinction within a fixed time. The critical requirement for using these models is understanding the underlying assumptions upon which they are based. One must be aware of the parameters used to drive the model and must have some familiarity with sensitivity of model outputs to variation in those parameters.

One example of the use of simulation models is the case of the northern spotted owl on the Olympic Peninsula of Washington (Holthausen et al. 1995). In this example, the U.S. Fish and Wildlife Service was interested in how alternative scenarios for retention of spotted owl habitat on non-federal lands (total land area = 606,800 ha) might influence the likelihood of owl persistence, given retention of habitat on federal lands (total = 619,800 ha) under the Northwest Forest Plan. Alternative scenarios were modeled by mapping habitat that might remain if all habitat on all ownerships were conserved, if habitat was conserved only on federal lands, if non-federal habitat were conserved under the current ESA take guidelines, and if non-federal habitat were conserved only on specially designated lands under the 4(d) provision of the ESA. A spatially-explicit population simulation model was run under three sets of assumptions about the relationships between habitat quality (percent habitat within fixed-area geographic blocks) and vital rates of owls (adult and juvenile survival, fecundity). Results of the simulations varied dramatically depending on which set of assumptions were used, but differences among scenarios were consistent. The results were used to assess the scenarios based on the simulated response of spotted owls using several criteria: extent of the peninsula that was projected to have high rates of occupancy by pairs of owls over the simulation run of 100 years, mean population size, and finite rate of population change.

Although simulation models can be very useful, and may be one of the only methods to evaluate population response to large-scale land management actions, users must understand the limitations of the models. Models are not reality. Results are dependent on the structure of the model, the assumptions used to parameterize the model, and the input data (including the representation of the land management action being evaluated). Unfortunately, good data on demographic characteristics exist for only a relatively few species, and information relating these characteristics to habitat is even scarcer. Such data are expensive and must be collected over a period of several to many years. If the quality of information used to link habitat quality and change with population responses is weak, model results must be interpreted with caution.

## 4.4 Expert Opinion Assessments

Because quantitative PVAs, and the data needed to conduct them, are scarce, management decisions have generally depended on information provided by qualitative assessments. Although these assessments may have little scientific standing (Boyce 1992), they often carry significant weight in management decision-making. It is important to discuss ways that such assessments can be as credible and informative as possible given the reality of scarce information (Ruggiero et al. 1994).

Expert opinion, gathered from panels of experts in a carefully structured process, has been used in several large-scale viability assessments (Thomas et al. 1993a, Thomas et al. 1993b). Guidelines for the use of such panels have been described by Cleaves (1994). These panels have been used most successfully in the context of judging likely outcomes for species resulting from the implementation of land management activities. For example, in a recent assessment of planning alternatives for the Columbia River Basin, Lehmkuhl et al. (1997) described the following five possible outcomes for species:

- *Outcome 1.* Populations are broadly distributed across the planning area with little or no limitation on population interactions.
- *Outcome 2.* Populations are broadly distributed across the planning area but gaps exist within this distribution. Disjunct populations are typically large enough and close enough to other populations to permit dispersal among populations and to allow species to interact as a metapopulation.
- *Outcome 3.* The species is distributed primarily as disjunct populations, some of which are small or isolated to the degree that species interactions are limited. Local subpopulations in most of the species' range interact as a metapopulation but some populations are so disjunct that they are essentially isolated from other populations.
- *Outcome 4.* Populations are typically distributed as isolated subpopulations, with strong limitation in interactions of subpopulations and limited opportunity for dispersal among patches. Some local populations may be extirpated and rate of recolonization of vacant habitat will likely be slow.
- *Outcome 5.* Populations are highly isolated throughout the area with little or no possibility of interactions among local populations, strong potential for extirpations, and little likelihood of recolonization of vacant habitat.

Expert panels were then led through a structured judgment process that allowed them to provide opinions about the likelihood that each of these outcomes would occur under several land management alternatives. The same set of outcomes was used to express judgment about both the historical condition of the species and its current status. Judgments were made for all species for which habitat modification was thought to be a potential concern. This process, further described in Lehmkuhl et al. (1997), has several strengths. First, the experts were asked to make judgments about biological conditions rather than judging legal interpretations of viability. Second, variability in judgments of individual experts and across experts was captured and expressed by the process. Finally, the process allowed for rapid assessment of a large number of taxa. A similar process (Thomas et al. 1993b), used to assess alternatives for the Northwest Federal Forest Plan, was found legally adequate as a means of implementing the NFMA viability regulation for most species. Additional information was required for several federally-listed species and for some local endemic species (USDA/USDI 1994).

The use of expert opinion assessment may also help solve another dilemma — deciding which species should be the subject of PVA. An initial assessment, addressing the broadest possible array of taxa, could be used to determine those species for which more detailed analysis is appropriate. Species whose habitats and populations were considered secure in the expert opinion assessment would require little additional attention, allowing the use of more time and resources to deal with those species for which experts expressed higher levels of risk to viability. This would be particularly helpful in large-scale assessments for areas where there has been no thorough review of the status of a broad array of species.

## 5 CASE STUDIES

Three case studies are presented below to illustrate the use of population viability analysis in the context of ecosystem management and planning. These include the northern spotted owl, the grizzly bear, and multiple species assessments conducted for the 1993 Pacific Northwest Forest Plan.

### 5.1 Northern Spotted Owl (*Strix occidentalis caurina*)

*Species status*

The northern spotted owl is listed as threatened by the U.S. Fish and Wildlife Service under the Endangered Species Act. Concern for viability of this species has been a major driver in litigation over management of late-successional and old-growth forests in the Pacific Northwest. A recovery plan was written (U.S. Department of the Interior 1992) but never formally adopted. The habitat provisions of this plan were subsumed within the Northwest Forest Plan which was designed, in part, to provide habitat conditions for a stable and well-distributed population of this species. PVA was used to judge the implications of the forest plan on viability of the owl in relation to a variety of alternative land management plans.

*Habitat and distribution*

The northern spotted owl occurs in late-successional coniferous forest in southwestern British Columbia, western Washington, western Oregon, and northwestern California. Pairs of owls nest and forage in large tracts of forest at lower elevations, with home ranges exceeding 2,000 ha throughout most of the species' range. Within each geographic area of the owl's range, home range size is smaller with increasing quality of habitat (higher prey density or greater density of favorable habitat). Forests selected by northern spotted owls typically have moderate to high canopy closure, a multilayered canopy with large overstory trees, numerous large snags and broken-topped live trees, large cavities, heavy accumulations of downed wood, and open space within and beneath the canopy.

*Population risk factors*

The primary risk to this species has been loss and fragmentation of habitat because of logging, development, and natural disturbance, primarily fire and wind. Other risks include competition with expanding populations of barred owls, mortality from predators, fluctuating prey availability, genetic isolation, and environmental fluctuation.

*Population viability and likelihood of persistence*

The northern spotted owl has been a focus of concern over forest management practices since the early 1970s. Early research indicated an association of the owl with older forest, and the rapid logging of that forest led to recognition that special habitat management provisions would be required to meet legal requirements of the NFMA. Various sets of habitat management guidelines were proposed and revised, leading to a major Forest Service analysis in 1985 in support of a supplemental EIS on spotted owl guidelines. This analysis included a formal PVA of the spotted owl in which estimates of vital rates were used in a population model to investigate the likelihood of species persistence (Marcot and Holthausen 1987). For the evaluation of land management alternatives, these demographic estimates were coupled with estimates of the amount and distribution of future habitat as well as estimates of genetic effects and habitat capability. Using a set of ranking criteria, likelihood of viability under each alternative was expressed in one of five categories ranging from very high to very low. The value of this analysis was the explicit treatment of each of the risk factors and a clear linkage between habitat quality and likelihood of persistence.

Over the past decade, increasingly sophisticated analyses have led to further understandings of the population viability of the owl. The Interagency Scientific Committee (ISC, Thomas et al. 1990) used predictions from simulation models to test the persistence of alternative theoretical sizes and spacings of clusters of owls (Lamberson et al. 1992, 1994). This work supported the idea that arranging suitable habitat into large blocks spaced closely together increased the occupancy rate of sites, which led to higher likelihoods of species persistence. Modeling since the ISC report has emphasized translation of results from theoretical landscapes to real ones using GIS maps of forest cover. These later models have enabled researchers and managers to investigate the effects of differing geometries of habitat resulting from alternative land management plans. The development of a spatially explicit population model (McKelvey et al. 1993) provided the tool to accomplish such analyses (see Noon et al., this volume).

The first application of this new model in support of a spotted owl PVA was done for the Bureau of Land Management (BLM) in Oregon (Dippon et al. 1992). A series of regression equations were used to relate birth and death rates to the amount of suitable habitat within home-range sized (1,000 ha) hexagonal cells throughout the analysis area. A GIS map of habitat was created, and this was updated at 10-year intervals under each of six management alternatives using timber harvest scheduling projections. The model was run for 100 years and yielded estimates of population trend and site-specific occupancy. These results provided criteria that managers could use to evaluate the likelihood of owl persistence into the future under each of the six alternatives. It was the relative ranking of the alternatives that was useful, not the absolute values of the projected owl population sizes and occupancy rates.

The BLM application applied only to parts of western Oregon. The first rangewide application was conducted in support of an evaluation of land management alternatives (USDA et al. 1994) developed by the Forest Ecosystem Management Assessment Team (Thomas et al. 1993b). For this analysis (Raphael et al. 1994, Raphael et al. in press) three of 10 alternatives were compared, including the proposed spotted owl recovery plan (Alternative 7, which allowed the most timber cutting), the preferred alternative (Alternative 9, intermediate timber cutting), and retention of most late-successional and old-growth forest (Alternative 1, least timber cutting). A fourth (control) scenario was included under which all existing spotted owl habitat was retained without harvest or growth for the full 100-year simulation. This analysis was run using a GIS map of spotted owl habitat on all federal lands throughout the owl's range in Washington, Oregon, and California. Timber cutting was modeled using harvest projections applied to those lands available for harvest according to land allocations under each alternative. Regrowth of currently unsuitable habitat was modeled assuming a simple linear growth model. A GIS map of current habitat was updated at 10-year intervals for any cutting and growth for 50 years; population simulations were extended to 100 years. As in the BLM analysis, parameters derived from ongoing demographic studies (Forsman et al. 1996a, Burnham et al. 1996) were assigned relating rates of birth and death to amounts of habitat within hexagonal cells. Three different sets of parameters were developed so that model results could be compared using differing assumptions about the relationships between demographic rates and habitat quality. Results of the simulations included projections of mean population size and estimated rates of occupancy of each cell under each alternative. Within any parameter set, the alternatives differed, with Alternative 1 supporting the largest population and more extensive areas of high occupancy. Alternative 7 supported the lowest population and occupancy rate and Alternative 9 had intermediate results (Fig. 1). This analysis was useful in portraying geographic areas where larger blocks of habitat might support more persistent populations and conversely where habitat is scarce and population persistence would likely be lower. As illustrated in Fig. 1, the western Cascade Range of Oregon was revealed as a stronghold for owl populations under each of the alternatives (although much weaker under Alternative 7).

The most recent and most sophisticated application of these models for PVA of the spotted owl was conducted in support of ongoing evaluations of the application of the ESA to nonfederal lands in the Olympic Peninsula of Washington (Holthausen et al. 1995). For this analysis, parameters were tuned to site-specific demographic data from the Olympic Peninsula (Forsman et al. 1996b) and hexagonal cell size was adjusted to match expected density of owls as estimated from existing data. A current habitat map covering all lands (both federal and nonfederal) was obtained and a habitat growth model was developed using the distribution of current seral stages to predict recovery of habitat over time on federal lands. The simulation model was used to evaluate a variety of scenarios for retention of habitat on nonfederal lands, coupled with retention of habitat on federal land under the existing forest plan. Scenarios ranged from complete retention of nonfederal habitat to complete

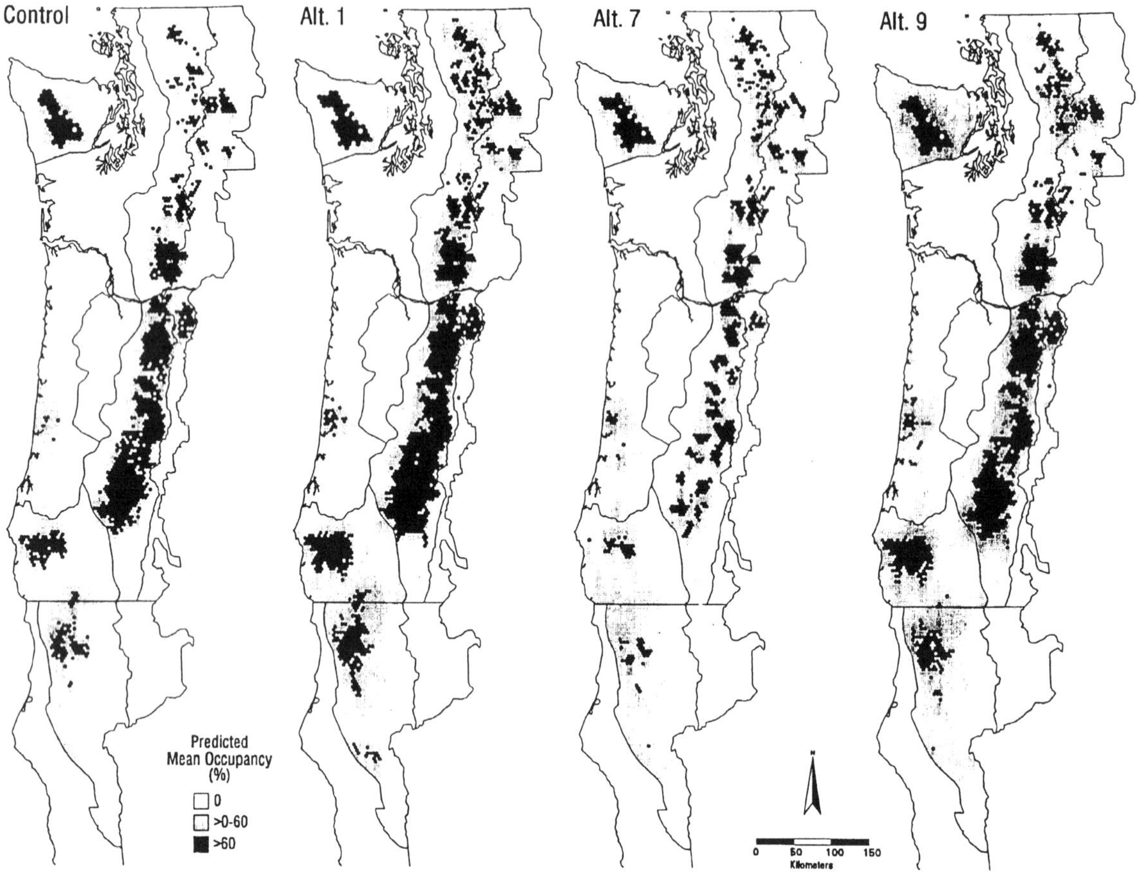


Fig. 1. Map of mean occupancy by pairs of Northern Spotted Owls over 10 replications of a 100-year simulation analysis of population dynamics under four scenarios of harvest and growth of suitable owl habitat. Control: no harvest or regrowth; Alternative 1: retain all current late-successional forest and regrow younger forest; Alternative 7: harvest in matrix following land allocations of the Final Draft Spotted Owl Recovery Plan and regrow younger forest; Alternative 9: harvest in matrix following land allocations of the Northwest Forest Plan and regrow younger forest. Simulations using Rule Set 2 (see Raphael et al 1994) are illustrated. Maps depict the range of the Northern Spotted Owl along the northwest boundary of the United States in western Washington, western Oregon, and northwestern California. Redrawn from Raphael et al. (in press).

removal of non-federal habitat and included retention of habitat within mapped "Special Emphasis Areas." Results of the simulations were expressed in a variety of ways, including mean population size, estimated rates of population change, and numbers of sites with high mean occupancy. Scenarios were ranked using these results, and managers were able to judge the relative value of each scenario to long-term persistence of the owl on the peninsula. In addition, specific model runs were used to test whether persistence of the owl on the isolated peninsula was enhanced if stepping stones of suitable habitat were designated between the peninsula and the nearest neighboring habitat in the Washington Cascades. The model results, in this case, showed negligible benefit from such a proposal. The model was also used to test for more efficient patterns of habitat retention than those of the prescribed scenarios. The analysis revealed that there are more efficient patterns of retention, that is, larger numbers of owls and larger areas of high occupancy could be obtained from fewer acres of nonfederal habitat than in the amounts of habitat proposed for retention under any of the alternatives considered. Subsequent optimization modeling (Hof and Raphael 1997) reinforced these findings.

## 5.2 Grizzly Bear (*Ursus arctos horribilus*)

### *Species status*

The grizzly bear was listed as a threatened species in July 1975 with an initial recovery plan approved in 1982 (U.S. Fish and Wildlife Service 1982). The 1982 plan called for the determination of occupied space and habitats in order to achieve recovery goals, and identified demographic targets including reproductive rate, average litter size, reproductive intervals, and annual total mortality. Of fundamental concern in the initial recovery effort was the concept of a minimum viable population. Shaffer (1978) and Shaffer and Samson (1985) developed a stochastic simulation model to determine the relation of population size to extinction probabilities for grizzly bears (Interagency Grizzly Bear Committee 1987).

The recovery plan for the grizzly bear was revised in 1993 incorporating several significant changes (U.S. Fish and Wildlife Service 1993). The concept of a recovery zone replaced occupied habitat, a small set of easily measured indicators was substituted for demographic targets, and the possible importance of linkages between grizzly bear ecosystems was included as a consideration of the recovery effort. Recovery zones are those areas within which grizzly bears and their habitats will be managed for recovery, population parameters will be monitored, and linkages will be addressed. The revised recovery plan further called for the development of a conservation strategy for each grizzly bear population prior to its delisting and the placement of one grizzly bear into the Yellowstone population every 10 years as an effort to maintain the genetic health of that population. Conservation strategies are to be developed through an interagency process.

### *Habitat and distribution*

Historically in the continental 48 states, the grizzly's range extended from the mid-Great Plains westward to the California coast and south into Texas and Mexico (Storer and Tevis 1955, Herrero 1972). The broad historical distribution suggests flexibility in habitat use and in food habitats. The morphology of the bear — crushing molars and length of the digestive system — are characteristic of a carnivore, but bears rely heavily on plant materials (roots, bulbs, tubers, fungi, seeds among others) during certain critical periods of the annual cycle, for example, denning and post-denning in late fall and early spring.

Distribution in the continental 48 states today includes only five areas in wilderness, national parks and mountainous areas of Idaho, Montana, Washington and Wyoming (U.S. Fish and Wildlife Service 1992). These grizzly bear ecosystems are called the Northern Cascades, Selkirks, Cabinet Yaak, Northern Continental Divide, and Yellowstone. Two additional areas, the Bitterroot Mountains in Idaho and the San Juan Mountains in Colorado, supported grizzly populations in the recent past. These two areas along with the five occupied areas are the seven grizzly bear ecosystems.

### *Population risk factors*

Today the primary risk factors to grizzly bear populations are human-bear conflicts, habitat degradation or loss because of human activities, and loss of habitat security (U.S. Fish and Wildlife Service 1992).

### *Population viability and likelihood of persistence*

The current understanding of population risk factors reflects nearly three decades of research and management effort. It combines insights gained from management with those gained from detailed modeling of demographic and genetic considerations.

Beginning in the mid 1970s, attempts to determine population viability and the likelihood of persistence have focused on the relationship between the amount of space available and the number of grizzly bears that can be sustained in an area (U.S. Fish and Wildlife Service 1992). Determination of the numbers that can be sustained within a specific bear ecosystem is based on research information relative to population density, individual habitat use, and home range when available for a particular ecosystem or, if unavailable, from reasonable extrapolations of information from other ecosystems. Based on these interpretations, the recovery plan calls for a minimum population goal of 90 bears in the two smaller ecosystems, the Selkirks and Cabinet Yaak, which are contiguous with Canada. The Northern Continental Divide and Yellowstone are larger and targeted for populations of 236 and 306 respectively, given the assumptions of human-caused mortality outlined in the recovery plan. A population in the range of 200 to 400 is considered to be recovered in the Northern Cascade Recovery Zone and as part of the proposed reintroduction of bears into the Bitterroot Mountains (U.S. Fish and Wildlife Service 1997). No current plans exist for reintroducing grizzly bears to the San Juan Mountains in Colorado.

The minimum viable population estimate of 90 grizzly bears is based on a computer simulation model and is for an isolated population with no immigration (Shaffer and Samson 1985, U.S. Fish and Wildlife Service 1992). This computer simulation model for grizzly bears incorporates the effects of inbreeding depression as a function of effective population size and the model is similar to that of Shaffer (1978) in that population structure, age and sex specific reproductive, and mortality rates as well as stochastic factors are considered (Interagency Grizzly Bear Committee 1987). There is no direct evidence of a negative impact because of the loss of genetic diversity — a factor often cited as contributing to population declines in grizzly bears (Allendorf and Servheen 1986) and other species. It is, however, considered to be sound management to introduce one grizzly bear every 10 years to an isolated ecosystem as is the case in the Yellowstone Recovery Zone to ensure minimal loss of genetic variation.

The current recovery plan recognizes that any increase in bear-human conflict or negative change in the quality and security of grizzly bear habitat could influence the viability of a bear population (U.S. Fish and Wildlife Service 1992). To remove the grizzly bear from threatened status in each ecosystem, the recovery plan suggests demographic goals and a step-down outline to establish the population objective for recovery and to identify limiting factors. The demographic recovery goals include a target number of females producing cubs over a period of time (usually 6 years) and an acceptable level of human-caused mortality during a particular period of time (usually 2 years). Where populations are known to be very low as in the Cabinet Yak and Selkirk ecosystems, the mortality goal is no known human-caused mortalities.

The step-down outline to achieve recovery goals and therefore a viable population involves eight steps (U.S. Fish and Wildlife Service 1992):

- Establish the population objective for recovery and identify limiting factors: determine the population conditions at which the species is viable and self-sustaining; determine current population conditions; and identify human-related population limiting factors if present populations differed from desired.
- Redress population limiting factors: manage sources of direct mortality; reduce accidental mortality; identify and reduce sources of indirect mortality; and coordinate, monitor and report activities relating to redressing population limiting factors and compliance with the Recovery Plan.
- Determine the habitat and space required for achievement of the grizzly bear population goal.
- Monitor populations and habitats before, during and after recovery.
- Manage populations and habitats.
- Develop and initiate appropriate information and education programs.
- Implement the recovery program through a coordinator.
- Revise appropriate federal and state regulations to reflect current situations and initiate international cooperation.

## 5.3 Assessment of Multiple Species for the Northwest Forest Plan

### *Background*

Throughout the 1980s, much of the controversy surrounding the management of old-growth forests in the Pacific Northwest revolved around northern spotted owls. However, concurrent with ongoing efforts to develop appropriate management for spotted owl habitat, a growing recognition of the broader concern for biodiversity in old-growth forests emerged. Early attempts to describe the biology of these forests included the work of Franklin et al. (1981) and the results of the Forest Service old-growth research program (Ruggiero et al. 1991). The first explicit assessment of land management impacts on biodiversity in the Pacific Northwest was included in a report prepared for Congress by Johnson et al. (1991). The findings of this report were included in the Forest Service's EIS on spotted owl management (USDA 1992), and were in part responsible for the court's finding that the EIS was inadequate. To address the court's concerns, the Forest Service formed the Scientific Analysis Team (Thomas et al. 1993a) to assess additional actions that should be taken to maintain biodiversity of late-successional and old-growth forests. This team attempted an analysis of effects of land management on viability of over 600 species closely associated with old-growth forests. Before the Forest Service could act on the team's recommendations, President Clinton chartered the Forest Ecosystem Management Assessment Team (FEMAT) to assess the impacts of alternative land management proposals on both the biology and economy of the Pacific Northwest. FEMAT's assessment was even more comprehensive than that of the Scientific Analysis Team.

### *Analysis*

FEMAT (Thomas et al. 1993b) assessed the effects of seven management alternatives on persistence of 82 terrestrial vertebrates, 21 groups of fish, 102 species of mollusks, 124 vascular plants, 157 species of lichens, 527 species of fungi, 106 species of bryophytes, and 15 functional groups of arthropods that may include 10,000 or more species. Assessments were done by 14 separate panels employing more than 70 species experts. These experts were presented with maps and data describing the proposed management options, and then instructed to use a likelihood voting methodology to express their opinion about the likely persistence of species under each of the options over a period of at least 100 years. Panelists assigned 100 "likelihood" points across four outcomes. The spread of these points was used to express scientific and personal uncertainty. The outcomes used in this process were:

- *Outcome A.* Habitat is of sufficient quality, distribution, and abundance to allow the species population to stabilize, well distributed across federal lands.
- *Outcome B.* Habitat is of sufficient quality, distribution, and abundance to allow the species population

to stabilize, but with significant gaps in the historical species distribution on federal land. These gaps cause some limitation in interactions among local populations. The significance of gaps must be judged relative to the species distribution, range, and life history, and the concept of metapopulations.

- *Outcome C.* Habitat only allows continued species existence in refugia, with strong limitations on interactions among local populations.
- *Outcome D.* Habitat conditions result in species extirpation from federal land.

It was not the objective of the process to achieve consensus among experts, so each expert's scores were recorded independently. Results brought forward by FEMAT were the means of the panelists' scores and indicated that the options would provide a greater likelihood of persistence for vertebrates than for the other taxonomic groups analyzed (Meslow et al. 1994). All but two of the options were judged to provide at least an 80 percent likelihood of Outcome A for virtually all of the vertebrates. Most vascular plants were judged to have high likelihoods of persistence under the options, although some options were clearly more beneficial than others. Those vascular plants that were not rated at 80 or higher likelihood of achieving Outcome A were generally rare or locally endemic species. Mitigation for these species is relatively straightforward.

Other taxonomic groups fared less well. Even under the option that would preserve the most late-successional forest, only about one fourth of the invertebrate and nonvascular species were judged to have an 80 percent likelihood of Outcome A. The inherently rare and/or locally endemic nature of these species caused the experts to judge that their habitat could be significantly disrupted even under alternatives that established large, well- distributed reserves of late-successional forests.

### *Use of the analysis to establish management direction*

The FEMAT report was used in the development of first a draft and then a final EIS for management of late-successional and old-growth habitat in the Pacific Northwest (USDA/USDI 1994). The analysis of species viability was considered by management in selecting the preferred alternative (which had been FEMAT Option 9) for the draft EIS. Many public comments on the draft expressed concern that a large number of species were judged to be at significant risk under Alternative 9, and that an alternative that provided higher assurance of viability for these species should be selected. Based on these concerns, a team of federal biologists (Holthausen et al. 1994) was chartered to review the FEMAT viability assessment and to recommend additional mitigation for species judged to be at risk under Alternative 9. This team reviewed the status of 490 species or species groups out of the 1,120 that had been initially analyzed by FEMAT. The review provided additional detail about the basis for the FEMAT ratings for each species. The large majority of species involved in this review were rare and/or locally endemic. Concerns about these species generally related to past declines and the likelihood for simple habitat disruption in the future. In some cases, the known locations of the species were in areas left open to timber harvest under draft Alternative 9, and no measures were in place to protect the sites. In other cases, species locations were not known, but it was suspected that they were present in small, remnant old-growth patches that had received little protection under draft Alternative 9. Finally, a significant number of the species that were judged to be at risk were riparian associates, and the degree of protection that would be provided for these species under draft Alternative 9 was unclear.

As a result of the additional analysis completed by Holthausen et al. (1994), a series of mitigation measures was recommended in the final EIS (USDA 1994). These included broadening riparian management zones, protecting a portion of the small, remnant old-growth patches, strengthening provisions for large tree, snag and log retention in harvest areas, and the use of surveys to locate and provide protection for sites of rare and endemic species.

Lawsuits were filed challenging the government's adoption of the Northwest Forest Plan. Several of these directly addressed the interpretation of the National Forest Management Act (NFMA) viability provision. The Ninth Circuit strongly endorsed the government's application of the diversity provision of NFMA and the viability provision in the NFMA regulations ((*Seattle Audubon Society v. Lyons*, 871 F. Supp. 1291 (W.D. Wash. 1994)):

> "There is similarly little or no support for the environmental plaintiffs' contention that the selected alternative violates the applicable viability standards. The district court correctly explained that the selection of an alternative with a higher likelihood of viability would preclude any multiple use compromises contrary to the overall mandate of the NFMA. See *SAS*, 871 F.Supp. at 1315–16; see also 16 U.S.C. S 1604 (g)(3)(B) (diversity is to be addressed in light of "overall multiple-use objectives"); 36 C.F.R. 219.27(a)(6) (habitat main-

tained and improved "to the degree consistent with multiple-use objectives"); 219.26 (provide for diversity consistent with multiple-use objectives); 219.27(a)(5) (forest plans should "maintain diversity of plant and animal communities to meet overall multiple-use objectives"). Here, the record demonstrates that the federal defendants considered the viability of plant and animal populations based on the current state of scientific knowledge. Because of the inherent flexibility of the NFMA, and because there is no showing that the federal defendants overlooked any relevant factors or made any clear errors of judgment, we conclude that their interpretation and application of the NFMA's viability regulations was reasonable. See *Batterson v. Francis*, 432 U.S. 416, 425-26 (1977) (the Secretary's interpretation of a statutory term is entitled to substantial deference)."

## 6 CONCLUSIONS

Consideration of individual species should remain a vital part of natural resource management under the paradigm of ecosystem management. Where the continued existence of species across their range is in question, formal analyses of population viability may be needed. Such analyses may be required by NFMA or ESA in some cases. In others, viability analysis is simply a necessary part of making informed management decisions. To be most useful, PVAs must be linked to factors under control of management, integrated with predictive models, conducted for an appropriate set of species at appropriate scales, and be carefully coordinated between science and management.

Viability analysis can be used by management in several different ways, as illustrated by examples presented in this chapter. The Interagency Scientific Committee used the component parts of a viability analysis to develop a management proposal for northern spotted owl habitat. Raphael and his colleagues (1994) used a more sophisticated analytical tool to assess long-term persistence of owls under alternative management scenarios. FEMAT used a simple expert opinion process to assess implications of alternative ecosystem management strategies on a broad array of species. The ultimate outcome of this assessment was adjustment of the selected management proposal to accommodate better the needs of multiple species. In multiple-species analysis in the Columbia River Basin, results of expert opinion assessments are to be used to help prioritize habitat restoration efforts. An additional use of viability analysis is identification of key parameters for monitoring. Finally, as illustrated by the FEMAT case study in this chapter, viability analysis can play a key role in legal defense of proposed management actions.

In two decades of implementation of the viability provision of NFMA, the Forest Service has had to deal with a series of significant policy dilemmas. The questions that have been raised are specific to NFMA, but are likely also to be encountered by anyone dealing with population viability within the context of ecosystem management. What species should be the subject of analytical and management attention? What is an appropriate level of assurance that species will continue to exist? What does "well distributed" mean for different taxa, and what is the relationship between distribution and persistence? How should locally endemic species be treated in large-scale assessments? What contribution should federal and non-federal lands make to species persistence? How should we use knowledge of the past and current conditions of a species to shape our goals for its future condition? How should we deal with species for which knowledge is extremely limited?

It is likely that views about these questions, which sit at the intersection of science and public policy, will continuously evolve. The challenge to professionals who undertake viability analysis within a management context is to bring the analysis forward in a way that allows managers to understand the implications of different policies. There are three keys to such a presentation. First, viability analysis should be conducted within the framework of risk analysis so that the implications of different potential actions can be fairly compared. Second, the analysis ought to be structured around the questions that have to be answered by management. For example, a recent analysis by Lehmkuhl and his colleagues (1997) looked separately at cumulative effects on viability, and the contribution of federal lands to those cumulative effects. This analysis also explicitly assessed historic, current, and projected future conditions. The third key to the successful use of viability analysis within a management context is simple communication between biologists performing the analysis and managers using it. Biologists should understand that managers must look at a broad array of options, and that the role of the analysis is to present unbiased information on the implications of all options. Managers should understand the uncertainty surrounding the analysis and the limitations imposed by lack of knowledge and limited ability to predict the future.

A variety of techniques have been used for viability analysis, ranging from simple rating systems to sophisticated models employing extensive demographic data. It is unlikely that data will ever be available to allow sophisticated analysis of most species of concern, so managers will continue to rely on simpler techniques that are less rigorous and possibly less reliable. Despite scientific criticism of these simpler processes, they have been upheld in important court decisions. Regardless of the technique used to assess viability, extensive documentation and the use of a logical process are necessary for credibility.

## REFERENCES


Akçakaya, H.R. 1992. Population viability analysis and risk assessment. In: McCollough, D.R. and R.H. Barrett (eds.), *Wildlife 2001: Populations*: 148–157. Elsevier Publishers, Oxford.

Akçakaya, H.R., M.A. McCarthy, and J. Pearce. 1995. Linking landscape data with population viability analysis: management options for the helmeted honeyeater. *Biological Conservation* 73: 169–176.

Allendorf, F.W., and C. Servheen. 1986. Genetics and the conservation of the grizzly bear. *Tree* 1: 88–89.

Bart, J. 1995. Amount of suitable habitat and viability of northern spotted owls. *Conservation Biology* 9: 943–946.

Bart, J., and E.D. Forsman. 1992. Dependence of northern spotted owls *Strix occidentalis caurina* on old-growth forest in the western USA. *Biological Conservation* 62: 95–100.

Boyce, M.S. 1992. Population viability analysis. *Annual Review of Ecology and Systematics* 23: 481–506.

Boyce, M.S., J.S. Meyer, and L. Irwin. 1994. Habitat-based PVA for the northern spotted owl. In: Fletcher, D.J. and B.F.J. Manly (eds.), *Statistics in Ecology and Environmental Monitoring*, 63–85. Otago Conf. Ser. 2, Univ. Otago Press, Dunedin, New Zealand.

Brown, E.R. (ed.). 1985. *Management of Wildlife and Fish Habitats in Forests of Western Oregon and Washington*. USDA Forest Service, Pacific Northwest Region, Portland, OR. 2 Vols.

Burnham, K.P., D.R. Anderson, and G.C. White. 1996. Meta-analysis of vital rates of the northern spotted owl. *Studies in Avian Biology* 17: 92–101.

Cleaves, D.A. 1994. *Assessing Uncertainty in Expert Judgments about Natural Resources*. USDA Forest Service, Southern Forest Experiment Station, New Orleans, LA. GTR SO-110.

Cole, C.A., and R.L. Smith. 1983. Habitat suitability indices for monitoring wildlife populations — an evaluation. *Transactions of the North American Wildlife and Natural Resources Conference*. 48: 367–375.

Conner, R.N. 1988. Wildlife populations: minimally viable or ecologically functional? *Wildlife Society Bulletin* 16: 80–84.

Conroy, M.J. 1993. Testing hypotheses about the relationship of habitat to animal survivorship. In: Lebreton, J.D. and P.M. North (eds), *Marked Individuals in the Study of Bird Populations*, pp. 331–342. Birkhäuser Verlag, Basel.

Dippon, D., C. Caldwell, J. Nighbert, and K. McKelvey. 1992. Linking a spatially-oriented northern spotted owl population dynamics model with the Western Oregon Digital Database. GIS 92 Symposium, Vancouver, British Columbia, Canada.

Fagen, R. 1988. Population effects of habitat change: a quantitative assessemnent. *Journal of Wildlife Management* 52: 41–46.

Forsman, E.D., S. DeStefano, M.G. Raphael, and R.J Gutiérrez. 1996a. Demography of the northern spotted owl. *Studies in Avian Biology* 17: 1–122.

Forsman, E.D., S.G. Sovern, E.D. Seaman, K.J. Maurice, M. Taylor, and J.J. Ziza. 1996b. Demography of the northern spotted owl on the Olympic Peninsula and Cle Elum Ranger District, Washington. *Studies in Avian Biology* 17: 21–30.

Frankel, O.H., and M.E. Soulé. 1981. *Conservation and Evolution*. Cambridge Univ. Press, Cambridge, UK.

Franklin, J.F., K. Cromack, Jr., W. Denison, A. McKee, C. Maser, J. Sedell, F. Swanson, and G. Juday. 1981. Ecological Characteristics of Old-growth Douglas-fir Forests. USDA Forest Service General Technical Report PNW-118. Portland, OR: Pacific Northwest Forest and Range Experiment Station.

Franklin, J.F. 1993. Preserving biodiversity: species, ecosystems, or landscapes? *Ecological Applications* 3: 202–205.

Gilpin, M.E. and M.E. Soule. 1986. Minimum viable populations: processes of species extinction. In: Soule, M.E. (ed.), *Conservation Biology: the Science of Scarcity and Diversity*, pp. 19–34. Sunderland, MA: Sinauer Associates.

Grubb, T.G. 1988. Pattern recognition — a simple model for evaluating wildlife habitat. USDA Forest Service, Rocky Mountain Forest and Range Experiment Station, Fort Collins, CO. Research Note RM-487.

Grumbine, R.E. 1994. What is ecosystem management? *Conservation Biology* 8: 27–38.

Herrero, S. 1972. Aspects of evolution and adaptation in American black bears (*Ursus americanus Pallas*) and brown and grizzly bears (*Urus arctos Linne*) of North America. In: Herrero, S (ed.) *Bears — Their Biology and Management*, pp. 221–233. IUCN Publ. New Series 23.

Hof, J., and M.G. Raphael. 1997. Optimization of habitat placement: a case study of the northern spotted owl in the Olympic Peninsula. *Ecological Applications* 7: 1160–1169.

Holthausen, R.S., R. Anthony, K. Aubry, K. Burnett, N. Fredricks, J. Furnish, R. Lesher, E.C. Meslow, M.G. Raphael, R. Rosentreter, and E.E. Starkey. 1994. Appendix J2: Results of Additional Species Analysis. In: U. S. Department of Agriculture; U.S. Department of the Interior; National Oceanic and Atmospheric Administration; U.S. Environmental Protection Agency. 1994. Final supplemental environmental impact statement on management of habitat for late-successional and old-growth forest related species within the range of the northern spotted owl. Portland, OR. 2 vols.

Holthausen, R.S., M.G. Raphael, K.S. McKelvey, E.D. Forsman, E.E. Starkey, and D.E Seaman. 1995. The Contribution of Federal and Non-federal Habitat to Persistence of the Northern Spotted Owl on the Olympic Peninsula, Washington: Report of the Reanalysis Team. USDA Forest Service Gen. Tech. Rep. PNW-GTR-352. Portland, OR: Pacific Northwest Forest and Range Experiment Station.

Hoover, R.L., and D.L. Wills (eds.). 1984. Managing Forested Lands for Wildlife. Colorado Division of Wildlife. In cooperation with USDA Forest Service Rocky Mountain Region, Denver, CO.

Hunter, M.L., Jr. 1990. *Wildlife, Forests, and Forestry*.

Englewood Cliffs, NJ: Prentice-Hall.

Interagency Grizzly Bear Committee. 1987. Washington, DC: The National Wildlife Federation.

Johnson, K.N., J.F. Franklin, J.W. Thomas, and J. Gordon. 1991. Alternatives for management of late-successional forests of the Pacific Northwest. A report to the Agriculture Committee and the Merchant Marine Committee of the U.S. House of Representatives.

Kangas, J., J. Karsikko, L. Laasonen, and T. Pukkala. 1993. A method for estimating suitability function of wildlife habitat for forest planning on the basis of expertise. *Silva Fennica* 27: 259–268.

Lamberson, R.H., McKelvey, R., B.R. Noon, and C. Voss. 1992. A dynamic analysis of northern spotted owl viability in a fragmented forest landscape. *Conservation Biology* 6: 505–512.

Lamberson, R.H., B.R. Noon, C. Voss, and K. McKelvey. 1994. Reserve design for territorial species: the effects of patch size and spacing on the viability of the northern spotted owl. *Conservation Biology* 8: 185–195.

Lande, R. 1987. Extinction thresholds in demographic models of territorial populations. *American Naturalist* 130: 624–635.

Lande, R. 1988. Demographic models of the northern spotted owl (*Strix occidentalis caurina*). *Oecologia* 75: 601–607.

Lande, R. 1993. Risks of population extinction from demographic and environmental stochasticity and random catastrophes. *American Naturalist* 142: 911–927.

Landres, P.B., J. Verner, and J.W. Thomas. 1988. Ecological uses of vertebrate indicator species: a critique. *Conservation Biology* 2: 316–328.

Lebreton, J.D., K.P. Burnham, J. Clobert, and D.R. Anderson. 1992. Modeling survival and testing biological hypotheses using marked animals: a unified approach with case studies. *Ecological Monographs* 62: 67–118.

Lee, D.C. In press. Assessing land-use impacts on bull trout using Bayesian belief networks. In: Ferson, S. (ed.), *Quantitative Methods in Conservation Biology*, New York: Springer-Verlag.

Lehmkuhl, J.F., M.G. Raphael, R.S. Holthausen, J.R. Hickenbottom, R. Naney, and J.S. Shelly. 1997. Chapter 4: Historical and current status of terrestrial species and the effects of proposed alternatives. In: Quigley, T.T., K.M. Lee, and S.J. Arbelbide (eds.), Evaluation of the Environmental Impact Statement Alternatives by the Science Integration Team. USDA Forest Service General Technical Report PNW-GTR-406. Portland, OR: Pacific Northwest Research Station. 2 volumes.

Lehmkuhl, J.F., and L.F. Ruggiero. 1991. Forest fragmentation in the Pacific Northwest and its potential effects on wildlife. In: Ruggiero, L.F. K.B. Aubry, A.B. Carey, and M.H. Huff (eds.) Wildlife and Vegetation of Unmanaged Douglas-fir Forests, 35–46. USDA Forest Service General Technical Report PNW-GTR-285. Portland, OR: Pacific Northwest Research Station.

Leopold, A. 1933. *Game Management*. Charles Scribner Sons, New York.

Mace, G.M., and N.J. Collar. 1995. Extinction risk assessment for birds through quantitative criteria. *Ibis* 137: 240–246.

Mace, G.M. and R. Lande. 1991. Assessing extinction threats: toward a reevaluation of IUCN threatened species categories. *Conservation Biology* 5: 148–157.

Manly, B.F.J., L.L. McDonald, and D. Thomas. 1993. *Resource Selection by Animals: Statistical Design and Analysis for Field Studies*. Chapman and Hall, New York.

Marcot, B.G. 1986. Use of expert systems in wildlife-habitat modeling. In: Verner, J., M.L. Morrison, and C.J. Ralph (Eds.), *Wildlife 2000: Modeling Habitat Relationships of Terrestrial Vertebrates*, pp. 145–150. University of Wisconsin Press, Madison.

Marcot, B.G., M. Castellano, J. Christy, L. Croft, J. Lehmkuhl, R. Naney, R. Rosentreter, R. Sandquist, and E. Zieroth. 1997. Chapter 5: Terrestrial ecology assessment. In: Quigley, T.T., S.J. Arbelbide, and S.F. McCool (eds.), An Assessment of Ecosystem Components in the Interior Columbia Basin and Portions of the Klamath and Great Basins. USDA Forest Service General Technical Report PNW-GTR-405. Portland, OR: Pacific Northwest Research Station.

Marcot, B.G., and R. Holthausen. 1987. Analyzing population viability of the spotted owl in the Pacific Northwest. *Transactions of the 52nd North American Wildlife and Natural Resource Conference*: 333–347.

McDonald, D.B., and H. Caswell. 1993. Matrix methods for avian demography. In: Power, D.M. (ed.), *Current Ornithology*, Vol. 10, pp. 139–185. New York: Plenum Press.

McKelvey K., B.R. Noon, and R H. Lamberson. 1993. Conservation planning for species occupying fragmented landscapes: the case of the northern spotted owl. In: Kareiva, P., J. Kingsolver, and R. Huey (eds.), *Biotic Interactions and Global Change*, pp. 424–450. Sinauer Assoc., Sunderland MA.

Meslow, E.C., R.S. Holthausen, and D.A. Cleaves. 1994. Assessment of terrestrial species and ecosystems. *Journal of Forestry* 92: 24–27.

Mills, L.S., M.E. Soulé, and D.D. Doak. 1993. The keystone-species concept in ecology and conservation. *BioScience* 43: 219–224.

Morrison, M.L., B.G. Marcot, and R.W. Mannan. 1992. *Wildlife-habitat Relationships: Concepts and Applications*. University of Wisconsin Press. Madison, WI.

Nelson, R. A., and H. Salwasser. 1982. The Forest Service wildlife and fish habitat relationships program. *Transactions of the North American Wildlife Natural Resources Conference* 47: 174–183.

O'Connell, M. 1992. Response to: six biological reasons why the endangered species act doesn't work and what to do about it. *Conservation Biology* 6: 140–145.

Patton, D.R. 1987. Is the use of management indicator species feasible? *Western Journal of Applied Forestry* 2: 33–34.

Pollack, K.H., J.D. Nichols, C.Brownie, and J.E. Hines. 1990. Statistical inference for capture-recapture experiments. *Wildlife Monographs* 107: 1–97.

Rabinowitz, D. 1981. Seven forms of rarity. In Synge, H. (Ed.), *The Biological Aspects of Rare Plant Conservation*, pp. 205–217. Wiley, New York.

Rabinowitz, D., S. Cairns, and T. Dillon. 1986. Seven forms of rarity and their frequency in the flora of the British Isles. In: Soule, M.E. (ed.), *Conservation Biology: the Science of Scarcity and Diversity*, 182–204. Sinauer Publishers, Sunderland, MA.

Raphael, M.G., R.G. Anthony, S. DeStefano, E.D. Forsman, A.B. Franklin, R. Holthausen, E.C. Meslow, and B.R. Noon. 1996. Use, interpretation, and implications of demographic analyses of northern spotted owl populations.

*Studies in Avian Biology* 17: 102–112.

Raphael, M.G., and B.G. Marcot. 1986. Validation of a wildlife–habitat relationships model: vertebrates in a Douglas-fir sere. In: Verner, J., M.L. Morrison, and C.J. Ralph (eds.), *Wildlife 2000: Modeling Habitat Relationships of Terrestrial Vertebrates*, pp. 129–148. University of Wisconsin Press, Madison.

Raphael, M.G., and B.G. Marcot. 1994. Key questions and issues — species and ecosystem viability. *Journal of Forestry* 92: 45–47.

Raphael, M.G., K.S. McKelvey, and B.M. Galleher. In press. Using geographic information systems and spatially explicit population models for avian conservation: a case study. In: Marzluff, J.M. and R. Salabanks (eds.), *Avian Conservation: Research and Management*, Island Press, Washington, DC.

Raphael, M.G., J. A. Young, K. McKelvey, B.M. Galleher, and K.C. Peeler. 1994. A simulation analysis of population dynamics of the northern spotted owl in relation to forest management alternatives. Final Environmental Impact Statement on Management of Habitat for Late-Successional and Old-Growth Forest Related Species Within the Range of the Northern Spotted Owl. Volume II, Appendix J-3.

Reed, J.M. 1992. A system for ranking conservation priorities for neotropical migrant birds based on relative susceptibility to extinction. In: Hagan, J.M. and D.W. Johnston (eds.), *Ecology and Conservation of Neotropical Migrant Landbirds*, pp. 524-536. Smithsonian Institution Press, Washington, DC.

Rohlf, D.J. 1991. Six biological reasons why the Endangered Species Act doesn't work — and what to do about it. *Conservation Biology* 5: 273–282.

Ruggiero, L.F., K.B. Aubry, A.B. Carey, and M.H. Huff. 1991. Wildlife and vegetation of unmanaged Douglas-fir forests. USDA Forest Service General Technical Report PNW-285. Portland, OR: Pacific Northwest Forest and Range Experiment Station.

Ruggiero, L.R., G.D. Hayward, and J.R. Squires. 1994. Viability analysis in biological evaluations: concepts of population viability analysis, biological population, and ecological scale. *Conservation Biology*, 8: 364–372.

Salwasser, H., S.P. Mealey, and K. Johnson. 1984. Wildlife population viability — a question of risk. *Transactions of the North American Wildlife and Natural Resources Conference*, 49: 421–439.

Samson, F.B., F. Perez-Trejo, H. Salwasser, L.F. Ruggiero, and M.L. Shaffer. 1985. On determining and managing minimum population size. *Wildlife Society Bulletin* 13: 425–433.

Schamberger, M., and W.B. Krohn. 1982. Status of the habitat evaluation procedures. *Transactions of the North American Wildlife and Natural Resources Conference* 47: 154–164.

Schonewald-Cox, C.M. 1983. Conclusions: guideline to management: a beginning attempt. In: Schonewald-Cox, C.M., S.M. Chambers, B. MacBryde, and W.L. Thomas (Eds.), *Genetics and Conservation: a Reference for Managing Wild Plant and Animal Populations*, pp. 414–445. Benjamin/Cummings, Menlo Park, CA.

Schonewald-Cox, C.M., S.M. Chambers, B. MacBryde, and W.L. Thomas. 1983. *Genetics and Conservation: a Reference for Managing Wild Plant and Animal Populations.* Benjamin/Cummings, Menlo Park, CA.

Shaffer, M.L. 1978. Determining minimum viable population size: a case study of the grizzly bear (*Ursus arctos*). Ph.D. Dissertation. School of Forestry and Environmental Studies, Duke University, Durham, NC.

Shaffer, M. 1981. Minimum population sizes for species conservation. *BioScience* 31: 131–134.

Shaffer, M.L. 1987. Minimum viable populations: coping with uncertainty. In: Soule, M.E. (ed.), *Viable Populations*, pp. 69–85. Cambridge University Press, New York.

Shaffer, M.L., and F.B. Samson. 1985. Population size and extinction: a note on determining critical population sizes. *American Naturalist* 125: 145–152.

Sidle, W.B., and L.H. Suring. 1986. *Wildlife and Fisheries Habitat Management Notes: Management Indicator Species for the National Forest Lands in Alaska.* Technical Publication R10-TP-2. USDA Forest Service, Alaska Region, Juneau, Alaska.

Skalski, J.R., A. Hoffman, and G.S. Smith. 1993. Testing the significance of individual- and cohort-level covariates in animal survival. In: Lebreton, J.D. and P.M. North (eds.), *Marked Individuals in the Study of Bird Populations*, pp. 9–28. Birkhäuser Verlag, Basel, Switzerland

Soulé, M.E. 1980. Thresholds for survival: maintaining fitness and evolutionary potential. In: Soulé, M.E. and B.A. Walkaways (eds.), *Conservation Biology: an Evolutionary-Ecological Perspective*, pp. 151–170. Sinauer Associates, Sunderland, MA.

Soulé, M.E. 1983. What do we really know about extinction? In: Schonewald-Cox, C.M., S.M. Chambers, B. MacBryde, and W.L. Thomas (eds.), *Genetics and Conservation: a Reference for Managing Wild Plant and Animal Populations.* Benjamin/Cummings. Menlo Park, CA.

Soulé, M.E. 1987 (ed.). *Viable Populations for Conservation.* Cambridge University Press, Cambridge.

Soulé, M.E., and B.A. Wilcox. 1980. *Conservation Biology: an Evolutionary-ecological Perspective.* Sinauer Associates, Inc., Sunderland, MA.

Storer, T.I., and L.P. Tevis. 1955. *California Grizzly.* University of Nebraska Press, Lincoln and London.

Taylor, B.L. 1995. The reliability of using population viability analysis for risk classification of species. *Conservation Biology* 9: 551–558.

Thomas, J.W. 1979. *Wildlife Habitats in Managed Forests of the Blue Mountains of Oregon and Washington.* USDA Forest Service Handbook No. 553. Portland, OR. 512 p.

Thomas, J.W., E.D. Forsman, J.B. Lint, E.C. Meslow, B.R. Noon, and J. Verner. 1990. A conservation strategy for the northern spotted owl. Interagency Scientific Committee to Address the Conservation of the Northern Spotted Owl. USDA Forest Service, USDI Bureau of Land Management, Fish and Wildlife Service, and National Park Service. Portland, OR. U.S. Government Printing Office 791-171/20026. Washington, D.C.

Thomas, J.W., M.G. Raphael, R.G. Anthony, E.D. Forsman, A.G. Gunderson, R.S. Holthausen, B.G. Marcot, G.H. Reeves, J.R. Sedell, and D.M. Solis. 1993a. Viability Assessments and Management Considerations for Species Associated with Late-successional and Old-growth Forests of the Pacific Northwest: the Report of the Scientific Analysis Team. USDA Forest Service, National Forest System, Forest Service Research. Portland, OR.

Thomas, J.W., and M.G. Raphael, R.G. Anthony, E.D.

Forsman, A.G. Gunderson, R.S. Holthausen, B.G. Marcot, G.H. Reeves, J.R. Sedell, and D.M. Solis. 1993b. Forest Ecosystem Management: an Ecological, Economic, and Social Assessment. Report of the Forest Ecosystem Management Assessment Team. USDA Forest Service, USDA Bureau of Land Management, and USDI Fish and Wildlife Service, Portland, OR.

Thomasma, L.E., T.D. Drummer, and R.O. Peterson. 1991. Testing the habitat suitability index model for the fisher. *Wildlife Society Bulletin*. 19: 291–297.

U.S. Department of Agriculture, Forest Service. 1975. Fisheries and Wildlife Habitat Management Handbook. Ottawa, Hiawatha, and Huron-manistee National Forests, Michigan. FSH 2609.23-R9. Milwaukee, WI.

U.S. Department of Agriculture. 1992. Final Supplement to the Environmental Impact Statement for an Amendment to the Pacific Northwest Regional Guide. Portland, OR. 2 vols.

U.S. Department of Agriculture; U.S. Department of the Interior; National Oceanic and Atmospheric Administration; U.S. Environmental Protection Agency. 1994. Final Supplemental Environmental Impact Statement on Management of Habitat for Late-successional and Old-growth Forest Related Species Within the Range of the Northern Spotted Owl. Portland, OR. 2 vols.

U.S. Department of the Interior. 1990. Endangered and threatened wildlife and plants: determination of threatened status for the Northern Spotted Owl. *Federal Register* 55: 26114–26194.

U.S. Department of the Interior. 1992. Final Draft Recovery Plan for the Northern Spotted Owl. Portland, OR. 2 vols. U.S. Department of Interior, Washington, DC.

U.S. Department of the Interior. 1994. Desert Tortoise (Mojave population) Recovery Plan. USDI Fish and Wildlife Service, Portland, OR.

U.S. Fish and Wildlife Service. 1982. Grizzly Bear Recovery Plan. U.S. Fish and Wildlife Service, Denver, CO.

U.S. Fish and Wildlife Service. 1993. Grizzly Bear Recovery Plan (revised). U.S. Fish and Wildlife Service, Denver, CO.

U.S. Fish and Wildlife Service. 1997. Grizzly Bear Recovery in the Bitterroot Ecosystem. U.S. Fish and Wildlife Service, Denver, CO.

Van Horne, B. 1983. Density as a misleading indicator of habitat quality. *Journal of Wildland Management* 47: 893–901.

Williams, G.L., D.R. Russell, and W.K. Seitz. 1977. Pattern recognition as a tool in the ecological analysis of habitat. In: Classification, Inventory, and Analysis of Fish and Wildlife Habitat, 521–531. U.S. Fish and Wildlife Service FWS/OBS-78/76.

Wilcove, D. 1993. Getting ahead of the extinction curve. *Ecological Applications* 3: 218–220.

Wisdom, M.J., and L.S. Mills. 1997. Sensitivity analysis to guide population recovery: prairie-chickens as an example. *Journal of Wildlife Management* 61: 302–312.

With, K.A., and T.O. Crist. 1995. Critical thresholds in species' response to landscape structure. *Ecology* 76: 2446–2459.

Zeiner, D.C., W.F. Laudenslayer, Jr., and K.E. Mayer (eds.). 1990. *California's Wildlife* (3 vols.). California Department of Fish and Game, Sacramento, CA.

## THE AUTHORS

**Richard S. Holthausen**
*US Forest Service*
*Rocky Mountain Research Station*
*2500 South Pine Knoll*
*Flagstaff, AZ 86001, USA*

**Martin G. Raphael**
*USDA Forest Service*
*Pacific Northwest Research Station*
*3625 93rd Ave. SW*
*Olympia, WA 98512, USA*

**Fred B. Samson**
*USDA Forest Service*
*Northern Region*
*Box 7669*
*Missoula, MT 59807, USA*

**Dan Ebert**
*USDA Forest Service*
*Boise National Forest*
*1750 Front Street*
*Boise, ID 83702, USA*

**Ronald Hiebert**
*National Park Service*
*1709 Jackson Street*
*Omaha, NE 68102, USA*

**Keith Menasco**
*USDA Forest Service*
*Kaibab National Forest*
*800 S. 6th Street*
*Williams, AZ 86046, USA*

# An Ecosystem Approach for Understanding Landscape Diversity

James R. Gosz, Jerry Asher, Barbara Holder, Richard Knight, Robert Naiman, Gary Raines, Peter Stine, and T.B. Wigley

**Key issues addressed in this chapter**

- *A broad review of factors responsible for landscape diversity*
- *Using an ecosystem approach for evaluating landscapes*
- *Analysis of landscape patterns and guidance on understanding pattern effects on management*
- *Anthropogenic factors, historical and current, and land-use effects on landscapes*
- *Roles of Keystone species and invasive exotic species*
- *Data management issues for measuring, analyzing, and managing landscape*

**Keywords: Fragmentation, GIS, riparian zones, scale, biological diversity, land-use planning**

## 1 INTRODUCTION

One has only to consider the ideas of Thoreau and the descriptions in the "Sand County Almanac" to realize that the concept of the landscape is not new (Risser 1987). The term "landscape ecology" was used by Carl Troll (1939) to describe his study of aerial photographs (ref. in Golley 1987) and arose out of European traditions of landscape-level geography and vegetation science. Concepts of landscape ecology have been developing around the world for years (see ref. in Forman and Godron 1986) and more recently in the United States (Forman 1995). As a separate field of scientific inquiry, landscape ecology considers the development and maintenance of spatial heterogeneity, interactions and exchanges across heterogeneous landscapes, the influence of heterogeneity on biotic and abiotic processes, and the management of that heterogeneity.

Forman (1995) defines a landscape as "a heterogeneous land area composed of a cluster of interacting components that is repeated in a similar format throughout". Hunter (1996) described landscape diversity as including all biotic and abiotic components of a landscape, defined as a mosaic of ecosystems. These definitions have a definite impression of what scale the landscape is; the habitats or community types that make up the mosaic. This also defines the scale of the ecosystem; however, this choice is an arbitrary one because the ecosystem is better defined as a concept that can be applied at any spatial scale. The landscape also could be the ecosystem with its internal habitats/communities representing components of the ecosystem's structure. Likewise, a subcomponent of a habitat in the landscape, such as a stand of young trees, could also be usefully identified as the ecosystem.

Figure 1 is a conceptual diagram of components nested within each other (habitats within landscape) from work by Shaver and colleagues in the arctic tundra in Alaska (Shaver et al. 1991). Their aim was to provide an example of the importance of spatial heterogeneity in element cycling at the landscape level. They viewed the landscape as a patchwork of internally homogeneous ecosystems linked by soil water transport. They focus on variation within the "watershed" or landscape in an attempt to identify which components of the landscape are most important as element sources or sinks and how they control flows of materials to downstream or aquatic systems. In their approach, each habitat was treated as the ecosystem. By defining the habitat as the ecosystem, they also imply that habitats are the primary controllers of process; i.e., most important. They also could have looked at individual plant groupings within the habitat as ecosystems and controllers of material flows or the integrated mosaic as a whole in how the ecosystem at this scale affects transport to surrounding environments. Although the habitat control over nutrient export may appear most significant, other factors (e.g., animal

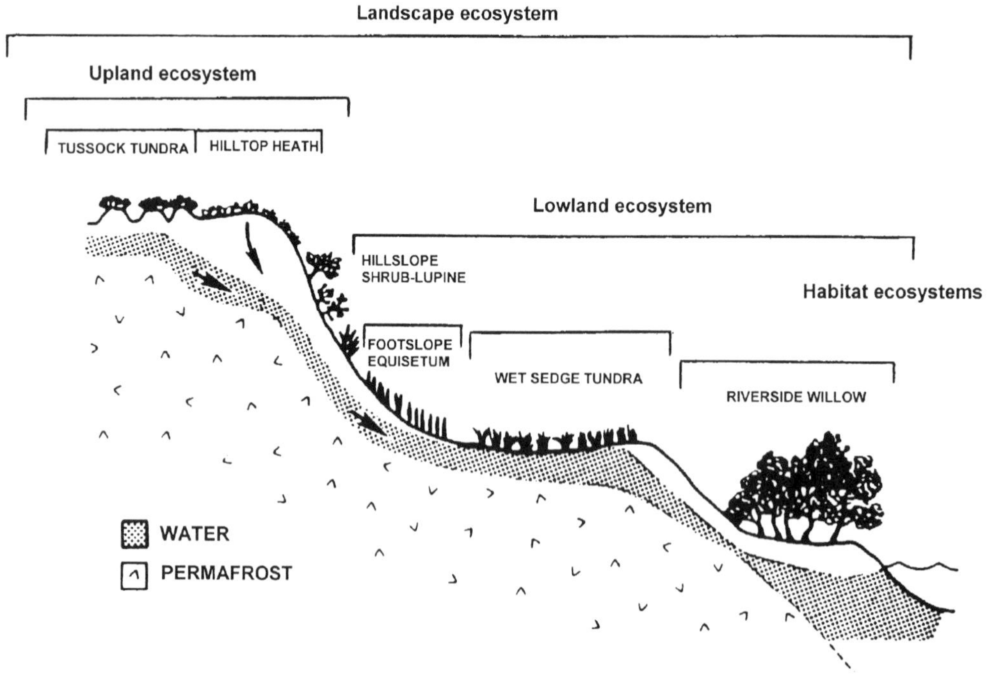


Fig. 1. Habitat to landscape hierarchy: the ecosystem concept may be applied at various levels in the landscape hierarchy to understand processes at different scales. The habitat may be the ecosystem for some questions, while other groupings (upland, lowland) may serve as ecosystems as well as the entire landscape (modified from Shaver et al. 1991).

movements, water) in this environment may be more significant at other scales or land-units (e.g., migration routes, watersheds).

Concannon et al. (this volume) provide an excellent example of management questions for an area that range from timber sales (hundreds of acres) to reserves for wide ranging species (e.g., wolverines, 100,000 acres). Other management goals also may dictate that other factors and scales be the primary focus. The important point is that an ecosystem approach can and should be used to evaluate each of these "hierarchial levels" as well as the interactions between them. The above definitions of Forman, Godron, and Hunter have a bias on what the ecosystem is that limits the use of the ecosystem concept in landscape ecology. We prefer to use the ecosystem concept at each scale as well as across the spatial scales represented within the landscape, including the entire landscape. *Different questions or management procedures often are used at each of these different scales.* We use the ecosystem in its original sense; a functional not a spatial concept, and it functions at all spatial scales in response to environmental conditions. The implications are important to the landscape mapping because functioning does not necessarily follow size rules (see also Lugo et al., this volume). Multiple-scale studies are critical in landscape ecology and an ecosystem approach is important in achieving cross-scale studies (Allen and Starr 1982).

## 1.1 Spatial Scales

For introducing the issue of scale, the following quote from S.A. Levin in the abstract from his Robert H. MacArthur's Award address is appropriate.

> "It is argued that the problem of pattern and scale is the central problem in ecology, unifying population biology and ecosystems science, and marrying basic and applied ecology. Applied challenges, such as the prediction of the ecological causes and consequences of global climate change, require the interfacing of phenomena that occur on very different scales of space, time and ecological organization. Furthermore, there is no single natural scale at which ecological phenomena should be studied; systems generally show characteristic variability on a range of spatial, temporal and organizational scales. The observer imposes a perceptual bias, a filter through which the system is viewed. This has fundamental evolutionary significance, since every organism is an 'observer' of the environment, and life history adaptations such as dispersal and dormancy alter the perceptual scales of the species, and the observed variability. It likewise has fundamental significance for our own study of ecological systems, since the patterns that are unique to any range of scales will have unique causes and biological consequences.
>
> The key to prediction and understanding lies in the elucidation of mechanisms underlying observed patterns. Typically, these mechanisms operate at different scales than those on which the patterns are observed; in some cases, the patterns must be understood as emerging from the collective behaviors of large ensembles of smaller scale units. In other cases, the pattern is imposed by larger scale constraints. Examination of such phenomena requires the study of how pattern and variability change with the scale of description, and the development of laws for simplification, aggregation, and scaling" — (Levin 1992).

Other chapters in this volume also deal with the influence of scale (see Caraher et al. and Haufler et al., this volume). Our goal is to demonstrate that landscape ecology cannot escape dealing with spatial analysis, spatial and temporal scales, and scale-change effects (Meentemeyer and Box 1987). A landscape may appear to be heterogeneous at one scale but quite homogeneous at another scale, making spatial scale inherent in definitions of landscape heterogeneity and diversity. This also then impacts what is called an ecosystem in the definitions of Forman, Godron, and Hunter. Something that looks homogeneous to one observer (e.g., a human) may be very heterogeneous to a different human observer, and certainly to a different species! These homogeneous patterns also are commonly determined by community dominants (e.g., dominant plants)

> The word scale has several meanings that complicate communication. In this paper, scale generally means spatial or temporal extent. Thus, small (fine) scale is a small area or short time span and large (broad) scale is a large area or long span of time. In a few places, scale is modified by the word map. Typically used to specify the units on a map, scale means a unitless fraction, such as 1 to 100,000 meaning 1 unit on the map equals 100,000 units on the ground. Note in the case of map scale, the meaning is reversed. Small scale, such as 1 to 500,000 (a small fraction) covers a large area with less detail, and large scale, such as 1 to 24,000 (a large fraction) covers a small area with more detail.
>
> Scales of space and time are very important concepts in understanding landscapes and the processes involved. So it is also important to recognize these contradictory meanings of the word scale and to clearly communicate the intended meaning.

which is another type of bias implied by the former definitions. All species do not recognize the same species as community dominants!

Landscapes are usually considered as items of interest at the spatial scale of tens to hundreds of kilometers; however, explanations of landscape characteristics involve processes on other scales (Risser 1987). As discussed by Golley (1987) and Allen and Starr (1982), all questions or management programs should involve three levels of attention (scales of study): the object of interest (the primary scale of interest), the components and functions within that object which explain its behavior (finer scales than the primary scale), and the larger system of which the object is a part and which establishes its significance (broader scale than the primary scale of interest). The influence of scale will be discussed in other sections of this chapter.

## 1.3 Temporal Scales

Different processes operate at different scales (Fortescue 1980). Landscape diversity also can be approached from an understanding of the interrelated factors of time, space, and adjustments to change (Bull 1991). Time involves duration but also rates of processes. An understanding of how landscapes changed in the past provides insights in future impacts of natural and human-induced change on storm patterns, flood frequency, landslides, stability of valley floors, and forest and agricultural productivity (Wright 1974, Bull 1991). Press (1994) suggested that the activities of humans will be seen to have a multiplying effect on the magnitude of disturbances. An example of this is the widespread flooding of the Mississippi in 1994. The magnitude of this event is most meaningful from a full understanding of the flood history before the last 300 years of development and urbanization of the Mississippi flood plain. Thus, an understanding of how the landscape has evolved since the end of the Pleistocene glaciation provides the necessary framework to evaluated current changes and potential evolutions.

The second aspect of time is the rate at which processes act, such as weathering, erosion, soil development, nutrient cycling, or the frequency of disturbances (e.g., fire, floods; Wright 1974). Each variable has its own rate that has to be understood. The variables, such as climate, total relief, and base level, have rates of change measured in hundreds to thousands of years, whereas human activities have rates of changes measured in a few to hundreds of years. Storm events, which can be critical events in shaping many environments, can have rates measured in hours to minutes. Another example of this perspective on rates is the question of extinctions. Based on geologic studies of number of species and extinctions, the mean annual rate of extinction is two to three species per average year (Sepkoski 1994). This number is, however, misleading. The fossil record shows huge variation around the mean and most short-term intervals (short in a geologic sense) contain little extinction (Sepkoski, 1994). Table 1 lists some examples of process rates.

Table 1. Examples of rates of processes. Taken from Goudie (1993) unless otherwise referenced.

| | |
|---|---|
| Sea level rise in last 100 years | 10–20 cm/100 year |
| Beach erosion (Committee on Engineering Implications of Changes to Relative Mean Sea Level, 1987) | 100 m/1 m sea level raise |
| Deltaic subsidence | 10–100 cm/100 year |
| Subsidence from oil and gas extraction | 50–216 cm/100 year |
| Tectonic uplift | 0–>200 cm/100 year |
| Change of vegetation belts for 3°C temperate change (Peters 1988) | 500 m in altitude |
| Salt marsh accretion | 51–1000 cm/100 year |
| Worldwide rate of erosion (Leopold et al. 1964) | 2.7 cm/1000 year |
| Erosion rates (Leopold et al. 1964) | 1–100 cm/ 1000 year |
| Average lowering of rivers (Sparks 1986) | 1–1350 mm/1000 year |
| Weathering rates (Bull 1991) | <1 cm/ 1000 year |
| Volcanic eruptions in western US (Dzurisin et al. 1994) | 2/100 year |
| Magnitude 8 earthquakes — Oregon coast (Yelin et al. 1994) | 1/600 year |
| Average extinction rate (Sepkoski 1994) | 2–3 species/year |
| Droughts (White 1979) | 1 in 5 year |

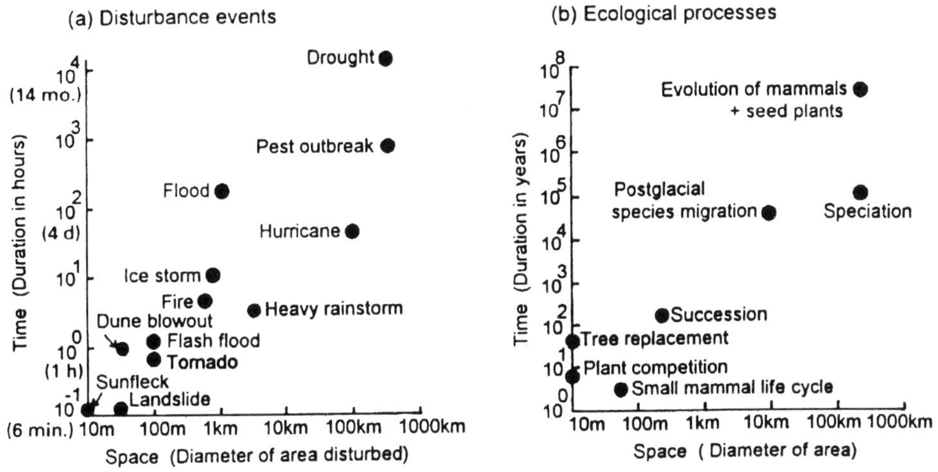


Fig. 2. Space-time principle expressed in a graph for environmental and biological changes. The principle generalizes that most short-duration changes affect small areas and most long-term changes affect large areas (from Forman 1995).

A general relationship exists between temporal and spatial scales as shown in Fig. 2 from Forman (1995). This generalized space-time principle suggests that most short-duration changes affect a small area, and most long-term changes affect a large area. This relationship has been observed for many biological responses and other ecological attributes. The principle also implies that phenomena at broad scales are more persistent or stable than those at fine scales. Fine-scale phenomena should be more variable in both time and space.

## 2 LANDSCAPE PATTERNS

Modern natural resource management has an appreciation for how landscape variability and patchiness affect wildlife populations and communities. This has occurred largely through the renewed awareness of how landscape patterns influence ecological processes, such as disturbances (fires, floods), nutrient cycling, and animal and plant distributions. Because species, such as wide-ranging animals, use more than one habitat patch type for different activities, such as breeding, resting, feeding, and dispersal, they may respond actively to the mosaic of habitats they encounter, avoiding some and using others, resulting in the surrounding matrix being as important as any particular habitat patch. Other species may be restricted to a single patch type and factors influencing that patch type directly influence those species. A simple analogy is that of a chessboard and how the game would change if the basic board (i.e., landscape) would change. The normal chessboard has a definite pattern of squares of two colors (i.e., habitats) and "figures" have particular abilities; e.g., pawns move 1 square, bishops move in diagonal directions, knights move 3 contiguous squares in one direction then 1 square in a 90° direction, etc. The movement of the "figures" can be thought of as the behaviors and movement capabilities of species (evolutionary characteristics). If the chessboard pattern is changed dramatically (like some landscapes), it is clear that the "game" has changed and movements of the various chess figures are constrained, enhanced, or even inhibited. As an example, Forman (1995) showed how various species are affected by the width of a road corridor (Fig. 3). Surface arthropods such as Wolf spiders and beetles almost never cross a lightly traveled, 6-m wide paved road. Small mammals cross lightly traveled roads of 6–15 m width, less than 10 percent of what is normal for movements within the adjacent habitat. Mid-sized animals crossed road corridors up to ca. 15 m wide but not 15–30 m. Large animals crossed most roads but the rate of crossing is typically lower than movement rates in the matrix. Habitat patterns are, therefore, critical to species movements and appropriate management of the landscape must consider how changing habitats will affect the ability of species to operate in ways necessary for their survival (evolutionarily determined). Forman (1995) cited the following changes in spatial patterns following fragmentation. There are increases in isolation, number of generalists, number of multihabitat species, number of edge species, number of exotic species, nest predation, and extinction rate. There are decreases in dispersal of interior specialists, large-home-range species and richness of interior species. Some variables increase, decrease, or remain unchanged, such as natural disturbance, hydrologic flows, wind movement, nutrient cycling, productivity, and gene flow.

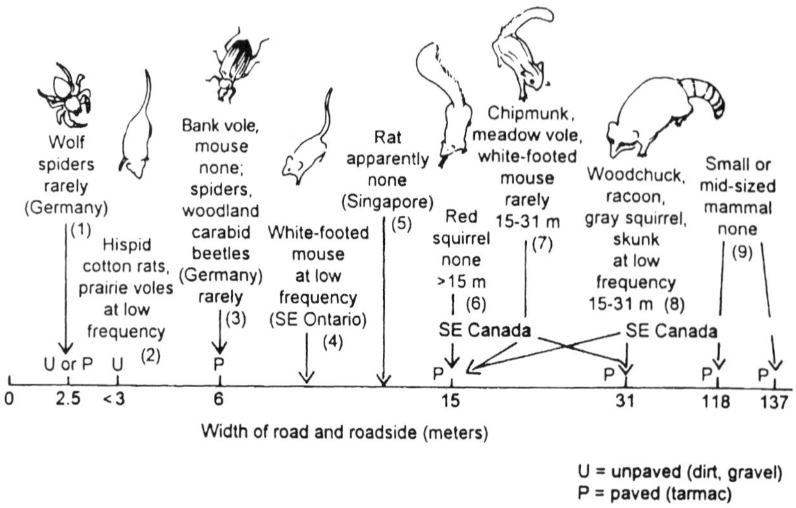


Fig. 3. Effect of road corridor width on animal crossings. The maximum corridor width at which animals crossed is given (from Forman 1995).

## 2.1 Landscape Terms

Before we go further, it is helpful to understand the variety of terms that relate to landscapes (see Forman and Godron 1986, Forman 1995):

1. *Patch* — a surface area differing in appearance from its surroundings.
2. *Matrix* — a landscape element surrounding a patch, also the most extensive and most connected landscape element present.
3. *Landscape structure* — is made up of elements as the basic building blocks of the mosaic of patches.
4. *Landscape boundaries* — spatial discontinuities in features of the soil and/or vegetation; abrupt transitions between woodlands and grasslands, between riparian and desert vegetation, riparian vegetation and water, or between an alluvial fan and adjacent plateaus are clear examples.
5. *Corridor* — a strip of a particular type that differs from the adjacent land on both sides.
6. *Network* — an interconnected system of corridors.

The first two of these terms, patch and matrix, can be measured in a variety of ways using spatial analysis, which is the process of describing and/or quantifying the relationship between ecological units (e.g., habitat patches) in two or three dimensions. Some of the more common measures used include:

1. *Patch size*
2. *Patch perimeter*
3. *Patch shape*
4. *Effective patch interior* — depends on depth of edge effects
5. *Distance-to-nearest-neighbor* — usually measured between adjacent patches: (a) *edge-to-edge* — distance of nearest edges between two patches; (b) *centroid-to-centroid* — distance of centers between two patches.
6. *Connectivity* — a measure of how connected or spatially continuous a corridor, network, or matrix is (i.e., the fewer the gaps, the higher the connectivity).

In addition to these descriptive terms for patches, other simple indices are available which provide useful information regarding a mosaic of patches.

1. *Richness* — the number of different patch types present.
2. *Diversity* — a measure of both the number of patch types present (richness) and the relative abundance (evenness) of the patches.
3. *Dominance* — based on number of patches or amount of area in a given patch type category.
4. *Juxtaposition* — what surrounds a given patch.
5. *Contagion* — aggregation or dispersion of patches
6. *Edge number* — number of horizontal and vertical interfaces

## 2.2 Understanding Landscape Pattern

Since the development of agriculture, the natural vegetative cover of every continent except Antarctica has been extensively modified by humans. Indeed, a cycle of human land-use and revegetation has been repeated throughout recorded history. For example, Plato described how ancient Greece was stripped of its topsoil following logging and grazing leaving "the mere skeleton of land." Habitat destruction, or conversion depending upon your persuasion, has done more than simply eliminate habitat; it reconfigures landscapes. One of the most significant realizations of this century, in regards to natural resource management, has been the appreciation that habitat alteration results in virtually all natural areas resembling oceanic islands, that is, habitats become smaller and more isolated and occur in a matrix other than the original one.

Over time, virtually every landscape with even a moderate level of human occupation has gone through a transition from a natural landscape matrix with human-altered habitat fragments embedded within it to a managed landscape matrix with natural habitat fragments embedded within it. Basically, everything humans do fragments, perforates, or shreds landscapes, whether building roads or transmission powerlines, clearing land for farming, logging, or housing developments. Over much of the world, conservation of regional biotas depends entirely on the retention and management of these habitat remnants. Fragmentation affects nearly all ecological patterns and processes, from genes to ecosystem functions (Forman 1995). Natural resource managers, therefore, are faced with the dual issues of whether the remnants have any practical conservation values, and if they do, of how they must be managed to retain these values.

A habitat fragment is defined as any patch of native vegetation around which most or all of the original vegetation has been removed. Habitat fragments have been the subject of considerable debate and research. Most of the research has centered on the equilibrium theory of island biogeography. This theory examined island species lists and put forth a series of curves which portrayed extinction and colonization rates based on island size and degree of isolation from the colonizing source. It was theorized that each island would have an equilibrium point where the extinction and immigration curves crossed. This would be the number of species which might be expected for that island's size and degree of isolation from the colonizing source.

Two major conclusions emerged from the island biogeography work: (1) the number of species found on an island will increase, all other things being equal, with decreasing distance to the colonization source; and (2) the number of species found on an island will increase, all other things being equal, with increasing island size. The intuitive power of this hypothesis was the reason for the tremendous amount of dogma which it spawned. Today, a number of limitations with the equilibrium theory of island biogeography are acknowledged. First, virtually no experimental studies have substantiated the major deductions from the theory. Second, the theory discusses only species, not relative abundances or persistence of species. And finally, the equilibrium theory bases species richness on the balance between extinction and colonization and offers no biological explanations for either. This is troubling because it has no explanation for the ultimate factors that explain species richness. At the same time that the island biogeography model was being developed, a British ornithologist, David Lack, was putting forth a biotic-interaction model (Lack 1976). This model offered ecological explanations for the species/area relationships seen on oceanic islands. It noted that species composition was regulated by area-dependent changes in the island's environment and incorporated the ideas of competition, predation, and parasitism. Also, Forman (1995) suggested that in a landscape mosaic on land, species isolation, a primary characteristic of island biogeography, is generally a minor variable. Because most patches in a mosaic contain internal habitat diversity, the area-per-se effects present are often difficult to demonstrate and probably only affect a minority of species. The evidence for island size effects on colonization rate is limited and the ability of early colonists to repel later arrivals may be more important than the area effect. Rather than assuming primary control of species number by immigration and extinction rates, species-area patterns seem more directly explained by the importance of the edge effect and the interior-to-edge ratio, also present on islands. Some proposed area-related extinctions may be due to boundary processes.

Which of these conflicting hypothesis is valid is not an esoteric issue. Because the equilibrium and biotic-interaction models of biogeography may lead to conflicting management strategies, it is important to be aware of them. For example, the equilibrium theory implies that corridors connecting islands are extremely important because they enhance colonization and decrease the chances of extinction. Lack's theory, on the other hand, suggests corridors connecting islands may be detrimental because they aid the spread of other species which may be effective predators, competitors, diseases, or parasites, i.e., biotic interactions. Ambuel and Temple (1983) examined these competing hypotheses using songbird communities in Wisconsin woodlots, and found support for the idea that area-dependent changes in biotic interactions explained the species diversity patterns present and provided no support for the equilibrium model. Since the equilibrium theory was developed, most ecologists have now altered their views on communities and assume they are non-equilibrium, with species populations and numbers normally in wide fluctuations and subject to biotic interactions.

Three primary characteristics are known for habitat fragments, which, to a large degree, explain the relative conservation value they may have.

*Area*: Area decreases with increased fragmentation and has led to the species/area relationship which states that species richness increases with increasing area. There are several explanations for this. First, as area increases the probability increases that dispersing species will encounter a patch. Second, as area increases, more different habitat types and topographic aspects are incorporated thereby increasing the number of available niches. Third, increasing area allows for species that have large area requirements, so-called area-sensitive species. Fourth, increasing area allows larger populations of individual species, and therefore, a greater likelihood of persistence. Fifth, as area increases, edge decreases and therefore minimizes the deleterious effects associated with edge (these will be discussed later).

*Shape*: As fragmentation increases, there is also likely to be an increase in the amount of habitat edge. Patch shape is of great importance in determining how much edge a patch has. This incorporates the concept of perimeter to habitat interior ratio. High ratios indicate a large amount of edge and therefore problems with edge-sensitive species, those species that are susceptible to edge generalist species. Edge generalists are species that often have broad life-history requirements in terms of food and habitat needs. Species that evolved in the midst of contiguous habitats traditionally had little contact with species along edges because the amount of edge in unfragmented landscapes was less overall than in fragmented landscapes. Habitat fragmentation increases the likelihood of interactions between these two types of species with the result that edge-sensitive species decline.

*Isolation*: Isolation between patches increases with increasing fragmentation. Isolation alters rates of successful colonization and may affect the biodiversity on habitat patches in three ways: time since isolation, distance from other remnants, and connectivity. Upon isolation, a fragment is likely to have more species than it can maintain and species will be lost over time. For example, area-sensitive and edge-sensitive species will begin to disappear. This process is known as "relaxation" and varies across taxa. Populations that are too small to be viable may still persist for long periods on patches simply because of the longevity of the individuals. For example, many tree species live for well over a century, although isolated patches may not allow colonization and, therefore, long-term persistence of these species. Additional discussion of this feature is in Noon et al. (this volume).

The ability to colonize a fragment depends to some degree on the distance of the fragment from other native vegetation. Species which are poor dispersors are called dispersal-sensitive species. Reduced dispersal ability may be due to physical, behavioral, or physiological reasons. For example, species may have characteristics which enhance dispersal such as the ability to fly or be wind dispersed. Sedentary species or those which have to walk may be dispersal sensitive. On the other hand, some species are physically capable of dispersing but are behaviorally constrained, they simply do not have the behavioral repertoire to cross an alien environment. Other species are physiologically constrained from dispersing across human altered environments. For example, eastern chipmunks cannot tolerate the heat extremes crossing from one forest patch to another. Finally, many species suffer none of these problems and attempt to disperse but suffer abnormally high mortality when dispersing through a fragmented landscape as opposed to through a contiguous habitat. Many examples are known of species, from Florida panthers to rattlesnakes and prairie dogs, that experience elevated death rates trying to move across human-dominated landscapes. This is the real-life example of the altered chessboard discussed earlier.

Associated with the effects of distance is the degree to which individual fragments are connected to adjacent suitable habitat. The role of corridors has been viewed as positive even though very little empirical data support this conclusion. The advantages attributed to corridors include enhancing movement and survival of dispersal-sensitive species, as habitat in their own right, and as movement paths during periods of rapid climate change. Some disadvantages attributed to corridors include the enhanced movement of predators, competitors, diseases and parasites, and exotic species, all of which may affect the fitness of species subject to conservation efforts. Regardless of these potential limitations, conservation biologists have taken the approach that corridors do have value for biotic movement and attempt to retain a good corridor network wherever possible.

Finally, to understand landscape-level patterns and processes, we must know the details of "boundaries and their dynamics" — what determines why a boundary is located where it is; how boundaries influence ecological processes within patches and over the larger landscape; how boundaries affect the exchanges or redistribution of materials, energy, and organisms between landscape elements; and how these transfers can in turn act to change the location and nature of boundaries. Wiens et al. (1985) originally drew the analogy of boundaries between elements in a landscape to membranes in organismal or physical systems. Like membranes, boundaries vary in their permeability or resistance to flows. This variation is a consequence of characteristics of the boundary itself (e.g., its thickness, the degree to which the separated patches differ) and of the responses of different materials, organisms, or abiotic factors to the boundary. The dynamics of boundaries will be discussed in Section 5.2.

## 3 FACTORS AFFECTING LANDSCAPE DIVERSITY

### 3.1 Geology and Geomorphology

Landscape diversity from a geologist's perspective is a large, complex subject. It involves primarily geomorphology and geochemistry and significantly overlaps with areas of ecology. This review is prepared for land managers involved in land stewardship and biological diversity, not professionals in geomorphology or geochemistry. The intent is not to review the literature, but to outline the breadth of the subject and provide some focus to developing an understanding of what earth science can contribute to managing biological resources.

The primary driving variables in landscape features are time, climate, geologic variables (lithology, geologic structure, and disturbance), and human activity. Climate is an independent variable but it is modified by geologic and human activities. Lithology, geologic structure, and disturbance are the sphere of earth science. Time from a geologic perspective also deals with longer time spans such as thousands to millions of years and the slow rates of many geologic processes.

#### 3.1.1 Physical Properties

Lithology and geologic structure are dominant controlling factors in the evolution of land forms and are reflected in them (modified from Thornbury 1969). Geomorphic processes leave their distinctive imprint upon land forms, and each geomorphic process develops its own characteristic assemblage of land forms (Thornbury 1969). The physical properties of rocks and minerals, that is the lithology and structure, interacting with the other independent variables, time, climate,

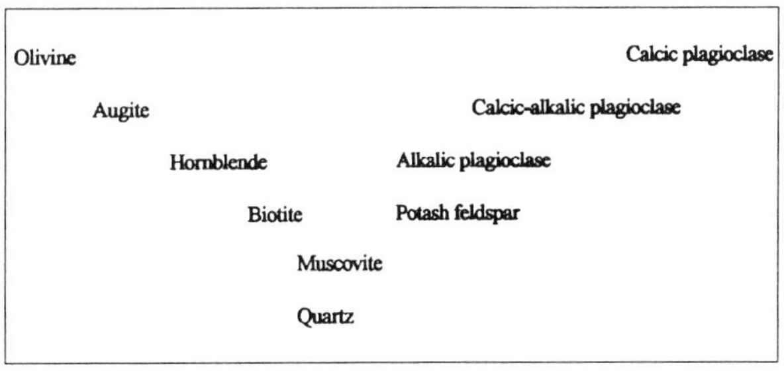


Fig. 4. Stability of minerals to weathering. The least stable minerals are at the top (Goldich 1938, Thornbury 1969).

and human activities, determine the form of the land. Rocks and minerals when freshly exposed to the Earth's surface begin to weather to new more stable forms. Goldich (1938) tabulated the relative stabilities of minerals (Fig. 4). The minerals in the top, middle, and bottom of this chart are characteristic of igneous rocks such as basalts, andesites, and rhyolites, respectively. Sedimentary rocks such as sandstones commonly contain more minerals from the bottom of the chart. Those minerals at the top of the chart generally weather to clay minerals, which form the spatially most variable rocks on land, shales. The weathering processes include physical disintegration, chemical disintegration (hydration, hydrolysis, oxidation and simple solution), and organic weathering processes (Birkland 1974, Rose et al. 1979). Factors affecting weathering processes include resistance of minerals in weathering, permeability of the rocks, climate, relief, and drainage (Birkland 1974, Rose et al. 1979). Thus, complex feedback mechanisms between the physical properties of rocks and minerals and the climate and the vegetation lead to the form of the land.

When rocks are brought to or near the Earth's surface, the chemical and mechanical properties of rocks and minerals interacting with climate (primarily precipitation and temperature) begins the disintegration of rocks, the first stage in the development of soil. At some point plants and animals begin to interact with these new materials. These interactions lead to further modifications of the chemical and mechanical processes of weathering. The form of the land then evolves and records in its form all of these materials and processes (Thornbury 1969). Recognizing this relationship of form and materials, soil scientists, for example, have long believed that certain soils are associated with particular land forms (Peterson 1981). Because the landforms are more readily observed, the first step in mapping soils is understanding landforms. Geologists, also, have long used vegetation as an aid in mapping lithology (Lehee 1961, Compton 1962, Brooks 1972, Raines and Canney 1980). Thus, these complex interactions and feedback loops create many diverse

Photo 1. This photo depicts two plant communities influenced by different geologies, limestone (foreground) and Precambrian granite (background). The person is standing at the fault line separating the two geologies The coarse textured soil plus rocky nature of the granite substrate allows better moisture relations resulting in higher plant diversity and more shrub biomass than the fine textured soils of the limestone area for this semiarid area. The limestone area is dominated by *Bouteloua* grasses and *Juniperus monosperma*. Many small animal species also show decided preferences for one or the other of the habitats. The area is on the Sevilleta National Wildlife Refuge and Long Term Ecological Research (LTER) site in central New Mexico (photo by James Gosz).

opportunities to use the physical form of the land to understand the landscape. For example, in semiarid areas this physical structure has profound influence on community composition and ecological processes. Different communities occur on either side of a fault line that separates Precambrian granite from limestone at the Sevilleta National Wildlife Refuge and Long Term Ecological Research (LTER) site in central New Mexico (Photo 1). The road lies directly on the fault line. The coarse textured soil plus rocky nature of the granite substrate allows better moisture relations than the fine textured soils of the limestone area resulting in the granite area having higher plant diversity, especially of shrub species. The limestone area is dominated by grasses and *Juniperus monosperma*. Many small animal species also show decided preferences for one or the other of the habitats; geology in this environment has a profound effect on its biology.

### 3.1.2 Geochemical Properties

Geochemical structure provides the framework and bounds on the nutrients that are required for plant and animal nutrition (Duvigneaud and Denaeyer-De Smet 1975). Mineral elements, as much as energy and water, are essential to maintain the continuity and stability of ecosystems. Mineral cycling appears to be one of the best and easiest ways to characterize the general metabolism and functioning of ecosystems (Duvigneaud and Denaeyer-De Smet 1975). An interesting perspective on lithology is that the same major elements (Al, Ca, Fe, K, Mg, Na, Si) important in plant and animal nutrition are the major elements used to classify minerals. The minerals and their form and texture are in turn the basis for classification of rocks. The major difference between the living chemistry of plants and animals and abiotic chemistry of rocks and minerals of the Earth's crust is that living chemistry uses these major elements as components of a carbon, hydrogen, nitrogen, and oxygen framework, whereas lithology uses these same major elements as components of a silica and oxygen framework. Additional information on the link between biology and geology can be found in Dobrovolsky (1994) and Connor and Shacklette (1975), and a brief summary on bioavailability is found in John and Leventhal (1995).

The chemical properties of rocks and minerals interacting with the other independent variables, time, climate, and human activities determine the chemical patterns of the landscape. Dramatic influences of these chemistries on plant communities are demonstrated by serpentine soils or gypsum soils. In central New Mexico, distinct communities of "gypsophiles," gypsum-loving plants, have abrupt boundaries with communities on non-gypsum soils (Photo 2). Powell and Turner (1977) demonstrated the uniqueness of gypsophily and found an interesting array of gypsophilic taxa inhabiting every island-like exposure in the Chihuahuan Desert. Gypsum exposures are found throughout much of the West; however, regardless of size, the island-like exposures always support communities (florulas) of plant species characteristic of gypsum soils. Specific fauna are also confined to these communities, such as the gypsum-specific grasshopper, an undescribed species most closely related to *Trimerotropis latifasciata* (D. Lightfoot, pers. comm.). This grasshopper lives only on gypsum outcrops, and occurs throughout the Chihuahuan Desert.

Lithology can also be used to indicate potential health conditions for species (and humans) ingesting plants grown on the soils from those lithologies. The health aspects are reviewed in Cannon and Hopps (1971) and Freeman (1995). Some examples for natural geochemical toxicity and health problems caused by ingesting specific elements are referenced in Table 2.

Understanding geologically relevant rates is a critical aspect of understanding landscape diversity and the relationship to biodiversity. One approach to monitoring geologic processes is the Hubbard Brook Ecosystem Study (Likens et al. 1977) in which products of processes were measured in a small watershed in the northeastern United States for a period of several

Photo 2. The chemical properties of rocks and minerals can dramatically influence plant communities as demonstrated by this photo of a gypsum substrate (foreground) in central New Mexico. Distinct communities of "gypsophiles", gypsum-loving plants, have abrupt boundaries with communities on non-gypsum soils. The community in the background is on a different sedimentary- derived soil and consists of creosote bush (*Larrea tridentata*) and *Bouteloua* and *Sporobolus* grasses. Gypsum exposures are found throughout much of the West; however, regardless of size, the island-like exposures always support plant communities that are characteristic of gypsum soils. Major plants of gypsum soils include *Coldenia hispidissima, Pseudoclappia arenaria, Enneapogon desvauxii, Sporobolus nealleyi, Anulocaulis gypsogenus, Anulocaulis leiosolenus, Frankenia jamesii, Mentzelia perennis, Mentzelia pumila* var. *procera, Nama carnosum, Nerisyrenia linearifolia, Oenothera hartwegii, Oenothera pallida, Phacelia integrifolia, Sartwellia flaveriae, Selinocarpus lanceolata* (Dick-Peddie 1993). There also are specific fauna confined to these communities such as the gypsum-specific grasshopper, an undescribed species most closely related to *Trimerotropis latifasciata* (Lightfoot, personal communication). This grasshopper lives only on gypsum outcrops, and occurs throughout the Chihuahuan Desert. This area is on the Sevilleta National Wildlife Refuge and Long Term Ecological Research (LTER) site (photo by James Gosz).

Table 2. Examples of natural geochemical toxicity and health problems.

| | |
|---|---|
| Arsenic | Reay (1972) |
| Calcium-carbonate hardness of water | Anderson (1971) |
| Cadmium | Perry (1971) |
| Chromium | Mertz (1971) |
| Copper | Cannon (1960); Chaney (1988) |
| Lead | Boggess (1977) |
| Mercury | Rytuba and Klein (1995); Freeman (1995) |
| Molybdenum | Neuman et al. (1987) |
| Selenium toxicity | Presser et al. (1990) |
| Serpentine Soils (Ca, Mg, Ni, Cr, Co) | Freeman (1995) |
| Zinc | Pories et al. (1971) |

decades. This study quantified weathering processes, nutrient cycling, water flows, and many interrelated processes. For the semi-arid and arid environments, Bull (1991), Chorley et al. (1984), and Sparks (1986) reviewed the processes and literature on the types of monitoring and resulting predictive models that can be developed. These are some aspects of monitoring that are necessary to understand how an ecosystem functions. They also explain the temporal and spatial scales at which such direct measurements are practical.

Geologic maps provide a tremendous wealth of information. Unfortunately for nongeologists, these maps contain an incredible amount of information that is generally encoded in the vocabulary of geology. Thus, it is not recognized, for example, that information about nutrients such as aluminum, calcium, iron, magnesium, sodium, and silica are buried in these maps (e.g. Raines et al. 1995). Related information is also available from stream sediment and water geochemistry programs routinely performed for geologic purposes (e.g., Hoffman et al. 1991). It is not recognized that information about animal habitats, such as caves for bats, is buried in these maps (e.g., Frost et al. 1995). Geomorphology, the study of the configuration of the Earth's surface, and geochemistry, the study of the distribution and amounts of chemical elements in earth materials, are critical tools routinely used by geologists to understand landscapes but not well understood by biologists who have needs to understand landscapes. Geologic and geochemical maps provide critical information for monitoring any landscape.

## 3.2 Aquatic Environments

The aquatic ecosystems of North America contain a rich diversity of flora and fauna that are products of the geologic and evolutionary processes on this continent (Williams and Neves 1992). The biological diversity of aquatic environments is under ever-increasing pressure as human population growth continues to change the landscape. As a consequence of the collective culturally-induced changes, there has been a 45 percent increase in the last decade in the number of rare freshwater fishes in North America, major losses of genetic diversity in West Coast salmon (*Oncorhynchus* spp.) stocks, and, in general, a much higher percentage of aquatic organisms (mollusks, crayfish, amphibians, fishes) at risk of extinction than terrestrial species (Naiman et al. 1995).

The physical factors that produced the great biological diversity now in such grave danger are well-known: geological processes shaping basin morphometry, edaphic factors resulting from geological and biological interactions, and disturbance regimes largely driven by interactions between regional geology and climate. Regional-scale characteristics of aquatic environments reflect variations in local geomorphology, sediments and soils, climatic gradients, spatial and temporal variations in disturbances, as well as the dynamic features of riparian vegetation acting to modify all of the above factors. Collectively, this results in a broad variety of aquatic environments that are modified by humans presenting substantial challenges for understanding and managing biodiversity.

### 3.2.1 Basin Morphometry

The morphology of aquatic basins has profound effects on nearly all physical, chemical, and biological characteristics occurring within those basins (Vannote et al. 1980, Wetzel 1983). The formation of aquatic basins (i.e., lakes, wetlands, streams, and subsurface waters) results from both catastrophic and gradual geological processes. Catastrophic processes include tectonic depressions formed by the displacement of the Earth's crust. For example, many lakes and streams, especially in the western half of North America, were formed by faulting movements or uplift. Likewise, aquatic basins result from volcanic activity. Small, deep maar lakes form in volcanic cones and large, deep caldera lakes form when emptied magmatic chambers collapse (e.g., Crater Lake, OR). In addition, temporary or permanent aquatic basins result from landslides into stream valleys. Finally, the erosional and depositional activities of glaciers were probably the most important geological agents of aquatic basin formation in North America. In eastern and northern regions outwash morainal deposits at the retreating edges of continental glaciers physically shaped the basins for many lakes, wetlands, and streams as buried blocks of ice in the morainal debris formed kettle lakes. Glaciers in mountainous regions scoured amphitheater-shaped depressions (e.g., cirque lakes) and water seepage into the permafrost formed cryogenic lakes and wetlands in arctic regions.

Other aquatic basins were formed by more gradual geological processes. The gradual dissolution of rock, such as limestone, along fissures and fractures resulted in solution basins (e.g., dolines). The erosional and depositional action of rivers isolated physical depressions to form plunge-pool and oxbow lakes. Wind erosion (in conjunction with animal activities, see later) formed shallow depressions that contain dune lakes or wetlands in sandy areas and playa lakes in arid or semiarid regions. Coastal aquatic basins often formed along irregularities in the shore line of the sea or large lakes where longshore currents deposited sediments that isolated an aquatic basin from the larger body of water.

Cultural activities also result in the formation of new aquatic environments and extensive modifications of existing ones. Reservoirs, both large and small, are now a dominant landscape feature. However, because of high sedimentation rates, many undergo rapid changes in morphometry as the basins fill. Likewise, modifications to natural aquatic basins are the norm rather than the exception for North America. Irrigation, power production, interbasin water transfers, pumping of ground waters, construction of flood control devices, dredging for navigation, and a host of other engineering achievements have resulted in substantial modifications to natural hydrological regimes.

### 3.2.2 Edaphic Factors

Interactions between the morphology of aquatic basins and the geological and soil-based substrates within the larger drainage basin (i.e., edaphic factors) strongly influence system productivity by regulating community composition, sediment–water exchanges, and material fluxes between the terrestrial and aquatic environments. Edaphic factors often can significantly modify morphometric controls on chemical constituents (i.e., nutrients and ions) to determine whether a water body will be productive or not. Edaphic characteristics of a drainage basin largely determine the potential amount of nitrogen, phosphorus, chloride, sodium, sulfate, potassium, carbonate, calcium, magnesium, and other elements for the aquatic environment. However, as we will see later, these potential fluxes can be modified by riparian vegetation at the land–water ecotone (Naiman and Décamps 1990).

## 3.3 Land–Water Interactions

Interactions between land and water systems take many forms: surface and subsurface material fluxes, transfer of information (i.e., sound, chemical communication), and local climate modification, to name a few. Historically, the interactions have been considered to be unidirectional from the land to aquatic systems as the landscape erodes, as human activities modify biogeochemical cycles, and as vegetative cover changes. However, it is now widely recognized that aquatic systems have several important influences on the land, such as creating and maintaining floodplain forests, controlling the demographics of riparian forests and many wildlife species, and local modification of climate (Forman 1995, Décamps 1996). Water movement is the principal physical agent controlling the intensity, frequency and duration of the land-water interactions but the nature of the interaction, as we will see in the following sections, can be substantially modified by vegetative communities at the land–water ecotone (Naiman and Décamps 1990).

### 3.3.1 Significance of Riparian Zones

Natural riparian communities are the most diverse, dynamic, and complex biophysical habitats on the terrestrial portion of the Earth. Riparian communities, as interfaces between terrestrial and aquatic systems, encompass sharp environmental gradients, ecological processes, and communities. Riparian communities are an unusually diverse mosaic of landforms, communities, and environments within the larger landscape. As such, we believe they serve as a framework for understanding the organization, diversity, and dynamics of communities associated with aquatic ecosystems (Gregory et al. 1991, Naiman et al. 1993).

The riparian community encompasses the stream channel and that portion of the terrestrial landscape from the high water mark toward the uplands where vegetation may be influenced by elevated water tables or flooding, and by the ability of soils to hold water. The width of the riparian community, the level of control that vegetation has on the aquatic environment, and the diversity of functional attributes (e.g., information flow, biogeochemical cycles) are related to the size of the aquatic system, the position of the aquatic system in the drainage network, the hydrologic regime, and the local geomorphology. For example, the riparian community is often small in the numerous headwater streams that are most completely embedded in the forest. In mid-sized streams the riparian community is larger, being represented by a distinct band of vegetation the width of which is determined by long-term (>50 year) channel dynamics and the annual discharge regime. Riparian communities on large streams are characterized by well-developed, geomorphically complex floodplains with long periods of seasonal flooding, lateral channel migration, oxbow lakes in old river channels, a diverse vegetative community, and moist soils (Salo and Cundy, 1987, Naiman et al. 1992).

Ecological investigations of riparian communities have demonstrated them to be a key landscape feature with substantial regulatory controls on environmental vitality (Naiman et al. 1992). For example, streams are non-equilibrium systems with strong effects on habitat formation and stability, on the attributes of riparian vegetation, on local geomorphology and microclimate, and on the diversity of ecological functions. The riparian community is frequently disturbed by floods and debris flows, creating a complex shifting mosaic of landforms over different spatial scales (Swanson et al. 1988). Consequently, plant species richness varies considerably in space and time along stream margins, and these variations have important influences on the in-stream biota and processes. It is well known that riparian vegetation regulates light and temperature regimes, provides nourishment to aquatic as well as terrestrial biota, acts as a source of large woody debris (which significantly influences sediment routing, channel morphology, and in-stream habitat), regulates the flow of water and nutrients from uplands to the stream, and maintains biodiversity by providing an unusually diverse array of habitat and ecological services (Naiman and Décamps 1990). Many of the ecological issues related to landuse and environmental quality could be ameliorated with effective riparian community management.

The term "biodiversity" encompasses the presence of species and ecological processes, and it may be more properly referred to or thought of as "ecological diversity." The maintenance of biodiversity requires a landscape perspective, especially for aquatic ecosystems whose drainage networks are embedded in the landscape. The riparian community may be viewed as the heart of the drainage basin because it may be the ecosystem-level component most sensitive to environmental change (Naiman et al. 1988, 1989). It is the delivery and routing of water, sediment, and woody debris that are the key processes regulating the ecological characteristics of the riparian community and it is the dynamics of these materials that are affected by alterations to the landscape. Available evidence suggests that ecologically diverse riparian communities are maintained by an active natural disturbance regime operating over a wide range of spatial and temporal scales. Ecologically diverse riparian communities are dependent on the nature of the disturbance (e.g., floods, fire, landslides, debris torrents, channel migration) and the ability of the biotic system to adjust to constantly changing conditions (Kalliola et al. 1992). The natural disturbance regime imparts considerable spatial heterogeneity and temporal variability to the biophysical components of the system. In turn, this is reflected in the life history strategies, productivity, and diversity of the ecological community. In support of these statements consider the following observations related to the biodiversity of riparian communities.

Even though the riparian community has been recognized for its high levels of biodiversity, it is still not known how many species are present for any system (Nilsson 1992). This is remarkable, considering that more than 80 percent of the riparian corridor area of North America and Europe has disappeared in the last 200 years. The general modification of this important habitat is continuing on a global scale, with little attention being paid to the ecological or human consequences of these changes. Biodiversity is best documented for vascular plants, even though nearly 70 percent of vertebrate species in a region will use riparian corridors in some significant way during their life cycle (Petts et al. 1985).

Studies of riparian vascular plants in Sweden (Nilsson 1986, 1992; Nilsson et al. 1989), in Finland (Kalliola and Puhakka 1988), in the Peruvian Amazon basin (Junk 1989, Kalliola et al. 1992), in southern France (Tabacchi et al. 1990, Décamps and Tabacchi 1994), and in the northwestern United States (Gregory et al. 1991, Pollock 1995), all demonstrate unusually high levels of biodiversity. For example, Nilsson (1992) reported that 13 percent (>260 species) of the entire Swedish flora of vascular plants occurred along a single river corridor; Junk (1989) reported that all periodically flooded forests in the Amazon basin may have >20 percent of the 4,000–5,000 estimated Amazonian tree species; and Tabacchi et al. (1990) reported over 900 taxa of vascular plants along the Adour River riparian corridor in France.

The reasons for the high diversity of vascular plants are thought to be related to (1) the intensity and frequency of floods, (2) small-scale variations in topography and soils as a result of lateral migration of river channels, (3) variations in climate as streams flow from high to low elevations or across biomes, and (4) disturbance regimes imposed on the riparian community by upland environments. The migration capacity of plants along riparian corridors is also an important factor explaining the high biodiversity observed along river courses. Collectively, these forces create a mosaic of habitats that allow a wide variety of species to co-exist. It is well known that environmental heterogeneity, productivity, and resource diversity have major effects on functional diversity and species richness (Solbrig 1991). Floods destroy older patches and create new patches, resulting in an annual redistribution and sorting of sediment sizes and new channel configurations.

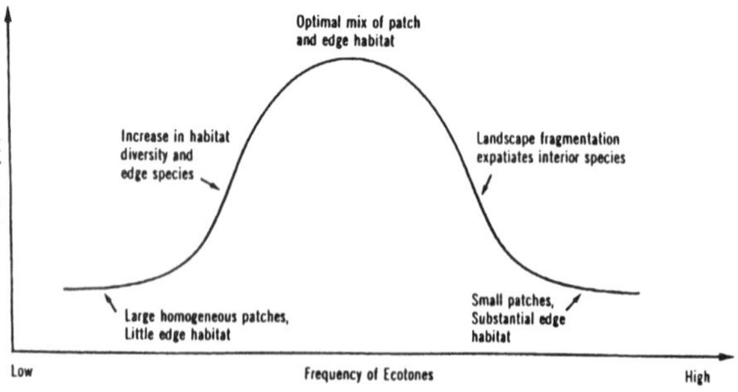


Fig. 5. Ratio of boundary edge to patch area. In highly fragmented landscapes the ratio of edge to patch area may be below some threshold value that causes local diversity to be lower than expected because of unsuitable habitats for non-boundary requiring species.

Elevation affects the overall climate, whereas individual vegetative patches influence the microclimate. Species from upland habitats add to the species diversity in the riparian community although, in many cases, they are relatively rare. Finally, riparian communities are productive systems because of the proximity of water and nutrients, but they are also subjected to regular and stochastic disturbance. These observations are consistent with the concepts of Huston (1979) and Solbrig (1991) related to the maintenance of high biodiversity in non-equilibrium situations where the process of competitive exclusion is retarded by periodic population reductions and environmental fluctuations.

### 3.3.2 Examples of Riparian Functions

Riparian forests act as edges in the landscape mosaic. One commonly noted phenomenon of biological communities is the "edge-effect" — the tendency for communities to be more dense and often more diverse in ecotonal situations (Leopold 1933). The edge-effect is a generality and probably does not apply to all taxa or to all boundaries. Nevertheless, in cases where the edge-effect does occur, biodiversity can be affected by the extent and quality of boundaries (Naiman et al. 1988, 1993). Within boundary communities, some species are characteristic of those areas whereas other species visit these communities only to perform activities essential for their survival. The abundance and survival of these species are related to the amount and quality of boundary space. The potential importance of boundaries for certain species, and the need for large unfragmented reserves for other species, creates a difficult management problem. Fragmented landscapes have an abundance of ecotonal space but may not provide suitable habitats for non-boundary requiring species (Harris 1984). Consequently, local diversity may not be maximized if the ratio of edge-to-patch area is reduced below some threshold value (Fig. 5). This would lead to a reduction in regional diversity because of a decrease in edge-avoiding species (Lovejoy et al. 1986). These factors have yet to be considered by those investigating biodiversity relationships in lotic systems. Yet, many species require more than one ecological system in which to complete their life cycle. Amphibians breed and lay eggs in water but live as adults on land; waterfowl and fish often feed in one ecological system but rest, nest, or hide from predators in another. Many species either pass through boundaries or require boundaries during critical periods of their life cycle. For example, salmonids use rivers for spawning whereas the newly emerged fry and young-of-the-year quickly move into flooded riparian zones or to stream edges for feeding and predator avoidance (Bustard and Narver

1975, Walsh et al. 1988). During this critical period salmonids use the stream edge boundary in a manner that significantly affects their abundance, growth, and survival. Other researchers have shown how the physical structure of riparian area can have positive effects on the diversity of birds and small mammals (Forman 1995), vegetation (Pollock 1995, Fetherston et al. 1995), and fish (Salo and Cundy 1987).

Another attribute of the ecotonal riparian community is the ability to modify material fluxes between water and land, thereby influencing not only the character of the riparian community but also the character of the community receiving the materials (Naiman et al. 1988). In a classic study, Peterjohn and Correll (1984) showed strong nutrient interactions between agricultural land, natural land, and the Rhode River, Maryland (Fig. 6). Agriculture provides the largest input of N and P to the watershed but harvests remove <50 percent of those inputs. Large amounts of nutrients leave the cropland and are either accumulated elsewhere in the landscape or are transformed to gases. Only 1 percent of the N and 7 percent of the P entering the watershed is discharged to the Rhode River; most nutrients are intercepted by adjacent riparian forests. These forests reduce total particulate concentrations in overland flow by 94 percent and reduce nitrate in groundwater by 85 percent. The nature and magnitude of the efficiency of riparian use is determined by the structural and functional properties of the boundary. The boundary, as a selective filter, acts to modify disturbances as well as the response of adjacent resource patches to that disturbance.

## 3.4 Disturbance

Disturbance is treated in detail in White et al. (this volume). Here we make some specific references and examples for influences on landscape diversity.

### 3.4.1 Geologic disturbance

Geologic disturbances or hazards are generally thought of as events or processes that are observed as single events occurring at a point in space and time. For the larger spatial and temporal scales, however, many types of disturbance are a single event in a long-term set of processes. For example, the eruption of Mt. Lassen in the early part of this century is simply one minor volcanic eruption in the long history of eruptions in the northwestern United States. Mt. Saint Helens is the latest eruption in this region, which is part of the larger area of volcanic eruptions surrounding the Pacific Ocean basin, the "Ring of Fire." White (1979) provided an excellent perspective on disturbance as a normal process of a landscape in a hierarchy of spatial and temporal scales, which is commonly observed as a short-term change in vegetation. Interestingly, White (1979) concluded that disturbance frequency, predictability, and magnitude are often correlated, and these three variables are also correlated with topographic form and lithology (Cline and Spurr 1942, Hayes 1942, Kessel 1976). Similarly, lithology influences the effects of climatic disturbance from wind (Cline and Spurr 1942; Webb 1958, Sauer 1962, Whitmore 1974), flooding (Bell 1974), and patterned

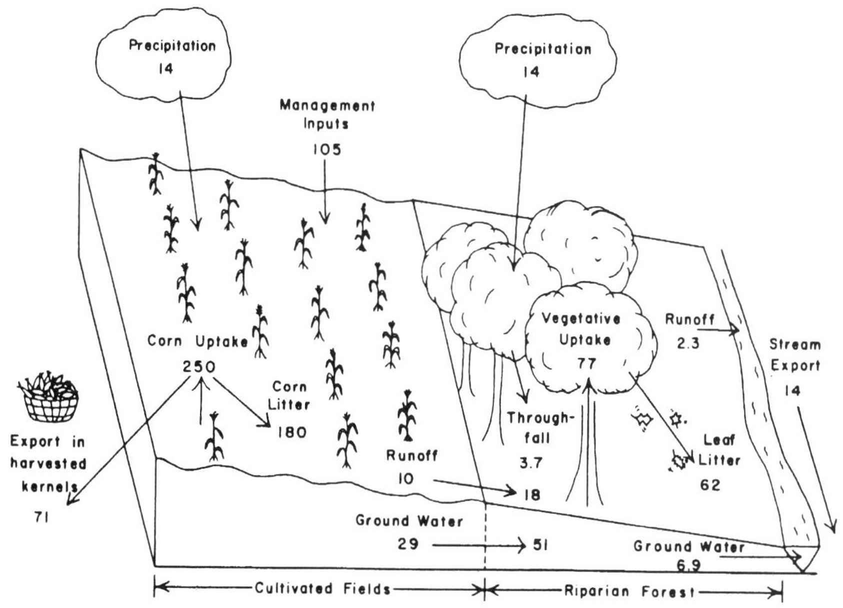


Fig. 6. Graphic of total nitrogen flux and cycling in a study watershed in units of kilograms per hectare. The habitats reflect cropland and riparian forest in a Rhode River watershed, Maryland. The forest removes large amounts of nitrogen from water entering the stream (from Peterjohn and Correll 1984).

Table 3. Examples of geologic disturbances. Modified from White (1979). See White (1979) for further discussion on each of these variables.

| |
|---|
| Cryogenic movement of soil |
| Freeze and thaw cycles |
| Extreme variation in stream flow |
| Stream erosion, deposition, and flooding |
| Ice push |
| Coastal erosion |
| Dune movement |
| Saltwater inundation |
| Landslides |
| Karst processes |
| Volcanic eruptions |
| Earthquakes |

ground (Watt et al. 1966). Nanson and Beach (1977) reported that stream erosion and other geomorphic processes associated with river meandering influence vegetation dynamics of patches. Other aspects of geologic disturbance are reviewed in White (1979) and listed in Table 3. White (1979) provided an extensive list of references for more details on these geologic disturbances.

Thus, geologic disturbances are the result of many types of processes that demonstrate the interaction of climate, lithology, geologic structure, and disturbance, that can be organized into a hierarchy of time and space. Geologic disturbance is thus an agent that shapes the landscape and the associated animal and vegetation communities in the complete spectrum of spatial and temporal scales.

### 3.4.2 Aquatic Disturbance Regimes

Disturbances, however defined, are fundamental for maintaining much of the variability associated with aquatic environments in North America. Disturbances can result from interactions between geological climatic processes, animal activities associated with modifying habitat and procuring food, and the type of aquatic environment (e.g., deep vs. shallow lakes, large vs. small streams). Geologic processes (as previously discussed) include volcanism, continental and alpine glaciation, and erosional processes. Climatic processes include the dominant form of precipitation (e.g., rain, snow or transition), the seasonal timing, and the hydrologic patterns. Animal influences (discussed later) include herbivory and physical habitat modifications such as dam building.

Aquatic systems are strongly affected by the type, frequency and intensity of disturbance, with physical disturbances having lasting effects (Niemi et al. 1990). The type of aquatic environment and its spatial position in the drainage network will determine much of the response to upstream geologic and climatic processes. Thus, a broad array of aquatic environments may be found within a relatively small area, with each having different biological characteristics and presenting different managerial challenges (Naiman et al. 1992).

Streams and rivers in the Pacific coastal ecoregion of North America provide good examples of how the disturbance regime and the spatial position of the aquatic systems in the drainage network affect biological characteristics (Naiman et al. 1992). Low-order (e.g., first and second order) stream segments represent more than 70 percent of the cumulative channel length in typical mountain watersheds. Hence, low-order streams are the principal conduits for water, sediment, and vegetative material routed from hill slopes to higher order rivers. First- and second-order streams are naturally prone to catastrophic erosion because steep slopes adjacent to steep channels favor landslides and debris flows. Filled with colluvium and large woody debris from lateral landslides, these channels are subject to devastating debris flows that scour channels to bedrock over long distances. The recurrence intervals for debris flows have been estimated to be 5–15 centuries with full recovery of the biological community estimated to be about 2–4 centuries.

Contrast that disturbance regime with regimes that occur farther downstream. In larger order channels the sediment supply is more steady in time and, as a result, the channel form (that which is dependent on the sediment supply rate, such as pools) is more uniform in space. In addition, extensive alluvial terraces and floodplains isolate the channel from direct contact with hill slopes and low-order streams, thereby limiting the direct influences of mass wasting (Benda et al. 1997). Lateral channel migration occurs gradually whereas multiple channels often are formed during floods. The recurrence intervals for these disturbances are years to decades as opposed to centuries for disturbances in the low order channels. In both cases, organisms with appropriate life history strategies comprise the biological communities and are well adapted to the dominant disturbance regimes (Naiman and Anderson 1997).

In North America and other regions, modification of aquatic environments has resulted in pervasive physical alterations to aquatic habitats and to the disturbance regimes that maintain them. This has caused extensive biological impoverishment as well as risks to human health and safety (Naiman et al. 1995). The biological impoverishment is manifested in the loss and fragmentation of habitat, acceleration of exotic invasions, overharvesting of fishery and wildlife resources, altered thermal regimes, nitrogen contamination, harmful algal blooms, and the spreading incidence of acid rain. Threats to human health and quality of life are manifested by the increasing impairment of ecological services (such as the absorption and detoxification of chemical contaminants), degradation of drinking water quality and quantity, contamination of aquatic food resources, spread of infectious disease, and deteriorating aesthetic and recreational opportunities.

### 3.4.3 Fire

Fire is a primary disturbance in most systems, either naturally or by human influences. Although this topic is covered more extensively in the Disturbance chapter (see White et al., this volume), we present some aspects of effects on landscape diversity. Fire regimes are often the major factors in determining landscape diversity as fire has a direct effect on stand structure and on plant and animal communities. Concannon et al. (this volume) states that knowledge of an areas's fire history is absolutely essential in all levels of planning and resource management because it affects both the production of goods and services as well as ecosystem structure, function, and composition. This information can be obtained through historical reports, paleoecological evidence of charcoal and plant species, photographic comparisons, tree ring analysis of scars and age classes, and understanding the mosaic of the current landscape. Perhaps the most obvious effect fire has on forest stands is on plant density regulation. Organized fire suppression in much of the West has lengthened the fire-free intervals of vegetation. Some stands of ponderosa pine (*Pinus ponderosa*) in the Klamath Province of Northern California and Southern Oregon have missed as many as 25 fire episodes. In the western United States, some ponderosa pine stands have more than 1,000 trees per acre today, whereas there were less than 100 trees per acre at the turn of the century (Tim Sexton, unpubl. data). Higher densities can lead to moisture and nutrient deficiencies, increased insect and disease susceptibility, species composition changes, increased wildfire hazard, and reduced biodiversity.

Changes in fire regimes can affect landscapes in various ways. In the absence of fire, site nutrients are immobilized in organic form much faster than decay can mineralize them. The net effect over many years is likely to lessen the available nutrients of the site. Frequent, low intensity fires increase soil pH and increase the ability of soil to retain mineralized nutrients.

Exclusion of fire in the Klamath Province has significantly changed ecosystem structure, composition, and function. In the Klamath Basin, interruption of natural fire regimes favors woody species over grasses, forb, and shade tolerant species. For example, east-side ponderosa pine forests are being replaced by Douglas-fir (*Psuedotsuga menziesii*), white fir (*Abies concolor*), or lodgepole pine (*Pinus contorta*) because of the absence of fire. Old aerial photos, from the 1940s, provide visible evidence of changes in stand structure and composition. Before intensive fire suppression, stands were more "park-like" and open, with less vegetative understory. Because of the heavy fuel accumulations and stand densities today, fires are resulting in a larger number of stand-replacing events with higher fire severity. These effects are highly visible when viewed from a landscape vantage point. Data from Pacific Southwest Range and Experiment Station in the Klamath Mountains Province have concluded that ranges of values for all vegetation variables measured were greater in 1944 compared to 1985, suggesting a more diverse landscape mosaic in 1944 than in 1985 (Skinner 1995).

The challenge today is how to bring these ecosystems back toward natural fire regime conditions in the face of many constraining policies and during a time of heightened public controversy over the management of public lands.

## 4 ANTHROPOGENIC FORCES AFFECTING LANDSCAPE DIVERSITY

Human activities have had profound effects on landscapes from the individual basin to continental scales (Delcourt and Delcourt 1991, Goudie, 1993). For North America, the impact of humans started approximately 10,000 years ago and continues with increasing intensity today. Many examples are known for geomorphologic phenomena in which the human role is evident: e.g., soil erosion, stream channelization, lake sedimentation (Table 4).

Information about cultural uses and transformations of aquatic environments is not well organized, allowing only an approximate description for North America. However, from the available information, the two most dramatic changes in this century have been the rapid increase in water consumption and in the amount of waste water produced. Changes to the physical habitat are equally pervasive but the data are less well organized. However, it is known that less than 2 percent of the total length of streams and rivers in the conterminous 48 states are in good condition (Benke 1990) and that nearly two-thirds of the organisms listed

Table 4. Examples of the human role in geomorphologic phenomena (Goudie 1993).

| | |
|---|---|
| *Aeolian* | |
| Dust storm generation | Goudie and Middleton (1992) |
| Wind erosion of soil | |
| Dune reactivation and stabilization | Watson (1990) |
| *Coastal* | |
| Salt marsh accretion | Adam (1990) |
| Delta retreat | Walker and others (1977) |
| Coral bleaching | Brown (1990) |
| Erosion and accretion | Hails (1977) |
| *Fluvial* | |
| Arroyo incision | Cooke and Reeves (1976) |
| Channelization | Brooks (1985) |
| Channel geometry change | Petts (1985) |
| Clear water erosion | Beckinsale (1972) |
| Soil erosion | Trimble (1988) |
| Sediment load change | Trimble (1974) |
| *Miscellaneous* | |
| Slope destabilization | Selby (1979) |
| Lake sedimentation | Jones et al. (1985) |
| Lake desiccation | Micklin (1972) |
| Hollow formation | Prince (1964) |
| Ground subsidence | Chi and Reilinger (1984) |
| Seismic disturbance | Meade (1991) |
| Thermokarst development | French (1976) |
| Accelerated salt weathering | Goudie (1977) |
| Lateritization | Gourou (1961) |
| Peat bog formation | Tallis (1985) |
| Tufa decline | Nicod (1986) |

as rare and endangered are aquatic (Naiman et al. 1995). The withdrawal of water for domestic, municipal, and industrial uses has caused major changes to runoff patterns, evaporation, and the volume and amount of pollutants returned to the environment. In addition, groundwater is being rapidly reduced by altered soil conditions and direct depletion of the aquifers. In the United States alone, water consumption is already excessive. Collectively, more than 8.7 billion cubic meters of municipal and industrial effluent are discharged annually. Residues of many of the more than 80,000 human-made chemicals in common use are present in this effluent, which is enriched further by nutrients from human and agricultural waste. This is occurring at a time when demands for clean water, recreation, and aquatic products are doubling every 10–20 years.

## 4.1 Land-Use Change

Land-use change in the United States represents an enormous uncontrolled experiment in the ways habitat changes influence plants and animals (Forman 1995). Land-use changes have been particularly profound since Europeans settled North America three centuries ago, with landscapes becoming mosaics of natural and human-influenced patches, and once-continuous natural habitats are becoming increasingly fragmented. Evidence is increasing to show that native Americans had important impacts on the landscape; however, rates of change increased markedly following European occupation.

The term "land-use change" has several meanings: changes in both land cover and land use. Land cover refers to the habitat or vegetation type present, such as forest, agriculture and grassland. Land-cover change describes differences in the area occupied by cover types through time; positive and negative. Land use is usually defined more strictly and refers to the way in which humans employ the land and its resources.

Many Americans assume that pre-colonial forests were ancient in age and vast and unbroken in size. This notion has been perpetuated to some degree by early American writers who sometimes described their surroundings with romantic terms such as "the forest primeval," e.g., H.W. Longfellow's poem *Evangeline* originally published in 1847. Although some were large and old, pre-settlement forests may not have been uniform or ancient. Dickson (1991) described pre-settlement forests as diverse, containing all age classes interspersed with openings. Johnson (1987) reported that pre-colonial southern forests contained a mixture of forest types with abundant old growth and substantial areas of openings and early successional forests interspersed. This view of pre-colonial forests as dynamic likely is accurate because fire, ice storms, tornadoes, hurricanes, insects, and diseases routinely disturb forest ecosystems.

For thousands of years, native Americans greatly influenced forests through their activities (e.g., Day 1953, Martin 1973, Hudson 1976). When constructing villages, Indians cleared forests for home sites and agriculture (Day 1953). They foraged long distances for plants and animals that were used for food, fiber, and medicine. They collected wood and bark for utensils, weapons, canoes, houses, and fuelwood. Native Americans also apparently used stone axes and fire to fell living trees for fuelwood and other purposes.

Indians frequently used fire for hunting and war, to improve visibility and traveling conditions, and other reasons (Day 1953, Martin 1973, Lewis 1980, Barrett 1980, Barrett and Arno 1982, Pyne 1983, Gruell 1985, Lewis 1985, Pyne et al. 1996). Even hardwood-dominated forests were burned with regularity (MacCleery 1992).

For three centuries following European settlement, most Americans were farmers. In 1800, 95 percent of Americans lived off the land (MacCleery 1992). Initially, settlers preferred to farm in clearings abandoned by Indians. However, iron tools and draft animals permitted them to convert large areas of older forest to other uses (e.g., row crops, pasture, urban development), thereby altering forest and landscape structure, composition, and processes. Railroads, cross-cut saws, and other technological advances rapidly facilitated even more efficient conversion.

Since 1600, American farmers have converted about 310 million acres of forest to other uses, mostly agriculture. Between 1850 and 1910, farmers cleared about 190 million acres, or about 13.5 square miles of forest per day (MacCleery 1992). Wood from forest conversion and other timber removals was used for ships, railroad ties, fences, buildings, and many other uses, such as fuel for heating, cooking, railroads, steamships, iron-making furnaces, and stationary engines. When forests were harvested, but not converted, regeneration was infrequent and wildfires were common. Some large-scale conversions have occurred in the 20th century. For example, about 80 percent of the floodplain forests in the Mississippi River alluvial valley (or about 52 percent of all southern bottomland forests) have been converted to agricultural or urban uses (Smith et al. 1993).

Early European settlers in some portions of the United States continued the Native American tradition of using fire to alter forest ecosystems. For example, until the late 19th century, burning was used extensively in the South to enhance foraging conditions for livestock and habitat for game species, such as northern bobwhites (*Colinus virginianus*) (Stoddard 1962). Large land-clearing and wildfires were not uncommon in many regions and were considered desirable because they reduced slash. For example, fires in Wisconsin and Michigan burned over 3 million acres during 1871, and fires in Minnesota burned over 500,000 acres between 1894 and 1918 (Johnson 1994).

Throughout much of the 20th century, ecological processes at the landscape scale have been dramatically altered by fire suppression. State forestry agencies, the USDA Forest Service, and the forest products industry initiated cooperative programs to fight wildfires. Examples of these programs include the

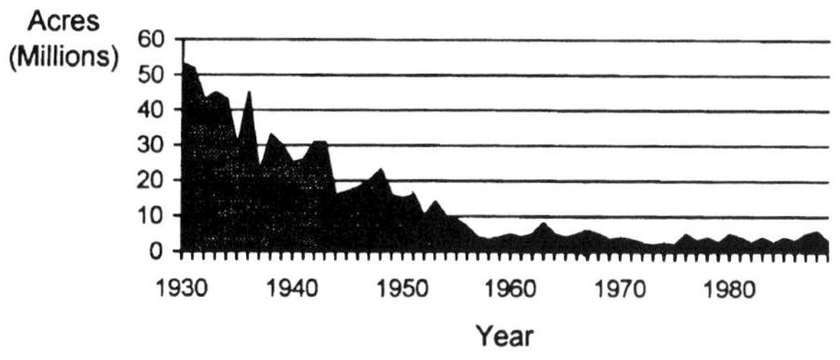


Fig. 7. Area burned in the United States, 1930–1989. Modified from MacCleery (1992).

"Smokey the Bear" program and the Southern Forestry Educational Project that functioned between 1927 and 1930. Through this project, landowners in Georgia, Florida, and Mississippi were informed that "fire destroys the natural breeding places of birds and animals" and that "so long as fire is kept out of the woods the community of trees and plants and animals have a chance to take care of themselves" (Komarek 1981). Because of suppression efforts, most fires today are small in size, and only about 3–5 million acres burn each year (Fig. 7) (MacCleery 1992).

Humans have grazed livestock in American forests since the 1600s. Early European settlers possessed large numbers of livestock. For example, MacCleery (1992) cited records from the Massachusetts Bay Colony in 1634 reporting a population of 4,000 people, 1,500 cattle, 4,000 goats, and "swine innumerable." Most of these livestock grazed in the forest; fences were used primarily to exclude livestock from gardens and crops rather than to confine them.

In the Southwestern United States, intensive livestock grazing also occurred following Spanish settlement. For example, by 1880 in the Rio Grande valley of New Mexico, there were over 2,000,000 sheep, over 150,000 cattle, and about 50,000 horses, mules and burros utilizing the rangelands and adjacent mesas and mountains (Burkholder 1928). Livestock grazing continues today throughout the nation affecting the structure and composition of forest and grassland habitats.

Early European settlers also influenced landscape and ecosystem diversity indirectly by altering wildlife communities. Market hunting reduced populations of some species, such as the passenger pigeon (*Ectopistes migratorius*) and white-tailed deer (*Odocoileus virginianus*), to extinction or to extremely low densities by the early 1900s. Forest succession, structure, and composition undoubtedly were influenced in many areas by altered abundances of important browsers, grazers, and seed dispersers.

Historical changes in land use and ecological processes have resulted in landscapes different in composition and structure from those that existed on this continent before settlement by Europeans (Palik and Pregitzer 1992). Because of conversion of forest to other uses, present-day landscapes can have less forest area than pre-settlement landscapes (e.g., Burdick et al. 1989). Present-day landscapes can be more diverse than pre-settlement forests and have more forest patches that are smaller and simpler shaped (Mladenoff et al. 1993). In addition, there often are fewer large patches, ecosystem juxtapositions sometimes have been changed, and the location of older stands may be restricted to sites bypassed by conversion or recent harvesting (Mladenoff et al. 1993).

However, in some landscapes, particularly in the East, fragmentation has actually declined and forest connectivity has increased relative to centuries ago. The amount of forest land actually has increased since 1900 in every state east of the Mississippi River (MacCleery 1992). In Georgia between the late 1930s and 1980s, the total amount of forest increased, the number of forest patches decreased, and in some areas of the state the amount of agricultural land declined (Turner 1990). Recent efforts to reforest marginal crop lands in formerly forested landscapes, such as the Mississippi Alluvial Valley, may yield similar results. This trend may again be reversing itself as more and more people are buying tracts of land and building year-round or seasonal houses on them. This requires roads and places homes in a landscape resulting in increasing fragmentation from roads and fences (Knight and Clark in press, Knight et al. 1995). Significantly, this reestablishment of forest environments as well as patchy, human-dominated landscapes, coupled with current wildlife management practices have caused very high densities of the white-tail deer and associated insects (e.g., ticks) and disease (Lyme disease) in areas of the Northeast.

Forman (1995) demonstrated this increase in forest type, a new land type in effect, with its attendant ecological characteristics. In the new land types, a somewhat different set of spatial processes emerge, such as patch appearance or proliferation, expansion, coalescence, aggregation, connection, and infilling. In Massachusetts, for example, fields were progressively abandoned from 1840–1985, regrew into forest, and the resultant forest patches gradually coalesced. The rate of forest spread was greatest in the middle of the period, and forest cover reached 90 percent by the end (Fig. 8). Boundary length of the forest type is greatest in the mid period, just as it is in a deforestation process where forest patches become increasingly dispersed.

Forman (1995) showed the dramatic effects of deforestation on landscape properties for a Pacific Northwest, Douglas-fir forest (Fig. 9). The transformation sequence has marked effects on many ecological factors, including windthrow, fire ignition, and interior species richness. Overall most of the rapid changes in ecological factors take place in the first 40 percent of transformation, and few rapid ecological changes occur in the last 40 percent (Fig. 9). Forman (1995) suggested that the most critical time for land planning and conservation appears to be when the landscape has 60–90 percent of its area in natural vegetation.

Because of changes in land use, some present-day management options have been largely foreclosed. For example, the option of maintaining pre-settlement conditions across large areas is not feasible across most of the East. Perhaps the only certainty regarding ecosystems is that most United States landscapes will continue to change and be heavily influenced by humans.

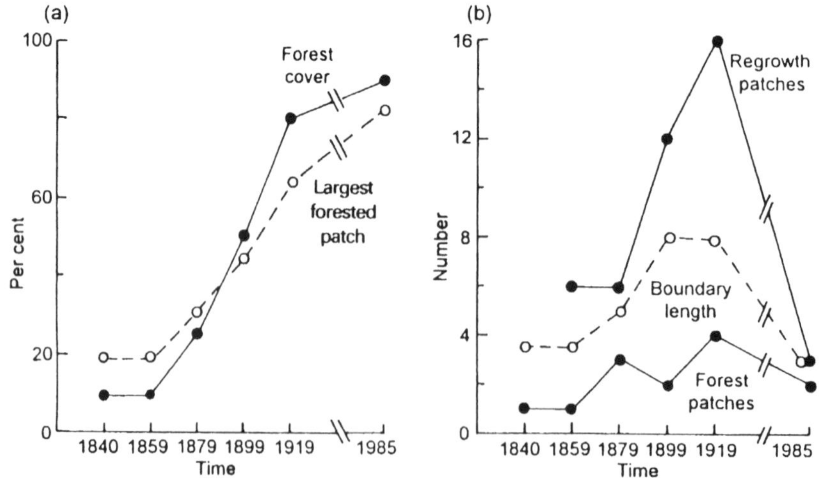


Fig. 8. Spatial changes during reforestation after field abandonment. (a) Forest cover and largest forest patch in 400-hectare landscape area. Relative patch size = diameter of largest circle of forest divided by diameter of largest circle in the landscape area. (b) Regrowth patches are of young trees on abandoned fields, and forest patches are of older trees with full canopy height. Boundary length is calculated by divided the total field-forested circle present. Harvard Forest in central Massachusetts (from Forman 1995).

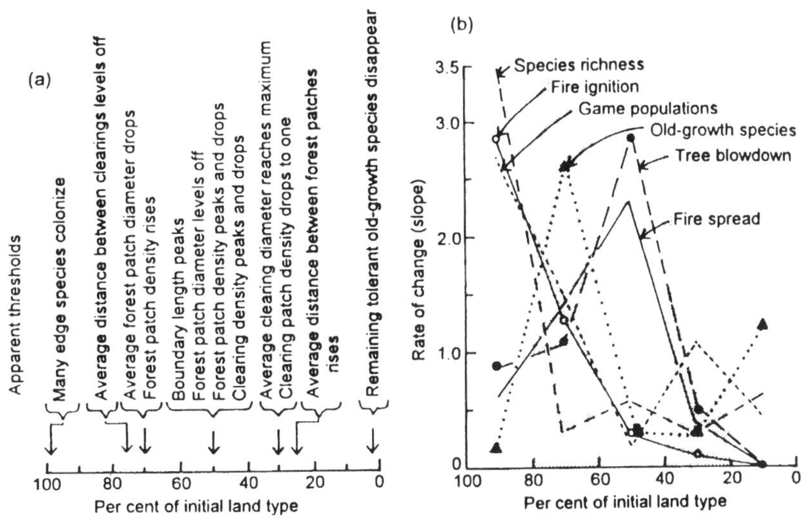


Fig. 9. Thresholds and periods of rapid change in a dispersed-patch deforestation process in a Pacific Northwest, Douglas-fir forest landscape. A. Apparent thresholds of spatial attributes based on a dispersed-patch or checkerboard model. B. The rate of change of six major ecological characteristics at different phases in the deforestation process (from Forman 1995).

## 4.2 Private Sector

Present-day influence of private landowners on landscape and ecosystem diversity is important because most land currently is owned not by federal, state, or local government, but by the private sector. About 66 percent of all land in the United States is owned by families and individuals; non-family corporations and partnerships own another 13 percent (Gustafson 1982). Much of all privately owned land is in small parcels and in parcels smaller than 500 acres (Gustafson 1982).

Like the pattern for all land types, most forest land also is owned by noncorporate individuals or families (Table 5). Nationally, private landowners control about 73 percent of timberland (forest capable of annually producing >20 $ft^3$ wood per acre), more than three times as much as the public owns through the federal government (Powell et al. 1993). Nationally, there are about 10 million nonindustrial, private owners of timberland, a number that has increased by about 20 percent since 1974 (Argow 1996).

Data presented in Birch (1996) revealed some important information about private landowners. Most (94 percent) private owners or ownership units (e.g., partnerships) possess fewer than 100 acres of forest land each (Table 6). However, these landowners control only about 32 percent of all private forest land. Thus, most (68 percent) private forest land is controlled by only 6 percent of owners who each have more than 100 acres.

Ownership objectives determine whether forest stands are converted to other uses, and if retained as forest, whether their structure, composition, and configuration remain natural. Because ownership objectives vary, ecological pattern and composition of private lands usually differs at the ownership level. Land-

Table 5. Percentage of timberland by landowner type and region (Powell et al. 1993).

| Landowner type | Region[1] | | | | | |
|---|---|---|---|---|---|---|
| | North-east | North Central | South | Rocky Mountain | Pacific Coast | United States |
| Federal | 3.2 | 12.4 | 8.0 | 63.5 | 43.1 | 19.7 |
| State | 6.5 | 15.3 | 1.8 | 4.1 | 11.3 | 5.6 |
| County/municipal | 1.3 | 7.1 | 0.4 | 0.2 | 0.5 | 1.5 |
| Forest products industry | 14.9 | 5.0 | 19.6 | 4.7 | 17.6 | 14.4 |
| Noncorporate private | 74.2 | 60.2 | 70.1 | 27.7 | 27.5 | 58.7 |

[1]Northeast includes CT, DE, ME, MD, MA, NH, NJ, NY, PA, RI, VT, and WV; North Central includes IL, IN, IA, MI, MN, MO, WI; South includes AL, AR, FL, GA, KY, LA, MS, NC, OK, SC, TN, TX, and VA; Rocky Mountains includes AZ, CO, ID, KS, MT, NE, ND, NV, NM, SD, UT, and WY; Pacific Coast includes AK, CA, HI, OR, and WA.

Table 6. Estimated number of private ownership units and acres of forest land, by size class of ownership, United States, 1994 (Birch 1996).

| Ownership size (acres) | Owners | | Acres | |
|---|---|---|---|---|
| | Thousands | % | Millions | % |
| 1–9 | 5795 | 58.6 | 16.6 | 4.3 |
| 10–49 | 2762 | 27.9 | 60.4 | 15.5 |
| 50–99 | 717 | 7.2 | 47.2 | 11.9 |
| 100–499 | 559 | 5.6 | 91.6 | 23.3 |
| 500–999 | 41 | 0.4 | 24.5 | 6.3 |
| 1000+ | 27 | 0.3 | 153.0 | 38.8 |
| Total | 9902 | 100.0 | 393.4 | 100.0 |

scapes composed of multiple private owners or a mixture of public and private owners usually are mosaics of forest types, ages, structures, and sizes.

Most landowners have multiple objectives for their land. For example, in Mississippi, Nabi et al. (1983) found that 63 percent of nonindustrial private landowners described "multiple-use" as the goal of owning their forest land. However, timber production was the most important of the uses, followed by wildlife, residence, and grazing. In New England, private landowners reported their forests were used for firewood cutting, open space, recreation, scenery, wildlife habitat, part of farm, hunting, and privacy (Alexander 1986).

Ownership objectives for private lands are constantly changing. Thus, management direction and ecological diversity can be expected to change temporally even within ownership. Annual turnover in ownership of private, noncorporate land is about 12 percent (Shaw 1981).

Most individual owners are "white collar" professionals (32 percent) or retired (29 percent) (Birch 1996). Most (39 percent) consider their forest land as part of their residence or farm (Table 7), and the primary benefit they most often expect to receive during the next 10 years is simply "enjoyment of owning" (Table 8). Nevertheless, timber harvesting is an important benefit in terms of the amount of area affected. Although most (97 percent) owners do not regard timber harvesting as their primary objective, these landowners control only 29 percent of private forest land. About 34 percent of owners say they never intend to harvest, but they own only 12 percent of private acreage. Regardless of primary objective, many owners (32 percent) expect to harvest within 1–10 years. These landowners control 63 percent of private forest land, thus presenting many opportunities to influence the future structure and composition of American forests.

A particularly large private owner and manager of forested land is the forest products industry, which owns about 70 million acres of timberland, about 75 percent as much as is managed by federal agencies. The

Table 7. Estimated number of private ownership units and acres of forest land, by primary reason for owning forest land, United States, 1994 (Birch 1996).

| Reason for owning | Owners | | Acres | |
|---|---|---|---|---|
| | Thousands | % | Millions | % |
| Land investment | 920.0 | 9.3 | 39.3 | 10.0 |
| Recreation | 874.5 | 8.8 | 37.9 | 9.5 |
| Timber production | 272.2 | 2.7 | 113.2 | 28.9 |
| Farm & domestic use | 816.4 | 8.3 | 35.8 | 9.1 |
| Enjoyment of owning | 1392.4 | 14.1 | 28.7 | 7.3 |
| Part of farm | 1189.8 | 12.0 | 38.6 | 9.8 |
| Part of residence | 2641.5 | 26.7 | 32.6 | 8.2 |
| Other | 1440.9 | 14.5 | 61.0 | 15.3 |
| No answer | 354.1 | 3.6 | 6.3 | 1.5 |
| Total | 9901.7 | 100.0 | 393.4 | 100.0 |

Table 8. Estimated number of private ownership units and acres of forest land, by primary benefit expected during the next 10 years from owning forest land, United States, 1994 (Birch 1996).

| Expected benefit | Owners | | Acres | |
|---|---|---|---|---|
| | Thousands | % | Millions | % |
| Land value increases | 2055.9 | 20.8 | 63.6 | 16.4 |
| Recreation | 959.4 | 9.8 | 42.8 | 11.0 |
| Timber production | 477.5 | 4.8 | 131.6 | 33.2 |
| Farm & domestic use | 1505.4 | 15.2 | 42.9 | 11.0 |
| Enjoyment of owning | 3356.2 | 33.9 | 63.3 | 16.0 |
| Firewood | 421.5 | 4.2 | 9.9 | 2.5 |
| Other | 478.9 | 4.8 | 27.3 | 6.6 |
| No answer | 646.9 | 6.5 | 12.0 | 2.8 |
| Total | 9901.7 | 100.0 | 393.4 | 100.0 |

forest products industry also is a diverse group of landowners. Forest products companies differ in size, type of ownership (e.g., family-, individual-, stockholder-owned), and the products that they make from timber they harvest. For example, they may emphasize glossy magazine paper; dimension products such as lumber or poles; paper cups, towels, and napkins; some combination of these products; or something entirely different. These differences also affect the management strategies of companies and resulting ecological and landscape diversity. For example, companies emphasizing paper products may require short rotations to be economically competitive, whereas other companies emphasizing lumber or other dimension products may favor longer rotations and different tree species.

The potential importance of private forest lands to landscape diversity is particularly salient in the eastern United States. In the Northeast and South, corporate and noncorporate private landowners control about 90 percent of all forest lands. Federal ownership is <4 percent in the Northeast, and <9 percent in the South. Thus, landscape and ecosystem diversity is largely determined by the objectives and actions of private landowners, and few efforts to manage forest ecosystems will be complete without involving them. The Departments of Interior and Agriculture recently acknowledged that "because they own the majority of the United States, the involvement of millions of private parties and landowners is critical to the overall success of conservation efforts" (Goklany 1992).

### 4.3 Federal Policies Affecting Landscape Diversity

Federal policies have had a profound influence on landscape patterns. In fact, ecosystem management policies are resulting in far greater interagency coordination, leading to regional and Province-wide cooperation to maintain biological diversity. One of the best examples is President Clinton's "Northwest Forest Plan for Management of Habitat for Late-Successional and Old-Growth Forest Related Species Within the Range of the Northern Spotted Owl." Large reserves have been allocated to late-successional species in a manner that allows connectivity and genetic interchange. The Forest Service and BLM utilize the same set of standards and guidelines, unique to each type of land allocation. This plan will have a visible effect, and is expected to have a major effect on landscape patterns and biological diversity. The Aquatic Conservation Strategy of the Pacific Northwest Plan provides another example. Its goal is to improve aquatic habitat conditions in order to reverse the decline in populations of anadromous salmon and steelhead. Its success is dependent on cooperative policies among federal agencies during implementing management actions. For example, the timing and quantity of flows from the Klamath River Project by the Bureau of Reclamation significantly influence the survival of fingerlings which result from improved habitat conditions under Forest Service and BLM habitat restoration programs. Concannon et al. (this volume) provide additional examples.

Turner et al. (in press) tested hypotheses about the influence of land ownership patterns on landscape structure by examining watersheds in the Olympic Peninsula, Washington, and the southern Appalachian highlands of western North Carolina. Two specific questions were addressed: (1) Does landscape pattern vary between federal, state and private lands? and (2) Do land cover changes differ among owners, and, if so, what variables explain the propensity of land to undergo change on federal, state and private lands? Their studies identified that in both regions, private lands contained less forest cover but a greater number of small forest patches than public lands. Lands that were actively managed for timber harvest, however, showed little difference in landscape pattern between ownerships.

## 5 BIOLOGICAL FORCES AFFECTING LANDSCAPE DIVERSITY

Previous sections have discussed several biological factors in association with physical and anthropogenic factors. This section identifies additional specific features of the biological world that have important roles at the level and scale of the landscape.

### 5.1 Roles of Individual Species/Populations

#### 5.1.1 Keystone Species

Since Paine (1969) first defined a keystone predator, the term "keystone" has been extended to species whose removal results in significant change in community structure or ecosystem functioning. Power et al. (1996) defined a keystone species as one whose effect is large, and disproportionately large relative to its abundance. When applied to landscapes or ecosystems, keystone species are those species whose contribution to ecosystem functioning is unique (Grimm 1995). This concept has raised the issue of the relative importance of keystone species versus functional redundancy in ecosystems (Jones and Lawton 1995). If an ecosystem could be divided into essential functions or processes, then organisms performing each function could be identified and a determination made as to whether or not there was a keystone species involved. Obviously, the more narrowly an individual function is defined, the more likely a keystone species is involved (Grimm 1995).

A more operational definition for keystone species involves defining the strength of the effect of a species on a community or ecosystem trait. Power et al. (1996) used a measure of community importance (*CI*); the change in a community or ecosystem trait per unit change in the abundance of the species; $CI = [d(\text{trait})/dp]\,[1/(\text{trait})]$, where $p$ is the proportional abundance of the species whose abundance is modified (in most cases, proportional biomass relative to the total biomass of all other species in the community). Trait refers to a quantitative trait of a community or ecosystem, (i.e., productivity, nutrient cycling, species richness, abundance of one or more functional groups of species of dominant species). Figure 10a provides general examples of total (collective) impact of a species (absolute value of community impact × proportional abundance of a species: $|CI\,i| \times p\,I$ versus its proportional abundance, $p\ I$. Points representing a species whose total impact is proportional to its abundance would fall along the diagonal line $X = Y$. Keystone species have effects that exceed their proportional abundances by some large factor will be above the line (Power et al., 1996). Figure 10b is a scenario of successional changes in the dominance and total impact of annual herbs (acting as keystone species) following fire in South African Savannah (Power et al., 1996). Immediately after fire, annuals sprout and make up most of the plant biomass (A early). Over time, woody shrubs and tree seedlings reinvade and make up increasing proportions of the total community biomass. The annuals at this stage (A middle) strongly determine sites at which the later successional plants can colonize, both positively (if annuals provide safe sites, such as more favorable microclimates for survival of seedlings) and negatively, if annuals compete with woody seedlings. During the third stage, annuals disappear (biomass becomes undetectable; A late).

Other examples include the abiotic and biotic influences of beaver and moose on the landscape which are spatially extensive and long-lasting (Johnston et al. 1993) and the ecological consequences of prairie dog disturbances. Prairie dogs (*Cynomys* spp.), like most fossorial rodents, alter grassland patch structure, nutrient cycling, and feeding site selection by other herbivores by selective grazing and burrowing (Whicker and Detling 1988). Weltzin et al. (1997) suggested that the broad-scale removal of prairie dogs from rangelands in the Southwest is a key to understanding the rapid expansion of desert shrubs into the grasslands of that region.

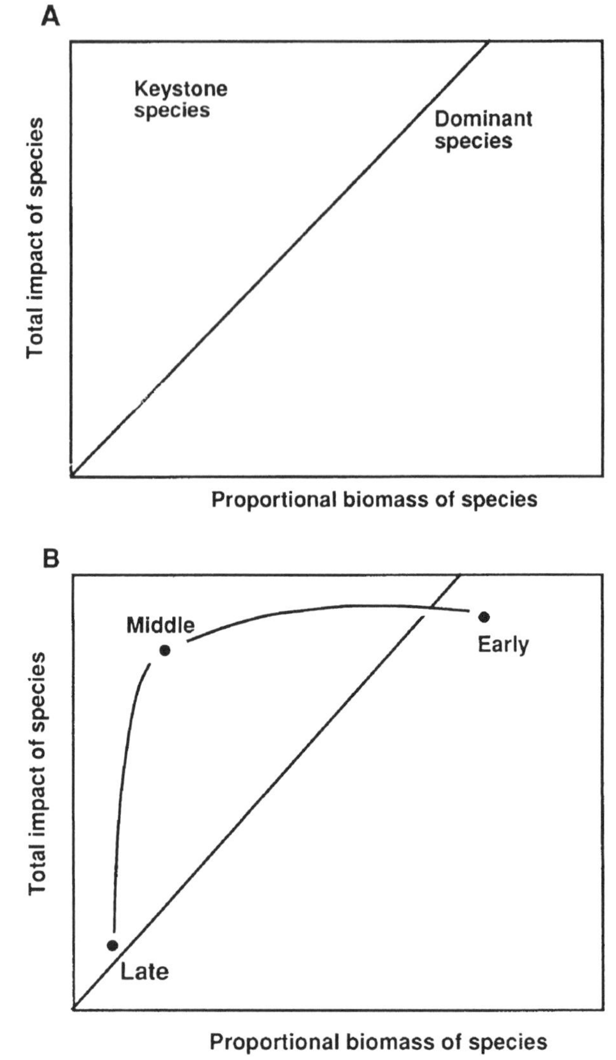


Fig. 10. Roles of keystone species. A. General examples of total (collective) impact of species (absolute value of community impact × proportional abundance of a species) versus its proportional abundance. Points representing a species whose total impact is proportional to its abundance would fall along the diagonal line ($X = Y$). Keystone species have effects that exceed their proportional abundances and fall above the line. B. Successional changes in the dominance and total impact of annual herbs following fire in South African Savannah: Immediately after fire, annuals sprout and make up most of the plant biomass ($A_{early}$). Over time, woody shrubs and tree seedlings reinvade and make up increasing proportions of the total community biomass. The annuals at this stage ($A_{middle}$) strongly determine sites at which the later successional plants can colonize, both positively (if annuals provide safe sites) and negatively (if annuals compete with woody seedlings). During the third stage, annuals disappear ($A_{late}$). From Power et al. 1996.

In general, animal–ecosystem interactions are complicated because population cycles of many keystone species occur over long periods (decades), alterations to the ecosystem (such as vegetation replacement or altered soil characteristics) are often subtle over short periods, and shifts in biogeochemical cycles or soil formation are often only detectable after a period of years. Nevertheless, alterations to the landscape by keystone species often result in patterns and processes that would not occur under the dominating influences of climate-driven or geological disturbances alone. If we can identify keystone species in various ecosystems, an important management policy could be to set aside critical areas and to manage them so as to maintain these keystones, instead of solely focusing on endangered local species or geographical hot spots of biodiversity (Power et al., 1996). If local keystones cannot be identified, the keystone concept points to the need for a cautious management strategy that takes into account potential surprises from small interventions or changes. For example, (1) the loss of some species of low abundance may have surprising dramatic effects; (2) the preservation of a species of concern may depend on the distribution and abundance of other species with which the target has no recognized interaction; and conversely, (3) the loss of a species, such as a top carnivore, may reverberate to affect members of seemingly disparate guilds, such as plants or decomposers (Power et al., 1996). Managers should carefully consider the consequences of the loss of species for which no obvious role in the ecosystem has been discovered. The keystone species concept indicates the need for a design with a wide margin of safety for managed lands to guard against the loss of those organisms with disproportionately high community importance values.

### 5.1.2 Exotic Species Invasions

More than anything else, biodiversity and ecosystem health on western federal wildlands depend on healthy, diverse, and resilient plant communities. Plants form the basic biological matrix of all communities, and the growth forms of plants are an important component of community structure. Unfortunately, these plant communities are under attack from invasive exotic plants.

Exotic plants arrived from other countries without their natural enemies that kept them in check in their country of origin. Consequently, these new plants are typically very successful and have the ability to dominate many locations with dramatic impacts to native plant communities. These undesirable plants are biological forces restructuring and changing processes that affect landscape diversity on a grand scale. Furthermore, invasive exotic plants are indicators of, and contributors to, desertification.

Invasive exotic plants move toward a near monoculture condition at the watershed and landscape scale. Obviously, the hundreds of very large, near monocultures that already exist on federal lands do not constitute a diverse landscape. With today's technology, it is either not practical or impossible to restore those sites. From an ecological standpoint the negative impacts from invasive plants may be categorized as: direct negative influences on native species, negative effects on soil (both erosion and soil quality), alteration of nutrient cycling, impacts on hydrology, and changes in disturbance regime.

A few examples will serve to demonstrate invasions that are underway. Spotted knapweed (*Centaurea maculosa* Lam.), first reported in Montana in 1920, has invaded nearly 5 million acres in that state alone. There are over 600,000 acres of leafy spurge (*Euphorbia esula* L.) in Montana and 1 million acres in North Dakota. In Idaho, rush skeleton weed (*Chondrilla juncea* L.) expanded from 40 acres in 1964 to over 4 million acres today! Since 1978, yellow starthistle (*Centaurea solstitialis* L.) has spread in Northern California from 1 million acres to over 10 million acres. Yellow starthistle is expanding rapidly in eastern Oregon and western Idaho as well. Weeds of the West (Whitson et al. 1996) lists many more examples.

Exotic plants increased on Bureau of Land Management lands from 2.5 million acres in 1985 to over 8 million acres in 1994. When exotic plant populations on the Forest Service, National Park Service, and the Fish and Wildlife Service land are included, that infested area nearly doubles. Recognizing that exotic plants typically spread about 14 percent per year if unchecked, the increase infestation rate on these federal lands is now about 4,600 acres per day — an "explosion in slow motion."

Next to grazing, invasive exotic weeds have probably caused the second largest changes in western arid and semiarid lands since pre-columbian times. Runoff and sediment yield were 56 percent and 192 percent higher respectively for spotted knapweed than for bunchgrass vegetation types. One of the five indicators for evaluating the susceptibility for desertification is exotics as a percent of total cover. The severe level of deterioration in four desertification classes is described in part as follows: "Undesirable forbs and shrubs have replaced desirable grasses or have spread to such an extent that they dominate the flora."

DeFerrari and Naiman (1994) described the landscape characteristics in the Olympic Peninsula of Washington that influenced the pattern of exotic invasions. Disturbance type and time since disturbance were major factors influencing invasibility. Of the plant species encountered in the watersheds studied, 23 percent were exotic, and the occurrence of these species was highly correlated with landscape patch type. Riparian areas are subjected to intense flooding disturbances, and these areas had the highest number and cover of exotics in the landscape. Although clearcutting is a major disturbance in upland plots, clearcuts had fewer exotics than riparian patches. Although clearcutting removes the forest overstory, it often leaves native understory plants relatively intact or able to resprout. Clearcutting promotes some invasion by exotic species, it is not equal in magnitude to a large flood that can remove most natives and competitively favor exotics, many of which are disturbance-oriented (see references in DeFerrari and Naiman 1994). Landscape patch size, position within watershed (i.e., distance from patch to human population centers, major highway, or river mouth), and environmental variables (slope, aspect, and elevation) were not important indicators of landscape patch invasibility. Riparian zones facilitated movement of exotic plants through landscapes, but did not appear to act as sources of exotics plants for undisturbed upland areas. The time since a landscape patch was disturbed is an important determinant of the extent of invasion (DeFerrari and Naiman 1994). The riparian and upland patch types sampled in their study represented successional stages, and the number and cover of exotics decreased with increasing successional stage or age.

Clearly, invasive exotic plants are having a dramatic effect on ecosystem and landscape diversity. Fortunately, 95 percent of the western federal wildlands are not yet seriously infested, and effective and economical measures are known to reduce future exotic infestations. Recognizing the long-term and often permanent damage to ecosystem health, invasive exotic plants should deserve at least as much attention as fire, insects, and disease.

## 5.2 Landscape Boundary Dynamics

### 5.2.1 Boundary Movement

Boundaries, even those of biomes, can change position (Gosz 1991). Land management and climate changes are two commonly cited reasons for the displacement of the transition zone associated with a boundary and its mosaic patterns of communities and species. Management can facilitate the extension of species' ranges into areas where climate has become favorable. One of the reasons for such management is the maintenance of biodiversity and ecosystem functional characteristics associated with these edges. Reducing biological inertia involves management techniques that allow species to migrate near the rate, and in the spatial direction, dictated by environmental change. Some suggestions from Gosz (1995) are: (1) Acquire or manage land that will become transitional. One example would be the predicted movement of species to higher elevations with presumed global warming. (2) Select and manage areas with heterogeneous topography and substrate to maximize the variety of microsites available for species establishment. The pattern of numerous, smaller and heterogeneous patches in boundary areas may be important for the formation/movement of the boundary. (3) Develop and maintain corridors to facilitate species movement. (4) Manage for (or anticipate) increased disturbance. Large and more intensive disturbance on new boundary positions will reduce biological inertia resulting from competition of previously established species. (5) Identify and manage for factors associated with successful plant establishment under new conditions (e.g., appropriate mycorrhizal fungal species, nutrient levels). (6) Plant with the goal of increasing intraspecies genetic variation. (7) Utilize threshold phenomena associated with climatic factors that cause the boundary movement (e.g., periodic wet or dry events). For example, the predictability of the El Nino Southern Oscillation can forecast wet and dry periods that can be used to increase establishment rates of new species and reduce competition from previously established species.

In some situations, management may want to prevent the normal pattern of species migration and boundary movement. Current boundaries may be instrumental in protecting existing land use, decreasing soil erosion or improving water quality. It may not be possible to change land use patterns in ways that follow environmental change and boundaries will have to be maintained in their current positions. Gosz (1995) suggested that biological inertia can be increased by: (1) Maintaining health/productivity of the current populations in boundary areas; (2) Decreasing stress through techniques such as thinning to reduce water stress, fertilization, and others. (3) Planting species or strains adapted to the new environment that maintain the current structure and function of the boundary area. (4) Managing in anticipation of, or fostering, disturbance that is fine-scale and low intensity. This can increase microsite heterogeneity that is a characteristic of the true boundary and will facilitate maintaining higher species variability. (5) Decreasing landscape connectivity in cases where other species can invade and compete with present species. (6) Managing to avoid threshold conditions that are likely to result in total community changes.

### 5.2.2 Boundary Permeability

To influence ecosystem properties or landscape patterns, vectors must actually transport materials (energy, nutrients, or organisms) or information in their movements. Directionalities and differentials in fluxes occur when vectors take up and release materials in a nonrandom spatial pattern. Different vectors have different probabilities of uptake and release of materials. In desert systems, wind is likely to transport fine sediment, detritus, and disseminules of organisms largely as a function of the mass of the material (Wiens et al. 1985). Depending on their characteristics, seeds may be transported by abiotic vectors such as wind or by biotic vectors (e.g., ants, rodents), which selectively glean them as they fall to the ground. Different outcomes have different effects on germination probabilities, and thus, on the overall demography and spatial distribution of the plant populations. These contribute importantly to the structure and function of local and landscape ecosystems. One way that boundary permeability relates to boundary dynamics is by deflecting or blocking vector movements. The degree to which a boundary deflects the movements of vectors may be expressed as boundary "permeability", and it may determine the degree to which the landscape element or patch can be viewed as a closed or an open system (Wiens et al. 1985).

Boundary permeability is a function of characteristics of the vectors and of features of the boundary itself. Abiotic vectors such as wind and water are strongly affected by features of the physical structure of the boundary. Among biotic sectors, characteristics of the animals themselves contribute to boundary permeability. Organisms with restricted movements such as termites are less likely to encounter a landscape boundary than are animals operating on broader scales, such as rodents or birds. Relatively sedentary forms may also be more likely than mobile vectors to recognize a boundary of a given degree of sharpness or difference as a strong spatial discontinuity and be deflected by it. The likelihood that a boundary will be crossed by a vector is also related to the density of animals within a patch. Increases in the density of the population within the patch produce a dramatic increase in the collective probability of boundary encounter and between-patch transport becomes more likely. Thus, there is likely to be a clear density-dependence to between-patch transfers in landscapes. Forman (1995) provides additional information on how various shapes of boundaries (e.g., concave, convex, undulating) influence plant and animal dispersal patterns.

## 5.3 Biological Lessons for Management

Most natural resource managers find themselves managing for biodiversity in a human-fragmented environment and one that, in all likelihood, will become more fragmented with time. From the material presented above, two generalizations can provide guidance to management. First, management needs to address the natural system or the internal dynamics of the fragments. This includes activities that occur within the patches such as natural disturbances (e.g., gap formation, fire, insect/plant disease epidemics, energy flow, nutrient cycling). Second, management needs to address the external influences of the system such as pollution, urbanization, spread of exotics, and alterations of water regimes. The degree that these two factors are prioritized depends on the size of the fragments. For large fragments, emphasis will be on the internal dynamics; for small fragments emphasis will be on the external influences. External factors, however, can be important regardless of the remnant size. Since most impacts on fragments originate from the surrounding landscape, there is clearly a need to depart from traditional notions of management and look instead toward integrated landscape management. Traditional natural resource management stopped at the reserve boundary; fluxes of water, pollutants, and organisms do not. Placing the conservation reserves firmly within the context of the surrounding landscape and attempting to develop complementary management strategies seems to be the only way to ensure the long term viability of remnant areas.

One goal of ecosystem management is the protection of biodiversity. Because a goal of conservation biology is to maintain species diversity, one approach to achieving this is to maintain representative examples of each ecosystem or community type present. To do this, it is necessary to know the distributions of species and communities and then select areas that represent them. There are at least two general approaches. First, managers may have a system that is about to be fragmented and are asked to design the ideal set of reserves for the area. Second, managers may already work in a fragmented landscape and need to make the most of it. Although most theories in conservation biology have dealt with the first, it is the second scenario that we most frequently have to confront. Some suggested guidelines for this approach include: (1) Determine the minimum subsets of the existing fragments that are required to represent the diversity of a given region. Determine the distribution of species and ecosystem types; (2) Manage the system to maintain the diversity of species or ecosystems. The question of whether management should be for single species or ecosystems is largely irrelevant because individual species require ecosystem functions to survive; (3) Establish priorities for management. There are more problems requiring attention than resources. Priority rankings should be established to ensure resources are deployed optimally; and (4) Continuous management is needed to maintain remnants in their desired state due to the constant pressure of altered internal dynamics and external influences.

Naiman (1996) provided recommendations for how landscape ecology can assist in solving problems of landscape change: (1) Focus on water. Fresh water is a strategic resource and understanding the abilities of freshwater systems to respond to human generated pressures and their limitations in adapting to such challenges is central to long term social and environmental vitality; (2) Provide alternative scenarios. The consequences of decisions are not always understood by policymakers. Developing modeling environments which link social, economic and environmental considerations are fundamental to providing policymakers with a more holistic understanding of their proposed actions; (3) Be cognizant of emerging technologies and political agreements that have the potential to make broad-scale landscape changes; (4) Ask and resolve questions at appropriately broad spatial and temporal scales. For example, where on the landscape are the biologically most important places for maintaining diversity, productivity, and resistance and resilience to disturbances? What are the optimal configurations of landscape elements that would promote a socio-environmental balance? (5) Become involved in the decision-making process by providing information. The research products of landscape ecologists are visual! — unlike statistics, they have impact on people! Rather than just documenting a changing landscape, landscape ecology can shape future landscapes based on sound ecological and social principles. The role of landscape ecology is central to balancing human and environmental needs (Naiman 1996).

## 6 DATA DEVELOPMENT AND DISSEMINATION

Ecosystem management is a process that most often applies to large, typically open-ended, geographic areas. These are areas generally vast enough to sustain viable populations of even large bodied species, and accommodate natural disturbance regimes (e.g., expected frequencies of fire, drought, diseases, and other stochastic events, Gregg 1994).

The information needed to support sound ecosystem management is diverse and extensive. To meet the challenges of ecosystem management we need to provide data that encompass these large areas as well as more traditional, site-specific kinds of ecological data. Landscape analyses purposefully attempt to address the broad perspective contained within large geographic areas. We use the term landscape analyses in the context of the National Hierarchical Framework of Ecological Units that envisions stratification of the earth into progressively smaller areas of increasingly uniform ecological potentials (McNab and Avers 1994). Thus, we need to provide information at several different resolutions (i.e., scales) to capture and understand ecological features and processes within the appropriate level of the hierarchy. It is this perspective that is perhaps the most meaningful and necessary contribution to ecosystem management.

Detailed analyses of landscapes, based on scientific methods, has been enabled only in recent years with the development of new technologies. Research and management activities at the landscape level (i.e. covering large geographic areas) depends largely on the availability of geographic information systems (GIS) and remotely sensed data, and the ability to link other non-spatial data sets to these baseline data sources. With the rapid growth and improvement in the reliability and utility of these data, landscape ecology and other similar disciplines have become major contributors to land assessment and management activities throughout the world. The management chapter for this topic (see Concannon et al., this volume) provides examples of how GIS information has been valuable for integrating different types of data and providing a common tool of use to many stakeholders.

### 6.1 Assess Data Needs

Data requirements for landscape analysis demand measurement of landscape features and processes that cover the spatial and temporal extremes, large geographic areas and long time frames. Scientists have addressed this conceptually for many decades, but only recently, with advances in technology, have we been able to actually collect and analyze suitable kinds of data.

An important concept in the consideration of data needs is the notion of a hierarchical relationship within ecosystems. Environmental heterogeneity is hierarchical and is controlled by different processes at different spatial and temporal scales (Turner et al. 1994). The large spatial extent and long time frames that landscape ecologists attempt to capture are part of a continuum of space/time resolutions that are useful and relevant to ecosystem management principles.

Landscape analysis also demands that relationships are understood between these broad, ecosystem features or processes that are measured over large geographic areas and more site-specific data that may address the immediate issues of concern. As determinations are made about what data are needed to provide the large context (landscape assessment), we must consider how those data and the results of any analyses should be linked to data of finer resolutions (see chapters on Genetic and Species Diversity, Population Viability, and Scale Phenomena, this volume).

### 6.1.1 Framework Data, National Spatial Data Infrastructure (NSDI)

Some of the most important baseline landscape level data come from (or will in the future come from) government agencies. Typically, though not exclusively, Federal and sometimes State agencies are the organizations that have both an interest and jurisdiction over areas large enough and appropriate for landscape level analysis. An important recent development is the concerted effort to establish the National Spatial Data Infrastructure. Representatives from local, State, and Federal agencies are developing a concept of a framework of geospatial data to meet the need of establishing a baseline of a few common themes of data. The National Spatial Data Infrastructure (NSDI) framework is a basic, consistent set of digital geospatial data and supporting services that will:

(a) provide a geospatial foundation to which an organization may add detail and attach attribute information;

(b) provide a base on which an organization can accurately register and compile other themes of data;

(c) orient and link the results of an application to the landscape.

The information content of the NSDI will initially include:

(1) geodectic control
(2) transportation (roads, trails, etc.)
(3) elevation data (Digital Elevation Models)
(4) digital orthorimagery (resolution from sub-meter to tens of meters)
(5) hydrography (EPA river reach file 3.0)
(6) governmental units (boundaries)
(7) cadastral reference systems

This effort, organized by the Federal Geographic Data Committee (FGDC), is targeted to be fully implemented by the year 2000.

Another major category of important baseline data, particularly the higher resolution, more detailed data, will come from local governments and the private sector. Local governments throughout the United States are devoting large financial resources towards development of baseline data, especially human/political GIS data such as digital parcel maps, general plans, etc.

### 6.1.2 Other Data Needs (Spatial and Non-Spatial)

Framework data should provide the foundation of geographic information needed for landscape analysis. A primary purpose of the framework data is to provide a base on which an organization can accurately register and compile other themes of data. Both spatial and non-spatial data, developed for supporting ecosystem management, can be built upon this foundation.

Landscape evaluation and ecosystem characterization are aimed at identifying the entire range of variability of the biotic and abiotic components of ecosystems as well as descriptions of the environmental variables and processes that control species, community, and ecosystem distributions (Bourgeron et al. 1994). The researcher or manager must determine what information is needed and how it will be developed. This component can often be the most expensive and time-consuming portion of a project. It is difficult and expensive to establish reliable ecological data sets that cover large geographic areas and particularly data sets that span long timeframes.

Some sources are beginning to provide the baseline of ecological information that can be useful to landscape analyses. The Gap Analysis program, under the guidance of the National Biological Service, is establishing statewide, and eventually, nationwide coverage of vegetation, land use/ownership, and in some cases vertebrate species distribution. Other data sets covering large geographic areas, either in preparation or already available, include soils maps (Natural Resources Conservation Service), historical climate averages (various sources), annual Christmas bird counts (National Audubon Society), and breeding bird censuses (National Biological Service).

## 6.2 Data Development Tools

The development of new technologies, especially computer technologies, has opened up new possibilities for landscape analyses. Most noteworthy are the many uses of remotely sensed data and geographic information systems (GIS) data.

### 6.2.1 Remotely Sensed Data

Land managers have made extensive uses of aerial photography for many decades. The development of improved optics and especially digital cameras/sensors have created many new sources of quality landscape data. These kinds of data include analog data in the form of aerial photography and digital data captured through satellite imagery and digital imagery captured from aircraft. Concurrent with emergence of these data sources is the development of computers capable of handling these data. Workstation computers and image processing software can now work with enormous volumes of data efficiently and quickly.

The obvious advantages of these sources of data are the large area covered and the relatively low costs per unit area of data. Thematic Mapper imagery, for example, costs between $4,000 and $5,000 per scene which covers 185 by 170 kilometers. This cost is less than 15 cents per square kilometer. Now it is possible for land managers to obtain data for enormous landscapes. Advances in remotely sensed imagery over the next few years, with the launching of new satellites and development of new airplane-mounted digital cameras, will likely make even greater advances in the development of landscape level data. Digital imagery will soon become an indispensable tool for virtually any land manager.

### 6.2.2 Geographic Information Systems (GIS)

The growth of GIS as a tool in land management over the last five years has been phenomenal. GIS makes possible, for the first time, truly interactive collaborations among resource managers, policysetters, scientists, and stakeholders. With GIS, researchers can attack the important questions involving larger spatial domains by synthesizing ecological data, developing concepts and displaying new findings (Franklin 1994). The spatial analytical capabilities of modern GIS are virtually limitless. Information, at any scale or resolution, about patches, edges, landscape matrices, connectivity, cumulative effects, and dispersed versus aggregated activities are just some of the kinds of analyses that GIS enables. Raster (cell-based) GIS offers the most versatile analytical capabilities.

All GIS software basically provides the same four types of data operations: (1) data input, (2) cartographic display, (3) data query, and (4) spatial analysis and modeling. It is the latter category that sets GIS apart as a new and valuable tool. Raster GIS has almost limitless opportunities to analyze the landscape and the spatial relationships of its component parts to create new insights, perspectives, and ultimately alternative scenarios of land use and preservation alternatives.

### 6.2.3 Shared Costs, Collaborative Programs, Partnerships

Developing dependable GIS and remotely sensed data can require as much as 90 percent of the financial resources available to a project. Many organizations, agencies, and research efforts cannot afford the costs required for the landscape data needed by their project. Many data sets could never be developed at all unless cooperating organizations are willing and able to share the costs and other facilities necessary to develop these data.

Cooperative efforts in data development are occurring more frequently, both out of need and directly due to the multi-organizational nature of ecosystem management. Most landscape level data sets transcend jurisdictional boundaries so there is incentive to pool resources for development of data sets of common interest.

## 6.3 Data Dissemination

The explosion of GIS and remotely sensed data offers opportunities to researchers and managers working in landscape analyses. One of the primary factors that constrain the use of this volume of data is the ability of any potential user to locate and access data of interest to that user. New tools have become available that are making the dissemination of digital data a simple matter, for example, World Wide Web, distributed networks (CERES, NBII, etc.), digital libraries, and massive storage capabilities of supercomputer centers. Advances in computer technologies in recent years have been phenomenal, particularly in the areas of storage and retrieval of digital information. Use of the Internet and the World Wide Web have increased tremendously since 1990 and access is now easily available to anyone. Two major efforts that exemplify the progress and future of distributed networks of environmental information include the National Biological Information Infrastructure (NBII), under the direction of the Biological Resources Division (BRD) of the U.S. Geological Survey, and the California Environmental Resources Evaluation System (CERES), under the direction of the State of California Resources Agency.

The NBII is an initiative of the BRD to foster the development of a distributed electronic "federation" of biological data and information, relying on a network of partners and cooperators to make the data they generate and/or maintain available to others through this federation. The objectives of the NBII are to make it easier for people to find the biological data and information they need, to integrate or combine data and information from different sources, and to apply data and information to actual resource management decisions. In addition to biological data and information, software tools will also be identified and available through the NBII, to help users to analyze, integrate, and display biological data and information. The NBII will also point to sources of biological expertise; people and organizations that users can contact to get advice and assistance on finding and understanding biological data.

### 6.3.1 Research and Development in Storage and Rapid Retrieval (Alexandria Project, Berkeley Digital Library Project, SDSC, etc.)

Within the past decade the number and kinds of digital information sources have proliferated. Computing system advances and the continuing networking and communications revolution have resulted in a remarkable expansion in the ability to generate, process and disseminate digital information. Together, these developments have made new forms of knowledge repositories and information delivery mechanisms feasible and economical.

Six research projects developing new technologies for digital libraries — storehouses of information available through the Internet — have been funded through a joint initiative of the National Science Foundation (NSF), the Department of Defense Advanced Research Projects Agency (ARPA), and the National Aeronautics and Space Administration (NASA). The projects' focus is to advance dramatically the means to collect, store, and organize information in digital forms, and make it available for searching, retrieval and processing via communication networks — all in user-friendly ways.

The projects are at Carnegie Mellon University, the University of California, Berkeley, the University of Michigan, the University of Illinois, the University of California, Santa Barbara and Stanford University. Each effort brings together researchers and users from the local university with those from other organizations including other academic institutions, libraries, museums, publishers, government laboratories, state agencies, secondary schools, and computer and communications companies.

### 6.3.2 Metadata Development (Spatial/Non-Spatial Data)

Metadata are "data about data." As such they provide information about the content, quality, structure, and other characteristics of data. Because metadata may be used to describe database structure (e.g., the definition of a class or a relation) as well as database content (e.g. old growth forest in the Sierra Nevada), they can play an important role in mediating between various trade-offs in data organization, access, and transfer (Kellogg 1994).

The Federal Geographic Data Committee (FGDC) has approved the "Content Standards for Digital Geospatial Metadata," to be used by all federal agencies with data sets they create and encouraged of all other entities creating spatial data. Suitable standards, similar in context, are in preparation for non-spatial data. The standards specify information that helps prospective users to determine what data exist, the fitness of these data for their applications, and the conditions for accessing these data. Metadata also aid the transfer of data to other users' systems.

The existing FGDC standard provides a common set of terminology and definitions for the documentation of geospatial data. The standard establishes the names of data elements and groups of data elements to be used for these purposes, the definitions of these data elements and groups, and information about the values that are to be provided for the data elements. Information about terms that are mandatory, mandatory under certain conditions, and optional (provided at the discretion of the data provider) also is provided by the standard.

### 6.4 Limitations of Landscape Data

The emergence of new analytical perspectives and the development of new technologies has made a tremendous contribution to ecosystem management. Future technological advances promise to enhance these potentials significantly. However, we must be careful always to demand sufficient quality from the underlying data and recognize the limitations of landscape analyses. The requirement for metadata to be attached to GIS data sets will provide information intended to advise users of the accuracy and precision limitations associated with the data. Data users must take responsibility for demanding quality data and determining the limitations of uses of any particular data set. A Pandora's box of pitfalls in uses of GIS and remotely sensed data include using tools wrongly, capturing data poorly, miscommunication of information, conveying incorrect results, and perhaps most importantly overselling the capabilities. Scientists must avoid these potential problems through solid project design, careful development and/or selection of data to be used in analyses, honest information portrayal, and constant communication.

## 7 CONCLUSIONS

Concepts of landscape ecology have been developing around the world for years. Common human perceptions are of landscapes composed of croplands, farmsteads, roads, woodlands, and streams; spatial heterogeneity that creates a diverse landscape. The challenge for landscape ecology is to integrate the ideas and research results of the many disciplines that provide information on these different habitat types into a coherent body of knowledge. As a separate field, landscape ecology considers the development and maintenance of spatial heterogeneity, interactions and exchanges across heterogeneous landscapes, the influence of heterogeneity on biotic and abiotic process, and the management of that heterogeneity.

The influence of scale (temporal and spatial) is paramount and landscape ecology cannot escape dealing with spatial analysis, spatial scale, and scale-change effects. A landscape may appear to be heterogeneous at one scale but quite homogeneous at another scale. An important concept is that landscapes change over time but not all landscape processes occur simultaneously or at the same rate. Three mechanisms are posed as causal processes for landscape heterogeneity: (1) specific geomorphologic processes occurring over long time periods; (2) colonization patterns of organisms occurring over short and long time scales; and (3) local disturbances of individual ecosystems occurring over a short time.

Recent changes in natural resources management demonstrate a new appreciation for how landscape variability and patchiness affect populations and communities. This has occurred largely through the renewed awareness of how landscape patterns influence ecological processes such as disturbances, nutrient cycling, and animal and plant distributions. Because species, such as wide-ranging animals, use more than one habitat patch type, many organisms respond actively to the mosaic of habitats they encounter, avoiding some and using others. Often the surrounding matrix is as important as the habitat patch itself. To understand landscape-level patterns and processes, we must know the details of "boundaries and their dynamics"; what determines why a boundary is located where it is, how boundaries influence ecological processes within patches and over the larger landscape, how boundaries affect the exchanges or redistribution of materials, energy, and organisms between landscape elements, and how these transfers can act to change the location and nature of boundaries. Land management and climate changes are two commonly cited reasons for the displacement of the transition zone associated with a boundary and its mosaic patterns of communities and species.

Over time, virtually every landscape with even a moderate level of human occupation has gone through a transition from a natural landscape matrix with human-altered habitat fragments embedded within it to a managed landscape matrix with natural habitat fragments embedded within it. Basically, human land use fragments, perforates, or shreds landscapes, whether building roads or transmission powerlines, clearing land for farming, logging, or housing developments. Historical changes in land use and ecological processes have resulted in landscapes different in composition and structure from those that existed on this continent before European settlement. Because of conversion of forest to other uses, present-day landscapes can have less forest area than pre-settlement landscapes. Present-day landscapes, however, can be more diverse than pre-settlement forests and have more forest patches that are smaller and more simple in shape. In addition, there often are fewer large patches, patch juxtapositions have changed and the location of older stands may be restricted to sites bypassed by conversion or recent harvesting. Because of changes in land use, some present-day management options have been largely foreclosed. Perhaps the only certainty regarding ecosystems is that most U.S. landscapes will continue to change and be heavily influenced by humans. An understanding of how landscapes changed in the past, naturally and as a result of human use, provides insights in future impacts of natural and human-induced change on storm patterns, flood frequency, landslides, stability of valley floors, and forest and agricultural productivity.

Disturbances, however defined, are fundamental for maintain much of the variability associated with terrestrial and aquatic environments. Disturbances result from interactions between geological climate processes, animal activities associated with modifying habitat and procuring food, and the type of environment. Geological processes include volcanism, continental and alpine glaciation, and erosion processes. Climatic processes include the dominant form of precipitation, the seasonal timing, and the hydrological patterns.

Perhaps special interest should be given to riparian zones. Natural riparian communities are the most diverse, dynamic, and complex biophysical habitats on the terrestrial portion of the Earth. Riparian communities, as interfaces between terrestrial and aquatic systems, encompass sharp environmental gradients, ecological processes, and communities. Riparian communities are an unusually diverse mosaic of landforms, communities and environments within the larger landscape. As such, they serve as a framework for understanding the organization, diversity, and dynamics of communities associated with aquatic ecosystems.

The roles of individual species, populations and communities in the properties and dynamics of landscapes are profound. Two key groups have been identified for special focus; keystone species and invasive exotic species. The term "keystone" has been extended to species whose removal results in significant change in community structure or ecosystem functioning. When applied to landscapes or ecosystems, keystone species are those species whose contribution to ecosystem functioning is unique. Animal–ecosystem interactions are complicated because population cycles of many keystone species occur over long periods (decades), alterations to the ecosystem are often subtle over short periods, and shifts in biogeochemical cycles or soil formation are often only detectable after a period of years. Nevertheless, alterations to the landscape by keystone species often result in patterns and processes that would not occur under the dominating influences of climate driven or geological disturbances alone.

Exotic species are important to landscape function because these species typically are very aggressive and have the ability to dominate many locations with dramatic impacts to native plant communities. Undesirable plants are biological forces restructuring and changing processes that affect landscape diversity on a grand scale. Invasive exotic plants move toward a near monoculture condition at the watershed and landscape scale.

There are several generalizations which can provide guidance to management to attain societal goals. First, management needs to address the natural system or the internal dynamics of the fragments. This includes things that occur within the patches such as natural disturbances. Second, management needs to address the external influences of the system such as pollution, urbanization, spread of exotics, alterations of water regimes, etc. The degree that these two factors are prioritized depends on the size of the fragments. For large fragments, emphasis will be on the internal dynamics; for small fragments emphasis will be on the external influences. External factors, however, can be important regardless of the remnant size. Since most impacts on fragments originate from the surrounding landscape, there is clearly a need to depart from traditional notions of management and look instead toward integrated landscape management. Traditional natural resource management stopped at the reserve boundary; fluxes of water, pollutants, and organisms do not. Placing important areas firmly within the context of the surrounding landscape and attempting to develop complementary management strategies seems to be the only way to ensure the long term viability of remnant areas.

The information needed to support sound ecosystem management is diverse and extensive. To meet the challenges of ecosystem management we need to provide data that can address these large areas as well as more traditional, site-specific kinds of ecological data. Landscape analyses purposefully attempt to address the broad perspective contained within large geographic areas. Information needs to be provided at several different resolutions to capture and understand ecological features and processes within the appropriate level of the hierarchy. Detailed analyses of landscapes, based on scientific methods, has been enabled only in recent years with the development of new technologies. Research and management activities at the landscape level (i.e., covering large geographic areas) depends largely on the availability of geographic information systems (GIS) and remotely sensed data, and the ability to link other non-spatial data sets to these baseline data sources. With the rapid growth and improvement in the reliability and utility of these data, landscape ecology and other similar disciplines have become major contributors to land assessment and management activities throughout the world.

## REFERENCES


Adam, P. 1990. *Saltmarsh Ecology.* Cambridge University Press, Cambridge.

Alexander, L. 1986. Timber-wildlife management from the forest landowner's perspective. In: J.A. Bissonette (ed.), *Is good forestry good wildlife management*? pp. 269–279. Maine Agric. Exp. Sta. Misc. Pub. No. 689.

Allen, T.F.H., and T.B. Starr. 1982. *Hierarchy: Perspectives for Ecological Complexity.* University of Chicago Press, Chicago.

Ambuel, B., and S.A. Temple. 1983. Area dependent changes in the bird communities and vegetation of southern Wisconsin. *Ecology* 64: 1057–1068.

Anderson, B.M. 1971. Calcium-carbonate hardness of public water supplies in the Conterminous United States. In: H.L. Cannon and H.C. Hopps (eds.), *Environmental Geochemistry in Health and Disease*, pp. 151–154. Geological Society of America, Memoir 123.

Argow, K.A. 1996. This land is their land: the potential and diversity of nonindustrial private forests. Journal of Forestry 94(2): 30–33.

Barrett, S.W. 1980. Indians and fire. *Western Wildlands* (Spring): 17–21.

Barrett, S.W., and S.F. Arno. 1982. Indian fires as an ecological influence in the northern Rockies. *Journal of Forestry* 80: 647–651.

Beckinsale, R.P. 1972. "The effect upon river channels of sudden changes in sediment load. *Acta Geographica Debrecina* 10: 181–186.

Bell, D.T. 1974. Tree stratum composition and distribution in a streamside forest. *Am. Midl. Nat.* 92: 35–46.

Benda, L., et al. 1997. Landscape Dynamics. In: R.J. Naiman and R.E. Bilby (eds.), *Ecology and Management of Streams and Rivers in the Pacific Northwest.* Springer-Verlag, New York (in press).

Benke, A. 1990. A perspective on America's vanishing streams. *Journal of the North American Benthological Society* 9: 77–88.

Berner, E.K., and R.A. Berner. 1996. *Global Environment.* Prentice Hall, New Jersey.

Birkland, P.W. 1974. Pedology, Weathering, and Geomorphological Research. Oxford University Press, New York.

Birch, T.W. 1996. Private forest-land owners of the United States, 1994. USDA For. Serv. Resour. Bull. NE-134.

Boggess, W.R. 1977. *Lead in the Environment*. National Science Foundation, Washington, DC.

Bourgeron, P.S., H.C. Humphries, and M.E. Jensen. 1994. General sampling design considerations for landscape evaluations. In: Volume II: Ecosystem Management: Principles and Applications, pp. 109–120. Gen. Tech. Rep. PNW-GTR-318. Portland, OR: U.S. Department of Agriculture, Forest Service, Pacific Northwest Research Station.

Brooks, A. 1985. River channelization — traditional engineering methods, physical consequences and alternative practices. *Prog. Phys. Geogr.* 9: 44–73.

Brooks, R.R. 1972. *Geobotany and Biogeochemistry in Mineral Exploration*. Harper and Row, New York.

Brown, B.E. 1990. Coral bleaching. *Coral Reefs* 8: 153–232.

Bull, W.B. 1991. *Geomorphic Responses to Climate Change*. Oxford University Press, New York.

Burdick, D.M., D. Cushman, R. Hamilton, and J.G. Gosselink. 1989. Faunal changes and bottomland hardwood forest loss in the Tensas Watershed, Louisiana. *Conservation Biology* 3: 282–292.

Burkholder, J.L. 1928. Report of the Chief Engineer. Middle Rio Grande Conservancy District, State of New Mexico, Albuquerque.

Bustard, D.R., and D.W. Narver. 1975. Aspects of the winter ecology juvenile coho salmon (*Oncorhynchus kisutch*) and steelhead trout (*Salmo gairdneri*). *Journal of the Fisheries Research Board of Canada* 32: 667–680.

Butler, D.R. 1995. *Zoogeomorphology*. Cambridge University Press, Cambridge, UK.

Cannon, H.L. 1960. Botanical prospecting for ore deposits: *Science* 132: 591–598.

Cannon, H.L., and H.C. Hopps. 1971. Environmental geochemistry in health and disease. *Geological Society of America*, Memoir 123.

Carpenter, S.R., and J. Kitchell (eds.). 1993. *The Trophic Cascade in Lakes*. Cambridge University Press, New York.

Chaney, R.L. 1988. Metal speciation and interaction among elements affect trace element transfer in agricultural and environmental food-chains. In: J.R. Kramer and H.E. Allen (eds.), *Metal Speciation — Theory, Analysis, and Application*, pp. 219–259. Lewis Publications, Boca Raton, FL.

Chi, S.C., and R.E. Reilinger. 1984. Geodetic evidence for subsidence due to groundwater withdrawal in many parts of the United States of America. *Journal of Hydrology* 67: 155–182.

Chorley, R.J., . A. Schumm, and D.E. Sugden. 1984. *Geomorphology*. Methuen, London.

Cline, A.C., and S.H. Spurr. 1942. The virgin upland forest of central New England — a study of old growth stands in the Pisgah Mountain section of southwestern New Hampshire. *Harv. Forest Bulletin* 21.

Committee on Engineering Implications of Changes to Relative Mean Sea Level. 1987. Responding to changes in sea level. National Academy Press, Washington, DC.

Compton, R.R. 1962. *Manual of Field Geology*. Wiley, New York.

Connor, J.J., and H.T. Shacklette. 1975. Background geochemistry of some soils, plants, and vegetation in the Conterminous United States. U.S. Geol. Survey Prof. Paper.

Cooke, R.U., and R.W. Reeves. 1976. *Arroyos and Environmental Change in the American South-west*. Clarendon Press, Oxford.

Décamps, H., and E. Tabacchi. 1994. Species richness in riparian vegetation along river margins. In: P. Giller, A. Hildrew, and D. Rafaelli (eds.). *Aquatic Ecology: Scale, Pattern, and Process*. Blackwell, Oxford.

Décamps, H. 1996. The renewal of floodplain forests along rivers: a landscape perspective. *Verh. Int. Verein. Limnol.* (in press).

Day, G. 1953. The Indian as an ecological factor in the northeastern forest. *Ecology* 34: 329–346.

DeFerrari, C.M., and R.J. Naiman. 1994. A multi-scale assessment of the occurrence of exotic plants on the Olympic Peninsula, Washington. *Journal of Vegetation Science* 5: 247–258.

Delcourt, H.R., and P.A. Delcourt. 1991. *Quaternary Ecology*. Chapman and Hall, New York.

Dick-Peddie, W.A. 1993. New Mexico Vegetation, Past, Present, and Future. University of New Mexico Press, Albuquerque, NM.

Dickson, J.G. 1991. Birds and mammals of pre-colonial southern old-growth forests. *Nat. Areas Journal* 11: 26–33.

Dobrovolsky, V.V. 1994. *Biogeochemistry of the World's Land*. CRC Press, Boca Raton, FL.

Duvigneaud, P., and S. Denaeyer-De Smet. 1975. Mineral cycling in terrestrial ecosystems, pp. 133–154. In: *National Academy of Sciences, Productivity of World Ecosystems*. National Academy of Sciences, Washington, DC.

Dzurisin, D., S.R. Brantley, and J.E. Costa. 1994. How should society prepare for the next eruption in the Cascades. Geol. Soc. Am. Program with Abstracts, Annual Meeting, p. A113.

Fetherston, K.L., R.J. Naiman, and R.E. Bilby. 1995. Large woody debris, physical process, and riparian forest development in montane river networks of the Pacific Northwest. *Journal of Geomorphology* 13: 133–144.

Forman, R.T.T. 1995. *Land Mosaics: The Ecology of Landscapes and Regions*. Cambridge University Press.

Forman, R.T.T., and M. Godron. 1986. Landscape Ecology. Wiley, New York.

Fortescue, J.A.C. 1980. *Environmental Geochemistry — A Holistic Approach*. Springer-Verlag, New York.

Franklin, J.F. 1994. Developing information essential to policy, planning, and management decision-making: the promise of GIS. In: V.A. Sample (ed.), *Remote Sensing and GIS in Ecosystem Management*, pp. 18–24. Island Press, Washington DC.

Freeman, B. 1995. *Environmental Ecology*, 2nd ed. Academic Press, San Diego, CA.

French, H.M. 1976. *The Periglacial Environment*. Longman, London.

Frost, T.P., Raines, G.L., Almquist, C. and Johnson, B.R. 1995. Digital map of potential habitat for cave-dwelling bats — a contribution to the Interior Columbia River Basin Ecosystem Management. U.S. Geological Survey Open-File Report 95.

Goklany, I. 1992. America's biodiversity strategy: actions to conserve species and habitats (fact sheet). USDI Office of Program Analysis and USDA Natur. Resour. and Manage. Washington, DC.

Goldich, S.S. 1938. A study of rock weathering. *Journal of Geology* 46: 17–58.

Golley, F.B. 1987. Introducing landscape ecology. *Landscape Ecology* 1: 1–3.

Gosz, J. R. 1991. Fundamental ecological characteristics of landscape boundaries. In: M.M. Holland, P.G. Risser, and R.J. Naiman (eds.), *Ecotones: The Role of Landscape Boundaries in the Management and Restoration of Changing Environments*. Chapman and Hall, New York.

Gosz, J.R. 1995. Edges and natural resource management: Future directions. *Ecology International* 22: 17–34.

Goudie, A.S. 1977. Sodium sulphate weathering and the disintegration of Mohenjo-Daro, Pakistan. *Earth Surface Processes* 2: 75–86.

Goudie, A.S. 1993. Human influence in geomorphology. *Geomorphology* 7: 37–59.

Goudie, A.S., and N.J. Middleton. 1992. The changing frequency of dust storms through time. *Climate Change* 20: 197–225.

Gourou, P. 1961. *The Tropical World*, 3rd ed. Longman, London.

Graf, W.L. (ed.). 1987. *Geomorphic Systems of North America.* Geological Society of America, Centenial Volume 2.

Gregg, W.P. 1994. Developing landscape-scale information to meet ecological, economic, and social needs. In: V. Alaric (ed). *Remote Sensing and GIS in Ecosystem Management*, pp. 13–17. Island Press, Washington, DC.

Gregory, S.V., F.J. Swanson, W.A. McKee, and K.W. Cummins. 1991. An ecosystem perspective of riparian zones. *BioScience* 41: 540–551.

Grimm, N.B. 1995. Why link species and ecosystems? A perspective from ecosystem ecology. In C.G. Jones, and J.H. Lawton (eds.), *Linking Species and Ecosystems*, pp. 5–15. Chapman and Hall, New York.

Gruell, G.E. 1985. Indian fires in the interior west: a widespread influence. In: Proc. Symp. and Workshop on Wilderness Fire, pp. 68–74. USDA For. Serv. Tech., Missoula, MT. Rept. INT-182.

Gustafson, G.C. 1982. Who owns the land? A state and regional summary of landownership in the United States. USDA Econ. Res. Serv. Staff Rep. No. AGES830405.

Hails, J.R., ed. 1977. *Applied Geomorphology*. Elsevier, Amsterdam.

Harris, L.D. 1984. *The Fragmented Forest.* University of Chicago Press, Chicago, IL.

Hayes, G.I. 1942. Difference in fire danger with altitude, aspect, and time of day. *Journal of Forestry* 40: 318–323.

Hoffman, J.D., G.B. Gunnels, and J.M. McNeal. 1991. National geochemical data base — national uranium resource evaluation data for the conterminous western United States. U.S. Geological Survey Digital Data Series DDS-1.

Hudson, C.M. 1976. *Southeastern Indians.* University of Tennessee Press, Knoxville.

Hunter, M.L., Jr. 1996. *Fundamentals of Conservation Biology.* Blackwell Science, Cambridge, MA.

Huston, M. 1979. A general hypothesis on species diversity. *American Naturalist* 113: 81–101.

John, D.A., and J.S. Leventhal, 1995. Bioavailability of metals. In: E.A. du Bray (ed.), 1995 Preliminary compilation of descriptive geoenvironmental mineral deposit models, pp. 10–18. In U.S. Geological Survey Open File Report 95–831.

Johnson, J.E. 1994. The Lake States region. In: J.W. Barrett (ed.), *Regional Silviculture of the United States*, 3rd ed., pp. 81–127. Wiley, New York.

Johnson, S.A. 1987. Pine plantations as wildlife habitat: a perspective. In: J.G. Dickson and O.E. Maughan (eds.), Managing southern forests for wildlife and fish, pp. 12–18. USDA For. Serv. Gen. Tech. Rep. SO-65.

Johnston, C.A., J. Pastor, and R.J. Naiman. 1993. Effects of beaver and moose on boreal forest landscapes. In: R. Haines-Young, D.R. Green, and S.H. Cousins (eds.), *Landscape Ecology and Geographic Information Systems*, pp. 237–254. Taylor and Francis, London.

Jones, C.G., and J.H. Lawton (eds.). 1995. *Linking Species and Ecosystems.* Chapman and Hall, New York.

Jones, R., K. Denson-Evens, and F.M. Chambers. 1985. Human influence upon sedimentation in Llangorse Lake, Wales. *Earth Surface Processes, Landforms* 10: 227–235.

Junk, W. 1989. Flood tolerance and tree distribution in central Amazonian floodplains. In: L.B. Holm-Nielsen, I.C. Nielson, and H. Balsley (eds.), Tropical Forests: Botanical Dynamics, Speciation, and Diversity, pp. 47–74. Academic Press, Orlando, FL.

Kalliola, R., and M. Puhakka. 1988. River dynamics and vegetation mosaicism: A case study of the River Kamajohka, northernmost Finland. *Journal of Biogeography* 15: 703–719.

Kalliola, R., J. Salo, M. Puhakka, and M. Rajasilta. 1992. New site formation and colonizing vegetation in primary succession on the western Amazon floodplains. *Journal of Ecology* 79: 877–901.

Kellogg, C. 1994. A brief overview of emerging metadata standards. Technical Report for the California Environmental Resources Evaluation System (CERES). The Resources Agency, Sacramento, CA.

Kessel, S.R. 1976. Gradient modeling — a new approach to fire modeling and wilderness resource management. *Environmental Management* 1: 123–133.

Knight, R.L. and T.W. Clark. (In press). Public-private land boundaries: Defining obstacles, finding solutions. Chapter 8. In: R.L. Knight and P. Landres (eds.), *Stewardship Across Boundaries.* Island Press, CA.

Knight, R.L., G.N. Wallace and W.E. Riebsame. 1995. Ranching the view: Subdivisions versus agriculture. *Conservation Biology* 9: 459–461.

Komarek, E.V. 1981. History of prescribed fire and controlled burning in wildlife management in the South. In: G.W. Wood (ed.), *Prescribed Fire and Wildlife in Southern Forests*, pp. 1–14. Belle W. Baruch Forest Science Institute, Clemson University, Georgetown, SC.

Lack, D.L. 1976. *Island Biology: Illustrated by the Land Birds of Jamaica.* University of California Press, Berkeley, CA.

Levin, S.A. 1992. The problem of pattern and scale in ecology. *Ecology* 73: 1942–1968.

Lehee, F.H. 1961. *Field Geology* (6th ed.), pp. 424, 434, 435. McGraw-Hill, New York.

Lewis, H.T. 1980. Indian fires of spring. *Natural History* 89: 76–83.

Lewis, H.T. 1985. Why Indians burned: specific versus general reasons. In: Proc. Symp. and Workshop on Wilderness Fire, pp. 75–80. USDA For. Serv., Missoula, MT. Tech. Rept. INT-82.

Leopold, A. 1933. *Game Management.* Scribner, New York.

Leopold, L.A., Wolman, M.G., and Miller, J.P. 1964. *Fluvial Processes in Geomorphology.* W.H. Freeman and Co., San Francisco.

Likens, G.E., Bormann, F.H., Pierce, R.S., Eaton, J.S., and

Johnson, N.M. 1977. *Biogeochemistry of a Forested Ecosystem*. Spinger-Verlag, New York.

Lovejoy, T.E., R.O. Bierragaard, A.B. Rylands, J. R. Malcolm, C.E. Quintela, L .H. Harper, K.S. Brown, A. H. Powell, G.V.N. Powell, H.O. R. Schubert, and M.H. Hays. 1986. Edge and other effects of isolation on Amazon forest fragments. In: MacArthur, R.H., and E.O. Wilson. 1967. *The Theory of Island Biogeography*, pp. 257–285 Princeton University Press, Princeton, NJ.

MacCleery, D.W. 1992. American forests: a history of resiliency and recovery: USDA For. Serv. FS-540.

Martin, C. 1973. Fire and forest structure in the aboriginal eastern forest. *The Indian Historian* 6: 38–42,54.

McNab, W.H., and P.E. Avers. 1994. Ecological subregions of the United States: section descriptions. USDA Forest Service, WO-WSA-5.

Meade, R.B. 1991. Reservoirs and earthquakes. *Eng. Geol.* 30: 245–262.

Meentemeyer, V., and E.O. Box. 1987. Scale effects in landscape studies. In: M.G. Turner (ed.), *Landscape Heterogeneity and Disturbance*, pp. 15–34. Springer-Verlag, New York.

Mertz, W. 1971. Health-related function of chromium in Cannon. In: H.L., and Hopps, H.C., Environmental geochemistry in health and disease, pp. 197–202. Geol. Soc. Am., Memoir 123.

Micklin, P.P. 1972. Dimensions of the Caspian Sea problem. *Sov. Geogr.* 13: 589–603.

Mladenoff, D.J., M.A. White, J. Pastor, and T.R. Crow. 1993. Comparing spatial pattern in unaltered old-growth and disturbed forest landscapes. *Ecol. Appl.* 3: 294–306.

Nabi, D.H., D.C. Guynn, Jr., T.B. Wigley, and S.P. Mott. 1983. Forest resource values of Mississippi nonindustrial private forest landowners. In: J.P. Royer and C.D. Risbrudt (eds.), *Nonindustrial Private forests: A Review of Economic and Policy Studies*, pp. 338–342. Duke Univ. Sch. For. and Env. Studies. Durham, NC.

Naiman, R.J. (compiler). 1988. How animals shape their ecosystems. *BioScience* 38: 750–800.

Naiman, R.J. 1996. Water, society and landscape ecology. *Landscape Ecology* 11: 193–196.

Naiman, R.J., and E.C. Anderson. 1997. Streams and rivers: their physical and biological variability. In: P. Schoonmaker, B. von Hagen, and E.C. Wolf (eds.), *The Rain Forests of Home: Profile of a North American Bioregion*, pp. 131–148. Island Press, Washington, DC.

Naiman, R.J. and H. Décamps (eds.). 1990. *The Ecology and Management of Aquatic-Terrestrial Ecotones*. Parthenon Publishing Group, Carnforth, UK and UNESCO, Paris.

Naiman, R.J. and H. Décamps. 1997. The ecology of interfaces — riparian zones. *Annual Review of Ecology and Systematics* (in press).

Naiman, R.J., H. Décamps, J. Pastor, and C.A. Johnston. 1988. The potential importance of boundaries to fluvial ecosystems. *Journal of the North American Benthological Society* 7: 289–306.

Naiman, R.J., H. Décamps, and F. Fournier. 1989. Role of land/inland water ecotones in landscape management and restoration: proposals for collaborative research. *MAB Digest* 4: 1–93.

Naiman, R.J., T.J. Beechie, L.E. Benda, D.R. Berg, P.A. Bisson, L.H. MacDonald, M.D. O'Connor, P.L. Olson, and E.A. Steel. 1992. Fundamental elements of ecologically healthy watersheds in the Pacific Northwest coastal ecoregion. In: R.J. Naiman (ed.), *Watershed Management: Balancing Sustainability and Environmental Change*, pp. 127–188. Springer-Verlag, New York.

Naiman, R.J., H. Décamps, and M. Pollock. 1993. The role of riparian corridors in maintaining regional biodiversity. *Ecological Applications* 3: 209–212.

Naiman, R.J., J.J. Magnuson, D.M. McKnight, and J. A. Stanford (eds.). 1995. *The Freshwater Imperative: A Research Agenda*. Island Press, Washington, DC.

Naiman, R.J. and K.H. Rogers. 1997. Large animals and the maintenance of ecosystem-level characteristics in river corridors. *BioScience* (in press).

Nanson, G.C., and Beach, H.F. 1977. Forest succession and sedimentation on a meandering river flood plain, northeast British Columbia, Canada. *J. Biogeogr.* 4: 229–252.

Neuman, D.R., Shrack, J.L, and Gough, L.P. 1987. Copper and molybdenum. pp. 215–232. In: Williams, R.D., and G.E. Shuman (eds.), Reclaiming mine soils and overburden in western U.S. Ankeny, IO, Soil Conservation Soc. Am.

Nicod, J. 1986. Facteurs physico-chimques de l'accumulation des formations travertineuses. *Mediteranee* 10: 161–164.

Niemi, G.J., et al. 1990. Overview of case studies on recovery of aquatic systems from disturbance. *Environmental Management* 14: 571–587.

Nilsson, C. 1986. Change in riparian plant community composition along two rivers in northern Sweden. *Canadian Journal of Botany* 64: 589–592.

Nilsson, C. 1992. Conservation management of riparian communities. In: L. Hansson (ed.), *Ecological Principles of Nature Conservation*, pp. 352–372. Elsevier, London.

Nilsson, C., G. Grelsson, M. Johansson, and U. Sperens. 1989. Patterns of plant species richness along riverbanks. *Ecology* 70: 77–84.

Office of Technology Assessment. 1987. Technologies to maintain biodiversity. OTA-F-330. U.S. Government Printing Office, Washington, D.C.

O'Hara, K.L., M.E. Jensen, L.J. Olsen, and J.W. Joy. 1994. Applying landscape ecology theory to integrated resource planning: two case studies. In: M.E. Jensen and P.W. Bourgeron (eds.), Vol. II. Ecosystem Management: Principles and Applications, pp. 225–236. Gen. Tech. Rep. PNW-GTR-318. USDA Forest Service, Pacific Northwest Research Station, Portland, OR.

O'Neill, R.V., D.L. DeAngelis, R.B. Waide, and T.F.H. Allen. 1986. *A Hierarchical Concept of Ecosystems*. Princeton University Press, Princeton, NJ.

O'Neill, R.V., A.R. Johnson, and A.W. King. 1989. A hierarchical framework for the analysis of scale. *Landscape Ecology* 3: 193–205.

Paine, R.T. 1969. A note on trophic complexity and community stability. *American Naturalist* 103: 91–93.

Palik, B.J., and K.S. Pregitzer. 1992. A comparison of presettlement and present-day forests on two bigtooth aspen-dominated landscapes in northern lower Michigan. *Amer. Midl. Nat.* 127: 327–338.

Perry, H.M., Jr. 1971. Trace-elements related to cardiovascular disease. In: H.L. Cannon and H. C. Hopps (eds.), *Environmental Geochemistry in Health and Disease*, pp. 179–196. Geol. Soc. Am., Memoir 123.

Peterjohn, W.T., and D.L. Correll. 1984. Nutrient dynamics in an agricultural watershed: observations of a riparian forest. *Ecology* 65: 1466–1475.

Peters, R.L. 1988. The effect of global climate change on natural communities. In: E.O. Wilson (ed.), *Biodiversity*, pp. 450–461. National Academy Press, Washington, DC.

Peterson, F.F. 1981. Landforms of the Basin and Range Province — defined for soil survey: Reno, Nevada Agricultural Experiment Station: University of Nevada, Reno, Technical Bull. 28.

Petts, G.E. 1985. *Impounded Rivers — Perspective from Ecological Management*. Wiley, Chichester.

Pollock, M.M. 1995. Patterns of plant species richness in emergent and forested wetlands of southeast Alaska. Dissertation. University of Washington, Seattle, WA.

Pollock, M.M., R.J. Naiman, H.E. Erickson, C.A. Johnston, J. Pastor, and G. Pinay. 1995. Beaver as engineers: Influences on biotic and abiotic characteristics of drainage basins. In: C.G. Jones and J.H. Lawton (eds.), *Linking Species and Ecosystems*, pp. 117–126. Chapman and Hall, New York.

Pories, W.J., W.H. Strain, and C.G. Rob. 1971. Zinc deficiency in delayed healing and chronic disease. In: Cannon, H.L., and Hopps, H.C., *Environmental Geochemistry in Health and Disease*, pp. 73–96. Geol. Soc. Am., Memoir 123.

Powell, A.M. and B.L. Turner. 1977. Aspects of the plant biology of the Gypsum outcrops of the Chihuahuan Desert. In: R.H. Wauer and D.H. Riskind (eds.), Transactions of the Symposium on the Biological Resources of the Chihuahuan Desert Region, United States and Mexico, pp. 315–325. U.S. Department of the Interior, National Park Service Transactions and Proceedings Series, Number 3.

Powell, D.S., J.L. Faulkner, D.R. Darr, Z. Zhu, and D.W. MacCleery. 1993. Forest resources of the United States, 1992. USDA For. Serv. Gen. Tech. Rep. RM-234.

Power, M.E., D. Tilman, J.A. Estes, B.A. Menge, W.J. Bond, L.S. Mills, G. Daily, J. C. Castilla, J. Lubchenco, and R.T. Paine. 1996. Challenges in the quest for keystones. *BioScience* 46: 609–620.

Press, F. 1994. Humankind and earth's natural systems. *Geotimes*, p. 4.

Presser, T.S, Swain, W.C., Tidball, R.R., and Severson, R.C. 1990. Geologic sources, mobilization, and transport of selenium from the California Coast Range to the Western San Juaquin Valley — a reconnaissance study: U.S. Geological Survey Water Resources Investigations Report 90-4070.

Prince, H.C. 1964. The origin of pits and depressions in Norfolk. *Geography* 49: 15–32.

Pyne, S.J. 1983. Indian fires. *Natural History* 92(2): 6–11.

Pyne, S.J., P.L. Andrews, and R.D. Laven. 1996. *Introduction to Wildland Fire*. Wiley, New York.

Raines, G.L., and Canney, F.C. 1980. Vegetation and geology. In: Siegal, B.S., and Gillespie, A.R. (eds.), *Remote Sensing in Geology*, pp. 365–380. Wiley, New York.

Raines, G.L., Johnson, B.R., Frost, T.P., and Zientek, M.L. 1995. Digital maps of major-element bedrock chemistry derived from 1: 500,000 scale geologic mapping in the Pacific Northwest — a contribution to the Interior Columbia River Basin Ecosystem Management Project. U.S. Geological Survey Open File Report 95-685.

Reay, R.F. 1972. The accumulation of arsenic from arsenic-rich natural waters by aquatic plants. *Journal of Applied Ecology* 9: 557–565.

Risser, P.G. 1987. Landscape ecology: state of the art. In: Turner, M.G. (ed.), *Landscape Heterogeneity and Disturbance*, pp. 3–14. Springer-Verlag, New York.

Rose, A.W., Hawkes, H.E., and Webb, J.S. 1979. *Geochemistry in Mineral Exploration*, 2nd ed. Academic Press, New York.

Rytuba, J.J., and Klein, D.P. 1995. Almaden Hg deposits. In: du Bray, E.A. (ed.), Preliminary compilation of descriptive geoenvironmental mineral deposit models, pp. 193–198. U.S. Geological Survey Open File Report 95-831.

Salo, E.O., and T.W. Cundy (eds.). 1987. Streamside Management: Forestry and Fishery Interactions. Contribution No. 57, College of Forest Resources, University of Washington, Seattle.

Sauer, J.D. 1962. Effects of recent tropical cyclones on the coastal vegetation of Mauritius. *Journal of Ecology* 50: 275–290.

Selby, M.J. 1979. Slopes and weathering. In: K.J. Gregory and D.E. Walling (eds.), *Man and Environmental Processes*, pp. 105–122. Dawson, Folkestone.

Sepkoski, J.J., Jr. 1994. Extinction and the fossil record. *Geotimes* 15–17.

Shaver, G.R., K.J. Nadelhoffer, and A.E. Giblin. 1991. Biogeochemical diversity and element transport in a heterogeneous landscape, the North Slope of Alaska. In: M.G. Turner and R.H. Gardner (eds.), *Quantitative Methods in Landscape Ecology*, pp. 105–125. Springer-Verlag, New York.

Shaw, S.P. 1981. Wildlife management on private nonindustrial forestlands. In: R.T. Dumke, G.V. Burger, and J.R. March (eds.), Proc. Symp. Wildl. Manage. on Private Lands, pp. 36–41. Wisc. Chapt. of The Wildl. Soc., Madison, WI.

Skinner, C.N. 1995. Change in spatial characteristics of forest openings in the Klamath Mountains of Northwestern California, USA. *Landscape Ecology* 10: 219–228.

Smith, W.P., P.B. Hamel, and R.P. Ford. 1993. Mississippi Alluvial Valley forest conversion: implications for eastern North American avifauna. *Proc. Annu. Conf. Southeast. Assoc. Fish and Wildl. Agency* 47: 460–469.

Solbrig, O.T. (ed.). 1991. From Genes to Ecosystems: A Research Agenda for Biodiversity. International Union of Biological Sciences, Paris, France.

Sparks, B.W. 1986. *Geomorphology* (3rd ed.). Longman Inc., New York.

Star, J., and J. Estes. 1990. *Geographic Information Systems: An Introduction*. Prentice Hall, Englewood Cliffs, NJ.

Steel, E.A., R.J. Naiman, and S. West. 1996. Woody debris piles: Habitat for birds and small mammals in the riparian zone. *Conservation Biology* (submitted).

Stine, P.A., F.W. Davis, B. Csuti, and J.M. Scott. 1996. Comparative utility of vegetation maps of different resolutions for conservation planning. In: R.C. Szaro and D.W. Johnston (eds.), *Biodiversity in Managed Landscapes, Theory and Practice*, pp. 210–220. Oxford University Press, Inc.

Stoddard, H.L. 1962. Use of fire in pine forest and game lands of the deep Southeast. Proc. Annu. Tall Timbers Fire Ecol. Conf., Tall Timbers Res. Sta., Tallahassee, FL 1: 31–42.

Swanson, F., T.J. Kratz, N. Caine, and R.G. Woodmansee. 1988. Landform effects on ecosystem patterns and processes. *BioScience* 32: 92–98.

Tabacchi, E., A.M. Planty-Tabacchi, and O. Décamps. 1990. Continuity and discontinuity of the riparian vegetation along a fluvial corridor. *Landscape Ecology* 5: 9–20.

Tallis, J.H. 1985. Mass movement and erosion of a southern Pennine blanket peat. *Journal of Ecology* 73: 283–315.

Thornbury, W.D. 1965. *Regional Geomorphology of the United States*. Wiley, New York.
Thornbury, W.D. 1969. *Principles of Geomorphology*, 2nd ed. Wiley, New York.
Trimble, S.W. 1974. *Man Induced Soil Erosion on the Southern Piedmont*. Soil Conservation Society of America, Madison, WI.
Trimble, S.W. 1988. The impact of organisms on overall erosion rates within catchments in temperate regions. In: H.A. Viles (ed.), *Biogeomorphology*, pp. 83–142. Basil Blackwell, Oxford.
Troll, C. 1939. Luftbildplan and Okologische Bodenforschung. *Z. Ges. Erdkunde* 7/8: 297.
Turner, M.G. 1990. Landscape changes in nine rural counties in Georgia. *Photo. Eng. Remote Sensing* 56: 379–386.
Turner, M.G., S.R. Carpenter, E.J. Gustafson, R.J. Naiman, and S.M. Pearson. Land Use. In: M.J. Mac, P.A. Opler, P. Doran, C. Haecker and L. Huckaby Stroh (eds.), Status and Trends of our Nations's Biological Resources. Volume 1. National Biological Service, Washington, DC (in press).
Turner, M.G., R.H. Gardner, R.V. O'Neill, and S.M. Pearson. 1994. Multiscale organization of landscape heterogeneity. In: M.E. Jensen and P.S. Bourgeron (eds.), Volume II: Ecosystem management: Principles and Applications, pp. 73–79. Gen. Tech. Rep. PNW-GTR-318. Portland, OR: U.S. Department of Agriculture, Forest Service, Pacific Northwest Research Station.
Turner, M.G., D. Wear, and R.O. Flamm. (in press). Influence of land ownership on land-cover change in the Southern Appalachian Highlands and the Olympic Penninsula. *Ecological Applications*.
Vannote, R.L., G.W. Minshall, K.W. Cummins, J.R. Sedell, and C.E. Cushing. 1980. The river continuum concept. *Canadian Journal of Fisheries and Aquatic Sciences* 37: 130–137.
Vitek, J.D., and Giardino, J.R. 1993. Geomorphology — the research frontier and beyond. *Geomorphology* 7(1–3).
Walker, H.J., Coleman, J.M., Roberts, H.H., and Tye, R.S. 1977. Wetland loss in Louisiana. *Geogr. Annaler* 69A: 189–200.
Walsh, G., R. Morin, and R.J. Naiman. 1988. Daily rations, diel feeding activity and distribution of age-0 brook charr (*Salvelinus fontinalis*) in two sub-arctic streams. *Environmental Biology of Fishes* 21: 195–205.
Watson, A. 1990. The control of blowing sand and mobile desert dunes. In: A.S. Goudie (ed.), Techniques for Desert Reclamation, pp. 35–85. Wiley, Chichester.
Watt, A.S., Perrin, R.M.S., and West, R.G. 1966. Patterned ground in Breckland — structure and composition. *Journal of Ecology* 64: 239–258.
Webb, L.J. 1958. Cyclones as an ecological factor in tropical lowland rain forest, north Queensland. Australia Journal of Botany 6: 220–228.
Weltzin, J.F., S. Archer, and R.K. Heitschmidt. 1997. Small-mammal regulation of vegetation structure in a temperate savanna. *Ecology* 78: 751–763.
Wetzel, R.G. 1983. *Limnology*, 2nd ed. Saunders, Philadelphia.
Whicker. A.D., and J.K. Detling. 1988. Ecological consequences of prairie dog disturbances. *BioScience* 38: 778–785.
White, P.S. 1979. Pattern, process, and natural disturbance in vegetation. *Botanical Review* 45: 229–299.
Whitmore, T.C. 1974. Change with time and the role of cyclones in tropical rain forest of Kolombangara, Solomon Islands. Univ. Oxford Commonwealth Forestry Institute, Institute Paper No. 46.
Whitson, T.D., L.C. Burrill, S.A. Dewey, D.W. Cudney, B.E. Nelson, R.D. Lee, and R. Parker. 1996. *Weeds of the West*. Pioneer of Jackson Hole, Jackson, WY.
Wiens, J.A. 1989. Spatial scaling in ecology. *Functional Ecology* 3: 385–397.
Wiens, J.A. 1990. *The Ecology of Bird Communities, Vol. II*. Cambridge Press, New York.
Wiens, J.A., C.S. Crawford and J.R. Gosz. 1985. Boundary dynamics: a conceptual framework for studying landscape eocsystems. *Oikos* 45: 421–427.
Williams J.E., and R.J. Neves. 1992. Introducing the elements of biological diversity in the aquatic environment. In J.E. Williams and R.J. Neves (eds.), *Biological Diversity in the Aquatic Environment*, pp. 345–354. Transactions of the 57th North American Wildlife and Natural Resources Conference. Wildlife Management Institute, Washington, D.C.
Wright, H.E., Jr. 1974. Landscape development, forest fires, and wilderness management. *Science* 186: 487–495.
Yelin, T.S., A.C. Tarr, J. A Michael, and C.S. Weaver. 1994. Washington and Oregon earthquake history and hazards. U.S. Geological Survey Open File Report 94-226B.

## THE AUTHORS

**James R. Gosz**
*Biology Department, University of New Mexico, Albuquerque, NM 87131, USA*

**Jerry Asher**
*Natural Resource Specialist, USDI BLM, P.O. Box 2965, Portland, OR 97208, USA*

**Barbara Holder**
*USDA Forest Service, Klamath National Forest, 1312 Fairlane Road, Yreka, CA 96097, USA*

**Richard Knight**
*Department of Fishery and Wildlife, Colorado State University, Fort Collins, CO 80523, USA*

**Robert Naiman**
*School of Fisheries, Box 357980, University of Washington, Seattle, WA 98195, USA*

**Gary Raines**
*USGS, MS 176, c/o Mackay School of Mines UNR, Laxalt Mineral Research, Room 366, Reno, NV 89557, USA*

**Peter Stine**
*California Science Center, Biological Resources Division, US Geological Survey, Room 1480 Chemistry Annex, University of California at Davis, Davis, CA 95616, USA*

**T.B. Wigley**
*Department of Aquaculture, Fisheries & Wildlife, G12H Lehotsky Hall, Clemson University, Clemson, SC 29634, USA*

# Describing Landscape Diversity: A Fundamental Tool for Ecosystem Management

Julie A. Concannon, Craig L. Shafer, Robert L. DeVelice,
Ray M. Sauvajot, Susan L. Boudreau, Thomas E. DeMeo,
and James Dryden

**Key questions addressed in this chapter**

- ♦ *What are the key concepts for defining landscape ecology, especially from a management point of view*
- ♦ *As managers, how can we assess scale of analysis for project areas?*
- ♦ *How can managers learn to identify and work around barriers to using the science and art of landscape ecology?*
- ♦ *Which resource areas can be most effectively managed using the tools of landscape ecology?*
- ♦ *Will we use these tools in the future or can we be creative enough to imagine even more sophisticated ways to assess landscapes?*

**Keywords: Landscape approach, habitat conservation, reserve design, disturbance**

## 1 INTRODUCTION

The diversity of life generally falls into three categories including genetic, species, and ecosystems (Noss and Cooperider 1994). Ecosystems described at the landscape level are generally characterized by the diversity of abiotic and biotic patterns throughout the landscape (Hunter 1996). The goal of this chapter is to illustrate the basic concepts of landscape level ecosystem analysis and show how that information can be used effectively by natural resource managers.

The terms "landscape" and "region" are used extensively throughout this chapter. A landscape "is a mosaic where the mix of local ecosystems or land use is repeated in similar form over a kilometers-wide area. In contrast, a region is a broad geographical area with a common macroclimate and space of human activity and interest" (Forman 1995). Managers generally describe landscape diversity at the regional level to account for variable management scenarios presented to the public.

A common question landscape ecologists are asked is when and where do we use the landscape approach? In an age of increasing information related from satellite imagery and high-tech photography, the public often has a visual idea of large landscapes. This technology using a landscape approach allows information to be recorded and monitored over time. Think about how much closer the moon became to earth when the first shots of Apollo 11 came back. Now when the Challenger aircraft/rocket leaves the Kennedy Space Station, we are more inclined to think about the flight mechanics than the moon's distance.

The analyses and management of ecosystems requires: (1) a knowledge of how landscape patterns reflect functional and structural elements of ecosystems — for each disciple of resource management, the elements vary — and (2) an understanding of the level of the landscape (based on hierarchical theory) at which these interactions take place. From there, most landscape ecologists recommend moving up one level on the spatial scale for assessment; however, the point is that the landscape goal will often dictate the level of analyses. For example, habitat conservation decisions should take into consideration patterns of natural diversity distributed on local, regional, and sometimes continental scales (Shafer 1990). Information on landscape patterns can be difficult to obtain. Such information varies from a plotted point on a 1:24,000 map (e.g. plotted points of rare plants and animals) to Geographic Information Systems (GIS) (e.g. 3-D plots of forest structure for large tracts of land). The U.S. Forest Service, State Natural Heritage Programs, The Nature Conservancy, the National Biological Service, and the National Park Service have access to such data.

Active landscape management activities are a daily challenge for public land managers. Whereas scientists define and examine the condition of ecosystems using science, public land managers must consider public values such as sustainability. Activities such as logging and passive recreation have different impacts. Land managers often consider "level of risk" to the ecosystems as well as the amount of reduction in landscape integrity with human activity. Public land managers are legally responsible for defining viability, integrity, health (condition), and sustainability of ecosystems within the context of federal laws (see Keiter, Vol. III of this book), such as the National Forest Management Act (NFMA), the National Environmental Protection Act (NEPA) and the Endangered Species Act (ESA). Intuitively, ecosystem management at the landscape level requires research, integration of science into management decisions, and public accountability.

Because so many variables can be described at the landscape level, it is important for the land manager to define the critical ones. One of the greatest challenges in ecosystem management is to integrate and use many types of information to describe the critical components of ecosystems in the landscape. Scientific interpretations of landscape patterns can produce a useful description of ecosystem function at the landscape scale. Although complex and often poorly understood, ecosystem science is generating better and more useful information for resource managers (Perry and Amaranthus 1996). This chapter provides methods to describe the landscape and interpret land patterns and variables for resource management decisions. Gosz (this volume) provides more detailed information on the scientific basis for describing and interpreting ecosystem diversity.

## 2 HISTORICAL CONTEXT

Managers of public lands have traditionally been responsible for large landscapes. Maps, aerial photography, and on-the-ground knowledge have long been used to provide a landscape perspective for decision makers. Today, the perspective required for many management issues is larger than it was just a decade ago. Resource managers once typically viewed their tasks at the scale of several hundred hectares. Today managers must often think about their tasks in a regional context that embraces thousands or even millions of hectares. In extreme examples, such as for neotropical migrant birds, population viability must be considered at both a hemispheric scale (e.g., nesting habitats in North America, wintering habitats in South or Central America, and migratory habitats in between) and in the immediate landscape. The survival of neotropical migrant birds depends on what happens to food and habitat in these widely separated places (Robinson 1996).

The "landscape approach" of describing physical patterns on the land has been recognized from the early 1900s (Troll 1939). However, the ecological interpretation of landscape diversity has taken longer to develop. The rapid evolution of landscape ecology in the early 1980s helped to correlate physical descriptions with ecological processes. In the late 1980s, public agencies began to combine descriptions of landscape heterogeneity with biological monitoring of wildlife populations and communities. Soule and Wilcox (1980), Shafer (1990), and the overview in Gosz (this volume) describe these developments in more detail.

Understanding the intrinsic affects of pattern and process on wildlife and communities within the landscape began in the 1930s to 1950s with microclimatic studies of edges and patches (Geiger 1959). Basic abiotic variables (e.g., windspeed, relative humidity, and air temperature) change with position in the landscape and so does biotic composition — a relationship shown dramatically when sophisticated abiotic modelling techniques are combined with on-the-ground biological data to quantify edge and interior conditions of landscape (Raynor 1971, Chen et al. 1995, Concannon 1995). Not surprisingly, species and communities react differently to changes in abiotic and biotic variables. A simple example is to monitor clearcut edges over space and time. South-facing forest edges have hotter, drier edges that extend further into the forest in summer than north-facing edges. Salamanders, for example, will have a harder time than grizzly bears in the summertime at a south-facing edge because of their moisture requirements.

Recently developed quantitative methods to define landscape heterogeneity provide land managers with useful tools to differentiate the variability of landscapes by linking spatial patterns with ecological processes operating at broad scales of space and time (Turner and Gardner 1991). For example, Johnson (1994) explored the relationship between increasing landslide frequency and increasing timber removals in a large watershed in Washington state over a 20-year period. Using successive aerial photographs, Johnson showed that an underlying geological process accelerated in relation to timber harvest intensity and caused increasing structural changes in a large landscape. This is an important concept because it suggests that changing structure at a large scale can provide an appropriate measure of regional ecological changes (see also O'Neill et al. 1988). It also shows that remote-sensing technologies can provide managers with a broader canvas for assessing landscape patterns related to biological diversity than ever before. The linkage of ground-based data (e.g., timber, wildlife, and soils) with geographic information systems (GIS) has provided the manager with the ability to think at the landscape level.

## 3 KEY CONCEPTS FOR MANAGERS TO CONSIDER

To define landscape diversity, two key concepts should be understood. The first is that spatial and time scales affect the patterns we observe. The second is that landscape diversity cannot be understood by vegetation patterns alone.

### 3.1 Landscapes in Time and Space

Time and geographic scales can significantly affect the interpretation of landscape patterns (see also Haulfer et al., and Caraher et al., this volume). Land managers need to monitor changes in landscape patterns because these patterns reflect changes in landscape structure and function. Changes in structure and function in turn, affect ecosystems and their organisms. Monitoring changes in landscape patterns at the scale of thousands to millions of hectares can be a daunting task. Monitoring changes in ecosystem processes that occur with changes in landscape patterns is even more challenging. Understanding what landscape changes mean for ecological processes often requires knowing the significance of those changes in the context of a large area or over a long period of time — or "scaling up" as ecologists (and others including geologists and hydrologists) refer to this challenge (Risser 1986). For example, a hydrologist may interpret a hundred-year flood in the Southwest as a rare event in a relatively undisturbed landscape. However, annual floods along forest streams on the east coast are predictable events that frequently disturb riparian zones across the landscape.

Knowledge of hierarchical theory is essential for interpreting landscape diversity. Urban et al. (1987) describes a nested forest hierarchy starting with tree gaps, then forest stands, watersheds, and landscapes (Fig. 1). Complexity and variation increases as spatial scale expands. For example, a storm's microburst can produce tree gap patterns at a very localized scale. At the landscape scale, patterns are driven by complex geological and climatic conditions that operate over decades, centuries, or longer to determine species distributions and mosaics of forest communities and seral stages. Patterns at the landscape scale are more resistant to change than patterns detected at the scale

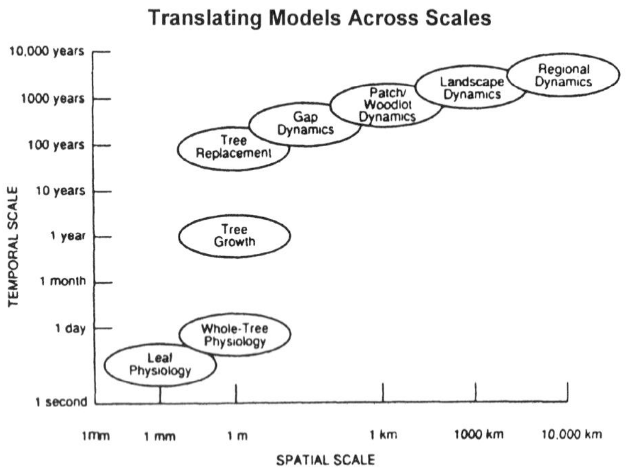


Fig. 1. A nested hierarchical representation of carbon/biomass dynamics in a forested landscape. The ellipses represent different scales of organization. Adapted from King et al. (1990) and Urban et al. (1987).

of tree gaps, forest stands, and watersheds. When large-scale change does occur on a landscape (e.g., large-scale fire), the recovery of landscape patterns takes much longer than the recovery of more localized tree gap or forest stand patterns. As a result, managers need to: (1) carefully define the spatial and time scale for collecting landscape data, and; (2) aggregate data from smaller scales (e.g., forest stands and watersheds) to define large landscapes.

The scale at which managers collect landscape information will need to vary depending on the decisions it is used to support. In some cases, a finer landscape scale may be sufficient, such as mapping soils in a valley bottom to assess the viability of a spatially-confined endemic plant species. In other cases, a broad scale approach is necessary, such as simultaneously determining timber harvest areas and the habitat requirements of grizzly bears in a national forest. To define the appropriate scale of analysis, it is essential that the manager or the researcher know what the long-term management goals are.

### 3.2 Beyond Vegetation Patterns

Recognizing the role of space and time in landscape patterns and ecological processes and functions is an important first step in using a landscape approach to help make management decisions. It is also important to recognize that landscape diversity is the product of different types of landscape patterns. Too often, vegetation patterns are considered to be synonymous with landscape or ecosystem diversity in the context of resource management. For example, timber inventories are sometimes generated by extrapolating plot surveys to maps of vegetation cover. Vegetation patterns, however, are only a subset of landscape patterns associated with ecological processes and functions, including those direct affecting timber growth and condition. Ecosystem approaches to natural resources management consider other patterns, such as geologic and microclimatic patterns, in relation to the management issue. The fact that various biological, physical, and human factors interact to create different vegetational patterns has been recognized for some time (e.g., Watt 1947). Today, ecologists are interested in how different landscape patterns including vegetation (e.g., patch size, shape, adjacency), geology (e.g., geomorphology of landslides), and microclimate (e.g., humidity, wind direction and speed) affect ecosystem processes and organisms (Forman and Godron (1986). Trying to understand these patterns and how management actions might affect them is a challenge that many of today's resource managers must confront. That is, resource managers today have responsibilities for maintaining biodiversity and ecosystem health that go beyond the timber (or fish, or elk, or cows, or recreation) concerns that once predominated.

Let us use a simple example. At the stand level, clearcut logging creates forest patches with sharp edges (Patton 1975, Thomas et al. 1979, Ranney et al. 1981). In southeastern Alaska, natural forest patches have gradual and transitional edges (e.g. muskeg and beach-front–forest edges; see Fig. 2) (Concannon et al. 1992, Concannon 1995). This difference in edge boundaries is ecologically significant. The microclimatic conditions typical of interior forest environments are found much closer to natural edges than they are to clearcut edges. The microclimatic conditions affect plant reproduction and growth and foraging quality for a number of interior dwelling small mammals and birds. Thus, two adjacent areas with the same forest type and similar patch sizes but different edge patterns can have diverging ecological conditions that affect species composition. Of course, the ecological consequences of these edges will not be the same everywhere. Newts in Alaska see edges very differently than salamanders in Appalachia. During the summertime, Florida black bears see hot dry edges differently than Alaskan brown bears. The lesson here is that management actions can affect not only vegetation patterns but underlying microclimatic patterns and ecological processes and functions as well.

## 4 BARRIERS TO USING A LANDSCAPE LEVEL APPROACH

Operating at the landscape scale can pose barriers to managers seeking to design and implement ecosystem

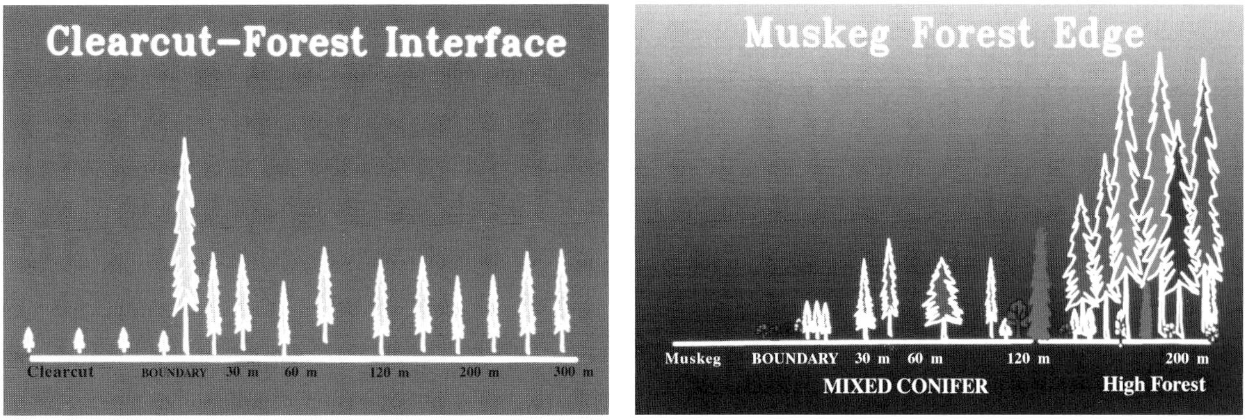


Fig. 2. Clearcut-forest edge structure differs markedly from muskeg-forest edge. The muskeg-forest transition is much more gentle than the sharp change along a newly-logged edge. Characteristic changes in microclimate correlate with these structure changes (Concannon 1995).

management programs. Barriers vary from agency to agency and from place to place, but they should be expected whether the area in question is managed for multiple-use purposes or as parks and protected areas. Regardless of the setting, obstacles to using landscape approaches can come in the form of technical, social, economic, and legal barriers. Anticipating such barriers is an important step in overcoming them and using landscape approaches effectively in natural resources management.

Simply deciding which scale to use is an example of a commonly encountered technical barrier. For example, a viability analysis should consider whatever geographic scale is appropriate for the organism involved (see also Holthausen et al., and Noon, this volume). But, the manager may face the dilemma of deciding which scale to use for a viability analysis when the technical answer does not necessarily jibe with their management jurisdiction ("Do I consider viability and distribution of a species on my planning area, throughout the national park or forest I work in, or throughout the range of the species?"). If the analyst selects too small a scale, the analysis is unlikely to be accurate and could lead to the wrong management decisions. If the appropriate scale goes beyond the manager's jurisdiction, he or she may need to plan on enlisting the cooperation of other managers and researchers to address the viability issue correctly. If not, valuable time and money could be wasted on an analysis that produces misleading information and, more importantly, leads to consideration of inappropriate management options.

Beyond technical barriers, economic barriers to using landscape approaches have become more prominent as public resource management agencies face declining budgets and staff. For example, the technology and staff for landscape assessments — particularly satellite imagery and GIS computer systems and the specialists who operate them — can be expensive. In a world of shrinking budgets, getting the best technology and expertise may not always be possible. It could also mean delaying or extending the project, or choosing not to do something else (e.g., incurring an opportunity cost). In a world of shrinking budgets, this makes it all the more important to carefully identify goals and objectives for managing the landscape and to set priorities for what needs to be done first (see Morrison and Marcot 1995).

Interactive Associates (1986) described a process illustrated in Fig. 3 to pose important problems and then set appropriate goals that respond to them. They suggested that management goals should be proposed

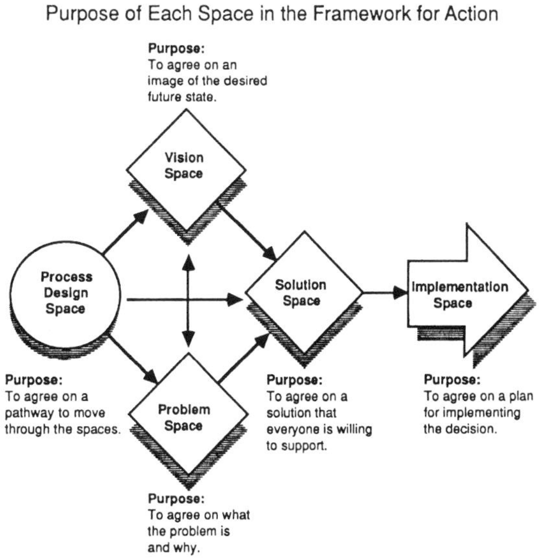


Fig. 3. The purpose of each space in the framework for action which is offered by Interaction Associates (1986) is to explore and produce an implementation plan for the landscape based on clear objectives.

and examined (e.g., desired future condition) first, followed by detailing the actions necessary to implement goals, with due consideration for essential data collection and analysis. Managers can use a similar process to: (a) define a landscape level problem; (b) identify the time and resources needed to respond to that problem, and (c) set realistic goals for managing landscape problems by prioritizing amongst possible responses.

The Great Lakes Ecological Assessment is among the better examples illustrating how managers have surmounted economic obstacles to carry out landscape assessments. In this case, managers of several national forests in the upper Midwest and staff in the U.S. Forest Service Region 9 office in Milwaukee pooled resources to catalyze a regional landscape assessment that eventually involved several federal agencies, departments of natural resources in three states, several large corporate landowners, and four universities (Padley 1996). Regional Forest Service staff started the effort because six national forests in the region were beginning forest plan revisions. Each of these national forests was located in the 98,000 square mile Laurentian Mixed Forest Province (Bailey 1996) that spans northern Minnesota, Wisconsin, and Michigan. They share similar resources and common problems, and each is surrounded by diverse ownership patterns. The individual forests were interested in cooperating because they could gain a more accurate assessment of the management issues they face than would be possible to develop on their own. The state agencies, universities, and private sector partners were motivated by different interests, but each wanted to use the integrated landscape databases the effort would produce.

The goal of the Great Lakes Assessment was to develop a comprehensive regional GIS database that would become an important tool for revising the individual forest plans. By the time the project was concluded, a wide variety of information had been mapped onto a common database encompassing the 98,000 square mile region. This included political boundaries and ownership boundaries between federal, state, county, and private lands, ecological unit maps (ecological units are described by soil, aspect, and vegetation type in map units—see also Grossman et al. and Carpenter et al., this volume), natural disturbance regimes (see also White et al. and Engstrom et al., this volume), historical vegetation, current vegetation, soils, elevation, climate, hydrology, wetlands, and species distributions.

Of course, the up-front cost of such assessments can be substantial (see also Lessard, Vol. III of this book). Engaging a large number of partners, each with skills, data, and financial resources of their own, however, can turn surmount economic obstacles and generate long term savings. One of the greatest benefits of this type of joint assessment is that resource managers and decision makers from across the region will be using the same information. That alone should save all of them considerable time and money. But, it may also avert debates and differences over data when shared decisions need to be made.

Finally, various legal obstacles can emerge to thwart landscape approaches to natural resources management. While it is true that the combined legal requirements of the Endangered Species Act (ESA), the National Forest Management Act (NFMA), and the National Environmental Policy Act (NEPA) have been major factors propelling federal agencies toward ecosystem management, it is also true that other laws can pose substantial and sometimes complex obstacles (see Keiter, Vol. III of this book). For example, the Sherman Antitrust Act may inhibit adjacent landowners from participating in landscape level planning and the Federal Advisory Committee Act (FACA) may inhibit federal resource managers from collaborating with citizens and interest groups.

## 5 HABITAT CONSERVATION

Landscape approaches have probably been used more extensively for habitat conservation planning than for any other purpose. This is not surprising given the fact that many species have wide ranging territories or are found in specific habitats widely separated across the landscape. Even in a relatively undisturbed landscape, landscape approaches are important tools for identifying conservation needs. For resource managers balancing their responsibilities to maintain biodiversity while also providing for commodities, recreation, or scenic preservation, landscape approaches are essential.

### 5.1 Fundamentals of Habitat Conservation at the Landscape Level

Conservation biologists have identified the types of habitat patterns that favor the retention of biodiversity (Noss, 1983; see Gosz, this volume). Their research, much of it based on the principles of island biogeography (MacArthur 1972), has shown that smaller and more isolated populations are more vulnerable to extinction than larger, adjacent populations. Thus, large blocks of habitat are generally superior to small blocks of habitat, proximity of habitat blocks is favored over distance, and habitats with large populations are preferable to habitats with small populations.

Five basic principles emerge from conservation biology that resource managers can use to retain habitat at the landscape and regional scale (drawn from Shafer 1990, Thomas et al. 1990, Wilcove and Murphy 1991, and Noss 1992). These principles include:

1. Minimizing the fragmentation of habitats across the landscape;
2. Conserving large blocks of habitat at the regional landscape scale;
3. Conserving blocks of habitat close together and in contiguous blocks;
4. Maintaining habitat corridors between large blocks of habitat, and;
5. Maintaining favorable habitat conditions for target species across their native range.

These principles, of course, have their exceptions. For example, some species (especially rare plants with very specific habitat requirements) may fare better with smaller but more numerous habitat blocks. Distance between blocks may sometimes have advantages over proximity (e.g., reducing the probability that a catastrophic disturbance such as a hurricane will affect most of a population). And, corridors between habitats in some cases may facilitate the spread of disease or invasive introduced species. Still, in most situations, conservation biologists favor large blocks of interconnected habitat to facilitate species dispersal and to minimize their vulnerability to catastrophic disturbances. These principles, therefore, should be seriously considered but not blindly applied. They are not a substitute for knowing the natural history of the species of concern.

Using a landscape or regional approach to select habitat conservation areas has a number of advantages over selecting conservation areas on a more localized basis. One reason is that it is easier to select a habitat large enough to protect a full range of natural communities and their seral stages when the assessment covers a large area (Noss 1983). Reserve areas that are chosen should collectively capture all or most of the species and community diversity in the landscape (Scott et al. 1988, Noss 1995).

Using a landscape approach also makes it possible to identify ecological processes, such as natural disturbance regimes, hydrologic processes, nutrient cycles, and biotic interactions essential for maintaining the natural variability of the landscape or regional biodiversity (Austin and Margules 1986, Noss 1992). These processes are not only essential to the near-term survival of some species, but they facilitate evolutionary processes which ultimately generate and maintain biodiversity (e.g. co-evolution of sympatic organisms). Thus, in addition to species and community diversity, reserve selection should consider the potential for these processes to operate.

In addition, a landscape or regional approach can identify areas that are least vulnerable to disruptive external influences such as pollution, introduced species, and creeping suburbanization. In short, an effective landscape approach to habitat conservation will help to design a reserve system that is resilient to short-term and long-term environmental change (Shafer 1990, Noss 1992).

Finally, a landscape approach to habitat conservation is important because protected areas alone will often be insufficient for the long-term maintenance of biodiversity. What happens on the surrounding landscape and where it occurs is a key factor in the survival of some species. For example, to maintain the viability of the grizzly bear population in Montana's Mission Mountain Wilderness Area, the U.S. Forest Service seasonally restricts logging and recreational use in the adjacent Seely-Swan valley. A landscape approach, therefore, is indispensable to determining how to manage the multiple-use matrix surrounding protected areas when biodiversity conservation is at stake (Brown, in Roberts 1988, Franklin 1993, Chen et al. 1996).

Of course, many land managers may not have the opportunity to design large reserves. Most of their daily work is managing smaller protected areas in a localized part of the landscape. But even here, a landscape perspective can be useful to ensure that small reserves accomplish their objectives (Shafer 1995, Shafer 1998b).

## 5.2 Information For Selecting Representative Habitats

So where does a manager obtain the information he or she will need to conserve habitats representative of the landscape or region? The answer will, of course, vary with the purpose, location, and scale at which the assessment is being conducted. In any case, the information needed will be diverse, cover a range of scales, and come from a variety of sources.

Assessing representative reserve design requires information on topics as diverse as soils, vegetation, species distribution, precipitation, landforms, and landuses. In many cases, there is value in going beyond the obvious vegetation and species distribution data to assess habitat conservation needs (Austin and Margules 1986). Environmental information on abiotic landscape factors can play a vital role in representative reserve design.

Mackey et al. (1989) provide an example of how a range of biotic and abiotic information should be

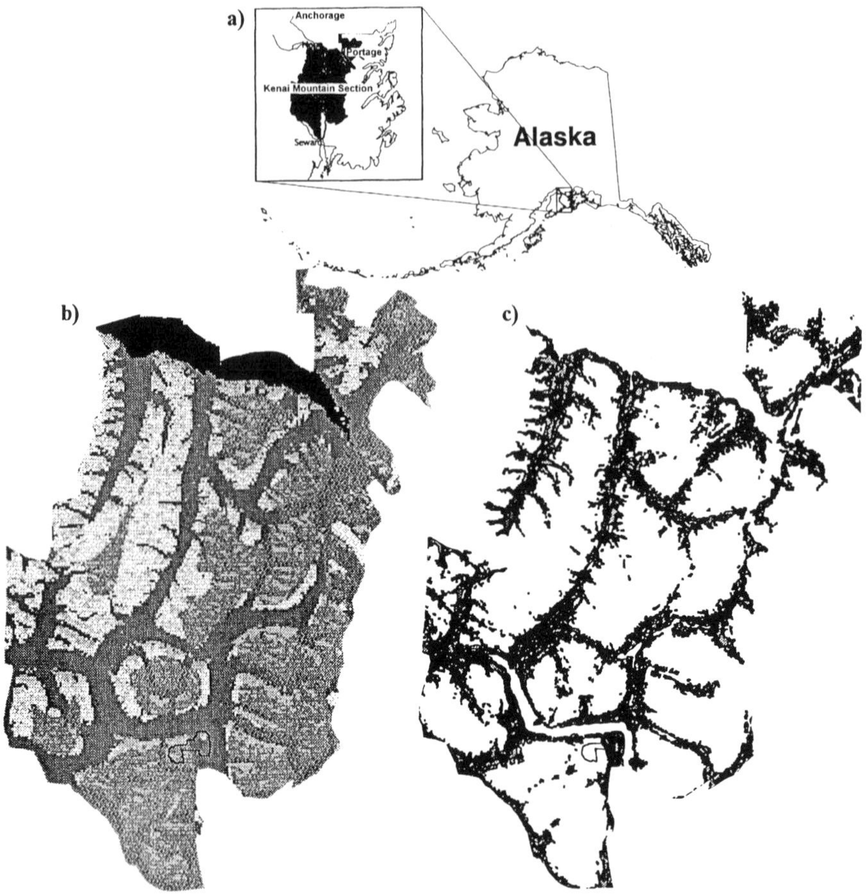


Fig. 4. Bioenvironmental features of the Kenai Mountain Ecological Section (sensu ECOMAP 1993) on the Chugach National Forest in southcentral Alaska. (a) The general location of the area in southcentral Alaska; (b) a plot of the 10 bioclimatic classes identify for the area by DeVelice and Hagenstein (1995); and (c) the distribution of forest vegetation in the area. The boundary of the Kenai Lake–Black Mountain candidate Research natural Area is shown in the lower portion of (b) and (c) (Imagery courtesy Chugach National Forest).

considered in establishing representative reserves. Their goal was to identify reserves to conserve lowland rainforest types in an extensively cleared landscape with remnant habitat patches in Australia. Since the goal of conserving these forest types could only be achieved by restoring habitat conditions on favorable sites, they needed to select information that indicated the potential for rainforest restoration. The information they used included GIS coverage of climate, topography, soils, surficial geology, and vegetation cover. The physical site data for the region were then grouped into biogeoclimatic classes using a statistical cluster analysis. This enabled them to generate potential vegetation maps and identify those areas most capable of restoring the desired lowland forest type. In the final step, existing reserves were overlaid on the potential vegetation maps and the most feasible options for expanding the reserve system to adequately conserve lowland rainforest types were identified. On the Chugach National Forest in Alaska, a classification of biologically relevant climatic conditions (i.e. bioclimates) was used to assess the degree to which research natural areas (RNAs) represent the range of bioclimatic variability on the forest (DeVelice and Hagenstein 1995). In this study, weather data from discrete stations were extrapolated across the entire national forest and classified into bioclimatic types. Of the 35 bioclimatic types identified forest-wide, on 22 are represented in a system of proposed RNA. Of these, only nine areas are represented by greater than 500 ha in the RNA system.

This kind of approach quantifies the adequacy of existing or proposed RNA in meeting representativeness of the area (Fig. 4). Currently, the approach used by DeVelice and Hagenstein (1995) is being investigated in conjunction with a computer-assisted reserve selection method by Bedward et. al. (1992) called Conservation Options and Decisions Analysis (i.e. CODA). The goal of this work is to identify a forest-wide network of areas that represent combinations of physical site and vegetation attributes.

Managers may not have access to the technology or statistical knowledge needed to carry out the sophisticated analysis used by Mackey et al. (1989). One alternative would be to group the data using non-statistical, "expert system" clustering techniques (DeVelice et al. 1994). The lesson illustrated here is that by describing various elements of the landscape in addition to vegetation, a more comprehensive summary of landscape diversity is possible for reserve design.

Managers should consider using information collected at a variety of scales. This may mean that relevant data comes from plotted points for the distribution of rare species on a 1:24,000 map, a GIS database on forest structure for an entire National Forest, and from statewide species distribution maps maintained by the State Natural Heritage Program. In any case, most relevant information has probably been collected at a scale different than the manager's planning area.

It is important, however, to recognize that the local scale of some data will have little value in landscape habitat analysis. For example, a National Forest planner considering the role of an 8,000 acre watershed in designing a reserve to maintain a viable population of wolverines is unlikely to get much help from a NEPA analysis for a 300 acre timber sale within that watershed. In this case, the planner would probably find it more useful to look the other way for information about surrounding watersheds — likely including some outside of the national forest. In general, assessing representation of ecosystems within a large landscape (>100,000 acres) has to rely heavily on the use of basic biophysical data (e.g., maps of macroclimate, geology and landform, vegetation, land use, and species distribution) because it is not possible to meaningfully aggregate much of the information collected at very local scales (Bourgeron et al. 1993).

Managers should not limit their search for information to data close at hand. Reliable sources of information for regional and landscape assessments include the Nature Conservancy, the National Biological Service, district and state offices of the Bureau of Land Management, U.S. Forest Service regional offices and research stations, U.S. Fish and Wildlife Service databases on threatened and endangered species, and frequently, state departments of natural resources. It may also be worth checking with area colleges and universities and non-governmental conservation groups.

### 5.3 Using Rarity Criteria For Reserve Designs

Representativeness is not the only criteria for selecting habitats for conservation. Another approach is to target species and communities that are rare on the landscape. Rarity criteria have been especially important in The Nature Conservancy's approach to selecting reserves (TNC 1987). In partnership with a network of Natural Heritage Programs managed by state governments, TNC maintains databases of rare species and rare community locations (known as "occurrences") that are used to evaluate alternative reserve sites (Noss 1987). Species and natural communities are called "elements" and they are ranked on a global scale ranging from extremely rare and imperiled (G1 with fewer than six occurrences) to globally secure (G5 with more than 100 occurrences). Areas with concentrations of these elements (i.e., rare species and community types) are identified and targeted for conservation actions by TNC including land acquisition, purchase of conservation easements, or management agreements with public land agencies. State Natural Heritage Programs use the same ranking system to rate the conservation status of species and natural communities resident in the state (S1 = critically imperiled in the state).

The TNC rarity approach, like all strategies for reserve selection, has both advantages and disadvantages. It has been used very successfully by TNC to identify several million acres of habitat for the rarest species and natural communities in the United States. The element/occurrence approach, however, does not consider the larger ecological context in which rare species and natural communities are located. A more holistic approach (e.g., Mackey et al. 1989) that details a range of biophysical factors may have more to offer. In fact, TNC has moved recently to identify conservation priorities on the basis of regional analyses designed to ensure that all ecological types are eventually represented at the ecoregional scale (see also Grossman et al., this volume). At many local levels this technique has not however, been readily adopted due to funding and expertise.

Methods to identify optimal reserve networks have evolved rapidly in recent years — some designed for use at the landscape scale, others for use at regional, national, or even continental scales. Johnson (1995) provides an overview of methods and experiences from around the world and contrasts the advantages and disadvantages of different approaches to identifying geographic conservation priorities.

## 6 DISTURBANCE AS A VARIABLE OF LANDSCAPE DIVERSITY

Disturbance is one of the most important factors determining how ecologically diverse a landscape is. Disturbance comes in all sizes and shapes and includes both natural events (e.g., a wind storm) and human actions (e.g., building a road across a wetland). White and Pickett (1985) defined disturbance in this context as "any relatively discrete event in time that disrupts an ecosystem, community, or population structure and changes resource and substrate availability, or the physical environment." Disturbance regimes are covered in more detail by White et al. (this volume) and Engstrom et al. (the volume). This section looks at how managers can describe disturbance to manage for an ecologically diverse and productive landscape.

## 6.1 Ecosystems and Disturbance

Most ecologists and natural resource managers today agree that: (a) ecological systems are regularly subjected to disturbances large (e.g., fire) and small (e.g. a fallen tree); (b) ecosystems respond in a variety of ways depending on the size and rate of the disturbance (Fig. 5), and; (c) some disturbance is a good thing (Oliver and Larson 1990, Perry 1995). For the manager, there is no question that understanding the role of disturbance is key to maintaining landscape diversity. In fact, much of the impetus for landscape pattern analyses came from assessing disturbance regimes on public lands (Pickett and Thompson 1985, Kaufman et al. 1994). The bottom line is that managers will have to deal with disturbance regimes on their landscape. Managers that seek to understand the role of disturbance might actually be able to manage both natural and human disturbances to maintain desirable landscape diversity patterns.

Although the general ecological role of disturbance is widely agreed upon, the details are not. In fact, the role of disturbance in maintaining landscape diversity is at the heart of some of the most heated debates in natural resource management today. Were pre-European settlement forest fires mostly natural or anthropogenic disturbances (see Bonnicksen et al. and Periman et al., this volume)? Has any landscape in the United States escaped human disturbance and if not, is there any such thing as a pristine environment? If some disturbance is a good thing, how much is too much? Is biodiversity intrinsically more vulnerable to human rather than natural disturbances (Noss and Cooperider 1994, Karieva 1994, Huston 1994, Perry 1995)? The manager should be forewarned: everyone agrees that disturbance is a necessary thing, but few agree on which forms of disturbance are preferable and how much.

In any case, a manager should seek to understand and describe the basic characteristics of natural disturbance events under which his or her landscape evolved. How frequent, how extensive, and how severe were fires?, floods?, droughts?, windstorms?, disease outbreaks? This understanding can help the manager to assess the potential impact of a new logging practice or regulation or anticipate the effects of a severe drought. It also helps to have a basic understanding of species sensitive to changes in disturbance regimes. For example, some forest interior nesting birds (e.g., certain warblers) depend on gap disturbances that favor insects they eat while shrikes thrive in open agricultural fields. Both are considered sensitive to changing conditions on the landscape. As the agricultural field is replaced by trees, shrikes begin to decline (Brooks and Temple, 1990). As older forest is replaced by younger stands, it shelters fewer interior forest birds. In this case, a landscape with expanding but younger forest has lost the disturbance patterns that these two sensitive species depend on. Perhaps the most useful maxim for considering the effects of disturbance comes from Noss and Cooperider (1994): "If we radically alter the natural regime that species are adapted to, there is the possibility that many will be unable to cope with the change."

## 6.2 Cumulative Effects of Disturbance

Most disturbance events take place very rapidly over a small area. A tree falls. A rock rolls down a scree slope into the forest. Lighting ignites a fire that roars across a hundred acres of grassland before a rain shower puts it out. In themselves, such events have little significance at the landscape scale. Over time, these events — and the occasional large event — add up and impact more and more of the landscape. To understand and successfully anticipate the effects of disturbance, managers should think about the additive or cumulative effect of separate disturbance events. Each clearcut considered on its own may have little impact on landscape diversity, but 100 clearcuts in 10 years would have some effect.

Disturbances can also be propagated across the landscape under certain conditions. One discrete event, such as a landslide at the head of a steep watershed, triggers another, then another, until, like a stack of dominoes, one little disturbance is transformed into one big one. Uniform landscapes are generally more vulnerable to the "amplification" of disturbances than heterogenous landscapes. Is such amplification characteristic of the disturbance regime the landscape evolved with or is it new? A natural even-aged forest of lodgepole pine in the northern Rockies and a large plantation of loblolly pine in a southern hardwood landscape share similar vulnerabilities to catastrophic fire and disease because of their uniformity. From the standpoint of a timber manager, fire or disease would spell disaster in both places. From an ecological perspective, the consequences of such disturbances might be quite different. The lodgepole ecosystem would recover more rapidly because the species inhabiting it are adapted to relatively frequent fire and the ecosystem is therefore better able to buffer and absorb the disturbance, both genetically and phenotypically.

Perry (1995) suggests managers keep two things in mind to reduce the probability of a catastrophic spread of disturbance. First, each situation must be evaluated independently. That is, identify the features of the ecosystem and its disturbance regime(s) that make it more or less vulnerable to widespread disturbance. Are natural disturbance regimes in the landscape low intensity or high intensity, widespread or localized, frequent or rare? What factors retard or promote the spread of common disturbances in the landscape? What are the most threatening disturbances? Which plant communities and age classes are most susceptible to disturbance, and which are not? In many ecosystems, knowledge of fire regimes can help assess ecosystem vulnerability to the spread of disturbance. Such information can be obtained through historic reports, paleo-ecological evidence (e.g. charcoal and plant species), photographic comparisons, tree ring analysis of scars and age classes, timber records (over decades), flight line photos, and GIS maps (see also Periman et al., Kenna, and Engstrom et al., this volume). Second, a general rule of thumb is that relatively uniform landscapes are more vulnerable to the rapid and possibly catastrophic spread of disturbance than relatively heterogenous landscapes.

Describing disturbance over a landscape in a cumulative manner requires both an ecological and mathematical approach (Franklin and Forman 1987, Turner and Gardner, 1991). The scale and rate of disturbance regimes are especially important in planning ecological restoration projects (Padley 1996). In restoring the Everglades ecosystem in south Florida, for example, managers have to know the timing, duration, rate, and intensity of inundation during the rainy season to provide just the right flooding regime for nesting of birds: too little too late and some species starve; too much too soon and other species get washed out.

While many native plant and animal communities co-evolved with natural disturbance and many thrive in its presence, human-caused disturbance can create serious problems in the same communities. For example, native grasslands in much of the western United States thrived in the presence of passing herds of native ungulates that may have contained thousands of animals. Today, native grasslands that are chronically and heavily grazed by cattle appear to be the most vulnerable to degradation and irreversible alteration by invasive weedy species (Noss and Cooperrider 1994; Vitousek 1986). Hobbs and Huenneke (1992) show clearly how changing the frequency and intensity of disturbance affects natural diversity and the susceptibility of an ecosystem to invasion by exotics (Fig. 5). As exotic species become established, they pose a growing threat to the composition and integrity of ecosystems (Box 1).

**NATURAL DISTURBANCE REGIMES**
**maintains native species diversity**

**(historical type, frequency, intensity of disturbance)**

| ***Decrease in frequency/ intensity*** | ***Change in type of disturbance*** | ***Increase in frequency/ intensity*** |
|---|---|---|
| Decreased diversity of natives **(dominance of competitively superior species)** | Elimination of natives; Enhancement of invasions **(direct damage to natives, creation of new microsites)** | Elimination of natives; Enhancement of invasions **(direct damage to natives, creation of new microsites)** |

Fig. 5. Any change in the historical disturbance regime of an ecosystem may alter species composition by reducing the importance of native species, by creating opportunities for invasion species or both (Hobbs and Huenneke 1992, Noss and Cooperrider 1994).

Human-caused disturbances are not necessarily better or worse for landscape diversity and ecosystem integrity than natural disturbances. Frequency, intensity, and spatial patterns of disturbance matter more than the source of that disturbance (Turner 1989). Landscape diversity created by clearcutting or suburban development differs from that created by fire or windthrow. If the natural ecosystem was characterized by infrequent fires and widely dispersed tree fall gaps, clearcuts are likely to dramatically alter landscape diversity, ecosystem processes, and ultimately species composition. Sharp-edged and right-angled patches created by dispersed clearcuts contrast sharply with the gradual and meandering edges of natural patch mosaics. Sharp edges create abrupt changes in interior microclimate that shift microclimatic patterns and set the stage for long-term changes in species composition (Thomas et al. 1979, Ranney et al. 1981, Chen et al. 1996). Human disturbances patterned on prevailing natural disturbances will tend to maintain similar landscape diversity patterns and species composition.

Anticipating the cumulative impacts of natural and human disturbance on landscape diversity can be a daunting challenge. Managers, however, are finding practical ways to analyze the cumulative effects of changes in disturbance (see Box 2).

**Box 1**
**Invasive Plants and Wildland Ecosystem Health**

Invasive plants are those noxious or exotic weeds that degrade wildland ecosystem health and productivity. The replacement of a native ecosystem by exotics is often permanent. The rate of replacement is often an important consideration because biodiversity, soil stability, land values, range values, and other important landscape variables are considerably reduced (USFS 1995). Successful weed management comes from understanding how, where, and at what scale these plants become so aggressive here in the United States (Mortensen et al. 1998).

Exotic plants arrived from other countries without the natural enemies that kept them in check in their country of origin (Radosevich and Holt 1984). These plants come to dominate many of our native communities. Along with physical presence, weeds reduce native regeneration, increase erosion, reduce water quality, diminish nutrient cycling (Radosevich and Holt 1984) and often alter the ecosystem of concern so irrevocably that the aesthetic and recreational value of the landscape is markedly reduced (Asher et al. 1997).

On a national scale, exotic plants increased on the Bureau of Land Management Lands from 2.5 million acres in 1985 to approximately 8.5 million acres in 1994. Add in the exotic weed acreage from National Forests, National Parks and Fish and Wildlife Service Reserves and the amount doubles. The infestation rate is 4600 acres per day — "an explosion in slow motion" (Asher et al. 1998). For example, exotic plants can invade a portion of a wilderness area following a fire (e.g. cheatgrass) or riparian area (e.g. tamarisk) following a flood. In a few years, a near monoculture can develop, drastically reducing the diversity of plants and wildlife habitat.

Satellite imagery classification has been used to identify spectral signatures for exotic plant movement. For instance, Fort Benning has identified the spectral imagery for Kudzu an aggressive weed in Georgia by identifying gullies and depositional areas for erosion (Legacy Project 1998). Erosion control management is being developed from this information to prevent the spread of the weed.

On a local scale, prevention, public and employee education, detection, and quick control of new/small infestations are very effective and economical first steps for implementing "integrated weed control". For example, In 1992 yellow starthistle was noticed in the Pueblo mountains in the Burns BLM District of Oregon. Seven hundred plants were immediately pulled. Subsequent handpulling included 2000 plants in 1993 and 6 in 1994 and 30 in 1995. At the same time, on a larger scale, susceptible habitat was identified (over 1.5 million acres).

As weeds know no boundaries, cooperation and commitment among county, state, and federal agencies, landowners, public land users, and conservation organizations is critical. Establishing cooperative weed management areas (CWMAs) is one way to do this (Asher et al. 1998). For instance, the Salmon River CWMA is responsible for weed management of the Salmon River Drainage in Idaho. This group facilitates effective treatment with coordinated efforts along logical geographic boundaries with similar landtypes, use patterns, and problem species. The goal is reduce extent and density of noxious weeds. Using landscape variables, they are able to alleviate local weed problems (Crabtree and Lake 1997).

The concept of integrated weed management requires knowledge of landscape variables, scientific knowledge of weed population ecology, detection and inventory, and the cooperation of landowners and managers to effectively apply sound landscape ecology principles for control of exotic weed movement in wildlands and native ecosystems.

*Source*: Adapted from Asher et al. 1997

## 6.3 Suppressing Disturbance Regimes

In many places, human suppression of disturbance regimes can also have major impacts on landscape diversity. Suppressing disturbance regimes that an ecosystem evolved with can profoundly alter landscape patterns and ecosystem processes. Fire suppression and dune stabilization are two notable examples where humans have suppressed ecosystem disturbances to protect lives and property, but with potentially serious impacts on ecosystem diversity and health.

### *Fire Suppression*

The effects of suppressing natural fire disturbance in National Parks has received wide attention (e.g., Harvey et al. 1980). Human activities outside the parks have altered natural fire regimes within the protected areas. As unnaturally high fuel loads built up, the potential for catastrophic fire increased. As fire disappeared, the species composition of forest types shifted away from those that prevailed for thousands of years prior to fire control. Now, controlled burning is being used in some National Parks to simulate natural fire intensity and periodicity and to reduce the potential for catastrophic fire near human habitation. The role of fire in National Parks was hotly debated in the wake of the 1988 Yellowstone fires, but prescribed and natural fires allowed to burn were recognized as important protected area management tools long before that — starting with the California national parks in 1968 (Stottlemyer 1981).

### *Dune Stabilization*

All along the East coast and the Gulf of Mexico, dunes have been "stabilized" to protect the narrow barrier islands and human dwellings behind them. Like fire,

**Box 2**
**Analyzing the Cumulative Impacts of Disturbance on the Payette National Forest**

In 1994, three wildland fires burned 290,000 acres on the Payette National Forest in Idaho in a variety of forest types from low elevation Ponderosa pine/savannah communities to high elevation spruce-fir (Boudreau and Maus 1996). To anticipate the broad impact of these fires and the management responses that followed, resource managers wanted to characterize pre- and post-fire vegetation and monitor environmental conditions. The goal was to create an information base that took the disturbance events into account when future management decisions are made.

Development of the monitoring system involved several steps. First, a broad-based system to monitor plant succession in the disturbed areas was designed instead of a more limited system focused only on timber conditions. To keep track of this information, the forest ecological staff developed a GIS database for both biological and physical environmental conditions. Next, in cooperation with the U.S. Forest Service Remote Sensing Applications Center in Salt Lake City, Utah, the staff used satellite imagery to develop a pre-and post-fire vegetation classification (Fig. 6). Another layer of information added to the database was a map of fire intensity and severity that was developed from LANDSAT thematic mapper imagery and high elevation photographs provided by NASA. Together, this information provides a baseline to evaluate the change in condition of areas affected by the 1994 fires.

The baseline database and monitoring system will enable managers to better assess the cumulative impacts of the fires, plant succession, and post- disturbance management actions. Some of the potential applications include: (1) tracking large-scale landscape patterns created by the 1994 fires and subsequent management actions; (2) continuous long-term monitoring of sensitive parts of the landscape; (3) detection of changes in plant succession and productivity; (4) identification of optimum sites for permanent vegetation monitoring plots (e.g. rare species) and association of that information with landscape patterns; (5) tracking range of variation in vegetation communities in response to fire severity and post-fire management actions; (6) monitoring post-disturbance tree mortality patterns over time; (7) modelling landslide prone areas over time, and; (8) identifying burn areas most suitable for possible salvage logging. More generally, the database allows forest staff to build a better picture of the structural features of mature and old-growth forest and keep track of changes in wildlife habitat conditions.

The 1994 fires on the Payette National Forest created an opportunity to build a database that allows forest staff to monitor and anticipate changes at the landscape level. Such a database would have been useful even without the extensive disturbance, but crisis affords opportunity and the staff took advantage of an opportunity to create a landscape database and monitoring system that otherwise would not have happened due to budget and personnel constraints. To use this database and monitoring system effectively, forest staff and decision makers have to develop and maintain new skills, especially the ability to interpret satellite imagery (see Correll et al. and Cooperrider in Vol. III of this book) and integrate patterns across time and space.

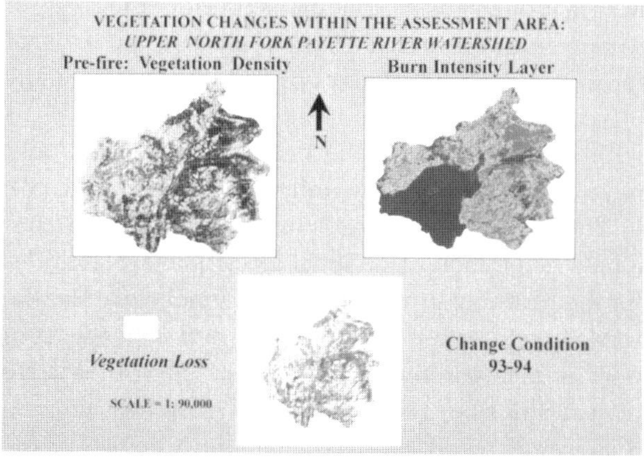


Fig. 6. Satellite imagery used to identify pre-fire vegetation and burn intensity of areas on the Payette National Forest in Idaho (Imagery courtesy the Payette National Forest).

the shifting sands and the occasional storm surge over barrier islands were viewed as destructive disturbances that needed to be controlled if not stopped. Even in natural areas, the National Park Service devoted a great amount of effort in attempts to stabilize coastal dunes on barrier islands (Fig. 7). Bulldozers piled berms of sand, sea oats and other dune grasses were planted, and miles of sand fences were erected to slow the natural inland movement of coastal dunes (Fig. 8).

Botanical and geological research on coastal dune dynamics suggested the control of beach dunes was not always beneficial — especially to native species and ecosystem processes. For example, the prevention of cross-island sediment transport starves soundside marshes of sediment (Dolan et al. 1978). Appreciation of how biotic communities are influenced by natural disturbance regimes grew rapidly in the 1970s and 1980s (Pickett and Thompson 1985, White 1987). It became clear that preventing coastal overwash and stabilizing dynamic dune structures rapidly altered natural communities in unanticipated ways. These processes joined a long list of other natural disturbances that should play an important role in the design and management of nature reserves (Baker 1989, Baker 1992, Shafer 1998a). By the late 1970s, enough was known about the dynamics of natural disturbances on barrier islands to develop a handbook (Leatherman 1979) that would help barrier island managers accommodate change rather than waste time and resources in a futile and misdirected effort to prevent change.

Fig. 7. Dunes at Cape Hatteras National Seashore, North Carolina, in the 1960s that illustrate revegetation with native dune grass with sand fences, an attempt to stop dune migration landward (Photo courtesy US National Park Service).

Fig. 8. A mechanical grass planter at work during the 1960s at Cape Hatteras National Seashore, North Carolina (photo courtesy of US National Park Service).

Starting in the 1973, the National Park Service's policy of dune stabilization was abandoned at Cape Lookout and Cape Hatteras National Seashores, North Carolina (Dolan 1972, Godfrey and Godfrey 1976). It has been replaced with a policy of allowing natural coastal dune processes to proceed unabated (Dolan et al. 1973). Similar policies now apply elsewhere including Cape Cod National Seashore, Massachusetts, Gulf Islands National Seashore, Mississippi, Assateague National Seashore, Maryland, and Padre Island National Seashore, Texas. There is no question, of course, that dune stabilization continues to play a vital role in protecting buildings and infrastructure on barrier islands. For example, historic structures, such as the Cape Hatteras Lighthouse, require extensive measures to protect them from the threat of inland migrating dunes and natural beach erosion. But, where natural features are to be maintained, the end of disturbance is often the beginning of trouble.

## 7 MULTIPLE RESOURCE MANAGEMENT

Most landscapes are managed for a variety of resource uses, usually including some combination of timber, grazing, wildlife and fisheries, recreation, water, and minerals. The ability to display and analyze a variety of resources at the landscape scale enables the resource manager and the public to see where problems and opportunities exist. Assessing the capability of ecological units to provide various resources (e.g., Pfister 1976) can provide a powerful tool for landscape level planning. Rapid growth in remote sensing and GIS technologies is making landscape level multiple resource assessments cheaper, more accurate, and more widely used.

### 7.1 Tools To Assess Multiple Resources

A wide range of tools are available to help managers assess multiple resources at an expanded geographic scale (see Cooperrider et al., Tolle et al., Maddox et al., Lessard et al., and Graham et al., in Vol. III of this book). This section looks at how landscape level, multiple resource assessments are being used by the U.S. Forest Service to develop forest plans. The Forest Service uses integrated inventories and databases to develop alternative forest plans for public review in accordance with the National Forest Management Act. To meet the requirements of NFMA (see also Keiter, in Vol. III of this book), forest planners must integrate a variety of information to ensure that plans consider the range of resources and resource uses that exist on the national forest. Geographic overlays of this information allow planners and the public to evaluate alternative plans with a better sense of the impacts each plan would have on the landscapes within the national forest.

The U.S. Forest Service is now beginning to develop a new generation of national forest plans. New rules for the planning process require planners to assess the capacity of a national forest's land and waters to produce or maintain multiple resources. Each alternative plan can then be evaluated according to how it impacts these resources. To help national forests develop these resource assessments, the U.S. Forest Service developed the National Ecological Hierarchical Framework (ECOMAP 1993). This framework is hierarchical because it delimits areas of different biological and physical resource capacity at any of eight different levels — from units less than 100 acres to regional areas in the tens or hundreds of thousands of square miles (see also Caraher et al., and Haufler et al., this volume). When smaller units are aggregated, the framework builds a biophysical picture of the ecosystem. At each scale, patterns can be detected in a number of ecological characteristics that may be related to resource capacity — from plant communities, soils, hydrologic functions, landform and topography to lithology, climate, air quality, energy and nutrient cycling. These variables can then be linked to aquatic and socio-economic frameworks to help determine possible management alternatives (see also Gosz, this volume).

In the Kisatchie National Forest in Louisiana, forest staff turned to the Nature Conservancy and the Louisiana Wildlife and Fisheries Department to help it simultaneously consider timber, endangered species, and other resources in its planning process. Together they developed an ecological classification (see also Grossman et al., and Carpenter et al., this volume) on land-type association (LTA). This classification uses geology, landform, climate, and other factors to identify potential natural vegetation — that is, the type of plant community that would be expected to grow there under natural conditions — at a landscape level. As in other parts of the south, conservation of the endangered red-cockaded woodpecker (*Picoides borealis*) has become a major management issue on the Kisatchie National Forest (Fig. 9). The landscape ecological classification enabled forest staff to identify longleaf pine LTAs that are prime habitat for the woodpecker. Based on this information, two red-cockaded woodpecker

Fig. 9. Red-cockaded woodpecker (*Picoides borealis*) in longleaf pine forest, a habitat which is necessary for the viability of the species (drawing courtesy the U.S. Fish and Wildlife Service).

reserve areas were established. This also facilitated timber resource management since it helped clarify which areas were sensitive habitat areas and which were not. Fort Benning in Georgia has run a similar analysis (Fig. 10) with parallel results to identify red-cockaded woodpecker habitat through remote sensing with spectral signatures for short- and longleaf pine.

One of the key goals for any national forest planning process is to develop management alternatives that can be readily understood by the public. In fact, most ecological classifications, and the resource assessments and management alternatives based on them, are

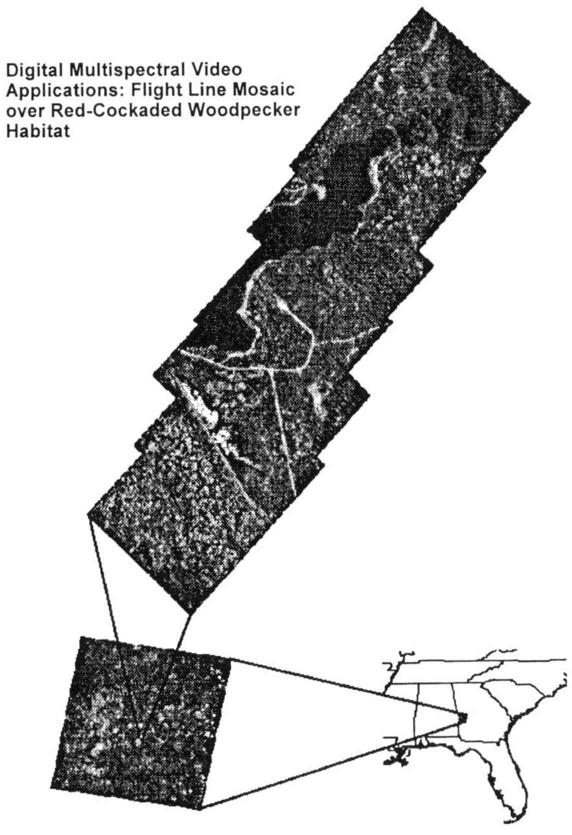
Digital Multispectral Video
Applications: Flight Line Mosaic
over Red-Cockaded Woodpecker
Habitat

Fig. 10. Remote sensing of a red-cockaded woodpecker habitat project considered from several different scales (imagery courtesy the Legacy Project).

difficult for both the public and resource managers to understand. Managers unfamiliar with the basic principles of an ecological classification system used in their forest's planning process cannot adequately explain to the public the basis for management alternatives. A confused public and a frustrated manager are often the result, increasing the likelihood that the planning process will take longer and generate greater disagreement than might otherwise have been the case.

On the Monongahela National Forest in West Virginia, staff ecologists, planners, and district rangers collaborated to develop a simplified ecological classification that would facilitate public understanding of forest management plans and decisions facing resource managers. In this case, the forest staff simplified a relatively complex ecological classification system with 26 LTAs into four distinct "ecozones" that encompass the entire national forest. The underlying LTAs are maintained for detailed planning and assessment, but the Ecozone Concept is used widely for a variety of planning and public information purposes. For example, there are several LTAs mapped at higher elevations on the Monongahela that share two distinctive characteristics — red spruce (*Picea rubens*) and frigid soils. The Ecozone Concept lumps these LTAs into one category called the "greater spruce zone". The public readily relates to this classification since it is similar to what they, their parents and grandparents, and local lore have called the high country for generations. While ecologists and resource managers might appreciate the distinctions between LTAs in the "greater spruce zone", they too find the Ecozone Concept useful for planning purposes. For example, the range of two of West Virginia's rarer forest species — the Virginia Northern Flying Squirrel (*Glaucomys sabrinus fuscus*) and the Cheat Mountain Salamander (*Plethodon nettingi*) — largely coincides with the greater spruce ecozone. The lesson here is that complex ecological classifications can be adapted to common experience and understanding. The result will be better public understanding of the choices that they and their public resource management agencies need to make. Landscape assessments — no matter how technically sound — must be easily understood by the public and resource managers if they are to have the intended effect. If done correctly, multiple resource assessments can become a powerful tool for compromise and negotiation.

## 8 DETECTING CHANGE OVER TIME

Different ecosystem elements require different sampling scales and times. If landscapes are sampled during the wrong time or over too narrow a time period, critical patterns may be missed (Morrison and Marcot 1995). Unfortunately, the time element in landscape variation is frequently overlooked. For example, the inventories used by many resource management agencies include organisms and ecosystem structure but do not capture ecosystem processes that operate over time as well as space.

The role of habitat fragmentation in the decline of some species is a case in point. Habitat fragmentation can reduce the population viability of species that require interior zones of their preferred habitats (Morrison et al. 1992). The relationship between habitat fragmentation and population viability is perhaps best known for some neotropical migratory bird species. Biologists, however, were able to associate habitat fragmentation with population declines in some neotropical migrant bird species only because they were able to link three decades of the Breeding Bird Survey (1966–1996) with changes in habitat conditions (Robinson 1996). A static snapshot of either variable would not have allowed researchers and managers to detect and manage for this problem. The longer projects are monitored, the more valuable the data become to successive managers.

Starting in 1994, the Monongahela National Forest in West Virginia has quantified forest fragmentation for large administrative areas (i.e. Opportunity Area) in the 1986 Forest Plan. Habitat fragmentation is analyzed with FRAGSTATS, a dynamic landscape fragmentation analysis program developed at Oregon State University (McGarigal and Marks 1995). When a timber cut is proposed within the administrative area, the use of FRAGSTATS allows forest staff to evaluate a range of management alternatives in an environmental assessment (EA) that helps determine whether and how to go forward with timber harvesting and provide a cumulative effects analysis for each successive action on the area (Fig. 11). Ideally, as Morrison et al. (1992) suggest, monitoring objectives for habitat fragmentation should include prevailing policies, land allocations, land use, environmental conditions, past management activities, and successional trends in the both within the target area and on surrounding ownerships or administrative areas.

## 9 CROSS-BOUNDARY CONCERNS AND THE VALUE OF A REGIONAL PERSPECTIVE

No management area — even one as large as Yellowstone National Park — is entirely self-contained when it comes to ecosystem functions and processes. Maintaining those functions and processes will be at least partially dependent on what happens on lands outside the management area. One of the most famous images

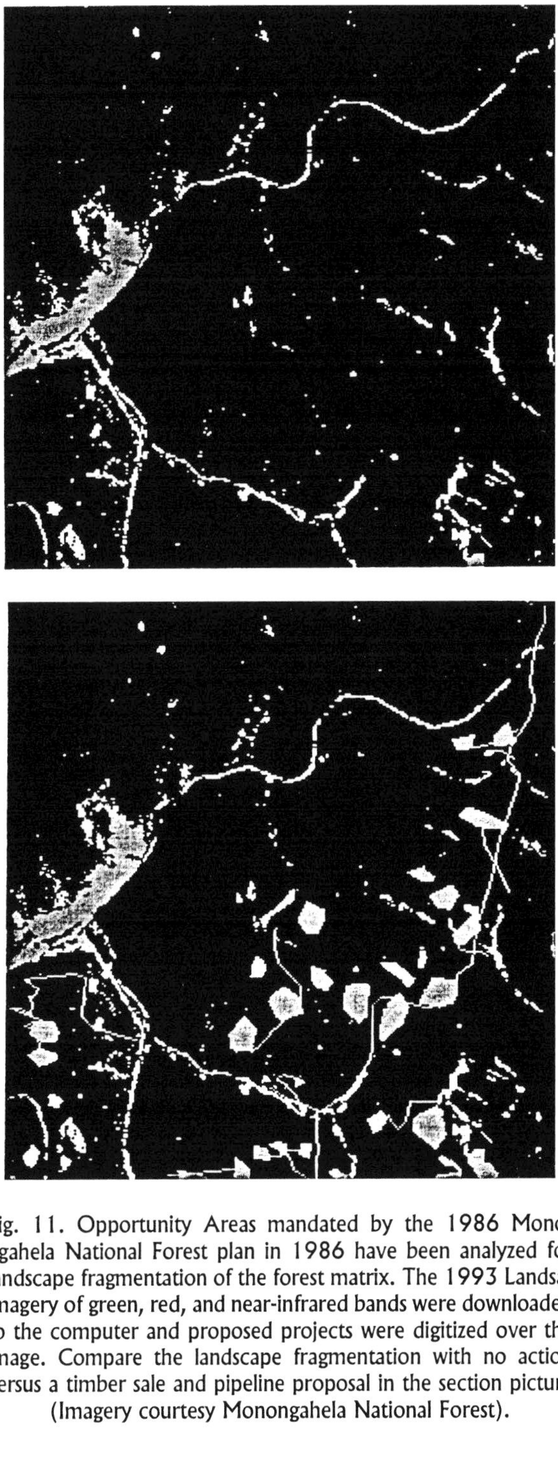

Fig. 11. Opportunity Areas mandated by the 1986 Monongahela National Forest plan in 1986 have been analyzed for landscape fragmentation of the forest matrix. The 1993 Landsat imagery of green, red, and near-infrared bands were downloaded to the computer and proposed projects were digitized over the image. Compare the landscape fragmentation with no action versus a timber sale and pipeline proposal in the section picture (Imagery courtesy Monongahela National Forest).

of this challenge is provided by the sharp contrast in forest conditions along the boundary between Yellowstone National Park and Gallatin National Forest, Idaho (Fig. 12). Clearly, the National Park Service and the U.S. Forest Service had very different management goals on the same landscape. Organisms and habitats, however, do not usually recognize the same boundaries that resource managers do. If conserving those species or habitats are within the manager's responsibilities, there is no substitute for good knowledge of the location and distribution of species and ecosystem diversity at a regional scale.

There is no question that satellite imagery, aerial photographs, and GIS technologies have enabled us to see more clearly the problems political and administrative boundaries can pose for biodiversity conservation. For example, the threat of dams both within and outside U.S. national parks first surfaced over 70 years ago in the wake of the Hetch Hetchy Dam's controversial impacts on Yosemite National Park (Cameron 1922). However, the issue did not really capture the general public's attention until publication of the "State of the Parks Report 1980" (NPS 1980). Today, it's widely recognized that habitats and species within protected areas can be profoundly affected by what happens outside of their boundaries. For example, air and water pollution generated hundreds of miles away can threaten park resources. And, most protected areas are too small to retain viable populations of some large mammals (especially carnivores) because human land uses restrict their movements and their ability to find prey, mates, or other resources needed to survive. The work of various scientists, most notably Schonewald-Cox and Bayless (1986), has helped establish the effects of protected area "boundary processes" in terms of maintaining ecosystem processes (see also Kushlan 1979) and viable wildlife populations (see also Salwasser et al. 1987).

Today, a "regional landscape" perspective has become indispensable to conserving biodiversity (e.g., Western and Pearl 1989). Three examples — from Chesapeake Bay, Acadia National Park, and San Diego County, California — illustrate the value of managing within a regional context.

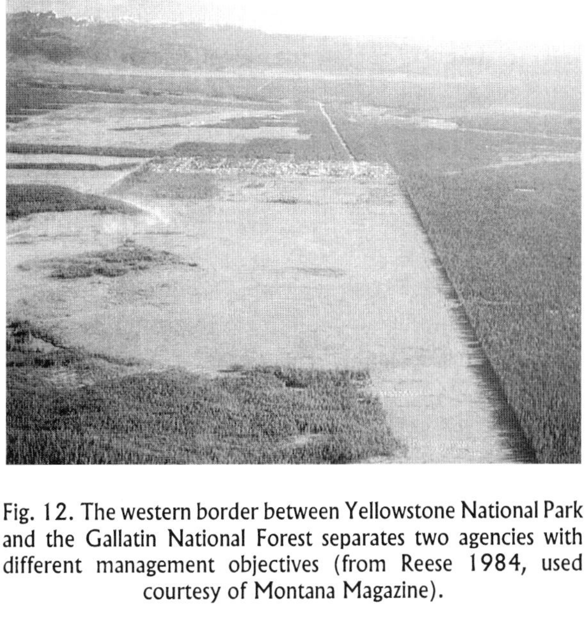

Fig. 12. The western border between Yellowstone National Park and the Gallatin National Forest separates two agencies with different management objectives (from Reese 1984, used courtesy of Montana Magazine).

## 9.1 Chesapeake Bay

The Chesapeake Bay is the largest estuary in the United States with a drainage basin of 64,000 square miles encompassing parts of Maryland, Virginia, West Virginia, Pennsylvania and New York — an area twenty-five times the size of the Bay itself. More than 40 significant rivers and thousands of streams drain into it— from central New York state in the north, from nearly to the North Carolina border to the south, and from the heart of West Virginia to the west (Horton 1993). Besides the Bay itself, the watershed includes marshes, swamps, streams, ponds, lakes, woodlands, farmlands and distant forested mountains (Fig. 13). The watershed also includes towns, cities, highways, reservoirs, golf courses, factories, military bases and a human population of over 20 million. For centuries, the Chesapeake has been widely known for its bountiful fish, oysters, crabs, waterfowl, shorebirds, and other species — many of them vital economic mainstays for hundreds of towns and cities along its shores. As a vital breeding and stopover habitat, a healthy Bay is also vital to thousands of migratory species including birds, fish, turtles, and a multitude of invertebrate species.

With so many people and so many intensive land uses within the watershed, it's no surprise that industrial pollution, sedimentation, over-fishing, over-hunting, and habitat fragmentation adversely impact the Bay's fisheries and wildlife. Industrial and municipal pollution from towns and cities and nutrient and sediment runoff from farms and development projects flow into the estuary, sometimes from sources hundreds of miles upstream. In short, the Chesapeake Bay is the classic example of a public "commons" (Hardin 1968) where distance diminishes responsibility for actions that degrade a resource that many depend on. Not only are many actors, such as factories, farmers, and city dwellers, remote from the Bay, but the governance of the watershed is fragmented between five states, several federal agencies, hundreds of counties, and thousands of townships, towns and cities. By the 1980s, the oyster fishery had collapsed, striped bass had nearly disappeared, and even the ubiquitous blue crab seemed to be in trouble. By then it was clear that isolated actions within the immediate confines of the Bay and its shoreline would do little to stem the Bay's accelerating environmental decline.

In 1983, the Federal government, Virginia, Maryland, Pennsylvania, the District of Columbia, mutually recognized their responsibilities to work cooperatively to "save the bay" and created the Chesapeake Bay Commission. The Commission catalyzed efforts by state and federal agencies to assess and monitor landscape diversity across the entire watershed with a focus on wetlands, submerged vegetation, and forests. Coordinated limits were set on commercial fishing, voluntary programs were established to encourage farmers to limit nutrient inflow, tributary streams were restored to allow annual fish migrations to move upstream, new wildlife refuges and habitat conservation areas were created, and sewage treatment plants were upgraded throughout the watershed (Rome 1991). By the early 1990s, this regional approach to conservation was beginning to show results: stripped bass populations rebounded, oyster populations started to slowly recover, and some nutrients declined and water quality improved (Horton 1993).

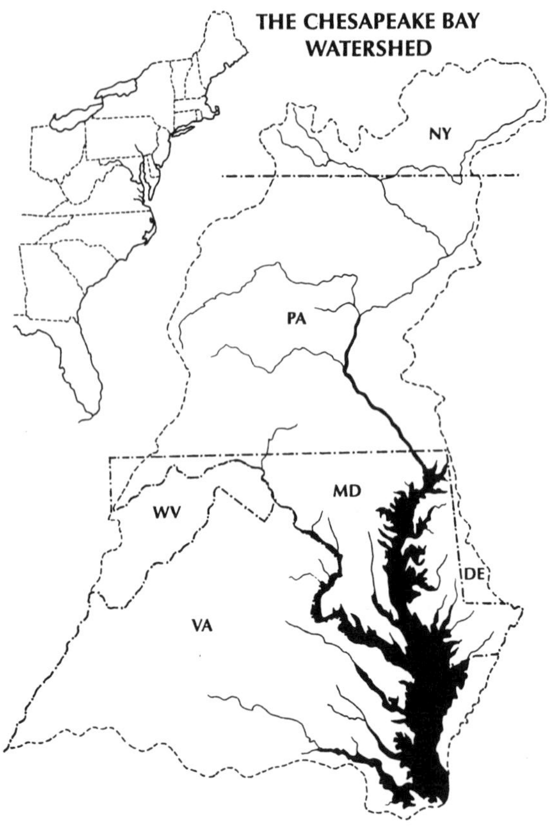


Fig. 13. The greater area of the Chesapeake Bay Project contains many large watersheds as well as surrounding urban areas (from Reshetiloff 1995).

In this case, reaching across political and ownership boundaries has worked on a grand scale to improve habitat conditions. Still, much remains to be done. An outbreak of the deadly *Pfisteria pescida* microbe in 1997 has been tentatively linked to unregulated nutrient runoff from rapidly expanding factory chicken farms from West Virginia to Delaware. Maryland's legislature has restricted development within 1000 feet of the shoreline, but such restrictions do not apply to nearby tributaries and rivers and Virginia has not regulated shoreline development at all. Meanwhile, thousands of people move into the watershed every year. If the Chesapeake is to show continued improvement, Chesapeake Bay Commission's efforts to coordinate the monitoring of the watershed's ecosystems at a regional landscape level must continue (Pendleton 1995). Just as important, expanded investments in education programs, new partnerships between government agencies and civic groups, and innovative voluntary and regulatory programs will be needed to keep up with social and economic changes in the watershed.

## 9.2 Acadia National Park

Acadia National Park on the Coast of Maine is one of most heavily visited national parks in the United States. It is also one of the smallest major national parks, and — like so many other parks — Acadia is increasingly surrounded by recreational and tourism development. If there was ever a good National Park candidate for a "buffer zone" to shield its natural features from adjacent development, Acadia is it. Unlike the western national parks where adjacent owners are often public agencies, Acadia, however, is surrounded by extremely expensive private land. Buying land or negotiating easements to buffer the park from adjacent development is an expensive proposition. To make the most of the money that is available for creating a buffer around Acadia demands considerable knowledge of the landscape diversity surrounding the park.

Since 1970, the Maine Coast Heritage Trust, a Brunswick-based non-governmental organization, has been securing easements around the park on Mount Desert Island and smaller nearby islands (Endicott 1993). In this case, a private conservation organization, in cooperation with the National Park Service and enabled by state legislation, has designed and implemented a buffer zone system around a publicly owned protected area. From the beginning, the Maine Coast Heritage Trust has gathered information on the environmental features of land near the park so it can identify priorities for making the most of its limited resources. Nearly thirty years after it began, the Maine Coast Heritage Trust continues to collect new information to build an increasingly sophisticated overview of the landscape of which Acadia is but a part.

The Acadia buffer zone is typical of buffer zone design and the importance of knowing landscape patterns to ensure the buffer adequately "filters" unwanted influences before they reach the boundaries of the park. The goal of buffers is to minimize the impact of surrounding human activities on the biota and ecosystems within a protected area (see Shafer 1998b for more details). A properly designed buffer zone can filter some, but not all, undesirable external influences. Buffer zones can be effective in controlling such problems as poaching, collecting, and harassment of wild species, incursion of exotic and domesticated animals, fragmentation of adjacent natural habitats, loss of migratory corridors to nearby habitats and protected areas, and protecting hydrological systems by controlling local wetland drainage, stream diversion, water pollution, and sedimentation. Buffer zones, on the other hand, may have little ability to protect against air pollution, many exotic species invasions (particularly weedy species), upstream dams, and upstream water pollution and nutrient loading.

A variety of land protection methods can be used to create buffer zones. These include outright land purchase (fee simple), conservation easements, zoning, land swaps, tax incentives for certain land management practices, and voluntary partnerships with adjacent landowners and civic groups (The Conservation Foundation 1985) (Brown, in Vol. III of this book). A buffer zone composed of these components may in turn be surrounded by a less restrictive "zone of cooperation," such as those described in the idealized model of a biosphere reserve (Gregg and McGean 1985). Buffer zones and outer "zones of cooperation" may rely entirely on voluntary cooperation, which is the general approach used most often by the National Park Service (National Park Service 1988). However, as habitat encroachment begins to abut park boundaries and land values increase, it may be advisable to use more secure forms of buffer protection such as easements and land purchases that are being used around Acadia National Park. In some areas, buffer zones can be designed to connect with habitat corridors that link natural habitats (Noss and Harris 1986).

## 9.3 San Diego and Los Angeles Counties

We usually think of a regional perspective in the context of relatively natural landscapes. But, semi-natural areas, even within rapidly growing urban and suburban areas, can play a vital role in biodiversity conservation. A regional perspective can be just as useful in this setting. San Diego County and Los Angeles Counties, California are two places that not only feature a large and rapidly growing human population, but still hosts pockets of habitat that are the last stronghold for a number of endangered species. During the past decade, San Diego County, the state of California, the U.S. Fish and Wildlife Service, and several conservation groups have mounted one of the most detailed nature reserve planning efforts anywhere in the United States (Boucher 1995; Holing 1997). Several studies have indicated that habitat

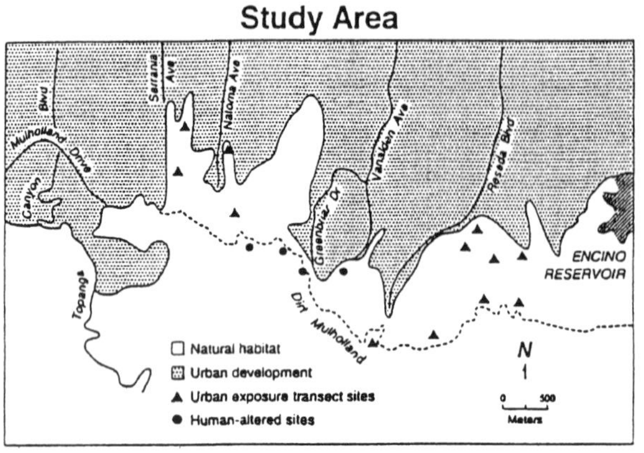


Fig. 14. Natural areas such as Topanga State Park of California require a buffering strategy from the surrounding urban encroachment. Subsequent habitat alternation from the encroachment has a direct negative impact on wildlife. Areas outside as well as within the city limit which could potentially serve as habitat corridors for landscapes fragmented by development (from Sauvajo and Buechner 1993).

alteration from urban encroachment has a direct impact on wildlife (e.g. Sauvajot and Buechner 1993) (Fig. 14). A regional perspective to provide reserves and a buffering capacity has been central to this ambitious program.

For instance, planners approached the challenge of biodiversity conservation in San Diego County by thinking of entire existing landscape in the southwestern part of the county as a *de facto* reserve system. They used a gap analysis methodology to survey the entire 235,387 ha area and identify what set of core reserves, habitat corridors, and buffer zones would do the most for their conservation objectives. Rather than focus on traditional rigidly bounded nature reserves, the San Diego County Plan (City of San Diego 1995) seeks to use an innovative combination of zoning, tax incentives, and other measures to protect key habitats (Reid and Murphy 1995, Manson 1994, Davis et al. 1995). In addition to natural undeveloped habitats, the planning process also examined how existing land uses such as water reservoirs, golf courses, large estates, and even cemeteries, could help meet conservation objectives. These small habitat patches can serve as habitat for naturally fragmented metapopulations, as stepping stones to other larger habitat patches, as sources of propagules for a species reintroduction elsewhere, and as safety nets in case of catastrophic disturbances in other areas (Shafer 1995). A growing focus on urban ecology in the last two decades demonstrates the value of using a regional perspective and integrating ecological concepts into urban reserve design (Gilbert 1991, Adams 1994, Adams and Dove 1989). For a review of designing reserves at the urban/rural interface, see Shafer (1998b).

## 10 CONCLUSIONS

Several themes are consistently woven throughout our discussion on the use of landscape approaches in natural resource management. These themes include the following:

1. Landscape ecology, which is the description and analysis of biotic and environmental landscape patterns, is vital to understanding ecosystem process and function. Many important ecosystem processes, such as natural fire disturbance regimes, operate at a broad geographic scale and cannot be accurately characterized unless patterns are examined at a landscape or regional scale ranging from thousands to millions of hectares.

2. There is no one correct scale for conducting a landscape analysis. The analysis should be dictated by the scale of the problem that managers are seeking to address. The scale used to address the conservation of a wide ranging mammal such as a wolverine will be different than the scale used to protect the habitat of a localized endemic plant. Determining which scale is most appropriate will depend in part on knowledge of relevant spatial and temporal patterns in a landscape. Depending on the hierarchy of time and space, a manager can make decisions at any scale, or move up and down scales as needed.

3. Managers will encounter barriers to analyzing and acting at a landscape level. Barriers may be technical, financial, legal, or political. Such barriers can be avoided or overcome, but they should not be ignored. It is important to acknowledge them when defining the analysis or management objectives for a landscape.

4. Landscape approaches are especially important for carrying out biodiversity and wildlife habitat conservation objectives. Habitat conservation strategies need to account for ecosystem patterns that change in space and time. Landscape or regional level analysis is vital for planning and maintaining a network of representative ecosystem reserves.

5. Ecosystems change over time and so do the disturbances that affect them. As a result, landscape level analysis, such as those conducted to characterize habitat fragmentation patterns, should be designed to detect cumulative changes.

6. In many cases, the development of a landscape level database will initially serve to provide benchmarks by which to measure and manage future

change. Landscape level monitoring and database development should be viewed as an ongoing process, rather than a one time event.

7. To make the most of landscape level information, managers need to have basic knowledge about landscape concepts and the skills to use the technologies (or at least understand how they work) that are vital to landscape analysis — especially GIS, remote sensing, and computer modeling.
8. Broad regional perspectives are indispensable to conservation and sustainable natural resource management in many situations. Experience from such efforts as the Chesapeake Bay Program show that success depends not just on the right technical approach and set of information, but on partnerships with a wide range of neighbors including other government agencies, landowners, and civic and conservation groups.

Finally, we urge resource managers to think at multiple scales on the landscape. This is no mere academic exercise. Moving from the site/unit/park level to the larger landscape level is the only way to deal with many critical issues facing managers today. Habitat fragmentation, maintaining viable populations of endangered species, managing watersheds for a range of conservation and resource objectives all demand an appreciation of how ecosystems and their components interact across the landscape. Landscape approaches to natural resources management not only requires new tools and techniques, but an expanded state of mind that encompasses large areas, new partners, and continual learning.

## REFERENCES


Adams, L.W. and L.E. Dove. 1989. *Wildlife Reserves and Corridors in he Urban Environment*. National Institute for Urban Wildlife, Columbia, Maryland.

Adams, L.W. 1994. *Urban Wildlife: A Landscape Perspective*. The University of Minnesota Press, Minneapolis.

Asher, J., S. Dewey, C. Johnson, J. Oliveraz, and R. Waller. 1997. Invasive exotic plants destroy wildland ecosystem health. Unpublished case study for Ecological Stewardship Workshop, Tucson, AZ.

Austin, M.P. and C.R. Margules. 1986. Assessing representativeness. pp. 45–67. In: M.B. Usher (ed.). *Wildlife Conservation Evaluations*, Chapman and Hall Ltd. London.

Bailey, R.G. 1996. *Ecosystem Geography*. Springer-Verlag, New York.

Baker, W.L. 1989. Landscape ecology and nature reserve design in the Boundary Waters Canoe Area, Minnesota. *Ecology* 70: 23–35.

Baker W.L. 1992. The landscape ecology of large disturbances in the design and management of nature reserves. *Landscape Ecology* 7: 181–194.

Bedward, M., R.L. Pressey, and D.A. Keith. 1992. A new approach for selection fully representative reserve networks: addressing efficiency, reserve design, and land suitability with an iterative analysis. *Biological Conservation* 62: 115–125.

Boucher, N. 1995. Species of the sprawl. *Wilderness* 58: 11–24.

Boudreau S.L. and P. Maus, 1996. An ecological approach to assess vegetation change after large scale fires on the Payette National Forest. In: *Proceedings of the Sixth Forest Service Remote Sensing Applications Conference, Remote Sensing: People in Partnerships with Technology*. American Society of Photogrammetry, Bethesda, MD. pp. 330–336.

Bougeron, P.S, H.C. Humphries, and M.E. Jensen. 1993. General sampling design considerations for landscape evaluations. pp. 119–130. In: M.E. Jensen and P.S. Bourgeron (eds.), *Eastside Forest Ecosystem Health Assessment — Volume II: Ecosystem Management Principles and Applications*. USDA Forest Service, Pacific Northwest
Research Station, Portland, OR.

Brooks, B.L. and S.A. Temple. 1990. Habitat availability and suitability for Loggerheads Shrikes in the upper Midwest. *American Midland Naturalist* 123: 75–83.

Cameron, J. 1922. *The National Park Service: Its History, Activities and Organization*. Appleton, New York.

Chen, J., J.F. Franklin, and T.A. Spies. 1995. Growing season microclimatic gradients extending into old-growth Douglas-fir forests from clearcut edges. *Ecological Applications* 5(1): 74–86.

Chen, J., J.F. Franklin, and J.S. Lowe. 1996. Comparison of abiotic and structurally defined patch patterns in a hypothetical forest landscape. *Conservation Biology* 10(3): 854–862.

City of San Diego. 1995. Multiple Species Conservation Program: MSCP Plan Executive Summary. City of San Diego, San Diego, California (draft memo).

Concannon, J.A. J.F. Franklin, and P.B. Alaback. 1992. Characterizing microclimate and forest structure along natural and newly-logged forest edges in southeast Alaska. Abstract in 77th Annual Ecological Society of America Meeting, Honolulu, Hawaii, Vol. 73:2 p. 145.

Concannon, J.A. 1995. Characterizing Structure, Microclimate and Decomposition of Peatland, Beachfront, and Newly-logged Forest Edges in Southeastern Alaska. Ph.D. Thesis. College of Forest Resources, Seattle, WA.

Davis, F.W., P.S. Stine, D.M. Stoms, M.I. Borchert, and A.D. Hollander. 1995. Gap analysis of the actual vegetation of California 1. The southwestern region. *Madrono* 42: 40–78.

DeVelice, R.L., G.J. Daumiller, P.S. Bourgeron, and J.O. Jarvie. 1994. Bioenvironmental representativeness of nature preserves: assessment using a combination of a GIS and a rule-based model. pp. 131–138. In: D.G. Despain (ed.), Plants and Their Environments:Proceedings of the First Biennial Scientific Conference the Greater Yellowstone Ecosystem. Technical Report NPS/NRYELL/NRTR-93 USDI National Park Service, Denver, CO.

DeVelice, R.L. and R.H. Hagenstein. 1995. Assessing the representativeness of ecological survey data and protected areas using a bioclimatic model. Final Report to USDA Forest Service, Pacific Northwest Forest and Range Experiment Station, Corvallis, OR. Dated March 15, 1995. On file at Chugach National Forest, Anchorage, AK.

Dolan, R. 1972. Barrier dune system along the Outer Banks of North Carolina. *Science* 176: 286–288.

Dolan, R., B. Hayden, J. Fisher, and P. Godfrey. 1973. A Strategy for Management of Marine and Lake Systems Within the National Park System. U.S. Department of the Interior, National Park Service, Dune Stabilization Study, Natural Science Report No.6. 40 pp.

Dolan, R. and B. Hayden. 1974. Management of highly dynamic coastal areas of the National Park Service. *Coastal Zone Management Journal* 1: 133–139.

Dolan, R., B.P. Hayden and G. Soucie. 1978. Environmental dynamics and resource management in the U.S. national parks. *Environmental Management* 2: 249–258.

ECOMAP. 1993. National hierarchical framework of ecological units. Unpublished administrative paper. USDA Forest Service, Washington D.C., 21 pp.

Endicott, E. and Contributors. 1993. Local partnerships with government. In: E. Endicott (ed.), *Land Conservation Through Public/Private Partnerships*, pp. 195–218. Island Press, Washington, D.C.

Engelking, L.D., H.C. Humphries, M.S. Reid, R.L. DeVelice, E.H. Muldavin, and P.S. Bourgeron. 1993. Regional conservation strategies: assessing the value of conservation areas at regional scales. pp. 219–235. In: M.E. Jensen and P.S. Bourgeron, Eds. *Eastside Forest Ecosystem Health Assessment—Volume II: Ecosystem Management: Principles and Applications*. USDA Forest Service, Pacific Northwest Research Station, Portland, Oregon.

Forman, R.T.T. 1995. *Land Mosaics:The Ecology Of Landscapes and Regions*. Cambridge University Press, New York. p. 13.

Forman, R.T.T. and M. Godron. 1986. *Landscape Ecology*. Wiley, New York.

Franklin, J.F. and R.T.T. Forman, 1987. Creating landscape patterns by forest consequences and principles. *Landscape Ecology* 1: 5–18.

Franklin, J.F. 1993. Preserving biodiversity: species, ecosystem, or landscapes. *Ecological Applications* 3: 202–205.

Geiger, 1959. *The Climate Near the Ground*. Harvard University Press. Cambridge, Mass.

Gilbert, O.L. 1991. *The Ecology of Urban Habitats*. Chapman and Hall, New York.

Godfrey, P.J. and M. M. Godfrey. 1976. *Barrier Island Ecology of Cape Lookout National Seashore and Vicinity, North Carolina*. National Park Service Scientific Monograph Series No. 9. U.S. Government Printing Office, Washington, D.C.

Gregg, W.P. and B.A. McGean. 1985. Biosphere reserves: their history and promise. *Orion* 4: 40–51.

Hardin, G. 1968. The tragedy of the commons. *Science* 162: 1243–1248.

Hobbs, R.J. and L.F. Huennecke. 1992. Disturbance, diversity, and invasion: implications for conservation. *Conservation Biology* 6(3): 324–337.

Holing, D. 1997. The coast sage scrub solution. *Nature Conservancy* 47: 16–24.

Horton, T. 1993. Chesapeake Bay: hanging in the balance. *National Geographic* 183: 2–35.

Huston, M.A. 1994. *Biological Diversity: the Coexistence of Species on Changing Landscapes*. Cambridge University Press, Cambridge.

Hunter, M.L., Jr. 1996. *Fundamentals of Conservation Biology*. Blackwell Science, Cambridge, MA, 482 pp.

Interactive Associates Inc. 1986. *Mastering Meetings For Results:The Interaction Methods, Changing the Way the World Meets*. San Francisco, CA.

Johnson, N.C. 1995. *Biodiversity in the Balance: Approaches to Setting Geographic Conservation Priorities*. Biodiversity Support Program, Washington, DC.

Kareiva, P. 1994. Diversity begets productivity. *Nature 368* (21): 686.

Kaufman, M.R.Graham, R.T., D.A. Jr. Boyce, W.H. Moir, L. Perry, R.T. Reynolds, R.L. Bassett, P. Mehlop, C.B. Edminster, W.M. Block, P.S. Corn. 1994. An Ecological Basis for Ecosystem Management. Gen. Tech. Rep. RM-246. Fort Collins, CO, U.S. Dept. of Ag., Forest Service, Rocky Mountain Forest and Range Exper. Station.

King, A.W., W.R. Emmanuel, and R.V. O'Neil. 1990. Linking mechanistic models of tree physiology with models of forest dynamics:problems of temporal scale. In: R.K. Dixon, R.S. Meldahl, G.A. Ruark, and W.G. Warren (eds.), *Process Modeling of Forest Growth Responses to Environmental Stress*, pp. 241–248. Timber Press, Portland, Oregon.

Kushlan, J.A. 1979. Design and management of continental wildlife reserves: lessons from the Everglades. *Biological Conservation* 15: 281–290.

Leatherman, S.P. 1979. *Barrier Island Handbook*. National Park Service Cooperative Research Unit, The Environmental Institute, University of Massachusetts, Amherst.

Legacy Project, 1998. Legacy Project: HyperSpectral Data and Identification of Red-cockaded Woodpecker Habitat. Internet topic. K. Slocum, kslocum@curly.tec.army.mil (last revision 8-21-96).

Mackey, B.G., H.A. Nix, J.A. Stein, S.E. Cork, and F.T. Bullen. 1989. Assessing the representativeness of the wet tropics of Queensland world heritage property. *Biological Conservation* 50: 279–303.

Manson, C. 1994. Natural communities conservation planning: California's new ecosystem approach to biodiversity. *Environmental Law* 24: 603–615.

McGarigal K. and B. Marks. 1995. Fragstats: Spatial Pattern Analysis Program for Quantifying Landscape Structure. USDA For. Serv. GTR PNW-GTR-351, Portland, OR, 122 pp.

McIntosh, R.P. 1985. *The Background of Ecology*. Cambridge University Press, Cambridge, UK.

Morrison, M.L. and B.G. Marcot. 1995. The evaluations of resource inventory and monitoring program used in National Forest planning. *Environmental Management* 19(1): 147–156.

Morrison, M.L., B.G. Marcot, and R.W. Mannan. 1992. *Wildlife–Habitat Relationships, Concepts and Applications*. University of Wisconsin Press. Madison, Wisconsin.

Morrison, M.L. and B.G. Marcot. 1995. An evaluation of resource inventory and monitoring program used in National Forest planning. *Environment Management* 19: 147–156.

Mortensen, P.A., L.G. Highley, J.A. Dieleman, and J.L. Lindquist. 1998. Ecological principles underlying integrated weed management systems. Abstracts: Integrated Weed Management, Weed Science Society of America Annual Meeting Abstract. Chicago, IL.

National Park Service. 1980. *State of the Parks–1980: A Report to the Congress*. U.S. Department of the Interior, Washington, D.C.

National Park Service. 1988. *Management Policies*. National Park Service. U.S. Department of Interior, Washington D.C.

National Park Service. 1996. Damaged and Threatened National Natural Landmarks 1995. National Park Service, U.S. Department of the Interior, Washington D.C. (Unpublished report to the U.S. Congress).

Noss, R.F. 1983. A regional landscape approach to maintain diversity. *BioScience* 33: 700–706.

Noss, R.F. 1987. From plant communities to landscapes in conservation inventories: a look at The Nature Conservancy (USA). *Biological Conservation* 41: 11–37.

Noss. R.F. 1990. Indicators for monitoring biodiversity: A Hierarchical approach. *Conservation Biology* 4: 355–364.

Noss, R.F. 1992. The Wildlands Project: Land conservation strategy. *Wild Earth* (Special Issue): 10–25.

Noss, R.F. 1995. Maintaining ecological integrity in representative reserve networks. Report to the World Wildlife Fund.

Noss, R.F. and A.Y. Cooperrider. 1994. *Saving Nature's Legacy: Protecting and Restoring Biodiversity*. Island Press. 416 pp.

Noss R.F. and L.D. Harris. 1986. Nodes, networks, and MUMs: Preserving diversity at all scales. *Environmental Management* 10: 299–309.

Oliver, C.D. and B.C. Larson 1990. *Forest Stand Dynamics*. McGraw-Hill, New York. 295 pp.

O'Neill, R.V. B.T. Milne, M.G. Turner, and R.H. Gardner. 1988. Resource utilization scales and landscape pattern. *Landscape Ecology* 2: 63–69.

Padley, E. 1996. Great Lakes Assessment Case Study. Ecological Stewardship Abstract, U.S. Forest Service Region 9 Office, Milwaukee, WI.

Patton, D.R. 1975. A diversity index for quantifying habitat edge. *Wildlife Society Bulletin* 3: 171–73.

Pendleton, E. 1995. Natural resources in the Chesapeake Bay watershed. In: E.T. LaRoe, G.S. Farris, C.E. Puckett, P.D. Doran and M.J. Mac (eds.), Our Living Resources: A Report to the Nation on the Distribution, Abundance, and Health of U.S. Plants, Animals, and Ecosystems, pp. 263–267. U.S. Government Printing Office, Washington, D.C.

Perry, A.D. 1994. *Forest Ecosystems*. The Johns Hopkins University Press, Baltimore and London. p. 649.

Perry A.D. and M.P. Amaranthus, 1996. Disturbance recovery, and stability. In: K.A. Kohn and J.F. Franklin (eds.). *Creating a Forestry for the 21st Century*. Island Press, Washington D.C. pp. 31–56.

Pfister, R.D. 1976. Land capability assessment by habitat types. In: *America's Renewable Resource Potential—1975: The Turning Point*, pp. 312–325. Proceedings of the 1975 National Convention of the Society of American Foresters, Washington, D.C.

Pickett, S.T.A. and J.N. Thompson (eds.). 1985. *The Ecology of Natural Disturbance and Patch Dynamics*. Academic Press, New York.

Radosevich, S.R. and J.S. Holt. 1984. *Weed Ecology*. Wiley, New York.

Ranney, J.W., M.C. Bruner, and J.B. Levenson. 1981. The importance of edge in the structure and dynamics of forest islands. In: F.L Berguss and D.M. Sharpe (eds.) In: *Forest Island Dynamics in Man-Dominated Landscapes*. Spring-Verlag, New York, pp. 67–95.

Raynor, G.S. 1971. Wind and temperature structure in a coniferous forest and a contiguous field. *Agricultural Meteorology*. 17: 3.

Reese, R. 1984. Greater Yellowstone: The national park and adjacent wildlands. *Montana*, Helena.

Reid, T.S. and D.D. Murphy. 1995. Providing a regional context for local conservation action. *BioScience* (Suppl.): S84–S90.

Reshetiloff, K. (ed.) 1995. *Cheseapeake Bay: Introduction to an Ecosystem*. U.S. Environmental Protection Agency, Washington, D.C.

Risser, P.G. 1986. Report of a Workshop on the Spatial and Temporal Variability of Biospheric and Geospheric Processes: Research Needed to Determine Interactions with Global Environmental Change, Oct. 28–Nov. 1, 1985. ICSU Press., St Petersburg, Fla. Paris.

Roberts L., 1988. Hard choices ahead on biodiversity. *Science* 241: 1759–1761.

Robinson, S. 1996. Nest losses, nest gains. *Natural History*. 105 (7): 40–47.

Rome, A. 1991. Protecting natural areas through the planning process: The Chesapeake Bay watershed. *Natural Areas Journal* 11: 199–202.

Salwasser, H.C., Schonewald-Cox, C., and R. Baker. 1987. The role of interagency cooperation in managing for viable populations. In: M. Soule (ed.), *Viable Populations for Conservation*, pp. 159–173. Cambridge University Press, New York.

Sauvajot, R.M. and M Buechner, 1993. Effects of urban encroachment on wildlife in the Santa Monica Mountains. In: J.E. Keeley (ed.), *Interface Between Ecology and Land Development in California*. Southern California Academy of Sciences, Los Angeles, CA.

Schonewald-Cox, C.M. and J.W. Bayless. 1986. The boundary model: a geographical analysis of design and conservation of nature reserves. *Biological Conservation* 38: 305–322.

Scott, J.M., B. Csuti, K.Smith, J.E. Estes, and S. Caicco. 1988. Beyond endangered species: an integrated conservation strategy for the preservation of biological diversity. *Endangered Species Update* 5: 43–48.

Shafer, C.L. 1990. *Nature Reserves:Island Theory and Conservation Practice*. Smithsonian Institution Press, Washington and London. 189 pp.

Shafer, C.L. 1995. Values and shortcomings of small reserves. *BioScience* 45: 80–88.

Shafer, C.L. 1997. Nature reserve design at the urban/rural interface. In: M.W. Schawartz (Ed.), *Conservation in Chronically Fragmented Landscapes*, pp. 345–378. Chapman and Hall, New York.

Shafer, C.L. 1998a. A history of selection and system planning for natural area U.S. national parks and monuments. *Beauty and Biology* (in press).

Shafer, C.L. 1998b. U.S. National Park buffer zones: Historical, scientific, social, and legal aspects. *Environmental Management* (in press).

Soule, M.E. and B.A. Wilcox (eds.). 1980. *Conservation Biology:An Evolutionary-Ecological Perspective*. Sinauer, Sunderland, MA.

Stottlemeyer, R. 1981. Evolution of management policy and research in the national parks. *Journal of Forestry* 79: 16–20.

The Conservation Foundation. 1985. *National Parks for a New Generation: Visions, Realities, Prospects*. The Conservation Foundation, Washington, D.C.

The Nature Conservancy (TNC). 1987. Preserve Selection and Design Operations Manual. TNC. Washington D.C.

Thomas, J.W., C. Maser and J.E. Rodiek. 1979. Edges. In: J.W.

Thomas (ed.) Wildlife Habitats. In: *Managed Forests: The Blue Mountains of Oregon and Washington*. USDA-FS Agricultural Handbook No. 553: 48–59.
Thomas, J.W., E.D. Forsman, J.B. Lint, E.C. Meslow, B.R. Noon, and J. Verner. 1990. A Conservation Strategy for the Spotted Owl. A report prepared by the Interagency Scientific Committee to address the conservation of the northern spotted owl. Portland, OR.
Troll, C. 1939. Luftbiodpan and okologische Bodenforschung. *Z. Ges. Erdkunde*: 241–98.
Turner, M.G. 1989. Landscape ecology: the effect of pattern on process. *Annual Review of Ecology and Systematics* 20: 171–197.
Turner, M.G. and R.H. Gardner. 1991. *Quantitative Methods in Landscape Ecology*. Ecological Studies 82. Springer-Verlag.
Urban, D.L. R.V. O'Neill, and H.H. Shugart. 1987. Landscape ecology. *BioScience* 37: 119–27.
U.S.D.A. National Forest Service, 1994. Interim Directive.
U.S.D.A. National Forest Service, 1995. Noxious Weed Management Interstate 90 Corridor" Environmental Assessment. Mt. Baker-Snoqualmie and Wenatchee National Forests, Washington. Internet Document.
Vitousek, P.M. 1986. Biological invasions and ecosystem properties: can species make a difference In: H.A. Mooney and J.A. Drake, (eds.) *Ecology of Biological Invasions of North America and Hawaii*, pp. 163–176. Springer-Verlag, New York.
Watt, A.S. 1947. Pattern and process in the plant community. *Journal of Ecology* 13: 27–73.
Western, D. and M.C. Pearl (eds.). 1989. *Conservation for the Twenty-First Century*. Oxford University Press, New York.
White, P.S. and S.T.A. Pickett. 1985. Natural disturbance and patch dynamics: an introduction. In: S.T.A. Picket and J.N. Thompson (ed). *The Ecology of Natural Disturbance and Patch Dynamics*, pp. 3–13. Academic Press, San Diego, CA.
White, P.S. 1987. Natural disturbance, patch dynamics, and landscape patterns in natural areas. *Natural Areas Journal* 7: 14–22.
Wilcove D. and D. Murphy. 1991. The spotted owl controversy and conservation biology. *Conservation Biology* 5: 261–262.

## THE AUTHORS

**Julie A. Concannon**
*USDI Fish & Wildlife Service*
*911 NE 11th Ave.*
*Portland, OR 97232-4181, USA*

**Craig L. Shafer**
*USDI National Park Service*
*1849 C St. NW*
*Washington, DC 20240-0001, USA*

**Robert L. DeVelice**
*USDA Forest Service*
*Chugach National Forest*
*3301 C Street, Suite 300*
*Anchorage, AK 99503-3998, USA*

**Ray M. Sauvajot**
*USDI Park Service*
*30401 Agoura Rd., Suite 100*
*Agoura Hills, CA 91301, USA*

**Susan L. Boudreau**
*USDA Forest Service*
*Payette National Forest*
*Box 1026, 800 W. Lakeside Ave.*
*McCall, ID 83638, USA*

**Thomas E. DeMeo**
*USDA Forest Service*
*Mt. Hood National Forest*
*16400 Champion Way*
*Sandy, OR 97055, USA*

**James Dryden**
*USDI Bureau of Land Management*
*Madison, WI, USA*

# Ecosystem Processes and Functioning

Ariel E. Lugo, Jill S. Baron, Thomas P. Frost,
Terrance W. Cundy, and Phillip Dittberner

**Key issues addressed in this chapter**

- ♦ *Understanding ecosystem processes and functioning is in the context of ecological stewardship.*
- ♦ *The importance of energy laws to ecological phenomena.*
- ♦ *Using a functional, as opposed to a geograpic, definition of ecosystems as a key to understanding ecosystem management.*
- ♦ *How natural and anthropogenic disturbances interact with ecosystems and how to incorporate them into management plans.*
- ♦ *How new paradigms of ecology provide flexibility and greater effectiveness to ecosystem management.*

**Keywords: Ecosystem processes, ecosystem functioning, biological hierarchies, geologic and tectonic processes, material cycles, sucession, disturbance, ecological space and functional interfaces, ecosystem concepts and ecosystem management**

# 1 INTRODUCTION

People obtain products and services from ecosystems. Lumber, mushrooms, fruits, flowers, chemicals, wildlife, and fuelwood are examples of products. Clean water and clean air, pollination, soil production, sights, and sounds suitable for recreation or meditation, climate control, and conservation of genetic diversity are examples of services. Ecosystem processes and functioning assure the continuous supply of these and many more products and services. Processes, such as those of the sedimentary cycle, support ecosystem functioning, and forest products result from the balance between the productive and respiratory functions of ecosystems. Services associated with clean water and air are byproducts of the cycling function of the ecosystem. The continuity of sights and sounds in a landscape results from the internal biotic organization of the ecosystem, which maintains itself through such functional attributes as feeding interactions and processes like succession. All ecosystem functions and processes contribute to conserving biodiversity, including genetic diversity.

This chapter provides information on how ecosystems function and shows how this information is related to ecosystem management. We introduce many concepts and ideas necessary as background to any intelligent manipulation of ecosystems. The chapter comprises three parts:

- Part I, Basic Knowledge, contains the necessary definitions and introductory information about ecosystem processes and functioning. We define the ecosystem as a functional rather than a geographic concept.
- Part II, Material Cycles, uses carbon and nitrogen to illustrate how matter circulates through and within ecosystems. Both cycles are fundamental to the functioning of any ecosystem, and they also reflect the effects of human activity on the biosphere. Because most readers are likely to be from the temperate zone where nitrogen is a limiting factor to ecosystem productivity (Vitousek and Howarth 1991), we provide an extended discussion on the nitrogen cycle. In tropical forests, phosphorus is more limiting than nitrogen (Vitousek 1982).
- Part III, Ecosystem Management Through a Functional Perspective, uses materials from the other two sections as well as new material on disturbances and ecological theory to arrive at new paradigms for managing ecosystems.

Throughout the text, we strive to relate scientific principles to management so readers can access any section of the chapter according to interest and understanding of the subject matter.

# 2 PART I: BASIC KNOWLEDGE

## 2.1 Definitions and Concepts

### 2.1.1 Processes

The definition of process as used here is close to that of the dictionary's definition: **a process** is a *sequence of events or states, one following from and dependent on another, which lead to some outcome.* The outcome itself may not be reached, or the process may be interrupted or replaced by another because of a disturbance, but the tendency toward an outcome is always realized. The outcome may be a step or state in some hierarchically broader scale process. Consider erosion, for example; it is a process influenced by regional climate, local climatic variations, bedrock type, topography, soil cover, vegetation cover, and management and disturbance history. The ultimate outcome of the processes of erosion and sedimentation is to form new sedimentary rocks which will, in turn, be uplifted and eroded, or buried under additional sediments to be metamorphosed, or melt to form entirely new volcanic or other rocks. The process of erosion and its rate are profoundly influenced by all the factors mentioned above, and erosion itself is a process that influences and is part of other processes: sedimentation and water and nutrient cycles, among others. The rate of erosion and its attendant processes may be influenced by management practices or naturally occurring disturbances, such as fire.

Processes that are critical in ecosystems include both one-way fluxes and cycles. The difference between these two terms is mostly a function of scale. *To a manager, erosion and the resulting sediment transport represent a flux, moving sediment from one place to another; in a global context, they become cycles.* For example, sediment is moved to a basin where it collects and may turn to stone; the resulting rock eventually will be uplifted and eroded, completing the cycle. Management activities or other disturbances affect the rates of some physical, biological, or closely linked processes, so considering broad-scale geological and climatological processes is important in more than just a contextual sense.

*Many "disturbances" in ecosystems are merely processes that have return frequencies that fall outside human planning horizons.* Fires and floods, outbreaks of insects and disease epidemics, very large storms, volcanoes, earthquakes, and landslides have all occurred continually throughout time. Inadequate understanding or human planning and management typically do not adequately allow for these processes. We use the term *disturbance* to mean a force that causes significant change in ecosystem state or trajectory. This force can be a naturally occurring process or a management-induced modification in processes. By this definition, the shift in Pacific Northwest forests from parklike stands of ponderosa pine to dense young Douglas-fir because fire was suppressed (Quigley 1997) represents a disturbance. Disturbances are discussed in more detail in Part III.

## 2.1.2 Function

*Function in an ecosystem is the role that any given process, species, population, or physical attribute plays in the interrelation between various ecosystem components or processes. For example, functions of carbonate rocks, in an ecological sense, are providing appropriate substrates for certain terrestrial species, critical macronutrients to aquatic species, and probably other functions as well.* A function of large dead snags in a forest is providing roosting or nesting sites, and a function of large woody debris in streams is breaking the flow into pools and riffles, and maintaining habitat. A function of coarse pebbly sediment in streams is to provide an appropriate substrate for fish spawning. Functional roles may be disturbed if fine sediment is deposited after fire, logging, or other events.

## 2.1.3 The Ecosystem Concept

Many definitions and views of what an ecosystem is have been published, and the literature about this concept is extensive (McIntosh 1985). O'Neill et al. (1986) make the distinction between studying ecosystems from a population-community perspective and from a process–functional perspective. The population–community approach emphasizes populations of species and the interactions among them, such as competition and predation. The process–functional approach, which we will take in this chapter, emphasizes the transfer and processing of matter and energy (O'Neill et al. 1986, King 1993).

The general use of the term "ecosystem" currently centers on the dynamics of a community with well-defined spatial boundaries, such as the spruce forest ecosystem or the watershed ecosystem. This use of the term recognizes the structural and functional aspects of plant communities or of particular physical entities such as watersheds. Traditionally, the study of plant communities focused on their structure, succession, and natural history with less attention to other functional attributes and the relations between structure and function. The term *ecosystem* is used to underscore the functional aspects of natural assemblages of organisms. Unfortunately, this general use of the term has led many to equate ecosystems with a particular place in spatial or biological hierarchies — for example, individual, population, community, ecosystems, and landscape. We disagree with the location of ecosystem in this hierarchy because it implies that communities and landscapes are not ecosystems. We use the term ecosystem to imply functioning at *any* scale of size, space, or biotic complexity (Box 1).

**Box 1**
**Hierarchies**

*Biological hierarchy* is defined as the progression of biotic systems along a scale of increasing biological complexity, from viruses to the Earth as a whole. A biological hierarchy can be displayed for convenience in a time and space scale as shown in Table 1 for temporal scales, but such scales are not the main organizational basis for constructing the biological hierarchy. A biological hierarchy is analogous to a time or space hierarchy, but instead of being organized around spatial or time concepts, it is organized around biological criteria. At each scale of time and space are increasingly complex biological entities that have functional continuity with each other (see Fig. 1). *Functional continuity* means that any system along the hierarchy is composed of systems located one level of complexity below, and it is part of a more complex system at one level of complexity above. Thus, a population is composed of individuals, and it forms associations. Associations are composed of populations, and they form life zones. Each level of the hierarchy is unique, but energy and matter flow unimpeded, seamlessly, and without discontinuity through the whole hierarchy. For this reason, the boundaries of ecosystems along the hierarchy can be chosen arbitrarily. *Ecosystem* does not appear anywhere in this hierarchy because ecosystems exist at all scales of organization and complexity: they are functional, not spatial units. Haufler et al. (see this volume) expand on the subject of hierarchies.

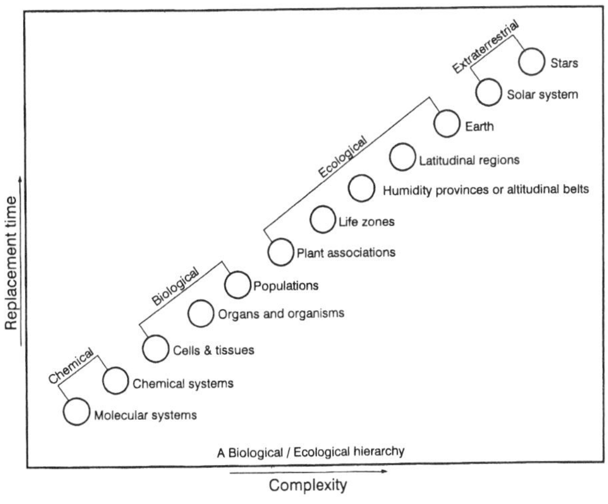


Fig. 1. A biological hierarchy with functional continuity along scales of complexity and turnover time. The hierarchy includes chemical, biological, ecological, and physical systems.

Table 1. Temporal scales with examples of biological phenomena that operate at these scales (modified from Magnuson 1990 and Swanson and Sparks 1990)

| Time (years) | Physical disturbance | Biological phenomena |
|---|---|---|
| $10^5$ (100 millennia) | | Evolution of species |
| $10^4$ (10 millennia) | Continental glaciation | Bog succession, forest community migration |
| $10^3$ (millennium) | Climate change, forest fires | Species invasion, forest succession |
| $10^2$ (century) | $CO_2$ climate warming | Cultural eutrophication |
| $10^1$ (decade) | Sun spot cycle | Hare population |
| | El Niño | Prairie succession |
| $10^0$ (year) | Prairie fires | Annual plants |
| | Lake turnover | Plankton succession |
| $10^{-1}$ (month) | Ocean upwelling | |
| | Storms | Algal blooms |
| $10^{-2}$ (day) | Diel light cycle | Diel migration |
| $10^{-3}$ (hour) | Tides | |

We will focus on the functional view of ecosystems as defined by Evans (1956) who, in agreeing with Tansley (1935, p. 1127) that the *ecosystem* is a basic unit of ecology, wrote

> "In its fundamental aspects, an ecosystem involves circulation, transformation, and accumulation of energy and matter through the medium of living things and activities."

This functional view of ecosystems is suited for ecosystem management approaches based on ecosystem science; it provides maximum flexibility for defining the boundaries and spatial scales of ecosystems as well as criteria for ecosystem management approaches. The definition also identifies the subjects needed to understand ecosystem functioning: energy flow, organic matter and nutrient dynamics, and biotic interactions (see Box 2 for an explanation of terminology to describe systems).

The definition also says that the term ecosystem represents a functional, not a spatial concept. Therefore, when the functional attributes of ecosystems or their management are described, no standard spatial scale is implied. Ecosystems function at *all scales* in which they are identified. Because ecosystems function at many scales of biological complexity, ecologists address the functional attributes relative to the scale at which each is being examined. For managers, addressing ecosystem scales that coincide with the particular scale at which management issues occur is important. Thus, one issue may be addressed at a small scale and another at a larger, more complex scale.

We selected four scales of ecological complexity to address ecosystem functioning: the stand, watershed, region or landscape, and global scales. We do not imply by these different scales that in every instance a size or area difference exists among them; they were selected to illustrate either scales for management or those influenced by activities at other scales. The four scales are defined as follows:

- *Stand* — a system homogeneous in its biological, edaphic, climatic, or hydrologic conditions (Matthews 1989). Much of the silvicultural treatment of forests is based on stand treatments. Stands, which can be of any size, are open systems (discussed below).
- *Watershed* — a well-defined hydrological unit (Likens and Bormann 1995). A watershed can be small or immense and can contain one or many stands. It can include a whole landscape or region or just part of one. The watershed is defined by hydrological and geomorphological conditions rather than by its biology, climate, geology, or soils. Watersheds provide well-defined boundaries for studying, managing, and conserving water, carbon, and nutrients (Likens and Bormann 1995).
- *Region* or *landscape* — three components define regions or landscapes: its visual or scenic component; its chronological component, that is, how its geomorphology developed; and the ecological component, that is, its ecological systems (Zonneveld 1990). Its boundaries are usually defined arbitrarily and usually represent a large area with a variety of open systems. A region or landscape may or may not have natural boundaries.
- *Global* — the global scale is defined by the boundaries of the planet and is the only natural system almost closed to carbon and nutrients (meteorites account for a small input of mass from space) but not to energy.

Managers address ecosystem functioning differently at the four scales just described; this difference can be illustrated with the productivity function. At the stand scale, the production function is addressed directly through silvicultural techniques, and management actions can be intensive. At the watershed scale, production is more complex to manage because of the heterogeneity of a watershed. A watershed may contain more than one stand, each with its own idiosyncrasies. Managers usually can not manipulate

**Box 2**
**Terminology Describing the Nature and Behavior of Systems**

System — a group of parts that are interacting according to some kind of process.
State variable — a quantity, stored in a system, that varies over time. It describes the condition of the system as it varies.
Flux — the amount of energy or materials in motion between a source (point of origin) and a sink (point of storage).
Steady state — when storages and patterns in open systems become constant with a balance of inflows and outflows in balance.
Balanced — inflows equal outflows. Used here to describe the inflows and outflows of primary productivity (P) and respiration (R), which result in a P/R of 1.
Homeostasis — self-regulation
Equilibrium — when storages become constant in a closed system. It means maximum entropy (disorder), or death. The term is confused with dynamic equilibrium or steady state, which apply to open systems. Dynamic equilibrium or steady state mean that the state variables of the system are constant (inputs equal outputs), and the system maintains a balance, with low entropy. Living systems are open systems not at equilibrium; they are at steady state or dynamic equilibrium. Chemical solutions in closed containers reach equilibrium when entropy is maximum. They require energy input to break the equilibrium.
Inertia — ability of an ecosystem to resist change applied by an external stressor.
Resilience — the degree, manner, and pace of recovery of the ecosystem to the predisturbance condition (Majer 1989a). The persistence of relations in a system and a measure of the ability of these systems to absorb changes in state variables, driving variables, and parameters, and still persist (Holling 1973).
Elasticity — the rate of recovery of an ecosystem property after disturbance.
Amplitude — the threshold of strain beyond which return to the original state is no longer possible.
Hysteresis — the degree to which the path of change in an ecosystem property under chronic stress differs from the path of succession on immediate removal of the stress.
Malleability — difference between the ecosystem's final recovery state and the predisturbance state.
Damping — degree and manner by which a path of restoration is altered by any forces that affect the restoring force.

Figure 2 illustrates the last seven terms.

*Source*: Odum (1971, 1983) and the interpretation of Westman (1986) by Majer (1989a).

production at regional and global scales, but they can influence the global and regional production process through the cumulative effects of what is done at stand and watershed scales. At the global scale, Vitousek et al. (1986) estimated that people consume or directly influence about 40% of the Earth's primary productivity.

## 2.2 Energy Basis of Ecosystem Functioning

Ecosystems have a finite input of energy available for all the work required for their maintenance and survival. The energy input in most ecosystems originates from the Sun and is based on area. One exception is that some deep ocean ecosystems are powered by chemosynthetic bacteria. The *solar constant*, measured at the top of the atmosphere, is the energy available to all ecosystems on Earth; it has a magnitude of about 2 cal/cm$^2$ per minute. Solar energy must somehow be effectively portioned out among all the vital functions necessary for the maintenance and survival of the biosphere. As a rule, maintaining complex ecosystems takes more energy than does maintaining less complex ecosystems because complexity itself requires energy for maintenance (cf. Odum 1970). Energy use must be efficient to accomplish as much work as possible within the limit set by the solar constant and the entropy tax of the second law of thermodynamics.

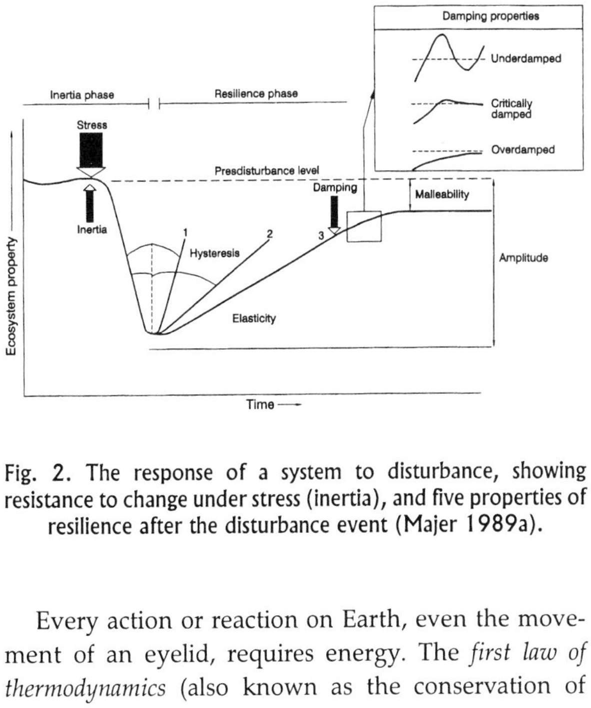


Fig. 2. The response of a system to disturbance, showing resistance to change under stress (inertia), and five properties of resilience after the disturbance event (Majer 1989a).

Every action or reaction on Earth, even the movement of an eyelid, requires energy. The *first law of thermodynamics* (also known as the conservation of energy law) states that the amount of energy in a system is constant and cannot be created nor destroyed. The *second law of thermodynamics* which exacts its toll of *entropy* (heat energy lost or unavailable to perform work) during every energy transformation in the biosphere assures that no process of energy transfer is 100 percent efficient. Thus, every time energy is used, the amount of energy available to do work by organisms and populations is reduced and dissipated to space as long wave heat.

The energy required to sustain activities of organisms is obtained through *respiration*, which is the oxidation of organic matter fixed by *photosynthesis*. Even communication among organisms and social organization of biotic activity demand energy because they require work in the form of sound, visual displays, or behavior. The more work organisms do, the more respiration must take place and the more organic matter from photosynthesis must be gathered and allocated for respiration. Logically, in areas where environmental stresses are high (for example with frost, fire, high soil salinity, or low concentrations of oxygen), ecosystems are less likely to become complex because they use their limited energy resources to overcome these stressors. In areas of low stress (for example, without extreme temperatures, high rainfall, or environmental fluctuations), energy is more available for growth, diversification, and — perhaps — greater evolutionary experimentation.

The processes and functioning of ecosystems are regulated by these two energy laws which underpin all biological and ecological activity. But how do physical processes create the conditions under which ecosystems must function? First, to help interpret the current landscape, we describe and give examples of geologic and tectonic processes that mold the Earth. We define five ecosystem processes and discuss how they can be addressed at the four scales identified. Then, we provide details of each of these ecosystem attributes, particularly as they relate to management, and end the paper with a discussion of the relations between disturbances, ecological space, and functional interfaces and how these are related to ecosystem resilience and management.

## 2.3 Geologic and Tectonic Processes

Geologic processes and their interactions with atmospheric and hydrologic processes, operating at many spatial and temporal scales (Hamblin 1985) have shaped, and continue to shape, the landscape (Summerfield 1991). Recognition of the role of geologic processes and their rates provides a counterpoint to the role of human-induced changes in any ecosystem. Recognition of the global scale and interactions of geologic, climatic, and biotic processes provides a framework within which human activities may be judged. Geologic processes are significant in ecosystem management where rates are high, where rates are closely linked to biological or other physical processes of concern, or where rates may be affected by management or disturbance (Jensen et al. 1997).

Most geologic processes operating today that directly affect current ecosystem structure, functioning, and potential development tend to be catastrophic or to be those whose local rates are high, or whose rates may be affected by management practice or disturbance. Such processes may occur infrequently but cause significant change in affected areas. On a human time scale, infrequent but rapid processes are thought of as disturbances; they recur so rarely that people tend not to prepare for them. Although these processes — including individual volcanic eruptions, earthquakes, floods, land movements, and their direct corollaries such as erosion and sedimentation — are infrequent they can cause significant change in affected areas. The processes themselves cannot be managed, but responses to them can be planned for and preparations made.

Efforts to manage rivers have sometimes had undesirable consequences in the extent and frequency of floods. The climatic conditions that resulted in the Upper Missouri River floods of 1993 clearly were beyond human control, but the efforts over the past century to manage the floodplain may have exacerbated the flow rates and other aspects of the flooding (SAST 1994). Human occupation of low ground, building dikes and channels, and other land uses and management activities all contributed to the flood's effects.

### 2.3.1 Tectonic Processes

A few examples of processes and their effects must suffice to introduce the role of plate tectonics and related processes in shaping the landscape. A more detailed discussion with pointers to significant references is presented in Jenson 1997. The distributions of continents and mountain ranges over the surface of the Earth directly result from plate tectonic processes operating at rates of from less than one to tens of centimeters per year. Processes deep within the Earth drive continental drift and sea-floor spreading. The Earth's crust is thin plates, no more than a hundred or so kilometers thick. Where these plates spread apart as at the mid-Atlantic ridge or the east-Pacific rise, basaltic volcanism on the sea floor creates new oceanic crust. Where crustal plates collide, as around much of the Pacific Ocean margin, the oceanic crust, which is heavy and dense, is subducted under the lighter, thicker, continental material.

#### *Volcanism*

Arc volcanoes form in response to subduction. Present day examples include the volcanic Cascades of British

Columbia, Oregon, Washington, and California; the Aleutian Islands; the Andes; and the archipelagos of Japan and the Caribbean. The eruption of Mount St. Helens in 1980 is merely a recent activity in the long, continuing process shaping the Pacific Northwest. During the last 4,000 years, volcanoes of the Cascade Range have erupted, on average, twice per century (Dzurizin et al. 1994). These return frequencies are similar to those for extremely large forest fire events (ICBEMP, in press). Even comparatively small eruptions like the 1980 Mount St. Helens eruption affect areas orders of magnitude larger than even the 1989 Yellowstone fires (Lipman and Mullineaux 1981). Eruptions may profoundly influence climatic conditions as well. Kerr (1993) documented a worldwide 0.6°C decrease in temperature after the eruption of Mount Pinatubo in the Philippines.

*Earthquakes*

Where crustal plates slide past one another, as along the San Andreas fault of California, or in more localized areas of the crust under regional stress, earthquakes mark breaks and slippage of crustal blocks past one another. Earthquakes may also occur in regions away from major crustal boundaries and, although less frequent, they may have major effects.

Effects of major earthquakes on people and social infrastructures are beyond the scope of this discussion. Apart from the social costs of large earthquakes, even moderate earthquakes may induce high demand for replacing infrastructure with consequent demands on natural resources. Less obvious is the likelihood for significant change in aquatic ecosystems by earthquakes, such as the landslide caused by an earthquake damming the Madison River forming Earthquake Lake in Yellowstone Park area. Although such spectacular effects are infrequent in human experience and earthquakes cannot now be predicted, the probabilities of earthquakes in parts of the United States are high.

## 2.3.2 Landscapes and Their Interpretations

Much of today's landscape is, in effect, a relict of the Ice Age (Pleistocene epoch, 1.5 million to about 10,000 years ago), when the rates of many of the geologic processes that shape the landscape were higher than they are today. Landscapes older than the Miocene (about 12 million years old) are rare (Summerfield 1991). Although these timeframes seem large, they are recent by geologic standards. Correct interpretations of the landscapes and the processes that shaped and continue to shape them are critical to managing them.

The Pleistocene was a time of many (several tens) cycles of major climate variation, ranging from ice ages to warm interglacial periods. The climax of the last cold period was 20,000 to 14,000 years ago, when average summer temperatures in the Pacific Northwest were 5°C to 7°C cooler and winter temperatures 10°C to 15°C less than today (Barry 1983). During these colder and effectively moister times, large ice sheets formed in the northern hemisphere, and covered most of Canada and the northern United States (Mickelson et al. 1983, Waitt and Thorson 1983). Lobes of the ice sheet originating in Canada advanced and retreated several times from the United States and excavated and molded valleys in the northern states (Waitt and Thorson 1983). Alpine glaciers carved many of the higher ranges and valleys in Washington, Oregon, California, Idaho, Montana, Wyoming, and Colorado (Porter et al. 1983). Much of these glaciated landscapes are now covered with a mantle of glacial till. Downstream, and in the wake of retreating glaciers, thick sedimentary sequences of silt, sand, and gravel outwash were left as valley fill and terraces flanking most rivers with glaciated headwaters. Lakes formed in the closed basins of the western United States, including the Bonneville (Salt Lake), Humboldt (Nevada), and many basins in southeastern Oregon during wet periods. These basins now have thick mantles of fine-grained lake deposits (Benson and Thompson 1987).

During the Pleistocene Ice Ages, silt and fine-sand outwash from alpine and continental glaciers and glacial outburst floods were blown by wind and deposited as thick blankets called loess. These loess deposits locally dominate the landscape; for example, the rolling hills of the Palouse in eastern Washington are entirely composed of loess deposited over the last 2 million years and locally over 75 m thick (Busacca 1991, Busacca and MacDonald 1994).

The changes in vegetation patterns and aquatic and terrestrial species since the end of the Ice Ages have been profound. Obviously, areas that were covered in ice have since been colonized by terrestrial and aquatic plants and animals, and many species not present in the basin on the past have moved in. For example, pollen data indicate that no ponderosa pine grew north of Flagstaff, Arizona, as recently as 11,000 years ago (Betancourt et al. 1990). Major changes in forest composition in the Columbia River basin may be even more recent because of plant colonization.

## 2.3.3 Weathering, Erosion, and Sedimentation

The three geologic processes of immediate concern to management are weathering, erosion, and sedimentation. They are closely linked, and changes in their rates may depend heavily on management practices or disturbance. *Weathering* is the chemical and physical change in rock and soil components resulting from exposure at the Earth's surface or *in situ* below the surface. *Erosion* includes the loosening, removal, and transport of the weathered material by gravity, wind, water, or ice. *Sedimentation* is the deposition of the transported sediment.

### Weathering

Weathering is continuous to materials whether they are exposed at the surface or not. Chemical weathering changes the constituent minerals and makes soluble components available for transport by solution. Chemical nutrients (or toxins) are released through chemical weathering and taken up by plants and animals. An example in which management practice has increased the rate of solution of toxic elements is in the Kesterson Reservoir area of California. Under normal climatic conditions, chemical weathering rates of the selenium-bearing sedimentary rocks were slow enough that selenium never reached toxic concentrations. With irrigation, however, enough selenium was dissolved from the rocks that it became toxic to some forms of life and profoundly affected breeding success among waterfowl in the area (Deverel et al. 1994).

Many mining districts provide examples for the release of toxic elements. In undisturbed areas, organisms adapt to the flux of metals and other drainage into the aquatic and terrestrial environment from concentrations of metals and minerals exposed at the surface. Mining, however, may expose large volumes of unweathered rock to physical and chemical modification and transport by air and water and increase the rates of introduction of metals, acid-drainage, or sediment to the environment (Kwong 1993). Because of changes in introduction rates of metals (lead, zinc, mercury, arsenic, antimony, cobalt, and others), acid-rock drainage, or sediment caused by mining or other disturbance, the aquatic or terrestrial habitat, ecosystem state, and processes can change in affected areas. Different mineral deposits have different ore, alteration, and host-rock mineralogies that lead to different environmental characteristics. Deposits high in pyrite and other acid-producing sulfide minerals and low in acid-buffering minerals, such as calcite, aragonite, or dolomite, are likely to produce acid-rock drainage (Kwong 1993, Plumlee et al. 1994, DuBray 1995). When these minerals are mined and exposed to air and water, weathering is accelerated and appropriate remediation technologies must be adopted to prevent acid-rock drainage problems (Plumlee et al. 1994). The amount of acid-rock drainage from a given deposit also depends on how much water is available to react with the minerals.

### Erosion

Erosion is subject to thresholds: material of a given size in a given setting will not move until some activation threshold is reached. Fine material, such as silt or clay, may be moved at low flows, but if no fine sediment is present in a stream or the fine sediments are sequestered in a low-flow regime portion of the stream or interbedded with coarser material, the stream will not transport any sediment. As flow rates increase, the carrying capacity, or *competence*, of the flow increases, and reach some threshold at which material will begin to move. The threshold depends on local flow rates, grain size, and other variables (Summerfield 1991).

Transport and introduction of volumes, grain sizes of materials, or both that are not typical to the system may constitute a disturbance. Transport of coarse material during a flood or high water will cause both erosion and redistribution of sediment as flows decrease during waning stages of a flood (Flint and Skinner 1974). Introduction of sediment sizes not common to the system may severely affect stream morphology, flow rates, and bottom characteristics. Influx of large volumes of fine sediment to a cobbled-bottom stream from increases in erosion rate as a consequence of logging, fire, grazing, or other surface disturbance uphill or upstream may decrease fishes' ability to spawn (Everest et al. 1987). Livestock grazing in rangelands may reduce vegetative cover to the point that all soil material is stripped by sheet wash or other flow (Flint and Skinner 1974). Changes in rates of transport and sedimentation are affected by management practices and need to be considered in developing management plans.

*Mass wasting*, the movement of weathered material downslope by gravity without influence of water, ice, or wind, includes landslides, debris flows, creep, solifluction, slope failures, and rock fall (Summerfield 1991). Mass wasting may be influenced by natural rock instability and may be triggered by earthquakes, saturation from increased precipitation, or other natural causes. Disturbance by human activities may also induce mass wasting, by removing material at the toes of slopes; increased loading from bridges, buildings, roads, or other structures; increased saturation from irrigation, waste disposal, or reservoir filling; or other activities (Flint and Skinner 1974, Summerfield 1991). In forested lands, logging, roadbuilding, and the resulting removal of vegetative cover and channelization of water may lead to mass wasting or gullying.

Wind-caused erosion may be significant and a local problem. Arid barren, agricultural, and range lands, or those with greatly reduced vegetation are particularly susceptible to wind erosion. Coarse (usually sand-sized) material is most commonly moved by bouncing along the ground surface as a traction load. Finer silt- and clay-size material typically moves as suspended load, carried along by air currents (Flint and Skinner 1974). Management practice may profoundly influence the susceptibility of a landscape to wind erosion.

#### Sedimentation

Sedimentation is deposition of materials carried by any transporting medium with insufficient energy to transport that material. Increased sedimentation is a natural consequence of increased erosion. Redistribution and deposition of material during waning floods in river systems is a natural and essential part of maintaining aquatic conditions. The recent human-caused flood in the Grand Canyon was an attempt to mimic natural, annual floods during spring runoff before Glen Canyon dam was built, and needed to flush sediment downstream and maintain beaches, pools, rapids, and aquatic habitat (Adler 1996). The Missouri River floods of 1993, although costly to infrastructure and human life, were exacerbated by the management practices and existing flood-control strategies (SAST 1994). Human settlement and use patterns should take into account the natural and recurring nature of floods.

Along many coasts, material supplied by rivers represents a significant proportion of the sediment budget. Dams built for flood control and power generation in these areas have reduced the supply of coarse sediment to the beaches, resulting in erosion and inland retreat of the coastline (Summerfield 1991). This example illustrates the manifold implications of poorly understanding the geological processes and their effects, which are often removed from the cause, in time and space.

## 2.4 Biotic Processes

### 2.4.1 Material Cycling

*Material cycling* is the movement of matter within and through the ecosystem, including cycles of water, carbon, and important mineral nutrients such as nitrogen, phosphorus, potassium, calcium, magnesium, and iron. Materials are exchanged between the physical environment (soil, water, or atmosphere) and organisms. Materials move within soil, water, the atmosphere, and between populations of organisms. Material cycling is also termed nutrient cycling, mineral cycling, or simply cycling — or recycling because matter can be used over and over in an ecosystem.

Stands, watersheds, and landscapes are open systems because the cycles of materials are not self-contained in any of these scales of organization. The Earth as a whole is large enough to internalize its mineral cycling (meteorites excepted). Earth is one large ecosystem, linked by global circulation of the atmosphere, oceans, and crustal movements. The Earth is a closed system where physical and biological processes are driven by the Sun's energy, with few other inputs or outputs. Schlesinger (1991) noted that meteorites are external inputs, and satellites, such as Galileo, are outputs. The ecosystem definition holds at smaller scales, those of stands, landscapes, or regions; at these smaller scales, the interaction of living things with their environments and each other still holds, but external inputs and outputs of materials and energy take on greater importance. The movement of air, water, and organisms keeps mineral cycles open. This concept of open vs. closed systems is discussed in more detail in Part III.

### 2.4.2 Primary Productivity and Respiration

Primary productivity is the transformation by organisms of solar or chemical energy into chemical energy suitable for respiration. Two groups of organisms participate in the primary productivity: one group is chemoautotrophic microorganisms that break up chemical bonds (usually sulfur) to obtain energy. The other group is organisms that convert solar energy, nutrients, and carbon dioxide into reduced chemical energy in the form of organic (carbon) matter and oxygen — a process called *photosynthesis*. A minor amount of water is used in primary productivity both as a raw material and a product. Organic matter becomes the fuel for respiration (below), as well as the building blocks of familiar structures such as leaves, roots, stems, soil organic matter, and animal tissue. Only photosynthetic plants contribute significantly to the primary productivity of terrestrial, freshwater, and most marine ecosystems. In vents of the deep ocean floor and in oxygen-free sediments of wetlands and freshwater ecosystems, organisms rely mostly on primary productivity by chemoautotrophs.

Respiration operates in the opposite direction from primary productivity. *Respiration* uses oxygen to oxidize organic matter into carbon dioxide and water, releasing nutrients and energy. Respiration makes energy from the Sun, originally stored by primary productivity, available to organisms. The organisms use the energy to do work in the form of maintaining their metabolism, reproducing, replacing worn-out tissues, and overcoming natural stressors such as frost, fire, salinity, or high temperature.

All organisms in an ecosystem continuously respire, thereby consuming organic matter. Because plants produce and respire simultaneously, gross primary productivity (GPP) must be differentiated from net primary productivity (NPP). Net primary productivity is the organic matter fixed in excess of plant respiration. Gross primary productivity is the total amount of organic matter produced by plants before they fuel their own respiration ($R_p$). In equation form:

$$NPP = GPP - R_p.$$

If plant respiration increases and the gross primary productivity is constant, net primary productivity decreases. Net primary productivity is measured by the increases in the mass of wood, leaves, and roots. Some plants may allocate more of the net production to roots and others may allocate more to wood or leaves, but the amount involved in these allocations is limited by the gross primary productivity and plant respiration.

Net ecosystem productivity (NEP) of whole ecosystems takes into account the respiration of all organisms. Net ecosystem productivity is what determines if a sector of the Earth is a carbon sink (NEP $> 0$), a carbon source (NEP $< 0$), or is neutral (NEP $= 0$) in its carbon exchange with the atmosphere. Biomass increment is essentially zero when NEP is 0. In equation form:

$$NEP = GPP - (R_p + R_a + R_m),$$

where $R_p$ is the total plant respiration, $R_a$ is the total animal respiration, and $R_m$ is the total microbial respiration. These three respiration rates equal total ecosystem respiration. Not included in this equation are possible inputs and outputs of organic matter, which can sometimes determine the dynamics of the ecosystem. For example, floodplains have high inputs and exports of organic matter. Caves lack GPP and depend exclusively on outside inputs of matter, usually from bat or other animal activity.

### 2.4.3 Feeding Interactions

The products of primary productivity are consumed by animals and transferred throughout the food web of plant, animal, and microbial communities. As these materials are transferred from one group of organisms to another — that is, from plant producers to *herbivores* (plant eaters), *omnivores* (both plant and animal eaters), *carnivores* (animal eaters), *top carnivores* (predator eaters), and parasites (feeding on other animals) — some fraction is consumed by each level of the food chain (known as *trophic levels*). Each organism consumes organic matter to maintain its respiration and to grow and reproduce. In the process, these organisms do ecological work, move about, and contribute to the functioning, products, and services of ecosystems.

When organisms or plant tissues die, the dead material is consumed by a complex food web of *detritivores*, the organisms that use dead organic matter to maintain their metabolism. At the bottom of this food web are bacteria and fungi, known as *decomposers*. These organisms eventually close the cycle of production and consumption, recycling the nutrients and carbon back to the plants responsible for primary productivity. The detritus food chains are responsible for about 90 percent of the energy flow in most ecosystems and the herbivorous food chains for the remaining 10 percent. These food chains are not isolated from each other and function as an interconnected food web.

### 2.4.4 Succession

*Succession* is the change in ecosystem structure and functioning through time. Succession typically is in response to a disturbance or disruption of ecosystem structure or function. External factors like fire, storms, wind, and other disturbances, and the organisms themselves change the environment and trigger succession.

Ecologists often describe specific stages or times in a successional sequence that can be recognized by particular characteristics. The stages of succession are named because the ecosystem remains in these stages for relatively long periods or passes through them according to a somewhat predictable pattern. An example would be the "old field stage" after a farm field is abandoned, or the "forest thicket stage" that appears after a mature forest is harvested and burned. Successional stages are not always predictable, however, because ecosystems can follow many paths as they age (Ewel 1980).

## 3 PART II: MATERIAL CYCLES

Basic understanding of biogeochemical fluxes is necessary for managing ecosystems because the materials moving around — such as water, carbon, phosphorus and nitrogen — are the building blocks of life and therefore of ecosystems. A system may have changed structurally, but functionally it may be intact (King 1993). Primary productivity remains the same as before acidification in many North American lakes, for example, even though the algae have shifted from highly diverse communities to ones with many fewer species. Here, ecosystem components, such as other species, can perform equivalent functions, so that changes may have little or only transient influences on ecosystem function (O'Neill et al. 1986, Vitousek and Howarth 1991, King 1993). All ecosystems have some capacity for *resilience* (the capacity to recover after disturbance, Box 2) and as long its capacity is not surpassed, these systems have integrity (Holling 1986).

Plant tissues, indeed all living tissues, are composed mainly of carbon, hydrogen, and oxygen. Nitrogen, phosphorus, potassium, calcium, and sulfur are additional elements necessary for plant growth and maintenance, as are up to 18 other elements. Elements must be available to the plant within a certain fairly narrow range of carbon to nitrogen (C:N) or carbon to phosphorus (C:P) ratios to have favorable conditions for growth. In aquatic ecosystems, these ratios are called Redfield ratios (Redfield 1958, Hecky et al. 1993). Trees and grasses have their own range of acceptable nutrient ratios (Parton et al. 1987, Vitousek et al. 1988).

When the availability of a nutrient is too low, plants show signs of stress in the short term, and will die over the long term and will be replaced by other species that can live under the existing nutrient conditions. When nutrients are too abundant, other consequences appear, including toxic responses by individual plants on a short-term, local scale, possibly leading to changing species composition and altered community dynamics over the mid-term and regional scales (Schindler 1974). Excessive concentrations of mobile nutrients can lead to export of that nutrient by streams or groundwater. As an example, excessive concentrations of nitrogen in groundwater have had adverse effects of human health (Tietema and Verstraeten 1991, Stoddard 1994) and consequences to the coastal waters so important to the global food supply (Valiela et al. 1990).

How an element or compound moves through plants, soils, the atmosphere, and the rock cycle defines the cycle of that element. We will look at cycles of carbon and nitrogen, and how changing availability of these substances influences plant health, regional vegetation dynamics, and global biogeochemistry.

Two important caveats are, first, that we must clearly define ecosystem boundaries as we begin to manage ecosystems. In evaluating cycles, we must define the boundaries of the ecosystem through which energy and matter flow. The determination of ecosystem boundaries set the amount or scale of complexity included in the management plan. Although boundaries are arbitrary, they should conform as closely as possible to the scale of complexity of the issue being addressed. For all cycles, the volume within these boundaries is used as an accounting tool to solve the conservation-of-mass equation:

input = output + change in storage.

**Box 3**
**Manipulating Vegetation Alters the Energy and Water Budgets**

Any manipulation of vegetation affects ecosystem energy, water, and nutrient budgets:

- Removing vegetation changes the albedo and energy balance, and thus it usually increases energy loss to the atmosphere, latent heat on soil surfaces, and decreases convective cooling and photosynthesis.
- Vegetation removal generally increases groundwater and streamflow because of reductions in interception and transpiration. The greatest increases in streamflow are in conjunction with highest precipitation. These streamflow increases will be observed annually.
- Severe disturbance or removal of the soil organic layer generally results in a shift from subsurface and groundwater flow to overland flow. During storms, this change results in greater streamflows (event scale). In between storms, streamflow may decrease. Annual streamflow will not necessarily change.
- Removing vegetation and the resulting changes in water and energy budgets can cause major shifts in ecosystem composition, as well as in the rates and timing of processes.

Second, all of the nutrient cycles are interconnected. The primary driver of all these cycles is energy from the Sun. This energy flux in turn, sets in motion the hydrologic cycle. The flow of water from the atmosphere to the land, into streams and oceans, and back to the atmosphere, is the primary mechanism by which chemicals are transported in the ecosystem. Some of the effects of vegetation management on the water cycle and energy balance are described in Box 3.

## 3.1 Carbon cycle

The atmosphere is the dominant carbon source for life on earth, and carbon dioxide ($CO_2$) makes up 0.03 percent of the atmosphere, on average.

### 3.1.1 Stand-scale Carbon Dynamics

The concentration of $CO_2$ in the atmosphere is rising because of combustion of fossil fuels and biomass burning for fuel or after deforestation. This change, discussed in greater detail below, raises the possibility that global primary productivity will increase because the $CO_2$ available to plants has been increased. In short-term laboratory and growth-chamber experiments, primary productivity does increase if sufficient amounts of other necessary nutrients and water are supplied. Natural stands may also have a short-term gain in biomass because of increased $CO_2$; but in the long term, however, plant growth rates could be constrained by limitations of other major nutrients, such as nitrogen or phosphorus (Likens et al. 1981, Peterson and Melillo 1985, Schlesinger 1991).

Carbon plays another important role in local vegetation dynamics, apart from being the major metabolic and structural component of all terrestrial plants. As plant tissue dies and falls to the ground, it accumulates to form organic matter. Organic matter is central to the cycling of plant nutrients and is a key factor in soil structure (Tisdall and Oades 1982, Parton et al. 1987). Soil organic matter enhances soil fertility by incorporating and stabilizing nutrients several ways, including enhancing soil water retention, producing organic acids that enhance mineral weathering, and retarding soil erosion. Soil organic-matter accumulation and retention in natural ecosystems is a balance between the rate of leaf and woody matter input and the rate of microbial decomposition and export of that material, and the rates are controlled by climate. The total amount of organic matter is a function of the rate of input and the rate of decomposition plus export.

Soil organic matter can be operationally divided into three components: an active fraction that is recycled within 1 to 5 years, a slow fraction that turns over at about 20- to 40-year intervals, and a passive, or recalcitrant, fraction with a recycle time between 200 and 1,500 years (Parton et al. 1987). The active fraction is comprised of live microbes, microbial products, and fresh plant remains. Slow-fraction organic matter can be physically protected by clay particles, and may also include chemically recalcitrant organic matter. The passive-matter fraction may also be physically protected and is often called *humus*.

Disturbance depletes soil organic matter, whether the agent is logging, plowing, or fire. Losses caused by disturbance from many soils are typically 20 to 30 percent (Schlesinger 1991). When soils are used for agriculture, organic matter loss is greatest in the first few years after cultivation, and levels off eventually to an amount that is in balance with the input from plant detritus and the greater rates of decomposition under disturbed ground (Jenkinson and Rayner 1977, Schlesinger 1991). Organic matter can recover after disturbance, although depending on conditions — decades to hundreds of years may be required (Schlesinger 1991, Lugo and Brown 1993).

### 3.1.2 Regional and Global Carbon Dynamics

Globally, most carbon on Earth is contained in the oceans (38,200 Gt) (Gt = $10^{15}$ g = $10^9$ t), followed by soils (1,580 Gt), the atmosphere (750 Gt), and terrestrial vegetation (610 Gt) (Schlesinger 1991, Houghton et al. 1994). The exchange of carbon between these pools is much more interesting than the pools themselves, particularly because in recent years, the flux rates have changed (Keeling and Whorf 1994).

Carbon dioxide in the atmosphere waxes and wanes seasonally, corresponding to the cycles of photosynthesis and respiration. Seasonality is most apparent in the northern hemisphere, reflecting the uptake of $CO_2$ by terrestrial vegetation of the northern continents. The seasonality, reversed in the southern hemisphere, is much less pronounced because so much less land area is south of the equator.

Because $CO_2$ is taken up by terrestrial vegetation, reducing atmospheric concentrations by increasing the amount of carbon taken up and stored in vegetation is theoretically possible. The terrestrial carbon storage pool may be increased by several positive-feedback processes between climate, mid- to high-latitude forest regrowth, and $CO_2$ and nitrogen fertilization, so that currently estimates suggest that up to 2.2 GtC/yr are being added to terrestrial ecosystems. All of these processes seem to contribute to the increased storage, but the uncertainties associated with each process are large (Houghton et al. 1994).

Climate influences carbon storage in terrestrial ecosystems through the direct effects of temperature, moisture, and solar radiation on carbon gain (net primary production) and carbon loss through respiration (Houghton et al. 1994, Schimel et al. 1994). Increased temperature may allow increased carbon uptake by trees because warming also increases soil microbial activity that may release nutrients from soil organic matter, making them available for plant uptake (Shaver et al. 1992). Enhanced microbial respiration reduces terrestrial carbon storage, however, so that boreal and tundra ecosystems — where large amounts of carbon are currently stored as organic matter — may have an increase in $CO_2$ efflux from soils as carbon is oxidized by microbes for energy (Oechel et al. 1993, Schimel et al. 1994).

A large proportion of forests in Europe, North America, and Northern Asia (Russia) deforested over the past several hundred years are now regenerating (Dixon et al. 1994, Houghton et al. 1994). The regeneration has been brought about by many complex social factors, such as abandonment of marginal farmlands, as in New England; large-scale establishment of tree plantations (FAO 1993); and increasing growth of urban centers and concurrent loss of rural, agrarian populations.

## 3.2 Nitrogen Cycle

Earth's atmosphere is 78 percent $N_2$ gas, the dominant source of nitrogen for plant growth. Atmospheric nitrogen becomes available for plant and microbial use in two ways: nitrogen can be fixed directly from the atmosphere by several types of bacteria and blue-green algae, and it can also be oxidized from $N_2$ gas to nitrogen oxides by combustion, including lightning and human use of fuels, allowing for deposition of NOx, including nitrate ($NO_3$) from the atmosphere to the biosphere. Although these atmospheric sources of nitrogen are the ultimate ones to the terrestrial biosphere, once the nitrogen is on the ground, retention and internal cycling account for most nitrogen available to ecosystem metabolism at any given time. In natural systems, internal cycling of nitrogen is 10 to 20 times greater than the amount received from outside the system; it arises from long-term accumulations of nitrogen in soil organic matter and plant biomass (Schlesinger 1991).

### 3.2.1 Stand-scale Nitrogen Cycling

Plants take up $NO_3$ and sometimes $NH_4$ from soil solutions and synthesize them to amino acids. Nitrogen is a component of chlorophyll so it can control rates of photosynthesis (Field and Mooney 1986). Nitrogen is also a component of enzymes, vitamins, and hormones, and it is essential for carbohydrate utilization (Stevenson 1986). In most ecosystems, nitrogen concentrations in solution are low, and plant uptake is actually enhanced by active transport by enzymes in root membranes (Ingestad 1982, Chapin 1988, Schlesinger 1991). Many, perhaps most plants also form *symbiotic* — mutually beneficial — associations with fungi or nitrogen-fixing bacteria. Fungal mycorrhizae can form a hyphal sheath around the fine roots of a plant, and extend additional hyphae — called *ectomycorrhizae* — into the soil. *Endomycorrhizae* actually penetrate into the plant root cells. The increased surface-area exposure to soil solutions brought about by mycorrhizal associations greatly increases the supply of nutrients to plant roots. The fungi obtain carbohydrates from the plant roots (Schlesinger 1991). Clearly plants gain through mycorrhizal association, but the cost also appears to be high. Vogt et al. (1982) found that mycorrhizal biomass was only 1 percent of the total biomass in a coniferous forest, but 15 percent of its net primary production was used by these fungi.

Symbiotic bacteria, such as *Rhizobium* and *Frankia*, live in root nodules of some plants and, with the enzyme nitrogenase convert atmospheric $N_2$ to $NH_4$. This process is energy-consuming, so nitrogen fixation is linked with the availability of organic carbon for respiration. Nitrogen fixation is inhibited by high available nitrogen, possibly also because of limitations of other nutrients, such as trace metals (like Mo, Co, and Fe); and by phosphorus (Schlesinger 1991). Conversely, nitrogen fixation enhances the ability of some primary succession plants, such as legumes and alders, to colonize new areas or recover after disturbance. Nitrogen-fixing exotic plants can rapidly gain a competitive advantage over non nitrogen-fixing natives. In Hawaii, invasion of *Myrica faya*, a nitrogen-fixing tree, has increased the availability of nitrogen to such an extent that not only is *Myrica* well established, but it has altered ecosystem dynamics and species composition in areas where it has invaded (Vitousek et al. 1987).

Asymbiotic bacteria and blue-green algae are important sources of nitrogen for some terrestrial ecosystems. Cryptogamic soil crusts have high rates of nitrogen fixation, annually delivering 1 to 5 kgN/ha to some desert ecosystems (Boring et al. 1988, Cushon and Feller 1989, Schlesinger 1991).

### 3.2.2 Nitrogen in Organic Matter

Nitrogen is retained in soils in the active, slow, and passive pools of soil organic matter, and is released or retained there by microbial activity. Nitrogen accumulation in soils begins with litter fall. Litter with high concentrations of nitrogen, such as some leaves, decompose rapidly, but litter with lower nutrient concentrations takes much longer to decompose and release nutrients (Fahey 1983, Schimel and Firestone 1989). Soil arthropods, nematodes, and earthworms perform an important function in churning and mixing fresh litter (Swift et al. 1979); however, the major breakdown of organic matter is by fungi and bacteria (Schlesinger 1991). *Mineralization* is the term used to describe conversion of organic matter — including nitrogen and other nutrients — from organic to inorganic (mineral) form. For nitrogen, this term usually refers to conversion from organic nitrogen to $NH_4$ to $NO_3$. *Immobilization*, the reverse process, is brought about by soil microbial consumption and sequestering of nitrogen and other nutrients in microbial biomass.

The carbon-to-nitrogen ratio (C:N) is often used as a measure of the quality of organic matter, with lower ratios being higher quality material. Woody material has a C:N ratio of about 150 to 160 because it is mostly structural carbon (Parton et al. 1987). Leaf litter is lower than wood, but it can differ by a factor of 4, depending on the nutrient availability of the site (Vitousek 1982). Plants, particularly on poor quality soils, resorb nitrogen and phosphorus from their leaves before the leaves fall; these nutrients, stored in roots and twigs, are used for the next year's growth (Vitousek 1982, Pastor et al. 1984, Chapin 1988). The C:N of microbial biomass has a low value of about 8, and unlike that in leaf litter, it has low variability (Parton et al. 1987). The slow and passive turnover pools of organic matter are also high in nitrogen, with C:N ratios of about 11 to 14 (Stevenson 1986, Parton et al. 1987).

### 3.2.3 Losses of Nitrogen from Ecosystems

Nitrogen in ecosystems is mobile and can be readily lost through several pathways, which undoubtedly contributes to nitrogen limitations in many systems. Nitrogen can be lost by bacterial denitrification, ammonia volatilization, leaching, and erosion. Bacteria denitrify flooded soils or sediments in the absence of oxygen and derive energy needed to break down sugars by converting $NO_3$ to $N_2O$ and $N_2$ gas. This process, widespread in terrestrial ecosystems, is stimulated by snowmelt and rain, flooding, and availability of $NO_3$, although it can be inhibited when concentrations of $NO_3$ are very high, and when it becomes carbon limited (Smith and Tiedje 1979, Davidson and Swank 1987, Matson et al. 1987, Keller et al. 1988, Schlesinger 1991). Losses of $N_2O$ from soils to the atmosphere have increased with increased application of commercial fertilizers to agricultural lands (Bremner and Blackmer 1978). Denitrification also increases in forests after burning or harvesting, and the combined increases in emissions are of concern because $N_2O$ is an important greenhouse gas that affects global climate change (Khalil and Rasmussen 1983, Keller et al. 1986, Matson and Vitousek 1987).

Ammonia volatilizes from soils that are alkaline, such as calcareous soils, and the rate increases with temperature. Losses are greatest in soils of low cation exchange capacity because clays and organic matter absorb $NH_4$ and immobilize it. Nitrogenous organic wastes, such as manures from heavily grazed pastures or feedlots, are concentrated sources of $NH_3$, that volatilizes to gaseous $NH_4$ (Stevenson 1986).

Significant amounts of nitrogen are not expected to leach from soils to surface waters in natural, undisturbed ecosystems that receive little atmospheric nitrogen (Hedin et al. 1995). In temperate ecosystems where vegetation goes through a dormant stage in winter, what little $NO_3$ is lost, leaches from the soil in winter and early spring, when snowmelt contributes $NO_3$ before vegetation begins to take it up. Soil microbial communities retain much of the available $NO_3$ (Zak et al. 1990), and the ephemeral spring plants in deciduous forest understories also act as "vernal dams" by taking advantage of the light, moisture, and nutrients available before canopies leaf out (Muller and Bormann 1976).

Leaching of soil $NO_3$ ($NH_4$ is nearly always retained in ecosystems because it is the most common nitrogen source for microbial organisms, which convert, or mineralize, $NH_4$ to $NO_3$ by a process known as *nitrification*) increases when terrestrial systems are disturbed, such as by fire, avalanche, hurricanes, blowdowns, or logging. Leaching increases because the processes of plant uptake are interrupted, and roots of the disturbed aboveground plant parts supply abundant carbon and nitrogen to microbial communities. Stottlemyer and Troendle (1995) found that nitrogen losses were still elevated 20 years after a coniferous forest in Colorado was harvested, suggesting that recovery from disturbance is not rapid under some conditions. Leaching can also increase when nitrogen inputs to ecosystems increase, such as from elevated nitrogen emissions (deposited wet or dry), or from excess nitrogen fertilizer. The nitrogen saturation hypothesis discussed in more detail under regional nitrogen processes, is based on the assumption that nitrogen deposition exceeds terrestrial plant and microbe demand, and the excess nitrogen leaches into aquatic systems (Aber et al. 1989, Stoddard 1994).

Erosion contributes to nitrogen losses from terrestrial ecosystems, particularly from soils predisposed to erode, such as disturbed soils. An estimated $4.5 \times 10^9$ kg of N is lost per year in erosion from agricultural soils, primarily as organic nitrogen (Stevenson 1986).

### 3.2.4 Regional or Landscape Scale to Global-scale Nitrogen Cycling

The chemical composition of leaves can serve as an index of site fertility and nutrient cycling rates in the plant, and broad regional patterns have been observed. Carbon-to-nitrogen and C:P ratios are higher in species from colder environments; they are lower in the tropics. In tropical forests, site fertility strongly influences content of nutrients in leaves (Vitousek and Sanford 1986, Vitousek et al. 1988). As with total plant biomass, nutrients in plant materials are distributed latitudinally, so that tropical > temperate > boreal in total nitrogen (Vitousek et al. 1988). In temperate biomes, the rates of nutrient circulation are much lower in conifers than in other forest types (Cole and Rapp 1981). Leaching losses are also lower in coniferous forests, and many evergreen species resorb more leaf nutrients during senescence than do deciduous species (Schlesinger 1991).

Before substantial human activity on Earth, the transfer of nitrogen from the atmosphere to the biosphere and back again was thought to be balanced at about 90 to 130 Tg N/y (Tg = $10^{12}$ g) (Delwiche 1970). In other words, the pools of $N_2$ in the atmosphere and the pools of organic nitrogen, nitrogen oxides such as $NO_3$, and ammonia-ammonium were stable. Human agricultural and energy production activities have changed this balance and much more N is being fixed from the atmosphere, which has regional and global consequences to ecosystem processes (Galloway et al. 1994). About an additional 140 Tg N/y is fixed from energy production (20 TgN/y), fertilizer production (80 TgN/y), and cultivation of legumes and rice (40 Tg N/y) (Galloway et al. 1994).

Most temperate and boreal terrestrial ecosystems are nitrogen limited, so that fertilizing them with nitrogen increases nitrogen storage, either in biomass or soil organic and microbial pools (Vitousek and Howarth 1991, Johnson 1992, Aber et al. 1993, Nadelhoffer et al. 1993). Dixon et al. (1994) estimated that temperate and other forests are accumulating about 7 Tg N/y, as a result of increased atmospheric $CO_2$ fertilization. Schindler and Bayley (1993) argue that increased nitrogen deposition will allow terrestrial vegetation to take up $CO_2$ more readily, thus increasing the amount of global carbon stored on land. For foresters interested in increasing timber yields, nitrogen deposition is a direct benefit from human activity (Kauppi et al. 1992).

Detrimental effects to natural ecosystems from nitrogen fertilization have also been reported, such as reducing species diversity and changing plant community composition (Van Vuren et al. 1992, Bowman et al. 1993, Tilman 1996). Over the long term, what appears to benefit plant growth may lead to losses in biodiversity of plant and aquatic ecosystems by *eutrophication* — the nutrient enrichment of an ecosystem (Galloway et al. 1994).

With excessive deposition, nitrogen becomes a pollutant, and Schulze et al. (1989) define annual critical loads for forest ecosystems of 3 to 14 kg/ha for forests on silicate soils and 3 to 48 kg/h for forests on calcareous soils. When plants and microbial biomass can no longer immobilize the nitrogen available, $NO_3$ leaches into ground and surface waters, carrying base cations along to maintain charge balance. Eventually, leaching depletes soils of Ca, Mg, Na, and K, leading to both cation limitation for plants and eventual acidification of surface waters. Although this theory of nitrogen saturation is still being debated, two lines of evidence point toward its validity: both gaseous and leaching losses of nitrogen are extremely low in ecosystems that receive low nitrogen inputs, but both leaching and gaseous losses increase with increased nitrogen inputs (Aber et al. 1989, Hedin et al. 1995, Stoddard 1994); and measurable loss of base saturation in soils from long-term inputs of strong acid anions (both $SO_4$ and $NO_3$) have been recorded in North America and Europe (Schulze et al. 1989, Dise and Wright 1995).

# 4 PART III: ECOSYSTEM MANAGEMENT THROUGH A FUNCTIONAL PERSPECTIVE

## 4.1 Succession in Terms of Primary Productivity, Respiration, and Cycling

In the absence of disturbance, ecosystems mature and pass through stages of succession with species replacements; species can be classified according to the stage of succession in which they appear. Eventually, a mature or near steady-state stage is reached, after which further structural or functional change is not apparent without detailed study. The chain of orderly events leading to the steady state assumes that no disturbance (natural or otherwise) will interrupt the succession. The effects of disturbances on succession will be discussed after we explain successional pathways in the absence of disturbance. Such pathways are intrinsic to ecosystems and operate most of the time because disturbances are usually short-term events that re-set ecosystems to earlier stages of succession and may change its direction dramatically.

Because of differences between inputs and outputs of an ecosystem, succession can deviate from, or never reach, a balanced condition between the production and respiratory functions of the ecosystem. We illustrate 14 successional trajectories in response to initial and terminal conditions of succession in the absence of disturbance (Fig. 1). The initial conditions are defined by the amounts of nutrients and organic matter in the ecosystem, and the ends are determined by the state of the ecosystem at maturity (balanced, autotrophic, heterotrophic, or equilibrium). All but two of the end stages of the successional trajectories are steady states. We will explain with a few examples.

Because primary productivity (P) and respiration (R) rates in the ecosystem are interconnected by the cycling of materials (Fig. 3), the natural tendency in succession is to balance these processes so that the ratio P/R approaches one (the photosynthesis rate is equal to the respiration rate). But an ecosystem can be balanced (that is, P/R = 1), only when it is open to energy but closed to matter. A closed terrarium and the Earth as a whole are examples of ecosystems that are closed to material cycles but open to solar energy and heat exchange. These conditions, in the absence of disturbance, lead to successions that end with P/R = 1.

Most natural ecosystems, however, are open to both matter and energy. Therefore, the materials required for the productive function (nutrients) and the respiratory functions (organic matter) can change and cause imbalances in their relative rates. Those that receive a predominance of materials that plants and animals can use in primary productivity more than they respire, are

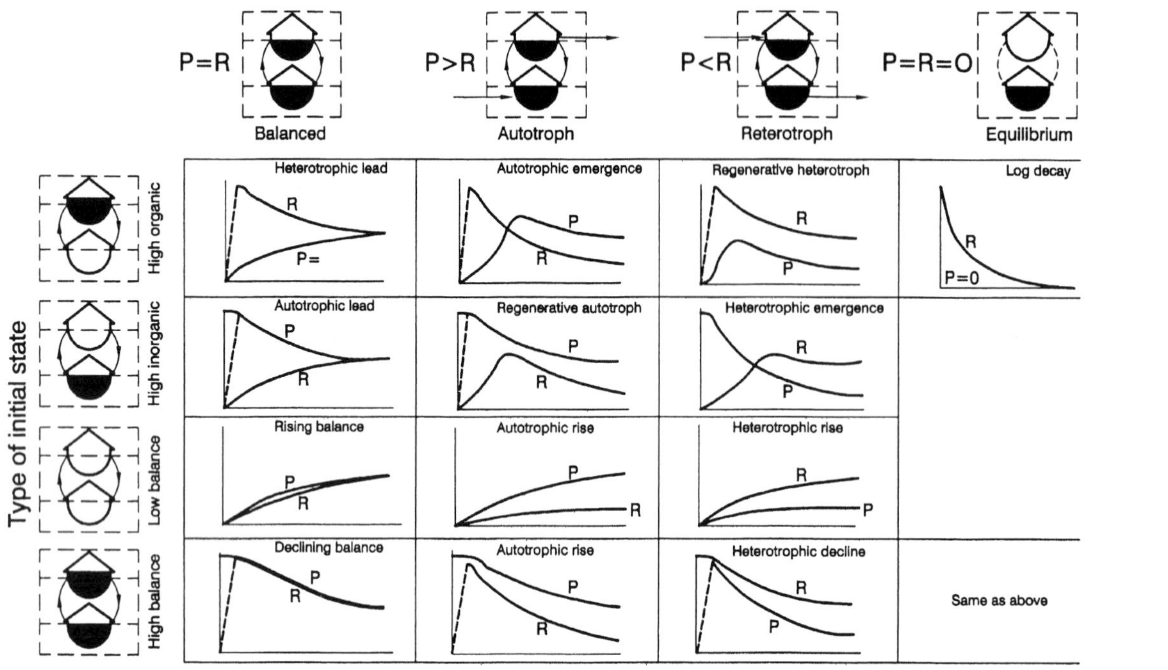


Fig. 3. Patterns of primary productivity and respiration in 14 types of succession uninterrupted by disturbance (from Odum 1971). The graphs show the successional trend in photosynthesis (P) and respiration (R). Graphs are arranged in a matrix with the initial conditions on the vertical axis and the terminal or climax conditions on the horizontal axis. The initial conditions shown are the nutrient and organic matter status of the ecosystem. These range from high and low nutrients and organic matter, to high nutrients but low organic matter, and low nutrients but high organic matter. The climax conditions include balanced P and R, greater R than P (autotrophic succession), P smaller than R (heterotrophic succession), and equilibrium (no light input).

termed *autotrophic* because their P/R is > 1. Agricultural systems and swamps are examples. Thus, successions in these systems tend to end with P/R > 1. If the exchange of materials across the ecosystem boundary favors respiration more than it does primary productivity, the system is termed *heterotrophic.* Estuaries, cities, caves, and sewage plants are examples. If light and organic matter input is blocked from an ecosystem — that is the system is closed to both matter *and* energy —food supplies are depleted and the organisms die off because no energy is input and regeneration and primary productivity are impossible. Once all of the food is consumed and decomposed, matter is dispersed into its components, $CO_2$ and water, and entropy is maximum. The resulting state is known as *equilibrium.* Equilibrium is the antithesis of life because no energy flows and no metabolic work is being done; in short, the system is dead. Many people confuse the term equilibrium with balance or steady state, but the meaning is erroneous in a thermodynamic sense (see Box 2). These types of succession do not end in steady state ecosystems.

The initial conditions of succession are also important because these conditions influence the successional trajectory. Ecosystems that initiate succession with a large storage of organic matter, have P/R ratios < 1 (more R than P); they change over time to systems with a greater contribution by plants and slowly approach a P/R of 1, as primary productivity catches up to respiration because the initial abundance of organic matter favors higher rates of respiration than of photosynthesis. An example would be succession that starts over organic sediments with high respiration rates. Ratios of P/R > 1 mean more production (P) relative to consumption (R), when storage of nutrients is high relative to the amount of organic matter. In this trajectory, primary production is higher than respiration during the initial stages of succession because the abundant nutrients stimulate the productive function of the ecosystem. An example is a young pine forest ecosystem growing on nutrient-rich, abandoned fields.

Mature forests are systems that have approached the terminal stages of succession either at P = R, P < R, or P > R. Most mature forests approach a balance between primary production and respiration (that is, $P/R \cong 1.0$). In these forests, most primary production is consumed by respiration, which maintains the system, leaving little net primary production for growth. Thus, the net ecosystem productivity approaches zero because most of the primary productivity is consumed by total ecosystem respiration (Part I).

When environmental resources are plentiful (that is, with ample, moisture, nutrients, and light) and temperatures are not extreme, biotic complexity and mechanisms that regulate the balance of the ecosystem (called *homeostatic* mechanisms) reach their maximum in mature ecosystems. The cost of maintaining high species diversity, complex biotic structures and interactions, and large biomass result in high respiration and slow growth of mature ecosystems. But the culmination of these trends is contingent on the absence of disturbances, which can reset ecosystem succession. To understand, conserve, and manage ecosystems better, we must understand their reactions to disturbances.

## 4.2 Disturbance and Succession

### 4.2.1 Disturbance

Successional changes are most obvious after a disturbance. A *disturbance* is a change to any state variable or flux of an ecosystem by any force external to the ecosystem. Disturbances drain or transfer mass and energy from one sector of the ecosystem to another. For example, the massive transfer of biomass and nutrients from the canopy to the forest floor after a storm (Frangi and Lugo 1991). A disturbance is a *stressor* as defined by Odum (1967a).

Disturbances can be natural — such as hurricanes, lighting-caused fires, and insect or disease outbreaks — or human-induced — such as arson, logging, artificial flooding, or drainage. If disturbances are repeated over long periods, natural selection may result in a genetic or evolutionary response in the affected organisms. *One difference between natural and human-induced disturbances is that the natural ones can have a periodicity (or predictability) but human-caused disturbances generally do not.* (Beissinger 1995). For example, oil spills, construction activities, and deforestation are haphazard events that affect particular ecosystems without warning or repeatability. *Some human-induced disturbances are chronic (for example, air pollution) and some are periodic* (such as regular draw-downs of wetlands or reservoirs) and result in genetic and community responses. Examples are mowing and grazing, which select for weeds of certain sizes and flowering phenologies, as well as certain densities of species and species composition (Grime 1973, Harper 1977).

*All disturbances have five attributes relevant to assessing ecosystem effects: severity, frequency of occurrence, duration, spatial scale, and point of interaction with the ecosystem* (see White et al., this volume). Here, we emphasize the point of interaction with the ecosystem because of its relevance to managing ecosystem functions.

The point of interaction with the ecosystem (that is, the structural or functional portions of the ecosystem with which the disturbance has initial or most critical interaction) varies for different types of disturbances

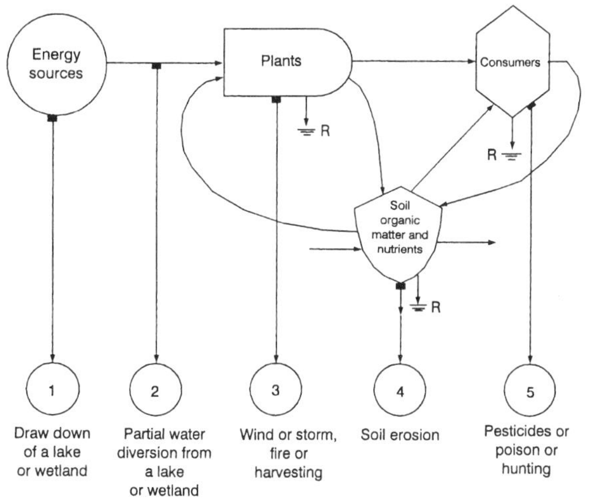


Fig. 4. Model of an ecosystem with five points of interaction with outside stressors shown as circles with numbers (Lugo 1978, Brown and Lugo 1994). The ecosystem consists of plants (capable of conducting photosynthesis as well as respiration [R]), consumers (animals and microbes capable only of respiration), and a storage of soil organic matter and soil nutrients. The soil compartment contains consumers as well, which is illustrated by the respiration rate. The ecosystem has external sources of energy (circle) as well as gains and losses of nutrients and organic matter (arrows to and from the storage symbol); that is, this system is open to both energy and materials. Lines with arrows show the direction and movement of matter and energy.

(Fig. 4). Some disturbances can affect more than one sector of the ecosystem (Lugo 1978). For example, some affect state variables (such as, plant biomass, soil, or populations). Others affect the influx of matter or energy (such as, reducing or increasing water supply or light), or affect internal fluxes (such as, interrupting herbivory, decomposition, or predation). Still others affect multiple loci of the ecosystem. For example, a hurricane transfers mass between sectors of a forest or between forest types along topographic gradients (Scatena and Lugo 1995), changes light inputs to the forest floor, and alters all internal processes of the ecosystem (Walker et al. 1991, 1996a).

We will now discuss succession after disturbance and ecological space, and then introduce the concept of interfaces to explain the functional responses that result from the interaction between external forces and the point of interaction with the ecosystem.

### 4.2.2 Succession After a Disturbance

A forest looks devastated after a major disturbance such as a fire, hurricane, or pest outbreak, but measurements suggest that actual tree mortality rarely exceeds 40% in the most devastated locations (Lugo and

Scatena 1996). An exception are landslides which can remove all the vegetation from a site (Walker et al. 1996b) or some fires that kill 100 percent of trees in a stand. Lugo and Scatena (1996) reviewed the rates of the tree mortality in tropical forests after major disturbances and differentiated between background (or normal) tree mortality with rates between 0 and 5%/yr, and sudden death catastrophic tree mortality, which usually has rates >5 percent and is fairly instantaneous (hours to weeks).

Catastrophic mortality is usually followed by one of two ecosystem states: (1) either the environment is radically different from what it was before the event or (2) the rainfall, temperature patterns, and soil conditions (but not the microclimate), that is, stand conditions — return to what they were before the event. Subtle environmental change is not as much a part of these phenomena as it is for background mortality. For example, after a landslide, debris flow, volcanic eruption, oil spill, massive flood, or lava flow, conditions on the affected sites are radically different from what they were moments earlier. Pre-event conditions are not likely to return to the site for a long time if ever (cf. Zarin 1993, Zarin and Johnson 1995a,b). After a fire, hurricane, wind storm, or a low-pressure atmospheric system (or tropical depression), climate and edaphic conditions return to pre-event conditions relatively quickly (cf. Silver 1992). The changes in these instances are microclimatic and attributed to residual effects of the disturbance: canopy destruction, massive transfer of mass and nutrients between ecosystem compartments, accumulation or pile-ups of organic debris at particular locations, and so on.

Regardless of the condition of the site, post-event ecological processes in ecosystems influenced by catastrophic tree mortality appear to change gears because succession is set back further than after background tree mortality, and ecological processes are either faster or slower. Succession is set back to early stages, even to primary succession with large landslides or lava and debris flows. The radical change in successional stage in such a short time is what differentiates catastrophic from background tree mortality.

Catastrophic tree mortality is associated with a different path of successional recovery than is background tree mortality which opens only small portions of the canopy or forest floor (Lieberman et al. 1985, Spies et al. 1990, Scatena and Lugo 1995), and its effects on light penetration, changes in microclimate, nutrient pools, and opportunity for species invasion or replacement are limited. Thus, that some studies show no changes in species composition after regeneration in forest gaps is not surprising (Lang and Knight 1983, Pérez Viera 1986).

In contrast, catastrophic tree mortality opens large areas of the canopy, changes the microclimate significantly (Fernández and Fetcher 1991), disrupts the forest floor, eliminates some species from stands, and allows new ones to invade the forest. Dittus (1985) found that after a cyclone struck a tropical forest in Sri Lanka, more canopy trees (46 percent) than subcanopy trees (29 percent) died. Furthermore, 22 percent of the upper canopy species were eliminated from the canopy and 12 percent were lost from the stand. Dittus pointed out that this differential damage leads to regional variability in species composition of the canopy and to a more uniform understory.

Dittus (1985) concluded that cyclones induce cyclic change in canopy composition. Similar responses have been reported for 1-ha patches of wet forests in Puerto Rico (Crow 1980, Doyle 1981, Weaver 1987, Lugo et al. 1995, Fu et al. 1996). In the absence of hurricanes, stands progressed toward states with lower tree species richness, a process that was reversed during the recovery period after hurricane disturbance. Similar cyclic successions are documented after catastrophic tree mortality from landslides (Guariguata 1989, 1990; Zarin 1993; Zarin and Johnson 1995 a,b; Myster and Fernández 1995).

Cyclic successions are also characteristic of naturally stressed, monospecific forests such as mangrove, and some temperate and boreal forests, but in these examples, catastrophic tree mortality does not usually induce changes in species composition (Lugo et al. 1995). Here, catastrophic tree mortality is followed by regeneration of the species present before the event (*autosuccession*), if site conditions remain unchanged. If stand conditions (such as salinity, hydrology, or geomorphology) change, a different group of species may dominate after the event (Lugo 1980, Jiménez et al. 1985). Regardless of latitude, however, human-caused disturbances provide opportunity for the invasion by exotic species into most ecosystems.

In summary, a catastrophic event or large and infrequent disturbance, including those induced by management activities, disrupts the ecosystem to such an extent that conditions for succession are radically different from when trees die at background rates. The canopy is opened, the forest is susceptible to species invasions, and changes in the direction of succession toward alternative states with different species composition are possible (Fig. 5). Processes can either be slowed so much that succession is arrested, or they can move at the fastest rates observable for particular locations. The first decade after any disturbance is a critical time to evaluate the effects of the event, the direction of succession, the needs for human intervention, and the alternatives available to the manager.

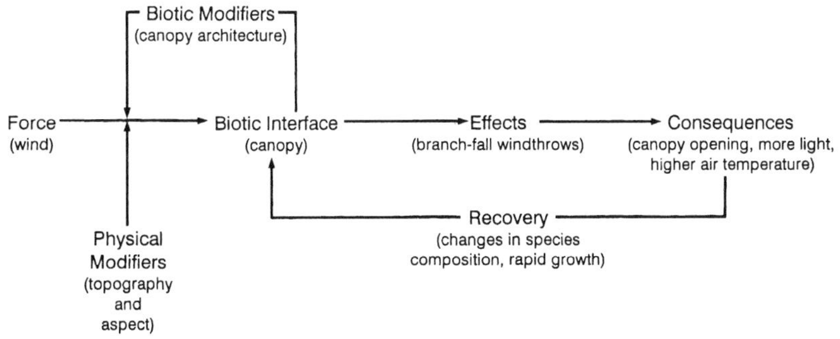


Fig. 5. Model of disturbance and succession. This diagram shows the action of disturbances and the response of ecosystems as a double feedback system. Examples are given in parenthesis. The disturbance is an external force affected by both physical and biotic modifiers. The biotic modifiers act as negative feedback on the disturbance (ecosystem resistance), and it originates from the biotic interface with which the disturbance interacts (point of interaction with the ecosystem). The result of the interaction between the disturbance and the biotic interface is shown as effects. Effects have consequences for ecosystem function, such as recovery or succession which act as another feedback into the interface. The recovery function is a measure of the resilience of the ecosystem. Because disturbances can change environmental conditions, exotic or other native species have an opportunity to invade the ecosystem.

## 4.3 Ecological Space and Functional Interfaces

The concept of ecological space (Hall et al. 1992) is based on the idea of the hypervolume of Hutchinson (1965). It relies on measuring the environmental gradients that affect the function of organisms. A *gradient* is defined by the change of any ecological parameter in space or time. We now have the technological means to be more inclusive and comprehensive in measuring environmental factors that affect ecological functioning. For this reason, estimating an annual mean of temperature, rainfall, or any other ecological parameter from a single location is no longer sufficient. Instead, we estimate how the environmental parameter changes over topography or temporally. For example, three generic sets of factors are shown in the bottom left panel of Fig. 6. Each of these generic factors is expanded in panels above and to the right. Gradients of these and other form the ecological context or *ecological space* in which organisms and ecosystems function.

Organisms and ecosystems perform differently along the range of values of each of the environmental factors that define ecological space. Organisms reproduce within their optimal range (Fig. 7). A broader range of conditions define their survival space, and the geographic range has an even wider range of values for each environmental factor.

Ecological space can be displayed with isopleths of organism performance within environmental conditions as shown in Fig. 8. An organism fixed in a given geographic space could walk through a wide variety of environmental conditions (dotted lines) that change within that space, particularly after a disturbance. At each point in ecological space, the performance of the population or the ecosystem differs (shown when it crosses the various isopleths).

Alternatively, the organism, population, or ecosystem can move geographically when ecological space changes drastically. Disturbance events, such as hurricanes or fires, have spatial and temporal patterns that shape the structure and functioning of ecosystems on a regional scale. They help define ecological space as well as ecosystems at larger spatial scales.

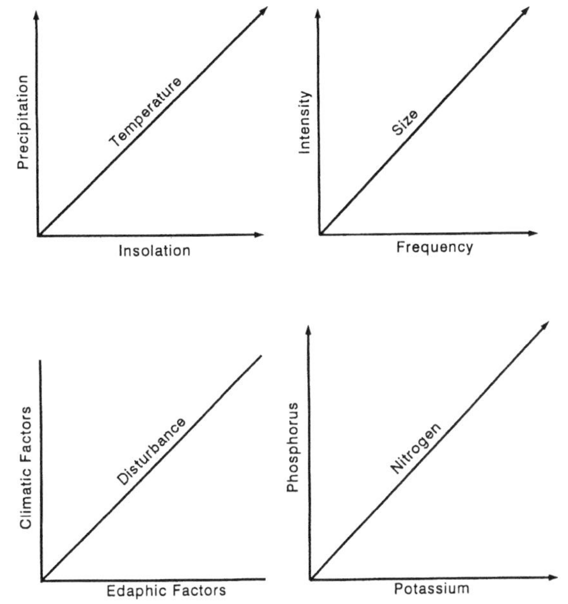


Fig. 6. Environmental gradients that affect organisms, populations, and ecosystems (from Hall et al. 1992). These gradients define the ecological space in which organisms, populations, and ecosystems function. The gradients are defined in the bottom left panel and include three types of environmental factors: climatic, disturbance, and edaphic (soil) factors. For each of these major groups of factors, a second set of gradients is illustrated. For example, disturbances are shown as having size, frequency, and intensity. Climatic factors are subdivided into precipitation, temperature, and insolation. Edaphic factors are subdivided into phosphorus, nitrogen, and potassium. Other sets of gradients can be shown for these and other environmental factors.

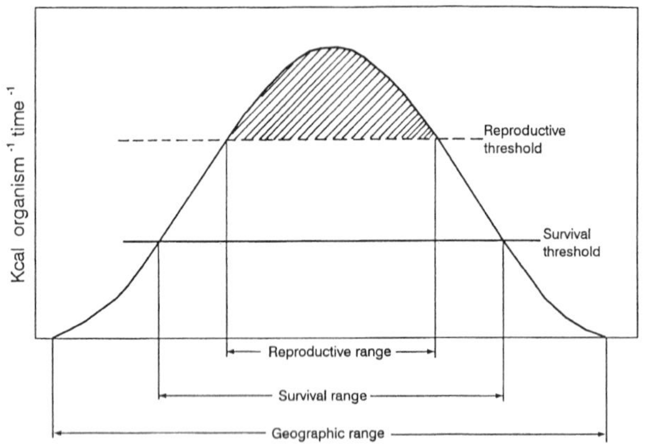


Fig. 7. Performance of organisms and populations along an environmental gradient (from Hall et al. 1992). For each environmental factor in ecological space, organisms and populations are expected to perform according to the graph shown; that is, they exhibit an optimum with decreasing performance towards both ends of the spectrum. Within this behavior, organisms exhibit maximum reproduction (fitness) at their optimum reproductive range, survive without major problems at a wider range of values (survival range), and can be found, albeit under stress, at the extremes (geographic range).

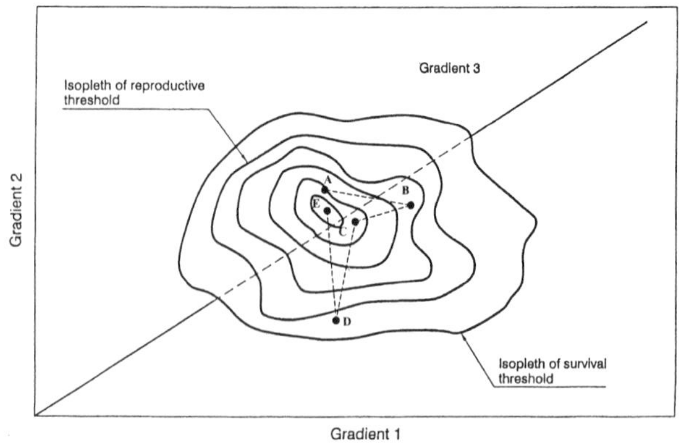


Fig. 8. Isopleths (or lines of equal value) of organism or population performance in relation to three environmental gradients that define their ecological space (from Hall et al. 1992). Letters A to E, connected by dotted lines, illustrate the movement of organisms and populations through ecological space. At point D, the population is outside its optimal ecological space for reproduction, while at point E, it is at its most optimal position.

Because of temporal and spatial variability, how we interpret environmental measurements must be reviewed. In Fig. 9, rainfall is presented in terms of intensity (cm/day) and frequency. Different amounts of rainfall affect different components of ecosystems, from individuals to landscapes. Such probabilistic analyses of environmental variables at particular places help to relate better the activities of organisms and ecosystems to ecological space.

The concept of ecological space is built in the Life Zone System of Holdridge (Fig. 10). Life zones are analogues of biomes but they are defined by climatic rather than taxonomic criteria. Life zones are bounded by gradients of rainfall, biotemperature (temperature corrected for frost and abnormally high temperatures), and potential evapotranspiration (cf. Holdridge 1967). Notice the use of logarithmic scales to define the life zone. Using climatic gradients as criteria to define life zones means that ecological space can be used to define the biotic organization and function at larger scales of complexity. Life zones are hierarchical; they can be subdivided into plant associations (communities) delimited by geomorphological conditions, and successional stages reflecting disturbance effects (human-caused or natural). They can be aggregated into humidity provinces (dry, moist, wet, and rain forest) and latitudinal (tropical, subtropical, temperate, boreal) and altitudinal (lowland, pre-montane, montane)

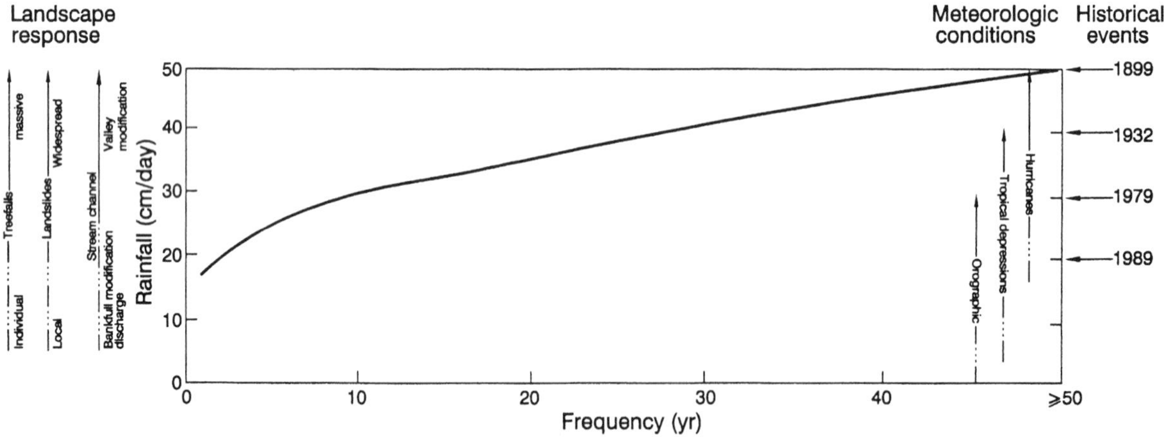


Fig. 9. Frequency and intensity of rainfall events and their effects on organisms, ecosystems, or the landscape at various geographical scales (from Lugo and Scatena 1995). These data are for a tropical climate with recurrent hurricanes (shown on the right of the figure). Notice that hurricanes are the terminus of a rainfall intensity gradient that starts with low-intensity orographic showers (showers caused by rising air masses). Landscape and biotic responses shown on the left, range in size according to the intensity of the rainfall.

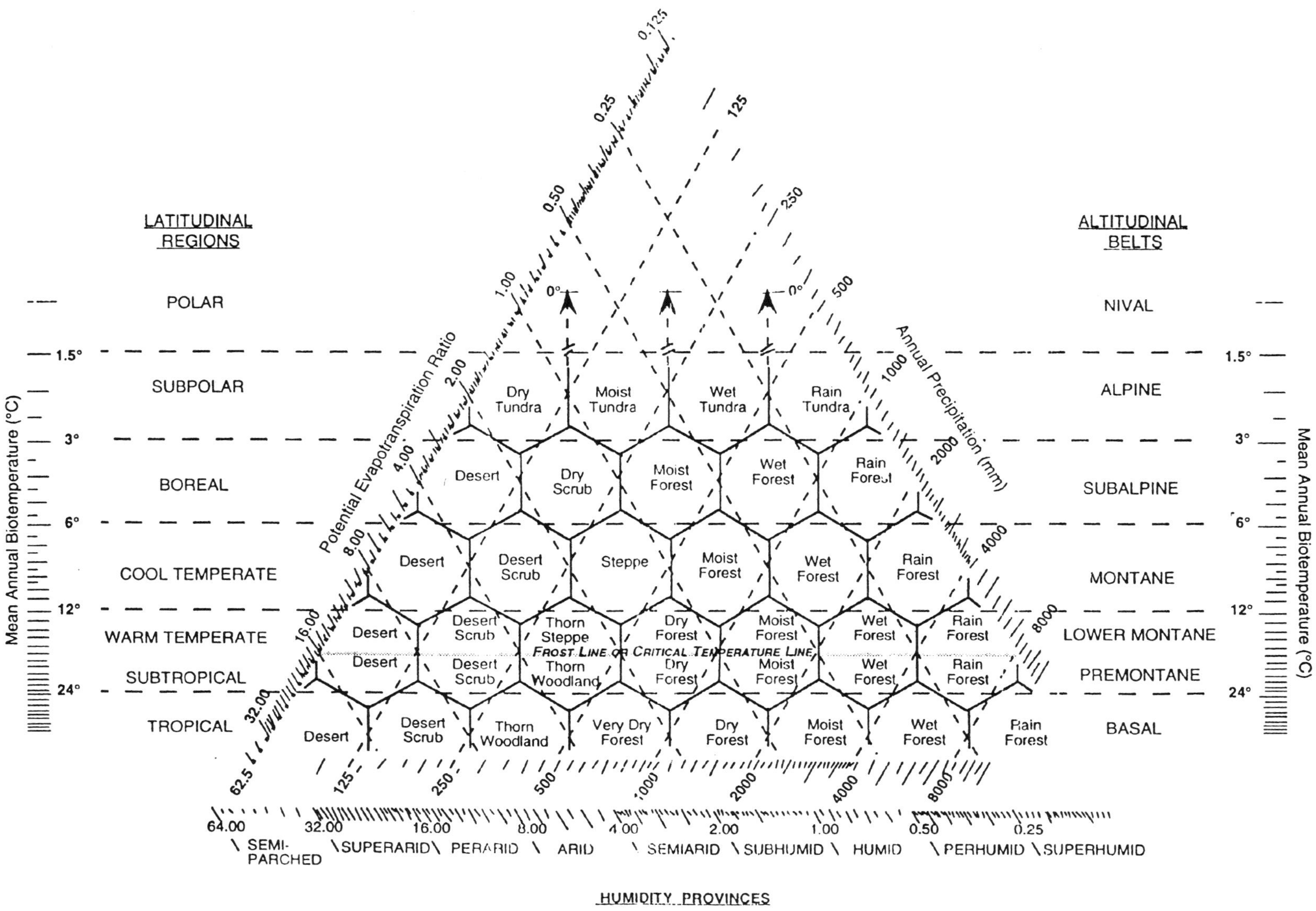


Fig. 10. The life-zone diagram for the classification of world life-zones or plant formations of Holdridge (1967). This system of ecosystem delimitation is based on annual rainfall, biotemperature, elevation, and potential evapotranspiration. Scales are logarithmic. Ecological space for life zones are delimited by the exagons, which represent points along climatic gradients. Biotemperature is the mean annual temperature (Celsius) from daily temperature data and assigning values of 0 to temperatures above 30°C and below 0°C.

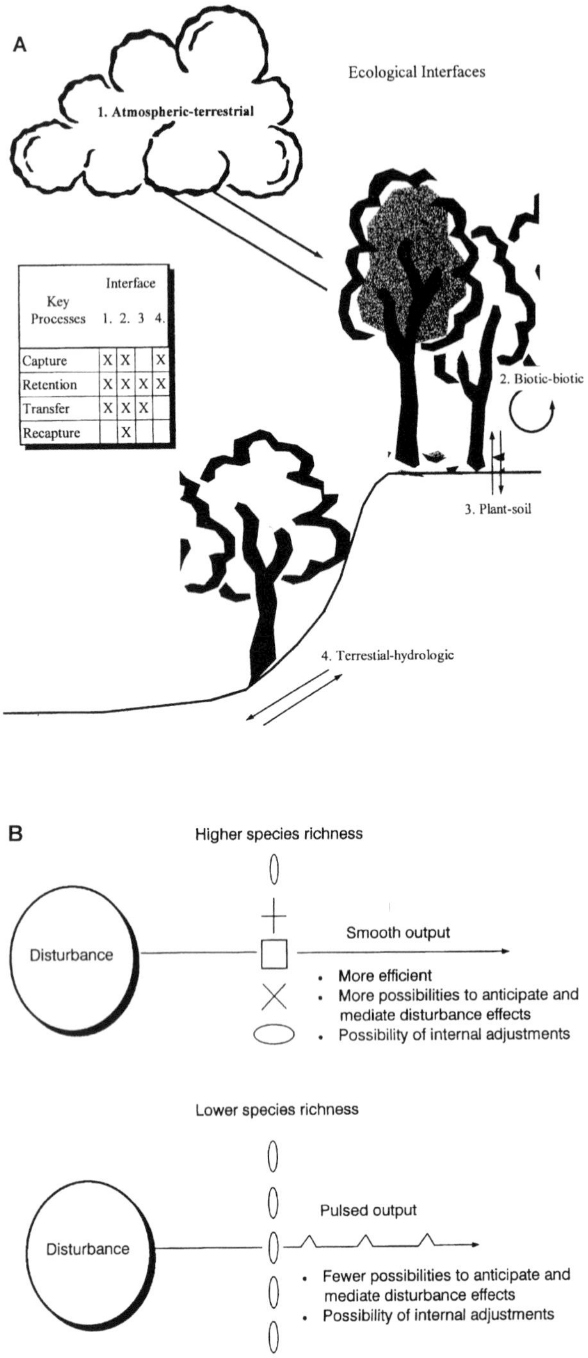


Fig. 11. Diagram of ecosystem interfaces (a) and an illustration of how species richness might affect ecosystem processes and functioning at an interface (b) (Silver et al. 1996). Four interfaces are illustrated in (a) and their function in terms of the key processes of capture, retention, transfer or recapture of nutrients is illustrated in the matrix. The number of species at any interface is probably responsible for the efficiency of nutrient use or response to disturbances (b).

groupings. Such a consistent and objective system of ecosystem classification provides functional continuity from the successional stage to the geomorphological association to the life zone to the humidity province to the altitudinal and latitudinal aggregate because systems are aggregated along natural gradients of the factors that drive their functioning (see Box 1).

We have defined ecological space as gradients of conditions that regulate the function of organisms and ecosystems regardless of their location in the hierarchy of biological and ecological complexity, i.e., from organisms to the whole Earth. Performance can range from optimal to below optimal along these ecological-space gradients. Ecosystem functioning is influenced by the gradients that define ecological space and can change dramatically after disturbances.

We now turn our attention to the point of interaction between the environmental factor and the ecosystem. We call the point of interaction between an environmental factor and the biota, the *interface.* We discussed how disturbance events dissipate energy at the interface or point of interaction with the ecosystem (Fig. 4), but not all external forces or events are destructive. Some disturbances are harnessed so that organisms do useful work at these interfaces. Examples of the interfaces where critical ecosystem functions occur (Fig. 11a,b) are the terrestrial–atmospheric, the plant–soil, terrestrial–hydrologic, and biotic–biotic interfaces (Silver et al. 1996). Functions at the various ecosystem interfaces are summarized in Table 2. A hypothesis of how organism diversity at the interface functions by influencing nutrient cycling is illustrated in Fig. 11b.

Individual species or particular life forms occupy interfaces in ecosystems; apparently, each species (Silver et al. 1996) or life form (Ewel and Bigelow 1996) has a particular role to play in ecosystem functioning. The ability of ecosystems to efficiently use all the resources available at a site and to support more or fewer species, depends on how effective species are in capturing resources at these interfaces (Haggar and Ewel 1997). But more species are present than the number of functions we can identify, which leads to equating functional diversity with species richness. Are they the same? Do ecosystems have functional redundancy? What is the significance of biodiversity to ecosystem function? No definitive answers to these questions have been found, and research on them is active. Species partition their functional roles along the gradients of ecological space, as described in Fig. 11b. This model means that the short list of ecological functions that we can identify is greatly multiplied by the huge range of conditions under which organisms live.

*Another point of view discussed in Ewel and Bigelow (1966) is that loss of species within a life form assemblage may not matter to ecosystem functioning because of the redundancy among them.* Grouping species by life forms provides a manageable number of functions to groups

Table 2. Examples of key functions in tropical ecosystems, the interface where they occur, the organisms or attributes that contribute to each key function and the mechanism through which they are accomplished (adapted from Silver et al. 1996)

| Ecological interface and key function | Responsible organism(s) or attribute | Mechanisms |
|---|---|---|
| *Atmospheric–terrestrial interface* | | |
| Energy capture | Foliage | Provide surface area for capture of sunlight. |
| Nutrient capture | Foliage and epiphytes | Increase surface area exposed to atmospheric nutrient inputs. |
| | N-fixing organisms | Fix atmospheric N in tissues. |
| Nutrient retention | Foliage, epiphytes, and wood | Immobilize nutrients in tissues and concentrate in stored water and arboreal soil; reduce the rate of water flow. |
| Nutrient transfer | Foliage, epiphytes, and wood | Cycle nutrients to terrestrial biota in litterfall; channel nutrients in stemflow and throughfall. |
| | N-fixing organisms | Transfer N to soils and plants in throughfall, litterfall, and decay. |
| *Biotic interface* | | |
| Energy retention | Live and dead plant tissues | Store energy in the form of biomass. |
| Energy transfer | Foliage and phloem | Transfer photosynthate from leaves to other plant tissues; transfer carbon-based products to herbivores. |
| Nutrient retention | Live plant tissues | Store nutrients in tissues; produce secondary chemicals to reduce herbivory. |
| Nutrient transfer | Live and dead plant tissues | Retranslocate nutrients within tissues to minimize nutrient losses; produce litter. |
| *Plant–soil interface* | | |
| Energy transfer | Litterfall | Transfer carbon to soil and forest floor. |
| Nutrient capture and nutrient and energy recapture | Rootmats and high fine-root biomass | Capture nutrients from rainfall, stemflow, throughfall; recapture nutrients and carbon before release to soil. |
| | Mycorrhizae and bacteria | Increase nutrient availability to plants by increasing exploitation of the rhizosphere; capture nutrients before release to soil. |
| Nutrient retention | Symbiotic N-fixers, rootmats, and high fine-root biomass | Fix soil atmospheric N and transfer directly to plant root; reduce the rate of water flow through the soil; store organic matter and nutrients in tissues. |
| Nutrient transfer | Fine roots and mycorrhizae; litterfall | Transfer nutrients from soil to plants; control decomposition rates by producing secondary compounds; and synchronize nutrient mineralization through control of litter quality and inputs. |
| *Terrestrial–hydrologic interface* | | |
| Nutrient capture and energy and nutrient recapture | Fine roots and soil microbes | Capture carbon and nutrients from stream water and alluvial sediments. |
| Energy and nutrient retention | Roots and coarse woody debris | Store carbon and nutrients in tissues, increase drainage; reduce the rate of surface flow of water and retain litter. |

of organisms: species are grouped by similar functions, and thus how serious the loss of a species might be can be evaluated. Maintaining all life forms is critical, and species become increasingly important when their loss represents the loss of a life form. For example, plants that only grow on other plants (epiphytes), vines, tall trees, and small trees appear to contribute uniquely to ecosystem functioning. Thus, the richness of species in a particular life-form group represents redundancy, and losing that whole life form would be more critical than losing a single, functionally redundant species. We present examples of how the biology of organisms viewed in a particular context leads to a better understanding of how biodiversity can contribute to ecological functioning at the interfaces of physical and biotic conditions.

Up to this point, we have:

- emphasized the functional aspects of ecosystems;
- clarified concepts and terminology to achieve consistency in their use;
- recognized ecosystems as functional units of ecology;
- advocated the use of environmental gradient analysis to define ecological functioning from organisms to life zones to the whole Earth;
- highlighted the use of environmental gradients to assure the continuity of ecosystem function when scales of biotic complexity are either expanded or reduced;
- advocated changing the analysis of ecosystems from geographic to ecological space;
- given examples of processes and functional attributes of ecosystems with a focus on energy, material cycles, succession, and biotic interactions;
- discussed how disturbances, like many management actions, shift ecological space and cause dramatic changes in the rates of ecosystem processes;
- suggested the need to match long-term, spatially specific environmental measurements to the proper scales of biotic activity; and
- identified interfaces as the places where ecological space interacts with ecosystem functioning.

This theoretical background is useful for guiding ecosystem management. Using the best theoretical framework available to support ecosystem management activities is most likely to produce effective ecosystem management because it best matches human actions to the forces that regulate the functioning of ecosystems.

## 4.4 A Functional Approach To Ecosystem Management

Ecosystem managers, be they foresters, agronomists, or horticulturists, use combinations of stress applications (disturbances) and stress removals (subsidies) to maintain ecosystem succession at a stage that matches their management objectives (Luken 1990). *Applying or removing stressors is synonymous with "management practices" because of the requisite costs, people continually seek to achieve management objectives at progressively lower costs.*

Succession can be hastened to achieve desired maturity by removing limiting factors, as is commonly done when poor soils are fertilized. Likewise, succession may be arrested or shifted backward by stressing the system in ways that prevent the potential energy of the system from being used to increase species diversity, biotic organization, or ecosystem complexity.

Pine forest managers in the southeastern United States commonly eradicate hardwoods, thus decreasing tree diversity and its associated ecosystem attributes. The primary productivity that would have been dissipated in maintaining (through respiration) the more complex hardwood systems is then concentrated as the woody biomass of the preferred pine forest. Thinning the stand has the same effect, by providing more space and resources to preferred trees and channeling a larger fraction of the stand's primary productivity toward a useful product. Natural disturbances and stressors achieve a similar sort of hardwood control, as is evident in sand pine forests, the sand hills, and the flatwoods of Florida. Disturbances or stressors, such as fire, maintain these ecosystems at earlier successional stages with fewer species than without the stressors.

The 14 successional pathways (Fig. 3) all have missing elements (that is, the disturbances), but one received more attention than the others in the 1970s: autotrophic succession starting with high nutrients and low organic matter and leading to a balanced steady state ecosystem. The classic example is the "old-field" succession in abandoned agricultural lands of the southeastern United States (Keever 1950), which led ecologists to emphasize the differences between mature and young systems (Odum 1969).

According to this vision, ecosystem succession culminated in mature systems with low net primary production, high gross primary production, high ecosystem respiration, closed and efficient nutrient cycles, maximum species diversity and biomass, and long-term stability (Fig. 12a). The resulting mature systems were considered fragile when subjected to human

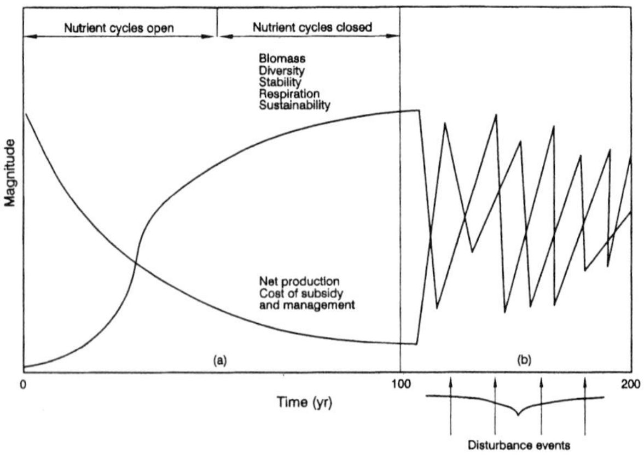


Fig. 12. Pattern of succession (a) from a young to a mature ecosystem, according to the paradigms of E.P. Odum (1969). This paradigm did not include disturbances and assumed that ecosystems had unlimited time to develop to maturity. Disturbances can set succession back at any time, however, creating the oscillations shown in (b). The outcomes of these two types of succession are quite different.

intervention because they had already reached a stage of succession where natural change was extremely slow. Environmental change was thus incompatible with maintaining such mature ecosystems because disturbances would set them back to an earlier stage of succession. Odum (1969) summarized this comparison by suggesting that mature systems were associated with protection (against disturbances), stability (few oscillations in time), and quality (chemical quality of tissues and types of species). In contrast, young systems were associated with production, growth, instability, and quantity.

This paradigm shows why managers must reduce the complexity of the ecosystem to maximize net production of commodities (for example, reducing species richness and using monocultures to increase wood or food production). For managers, high species diversity is incompatible with high net primary productivity of desirable species. Achieving high net production through management may also require a high energy subsidy because the natural tendency of ecosystems is to evolve complexity and exhibit low net production. Also, the nutrient cycles of low-diversity and high net-production ecosystems were considered open and not as efficient as those cycles in the mature forest considered closed. In short, this paradigm has managers spending energy and effort to work against nature and succession; affecting the fragility of mature ecosystems; and moving away from quality, protection, and stability.

A missing element from this concept of ecosystem management was the role of natural disturbances, with their various degrees of intensity and periodicity, and their ability to disrupt the steady maturation of ecosystems and maintain their oscillating states. Disturbances frequently prevent the establishment of a mature system or reduce the time the ecosystem can remain in a mature state (Fig. 12b). Also, field studies showed that the Odum paradigm was flawed in several ways. For example, ecosystem development is not well characterized by an asymptotic curve; rather, biomass accumulation reaches a maximum about the midpoint in ecosystem development and then declines in maturity (Bormann 1996). Also, nutrient cycles of all ecosystems are open, and the nutrient cycles of the successional systems are not necessarily less efficient than those of mature forests (Vitousek and Reiners 1975).

Moreover, ecologists have realized that comparing young ecosystems with mature ones, although of great ecological and academic interest, was the wrong comparison to assess the differences between managed and unmanaged ecosystems. Mature and young systems are different in part because of age differences, but many similarities emerge when comparisons are constrained to the same age and environmental conditions (Lugo 1992a). The differences depend on the intensity and quality of management. Ecosystems can be managed to conserve nutrients, preserve biological diversity, maintain options for natural succession, and achieve management objectives. Furthermore, many mature, natural ecosystems may not be as fragile as was once thought; the focus today is on ecosystem resilience (Box 4). But ecosystems are *always* vulnerable to human misuse and destruction.

**Box 4**
**Ecosystem Resilience**

"Some stress [or disturbance] is a regular part of all natural ecological systems, some more than others" (Odum 1967a).

If Odum is right, and we believe he is, ecosystems must have self-regulated processes to deal with stress and disturbance. Through adaptation, organisms contribute to the resilience of the ecosystems they form. Ecosystems exhibit resilience after a disturbance when environmental changes do not exceed the capacity of organisms to restore initial conditions. This capacity, in turn, depends on biological adaptations, and the availability of material and energy resources to repair the damage. Box 2 defined the terminology associated with ecosystem resilience and Fig. 2 illustrates the resilience phase of ecosystem response to disturbance as developed by Majer (1989a). Majer (1989b) edited a book with illustrative examples of the role of animals in ecosystem resilience during primary succession. A growing literature documents resilience mechanisms of ecosystems: for example, tropical forests-long believed to be the most fragile ecosystems of the world (Lugo 1995), lakes (Carpenter and Cottingham 1997), and the implications to ecosystem integrity (De Leo and Levin 1997). The resilience function of ecosystems is a major item in the tool kit of managers, one that has yet to be fully understood, but which provides flexibility to management when properly used.

Modern notions of ecosystem dynamics, ecosystem resiliency, and ecosystem restoration and rehabilitation are now leading to different ecological paradigms and a more flexible management environment (Holling 1973, 1978, 1986, 1992; Lugo 1988; Brown and Lugo 1994). Box 5 summarizes the paradigms now leading the way for ecosystem management. Before elaborating more fully on ecosystem management, we introduce the concept of the natural ecosystem manager.

## 4.5 Natural Ecosystem Managers

The natural world is often regarded as a place where creatures of all kinds live in a continuous struggle for survival. Although eat or be eaten appears to be the rule of nature, predation is not necessarily bad nor is nature in constant trauma (Boucher 1985). Careful

**Box 5**
**Paradigms of Ecosystem Management**

Advancement in ecosystem science has led to new paradigms in resource management. These are listed below with the resource management paradigms they replace.

- A non-steady-state concept of the ecosystem (as opposed to a focus on climax or balanced ecosystems).
- Managing from the perspective of resilience (as opposed to stability).
- Considering disturbance an integral part of ecosystems, required to maintain them (as opposed to suppressing or ignoring catastrophic factors).
- Considering past land-use legacies, all species, and the dead mass (necromass) of ecosystems, (as opposed to focusing only on the present state of the site, and the live component of a few species and populations).
- Focusing greater attention on the connections within and between ecosystems, particularly the interfaces of land, water, and atmosphere (as opposed to focusing only on the site under management).
- Considering all time and spatial scales (as opposed to focusing only on short-term and small geographic scales).
- Maintaining a global and long-term perspective even when managing at small scales (as opposed to a short-term, local perspective).
- Restoring whole ecosystems (as opposed to only rehabilitating land and water productivity).

interpretation of natural phenomena shows that relations among species in natural ecosystems can often assist in the functioning of the ecosystem that shelters them, as has been demonstrated for many animal species (Box 6).

Natural ecosystems contain — in their complex biological network —organisms that have been called natural ecosystem managers (Odum 1967b). *The natural ecosystem manager is an organism that uses a small fraction of the total energy budget of the ecosystem and provides a key function that pushes the ecosystem in a certain direction.* The functional outcome of the activities of ecosystem managers can be higher primary productivity, more efficient nutrient cycling, a service to other members of the ecosystem, or maintenance of the ecosystem in a particular state.

Interactions of natural ecosystem managers appear to be timed to the right moment and the right place. The natural ecosystem manager may be inconspicuous and seemingly unimportant (for example, earthworms), or dominant and destructive (for example, elephants), or its actions apparently aimed at a point far removed from the results of its actions (for example, predators that regulate the brouseline of forests by regulating herbivores). Organisms are effective ecosystem managers because they have evolved with the system they live in and perform a particular role in the ecosystem.

The concept of *keystone plant resources* was developed in animal studies for plant species that regularly produce edible fruits, seeds, or nectar during seasons when the food of birds and mammals are scarce (Terborgh 1983, 1986; Howe 1984; Mills et al. 1993). The plant species are considered keystones because, although they represent a tiny fraction of the ecosystem, nonmigrating animals depend on them. Ecosystem managers are also keystone species. Pollinators, seed dispersers, wood decomposers, and predators are natural ecosystem managers with keystone functions. These concepts developed independently for plants and animals as a result of similar types of observations of the importance of plants and animal functions in ecosystems.

All organisms have functions or roles in their ecosystems; when these functions are studied in detail at different scales of time and space, ecologists find that the web of nature is highly integrated and that all

**Box 6**
**Animals as Natural Ecosystem Managers**

Animal activity is responsible for multitudes of services to plants, microbes, and other animal species in the ecosystem. Many insects, birds, and mammals pollinate flowers, disperse seeds, trigger seed germination, and even plant seeds. Soil aeration, fertility, organic matter content, and structure in many ecosystems depend on animal activity. Notable examples are earthworms and other burrowing organisms. Decomposition of coarse woody debris and leaf litter is mediated and accelerated by animal activity (for example, termites, boring insects, crabs), and nutrient recycling can be accelerated by the activity of animals (such as leaf cutter ants). Herbivores control plant growth and even internal physiological states of plants. Predators regulate the population size of herbivore prey. Many animal species like alligators and elephants direct ecosystem succession or are directly responsible for the survival of other species during periods of drought or other stressful situations.

All of the activities have been documented by research that is the basis for terming animals "natural ecosystem managers." Termites, earthworms, ants, birds, ungulates, elephants, alligators, sea otters, and rabbits are a few examples of animal ecosystem managers that have been studied in some detail; summaries of this research can be found in the following books: Lee and Wood (1971), Odum (1971), Wilson (1971, 1975), Zlotin and Khodashova (1980), Satchell (1983), Majer (1989a), Luken (1990), Hendrix (1995).

**Box 7**
**Some Principles of Ecosystem Management**

The following suggestions for implementing ecosystem management are adapted from USDA Forest Service 1992.

- Follow nature's lead. Mimic natural disturbance patterns and recovery trends in your area.
- Think big. Manage for landscape diversity as well as within-stand or patch diversity, and maintain the largest possible contiguous patches of ecosystems of concern.
- Don't throw out any of the pieces. Maintain a diverse mix of genes, species, biological communities, and regional ecosystems.
- Side with the underdogs. Set priorities in favor of the species, communities, or processes that are threatened or otherwise warrant special attention.
- Try a different tool. Diversify management approaches and reduce the emphasis on complete conversion of ecosystems.
- Keep your options open. Use existing infrastructure or resources wherever possible.
- No ecosystem should be an island. Minimize fragmentation of continuous ecosystems by exploiting areas near existing clearings and by nibbling away at the edge instead of creating a new hole.
- Encourage free travel. Create a web of connected habitats. Leave broad travel connectors for plants and animals, especially along streams and ridgetops.
- Leave biological legacies. Select what to leave behind as carefully as what to take out; specifically, leave standing live and dead trees and fallen trees in managing forests.
- Leave it as nature would. Leave a mixture of tree sizes and species on the site. Restore naturally diverse forests after harvest.
- Be an information hound. Use the latest studies and state-of-the-art technologies to design, monitor, and evaluate new approaches.
- Be a critical thinker. Use only the scientific findings that make sense for your region and social setting.
- Monitor, monitor, monitor. Monitoring is the only sure way to tell if you are really conserving biological diversity.

organisms contribute in some way to the survival and functioning of the whole. All organisms play some role though we may not yet recognize what it is. How much functional redundancy exists in the web of nature is still unknown. Are all species needed to maintain ecosystem function? We have presented two contrasting points of view, both of which are speculations and the subject of active research and discussion (Schulze and Mooney 1993, Ewel and Bigelow 1996, Silver et al. 1996).

The lesson from natural ecosystem managers and keystone species is that complex ecosystems can be nudged one way or another, when low-energy interventions are applied at the right place and time. This lesson is the essence and goal of ecosystem management. In a world of declining fossil fuels, acting like natural ecosystem managers or keystone species suggests a strategy that ecosystem managers can emulate.

## 4.6 People as Ecosystem Managers

People are ecosystem managers, but our degree of integration with the natural world has varied through the history of our species with the intensity of our activities and their effects on ecosystems of all types. Our capacity to do things and influence ecosystems has recently increased as our energy resources changed from the food we could hunt or gather to making charcoal to harnessing rivers, estuaries, and winds and to using coal, petroleum, and nuclear power. Some current, less-energy-intensive, agriculture systems with ecological know-how reflecting the local circumstance are apparently sustainable. But the activities of early people were fundamental to the evolution of present ecosystems: numerous extinctions of species particularly on islands (Steadman 1991, 1995; Stoddart 1992), influences on landscapes through the use of fire (Bush et al. 1992), and changes in species composition in very large areas of the planet (Gómez-Pompa 1987a,b; Gómez-Pompa et al. 1987).

Today, people are modifying most of the biosphere and introducing new species and new ecosystems to landscapes (Burgess and Sharpe 1981, Odum and Turner 1990, Zonneveld and Forman 1990). Learning to follow the strategy of a natural ecosystem manager is a challenge that faces modern humanity. Odum (1962) summarized this challenge when he defined the field of ecological engineering:

> "...we suggest the term ecological engineering for those cases in which the energy supplied by man is small relative to the natural sources, but sufficient to produce large effects in the resulting pattern and processes."

Whether called ecological engineering or ecosystem management, this approach is clearly our best hope for

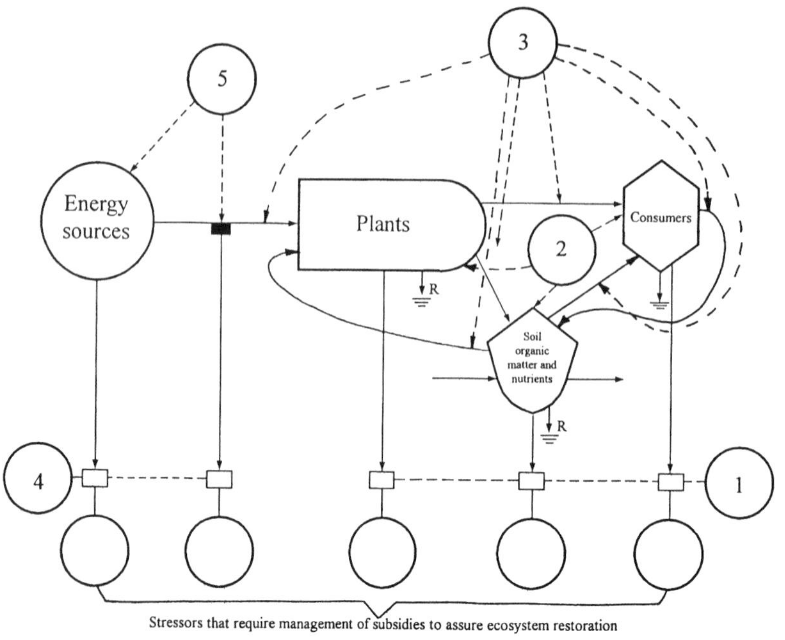


Fig. 13. Model of an ecosystem identical to the one shown in Fig. 3 to illustrate the interaction with five stressors, but adding the use of five types of subsidies to overcome the stressors and thus rehabilitate the ecosystem (based on Brown and Lugo 1994). Subsidies numbered 1 and 4 can be used to suppress all five stressors such as fire, frost, salinity, flooding, or drought. Type 2 subsidies can be used to add or subtract species, individuals, nutrients, or any other ecosystem structural component. Type 3 subsidies accelerate or decelerate ecosystem fluxes such as herbivory, decomposition, productivity, and respiration. Type 5 subsidies can be used to modify energy sources and inputs of the ecosystem. This type of analysis builds on the idea of ecosystem interfaces to focus management subsidies to particular target areas of ecosystem structures or functions.

emulating natural ecosystem managers and thus achieving our goals within the constraints of the natural laws that govern our world (Odum 1996) (Box 7).

We have asserted that management is a form of ecosystem disturbance. It differs from natural disturbances in that managing is a directed action with a purpose. Thus, the success of management depends on how well the disturbance is directed so that its purpose is achieved. The same ecosystem stress model discussed in Figure 4 appears in Figure 13 to illustrate the points of disturbance and stress interactions with ecosystems. We have added management actions that can suppress stressors (actions Types 1 and 4), as well as actions that manipulate ecosystems. Type 2 actions are directed at ecosystem reservoirs or populations. Type 3 actions are directed at ecosystem fluxes. Type 5 actions are directed at modifying the energy sources of the ecosystem. Management costs increase from actions 1 to 5, but the multiplier effect of actions and the unpredictability of their effects also increases from 1 to 5. One of the reasons for adaptive management, monitoring, and the need to maintain close links with research is the uncertainty in manipulating complex ecological systems. Many times, predicting how a particular sector of the ecosystem is going to respond to a given management action is impossible.

## 4.7 Combining Ecosystem Concepts with Ecosystem Management

Here we provide a seven-step process for applying ecosystem management criteria in the field (see Fig. 14). (For additional management considerations, see Paustian et al., this volume.) The macroscope in the figure helps visualize field situations holistically, an approach that requires a "detail eliminator" to focus properly on large scale rather than small-scale phenomena. What is not included below is the necessary first step of justifying management intervention by identifying the needs (products or services) sought, and the desired conditions of managed sites. These steps have been assumed to be well defined when the analysis begins.

1. *Identify ecosystems on the ground.* We made the point that ecosystem management uses ecosystem classifications based on a biological or ecological hierarchy as opposed to hierarchies of size because management actions are more successful if they are biologically correct. A single size-based scale can include more than one scale of biological or ecological complexity within its size units. Thus, each size unit cannot be managed uniformly because it can require a mix of treatments or strategies, each geared to a different biotic scale.

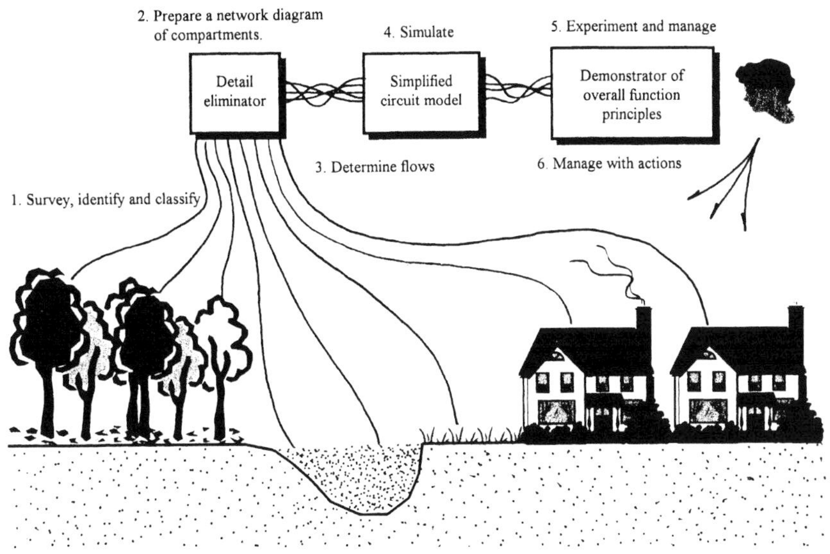


Fig. 14. The macroscope; the detail eliminator simplifies by grouping parts into compartments of similar function (from H.T. Odum 1971).

Regardless of theory, the fact is that managers, scientists, and conservationists deal with ecosystems defined by artificial boundaries, often based on size or political reality as opposed to biology or ecology. Clearly, managers must learn to deal with size-based hierarchies when organizing management or conservation actions, with the full understanding of the biotic complexity within the size boundaries selected. For example, a watershed provides convenient hydrogeochemical delimitation of ecosystems, but within the watershed boundary and depending on the size of the watershed, many scales of biological complexity can and must be identified. The study, management, and conservation of watersheds require identifying the contrasting ecological systems within their boundaries. The same requirement exists for identifying ecological systems within socioeconomic systems, such as county or state boundaries.

2. *Identify and quantify all demands to be placed on the resource.* Often, the demands on ecosystems include conflicting uses; if conflicts are not properly resolved, the resource can be damaged. Conflicting uses require a high degree of leadership and care on the part of the manager or conservationist. Having information about what demands are compatible with what ecosystem types is probably the most important contribution to ecosystem management and conservation from both scientists and experienced managers. Managers and conservationists can and must be responsive to people's needs without endangering the resources, including biodiversity, trusted to them.

Some of the most obvious demands on forests are wood, wildlife (to watch or hunt), preserving historical and unique sites, high-intensity recreation, vistas, a quiet place to meditate, controlling air and water quality, raw inorganic materials (such as, sand), or a place to grow food or graze cattle. How do we satisfy all these requests in a finite area? How do we prevent deterioration of the whole system? Where and how do we allocate resources for each activity? These questions must be answered by considering the capacity of each ecosystem to meet peoples needs and wants public demands. Figure 15 provides a framework for a decision process that laces together societal values and the ecological capacity of the ecosystem (see Bormann et al. 1994 for an elaboration of this framework).

3. *Describe and quantify the resource.* To begin to answer the questions just posed, we need to know what kinds of ecosystems are to be managed, which requires identifying all its hierarchies, from plant species associations to groups of life zones and their spatial distribution. The first step is thus a life-zone map *sensu* Holdridge (1967). Such a life-zone map is now available for the United States from A.E. Lugo. For smaller-scale analysis, the life-zone map can be supplemented with a vegetation map (at plant association scale) based on field surveys and remote sensing. Topographic, geologic, hydrologic, political, and climatic maps, as overlays or a geographic information system, will help in analyzing the resource under study. In examining the various ecosystems, managers are interested in the general physiological and structural adaptations to the local environment. The key to management success is identifying how best to use these adaptations and minimize problems of constantly subsidizing maintenance.

4. *List the opportunities and constraints for management.* Many factors can influence the success of a man-

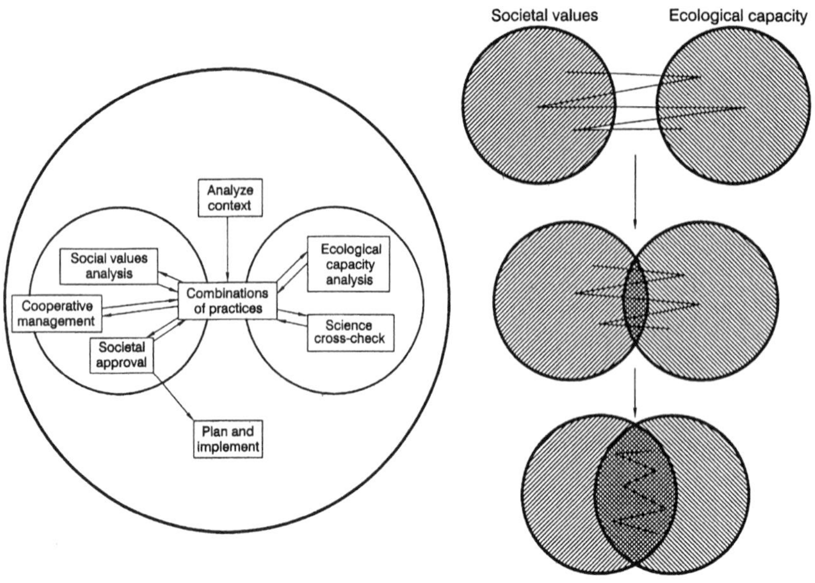


Fig. 15. An iterative decision process for any geographic scale to "lace" together societal values and ecological capacity of the ecosystem. Success in this process will produce overlap as shown on the bottom right (Bormann et al. 1994).

agement or conservation action. We have discussed the role and importance of disturbances in determining ecosystem states, structure, and functioning. Clearly any restoration or management scheme has to take into account the nature and importance of disturbances and stressors of the ecosystem under consideration. Failure to consider these components can ruin the best plans. In Jamaica, for example, extensive plantations of pine as part of a costly but well-executed plan to support a wood industry, were destroyed by a hurricane. These lands are now being converted to sun-grown coffee.

Social and cultural forces, bureaucratic constraints, and many other factors can influence ecosystem management objectives. Managing ecosystems will always yield some surprises; ecosystems are far too complex to be 100 percent predictable. And ecosystems are continuously changing because of changing global conditions. Dealing with global change and other surprises requires great flexibility (Holling 1986) and anticipatory management strategies (Lugo 1992b).

5. *Formulate a simulation model with alternative management options.* Advances in computer technology and understanding of ecosystem function allows ecosystems to be modeled in space and time; that is, before management actions are put to the test on the ground, they can be tested in the computer. Computer technology allows modern managers to test ideas and procedures electronically before a square meter of land is disturbed.

6. ***Determine the timing and application of management.*** Ecosystems have many rhythms that are expressed in a variety of cycles ranging from diurnal fluctuations to seasonal, annual, and longer cycles (such as, drought–flood cycles). These rhythms are often predictable to a degree and coupled to ecosystem function. For example, soil microbial biomass and nutrient uptake both increase in temperate forests in winter when trees are photosynthetically inactive. The high rates of spring growth of trees is supported by the mineralization of these nutrients. The processes of production and consumption are coupled, and a mechanism of nutrient conservation (increase in microbial biomass) is provided when tree nutrient uptake is inactive (Groffman et al. 1993). We should be cognizant of such ecosystem rhythms and periodicity so that management actions may be timed to the precise moment when such action is most needed, so the treated system is more likely to be able to continue its natural functions.

7. ***Adaptive management and self-design*** Adaptive management involves establishing monitoring systems in the ecosystems under management, and mechanisms to allow managers to adjust management actions in response to monitoring results. When properly implemented, adaptive management allows managers to maintain awareness on how ecosystems are functioning and they can adjust management strategies according to the results of previous actions. Bormann et al. (1994, and Vol. III of this book) give a detailed explanation of adaptive management, and Paustian et al. (this volume) discuss adaptive management in the context of managing ecosystem processes and functions. The ecosystem attribute most relevant to adaptive management is that of self-organization

(Odum 1988), or the capacity of ecosystems to develop new combinations of organisms and species in response to environmental gradients. Self-organization is the natural mechanism for developing complexity and functions that provide resilience and resistance to ecosystems. Managing with nature means allowing as much natural expression of self-organization as is possible without compromising the management objectives. Management costs decrease because natural processes take over much of the work of ecosystem maintenance and resilience.

## 5 CONCLUSIONS

Ecosystems are functional units of ecology. Ecosystem processes and functioning offer a powerful approach to ecosystem management. For best results, functional and process-oriented approaches to ecosystem management require a hierarchical and functional classification of ecosystems. Such classification of ecosystems is best achieved by using environmental gradients as the scales along which organisms, populations, association, and life zones function. These environmental gradients form multidimensional ecological spaces that influence rates and direction of processes and functioning of ecosystems. The biota is the medium through which ecosystems harness, distribute, and accumulate energy and materials. Energy and materials flow through the complex web of feeding interactions within ecosystems and enter the ecosystem at critical interfaces with the atmosphere, soil, water, or other organisms. Disturbances, whether natural or human-induced, influence ecosystem states over long periods. Ecosystem resilience in the face of disturbances, resistance to change, productive capacity, and ability to sustain productivity and complexity, *all* depend on maintaining ecosystem functions and processes. Some organisms perform critical functions in the ecosystem and are called keystone species or natural ecosystem managers. Other organisms that do not appear to have critical roles in the functioning of the ecosystem, are said to have functions redundant with those of other organisms. When disturbed, ecosystems undergo succession or change that results in self-organization and adjustment to the new conditions caused by the disturbance. Ecosystem management uses knowledge about ecosystem function and processes to obtain yields or services from ecosystems without irreversibly affecting their resilience and natural resistance to disturbances. Such management is termed sustainable: when it is conducted in an experimental mode with monitoring and re-adjustment, it is termed adaptive management.

## ACKNOWLEDGMENTS

We acknowledge the assistance of the following people: Mildred Alayón, Martha Brookes, Gisel Reyes, Carlos Domínguez Cristóbal, Frank H. Wadsworth, Nels Johnson, Gary K. Meffe, and the folks in the World Resources Institute.

## REFERENCES

Aber, J.D., A. McGill, R. Boone, J.M. Melillo, P. Steudler, and R. Bowden. 1993. Plant and soil responses to chronic nitrogen additions at the Harvard Forest, Massachusetts. *Ecological Applications* 3: 156–166.

Aber, J.D., K.J. Nadelhoffer, P. Steudler, and J.M. Melillo. 1989. Nitrogen saturation in northern forest ecosystems. *BioScience* 39: 378–386.

Adler, T. 1996. Healing waters: flooding rivers to repent for the damage done by dams. *Science News* 150: 188–189.

Barry, R.G. 1983. Late-Pleistocene climatology. pp. 390–407. In: H. E. Wright, Jr., and S. C. Porter (eds.), *The Late Pleistocene. Late-Quaternary Environments of the United States.* Vol. 1. University of Minnesota Press, Minneapolis, MN.

Beissinger, S.E. 1995. Modeling extinction in periodic environments: Snail kite population viability and Everglades water levels. *Ecological Applications* 5: 618–631.

Benson, L.V., and R.S. Thompson. 1987. The physical record of lakes in the Great Basin. pp. 241–260. In: W.F. Ruddiman and H.E. Wright, Jr. (eds.), *North America and Adjacent Oceans During the Last Deglaciation.* The Geology of North America, Centennial Special Volume K-3. Geological Society of America, Boulder, CO.

Betancourt, J.L., T.R. Vandever, and P.S. Martin (eds.). 1990. *Pack Rat Middens: The Last 40,000 Years of Biotic Change.* University of Arizona Press, Tucson, AZ.

Boring, L.R., W.T. Swank, J.B. Waide, and G.S. Henderson. 1988. Sources, fates, and impacts of nitrogen inputs to terrestrial ecosystems: review and synthesis. *Biogeochemistry* 6: 119–159.

Bormann, B.T., M.H. Brookes, E.D. Ford, A.R. Kiester, C.D. Oliver, and J.F. Weigand. 1994. Volume V: A Framework for Sustainable-Ecosystem Management. U.S.D.A. Forest Service General Technical Report PNW-GTR-331. U.S. Forest Service, Pacific Northwest Research Station, Portland, OR.

Bormann, F.H. 1996. Ecology: a personal history. *Annual Review of Energy and Environment* 21: 1–29.

Boucher, D.H. (ed.). 1985. *The Biology of Mutualism.* Oxford University Press, New York.

Bowman, W.D., T.A. Theodose, J.C. Schardt, and R.T. Conant. 1993. Constraints of nutrient availability on primary production in two alpine tundra communities. *Ecology* 74: 2085–2097.

Bremner, J.M., and A.M. Blackmer. 1978. Nitrous oxide: emission from soils during nitrification of fertilizer nitrogen. *Science* 199: 295–296.

Brown, S., and A.E. Lugo. 1994. Rehabilitation of tropical lands: a key to sustaining development. *Restoration Ecology* 2(2): 97–111.

Burgess, R.L., and D.M. Sharpe (eds.). 1981. *Forest Island Dynamics in Man-Dominated Landscapes.* Ecological Studies 41. Springer Verlag, New York.

Busacca, A.J. 1991. Loess deposits and soils of the Palouse and vicinity. pp. 216–228. In: R.B. Morrison (ed.), *Quaternary Nonglacial Geology, Conterminous U.S.* The Geology of North America, vol. K-2. Geological Society of America, Boulder, CO.

Busacca, A.J., and E.V. McDonald. 1994. Regional sedimentation of late Quaternary loess on the Columbia Plateau: sediment source areas and loess distribution patterns. pp. 181–190. In: R. Lasmanis and E.S. Cheney (eds.), *Regional Geology of Washington State.* Washington Division of Geology and Earth Resources, Bulletin 80. Olympia, WA.

Bush, M.B., D.R. Piperno, P.A. Colinvaux, P.E. de Oliveira, L.A. Krissek, M.C. Miller, and W.E. Rowe. 1992. A 14,300-yr paleoecological profile of a lowland tropical lake in Panama. *Ecological Monographs* 62: 251–275.

Carpenter, S.R., and K.L. Cottingham. 1997. Resilience and restoration of lakes. *Conservation Ecology* [online] 1(1): 2.

Chapin, F.S. 1988. Ecological aspects of plant mineral nutrition. *Advances in Mineral Nutrition* 3: 161–191.

Cole, D.W., and M. Rapp. 1981. Element cycling in forest ecosystems. pp. 341–409. In: R.E. Reichle (ed.), *Dynamic Properties of Forest Ecosystems.* Cambridge University Press, Cambridge, U.K.

Crow, T.R. 1980. A rain forest chronicle: A thirty-year record of change in structure and composition at El Verde, Puerto Rico. *Biotropica* 12: 42–45.

Cushon, G.H., and M.C. Feller. 1989. Asymbiotic nitrogen fixation and denitrification in a mature forest in coastal British Columbia. *Canadian Journal of Forest Research* 19: 1194–1200.

Davidson, E.A., and W.T. Swank. 1987. Factors limiting denitrification in soils from mature and disturbed southeastern hardwood forests. *Forest Science* 33: 135–144.

De Leo, G.A., and S. Levin. 1997. Multifaceted aspects of ecosystem integrity. *Conservation Ecology* [online] 1(1): 3.

Delwiche, C.C. 1970. The nitrogen cycle. *Scientific American* 223(3): 136–146.

Deverel, S.J., J.L. Fio, and N.M. Dubrovsky. 1994. Distribution and mobility of selenium in groundwater in the western San Joaquin Valley of California. pp. 157–183. In: W.T. Frankenberger, Jr., and S. Benson (eds.), *Selenium and the Environment.* Marcel Dekker, New York.

Dise, N.B., and R.F. Wright. 1995. Nitrogen leaching from European forests in relation to nitrogen deposition. *Forest Ecology and Management* 71: 153–162.

Dittus, W.P.J. 1985. The influence of cyclones on the dry evergreen forest of Sri Lanka. *Biotropica* 17: 1–14.

Dixon, R.K., S.A. Brown, R.A. Houghton, A.M. Solomon, M.C. Trexler, and J. Wisniewski. 1994. Carbon pools and flux of global forest ecosystems. *Science* 263: 185–190.

Doyle, T.W. 1981. The role of disturbance in the gap dynamics of a montane rain forest: An application of a tropical succession model. pp. 56–73. In: D.C. West, H.H. Shugart, and D.B. Botkin (eds.), *Forest succession concepts and application.* Springer-Verlag, New York.

DuBray, E.A. (ed.) 1995. Preliminary compilation of descriptive geoenvironmental mineral deposit models. USGS Open File Report 95–831.

Dzurizin, D., S.R. Brantley, and J.E. Costa. 1994. How should society prepare for the next eruption in the Cascades? Abstracts with Programs, Geological Society of America, 1994 Annual meeting. Seattle, WA.

Evans, F.C. 1956. Ecosystem as the basic unit in ecology. *Science* 123: 1127–1128.

Everest, F.H., R.L. Beschta, J.C. Scrivener, K.V. Koski, J.R. Sedell, and C.J. Cederholm. 1987. Fine sediment and salmonoid production: A paradox. pp. 98–142. In: E.O. Salo and T.W. Cundy, *Streamside management, forestry and fishery interactions.* Contribution 57. College of Forest Resources, University of Washington, Seattle, WA.

Ewel, J.J. 1980. Tropical succession: manifold routes to maturity. *Biotropica* 12 (suppl.): 2–7.

Ewel, J.J., and S.W. Bigelow. 1996. Plant life forms and tropical ecosystem functioning. pp. 101–126. In: G. Orians, R. Dirzo, and J. H. Cushman (eds.), *Biodiversity and Ecosystem Processes in Tropical Forests.* Springer Verlag, Berlin.

Fahey, T.J. 1983. Nutrient dynamics of aboveground detritus in lodgepole pine (*Pinus contorta* ssp. *latifolia*) ecosystems, southeastern Wyoming. *Ecological Monographs* 53: 51–72.

FAO. 1993. *Forest Resources Assessment 1990.* Tropical countries. FAO Forestry Paper 112, Rome.

Fernández, D.S., and N. Fetcher. 1991. Changes in light availability following Hurricane Hugo in a subtropical montane forest in Puerto Rico. *Biotropica* 23: 393–399.

Field, C., and H.A. Mooney. 1986. The photosynthesis–nitrogen relationship in wild plants. pp. 25–55. In: T.J. Givnish (ed.), *On the Economy of Plant Form and Function.* Cambridge University Press, Cambridge, U.K.

Flint, R.F., and B.J. Skinner. 1974. *Physical Geology.* Wiley, New York.

Frangi, J.L., and A.E. Lugo. 1991. Hurricane damage to a flood plain forest in the Luquillo Mountains of Puerto Rico. *Biotropica* 23: 324–335.

Fu, S., C. Rodríguez Pedraza, and A.E. Lugo. 1996. A twelve-year comparison of stand changes in a mahogany plantation and a paired natural forest of similar age. *Biotropica* 28: 515–524.

Galloway, J.N., H. Levy II, and P.S. Kasibhatla. 1994. Year 2020: Consequences of population growth and development on deposition of oxidized nitrogen. *Ambio* 23: 120–123.

Gómez-Pompa, A. 1987a. Tropical deforestation and Maya silviculture: An ecological paradox. *Tulane Studies in Zoology and Botany* 26: 19–37.

Gómez-Pompa, A. 1987b. On Maya silviculture. *Mexican Studies* 3(1): 1–17. Regents of the University of California.

Gómez-Pompa, A., J.S. Flores, and V. Sosa. 1987. The 'Pet Kot': A man-made tropical forest of the Maya. *Interciencia* 12: 10–15.

Grime, J.P. 1973. Control of species density in herbaceous vegetation. *Journal of Environmental Management* 1: 151–167.

Groffman, P.M., D.R. Zak, S. Christensen, A. Mosier, and J.M. Tiedje. 1993. Early spring nitrogen dynamics in a temperate forest landscape. *Ecology* 74: 1579–1585.

Guariguata, M.R. 1989. *Landslide natural disturbance and forest regeneration in the Luquillo Mountains of Puerto Rico.* M.S. Thesis. University of Florida, Gainesville, FL.

Guariguata, M.R. 1990. Landslide disturbance and forest regeneration in the upper Luquillo Mountains of Puerto Rico. *Journal of Ecology* 78: 814–832.

Haggar, J.P., and J.J. Ewel. 1997. Primary productivity and resource partitioning in model tropical ecosystems. *Ecology* 78: 1211–1221.

Hall, C.A.S., J.A. Stanford, and F.R. Hauer. 1992. The distribution and abundance of organisms as a consequence of energy balances along multiple environmental gradients. *Oikos* 65: 377–390.

Hamblin, W.K. 1985. *The Earth's Dynamic Systems*. Burgess Publishing Co., Minneapolis, MN.

Harper, J.L. 1977. *Population Biology of Plants*. Academic Press, New York.

Hecky, R.E., P. Campbell, and L.L. Hendzel. 1993. The composition of precipitation in remote areas of the world. *Journal of Geophysical Research* 87: 8771–8776.

Hedin, L.O., J.J. Armesto, and A.H. Johnson. 1995. Patterns of nutrient loss from unpolluted, old-growth temperate forests: evaluation of biogeochemical theory. *Ecology* 76: 493–509.

Hendrix, P.F. (ed.) 1995. *Earthworm Ecology and Biogeography in North America*. Lewis Publishers, Boca Raton, FL.

Holdridge, L.R. 1967. *Life Zone Ecology*. Tropical Science Center, San Jose, Costa Rica.

Holling, C.S. 1973. Resiliency and stability of ecological systems. *Annual Review of Ecology and Systematics* 4: 1–23.

Holling, C.S. 1978. *Adaptive Environmental Assessment and Management*. Wiley, New York.

Holling, C.S. 1986. Resiliency of ecosystems; local surprise and global change. pp. 292–317. In: W.C. Clark and R.E. Munn (eds.), *Sustainable Development of the Biosphere*. Cambridge University Press, Cambridge, U.K.

Holling, C.S. 1992. Cross-scale morphology, geometry, and dynamics of ecosystems. *Ecological Monographs* 62: 447–502.

Houghton, J.T., L.G. Meira Filho, J. Bruce, H. Lee, B.A. Callander, E. Haites, N. Harris, and D. Maskell (eds.) 1994. *Radiative Forcing of Climate Change and an Evaluation of the IPCC IS92 Emissions Scenarios*. Intergovernmental Panel on Climate Change. Cambridge University Press, Cambridge, U.K.

Howe, H.F. 1984. Implications of seed dispersal by animals for tropical reserve management. *Biological Conservation* 30: 261–281.

Hutchinson, G.E. 1965. The niche: an abstractly inhabited hypervolume. pp. 26–78. In: *The Ecological Theater and the Evolutionary Play*. Yale University Press, New Haven, CT.

Ingestad, T. 1982. Relative addition rate and external concentration: Driving variables used in plant nutrition research. *Plant, Cell and Environment* 5: 443–453.

Jenkinson, D.S., and J.H. Rayner. 1977. The turnover of soil organic matter in some of the Rothamsted classical experiments. *Soil Science* 123: 298–305.

Jenson, M., I. Goodman, K. Brewer, T. Frost, G. Ford, J. Nesser. 1997. Biophysical Environments of the Basin. In: T.M. Quigley and S.J. Arbrelbide (eds.), An Assessment of Ecosystem Components in the Interior Columbia Basin and Portions of the Klamath and Great Basins. Part I. Biophysical environment. U.S.D.A. Forest Service General Technical Report. PNW-GTR-408, pp. 99–307.

Jiménez, J.A., A.E. Lugo, and G. Cintrón. 1985. Tree mortality in mangrove forests. *Biotropica* 17: 177–185.

Johnson, D.W. 1992. Nitrogen retention in forest soils. *Journal of Environmental Quality* 21: 1–12.

Kauppi, P.E., K. Mielikäinen, and K. Kuusela. 1992. Biomass and carbon budget of European forests, 1971–1990. *Science* 256: 70–74.

Keeling, C.D., and T.P. Whorf. 1994. Decadal oscillations in global temperature and atmospheric carbon dioxide. In: W.A. Sprigg (ed.), *Natural Variability of Climate in Decade-to-Century Time*. National Academy of Sciences, Washington, DC, 132 p.

Keever, C. 1950. Causes of succession on old fields of the Piedmont, North Carolina. *Ecological Monographs* 20: 229–250.

Keller, M., W.A. Kaplan, and S.C. Wofsy. 1986. Emissions of $N_2O$, $CH_4$ and $CO_2$ from tropical forest soils. *Journal of Geophysical Research* 91: 11791–11802.

Keller, M., W.A. Kaplan, S.C. Wofsy, and J.M. DaCosta. 1988. Emission of $N_2O$ from tropical forest soils: Response to fertilization with $NH_4^+$, $NO_3^-$ and $PO_4^{3-}$. *Journal of Geophysical Research* 93: 1600–1604.

Kerr, R.A. 1993. Pinatubo global cooling on target. *Science* 259: 594.

Khalil, M.A.K., and R.A. Rasmussen. 1983. Increase and seasonal cycles of nitrous oxide in the Earth's atmosphere. *Tellus* 35B: 161–169.

King, A.W. 1993. Considerations of scale and hierarchy. pp. 19–46. In: S. Woodley, J. Kay, and G. Francis (eds.), *Ecology Integrity and the Management of Ecosystems*. St. Lucie Press.

Kwong, Y.T.J. 1993. Prediction and Prevention of Acid Rock Drainage from a Geological and Mineralogical Perspective. MEND Project 1.32.1. Ottawa, Canada: CANMET.

Lang, G.E., and D.H. Knight. 1983. Tree growth, mortality, recruitment, and canopy gap formation during a 10-year period in a tropical moist forest. *Ecology* 64: 1075–1080.

Lee, K.E., and T.G. Wood. 1971. *Termites and Soils*. Academic Pressm New York.

Lieberman, D., M. Lieberman, R. Peralta, and G.S. Hartshorn. 1985. Mortality patterns and stand turnover rates in a wet tropical forest in Costa Rica. *Journal of Ecology* 73: 15–924.

Likens, G.E., and F.H. Bormann. 1995. *Biogeochemistry of a Forested Ecosystem*. Springer Verlag, New York, 159 p.

Likens, G.E., F.H. Bormann, and N.M. Johnson. 1981. Interactions between major biogeochemical cycles in terrestrial ecosystems. pp. 93–112. In: G.E. Likens (ed.), *Some Perspectives of the Major Biogeochemical Cycles*. Wiley, New York.

Lipman, P.E., and D.R. Mullineaux (eds.) 1981. The 1980 Eruptions of Mt. St. Helens, *Washington*, Washington, D.C. U.S. Geological Survey Professional Paper 1250. pp. 347–377.

Lugo, A.E. 1978. Stress and ecosystems. pp. 62–101. In: J.H. Thorp and J.W. Gibbons (eds.), *Energy and Environmental Stress in Aquatic Systems*, U.S. Department of Energy Symposium Series CONF 77114. National Technical Information Services, Springfield, VA.

Lugo, A.E. 1980. Mangrove ecosystems: successional or steady state? *Biotropica* 12 (suppl.): 65–72.

Lugo, A.E. 1988. The future of the forest. Ecosystem rehabilitation in the tropics. *Environment* 30(7): 16–20, 41–45.

Lugo, A.E. 1992a. Comparison of tropical tree plantations with secondary forests of similar age. *Ecological Monographs* 62(1): 1–41.

Lugo, A.E. 1992b. Managing tropical forests in a time of climate change. pp. 336–344. In: A. Qureshi (ed.), *Forests in a Changing Climate*. Climate Institute, Washington, DC.

Lugo, A.E. 1995. Management of tropical biodiversity. *Ecological Applications* 5: 956–961.

Lugo, A.E., A. Bokkestijn, and F.N. Scatena. 1995. Structure, succession, and soil chemistry of palm forests in the Luquillo Experimental Forest. pp. 142–177. In: A.E. Lugo and C. Lowe (eds.), *Tropical Forests: Management and Ecology*. Springer-Verlag, New York.

Lugo, A.E., and S. Brown. 1993. Management of tropical soils as sinks or sources of atmospheric carbon. *Plant and Soil* 149: 27–41.

Lugo, A.E., J. Figueroa Colón, and F.N. Scatena. The Caribbean. In: D. Billings and M. Barbour (eds.), *Vegetation of North America*. Cambridge University Press, Cambridge, U.K., in press.

Lugo, A.E., and F.N. Scatena. 1995. Ecosystem-level properties of the Luquillo Experimental Forest with emphasis on the tabonuco forest. pp. 59–108. In: A.E. Lugo and C. Lowe (eds.), *Tropical Forests: Management and Ecology*. Springer Verlag, New York.

Lugo, A.E., and F.N. Scatena. 1996. Background and Catastrophic Tree Mortality. *Biotropica* 28(4a): 585–599.

Luken, J.O. 1990. *Directing Ecological Succession*. Chapman and Hall, New York.

Magnuson, J.J. 1990. Long-term ecological research and the invisible present. *BioScience* 40(7): 495–501.

Majer, J.D. 1989a. Long-term colonization of fauna in reclaimed lands. pp. 143–174. In: J.D. Majer (ed.), *Animals in Primary Succession: The Role of Fauna in Reclaimed Lands*. Cambridge University Press, Cambridge, U.K.

Majer, J.D. (ed.) 1989b. *Animals in Primary Succession: The Role of Fauna in Reclaimed Lands*. Cambridge University Press, Cambridge, U.K.

Matson, P.A., and P.M. Vitousek. 1987. Cross-system comparisons of soil nitrogen transformations and nitrous oxide flux in tropical forest ecosystems. *Global Biogeochemical Cycles* 1: 163–672.

Matson, P.A., P.M. Vitousek, J.J. Ewel, M.J. Mazzarino, and G.P. Robertson. 1987. Nitrogen transformations following tropical forest felling and burning on a volcanic soil. *Ecology* 68: 491–502.

Matthews, J.D. 1989. *Silvicultural Systems*. Clarendon Press, London.

McIntosh, R.P. 1985. *The Background of Ecology; Concept and Theory*. Cambridge University Press, Cambridge, U.K.

Mickelson, D.M., L. Clayton, D.S. Fullerton, and H.W. Borns, Jr. 1983. The late Wisconsin glacial record of the Laurentide ice sheet in the United States. pp. 3–37. In: H.E. Wright, Jr., and S.C. Porter (eds.), *The Late Pleistocene. Late-Quaternary Environments of the United States*. Volume 1. Minneapolis, MN: University of Minnesota Press.

Mills, L.S., M.E. Soulé, and D.F. Doak. 1993. The keystone species concept in ecology and conservation biology. *BioScience* 43: 219–224.

Muller, R.N., and F.H. Bormann. 1976. The role of *Erythronium americanum* Ker. in energy flow and nutrient dynamics of a northern hardwood forest ecosystem. *Science* 193: 1126–1128.

Myster, R.W., and D.S. Fernández. 1995. Spatial gradients and patch structure on two Puerto Rican landslides. *Biotropica* 27: 149–159.

Nadelhoffer, K.J., M. Downs, B. Fry, J.D. Aber, A.H. Magill, and J.M. Melillo. 1993. Biological sinks for nitrogen additions to a forested catchment. pp. 22–44. In: Lajtha, and Michener (eds.), *Stable Isotope Studies in Ecology and Environmental Science*. Blackwell, Oxford, U.K.

Odum, E.P. 1969. The strategy of ecosystem development. *Science* 164: 262–270.

Odum, E.P., and M.G. Turner. 1990. The Georgia landscape: a changing resource. pp. 137–164. In: I.S. Zonneveld and R.T.T. Forman (eds.), *Changing Landscapes: an Ecological Perspective*. Springer Verlag, New York.

Odum, H.T. 1962. Ecological tools and their use. Man and the ecosystem. pp. 57–75. In: P.E. Waggoner and J.D. Ovington (eds.), *Proceedings of the Lockwood Conference on the Suburban Forest and Ecology*, Bulletin 652. New Haven Connecticut Agricultural Experiment Station, New Haven, CT.

Odum, H.T. 1967a. Work circuits and systems stress. pp. 81–138. In: H.E. Young (ed.), *Symposium Primary Productivity and Mineral Cycling in Natural Ecosystems*. University of Maine Press.

Odum, H.T. 1967b. Energetics of food production. pp. 55–94. In: *The World Food Problem, Report of the President's Science Advisory Committee, Panel on World Food Supply*. The White House, Washington, DC.

Odum, H.T. 1970. Summary: an emerging view of the ecological system at El Verde. In: H.T. Odum and R. F. Pigeon (eds.), *A Tropical Rain Forest*. Chapters 1–10. National Technical Information Service, Springfield, VA.

Odum, H.T. 1971. *Environment, Power, and Society*. Wiley Interscience, New York.

Odum, H.T. 1983. *Systems Ecology: An Introduction*. Wiley, New York.

Odum, H.T. 1988. Self-organization, transformity, and information. *Science* 241: 1132–1139.

Odum, H.T. 1996. Scales of ecological engineering. *Ecological Engineering* 6: 7–19.

Oechel, W.C., S.J. Hastings, G. Vourlitis, M. Jenkins, G. Riechers, and N. Grulke. 1993. Recent change of Arctic tundra ecosystems from a net carbon dioxide sink to a source. *Nature* 361: 520–523.

O'Neill, R.V., D.L. DeAngelis, J.B. Waide, and T.F.H. Allen. 1986. *A Hierarchical Concept of Ecosystems*. Princeton University Press, Princeton, NJ.

Parton, W.J., D.S. Schimel, C.V. Cole, and D.S. Ojima. 1987. Analysis of factors controlling soil organic matter levels in Great Plains grasslands. *Soil Science Society of America Journal* 51: 1173–1179.

Pastor, J., J.D. Aber, C.A. McClaugherty, and J.M. Melillo. 1984. Aboveground production and N and P cycling along a nitrogen mineralization gradient on Blackhawk Island, Wisconsin. *Ecology* 65: 256–268.

Pérez Viera, I.E. 1986. *Tree Regeneration in Two Tropical Rain Forests*. M.S. Thesis. Department of Biology, University of Puerto Rico, Río Piedras, PR.

Peterson, B.J., and J.M. Melillo. 1985. The potential storage of carbon caused by eutrophication of the biosphere. *Tellus* 37B: 117–127.

Plumlee, G.S., K.S. Smith, and W.H. Ficklin. 1994. Geoenvironmental models of mineral deposits, and geology-based mineral-environmental assessments of public lands. U.S. Geological Survey Open-file Report 94-230. U.S. Government Printing Office, Washington, DC.

Porter, S.C., K.L. Pierce, and T.D. Hamilton. 1983. Late Wisconsin mountain glaciation in the western United States. pp. 71–111. In: S.C. Porter and H.E. Wright (eds.),

*Late-Quaternary Environments of the United States, the Late Pleistocene.* Volume 1. University of Minnesota Press, Minneapolis, MN.

Quigley, T.M., and S.J. Arbelbide (eds.) *An Assessment of Ecosystem Components in the Interior Columbia Basin and Portions of the Klamath and Great Basins: Vols. I and II.* Portland, OR

Redfield, A.C. 1958. The biological control of chemical factors in the environment. *American Scientist* 46: 206–226.

SAST. 1994. Science for floodplain management into the 21st century, part V. John A. Kelmelis, director. Preliminary report of the Scientific Assessment and Strategy Team. Report of the Interagency Floodplain Management Review Committee to the Administration Floodplain Management Task Force. U.S. Government Printing Office, Washington, DC.

Satchell, J.E. (ed.) 1983. *Earthworm Ecology from Darwin to Vermiculture.* Chapman and Hall, London, U.K.

Scatena, F.N., and A.E. Lugo. 1995. Geomorphology, disturbance, and the soil and vegetation of two subtropical wet steepland watersheds of Puerto Rico. *Geomorphology* 13: 199–213.

Schimel, D.S., B.H. Braswell, Jr., E.A. Holland, R. McKeown, D.S. Ojima, T.H. Painter, W.J. Parton, and A.R. Townsend. 1994. Climatic, edaphic and biotic controls over storage and turnover of carbon in soils. *Global Biogeochemical Cycles* 8: 279–293.

Schimel, J.P., and M.K. Firestone. 1989. Nitrogen incorporation and flow through a coniferous forest soil profile. *Soil Science Society of America Journal* 53: 779–784.

Schindler, D.W. 1974. Eutrophication and recovery in experimental lakes: Implications for lake management. *Science* 184: 897–899.

Schindler, D.W., and S.E. Bayley. 1993. The biosphere as an increasing sink for atmospheric carbon: Estimates from increased nitrogen deposition. *Global Biogeochemical Cycles* 7: 717–734.

Schlesinger, W.H. 1991. *Biogeochemistry: An Analysis of Global Change.* Harcourt Brace Jovanovich Publishers, San Diego, CA: Academic Press.

Schulze, E.D., and H.A. Mooney (eds.) 1993. *Biodiversity and Ecosystem Function.* Springer Verlag, New York.

Schulze, E.D., W. De Vries, M. Hauhs, K. Rosen, L. Rasmussen, O.C. Tann, and J. Nilsson. 1989. Critical loads for nitrogen deposition in forest ecosystems. *Water, Air and Soil Pollution* 48: 433–441.

Shaver, G.R., W.K. Billings, F.S. Chapin III, A.E. Giblin, K.J. Nadelhoffer, W.C. Oechel, and E.B. Rastetter. 1992. Global change and the carbon balance of Arctic ecosystems. *BioScience* 42: 433–441.

Silver, W. 1992. Effects of small-scale and catastrophic disturbance on carbon and nutrient cycling in a lower montane subtropical wet forest in Puerto Rico. Ph.D. Diss. Yale School of Forestry and the Environment, New Haven, CT.

Silver, W.L., S. Brown, and A.E. Lugo. 1996. Effects of changes in biodiversity on ecosystem function in tropical forests. *Conservation Biology* 10: 17–24.

Smith, M.S., and J.M. Tiedje. 1979. Phases of denitrification following oxygen depletion in soil. *Soil Biology and Biochemistry* 11: 261–267.

Spies, T.A., J.F. Franklin, and M. Klopsch. 1990. Canopy gaps in Douglas-fir forests of the Cascade Mountains. *Canadian Journal of Forestry Research* 20: 649–658.

Steadman, D.W. 1991. Extinct and extirpated birds from Aitutaki and Atiu, southern Cook Islands. *Pacific Science* 45: 325–347.

Steadman, D.W. 1995. Prehistoric extinctions of Pacific island birds: Biodiversity meets zooarchaeology. *Science* 267: 1123–1131.

Stevenson, F.J. 1986. *Cycles of Soil: Carbon, Nitrogen, Phosphorus, Sulfur, and Micronutrients.* Wiley-Interscience Publication, Wiley, New York.

Stoddard, J.L. 1994. Long-term changes in watershed retention of nitrogen: Its causes and aquatic consequences. pp. 224–284. In: L.A. Baker (ed.), *Environmental Chemistry of Lakes and Reservoirs, Advances in Chemistry.* American Chemical Society, Washington, DC.

Stoddart, D.R. 1992. Biogeography of the tropical Pacific. *Pacific Science* 46: 276–293.

Stottlemyer, R., and C.A. Troendle. 1995. Surface water chemistry and chemical budgets, alpine and subalpine watersheds, Fraser Experimental Forest, Colorado. pp. 321–327. In: K.A. Tonnessen, M.W. Williams, and M. Tranter (eds.), *Biogeochemistry of Seasonally Snow-Covered Catchments.* IAHS Publication 228. Oxfordshire, U.K.

Summerfield, M.A. 1991. *Global Geomorphology.* Longman House, Essex, U.K.

Swanson, F.J., and R.E. Sparks. 1990. Long-term ecological research and the invisible place. *BioScience* 40(7): 502–508.

Swift, M.J., O.W. Heal, and J.M. Anderson. 1979. *Decomposition in Terrestrial Ecosystems.* University of California Press, Berkeley, CA.

Tansley, A.G. 1935. The use and abuse of vegetational concepts and terms. *Ecology* 16: 284–307.

Terborgh, J. 1983. *Five New World Primates: a Study in Comparative Ecology.* Princeton University Press, Princeton, NJ.

Terborgh, J. 1986. Keystone plant resources in the tropical forest. pp. 1330–1334. In: M.E. Soulé (ed.), *Conservation Biology: the Science of Scarcity and Diversity.* Sinauer Associates, Sunderland, MA.

Tietema, A., and J.M. Verstraeten. 1991. Nitrogen cycling in an acid forest ecosystem in the Netherlands under increased atmospheric nitrogen input: the nitrogen budget and the effect of nitrogen transformations on the proton budget. *Biogeochemistry* 15: 21–46.

Tilman, D. 1996. Biodiversity: population versus ecosystem stability. *Ecology* 77: 350–363.

Tisdall, J.M., and J.M. Oades. 1982. Organic matter and water-stable aggregates in soils. *Journal of Soil Science* 33: 141–163.

Valiela, I., J. Costa, K. Foreman, J.M. Teal, B. Howes, and D. Aubrey. 1990. Transport of groundwater-borne nutrients from watersheds and their effects on coastal waters. *Biogeochemistry* 10: 177–197.

Van Vuren, M.M.I., R. Aerts, F. Berendse, and W. De Visser. 1992. Nitrogen mineralization in heathland ecosystems dominated by different plant species. *Biogeochemistry* 16: 151–166.

Vitousek, P.M. 1982. Nutrient cycling and nutrient-use efficiency. *American Naturalist* 119: 553–572.

Vitousek, P.M., and R.W. Howarth. 1991. Nitrogen limitation on land and in the sea: How can it occur? *Biogeochemistry* 13: 87–115.

Vitousek, P.M., and W.A. Reiners. 1975. Ecosystem succession and nutrient retention: A hypothesis. *BioScience* 25: 376–381.

Vitousek, P.M., and R.L. Sanford. 1986. Nutrient cycling in moist tropical forest. *Annual Review of Ecology and Systematics* 17: 137–167.

Vitousek, P.M., L.R. Walker, L.D. Whiteaker, D. Mueller-Dombois, and P.A. Matson. 1987. Biological invasion of *Myrica faya* alters ecosystem development in Hawaii. *Science* 238: 802–804.

Vitousek, P.M., P.R. Ehrlich, A.H. Ehrlich, and P.A. Matson. 1986. Human appropriation of the products of photosynthesis. *BioScience* 36: 368–373.

Vitousek, P.M., T. Fahey, D.W. Johnson, and M.J. Swift. 1988. Element interactions in forest ecosystems: Succession, allometry and input–output budgets. *Biogeochemistry* 5: 7–34.

Vogt, K.A., C.C. Grier, C.E. Meier, and R.L. Edmonds. 1982. Mycorrhizal role in net primary production and nutrient cycling in *Abies amabilis* ecosystems in western Washington. *Ecology* 63: 370–380.

Waitt, R.B., and R.M. Thorson. 1983. The Cordilleran Ice Sheet in Washington, Idaho, and Montana. pp. 53–70. In: S.C. Porter and H. E. Wright (eds.), *Late-Quaternary Environments of the United States, The Late Pleistocene.* Volume 1. University of Minnesota Press, Minneapolis, MN.

Walker, L.R., N.V.L. Brokaw, D.J. Lodge, and R.B. Waide (eds.) 1991. Ecosystem, plant, and animal responses to hurricanes in the Caribbean. *Biotropica* 23: 313–521.

Walker, L.R., M. Willig, W. Silver, and J.K. Zimmerman (eds.) 1996a. Long-term responses of Caribbean ecosystems to disturbance. *Biotropica* 28(4a): 414–613.

Walker, L.R., D.J. Zarin, N. Fetcher, R.W. Myster, and A.H. Johnson. 1996b. Ecosystem development and plant succession on landslides in the Caribbean. *Biotropica* 28(4a): 566–576.

Weaver, P.L. 1987. *Structure and Dynamics in the Colorado Forest of the Luquillo Mountains of Puerto Rico.* Ph.D. Diss. Michigan State University, East Lansing, MI.

Westman, W.E. 1986. Resilience: concepts and measures. pp. 5–19. In: B. Dell, A.J.M. Hopkins, and B.B. Lamont (eds.), *Resilience in Mediterranean Ecosystems.* Dr. W. Junk, Dordrecht, the Netherlands.

Wilson, E.O. 1971. *The Insect Societies.* Harvard University Press, Cambridge, MA.

Wilson, E.O. 1975. *Sociobiology.* Belknap Press, Harvard University Press, Cambridge, MA.

Zak, D., P.R. Groffman, D.S. Pregitzer, S. Christensen, and J.M. Tiedje. 1990. The vernal dam: plant-microbe competition for nitrogen in northern hardwood forests. *Ecology* 71: 651–656.

Zarin, D.J. 1993. *Nutrient Accumulation During Succession in Subtropical Lower Montane Wet Forests, Puerto Rico.* Ph.D. Diss. University of Pennsylvania.

Zarin, D.J., and A.H. Johnson. 1995a. Base saturation, nutrient cation, and organic matter increases during early pedogenesis on landslide scars in the Luquillo Experimental Forest, Puerto Rico. *Geoderma* 65: 317–330.

Zarin, D.J., and A.H. Johnson. 1995b. Nutrient accumulation during primary succession in a montane tropical forest, Puerto Rico. *Journal of the Soil Science Society of America* 59: 144–1452.

Zlotin, R.I., and K.S. Khodashova. 1980. *The Role of Animals in Biological Cycling of Forest-steppe Ecosystems.* Dowden, Hutchinson and Ross, Stroudsburg, PA.

Zonneveld, I.S. 1990. Scope and concepts of landscape ecology as an emerging science. pp. 3–20. In: I.S. Zonneveld and R.T.T. Forman (eds.), *Changing Landscapes: An Ecological Perspective.* Springer Verlag, New York.

Zonneveld, I.S., and R.T.T. Forman (eds.). 1990. Changing Landscapes: An Ecological Perspective. Springer Verlag, New York.

## THE AUTHORS

**Ariel E. Lugo**
*International Institute of Tropical Forestry, USDA Forest Service, P.O. Box 25000, Río Piedras, Puerto Rico 00928-5000, USA*

**Jill S. Baron**
*U.S. Geological Survey, U.S. Department of Interior, Natural Resources Ecology Laboratory, Colorado State University, Fort Collins, CO 80523, USA*

**Thomas P. Frost**
*Geological Survey, U.S. Department of Interior, W904 Riverside Avenue, Room 202, Spokane, WA 99201-1087, USA*

**Terrance W. Cundy**
*Potlach Corporation, Wood Products, Western Division, P.O. Box 1016, Lewiston, ID 83501-1016, USA*

**Phillip Dittberner**
*Bureau of Land Management, SC-213, U.S. Department of Interior, P.O.Box 25047, Denver, Co 80225-0047, USA*

# Ecosystem Processes and Functions: Management Considerations

Steven J. Paustian, Miles Hemstrom, John G. Dennis,
Phillip Dittberner, and Martha H. Brookes

**Key questions addressed in this chapter**

- ♦ *How can managers work with rather than against ecological processes to maintain key functions and sustainable ecosystems?*
- ♦ *How can managers deal with complex relationships between ecological processes that operate over a range of temporal and spacial scales?*
- ♦ *How can managers identify and characterize key ecological processes?*
- ♦ *How can managers monitor and evaluate the effects of management activities on ecosystem processes and functions?*

**Keywords: Ecosystem processes, Everglades restoration, western rangeland riparian ecosystems, western Cascade landscapes, interior Columbia Basin disturbance, watershed analysis process, multiple scale monitoring, long-term monitoring**

# 1 INTRODUCTION

Managers monitor and evaluate the effects of management activities on key ecological processes and functions in almost as many ways as there are managers. Many of them try to be precise and quantitative; they collect detailed data on variables and trends related to the decisions they need to make. Other managers are qualitative and base resource monitoring on observations and estimates. Between these two opposite approaches are all gradations of approaches to the whole spectrum of resource management issues.

The key objective in ecosystem management is maintaining ecosystem processes within the given range of natural variation or, in extreme cases, at least within acceptable thresholds that maintain ecosystem processes and functions critical to maintaining a functioning and sustainable ecosystem.

Interactions among ecological processes and functions are complex and account for a wide range of ecosystem variability, resilience, and adaptations. Approaches for monitoring ecosystem condition or health must be able to account for the wide range of variability in temporal and spatial changes in the system. Management must allow for that variability, and management decisions and implementation strategies must maintain variability within historical and acceptable ranges.

The linkages between ecological processes must be understood if ecosystem management is to be successfully implemented. The processes and linkages are complex and often have not been identified, much less understood. The implications of this lack of understanding for sustainable resource use are profound. Resource managers cannot afford to stand by and do nothing, however, and society cannot continue to manage, harvest, and develop resources at the same rates as in the recent past. Adaptive management is an approach that holds a great deal of promise for dealing with system complexity and unknowns. It also has tremendous potential for improving resource management and maintaining sustainable ecosystems if the concepts are understood and used effectively.

Managers also should assess ecological processes at scales (both temporal and spatial) higher and lower than in the past to provide context and linkages with other systems and variables. Knowledge about these linkages is often incomplete, and managers tend to want to put boundaries that are too restrictive around the "ecosystems" being managed. All systems and all variables are connected and interrelated in one way or another. Ecosystem management goals should include identifying biologically important linkages and variables, and conditions where the processes and functioning of the system being managed might be negatively affected.

## 1.1 Objectives

Our goals in this chapter are to illustrate the effects of past resource management on a range of ecological processes that influence ecosystem functions. We outline a management process and a variety of assessment approaches that can help managers work more in concert with key ecological processes (productivity, diversity, and resilience (Lugo et al. this volume).

Our specific objectives are to:

- Identify reference conditions for assessing ecosystem health.
- Identify appropriate spatial and temporal scales for addressing key processes and functions of an ecosystem, realizing that the scales may not be the same for all ecosystems.
- Identify indicators for monitoring trends in ecosystem processes and functions.
- Illustrate each of the above objectives with examples that will provide a good contextual setting so a resource manager can draw parallel comparisons and extrapolations.
- Outline a management process and assessment procedures that will help managers address complex issues associated with maintaining ecological processes and sustaining ecological functions.

## 1.2 Key Ecosystem Processes

Lugo et al. (this volume) identify the following ecological processes and functions as having the highest susceptibility to change or disruption as a result of resource management: *geologic processes*, *material (nutrient) cycling*, *energy transfer*, and *succession*. We believe that these general categories and the subset of key ecosystem processes and functions listed in Table 1 are critically important for resource managers to understand. Several case studies were selected to illustrate management issues and applications in several different ecosystems over a broad range of scales (Table 2). These case studies also demonstrate the influences of important ecosystem processes and functions on the outcomes of management actions.

# 2 HISTORICAL CONTEXT

During the many millennia through which people have evolved from nomadic hunters and gatherers to modern urban dwellers, human relations to the ecological processes of the ecosystems that supported

Table 1. Ecological processes commonly affected by resource management (adapted from Lugo et al., this volume).

| Geologic processes | Material and biogeochemical cycling | Energy transfer | Succession and disturbances |
|---|---|---|---|
| Fluvial erosion | Water cycle | Radiation transfer | Fire |
| Sedimentation | Carbon cycle | Heat transfer | Flooding |
| Mass wasting | Nutrient cycling | Photosynthesis | Wind,Insects,Diseases |
| Physical weathering | Chemical weathering | Consumption | Exotic species |

Table 2. Ecological processes and scales represented in case studies.

| Processes | Case studies | | | | | |
|---|---|---|---|---|---|---|
| | Everglades restoration | Range riparian mgmt. | Western Cascades forest structure | Columbia Basin fire ecology | Alaska watershed analysis | Smoky Mts. Acid rock drainage |
| Flooding and drainage | X | X | | X | X | |
| Erosion and sedimentation | | X | X | X | X | |
| Radiation/ heat transfer | | X | X | | X | |
| Water cycle | X | X | X | | X | |
| Nutrient cycles | X | X | X | X | X | |
| Carbon cycle | X | | X | X | | |
| Succession | X | X | X | X | | |
| Exotic species | X | | | | | |
| Insect/disease | | | X | X | | |
| Biochemical weathering | X | | | | | X |
| Primary scale | regional | watershed | site-stand | regional | watershed | watershed |

them have changed. Early in their evolution, people were at the mercy of environmental events. Developing the use of fire, shelter, and agriculture gave people some control over at least the local and short-term expression of ecological processes. Today, with large-scale technologies, people exert considerable control over many ecosystem processes by such actions as widespread converting from natural vegetation to crops, constructing reservoirs, channelizing rivers, stabilizing shorelines, applying pesticides and fertilizer, suppressing fire, using the energy from fossil fuels, and eliminating natural barriers to biological dispersion. People not only control ecosystem processes in desirable ways, but also exert influence in undesirable and often unknown ways. Society is beginning to realize that beneficial or desired control of processes at one scale of time and space may interfere with processes at another scale in time or space with detrimental — and sometimes catastrophic — consequences.

## 2.1 Human Resource Exploitation and Ecosystem Sustainability

People have experienced many examples of change where one or the sole agent of those changes was human intervention. Unknowing interference with species-population interactions has led to extinction or near extinction of North American biota — such as the passenger pigeon, the bison, and perhaps many of the large animals that disappeared after the last ice age. In recent times, people have triggered loss of additional species not directly through harvest but indirectly through changes in ecological processes. Habitat of the red-cockaded woodpecker, for example, was eliminated by declines in late-succession pine stands. Similarly, native ground-nesting bird species in Hawaii have been eliminated by changes in predation rates caused by introduction of foraging pigs and mosquitoes that spread avian malaria.

> Sustainability of important ecosystem processes and functions is dependent upon the interaction between societal goals and values, and the ecological capacity of the ecosystem.

Early resource exploitation in the West by trappers, farmers, ranchers, miners, and loggers resulted in dramatic changes to streams and associated riparian and aquatic communities. Resource use and disturbances were focused on riparian areas from the time of the early settlers because they provided travel corridors and sources for water. Fur trapping virtually eliminated beaver populations in numerous rivers and streams across the West. In many of these riparian ecosystems, beavers are a keystone species influencing hydrology, water temperature, nutrient cycles, vegetation structure, and aquatic habitats (Elmore and Kauffman 1994). The extirpation of beaver in these watersheds resulted in loss and simplification of extensive riparian and wetland habitats.

More recently, the construction of the Glen Canyon dam has resulted in major changes to the lower Colorado River ecosystem. Changes in water quality from warm, sediment-rich water to cooler water with low turbidity has favored populations of introduced aquatic species and greatly reduced the river's capacity to support many native fishes. Changes in flow regime, by eliminating annual spring floods, has resulted in widespread changes in riparian vegetation communities below the dam.

Sustainability of important ecosystem processes and functions depends on the interaction between societal goals and values, and the ecological capacity of the ecosystem (see Lugo et al., this volume; Bormann et al. 1994). Human development and manipulation of ecosystem processes in the Everglades ecosystem well illustrates loss of ecological capacity and integrity, and the high cost associated with restoration of key ecosystem functions.

## 2.2 Example — Everglades Restoration Project

### 2.2.1 Background

Important ecosystem functions and processes associated with wetlands include water cycling, energy capture, nutrient capture, nutrient transfer, and material transfer (Lugo et al. 1990, Brinson 1993). Wetlands are major primary and secondary producers and exporters of organic carbon. They play a major role in storing flood waters and serving as water donors during droughts. Wetlands are also a catalyst for biogeochemical processes including nutrient cycling and sinks for metals and toxins (Lugo et al. 1990).

### 2.2.2 1 Management Consequences

The Everglades offers a vivid example of how human interference with wetland ecosystem processes and function can result in catastrophic resource consequences on a regional scale. In the late 1800s, South Florida was a mosaic of freshwater and intertidal marshlands encompassing about 18,000 square miles (Williams 1994). The historical Everglades ecosystem was characterized as a 50-mile-wide river of grass stretching from Orlando to Lake Okeechobee and into Florida Bay. From 1900 through 1970, a series of flood-control projects was developed, including straightening the Kissimmee River, building thousands of levees, and constructing more than 1,000 miles of water-diversion canals. Reclaimed wetlands were converted to urban and agricultural uses resulting in water quality problems that further stressed Everglades ecosystem functions.

The effects of these alterations in the Everglades ecosystem have been dramatic. Alterations in groundwater transfer processes have produced many undesirable effects. Groundwater flow regimes in limestone quarries have changed and now provide open conduits for contaminating aquifers. Damage to offshore reefs have been linked to the influx of groundwater contaminated by agricultural activities. And drinking water sources for major urban centers are threatened by saltwater intrusion into Everglades aquifers (Finkl 1995).

Patterns of surface water movement and the hydroperiod — the duration and periodicity of wetland flooding — have been altered for the entire Everglades watershed. Changes to flooding extent and hydroperiod have had many detrimental effects on Everglades flora and fauna: wading bird populations have dropped 90%, numbers of all vertebrate species have declined by 75 to 90%, and 55 threatened or endangered species are at risk (SFERWG 1994, Finkl 1995). Extreme high water has flooded alligator nests and reduce food supplies to wading birds by altering wetland plant communities (David 1996). Increased drought frequency in other areas has increased the number of severe fires, resulting in rapid loss of organic soils and the invasion of exotic woody species into wet prairie communities. Surface water diversions along the west Florida coast have greatly increased salinity in coastal marshes and Florida Bay, causing anoxic conditions in critical fish-nursery areas (SFERWG 1994).

Changes to nutrient cycling processes have similarly had dramatic effects on water quality and plant succession in much of the Everglades. The historical Everglades was nutrient poor, as indicated by the predominance of macrophyte and periphyton plant communities (Finkl 1995). Phosphorus loading (up to 200 tons per year), primarily from sugar cane fields, is believed to be a major factor affecting wetland plant community composition and diversity in the Everglades (Williams 1994). Increased nutrient loading from agricultural and urban sources gives a competitive advantage to species such as cattail, which is rapidly displacing native sawgrass in parts of the Everglades wetlands (SFERWG 1994, David 1996).

Elevated mercury concentrations in high trophic level species including fish, wading birds, alligators, and raccoons is another vexing problem recently identified in the Everglades ecosystem (Williams 1994). Changes in geochemical soil weathering associated with agricultural tillage methods and soil subsidence associated with draining organic soils are likely to be factors in the release of reactive mercury compounds into the Everglades (Finkl 1995).

### 2.2.3 Future Management Implications

The Everglades is an excellent example of the need for regional, multi-agency planning and analysis to sustain critical ecosystem processes and functions across watershed and landscape scales. The developers that drained and cultivated the agricultural areas, and the engineers charged with replumbing the Kissimmee River and Lake Okeechobee region were probably aware of some of the obvious local consequences of their projects, but they clearly did not understand the hydrologic linkages in the Everglades ecosystem as a whole. Even with a more complete understanding of key ecosystem processes and functions and the availability of modern technologies, the task of restoring the Everglades will be monumental. A major wetlands restoration project is currently underway to revive ecological processes and functions for the entire South Florida Everglades ecosystem. This 15–20-year, 2-billion-dollar restoration program is unprecedented in its scope and complexity (SFERWG 1994).

South Florida Interagency Working Group (the Corps of Engineers, the National Park Service and the South Florida Water Management District) managers and planners acknowledge that a holistic approach will be needed to attain even modest restoration goals for the Everglades. The South Florida Restoration Project has adopted a comprehensive watershed management approach that requires the cooperation and participation of many land owners, state and federal regulatory and land management agencies, and the public at large. Cooperation and acceptance of restoration plans from such a diverse group of competing interests will be difficult to obtain.

A better understanding of the interactions between hydrologic, biogeochemical, and food-web processes will be needed to establish detailed management objectives and to design strategies for restoring wetland functions (SFERWG 1994, Brinson 1993). The Working Group is currently developing a comprehensive regional monitoring program focusing on hydrologic processes and ecological responses to restoration programs (SFERWG 1994). Because pre-development information on hydroperiod, flooding extent, and natural processes in the Everglades is lacking, several "Natural Systems Models" are being used to simulate hydrologic and other characteristics of the natural Everglades ecosystem. Project managers also realize, however, that a large amount of scientific uncertainty must be accepted in designing and implementing restoration projects. An adaptive management approach will help bridge the science gaps, reduce the potential for wasting funds, and limit the risk of creating further damage to wetland function. Adaptive management will be used to refine management procedures based on the results of iterative monitoring, research, and modeling.

### 2.2.4 Summary

The Everglades is an example of a complex ecosystem where human development has altered ecosystem processes and functions to a degree that the system is approaching collapse. A lack of comprehensive inventory and research data on ecosystem condition has made it difficult to establish a clear set of options for restoring proper conditions and functions in the Everglades wetland ecosystem. Resuscitating the Everglades will be very expensive and require cooperation between numerous management agencies and stakeholders.

## 2.3 Understanding the Role of Ecological Processes in Management

Native Americans understood the results of a variety of ecological processes well in advance of the current science of ecology and practice of ecosystem management. They mitigated soil nutrient depletion by using fish to fertilize mounds of corn. Their use of fire to drive game animals, clear vegetation, and improve food for game animals and themselves altered plant succession and nutrient cycling in ways beneficial to the people.

Agrarian societies learned over the years that bare soil cropping of single plant species accelerated soil

By understanding some of the ecological processes involved, working with rather than against those processes, managers can develop approaches that are effective in meeting many human needs and are less disruptive to ecosystem integrity.

erosion and soil nutrient depletion. They discovered through a variety of techniques that changes in tillage methods and cropping patterns could reduce the influence of these processes on crop production and provide for sustainable agriculture.

Wildland resource managers have been exploring and developing techniques for manipulating ecological processes to achieve management goals for many decades. Using selective harvest in some forests and clearcutting in others may mimic some aspects of natural regeneration and succession. Undesired changes in physical and biological characteristics of streams and rivers associated with riparian timber harvest and unfettered grazing along streams have led to an awareness of important riparian processes and functions such as mitigating floods, recharging groundwater, cycling nutrients and other materials, and transferring energy.

Resource managers currently observe changes in many ecological processes and functions that result from their actions. Examining causes of these changes is leading to new management approaches and policies designed to prevent or mitigate human interference with natural processes. By understanding some of the ecological processes involved, working with rather than against those processes, managers can develop approaches that are effective in meeting many human needs and are less disruptive to ecosystem integrity. The following example of Western rangeland riparian management illustrates a successful management approach for understanding the role of ecosystem processes and functions.

## 2.4 Example — Western Rangeland Riparian Management

### 2.4.1 Background

Riparian ecosystems provide a good example of basic ecosystem functions including nutrient capture, nutrient transfer, and material transfer (Meehan et al. 1977, Naiman 1992). The key to understanding riparian ecosystem functions is knowledge of how fluvial processes (erosion and sedimentation) influence the transfer of energy and materials in a watershed and, in turn, shape streams and adjacent riparian vegetation communities. Fluvial erosion is the dominant process for transfer of sediments from headwaters to valley stream segments. Vegetation in headwater zones acts to mediate rain splash and surface erosion, and it facilitates infiltration of precipitation that feeds shallow groundwater aquifers and is ultimately released into springs and surface-water bodies. Riparian areas are influenced to large degree by sediment deposition during over-bank floods. Floodplain deposition zones act as sinks for both sediments and nutrients. These deposits also slow the release of groundwater from stream terraces that sustain stream flow during droughts. The availability of nutrient-rich soils and water make productive sites for vegetation adapted to periodic flood disturbances. The roots of riparian vegetation also aid in stabilizing stream banks and create cover from overhanging banks. Riparian vegetation has another important function in trapping new sediments and reducing streambank erosion during over bank flood events. Extreme flood events recycle soils through floodplain erosion and transfer organic and inorganic sediment to downstream deposition sites.

### 2.4.2 Lessons learned from past riparian management

Rangeland riparian areas represent a relatively small portion of the landscape in the West, usually less than 1% of the land area (Elmore and Beschta 1987). In eastern Oregon, the interaction of trapping beaver and livestock grazing have altered basic ecosystem functions by changing the rates of sediment routing and storage in streams, reducing nutrient capture and transfer in riparian and aquatic ecosystems, and changing the hydrology of many stream systems. Concentrated, year-round foraging by domestic livestock has caused expanded gully networks, caused ephemeral flow regimes in streams that once had perennial flow, simplified stream banks and channels, and changed willow and wet meadow riparian communities to more xeric plant types. Grazing has also induced changes to riparian areas, which intensified the effects of natural disturbances and, in many instances, inhibited recovery from extreme flood events.

### 2.4.3 Successful management actions that have restored natural processes and improved rangeland riparian functions

The results of long-term rangeland riparian restoration projects in eastern Oregon are summarized in Elmore and Beschta (1987). Management actions included changes to grazing patterns in degraded watersheds and long-term exclusion of grazing in sensitive riparian

Fig. 1. Changes in grazing management. Left: Gordon Canyon Arizona 1987 with season long grazing. Right: Gordon Canyon 1992 after five years of grazing exclusion (photo credits: Wayne Elmore).

areas. Grazing restrictions for periods of 5–20 years resulted in dramatic improvements in riparian vegetation health and diversity (Fig. 1).

Elmore and Beschta, and Kauffman and others (1997) found that vegetation along stream channels responded rapidly to the new grazing regime. Rejuvenated riparian vegetation established a positive feedback mechanism by trapping nutrient-rich sediments and helping to bind sensitive stream banks. With increased infiltration and more extensive deposition of sediment along flood plains, water tables rose and expanded, resulting in perennial stream flow in streams that were previously intermittent.

A major benefit of this "ecological restoration" approach to riparian management is that it allows natural processes to restore riparian functions and desired conditions. Riparian ecosystems are dynamic, and working with, rather than being constrained by, natural processes is a critical consideration in managing these systems. Structural treatments, such as gully check dams and gabions, in a rangeland riparian setting have had limited success, primarily because this approach attempts to constrain natural processes of channel erosion and floodplain deposition. Beschta et al. (1994) note that projects relying on channel structures often sever the ecological linkages between terrestrial, riparian, and aquatic ecosystems.

Riparian restoration should involve a watershed-wide approach (Minshall 1994). Adjustments in land-use practices and management for an entire watershed may be needed to reduce surface erosion to rates that riparian areas can process. Riparian restoration projects that ignore upland conditions often fail to meet management objectives (Elmore and Beschta 1987, National Research Council 1994).

Restoring riparian and aquatic ecosystem processes and function is a long-term proposition. Improved riparian condition is quickly reversed if poor land management practices are reinstated (Elmore and Beschta 1987).

### 2.4.4 Summary

Integrated rangeland riparian management should consider key ecological processes that control ecosystem function, the degree of management and natural stress that affect these processes, and the ability of the system to respond positively to treatments designed to restore ecosystem processes. Passive restoration —stopping the human activities that have compromised ecosystem integrity — is a critical element in successful riparian restoration efforts (Kauffman et al. 1997).

## 3 INCORPORATING ECOSYSTEM PROCESSES INTO MANAGEMENT

The ultimate goal of ecosystem management is "to meet human needs while maintaining the health, diversity, and productivity of ecosystems" (Thomas and Huke 1996). To attain this goal of ecosystem health and productivity, the integrity of key ecosystem processes and functions must be sustained. In this section, we introduce a general management process that incorporates approaches to help managers integrate natural ecosystem process and function considerations into resource planning and project implementation. We place primary emphasis on tools and approaches to

> Managers must be aware that management can alter basic ecosystem processes, functions, and cycles and that the effects of these alterations can be dramatic.

evaluate ecosystem condition and trend with respect to the key process and function elements introduced earlier. We also emphasize the need for managers to consider what scales are most appropriate for analyzing potential effects of management practices for an array of ecosystem processes and functions.

## 3.1 A Management Process for Sustaining Ecosystem Integrity

How is a manager to find sound footing in the swamp of difficulties presented by often counter-intuitive relations among ecological processes and functions, and the burgeoning demands of human populations for natural resources? A general management process we believe will help resource managers find this sound footing in *developing sustainable ecosystem management practices* is shown in Fig. 2. The process has four steps: define *objectives*, *inventory* the ecosystem, *analyze* information, and *implement* projects by using adaptive management principles. The process is mediated by feedback loops between each of the steps.

The foundation for establishing management objectives requires identifying key resource management issues and questions, and describing a desired condition for the ecosystem being managed. Identifying management issues and questions is critical in establishing sideboards for inventorying key ecosystem components and analyzing important processes and functions (Regional Ecosystem Office 1994). Determining the appropriate scale to address specific ecosystem management questions is another important consideration. We recommend an approach that uses natural reference conditions to describe desired ecosystem conditions. Maintaining ecosystems in the range of natural variability has the best chance of sustaining ecosystem integrity by maintaining processes and functions that control ecosystem conditions (see Section 3.6).

Inventory and research data for analyzing management effects on ecosystem processes and functions will always be limited. Information ranging from local knowledge of managers and landowners to long-term research results should be used to fill information gaps. Our ability to gather more-comprehensive data for large-scale ecosystems, however, has expanded dramatically because of recent advancements in remote sensing technology (see Section 3.5). Ongoing development of hierarchical ecological classification and mapping approaches provide an effective tool to aggregate and extrapolate resource information across a broad range of scales, and to display linkages between ecosystem components and their functions (see Grossman et al. and Carpenter et al. this volume).

A wide array of analytical tools are also becoming more accessible to managers. Geographic information systems, along with an array of predictive models, risk-assessment procedures, and other decision-

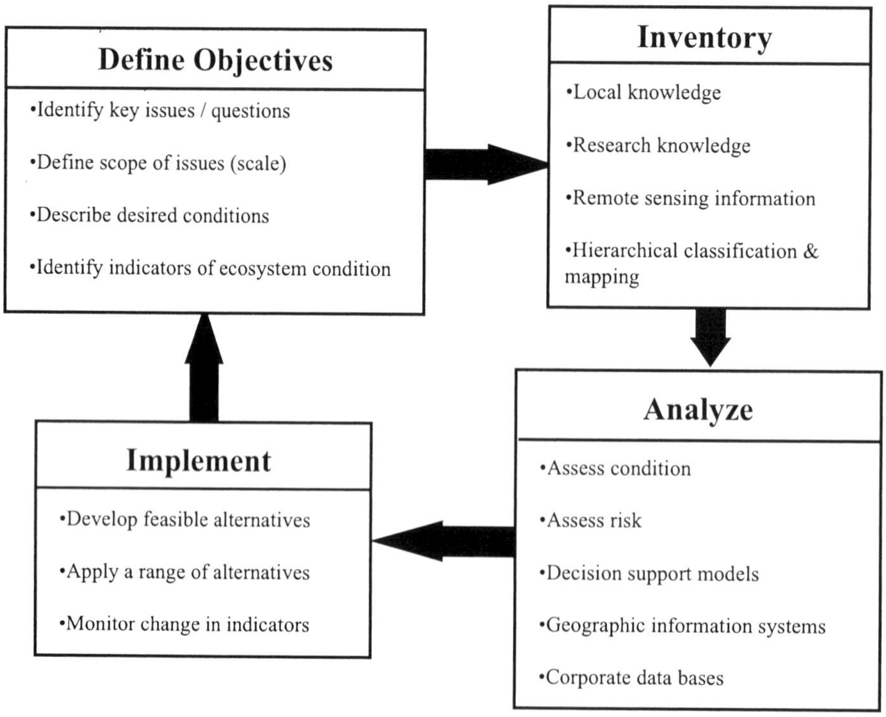


Fig. 2. A management process for sustaining ecosystem integrity (source: Elements of this framework were adapted from the USDA Forest Service, Ecosystem Characterization and Analysis Project and Ecosystem Analysis at the Watershed Scale (Regional Ecosystem Office 1994)).

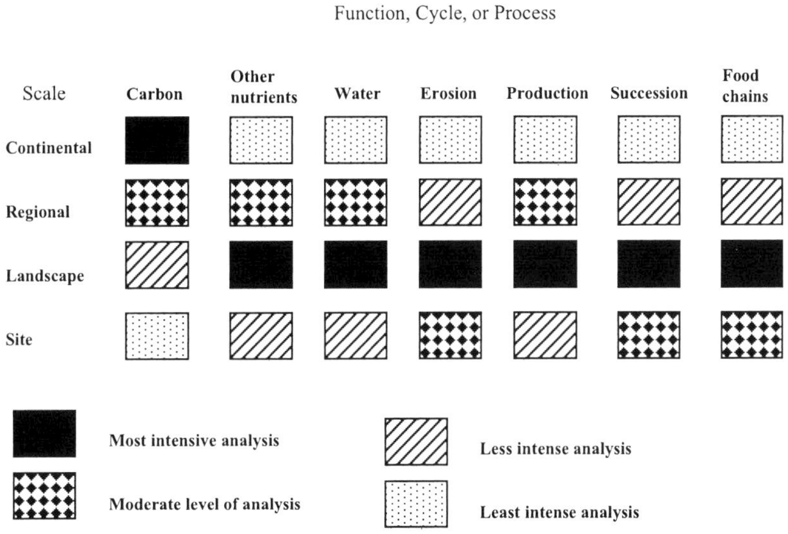


Fig. 3. Guidance for analysis of ecosystem processes and functions at various spatial scales.

support software provide managers with an improved capacity to assess complex relations between ecosystem processes and functions (see Oliver and Twery, and Tolle et al., in Vol. III of this book). Developing corporate, relational data bases that contain consistent data for key indices of ecosystem condition is a critical element for applying any of these analytical techniques (see Cooperrider et al., in Vol. III of this book).

Implementing management activities under the framework in Fig. 2 uses an adaptive management approach (see Bormann et al., in Vol. III of this book). Adaptive management entails designing feasible alternatives that can meet management objectives (to attain a desired condition), applying a range of promising alternatives, and monitoring the effects of each alternative on selected indicators of ecosystem condition to learn how well the alternative meets management objectives.

This general management process provides a conceptual framework for the following examples that illustrate important principles and effective management techniques that we believe can lead to more sustainable resource development practices in the future. This framework will also help managers assimilate many important concepts and approaches discussed throughout this volume.

## 3.2 Ecosystem Processes and Scale

The scale an observer adopts depends on the process of interest and determines which processes appear to be external, constraining the system, and which appear to be internal, part of and affected by the system. If a manager is interested in nitrogen fixation and productivity in a stand of trees, a site or landscape scale would be appropriate for investigation. If global warming and the influence of warming on carbon stored in forests are of interest, analysis would probably focus on regional, continental, and global scales (see Lugo et al., this volume).

Although management effects ultimately happen at the site scale for many functions, analysis should most often be done at landscape and higher scales. The intensity of analysis and, in fact, which ecosystem functions appear to act in the system, are functions of scale (Fig. 3). Consider each of the combinations of scale and function or process in Fig. 3. Concentrations of carbon dioxide in the atmosphere may contribute significantly to global climate change. The amount of atmospheric carbon dioxide is, in turn, influenced by storage in biological compounds. Management at the stand or site scale can change carbon stored in decomposition-resistant forms like large tree trunks. Land managers might well be interested in reducing atmospheric carbon by increasing the amount of carbon stored in large wood. Analysis at the site scale, although revealing amounts of large wood on a site and how long it is likely to be there, does not say much about changes to atmospheric carbon. Analysis of carbon storage, as it relates to climate change, is best done at regional or higher scales, which should set standards for stand management at lower scales. Analysis at the site scale determines if carbon storage assumptions made at the higher scale are valid and if carbon storage is accumulating at desired rates. Analysis at the site scale is fairly simple: how much wood is present, how much is accumulating, and has soil carbon (organic matter) been reduced by fire or other events? These analyses can be done by using a stratification and subsampling scheme to maximize the efficiency of field measurements.

Managers must be aware that management can alter basic ecosystem processes, functions, and cycles and that the effects of these alterations can be dramatic. From a pragmatic perspective, however, managing these ecosystem attributes will usually be indirect, through manipulating vegetation composition, structure, and pattern. This indirect management is especially likely at the site or project scale, where expertise, time, and funds are unlikely to be available for examining ecosystem effects of each project. Cumulative effects of land management are most realistically analyzed at higher organizational and geographic scales — regional and landscape scales in Fig. 3 — where standards, guidelines, and descriptions of desired conditions can be developed for implementing and validating at the site or project scale. Site-scale consideration of ecosystem processes should, for the most part, emphasize implementing standards, monitoring results, and testing assumptions made at higher scales (see Lugo et al., Haufler et al., and Caraher et al., this volume, for additional discussion of scaler considerations in evaluating ecosystem processes and function, and the implications for ecosystem management).

## 3.3 Indicators of Ecosystem Condition

Although scientific understanding of ecosystems has increased dramatically in the last several decades, only a few areas and systems have been closely studied. An approach to land management driven by current ecosystem theory would be fueled by gross overgeneralizations and often arrive at unexpected results.

Resource managers often do not have the staff and budget to analyze ecosystem processes and functions in depth. Given the difficulty inherent in direct analysis of ecosystem functions and processes, most management analysis will use indicators. Several basic indicators of ecosystem process and function may be useful in gauging the general effects of management activities on process and function. Some of these indicators require a baseline condition (temporal context) for comparison; some do not. The following general list is not necessarily all-inclusive nor may all indicators work in all ecosystems (see rangeland health indicators in National Research Council 1994).

Does the proposed management activity, especially when added to previous and planned future activities, significantly change any of the following indicators?

- The quality and quantity of soil organic matter compared to the historical baseline or to desired conditions
- Other soil properties, such as bulk density or the existence of surface soil horizons
- The processes of nutrient recharge (elimination of nitrogen-fixing species, for example)
- The disturbance factors that operate on the site (such as fire, insects, disease, wind) over a long period
- Hydrologic function by altering the movement of water or sediment from or through the site
- The structure or composition of flora and fauna on the site over the long term, considering baseline variability

Case studies from the western Cascade Range and the Columbia River basin in Oregon and Washington provide a comprehensive list of ecological structure and composition indicators that link to primary ecosystem process and function elements at the stand (or site) and landscape scales. The Columbia River basin fire-ecology case study demonstrates the linkage between changes in fire disturbance and changes in landscape-scale vegetation composition and structure. These changes along with other indirect effects from management activities, have reinforced other ecological process changes — such as insect and disease outbreaks — that have caused widespread forest health concerns in the region.

## 3.4 Example — Ecological Process and Function Indicators for Forest Stands and Landscapes in the Western Cascades of the Pacific Northwest

### 3.4.1 Issues on forest management

Controversy on managing federal forest lands in western Oregon, western Washington, and northern California had its genesis in the conflict between intensive timber management practices and concerns about biological diversity and ecosystem processes. Ultimately, the controversy led to the Northwest Forest Plan and the adoption of a Record of Decision to it (USDA/USDI 1994b). The driving issues in the Plan were maintaining population viability of northern spotted owls and marbled murrelets, conserving late-successional and old-growth forests, managing aquatic and riparian resources, and determining the status of hundreds of poorly understood species which might depend on late-successional and old-growth forests. The analysis behind the Plan assumes that ecosystem process and function must be maintained and that habitat and species viability relate to ecosystem process and function (FEMAT 1993, USDA/USDI 1994a).

Table 3. Composition and structure indicators of ecosystem process and function at the stand scale in the Pacific Northwest.

| Process — function element | Structure — composition indicators | Reference |
|---|---|---|
| Nutrient, water, and energy cycling — fixation, interception, release | Forest floor — fine roots, coarse | Amaranthus et al. 1993 |
| | Vegetation species — nitrogen fixers, litter quality, mycorrhizal host species | Edmonds et al. (1989) |
| | Live trees — foliage amount and distribution, stem size and per area distribution | Franklin et al. (1981) |
| | Standing dead wood — size distribution and per area density | Franklin et al. (1989) |
| | Down dead wood — size distribution and per area density | Franklin and Spies (1991) |
| | Soil characteristics — volume, porosity, water availability, chemistry, organic matter | |
| Disturbance | Fire — intensity, severity, extent | Amaranthus et al (1993) |
| | Logging — intensity, severity, extent | |
| Erosion | Road density | Swanson et al. (1989) |
| | Debris slides | |
| | Surface erosion | |
| | Vegetation cover and root strength | |
| Vegetative development — succession, regeneration | Structure —composition stages | Spies (1997) |
| | | Oliver and Larson (1990) |

### 3.4.2 Significant findings that relate process and function to composition and structure

The link between the composition and structure of upland ecosystems and their processes and functions is reasonably well documented at the stand scale in the Pacific Northwest. Although an exhaustive literature synopsis is not the purpose of this example, the literature does indicate composition and structure elements that are linked to process and function (Table 3). A more comprehensive discussion can be found in Kohm and Franklin (1997) and Perry et al. (1989), who provide syntheses of the relation between composition and structure and process and function. Ruggiero et al. (1991) link composition and structure to wildlife populations and habitat.

*Key indicators at the stand scale*

Several composition and structure indicators of stand-scale process and function emerge from research in the Pacific Northwest. The following are at least an initial set:

- Live trees — size, density, species composition, canopy cover, leaf area
- Standing dead trees — size density, species, decay class
- Down trees — size, density, decay class
- Stand structure and composition stage
- Soil porosity, compaction, and bulk density
- Soil organic matter
- Soil erosion
- Soil mass wasting
- Disturbance kind, intensity, and severity — logging, windthrow, and fire.

### 3.4.3 Landscape scale

The relation between composition and structure and ecosystem process and function is not as well documented at large landscape scales. A brief review of the literature provides an initial set of indicators that should be useful (Table 4).

Table 4. Composition and structure indicators of ecosystem process and function at the large landscape scale in the Pacific Northwest.

| Process — function element | Structure — composition indicators | Reference |
|---|---|---|
| Disturbance | Patch size | Agee (1993) |
| | Patch shape | Morrison and Swanson (1990) |
| | Patch type | Turner et al. (1989) |
| | Patch amount | Turner (1989) |
| | Patch juxtaposition | Forman and Godron (1986) |
| | Patch topographic position | |
| | Climate | |
| | Topographic position | |
| Energy and nutrient cycling | Patch type — composition, leaf area index, structure | Harmon et al. (1986) |
| | Patch amount | Forman and Godron (1986) |
| | Disturbance history — soil and legacy organic matter | Waring and Schlesinger (1985) |
| Erosion | Road density | Swanson et al (1989) |
| | Debris slides | |
| | Surface erosion | |
| | Timber harvest — cover and root strength | |
| Vegetative development | Patch — structure and composition, stage | Oliver and Larson (1990) |
| | | Spies (1997) |

## 3.5 Example—Changes in Interior Columbia Basin Vegetation Structure and Composition Resulting from Changes in Disturbance Processes (Fire, Insects, and Diseases)

### 3.5.1 Background

A complex mosaic of plant communities with various patch sizes, structure, and species have evolved under the influence of lightning-induced fire regimes in the interior Columbia River basin ecosystem (Johnson et al. 1994). In these forest landscapes, fire is a major disturbance factor influencing basic ecosystem processes, such as energy transfer, fluvial processes, plant succession, and nutrient and material transfer. Fire-related processes are particularly important for carbon and nutrient cycling (Agee 1994). Low-intensity fire accelerates decomposition of forest litter and provides a periodic flush of nutrients that would otherwise be tied up in soil litter, down woody debris, and understory vegetation. Another benefit of light ground fires is that the dark soil surface can increase soil temperature and stimulate soil microbial activity (Oliver and Larson 1990). High-intensity fire often has more dramatic effects, such as loss of soil organic matter, volatilization of nitrogen, destruction of soil microbes, and loss of soil structure and infiltration capacity (Harvey et al. 1994). The effects of high-intensity fire on forest litter and soil characteristics also influence runoff and erosion in interior Columbia Basin watersheds. Large concentrations of sediment and nutrients can be transported from uplands to streams through surface erosion, mass soil and debris movement, or leaching (Swanston 1991). Studies of fire-denuded watersheds in Washington and Idaho showed a doubling of peak stream flow, which, coupled with two to seven times greater sediment discharge, have affected riparian and channel condition, fish habitat, and water quality in many watersheds (Swanston 1991, Quigley et al. 1996).

Forests in the interior Columbia River basin of eastern Oregon and eastern Washington have undergone substantial change in the last 40 to 50 years (Quigley et al. 1996). These include increased fragmentation in intensively managed landscapes and decreased fragmentation in wilderness and roadless landscapes; decreased abundance of early seral, late seral, and climax stands and increased abundance of mid-seral stands; changed insect and disease hazard in some watersheds; increased tree densities, fuel loads, fuel continuities, and fire hazards in some areas and decreased them in others; adversely affected riparian vegetation and associated fish habitat in many watersheds; and fire disturbance regimes have been altered through fire suppression, especially on sites adapted to frequent, low, and moderate-severity fires.

These changes result from altering forest processes directly (especially fire suppression) and from changes in composition and structure that influence ecosystem process and function in a feedback loop. This example illustrates how management activities can influence ecosystem process and function through direct manipulation and indirectly through changes in composition and structure.

### 3.5.2 Direct effects on ecosystem process: fire suppression

Fire has been actively suppressed in many places in eastern Oregon and Washington for 50 or more years

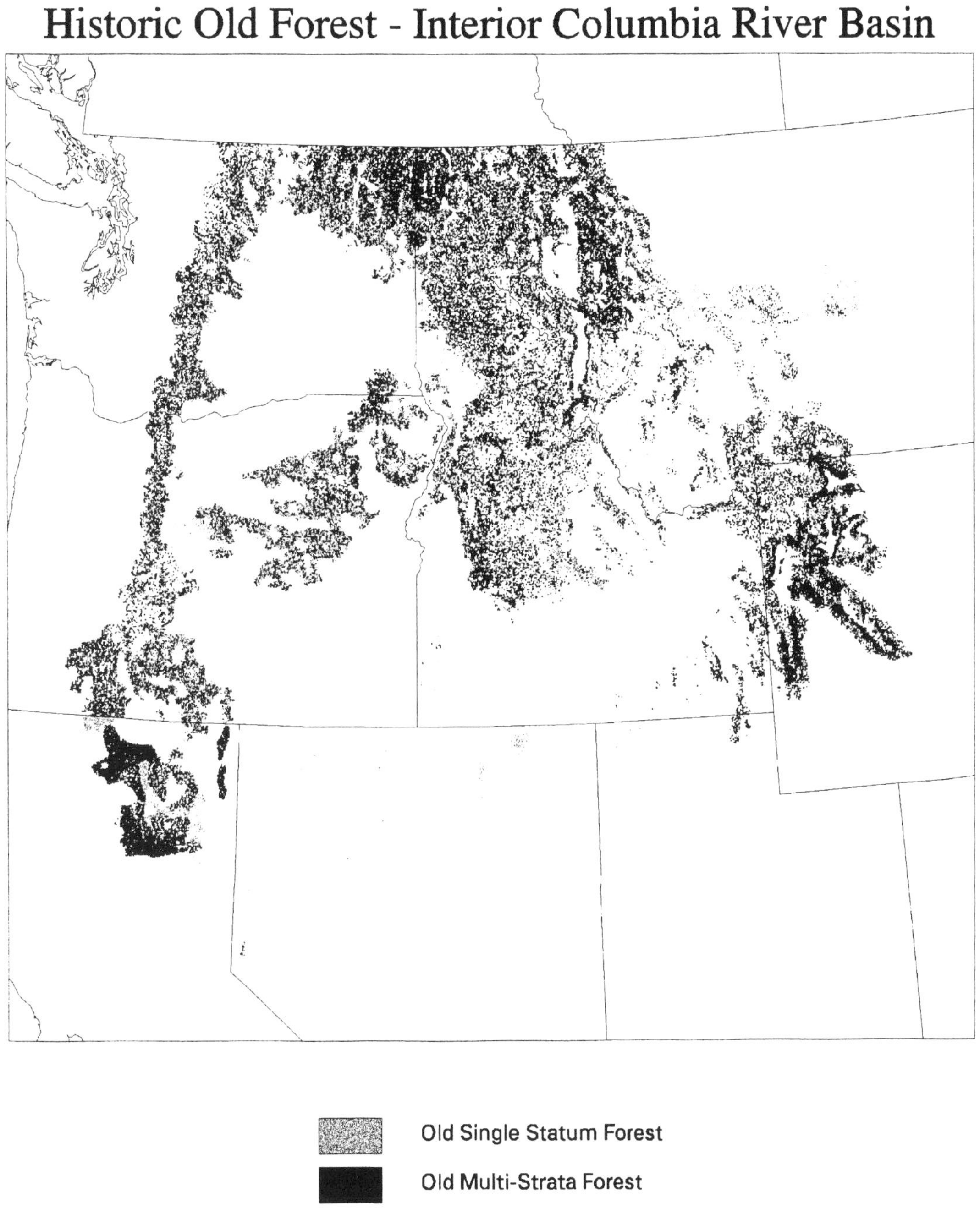


Fig. 4. Historic old forest stand types in the interior Columbia River Basin (source: unpublished Interior Columbia Basin Ecosystem Management Project map, USDA Forest Service, Portland, Oregon).

(Agee 1993, Quigley et al. 1996). Effects of fire suppression vary by vegetation type and environment (Quigley et al. 1996). For example, in the ponderosa pine and grand fir zones of eastern Oregon, frequent, low-intensity underburns historically killed small trees but not the large, thick barked pine and Douglas-fir. Historical stands typically consisted of large, widely spaced ponderosa pine and Douglas-fir. Fire suppression has allowed regeneration of relatively shade-tolerant tree species under an overstory of fire-resistant, shade-intolerant species. Understory regeneration of relatively shade-tolerant grand fir and Douglas-fir has altered fuel continuity and density. Fires are less frequent because of fire suppression efforts and more lethal due to successional changes in forest understory. As a result, late-successional stands of large trees have declined across the interior Columbia River Basin and open stands of large, fire-resistant trees have declined dramatically (Figs. 4 and 5).

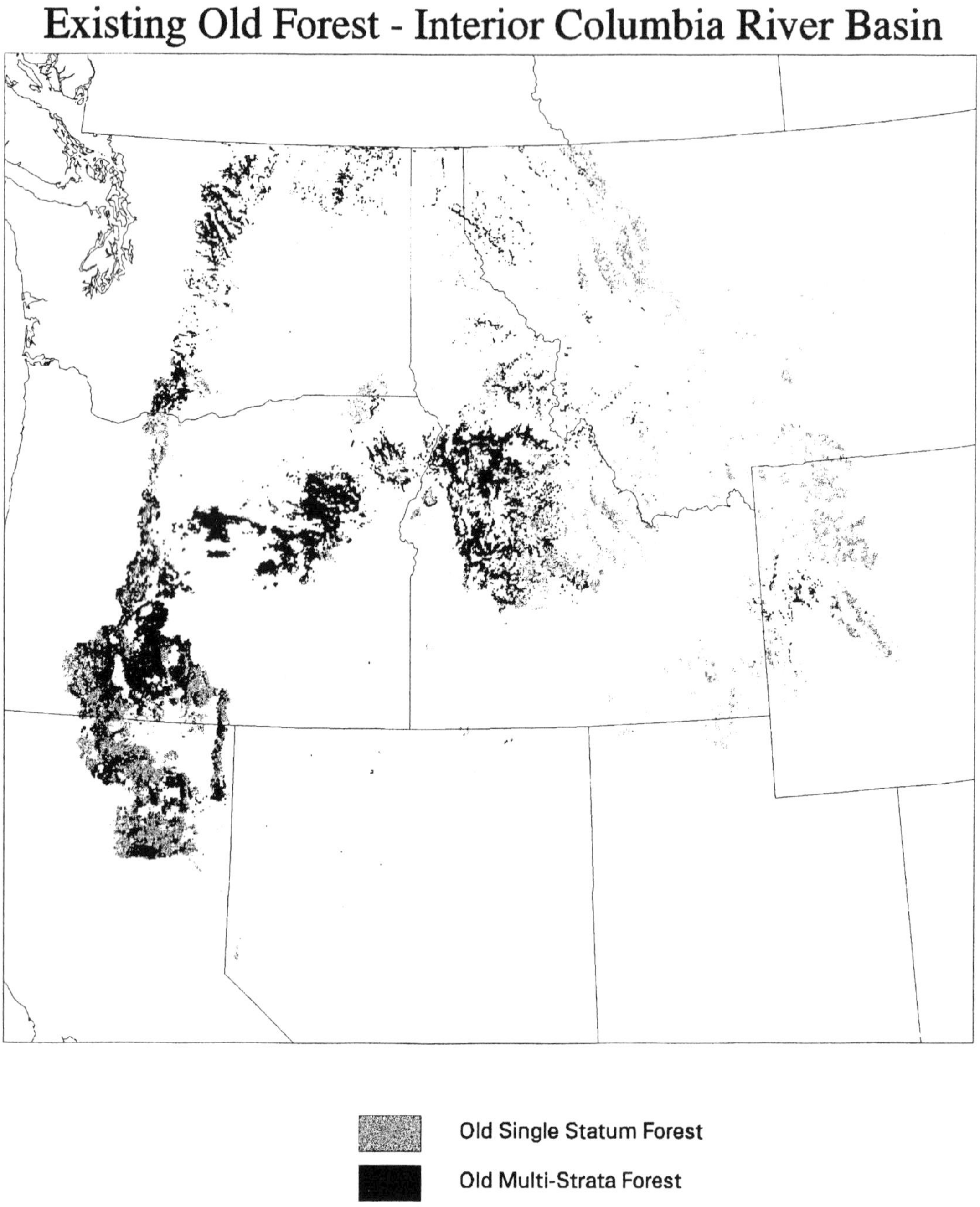


Fig. 5. Existing old forest stand types in the interior Columbia River Basin (source: unpublished Interior Columbia Basin Ecosystem Management Project map, USDA Forest Service, Portland, Oregon).

### 3.5.3 Indirect effects on ecosystem processes: managing vegetation

Fire suppression is not the only change in these forests. Decades of timber management, domestic livestock grazing, and other activities have also altered vegetation structure and composition (Quigley et al. 1996). Managed stands were typically clearcut planted with seedlings, and thinned to maintain relatively high stand density and uniform composition. The resulting forests lack large trees of early seral, fire-resistant species and are relatively uniform in density and composition. As a result, their crowns are dense and consistent fuels for wildfire. These stand-structure alterations reinforce direct changes in fire regime initiated by fire suppression. As a result of human actions that have changed vegetation composition and structure both directly and indirectly, forests are more susceptible to stand-replacing fire, again reinforcing changed ecosystem processes.

### 3.5.4 Management implications

Concerns about the current condition and health of Columbia Basin ecosystems provided the impetus for the Interior Columbia River Basin Ecosystem Management Project. The Integrated Science Assessment (Quigley et al. 1996) recommends a shift in management emphasis for Forest Service and Bureau of Land Management resources to restoration. It's cornerstone is to restore landscape "patterns, structures, and vegetation types to be more consistent with those occurring under disturbance regimes more typical of biophysical environments" (Quigley et al. 1996). A comprehensive discussion about practical applications of disturbance ecology and examples of approaches for restoring altered disturbance regimes to more natural conditions can be found in Engstrom et al. (this volume).

### 3.5.5 Summary

As a result of changing a key ecosystem process — the fire disturbance regime — vegetation composition and structure have been altered in forest landscapes across much of the interior Columbia Basin Ecoregion. This change along with other development activities over the past 50 years have reinforced the altered state of forest stands and have affected many other ecosystem processes and functions (Quigley et al. 1996).

## 3.6 Reference Conditions

Evaluating change in an indicator of ecosystem condition requires some frame of reference. If the indicator is the frequency distribution of patch sizes in a landscape, for example, a small change in average patch size may mean little or may be important. The "so-what?" question is a good test to use with most indices or indicators. One possible approach is based on understanding key elements of ecosystem function — for example, nutrient cycling. Soil organic matter content was discussed earlier as being a good indicator of nutrient status. Using this approach, however, is limited by our ability to obtain sufficient soils data to characterize conditions across a landscape, and by our ability to understand the complex linkages between this index and system structure, processes, and functions.

> The composition, structure, and landscape pattern of vegetation that exist in "natural" ecosystems or existed in the recent past provide one set of templates that have at least a somewhat predictable influence of ecosystem process and function.

Another approach to establishing a reference examines "natural" ecosystems, past or present, for clues to the future. The composition, structure, and landscape pattern of vegetation that exist in "natural" ecosystems or existed in the recent past provide one set of templates that have at least a somewhat predictable influence on ecosystem process and function (Morgan et al. 1993, Swanson et al. 1993). This approach requires studying the ranges in composition, structure, and pattern that characterized "natural" systems, present or past. This historical range of conditions, over sufficient time, at least maintained background rates of change in ecosystem process and species diversity. This approach suffers two difficulties: defining "natural" and assuming similar broad scale constraints, especially climate.

The utility of lessons learned from "natural" systems in predicting the behavior of managed systems depends, to some degree, on how far the conditions in the managed landscape diverge from the "natural" system studied and for how long. The more divergent the managed landscape, the more difficult applying templates from natural systems becomes. Small remnant parcels of native forest surrounded by highly developed land, for example, might not respond to managing disturbance according to lessons from similar large ecological systems. Similarly, land managed under intensive agriculture for many decades might not respond the same to fire as would a native prairie, because of changes in soil structure and chemistry.

Even so, using natural or historical conditions is an essential component in formulating sustainable resource-management objectives (Swanson et al. 1993). Determining desired conditions based on the historical range of variability can be used to establish acceptable limits of change in ecosystem processes and functions. This approach also provides essential insights for predicting behavior of both natural and altered ecosystems (Morgan et al. 1993).

## 3.7 Example — Ecosystem Analysis at the Watershed Scale, Game Creek, Southeast Alaska

### 3.7.1 Watershed analysis process

The Interagency Watershed Analysis procedure is being widely implemented on federal lands across the Pacific Northwest Coastal Ecosystem. We chose this example because it ties together many of the important concepts and methods for incorporating ecosystem process and function considerations into management strategies introduced at the beginning of this section. *Ecosystem Analysis at the Watershed Scale* (Regional

Ecosystem Office 1994) has been adopted for the Northwest Forest Plan (USDA/USDI 1994b) and recently for the Tongass National Forest Plan (USDA 1997). This process is designed to guide ecosystem analysis for watersheds of about 20 to 200 square miles and consists of six steps designed to assess the dominant ecological processes, functions, conditions, and uses of a watershed. The analysis should result in a clear list of management recommendations, based on the goal of sustaining key ecosystem processes and functions. The analysis steps are simple, direct, and useful for a variety of scales and purposes. The six steps include

- **Characterizing the watershed.** "Identify the dominant physical, biological, and human processes of features that affect ecosystem functions or conditions."
- **Identifying issues and key questions.** "Surface the key elements (indicators) of the ecosystem that are most relevant to management questions and objectives, human values, or resource conditions within the watershed."
- **Describing current conditions**. "Develop information relevant to the issues and key questions identified in step 2."
- **Describing reference conditions.** "Explain how ecological conditions have changed over time as a result of human influences and natural disturbances."
- **Synthesizing and interpreting information.** "Compare existing and reference conditions of specific ecosystem elements and explain significant differences, similarities, or trends and their causes."
- **Formulating recommendations.** "Bring the results to conclusion, focusing on management recommendations that are responsive to watershed processes identified in the analysis."

### 3.7.2 Background

Naiman (1992) identified the fundamental components for sustaining healthy watersheds in the Pacific Northwest Coastal Ecoregion. Two of the most important components identified were natural geomorphic processes and riparian vegetation. Disturbance regimes associated with episodic fire, flood, and mass-wasting events are primary factors influencing the delivery and routing of sediment in stream networks in this ecoregion. These geomorphic processes play a major role in shaping channel morphology and aquatic habitat structure and distribution (Reeves et al. 1995). Riparian vegetation influences energy transfer and nutrient cycling in the aquatic system through shading and input of organic materials. Large woody debris derived from riparian forests are a key element shaping stream channels and fish habitat structure. Riparian vegetation also regulates fluvial processes including stream bank erosion and floodplain deposition (Meehan et al. 1977, Naiman 1992). The convergence of surface and subsurface water flow in floodplain and wetland areas control nutrient, temperature, and oxygen exchange in these aquatic ecosystems (Naiman 1992).

Timber harvest in the Northwest Coastal Forest Ecoregion have affected watershed health and condition in many areas. Major effects of timber management on watershed and riparian processes include increases in size or frequency of peak stream flow, substantial increases in sediment supply, streambank destabilization, and loss of stable in-stream large woody debris, all of which, are critical to maintaining aquatic habitat diversity (Chamberlin et al. 1991). Timber harvesting can also alter seasonal flow and temperature regimes, which are important factors regulating feeding and reproductive behavior in aquatic organisms. Forest management practices combined with other kinds of development, such as hydroelectric dams and agriculture, have led to extensive habitat loss and the listing of 400 stocks of Pacific salmon as threatened or endangered species in Washington, California, Oregon, and Idaho (Nehlsen et al. 1991). Resource managers' approaches to managing these ecosystems have often been concerned more with mitigating past problems than with preventing future ones (Reeves et al. 1991). Unfortunately, a lack of basic understanding of key watershed and riparian processes has resulted in many stream restoration approaches being ineffective or even counterproductive (Beschta et al. 1994).

### 3.7.3 Management issues in the Game Creek watershed, Southeast Alaska

In contrast to other parts of the Northwest Coastal Ecoregion, the watershed function and processes in most watersheds in coastal Alaska are unimpaired (Fig. 6). The recently completed Anadromous Fish Habitat Assessment for Southeast Alaska (AFHA 1995) found, however, that expanded development of watersheds for timber production may eventually result in habitat degradation and damage to fish stocks unless forest management is improved. The Game Creek watershed analysis illustrates how a detailed blueprint of riparian areas, along with an understanding of ecosystem processes, can be used to develop management recommendations that will help sustain important riparian functions.

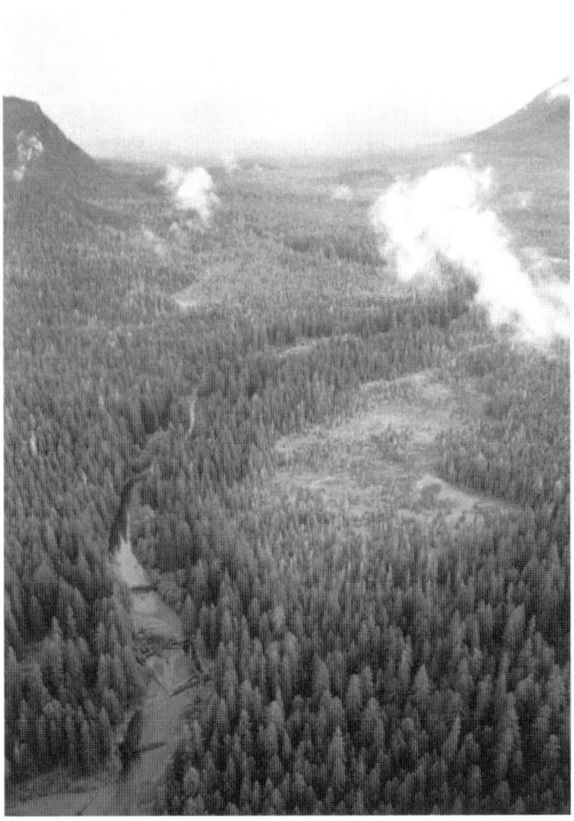

Fig. 6. Game Creek watershed, Southeast Alaska; an example of healthy watershed and riparian conditions in the Pacific Northwest Coastal Ecoregion (photo credits: Steve Paustian).

### 3.7.4 Watershed characterization

The design of the Game Creek riparian conservation areas incorporates three major components that provide a functional linkage between aquatic habitats, stream riparian areas, and adjacent hillslopes (Paustian et al. 1995). The delineation of these three components — riparian areas, sediment source areas and fens — is tied to the fundamental components, hydrogeomorphic and riparian processes identified by Naiman (1992). Streamside riparian area was determined by the extent of vegetation indicating shallow groundwater table or periodic flooding. For constrained or entrenched stream reaches, the first significant slope break above the bank-full stream-stage, or the height of dominant riparian trees define the riparian area. Average riparian area widths are tabulated by channel-type classification (Paustian 1992). An example of GIS stream riparian polygons for a portion of Game Creek is shown in Fig. 7.

The second key component of the riparian conservation area is the sediment source area. Mass wasting events — particularly snow and debris avalanches — are major natural disturbance factors in the Game Creek watershed. These unstable areas are especially sensitive to road construction and timber harvest, and they are a special concern because of accelerated erosion, sediment delivery, and indirect effects on downstream aquatic and riparian habitats (Swanston 1991). Sediment source areas in the Game Creek watershed were defined by using landform and soils criteria (Paustian et al. 1995). These delineations encompass steep mountain slopes (70% gradient), frequently dissected by small headwater drainages, and colluvial hollows with high potential for sediment delivery to perennial stream channels (Fig. 8).

A third key component of the Game Creek riparian conservation area is the rich fen wetlands. These wetlands are along valley margins below coarse alluvial–colluvial toe slopes (Fig. 9). Rich fens are the hydrologic link between resurgent alpine streams, hill slope aquifer and alluvial valley stream. Fens support a diverse, wet meadow vegetation community and are laced with small streams fed by seeps and springs. These donor wetlands provide mineral-rich groundwater at nearly

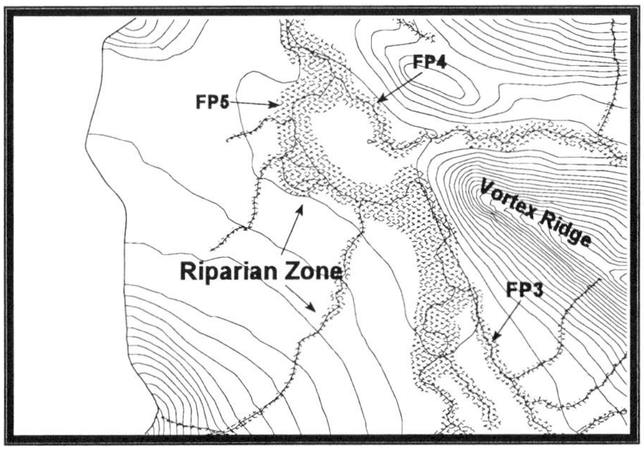


Fig. 7. Riparian conservation areas in a portion of the Game Creek watershed, Southeast Alaska; different riparian zone widths correspond to variations between channel type (e.g. FP3, FP4, FP5) riparian characteristics (source: Paustian et al. 1995).

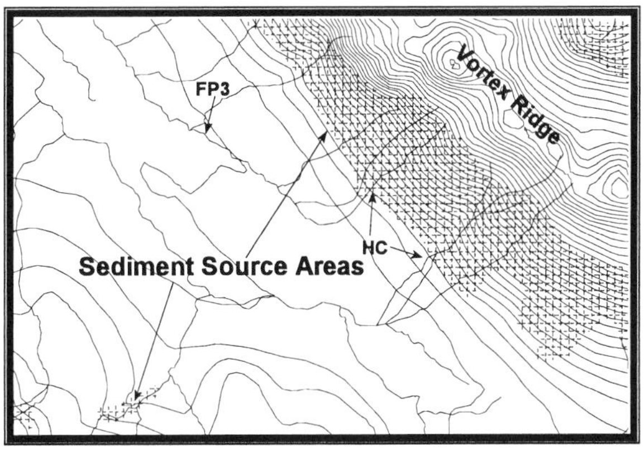


Fig. 8. Sediment source areas in a portion of the Game Creek watershed, Southeast Alaska (source: Paustian et al. 1995).

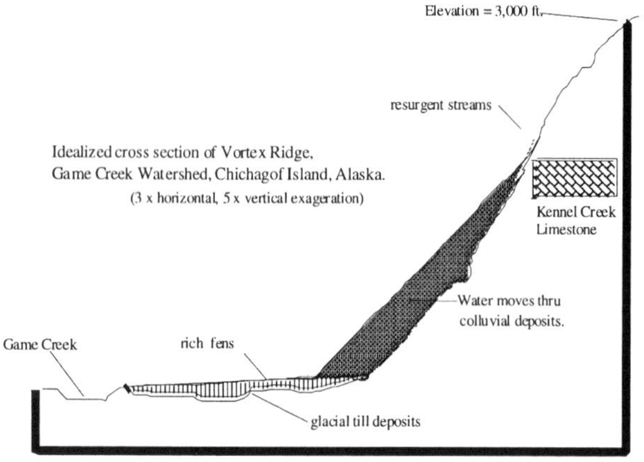


Fig. 9. Relations of rich fen wetlands to watershed morphology, Game Creek, Southeast Alaska (source: Paustian et al. 1995).

constant temperature and flow rate to adjacent streams. Fen complexes also provide critical over-winter and summer low-flow refuge habitat for juvenile salmonids in the Game Creek watershed.

### 3.7.5 Assessing watershed condition

The Game Creek analysis focused on the relative condition of the riparian conservation areas described above along with quantitative measurements of aquatic habitat indices to address the issue of existing watershed health. Recent timber-harvest activities currently affect about 4% of the 13,560-ha watershed. Harvest units and roads affect about 2.5% of the total riparian conservation area delineated in Game Creek. The analysis showed that key riparian areas, sensitive soil, and wetland areas were largely intact. Stream buffer areas had been established along all fish streams: therefore, most potential effects from riparian harvest activities in Game Creek were indirect.

In the Game Creek watershed, 20% of the fish habitat was randomly sampled to determine the condition of aquatic habitats. Interim Region 10 fish habitat objectives based on key stream morphology and habitat indicators were used as reference conditions in the Game Creek analysis (USDA 1995). These fish-habitat objectives established a range of natural variability for three indices — large woody debris, pool area, and channel width-to-depth ratio — derived from data sets collected in pristine watersheds. These metrics are defined for specific channel-type stream classification (Paustian 1992) categories to group habitat data by stream segments that are influenced by similar geomorphic and riparian processes. Fish habitat objectives are expressed as a range of values based on the 25th, 50th (median), and 75th percentiles for benchmark data collected from a given channel type or process group. Results from the Game Creek analysis are compared to interim Region 10 habitat objectives displayed in Table 5. Large woody debris values for Game Creek are consistently at or above the 75th percentile for this habitat objective, indicating excellent habitat conditions related for large wood. Most values for pool area are at or above the 50th percentile, indicating above-average habitat associated with pools. One channel type segment — MM1 — had poor pool habitat relative to the reference condition. Width-to-depth measurements for the Game Creek flood plain (FP) channel categories were near the 50th percentile of the habitat objective indicating relatively stable channel condition and balanced sediment transport regime. Width-to-depth values for moderate-gradient-mixed-control (MM) channels are at the 25th percentile for reference conditions. The Game Creek channel segments in this category are narrow and deeper than similar channels of the same type, but this condition probably has little significance to the geomorphic processes in these streams.

Table 5. Comparison between Game Creek watershed-analysis habitat measurements and Interim Alaska (Region 10) habitat standards for pools, large woody debris and channel width-to-depth ratio for different process group/channel type classes (source: USDA 1995).

| Fish habitat objective | Process group/ channel type | Interim Region 10 habitat standards (percentiles) | | | Game Ck. habitat values |
|---|---|---|---|---|---|
| | | 25th | 50th | 75th | |
| Large woody debris (pieces/ 1,000 m$^2$) | FP3 | 10 | 32 | 54 | 96 |
| | FP4 | 8 | 24 | 34 | 46 |
| | FP5 | 4 | 5 | 6 | 21 |
| | LC & MC | 6 | 15 | 22 | 20 |
| | MM1 | 27 | 45 | 82 | 96 |
| | MM2 | 33 | 35 | 44 | 42 |
| Pool area (%) | FP3 | 20 | 53 | 76 | 45 |
| | FP4 | 35 | 47 | 59 | 50 |
| | FP5 | 47 | 51 | 60 | 37 |
| | LC | 8 | 20 | 27 | 39 |
| | MM1 | 28 | 40 | 52 | 18 |
| | MM2 | 2 | 22 | 39 | 33 |
| Stream channel width-to-depth ratio (dimensionless) | FP3 | 8 | 13 | 18 | 11 |
| | FP4 | 16 | 25 | 35 | 20 |
| | FP5 | 30 | 45 | 70 | 32 |
| | MM1 | 9 | 12 | 18 | 7 |
| | MM2 | 17 | 24 | 33 | 15 |

The general conclusion from the assessment of riparian conservation area and fish-habitat conditions in the Game Creek Watershed Analysis is that the watershed is in good health. Stream survey data used to assess fish habitat objectives for various channel types in the watershed will provide a basis for future analysis of watershed and riparian-condition trends.

This quantitative, objective approach is a very effective technique for assessing ecosystem condition but it does require more time and expense compared to more qualitative assessment procedures. An alternative approach to assessing riparian condition can be found in *Riparian Area Management: Process for Assessing Proper Functioning and Condition* (USDI 1995).

### 3.7.6 Management recommendations

Game Creek riparian conservation areas, as defined by stream riparian areas, sediment source areas, and rich fens, require special management consideration. Intensive timber harvest activities should be minimized in riparian conservation areas. Silvicultural practices in these areas should be designed to avoid risks to desired riparian and aquatic ecosystem processes and functions. Management objectives for riparian areas should focus on maintaining existing healthy riparian and aquatic habitat conditions. Future road construction in riparian conservation areas should be limited to roads providing essential access. Roads should be located and designed to limit the number of stream crossings, maintain unrestricted fish migration, and maintain stream channel integrity and function. Timber harvest and road construction should be prohibited in sensitive soil areas because of the high risk of mass wasting and potential effects on down-slope riparian resources. Ground-disturbing activities, such as road construction, should also be discouraged in rich fen areas. Special mitigation measures are required when activities in fens cannot be avoided, to minimize potential effects on surface and subsurface hydrologic function. These measures may include use of geotextile matting below the road fill, to reduce compaction of organic soils and the use of porous rock fill for road subgrades, with numerous cross drain culverts, to maintain natural surface runoff. Road drainage structures should be regularly maintained to minimize the potential for indirect effects on riparian areas from surface erosion and sedimentation. If roads cannot be maintained, they should be put into storage or obliterated to restore the natural hydrologic functions of the area.

Watershed analysis steps are simple, direct and useful for a variety of scale and purposes:

- Characterizing the watershed
- Identifying issues and key questions
- Describing current conditions
- Describing reference conditions
- Synthesizing and interpreting information
- Formulating recommendations

### 3.7.7 Summary

Watershed analysis is a simple, direct approach for assessing key ecological processes, functions, and conditions. Delineation of riparian conservation areas is a useful technique for identifying key or sensitive areas in a watershed, which need to be managed in a manner that sustains important ecological processes and functions. Use of key indicators, such as large woody debris and pool habitat characteristics, provide an unbiased, ecologically sound frame of reference for judging the health of watersheds and riparian ecosystems. Recommendations resulting from watershed analysis focuses managers' attention on management practices and opportunities responsive to the goal of maintaining healthy watershed and riparian conditions.

## 4 ASSESSING CHANGE IN ECOSYSTEM PROCESSES

Most of our management actions focus on components of ecosystems rather than on ecosystem processes. Yet many of the outcomes managers seek from those actions depend on one or more ecological processes operating as expected. Monitoring provides the means to verify whether their expectations are being met. Monitoring of existing human effects can be based on knowing about the past by using historical records, concurrently comparing managed to similar unmanaged sites, measuring easily observed ecosystem components believed to reflect the status of hard-to-measure components, recurrently measuring at the same sites to reveal long term-trends, and using ongoing observations of a selected range of management options to reduce the risk and uncertainty associated with management activities that have unknown outcomes (adaptive management). The topics of monitoring and adaptive management are discussed in detail by Powell et al. and Bormann et al. in Vol. III of this book. In the following discussion, our intent is not to duplicate that information but rather to focus attention on major issues pertaining to assessing change in ecological processes.

## 4.1 Monitoring

The long-term effects on other ecosystem processes of managing a specific ecosystem process are often difficult to quantify and may be counter-intuitive. Managing woody debris in anadromous fish streams in the Pacific Northwest provides a prime example. In the 1960s, managers routinely removed large wood from streams because they presumed it impeded fish migration. Removing it would, in theory, improve fish passage, spawning, and run size. The large wood was often sound and saleable, and more fish would provide more recreation and commercial fishing. Unfortunately, large woody debris proved to be critical in dissipating stream energy, capturing sediment, and providing nutrient capital and habitat for fish. Managers are now replacing wood where it was formerly removed. Even well-thought-out programs backfire. Failure to consider ecosystem function can prove disastrous (as in the Everglades example).

To address these complex relations, ecosystems monitoring efforts must consider both the broad spatial scales and the long-term temporal scales in which these processes operate. Working at these scales is very difficult, however, because of their great variability and the difficulty in maintaining uniform long-term data collection and determining the natural variability in the systems. Infrequent catastrophic events from within and outside the system are also very difficult or impossible to account for.

Ecological processes may influence ecosystem development at landscape and regional scales as well as at the site scale. Because individual site management programs may influence or be influenced by these broader processes, managers must collaborate to coordinate monitoring at the landscape scale. Periodic droughts, acid precipitation, insect and disease outbreaks, wildfires, seasonal animal migrations, and spread of exotic species can generate changes to ecological processes that transcend local management boundaries. The purpose of coordinated monitoring is to provide early warning about landscape trends that may influence the freedom or appropriateness of individual actions. A positive outcome at a given site might be negated by landscape-scale processes, or effects from a project might compound unwanted outcomes to another manager's programs.

> Monitoring and adaptive management programs should help determine whether management is resulting in desired outcomes, providing a factual basis for changing management objectives, and providing consistent data sets applicable across broad areas and jurisdictional boundaries.

## 4.2 Example — Monitoring Ecosystem Composition and Structure Across Multiple Scales in the Pacific Northwest

A conceptual model (Fig. 10) for the relations of process and function to vegetation structure and composition displays indicators that could be used in monitoring the effects of vegetation management on processes and functions across multiple scales. This model (Tom Spies, USDA Forestry Science Laboratory, Corvallis, OR, personal communication) assumes that ecological processes and functions drive composition and structure, and conversely, composition and structure mediate processes and functions (see "Indicators" Sections 3.3–3.4, this volume). Fire, for example, is an ecological process that initiates stand regeneration, influences nutrient cycling, carbon storage, hydrology, and other processes. Fire, in turn, is influenced by the composition and structure of stands and the distribution of stands across a landscape. The model also assumes a relation between biological diversity and the composition and structure of vegetation, essentially in the form of habitat relations.

Use of this conceptual model allows identification of key indicators which can be relatively easily tracked at landscape and site scales. Data on patch sizes, shapes, types, and landscape distribution can be derived from remotely sensed imagery. Tree size distribution, dead wood, understory vegetation, and forest floor information necessarily come from field data. A sampling scheme can be developed, one that uses field plots stratified by classes of stands from remotely sensed images.

The model (Fig. 10) also specifies assumed relations between *structure* and *processes*, such as fire (intensity,

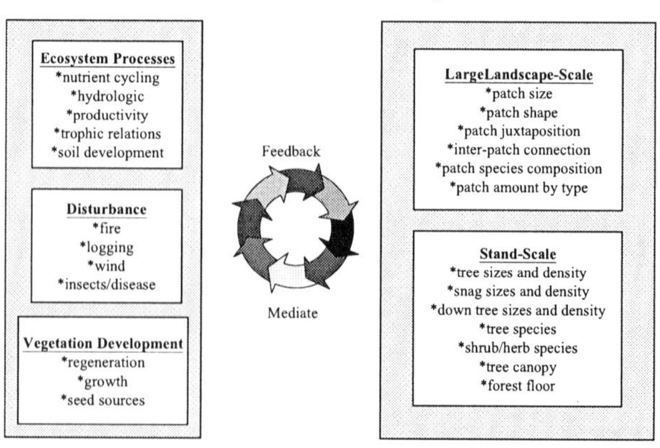


Fig. 10. Conceptual model for monitoring relations of vegetation composition and structure to ecosystem processes and functions, at multiple spatial scales in the Pacific Northwest.

frequency), tree mortality, gap regeneration, and others; *functions* such as nutrient cycling, primary production, and others; to *biological diversity* (richness, evenness, and so on), all of which can be tested in a rigorous research design.

## 4.3 Long-Term Monitoring of Ecosystem Processes —Example: Acid Rock Drainage, Great Smoky Mountains National Park

Long-term monitoring is needed to assess subtle cumulative changes in ecological processes. The need for long-term monitoring varies, depending on the processes of concern and the manager's understanding of potential consequences associated with altering a given process or processes. For example, long-term grazing of rangeland may result in cumulative changes in nutrient cycling and soil erosion processes more quickly than effects on soil productivity associated with clearcutting in a temperate forest. Some management practices may have fairly immediate direct effects on ecological processes, but indirect effects from the activity may not appear for decades. Timber harvest along forest streams can result in changes to energy transfer processes that immediately affect fish populations through increases in water temperature. Riparian timber harvest can indirectly affect fish habitat by reducing recruitment of pool-forming woody debris. These indirect effects on pool habitat may not become evident for several decades until the residual woody debris derived from the old-growth forest decays and is flushed out of the stream system.

### 4.3.1 Background

Water-quality problems associated with biogeochemical weathering in Great Smoky Mountains National Park, North Carolina–Tennessee, illustrate the importance of long-term monitoring to assess ecosystem condition and evaluate restoration effectiveness. Lugo et al. (see this volume) discuss the mechanisms of biochemical weathering and production of acid rock drainage. Acid drainage problems are most commonly associated with mining activities and often have serious long-term effects on water quality and aquatic ecosystems. The Great Smoky Mountains National Park case study illustrates that acid rock drainage can be a serious management concern with any significant ground-disturbing activity where climatic and geologic conditions promote the oxidation of sulfide-bearing rock formations.

### 4.3.2 Management effects

Reconstruction of the Newfoundland Gap section of U.S. Highway 441 in 1964 between Gatlinburg, Tennessee, and Cherokee, North Carolina, exposed pyritic slate and phyllite rock known as the Anakeesta Formation. Subsequent landslides have periodically exposed fresh rock above the road cutslopes. Large volumes of slide debris have also been sidecast into small ravines and hollows that enter Beech Flats Creek, which flows parallel to the highway.

### 4.3.3 Water quality monitoring

Significant changes in stream chemistry were first documented in 1975 in Beech Flat Creek below Highway 441. Acid drainage from the highway construction decreased pH and increased concentrations of heavy metals including iron, zinc, manganese, and aluminum along a 3-mile section of Beech Flats Creek. Once a productive trout fishery, the creek became virtually devoid of aquatic organisms. Additional water-quality monitoring between 1988 and 1990 showed no improvement in stream chemistry and significant increases in toxic metal concentrations at many sites. The later monitoring effort identified seeps and ravines that were acid drainage hot-spots.

### 4.3.4 Restoration

Mitigation of acid rock drainage requires increased acid buffering and limiting the input of oxygen and water to waste rock deposits. Traditional mitigation techniques for acid rock drainage include layering waste rock with limestone, capping the material with an impervious layer of clay or compacted soil, installing finger drains at the bottom of waste deposits, and establishing a dense vegetative cover on the waste rock deposit to limit water infiltration and erosion of the soil cap. The site conditions and lack of remediation in the original project design will make restoration of the Great Smoky Mountains site extremely difficult. Test plots in conjunction with additional water-quality monitoring will determine whether soil capping and revegetation are effective in reducing acid rock drainage into Beach Flats Creek.

## 4.4 Adaptive Management

Application of adaptive management principles (see Bormann et al., in Vol. III of this book) is particularly useful in assessing ecological processes and functions. With the current limited understanding of complex ecological processes and interrelations, the outcomes of management actions are often uncertain. Indeed, all of the effects of any one management action are unlikely ever to be known. Widely implementing a single management strategy can severely limit a manager's future options to respond to unanticipated outcomes. Under an active, adaptive management program, citizens, managers, and scientists collaborate to define learning objectives, to predict outcomes and monitor appropriate variables to determine whether the predictions were reasonably close to the actual outcomes and then adapting their management practices in response to the new information. Several alternative management strategies for meeting these objectives are developed, and factual outcomes are compared by using statistical techniques such as random treatments and replicated treatments. This approach helps managers deal with the inherent risks and uncertainties associated with managing complex ecological processes.

Evaluating change in ecological processes and functions associated with management activities is a major challenge. Some outcomes are apparent at the site of management; others appear some distance away. Some outcomes are apparent within a short time after the treatment; others appear years or decades later. Because outcomes appear at different times and places than management treatments, learning approaches must be designed to assess the timing and effect of processes through the outcomes and not the treatments. In meeting this function, monitoring and adaptive management programs should help determine whether management is resulting in desired outcomes, providing a factual basis for changing management objectives, and providing consistent data sets applicable across broad areas and jurisdictional boundaries. Monitoring ecological processes needs to be considered as a long-term management program. A sound monitoring program must have a consistent funding and staffing base, standard definitions and measurement criteria to collect consistent data, statistically valid sampling designs, and periodic sample regimes to detect trends (National Research Council 1994).

## 5 CONCLUSIONS

Ecosystem management should strive to maintain ecological processes within acceptable thresholds and to sustain diverse and functioning ecosystems to meet long-term human needs. Understanding natural ecosystem processes and functions is our only option for maintaining ecosystem sustainability, productivity, and long-term integrity.

Key findings and conclusions include:

- By working with rather than against ecological processes, resource managers can develop effective approaches to meet many human needs that are less disruptive to ecosystem functions and integrity. A passive management approach may sometimes be best.
- In defining management objectives, inventorying, analyzing, manipulating and monitoring ecosystems — and their processes and functions — managers must determine the appropriate scales, indicators, and reference conditions.
- Analyses should be focused at the watershed and landscape scales for ecosystem process and function issues that are of principal importance to land managers.
- The composition, structure, and landscape pattern of vegetation or other fundamental components of "natural" ecosystems provide a set of templates or reference points that have somewhat predictable influence on ecosystem process and function.
- Managing within ecosystem constraints, and using new management approaches — such as watershed analysis and adaptive management — and powerful tools for resource information gathering and analysis such as remote sensing and GIS — will increase the likelihood of attaining the goal of meeting human needs and maintaining healthy ecosystems.

## ACKNOWLEDGMENTS

We wish to thank Gary Meffe, David Perry, Donald Imm, Ariel Lugo, and two anonymous reviewers for their helpful comments on earlier drafts of this paper. We also wish to thank Wayne Elmore and Cindy Correll for providing photographs of rangeland riparian rehab projects.

## REFERENCES

Agee, J.K. 1993. *Fire ecology of Pacific Northwest Forests*. Island Press, Washington, DC.

Agee, J.K. 1994. Fire and Weather Disturbances in Terrestrial Ecosystems of the Eastern Cascades. USDA Forest Service General Technical Report PNW-GTR-320. USDA Forest Service, Pacific Northwest Research Station, Portland, OR.

Amaranthus, M.P., J.M. Trappe, and D.A. Perry. 1993. Soil moisture, native revegetation, and *Pinus lambertiana* seedling survival, growth, and mycorrhiza formation following wildfire and grass seeding. *Restoration Ecology* 1(3): 188–195.

Beschta, R.L., W.S. Platts, J.B. Kauffman, and M.T. Hill. 1994. Artificial stream restoration: money well spent or an expensive failure? *Proceedings of Environmental Restoration.* Universities Council of Water Resources, 1994 Annual Meeting, August 2–5, 1994. Universities Council of Water Resources, Big Sky, MT.

Bormann, B.T., M.H. Brookes, E.D. Ford, A.R. Kiester, C.D. Oliver, and J.F. Weigan. 1994. A Framework for Ecosystem Management. USDA Forest Service General Technical Report PNW-GTR-331. USDA Forest Service, Pacific Northwest Research Station, Portland, OR.

Brinson, M.M. 1993. A Hydrogeomorphic Classification for Wetlands. U.S. Army Corps of Engineers, Wetlands Research Program Technical Report, WRP-DE-4. U.S. Corps of Engineers, Vicksburg, MS.

Chamberlin, T.W., R.D. Harr, and F.H. Everest. 1991. Timber harvesting, silviculture, and watershed processes. pp. 181–205. In: W.R. Meehan (ed.), *Influences of Forest and Rangeland Management on Salmonid Fishes and their Habitats,* American Fisheries Society Special Publication 19. American Fisheries Society, Bethesda, MD.

David, D.G. 1996. Changes in plant communities relative to hydrologic conditions in the Florida Everglades. *Wetlands* 16 (1): 15–23.

Edmonds, R.L., D. Binkley, M.C. Feller, P. Sollins, A. Abee, and D.D. Myrold. 1989. Nutrient cycling: effects on productivity of Northwest forests. pp. 17–35. In: D.A. Perry, R. Meurisse, B. Thomas, R. Miller, J. Boyle, J. Means, C.R. Perry, and R.F. Powers (eds.), *Maintaining the Long-Term Productivity of Pacific Northwest Forest Ecosystems.* Timber Press, Portland, OR.

Elmore, W., and R.L. Beschta. 1987. Riparian areas: perceptions in management. *Rangelands* 9(6): 260–265.

Elmore, W., and J.B. Kauffman. 1994. Riparian and watershed systems: degradation and restoration. pp. 211–232. In: M. Vavra, W.A. Laycock, and R.D. Piper (eds.), *Ecological Implications of Livestock Herbivory in the West.* Society of Range Management, Denver, CO.

FEMAT. 1993. Forest Ecosystem Management. An Ecological, Economic and Social Assessment. Report of the Forest Ecosystem Management Assessment Team. Portland, OR: USDA Forest Service, USDC National Marine Fisheries Service, USDI Bureau of Land Management, USDI National Park Service, and USEPA.

Finkl, C.W. 1995. Water resource management in the Florida Everglades: are lessons from experience a prognosis for conservation in the future? *Journal of Soil and Water Conservation,* 50(6): 592–611.

Forman, R.T.T., and M. Godron. 1986. *Landscape Ecology.* Wiley, New York.

Franklin, J.F., K. Cromack, Jr., W. Dennison, A. McKee, C. Maser, J. Sedell, F. Swanson, and G. Juday. 1981. Ecological Characteristics of Old-Growth Douglas-Fir Forests. Gen. Tech. Rept. PNW-118. USDA Forest Service, Pacific Northwest Research Station, Portland, OR.

Franklin, J.F., D.A. Perry, T.D. Schowalter, M.E. Harmon, A. McKee, and T.A. Spies. 1989. Importance of ecological diversity in maintaining long-term site productivity. pp. 82–97. In: D.A. Perry, R. Meurisse, B. Thomas, R. Miller, J. Boyle, J. Means, C.R. Perry, and R.F. Powers (eds.), *Maintaining the Long-Term Productivity of Pacific Northwest Forest Ecosystems.* Timber Press, Portland, OR.

Franklin, J.F., and T.A. Spies. 1991. Ecological definitions of old-growth Douglas-fir forests. pp. 61–69. In: L.F. Ruggiero, K.B. Aubry, A. B. Carey, and M. Huff (tech. coords.), Wildlife and Vegetation of Unmanaged Douglas-Fir Forests. Gen. Tech. Rept. PNW-GTR-285. USDA Forest Service, Pacific Northwest Research Station, Portland, OR.

Harmon, M.E., J.F. Franklin, F.J. Swanson, P. Sollins, S.V. Gregory, J.D. Lattin, N.H. Anderson, S.P. Cline, N.G. Aumen, J.R. Sedell, G.W. Lienkeamper, K. Cromack, Jr., and K.W. Cummins. 1986. Ecology of coarse woody debris in temperate ecosystems. *Advances in Ecological Research* 11: 155–167.

Harvey, A.E., J.M. Geist, G.I. McDonald, M.F. Jurgensen, P.H. Cochran, D. Zabowski, and R.T. Meurisse. 1994. Biotic and Abiotic Processes in Eastside Ecosystems: The Effects of Management on Soil Properties, Processes, and Productivity. USDA Forest Service General Technical Report PNW-GTR-323. USDA Forest Service, Pacific Northwest Research Station, Portland, OR.

Johnson, C.G., R.R. Clausnitzer, P.J. Mehringer, and C.D. Oliver. 1994. Biotic and Abiotic Processes in Eastside Ecosystems: The Effects of Management on Plant and Community Ecology, and on Stand and Landscape Vegetation Dynamics. USDA Forest Service General Technical Report PNW-GTR-322. USDA Forest Service, Pacific Northwest Research Station, Portland, OR.

Kauffman, J.B., R.L. Beschta, N. Otting, and D. Lytjen. 1997. An ecological perspective of riparian and stream restoration in the western United States. *Fisheries* 22 (5): 12–24.

Kohm, K.A., and J.F. Franklin. 1997. *Creating a Forestry for the 21st Century.* Island Press, Washington, DC.

Lugo, A.E., S. Brown, and M.M. Brinson. 1990. Concepts in wetland ecology. pp. 53–85. In: A.E. Lugo, S. Brown, and M.M. Brinson (eds.), *Forested Wetlands, Ecosystems of the World.* Elsevier, Amsterdam, The Netherlands.

Meehan, W.R., F.J. Swanson, and J.R. Sedell. 1977. Influences of riparian vegetation on aquatic ecosystems with particular reference to salmonid fishes and their food supply. *Proceedings of Importance, Preservation and Management of Riparian Habitat: A Symposium, July 9, 1977.* USDA Forest Service, Tucson, AZ.

Minshall, G.W. 1994. Stream-riparian ecosystems: rationale and methods for basin-level assessments of management effects. pp. 143–167. In: M.E. Jensen and P.S. Bourgeron (eds.), Volume II: Ecosystem Management: Principles and Application. USDA Forest Service General Technical Report PNW-GTR-318. Portland, OR: USDA Forest Service, Pacific Northwest Research Station.

Morgan, P., G.H. Aplet, J.B. Haufler, H.C. Humphries, M.M. Moore, and W.D. Wilson. 1993. Historical range of variability: a useful tool for evaluating ecosystem change. *Journal of Sustainable Forestry* 2(1/2): 87–111.

Morrison, P.H., and F.J. Swanson. 1990. Fire History and Pattern in a Cascade Range Landscape. USDA Forest Service, Pacific Northwest Research Station, General Technical Report PNW GTR-254. USDA Forest Service, Pacific Northwest Research Station, Portland, OR.

Naiman, R.J. (ed.). 1992. *Watershed Management: Balancing Sustainability and Environmental Change.* Springer Verlag, New York.

National Research Council. 1994. *Rangeland Health: New*

*Methods to Classify, Inventory, and Monitor Rangelands.* National Academy Press, Washington, DC.

Nehlsen, C.K., J.E. Williams, and J.A. Lichatowich. 1991. Pacific salmon at the crossroads: Stocks at risk from California, Oregon, Idaho, and Washington. *Fisheries* 16(2): 4–21.

Oliver, C.D., and B.C. Larson. 1990. *Forest Stand Dynamics.* McGraw-Hill, New York.

Paustian, S.J. (ed.). 1992. *A Channel Type Users Guide for the Tongass National Forest, Southeast Alaska.* USDA Forest Service, Alaska Region, R10-TP-26, Juneau, AK.

Paustian, S.J., M.E. Shephard, J.R. Rickers, D.F. Kelliher, and S. Kessler. 1995. Watershed analysis of the Game Creek basin in Southeast Alaska utilizing GIS. pp. 7–21. In: R. Noll (ed.), *Proceedings of Alaska Water Issues.* 1995 Annual Conference Alaska Section. American Water Resources Association, Water Research Center, Institute of Northern Engineering, Fairbanks, AK.

Perry, D.A., R. Meurisse, B. Thomas, R. Miller, J. Boyle, J. Means, C.R. Perry, and R.F. Powers (eds.). 1989. *Maintaining the Long-Term Productivity of Pacific Northwest Forest Ecosystems.* Timber Press, Portland, OR.

Quigley, T.M., R.W. Haynes, and R.T. Graham (tech. eds.). 1996. Integrated Scientific Assessment for Ecosystem Management in the Interior Columbia Basin. USDA Forest Service, Pacific Northwest Research Station, Gen. Tech. Rept. PNW GTR-382. U.S. Forest Service, Pacific Northwest Research Station, Portland, OR.

Reeves, G.H., L.E. Benda, K.M. Burnett, P.A. Bisson, and J.R. Sedell. 1995. A disturbance-based ecosystem approach to maintaining and restoring freshwater habitats of evolutionarily significant units of anadromous salmonids in the Pacific Northwest. *American Fisheries Society Symposium* 17: 334–349.

Reeves, G.H., J.D. Hall, T.D. Roelofs, T.L. Hickman, and C.O. Baker. 1991. Rehabilitating and modifying stream habitats. pp. 519–557. In: W.R. Meehan (ed.), *Influences of Forest and Rangeland Management on Salmonid Fishes and Their Habitats.* Special Publication 19. American Fisheries Society, Bethesda, MD.

Regional Ecosystem Office. 1994. Ecosystem Analysis at the Watershed Scale. Revised August 1995, version 2.2. On file at the Region Ecosystem Office, Portland, OR.

Ruggiero, L.F., K.B. Aubry, A.B. Carey, and M.H. Huff. 1991. Wildlife and Vegetation of Unmanaged Douglas-Fir Forests. USDA Forest Service, Pacific Northwest Research Station, Gen. Tech. Rept. PNW GTR-285. USDA Forest Service, Pacific Northwest Research Station, Portland, OR.

SFERWG. 1994. South Florida Ecosystem Restoration Working Group 1994 Annual Report. National Park Service, Homestead, FL.

Spies, T.A. 1997. Forest stand structure, composition and function. In: K.A. Kohm and J.F. Franklin (eds.), *Creating a Forestry for the 21st Century.* Island Press, Washington, DC.

Swanson, F.J., J.L. Clayton, W.F. Megahan, and G. Bush. 1989. Erosional processes and long-term site productivity. pp. 67–81. In: D.A. Perry, R. Meurisse, B. Thomas, R. Miller, J. Boyle, J. Means, C.R. Perry, and R.F. Powers (eds.), *Maintaining the Long-Term Productivity of Pacific Northwest Forest Ecosystems.* Timber Press, Portland, OR.

Swanson, F.J., J.A. Jones, D.O. Wallin, and J.H. Cissel. 1993. Natural variability — Implications for ecosystem management. pp. 89–103. In: M.E. Jensen and P.S. Bourgeron, *Ecosystem management; principles and applications.* Vol. II. *Eastside Forest Ecosystem Health Assessment.* USDA Forest Service, Pacific Northwest Research Station, Portland, OR.

Swanston, D.N. 1991. Natural processes. pp. 139–179. In: W.R. Meehan (ed.), *Influences of Forest and Rangeland Management on Salmonid Fishes and their Habitats.* American Fisheries Society Special Publication 19. American Fisheries Society, Bethesda, MD.

Thomas, . W., and S. Huke. 1996. The Forest Service approach to healthy ecosystems. *Journal of Forestry* 94(8): 14–18.

Turner, M.G. 1989. Landscape ecology: the effect of pattern on process. *Annu. Rev. Ecol. Syst.* 20: 171–197.

Turner, M.G., R.H. Gardner, V.H. Dale, and R.V. O'Neil. 1989. Predicting the spread of disturbance across heterogeneous landscapes. *Oikos* 55:121–129.

USDA. 1995. Anadromous Fish Habitat Assessment, Report to Congress. Juneau, AK: USDA Forest Service, Pacific Northwest Research Station and Alaska Region, R10-MB-279.

USDA Forest Service. 1997. Land and Resource Management Plan: Tongass National Forest. Alaska Region, R10-MB-338dd. Juneau, AK.

## THE AUTHORS

**Steven J. Paustian**
*USDA Forest Service,*
*Tongass National Forest,*
*204 Siginaka Way,*
*Sitka, Alaska 99835, USA*

**Miles Hemstrom**
*USDA Forest Service,*
*Pacific Northwest Region,*
*P.O. Box 3624,*
*Portland, OR 97208, USA*

**John G. Dennis**
*USDI National Park Service,*
*P.O. Box 37127,*
*Washington, DC 20013-7127, USA*

**Phillip Dittberner**
*USDI Bureau of Land Management,*
*P.O. Box 25047,*
*Denver, CO 80225-0047, USA*

**Martha H. Brookes**
*USDA Forest Service,*
*Pacific Northwest Regional Station,*
*3200 S.W. Jefferson Way,*
*Corvallis, OR 97331, USA*

USDA/USDI. 1994a. *Final Supplemental Environmental Impact Statement on Management of Habitat for Late-Successional and Old-Growth Related Species with the Range of the Northern Spotted Owl.* USDA Forest Service and USDI Bureau of Land Management, Portland, OR.

USDA/USDI. 1994b. *Record of Decision for Amendments to Forest Service and Bureau of Land Management Planning Documents Within the Range of the Northern Spotted Owl, and Standards and Guidelines for Management of Habitat for Late-Successional and Old-Growth Forest Related Species within the Range of the Northern Spotted Owl.* USDA Forest Service and USDI Bureau of Land Management, Portland, OR.

USDI Bureau of Land Management. 1995. *Riparian Area Management: Process for Assessing Proper Functioning Condition,* TR 1737–11. Bureau of Land Management Service Center, Denver, CO.

Waring, R.H., and W.H. Schlesinger. 1985. *Forest Ecosystems Concepts and Management.* Academic Press, Orlando, FL.

Williams, J. 1994. The imaginary Everglades. *Outside.* January 1994.

# Disturbance and Temporal Dynamics

Peter S. White, Jonathan Harrod, William H. Romme, and Julio Betancourt

## Key topics addressed in this chapter

- ♦ *A review of scientific findings about the role of disturbance and other kinds of temporal dynamics in ecosystems*
- ♦ *A definition of disturbance and a review of descriptors of disturbance regime*
- ♦ *An overview of the kinds of natural disturbance in North American ecosystems*
- ♦ *A discussion of interactions and feedbacks among disturbances, the influence of landscape pattern on the process of disturbance, the concept of equilibrium with regard to disturbance, and ecosystem responses to disturbance*
- ♦ *An identification of emerging issues in disturbance ecology, including the relationship of disturbance and climate, Native American influences on disturbance rate, the human imposition of new scales on ecosystems through habitat fragmentation, the invasion of exotic species, and the relationship of ecological variation and resilience*

**Keywords: Disturbance, temporal dynamics, feedbacks and interactions, landscape mosaics and patterns, dynamic equilibrium, managing disturbance, habitat fragmentation, ecosystem dynamics**

## 1 INTRODUCTION

All ecosystems are dynamic. Relatively sudden and dramatic changes result from natural disturbances like fire, windstorm, flooding, catastrophic drought, avalanche, coastal erosion, insects, and pathogens (White 1979). Ecosystems also undergo gradual changes due to succession (Olson 1958), climate variation (Davis 1981, Clark 1988), and geomorphic processes (Swanston and Swanson 1976, Swanson 1981). Change is intrinsic and inevitable; ecosystem management must be based on an understanding of this change, whether an ecosystem is managed for harvest of natural resources or preservation (White and Bratton 1980).

While disturbances characterized the evolutionary setting of organisms before the human era, humans have also influenced disturbance regimes and introduced new forms of disturbance. Most management actions involve intentionally disturbing ecosystems (e.g., logging, prescribed fire) or suppressing disturbance (e.g., fire, flood, and insect control). Human activities like logging and livestock grazing may superficially resemble natural disturbances but may differ in important ways (Hansen et al. 1991). In addition to these direct effects, humans have indirectly altered the propagation of disturbances by changing the spatial structure of landscapes (Turner et al. 1989, 1993). Even when disturbances are not under human control (e.g., hurricanes, earthquakes, volcanic eruptions), management actions which alter landscape pattern and successional state may influence ecosystem response. Natural and human-caused disturbances have social and economic consequences, affecting natural resources like timber and fisheries and non-resource values like aesthetics and biodiversity. Although public perceptions focus on negative aspects of large "natural disasters" (e.g., fires, floods, and hurricanes), disturbances often play a crucial positive role in maintaining ecosystem variability and biological diversity (Christensen et al. 1989). The suppression of disturbances leads to the loss of biological diversity and may contribute to larger and more severe disturbance events later.

In this paper, we review principles of disturbance and ecosystem dynamics. Because ecosystem structure and productivity depend largely on primary producers, our focus is on vegetation. We begin by reviewing definitions and characteristics of disturbance, the kinds of disturbance as they vary with climate and site, and the concept of disturbance regime.

We then discuss disturbance interactions and feedbacks, effects of landscape-level patterns on disturbance processes, concepts of equilibrium, and species and community responses to disturbance. We then turn to five emerging issues that will affect how we incorporate disturbance into ecosystem management: climate variability and disturbance regimes, Native American disturbance, habitat fragmentation and the human imposition of new scales on management, exotic species invasions, and the restoration of ecological variation.

## 2 HISTORICAL CONTEXT

Ecologists have long recognized disturbance as a factor shaping ecological communities. For example, Darwin (1859) noted that when mowing of a meadow ceased, plant diversity declined. During the late 19th and early 20th centuries, much research focused on succession (Cowles 1899, Clements 1916). Disturbance was viewed primarily as a force moving systems away from a stable late-successional condition in which climatic, topographic, and soils determine composition and structure. However, a few early workers emphasized the importance of disturbance itself in shaping ecosystems (Cooper 1926, Raup 1941) or argued that successional concepts of the day did not apply well to vegetation with frequent disturbance (Churchill and Hanson 1958). The work of Watt (1947) drew attention to small-scale disturbances such as treefall gaps in mature forests and suggested that understanding patterns in plant communities required an understanding of dynamic processes, including disturbances. Watt's ideas have been extrapolated through computer simulations (Shugart 1984) and empirical studies (Bormann and Likens 1979, Christensen and Peet 1984, Peet and Christensen 1987) and form the basis for much of modern successional theory.

In the 1970s, attention focused on describing disturbances and documenting their effects, and evidence accumulated that disturbances play an important role in determining the structure of many communities, landscapes, and ecosystems (Dayton 1971, Heinselman 1973, Bormann and Likens 1979, White 1979, Runkle 1982). Empirical and conceptual studies suggested that disturbances may maintain species diversity (Connell 1978, Huston 1979), and increasing awareness of natural disturbances prompted interest in the effects of fire suppression on community and ecosystem structure (Kilgore and Taylor 1979, Harmon 1984). In the 1980s, the emerging discipline of landscape ecology (Forman and Godron 1986, Turner 1989) turned Watt's formulation around, examining ways in which spatial patterns (particularly coarse-scale patterns) influence disturbance processes. Research also focused on the role of residual structures such as logs and snags in post-disturbance recovery (Harmon et al. 1986, Franklin 1989). The 1988 fires in the Yellowstone area and the debate over logging practices in the Pacific Northwest drew public attention to disturbance ecology. Emerging topics in the 1990s include effects of climate on disturbance regimes (Swetnam and Betancourt 1990), influence of disturbance history on the occurrence and outcome of subsequent disturbances (Schowalter and Filip 1993), and approaches to integrating natural disturbances and management activities (Swanson and Franklin 1992, Christensen et al. 1996).

## 3 CHARACTERIZING DISTURBANCE AND DISTURBANCE REGIMES

We define a disturbance as a relatively discrete event in time that disrupts ecosystem, community, or population structure and changes resources, substrate availability or the physical environment (White and Pickett 1985). This "absolute" definition of disturbance stresses disturbance as a measurable physical event and suggests the need for mechanistic studies of disturbance effects and ecosystem recovery. A Forest Service definition is similar but somewhat less specific: disturbance is "a discrete event, either natural or human induced, that causes a change in the existing condition of an ecological system" (Kaufmann et al. 1994).

The alternative to this definition is the "relative" definition of disturbance: disturbance as a departure from the "normal" range of conditions. However, the applicability of this relative definition is limited by problems in defining "normal" conditions. Some disturbance regimes are unstable, and many are poorly known. Even disturbance regimes that appear stable in the short-term vary over longer time periods or in the face of changing climates. Our understanding of "normal" conditions is further complicated by variable and incompletely known histories of human influence. For these reasons, a physical and absolute definition of disturbance provides a better basis for understanding, prediction, and management. As we will argue below, even when the goal is to quantify the range of variation within an ecosystem, we are better off with an absolute measure which stresses the physical characteristics of disturbance and the mechanisms of ecosystem response than an approach that focuses solely on the bounds of variation.

### 3.1 Kinds of Disturbance in North America

The kinds of natural disturbances that are important vary with climate, topographic position, substrate, and successional age (Table 1, White 1979). Some disturbances are endemic to particular climates. Examples are cryogenesis in arctic and alpine tundra soils, freeze damage in subtropical and warm temperate vegetation, ice storms in temperate areas with continental climates and high precipitation, ice battering on shores, and flash floods after intense rain storms. Fire is important in climates with ignition sources, sufficient biomass to carry a fire, and long enough dry periods to permit burning. Dry sites in humid areas (e.g., pine barrens on sand deposits and pine stands on well drained ridges in humid mountains) also permit fires which may be severe in drought years. Other disturbances occur in a variety of climates but are specific to particular topographic settings: landslide and avalanche in mountainous areas, alluvial erosion, deposition, and flooding, wave battering of shores, water level fluctuation in basins, and salinity encroachments in coastal rivers. In general, disturbance varies along topographic gradients (Fig. 1) (Romme and Knight 1981, Harmon et al. 1983, White 1994), as do other physical factors like insolation, temperature, and precipitation. Some disturbances are associated with particular geological settings and substrates; these include volcanic eruption, earthquake, sand dune dynamics, and coastal erosion and deposition. Some disturbances are biological in origin; examples are the activities of burrowing mammals, grazers, and ants in the prairie, beaver activity along streams, and insect and pathogen outbreaks in forests. Most ecosystem types experience not only several kinds of disturbance, but a range of disturbance impacts within each kind (Fig. 2) (Harmon et al. 1983, Lang 1985).

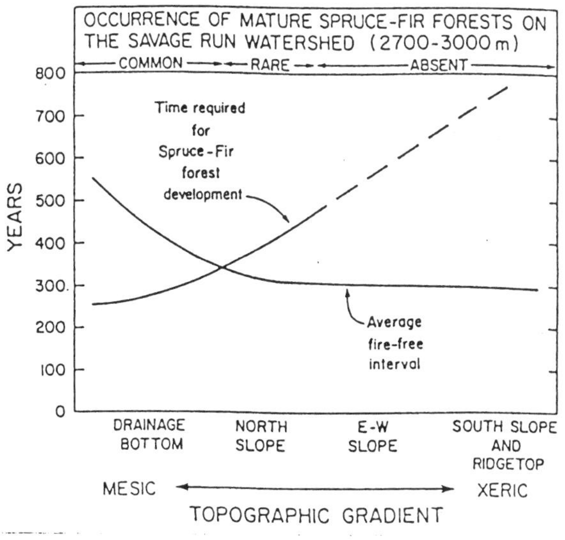


Fig. 1. The average fire free interval decreases from moist to dry sites, while the time for succession to spruce-fir forest increases along the same gradient. As a result, spruce-fir forests will probably never develop on the driest slope positions (from Romme and Knight 1981).

Table 1. Natural disturbances in North America. The kinds of disturbances vary geographically, and by topographical position and substrate.

| | |
|---|---|
| Eastern mixed and deciduous forests | Gap dynamics: Runkle 1982, 1985; Forcier 1975; Lorimer 1980 |
| | Hurricane, catastrophic wind: Foster 1988, Foster and Boose 1992 |
| | Fire: Komarek 1974, Harmon 1982, 1984; Abrams 1992; Clark and Royall 1996 |
| | Landslide: Hupp 1983 |
| | Insects and pathogens: Schowalter 1985, Harmon et al. 1983, Daughtry and Hibben 1994 |
| | Ice storm: Lemon 1961, Whitney and Johnson 1984 |
| | Catastrophic drought: Hough and Forbes 1943 |
| Southeastern pine forests | Fire, beetles: Komarek 1974, Rykiel et al. 1988, Frost 1993 |
| Appalachian spruce-fir forests | Gap dynamics, wind: White et al. 1985a,b; Sprugel 1976 |
| | Debris avalanche: Flaccus 1958 |
| Central grasslands | Fire, grazing, burrowing animals: Vogl 1974, Collins 1987, Hobbs et al. 1991, Vinton et al. 1993 |
| | Catastrophic drought: Weaver 1968 |
| Deserts | Rare rain storms, flash floods: Zedler 1981 |
| Western conifer forests | |
| Rocky Mountains | Fire, insects: Knight 1987, Romme and Knight 1981, Romme 1982, Romme and Despain 1989, Veblen et al. 1994 |
| | Cryogensis in alpine communities: Johnson and Billings 1962 |
| Sierra Mountains | Fire: Kilgore and Taylor 1979, Stephenson et al. 1991, Stephenson 1996, Swetnam 1993 |
| Pacific Northwest | Fire, windstorm: Stewart 1986, Franklin and Forman 1987, Hansen et al. 1991 |
| | Landslides: Swanson and Dyrness 1975 |
| | Volcanic eruption: Franklin et al. 1985 |
| Western shrublands | Fire: Biswell 1974, Minnich 1983, Christensen 1985 |
| | Debris flows: Biswell 1974 |
| Boreal forest | Fire, insects: Heinselman 1973, 1981; Dansereau and Bergeron 1993 |
| Arctic tundra | Cryogenesis: Churchill and Hanson 1958 |
| Subtropical areas | Freeze damage: Silberbauer-Gottsberger et al. 1977 |
| Lakes | Fluctuating water levels: Shipley et al. 1991 |
| | Ice battering on shorelines: Raup 1975 |
| Streams | Floods and erosion: Hemphill and Cooper 1983, Resh et al. 1988, Pringle et al. 1988 |
| | Beaver: Ives 1942 |
| | Debris flows: Lamberti et al. 1991 |
| Coastal areas | Dune movement: Schroeder et al. 1976 |
| | Hurricanes and other storms: Chabrek and Palmisano 1973 |
| | Salinity changes: Chabrek and Palmisano 1973 |
| Rocky intertidal communities | Wave action, storms, predation, dessication, drift log battering: Paine and Levin 1981; Sousa 1984, 1985; Dayton 1971 |
| Mangroves | Hurricanes, salinity changes: Thom 1967 |

Disturbances interact with each other and are imposed on more gradually acting sources of ecosystem change, such as soil development, geomorphological changes, and climate change. For example, the distribution and availability of phosphorus on Australian sand dunes shifts dramatically over the course of long-term soil development (Walker and Syers 1976, Vitousek and White 1981). Initially, phosphorus is relatively abundant in the mineral soil; over millennia, availability declines and the element becomes largely restricted to soil organic matter. Walker and Syers (1976) argue that these changes in soil chemistry will lead to changes in ecosystem response to disturbance; nitrogen fixers, which require relatively high phosphorus levels, will respond more strongly to fires on young soils than on older ones. In Everglades National Park, fire frequency varies with topographic position relative to the water table (White 1994). The water table

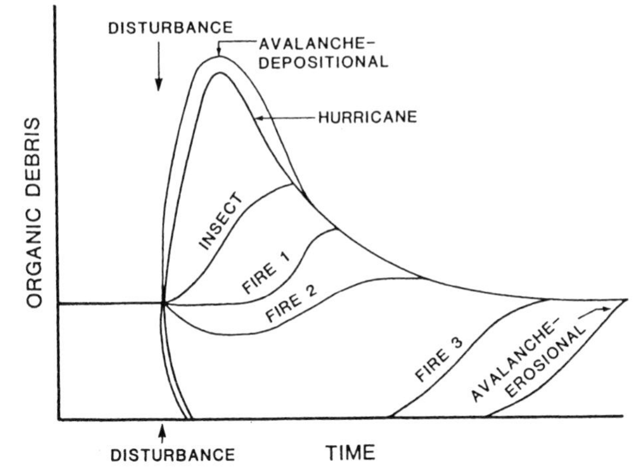


Fig. 2. Within a single ecosystem (in this example, northern Appalachian fir forests), different disturbances produce widely different effects on organic debris (redrawn from Lang 1985).

rises as the sea level rises (which it has been doing over the last few thousand years) and is also influenced by impoundments upstream from the National Park. Topography in this low elevation landscape is partly controlled by the amount of organic matter present. Intense fire removes peat and lowers topographic position; lowered water table (because of impoundments, droughts, or the high evapotranspiration of an introduced tree, *Melaleuca*) leads to a higher decomposition rate and lowers the topography. The incidence of fire is one of many interacting sources of change in this ecosystem.

## 3.2 Disturbance and disturbance regime descriptors

Not all disturbances are equivalent. Disturbances differ in six categories of descriptors (Table 2): kind, spatial characteristics, temporal characteristics, specificity, magnitude, and synergisms (Sousa 1984; White and Pickett 1985; Runkle 1985; White and Harrod 1997). Each of these categories is described below. Taken together, the attributes of all the disturbances occurring in a system, the interactions between them, and their linkages with biotic and abiotic factors, define the disturbance regime.

Disturbances not only affect the sites where they occur, but they also can affect nearby ecosystems. For example, fire in the upper part of a forest watershed can affect nutrients and siltation downstream. Romme and Knight (1981) speculated that a recent downward trend in fish populations in a watershed in Yellowstone National Park was the result of long absence of fire. Immediately after fire, they suggested that nutrient inputs to streams would be high, causing relatively high aquatic productivity and higher fish populations. Fire also influences wildlife movement patterns, thus changing the level of herbivory at sites that were not burned. Although offsite effects like these are important, in this section we focus on onsite effects.

Table 2. Parameters of disturbance regimes (from Sousa 1984; White and Pickett 1985; Runkle 1985; White and Harrod 1997).

| | |
|---|---|
| Kind | |
| Spatial characteristics | Size: patch size, area per event, area per time period, area per event per time period, total area per disturbance per time period |
| | Shape |
| | Distribution: spatial distribution including relationship to geographic, topographic, environmental and community gradients |
| | Landscape context: patch dispersion, contiguity, matrix |
| Temporal characteristics | Frequency: Number of events per time period |
| | Rotation period: Time needed to disturb an area equivalent to the study area |
| | Return interval, cycle, or turnover time: Interval between disturbance events |
| | Predictability: A scaled inverse function of the variance in return interval |
| | Contagion: Rate and probability of spread |
| | Seasonality: Seasonal distribution |
| Specificity | To species: Probability of disturbance by species |
| | To age or size classes: Probability of disturbance by age or size classes and feedback between community state and disturbance rate |
| | To landforms: Probability of disturbance by landform element |
| Magnitude | Intensity: Physical force of the event per area per time |
| | Severity: Impact on the organism, community, or ecosystem |
| | Ecosystem effects: |
| | Internal heterogeneity: Degree of internal patchiness within disturbed areas |
| | Ecosystem legacies: Structures, dead, and living biomass remaining |
| Synergisms | Interactions between disturbances |
| | Feedbacks through successional state |
| | Coupling with climate |

### 3.2.1 Kind

The types of disturbance that occur within ecosystems, landscapes, or regions vary with climate, topography, substrate, and biota (Table 1).

### 3.2.2 Spatial characteristics

Disturbances differ in size (patch size, area per event, area per time period, area per event per time period, total area per disturbance type per time period), in distribution (on geographic, topographic, environmental and community gradients), and landscape pattern (patch shape and dispersion, contiguity, and relationship to the surrounding matrix). The size of individual disturbances (few large versus several small disturbances) affects amount of edge, contiguity, and other spatial parameters. Size may also affect the nature of subsequent colonization and succession (lateral expansion versus vertical growth; shade tolerant versus intolerant species; advance regeneration versus the establishment of new individuals) in both terrestrial and marine systems (Fig. 3) (Runkle 1985, Sousa 1985).

The shape of disturbance patches can also be important. The relationship between length of edge and interior area has implications for wildlife and vegetation. Circular or square patches have smaller edge/area ratios than elongate or convoluted patches. The shape and orientation of gaps may affect levels of incident light, particularly in higher latitudes. The distribution of disturbed patches across landscapes, geographic and environmental gradients, and community types is also important. Disturbed patches occur in the context of a larger landscape which may act as a source of colonists and as a refuge for disturbance-sensitive organisms. The composition and spatial structure of the surrounding landscape will affect post-disturbance recovery and the persistence of biological diversity.

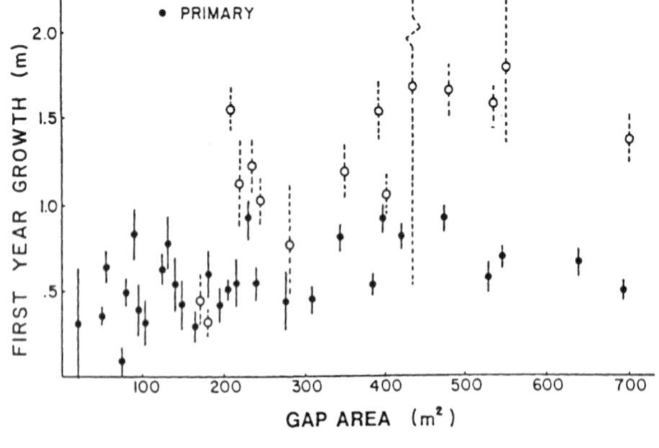


Fig. 3. Height growth (mean and one standard error of annual height increment for stems greater than 1 m tall) of pioneer (open circles) and primary tree species (closed circles) as a function of gap size (from Brokaw 1985). Primary forest species show some response to increased gap size and occur at a wide range of gaps sizes. Pioneer species have the highest growth rates but appear only in the highest light conditions (largest gaps).

### 3.2.3 Temporal characteristics

Disturbances differ in frequency (number of events per time period), rotation period (time needed to disturb an area equivalent to the study area), return interval, cycle, and turnover time, predictability, regularity, and stochasticity, contagion (rate and probability of spread), and seasonality. The concept of "predictability" has been used in the context of morphological and life-history adaptations to disturbance. From a community or management perspective, "regularity" or "stochasticity" are more useful terms. Temporal stochasticity will contribute to variation between sites in a landscape and may thus increase the diversity of successional states. Management plans can incorporate stochastic factors. For example, when conditions are suitable for prescribed fire, burn/no burn decisions for particular units may be made by applying a probability (e.g., rolling a die or using a random number table). The resulting return interval will reflect a statistical distribution rather than a single value. Periodicity may be driven by endogenous feedback mechanisms (e.g., increasing flammability with stand age) or by exogenous climate factors (e.g., the Southern Oscillation).

Fires, hurricanes, ice storms, and insect outbreaks are among the disturbances which exhibit marked seasonality in their occurrence. Season of disturbance may affect availability of propagules from outside the disturbed area and physiological/phenological response of species within the patch. For example, saplings may be killed outright by a growing season fire, but may resprout if fire occurs during the dormant season; some species like wiregrass on the southeastern coastal plain may only flower in response to properly timed burns. Sousa (1985) reviewed the effects of disturbance timing and seasonality on marine intertidal organisms.

### 3.2.4 Specificity

The susceptibility of organisms to disturbance may vary with species, age or size class, successional time, community state, and landscape position. Some physical disturbances such as lava flows and catastrophic debris slides may obliterate all organisms in their path. Other disturbances, particularly biotic disturbances like insect parasites and fungal pathogens, impact one or very few species. Wind, fire, and vertebrate grazers tend to be intermediate in specificity. Wind and fire effects vary with species and size/age class (Harmon 1984, Foster 1988). Because disturbance effects are often species- and age class-specific, descriptions of disturbance-induced damage and mortality should be described relative to species and age classes.

### 3.2.5 Magnitude

Disturbances vary in intensity (the physical force per event per area per time) and severity (the impact on organisms and ecosystem structure and composition). This variation is reflected in the percentage of living biomass killed and the amount of dead biomass added to or removed from a patch. Intensity and severity affect resource levels, structural heterogeneity, and mechanisms of recovery. Disturbances rarely remove all biota and organic matter; the amount and nature of living and dead material left in a patch after disturbance may play a key role in determining community and ecosystem response. These residual structures (dubbed the "biological legacy;" Franklin 1989) may include standing live trees, trees knocked over but still alive, seedlings, saplings, herbs, shrubs, buried seeds, standing snags, logs and other coarse woody debris, humus layer and soil biota, including mycorrhizal fungi. These residual organisms and structures may help maintain ecosystem function, moderate fluctuations in temperature and humidity, and restrict nutrient and sediment loss during early stages of post-disturbance recovery (Marks 1974, Franklin 1989). They may also serve as refugia and corridors for disturbance-sensitive species (Franklin 1989) and foci for seed dispersal (McDonnell and Stiles 1983) and contribute to structural heterogeneity and habitat diversity in young, aggrading systems (Hansen et al. 1991). The importance of biological legacies is one reason that we should be careful in accepting management treatments (e.g., logging) as analogs for natural disturbances (e.g., windstorm).

Historic disturbance regimes typically include events of various magnitudes (Lorimer 1980, Barrett et al. 1991). Both high and low intensity disturbances may play roles in maintaining ecosystem structure. On xeric low elevation sites in the southern Appalachians, rare crown fires create canopy and soil conditions necessary for the vigorous growth of pine seedlings; low-intensity surface fires, though more frequent, lead to little pine regeneration (Barden and Woods 1976). However, low-intensity fires help to maintain pine dominance and historic structure by top-killing hardwood seedlings and saplings (Harmon 1984). Intense fires may also be important in sequoia-mixed conifer forests, often termed low-intensity fire systems (Stephenson et al. 1991). Full characterization of disturbance regimes requires assessment of the range of disturbance magnitudes.

### 3.2.6 Synergisms

At the levels of individuals, stands, and landscapes, the occurrence and outcome of disturbances depend, to some extent, on the history of past disturbance. For example, fire scars may make pine trees more vulnerable to bark beetle attack (Geiszler et al. 1980). As we will argue below, such synergisms are widespread and have important implications for community and ecosystem dynamics.

## 3.3 Characterizing Disturbance Regimes: Approaches and Challenges

Efforts to characterize disturbances and disturbance regimes involve four basic approaches. The historical approach involves documenting past events and ecosystem states through fossil pollen and charcoal (Foster and Zebryk 1993, Clark and Royall 1996), stand origin dates and regeneration patterns (Heinselman 1973, Romme 1982), fire scar analyses (Harmon 1982), detailed reconstructions of stand structure and patterns of release (Henry and Swan 1972, Lorimer 1980), and historical survey records, narratives, and photographs (Seischab and Orwig 1991, Motzkin et al. 1996). Additional information and references on historical methods can be found in the companion management paper (see Engstrom et al., this volume, Section 2.2.1), Agee (1993), and Lorimer (1985; see also Lorimer and Frelich 1989). The observational approach involves description of present-day disturbances, conditions, and responses (Dayton 1971, Runkle 1982, Harmon 1984, Hansen et al. 1991). The experimental approach involves deliberate disturbance of an ecosystem followed by monitoring of disturbance effects (Bormann and Likens 1979, Collins 1987). The simulation approach involves the use of models to examine disturbance behavior and the effects of changes in disturbance regime (Shugart 1984, Franklin and Forman 1987, Turner et al. 1989, Keane et al. 1990, Covington and Moore 1994). These four approaches differ in their spatial and temporal resolution, accuracy, and scope; research programs which integrate multiple approaches will provide the most useful information.

Disturbance regimes can vary considerably between areas with similar vegetation. For example, presettlement lodgepole pine forests supported a range of fire patterns. In the northern Rockies, lower elevation sites burned every 25–150 years; severity ranged from underburns causing little canopy mortality to stand-replacing fires (Barrett 1994). Higher elevation sites experienced mostly high-severity fires; return intervals ranged from about 200 years on productive andesitic soils (Barrett 1994) to 300–400 years on less fertile rhyolite (Romme 1982, Romme and Despain 1989). In lodgepole pine forests in the Pacific Northwest, fires of variable severity have burned at intervals of 60–80 years (Agee 1993). Although characterizing disturbance regimes involves considerable time, effort, and expense, ecologically sound management requires site-specific information.

### 3.3.1 The historical or natural range of variation

Ecosystem scientists and managers have attempted to use the range of variation that characterizes ecosystems as a guide to understanding and management (Landres 1992, Swanson and Franklin 1992, Hunter 1993, Morgan et al. 1994, Swanson et al. 1994, Landres et al., in press). One might, for example, seek to quantify the "natural range of variation" in biomass or population density over several generations of the dominant organisms. Although this approach may provide valuable information, we cannot assume that all ecosystems have well-defined bounds of variation. The farther back in time we look, the more variation we will see. In addition, recent studies have shown that humans have influenced some ecosystems that were once considered pristine. The "historic range of variation" (Swetnam 1993, Morgan et al. 1994, Wright et al. 1995) is an attractive phrase because it makes no assumption about naturalness or normalcy and accepts the arbitrary and variable duration of the historical record. Documenting historical variation in ecosystems will help us to understand better both disturbance effects and the influences of climate and human activity. The recent history (decades to millennia) tells us about past behavior of the ecosystem and the natural processes with which management actions will have to interact.

## 4 SYNERGISMS: FEEDBACKS AND INTERACTIONS

Here we discuss the potential effects of disturbance history on subsequent disturbance events and include both natural and human caused disturbance. White (1987) identifies two types of synergisms, feedbacks and interactions. A feedback is a situation in which a disturbance influenced subsequent disturbances of the same type. For example, flammability of chaparral may be low immediately after a fire and increase as a stand matures. An interaction is a situation in which a disturbance influences subsequent disturbances of a different type. For example, fires alter chaparral soils and increase likelihood of landslides on steep slopes. Feedbacks and interactions have been documented in many systems and occur at a range of scales (Table 3).

Table 3. Disturbance feedbacks and interactions by biome.

| | |
|---|---|
| Eastern mixed and deciduous forests | fire-fire: Harmon 1984 |
| | fire-fungi-wind: Matlack et al. 1993 |
| | lightning-fire-fungi-bark beetle: Schowalter 1985, Schowalter et al. 1981, Rykiel et al. 1988, Flamm et al. 1993 |
| | agriculture-wind: Foster 1988 |
| Central grasslands | fire-grazing: Collins 1987, Hobbs et al. 1991, Vinton et al. 1993 |
| | prairie dog activity-grazing: Coppock et al. 1983 |
| Western conifer forests | fire-fire: Kilgore and Taylor 1979, Agee and Huff 1987, Romme and Despain 1989, Covington and Moore 1994 |
| | lightning-fire-fungi-bark beetle: Geiszler et al. 1980, Knight 1987, Paine and Baker 1993, Schowalter and Filip 1993, Hagle and Schmitz 1993 |
| | fire-parasitic plants: Knight 1987 |
| | fire-weather-large mammal mortality: Turner et al. 1994 |
| | avalanche-fire-bark beetle: Veblen et al. 1994 |
| | fire-grazing: Covington and Moore 1994 |
| | logging-wind: Franklin and Forman 1987 |
| | logging-fire: Franklin and Forman 1987 |
| | logging-landslides: Swanson and Dyrness 1975 |
| | logging-pathogens: Paine and Baker 1993, Hagle and Schmitz 1993 |
| Western shrublands | fire-fire: Minnich 1983, Christensen 1985 |
| | fire-debris slides: Biswell 1974 |

Individual-level feedbacks and interactions can occur whenever disturbances leave damaged survivors. Wounds caused by fire, lightning, or human activity predispose trees to fungal infection and insect attack. Fungal infection can increase the likelihood of other disturbances. For example, in the New Jersey pine barrens, trees with extensive fungal rot suffered higher rates of wind breakage than sound trees (Matlack et al. 1993). Fungal rot is most common in trees with basal fire scars. Vulnerability to scarring varies with tree age at time of fire; young trees are particularly vulnerable. Thus, susceptibility of individual trees to wind damage may depend on date of recruitment relative to fire events several decades in the past.

Many other examples of individual level interactions and feedbacks are known. Bark beetles transmit pathogens between trees, and pathogens reduce trees' ability to resist beetles. In general, trees weakened by mechanical injury, disease, or herbivory have fewer resources for growth, maintenance, and defense and are more susceptible to subsequent disturbance. Damage to an individual is often cumulative; a single defoliation by gypsy moths rarely kills an oak, but repeated episodes cause high rates of mortality. Some plants respond to herbivory by increasing toxin levels or reducing nutritional quality of leaves. Such cases provide examples of negative feedback; one defoliation makes another defoliation less likely.

At the stand level, disturbances alter species composition, canopy structure, and fuel levels in ways which affect susceptibility to subsequent disturbances. For example, the distribution of hurricane damage in central New England is largely a function of the history of agricultural disturbance (Foster 1988). Stand susceptibility to wind varies with age and species composition; pine stands over 30 years old are particularly susceptible. The extensive damage caused by the 1938 hurricane can be explained largely by the abundance at that time of 30–100 year old white pine on abandoned agricultural fields.

Fire likelihood and intensity are subject to stand-level feedbacks. In systems in which fires consume most fine fuels, stand flammability is low shortly after a fire, increases in a developing stand, and levels off as the stand matures. The cycle may take 30–50 years in California chaparral or 200–400 years in higher elevation Rocky Mountain lodgepole pine forests. When a fire kills trees without consuming them, post-fire fuel levels and flammability may be high. In western hemlock–Douglas-fir forests in Washington, flammability is highest in the first 20 years after a fire, drops to a low level in ~100-year-old stands, and increases thereafter (Fig. 4) (Agee and Huff 1987). In ponderosa pine/bunchgrass woodlands in the Southwest, frequent surface fires maintain an open stand structure and grassy ground layer and prevent accumulation of woody debris. Surface fires burn grasses and pine needles, but the lack of larger fuels makes crown fires unlikely. Fire suppression results in increases in stand density and woody fuels and decreases in grass abundance. For example, since the onset on fire suppression, fuel loads in forests near Flagstaff, Arizona have increased by 20-fold, while grass and forb production has fallen by 90 percent (Covington and Moore 1994). As grasses decline and woody fuels accumulate, the potential for low-intensity surface fires decreases and catastrophic crown fires become more likely. Fire suppression may produce even more dramatic effects in ponderosa pine communities in the inland Northwest; there, increases in the densities of fire-sensitive Douglas-fir, grand fir, and white fir have contributed to high-intensity fires and outbreaks of spruce budworms and pathogens (Anderson et al. 1987, Keane et al. 1990, Arno et al. 1995).

Disturbance feedbacks and interactions also include the interplay of landscape pattern and process as discussed below. Interactions may propagate individual and stand-level phenomena to larger areas. For example, a pine beetle infestation may spread from a lightning-damaged tree through a stand to other stands in the landscape (Rykiel et al. 1988).

Stand- and landscape-level feedbacks and interactions between insects, wildfire, and plant pathogens have become management issues on public lands. In stands dominated by susceptible species, insect and pathogen outbreaks may produce large quantities of dead woody fuel. But the effects of insects and pathogens on fire regimes are complex and incompletely understood. In some forests, fire risk actually decreases in the first few decades following a beetle outbreak (Knight 1987). Moderate levels of insect and pathogen activity may reduce the risk of catastrophic fire by thinning stands and preventing excessive fuel buildups; insects and pathogens may also promote forest health by gradually culling weakened trees (Schowalter and Filip 1993).

The stand- and landscape-level effects of fires on insects and pathogens are also complex. The presence of fire-damaged trees may allow bark beetles to persist

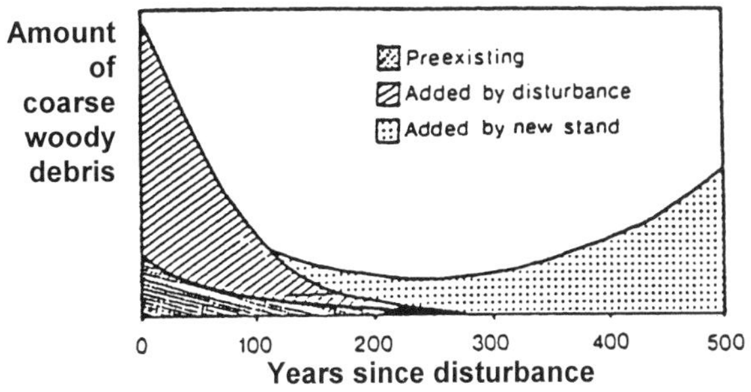


Fig. 4. A general model for changes in coarse woody debris after disturbance (from Agee and Huff 1987). Some coarse woody debris survives the disturbance ("Preexisting"). Some disturbances (including many fires) actually create woody debris through heavy tree mortality ("Added by disturbance"). Decomposition reduces these pools of debris, but succession gradually reestablishes predisturbance levels. During the thinning phase of forest succession, woody debris may surpass predisturbance levels, but larger size classes of logs would be absent.

at low levels until undamaged trees are made vulnerable by competition or drought (Schowalter 1985). But a regime of low-intensity surface fires may also maintain an open stand structure; with lower densities and more vigorously growing trees, open stands may be less susceptible to insect outbreaks. Hotter fires initiate patches of young trees which do not become vulnerable to bark beetles until they are several decades old (Schowalter 1985, Veblen et al. 1994). As components of a larger landscape, these non-susceptible patches may limit spread of insect outbreak.

Forest management activities also influence insects and pathogens. In parts of the southeastern and western United States, fire suppression and the conversion of natural forests to plantations have led to dense stands with low diversity and poor vigor which are susceptible to severe insect and pathogen outbreaks (Anderson et al. 1987, Schowalter and Filip 1993, Hagle and Schmitz 1993). Thinning treatments may reduce crowding and improve vigor but may also promote insect and disease spread by wounding trees and damaging roots (Paine and Baker 1993). Airborne spores of some root pathogens infect cut stumps; the pathogens may then spread to adjacent trees. The appropriate management response will vary with stand characteristics and the species of insects and pathogens involved. Hagle and Schmitz (1993) discussed options for managing insect–pathogen interactions.

Disturbance interactions play a role in water, nutrient, and sediment dynamics. In many forests, ecosystem function recovers rapidly after a single disturbance. Even though fires, windstorms, insect outbreaks, and even logging may damage or kill large numbers of organisms, biological legacies — the trees, shrubs, and herbs which survive —continue to transpire water, cycle nutrients, and stabilize soils. But if a second disturbance such as salvage logging, herbicide application, or mechanical site preparation compromises the system's ability to compensate, nutrient and sediment losses will increase. In some ecosystems, disturbance interactions lead to conservation of nutrients. In tallgrass prairie, grazing prevents nitrogen losses from the burning of plant biomass (Hobbs et al. 1991). Grazers reduce the amount of biomass available to be burned and return nitrogen to the soil as waste.

Disturbance interactions may also play a role in maintaining biological diversity. For example, in tallgrass prairie in Oklahoma (Collins 1987), fire and grazing create four patch types: undisturbed, grazed, burned, and both burned and grazed. These patch types differ in vegetation structure and species composition. Burning stimulates growth of many grass and forb species. But unless burning is followed by grazing, fast-growing grasses such as big bluestem crowd out other species, and diversity declines. Cattle and bison prefer these grasses; grazing keeps the grasses in check and allows other species to persist. Thus, while burned, ungrazed patches are least diverse, burned, grazed patches are most. Fire and grazing act together to maintain variety of patch types in the landscape and high levels of diversity within patches.

Feedbacks and interactions are important but poorly documented aspects of disturbance regimes. Although these synergisms occur in many vegetation types, they have been explored in detail in only a few. In most systems, additional research on both mechanisms and long-term consequences is needed before management recommendations can be made. Managers should be aware of the potential for feedbacks and interactions; present activities may have unintended consequences decades in the future. Multiple, interacting disturbances may play an important role in maintaining vegetation structure, ecosystem health, and species diversity. Actions which simplify the disturbance regime may compromise biological integrity. The widespread occurrence of feedbacks and interactions suggests that disturbances should not be studied or managed as independent events. Rather, it argues for a historical, synthetic approach to ecosystem dynamics.

## 5 THE LANDSCAPE MOSAIC AND THE INFLUENCE OF PATTERN ON PROCESS

The early literature on patch dynamics emphasized the effects of processes on compositional and structural patterns. Over the past 20 years, ecologists and land managers have become increasingly interested in the effects that spatial patterns (particularly the size, shape, and arrangement of patches) exert on ecological processes. Among the processes influenced by landscape pattern are seed dispersal, exotic species invasions, and the propagation of fires and insect outbreaks. This recognition of the importance of spatial pattern, and the accompanying focus on phenomena occurring over large areas, have given rise to the discipline of landscape ecology (Turner 1989). The rapid development of landscape ecology has been facilitated by new technology, including geographic information systems (GIS) and high-resolution satellite images. The landscape perspective provides new insight into the dynamics of ecosystems and the impacts of management activities.

A landscape can be envisioned as a mosaic of patches which differ in history, environment, and species composition. A central paradigm in landscape ecology is that processes create landscape patterns, which, in turn, control subsequent processes. For example, the practice of "checkerboard" cutting in the forests of the Pacific Northwest produces a pattern of small (10–15 ha) clearcuts dispersed across the landscape (Franklin and Forman 1987). Windthrow damage is concentrated along the edges of clearcut patches. Compared with a landscape with a few large patches, a landscape with many small patches has a greater length of edge per area of clearcut. Thus, dispersed cutting creates a pattern (many small patches with high total edge length) which promotes a disturbance process (windthrow along forest edges).

Natural disturbances create new patches, modifying the existing landscape pattern. Landslides in the Pacific Northwest tend to occur on unstable soils, forming wedge- or bullet-shaped scars. Growing pine beetle infestations in the coastal plain of the southeastern United States often form circular "spots" of dying trees. Wildfires in the Rockies may burn over several square kilometers, following prevailing winds, topography, and fuels. A single fire may create patches of several types, consuming some stands in intense crown fires, burning others with cooler surface fires, and leaving unburned islands within burned areas. The effects on landscape pattern will depend on disturbance type, intensity, size, shape, relationship to other patches, and position along environmental gradients. Human activities also shape landscape patterns. Logging, mining, agriculture, road building, and construction fragment existing patches and create new ones. Although natural disturbances usually produce irregular patterns, human activities often create geometric patches with straight boundaries.

Landscape patterns also change through succession. As a clearcut or burned stand matures, it becomes more similar to the surrounding undisturbed forest. Eventually, it may no longer appear as a distinct patch. Landscapes dominated by stand-destroying fires are made up of even-aged stands, each of which develops through a sequence of successional stages. The proportion of the landscape in each stage may change through time; during periods when little area burns, more of the landscape passes into older stages. Figure 5 shows fluctuations in the composition of a landscape in Yellowstone National Park over the past 250 years (Romme and Despain 1989).

Landscapes can be characterized by the number of patch types and the proportion of the total area in patches of each type. The spatial arrangement of patches may also be important. Is the landscape made up of many small patches, or a few large ones? How remote are patches of the same type from each other? What patch types occur together? Are patches compact or elongate in shape? Are their boundaries simple or complex? What are the dimensions of patch edges?

Increasingly, ecologists and resource managers are using geographic information systems (GIS) to address such questions. GIS allows a researcher to analyze a digital map and quantify aspects of landscape structure. Used properly, GIS provides a powerful tool for evaluating management alternatives. However, caution is required in interpreting results. Not all patch boundaries are equal in their ecological significance. For example, adjacent stands of pine and hardwood show far smaller contrasts in light, temperature, humidity and wind speed than either stand does with a clearcut, yet the GIS may show both kinds of boundaries with equal clarity. Landscape attributes will depend on the way in which the landscape is classified. If forest patches are classified simply as "hardwood" or "pine," a landscape may appear homogeneous, with large, continuous tracts of each type. Using a finer scheme which differentiates oak-hickory, beech, and young and mature pine forests, the same landscape will appear more complex. Different management questions warrant different levels of classification.

Both the nature of the pattern and the processes which create it will differ with scale. As one examines areas of increasing size, details are lost but broader patterns emerge. At the scale of a few hundred square meters, the spatial pattern of tree crowns in a southern Appalachian forest depends on the birth and death of individual trees. At the scale of many hectares, single trees are no longer visible; the pattern of stands reflects minor landforms and stand-level disturbance history. At the scale of tens of square kilometers, individual stands merge into larger land-cover units, and vegetation patterns follow broader physiographic gradients

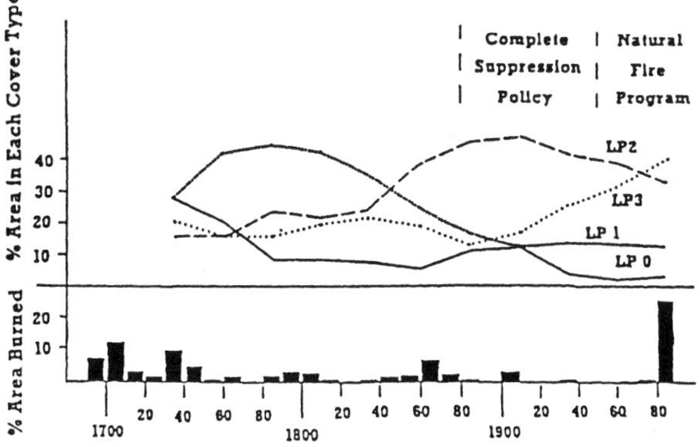


Fig. 5. Disturbance and succession on a 129,600 ha study area in Yellowstone National Park (Romme and Despain 1989). Changes in the percent of the study area in each of four successional types (top) and the percent of the study area burned each decade (bottom) over the past three centuries. LP0 are lodgepole pine stands less than 40 years old; LP1 are even-aged stands 40–150 years old; LP2 are even-aged stands 150–300 years old; and LP3 are mixed-age pine/fir/spruce stands.

and boundaries between public and private land. Many factors act simultaneously to shape the landscape; the scale of observation determines which appear most prominent. No one scale is best for all management applications; the scale used will depend on the questions asked and the resolution of the available data.

Dynamic processes can homogenize a landscape or make it more complex. For example, cattle in unburned tallgrass prairie create a mosaic of small grazed patches (Hobbs et al. 1991). Cattle prefer to graze on the young grass in these patches and thus maintain the initial pattern through time. Prairie fires erase the fine-scale pattern of grazed and ungrazed patches and produce a more uniform landscape. In other systems, the vagaries of fire or wind behavior may increase landscape heterogeneity. Successional processes may in some cases erase the patterns created by disturbance. In other cases, landscape heterogeneity will increase with succession as mature communities develop to reflect local site conditions (Werner and Platt 1976, Christensen and Peet 1984).

Rates of disturbance and succession vary along environmental gradients. These dynamic processes may create and maintain associations between vegetation patterns and features of the physical environment. For example, in southeastern Wyoming (Romme and Knight 1981), lower elevation spruce-fir forests are restricted to ravines and valley bottoms, while lodgepole pine forests dominate slopes and ridges. Lodgepole pine, an early-successional species, typically establishes after fire and forms even-aged stands. Given sufficient time, the later-successional spruce and fir can replace pine on both upland and lowland sites. However, upper slopes and ridges have shorter fire-return intervals and slower rates of succession. On these sites, fires destroy stands before spruce-fir forests can develop. The landscape pattern of pine forests on the uplands and spruce-fir in the valleys and ravines results from a dynamic tension between disturbance and succession.

How do landscape patterns shape processes? In some cases, the composition of an individual patch will be the major factor determining the changes that patch undergoes. In the North Carolina Piedmont, abandoned agricultural fields pass through a successional sequence from grasses and forbs to pines to mature oak-hickory forest. Successional state depends on time since abandonment; landscape changes are largely a function of the proportion of area in each age class. But, even in this case, the size, shape, and arrangement of patches can influence processes. For example, low rates of seed dispersal and poor seedling establishment lead to slower pine invasion on larger fields (Pinder et al. 1995).

In some cases, the dynamics of one patch may be dominated by its relationship with other patches. Some disturbances tend to originate at patch boundaries. In forests of the Pacific Northwest, windthrow may be most severe in forests adjacent to clearcuts (Franklin and Forman 1987), and a disproportionate number of landslides occur near roads (Swanson and Dyrness 1975). Human-caused fires often start near roads or campsites and spread into wildlands; when these patch types are juxtaposed, the likelihood of ignition increases. In the southeastern United States, some exotic species, including Japanese honeysuckle and kudzu, thrive along forest edges but do poorly in the shade of the forest interior. Activities such as dispersed-patch clearcutting and road construction, which fragment the landscape, magnify such edge effects.

In other cases, large, continuous patches of susceptible habitat may promote disturbance spread. The disturbance regime of California chaparral is characterized by stand-destroying fires. Stands accumulate woody fuels as they mature and reach extreme flammability at 30–50 years of age. Minnich (1983) compared fire size and frequency in southern California, where there are active fire control efforts, with nearby Baja California, where there are not. He found that while there were fewer fires in California, the total area burned in both regions is similar. The suppression of small fires appears to lead to large patches of highly flammable older chaparral and larger, more difficult to control fires.

The effects of fire suppression on fire size and intensity have become management issues in many western landscapes. In 1988, fires burned nearly 3,000 square kilometers in the Greater Yellowstone Area of Wyoming, Idaho, and Montana. To gain a historical perspective on the 1988 fires, Romme and Despain (1989) reconstructed fire history and past landscape composition for a portion of Yellowstone National Park (Fig. 5). They found that fires comparable to those in 1988 had occurred in the early 1700s. For the next 200 years, the landscape was dominated by relatively nonflammable early- and mid-successional forests. Although ignitions occurred every decade, the size of burned patches remained small. By the mid 20th century, most of the landscape had developed into highly flammable older forest. The 1988 fires occurred in unusually dry and windy weather, but also in a landscape which was particularly susceptible to fire spread. With or without fire suppression, large fires probably would have occurred in the Greater Yellowstone Area in the mid or late 20th century. The period of effective fire exclusion (mid 1940s–mid 1970s) was short relative to the ~300 year cycle of forest development. Fire suppression may thus have delayed the burning of some areas and led to one year with particularly extensive fires, but in the long run probably had little effect on the total area burned.

Studies of lodgepole pine forests in the Yellowstone area (Romme 1982, Barrett 1994) suggest that in systems in which infrequent crown fires dominated the presettlement disturbance regime, recent fire suppression has not fundamentally altered landscape dynamics. Other systems, such as Douglas-fir woodlands in Yellowstone, sequoia-mixed conifer forests in California, and ponderosa pine woodlands throughout the West, were maintained historically by frequent low-intensity surface fires. Fire suppression in these systems has led to increases in stand density and fuel loads, creating the conditions for high intensity crown fires (Kilgore and Taylor 1979, Barrett 1994, Covington and Moore 1994, Arno et al. 1995).

Large, homogeneous tracts dominated by a single species may also facilitate insect outbreaks. Several hundred years ago, the coastal plain of the southeastern United States was a complex mosaic of hardwood forests, shrub bogs, and open pine savannas (Frost 1993). Over large areas, that mosaic has now been replaced by commercial pine plantations in which high densities of physiologically stressed trees create ideal conditions for the spread of the southern pine beetle. In the boreal zone of Alaska and Canada, where species-poor spruce-fir forests dominate the landscape, large outbreaks of spruce budworm are common.

Simple computer models suggest that the proportion of a landscape susceptible to a disturbance such as fire or insect outbreak influences the nature of disturbance propagation (Turner et al. 1989). Modeling landscapes with a random pattern, Turner et al. found that when less than about 60 percent of the landscape is made up of susceptible cells, susceptible patches are disjunct, and disturbance spread is limited by patch boundaries. In such a landscape, the amount of area disturbed is sensitive to the number of disturbance initiations (e.g., lightning strikes). When the proportion of susceptible cells exceeds 60 percent, susceptible areas coalesce into a few large patches. Because a single disturbance can then spread across much of the landscape, the number of initiations becomes less important. The extent to which susceptible patches are connected with each other appears to be a critical factor in disturbance spread. Real-world landscapes tend to have higher connectivity than random ones; thus, the threshold at which actual disturbances can spread over large areas is probably lower than 60 percent.

Both patches of non-susceptible vegetation and geomorphic features may act as barriers to disturbance spread. In the Boundary Waters Canoe Area in Minnesota, large fires tend to burn from west to east, following prevailing winds; lakes, particularly large lakes with north-south orientation, interrupt the spread of fires (Heinselman 1973). Fire regimes on islands in boreal forest lakes differ from that of the surrounding mainland (Bergeron and Brisson 1990). On the mainland, a fire started from a single lightning strike can spread across large areas of forest. Islands, isolated from fire spread, burn far less frequently.

In subalpine forests in the Rocky Mountains, spruce beetle outbreaks, fires, and snow avalanches shape and are shaped by the pattern of susceptible patches and barriers to disturbance spread (Veblen et al. 1994). Spruce do not become susceptible to beetles until about 70 years of age. Thus, sites of severe fires and beetle outbreaks are unlikely to support subsequent infestations for several decades. Avalanche paths lack the mature spruce attacked by beetles. They also have little fuel accumulation and thus check fire spread.

Landscape pattern influences the dynamics of wildlife populations. Some species thrive along edges of disturbed patches, where a mix of habitats provides forage and cover. As a landscape becomes increasingly fragmented, these species increase. Others prefer the moist, shady conditions found in the interior of mature forest patches. In western coniferous forests, the spotted owl, varied thrush, and red-backed vole are among the species more common in interior habitats. For these species, the suitability of forest patches depends on their size. Because edge influences on microclimate may extend two to three tree heights into the forest, patches below a certain size are, in effect, all edge (Franklin 1989). Forest fragmentation brings increased contact with predators and parasites with which interior species have little evolutionary experience. For example, brown-headed cowbirds, brood parasites which lay their eggs in the nests of songbirds, are more active near forest edges than in the interior of large forest patches.

The density and placement of roads may affect wildlife. Roadkill is a major source of mortality for some species (Schoenwald-Cox and Buechner 1992). Roads also increase access for legal and illegal hunting and off-road vehicle use. The presence of roads may reduce the quality of otherwise suitable habitat. Grizzly bears in the northern Rockies avoid areas within 100 m of roads. Because the roads often run along streams in valley bottoms, they limit the bears' use of those productive areas (McLellan and Shackleton 1988).

Some conservation biologists have advocated wildlife corridors, strips of suitable or semisuitable habitat which connect larger patches. Corridors may allow species to recolonize areas from which they have disappeared and facilitate movement of animals that require large areas or multiple kinds of patches. However, the value of corridors remains controversial (Mann and Plummer 1995). The extent to which most species will use them is unknown. High ratios of edge to interior may promote invasion by exotic species and bring wildlife into increased contact with humans and natural predators and parasites. Perhaps the best case for corridors can be made when they provide multiple benefits. For example, a strip of undisturbed vegetation along a stream will protect fish habitat and water quality whether or not it is used by terrestrial wildlife.

Studies of landscape-level phenomena have drawn attention to functional linkages across large areas. The size, shape, and spatial configuration of patches may influence the initiation and spread of disturbance and the abundance of wildlife species. Logging, road building, and fire control may have effects in areas far removed from the site of activity. When evaluating alternatives, managers need to consider implications for landscape-scale patterns and processes.

## 6 DISTURBANCE AND THE DYNAMIC EQUILIBRIUM

One of the most important questions for understanding and managing ecosystems is whether processes, such as disturbance and succession, are consistent with and can sustain observed patterns (the frequency of various compositional and structural states) (White 1987). Consequences for management are clear: if process and pattern are in equilibrium (we introduce several definitions of equilibrium below), structure, process, and composition will persist across many locally fluctuating patches. Shugart (1984) used simulation models to suggest that biomass would be in equilibrium across patches if patch size was 1/50th or less of landscape size, i.e., if 50 or more patches with independent dynamics comprised the simulated landscape (Fig. 6); see also below for the model of Turner et al. (1993) that incorporates both disturbance and recovery rate.

A steady state equilibrium exists most likely where patch size is small relative to landscape size, where recovery time is much less than average recurrence interval, and where disturbance regimes are stable (White and Pickett 1985). If the probability of disturbance increases with patch age, and these other conditions are met, a steady state equilibrium is also more likely. An exception to this statement may be contagious disturbances such as fire and insect outbreak. In a landscape with a high proportion of susceptible patches, such disturbances may spread over large areas. A large disturbance will create a cohort of patches which will increase in susceptibility as they mature. The result may be long periods during which little disturbance activity occurs punctuated by shorter periods of widespread disturbance.

Whether or not landscapes and regions ever had equilibrium disturbance regimes, human management has drastically changed the scale of many ecosystems. The natural or original patch size of disturbance may exceed the scale of the remnant natural area. The result may be the complete disturbance of the area in a single disturbance event. If there are no refuges for species vulnerable to disturbance, disturbance will cause an immediate loss of diversity even though it is essential to the persistence of other biota (Pickett and Thompson 1978, Baker 1992a,b).

Ideas of stability, constancy, and steady-state are deeply ingrained in our concepts of nature (Botkin 1990), and we have a tendency to expect equilibrium within the wildlands that we manage. However, ecological research during the last two decades has shown that equilibrium is not a simple concept; in many situations we cannot unambiguously determine that an ecological system is or is not in an equilibrium state (DeAngelis and Waterhouse 1987). There are at least three reasons for this uncertainty: (i) multiple definitions of equilibrium, (ii) multiple parameters for measuring equilibrium, and (iii) effects of spatial and temporal scale. Nevertheless, the idea of equilibrium is useful in characterizing and understanding the long-term dynamics of these landscapes. In the next section we review the various ways in which the concept of equilibrium can be fruitfully applied.

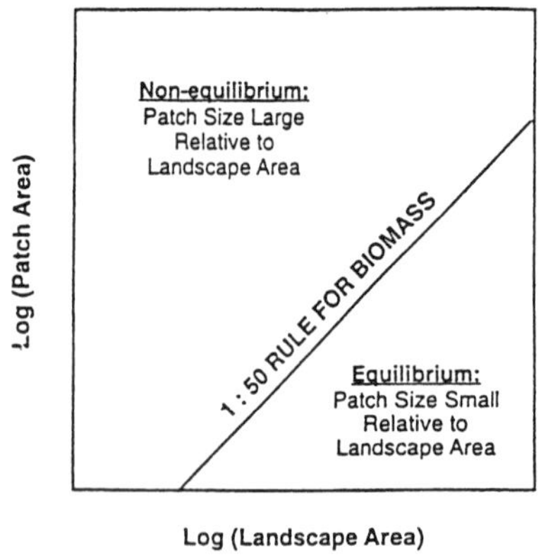


Fig. 6. Shugart's (1984) depiction of the effects of patch size and total landscape area on the nature of dynamic equilibrium in ecosystems. In his simulation models, patch sizes which were less than one-fiftieth of landscape area produced an overall equilibrium of biomass values at the landscape scale because the independent dynamics of individual patches produced a stable average.

## 6.1 Equilibrium, Change, and Scale

Scale refers to the spatial or temporal dimensions of an observation. The components of scale include grain, which is the level of resolution or smallest entity that can be recognized within a data set, and extent, which is the size of the study area or the time frame being considered (see Caraher et al. and Haufler et al., this volume; Allen and Starr 1982; O'Neill et al. 1986; Turner et al. 1989, 1993).

Some controversies over equilibrium versus non-equilibrium characteristics can be resolved simply by specifying the relevant spatial and temporal scales and ecological parameters. Consider a 100 $m^2$ plot within the deciduous forest of Great Smoky Mountains National Park. Such a plot is about the median size of treefall gaps in this ecosystem, and treefalls are likely to occur about every 100 years (Runkle 1982). If we monitored above-ground living biomass, biomass of large dead boles, or canopy cover for 200 years, we would conclude that this plot was a non-equilibrium system, because all of these parameters would change abruptly when a treefall occurred and continuously as smaller trees grew into the gap during the next 100 years. However, our perceptions change when we view this situation from a different scale. Treefall gaps affect about 1 percent of entire park annually (Runkle 1982). Living biomass, coarse woody debris, and canopy cover, averaged over the larger area, remain nearly constant even though dramatic changes are occurring within the smaller units; thus, the larger system appears to be in equilibrium. If we reduce the temporal scale within the original 75 $m^2$ plot to a single decade and if a tree-fall does not happen to occur during this short time period, the system may appear to be in equilibrium during the time of observation. Finally, consider another parameter of the system, richness of vascular plants on the forest floor, which may remain essentially constant at all of the scales of time and space described above, or may vary in response to other factors. The first step in any assessment of equilibrium/nonequilibrium dynamics is to specify the spatial and temporal scales of analysis and the parameters of the system that are being considered.

The equilibrium concept probably is most useful at intermediate scales. We would not expect to find equilibrium at the scale of a few plants, nor would this kind of equilibrium provide much insight for managers responsible for areas of hectares or square kilometers. Yet if we expand the spatial scale too far, we begin to encompass regions having fundamentally different climate and soil properties, in which different disturbance regimes and ecological dynamics prevail. Clearly, the equilibrium concept is not particularly meaningful at a scale that crosses biomes. Generally, a landscape scale, i.e., a spatial extent of tens to hundreds of kilometers, is most useful for consideration of equilibrium disturbance regimes (Risser 1987, Forman 1990). Similarly, equilibrium is not very meaningful to most managers within time frames shorter than several decades, and if we try to deal with many thousand of years, we encounter profound changes in climate and soils related to glacial cycles and long-term geologic processes. Thus, the appropriate temporal scale for assessing equilibrium generally is in the range of decades to centuries, though it may be appropriate to use millennia in very long lived trees (Stephenson 1996).

## 6.2 Concepts of Equilibrium

Various definitions of equilibrium have been used to assess ecological systems. Let us first dispense with two concepts that are not particularly useful to wildland managers. One of these is the idea of equilibrium as complete absence of change; clearly this notion is not applicable to ecosystems. The other is the idea that a system is in equilibrium only when it returns to its original set point after being disturbed. This definition assumes that stasis is the norm and that disturbance necessarily moves the system into an "unnatural" or otherwise inappropriate state. Moving beyond these overly simplistic concepts, however, we can identify four general classes of definitions of equilibrium that provide useful insights into the workings of ecological systems:

(1) *Persistence or qualitative equilibrium* — A system may be regarded as in equilibrium if none of its characteristic elements becomes extinct during a specified period of time (DeAngelis and Waterhouse 1987, DeAngelis and White 1994). These elements might be species, successional stages, surface water, or any other ecological parameter of interest. This is the least rigorous of the four concepts of equilibrium discussed here, but it is useful nonetheless: at a bare minimum, an equilibrium system must not lose any of its components, and a system that does lose components is clearly not in equilibrium, at least with respect to those components. This definition allows much local fluctuation in abundance, as long as elements persist.

(2) *Shifting mosaic, steady-state, or quantitative equilibrium* — A system may be regarded as in equilibrium if the abundance of specific elements or the rates of specific processes remain more or less constant throughout a specified time period. Some temporal variation is allowed, but it must be small, and the levels of the parameters of interest must remain close to some average value. For example, Bormann and Likens (1979) suggested that total biomass in watersheds or

landscapes dominated by northern hardwoods forests varied only slightly during the period before European settlement, despite large fluctuations within smaller units of the landscape. Bormann and Likens described this situation as a shifting mosaic steady-state. Similar equilibria, defined as constancy of biomass or of proportions of the landscape occupied by each successional stage, have been suggested for wave-regenerated fir forests in New England and elsewhere (Sprugel 1976), for riparian woodlands disturbed by recurrent floods (Everitt 1968; cited in Baker 1989a), and for fire-regenerated boreal forests in northern Sweden (Zachrisson 1977) and Isle Royale in Lake Superior (Cooper 1913).

This type of equilibrium is strongly dependent on the spatial and temporal characteristics of the disturbance regime. Based on simulation studies, Shugart (1984) suggested that a quasi-steady-state landscape was likely only in situations where the total extent of the landscape was at least fifty times the average size of a disturbance event. Zedler and Goff (1973), Connell and Sousa (1983), and DeAngelis and Waterhouse (1987) also suggested that stable mosaics of successional stages are more likely to occur when the landscape is large relative to the size of disturbed patches.

Shifting mosaic steady-state systems may be fairly rare because large, infrequent disturbances are a feature of many ecosystems. Baker (1989a) did not detect equilibrium in the mosaic of successional stages in the Boundary Waters Canoe Area in northern Minnesota even at a spatial scale 87 times as large as the average disturbance-patch. The relative proportions of post-fire successional stages in subalpine forests of Yellowstone National Park have not exhibited constancy during the last 250 years at any spatial scale up to 130,000 ha because of the persistent effects of infrequent large fires (Fig. 5) (Romme 1982, Romme and Despain 1989).

(3) *Stable trajectory or stationary-dynamic equilibrium* — Loucks (1970) suggested that although individual communities at particular locations are continually changing through time as a result of disturbance, the long-term process of disturbance and recovery constitutes a stable system because the same successional sequence occurs after each disturbance event. The system never reaches any final, undisturbed state, but it does return along the same trajectory of change, i.e., the same dynamics occur after each disturbance (O'Neill et al. 1986).

(4) *Statistical equilibrium* — A system or subsystem can be regarded as in equilibrium if the distribution of individual disturbance events does not deviate significantly from the expected statistical distribution, and if a normal successional sequence follows each disturbance. If disturbance intervals, intensities, sizes or responses fall outside the expected distribution, then the system may be out of equilibrium or it may have shifted to a new equilibrium. Probability density functions of disturbance intervals and disturbance sizes in fire-dominated systems have been described quantitatively using the Weibull model (Johnson and Van Wagner 1985, Baker 1989b, Johnson and Gutsell 1994).

The statistical distribution of disturbance intervals is sensitive to the spatial and temporal scales at which the disturbance regime is described. For example, the Weibull model assumes that the largest individual disturbances are much smaller than the total extent of the study area (Johnson and Van Wagner 1985); where this assumption is not true, the Weibull distribution may be an inappropriate model for landscape dynamics (Baker 1989b). Statistical distributions of disturbance intervals also may be complicated by the presence of subregions, having somewhat different disturbance regimes, that lie within the larger landscape unit being analyzed. Subtle spatial heterogeneity in disturbance regimes may not be apparent until it is teased out by means of cluster analysis or some other analytical method (Baker 1989b).

## 6.3 Management Implications of Equilibrium/Nonequilibrium Disturbance Regimes

Given the complexity of the equilibrium concept as applied to ecological systems, few managers will be able to determine unambiguously whether their particular parcel of land is an equilibrium or nonequilibrium system. Fortunately, it probably is not necessary to categorize specific areas as equilibrium or nonequilibrium with respect to every ecological parameter. However, it is important for managers to explicitly recognize the spatial and temporal dynamics of disturbance regimes and to be aware of the scale of the landscape processes that characterize their systems.

For example, it is helpful to understand that most crown-fire ecosystems are not in quantitative equilibrium, although such landscapes may meet the criteria for qualitative or stable-trajectory equilibrium (Baker 1989b, 1992b; Turner and Romme 1994). The proportion of successional stages in Yellowstone National Park was drastically altered in 1988. Prior to the fires, the landscape was dominated by middle and late successional stages, and early successional habitat was rare. The fires transformed nearly 30 percent of the landscape to early successional stages and reduced middle and late stages correspondingly. A change of this magnitude would be regarded as a catastrophe if the expectation was for a quantitative equilibrium. However, the characteristic spatial and temporal scales of Yellowstone's disturbance regime indicate that fires like those in 1988 are infrequent but expected events in this kind of ecosystem. Although proportions of successional stages were altered by the 1988 fires, no landscape elements were eliminated entirely, and historical landscape reconstructions indicate that qualitatively similar disturbances and recovery have occurred in the past (Romme and Despain 1989). From a historical perspective, the 1988 fires were not a unprecedented catastrophe but a normal feature of the area's disturbance regime.

We need a practical method by which managers can assess whether the landscapes for which they are responsible are characterized by equilibrium vs. nonequilibrium or stable vs. unstable dynamics. Turner et al. (1993) have developed a model that can help to answer this question. A state-space diagram (Fig. 7) depicts qualitatively different kinds of landscape dynamics in relation to spatial and temporal scales of disturbance. The ecological parameter upon which this model is based is the relative proportion of the landscape occupied by each successional stage. If we have reasonably reliable empirical data with which to characterize the disturbance regime of a particular landscape, then we can locate that system within the state-space shown in Fig. 7.

The horizontal axis (S) represents the spatial scale, i.e., the ratio of disturbance extent to landscape extent.

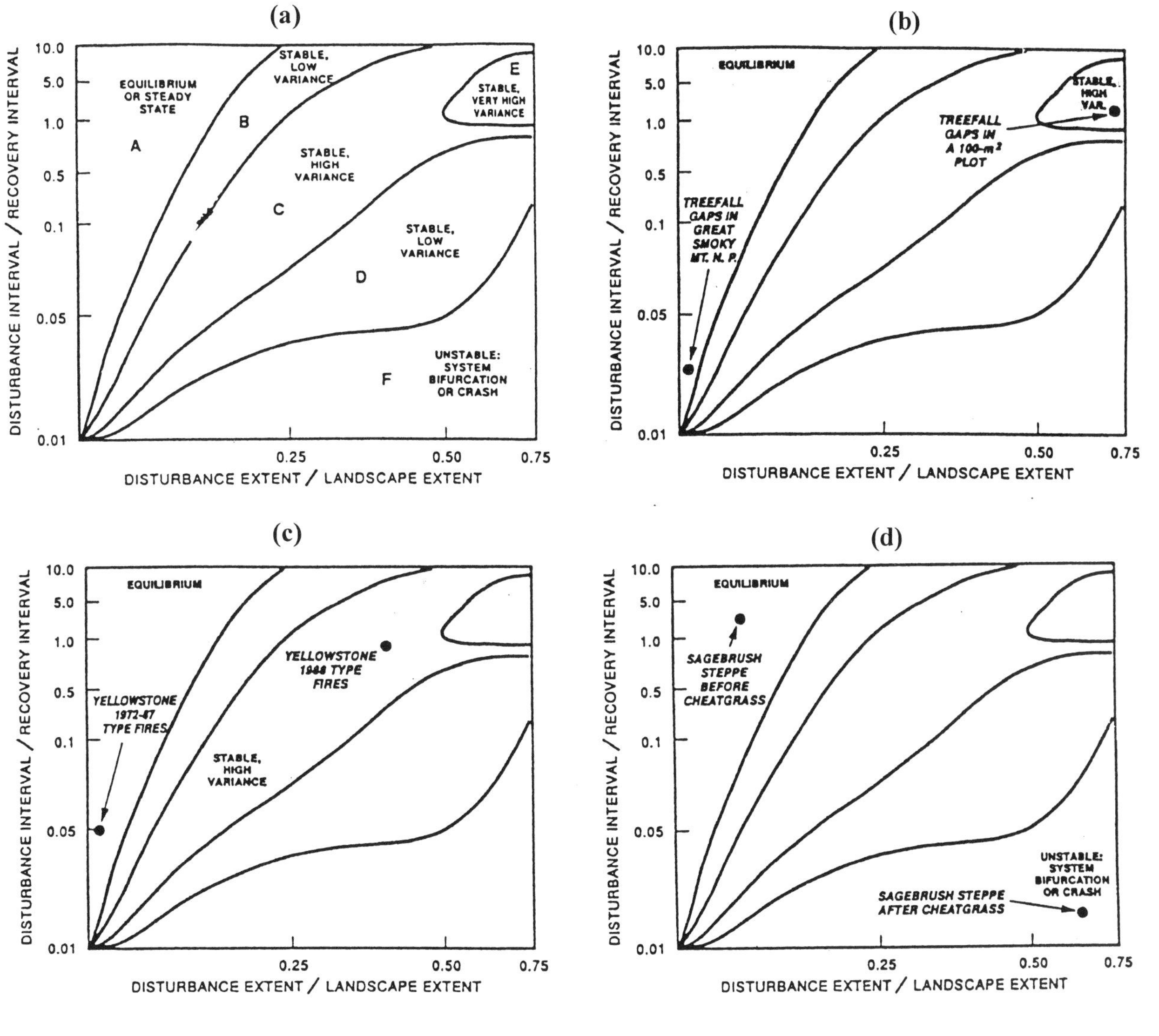


Fig. 7. State-space diagram of qualitatively distinctive types of landscapes in relation to spatial and temporal scales of disturbance (adapted from Turner et al. 1993). (a) The regions of six different kinds of landscape dynamics (see text). (b) Effects of scale on landscape dynamics of mesic forests in Great Smoky Mountains National Park (adapted from Runkle 1982). (c) Two types of disturbance regimes in Yellowstone National Park (adapted from Romme and Despain 1989). (d) Effects of introduction of exotic cheatgrass in sagebrush-steppe landscapes of western North America (adapted from Whisenant 1990). See text for further explanation.

Disturbance extent refers to the mean, median, or some other measure of the size of individual disturbances (for some landscapes the maximum likely disturbance extent might be most appropriate here). There are two qualitatively different regions along this axis: regions where disturbances are small relative to the size of the landscape (i.e., low ratios) and regions where disturbances are large relative to the landscape (i.e., high ratios).

The vertical axis (T) represents the temporal scale, i.e., the ratio of disturbance interval to recovery interval. Disturbance interval might be expressed as the mean, median, or some other measure of the time between successive disturbance events (the expected interval between large disturbances might be most appropriate for some systems). Recovery interval refers to the time required for succession to reach some late or "climax" stage of development if there were no further disturbance. There are three qualitatively different regions along the vertical axis: regions where the interval between successive disturbances is longer than the time required for recovery ($T > 1$); regions where disturbance and recovery intervals are equal ($T = 1$); and regions where the disturbance interval is shorter than the time required for recovery ($T < 1$).

The region in the upper left-hand corner (A) is characterized by disturbances that are small, relative to the total extent of the landscape, and that occur at intervals longer than the time required for recovery. This is the region in which we find systems traditionally regarded as equilibrium or steady-state (e.g., the northern hardwood forest or the Great Smoky Mountains landscape). To the right of the equilibrium zone is a region where disturbances are a little too large and too frequent to allow a true equilibrium state as defined by the constancy concept described above, but where the variation in proportion of successional stages over time is relatively low (region B). This is a stable landscape state, where all elements persist, but where there is a small amount of fluctuation over time. The landscape in region B is dominated by late successional stages. Region C extends through the center of the Fig. 7, and is a state in which the landscape is stable, in the sense that all elements persist, but in which relative proportions of landscape elements fluctuate strongly over time as a result of relatively infrequent but large disturbance events. Many crown-fire ecosystems fall into this kind of landscape (e.g., the Boundary Waters Canoe Area and Yellowstone National Park). Below and to the right of region C is an area where variance is again low, and all elements persist, but this kind of landscape (region D) is dominated by early successional stages because disturbances tend to be both large and frequent. In the upper right-hand corner (region E) is a landscape characterized by very great fluctuation in landscape structure, resulting from very large disturbance events (affecting > 50% of the landscape). Nevertheless, this kind of landscape does not lose any of its elements because these large disturbances occur only at very long intervals. Finally, we see the region where disturbances are both large and frequent (region F). This is an unstable kind of landscape, susceptible to irretrievable loss of one or more of its elements (e.g., late successional stages). Such a landscape is likely to "crash" or "bifurcate," i.e., to shift to a qualitatively different ecological state or developmental trajectory. We do not see many systems in region F because they do not persist for very long. However, it is useful to recognize that alterations of the spatial and temporal dimensions of a disturbance regime can cause profound and probably irreversible changes in the structure and dynamics of an ecosystem. For example, chronic air pollution in Copper Basin, Tennessee, transformed a deciduous forest landscape into a sparsely vegetated landscape that has not recovered even after the pollution sources were eliminated (Turner et al. 1993).

Long lasting effects of infrequent, large disturbances may overshadow the effects of more frequent small disturbances (Turner and Romme 1994). A possible reason why Baker (1989a) found no evidence of a steady-state patch mosaic in the Boundary Waters Canoe Area, despite a study area 87 times the size of an average fire-created patch, was because the largest fire in the 141-year period of study burned nearly 1/5 of the 400,000 ha study area. In the Yellowstone landscape before 1988, the most abundant elements were middle and late successional stages that had developed following extensive fires around 1700 A.D. (Romme and Despain 1989). Even though smaller fires occurred in every decade thereafter, the imprint of those early large fires was still evident. The fires of 1988 created a new landscape pattern that will persist until the next extensive fire event. Romme (1982) simulated landscape changes in Yellowstone during the last 200 years under three scenarios: (i) actual fire history (documented from fire scars and current stand ages), (ii) all fires excluded, and (iii) large fires excluded but small fires allowed to occur. Interestingly, the landscape patterns generated in the second and third scenarios were almost indistinguishable, and both were strikingly different from the patterns produced by the actual fire history. Based on these considerations, the location of a landscape in the state-space diagram of Fig. 7 probably should be based on the size of the largest disturbance events rather than on mean, median, or modal disturbance patch sizes. The Weibull distribution can be used to estimate the occurrence of very large or infrequent disturbance events.

Landscapes having disturbance regimes that place them in the upper right-hand portion of the state-space diagram shown in Fig. 7 pose special challenges to managers. These are systems that require large land areas to retain all of their ecological elements because the spatial scale of disturbances is so great. In a very small management unit having this kind of disturbance regime, a single large disturbance could affect the entire landscape and eliminate all but the earliest successional stages. The result would be complete loss of the species and ecological processes that characterize later successional stages. Even in a very large management unit, we cannot be 100 percent certain that a single disturbance event will not affect all or most of the area. Large wilderness areas and national parks in western North America contain some of the most extensive remaining tracts of old-growth forest, but many of these protected areas are crown-fire ecosystems in which a single fire event could eliminate most of the old forests. In such systems, it is important to maintain or restore mature forests in surrounding multiple use lands as well as in designated nature reserves. In contrast, even a small tract of land may support equilibrium conditions if it has a disturbance regime that places it in the upper left-hand corner of the state-space diagram.

Managers should seek to avoid situations in which a system is shifted into a region of the state-space diagram dramatically different from its natural or historical norm. This may be difficult in the world of the early 21st Century, given the variety and magnitude of stresses being imposed on our wildland ecosystems. For example, consider the semi-arid sagebrush steppes, which cover thousands of square kilometers in the Great Basin of western North America (West 1988). Prior to the late 1800s, this region was characterized by infrequent, small fires, placing it in the upper left-hand corner of the state-space diagram (Fig. 7). Fires, whether ignited by lightning or by humans, rarely spread over large areas because the scattered shrubs and bunchgrasses, separated by patches of bare ground, formed a discontinuous fuel bed (Whisenant 1990). In the early 1900s this region was invaded by the Eurasian cheatgrass, which spread with alarming speed through plant communities already stressed by heavy livestock grazing (Leopold 1966, Mack 1981, Billings 1990). A winter annual, cheatgrass grows quickly in the early spring then dies and becomes dry by early summer, creating a highly flammable, continuous fuel bed. Cheatgrass is not injured by fire; on the contrary, fire creates an ideal seedbed for cheatgrass germination. Since its arrival, fires have become more frequent and have begun to burn over thousands of hectares. Changes in the spatial and temporal parameters of the fire regime have moved the system into the lower right-hand corner of the state-space diagram (Fig. 7). Neither native shrubs nor bunchgrasses can withstand such frequent burning; they have disappeared over large areas and been replaced by cheatgrass and a few other fire-tolerant species (Whisenant 1990). The original sagebrush-steppe system has "crashed" following the change in disturbance regime, and the original post-fire succession has been replaced by a new trajectory dominated by cheatgrass.

In addition to the stresses posed by biological invasion and legacies of unsustainable land- use practices in the past, managers of the 21st century will have to contend with the impact of elevated atmospheric $CO_2$, altered global nitrogen dynamics, and possible climate change (Vitousek 1994). It is therefore important not only to document past disturbance regimes in wildland ecosystems, but also to monitor current and future disturbance regimes in order to detect changes in intensity, kind, frequency, or size of disturbance — changes that could push a system into an unstable region of the state-space diagram. Even if future changes in disturbance regimes do not cause systems to crash, they may cause quantitative or qualitative changes in landscape dynamics (Baker 1995). For example, warmer and drier conditions will likely be associated with more frequent fires (Romme and Turner 1991), which could shift the Yellowstone landscape from a stable system with high variability in proportions of successional stages (region C in the state-space diagram) to a stable system with low variability dominated by early successional stages (region D). Such a change would have obvious implications for long-term persistence of species associated with old forests, such as pine marten and goshawk (Romme and Turner 1991). Flannigan and Van Wagner (1991) coupled global predictions for precipitation and temperature under conditions of doubled atmospheric $CO_2$ with the Canadian Forest Fire Weather Index System, which provides an index of fire intensity. Their results suggest that a doubling of atmospheric $CO_2$ could lead to a 40 percent increase in the annual area burned in Canada.

Spatial and temporal parameters of disturbance regimes have important implications for the existence and nature of equilibrium. Steady-state equilibrium is most likely to occur where disturbances are infrequent and small relative to recovery time and the total extent of the landscape. Where disturbances tend to be frequent or large, other kinds of landscape dynamics will occur. Although these landscapes are not in quantitative equilibrium, they may nonetheless exhibit qualitative and stable-trajectory equilibrium. An increase in disturbance size or frequently resulting from biological invasion, global climatic and atmospheric changes, or unsustainable land use practices may cause a qualitative shift in landscape structure and dynamics. Whether a particular land unit is in equilibrium or not, managers need to document past disturbance regimes and monitor potential future changes in the scale, intensity, and kinds of disturbances that shape the structure and dynamics of the landscape.

## 7 ECOSYSTEM RESPONSES TO DISTURBANCE

By definition, disturbances alter the physical environment or the availability of resources or space. Levels of available light, moisture, and nutrients often increase as a result of reduced uptake and enhanced decomposition (Vitousek 1984). For example, treefall in most temperate and tropical forests increases understory light levels (Canham et al. 1990), and fires in tallgrass prairie release phosphorus sequestered in dead biomass (Knapp and Seastedt 1986). Disturbances may also make available a range of substrates such as logs, tip-up mounds, and mineral soil. Conversely, the physical effects of disturbance may reduce resource availability. Overall moisture availability may decrease in large forest gaps as increased wind speeds and temperatures offset reduced water uptake by vegetation. Disturbances that remove soil, leaf litter, or coarse woody debris reduce the availability of these resources to species which require them (Hansen et al. 1991). In addition to altering absolute levels, disturbances may change the distribution of resources and increase or decrease spatial heterogeneity (see section on pattern and process above).

Response to disturbance involves several mechanisms. In forests, these include expansion of crowns of surviving canopy trees, release of suppressed individuals in the understory (advance regeneration), sprouting from roots and stumps of damaged individuals, establishment of new seedlings, and suppression of trees by vigorously growing shrubs, herbs, and vines (Woods and Shanks 1959, Runkle 1985, Everham and Brokaw 1996). Which of these mechanisms dominates depends on the characteristics of the disturbance (particularly size, specificity and severity) and the structure and composition of the community. For example, small gaps in eastern deciduous forests are typically filled by crown expansion and advance regeneration (Runkle 1985). Because light levels are fairly low and gap closure is rapid, new seedlings have little chance of reaching the canopy. In larger, more persistent gaps, new seedlings of fast-growing, shade-intolerant species such as tulip-poplar can compete for space in the canopy. A dense understory of shrubs such as rhododendron and mountain laurel may inhibit tree regeneration (Clinton et al. 1994) and slow gap closure.

Longer term changes following large disturbances have been described using a four-phase model (Oliver 1980, Peet and Christensen 1987). The establishment phase, which immediately follows disturbance, is characterized by open conditions and the germination and rapid growth of herbs, shrubs, and tree seedlings. Resource availability often exceeds uptake; the result may be a pulse of nutrient runoff in ground water. During the establishment phase, competition is reduced, and fine-scale species diversity is often high.

As seedlings grow up to form a closed canopy, the stand enters the thinning phase. Biomass continues to accumulate, but growth of the more vigorous trees is offset by death of weaker competitors. Most of the trees which die have been overtopped by stronger competitors and thus do not create gaps in the canopy. The few small gaps which form are filled rapidly by expansion of adjacent trees. Competition for light and nutrients is intense, and nutrient runoff drops to very low levels. The initial, even-aged canopy is typically dominated by fast-growing, shade-intolerant species. After the canopy closes, opportunities for new establishment are limited, and only species tolerant of low resource levels persist in the understory. As a result, fine-scale species richness is often low. Living biomass typically peaks toward the end of the thinning stage.

After several decades, canopy trees become large enough to form gaps which cannot be filled by lateral expansion, and the stand enters the transition phase. At this point, death of canopy trees creates patches of resource availability which promote seedling and sapling growth. As the initial even-aged canopy breaks up, the stand begins to develop a multi-age size structure. Although individual trees continue to increase in size, total biomass may decline. Gap formation increases structural diversity, and reduced competition in gaps may lead to an increase in fine-scale species diversity.

In the final, steady-state phase, the breakup of the initial even-aged canopy is complete, and the stand has developed an all-aged, all-sized structure. Barring subsequent large disturbance, formation of small gaps will be balanced by germination and growth. At a scale many times that of individual gaps, structure, composition, and biomass will be in rough equilibrium. The model also suggests that nutrient runoff during this phase will be balanced by inputs from atmospheric and geologic sources. Although it is often assumed that biomass and productivity in this phase will converge on pre-disturbance values, other relationships are possible (Reiners 1983). For example, erosion or nutrient losses immediately following a disturbance may lead to long-term reductions in productivity.

The four-phase model, originally developed to explain secondary succession in eastern forests following clearcutting or agricultural abandonment, applies with minor modifications to secondary forest succession following other large, severe disturbances such as crownfires in Rocky Mountain forests (Peet 1992). The model suggests that stands of different ages will differ fundamentally in structure, dynamics, and diversity; thus, observed effects of disturbance will vary with successional age. In addition, the model provides insights into mechanisms by which disturbances alter community composition. For example, Meier et al. (1995) suggest that reduced densities of some vernal herb species in logged southern Appalachian forests relative to unlogged sites may reflect mortality incurred during logging itself, inability of these species to tolerate the open, high-light environment of the establishment phase, or intense competition, low resource levels, and lack of gaps during the thinning phase.

The four-phase model can be generalized to explain forest response to disturbances of low and medium severity. Disturbances which remove light-demanding canopy trees but leave shade-tolerant advance regeneration more or less intact move can stands toward steady-state composition; examples of such "accelerated succession" include logging in oak forests (Abrams and Nowacki 1992), beetle outbreaks in southern Appalachian pine stands (Kuykendall 1978), and hurricane damage on old agricultural fields in New England (Foster 1988). Systems in which frequent, low-intensity disturbances such as surface fires maintain open conditions and create opportunities for regeneration of light-demanding species may never leave the establishment phase. In such systems (e.g., longleaf pine savannas), fine-scale species diversity is often very high (Walker and Peet 1983, Peet et al. 1983).

Understanding the role of disturbance in maintaining biological diversity requires knowledge of the characteristics of individual species. Species differ in dispersal ability, mode of reproduction, and ability to survive, grow, and reproduce under various conditions. Most ecosystems contain species which require the open conditions and resources made available by disturbance. Some, such as fireweed, flourish briefly following disturbance and disappear as succession proceeds (Agee 1993). Others, such as tulip-poplars and giant sequoias, require disturbance for initial establishment but persist for centuries. In ecosystems throughout the United States, the suppression of natural fires and floods has led to declines of open, early successional habitats and the species that require them (see Engstrom, this volume, Sections 9.01–9.09). Most systems also contain species that can establish and survive under both early and late-successional conditions. Although the abundance of such species may change through successional time, they occur on both recently disturbed and undisturbed sites. The existence of species which require late-successional habitats is less well documented. The best examples come from the Pacific Northwest, where concern over the loss of mature forests has prompted extensive study (Hansen et al. 1991).

Determining species' successional requirements is not always straightforward. Some species which require disturbance for establishment or survival over much of their ecological range may be able to persist on extreme sites in the absence of disturbance (Barden 1988). Some species require microenvironments or structures such as large snags rather than particular successional states (Hansen et al. 1991); even though such species may thrive in both mature stands and stands recently disturbed by natural fire or windthrow, they may disappear from clearcut stands which lack those structures. And populations of some species which occur in both early and late successional managed stands may gradually decline if logging rotations are too short to allow recovery to pre-disturbance levels.

The relationship between species diversity and the frequency and intensity of disturbance is the subject of the intermediate disturbance hypothesis ("IDH", below; Connell 1978), which suggests that diversity peaks at some intermediate frequency and intensity of disturbance. Huston (1979) made a similar prediction, proposing that diversity is highest at intermediate disturbance frequencies and population growth rates. Intermediate levels of disturbance create opportunities for species which require early-successional conditions without eliminating disturbance-sensitive species. Thus, the IDH should be true in a qualitative sense, at least in systems which contain both groups of species. However, in systems with no history of a particular disturbance, all species might be sensitive, and in systems with a very high frequency of disturbance, there might be no species which require late-successional conditions. Even in systems in which the IDH applies, the model lacks much predictive value. Does diversity show a distinct peak or a broad plateau with increasing frequency or intensity of disturbance? And at what levels is diversity maximized? For most systems, the answers to these questions are not known. The hypothesis that diversity is highest at the disturbance levels which characterized the system over its ecological and evolutionary history has received some anecdotal support (Peet et al. 1983).

The effects of disturbance on species diversity also vary with scale (Palmer and White 1994, White and

Harrod 1997) and with simple changes in stem density (Busing and White 1997). The number of species in a single square meter is often much higher in a recent clearcut than in an intact forest. But if all the species that establish in the clearcut were already present in the watershed before clearcutting, the total number of species in the watershed remains unchanged. And if clearcutting in the watershed leads to loss of mature forest species, increases in the number of species at fine scales will be accompanied by a decline in species numbers at coarser scales. For these reasons, statements on the relationship between disturbance and diversity need to specify the scale of observation.

## 8 MANAGING DISTURBANCE: EMERGING ISSUES

Because species show a wide range of responses to disturbance, and because historical composition and structure are themselves products of past disturbance, changes in disturbance regimes often threaten biological diversity and ecosystem function. The two natural disturbances most manipulated by people are fire and hydrology. In North America, fire frequency has been greatly reduced in the 20th century (Kilgore and Taylor 1979, Covington and Moore 1994). Channelization, damming, and draining have affected many river systems. Reintroducing fire and hydrologic fluctuation are key to the restoration of many ecosystems (Baker 1994, Dahm et al. 1995). Where suppression has led to major changes, it will often be necessary to modify ecosystem structure (e.g., thinning treatments or fuel reduction burns) before reintroducing natural processes. Restoration of natural structure and dynamics may occur only gradually over the course of several disturbance cycles.

To what extent do management activities mimic natural events? Human-caused disturbances resemble natural disturbances in some ways, but differ in others. Clearcutting might resemble windthrow in patch size and return interval, but differ in its effects on woody debris, soils, and salamander populations. Prescribed fire might mimic the seasonality and average intensity of natural fires, but differ in the patchiness of effects. Meaningful comparisons require accurate, detailed information on both natural and human-caused disturbances. They also require us to specify which aspects of the disturbances are being compared.

Human-introduced disturbances, including logging and grazing, may also threaten species and ecosystems. The study of natural disturbances suggests ways to modify management activities to balance commodity production with ecosystem health and diversity. The restoration of ponderosa pine forests in western Montana provides one example of the application of disturbance ecology in ecosystem management (Arno et al. 1995). There, managers are using prescribed fire and thinning treatments (including commercially-oriented harvests) to restore open stand structure, enhance canopy and understory vigor, and reduce wildfire and insect hazards. Disturbance ecology also forms a foundation for "New Forestry" in Douglas-fir forests of the Pacific Northwest (Franklin 1989, Swanson and Franklin 1992). To mimic the effects of natural disturbances more closely, managers have modified clearcutting practices to retain more coarse woody debris and standing live trees. Managers are also modifying the traditional pattern of dispersed 10–15-ha cutting units to reduce fragmentation of mature forest tracts and create patches of variable size. In both regions, management actions have been informed by historical, observational, and simulation studies and are treated as ongoing experiments subject to monitoring and evaluation.

Application of disturbance ecology to management of ecosystems is complicated by several factors. In the next sections we summarize five major challenges: the relation of disturbance regime to climate variability and climate change, the influence of Native Americans, habitat fragmentation and the human imposition of new scales on ecosystems, exotic species invasions, and documenting and understanding the role of ecological variation.

### 8.1 Climate Variability, Regional Scales, and Disturbance Regimes

Documenting the connection between climate and disturbance will improve our ability to predict ecosystem dynamics and assess the potential impact of climate change. Natural climatic fluctuations occurring at time scales of decades may have broad impacts on terrestrial ecosystems; examples of such events include multiyear droughts in the 1930s in the Great Plains and the 1950s in the North American subtropics (Betancourt et al. 1993). These events may affect disturbance and recovery rates and suppress or amplify human impacts on terrestrial ecosystems. For example, range conditions in much of the American Southwest have improved over the last 20 years. Does this improvement reflect better management or simply rebounding precipitation levels following the 1950s drought? And to what extent does interdecadal variability explain changes in flood regimes (Webb and Betancourt 1992), fire behavior (Fig. 8) (Swetnam and Betancourt 1990, Swetnam 1993) and outbreaks of forest pests (Swetnam and Lynch 1993)?

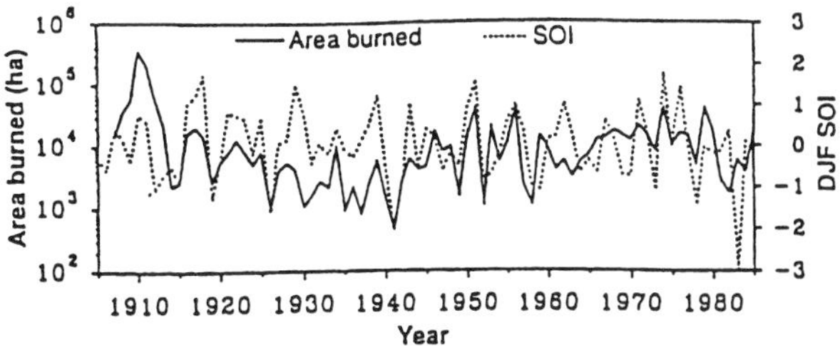


Fig. 8. The association between area burned in Arizona and New Mexico between 1905 and 1985 and an index of the Southern Oscillation (SOI) for the December to February (from Swetnam and Betancourt 1990). High values of SOI are associated with droughts and higher fire incidence across large areas.

Documenting effects of interdecadal variability on the demography of trees presents several challenges. Data on tree recruitment and mortality must be regional in geographic extent and both long-term (much longer than 100 years) and high resolution (preferably annual) in temporal coverage. Few such studies have been attempted (e.g., Villalba 1995), in part because the necessary resolution and sample size are often prohibitive.

The extent to which climate synchronizes ecological processes at regional scales has implications for management. For example, in years of severe drought, intense fires may affect many units and land-managing agencies and strain management resources. In addition, regional synchronization of the dynamics of wildlife populations would make rare species more vulnerable to extinction. The spatial scale of this synchronization is likely to vary from one part of North America to another. Documenting responses to decade-scale climatic variation may also help to predict the ecological effects of human-induced climate change.

## 8.2 Native American Disturbance

Ecologists and land managers once assumed that ecosystem dynamics prior to European settlement of North America were influenced only locally or in minor ways by Native Americans. It is now becoming clear that at some times and places, Native American populations had important direct and indirect impacts on ecosystem processes. These impacts include cultivation, the use of fire, and the direct harvest of wild plants and animals. Impacts on disturbance regimes and temporal dynamics varied with the size, location, and evolving cultural practices of these populations. In some areas these influences were relatively minor (e.g., occasional hunting and gathering activities), whereas in others they were significant (e.g., frequent burning and large agricultural areas; Delcourt et al. 1986). Evidence of widespread Native American influence suggests that we re-evaluate the "naturalness" of pre-Columbian ecosystems (see also Bonnicksen et al., this volume.)

Understanding the effects of Native American activities may also provide insights into ongoing ecological change. In the 1950s through 1970s, chaining of southwestern pinyon-juniper woodlands to improve forage and water yield proceeded on the assumption that pinyon and juniper had recently invaded grasslands as a result of overgrazing and fire suppression. However, midden records suggest that at least some of the areas chained were impacted by intensive prehistoric fuel harvesting (Betancourt and Van Devender 1981). These so-called invasions may in fact represent recovery from prehistoric human impact (Samuels and Betancourt 1982).

## 8.3 Habitat Fragmentation and the Human Imposition of New Scales

Managers must be concerned not only with the characterization of disturbances, but also with the scale of managed systems. The size of disturbance events may exceed that of management units, and the boundaries of management units may not necessarily correspond to natural boundaries. In areas where several agencies oversee adjacent units with similar ecology, ecosystem management on a scale appropriate to the disturbance regime may require coordination among agencies. In other situations, we may need to ask if management can reduce disturbance size and still maintain the desired patch mosaic.

Scale-dependence of ecological variables such as species richness has important consequences for how we observe and manage ecosystems. Different ecosystem parameters (e.g., leaf area, biomass, downed woody debris, species richness) change at different rates with scale. As a consequence, there is no single optimum scale of observation and management for an ecosystem, although particular management goals may be best addressed at particular scales. Although finer scales (e.g., patches, stands and watersheds) have received more attention in the past, coarser scales (e.g., landscapes and regions) may be critical for managing biological diversity (Swanson and Franklin 1992).

Reduction in the size of natural areas because of habitat loss and fragmentation may threaten the persistence of both early and late successional species. Early-successional species that disappear from older patches can persist in a landscape only if newly disturbed patches occur within dispersal distance (Pickett and Thompson 1978; Baker 1992a, b). The alteration of spatial context may thus impair response to disturbance itself. Similarly, the loss of late-successional refuges for disturbance-sensitive species may jeopardize their persistence. Human alteration of the landscape matrix will also affect the propagation of disturbances. Changes in the size and spatial context of natural areas represent an important challenge for the management of biological diversity in the future.

## 8.4 Exotic Species Invasions and Disturbance Regimes

One of the most troubling problems in the management of ecosystems is the invasion of exotic species, sometimes called the least reversible of human impacts (Coblentz 1990). Humanity is now engaged in a massive experiment involving purposeful and accidental transport of species across natural barriers to their dispersal (Drake et al. 1989, Groves and Burdon 1986, Mooney and Drake 1986, Pycek et al. 1995). Exotics may exert direct impacts on native species through competition, predation, and pathogenesis, and may affect communities indirectly by altering habitat structure, resource availability (e.g., the invasion of nitrogen fixers in Hawaii, Vitousek 1990), hydrology, and ecosystem function. Dogwood anthracnose, an exotic fungus, has devastated populations of flowering dogwoods in eastern North America (Daughtrey and Hibben 1994). In doing so, it has removed an major component of the small tree stratum and an important source of fruit for migratory songbirds. Because flowering dogwoods act as "calcium pumps," extracting the nutrient from the subsoil, anthracnose may alter soil nutrient status and thus the abundance of other plant species.

Exotic species invasions change disturbance regimes. Exotic insects and pathogens such as the gypsy moth, hemlock aphid, and chestnut blight kill or damage trees and thus represent a new source of disturbance to the forest canopy. The introduction of Eurasian cheatgrass to semi-arid sagebrush steppes, discussed above, has caused a fundamental shift in disturbance regime, increasing the size and frequency of fires and threatening native species (Billings 1990). In south Florida, the exotic tree *Melaleuca* transpires so much water that it lowers the water table; the result is an increase in fire intensity (Bodle et al. 1994).

Rates of exotic species invasion can be hastened by both human and natural disturbances. For example, many exotic plants common in the eastern United States, such as Japanese honeysuckle, princess tree, and kudzu, thrive in open and edge habitats but are slow to invade intact forest. Human-caused habitat fragmentation and natural disturbances such as hurricanes allow these species to spread by creating gaps and edges. The invasion of *Potamocorbula amurensis*, an Asian filter-feeding clam, in San Francisco Bay provides an example of the effects of disturbance on exotic species invasions (Carlton et al. 1990, Nichols et al. 1990). The 1986 introduction of this clam in ballast water coincided with a major flood which disrupted benthic communities. By 1988, the clam was the dominant organism in the north section of the bay and had begun spreading into the southern section.

The spread of exotic species will affect our ability to restore historical processes in certain landscapes. Management of exotic species may involve direct removal, reduction of human activities which promote their spread, and testing and release of biological control agents. Biological control agents which are themselves exotics must be introduced with caution; managers must seek to minimize the risk that new exotic pests will be released.

## 8.5 Ecological Variation, Scale Dependence, and Resilience

In many ecosystems, disturbance maintains ecological variability. However, we can also turn this formulation around: ecosystem response to disturbance is itself be a function of the ecological variability present. For example, some species and patch types require disturbance to persist, and some of these disturbance-dependent elements are critical to ecosystem function (Marks 1974, Hansen et al. 1991). The interrelationship of disturbance and variability thus has implications for ecosystem resilience. Effects of ecological diversity on resilience have only recently attracted rigorous empirical study (e.g. Tilman 1996, Tilman et al. 1996, Huston 1997). Relationships between disturbance, diversity, and ecosystem dynamics represent an important research frontier.

Results of ecological studies depend, sometimes critically, on spatial and temporal scales of observation (Busing and White 1993, Reed et al. 1993, Zedler and Goff 1973). One implication of scale dependence in a patchy ecosystem is that small samples may contain only a subset of ecosystem states. For example, studies which position relatively small plots in areas of large trees typically overestimate biomass in southern Appalachian old growth forests (Busing et al. 1993). If the time span of a historical study is too short, extremely large and destructive disturbance events which are an infrequent part of a system's dynamics may appear unprecedented. Rather than relying on a narrow range of states defined by a few small samples or reference points, managers should seek to characterize and understand the full range of spatial and temporal variation in their systems (White and Walker, in press).

## 9 CONCLUSIONS AND KEY POINTS

1. A disturbance is a relatively discrete event in time that disrupts ecosystem, community, or population structure and changes resources, substrate availability or the physical environment. Natural disturbances occur in all ecosystems, and in many cases, play an important role in maintaining species diversity and community and ecosystem structure. Describing disturbance regimes is an important step towards understanding the natural dynamics of ecosystems.
2. Most management activities involve disturbing ecosystems (e.g., logging, prescribed fire) or suppressing disturbance (e.g., fire and pest control). These alterations of the disturbance regime may affect ecosystem health and productivity, biological diversity, and aesthetics.
3. Disturbances are not all equivalent, but differ in spatial parameters (e.g., size and shape), temporal parameters (e.g., return interval and seasonality), magnitude (e.g., proportion of living biomass removed), and specificity (relative impacts on different species and size classes). Historical, observational, experimental, and simulation approaches have been used to document disturbance regimes and effects. Considerable geographic variation in disturbance regimes, even within single vegetation types, suggests the need for site-specific data.
4. Climate change, the pervasive influence of human cultures, and low-frequency events make it difficult to fully document disturbance regimes. Because disturbance regimes are ultimately coupled to climate, differences between past and present climates may limit the application of historic disturbance data to present situations. Human-induced changes in climate and atmospheric chemistry are likely to further complicate the situation in the future. Historical and prehistorical land-use have affected disturbance regimes in both direct (burning and clearing) and indirect (e.g., changes in landscape structure) ways. Disturbance regimes are not necessarily stable, and any demonstration of stability is contingent on temporal scale. For these reasons, we should be careful in our interpretation of and reliance on "natural," "original," or historical (e.g., presettlement) disturbance regimes.
5. Community and ecosystem responses to disturbance reflect the impacts of the disturbance on resource availability and the structure of the post-disturbance community. "Biological legacies" (e.g., standing live trees, coarse woody debris, and buried seeds which remain following disturbance) often play an important role in response. Patch structure and dynamics change as a function of successional time.
6. Species show a range of responses to disturbance. Some require open, early-successional conditions, some require mature communities, and some can survive in both early and late-successional habitats. Increases or decreases in disturbance levels relative to historical conditions may threaten biological diversity.
7. In many ecosystems, the occurrence and outcome of disturbance depend on the history of previous disturbances. These synergisms occur at the levels of individual organisms, patches, and landscapes and may profoundly influence parameters such as species richness, community composition, and the likelihood and intensity of fires and insect outbreaks.
8. The effects of 20th-century fire suppression vary considerably between ecosystems. In vegetation types with historical disturbance regimes dominated by high-intensity, stand-regenerating fires, fire suppression may lead to an increase in fire size but otherwise have little effect on ecosystem dynamics. In some vegetation types (e.g., ponderosa pine) in which the historical disturbance regime was characterized by frequent, low intensity fires, fire suppression may lead to increases in stand density and fuel levels; as a result, likelihood of high-intensity fire may increase. In other frequent-fire systems, fire suppression may lead to replacement of fire-prone vegetation (e.g., longleaf pine and wiregrass) with a fire-resistant type (closed canopy mesic forest).
9. Disturbance interacts with climate, topography, geology, and successional processes to create landscape-level patterns. The size, shape, and spatial configuration of patches in a landscape, in turn, influence the initiation and spread of disturbance and the dynamics of ecosystem response. Disturbances may increase or decrease landscape heterogeneity. Large, continuous tracts of susceptible vegetation promote the spread of fires and insect outbreaks. Other disturbances, such as windthrow, may increase in more heterogeneous landscapes. Landscape patterns influence wildlife populations (e.g., area-sensitive forest interior species, "edge" species) and ecological processes (e.g., seed dispersal, spread of exotics).

10. The spatial and temporal characteristics of a disturbance regime have important implications for the existence and nature of equilibrium. Steady-state equilibrium is most likely to occur where disturbances are small and infrequent relative to recovery time and total landscape size. While landscapes with large or frequent disturbances may not be in quantitative equilibrium, they may nonetheless exhibit qualitative or stable-trajectory equilibrium. Habitat loss and fragmentation alter the scales of landscapes, making it more difficult to preserve species and maintain historical disturbance regimes.

11. Applying disturbance ecology to ecosystem management will require investigation of the relation between climate and disturbance regime, the effects of Native Americans, the influence of habitat fragmentation on the spatial and temporal characteristics of disturbance, the relationship between exotic species invasions and disturbance, and the relation of ecological variation and resilience.

12. The study of natural disturbances provides guidelines for balancing commodity production with the maintenance of ecosystem health and biological diversity. However, our ability to incorporate disturbance ecology into ecosystem management is currently restricted by the limitations of historic data, a lack of site-specific information, and the inherently stochastic nature of disturbance occurrence and response. The current gaps in our knowledge about disturbance suggest that we treat human management as an iterative, adaptive process linked to long-term monitoring and research.

## ACKNOWLEDGMENTS

We wish to thank Nate Stephenson and Jimmie Chew, who participated in the workshop presentation on which this paper is based. We also thank the following colleagues for useful discussions of this subject: Tom Swetnam, Greg Aplet, Beverly Collins, Dave Parsons, Peter Landres, and Joan Walker.

## REFERENCES

Abrams, M.D. 1992. Fire and the development of oak forests. *BioScience* 42: 346–353.

Abrams, M.D. and G.J. Nowacki. 1992. Historical variation in fire, oak recruitment, and post-logging accelerated succession in central Pennsylvania. *Bulletin of the Torrey Botanical Club* 119: 19–28

Agee, J.K. 1993. *Fire Ecology of Pacific Northwest Forests.* Island Press, Washington, D.C.

Agee, J.K., and M.H. Huff. 1987. Fuel succession in a western hemlock/Douglas-fir forest. *Canadian Journal of Forest Research* 17: 697–704.

Allen, T.F.H., and T.B. Starr. 1982. *Hierarchy: Perspectives for Ecological Complexity.* University of Chicago Press, Chicago, IL.

Anderson, L., C.E. Carlson, and R.H. Wakimoto. 1987. Forest fire frequency and western spruce budworm outbreaks in western Montana. *Forest Ecology and Management* 22: 251–260.

Arno, S.F., M.G. Harrington, C.E. Fiedler, and C.E. Carlson. 1995. Restoring fire-dependent ponderosa pine forests in western Montana. *Restoration & Management Notes* 13: 32–36.

Baker, W.L. 1989a. Landscape ecology and nature reserve design in the Boundary Waters Canoe Area, Minnesota. *Ecology* 70: 23–35.

Baker, W.L. 1989b. Effects of scale and spatial heterogeneity on fire-interval distributions. *Canadian Journal of Forest Research* 19: 700–706.

Baker, W.L. 1992a. Effects of settlement and fire suppression on landscape structure. *Ecology* 73: 1879–1887.

Baker, W.L. 1992b. The landscape ecology of large disturbances in the design and management of nature reserves. *Landscape Ecology* 7: 181–194.

Baker, W.L. 1994. Restoration of landscape structure altered by fire suppression. *Conservation Biology* 8: 763–769.

Baker, W.L. 1995. Longterm response of disturbance landscapes to human intervention and global change. *Landscape Ecology* 10: 143–159.

Barden, L.S. 1988. Recruitment and survival of a *Pinus pungens* population: equilibrium or non-equilibrium? *American Midlands Naturalist* 119: 253–257.

Barden, L.S., and F.W. Woods. 1976. Effects of fire on pine and pine-hardwood forests in the southern Appalachians. *Forest Science* 22: 399–403.

Barrett, S.W. 1994. Fire regimes on andesitic mountain terrain in northeastern Yellowstone National Park, Wyoming. *International Journal of Wildland Fire* 4: 65–76.

Barrett, S.W., S.F. Arno, and C.H. Key. 1991. Fire regimes of western larch-lodgepole pine forests in Glacier National Park, Montana. *Canadian Journal of Forest Research* 21: 1711–1720.

Bergeron, Y., and J. Brisson. 1990. Fire regime in red pine stands at the northern limit of the species' range. *Ecology* 71: 1352–1364.

Betancourt, J.L., E.A. Pierson, K. Aasen-Rylander, J.A. Fairchild-Parks, and J.S. Dean. 1993. Influence of history and climate on New Mexico pinyon-juniper woodlands. In: Aldon, E.F. and Shaw, D.W. (eds.), Managing Pinyon–Juniper Ecosystems for Sustainability and Social Needs. Proceedings of the symposium April 26–30, Santa Fe, New Mexico. Gen. Tech. Rep. RM-236. USDA Forest Service, Rocky Mountain Forest & Range Exp. Station, Fort Collins, CO, pp. 42–62.

Betancourt, J.L., and T.R. Van Devender. 1981. Holocene vegetation in Chaco Canyon, New Mexico. *Science* 214: 658–660.

Billings, W.D. 1990. *Bromus tectorum*, a biotic cause of ecosystem impoverishment in the Great Basin. pp. 301–322. In:

G.M. Woodwell (ed.), *The Earth in Transition.* Cambridge University Press, Cambridge.

Biswell, H.H. 1974. Effects of fire on chaparral. pp. 321–364. In: T.T. Kozlowski and C.E. Ahlgren (eds.), *Fire and Ecosystems.* Academic Press, New York.

Bodle, M.J., A.P. Ferriter, and D.D. Thayer. 1994. The biology, distribution, and ecological consequences of *Melaleuca quinquenervia* in the Everglades. pp. 341–355. In: S. Davis and J. Ogden (eds.), *Everglades: The Ecosystem and its Restoration.* St. Lucia Press.

Bormann, F.H., and G.E. Likens. 1979. Catastrophic disturbance and the steady state in northern hardwood forests. *American Scientist* 67: 660–669.

Botkin, D.B. 1990. *Discordant Harmonies: A New Ecology for the Twenty-first Century.* Oxford University Press, Oxford.

Brokaw, N.V.L. 1985. Gap-phase regeneration in a tropical forest. *Ecology* 66: 682–687.

Busing, R.T., and P.S. White. 1993. Effects of area on old-growth forest attributes: implications for the equilibrium landscape concept. *Landscape Ecology* 8: 119–126.

Busing, R.T., and P.S. White. 1997. Species diversity and small-scale disturbance in an old-growth temperate forest: a consideration of gap partitioning concepts. *Oikos* 78: 562–568.

Busing, R.T., E.E.C. Clebsch, and P.S. White. 1993. Biomass and production of southern Appalachian cove forests reexamined. *Canadian Journal of Forest Research* 23: 760–765.

Canham, C.D., J.S. Denslow, W.J. Platt, J.R. Runkle, T.A. Spies, and P.S. White. 1990. Light regimes beneath closed canopies and treefall gaps in temperate and tropical forests. *Canadian Journal of Forest Research* 20: 620–631.

Carlton, J.T., J.K. Thompson, L.E. Schemel, and F.H. Nichols. 1990. The remarkable invasion of San Francisco Bay (California, USA) by the Asian clam Potamorcorbula amurensis, I. Introduction and dispersal. *Marine Ecology Progress Series* Vol. 66, pp. 81–94.

Chabrek, R.H., and A.W. Palmisano. 1973. The effects of Hurricane Camille on the marshes of the Mississippi River delta. *Ecology* 54: 1118–1123.

Christensen, N.L. 1985. Shrubland fire regimes and their evolutionary consequences. pp. 86–100. In: S.T.A. Pickett and P.S. White (eds.), *The Ecology of Natural Disturbance and Patch Dynamics.* Academic Press, New York.

Christensen, N.L., J.K. Agee, P.F. Brussard, J. Hughes, D.H. Knight, G.W. Minshall, J.M. Peek, S.J. Pyne, F.J. Swanson, J.W. Thomas, S. Wells, S.E. Williams, and H.A. Wright. 1989. Interpreting the Yellowstone Fires of 1988. *Bioscience* 39: 678–685.

Christensen, N.L., A.M. Bartuska, J.H. Brown, S. Carpenter, C. D'Antonio, R. Francis, J.F. Franklin, J.A. MacMahon, R.F. Noss, D.J. Parsons, C.H. Peterson, M.G. Turner, and R.G. Woodmansee. 1996. The report of the Ecological Society of America committee on the scientific basis for ecosystem management. *Ecological Applications* 6: 665–691.

Christensen, N.L. and R.K. Peet. 1984. Convergence during secondary succession. *Journal of Ecology* 72: 25–36.

Churchill, E.D., and H.C. Hanson. 1958. The concept of climax in arctic and alpine vegetation. *Botanical Reviews* 24: 127–191.

Clark, J.S. 1988. Effect of climate change on fire regimes in northwestern Minnesota. *Nature* 334: 233–235.

Clark, J.S., and P.D. Royall. 1996. Local and regional sediment charcoal evidence for fire regimes in presettlement north-eastern North America. *Journal of Ecology* 84: 365–382.

Clements, F.E. 1916. *Plant Succession: An Analysis of the Development of Vegetation.* Carnegie Institute of Washington, Washington, D.C., 512 pp.

Clinton, B.D., L.R. Boring, and W.T. Swank. 1994. Regeneration patterns in canopy gaps of mixed-oak forests in the southern Appalachians: Influences of topographic position and evergreen understory. *American Midland Naturalist* 132: 308–319.

Coblentz, B.E. 1990. Exotic organisms: a dilemma for conservation biology. *Conservation Biology* 4: 261–265.

Collins, S.L. 1987. Interactions of disturbance in tallgrass prairie: A field experiment. *Ecology* 68: 1243–1250.

Connell, J.H. 1978. Diversity in tropical rain forests and coral reefs. *Science* 199: 1302–1310.

Connell, J.H., and M.J. Keough. 1985. Disturbance and patch dynamics of subtidal marine animals on hard substrata. pp. 125–151. In: S.T.A. Pickett and P.S. White (eds.), *The Ecology of Natural Disturbance and Patch Dynamics.* Academic Press, New York.

Connell, J.H., and W.P. Sousa. 1983. On the evidence needed to judge ecological stability or persistence. *American Naturalist* 121: 789–824.

Cooper, W.S. 1913. The climax forest of Isle Royale, Lake Superior, and its development. *Botanical Gazette* 55: 1–44.

Cooper, W.S. 1926. The fundamentals of vegetation change. *Ecology* 7: 391–413.

Coppock, D.L., J.E. Ellis, J.K. Detling, and M.I. Dyer. 1983. Plant–herbivore interactions in a North American mixed-grass prairie II: Responses of bison to modification of vegetation by prairie dogs. *Oecologia* 56: 10–15.

Covington, W.W., and M.M. Moore. 1994. Southwestern ponderosa pine forest structure: Changes since Euro-American settlement. *Journal of Forestry* 92: 39–47.

Cowles, H.C. 1899. The ecological relations of the vegetation on the sand dunes of Lake Michigan. *Botanical Gazette* 27: 95–117; 167–201; 281–308; 361–391.

Dahm, C.N., K.W. Cummins, H.M. Valett, and R.L. Coleman. 1995. An ecosystem view of the restoration of the Kissimmee River. *Restoration Ecology* 3: 225–238.

Dansereau, P. and Y. Bergeron. 1993. Fire history in the southern boreal forest of northwestern Quebec. *Canadian Journal of Forest Research* 23: 25–32.

Darwin, C. 1859. *On the Origin of Species by Means of Natural Selection.* John Murray, London.

Daughtrey, M.L., and C.R. Hibben. 1994. Dogwood anthracnose: a new disease threatens two native *Cornus* species. *Annual Review of Phytopathology* 32: 61–73.

Davis, M.B. 1981. Quaternary history and the stability of forest communities. pp. 132–153. In: D.C. West, H.H. Shugart, and D.B. Botkin (eds.), *Forest Succession: Concepts and Application.* Springer-Verlag, New York.

Dayton, P.K. 1971. Competition, disturbance, and community organization: the provision and subsequent utilization of space in a rocky intertidal community. *Ecological Monographs* 41: 351–389.

DeAngelis, D.L., and J.C. Waterhouse. 1987. Equilibrium and nonequilibrium concepts in ecological models. *Ecological Monographs* 57: 1–21.

DeAngelis, D.L., and P.S. White. 1994. Ecosystems as products of spatially and temporally varying driving forces, ecological processes, and landscapes — a theoretical perspective. Chapter 2, pp. 9–28. In: S. Davis and J. Ogden (eds.), *Everglades: The Ecosystem and its Restoration*. St. Lucia Press.

Delcourt, P.A., H.R. Delcourt, P.A. Cridlebaugh, and J. Chapman. 1986. Holocene ethnobotanical and paleoecological record of human impact on vegetation in the Little Tennessee River Valley, Tennessee. *Quaternary Research* 25: 330–349.

Drake, J.A., H.A. Mooney, F. di Castri, R.H. Groves, F.J. Kruger, M. Rejmanek, and M. Williamson (eds.) 1989. *Biological Invasions: A Global Perspective. SCOPE 37.* Wiley, New York.

Everham III, E.M. and N.V.L. Brokaw. 1996. Forest damage and recovery from catastrophic wind. *The Botanical Review* 62: 113–185.

Everitt, B.L. 1968. Use of the cottonwood in an investigation of the recent history of a flood plain. *American Journal of Science* 266: 417–439.

Flaccus, E. 1959. Revegetation of landslides in the White Mountains of New Hampshire. *Ecology* 40: 692–703.

Flamm, R.O., P.E. Pulley, and R.N. Coulson. 1993. Colonization of disturbed trees by the southern pine bark beetle guild (Coleoptera: Scolytidae). *Environmental Entomology* 22: 62–70.

Flannigan, M.D. and C.E. Van Wagner. 1991. Climate change and wildfire in Canada. *Canadian Journal of Forest Research* 21: 66–72.

Forcier, L.K. 1975. Reproductive strategies and the co-occurrence of climax tree species. *Science* 189: 808–810.

Forman, R.T.T. 1990. Ecologically sustainable landscapes: the role of spatial configuration. pp. 261–278. In: Zonneveld, I.S., and R.T.T. Forman (eds.), *Changing Landscapes: An Ecological Perspective*. Springer-Verlag, New York.

Forman, R.T.T. and M. Godron. 1986. *Landscape Ecology*. Wiley, New York, 619 pp.

Foster, D.R. 1988. Species and stand response to catastrophic wind in central New England, U.S.A. *Journal of Ecology* 76: 135–151.

Foster, D.R., and E. Boose. 1992. Patterns of forest damage from catastrophic wind in central New England. *Journal of Ecology* 80: 79–98.

Foster D.R. and T.M. Zebryk. 1993. Long-term vegetation dynamics and disturbance history of a Tsuga-dominated forest in New England. *Ecology* 74: 982–998.

Franklin, J.F. 1989. Towards a new forestry. *American Forests* November/December: 37–44.

Franklin, J.F. and R.T.T. Forman. 1987. Creating landscape patterns by forest cutting: Ecological consequences and principles. *Landscape Ecology* 1: 5–18.

Franklin, J.F., J.A. MacMahon, F.J. Swanson, and J.R. Sedell. 1985. Ecosystem responses to catastrophic disturbances: lessons from Mount St. Helens. *Nat. Geogr. Research* 1: 198–216.

Frost, C.C. 1993. Four centuries of changing landscape patterns in the longleaf pine ecosystem. *Proceedings of the Tall Timbers Fire Ecology Conference* 18: 17–43.

Geiszler, D.R., R.I. Gara, C.H. Driver, V.F. Gallucci, and R.E. Martin. 1980. Fire, fungi, and beetle influences on a lodgepole pine ecosystem of south-central Oregon. *Oecologia* 46: 239–243.

Graham, N.E. 1995. Simulation of recent global temperature trends. *Science* 267: 666–671.

Groves, R.H., and J.J. Burdon (eds.). 1986. *Ecology of Biological Invasions*. Cambridge University Press, London.

Hagle, S. and R. Schmitz. 1993. Managing root disease and bark beetles. pp. 209–228. In: T.D. Schowalter and G.M. Filip (eds.), *Beetle-Pathogen Interactions in Conifer Forests*. Academic Press, San Diego.

Hansen, A.J., T.A. Spies, F.J. Swanson, and J.L. Ohman. 1991. Conserving biodiversity in managed forests: lessons from natural forests. *Bioscience* 41: 382–392.

Harmon, M.E. 1982. The fire history of the westernmost portion of Great Smoky Mountains National Park. *Bulletin of the Torrey Botanical Club* 109: 74–79.

Harmon, M.E. 1984. Survival of trees after low-intensity surface fires in Great Smoky Mountains National Park. *Ecology* 65: 796–802.

Harmon, M.E., S.P. Bratton, and P. S.White. 1983. Disturbance and vegetation response in relation to environmental gradients in the Great Smoky Mountains. *Vegetatio* 55: 129–139.

Harmon, M.E., J.F. Franklin, F.J. Swanson, P. Sollins, S.V. Gregory, J.D. Lattin, N.H. Anderson, S.P. Cline, N.G. Aumen, J.R. Sedell, G.W. Lienkaemper, K. Cromack, Jr., and K.W. Cummins. 1986. Ecology of coarse woody debris in temperate ecosystems. *Advances in Ecological Research* 15: 133–302.

Heinselman, M.L. 1973. Fire in the virgin forests of the Boundary Waters Canoe Area, Minnesota. *Quaternary Research* 3: 329–382.

Heinselman, M.L. 1981. Fire and succession in the conifer forests of northern North America. pp. 374–405. In: D.C. West, H.H. Shugart, and D.B. Botkin (eds.), *Forest Succession: Concepts and Application*. Springer-Verlag, New York.

Hemphill, N., and S.D. Cooper. 1983. The effect of physical disturbance on the relative abundances of two filter-feeding insects in a small stream. *Oecologia* 58: 378–382.

Henry, J.D., and J.M.A. Swan. 1972. Reconstructing forest history from live and dead plant material — an approach to the study of forest succession in southwest New Hampshire. *Ecology* 55: 772–783.

Hobbs, N.T., D.S. Schimel, C.O. Owensby, and D.S. Ojima. 1991. Fire and grazing in the tallgrass prairie. Contingent effects on nitrogen budgets. *Ecology* 72: 1374–1382.

Hough, A.F., and R.D. Forbes. 1943. The ecology and silvics of trees in the high plateaus of Pennsylvania. *Ecological Monographs* 13: 299–320.

Hunter, M.L. 1993. Natural fire regimes as spatial models for managed boreal forests. *Biological Conservation* 65: 115–120.

Hupp, C.R. 1983. Seedling establishment on a landslide site. *Castanea* 48: 89–98.

Huston, M. 1979. A general hypothesis of species diversity. *The American Naturalist* 113: 81–101.

Huston, M.A. 1997. Hidden treatments in ecological experiments: re-evaluating the ecosystem function of biodiversity. *Oecologia* 110: 449–460.

Ives, R.L. 1942. The beaver-meadow complex. *Journal of Geomorphology* 5: 191–203.

Johnson, E.A., and C.E. Van Wagner. 1985. The theory and use of two fire history models. *Canadian Journal of Forest Research* 15: 214–220.

Johnson, E.A., and S.L. Gutsell. 1994. Fire frequency models, methods and interpretations. *Advances in Ecological Research* 25: 239–287.

Johnson, P.L., and W.D. Billings. 1962. The alpine vegetation of the Beartooth Plateau in relation to cryopedogenic processes and patterns. *Ecological Monographs* 32: 102–135.

Kaufmann, M.R., R.T. Graham, D.A. Boyce, Jr., W.H. Moir, L. Perry, R.T. Reynolds, R.L. Bassett, P. Melhop, C.B. Edminster, W.M. Block, and P.S. Corn. 1994. An ecological basis for ecosystem management. USDA Forest Service, General Technical Report RM-246.

Keane, R.E., S.F. Arno, and J.K. Brown. 1990. Simulating cumulative fire effects in ponderosa pine/Douglas-fir forests. *Ecology* 7: 189–203.

Kilgore, B.M. and D. Taylor. 1979. Fire history of a seqoia-mixed conifer forest. *Ecology* 60: 129–142.

Knapp, A.K., and T.R. Seastedt. 1986. Detritus accumulation limits productivity of tallgrass prairie. *Bioscience* 36: 662–668.

Knight, D.H. 1987. Parasites, lightning, and the vegetation mosaic in wilderness landscapes. pp. 61–80. In: M.G. Turner (ed.), *Landscape Heterogeneity and Disturbance*. Springer-Verlag, New York.

Komarek, E.V. 1974. Effects of fire on temperate forests and related ecosystems: southeastern United States. pp. 251–277. In: T.T. Kozlowski and C.E. Ahlgren (eds.), *Fire and Ecosystems*. Academic Press, New York.

Kuykendall, N.W., III. 1978. Composition and structure of replacement forest stands following southern pine beetle infestations as related to selected site variables in the Great Smoky Mountains. M.S. Thesis, University of Tennessee, Knoxville. 122 pp.

Lamberti, G.A., S.V. Gregory, L.R. Ashkenas, R.C. Wildman, and K.M.S. Moore. 1991. Stream ecosystem recovery following a catastrophic debris flow. *Canadian Journal Fish Aquat. Science* 48: 196–208.

Landres, P.B. 1992. Temporal scale perspectives in managing biological diversity. *Transactions of the North American Wildlife and Natural Resources Conference* 57: 292–307.

Landres, P.B., P.S. White, G. Aplet, and A. Zimmerman. Naturalness and natural variability: definitions, concepts, and strategies for wilderness management. *Proceedings of the Eastern Wilderness Conference*, June, 1996, Gatlinburg, TN, in press.

Lang, G.E. 1985. Forest turnover and the dynamics of bole wood litter in subalpine balsam fir forest. *Canadian Journal of Forest Research* 15: 262–268.

Lemon, P.C. 1961. The forest ecology of ice storms. *Bulletin of the Torrey Botanical Club* 88: 21–29.

Leopold, A. 1966. Cheat takes over. pp. 164–168. In: *A Sand County Almanac*, with Essays on Conservation from Round River. Ballantine Books, New York.

Lorimer, C.G. 1980. Age structure and disturbance history of a southern Appalachian virgin forest. *Ecology* 61: 1169–1184.

Lorimer, C.G. 1985. Methodological considerations in the analysis of forest disturbance history. *Canadian Journal of Forest Research* 15: 200–213.

Lorimer, C.G. and L.E. Frelich. 1989. A methodology for estimating canopy disturbance frequency and intensity in dense temperate forests. *Canadian Journal of Forest Research* 19: 651–663.

Loucks, O.L. 1970. Evolution of diversity, efficiency, and community stability. *American Zoologist* 10: 17–25.

Mack, R.N. 1981. Invasion of Bromus tectorum L. into western North America: an ecological chronicle. *Agro-Ecosystems* 7: 145–165.

Mann, C.C., and M.L. Plummer. 1995. Are wildlife corridors the right path? *Science* 270: 1428–1430.

Marks, P.L. 1974. The role of pin cherry (*Prunus pensylvanica* L.) in the maintenance of stability in northern hardwood ecosystems. *Ecological Monographs* 44: 73–88.

Matlack, G.R., S.K. Gleeson, and R.E. Good. 1993. Treefall in a mixed oak-pine coastal plain forest: Immediate and historical causation. *Ecology* 74: 1559–1566.

McDonnell, M.J., and E.W. Stiles. 1983. The structural complexity of old field vegetation and the recruitment of bird-dispersed plant species. *Oecologia* 56: 109–116.

McLellan, B.N. and D.M. Shackleton. 1988. Grizzly bears and resource-extraction industries: Effects of roads on behaviour, habitat use and demography. *Journal of Applied Ecology* 25: 451–460.

Meier, A.J., S.P. Bratton, and D.C. Duffy. 1995. Possible ecological mechanisms for loss of vernal-herb diversity in logged eastern deciduous forests. *Ecological Applications* 5: 935–946.

Minnich, R.A. 1983. Fire mosaics in southern California and northern Baja California. *Science* 219: 1287–1294.

Mooney, H.A., and J.A. Drake (eds.) 1986. *Ecology of biological invasions of North America and Hawaii*. Ecological Studies 58. Springer-Verlag, New York.

Morgan, P., G.H. Aplet, J.B. Haufler, H.C. Humfries, M.M. Moore, and W.D. Wilson. 1994. Historical range of variability: A useful tool for evaluating ecosystem change. pp. 87–111. In: R.N. Sampson and D.L. Adams (eds.), *Assessing Forest Ecosystem Health in the Inland West*. The Haworth Press.

Motzkin, G., D. Foster, A. Allen, J. Harrod, and R. Boone. 1996. Controlling site to evaluate history: Vegetation patterns of a New England sand plain. *Ecological Monographs* 66: 345–365.

Nichols, F.H., J.K. Thompson, and L.E. Schemel. 1990. The remakable invasionof San Francisco Bay (California, USA) by the Asian clam Potamorcorbula amurensis, II. Displacement of a former community. *Marine Ecology Progress Series*, v. 66, pp. 95–101.

Oliver, C.D. 1980. Forest development in North America following major disturbances. *Forest Ecology and Management* 3: 157–168.

Olson, J.S. 1958. Rates of succession and soil changes on southern Lake Michigan sand dunes. *Botanical Gazette* 119: 125–170.

O'Neill, R.V., D.L. DeAngelis, J.B. Waide, and T.F.S. Allen. 1986. *A Hierarchical Concept of Ecosystems*. Princeton University Press, Princeton, NJ.

Overpeck, J.T., D. Rind, and R. Goldberg. 1990. Climate-induced changes in forest disturbance and vegetation. *Nature* 343: 51–53.

Paine, R.D., and S.A. Levin. 1981. Intertidal landscapes: disturbances and the dynamics of pattern. *Ecological Monographs* 51: 145–178.

Paine, T.D. and F.A. Baker. 1993. Abiotic and biotic predisposition. pp. 61–79. In: T.D. Schowalter and G.M. Filip (eds.), *Beetle–Pathogen Interactions in Conifer Forests*. Academic Press, San Diego.

Palmer, M.W., and P.S. White. 1994. Scale dependence and the species-area relationship. *The American Naturalist* 144: 717–740.

Peet, R.K. 1992. Community structure and ecosystem function. pp. 103–151. In: D.C. Glenn-Lewin, R.K. Peet, and

T.T. Veblen (eds.), *Plant Succession: Theory and Prediction.* Chapman and Hall, London.

Peet, R.K. and N.L. Christensen. 1987. Competition and tree death. *Bioscience* 37: 586–595.

Peet, R.K., D.C. Glenn-Lewin, and J.W. Wolf. 1983. Prediction of man's impact on plant species diversity: A challenge for vegetation science. In: W. Holzner, M.J.A. Werger, and I. Ikusima (eds.), *Man's Impact on Vegetation.* Dr. W. Junk, Boston.

Pickett, S.T.A., and J.N. Thompson. 1978. Patch dynamics and the design of nature reserves. *Biological Conservationist* 13: 27–37.

Pickett, S.T.A., and P.S. White, (eds.). 1985. *The Ecology of Natural Disturbance and Patch Dynamics.* Academic Press, New York. 496 p.

Pinder, J.E. III, F.B. Golley, and R.F. Lide. 1995. Factors affecting limited reproduction by loblolly pine in a large old field. *Bulletin of the Torrey Botanical Club* 122: 306–311.

Pringle, C.M., R.J. Naiman, G. Bretschko, J.R. Karr, M.W. Oswood, J.R. Webster, R.L. Welcomme, and M.J. Winterbourn. 1988. Patch dynamics in lotic systems: the stream as a mosaic. *Journal of the North American Benthological Society* 7: 503–524.

Pycek, P., K. Prach, M. Rejmanek, and M. Wade (eds.). 1995. *Plant Invasions: General Aspects and Special Problems.* SPB Academic, Amsterdam.

Raup, H.M. 1941. Botanical problems in boreal America. *Botanical Review* 7: 147–248.

Raup, H.M. 1975. Species versatility in shore habitats. *Journal Arnold Arboretum Harvard University* 55: 126–165.

Reed, R.A., R.K. Peet, M.W. Palmer, and P.S. White (1993). Scale dependence of vegetation-environment correlations: a case study of a North Carolina piedmont woodland. *Journal Vegetation Science* 4: 329–340.

Reiners, W.A. 1983. Disturbance and basic properties of ecosystem energetics. pp. 83–98. In: H.A. Mooney and M. Godrop (eds.), *Disturbance and Ecosystems: Components of Response.* Springer, New York.

Resh, V.H., A.V. Brown, A.P. Covich, M.E. Gurtz, H.W. Li, G.W. Minshall, S.R. Reice, A.L. Sheldon, J.B. Wallace, and R.C. Wissmar. 1988. The role of disturbance in stream ecology. *Journal of the North American Benthological Society* 7: 433–455.

Risser, P.G. 1987. Landscape ecology: state of the art. pp. 3–14. In: Turner, M.G. (editor), *Landscape Heterogeneity and Disturbance.* Springer-Verlag, New York.

Romme, W.H. 1982. Fire and landscape diversity in subalpine forests of Yellowstone National Park. *Ecological Monographs* 52: 199–221.

Romme, W.H., and D.G. Despain. 1989. Historical perspective on the Yellowstone fires of 1988. *BioScience* 39: 695–699.

Romme, W.H. and D.H. Knight. 1981. Fire frequency and subalpine forest succession along a topographic gradient in Wyoming. *Ecology* 62: 319–326.

Romme, W.H., and M.G. Turner. 1991. Implications of global climate change for biogeographic patterns in the Greater Yellowstone ecosystem. *Conservation Biology* 5: 373–386.

Runkle, J.R. 1982. Patterns of disturbance in some old-growth mesic forests of eastern North America. *Ecology* 63: 1533–1546.

Runkle, J.R. 1985. Disturbance regimes in temperate forests. pp. 17–34. In: S.T.A. Pickett and P.S. White (eds.), *The Ecology of Natural Disturbance and Patch Dynamics.* Academic Press. New York.

Rykiel, E.J. Jr., R.N. Coulson, P.J.H. Sharpe, T.H.F. Allen, and R.O. Flamm, 1988. Disturbance propagation by beetles as an episodic landscape phenomenon. *Landscape Ecology* 1: 129–139.

Samuels, M., and J.L. Betancourt. 1982. Modeling the long-term effects of fuelwood harvests on pinyon–juniper woodlands. *Environmental Management* 6: 505–515.

Schoenwald-Cox, C. and M. Buechner. 1992. Park protection and public roads. pp. 373–395. In: P.L. Fiedler and S.K. Jain (eds.), *Conservation Biology.* Chapman and Hall, New York.

Schowalter, T.D. 1985. Adaptations of insects to disturbance. pp 235–252. In: S.T.A. Pickett and P.S. White (eds.), *The Ecology of Natural Disturbance and Patch Dynamics.* Academic Press, New York.

Schowalter, T.D., R.N. Coulson, and D.A. Crossly, Jr. 1981. Role of southern pine beetle and fire in maintenance and structure of the southeastern coniferous forest. *Environmental Entomology* 10: 821–825.

Schowalter, T.D., and G.M. Filip. 1993. Bark beetle–pathogen–conifer interactions: An overview. pp. 3–19. In: T.D. Schowalter and G.M. Filip (eds.), *Beetle–Pathogen Interactions in Conifer Forests.* Academic Press, San Diego.

Schroeder, P.M., R. Dular, and B.P. Hayden. 1976. Vegetation changes associated with barrier dune construction on the Outer Banks of North Carolina. *Environmental Management* 1: 105–114.

Seischab, F.K. and D. Orwig.1991. Catastrophic disturbances in the presettlement forests of western New York. *Bulletin of the Torrey Botanical Club* 118: 117–122.

Shipley, B., P.A. Keddy, C. Gaudet, and D.R.J. Moore. 1991. A model of species density in shoreline vegetation. *Ecology* 72: 1658–1667.

Shugart, H.H. 1984. *A Theory of Forest Dynamics.* Springer, New York. Silberbauer-Gottsberger, I., W. Morawetz, and G. Gottsberger. 1977. Frost damage of Carrado plants in Botucatu, Brazil, as related to the geographic distribution of species. *Biotropica* 9: 253–261.

Sousa, W.P. 1984. The role of disturbance in natural communities. *Annual Review of Ecology Systematics* 15: 353–391.

Sousa, W.P. 1985. Disturbance and patch dynamics on rocky intertidal shores. pp. 101–124. In: S.T.A. Pickett and P.S. White, *The ecology of Natural Disturbance and Patch Dynamics.* Academic Press, New York.

Sprugel, D.G. 1976. Dynamic structure of wave-regenerated Abies balsamea forests in the north-eastern United States. *Journal of Ecology* 64: 889–911.

Stephenson, N.L. 1996. Ecology and management of giant sequoia groves. pp. 1431–1467. In: Sierra Nevada Ecosystem Project: Final Report to Congress. University of California, Davis, California. Centers for Water and Wildland Resources.

Stephenson, N.L., D.J. Parsons, and T.W. Swetnam. 1991. *Restoring natural fire to the sequoia-mixed conifer forest: should intense fire play a role?* Proc. 17th Tall Timbers Fire Ecology Conference: pp. 321–337.

Stewart, G.H. 1986. Population dynamics of a montane conifer forest, western Cascade Range, Oregon, USA. *Vegetatio* 76: 79–88.

Swanson, F.J. 1981. Fire and geomorphic processes. pp. 401–420. In: H.A. Mooney, T.M. Bonnicksen, N.L. Christensen,

J.E. Lotan, and W.A. Reiners (eds.), Fire regimes and ecosystem properties. USDA Forest Service, Gen. Tech. Rept. WO-26.

Swanson, F.J. and C.T. Dyrness. 1975. Impact of clear–cutting and road construction on soil erosion by landslides in the western cascade range, Oregon. *Geology* 10: 393–396.

Swanson, F.J., and J.F. Franklin. 1992. New forestry principles from ecosystem analysis of Pacific Northwest forests. *Ecological Applications* 2: 262–274.

Swanson, F.J., J.A. Jones, D.O. Wallin, and J.H. Cissel. 1994. Natural variability — implications for ecosystem management. pp. 80–93. In: M.E. Jensen and P.S. Bourgeron (eds.), Volume II: Ecosystem management: principles and applications. USDA Forest Service, Pacific Northwest Research Station, General Technical Report PNW-GTR-318.

Swanston, D.N., and F.J. Swanson. 1976. Timber-harvesting, mass erosion, and steep land form geomorphology in the Pacific Northwest. pp. 199–221. In: D.R. Croats (ed.), *Geomorphology and Engineering*. Dowden, Hutchinson, and Ross, Inc., Stroudsburg, PA.

Swetnam, T.W. 1993. Fire history and climate change in giant sequoia groves. *Science* 262: 885–889.

Swetnam, T.W., and J.L. Betancourt. 1990. Fire-southern oscillation relations in the southwestern United States. *Science* 249: 1017–1020.

Swetnam, T.W. and A.M. Lynch. 1993. Multicentury, regional-scale patterns of western spruce budworm outbreaks. *Ecological Monographs* 63: 299–424.

Thom, B.G. 1967. Mangrove ecology and geomorphology: Tabasco, Mexico. *Journal of Ecology* 55: 301–343.

Tilman, D. 1996. Biodiversity: Population versus ecosystem stability. *Ecology* 77: 350–363.

Tilman, D., D. Wedin and J. Knops. 1996. Productivity and sustainability influenced by biodiversity in grassland ecosystems. *Nature* 379: 718–720.

Turner, M.G. 1989. Landscape ecology: the effect of pattern on process. *Annual Review of Ecology and Systematics* 20: 171–197.

Turner, M.G., R.H. Gardner, V.H. Dale, and R.V. O'Neill. 1989. Predicting the spread of disturbance across heterogeneous landscapes. *Oikos* 55: 121–129.

Turner, M.G., R.V. O'Neill, R.H. Gardner, and B.T. Milne. 1989. Effects of changing spatial scale on the analysis of landscape pattern. *Landscape Ecology* 3: 153–162.

Turner, M.G., and W.H. Romme. 1994. Landscape dynamics in crown fire ecosystems. *Landscape Ecology* 9: 59–77.

Turner, M.G., W.H. Romme, R.H. Gardner, R.V. O'Neill, and T.K. Kratz. 1993. A revised concept of landscape equilibrium: disturbance and stability on scaled landscapes. *Landscape Ecology* 8: 213–227.

Turner, M.G., Y. Wu, L.L. Wallace, W.H. Romme, and A. Brenkert. 1994. Simulating winter interactions among ungulates, vegetation, and fire in northern Yellowstone Park. *Ecological Application* 4: 472–496.

Veblen, T.T., K.S. Hadley, E.M. Nel, T. Kitzenberger, M. Reid, and R. Villalba. 1994. Disturbance regime and disturbance interactions in a Rocky Mountain subalpine forest. *Journal of Ecology* 82: 125–135.

Vinton, M.A., D.C. Hartnett, E.J. Finck, and J.M. Briggs. 1993. Interactive effects of fire, bison grazing and plant community composition in tallgrass prairie. *American Midland Naturalist* 129: 10–18.

Villalba, R. 1995. Climatic influences on forest dynamics along the forest-steppe ecotone in northern Patagonia. Ph.D. dissertation, University of Colorado, Boulder, 288 p.

Vitousek, P.M. 1984. A general theory of forest nutrient dynamics. pp. 121–135. In: G.I. Agren (ed.), State and change of forest ecosystems—indicators in current research. Swedish University of Agricultural Science, Report Number 13.

Vitousek, P.M. 1990. Biological invasions and ecosystem processes: towards an integration of population biology and ecosystem studies. *Oikos* 57: 7–13.

Vitousek, P.M. 1994. Beyond global warming: ecology and global change. *Ecology* 75: 1861–1876.

Vitousek, P.M., and P.S. White. 1981. Process studies in succession. pp. 262–276. In: D.C. West, H. Shugart, and D. Botkin (eds.), *Forest Succession*. Springer-Verlag. New York.

Vogl, R.J. 1974. Effects of fire on grasslands. pp. 139–194. In: T. T. Kozlowski and C.E. Ahlgren (eds.), *Fire and Ecosystems*. Academic Press. New York.

Walker, J., and R.K. Peet. 1983. Composition and species diversity of pine-wiregrass savannas of the Green Swamp, North Carolina. *Vegetatio* 55: 163–179.

Walker, T.W., and J.K. Syers. 1976. The fate of phosphorous during pedogenesis. *Geoderma* 15: 1–19.

Watt, A.S. 1947. Pattern and process in the plant community. *Journal Ecology* 35: 1–22.

Weaver, J.E. 1968. *Prairie Plants and Their Environment: A 50 Year Study in the Midwest*. University of Nebraska Press, Lincoln.

Webb, R.H. and J.L. Betancourt. 1992. Climatic variability and flood frequency of the Santa Cruz River, Pima County, Arizona: U.S. Geological Survey Water-Supply Paper 2379.

Werner, P.A., and W.J. Platt. 1976. Ecological relationships of co-occurring goldenrods (*Solidago*: Compositae). *American Naturalist* 110: 959–971.

West, N.E. 1988. Intermountain deserts, shrubsteppes and woodlands. pp. 209–230. In: Barbour, M.G., and W.D. Billings (eds.), *North American Terrestrial Vegetation*. Cambridge University Press, Cambridge.

Whisenant, S.G. 1990. Changing fire frequencies on Idaho's Snake River Plains: ecological and management implications. pp. 4–10. In: Proceedings — Symposium on cheatgrass invasion, shrub die-off, and other aspects of shrub biology and management. USDA Forest Service General Technical Report INT-276.

White, P.S. 1979. Pattern, process, and natural disturbance in vegetation. *Botanical Reviews* 45: 229–299.

White, P.S. 1987. Natural disturbance, patch dynamics, and landscape pattern in natural areas. *Natural Areas Journal* 7(1): 14–22.

White, P.S. 1994. Synthesis: vegetation pattern and process in the Everglades ecosystem. Chapter 18, pp. 445–460. In: S. Davis and J. Ogden (eds.), *Everglades: The Ecosystem and its Restoration*. St. Lucia Press.

White, P.S., and S.P. Bratton. 1980. After preservation: the philosophical and practical problems of change. *Biological Conservation* 18: 241–255.

White, P.S., and J. Harrod. 1997. Disturbance and diversity in a landscape context. pp. 128–159. In: J.A. Bissonette (ed.), *A Primer in Landscape Ecology*. Springer-Verlag, New York.

White, P.S., M.D. MacKenzie, and R.T. Busing. 1985a. Natural disturbance and gap phase dynamics in southern Appala-

chian spruce-fir. *Canadian Journal of Forest Research* 15: 233–240.
White, P.S., M.D. MacKenzie, and R.T. Busing. 1985b. A critique of overstory/understory comparisons based on transition probability analysis of an old growth spruce-fir stand in the Appalachians. *Vegetatio* 64: 37–45.
White, P.S. and S.T.A. Pickett. 1985. Natural disturbance and patch dynamics, an introduction. pp. 3–13. In: S.T.A. Pickett and P.S. White (eds.), *The Ecology of Natural Disturbance and Patch Dynamics*. Academic Press. New York.
White, P.S., and J.L. Walker. Approximating nature's variation: selecting and using reference sites and reference information in restoration ecology. *Restoration Ecology*, in press.
Whitney, H.E., and W.C. Johnson. 1984. Ice storms and forest succession in southwest Virginia. *Bulletin of the Torrey Botanical Club* 111: 429–437.
Woods, F.W., and R.E. Shanks. 1959. Natural replacement of chestnut by other species in the Great Smoky Mountains National Park. *Ecology* 40: 349–361.
Wright, K.A., L.M. Chapman, and T.M. Jimerson. 1995. Using historic range of vegetation variability to develop desired conditions and model forest plant alternatives. pp. 258–266. In: Analysis in support of ecosystem management: analysis workshop III. USDA Forest Service, Ecosystem Management Analysis Center, Washington, D.C.
Zachrisson, O. 1977. Influence of forest fires on the North Swedish boreal forest. *Oikos* 29: 22–32.
Zedler, P. 1981. Vegetation change in chaparral and desert communities in San Diego County, California. pp. 406–430. In: D.C. West, H.H. Shugart, and D.B. Botkin (eds.), *Forest succession: concepts and application*. Springer-Verlag, New York.
Zedler, P., and F.G. Goff. 1973. Size-association analysis of forest successional trends in Wisconsin. *Ecological Monographs* 43: 79–94.

## THE AUTHORS

**Peter S. White**
*CB# 3280,*
*Department of Biology,*
*University of North Carolina,*
*Chapel Hill, NC 27599-3280, USA*

**Jonathan Harrod**
*CB# 3280, Curriculum in Ecology,*
*University of North Carolina,*
*Chapel Hill, NC 27599-3280, USA*

**William H. Romme**
*Department of Biology,*
*Fort Lewis College,*
*Durango, CO 81301, USA*

**Julio Betancourt**
*US Geological Survey,*
*Office of Regional Hydrologists,*
*1671 W. Anklam Road,*
*Tucson, AZ 85745, USA*

# Practical Applications of Disturbance Ecology to Natural Resource Management

R. Todd Engstrom, Sam Gilbert, Malcolm L. Hunter, Jr., David Merriwether, Gregory J. Nowacki, and Page Spencer

**Key issues addressed in this chapter**

- ♦ *Disturbance ecology furnishes a valuable conceptual framework for natural resource management.*
- ♦ *Numerous techniques exist for documenting past disturbance regimes and the historic range of variability of key disturbances.*
- ♦ *Management goals should be viewed as motion pictures instead of static snapshots because natural disturbances and a changing environment constantly interact with management.*
- ♦ *Monitoring within an adaptive management model is essential to understanding the effects of disturbance on natural resources.*
- ♦ *Human-initiated disturbances can be designed to mimic aspects of natural disturbance even though they often differ in fundamental ways.*
- ♦ *Although disturbance ecology concepts are relatively new in natural resource management, a wide variety of applications are in progress. These applications provide a fertile source of case studies that should be consulted for development of management plans.*

**Keywords: Disturbance ecology model, disturbance regimes, reference conditions, desired conditions, setting goals, old-field pineland, Glen Canyon Dam, Ouachita mountains, Chattooga Project, Denali mine reclamation**

## 1 INTRODUCTION

Disturbance ecology provides an important conceptual framework for resource management because disturbances are agents of change, and change is intrinsic to the natural world. Natural disturbances (discrete events that disrupt ecosystem structure and change resource availability) alter ecosystem characteristics that matter to managers, including species composition, habitat structure, biological diversity, resource productivity, and incidence of disease (White et al., this volume). Managers respond to and introduce disturbances all the time. Hurricanes, floods, landslides, and fires are inevitable natural disturbances. Timber harvest, prescribed fire, water flow alteration, grazing, road and trail building, and plowing are anthropogenic disturbances that have ecological effects at various temporal and spatial scales. Global alterations of the environment, such as air chemistry or temperature, although not disturbances (see White et al., this volume), can interact with disturbances and profoundly affect land management in ways that are extremely difficult or impossible for a local manager to address.

In addition to their ecological aspects, dramatic disturbances have important human dimensions. Society's responses to disturbances (e.g., public perceptions of the 1998 Yellowstone fires) may have indirect, but critical feedback on management. Public demand for resources — water, timber, public access — can shape land management regardless of ecological considerations. Likewise, public attitudes can alter the harvest or use of natural resources that have variable economic repercussions. For example, reduction of timber harvest in the Pacific Northwest put pressure on timber prices, which caused timber harvest in the Southeast to rise.

This paper is written for all land managers, but particularly those working on public lands. We provide case studies of how knowledge of ecological disturbance has been used by managers for a variety of purposes. These case studies can be used conceptually or for establishing contacts with those involved in the individual examples. Our goals for the paper are: (1) to establish the importance of the ecology of disturbance for ecological stewardship, (2) to provide a conceptual framework and tools for maintaining an appropriate disturbance regime, and (3) to give examples of how principles of disturbance ecology have been used by natural resource managers.

Disturbance ecology is rooted in an old debate over what factors determine community structure and composition and the issues of succession and equilibrium (McIntosh 1992, Reice 1994, Rogers 1996). Disturbance has long been recognized in studies of ecological communities, but often as a setback to succession that moves communities toward equilibrium conditions (Pickett and White 1985). Integration of disturbance ecology into resource management — designing management activities based on a natural disturbance regime — is a recent development. The attractiveness of disturbance ecology to resource management has been enhanced, in part, by the accumulation of examples of undesirable ecological consequences — particularly loss of biological diversity and productivity — caused by some management actions. Use of disturbance ecology in natural resource management is grounded in the idea that an indigenous biota is adapted to disturbance conditions that developed over thousands of years, and increased deviation from those conditions risks species loss and other undesirable ecological change (Swanson et al. 1994).

A steady stream of examples of failed human attempts to control ecosystems has forced recognition of the inherent variability within ecosystems (Holling and Meffe 1996). Efforts to control water flow in the Everglades contributed to declines in wading bird populations, hypersalinity of Florida Bay, phosphorus enrichment, and invasion of exotic plants (Davis and Ogden 1994, Light et al. 1995). Fire suppression in the western United States has led to tree invasion of grasslands, shifts in species composition and structure, and increased susceptibility to insect infestations and catastrophic fire (Arno and Gruell 1986, Hoff and Hagle 1990, Covington and Moore 1994). Loss of biological diversity is a common result of reduction of ecosystem variability. It has generated interest in models, such as coarse- and fine-filter approaches, for natural resource management that maintain diversity in an efficient manner (Noss 1987, Hunter 1990). Knowledge of how disturbance shapes ecosystems can be used to understand better the limits of an ecosystem.

## 2 THE DISTURBANCE ECOLOGY MODEL

A model for using disturbance ecology in natural resource management is emerging. A key step in this model is to understand the disturbance regime (including human influences) that interacted with and helped shape the ecosystem. For managers this may be complicated because multiple types and intensities of disturbance may generate complex patterns in time and space. Frequently, reference conditions and the historical range of variability are used to provide targets and boundaries, respectively, for management. Implicit in range of variability are the crucial roles of the temporal and spatial scales of ecosystems and the idea that an ecosystem has limits.

Disturbances affect ecosystems over various scales in time and space, and these scales can be arranged hierarchically. Holling (1995) used a simple graphical display to contrast time and space dimensions of structure and disturbance of a boreal forest (Fig. 1). Climate change can alter a 1,000 $km^2$ biome completely over 10,000 years. In contrast, a lightning strike occurs in a fraction of a second and has a direct effect on 1 to 10 $m^2$. This graph, coupled with a third axis of frequency, could be used to place management actions within the context of different types of natural disturbance. Special care should be taken to "adjust the scales of management to those of natural processes, insofar as economic, social, and political constraints permit" (Wiens 1997). This provides the ecological basis for a conservative approach to land management (Attiwill 1994).

## 2.1 Ecosystem Limits

The concept that an ecosystem has limits or thresholds is based in complexity theory. This theory holds that ecosystems, like self-organizing systems, may "change rapidly [into a new state] when their bounds of adaptability are exceeded" (Perry and Amaranthus 1997). Crossing the threshold has the critical feature that the new state is self-reinforcing, which means that return to the old state may be difficult or impossible. Alterations of the scale, intensity, frequency, or novelty of disturbances are catalysts for exceeding limits. An example of an ecosystem that has been altered past a threshold is the old-field pine forest of the southeastern United States.

### 2.1.1 Old-field Pineland

Land-use history of Tall Timbers Research Station (1,600 ha) north of Tallahassee, Florida, followed a pattern similar to many lands in the southeastern United States: Native American intensive cotton agriculture in the first half of the 19th century, and smaller scale share-crop farming during the depressed economic conditions following the Civil War (Paisley 1990).

The landscape distribution of relatively undisturbed vegetation types in this region is characterized by longleaf pine (*Pinus palustris*) on the uplands with a diverse ground flora well-adapted to frequent fire, grading to a hardwood-pine mixture along moister slopes, and a hardwood hammock rich in tree species in relatively fire-protected areas (Myers 1991). The disturbance regime for most upland habitats consisted of frequent (two-to-five-year return interval), low-intensity, lightning-started fires during the growing season (Robbins and Myers 1992), small gaps created by lightning and insect-caused canopy tree mortality, and infrequent but more extensive hurricane damage (Myers 1991, Palik and Pederson 1996). The vegetation of Tall Timbers before cultivation most likely had a greater abundance of longleaf pine on the uplands, but the actual characterization of the natural or presettlement vegetation types is unclear. Drainages dissect the landscape creating a mosaic of pine ridges and hardwood forest.

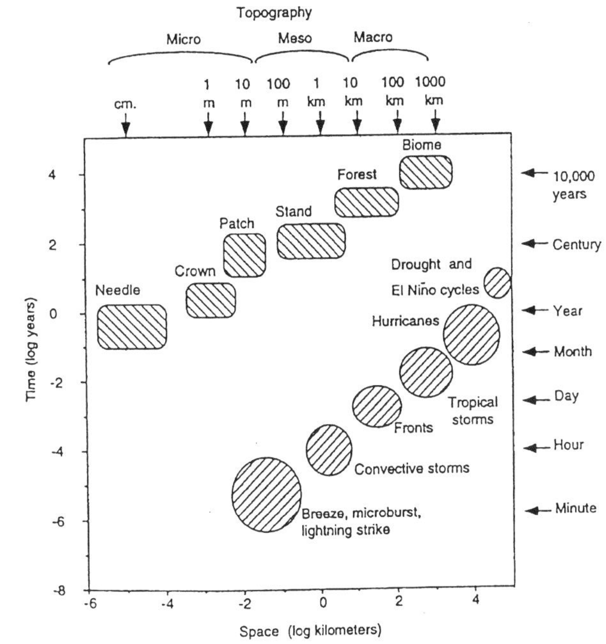


Fig. 1. Space-time hierarchy of forest structure and disturbances of the boreal forest (reprinted from Holling 1995).

Intensive agriculture on the uplands of Tall Timbers completely changed groundcover composition and radically altered tree species composition. Longleaf pine was greatly reduced in abundance and wiregrass (*Aristida beyrichiana*), a highly flammable groundcover dominant, was completely eliminated. After the agricultural fields were abandoned, gullies developed and old-field pines, loblolly (*P. taeda*) and shortleaf (*P. echinata*) — species less fire-adapted — colonized the upland to replace longleaf pine. The groundcover community following cultivation was composed of less-flammable broad-leaved annuals.

Disruption of the vegetation via cultivation strongly influences fire. First, it changed the nature of the fine fuels: groundcover plants and pine needles. Groundcover species of the disturbed community tend to have a narrower seasonal window when they readily burn — the dormant season (late winter–early spring) when the broad-leaved species were dead and dry. Also, old-field pine needles are lower quality fuel than longleaf needles. Second, old-field pine species are less well-adapted to fire. Loblolly and shortleaf pine seedlings and saplings are fire-sensitive and must be protected from fire for several years. Third, hardwood species readily invade highly disturbed upland pinelands, which further disrupts fire. Hardwoods can be killed by growing season fire, but tend to be topkilled if burned in the dormant season. Root crowns develop and readily resprout under these conditions (Lewis and Harshbarger 1976).

These interrelated factors create difficult conditions for ecological management. Lengthening the interval between prescribed fire allows pine regeneration, but it releases hardwoods. Hardwoods are difficult to burn, because they supply poor quality fuel and shade out grasses that make up the finer fuels. Also, the increasing abundance of relatively poor fuels makes use of fire difficult during the growing season when fire would cause the highest mortality of hardwoods.

Old-field vegetation at Tall Timbers provides an example of a severe disturbance (cultivation) that radically altered ground vegetation composition to the point that fire, a crucial part of the natural disturbance regime, does not function within a historical pattern. Maintaining this disturbed ecosystem to provide habitat for animal species adapted to the longleaf pine-wiregrass ecosystem, such as Northern Bobwhite (*Colinus virginianus*) and Red-cockaded Woodpecker (*Picoides borealis*) now requires intensive management. Given the management objective to retain an open-structured pine-dominated forest ecosystem, a long-term strategy has been developed that includes aggressive attempts to shift the season of fire to control hardwoods and restoration of longleaf pine. Restoration of fire-adapted groundcover species has also been used on a limited basis, but it is an expensive and slow process.

## 2.2 Disturbance Regimes

Reconstructing past disturbance regimes is essential for understanding the evolutionary context within which ecosystems have developed. A variety of techniques are available to document the disturbance history, and historical structure and composition of ecosystems (Table 1), including (1) written and oral histories (Day 1953, Cronon 1983, Denevan 1992, Russell 1983), (2) paleoecological (pollen) studies (Grimm 1983, Clark 1990; Foster and Zebryk 1993, Russell et al. 1993), (3) historical photos (Progulske 1974, Gruell 1983, Skovlin and Thomas 1995), (4) land survey records (Bourdo 1956, Lorimer 1977, Rogers and Anderson 1979, Seischab and Orwig 1991), (5) weather records (Curtis 1943, Boose et al. 1994), (6) fire scars on tree boles (Guyette and McGinnes 1982), (7) forest age-structure patterns (Heinselman 1973, McClaran and Bartolome 1989, Nowacki and Abrams 1997), (8) dendroecological analyses (Lorimer and Frelich 1989), (9) vegetation reconstructions using on-site evidence (Henry and Swan 1974), and (10) soil phytolyths (Birkeland 1974). A variety of forest age-structure and land survey records can be used to reconstruct disturbance regimes like that of conifer–northern hardwood forests that lie along the interface of boreal and temperate broadleaf biomes in eastern North America (Box 1).

Table 1. Tools for defining reference conditions. (Modified from Kaufmann et al. 1994).

| Analyses and approaches | Information provided |
|---|---|
| Historical records | Past conditions |
| Tree rings | Climate, disturbance regimes |
| Palynology | Past vegetation and climates |
| Packrat middens | Past vegetation |
| Natural areas | Natural variation, minimum human disturbance |
| Archival literature and photographs | Vegetation change in last century |
| Potential and existing natural vegetation | Estimates of future and former vegetation, natural variation |
| Predictive models | Future vegetation |

**Box 1**
**Conifer-Northern Hardwood Forest of Eastern North America**

Conifer-northern hardwood forests comprise various mixtures of hemlock (*Tsuga canadensis*), eastern white pine (*Pinus strobus*), red spruce (*Picea rubens*), sugar maple (*Acer saccharum*), beech (*Fagus grandifolia*), and yellow birch (*Betula alleghaniensis*). Wind is the predominant disturbance factor on mesic sites at mid-latitudes, frequently creating single- or small multiple-tree canopy gaps (Frelich and Graumlich 1994). Small-scale patch dynamics usually prevail over extended time periods (≈1200 years) (Lorimer 1977, Canham and Loucks 1984, Whitney 1986), being occasionally interrupted by catastrophic disturbances. The effects of large-scale disturbances is dependent on storm type (e.g., tornadoes, downbursts) and intensity, and can result in either large contiguous blowdowns (Dunn et al. 1983) or, more typically, a broadcast scattering of small- and mid-sized openings (Frelich and Lorimer 1991, Frelich and Graumlich 1994). The long-term stability exhibited by this forest type has resulted in many co-evolved and specialized species.

European logging and subsequent fires have had major effects on conifer-northern hardwood structure at stand and landscape levels (Mladenoff et al. 1993). The conifer component has been largely reduced in the process, whereas hardwoods such as maple, oak and aspen have largely benefitted (Elliott 1953; Whitney 1986, 1987; Nowacki et al. 1990). In addition, European alterations have indirectly stifled the post-logging recovery of highly-palatable hemlock and yew (*Taxus canadensis*) this century by elevating white-tailed deer (*Odocoileus virginianus*) populations (predator elimination; increased edge and agro-based habitats); over-protective game laws (Hough 1965, Anderson and Loucks 1979, Alverson et al. 1988, Frelich and Lorimer 1985). At artificially high levels, animal populations can have deleterious effects on native flora.

**Box 2**
**Former Pine Forests of the Upper Great Lakes Region of North America**

Fire was the premier force governing the former pine forests of the upper Great Lakes Region of North America (Wright and Bailey 1982). Forests of eastern white pine (*Pinus strobus*), red pine (*P. resinosa*), and jack pine (*P. banksiana*) were located on coarse-textured, dry, glacial-outwash soils. Although these sites were inherently prone to burn, the flammability of pine needles (high terpene and resin levels and low foliage moisture) further enhanced the fire regime. Jack pine flourished where fires were frequent (83–167 year return time), whereas longer fire intervals encouraged stately red and white pine stands to develop (129–344 years, Whitney 1986; also see Heinselman 1973). Extensive logging by European settlers and post-harvest fire conditions in tandem, caused the rapid demise of this ecosystem (Maissurow 1935, Chapman 1952, Whitney 1987). By concentrating on large-girth pines, logging operations invariably targeted individuals with the greatest seed-bearing potential, substantially reducing their number. The vast quantities of post-harvest slash coupled with newly added ignition sources (railroads, farmers) led to uncontrolled fires over the land. Successive waves of fires ultimately eliminated the "pinery" by consuming pine regeneration and most of the remaining seed-bearing trees (Whitney 1987). Oaks (*Quercus* spp.) and more shade tolerant hardwoods that were formerly relegated to understory positions, had the ability to resprout and maintain themselves during this time until recurring fires were successfully suppressed in the 1920s through 1950s. Today, forests of oak and red maple (*Acer rubrum*) exist where impressive pines formerly resided (Elliott 1953, Nowacki et al. 1990). Prescribed burning is an essential component in reestablishing pine forests over portions of its former range. Reintroducing fire is a difficult prospect since much of the public suffers from what can be called "Smoky the Bear" Syndrome (i.e., perception that all fire is bad). Success of reintroducing fire into pyrogenic-based systems hinges on public education and acceptance.

To a large degree, interactions among climate, soils, and the environmental attributes of disturbance shape key habitat conditions that control species distributions. Understanding these linkages allows managers to formulate strategies for achieving objectives such as maintaining biological diversity or function of ecosystems (Murray 1996). For example, shade-adapted trees prevail in forests where gap-phase dynamics occur (e.g., mixed mesophytic and northern hardwood forests; Runkle 1981, 1982; Barden 1980, 1981; Frelich and Lorimer 1991). In contrast to gap-phase forests, opportunistic, light-demanding species tend to dominate systems that are subject to frequent or chronic, catastrophic disturbance. Fire-shaped landscapes are most illustrative (Wright and Bailey 1982). Here, past fire behavior directly determines vegetation composition, structure, and patterning. Often, plant communities change from open grasslands through savannas to closed forests along a fire gradient from high frequency/intensity to low (Cole et al. 1992). Disrupting historical fire cycles can lead to dangerous fuel buildups with catastrophic consequences (Baker 1994). Prescribed burning is an integral management tool ensuring the maintenance of pyrogenic systems. Species loss through competitive displacement is often an inevitable consequence of fire suppression in fire-dominated ecosystems (Mehlman 1992, McClain et al. 1993, Grigore and Tramer 1996). The demise of the former pine forests of the upper Great Lakes region provides an excellent example of deleterious effects caused by the disruption of a fire regime by logging (Box 2).

Disturbance regimes may change with climatic and physiographic settings (Hadley 1994), thus making it difficult to make generalized statements and universal models. For example, the importance of a single component of a disturbance regime (e.g., an ice storm) in shaping an ecosystem may vary geographically. Because disturbance history is so fundamental in ecosystem development and function, alterations to the prevailing disturbance regime can have tremendous ramifications. Any change in disturbance type, frequency, and intensity may have environmental consequences to the system. Eliminating (or severely curtailing) historically important disturbances can have as large an impact as adding "new" (anthropogenic) disturbances to an ecosystem. The cessation of fire in many North American ecosystems is a case in point. Fire suppression can lead to substantial changes in composition (successional shifts from fire-adapted to fire-sensitive species), a change in structure from open to closed canopy conditions, microclimate, and ecological processes (e.g., nutrient cycling). Concurrent changes in habitat conditions have major repercussions on faunal assemblages (Engstrom et al. 1984, Herkert 1994) and hydrologic processes.

## 2.3 Range of Variation

Range of variation (also called historical variation or natural variability) is an important practical concept that can be used in model development for ecosystem management (Kaufmann et al. 1994, Morgan et al. 1994, Swanson et al. 1994, White et al. this volume). One rationale for using the range of variation approach is that some management practices tend to eliminate or severely alter natural variation in ecosystems (Holling and Meffe 1996, Poff et al. 1997, Sidle and Faanes 1997), and this has negative repercussions on native plants

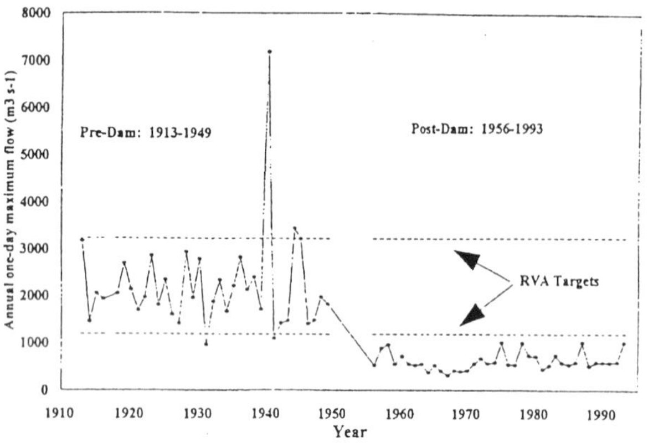


Fig. 2. This plot of one-day maxima streamflows ($m^3 s^{-1}$) reveals the effects of dam construction on the Roanoke River in North Carolina (reprinted from Richter et al. 1997). Dashed horizontal lines denote the mean and $\pm$ 1 SD of the pre-dam variation and could be used for Range of Variability Approach targets.

and animals. Using the range of variation approach, management guidelines can be derived from measurements of spatial and temporal variation within disturbance regimes.

Swanson and others (1994) identified some problems with using the range of variation approach. First, our ability to interpret past ecosystem variability is limited. For example, Shinneman and Baker (1997) claimed that stand-replacing fires that produced large patches of dense ponderosa pine (*Pinus ponderosa*), not low-intensity surface fires, are more important to maintain the historic range of variability in the Black Hills of South Dakota and Wyoming. The relative influence of large, infrequent disturbances versus numerous small disturbances is unresolved (Turner et al. 1997). Second, change in environmental conditions, such as climate change or altered species composition, may move disturbances within ecosystems outside the range of historic variability (D'Antonio and Vitousek 1992) (Section 2.1.1 above). Third, the scientific basis for establishing natural variability as a component of ecosystem management must be broadly accepted by the public. Frequently, there are few opportunities for public discussion of natural variability of ecosystems and its relevance to natural resource management.

Although applications of the range of variation approach are emerging for terrestrial ecosystems (Morgan et al. 1994, Swanson et al. 1994), Richter and others (1997) have formalized the concept into the "Range of Variability Approach" (RVA) to provide guidelines for managing rivers. RVA relies on statistical characterization of ecologically relevant physical parameters that describe ecosystem characteristics. Thirty-two parameters were used to characterize the magnitude, duration, timing, and extreme conditions of river flow. For example, the mean $\pm$ 1 SD 1-day maxima streamflow in the Roanoke River over a 45-year period before a dam was constructed can be used to guide the variability introduced by management into the system (Fig. 2). A recent example of attempts to reintroduce more variation into water flow is the Glen Canyon Dam.

### 2.3.1 Glen Canyon Dam

Closing the gates of the Glen Canyon Dam on the Colorado River in 1963 precipitated major ecological changes downstream in the Grand Canyon. Average peak flow before 1963 exceeded 2400 $m^2$/s; after dam completion, in order to control the release of water for power generation, peak flow water release was reduced to 500 $m^2$/s (Collier et al. 1997). The resulting reduction in peak flows has had several impacts on the river channels and associated sandbars in the Grand Canyon. Under normal flood regimes, sandbars along the Colorado River were actively eroded, deposited, and reshaped. With post-dam flows the sandbars eroded, which reduced them in size and firmly established vegetation on the upper levels of some bars. This has resulted in a net loss of camp sites along the river for recreationists traveling in the river corridor of the Grand Canyon. Low volume and velocity water also deposited the eroded sediments in the main channel and backwater sloughs, thus reducing habitat for native fish species. Additionally, the water released from Lake Powell is colder and has less sediment than the original Colorado River water, which has encouraged cold-water fish such as trout and reduced populations of native warm-water fish.

Even before the dam was constructed, numerous scientists and river lovers were concerned about these and other impacts to the Colorado River ecosystem. Ongoing studies by government agencies and universities have documented the changes listed above. Much of this work was synthesized in an environmental impact statement for a proposal to generate a test flood from the Glen Canyon Dam in the Colorado River (USDI Bureau of Reclamation 1995). A test flood flow of 1275 $m^3$/s was released for six days in spring 1996. The objectives of the controlled release were to rebuild sandbars with sediment from river pools and back water channels, to clean vegetation from high zones of sandbars to establish more beaches for camping, to scour slack water channels and marshes to increase habitat for native wildlife, and to clean out debris fans that cause rapids. Extensive sampling and modeling prior to the flood had led to the expectation that these objectives could be achieved with the controlled flood level.

Preliminary observations and data indicate dramatic changes in the Colorado River floodplain below the Glen Canyon Dam (Collier et al. 1997). Most of the sand was moved in the first 48 hours of the flood. Pre- and post-flood transects of the river show that most of the sand which was deposited on the sandbars came from the bed of the river channel and pools. Changes in the back channel marshes were variable: some areas were scoured out and others were filled in with sediment. Debris fans were cleaned of fine-grained materials and some boulders up to 1-m diameter moved downstream. Overall, the models were excellent predictors of sediment movement and deposition.

Additionally, there were no loss of cultural features or marsh habitat, and populations of leopard frogs (*Rana pipiens*) and trout below Glen Canyon Dam survived. A substantial amount of woody debris stranded at high elevations on sandbars was moved back into the river and redeposited near normal water levels in the sandbars. Most shrubs stayed in place on the sandbars with the controlled flood level. Continued monitoring is needed to evaluate the overall effects on small bird species and the humpback chub (*Gila cypha*) population at the mouth of the Little Colorado.

Initial barriers to the controlled flood test were institutional mind-sets and "turf battles" between agencies that had differing objectives, although they managed interdependent resources. For example, the National Park Service (NPS) and the Bureau of Reclamation (BOR) are both Department of Interior agencies, the NPS manages to "preserve and protect" natural and cultural resources while the BOR manages water flows for economic interests, including power generation.

Several factors contributed to the overall success of the controlled flood in the Grand Canyon. Over 10 years of concentrated, careful scientific study formed the backbone of the proposed flood event. Models were built and tested to forecast effects of a spike flow on a river system that had reached an uncomfortable equilibrium with low flows controlled by the need for power, rather than by natural floods. An equally important factor was the cooperation of a varied number of individuals and organizations to finally bring a test flood to reality. Federal agencies included BOR, NPS, U.S. Geological Survey, and U.S. Fish and Wildlife Service. Several Native American tribes were represented, as well as the power interests, water institutions, the State of Arizona, the environmental community, universities, and river guide associations. Although not without controversy, the long process of building cooperation eventually led to a successful test. A critical crossover occurred as people with economic interests in power generation began to understand that the American people put very high value on the natural, cultural, and recreational resources of the Grand Canyon.

The test-controlled flood release from the Glen Canyon Dam was a great success from both scientific and management perspectives. A long-term monitoring and research center will be set up to continue the scientific studies and provide information and recommendations to the management group on the timing, size, and other parameters of future flows through the Grand Canyon. The success of this test flood will probably be extended to other western river systems through adaptive management.

## 2.4 Reference Conditions

Reference conditions are current examples or descriptions of historical circumstances that characterize desirable ecosystem conditions (Kaufmann et al. 1994). Reference conditions need not be described as "normal" or "natural" conditions (Sprugel 1991, White et al. this volume). Because change is inevitable for any conditions, they should be viewed as part of a moving picture instead of a snapshot. Use of reference conditions must be put within the context of achieving ecosystem objectives regardless of whether or not those conditions are natural (Agee 1993).

Although not without controversy (see Hoerr 1993), conditions prior to European settlement of North America are often used as reference benchmarks. If "presettlement" conditions are used as reference conditions, it is critical to recognize that humans have been part of the environment of North America for thousands of years, and have had a profound influence on disturbance regimes (Hoerr 1993). Where prehistoric human populations resided, fire frequencies were often elevated over what would have occurred naturally (Stewart 1956, Sauer 1975, Pyne 1982). Furthermore, climatic controls on fire were often overridden by humans, which resulted in fire-dependent communities in locales where they would not exist in the absence of humans. The Prairie Peninsula, which formerly extended through Illinois, Indiana, and Ohio is a classic example (Transeau 1935). On the other hand, 10,000 years of human presence in the western hemisphere only represent 1 percent of the evolutionary history of a typical species (Hunter 1996).

Anthropogenic burns occurred for a variety of reasons: encouraging mast and berry production, clearing vegetation for villages and agricultural fields, reducing understories for overland transportation and security, for warfare, promoting grasses for ungulates, and reducing insects and fuel hazards. Overall, humans clearly alter ecosystems and should not be overlooked when assessing past disturbance histories.

## 2.5 Managing for Desired Conditions

Acquiring information on present-day community composition and structure is a key to evaluating change and serves as a starting point for predicting future trends. A variety of techniques are used to document current conditions based on location (terrestrial, aquatic), lifeform (vegetation, faunal) and scale of interest (species, community, landscape, regional). To provide for valid comparison, attempts should be made to match data collection protocols of present-day data with those past surveys where practicable.

On-site evidence coupled with available management records or other information sources can be used to identify important disturbance events that have affected the area of interest. For example, charred debris within the forest floor or on tree boles are tangible indications of past fire. Skid trails, stumps, and even-aged conditions are indicative of past logging activities. Those disturbance events that have led to current communities can be compared and contrasted to historic variation in conditions.

Reasonable predictions about the future condition of ecosystems can be made by putting current ecological conditions (community composition, size class distribution, etc.) within the framework of the disturbance regimes and range of variability. If disturbance events leading to the current community are projected forward, ecological changes are assumed to continue (e.g., succession). However, it must be understood that there are limitations because of unpredictable events. Unforeseen shifts in the climate, for instance, may easily dismantle predicted trends. Predicting ecological trajectories based on current disturbance regimes invariably involves uncertainty.

## 2.6 Disturbances as Templates for Natural Resource Management

Human-induced disturbances are often more successful if the tools match the natural disturbance regime in spatial and temporal scale. Here a thorough understanding of the cyclic patterns of an ecosystem should be incorporated into the project design. For example, when designing floodplain capacity and configurations, the engineer needs to allow for 100-year or 500-year flood events, even though these time frames may exceed the historic record of the region. When considering size of logging or prescribed fire patches, the original heterogeneity of the landscape should be approximated.

Many anthropogenic disturbances have analogues, to some degree, in natural disturbances (Smith et al. 1997). But natural and anthropogenic disturbances also have critical differences that must be considered by managers (Table 2) (Hansen et al. 1991). Understanding of these differences can "make-or-break" the effectiveness of using natural disturbances as a model for management activities. Forest managers, in particular, are rapidly developing models of natural disturbance as templates for management activities.

Table 2. Comparison of human-induced and natural disturbances.

| Human-induced disturbances | Natural disturbances | Differences |
|---|---|---|
| Prescribed fire/fire suppression | lightning fires | season, intensity, ignition pattern |
| Logging | wind throw, insect infestation, fire, flood | biological legacy, spatial pattern |
| Alteration of water flow | drought and flood | season, frequency, magnitude |
| Dams | landslides, avalanches, glaciers, earthquakes | water flow regulation, permanence |
| Pest management | cycles of insect infestations | pesticides, introduction of exotic biological control organisms |
| Roads | linear disturbances, game trails | persistence, chemical characteristics, traffic |
| Herbicides | pathogen outbreak | plant specificity, persistence |

The nucleus of innovation in forestry (i.e. the New Forestry [Franklin 1989]) has been in the Pacific Northwest, particularly the H.J. Andrews Ecosystem Research Group. The shift in focus of New Forestry is toward recognition of the complexity of forest ecosystems, the importance of biological legacies, and maintenance of viable landscapes (Franklin 1989, Hunter 1990, Kohm and Franklin 1997). Swanson et al. (1994) describe the Augusta Project from the Willamette National Forest in which spatial and temporal aspects of silvicultural treatments were designed partially to emulate the natural disturbance regime, particularly the size and frequency of fires. Care was taken to retain all vegetation seral stages across the landscape and to tailor stands to topographic locations (e.g., cutting units were matched to the size of the watershed).

Fire, wind, and lightning regimes can be used as templates for design of forest management activities.

Size of the burned area (Hunter 1993), consideration of natural stand-replacement dynamics (Bergeron and Harvey 1997), spatial distribution of harvest areas (Delong and Tanner 1996, Fries et al. 1997) are all considerations in management. As an example, Palik and Pederson (1996) characterized disturbance frequency and size in longleaf pine forest over a five-year period. Most of the canopy tree mortality was caused by lightning strikes of single trees with occasional strikes that killed 15–30 trees. A combination of thinning (single-tree death by lightning), small gap formation (group lightning strike), and larger gaps with some live tree remaining (wind event) could be used in silviculture to mimic natural patterns of disturbance.

By mimicking this disturbance regime through silvicultural practices (i.e., selection harvesting in this case), wood products can be removed from these sites while minimizing ecological disruptions to the system (e.g., continuation of favorable regeneration conditions, prevention of forest type conversion, and biodiversity loss) (Kimball et al. 1995).

### 2.7 Integration of Natural Disturbances into Management

Natural disturbances happen. Incorporating natural disturbances into management plans is a challenge because their timing, scale, and intensity is unpredictable. In some cases natural disturbances can be anticipated and used for management purposes. A management decision to permit a natural fire to burn if it is within prescription (i.e., the "let-it-burn" policy) is an obvious example. But many large disturbances (hurricanes or major floods) will be uncontrollable. Anticipation of the ecological and social consequences of natural disturbances should be included in management plans. For example, Wunderle and Wiley (1996) list several management strategies to lessen the impact of hurricanes on wildlife populations.

## 3 SETTING GOALS

A rational decision-making process requires setting goals, identifying problems and opportunities, forming objectives and standards, developing management action alternatives, and evaluating and implementing alternatives (Culhane and Friesema 1979). In the context of ecosystem management, goals represent the desired end state and reflect policies and values of the organizations and individuals which define them. Goal-setting guides the actions of ecosystem management. For example, a major goal for management of sites along the Oregon Trail is restoration of conditions encountered by wagon train pioneers traveling west. Selection of this goal must be achieved within the mission of an organization and, in the case of public agencies, societal support.

We set natural resource goals to meet human desires and expectations. However, the concept of setting goals may appear futile in an ever-changing system. How can we be assured that our goals for natural resource conditions and uses are attainable and sustainable? How can we reduce the risk of failure in achieving goals given the unpredictable nature of disturbance events and other dynamics? As Reice (1994) wrote, "...most biological communities are always recovering from the last disturbance." Disturbance ecology would seem anathema to setting goals. These questions can be addressed by setting goals with a historical perspective of inherent variability within the systems, and within the context of current dynamics that continue to shape natural resources. The range of variability and the nature of the current forces that affect ecosystems help us to define ecological limits that determine the range of conditions around which human desires and expectations should be shaped.

Goals for management of natural resources are often expressed in terms of outputs, desired products, and uses. Goals for agricultural systems, for example, may be expressed in terms of crop yields or acres under management. Goals for complex ecosystems or shared resources, such as water, are often much more difficult to describe by simple measures. When numerous competing desires for products and uses exist, natural resources may be taxed beyond the limits of sustainability, many goals may not be attained, and substantial financial, human, and natural resources can be at risk. Setting goals for natural resource utilization in such a complex and competitive situation must consider the limits of natural ecosystems both to sustain themselves and to produce desired outputs. Another way to describe goals, especially useful in such a complex situation, is in terms of the conditions that must be maintained to meet all expectations. Desired conditions include biological, physical, and human elements and the forces that cause these conditions to be dynamic.

When trying to describe desired conditions, an important element is an understanding of the inherent limits and thresholds of ecosystems. As previously described, an understanding of past disturbances regimes helps to frame these limits. If planned outputs are close to the system's capability to produce these outputs then a disturbance event may have a significant effect on desired conditions (Averill et al. 1995). A variety of natural resource uses, from wilderness to commodity production to urban development, can be addressed with this approach.

Initiation of major projects to apply concepts of ecosystem management is often coincidental with investigations of disturbance histories. For example, the Chattooga Project (Section 5.1) included an assessment of natural and anthropogenic disturbance to provide information needed to characterize vegetative composition and structure. Knowledge gained regarding climate changes, wind effects, lightning and human-induced fire contributed to the identification of desired vegetation conditions for parts of the project area. An ecosystem management initiative in the Ouachita Mountains addressed historical conditions of forests and the interaction of the physical environment and disturbances that shaped the natural communities. This knowledge was used to describe the desired vegetation and associated wildlife species to be managed on public lands.

## 3.1 Ouachita Mountains

Disturbance events operate at many temporal and spatial scales in the temperate forests in the Ouachita Mountains of Arkansas and eastern Oklahoma. We know from fossil pollen analyses that boreal forests of jack pine and spruce dominated Ouachita Mountain landscapes during the period from about 20,000 to 15,000 years before present (BP). These boreal species were replaced by forests of oak, hickory, and ash (12,000 to 8,000 years BP), which gave way to prairies or oak-grass savannas (8,000 to 4,000 years BP), which were replaced by forests dominated by shortleaf pine and oaks (4,000 BP to present). Long-term climatic changes caused these dynamics in vegetation.

Historical accounts and photographs indicate that open pine forests dominated by shortleaf pine and a lush herbaceous understory comprising bluestem grasses and forbs were once common in the Ouachita Mountains. Fires, started by lightning or set by Native Americans, and windstorms played major roles in shaping the current vegetation of the forest historically. These forces resulted in a forest of varying age classes and of varying patch sizes. Historic accounts describe the forests as similar to, but typically more open than, those of the present (Jansma and Jansma 1991) with tree densities much lower (16–20 trees/ha) than today's forest (Foti and Glenn 1991, Masters et al. 1993, Kreiter 1995).

Shortleaf pine/bluestem communities have become rare due to the elimination of old-growth pine and active fire prevention and suppression efforts during the 20th century. Encroachment of dense midstories of pine and hardwood trees reduced or eliminated the formerly open condition. Current inventories of mature trees show densities of 50–60/ha with an 84% average canopy cover.

Renewal and restoration in the shortleaf pine/bluestem grass ecosystem is needed for perpetuation of a diverse biota that has declined throughout this century. A key species in this historical habitat, the endangered Red-cockaded Woodpecker once occurred widely in the southeastern United States, including the Ouachita mountains. Human intervention is essential, given the inability of natural processes such as wildfire to function at pre-settlement landscape scales in a fragmented modern landscape. Based on the knowledge of changes in forest composition, historical variation of ecosystem characteristics, and disturbance processes, the Ouachita National Forest is preparing to change management to restore the historic shortleaf pine/bluestem grass ecosystem and recover the Red-cockaded Woodpecker.

Restoration and management of the historical shortleaf pine/bluestem grass ecosystem can be achieved by a combination of natural events such as fires, wind storms, and insect outbreaks and by the use of prescribed fire and timber harvests. Key elements of this approach are: (1) using fire and tree cutting to simulate natural disturbance patterns, (2) using natural events such as insect outbreaks and wind storms to provide open conditions and opportunities for natural forest regeneration, (3) increasing the minimum time between regeneration cutting from 70 years to 120 years, (which would allow for development of older trees required by Red-cockaded Woodpeckers and other cavity-dependent species), (4) maintaining mixtures of native pines and hardwoods, (5) developing and maintaining forested linkages between mature forest stands, (6) minimizing ecotonal differences between contiguous stands, and (7) reducing forest fragmentation.

Strong collaborative analysis and planning among federal and state agencies, academia, and private organizations and individuals has made this a successful planning activity to date. Reliance on science to support concepts of desired conditions has helped to provide a foundation for a management plan that is practical and achievable.

# 4 MONITORING

Monitoring disturbance events and effects is integral to natural resource stewardship. Regardless of the type of action, from "let nature take its course" management to a timber sale or reclamation project, monitoring is necessary to evaluate the effectiveness of the action and to guide future management. Historically this phase of work has often been ignored due to time constraints, lack of understanding of the importance of monitoring, or funding shortages. However, monitoring and evaluating the short and long-term outcomes are just as critical to a successful project as setting goals and effective planning.

Why monitor at all? The substantial outlay of funds and personnel to track and interpret effects over time often has no short-term return. The true benefits of monitoring a project are realized over longer time periods in improved relations between the public and the management agency and savings of time, personnel, and finances as lessons learned are applied to future work in similar areas.

Monitoring is essential to develop and maintain agency credibility with various public groups and for these individuals to have confidence in the management actions of the agency. High public expectations, locally and nationally, may influence management action. Stringent measurements and honest interpretations help gain public confidence, even if the results are not exactly as predicted. Over the longer term, the public often provides strong support for additional project work or even augmented budgets.

One of the major reasons for monitoring is to determine if a management project is achieving the desired goal: Is it a success, a failure, or a learning experience? The human-induced disturbances often inherent in management actions interact with other natural disturbances on a site, as well as the legacy of past disturbances and the imported effects of off-site disturbances to create a new mix of results that may be a surprise to the manager. Remember that the definitions of "success" and "failure" and even the ultimate goals for an area may change over time with changing management mandates and political/social imperatives.

Monitoring allows a manager to evaluate the direction and progress of change in an ecosystem and make course corrections while the project or plan is being implemented. This is the core of adaptive management. Often this results in direct budget and time savings over the term of a management action. Occasionally, there is a need for additional interagency coordination or public education and input to take advantage of opportunities revealed by monitoring.

The longer term profits of monitoring are to learn for future work and to increase our understanding of natural disturbance processes in an ecosystem. The data collected during monitoring build a new baseline, providing a reference condition for the future. Great care must be exercised when extending management techniques to different ecosystems or even similar types which may have different histories. In these new areas monitoring should also be continued to ensure that management is achieving the predicted results here as well.

Monitoring has to be built into a project plan from the beginning with a commitment for adequate funding and sufficient time frames to evaluate long-term impacts. Often there is a rush at the end of implementing a project to write a report and move onto the next pressing project. Valuable data are never collected, or if collected, not adequately analyzed and interpreted. Resource specialists and scientists need to have a commitment to actual data collection, analysis, and interpretation over the entire time frame that the results of a management action are coming into equilibrium with the rest of the local ecosystem.

## 4.1 Techniques

Two major types of tools are available to track progress of an ecosystem: quantitative (numeric/statistical) and qualitative (non-numerical observations). Understanding of ecosystem dynamics requires the integration of both types of data and interpretation in light of management mandates for the area.

Quantitative techniques provide rigorous tools that attempt to minimize observer bias. Numerous sampling regimes and statistical analysis tools are available for the various components of an ecosystem, and there is a vast literature describing the sampling and statistical analysis techniques.

The qualitative tools are more difficult to define, but there is no substitute for a thorough, intuitive understanding of the natural ecosystem and the various disturbance forces on it. This can be gathered by knowledgeable resource managers who take time to walk on the land, fly over it, sit by the streams, and talk with knowledgeable local people who have a long and deep understanding of the area.

All the usual caveats about using these tools in research also apply to the interpretation of monitoring results. The critical cautions are that the quantitative results need to be melded with general knowledge of the land and nearby areas. Despite the most rigorous techniques it is possible for statistical sampling to miss parts of an ecosystem, or to indicate trends that do not match reality.

When designing the statistical sampling part of the monitoring, we must be sure to ask the core questions. Sometimes one or two key parameters that are reliable indicators will integrate several processes. We must balance the need to have data which correspond with other agencies or previous investigators with the need to answer the questions at hand.

Another important consideration is that cause-and-effect relationships in ecosystem disturbance dynamics are seldom simple or fully understood. The scale of the question and the measurements being collected must match the scale of the disturbance or landscape being addressed. A final caveat involves paying attention to the entire ecosystem under consideration, not just the easily observed macro floral and faunal components.

A variety of tools is now available to help us with our planning and decision-making processes. These include GIS capability to display information, social assessments to determine the desires of stakeholders, economic analysis tools to compare alternatives, and computer models to predict ecological changes. In the latter case, models are available to project vegetative development over time. These are now linked to various disturbance models such as insects, disease, and fire to project a probable scenario of changes that would occur beyond what would be expected with just successional development. Some models have the capability to include the influence of adjacent sites.

Watershed models are available to predict the consequences of vegetative changes on water yield, sediment production, and fisheries (fry survival, food production, temperature, etc.). Models have also been developed to display the effects of vegetation changes on populations of various animal species. Fire behavior models (e.g., BEHAVE) can predict the pattern and intensity of fire under given conditions of fuel and climate. This type of model has been very useful to fire suppression efforts, but it has equal value in analyzing current patch sizes and shapes and vegetative structures to predict what the range of climatic and fuel conditions might have been at the time the fire burned.

To date, most of these models have been used to predict something that is likely to occur in the future. It appears that many of these models (perhaps with modifications), could be equally beneficial in evaluating the conditions of the past and what the effects of the various disturbances might have been on the resources.

## 5 APPLICATIONS

We describe two case studies of the use of disturbance ecology to natural resources management from widely different parts of the country. The first is a description of the Chattooga Project, which was initiated by the Southern Region of the U.S. Forest Service across a three-state watershed in 1993. This project, considered a model for national forest management, took novel approaches to develop tools and information necessary for land managers to implement ecosystem management. The use of disturbance ecology to describe the disturbance regime, identify reference conditions, and to design management techniques seems to be common to many of the case studies that we have been able to identify. Our second application is the restoration of an old mine site in Denali National Park. Understanding the range of variation in Glen Creek was essential to design effectively its restoration.

### 5.1 The Chattooga Project

The Chattooga River, a National Wild and Scenic River, flows 91 km from the Appalachian Mountains in North Carolina along the boundary between Georgia and South Carolina. Two-thirds of the 72,850-ha Chattooga watershed is in the Chattahoochee (Georgia), Sumter (South Carolina), and Nantahala (North Carolina) National Forests. The broad range of annual precipitation and elevation along the watershed supports a tremendous diversity of plants and animals. Before federal acquisition in 1916, the watershed was intensively logged, farmed, and grazed, and typified "land that nobody wanted." The present vegetation is a pine–hardwood mixture, and the area is used for hunting, fishing, sightseeing, rafting, and timber production.

A key objective for the project was to develop and test a process to inventory and assess biological diversity in National Forests in the Southern Blue Ridge area. The process used coarse-filter (looking at landscape level processes and disturbances to define the natural range of variation) and fine-filter approaches (identifying and inventorying rare communities and species). A team of researchers simultaneously surveyed the "knowledge base" and implemented on-the-ground inventories. An annotated bibliography summarized previous and ongoing studies (Rundle 1994). Obvious information gaps on topics such as old-growth forest, waterfall spray zones, flood plains, bogs, wetlands, aquatic and terrestrial mollusks, reptiles and amphibians, small mammals, neotropical migratory birds, and rare communities (e.g., table mountain pine), were filled through cooperative efforts between the USDA Forest Service, universities, other agencies and individual scientists.

The coarse-filter approach included an analysis of past disturbance patterns and processes (Alger 1995), such as logging, agriculture, and mining in the watershed. Natural disturbances — wind, fire, insects, disease, snow, and ice —and the temporal and spatial range of variation were described by Bratton and Meier (1995). Information on past species composition of the landscape was gathered through a pollen/charcoal analysis (Delcourt and Delcourt 1996). The objective of this approach was to design practices to provide habitat for plants and animals that would be expected in a natural forest.

The study of European disturbance showed that disturbance was much more intense in local areas than expected (Alger 1995). Up to five entries for logging was noted in some areas. Past mining was intense, and agricultural areas suffered a loss of over 30 cm of top soil, which severely affected soil productivity. Analysis of wind disturbance showed standard gap dynamics: about one percent openings per year. Sampling methodology was not suitable for analyzing infrequent large disturbances or mico-bursts, which occur in the Chattooga but were not observed in the study.

Fire was prevalent nearly everywhere. About 90 percent of the ignitions were estimated to be anthropogenic. Native Americans and white settlers tended to run fire from valley bottoms to ridge tops in the fall and spring. Lightning ignitions were usually top or ridge down during summers of drought years. An early surveyor of the area, Andrew Ellicott, noted in 1795 that it was nearly impossible to survey that fall because of all the smoke from the Indian fires. Fire frequency increased in lower elevations. Bratton and Meier (1995) felt that the current fire suppression policy was causing a shift in species composition from oak-hickory to red maple-black gum.

Major changes in vegetative composition in the area have occurred over the past 4000 years. Pollen and charcoal analysis of a peat core removed from Horse Cove Bog dated to 3550 years BP ±10 years (Delcourt and Delcourt 1996). Recent species declines in the area are credited to be a result of human intervention, especially fire suppression over the past century. Results indicate that Native Americans played a key part in shaping the vegetation of the Southern Appalachian forests by their use of fire and agriculture. Because precipitation is relatively high and the incidence of lightning-caused fire has been very low, fire tolerant species increased over much of this period. Recent declines in fire-adapted and tolerant species were noted.

Accomplishments of the project include: development of a geographic information system; an ecological classification based on mapping and inventory system that describes ecological potentials of National Forest lands in 2 to 40 ha units; biodiversity and water quality assessment; desired condition scenarios developed with public involvement; and land acquisition and exchange criteria. This information is being used to implement new Forest Service projects in an adaptive management process. Mitigation of water quality problems, timber harvest systems that better reflect natural disturbance patterns, and implementation of landscape approaches in managing for both outcomes and outputs in the watershed are some of the positive results to date.

## 5.2 Denali Mine Reclamation

Streams in the Kantishna Hills on the north side of Denali National Park and Preserve in Alaska were placer-mined during the early 1920s to 1940s and again during the late 1970s–early 1980s. The recent mining activities were often conducted with heavy equipment —bulldozers and wash plants — with little to no attention to reclamation of the watershed. A block of unpatented claims in lower Glen Creek, a drainage on the south side of the Kantishna Hills, reverted to the National Park Service. The National Park Service decided to use the area as a prototype reclamation study to test a variety of reclamation techniques and to slow the deterioration of the watershed (Densmore 1994).

Approximately 5 km of the stream and surrounding watershed had been intensively mined, leaving the stream channelized with no functioning floodplain. Tailings were strewn around in steep piles of gravel. As all the fine-grained materials and organic soils had been washed downstream during gold processing, very little revegetation has occurred in the 10–15 years since mining. A mining camp of old travel trailers and plywood buildings was rapidly deteriorating; debris and broken machinery were scattered through the drainage. Approximately 12 ha were devoid of vegetation.

The first step was mapping the existing condition from aerial photos. Reconnaissance field work in nearby streams identified pre-mining conditions of the stream, the floodplain, and the vegetated uplands. Field work in Glen Creek located piles of overburden, with some fine-grained materials, and pure gravel tailing piles. A project plan and reclamation design were prepared and implemented which included debris removal, reshaping of the tailings piles, respreading available overburden materials, a complex research project to evaluate the potential of restoring several plant species. In a later phase, several reaches of the stream were rebuilt with 2- and 100-year floodplains, and one reach was rerouted. Floodplains were stabilized with several techniques including half-buried alder branch bundles, planted willow shoots, large rocks, micro-topography (dozer track spins) and commercial mats and fiber logs. The project was implemented over three summer seasons, and monitoring continues for several stream, floodplain, and vegetation components. Concurrent monitoring is also conducted in a nearby control watershed.

The Glen Creek reclamation project has been a model of several techniques that have been used by other agencies on other streams in interior Alaska. Results of this research help guide reclamation standards in other parks. The reclamation work has given a "jump start" to normal succession of the vegetation and riparian systems in the watershed. A 50–100 year flood event occurred within a week of finishing the dirt work on a stretch of floodplain rebuilding. Some of the floodplain was already stabilized with alder bundles.

The stream rerouted itself within the floodplain, but the floodplain performed beautifully.

Factors that contributed to the success of the project included: (1) using an adequate time frame that permitted changes to the project design and adaptive management of the floodplain rebuilding effort, (2) use of several species in the revegetation effort to better duplicate natural succession patterns, (3) direct involvement of a qualified revegetation scientist and a innovative hydrologic engineer, who worked together throughout the project and continue monitoring and integration of the results.

## 5 CONCLUSIONS

Ecosystems are routinely subject to disturbances (discrete events that disrupt ecosystem structure and change resource availability) and thus an understanding of disturbance regimes is an essential foundation for ecosystem management. In particular, staying within the boundaries of a natural disturbance regime provides a practical, conservative framework for sustainable use of ecosystems in the absence of detailed knowledge. For example, modeling timber harvesting activities after the types of natural disturbances that shape a particular forest ecosystem is likely to mitigate the effects of logging on ecosystem structure. The effort to understand disturbance regimes in an ecosystem management context can be organized around four key questions: (1) What are the desired reference conditions (e.g., pre-European settlement)?; (2) What is it like today?; (3) What happened in between?; and (4) What is the likely future based on postsettlement trends? With answers to these four questions one will have a framework for setting realistic goals for ecosystem management. These goals need to reflect both the policies and values of organizations and individuals as well as the limitations imposed by the ecosystem.

Disturbance regimes are key factors that shape ecosystem limits; for example one cannot realistically have a goal of maintaining 500-year-old trees in an ecosystem that experiences severe windstorms every few decades. Furthermore, disturbance regimes make ecosystems dynamic and this dynamism must also be considered when setting realistic goals, especially because disturbance events are often unpredictable.

Disturbance regimes need to be considered in all aspects of ecosystem management planning and implementation, not just the goal-setting stage. Specifically, knowledge of disturbance regimes needs to inform: (1) the cross-validation between local management plans and higher order (agency-level) plans, (2) the identification of opportunities for effective action, (3) evaluations of alternative actions, (4) assessments of potential risks, (5) the selection of appropriate planning tools such as computer models, and (6) the selection of implementation tools such as prescribed fire or logging.

Monitoring disturbance events and their impact is essential whether the disturbances were initiated as part of an ecosystem management plan, or were a natural or accidental event. Without rigorously designed and executed monitoring there is no way to evaluate the efficacy of management and take corrective action if necessary. Over the long term monitoring will lead to a far deeper understanding of ecosystem dynamics and more effective ecosystem management.

## ACKNOWLEDGMENTS

We thank Gary Meffe and three anonymous reviewers for comments on the manuscript. We thank Columbia University Press and Blackwell Science Ltd. for permission to reproduce Figures 1 and 2, respectively.

## REFERENCES

Agee, J.K. 1993. *Fire Ecology of Pacific Northwest Forests.* Island Press, Washington, DC.

Alger, J. 1995. Anthropogenic Disturbance of the Chattooga Watershed. Contract for the Chattooga project.

Alverson, W.S., D.M. Waller, S.L. Solheim. 1988. Forests too Deer: Edge Effects in Northern Wisconsin. *Conservation Biology* 2: 348–358.

Anderson, R.C. and O.L. Loucks. 1979. White-tail Deer (*Odocoileus virginianus*) influence on structure and composition of *Tsuga canadensis* forests. *Journal of Applied Ecology* 16: 855–861.

Arno, S.F., and G.E. Gruell. 1986. Douglas-fir encroachment into mountain grasslands in Southwestern Montana. *Journal of Range Management* 39: 272–276.

Attiwill, P.M. 1994. The disturbance of forest ecosystems ecological basis for conservative management. *Forest Ecology and Management* 63: 247–300.

Averill, R.D., L. Larson, J. Saveland, P. Wargo, and J. Williams. 1995. *Disturbance Process and Ecosystem Management.* USDA Forest Service, white paper.

Baker, W.L. 1994. Restoration of landscape structure altered by fire suppression. *Conservation Biology* 8: 763–769.

Barden, L.S. 1980. Tree replacement in a cove hardwood forest of the Southern Appalachians. *Oikos* 35: 16–19.

Barden, L.S. 1981. Forest development in canopy gaps of a diverse hardwood forest of the Southern Appalachian mountains. *Oikos* 37: 205–209.

Bergeron, Y., and B. Harvey. 1997. Basing silviculture on natural ecosystem dynamics: an approach applied to the southern boreal mixedwood forest of Quebec. *Forest Ecology and Management* 92: 235–242

Birkeland, P.W. 1974. *Pedology, Weathering, and Geomorpho-*

*logical Research*. Oxford University Press, New York, NY.

Boose, E.R., D.R. Foster, and M. Fluet. 1994. Hurricane impacts to tropical and temperate forest landscapes. *Ecological Monographs* 64: 369–400.

Bourdo, E.A. 1956. A review of the General Land Office Survey and of its use in quantitative studies of former forests. *Ecology* 37: 754–768.

Bratton, S.P., and A.J. Meier. 1995. The Natural Disturbance History of the Chattooga Watershed: Written Records. Report submitted to the USDA Forest Service, Chattooga River Ecosystem Management Demonstration Project, Clemson, SC.

Canham, C.D. and O.L. Loucks. 1984. Catastrophic windthrow in the presettlement forests of Wisconsin. *Ecology* 65: 803–809.

Chapman, H.H. 1952. The place of fire in the ecology of pines. *Bartonia* 26: 39–44.

Clark, J.S. 1990. Fire and climate change during the last 750 years in northwestern Minnesota. *Ecological Monographs* 60: 135–159.

Cole, K.L., K.F. Klick, and N.B. Pavolvic. 1992. Fire temperature monitoring during experimental burns at Indiana Dunes National Lakeshore. *Natural Areas Journal* 12: 177–183.

Collier, M.P., R.H. Webb, and E.D. Andrews. 1997. Experimental flooding in the Grand Canyon. *Scientific American* 276: 82–89.

Covington, W.W., and M.M. Moore. 1994. Southwestern Ponderosa pine forest structure: Changes since Euro-American settlement. *Journal of Forestry* 92: 39–47.

Cronon, W. 1983. *Changes in the Land: Indians, Colonists, and the Ecology of New England*. Hill and Wang, New York, NY, 241 pp.

Culhane, P.J. and H.P. Friesema. 1979. Land use planning for the public lands. *Natural Resources Journal* 1: 43–74.

Curtis, J.D. 1943. Some observations on wind damage. *Journal of Forestry* 41: 877–882.

D'Antonio, C.M., and P.M. Vitousek. 1992. Biological invasions by exotic grasses, the grass/fire cycle, and global change. *Annual Review of Ecology and Systematics* 23: 63–88.

Davis, S.M., and J.C. Ogden (eds.). 1994. *Everglades: The Ecosystem and Its Restoration*. St. Lucie Press, Delray Beach, FL, 826 pp.

Day, G.M. 1953. The Indian as an ecological factor in the Northeastern Forest. *Ecology* 34: 329–346.

Delcourt, P.A., and H.R. Delcourt. 1996. Holocene Vegetation History of the Northern Chattooga Basin, North Carolina. Report submitted to the Tennessee Valley Authority.

Delong, S.C., and D. Tanner. 1996. Managing the pattern of forest harvest: lessons from wildfire. *Biodiversity and Conservation* 5: 1191–1205.

Denevan, W.M. 1992. The pristine myth: The landscape of the Americas in 1492. *Annals of the Association of American Geographers* 82: 369–385.

Densmore, R.V. 1994. Succession on regraded placer mine spoil in Alaska, U.S.A., in relation to initial site characteristics. *Arctic and Alpine Research* 26: 354–363.

Dunn, C.P., G.R. Guntenspergen, and J.R. Dorney. 1983. Catastrophic wind disturbance in an old-growth hemlock-hardwood forest, Wisconsin. *Canadian Journal of Botany* 61: 211–217.

Elliott, J.C. 1953. Composition of upland second growth hardwood stands in the tension zone of Michigan as affected by soils and man. *Ecological Monographs* 23: 271–288.

Engstrom, R.T., R.L. Crawford, and W.W. Baker. 1984. Breeding bird populations in relation to changing forest structure following fire exclusion: A 15-year study. *Wilson Bulletin* 96: 437–450.

Foster, D.R., and T.M. Zebryk. 1993. Long-term vegetation dynamics and disturbance history of a tsuga-dominated forest in New England. *Ecology* 74: 982–998.

Foti, T.L., and S.M. Glenn. 1991. The Ouachita Mountain landscape at the time of settlement. pp. 49–66. In: D. Henderson and L.D. Hedrick (eds.), *Restoration of Old Growth Forests in the Interior Highlands of Arkansas and Oklahoma*. Proceedings of the Conference, September 19–20, 1990, Morrilton, AR. Ouachita National Forest and Winrock International Institute.

Franklin, J.F. 1989. Toward a new forestry. *American Forests*, November/December: 37–44.

Frelich, L.E., and L.J. Graumlich. 1994. Age-class distribution and spatial patterns in an old-growth hemlock-hardwood forest. *Canadian Journal of Forest Research* 24: 1939–1947.

Frelich, L.E., and C.G. Lorimer. 1985. Current and predicted long-term effects of deer browsing in hemlock forest in Michigan, USA. *Biological Conservation* 34: 99–120.

Frelich, L.E., and C.G. Lorimer. 1991. Natural disturbance regimes in hemlock-hardwood forests of the Upper Great Lakes Region. *Ecological Monographs* 61: 145–164.

Fries, C., O. Johansson, B. Petterson, and P. Simonsson. 1997. Silvicultural models to maintain and restore natural stand structure in Swedish Boreal Forests. *Forest Ecology and Management* 94: 89–103.

Grigore, M.T., and E.J. Tramer. 1996. The short-term effect of fire on *Lupines paranoias* (L.). *Natural Areas Journal* 16: 41–48.

Grimm, E.C. 1983. Chronology and dynamics of vegetation change in the prairie-woodland region of southern Minnesota, USA. *The New Phytologist* 93: 311–350.

Gruell, G.E. 1983. Fire and vegetative trends in the Northern Rockies: Interpretations from 1871–1982 photographs. USDA Forest Service General Technical Report INT-158.

Guyette, R., and E.A. McGinnes, Jr. 1982. Fire history of an Ozark Glade in Missouri. *Transactions of the Missouri Academy of Science* 16: 85– 93.

Hadley, K.S. 1994. The role of disturbance, topography, and forest structure in the development of a montane forest landscape. *The Bulletin of the Torrey Botanical Club* 121: 47–61.

Hansen, A.J., T.A. Spies, F.J. Swanson, and J.L. Ohmann. 1991. Conserving biodiversity in managed forests. *Bioscience* 41: 382–392.

Heinselman, M.L. 1973. Fire in the virgin forests of the Boundary Waters Canoe Area, Minnesota. *Quaternary Research* 3: 329–382.

Henry, J.D., and J.M.A. Swan. 1974. Reconstructing forest history from live and dead plant material — an approach to the study of forest succession in southwest New Hampshire. *Ecology* 55: 772–783.

Herkert, J.R. 1994. Breeding bird communities of midwestern prairie fragments: The effects of prescribed burning and habitat-area. *Natural Areas Journal* 14: 128–135.

Hoerr, W. 1993. The concept of naturalness in environmental discourse. *Natural Areas Journal* 13: 29–32.

Hoff, R., and S. Hagle. 1990. Diseases of Whitebark Pine with special emphasis on White Pine Blister Rust. pp. 179–190. In: W.C. Schmidt and K.J. McDonald (compilers). Proceedings — Symposium on Whitebark Pine Ecosystems: Ecology and Management of a High-mountain Resource. USDA Forest Service General Technical Report INT-270.

Holling, C.S. 1995. What barriers? What bridges? pp. 3–36. In: L.H. Gunderson, C.S. Holling, and S.S. Light (eds.), *Barriers and Bridges to the Renewal of Ecosystems and Institutions*. Columbia University Press, New York, NY.

Holling, C.S., and G.K. Meffe. 1996. Command and control and the pathology of natural resource management. *Conservation Biology* 10: 328–337.

Hough, A.F. 1965. A twenty-year record of understory vegetational change in a virgin Pennsylvania forest. *Ecology* 46: 370–373.

Hunter, M.L. Jr. 1990. *Wildlife, Forests, and Forestry: Principles of Managing Forests for Biological Diversity.* Prentice Hall, Englewood Cliffs, NJ, 370 pp.

Hunter, M.L. Jr. 1993. Natural fire regimes as spatial models for managing boreal forests. *Biological Conservation* 65: 115–120.

Hunter, M.L. Jr. 1996. Benchmarks for managing ecosystems: Are human activities natural? *Conservation Biology* 10: 695–697.

Jansma, J., and H.J. Jansma. 1991. George Englemann in the Arkansas Territory. *Arkansas Historical Quarterly.* Vol. L: 225–248.

Kaufmann, M.R., R.T. Graham, D.A. Boyce, Jr., W.H. Moir, L. Perry, R.T. Reynolds, R.L. Bassett, P. Mehlhop, C.B. Edminster, W.M. Block, and P.S. Corn. 1994. An ecological basis for ecosystem management. USDA Forest Service General Technical Report RM-246.

Kimball, A.J., J.W. Witham, J.L. Rudnicky, A.S. White, and M.L. Hunter, Jr. 1995. Harvest-created and natural canopy gaps in an oak-pine forest in Maine. *The Bulletin of the Torrey Botanical Club* 122: 115–123.

Kohm, K.A., and J.F. Franklin. 1997. *Creating a Forestry for the 21st Century.* Island Press, Washington, DC.

Kreiter, S.D. 1995. Dynamics and spatial pattern of a virgin old-growth hardwood-pine forest in the Ouachita Mountains, Oklahoma, from 1896 to 1994. M.S. Thesis. Oklahoma State University.

Lewis, C.E., and T.J. Harshbarger. 1976. Shrub and herbaceous vegetation after 20 years of prescribed burning in the South Carolina Coastal Plain. *Journal of Range Management* 29: 13–18.

Light, S.S., L.H. Gunderson, and C.S. Holling. 1995. The Everglades: Evolution of management in a turbulent ecosystem. pp. 103–168. In: L.H. Gunderson, C.S. Holling, and S.S. Light (eds.), *Barriers and Bridges to the Renewal of Ecosystems and Institutions.* Columbia University Press, New York.

Lorimer, C.G. 1977. The presettlement forest and natural disturbance cycle of northeastern Maine. *Ecology* 58: 139–148.

Lorimer, C.G., and L.E. Frelich. 1989. A methodology for estimating canopy disturbance frequency and intensity in dense temperate forests. *Canadian Journal of Forest Research* 19: 651–663.

Maissurow, D.K. 1935. Fire as a necessary factor in the perpetuation of white pine. *Journal of Forestry* 33: 373–378.

Masters, R.E., J.E. Skeen, and J. Whitehead. 1993. Preliminary fire history of McCurtain County Wilderness Area and implications for Red-cockaded Woodpecker management. pp. 290–302. In: R. Costa, D.L. Kulhavy, and R.G. Hooper (eds.). *Red-cockaded Woodpecker Symposium III: Species Recovery, Ecology and Management.*

McClain, W.E., M.A. Jenkins, S.E. Jenkins, and J.E. Ebinger. 1993. Changes in the woody vegetation of a Bur Oak savanna remnant in central Illinois. *Natural Areas Journal* 13: 108–114.

McClaran, M.P., and J.W. Bartolome. 1989. Fire-related recruitment in stagnant *Quercus douglasii* populations. *Canadian Journal of Forest Research* 19: 580–585.

McIntosh, R.P. 1992. Succession and ecological theory. pp. 10–23. In: D.C. West, H.H. Shugart, and D.B. Botkin (eds.). *Forest Succession: Concepts and Application.* Springer-Verlag, New York, NY.

Mehlman, D.W. 1992. Effects of fire on plant community composition of north Florida second growth pineland. *The Bulletin of the Torrey Botanical Club* 199: 376–383.

Mladenoff, D.J., M.A. White, J. Pastor, and T.R. Crow. 1993. Comparing spatial pattern in unaltered old-growth and disturbed forest landscapes. *Ecological Applications* 3: 294–306.

Morgan, P., G.H. Aplet, J.B. Haufler, H.C. Humphries, M.M. Moore, and W.D. Wilson. 1994. Historical range of variability: A useful tool for evaluating ecosystem change. *Journal of Sustainable Forestry* 2: 87–111.

Murray, M.P. 1996. Natural processes: Wilderness management unrealized. *Natural Areas Journal* 16: 55–61.

Myers, R.L. 1991. High pine and scrub. pp. 150–193. In: R.L. Myers and J.J. Ewel (eds.), *Ecosystems of Florida*, University of Central Florida Press, Orlando, FL.

Noss, R.F. 1987. From plant communities to landscapes in conservative inventories: A look at the Nature Conservancy (USA). *Biological Conservation* 41: 11–37.

Nowacki, G.J., and M.D. Abrams. 1997. Radial-growth averaging criteria for reconstructing disturbance histories from presettlement-origin oaks. *Ecological Monographs* 67: 225–249.

Nowacki, G.J., M.D. Abrams, and C.G. Lorimer. 1990. Composition, structure, and historical development of northern red oak stands along an edaphic gradient in north-central Wisconsin. *Forestry Science* 36: 276–292.

Palik, B., and N. Pederson. 1996. Overstory mortality and canopy disturbances in longleaf pine ecosystems. *Canadian Journal of Forestry Research* 26: 2035–2047.

Paisley, C. 1990. *The Red Hills of Florida, 1528–1865.* University of Alabama Press, Tuscaloosa, AL.

Perry, D.A., and M.P. Amaranthus. 1997. Disturbance, recovery, and stability. pp. 31–56. In: K.A. Kohm and J.F. Franklin (eds.), *Creating a Forestry for the 21st Century.* Island Press, Washington, DC.

Pickett, S.T.A. and P.S. White. 1985. Patch dynamics: A synthesis. pp. 371–384. In: S.T.A. Pickett and P.S. White (eds.), *The Ecology of Natural Disturbance and Patch Dynamics.* Academic Press, Orlando, FL.

Poff, N.L., J.D. Allan, M.B. Bain, J.R. Karr, K.L. Prestegaard, B.D. Richter, R.E. Sparks, and J.C. Stromberg. 1997. The natural flow regime: a paradigm for river conservation and restoration. *BioScience* 47: 769–784.

Progulske, D.R. 1974. *Yellow Ore, Yellow Hair, Yellow Pine — A Photographic Study of a Century of Forest Ecology.* South Dakota State University Agricultural Experiment Station Bulletin 616.

Pyne, S.J. 1982. *Fire in America: A Cultural History of Wildland and Rural Fire*. Princeton University Press, Princeton, NJ, 654 pp.

Reice, S.R. 1994. Non-equilibrium determinants of biological community structure. *American Scientist* 82: 424–435.

Richter, B.D., J.V. Baumgartner, R. Wiginton, and D.P. Braun. 1997. How much water does a river need? *Freshwater Biology* 37: 231–249.

Robbins, L.E., and R.L. Myers. 1992. *Seasonal Effects of Prescribed Burning in Florida: A Review*. Tall Timbers Research Station Miscellaneous Publication No. 8.

Rogers, C.S., and R.C. Anderson. 1979. Presettlement vegetation of two prairie peninsula counties. *Botanical Gazette* 140: 232–240.

Rogers, P. 1996. Disturbance ecology and forest management: A review of the literature. USDA Forest Service General Technical Report INT-GTR-336.

Rundle, T. 1994. *An Annotated Bibliography of Research in the Chattooga Watershed and Adjacent Watersheds*. Biology Department, Western Carolina University, Cullowhee, NC.

Runkle, J.R. 1981. Gap regeneration in some old-growth forests of the eastern United States. *Ecology* 62: 1041–1051.

Runkle, J.R. 1982. Patterns of disturbance in some old-growth mesic forests of eastern North America. *Ecology* 63: 1533–1546.

Russell, E.W.B. 1983. Indian-set fires in the forests of the northeastern United States. *Ecology* 64: 78–88.

Russell, E.W.B., R.B. Davis, R.S. Anderson, T.E. Rhodes, and D.S. Anderson. 1993. Recent centuries of vegetational change in the glaciated north-eastern United States. *Journal of Ecology* 81: 647–664.

Sauer, C.O. 1975. Man's dominance by use of fire. *Geoscience and Man* 10: 1–13.

Seischab, F.K., and D. Orwig. 1991. Catastrophic disturbances in the presettlement forests of Western wew York. *The Bulletin of the Torrey Botanical Club* 118: 117–122.

Shinneman, D.J. and W.L. Baker. 1997. Nonequilibrium dynamics between catastrophic disturbances and old-growth forests in ponderosa pine landscapes in the Black Hills. *Conservation Biology* 1276–1288.

Sidle, J.G., and C.A. Faanes. 1997. Platte River Ecosystem resources and management, with emphasis on the Big Bend Reach in Nebraska. U.S. Fish and Wildlife Service, Grand Island, Nebraska. Jamestown, ND: Northern Prairie Wildlife Research Center Home Page. http://www.npsc.nbs.gov/resource/othrdata/platte2/platte2.htm

Skovlin, J.M., and J.W. Thomas. 1995. Interpreting Long-term Trends in Blue Mountain Ecosystems From Repeat Photography. USDA Forest Service General Technical Report PNW-GTR-315.

Smith, D.M., B.C. Larson, M.J. Kelty, P.M.S. Ashton. 1997. *The Practice of Silviculture: Applied Forest Ecology* (9th ed.), Wiley, New York, NY.

Sprugel, D.G. 1991. Disturbance, equilibrium, and environmental variability: What is 'natural' vegetation in a changing environment? *Biological Conservation* 58: 1–18.

Stewart, O.C. 1956. Fire as the first great force employed by man. pp. 115–133. In: W.H. Thomas (ed.), *Man's Role in Changing the Face of the Earth*. University of Chicago Press, Chicago, IL.

Swanson, F.J., J.A. Jones, D.O. Wallin, and J.H. Cissel. 1994. Natural variability —Implications for ecosystem management. pp. 80–94. In: M.E. Jensen and P.S. Bourgeron (Tech. eds.), Volume II: Ecosystem Management: Principles and Applications. USDA Forest Service, General Technical Report PNW-GTR-318, Pacific Northwest Research Station.

Transeau, E.N. 1935. The Prairie Peninsula. *Ecology* 16: 423–437.

Turner, M.G., V.H. Dale, E.H. Everham III. 1997. Fires, hurricanes, and volcanoes: Comparing large disturbances. *BioScience* 47: 758–768.

USDI Bureau of Reclamation. 1995. Operation of Glen Canyon Dam, Colorado River Storage Project, Arizona — Final Environmental Impact Statement. 425 pp.

Whitney, G.G. 1986. Relation of Michigan's presettlement pine forests to substrate and disturbance history. *Ecology* 67: 1548–1559.

Whitney, G.G. 1987. An ecological history of the Great Lakes forest of Michigan. *Journal of Ecology* 75: 667–684.

## THE AUTHORS

**R. Todd Engstrom**
*Tall Timbers Research Station*
*Route 1, Box 678*
*Tallahassee, FL 32312, USA*

**Sam Gilbert**
*USDA Forest Service,*
*Helena National Forest,*
*2880 Skyway Drive,*
*Helena, MT 59601, USA*

**Malcolm Hunter**
*University of Maine,*
*Department of Wildlife,*
*College of Forest Resources,*
*Orono, ME 04469-0125*

**David Merriwether**
*USDA Forest Service,*
*Southern Region,*
*1720 Peachtree Road, NW,*
*Atlanta, GA 30367, USA*

**Greg Nowacki**
*USDA Forest Service,*
*Region 10 Regional Office,*
*Federal Office Building, Box 21628,*
*Juneau, Alaska 99802-1638, USA*

**Page Spencer**
*NPS Alaska SSO,*
*Reclamation Specialist,*
*2525 Gambell Street,*
*Anchorage, Alaska 99503, USA*

Wiens, J.A. 1997. Scientific responsibility and responsible ecology. *Conservation Ecology* [online] 1: 16. Available from the Internet. http: //www.consecol.org/vol1/iss1/art16

Wright, H.A., and A.W. Bailey.1982. *Fire Ecology: United States and Southern Canada*. Wiley, New York, NY. 501 pp.

Wunderle, J.M., Jr., and J.W. Wiley. 1996. Effects of hurricanes on wildlife: implications and strategies for management. pp. 253–264. In: R.M. DeGraaf and R.I. Miller (eds.), *Conservation of Faunal Diversity in Forested Landscapes*. Chapman and Hall, London.

# Scale Considerations for Ecosystem Management

Jonathan B. Haufler, Thomas Crow, and David Wilcove

**Key questions addressed in this chapter**

- *Spatial and temporal components of scale that are important to ecosystem management.*
- *Why careful consideration of scale is critical to ecosystem management.*
- *Criteria and suggestions for determining the extent of planning landscapes.*
- *Considerations in identifying appropriate resolution of mapping or data.*
- *Time-spans for ecosystem management planning*
- *Time-spans for historical perspectives.*

**Keywords: Landscape planning, spatial scale, temporal scale, mapping**

## 1 INTRODUCTION

One of the difficult challenges facing ecosystem management is the determination of appropriate spatial and temporal scales to use. Scale in a spatial sense includes considerations of both the size area or extent of an ecosystem management activity, as well as the degree of resolution of mapped or measured data. In the temporal sense, scale concerns the duration of both natural and human induced disturbances, duration and time intervals of successional trajectories, the appropriate planning horizon of future activities, and the length of any historical perspective.

A review of literature relating to scale as a component of ecosystem or natural resource management reveals how recently identified this topic is, with considerable focus of attention only within the last two decades. Early plant ecologists such as Clements (1916) and Tansley (1924) addressed questions about plant dynamics at a stand (homogeneous group of plants) level. This focus dominated debates concerning species composition, stand structural relationships, and successional trajectories through the next 50 years. Similarly, most animal ecologists focused their attention on the relationships among species, or in describing stand-level habitat conditions required by different species. Leopold (1933) contributed a broader view of wildlife–habitat relationships by introducing the multi-stand concepts of edge, interspersion, and juxtaposition. The importance of integrating regional geography and vegetation science was first termed landscape ecology by Troll (1939), as discussed by Turner and Gardner (1991). Greig-Smith (1952) discussed the importance of scale in evaluating the distributional patterns of plants. However, it was not until the 1980s that scale issues relative to resource management became a major component in the ecological literature. Schneider (1994:2) stated "In reading the ecological literature prior to 1980, I gained the impression that nearly all papers before 1980 treat scale either implicitly, or not at all." He attributed this partially to the relatively recent technological advances in computers, geographical information systems (GIS), and remote sensing tools that have facilitated many new types of scale analyses.

Ecosystem management has emerged as a way of addressing increasingly complex management planning needs. It can be defined in various ways, but regardless of a specific definition, it generally requires management decisions over a large geographic area (Gregg 1994). Effective ecosystem management will need to utilize many of the tools developed by the expanding field of landscape ecology. As Wiens (1992) noted, scale issues are one of the largest future challenges to ecologists.

In this paper on scale considerations, we address the following key issues:

- Spatial and temporal components of scale that are important to ecosystem management.
- Why careful consideration of scale is critical to ecosystem management.
- Criteria and suggestions for determining the extent of planning landscapes.
- Considerations in identifying appropriate resolution of mapping or data.
- Time-spans for ecosystem management planning
- Times-spans for historical perspectives.

Scale considerations are integrally linked to the definition and objectives of ecosystem management. Ecosystem management generally involves the consideration of ecological, social, and economic objectives, each of which will require different scale considerations. Although all three of these are important objectives for successful ecosystem management, this paper will emphasize the consideration of scale issues primarily for ecological objectives.

## 2 BACKGROUND FOR SCALE CONSIDERATIONS

Scale is defined by the size and extent of the observations in time and space as well as by the resolution (i.e., pixel size or grain) of the measurements. Scale is relative because it is either large or small compared to some reference generally defined by an observer (Hoeskstra et al. 1991). The discussion of scale relative to ecosystem management requires the use of numerous terms, some of which often have different meanings. The use of scale terminology has differed somewhat between ecologists and geographers. Geographers use the terms large and small scale to describe the scale of a map, with a large-scale map depicting less land area per cm of map than a small-scale map. Ecologists have generally referred to large scale as a description for an analysis of a large-sized area. In this paper, we define terms in the following ways:

- *size or extent*: the amount of area or length of time contained in a delineated landscape or time-span, or a measure of its breadth and width or duration;
- *stand*: an identified area with relatively homogeneous structure and composition of vegetation;
- *resolution*: the level of detail, such as pixel size or graininess, that is incorporated into the mapping of an area or in the collection of data;

- *coarse filter*: an approach to ecosystem management that involves providing for an appropriate mix of ecological communities across a planning landscape;
- *fine filter*: an approach to ecosystem management that involves a focus on the needs of individual species or groupings of species as a basis for landscape planning;
- *coarse scale*: a level of resolution or grain size used in mapping or measuring data based on units such as large pixel sizes, large grain, broad categories, etc.;
- *fine scale*: a level of resolution or grain size used in mapping or measuring data based on units such as small pixel sizes, small grain, detailed data, etc.;
- *broad scale*: an area of analysis or management with a large extent, containing a relatively large amount of acreage or a long duration;
- *small scale*: an area of analysis or management with a small extent, containing a relatively small amount of acreage or a short duration.

How we perceive an object or a phenomenon is greatly influenced by the scale, both in space and time, at which it is viewed. This rather obvious fact has important implications for both science and resource management (Hoekstra et al. 1991). In many published studies, there is no recognition of the sensitivity of the results to the scales at which they are conducted. Indeed, it is not unusual for the same question to be studied at many different spatial and temporal scales. In ecosystem management, the approach used and objectives being addressed will have a direct bearing on the appropriate scale to use. If a fine filter approach is being used, then the planning environment should consider the needs of the specific species of interest. Typically resource managers have mapped landscapes into stands on the basis of what they observe to be homogeneous conditions relative to their land management objectives. However, from the vantage point of a small mammal, the important components of a stand might look completely different, and scale related issues would be very different. Instead of looking at the composition and structure of vegetation in a forest in terms of the overstory of trees, the critical scale might be the arrangement and patchiness of the understory herbaceous vegetation. Addressing a question at the wrong scale often leads to a failure of explanation and to the wrong conclusions (Wiens 1989, Turner 1990).

Turner and Gardner (1991) distinguished between scale and level of organization. They defined scale as "the spatial or temporal dimension" (1991: 6), whereas level of organization was defined as "the place within

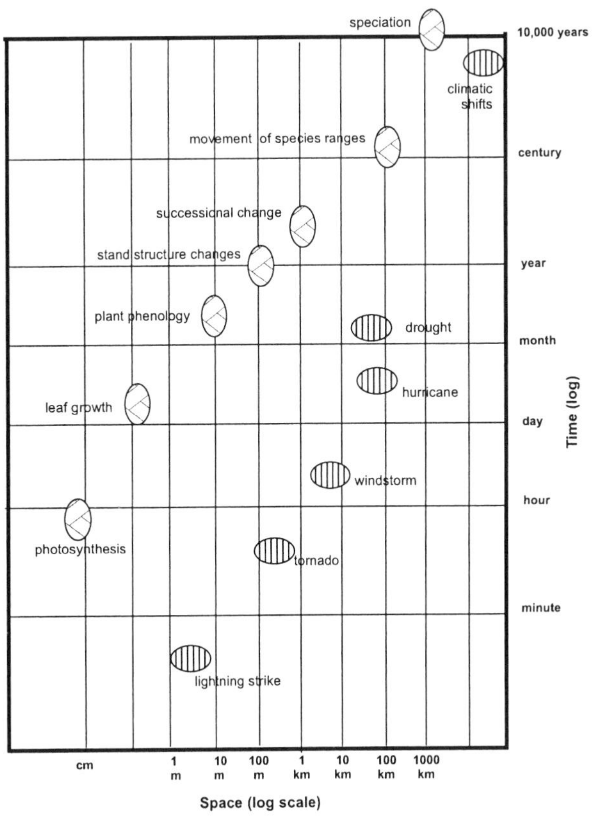


Fig. 1. Space/time hierarchy (after Holling 1995) for plant biotic processes and climatic disturbance events.

some biotic hierarchy" (1991: 6). Others have described scale as operating in a hierarchical fashion. Spatial hierarchies have been described for ecological diversity from niches to biospheres (Miller 1996), for ecological classification from sites to domains (Ecomap 1993), for disturbance activities from small mammal influences to major floods (Bourgeron and Jensen 1994), and for aquatic communities from channel units to river basins (Maxwell et al. 1995). Temporal hierarchies have been described for aquatic systems (Maxwell et al. 1995) and disturbance regimes (Bourgeron and Jensen 1994). Holling (1995) (Fig. 1) depicted a combined spatial and temporal hierarchy. Hierarchy theory provides an organizing framework to search for common properties across broad classes of complex systems, including physical, biological, social, and artificial systems. It is important to recognize these hierarchical levels of organization, and their potential influences on defining appropriate scales.

How do we take fundamental information about ecological processes obtained at fine scales (e.g., physiological response of leaves to elevated levels of ozone) and apply it to responses observed at broad scales (e.g., a landscape, a region, or even the entire ecosphere)? Understanding the properties of complex hierarchical organizations is useful in addressing this question (Pattee 1973, Simon 1973). For example, the transfer of information from one level of ecological organization to another is not a simple additive process. Vital information from lower levels in a hierarchy may be extraneous information at higher levels of organization. Furthermore, levels in an organizational hierarchy can be isolated from one another because they operate at distinctly different rates. A leaf, for example, is sensitive to changes in light conditions that can be measured in seconds and minutes. A forest stand integrates this information over weeks, months, and growing seasons, and its growth responses are measured in these timeframes. Successional change influences light conditions over years and decades. At each level in the organizational hierarchy, new organizing principles apply and new properties emerge, so properties of whole systems cannot effectively be predicted from the properties of simpler subsystems (Allen and Starr 1982).

What is an appropriate area (extent) for ecosystem management and planning? In determining extent, numerous factors should be considered, including ecological, economic, legal, political, and other social considerations, all of which are influenced by different factors operating at potentially different scales. This determination must consider the minimum-size area that is capable of addressing the stated objectives of the initiative as well as the maximum-size area that can feasibly be included from the standpoint of data collection, data storage and analysis, collaborative partnerships, and the resources available to do the job. The extent of the area can also influence the resolution to be used in mapping or data collection. The data must have the necessary or required precision to achieve the desired objectives, but also be feasible to collect, store, and analyze. No one scale of extent or resolution will meet all of the objectives of ecosystem management, but instead, multiple spatial scales are likely to be incorporated into the decision-making process. Project planning can involve considering a range of spatial scales from a few hectares to millions of hectares. In reality, however, decisions are most often made at local levels by people working on the ground. It is important that these decisions be made within the broader context provided by a landscape or regional assessment or a regional planning framework (Weintraub and Cholacky 1991, Crow and Gustafson 1996).

Time-frames to be used in ecosystem management also require careful considerations. The planning time-frame for ecosystem management must consider the practical realities of legal, political, and legislative constraints, but also the relevance and requirements of ecological parameters. Time-frames for historical perspectives must balance the availability of information, the influences of industrial societies on disturbance regimes, as well as factors such as global climate changes (e.g., ice ages), shifts in species' ranges and interactions, and evolutionary processes.

## 3 RELEVANCE OF SCALE TO ECOSYSTEM MANAGEMENT

Ecosystems are not closed, self-supporting systems, but rather are parts of larger interacting systems. Significant problems arise if ecosystems are treated as an isolated "entity." Rather, the ecosystems in an area must be viewed in the context created of the broader surrounding landscape. For example, providing habitat for forest interior birds often involves identifying within-stand areas of sufficient distance from an edge to provide protection from parasitism from brown-headed cowbirds (*Molothrus ater*) (Brittingham and Temple 1983, Reese and Ratti 1988). However, Robinson (1990) reported that the Shawnee National Forest was so saturated with cowbirds from the surrounding agricultural landscape that wood thrush (*Hylocichla mustelina*) nests were just as heavily parasitized (90% parasitism) at 400 m from an edge as at the edge. In contrast, Stribley (1993) found that in the primarily forested area of northern Michigan, cowbird parasitism was not a problem for nests at any proximity to edges, if agricultural lands or other cowbird foraging areas were located farther than 3 km from the site. Thus, surrounding lands can have a significant influence on relationships occurring within similar stand types in two different landscapes, and can significantly influence the character of a habitat patch (Janzen 1986).

Maintaining biological diversity, including genetic, species, and ecosystem levels, is one generally identified objective of ecosystem management. Attainment of ecological objectives, such as maintaining biological diversity, requires at least some minimum spatial area and time span. A fine filter approach is typically used for maintaining viable populations with sufficient genetic diversity and interchange to avoid inbreeding concerns and to provide sufficient resilience against demographic or environmental stochasticities. In contrast, a coarse filter approach, strives to provide the ecological communities necessary to maintain ecosystem function and integrity, and thus provide for viable populations of species within these ecological communities over an appropriate time span.

The area required to meet these ecological objectives can vary greatly, especially depending on how temporal considerations are factored into the planning effort. For example, maintenance of population viability is an essential component of ecological objectives relating to biodiversity. Population viability analysis attempts to predict the probability of persistence of a plant or animal population over a specified time period under various habitat conditions (see Noon et al. and Holthausen et al., this volume, ). Viable populations of different species will have different area requirements that need to be factored into the planning effort. Roloff and Haufler (1997) described the different home range requirements of snowshoe hare (*Lepus americanus*) and Canadian lynx (*Felis lynx*) and how they relate to population viability. Population viability of snowshoe hare may be achieved in a much smaller area than Canadian lynx. Lynx depend upon the distribution and population sizes of snowshoe hare over a larger area. Thus, assuring adequate consideration for population viability of lynx requires that population distribution and relative size for snowshoe hare be determined first. Snowshoe hare habitat requirements must be addressed at scales appropriate to this species. Data on hare relative abundance are then aggregated into larger spatial scales that more closely approximate the areas required by lynx populations. Viability analyses for both of these species at appropriate scales can then be factored into analyses of ecosystem diversity that can occur for landscapes of an even larger extent.

Maintaining population viability of species must also address important temporal questions. Managers must select a time-span for assessing population viability, a decision that typically reflects a blend of ecological and political concerns as well as realistic long-term planning capabilities. In the Pacific Northwest, for example, federal land management agencies chose a 100-year time-span for predicting the impacts of different timber-harvest levels on wildlife associated with late-successional forests (Forest Ecosystem Management Assessment Team 1993). This time-span might seem long relative to actual applicability of forest plans, which seldom stay consistent for even 10 years. However, 100 years is a relatively short time-span when viewed from the context of the average "life span" of a species (i.e., the time a species survives before it goes extinct or evolves into new species). Based on fossil records, Ehrlich and Wilson (1991) reported that the life span of most species has ranged from one to ten million years. Species exist as aggregates of discrete populations. Individual populations of a species may exist for a much shorter period of time than the life span of the species, and it is only when the last population has expired that the species becomes extinct. By some estimates, because of human activities, the extinction rate of species today may be nearly 100 times the average level of the past 50 million years (Ehrlich 1986). When viewed from this temporal perspective, the importance of populations of a species as interim contributors to an evolving species continuum becomes more apparent.

Another temporal concern is the persistence or rates of successional change within natural communities. Some ecosystems, such as prairies or oak (*Quercus* spp.) savannas, were historically maintained by frequent (1–25 year interval) low intensity ground fires (Noss and Cooperrider 1994). Other systems, such as the Sitka spruce (*Picea sitchensis*) forests along the Washington coast may have fire-return intervals of up to 1,100 years (Fahnestock and Agee 1983), with wind events as a disturbance factor occurring on a more frequent and persistent basis (Ruth and Harris 1979, Agee 1993). Thousand-year disturbance regimes challenge the capabilities of our planning time-spans (see White et al., Engstrom et al., this volume)..

The level of detail or resolution of data used in making ecosystem management decisions can also vary depending on the approach being used. Using a fine filter approach or addressing the needs of one or more species may require fine scale data to describe adequately specific habitat requirements. Using a coarse filter approach to identify the diverse communities necessary to provide for ecosystem diversity may require a lower resolution, or coarser scale data, although finer scale data should work equally well assuming data handling capabilities are sufficient. Thus, scale considerations, both spatial and temporal, are significant concerns in planning efforts that address biodiversity objectives.

Ecosystem analyses must span several scales whenever the management actions being considered affect ecological processes that operate at different scales. For example, a timber harvesting activity has a direct effect on the process of regeneration at the stand level. But the same activity can, through cumulative effects, change sediment deposition and insect population dynamics at the watershed level and economic outcomes at a local community level. However, if one were to consider even all the ecological processes that could be potentially affected, the analysis would quickly get out of hand. Obviously, some practical bounds are essential. Some practical tools for ecosystem management are also needed. Decision-support models such as the northeast Decision (Kollasch and Twery 1995) and spatial models such as the allocation model HARVEST (Gustafson and Crow 1996), or the forest succession and landscape management model LANDIS (Mladenoff et al. 1996) are helpful for evaluating multiple factors and interactions at several spatial scales.

## 4 SCALE CONSIDERATIONS

Scale considerations that should be addressed in any ecosystem management process include:

- the extent of the planning landscape;

- the appropriate resolution for mapping or data collection;
- the planning time-span; and
- the time-span for an historical perspective.

## 4.1 Extent of Planning Landscape

One of the first steps in ecosystem management, following an initial identification of the participants and management objectives, is determining the extent and boundary(ies) of the planning landscape (Haufler et al. 1996). This involves considering the appropriate size of the landscape, and the boundary criteria. Haufler et al. (1996: 201) listed the following as criteria for delineating ecological boundaries:

- Similar biogeoclimatic conditions that influence site potentials.
- Similar historical disturbance regimes that influence vegetation structures and species compositions.
- Adequately sized landscape to provide sufficient ranges of habitat conditions to assure population maintenance of the majority of native species that historically occurred in the planning landscape, excluding certain species such as megafauna. Megafauna or species with low population densities will require analyses at broader scales where contributions from landscapes are aggregated to address population maintenance of these species.
- Recognition of maximum size to avoid practical operational limitations in terms of data management, implementation restrictions, and number of cooperating landowners necessary for successful plans.

Ecomap (1993) and Maxwell et al. (1995) provide a hierarchical classification for boundary determination that can be used to define a planning landscape. Haufler et al. (1996), using the above criteria and classification, advocated using the section or aggregates of subsections level (equivalent to Bailey's subregions (Bailey 1995, 1996)), to provide an appropriate landscape size to meet ecological objectives while being operationally functional. In Idaho, Haufler et al. (1996) described the Idaho Southern Batholith landscape comprised of an aggregation of subsections containing 2.3 million ha within the Idaho Batholith Section (McNab and Avers 1994). In Washington State, an aggregation of subsections was selected to delineate the 1.1 million ha Central East Cascade planning landscape (D. Volsen, Boise Cascade Corp., pers. comm.) In Minnesota, the entire Northern Minnesota/Ontario Peatlands Section was identified as a planning landscape for an ecosystem management initiative in that state (B. Kernohan, Boise Cascade Corp., pers. comm.). Within each of these planning landscapes, an ecosystem diversity matrix (Haufler 1994, 1995, Haufler et al. 1996) was then developed to characterize and quantify the forested ecosystems. This matrix provides for the quantification of ecological land units within the planning landscape. An additional criteria for delineating planning landscapes is that ecological land units within the landscape should be consistent enough throughout the landscape that they have an acceptable level of variability for key habitat variables or characteristics (Roloff 1994). In other words, all stands that comprise an ecological land unit should have similar enough composition, structure, or other characteristics so that habitat variables that describe the unit have small enough variance for determining habitat quality or quantity for a species.

Other ecosystem management initiatives, with different objectives or organizational structures, have used different scales. The Federal government delineated the Interior Columbia Basin Ecosystem Management Project as a planning landscape for an ecosystem management assessment, focusing on the Federal land holdings (Quigley et al. 1996). The assessment area included more than 58 million ha. At the other end of the scale gradient, the Watershed Analysis Coordination Team (1995) recommended conducting ecosystem analyses based on watersheds of 4500–52,000 ha in size.

No single scale exists for describing ecosystem patterns or diversity (Levin 1992, Noss 1990). The inherent complexity of landscapes results in a mozaic of both micro- and macro-site conditions that provide for the range of patterns apparent at different scales. Turner et al. (1994: 76) stated "View the landscape as a whole and use landscape-level indices to measure pattern at multiple scales. Do not focus solely on single, simple concepts like patches and corridors, and recognize that these concepts are scale-dependent." These views are not inconsistent with a hierarchically based delineation of landscapes for planning purposes. When a planning landscape is delineated, it should contain various descriptors of ecosystem complexity within the planning landscape, and allow interpretation of this complexity into larger-size areas in the hierarchical classification.

Most past management planning has utilized legal, political, and ownership boundaries as the basis for decisions. Although these boundaries remain critical, ecosystem management has added new ecological criteria to land management planning. These new criteria require the consideration and identification of ecological boundaries and scales. The social and economic objectives of an ecosystem management initiative must incorporate the ecological boundaries of the landscape planning unit, but will usually reflect economic markets, political structures, and social influences. All of these operate at multiple-spatial scales. For example, a local community will have many components of its quality of life influenced by the surrounding landscape. The surrounding landscape will largely determine the scenery, recreational opportunities, opportunities for firewood cutting or mushroom picking, as well as commercial extraction to support local commodity-based industries. This same landscape will have additional objectives placed on it by state authorities to meet objectives such as water quality standards. Additional national priorities for wilderness, mineral exploration, or other objectives may override local objectives. Desires at the local level for commodity-based industry are dependent on the economics of global markets. Conversely, there is the possibility that local restrictions on commodity extraction may raise costs of a local supply of commodities to high enough levels that supply is shifted to more distant sources. Whether or not this happens depends upon a complex set of economic factors, but it has the potential to "export" environmental problems to other places with less stringent environmental safeguards.

Meshing the social and economic with the biological and physical worlds remains a major challenge facing resource managers. Significant advances have been made in integrating these disciplines under the general rubric of ecological economics (Constanza et al. 1991). In this emerging science, great importance is attached to the interaction of environment and economics and to themes common to our chapter such as multi-scale synthesis, hierarchical theory, and interconnections.

## 4.2 Resolution Issues for Mapping or Data Collection

The resolution of the mapping units and data used in ecosystem management can have a significant influence on the conclusions of an ecosystem analysis. For example, Gap Analysis has used a fairly coarse scale (1 km pixel) in some of the state analyses. This means that each pixel can be assigned only a single vegetation or ecosystem characteristic. At this resolution, plant communities that typically occur in relatively small patches; for example willow (*Salix* spp.) along riparian zones, will never occur on a map of vegetation types. Species dependent on such plant communities, such as the yellow warbler (*Dendroica petechia*) in much of the Western U.S. will not be recognized as having any available habitat, even though relatively significant amounts of habitat may exist. Thus, using a very coarse scale of analysis, yellow warblers might be identified as a species of concern because of a lack of sufficient resolution to identify the presence of suitable habitat.

In a similar example, Capen et al. (1994) compared the mapping resolution used by a Gap analysis project in Vermont (100 ha pixels) to a finer scale resolution that mapped stands to an average size of 9.5 ha. They found that with the finer scale resolution there were 68 community types on a 62,000-ha study area that were analyzed with species habitat models to support 98 bird species. Using a similar analysis with the Gap data, 56 of the 68 community types were "lost" and only 67 of the 98 bird species were retained. Thus, too coarse a scale of resolution can lead to different and often misleading results of both available ecosystems and associated species.

On the other hand, using data at too fine a scale can overwhelm data storage and analysis capabilities for most planners. One landsat thematic mapper scene (185 km × 170 km) for 7 bands at a 30 m pixel resolution requires 244.3 megabytes of computer memory (C. Campbell, Boise Cascade Corp., pers. comm.). With current technologies leading to capabilities of a 1-m pixel resolution, with potentially 900 times the data generation of a 30-m pixel, the data support needs can become staggering. As computer speed and data handling and storage capabilities expand, these barriers may disappear. At the present, however, real limitations of hardware and software relative to the planning landscape area exist and must be recognized.

Another example of resolution delineation was discussed by Schneider (1994: 27). He stated "The length of the seacoast as measured on a map will differ from that measured by pacing along the beach because the map measurements are at a much coarser scale than pacing. The customary view of this difference is that the beach has a true length and that measurement with a meter stick is closer to the true value than measurement with a larger unit, such as a kilometer stick. But how far do we take this? Should we say that measurement with a meter stick is also inaccurate, and that a centimeter stick must be used instead? How small a stick is necessary to obtain the "true" length?"

For stand delineation or measurements, similar decisions must be made. Fine-scale data, such as a 1-m pixel resolution, theoretically can distinguish a 1-m gap in canopy coverage in a stand of trees. Should all of these gaps be mapped out as separate stands? For most purposes, this would present too fine a resolution of a landscape to interpret relative to mapping of "homogeneous" stands. At some fine scale of resolution, additional precision may be greater than that discerned by a species of interest, if a fine filter assessment is being utilized. For example, it is very unlikely that an elk (*Cervus elaphus*) would respond to a 1-m gap in canopy coverage. However, a vole (*Microtus* spp.) might select a 10-m patch of grass in an understory. What then is appropriate, mapping of 10-m gaps, 50-m gaps, 1-ha openings, or 5-ha openings? The larger the discontinuity in the vegetation that is accepted, the greater will be the variance around parameters for descriptions of stands. The finer the resolution, the more homogeneous the overall stand delineations, up to a point where additional precision is actually sampling variation within otherwise homogeneous stand conditions.

The concept of minimal area of a stand (Mueller-Dombois and Ellenberg 1974, Barbour et al. 1980) has considerable relevancy to this resolution question. Thus, resolution of data and mapping precision are critical considerations in classifying and describing ecosystem management landscapes. The specific use of the classification will help to identify an appropriate resolution. The resolution of data and mapping is usually set by what is available within a planning budget, with little consideration given to the assessment or consequences of using the selected scale.

Resolution considerations also operate for temporal components of ecosystem management. In monitoring or data collection, the sampling intensity or interval should consider the periodicity of the process or phenomena being sampled or monitored. For example, measurement of humidity in riparian zones in the Western United States reveals more variability on a hourly basis (Fig. 2) throughout a day, than variation from different locations at any one point in time (R. Danehy, Boise Cascade Corporation, pers. comm.). Thus, investigations of factors influencing humidity in riparian zones would need to account for the temporal changes in humidity throughout a day in any meaningful analysis of site effects. Similarly, habitat use by many animals is strongly influenced by daily activity patterns. Beyer and Haufler (1994) displayed how failure to monitor habitat use throughout a 24-hour

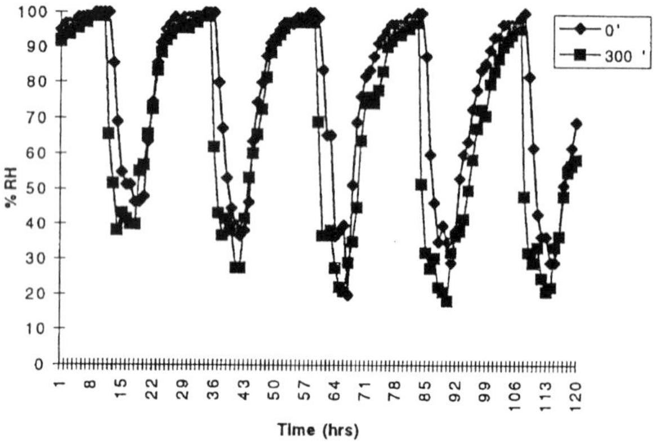


Fig. 2. Graph of relative humidity (RH) recorded hourly at two distances from a stream in Western Oregon. R. Danehy, Boise Cascade Corporation, unpublished data.

time-span for animals that are both diurnally and nocturnally active can lead to inaccurate descriptions of habitat use and importance. At a different temporal scale, vegetation structure, composition, nutrient status, and other factors change seasonally, and even within a season, as plants grow and senesce. These temporal patterns may not be consistent even in fairly local environments depending on influences such as elevation, aspect, shade, and soil moisture. At a longer temporal scale, the vegetation being sampled will change annually with growth and maturation of stands, and successional change. At longer time scales, differences have been noted between ecological time and evolutionary time. Schneider (1994: 28) distinguished these as "Evolutionary time operates on a longer time scale, over which changes in gene frequency can be described as trends, rather than a noisy coming and going of alleles. Ecological time operates on a shorter time scale, over which changes in population size occur with little or no change in gene frequency." The significance of all of these temporal effects on the information being collected by the sampling or monitoring should be considered in sampling designs.

## 4.3 Time-spans for Future Planning

Maintaining biological diversity and ecological processes involves considering time frames that are often far beyond traditional planning horizons. Genetic components of biological diversity may involve the analysis and planning for multiple generations of a species to assure that adequate heterozygosity of gene pools are maintained. Such time frames are beyond the practical realm of resource management decisions, yet ecosystem managers must factor these long-term concerns into the planning process.

Another example of the importance of temporal scales in ecosystem management involves the exchange of genetic information and provisions for demographic and environmental stochasticity among metapopulations of a species. Noss and Cooperrider (1994) and Harris and Gallagher (1989) discussed the importance of maintaining corridors that allow connectivity of similar habitat, such as old growth, to facilitate dispersal and genetic interchange. They felt that this was important to avoid problems with inbreeding or stochastic population fluctuations that could disrupt or extinguish metapopulations. This concept views the landscape as a static condition, where metapopulations are continuously linked by corridors of similar habitat condition. Old growth corridors would provide connectivity among late successional stands in a landscape, assuming animals used the established corridors, but they would also isolate earlier successional stands in the landscape, as it is impossible to provide connectivity of all late successional dependent metapopulations and early successional metapopulations at the same time. While dispersal capabilities of old growth dependent versus early successional dependent species have not been extensively examined, metapopulation considerations point to the importance of addressing all connectivity concerns. By incorporating a temporal scale into the plan, connectivity of diverse populations at appropriate time intervals to protect against inbreeding concerns can be factored into the plan. Camp et al. (1997) and Oliver et al. (1997) discussed a landscape analysis that showed dynamic shifts over time of refugia for early or late successional species, and that through changing landscape configurations, connectivity of all metapopulations can be achieved. Thus, temporal considerations can provide critical linkages in landscape planning designs.

Another consideration for the planning time-span is the duration of the planning horizon relative to the duration of the disturbance regimes of the primary ecosystems in the planning landscape. As mentioned previously, return-intervals for fire disturbance can range from less than 25 years to over 1000 years (Agee 1993). Tides cycle more frequently than daily. Major flood events may occur at very irregular cycles, but aquatic ecosystems may take hundreds of years to recover. Planning for 1000 years is unrealistic from a pragmatic management standpoint. Yet, the significance of 1000-year disturbance intervals should be factored into ecosystem management plans for such ecosystems, without overlooking the importance of the more frequent, often less severe, disturbances.

At very long time-spans (i.e., >1000 years), the effects of shifts in species ranges, and climatic events such as global warming or cooling, could produce changes in species interactions and competitive advantages, and change the basic composition and structure of ecological communities. Some would use such information to argue that any attempts to characterize or classify ecosystem compositions, structures, or distributions are inappropriate. However, the necessity of providing a reasonable projection of ecological outputs from a planning activity overrides these broader temporal scale views of ecosystem dynamics.

### 4.4 Historical Time-span Considerations

Perhaps less ominous but equally challenging is defining an appropriate historical perspective to use in ecosystem management. To meet ecological objectives, understanding historical ranges of variability can provide important reference points. However, what range of variability is being defined? Human alteration of natural disturbance regimes has been dramatic over the last 100–200 years. Even prior to this, indigenous populations were significantly influencing ecosystems to varying degrees. With some disturbances occurring at intervals of 500+ years, range of variability can become intermeshed with shifting species distributions and climatic changes, as discussed above.

Ecosystem management must factor in an understanding of the influences of historical disturbances on the occurrence, structure, and functioning of component ecosystems in the landscape, and understand the relationship of these historical disturbances with recent anthropogenic disturbance. Each landscape must be evaluated as to the appropriate reference timeframe for an understanding of historical disturbance regimes. In the Western United States, an historical perspective might focus on a time-span of 100–400 years ago (Steele 1994). In the Eastern United States, because of the earlier extent of dramatic anthropogenic influences, the appropriate time-span for analysis and quantification may be from 200–400 years ago.

Incorporating historical perspectives into ecosystem management has severe restrictions from availability of data. Typically, forests in drier landscapes have preserved more stumps and logs that can be used for dendrochronology studies, but in wetter environments, sources of data may be very restricted. Morgan et al. (1994) discussed sources of information on historical disturbance regimes.

## 5 SCALE RECOMMENDATIONS

All efforts at ecosystem management at a landscape scale are new approaches, and as such, should be viewed as experimental and adaptive programs. This is not to imply that landscape approaches are not based on the best available information, or that they are in any way inappropriate management directions, but rather, that their results should be monitored and adjusted as needed. Scale recommendations fall directly into this category. Little empirical data exist as a direct basis for designating appropriate scales. However, various recommendations can be proposed that may serve as initial targets for adaptive management.

The extent of an ecosystem management initiative needs to balance the various objectives and constraints, as discussed previously. The section level or aggregates of sub-sections within Ecomap (1993) seems to be a reasonable balance of minimum size to meet ecological objectives with a maximum size to maintain an acceptable amount of variability in ecological communities and to involve partnerships and data compilation. Terrestrial components, such as cells within an ecosystem diversity matrix (Haufler et al. 1996), should have acceptable levels of variance at this scale, with most of the biodiversity of the landscape occurring in maintainable populations. Aquatic components should also be able to aggregate to this scale, with watersheds as one level of identifiable units that can aggregate up to the landscape level (Bailey 1996). This extent of a landscape should also allow data to be collected and analyzed at resolutions that are sufficiently fine-grained to meet the needs of fine filter assessments.

Resolution of maps and data collection should be detailed enough to meet the desired objectives, but allow for compatibility with available storage and analysis systems. Landsat imagery at 30-m pixels can meet the needs of many objectives, but may not be detailed enough for some fine filter assessments. It also lacks the ability to discern many kinds of information that may be needed in ecosystem management including understory characteristics, soil typing, and stream classifications. Stands should be mapped at resolutions that provide sufficient description of community and habitat features to allow for all components of biodiversity to be accounted for. If resolutions are too coarse, then components of biodiversity can slip through the heterogeneity of stand conditions in the mapping, and ecological objectives may not be met. Even mapping resolutions of 5–10 ha will not typically identify many small or linear habitats, such as narrow bands of riparian vegetation. Such linear habitat may be the primary habitat of species such as the previously mentioned yellow warbler. Too detailed a resolution may not allow functional homogeneous units to be mapped in a landscape. The appropriate mapping resolution for a planning activity must be carefully evaluated relative to the full suite of management objectives. If the desired level of resolution cannot be obtained at present due to budget or technological limitations, then the potential implications of this should be clearly identified and stated.

Planning time-spans should recognize degrees of expectations at increasing time intervals. Relatively detailed plans are often expected for the immediate future (e.g., 10 years). Less detailed plans, that still prescribe specific targeted conditions and their locations on the ground, might be expected for a 20–50-year horizon. Plans for longer than 50 years tend to portray trends for the conditions that need to be present, and descriptions of how they will be provided, but not to the same degree of specificity as for the shorter time frames. The realization that demands, knowledge, and technologies will undoubtedly change dramatically over the next 20 years makes unrealistic the expectation that detailed plans will remain in effect for long time periods. For longer time periods, it is probably more reasonable to target ecological conditions that are expected to be needed to meet ecological objectives, and assure that short-term plans provide for the capability of providing these conditions in the future. Thus, plans should strive to meet short-term specific objectives, and assure that longer-term objectives are not precluded.

Historical perspectives of ecosystem management are well summarized by Morgan et al. (1994). They felt that the historical ranges of variability of significance to ecosystem management should be "assessed over a time period characterized by relatively consistent climatic, edaphic, topographic, and biogeographic conditions" (1994: 94). For inland forests in the Western United States, Steele (1994) recommended a time interval of 100–400 years for defining historical disturbance regimes. In other areas this time interval may need to start further back to factor in the earlier influences of European settlement.

## 6 CONCLUSIONS

1. Scale considerations are a critical component in all ecosystem management efforts, and include the extent of the planning landscape, the resolution of mapping and data collection, the time-span for the planning horizon, and the time-span for an historical perspective.
2. No one scale will meet all objectives of ecosystem management. Rather, appropriate scales must be selected for the various objectives, and linkages among these scales identified.
3. Analyses at different scales, or at the same scale but in different landscapes, can lead to significantly different conclusions. Thus, many ecosystem management relationships are scale and/or landscape dependent.
4. The spatial extent of planning landscapes must be large enough to address adequately population viability, biodiversity, and other such components of ecological objectives, but not be so large as to cause either too much variance in delineated ecological communities within the landscape, or make infeasible the building of collaborative partnerships or databases. The section level of Ecomap (1993), or aggregates of subsections, may be an example of this balance.
5. The resolution of mapping and data should be detailed enough to allow for the identification of landscape mapping units (e.g., stands, stream reaches) that can provide descriptions of the habitat requirements of species, but allow for a reason-

able identification of homogeneous conditions for planning. Pixel sizes of 30 m, or mapping resolutions of approximately 2 ha may balance these needs for many landscapes, although all management objectives need to be evaluated relative to the planned resolution. Data and budgeting restrictions may preclude a desired level of detail, but desired levels of detail may be targeted for future efforts.

6. Planning time-spans should focus on providing detailed actions for short-term objectives, while also providing for conditions to be produced or maintained to meet long-term objectives. The duration of successional trajectories and disturbance regimes must be factored into planning time-spans.

7. Historical perspectives must address time-spans that allow for historical disturbance regimes to be considered prior to dramatic anthropogenic alteration, but balance this with the length of time that data on these disturbances can be generated. In the inland forests of the Western United States, a 100–400-year perspective may be appropriate (Steele 1994).

## ACKNOWLEDGEMENTS

A number of people have provided useful guidance and input in the development, writing, or editing of this paper. We wish to acknowledge and thank the assistance of Gary Roloff, Rique Campa, Dean Beyer, Brian Kernohan, Carolyn Mehl, Mike Scott, Gary Meffe, Bob Szaro, Jan Zendzimer, and Dave Carothers.

## REFERENCES

Agee, J.K. 1993. *Fire Ecology of Pacific Northwest Forests*. Island Press, Washington, D.C.

Allen, T.F.H., and T.B. Starr. 1982. *Hierarchy, perspectives for ecological complexity*. University of Chicago Press, Chicago, IL.

Bailey, R.G. 1995. *Descriptions of the ecoregions of the United States*. USDA For. Serv. Misc. Publ. 1391.

Bailey, R.G. 1996. *Ecosystem Geography*. Springer, New York.

Barbour, M.G., J.H. Burk, and W.D. Pitts. 1980. *Terrestrial Plant Ecology*. Benjamin/Cummings. Menlo Park, CA.

Beyer, D.E., Jr., and J.B. Haufler. 1994. Diurnal versus 24-hour sampling of habitat use. *Journal of Wildlife Management* 58: 178–180.

Bourgeron, P.S., and M.E. Jensen. 1994. An overview of ecological principles for ecosystem management. pp. 45–57. In: M.E. Jensen, and P.S. Bourgeron (tech. eds.) Volume II: Ecosystem management: principles and applications. USDA For. Serv. Gen Tech. Rep. PNW-GTR-318.

Brittingham, M.C., and S.A. Temple. 1983. Have cowbirds caused forest songbirds to decline? *Bioscience* 33: 31–35.

Camp, A., C. Oliver, P. Hessburg, and R. Everett. 1997. Predicting late successional fire refugia predating European settlement in the Wenatchee Mountains. *Forest Ecology and Management* 95: 63–77.

Capen, D.E., D.R. Coker, A.B. Cumming, and Y.K. Ortega. 1994. *Habitat models for predicting avian diversity: importance of the minimum mapping unit*. Poster presented at the Annual Meeting, The Wildl. Soc., Albuquerque, N.M.

Clements, F.E. 1916. *Plant succession: an analysis of the development of vegetation*. Carnegie Inst. Pub. 242. Washington, D.C.

Costanza, R., H.E. Daly, and J.A. Bartholomew. 1991. Goals, agenda, and policy recommendations for ecological economics. pp. 1–20. In: R. Constanza (ed.), *Ecological Economics, The Science and Management of Sustainability*. Columbia Univ. Press, New York.

Crow, T.R., and E. Gustafson. 1996. Ecosystem management: managing natural resources in time and space. In: K.A. Kohm and J.F. Franklin (eds.), *Creating a Forestry for the 21st Century: The Science of Ecosystem Management*. Island Press, Washington, D.C.

Ecomap. 1993. *National hierarchical framework of ecological units*. USDA For. Serv., Washington, D.C.

Ehrlich, P.R. 1986. Extinction: what is happening now and what needs to be done. pp. 157–164. In: D.K. Elliott (ed.), *Dynamics of Extinction*. Wiley, New York.

Ehrlich, P.R., and E.O. Wilson. 1991. Biodiversity studies: science and policy. *Science* 253: 758–762.

Fahnestock, G.R., and J.K. Agee. 1983. Biomass consumption and smoke production by prehistoric and modern forest fires in western Washington. *Journal of Forestry* 81: 653–657.

Forest Ecosystem Management Assessment Team. 1993. *Forest ecosystem management: an ecological, economic, and social assessment*. USDA For. Serv., USDI Fish and Wildl. Serv., USDI Nat. Park Serv., USDI Bureau Land Mgmt., USDC Nat. Marine Fish. Serv., and Environ. Protection Agency, Washington D.C.

Gregg, W.P., Jr. 1994. Developing landscape-scale information to meet ecological, economic, and social needs. pp. 13–17. In: V.A. Sample (ed.), *Remote Sensing and GIS in Ecosystem Management*. Island Press, Washington, D.C.

Greig-Smith, P. 1952. The use of random and contiguous quadrats in the study of the structure of plant communities. *Annals of Botany* 16: 293–316.

Gustafson, E.J., and T.R. Crow. 1996. Simulating the effects of alternative forest management strategies on landscape structure. *Journal of Environmental Management* 46: 77–94.

Harris, L.D., and P.B. Gallagher. 1989. New initiatives for wildlife conservation: The need for movement corridors. pp. 11–34. In: G. Mackintosh (ed.), *Preserving communities and corridors*. Defenders of Wildlife, Washington, D.C.

Haufler, J.B. 1994. An ecological framework for forest planning for forest health. *Journal of Sustainable Forestry* 2: 307–316.

Haufler, J.B. 1995. Forest industry partnerships for ecosystem management. *Transactions of the North American Wildlife and Natural Resources Conference* 60: 422–432.

Haufler, J.B., C.A. Mehl, and G.J. Roloff. 1996. Using a coarse filter approach with a species assessment for ecosystem management. *Wildlife Society Bulletin* 24: 200–208.

Hoekstra, T.W., T.F.H. Allen, and C.H. Flather. 1991. Implicit scaling in ecological research. *Bioscience* 41: 148–154.

Holling, C.S. 1995. What barriers? What bridges? pp. 3–34. In: Gunderson, L.H., C.S. Holling, and S.S. Light (eds.), *Barriers and Bridges to the Renewal of Ecosystems and Institutions.* Columbia University Press. New York.

Janzen, D.H. 1986. The eternal external threat. pp. 286–303. In: M.E. Soule (ed.), *Conservation Biology, The Science of Scarcity and Diversity.* Sinauer Associates, Sunderland, Mass.

Kollasch, R.P., and M.J. Twery. 1995. Object-oriented system design for natural resource decision support: the northeast decision model. *AI Applications* 9: 73–84.

Leopold, A. 1933. *Game Management.* Charles Scribner's Sons, New York.

Levin, S.A. 1992. The problem of pattern and scale in ecology. *Ecology* 73: 1942–1968.

McNab, W.H., and P.E. Avers. 1994. *Ecological subregions of the United States: Section descriptions.* USDA For. Serv. Admin. Publ. WO-WSA-5.

Maxwell, J.R., C.J. Edwards, M.E. Jensen, S.J. Paustian, H. Parrott, and D.M. Hill. 1995. A hierarchical framework of aquatic ecological units in North America (Nearctic zone). USDA For. Serv. Gen Tech. Rep. NC-176.

Miller, K.R. 1996. *Balancing the scales: guidelines for increasing biodiversity's chances trough bioregional management.* World Resources Institute, Washington, D.C.

Mladenoff, D.J., G.E. Host, J. Boeder, and T.R. Crow. 1996. LANDIS: a spatial model of forest landscape disturbance, succession, and management. pp. 175–179. In: M.F. Goodchild, L.T. Steyaert, B.O. Parks, C. Johnston, D. Maidment, M. Crane, and S. Glendinning (eds.), *GIS and Environmental Modeling; Progress and Research Issues.* GIS World Books, Fort Collins, CO.

Morgan, P., G.H. Aplet, J.B. Haufler, H.C. Humphries, M.M. Moore, and W.D. Wilson. 1994. Historical range of variability: a useful tool for evaluating ecosystem change. *Journal of Sustainable Forestry* 2: 87–111.

Mueller-Dombois, D., and H. Ellenberg. 1974. *Aims and Methods of Vegetation Ecology.* Wiley, New York.

Noss, R.F. 1990. Indicators for monitoring biodiversity: a hierarchical approach. *Conservation Biology* 4: 355–364.

Noss, R.F., and A.Y. Cooperrider. 1994. *Saving Nature's Legacy: Protecting and Restoring Biodiversity.* Island Press, Washington, D.C.

Oliver, C.D., A. Osawa, and A. Camp. 1997. Forest dynamics and resulting animal and plant population changes at the stand and landscape levels. *Journal of Sustainable Forestry* 6 (3/4): 281–312.

Pattee, H.H. 1973. the physical basis and origin of hierarchical control. pp. 73–108. In: H.H. Pattee (ed.), *Hierarchy Theory, The Challenge of Complex Systems.* George Braziller, New York.

Quigley, T.M., R.W. Haynes, and R.T. Graham (tech. eds.). 1996. Integrating scientific assessment for ecosystem management in the Interior Columbia Basin and portions of the Klamath and Great Basins. USDA For. Serv. Gen. Tech. Rep. PNW-GTR-382.

Reese, K.P., and J.T. Ratti. 1988. Edge effect: a concept under scrutiny. *Trans. N. Am. Wildl. and Nat. Res. Conf.* 53: 127–136.

Robinson, S.K. 1990. Effects of forest fragmentation on nesting songbirds. *Illinois Natural History Reports* 296: 1–2.

Roloff, G.J. 1994. Using an ecological classification system and wildlife habitat models in forest planning. Ph.D. Thesis. Mich. State Univ., East Lansing.

Roloff, G.J., and J.B. Haufler. 1997. Establishing population viability planning objectives based on habitat potentials. *Wildlife Society Bulletin* 25: 815–904.

Ruth, R.H., and A.S. Harris. 1979. Management of western hemlock-Sitka spruce forests for timber production. USDA For. Serv. Gen. Tech. Rept. PNW-88.

Schneider, D.C. 1994. *Quantitative Ecology: Spatial and Temporal Scaling.* Academic Press, San Diego, CA.

Simon, H.A. 1973. The organization of complex systems. pp. 3–27. In: H.H. Pattee (ed.), *Hierarchy Theory, The Challenge of Complex Systems.* George Braziller, New York.

Steele, R. 1994. The role of succession in forest health. *Journal of Sustainable Forestry* 2: 183–190.

Stribley, J.M. 1993. Factors influencing cowbird distributions in forested landscapes of northern Michigan. M. S. Thesis, Michigan State Univ., East Lansing.

Tansley, A.G. 1924. The classification of vegetation and the concept of development. *Journal of Ecology* 8: 118–149.

Troll, C. 1939. Luftbildplan and okologische Bodenforschung. *Z. Ges. Erdkunde*: 241–298.

Turner, M.G. 1990. Spatial and temporal analysis of landscape patterns. *Landscape Ecology* 4: 21–30.

Turner, M.G., and R.H. Gardner. 1991. Quantitative methods in landscape ecology: an introduction. pp. 3–14. In: M.G. Turner, and R.H. Gardner (eds.), *Quantitative Methods in Landscape Ecology: The Analysis and Interpretation of Landscape Heterogeneity.* Springer, New York.

Turner, M.G., R.H. Gardner, R.V. O'Neill, and S.M. Pearson. 1994. Multiscale organization of landscape heterogeneity. pp. 73–79. In: M.E. Jensen, and P.S. Bourgeron (tech. eds.), Volume II: Ecosystem management: principles and applications. Gen. Tech. Rep. PNW-GTR-318.

Watershed Analysis Coordination Team. 1995. Ecosystem analysis at the watershed scale: the revised federal guide for watershed analysis, Version 2.1. Review Draft. Regional Ecosystem Office. Portland, OR.

Weintraub, A., and A. Cholacky. 1991. A hierarchical approach to forest planning. *Forest. Science* 37: 439–460.

Wiens, J.A. 1989. Spatial scaling in ecology. *Functional Ecology* 3: 385–397.

Wiens, J.A. 1992. Ecology 2000: an essay on future directions in ecology. *Bulletin of the Ecology Society of America* 73: 155–164.

## THE AUTHORS

**Jonathan B. Haufler**
*Boise Cascade Corporation,*
*P.O. Box 50, Boise, ID 83728, USA*

**Thomas Crow**
*USDA Forest Service,*
*North Central Forest Experiment Station,*
*Forestry Sciences Laboratory,*
*5985 Highway K,*
*Rhinelander, WI 54501, USA*

**David Wilcove**
*Environmental Defense Fund,*
*1875 Connecticut Avenue, Suite 1016,*
*Washington, DC 20009*

# Scales and Ecosystem Analysis

David L. Caraher, Arthur C. Zack, and Albert R. Stage

**Key questions addressed in this chapter**

- ♦ *Selecting appropriate spatial scales for analyses*
- ♦ *Identifying appropriate scales of information within an analysis*
- ♦ *Integrating information from a variety of hierarchies and scales*

**Keywords: Spatial scale, temporal scale, watershed**

## 1 INTRODUCTION

Land managers who become involved with ecosystem analyses are likely to encounter a number of problems related to scale. Some of these problems are rooted in the nature of ecosystems as indistinct entities with vague boundaries that shift through space and time. Other problems stem from the information we develop and use to describe ecosystems, their attributes and phenomena, with its inherent limitations in the space and time to which it applies. Still other problems arise out of the nature of human purposes and interests for ecosystem management and ecosystem analysis, and the way they shift from place to place and time to time. The purpose of this paper is to help resource managers recognize and avoid common problems related to scale.

The existence of various hierarchies and scales for ecosystems, along with the nature of ecosystems, of information about ecosystems, and of human purposes and interests, suggests several key questions for resource managers.

- How can resource managers select appropriate spatial scales for their analyses?
- Within an analysis, what are the appropriate scales of information?
- How can managers integrate information from a variety of hierarchies and scales?

## 2 SCALE PROBLEMS

The nature of ecosystems is such that they exist and function at a variety of scales (Allen and Starr 1982), from a seasonal pocket of water in surface rock to an ancient tropical rainforest spanning an entire continent, from the canopy of a single tree to the canopy of an entire forest. And within any one ecosystem at any scale, components and processes occur and function at different scales as well. As a result, ecosystems and their components and phenomena can be viewed within space/time hierarchies where each scale has its own spatial and temporal attributes (Holling 1995) (Fig. 1). Thus, a view at any scale reveals some information while at the same time, conceals other information. Consequently, when land managers plan and conduct ecosystem analyses, they not only need to select scales that are likely to yield the information they seek, they also need to consider what kind of information might be excluded from view at that scale.

The nature of information that land managers use in conducting their analyses, such as maps and photos and data from surveys and inventories, is also

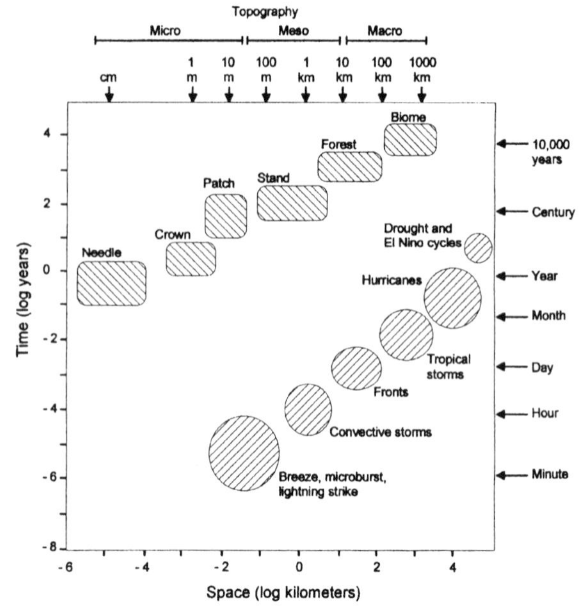


Fig. 1. An example of two hierarchies, each with its own set of distinct scales in space and time (after Holling 1995).

problematic. Such information tends to be scale-specific, that is, it represents limited spatial extents, periods of time, and levels of detail. For example, information represented by an aerial photograph is limited to the spatial extent of that image, and a photograph of the same area 20 years later will reveal different information. Furthermore, if the new photograph has been taken with newer equipment capable of higher resolution, it will reveal more detail than the original. These considerations apply as well to maps, inventories, and surveys. Managers should understand these limitations so that they can (1) select information from appropriate scales to meet their needs and (2) understand the spatial and temporal limitations of the information they use.

Finally, the nature of human interests and purposes for ecosystem management, and particularly for a given ecosystem analysis also give rise to problems with scale. When land managers plan or undertake an ecosystem analysis, they do so with either implied or express interests or purposes in mind. They might be trying to solve a resource management problem, trying to build the foundation for ecosystem management on a given ownership or administrative area, trying to determine the effects of management and human activity, or trying to evaluate alternative management proposals. These and other human interests and purposes for land and ecosystem management imply corresponding spatial scales for analysis. Unfortunately, the scales represented by human interests and purposes do not often fit the scales represented by ecosystems. Property, administrative, and jurisdictional boundaries rarely conform to ecosystem boundaries.

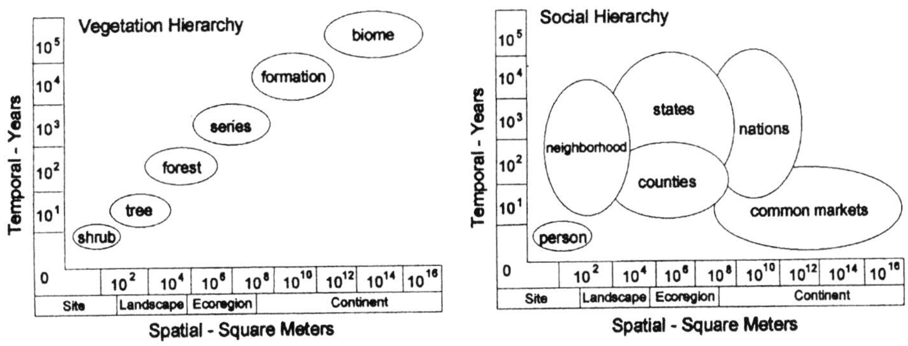


Fig. 2. Comparing biological and social hierarchies often reveals considerable disparity between their spatial and temporal scales (Quigley et al. 1996).

Nor does the duration of a human or social endeavors such as an annual budgets, durations of ownership, or planning horizons conform to the time span of ecological processes (Fig. 2). These problems are confounded when human interests and purposes for ecosystem management shift over time. Yesterday's analysis might have been adequate for yesterday's needs, but could well become inadequate for tomorrow's needs.

## 3 TERMINOLOGY

The terms used to describe and discuss scales for resource management and ecosystem analysis are less intuitive than first appears. Inconsistent and overlapping common use of related terms such as scale, hierarchy, resolution, grain, and extent can be confusing. In this paper, we use the following terms and definitions (see also Haufler, this volume), consistent with the definitions used by Turner and Gardner (1991) and Allen and Hoekstra (1992).

*Scale*: spatial or temporal size. "Scale" has been variously defined as the "relative size and extent" (Barnhart and Barnhart 1988), "the level of spatial resolution perceived or considered" (Forman and Godron 1986), something that "pertains to size in both time and space" (Allen and Hoekstra 1992), and "the spatial or temporal dimension of an object or process, characterized by both grain and extent" (Turner and Gardner 1991). "Scale" can be synonymous with "scope" because both can refer to "extent or range of view." But in the context of resource and ecosystem management, scale alone embodies concerns for not only the extent, but also the amount of detail in view. Large-scale and small-scale are relative terms, and their meaning in a specific instance depends on the boundaries of the space or time being discussed.

*Grain:* the finest spatial or temporal entity discernible at a given scale of analysis (Turner and Gardner 1991, Allen and Hoekstra 1992). Grain determines the lowest or smallest visible level in a hierarchy.

*Resolution:* the precision of measurement, usually defined by specifying grain size (Turner and Gardner 1991).

*Extent:* the scope or total size of the area or span of time being considered of (Turner and Gardner 1991; Allen and Hoekstra 1992). "Extent" essentially determines the largest or highest level in a hierarchy that is entirely visible. Generally, larger extents are represented by larger minimum grain sizes and coarser resolutions in order to have a manageable number of units and relationships to analyze. For example, the large-scale (144 million acres) Columbia River Basin Assessment (Quigley et al. 1996) characterized vegetation using 247-acre pixels, meaning that vegetation units smaller than that are not discernible.

*Large-scale:* a broad area or amount of space (which is opposite of the definition used in cartography, where "large-scale" refers to a representation of a relatively small area) (Allen and Hoekstra 1992). Although large-scales often include a relatively coarse grain size, this is not true by definition. When a question involves structures and processes with a coarse grain size, a large-scale perspective is usually required simply to encompass all the relevant variables. In large-scale analyses with relatively coarse-grained variables, gathering data and conducting the analysis at too fine of a grain size may actually mask the processes relevant to the questions at hand. For example, if changes in wildfire regimes of North American forests are a function of climate change over a scale of decades to centuries (Swetnam and Betancourt 1990, Swetnam 1993), then gathering hourly or daily fire weather data might actually obscure the larger and longer climatic patterns.

*Small-scale:* a relatively confined area or limited extent (Turner and Gardner 1991, Allen and Hoekstra 1992). By necessity, a small-scale view requires a relatively fine-grain size to reveal features or processes within it's extent.

*Hierarchy*: A hierarchy is a formally ranked system of a number of related scales, graded from small to large.

"Hierarchy" has been variously defined as "an organization of things arranged into higher and lower ranks, classes, or grades" (Barnhart and Barnhart 1988); "a sequence of sets composed of smaller subsets" (Forman and Godron 1986); and "a formal approach to the relationship between upper-level control over lower-level possibilities" (Allen and Hoekstra 1992). An example of a spatial hierarchy for natural resource planning based on physical aspects of ecosystems might consist of something like Robert Bailey's ecoregions scaled from land-types, to land-type associations, to subsections, to sections, to provinces, to divisions, and finally to domains (Bailey 1995). In this system, divisions are large-scale entities and land-types are small-scale entities.

*Analysis and Assessment: Webster's New Collegiate Dictionary* defines "analysis" as "(1) separation of a whole into its component parts; (2a) an examination of a complex, its elements, and their relations; (2b) a statement of such an analysis." The same dictionary also defines "assessment as (1) the act or an instance of assessing; (2) the amount assessed." In the field of natural resources, "analysis" and "assessment" seem to be used interchangeably so that both can refer to either *a process of inquiry* ("conduct an analysis" or "conduct an assessment") or the *results of an inquiry* ("the analysis reveals" or "the assessment reveals").

## 4 HISTORICAL PERSPECTIVES

Land and resource managers have probably always conducted their work at various scales, designing projects for sites consisting of less than an acre to developing plans encompassing thousands of acres. Throughout, many have undoubtedly encountered problems related to scale. Those problems are probably universal to managers of both private and public lands, but the experiences of those working for the USDA Forest Service offer typical examples.

In the 1950s, the Forest Service based its plans for managing timber on inventories that encompassed entire National Forests often consisting of a million acres or more. Each inventory contained enough field samples to enable broad, long-term planning, but not enough to locate specific harvest units nor enough to identify potential environmental effects of alternative harvest levels.

To meet their needs for more specific information, Forest Service managers shifted to smaller scale inventories. They subdivided National Forests into land units called "Ecological Planning Units." A single National Forest of well over a million acres might contain several dozen of these units, and each unit was analyzed separately. But with this approach resource managers were unable to determine how management activities within one unit affected the resources of other units, even though they were within the same National Forest. Forest Service managers compensated by attempting to meet objectives for all resources within each planning unit.

The National Forest Management Act may have been partly a response to the short-comings of small scale planning. The act required the Forest Service to develop a resource management plan for each National Forest. These plans allow the public and the Forest Service to allocate broad, homogenous portions of a landscape to specific uses, activities, and management guidelines. But to design management activities and projects, Forest Service resource managers still have to conduct analyses at smaller, project level scales.

On both public and private lands, three recent developments are presenting resource managers with new problems of scale. First, the need to consider the cumulative effects of management activities leads resource managers to consider several past and potential future project sites simultaneously. Second, efforts to save endangered species require biologists and resource managers to expand their analytical horizons to much larger scales, often entire regions encompassing millions of acres. Finally, and most significantly, resource management, especially on public lands, is shifting to ecosystem management, and ecosystems are thought to occur and function at a variety of scales.

## 5 SELECTING SPATIAL SCALES FOR ANALYSIS

When landowners and resource managers plan to analyze an ecosystem or part of an ecosystem, they can choose from a wide range of spatial scales, from less than an acre to millions of acres. They may chose a scale to match a physical boundary such as that for an ownership, jurisdiction, or watershed, or one that matches the implied boundaries of a subject or topic, such as bald eagles, water quality, or ecosystem health. Ultimately, the scale of their analyses should match the scales of their purposes and interests (Table 1). But when landowners and resource managers are confronted with multiple purposes and interests, or when those purposes and interests change, they encounter problems with scales.

### 5.1 Problems With Analytical Scales

Landowners and managers can usually identify an appropriate scale of analysis to meet a specific purpose without difficulty. For project designs and site problems they intuitively conduct small-scale analyses, and

Table 1. Analytical considerations for three broad spatial and temporal scales.

| Analytical Considerations | Relative Scale | | |
|---|---|---|---|
| | Large | Mid | Small |
| Purpose or Interest | Develop policies, laws, standards, and practices | Develop management strategies and programs | Design projects |
| Spatial Scale | >2,000,000 acres | 2,000–2,000,000 acres | < 2,000 acres |
| Temporal Scale | > 100 years | 10–100 years | 1–10 years |
| Focus | Patterns of ecosystems | Patterns of ecological features and components | Individual ecological features and components |
| Character of information | Mostly qualitative | Qualitative and quantitative | Mostly quantitative |

for setting project priorities or developing plans, they move to larger scales. But they encounter problems when unforeseen needs arise after an analysis has begun or has been completed and they discover that the analysis was conducted at a scale either too small or too large to meet those needs. For example, a manager might have an interest in developing an understanding of the ecosystem represented by a 30,000-acre watershed. But that interest can shift to larger scales if the analysis reveals that the watershed provides summer habitat for a species of migrating wildlife, or to smaller scales if the analysis reveals several nesting locations for an endangered species.

Unforeseen needs often emerge from social interests after an analysis has been completed. Members of the public often challenge a management proposal, even on private lands, on the grounds that an analysis is inadequate. Those challenges often introduce new purposes and interests which in turn represent new scales of analysis. To avoid such problems, managers of public lands often invite the public to participate during various stages of an analysis, but can still have difficulty when a particular purpose or interest is either beyond the scope or too specific for the analysis in question.

## 5.2 Guidelines for Selecting Spatial Scales of Analysis

The following guidelines can help managers select scales of analyses to match their interests and purposes.

### 5.2.1 To Support General Ecosystem Management

The following three guidelines apply to those analyses managers conduct to provide a general ecological foundation and background for resource management.

1. *Develop explicit statements of interest or purpose for the analysis.* If a landowner or administrator proposes simply to "conduct an ecosystem analysis," the scale for the analysis and the utility of its results are open to question. But when landowners and administrators express specific statements of interest or purpose, those who conduct the analysis can select appropriate scales to meet them. For example, the Federal Guide for Watershed Analysis, *Ecosystem analysis at the Watershed Scale* (Regional Ecosystem Office 1995) declares that the purpose of ecosystem analysis at the watershed scale is to develop an understanding of the processes and interactions occurring within a watershed. On the other hand, a request by U.S. Senator Hatfield and U.S. Congressman Foley for "a scientific evaluation of the effects of Forest Service management practices on the sustainability of eastern Oregon and Washington ecosystems" (Everett et al. 1993) suggests an analysis at a much larger scale.

2. *Identify the terms of analysis.* The purposes of the analysis will either imply or specify topics of interest, such as water quality, sediment, erosion, or wildlife habitat. But these topics can be represented on landscapes by a wide range of attributes and phenomena. Wildlife habitat could be analyzed in terms of nest sites, cover types, structure types, habitat types, habitat patterns, amount of edge, patch size, habitat configuration, connectivity, travel corridors, and so on, with each term referring to a particular attribute or phenomenon and suggesting a particular scale of analysis.

3. *Consider expanding the analysis one scale above and below those represented by the immediate scale of interest.* Managers can sometimes expand a scale of analysis without adding substantially to its cost. They might, for example, shift from an analysis to sup-

port annual project plans to one that will support multi-year project planning horizons, from an analysis to support a management plan for an ownership to one that will support plans for a watershed, or from an analysis of one habitat type to one that includes a mosaic of habitats.

The primary spatial scale of an analysis can also be subdivided to develop information about attributes and phenomena at the next smaller scale, although this presents drawbacks. The primary scale of interest may need to be stratified in some way first, and smaller-scale sampling can add to the time and cost of an analysis. Still, by expanding and subdividing the scale of analysis, managers may be able to meet some unanticipated or future needs that appear with new broader and narrower scales of interest.

### 5.2.2 To Support Decision-Making

The following guidelines apply to ecosystem analyses that are conducted to support a decision-making process, especially on public lands. On federal lands, these analyses often are part of an environmental impact statement and include procedures for including the public in comparing and choosing among alternative courses of action. Spatial scales for these analyses should be defined and bounded by three elements: potential management actions, ecological indicators, and ecological processes. Taken together, these elements determine the appropriate geographic extent and temporal span of such an analysis, and success of the analysis can depend on the forum and sequence in which these elements are defined.

1. *Identify actions that are to be considered in the alternatives.* When the alternatives involve policy decisions, the actions that would be implemented in each alternative should be listed along with the conditions under which the actions are deemed appropriate. For example, if a policy alternative being considered is to use prescribed fire to reduce fuel loading, then a specific action might be to permit natural fires in mature forest to burn if the fire danger is low and fuel-ladders are not present within the affected area. When the alternatives involve specific project decisions, then alternative actions being considered might include the use of heavy equipment to construct fire-lines, the use of hand tools to build fire-lines, burn in the spring, or burn in the fall. Identifying the list of potential actions is primarily the role of resource managers.

2. *Select attributes that will indicate how the ecosystem will look and function in the future.* These attributes, which are the essence of communication between interested parties, the analysts, and ultimately the decision-maker, are the basis for evaluating the trade-offs between alternative courses of action. They may be communicated as pictorial displays of the forest as it changes through time, or as graphical displays of the time-trends of quantitative indicators. The list of indicators might include both quantities of resource yields as well as descriptions of specific landscape attributes, such as areas in different structural classes of vegetation, length of roads, or areas of disturbance. Different types of indicators emerge and are discernable at different scales. These indicators reflect properties of the ecosystem at this scale which may not be observable by aggregating data from lower levels. These are the properties that provide the necessary context for analysis and it may be both necessary and more efficient to make observations of these properties at broader scales.

3. *Identify the ecological processes that would be influenced by the proposed management actions.* These biophysical and socio-economic mechanisms control the functioning of the ecosystem and cause changes in indicator trends. The spatial and temporal extent at which these processes function may be substantially different than that of the actions being considered. The specialists who conduct an analysis can develop hypotheses about how potential management actions might affect ecological attributes (Fig. 3). Understanding the potential effects of management activities on ecological attributes helps define the scale of analysis.

4. *Define the spatial and temporal scales of the analysis to encompass the maximum space and time over which the actions and processes operate.* These bounds are implied by the choice of actions and indicators. Processes and actions that operate beyond these bounds can be represented as driving variables. Within the outer bounds of space and time, effects on small-scale processes must also be represented (Allen and Starr 1982). Short-term disturbances with rapid recovery can appear as "noise" over longer time periods, while small, persistent, cumulative effects that appear insignificant in the short term can have major long-term significance.

## 6 SELECTING SCALES OF INFORMATION

Although not always readily apparent, maps, photographs, data, and other sources of information reflect inherent limitations in the spatial extent, span of time, and level of detail (grain and resolution) that they

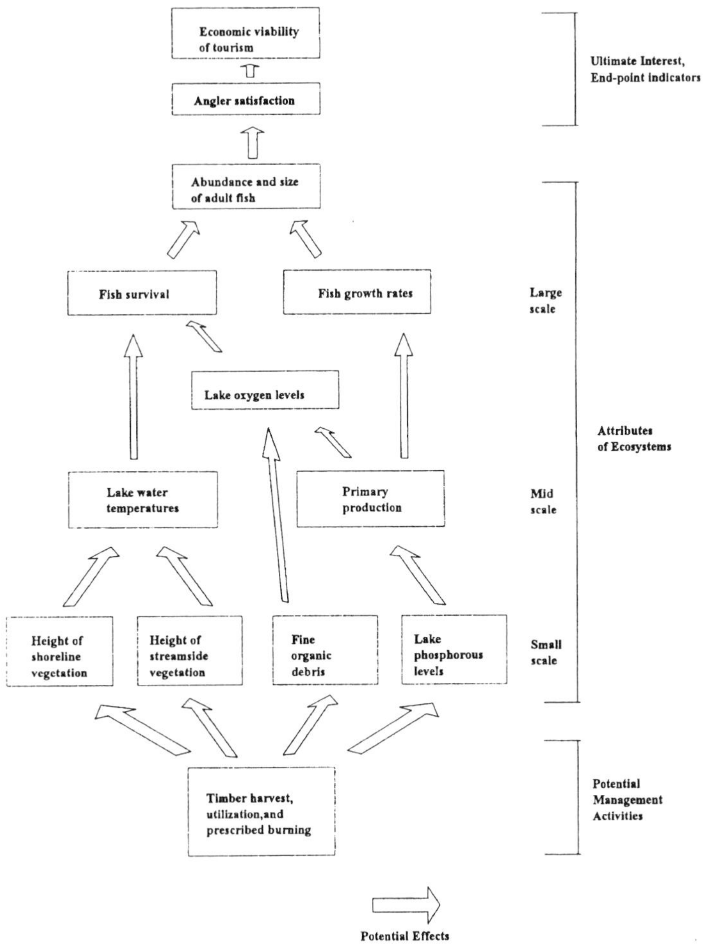


Fig. 3. Identifying the potential effects of management activities on ecological attributes can help define an appropriate scale of analysis.

represent. It is tempting to believe that a given spatial scale of analysis imposes a given scale of information, but the two are separate. An analysis that focuses on a particular spatial scale can, and often does, incorporate information from a wide range of larger and smaller scales. For example, an analysis might rely on long-term, large-scale regional data to describe the climate, while using short-term, small-scale inventories to describe fish habitat. Information from both larger and smaller scales does have meaning and utility at the primary scale of interest. Allen and Hoekstra (1992, p. 9) point out that "...it is necessary to look to both larger scales to understand context and to smaller scales to understand mechanism; anything else would be incomplete." But information from larger and smaller scales, relative to the primary scale of interest, can also cause problems.

## 6.1 Problems With Scales of Information

Resource managers encounter two kinds of problems with scales of information. The first occurs during an analysis when they attempt to describe attributes or phenomena at one scale by either aggregating information from smaller scales or dis-aggregating information from larger scales. The second occurs after an analysis has been completed at a given scale, and resource managers try to extrapolate results to larger scales or interpolate them to smaller scales.

*During an Analysis:* Information derived from data, maps, photographs, inventories, surveys, and reports is scale-specific in two regards. First, it applies to an area that has limited spatial extent and is valid for a limited span of time. Second, because ecosystem attributes and phenomena change with scale, the kind of information that can be used to describe them — the terms of our analysis — also changes with scale. A forest, for example might be described in terms of species composition, age and size classes of the species represented, and dominant understory vegetation. At a smaller scale, an individual tree would be described in terms of its species, age, height, form, and diameter. But foresters may attempt to aggregate descriptions of individual trees to characterize the large-scale distribution and pattern of a forest, perhaps without realizing that the large-scale distribution of forest patches cannot be discerned from samples (and descriptions) of individual trees.

Managers often encounter difficulty when they attempt to fill gaps in information at a given scale by either aggregating it from smaller scales or disaggregating it from larger scales. Perhaps the most common current problems between scales of analysis and scales of information occur with mid-scale ecosystem analyses, those conducted at watershed, landscape, and sub-basin scales. These analyses are being conducting throughout the Pacific Northwest on both private and public lands so that landowners, citizens, and resource managers can learn more about the ecosystems that function across property and jurisdictional boundaries and so that they can incorporate multiple issues into a single analysis. These are mid-scale interests, but most of the analyses rely heavily on small-scale information because it is readily available and because methods for developing mid-scale information has not been well developed for most natural resource disciplines. The proliferation of small-scale information in mid-scale analyses either masks the information that describes mid-scale attributes and phenomena, or replaces it altogether. As a result, managers can see the trees but not the forest — or ecosystem — they were looking for.

For mid and large-scale analyses, the spatial scale chosen for the analysis may correlate well with the manager's scales of interest, but the information developed during the analysis often turns out to be far too specific to satisfy those interests.

*After an Analysis:* Managers encounter other problems when they attempt to extrapolate information from one scale to another. They may, for example, have data from a small watershed indicating that removing more than 15% of forest canopy produces measurable responses in water yields and in stream channel dynamics. But they should not conclude that the same relationship between canopy removal and water yield applies at either larger scales encompassing many such small watersheds or at smaller scales within the watershed. As the spatial and temporal scales change, the context, the way the system responds, and even the relevant principles are likely to also change. A principle appropriate at one scale is frequently inappropriate if carried too far to other scales. The same problem of scaling occurs when models are used to predict ecological effects of actions.

## 6.2 Guidelines for Selecting Scales of Information

1. *Translate the terms of analysis to scales of information.* The terms of analysis imply not only a corresponding scale of analysis, but corresponding scales of information as well. Terms such as habitat type, nest locations, and habitat structure suggest a scale of information that relates to attributes and phenomena at individual sites. Terms such as "habitat patterns, connectivity, and migration routes" suggest a scale of information that relates to attributes and phenomena for a collection of sites.

2. *What can you see from here?* Each scale of analysis will reveal a unique set of attributes and phenomena. "Stepping back" (to include a larger view) reveals the context in which they occur and function, and "moving in" (for a closer view) will reveal the mechanisms by which they function. When defoliating insects attack a forest, a single aerial photo can reveal which trees are affected within the scope of the photo, but only a broader view, perhaps from a satellite image or a mosaic of aerial photos can reveal the total extent and distribution of the infestation. And only close examination on the ground will reveal the species of insects responsible. Resource managers and specialists, should, as noted earlier, include larger and smaller scales in their analyses. But in so doing, they need to be careful not to inadvertently abandon the central focus of their analysis and risk being unable to satisfy their original purposes or interests. Managers and resource specialists can avoid inadvertent shifts in scale by consciously maintaining their central focus on information represented by "what you can see from here," the chosen scale of analysis. Our case study, "Restoring Ecosystems in the Blue Mountains" (see Box 1) describes one successful effort where the scale of information corresponded well with the scale set for the analysis.

3. *During small-scale analyses, be alert to limitations of large-scale information.* Small-scale analyses, such as those conducted within project sites consisting of a few acres, often need to incorporate information that has been developed only at larger scales, such as frost-free days per year, or dominant plant community. But large-scale information often relates to attributes and phenomena that can vary significantly among locations at smaller scales. Resource managers can avoid mis-application of large-scale information by being alert to its limitations.

4. *Conversely, during mid and large-scale analyses, be alert to the limited applicability of small-scale information.* During a mid-scale (watershed or landscape) analysis, small-scale information, such as that derived from field measurements and plot data, can, as noted earlier, identify and describe the mecha-

**Box 1**
**Restoring Ecosystems in the Blue Mountains**

In the Blue Mountains of Eastern Oregon, ecosystems have been widely affected by fire, insects, and diseases for the past twenty years. The damage now affects more than 3 million acres, and exceeds traditional, smaller scales of resource management on both private and public lands. In July 1992, a panel of Forest Service scientists undertook a brief, large-scale assessment of the problem.

Assessment Attributes

1. Purpose or interest: establish long-term objectives, and identify priorities for restoring ecosystem health on National Forest lands in the Blue Mountains.
2. Scale of analysis: 6 million acres
3. Terms of analysis: percent cover of (a) various timber stand conditions, (b) standing dead and down trees, (c) juniper colonization, and (d) riparian shrubs, and percent of stable streambanks.
4. Scale of information: river basins covering from a half to one million acres.
5. Assessment time: 3 months

Analytical Procedure

The panel used their collective, best professional judgement to establish ranges of *natural or expected* variability of percent cover for each indicator in each of three physiographic zones represented in the study area. Then the panel employed local resource specialists with surveys and inventories to identify ranges of *current* variability for each indicator in each river basin. The panel concluded by comparing ranges of current conditions with those for natural conditions of each river basin in the study area.

Results

The panel was able to distinguish between river basins where indicators appeared to be either *far outside*, *somewhat outside*, or *within* their ranges of natural variability. The panel was also able to identify the highest priority corrective objectives and measures: (1) reduce fuels to lower the risk of catastrophic fires, (2) establish forest conditions to minimize insect epidemics, (3) restore ponderosa pine on sites where the species is adapted, (4) restore riparian vegetation, (5) reduce sediment from the road system, and (6) manage livestock and big game to improve water quality and fish habitat. Finally, the assessment highlighted the most sensitive locations within poor-condition river basins: watersheds which (1) contain threatened and endangered fish species, (2) provide municipal water supply, (3) interface with urban areas, and (4) provide habitat for deer and elk.

As a result of this assessment, resource managers were able to develop a large-scale restoration strategy encompassing three National Forests. The strategy helps assure that projects developed at smaller scales that will contribute to ecosystem health at larger scales.

For more information about the Blue Mountain ecosystem assessment, contact the Director of Natural Resources, USDA Forest Service, P.O. Box 3623, Portland, OR 97208.

nisms for mid-scale attributes and phenomena. But over-reliance on small-scale information can keep managers from detecting essential or critical mid-scale attributes and phenomena. A stream survey, for example, conducted reach-by-reach, can reveal not only stream temperature, sediment, and fish-passage problems, but also identify the type of channel according to some classification, its width, depth, meander length, pool to riffle ratio, and so on. But such small-scale information, even when it is available for a hundred locations within a watershed, will not satisfy such mid-scale interests as the relative density of streams within the watershed, their pattern of distribution, nor their relationship to upland features and patterns.

## 7 SUMMARY

Landowners and managers who plan, conduct, or use ecosystem analyses are likely to encounter problems related to scale. Those problems arise out of the variable nature of ecosystems, the limitations embodied by information about ecosystems, and the variety of human interests and purposes for ecosystem management.

Ecosystems change over space and time. Attributes and phenomena that are discernable at one scale can be invisible at others.

When land managers specify their interests and purposes for an ecosystem analysis, they need to understand that those interests and purposes relate to particular attributes and phenomena, and therefore imply a particular spatial scale of analysis.

The spatial and temporal scale selected for an analysis should correspond with the spatial and temporal scales that are represented by management interests and purposes.

It is important to distinguish between scale of analysis and scale of information. Scale of analysis refers to the spatial extent chosen for an inquiry. Scale of information refers to the spatial and temporal extents, and the levels of detail represented in photographs, maps, and data that are actually used during the analysis or inquiry.

Most ecosystem analyses employ information from a variety of scales, but to fulfil their purposes, they must focus on information that relates to the attributes and phenomena that are operating at the spatial scale set for the analysis.

## REFERENCES

Allen, T.F.H. and T.W. Hoekstra. 1992. *Toward a Unified Ecology*. Columbia University Press, New York.

Allen, T.F.H., and T.B. Starr. 1982. *Hierarchy: Perspectives for Ecological Complexity*. University of Chicago Press, Chicago, IL.

Bailey, R.G. 1995. *Descriptions of the Ecoregions of the United States*. USDA Forest Service, Misc. Pub. No. 1391.

Barnhart, C.L., and R.K. Barnhart (eds.). 1988. *The World Book Dictionary*. World Book, Chicago, IL.

Everett, R., et al. 1993. *Eastside Forest Ecosystem Health Assessment*. USDA Forest Service, Pacific Northwest Research Station, Portland, OR.

Forman, R.T.T., and M. Godron. 1986. *Landscape Ecology*. Wiley, New York.

Holling, C.S. 1995. What Barriers? What bridges? pp. 3–34. In: L.H. Gunderson, C.S. Holling, and S.S. Light (eds.), *Barriers and Bridges to the Renewal of Ecosystems and Institutions*. Columbia University Press, New York.

Perry, D.A. 1994. *Forest Ecosystems*. The Johns Hopkins University Press, Baltimore, MD.

Quigley, T.M., et al. 1996. Integrated Scientific Assessment for Ecosystem Management in the Interior Columbia Basin. USDA Forest Service General Technical Report, PNW-GTR 382. Portland, OR: Pacific Northwest Research Station.

Regional Ecosystem Office. 1995. Ecosystem Analysis at the Watershed Scale: Federal Guide for Watershed Analysis. Regional Ecosystem Office, P.O. Box 3623, Portland, OR.

Swetnam, T.W. 1993. Fire history and climate change in giant sequoia groves. *Science* 262: 885–889.

Swetnam, T.W., and J.L. Betancourt. 1990. Fire—Southern oscillation relations in the southwestern United States. *Science* 249: 1017–1020.

Turner, M.G., and R.H. Gardner. 1991. pp. 3–14. In: M.G. Turner and R.H. Gardner (eds.), *Quantitative Methods in Landscape Ecology*. Springer-Verlag, New York.

## THE AUTHORS

**David L. Caraher**
*4388 Kenthorpe Way*
*West Linn, OR 97068, USA*

**Arthur C. Zack**
*USDA Forest Service,*
*3815 Schreiber Way,*
*Coeur d'Alene, ID 83814, USA*

**Albert R. Stage**
*USDA Forest Service,*
*1221 S. Main Street,*
*Moscow, ID 83843, USA*

# Principles for Ecological Classification

Dennis H. Grossman, Patrick Bourgeron, Wolf-Dieter N. Busch, David Cleland, William Platts, G. Carleton Ray, C. Richard Robins, and Gary Roloff

**Key questions addressed in this chapter**

- *Review of historic and current approaches to ecological classification*
- *Relationship between ecological classifications and management objectives*
- *Strengths and weaknesses in the applications of current ecological classification systems*
- *Role of data and quantitative methods for the development of ecological classifications*
- *Conceptual and practical differences in the development and application of ecological classifications for terrestrial, freshwater and marine ecosystems*
- *Challenges for the future development of integrated ecological classification approaches*

**Keywords: Biogeography, ecoregions, vegetation, watersheds, climate, geology**

## 1 INTRODUCTION

The principal purpose of any classification is to relate common properties among different entities to facilitate understanding of evolutionary and adaptive processes. In the context of this volume, it is to facilitate ecosystem stewardship, i.e., to help support ecosystem conservation and management objectives.

This chapter has three purposes. The first is to provide a broad, scientific overview of the theory and methods of ecological classification. The second is to review past and current efforts to shed light on the characteristics of current classifications and how they have evolved. The third is to provide the scientific foundation for applying ecological classification to resource planning and management in the ecosystem context, with a primary emphasis on new paradigms for classification. In all three aspects, the focus is on the United States because the intent is to help meet the goal of improved ecological stewardship of federal lands and waters.

It is now widely accepted that federal management activities should take an ecosystem approach. The goal of this approach to management has been defined by the IEMTF (1995) as follows: "to restore and sustain the health, productivity, and biological diversity of ecosystems and the overall quality of life through a natural resource management approach that is fully integrated with social and economic goals. This is essential to maintain the air we breathe, the water we drink, the food we eat, and to sustain natural resources for future populations."

To this somewhat anthropocentric definition, one may add the importance of maintaining entire landscapes (e.g., Franklin 1993), the conservation of evolutionary and ecological processes, and the protection of species and ecosystems in protected areas (e.g., Grumbine 1994). Virtually all of the approaches have explicit or implicit requirements in common for the application of the best of theoretical science, and for clearly articulated conservation and management goals. Historically, these two requirements have not necessarily been congruent; in fact, they have often been at odds. Yet only through the alignment of ecosystem concepts with realistic management goals may the necessary research and monitoring activities emerge to meet the challenges involved in sustaining productive ecosystems, with their full complement of biological diversity.

Solutions to problems related to the common interests of science and management directly affect ecological classification. In the past, problems were expressed narrowly according to site- or resource-specific goals. This led to narrowly defined classifications specific to those problems. The ecosystem approach is broad and integrative, multi-factorial, and synthetic. Consequently, it is clear that a paradigm shift is required in how classifications may become ecosystem-oriented, dynamic, and process-oriented. Also, they must evolve according to increases in knowledge and to the evolving needs of ecosystem management.

A fundamental question is whether a common classification system can be developed that would meet the needs of all ecosystem managers. The answer seems to lie in the ecosystem concept and the clarification of what is "common" to all ecologically based classifications. For the complex variety of conservation and management questions that will always exist it would appear that a variety of biophysical classifications will always by needed.

This chapter attempts to respond to the challenge of developing common ecosystem properties for ecological classifications. It is structured in five sections. After providing a definition of ecological classification, below, Section 2 describes the theory, conceptual approaches, and methods that have been developed, and that continue to be the basis for, ecological classification. Section 3 is a summary of existing classifications for the major ecological realms — terrestrial, freshwater, and coastal-marine. It will become apparent that each of these realms has been historically treated quite differently, conceptually, and methodologically. In some cases, management applications are made difficult by these contrasting approaches, as for example when aquatic systems are subsumed within terrestrial, floristic provinces.

Section 4 describes how classifications may be derived by means of multi-factor, integrated methods. This approach is required for ecosystems, which are complex, hierarchical, integrated systems involving both biotic and abiotic elements. Section 5 summarizes major existing classification systems and their applications. Section 6 presents conclusions and future needs.

This chapter is written in recognition that no ecosystem classification system yet exists that truly integrates all ecosystem attributes, nor that can suffice for all plants and animals, nor for all lands and waters, nor that can fully describe or predict ecological change. Notwithstanding these limitations, our intent has been to cull out the essential elements of the subject with future needs in mind, trusting that the reader will refer to the literature cited for further insight and information.

### 1.1 What Is Ecological Classification?

The objective of any classification is to group together sets of observational units based on their common attributes (Kent and Coker 1992). Thus, ecological classification refers to the development and characterization of a set of units that are assigned by analyzing environmental and biological variables. The variables that can be used for classification are determined by available data and other resources. The end product of a classification is a set of groups derived from the units of observation where units within a group share more attributes with one another than with units in other groups. The ecological stewardship implications are that all representatives of an ecological unit should respond predictably to both natural changes and resource management practices (Driscoll et al. 1984).

Four important historical facts have set the stage for present-day ecological classification. First, classification in the past has been dominated by static approaches; ecological dynamics were poorly understood and environmental change was generally not considered. This is especially apparent on maps, where the polygons do not represent spatial, environmental and temporal dynamics. Second, by far the majority of work on ecological classification has been land-based, resulting in the omission of aquatic features. Third, aquatic systems have been classified generally by means of physical features, such as temperature, salinity, topographic features that define watersheds, the dendritic structures of river systems, estuarine structure, and the like. This is in strong contrast to the more biotic approach of land systems. These segregated treatments of terrestrial and aquatic systems defy attempts at integrating these ecosystem classification approaches, as is obvious in the case of omitting rivers from terrestrial biotic provinces or not considering land-sea interactions when describing coastal classifications. (An assumption for the land has been that vegetation is a surrogate for other biota. An assumption for the sea is that zoogeography or physical features can be used to define oceanographic regions, or vice versa. These assumptions have met with uncertain results.)

Fourth, a host of classifications has resulted from both the fragmented approach of management and the disciplinary approach of science. As a result, some classifications are defined at a single level, whereas others contain multiple levels, which may or may not be hierarchical. Most classification systems, and levels within classification systems, are associated with a particular geographic area and with specific spatial scales for implementation.

It is now abundantly clear that biotic–abiotic, terrestrial–aquatic integration is required for an ecosystem approach to classification. If ecological classification systems are generally to be used to delineate sets of biophysical units across landscapes, they must go beyond climate, floristics, zoogeography or physical features, as sole descriptors for those units. If coastal systems are to be classified, they must be fully based on the interactions of land, sea, and atmosphere, which define the coastal zone.

## 1.2 Classification Must be Driven by Objective

Different approaches to ecological classification have been developed to address specific objectives. As reviewed in this chapter, numerous ecological classification systems have been applied in many different areas over the years. There is no one correct way to classify ecosystems; the success of an approach is measured by its ability to meet management and/or scientific objectives, preferably both when they exist. Classifications have primarily been constructed for three purposes: (1) to help develop and represent the science of biogeography itself; (2) to help identify "representative" ecosystems for protection, restoration production, management, research, monitoring, and inventory, most notably of living resources; and (3) conservation of biological diversity. With respect to management itself, many possible applications exist for classifications, of which Frayer et al. (1978) have identified four: (1) a basis for cataloging the status of current resources; (2) a means of transferring experience and knowledge of a studied area to a similar but unstudied area; (3) a framework for assessing local management opportunities and predicting the outcomes of treatments or actions; and (4) a vocabulary for communication between managers, between managers and researchers, and between managers and the public.

The principal factor that controls the relationship between classification and objectives is scale. For example, ecological classification requirements for songbirds (functioning at the vegetation stand level) are different than those for amphibians (functioning at the micro-habitat level). However, under a properly designed system, ecological units that are appropriate for management objectives related to amphibians can be aggregated into or nested within units meaningful for songbird management. Thus, the first step in ecological classification is to determine the spatial scales needed to address management objectives.

Historically, management objectives have emphasized single resources as separate entities in the landscape or seascape. Contemporary land managers, however, are asking new sets of questions that require integration of multiple factors. These new questions include:

- What are the cumulative effects of a management activity on multiple components of a system?

- Will a management activity disrupt the inherent ecological processes and functions of the system?
- What level of a management activity will direct the system into an aberrant ecological trajectory?

Such questions are the essence of ecosystem stewardship because they place resource management in the context of multifaceted, hierarchical, ecological relationships. This is a particular challenge for the integration of science and management, as science is not able to give answers with high degrees of certainty to such questions. An instructive example concerns fisheries management, which has been treated from the point of view of simple "yield" models, out of context of ecosystem properties. A growing concern is that of sustained ecosystem functions and processes over time in light of management actions (question 3). Orians (1975) defines the following measures of ecosystem "stability": (1) constancy, (2) persistence, (3) inertia, (4) elasticity, (5) amplitude, (6) cyclical stability, and (7) trajectory stability. Which measures are operational can be very difficult to identify, but these terms define a way to approach ecological classification from a functional-process viewpoint.

## 2 HISTORICAL OVERVIEW OF ECOLOGICAL CLASSIFICATION

Biogeographic classification has a very long history, with strong links to biological diversity and biogeography. Since the time of the Greek scholars Aristotle and Theophrastus during the third century B.C., and of the Latin scholar Pliny the Elder during the first century A.D., descriptions of species have been accompanied by classifications, however rudimentary, and data on site locations. Over subsequent centuries to the present, many questions concerning the characterization of biological species and communities, including their relationships to environmental and latitudinal gradients, have emerged. As a result of rapid discoveries and the increasingly disciplinary nature of the inquiry, many different approaches to biological and ecological classification have been developed, along with a massive literature.

Historical approaches to ecological classifications may be broadly characterized as being either locally specific or regionally simplified. For local classifications, there has generally been a highly focused objective and a large quantity of field information; this has resulted in classifications and maps that are specific and detailed. In contrast, regional treatments have tended to generalize data from few points over large areas and, thus, had limited use at a local level. The most important contemporary advances have utilized the greater availability of biotic and abiotic data and improved spatial technologies to create multiple hierarchical classifications that can be applied to different objectives at multiple scales. What may be described as "local" vs. "regional" differs markedly among various systems. For the land, the difference is apparent. But for rivers, the watershed is the usual "regional" frame of reference, with "local" referring to third- or fourth-order tributaries. For coastal systems, "regional" is the biotic province, usually defined by the ranges of endemic fauna; "local" refers to physical subunits such as estuaries, bays, tributaries, and coastal formations. For open marine waters, including both coastal oceans and ocean basins, "regional" may refer to zoogeographic provinces or water masses; "local" is usually referred to as "sample sites" or "patchiness."

Ecological classifications have historically been based on abiotic, biotic, or integrated approaches. Abiotic classifications have defined sites and mapping units by singular or multiple components of the abiotic environment. These components include land forms (Hammond 1964), physiographic provinces (Fenneman 1928), and climate (Trewartha 1968), along with other variables such as geology, watersheds, and elevation. Biotic classifications have primarily been developed to portray vegetation classes and animal habitats.

Examples of terrestrial biotic classifications include potential natural vegetation (Küchler 1964), vegetation physiognomy (UNESCO 1973, Grossman et al. 1994b), Society of American Foresters cover types (Eyre 1980), plant communities (Grossman et al. 1994b), and natural vegetation of the Southwest (Brown et al. 1980). Hybrid approaches have integrated abiotic and biotic components to classify and map actual or potential land units. Examples of integrated approaches include habitat types (Krajina 1965, Pfister et al. 1977), ecoregions (Bailey 1976, Omernik 1987), forest regions (Braun 1950, Rowe 1972), ecological land units (Haufler 1994), and wetlands (Cowardin et al. 1979).

In Europe, land use planning assessments prompted the development of ecological classification in the early 1930s. Ecological, as opposed to purely biotic, classifications attempt to interpret the ecological meaning of the distribution of biotic types by relating this distribution to one or more features of the environment. Tools such as "ecological vegetation maps," which noted the environmental features believed to be indicated by the vegetation, were used for large-scale assessments of the suitability of sites for particular uses (Zonneveld 1988). Similarly, the North American concept of "ecosystem" became the basis for evaluating suitability in that site characteristics for a given mapping unit were used to determine appropriate land use categories and for other resource management needs. In the United States from 1900 to the late 1960s, resource management was mainly concerned with protection from catastrophic events (e.g., fire, floods), and with the efficient extraction of commodities. Throughout the 1970s and 1980s, however, an increased environmental awareness on the part of the public and within government confirmed and strengthened the need for structured, ecological data assessment tools. Single-factor classification systems were ineffective at addressing these "new" issues and did not capture land functions and processes. As a result, multi-factor ecological classification systems that recognized system inputs, outputs, complexity, and inter-connectedness were developed to address multiple-use and biodiversity issues.

Terrestrial classification systems have historically been given more attention and are considerably more advanced than other classification systems. As a result, aquatic systems tend to be included within terrestrial systems. For example, Bailey (1995) described "ecoregions" on the basis of land surface, climate, vegetation, soils, and fauna. This approach neglects watersheds and their dynamics and even the existence of the coastal zone, which includes a major portion of the land surface. Additionally, Cowardin et al. (1979) described wetland and deepwater habitats in traditional, typological ways that did not take into account the dynamic approaches employed by coastal biologists and oceanographers.

Coastal-marine physiographic and biogeographic classification has its own long history, perhaps dating from the scientific voyages of the 19th century. Modern zoogeography was initiated by Ekman (1953) and Hedgpeth (1957), who took zoogeographic and physical-ecological approaches to sub-dividing the near-shore seas and oceans into now-familiar zones (from littoral to abyssal). Somewhat later, Briggs (1974) re-evaluated the zoogeography and proposed a system of coastal biogeographic provinces primarily based on endemic species. This method allows clear identification of province divisions, but does not create a relational basis among provinces for evaluation, as all provinces defined by endemism become, by definition, unique. Nonetheless, Briggs's approach does confirm divisions that have long been apparent among coastal biotic provinces (e.g., as given for North America in Robins and Ray 1986, Robins 1992).

A major, relatively recent development has been the recognition of the "coastal zone" as a distinct earth realm where land, sea, and atmosphere uniquely interact. Ketchum (1972) essentially defined this zone to include the entirety of the coastal plains and continental shelves of the world. This zone is now estimated to include sea level plus and minus 200 m. It encompasses 18 percent of the earth's land surface and 8 percent of the oceans' surface (but less than 0.5 percent of the oceans' volume). However, its importance is disproportionate to its size: it supports 60 percent of the human population, and accounts for a quarter of the global primary productivity and more than 90 percent of fisheries (LOICZ 1996). The recognition of the coastal zone warrants a new ecological classification that not only transgresses traditional land–sea boundaries, but also functionally integrates terrestrial, marine, and atmospheric processes.

The need for a global classification of coastal and marine systems became especially clear when IUCN and UNESCO recognized during the 1970s that there was no comprehensive, global classification for conservation purposes. Udvardy's (1975) "world" classification was intended to meet that need for the land, but this "world" did not include freshwater and marine systems. Consequently, IUCN and UNESCO supported an effort to develop a matching, coastal-marine classification. The result was Hayden et al. (1984), which reviewed the "state-of-the-art" and presented a tripartite scheme for comparing coastal geomorphological provinces, coastal biotic provinces, and marine realms. It was recognized that this scheme needed significant extensions, particularly into the third dimension of deeper ocean waters and into smaller scales of interaction. However, to date, this global treatment has not been expanded.

In summary, most work on ecological classification and the related field of biogeography has focused on specific systems or elements of systems. There are many examples of classifications that represent specific taxa, landforms, vegetation and climate, physical geography, the phytoplankton of oceanic waters, estuaries and habitats of continental shelves, coral reefs, and other factors. It has become increasingly clear that many of the challenges presented by management of whole ecosystems cannot be addressed by these fragmented classification approaches.

## 3 ECOLOGICAL CLASSIFICATION: THEORY AND METHODS

From this brief history, it is clear that ecosystem management requires a reversal of trends, in both science and management, away from fragmentation and reductionism. The Ecological Society of America (ESA) has led attempts to define concepts for holistic management application, e.g., for "sustainability" (Lubchenco et al. 1991). More recently, the ESA has reported on the scientific basis for ecosystem management, defined by: "...explicit goals, executed by policies, protocols, and practices, and made adaptable by monitoring and research based on our best understanding of the ecological interactions and processes necessary top sustain ecosystem composition, structure, and function" (Christensen et al.1996). This amplifies the definition of the IETMF (1995) to include specific scientific goals, as well as social ones. Among the elements included in ecosystem management, as defined by this ESA report, are: sustainability, ecological modeling, adaptability and accountability, spatial and temporal scale, ecosystem function and dynamics, ecosystem integrity, and the uncertainties of our knowledge.

## 3.1 Theoretical Basis for Ecological Classification

In this context, it is abundantly clear that most present classifications are incomplete and possibly even misleading in cases where the approaches are based on narrowly defined ecosystem components. To address future ecosystem-oriented needs, seven areas of ecological theory need to be considered: (1) pattern recognition; (2) system dynamics; (3) hierarchy theory; (4) systems limits; (5) the biotic component of ecosystems; (6) the abiotic component of ecosystems; and (7) ecosystem characterization. Each of these areas is described in greater detail below. The treatment of these issues affects the selection of the concepts guiding ecological classifications, the most appropriate data for such classification, the best techniques for classifying ecosystems and their components (e.g., assemblages of species, landscapes, seascapes), and the best methods for predicting ecosystem properties and their responses to various management scenarios.

### 3.1.1 Pattern Recognition

To describe ecological systems, pattern recognition techniques are often employed. But because ecosystems are complex systems, defined by exchanges of energy, materials, and information, which may be described at many different scales (Levin 1992), their boundaries may be difficult to define and may even be considered arbitrary. Nevertheless, boundaries are determinable, for example, by recognizing gradients or "ecotones," which also provide a context for understanding ecological function (Hansen and di Castri 1992).

Ecological classification requires that spatial and temporal relationships need to be defined clearly (Bourgeron and Jensen 1993, Jensen et al. 1996). Furthermore, ecosystem management requires making predictions about ecological systems, that is, determining relationships between patterns and hypothesized causal factors (Urban et al. 1987, Bourgeron and Jensen 1993, Jensen et al. 1996). Once a correlation or a cause–effect relation between pattern and process is determined, predictions are made using the summary of data and information performed during the classification process and/or the generation of maps, statistical or simulation ecological models, and/or a combination of the first two approaches. Such interpretations are reasonably well developed for terrestrial systems, but are only at a primitive stage of development for coastal (Ray 1991) and marine (Steele 1989, Ogden et al. 1994) systems.

### 3.1.2 System Dynamics

Most existing ecological classifications are concerned with patterns or static structures. However, ecosystems may exhibit several trajectories (Orians 1975), and future ecosystem classifications cannot ignore these dynamics. A new approach is that of dynamic biogeography, which merges the large-scale approaches of traditional biogeography with smaller scale approaches of ecology. Dynamic biogeography concerns the study of biological patterns and processes on broad geographical and time scales, and represents spatial patterns at different scales of variation (Hengeveld 1990).

Ecological classifications are used for many purposes, such as assessing spatial relations and comparative features of ecological systems. However, ecological systems exhibit temporal changes along various developmental pathways that result in different types of organization. Many conceptual models used for planning purposes assume that all ecosystems reach a state of equilibrium or quasi-equilibrium (e.g., the assumption that succession leads to climax). However, we now know that ecosystems are rarely in equilibrium. Although it is convenient for scientific (e.g., modeling) and planning purposes to assume equilibrium at a given scale, we must recognize that ecological systems are fully dynamic (e.g., Kay 1991, Constanza et al. 1993). Thus, ecological classifications at all scales need to be designed to account for change, which may be slow or rapid, in terrestrial and aquatic systems.

### 3.1.3 Hierarchical Organization

Complex ecosystem patterns, and the multitude of processes that form them, exist within a hierarchical framework (Allen and Starr 1982, Allen et al. 1984, O'Neill et al. 1986), as is readily observed in landscapes or seascapes. In recent years, considerable attention has been directed to describing this hierarchical organization. Hierarchy theory (Allen and Starr 1982, O'Neill et al. 1986) views multi-scaled systems as a series of constraints in which higher levels of organization provide the environment in which the lower levels function. However, constraints are not necessarily "top-down," as Holling et al. (1993) have described. Ecosystems at any level may cycle through functional states of "exploitation," "conservation," "release," and "reorganization," in which both "top-down" and "bottom-up" interactions are evident.

Four tenets of hierarchy theory are critical to understanding landscape patterns and their dynamics, and to using that knowledge for ecosystem management (Allen et al. 1984; O'Neill et al. 1986).

1. Every component of a system, ecological or otherwise, is both a whole and a part at the same time. This concept is called "whole/part duality." For example, a forest (a whole) is made up of trees (the parts), each of which would consist of multiple communities for many organisms. At a larger spatial scales, the forest is part of a larger scale regional landscape.
2. Patterns, processes and their interactions can be defined at multiple spatial and temporal scales. These scales need to be identified clearly, according to the question being asked or the function being examined.
3. There is no single scale of ecological organization that can be used for all purposes. This is important to consider because ecological systems are often interpreted at a single scale or at a limited number of scales.
4. The definition of the component patterns and processes of any particular ecological hierarchy is dictated by the objectives of the study or by the objectives of management.

Most classifications are static and "taxonomic," whereas hierarchy theory challenges them to become dynamic and integrative. The "whole/part duality" of ecosystems (Koestler 1967, Allen and Starr 1982, Allen et al. 1984) clarifies ecological classifications because hierarchy theory expresses how different levels of ecological organization are linked. The findings at any one level may assist the understanding of another level, but can never fully explain phenomena occurring at other levels (Odum 1971). That is, each level of a hierarchy has characteristics that can only be explained with knowledge of other levels.

### 3.1.4 System Limits

As noted above, ecological systems may follow different trajectories of change. Much attention has been given to whether natural change is limited to approaches to a "climax" state in which an ecosystem functions at an "optimum." For example, it has been assumed that, eventually, ecological succession continues until species stop replacing each other, at which point the processes that lead to change balance the processes that lead to ecological organization. At this stage, called "climax," the ecosystem presumably would be operating at "optimum." Yet actual field observations reveal that this point does not remain constant. Environmental conditions may change (e.g., because of local, regional, or global climate change), new species may appear, and, even in the absence of major disturbances such as regional fires or coastal storms, the "climax" will change to a new community. The result is an "altered ecosystem state" with different optimal conditions (Hayden et al. 1991).

Obviously, the concept that optimal limits exist for any ecological system deserves re-examination. The observed state of an ecosystem may be only one of several possible and, hence, that the concept of an "optimum operating point" at "climax" may be misleading. It has been assumed that if an ecological system follows one path rather than another in response to disturbance or environmental change, it may operate far below its optimal level. An example is the maintenance of prairie grasslands under the natural disturbance regime of low intensity, high frequency fires. When such fires are suppressed, succession leads to the development of a forest stage (e.g., Arno and Gruell 1985). Thus, the conclusion was that the ecosystem functioned at suboptimal levels under presettlement disturbance regimes. On the other hand, disturbance is known to increase the biological diversity of some systems, notably tropical rain forests and coral reefs (Connell 1978), and recent research has shown that an increase in biological diversity may increase the productivity of ecosystems (Naeem et al. 1994).

Ecological classifications need to be designed to reflect these factors, especially if they are to be used in ecosystem management or assessments. This is especially important because many of the ecological units that require protection and management do not represent late successional or "climax" stages and/or are not in a state of equilibrium.

### 3.1.5 Biotic Component

Species patterns over time and space may be used to define the biotic component of ecosystems in two ways: by means of the continuum or community concepts. The continuum concept states that species assemblages are temporary and fluctuating phenomena along regional gradients. In contrast, the community concept maintains that repeatable assemblages of species occur in discrete habitats with characteristic properties. The conflict between these two approaches has far-reaching implications for ecological classification. For example, if the continuum concept is correct, the biotic component of ecosystems cannot be classified because species respond individualistically to environmental change over space and time. Most researchers implicitly accept the continuum concept, even avoiding the term community and referring to the more neutral term "assemblage" (Austin 1991). The same researchers, cartographers, and other practitioners, however, continue to recognize homogeneous units, hence implicitly using the community concept for pragmatic management purposes.

In defining the biotic component of ecosystems, finding a practical approach has faced major limitations. Weak and ambiguous classifications have resulted from unspecified or inconsistent criteria, vague definitions of key concepts, unspecified minimal areas of reference, and undocumented sorting strategies (Whittaker 1978, Küchler and Zonneveld 1988). In addition, existing data have not supported rigorous testing among the various continuum alternatives (Austin 1991). Nevertheless, various aspects of both the continuum and the community views appear to complement rather than exclude each other (Westhoff and van der Maarel 1978, Whittaker 1980, Austin 1991). It has been shown, at least for terrestrial vegetation, that species can be individually distributed along gradients (Austin 1987, Austin and Smith 1989) *and* that the distribution pattern of controlling environmental factors constrains the pattern of species combinations, their distribution in the landscape, and their frequency.

Even if some form of the community concept is accepted, the problem of identifying the full set of environmental factors shaping the composition of the biotic community remains. For terrestrial classification, the concept of a floristic or biotic province is defined within the context of a region with a distinct pattern of climate and landscape characteristics (Bailey et al. 1994). This is because such classifications are useful for defining regional ecosystem management guidelines only if vegetation units, defined as biotic or floristic provinces, or any variant (e.g., Küchler 1967, Brown et al. 1979) are modified to correlate with climatic/landscape regions (Burger 1976, Rowe 1980, Bailey and Hogg 1986, Küchler 1988). A primary purpose of such regionally defined ecosystems is to serve as a reporting structure for information about regional resources and environment (Bailey and Hogg 1986, Bailey et al. 1993). Another purpose is to define homogeneous regions within which to characterize finer scale ecosystems and/or landscape properties (Westhoff and van Der Maarel 1978, Austin and Smith 1989). Similarly, attempts to characterize landscape level ecosystems and conduct landscape assessment using such communities can be successful only if the intra-community pattern of gradual change is correlated with gradual changes in the environment. Therefore, effective landscape surveys need to take into account this dependence of biotic patterns on abiotic variables (Bourgeron et al. 1993).

The extent to which regional climatic/floristic classifications may apply to the fauna or to aquatic systems is highly dependent upon the strengths of the ecological linkages or dependencies among these different components. It would be incorrect to assume, for example, that different taxa respond in the same way to a particular set of environmental variables. Bibby et al. (1992) reported that areas of high endemism for birds are not the same as those for other vertebrates. Freshwater fishes are distributed according to the confines of watersheds, not floristics, and aquatic-coastal biotic provinces (Hayden et al. 1984) appear only roughly related to adjacent terrestrial provinces (Bailey 1995).

### 3.1.6 Abiotic Component

Two major categories of abiotic environmental variables have been useful for ecological classification (Austin et al. 1984, Austin 1985, Austin and Smith 1989):

- indirect factors — known to be correlated to direct factors that exert physiological influence on species (e.g., elevation, bathymetry);
- direct factors — known to have a direct physiological influence on species (e.g., temperature, pH, nutrients).Both of these have been variously been used for classification of terrestrial, freshwater, and marine systems.

Numerous assumptions are made in developing ecological classifications that involve the abiotic components of ecosystems. The first maintains that biological communities are surrogates for the environment. This approach, for example, uses the vegetation as a surrogate for the environment as a whole, based on the assumption that vegetation is a faithful expression of site characteristics (Troll 1941, 1943, 1955, 1956; Küchler 1988). Therefore, vegetation maps have been central to describing ecological patterns (Whittaker 1980). Similarly, for deep oceanic systems, plankton diversity has been used for decades as an indicator of environmental conditions for the delineation of ocean zones (McGowan 1971, Dunbar 1979).

The second assumption is that ecological units contain recurrent patterns and characteristics that can be delineated. This approach uses broad environmental patterns alone to describe and delineate abiotic elements in both terrestrial (Rowe and Sheard 1981,

Zonneveld 1989) and coastal systems (Hayden and Dolan 1976).

The third assumption is that environmental patterns may be integrated with biotic patterns to describe and delineate habitats of both plant and animal communities. For terrestrial systems, classifications have been developed using climatic attributes either alone or in conjunction with other attributes (e.g., Bailey 1976, Austin and Yapp 1978, Walter 1985, Omernik 1987, Bailey et al. 1994). For aquatic systems, very different approaches are used for freshwater, estuarine, and marine systems (Section 3.2).

A fourth assumption is based on the argument that, to be meaningful, ecological evaluation should be based on species' niche–habitat relationships (e.g., Hutchinson 1959, Whittaker 1972, Nix 1982, Brown 1984). The aim is to summarize environmental variability, identify the distribution of major environmental gradients, and indicate where significant shifts in ecological variability might occur (Mackey et al. 1988, 1989). To accomplish this, species' responses to a limited set of dominant environmental variables that comprise primary niche dimensions must be estimated (Nix 1982; Mackey et al. 1988, 1989). Site-specific data are used to generate classes of sites sharing similar ranges of values of the environmental variables. A map of these classes, or bioenvironments, can be used alone in the ecological assessment stage of an area for given purposes (DeVelice et al. 1993), and/or in conjunction with vegetation data for quantifying biotic–abiotic correlations (Mackey et al. 1989).

In all cases, the classification units arguably represent natural levels of ecosystem integration with respect to environmental regimes and key processes - and there is evidence that this is the case (Swanson et al. 1988). For example, geomorphic pattern, through erosional/sedimentation processes, controls carbon, nitrogen, and phosphorus cycles in soils of riparian forests in southern France (Pinay et al. 1992).

### 3.1.7 Ecosystem Characterization

In the field, various ecological classifications are often used together to delineate the boundaries of ecological systems. For example, a classification of the vegetation may be used with classifications of land forms and disturbances. The process of using ecological classifications for matching patterns and processes is called ecosystem characterization (Levin 1992). This process is used for mapping (e.g., Avers et al. 1994, Zonneveld 1989) and landscape evaluation from the site to the continent level. In defining ecological land types (e.g., Wertz and Arnold 1972), the idea is to draw boundaries around ecological systems. Relevant patterns and agents of pattern formation at all appropriate ecological and planning scales may then be defined. The proper match of patterns and agents of pattern formation is important because any incorrect coupling of pattern to agents impedes the ability to make predictions about the future state of the ecological system. As a result, the value of conservation and management actions and prescriptions is limited.

## 3.2 Goals and Methods of Multivariate Classification

The introduction to this chapter defined the purpose of ecological classification in the context of meeting the goal of ecosystem management. Sections 2.1 and 2.2 described past approaches and presented some aspects of modern ecosystem theory as a setting for future development. Clearly, classifications are now challenged to become integrative, dynamic, and adaptive. Accordingly, this section reviews some multi-factor methods and their applications.

Developing ecological classification systems is an iterative process involving field observation, statistics, and numerical modeling. At fine spatial scales, ecological units are often identified in the field, based on such things as flora, soils, and physiography (Host et al. 1988). At coarser scales, data from remote sensing or other sources are commonly analyzed (Denton and Barnes 1988). Classification units are refined as additional sampling and analyses are completed, and new relationships among components are identified. Additionally, sampling of ecosystem components aids in the recognition of such patterns as soil–vegetation correlations. This information is then integrated into the classified units.

Once large data sets are on hand, a variety of multivariate methods may be employed to identify patterns and to reduce the number of variables. Multivariate methods have been used to detect patterns in different types of data, including overstory basal areas, ordinal ground flora coverages, and nominal soil texture classes (Cleland et al. 1993). Also, statistical methods can be employed to simplify data sets by extracting their "principal components," as has been done for estuaries by Ray et al. (1997) through the analysis of NOAA's (1985) large data set. Several methods have been widely applied in recent years to the analysis of flora, fauna, plankton, abiotic features, and whole environments (e.g., estuaries), especially where it is necessary to condense and summarize extremely large data matrices.

Multivariate ecological classifications for terrestrial and aquatic environments have changed dramatically over the past few decades. Historically, analysis was carried out through the tabular analysis of plot data. Braun-Blanquet's procedure of 1921 used tabular methods in successive approximations to identify groups of species occurring in similar samples, and to identify samples with similar species composition. These early tabular classification techniques were informal and inherently subjective (Whittaker 1962, Mueller-Dombois and Ellenberg 1974). As a result, recognizing different species groups as well as groups of similar samples depended heavily on the individual investigator's understanding of species-species and species–environmental relationships within a study area. The results have been variously expressed in tables, dendrograms, and maps.

The past few decades have seen the proliferation on more objective procedures and refined multivariate analyses, in some cases leading to models expressive of ecosystem function (Mueller-Dombois and Ellenberg 1974, Gauch 1982, Spies and Barnes 1985, Denton and Barnes 1988, Hix 1988, Host and Pregitzer 1991, Bulger et al. 1993). These procedures include explorative data analyses involving descriptive statistics and graphical displays that are used to: (1) detect intercorrelations among variables, thus reducing in the number of variables that need to be considered; (2) check assumptions about the data structure underlying particular analyses; (3) suggest appropriate transformations; and (4) identify sample outliers.

### 3.2.1 Ordination

Ordination is a powerful technique used to identify important variables, to summarize data, and to reduce the complexity of the data set. The results may be displayed diagrammatically or mapped. In ecological studies, ordination is also used to discover the latent structure of data by analyzing species' responses to underlying environmental gradients (Prentice 1977, Bulger et al. 1993).

Principal component analysis (Gauch 1982, Morrison 1976), correspondence analysis (Hill 1974, Greenacre 1984), and detrended correspondence analysis are among the most commonly used ordination techniques in modern ecological studies. They employ different methods to account for the variance in the data.

Ordination is often used in an exploratory sense to detect trends as well as outliers, to screen variables, to reduce dimensionality, and to summarize community and environmental patterns. Ordination is often accompanied by clustering procedures to clarify natural groupings of sampled data. Results of ordination and clustering may be compared, and subsets of data may be interrogated to elucidate relationships further and to develop hypotheses. Several complementary analysis techniques may be applied to the same data set. The communication of results is promoted by employing a moderate number of commonly used, relatively standardized methods (Pielou 1977).

### 3.2.2 Clustering

The objective of clustering is to identify naturally occurring groups based on all variables in a data set. Both the process and the choice among techniques are more complex and more subjective than those of ordination. The most commonly used clustering methods are: (1) non-hierarchical, (2) polythetic hierarchical agglomerative, and (3) polythetic hierarchical divisive. Non-hierarchical clustering assigns data to clusters, placing similar samples or species together. This is an excellent way to handle redundancy and outliers, but is limited in its ability to analyze relationships. Hierarchical clustering also groups similar entities together into classes, but additionally arranges these within a hierarchy such that a single analysis may be viewed on several levels, with relationships expressed among the entities classified. The utility of a given technique is judged in relation to others, and often several classification techniques are applied to the same data sets with results compared afterwards.

The term "polythetic" means that information on all variables is used to assign observations to a cluster, as opposed to earlier monothetic methods that used single variables in a non-multivariate analyses. Polythetic agglomerative clustering has two steps: (1) the samples-by-species data matrix is used to compute a samples-by-species dissimilarity matrix using any of several distance measures such as Euclidean distance or percent dissimilarity; (2) an agglomeration procedure is applied successively to build up a hierarchy of increasingly large clusters, starting with clusters consisting of a single member, and agglomerating these hierarchically until a single cluster contains all the samples or species. The polythetic hierarchical divisive method similarly computes the dissimilarity matrix in the first step, and then successively applies a divisive procedure to a single large cluster to create a hierarchy of individual members.

The simplest polythetic-divisive classification is ordination space partitioning (e.g., Noy-Meir 1973, Peet 1980). Two-way indicator species analysis (TWINSPAN) (Hill 1979) is another polythetic divisive clustering technique. TWINSPAN begins with all species or samples (depending on the objectives) in a single cluster and divides these into smaller clusters by first ordinating data by reciprocal averaging. The reciprocal averaging procedure is repeated until each cluster has no more than a chosen minimum number of members.

### 3.2.3 Summary

Presently, ecological classification is often approached in an combined program of ordination, clustering, direct gradient analysis, and tabular synthesis of results. A five-step procedure has become routine, especially for terrestrial systems where procedures are most advanced. The analysis progresses by successive stages of refinement (Cleland 1996). First, exploratory data analyses are conducted to ensure that the assumptions underlying particular methods are met, and to reduce the number of variables. Second, ordination is used to summarize community and environmental patterns. Third, clustering is used to identify groupings of samples and variables, and to corroborate patterns detected through ordination. Fourth, community patterns are compared with environmental information to elucidate congruent changes and to produce an integrated interpretation of the ordination and clustering results. Finally, hypothesis testing methods may be used *a posteriori* to assess the relationship between classification and mapping as well as the ecosystem-level differences in structure and function. Such differences might include biomass production and productivity (Host et al. 1988), successional pathways (Host et al. 1987, Johnson 1992), and nitrogen dynamics (Zak et al. 1986, 1989).

Thus, multivariate methods are extremely useful in ecological classification because of the large number of variables and observations commonly analyzed, the difficulty in detecting patterns due to the complex interrelationships involved, and the need to verify the taxonomic and spatial hypotheses represented by classification and mapping. Multivariate analyses alone, however, are insufficient to develop classification systems. Even though the analysis methods themselves may be relatively objective, the selection of particular multivariate methods, decisions about how to standardize data, the selection of important variables or the removal of superfluous, redundant, or rare variables are subjective choices that can influence the resulting classification. In addition, the experience gained through reconnaissance, observation, and field sampling, and through the thought processes required to develop an integrated ecosystem classification, is valuable and *should be* used to augment strictly mechanical methods. The role of multivariate analysis, therefore, is that of a tool to be used in conjunction with knowledge of ecological relationships gained in the field.

## 4 ECOLOGICAL CLASSIFICATION SYSTEMS

Various types of classifications are summarized in this section, to illustrate their objectives and priorities. All classifications have the goal of determining the relative degree of similarity and difference among units. However, it is apparent that none comprehensively addresses the holistic character of ecosystems. This is not to say that they are not useful, but that integrated, ecological classification still lies ahead. It is also apparent that a classification is most useful to management when it can be mapped and related to biogeography (e.g., Pielou 1979, Brown and Gibson 1983).

### 4.1 Terrestrial Classifications

Most terrestrial classifications are biotic in nature, although many incorporate references to physiognomic regions and climatic zones. Terrestrial classifications are often simpler than aquatic classifications because of the comparably static nature of landforms and vegetation. Also, more research has been completed in terrestrial systems than in aquatic systems. As a result, terrestrial classifications are more advanced in theory and in practice than are aquatic classifications.

### 4.1.1 Vegetation Classification

Most terrestrial, biotic classifications have been based on vegetation. Beginning with the viewpoint of Gleason (1917, 1926) and extended by others (e.g., Whittaker 1956, 1962; Curtis 1959), it was generally believed that vegetation units could not be defined. The approaches taken often became polarized between the "continuum" and the "community unit" concepts, described above. But despite differing viewpoints, several features became widely recognized (Mueller-Dombois and Ellenberg 1974): (1) similar species combinations recur; (2) no two sampling units are exactly alike; and (3) species assemblages change more or less continuously if a geographically widespread community is sampled throughout its range. Thus, recurring species combinations may be correlated with their environments, and these combinations shift geographically, meaning that a limited degree of ordering *is* possible.

#### *Potential Versus Existing Vegetation*

Two differing approaches to the classification of vegetation are the portrayal of existing versus potential natural vegetation (PNV). Classifications emphasizing existing vegetation determine vegetation units from the current characteristics of the vegetation. Classifications emphasizing potential natural vegetation use characteristics that represent the most mature conditions of vegetation development (Tüxen 1956, Küchler 1964). It is, however, difficult to model the relationships between existing vegetation and potential natural vegetation. This modeling is limited by the current knowledge of vegetation-site relationships, and the ability of the observer to infer these relationships (Cook 1995). These models also emphasize hypothesized climax vegetation, a concept that is fraught with theoretical and practical difficulties. For example, most existing schemes assume that exotic species will be replaced by native species when, in fact, the behavior of exotics is notoriously complicated to predict.

### *Basic Classification Approaches*

Three vegetation classification systems have gained widespread acceptance: physiognomic classifications, floristic classifications, and combined physiognomic–floristic classifications (Howard and Mitchell 1985). All three approaches provide a systematic ordering of vegetation or ecosystem pattern and relate these patterns to ecological processes. Beginning in the 19th century with the work of plant geographers such as Humboldt, Warming, and Grisebach, vegetation classification focused on the physiognomy of the vegetation. Broadly speaking, physiognomy refers to the structure (height and spacing) of the vegetation and to the life forms of the dominant species (i.e., gross morphology and growth aspect of the plants). In addition, physiognomy refers to characters of seasonality, leaf shape, phenology, duration. These features are relatively easy to recognize in the field with limited knowledge of the flora. Physiognomy provides a fast, efficient way to categorize vegetation and can often be linked to remote sensing signatures. These characters are also useful for initial reconnaissance of areas to be surveyed. In addition, they permit generalizations about the vegetation at a coarse, often worldwide, scale.

The principles underlying physiognomic classification are that each specific life form represents a strategy (Stearns 1976), and that the composition of life forms in a vegetation type is governed by these strategies (Raunkier 1937, Monsi 1960, Walter 1973, Whittaker 1975). The predominance of certain physiognomic types in a region tends to correspond to major climatic zones. Thus, physiognomic categories are often expressions of macroclimate, soils, and vegetation (Holdridge 1947, Walter 1985, Howard and Mitchell 1985).

The basic unit of physiognomic classifications is the "formation," i.e., a "community type defined by dominance of a given growth form in the uppermost stratum ... of the community, or by a combination of dominant growth forms" (Whittaker 1962). These classifications emphasize a "top-down," divisive approach, which in practice, defines formations defined by varied, conventionally accepted combinations of growth-form dominance and characteristics of the environment.

Floristic methods characterize the species themselves. Early 20th century ecologists who favored a strict floristic system included members of what has been termed the Zurich–Montpellier Tradition in central Europe (Shimwell 1971). The most well known among them is that of Braun-Blanquet (1928), who called plant associations with common diagnostic species "alliances." Nowadays, the most commonly defined floristic unit is known as the "association," defined by Flahault and Schroter (1910) as "a plant community ... presenting a uniform composition and physiognomy, and growing in uniform habitat conditions." This definition implies that associations sharing a certain physiognomy would be grouped together into the same formation. Some floristic methods focus on species that occur constantly throughout a set of stands; others emphasize indicator or diagnostic species, which are dominant in or restricted to these stands. Floristic methods require intensive field sampling, detailed knowledge of the flora, and careful tabular analysis of stand data to determine diagnostic species groups. These methods reflect local and regional patterns of vegetation and are more detailed than physiognomic methods. They also provide detailed descriptions of biotic communities regardless of their successional stage or origin. As such, they are typified by an agglomerative, "bottom-up" approach. Thus, floristic units have been used frequently as indicators of ecosystem processes and are a useful component of ecosystem classifications (Mueller-Dombois and Ellenberg 1974, Rowe 1984, Strong et al. 1990).

Combinations of these two approaches have also been developed on the basis that vegetation is most thoroughly described by both structure and floristic composition. As stated previously, physiognomic systems are easily recognized in the field, can be applied with limited knowledge of the flora, permit generalizations of vegetation patterns over large areas, and can be linked to remotely sensed data to facilitate vegetation mapping. In contrast, floristic information is almost always used for detailed site analyses, whether for studying environmental gradients, ecological site factors, or describing and forming classification units. Furthermore, studies have shown that a very good fit exists between floristic and physiognomic classifications because both types of attributes are borne by individual species (e.g., Rübel 1930, Westhoff and Held 1969, Webb et al. 1970, Wergner and Sprangers 1982, Borhidi 1991). The Nature Conservancy (Grossman et al. 1994b; Grossman et al. 1998) have refined the

**CLASSIFICATION HIERARCHY**

**SYSTEM — Terrestrial**

**PHYSIOGNOMIC CLASS — Woodland**

**PHYSIOGNOMIC SUBCLASS — Mainly evergreen woodland**

**PHYSIOGNOMIC GROUP — Temperate evergreen needle-leaved woodland**

**FORMATION — Needle-leaved evergreen seasonally flooded/saturated woodland with rounded crowns**

**ALLIANCE — *Pseudotsuga menziesii* woodland**

**ASSOCIATION — *Pseudotsuga menziesii/Festuca idahoensis* woodland**

Fig. 1. An example of the TNC terrestrial physiognomic–floristic hierarchy (Grossman et al. 1994b).

physiognomic classifications of UNESCO (1973) and Driscoll et al. (1984) and combined this with two floristic levels to create an hierarchical, physiognomic–floristic approach (Fig. 1) which is now widely used for conservation and resource planning across the United States.

The common question for terrestrial classification approaches is whether the objective is to portray existing vegetation, potential natural vegetation, or both. Classifications emphasizing existing vegetation determine vegetation units based on existing structure, composition, and successional status (Haufler 1994). Classifications emphasizing potential natural vegetation use vegetation characteristics that represent the most mature conditions of vegetation development (Küchler 1964). In the words of Tüxen (1956, in Mueller-Dombois and Ellenberg 1974), potential vegetation becomes "the vegetation structure that would become established if all successional sequences were completed without interference by man under the present climatic and edaphic conditions."

### 4.1.2 Zoogeographic Classification

Zoogeography is the scientific study of the distribution of animals on the earth and the mutual influence of environment and animals on each other (Allee and Schmidt 1937). It is more difficult to summarize and represent than plant geography, simply because animals are much less easily observed and are highly mobile. Approaches to zoogeography include: (1) the faunal compositions of landscapes and regions, (2) the evolutionary dynamics of the geographical ranges of animals, and (3) the mutual relationships of geographic ranges with humankind (Allee and Schmidt 1937, Muller 1974).

Classifying animals with reference to their geographic locations dates back to the first century A.D. when faunal lists were created for specific areas (cf. Allee and Schmidt 1937). Faunal classifications have focused on numerous factors, including morphological traits, geologic influences (e.g., continental drift), and ecological relationships (e.g., animal distributions seen as dependent upon localized environments, as opposed to geographic distribution and range). Regardless of emphasis, all zoogeographic approaches recognize the link between habitat and animal distributions.

As is the case for plants, animal life is unequally distributed within any zoogeographic area. Primary factors influencing the distribution of animals include (1) the means of dispersal; (2) inherent or introduced landscape barriers; (3) the degree of adaptability exhibited by the species (Allee and Schmidt 1937); and (4) competition with other species. Animal distributions are further complicated by the fact that many species (e.g., most insects, crustaceans and fishes) have several life history stages, each with its own discrete habitat. Another major complicating factor is that different taxonomic animal groups have strikingly different geographical affinities. It is possible to construct classifications for birds, mammals, and invertebrates that have only a superficial resemblance in geographic pattern of distribution to one another. Even when "hot spots" for endemic species with limited ranges are identified for conservation purposes, the resulting maps for different taxonomic groups are not well aligned (see the case of birds vs. amphibians and mammals in Bibby et al. 1992). Thus, the concept of a "formation" or "association" does not apply to fauna, except for species that are highly restricted to specific plant units at small scales. In these cases, the term "community" may be used. Additionally, the habitat requirements of most species are multi-scale in accord with differing life-history requirements, meaning that the distributions of most animals are best viewed in an hierarchical framework in which distributions at different life-history stages are dependent on sets of habitat requirements defined at those scales. For many species, the scales may vary from inter-continental during migration to highly local during reproduction.

Lawton et al. (1993) explored numerous hypotheses to describe the relationships between range sizes or distribution limits and body size, latitudinal patterns, and metapopulation theories. Multiple approaches to coarse-scale zoogeographic classification have also taken place, for example, zoogeographic realms (Muller 1974) and life zones (Merriam 1898). However, these coarse-scale classifications are not very useful for operational land managers. Scott et al. (1993) developed a Gap Analysis approach more conducive to operational planning; it depicts species distributions based on vegetation alliances using small-scale satellite imagery. Even in this case, however, significant scale-related problems emerge.

## 4.2 Aquatic Classifications

As compared with terrestrial systems, aquatic systems are highly dynamic. Classification is facilitated more by abiotic than by biotic features. The water's surface also presents a screen to direct viewing by visible light. As a result, aquatic classification is much less advanced than terrestrial classification. It is apparent that aquatic classification systems should not be assumed to nest easily, if at all, into terrestrial classification systems.

As aquatic systems in general are highly dynamic and subject to intense natural perturbations, the concept of a stable "climax" has not been useful. Coastal-marine systems shed some light on the question whether classifications of existing biology or biotic potential apply to other than terrestrial systems. Valiela (1984) devoted an entire chapter to colonization and succession in marine communities; it is significant that the concept of "climax" does not appear. In fact, Peters (1976) pointed out that many properties associated with succession are tautological, creating doubt on the validity of the concept of "potential" ecosystem states.

Indeed, as any recent marine biology or oceanography text will illustrate, marine ecology is dominated by process concepts that emphasize change. Succession is seen as the composite result of a complex of factors. Perhaps it is for this reason, that few "potential" or "climax" maps equivalent to those for terrestrial vegetation exist. An exception is the work of Hayden et al. (1984), which brought together mapped classifications of coastal–biogeographic, coastal–physical, and oceanographic environments. Although there was some concurrence in the boundaries, enough differences were present to illustrate that these three approaches express different attributes.

### 4.2.1 Freshwater Classifications — Rivers

Approaches to classifying freshwater ecosystems range from those focusing on single physical, chemical, or biological variables to efforts using complex combinations of all three types of variables (Naiman et al. 1992, Hudson et al. 1992). However, despite many recent advances, issues related to parameterizing the many variables and physical processes involved in watershed modeling remain a major challenge for hydrologists. These include the need to deal with scaling and the linkages among hydrology, geochemistry, environmental biology, meteorology, and climatology (Hornberger and Boyer 1995).

The development of freshwater classification has occurred in phases. Early efforts to classify river systems identified whole river system types (Davis 1890, Shelford 1911). The recognition of biological zones led to a second phase of classification (Carpenter 1928, Ricker 1934). The utility of this approach was limited, however, because the classes were only applicable in basins where zoogeography, geology, and climate were held constant; biological zones could not be compared across regions (Naiman et al. 1992). The third phase, which continues to the present day, attempts to describe systems in terms of general ecosystem processes (Illies 1961). Concurrently, geomorphologists worked to describe the physics of channel formation (Horton 1945, Leopold and Wolman 1957, Strahler 1957); their findings led to a recognition of the need to consider geologic and climatic processes (Naiman et al. 1992).

The most recent efforts classify river systems within surrounding landscapes to account for the dynamics of ecosystem processes. Building on earlier work (Warren 1979, Lotspeich 1980, Lotspeich and Platts 1982), Frissell et al. (1986) applied principles of hierarchy theory to describe stream ecosystems in a watershed context. This model has two important elements. First, ecosystem processes are regionally scaled (i.e., whole stream systems are formed by climatic and geologic events), whereas variation in channel form results from more local events, such as storms and landslides. Second, large-scale processes constrain the development of small-scale stream features. In other words, the genesis of the stream system limits the range of microhabitat systems.

Much testing of hierarchical models has focused on the constraint of the upper level ecological processes on stream systems, particularly their biological composition. For example, Larsen et al. (1986) found that ecoregions corresponded to the distribution of fish assemblages in Ohio. Similar studies were conducted in Kansas (Hawkes et al. 1986), Arkansas (Rhom et al. 1987), Oregon (Hughes et al. 1987, Whittier et al. 1988), and Wisconsin (Lyons 1989). At a coarse scale, these studies showed that ecoregional classification provides a useful means to stratify variability in fish distributions (Hughes et al. 1994). However, prediction of fish species occurrence at a given site will require a classification system that accounts for local-scale habitat features that are not apparent at the regional level (Lyons 1989).

### 4.2.2 Freshwater Classifications — Lakes

The development of lake classification has followed a path similar to that for rivers. Typically, they have been based on single or multiple variables (Leach and Herron 1992) that generally fall into four groups: trophic or productivity levels (e.g., Hooper 1969; Shannon and Brezonik 1972), chemical and/or biological characteristics (e.g., Moyle 1956, Bright 1968), basin origin and physiographic characteristics (e.g., Hutchinson 1957, Winter 1977), and hydrologic setting (Winter 1977).

Current efforts to classify lakes relate biological composition to large-scale ecosystem processes (Tonn and Magnuson 1982; Tonn 1990; Schupp 1992). Tonn (1990) described the environmental and historical factors that determine patterns of fish biogeography and community structure in lakes as a series of filters that operate from continental to local scales. This incorporates appropriate hierarchical levels for aquatic ecosystem processes and underscores the importance of considering the interaction of zoogeography and geoclimatic factors at the higher levels of the hierarchy. Sly and Busch (1992) suggested an approach for classifying large aquatic systems based on multidimensional, geophysical parameters. This approach allows the consideration of productivity or functional performance criteria within geophysical boundaries. It also allows flexibility in identifying three-dimensional boundaries.

### 4.2.3 Wetland Classifications

Wetland classification became an issue in the early 20th century through efforts related to the identification of peatlands (Mitsch and Gosselink 1986). More recently, scientists and environmental regulators have developed various wetland classifications to identify and quantify areas for management and protection purposes. The goal, as stated in Cowardin et al. (1979), is to create "boundaries on natural ecosystems for the purpose of inventory, evaluation and management." Nonetheless, this classification system is a generalized "taxonomy" that is not georeferenced and that lacks three-dimensional boundaries — a much-needed extension and refinement.

Water levels dictate the type of wetland community. Wetlands usually have their water table near or above the terrestrial surface, so that the area is seasonally or permanently covered by shallow water, in which hydrophytes dominate. The water source and duration often differs for coastal versus inland wetlands. Coastal wetlands are usually located on gentle, sloping topography, and are exposed to regular, seasonal water-level fluctuations. However, abrupt, short-term changes caused by winds are common. These dynamics encourage more distinctive zonation and prevent the aging process usually found in inland wetlands (Herdendorf et al. 1986). Inland wetlands respond to rapid water-level increases from runoff, and groundwater contribution can be significant. During hot and dry periods, the survival of some inland wetland plant communities may depend on groundwater inflow. Groundwater is of lesser significance for coastal wetlands.

Wetlands are often classified according to their plant communities. For example, coastal wetlands may be referred to by a major plant species, e.g., mangrove, *Spartina*, or *Juncus*. The approach for this classification is floristic, as described above for terrestrial vegetation.

### 4.2.4 Estuarine Classifications

Estuaries represent important mixing zones of fresh and marine waters. They have been classified in several ways, largely by combinations of geomorphology, hydrodynamics, and geographic (climate). Pritchard (1967) recognizes five geomorphologic categories: (1) drowned river mouths, e.g., Chesapeake Bay, (2) tectonically produced, e.g., San Francisco Bay, (3) bar-built, e.g., coastal-lagoon systems of the southeastern and Gulf coasts, and (4) fjord-like, e.g., glacially overdeepened valleys along the Alaskan coast. Birdsfoot river delta or everted river mouths, e.g., the Mississippi River Delta, may be a fifth category.

The most simplistic estuarine classification is the so-called "Venice System," which classifies estuaries solely on the basis of salinity regimes (Anonymous 1959). This has long been considered an oversimplification. In contrast, estuaries may be classified hydrodynamically as: (1) stratified, (2) partially stratified or salt-wedge, and (3) completely mixed. In stratified estuaries the freshwater floats over saltwater; river influence is greater than tidal influence. Little mixing occurs, and the wedge may move far inland. In partially stratified estuaries, freshwater flow equals tidal flow as an agent of turbulence; the resultant layering may be complex. Completely mixed estuaries are vertically homogeneous, with tidal currents dominating the freshwater inflow. All these classifications are entirely physical. Bulger et al. (1993) used a principal components analysis to demonstrate that biologically relevant salinity regimes can be derived.

Geographic (climatic) classifications represent attempts to categorize seasonal effects of climate on run-off, temperature, and daylight, all of which trigger biological responses, such as migration and spawning. Runoff is especially subject to seasonal shifts. For example, in the tropics too little rainfall occurs in the dry season for any runoff to reach coastal areas. Lagoons may become sealed by bars formed by long-shore currents and quickly become hypersaline. With the onset of the rainy season, freshets break through the bars, and flush the lagoon. As a result, the lagoon may become entirely fresh. Such dynamics are important considerations for estuarine classification. Roy (1984) recognized three main estuarine types for New South Wales, Australia: (1) drowned river valley, (2) barrier, and (3) saline coastal lake, all of which evolve by processes of infilling. He documented unique biotic assemblages in each estuarine type.

### 4.2.5 Coastal-Marine Classifications

It has been recognized for some time that a useful ecological integrator for aquatic systems is the watershed (e.g., Sheldon 1972, Lotspeich 1980, Seaber et al. 1984). This approach may be extended into the sea as a "seashed." The advantage of using this approach is that watersheds reflect ecological properties across terrestrial–aquatic boundaries, and thus act as controls on the distribution of biological diversity. This approach is driven by multi-scale ecosystem factors, as exemplified by Lotspeich (1980); i.e., large-scale controlling forces are climate and geology, reacting forces are soils and vegetation, and the streams themselves respond to all factors of the system. Although watershed and seashed models are relatively well-developed, they are also limited by the lack of available information on ecological interactions of the coastal zone. Significant research is needed to establish ecological relationships that are useful for coastal-marine ecological classification.

Ray and Hayden (1992) adopted concepts from Seaber et al. (1984) to develop a watershed- or seashed-based classification that recognized five coastal-zone subdivisions: uplands, coastal plain, and tidelands on the terrestrial side and the shoreface entrainment volume and offshore entrainment volume over the continental shelf. Hydrological interactions are key to this system, which is specifically intended as a comparative-hierarchical classification that is adaptable to both scientific and conservation purposes. This hierarchical approach is necessary to address the many scales that are involved. For example, the ability of any location to provide suitable habitat for a diversity of species depends on an array of physical and biotic factors that vary at different temporal and spatial scales from whole ocean basins to local estuarine and freshwater tributaries. (Regional distributions of fishes have been described by Robins [1971], Robins and Ray [1986], and Robins [1992]; the multi-scale dynamics of these relationships have been examined by Ray et al. [1997].)

### 4.2.6 Hierarchical Approaches

The Nature Conservancy is developing an aquatic classification system to provide a framework for the inventory, identification, and characterization of the freshwater biodiversity of North America (Higgins et al. 1996). This classification is hierarchical and allows for the characterization of aquatic communities on both abiotic and biotic levels at multiple scales (Fig. 2). This approach captures information on the ecological context of community types, including physical processes. It also characterizes the physical environment, which provides an indirect way to identify community types where biological information is not sufficient (Angermeier and Schlosser 1995).

Another recent development is that of Maxwell et al. (1995), which calls for an hierarchical framework of aquatic units (Fig. 3). This system is based on the aggregation of hydrologic units and the geographic distribution of aquatic biota (particularly fish species). The system is therefore not merely typological or descriptive, but ecological. It provides insights into system dynamics and how these units may be used for management and other purposes.

## 4.3 Integrated Classification Approaches

The introduction to this chapter presented the overall purposes of ecological classification. From previous sections, as well as Tables 4 and 5, it is apparent that many different classifications exist. As has been emphasized previously, terrestrial classifications have dominated the field. The classifications have many properties in common, namely the organization of data into categories, which facilitates retrieval and analysis and allows comparisons among systems. Most of these systems do not share a common approach to

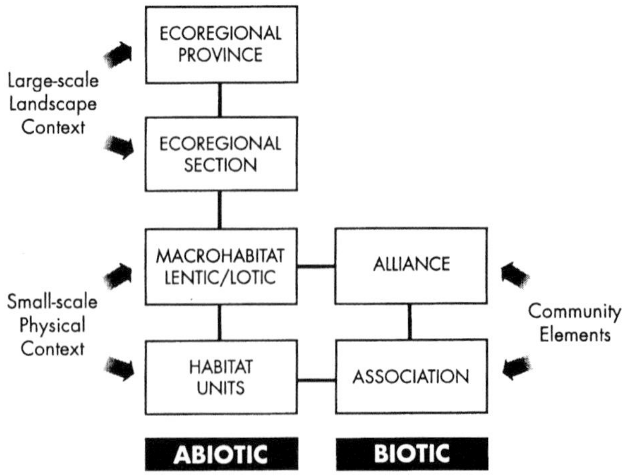


Fig. 2. TNC Aquatic Classification Hierarchy (Higgins et al. 1996).

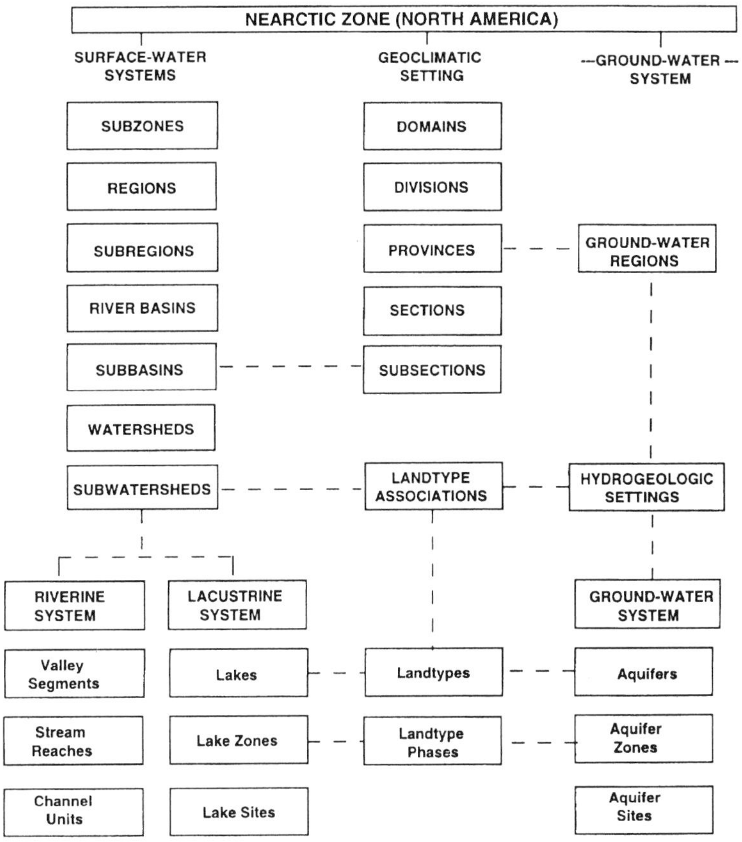


Fig. 3. US Forest Service Aquatic Ecological Unit Hierarchy (Maxwell et al. 1995).

classification, and this is of utmost importance for ecosystem management. Nor is it clear, at this point, if the different classifications may be reconciled into or around a common classification approach.

### 4.3.1 Site Classifications

*Single Factor*

Single-factor site classifications generally assume that the chosen characteristic best reflects spatial patterns of the environmental resources of interest (Omernik and Gallant 1989). For terrestrial sites, these classifications are primarily intended to reflect plant and animal potentials. Several site classification systems have used only vegetation to determine site potentials, usually with reference to successional trends or productivity. In this sense, these systems focus on potential natural vegetation (e.g., Fenneman 1946; Küchler 1967; Udvardy 1975; Brown et al. 1979, 1980). A widespread approach to site classification using vegetation is the habitat type classification system (Daubenmire 1952, Pfister and Arno 1980, Kotar et al. 1988). This system focuses on natural climax or near climax vegetation, and recognizes all understory species as a reflection of site characteristics. Relationships between vegetation and the soils or landform factors are established during and after the classification process, but these factors are not used to define the vegetation units (Komarkova 1983). The units described are natural ones, but emphasis is placed on determining vegetation units that represent "ecologically equivalent landscapes" (Kotar et al. 1988). Insofar as they describe the floristic composition of part of the natural vegetation, namely climax stands, the units of the habitat type are fairly equivalent to the plant association concept used in the western United States (Komarkova 1983). The intent is to use these descriptions to classify sites that are not at climax and, by examining their understory composition, to infer their ecological potential.

Somewhat different from the habitat type approach is the ecological species group approach, which maintains that species show similar "ecological behavior." Generally, these species belong to the same layer of vegetation (e.g., the herb layer, nonvascular layer, or shrub layer). The method presumes that communities are combinations of plant species whose composition depends on the local environment (Mueller-Dombois and Ellenberg 1974). The community unit identified can, at times, be very similar to the plant association, whereby the ecological species groups are the diagnostic species for the association. However, it is also possible that the same association could contain several ecological species groups (Mueller-Dombois and Ellenberg 1974). The ecological species group information can either be used by itself to indicate site characteristics, in which case the system partially resembles the habitat type system, or it can be integrated with other measured site factors as part of an ecosystem classification (Pregitzer and Barnes 1982, Cleland et al. 1994).

*Multiple factor*

Multi-factor classifications are based on the idea that ecosystems and their components display regional patterns that are reflected in a combination of causal and integrating factors (Bailey 1983; Austin and Margules 1986; Omernik and Gallant 1987). Existing multi-characteristic frameworks include, but are not limited to, the British Columbia biogeoclimatic system (Pojar et al. 1987), Cajender's (1926) forest types, Daubenmire's (1968) habitat types, the ecoregions of the conterminous United States (Omernik 1987, Gallant et al. 1989, Omernik and Gallant 1989), Bailey's (1976, 1980) ecoregions of the United States, and the major land resource areas (USDA Soil Conservation Service 1981). Many similarities are found among these classification systems. For example, site classifications include floristic information that is collected in the same way as that which would be used for vegetation classification (Mueller-Dombois and Ellenberg 1974, Pregitzer and Barnes 1982). Similarly, habitat type classifications define plant associations in a similar manner to that of the floristic system of Braun-Blanquet. Furthermore, site classifications that bring together independent vegetation, soil, and landform classifications rely on the independent classification of these variables as their starting point (Jones et al. 1983, Sims et al. 1989, Host 1993). Site classification systems that use multiple factors subdivide land into major and minor land types or landscape ecosystems. They have been developed primarily for land managers who need to integrate resource management, biological conservation, and restoration planning. These systems are most appropriate for classifying ecosystems, which are defined by the dynamic interactions of the biotic and physical components. As with vegetation classifications, emphasis is placed on units that are more or less homogeneous both as to form and structure, but in this case with respect to all factors of the land and the vegetation (Rowe 1961).

An ecosystem approach to classification, namely that the plant community is considered together with its environment, was implicit in Clements work (1916), but was defined explicitly by Tansley (1935) and similarly by Sukachev (1945) as "biogeocoenosis." Central to the application of the approach is that all parts of the ecosystem are included. In some ecosystems, each part — vegetation, soils, climate and landform — is first studied independently and then combined (Jones et al. 1983, Driscoll et al. 1984, Sims et al. 1989). In others, the parts are combined at the outset because it is their joint interactions on the landscape that define the units. Because it is difficult to integrate multiple factors and understand their interactions beyond the local level, the units are usually considered hypotheses in need of further testing (Albert et al. 1986). Mapping is a key product from this process (Rowe 1984, Zonneveld 1989). Bailey's ecoregional map of the United States (1976, 1995) is more similar to the independent approach because Bailey relies heavily on separate climatic, physiographic, and vegetation maps and then reconciles their boundaries. The works of Albert et al. (1986) and Cleland et al. (1994) represent the combined approach.

The biogeoclimatic zone system of Krajina (1965, in Mueller-Dombois and Ellenberg 1974) is another system in which vegetation is emphasized when defining landscape or ecosystem units. These zones are defined as geographic areas that are predominantly controlled by the same macroclimate and contain similar soils and (climatic climax) vegetation. The definition of the zones at lower scales uses vegetation units that are defined by the plant association concept (Pojar et al. 1987). At higher levels, climatic zones and topographic position are used to group vegetation units into biogeoclimatic zones.

### 4.3.2 Ecoregional Classifications

Ecoregions are defined as geographical zones that represent groups or associations of ecosystems that function in a similar way. Ecoregions are assumed to be highly uniform with respect to their abiotic and biotic factors, the processes controlling their ecosystems, and their limits and prospects for human exploitation. Consequently, classifying the world into an organized geographic system and subdividing the earth's surface into consistent regions is difficult.

In general, two distinct concepts have been used. The first is the "controlling factor(s)" concept (Bailey 1983, 1996), in which it is assumed that one or more environmental factors act as primary controls or limitations for a particular ecoregion as defined at a particular scale. The second is the "synthesis approach" (Omernik 1987), which assumes that a holistic integration of all predominant and stable components must be considered in the regionalization. This approach defines ecoregions by perceived patterns in a combination of causal and integrative factors, including land use, land surface form, potential natural vegetation, soils, and so on. The distinction between these approaches is obscured when they are mapped because mapping always requires a reduced set of differential and associated characteristics.

Through over 100 years of development of biogeography, many ecoregion systems have been produced at various scales from global to local. Many approaches and applications have been developed, including some which integrate soils, land forms, and vegetation (see Herbertson 1905, Austin 1972, Udvardy 1975, Bailey 1976, Omernik 1987, Bailey 1995). Below is a brief description of four ecoregional classification systems that are presently used in North America — two for Canada and two for the United States. Also included is a brief description of a new ecoregional classification system that was developed by the World Wildlife Fund to evaluate conservation priorities at an ecoregional level in Latin America and the Caribbean.

### Controlling Factor Approaches

*Ecoclimatic regions of Canada*: Environment Canada has developed the "Ecoclimatic Regions of Canada, First Approximation" (1989). This is a traditional "control factor(s)" approach to creating a hierarchy of regionalizations. In order of decreasing size and generality, the map includes ecozone, ecoprovince, ecoregion, and ecodistrict levels. Climate guides the development of the ecoprovinces, although physiography, such as large mountain ranges, has major modifying influences. Ecoregions were subdivided based on the effects of macroclimate on the vegetation. Ecoclimatic regions were defined as broad areas on the earth's surface characterized by distinctive ecological responses to climate as expressed by vegetation and reflected in soils, wildlife, and water. Within ecoclimatic regions, similar trends in vegetation succession will be found on similar soils occurring on similar parent materials and positions on the landscape. Ten ecoclimatic provinces and 77 ecoclimatic regions were identified in Canada.

*Ecoregions of the United States*: Bailey (1976, 1980, 1983) developed a national regionalization of ecosystem units for the U.S. Forest Service. It is a hierarchical system, and similar in concept to the Environment Canada system. Four broad ecological levels (domain, division, province, section) are distinguished, based on climate and biogeography. Domains are identified by broad climatic similarity, such as lands having dry climates at the subcontinental scale. The domains are quite heterogeneous and are further subdivided into divisions through additional climatic criteria. Divisions are subdivided into provinces on the basis of the climax plant formation that geographically dominates the upland area of the province. Provinces are further subdivided into sections by differences in the composition of the climax vegetation type. The sections correspond generally to the potential natural vegetation types of Küchler (1964). As an extension of this system, an Ecological Classification and Mapping Task Team (ECOMAP), was formed in the U.S. Forest Service to develop a consistent approach to ecosystem classification and mapping at multiple geographic scales. A national hierarchical framework of ecological units was developed by the ECOMAP team with four new levels under Bailey's section: subsection, landtype association, landtype, and landtype phase (McNab and Avers 1994). ECOMAP is described in greater detail in the "Synthesis Approaches" section.

### Synthesis Approaches

*Terrestrial ecozones of Canada*: Using the synthesis approach to integrate a different set of assumptions in mapping and characterizing regional land units, Wiken (1986) developed "Terrestrial Ecozones of Canada." Although its nomenclature is similar to that of the Ecoclimatic Regions of Canada, this system is based on predominant and stable biophysical characteristics, rather than a singular controlling mechanism. Ecozones (15 units) are areas of the earth's surface representing large and very generalized ecological units, characterized by various abiotic and biotic factors. Ecoprovinces (45 units) are parts of ecozones characterized by major assemblages of structural or surface forms, faunal realms, and vegetation, soil, hydrological and climatic zones. Ecoregions (177 units) are parts of ecoprovinces characterized by distinctive regional ecological responses to climate, as expressed by the development of vegetation, soil, water, and fauna.

*Ecoregions of the conterminous United States*: Omernik has produced a map entitled "Ecoregions of the Conterminous United States" (Omernik 1986, 1987) that shows 76 ecoregions at a scale of 1:7,500,000. Omernik's method grew out of an effort to create a stream classification system for the U.S. Environmental Protection Agency to plan water resource management. The premise behind Omernik's approach is that ecological regions can be identified by analyzing the patterns and composition of biotic and abiotic phenomena that reflect differences in ecosystem quality and integrity (Wiken 1986; Omernik 1987, 1995). These phenomena include geology, physiography, vegetation, climate, soils, land use, wildlife, and hydrology. The relative importance of each characteristic varies from one ecological region to another, regardless of the hierarchical level.

## ECOMAP

Until the early 1990s, no single system had been developed with the structure and flexibility needed to develop ecological units from continental to local scales.

The National Hierarchical Framework of Ecological Units (Avers et al. 1994) was developed by the U.S. Forest Service to provide a consistent, hierarchical framework for application throughout the United States. ECOMAP uses a single factor approach at some levels, and a synthesis approach at others. Because ecological variables and processes operate at different spatial spaces, each level of the hierarchy is associated with particular design criteria for its scale (see Table 1).

At the ecoregion scale, the domains, divisions, and provinces are adapted from Bailey (1980, 1995), and are recognized by differences in global, continental, and regional climatic regimes and gross physiography. These upper levels are based on regional climatic types of Köppen (1931), as modified by Trewartha (1968). In addition to climate and physiography, the units at the

Table 1. ECOMAP: Principal map units and design criteria (Avers et al. 1994).

| Scale | Ecological Unit | Principal Map Unit Design Criteria* | Map Scale Range | General Polygon Size |
|---|---|---|---|---|
| Ecoregional | Domain | • Broad climatic zones or groups (e.g. dry, humid tropical). | 1:30,000,000 or smaller | 1,000,000's of square miles |
| | Division | • Regional climatic types (Koppen 1931, Trewatha 1968).<br>• Vegetational affinities (e.g. prairie or forest).<br>• Soil order. | 1:30,000,000 to 1:7,500,000 | 100,000's of square miles |
| | Province | • Dominant potential natural vegetation (Kuchler 1964).<br>• Highlands or mountains with complex vertical climate–vegetation–soil zonation. | 1:15,000,000 to 1:5,000,000 | 10,000's of square miles |
| Subregion | Section | • Geomorphic province, geologic age, stratigraphy, lithology.<br>• Regional climatic data.<br>• Phases of soil orders, suborders or great groups.<br>• Potential natural vegetation.<br>• Potential natural communities (PNC) (FSH 2090). | 1:7,500,000 to 1:3,500,000 | 1,000's of square miles |
| | Subsection | • Geomorphic process, surficial geology, lithology.<br>• Phases of soil orders, suborders, or great groups.<br>• Subregional climatic data.<br>• PNC—formation or series. | 1:3,500,000 to 1:250,000 | 10's to low 1,000's of square miles |
| Landscape | Landtype Association | • Geomorphic process, geologic formation, surficial geology, and elevation.<br>• Phases of soil suborders, families, or series.<br>• Local climate.<br>• PNC—series, subseries, plant associations. | 1:250,000 to 1:60,000 | 100's to 1,000's of acres |
| Land unit | Landtype | • Landform and topography (elevation, aspect, slope gradient, and position).<br>• Phases of soil subgroups, families, or series.<br>• Rock type, geomorphic process.<br>• PNC–plant associations. | 1:60,000 to 1:24,000 | 10's to 100's of acres |
| | Landtype Phase | • Phases of soil families or series.<br>• Landform and slope position.<br>• PNC—plant associations or phases. | 1:24,000 or larger | < 100 acres |

*Note: Criteria listed are broad categories of environmental and landscape components. The actual classes of components chosen for designing map units depend on the objectives for the map.

Table 2. Examples of subsection descriptions (Keys et al. 1996)

MAP UNIT TABLES: ECOLOGICAL UNITS OF THE EASTERN UNITED STATES -- FIRST APPROXIMATION

| Subsection | Geomorphology; Elevation | Quaternary geology; Stratig. & lithology | Soil taxa; Temp. & moisture regimes | Climate P(in) | Climate T(F.) | Climate Gs(d) | Potential Vegetation | Surface water characteristics | Human use |
|---|---|---|---|---|---|---|---|---|---|
| **212A AROOSTOOK HILL AND LOWLANDS SECTION** | | | | | | | | | |
| 212Aa Aroostook Hills and Lowlands | Open low mountains; 200-2500 ft. | Wis. loamy-sandy loamy till; Paleozoic pelite-sandstone-conglomerate | Haplorthods, Epiaquepts, Endoaquepts; Frigid, udic, aquic | 36 | 39 | 109 | Sugar Maple-Birch-Beech, Red Spruce-Balsam Fir Forest, Pine-Heath Woodland | Few lakes, ponds, rivers, numerous streams | Forestry Agriculture |
| 212Ab Aroostook Lowlands | Plains with high hills; 100-1000 ft. | Wis. loamy-sandy loamy till; Silurian pelite-sandstone-limestone | Haplaquents, Haplorthods; Frigid, udic, aquic | 38 | 40 | 110 | Sugar Maple-Birch-Beech, Red Spruce-Balsam Fir Forest, N. Red Oak-White Pine Forest | Numerous lakes, rivers, streams, large wetlands | Forestry Agriculture |
| **212B MAINE-NEW BRUNSWICK FOOTHILLS AND LOWLANDS SECTION** | | | | | | | | | |
| 212Ba Central Maine Foothills | Low mountains; 200-1200 ft. | Wis. sandy loam till, marine deposits; Paleozoic calcareous sandstone, limestone, pelite | Borofolists, Haplorthods, Epiaquents; Frigid, udic, aquic | 41 | 42 | 131 | Sugar Maple-Birch-Beech, Red Spruce-Balsam Fir Forest and Wetlands, N. Red Oak-White Pine Forest | Numerous large lakes, rivers and streams, few wetlands | Forestry Agriculture |
| 212Bb Maine/New Brunswick Lowlands | Open high hills; 100-800 ft. | Wis. loamy-sandy till, marine silt-clay; Devnian calcareous sandstone, limestone, monzonite | Epiaquents, Haplorthods; Frigid, udic, aquic | 42 | 42 | 141 | Sugar Maple-Birch-Beech, Red Spruce-Balsam Fir Forest, N. Red Oak-White Pine Forests, N. Cedar Limestone Woodland | Common lakes, ponds rivers, stream, and large wetlands | Forestry Agriculture |
| **212C FUNDY COASTAL AND INTERIOR SECTION** | | | | | | | | | |
| 212Ca Maine Eastern Interior | Open hills; 100-800 ft. | Wis. loamy-sandy till, Devonian alkali feldspar syenite-biotite granite | Haplorthods, Epiaquents, Borosaprists; Frigid, udic, aquic | 44 | 42 | 139 | Sugar Maple-Birch-Beech, Maritime Red Spruce-Balsam Fir, N. Red Oak-White Pine Forests, N. Red Oak Summit Woodland | Common lakes, ponds and streams, few rivers; few, large wetlands | Forestry Agriculture Rural |
| 212Cb Maine Eastern Coastal | Plains with hills: 100-300 ft. | Quaternary marine silt-clay, sandy loam till; Devonian gabbro diorite, monzonite diorite, quartz monzonite | Haplorthods, Epiaquepts, Endoaquods, Borofolists; Frigid, udic, aquic | 47 | 43 | 152 | Sugar Maple-Birch-Beech, Maritime Red Spruce-Balsam Fir, N. Red Oak-White Pine Forests, Freshwater Tidal Marsh | Coastal, saltmarsh, few lakes, ponds, rivers, streams | Rural Agriculture Recreation |

Table 3. ECOMAP: National hierarchy of ecological units associated with purpose.

| Planning and Analysis Scale | Ecological Units | Purpose, Objectives, and General Use |
|---|---|---|
| Ecoregions | | |
| Global | Domain | Broad applicability for modeling and sampling. RPA assessment. International planning. |
| Continental | Division | |
| Regional | Province | |
| Subregions | Sections<br>Subsections | RPA planning. Multi-forest, statewide, and multi-agency analysis and assessment. |
| Landscape | Landtype Association | Forest or area-wide planning, and watershed analysis. |
| Land Unit | Landtype<br>Landtype Phase | Project and management area planning and analysis. |

province level are further characterized and classified by soil orders and potential natural communities (from Küchler 1964).

At the subregion scale, sections and subsections are characterized by combinations of climate, geomorphic process, topography, and stratigraphy. These factors influence moisture availability and exposure to radiant solar energy, which in turn directly control hydrological function, soil-forming processes, and potential plant community distributions (Avers et al. 1994). The classification and descriptions have now been completed throughout the United States at the section level (McNab and Avers 1994). Numerous regions (the eastern United States, California, and the Columbia Basin) have completed classification, characterization, and mapping at the subsection level (see Keys et al., in press). An example of the level of detail for the map units is shown in Table 2. In the upper midwest (Minnesota, Wisconsin, and Michigan) and other areas this work has been completed to the subsubsection level (see Albert 1994).

At the landscape scale, the landtype association is defined by general topography, geomorphic process, surficial geology, soil, potential natural community patterns, and local climate (Formann and Godron 1986; Avers et al. 1994). At this level, terrestrial features and processes may also have a strong influence on ecological characteristics of aquatic habitats (Platts 1979, Ebert et al. 1991). At the land unit scale, landtypes and landtype phases are designed and mapped in the field based on properties of local topography, rock types, soils, and vegetation. These factors influence the structure and composition of plant communities, hydrologic function, and basic land-use capability (Avers et al. 1994).

*Taxonomic and geographic scales: implications for classifications and their integration*: The different scales inherently associated with the levels of the National Hierarchical Framework of Ecological Units have implications for the specific types of planning objectives for which each can be used. Table 3 shows the general utility of each ECOMAP level.

### *Biodiversity Pattern Approaches*

*Terrestrial ecoregions of Latin America and the Caribbean*: With the objective of setting conservation priorities at an ecoregional level, the World Wildlife Fund developed a hierarchical, terrestrial ecoregion system for Latin America and the Caribbean (Dinerstein et al. 1995). This system is based on an estimation of the "original," pre-Colombian distribution of habitats in this area, and differs significantly from the approaches used in the United States and Canada. Three hierarchical levels were identified: major ecosystem types, major habitat types, and ecoregions. Major ecosystem types (5 units) are defined primarily by dynamic properties and spatial patterns of biodiversity, not wholly by vegetation structure. Major ecosystem types are subdivided into major habitat types (11 units) based on general habitat structure, climatic regimes, and major ecological processes. The level of species turnover with distance (beta diversity) is also considered, with flora and fauna showing similar guild structures and life histories. Nested within the major habitat types, ecoregions are identified to represent geographically distinct assemblages of natural communities that share a large majority of their species, ecological dynamics, and similar environmental conditions. The ecological interactions critical for the long-term persistence of these communities are also considered.

### 4.3.3 Land Cover Classifications

Land cover classifications are primarily intended for land management or resource planning. They emphasize conspicuous features of the land surface, and can be combined with land-use maps to convey an overall perspective of what is visually perceived on the land. Many classification systems derive land cover units from general structure and composition of the vegetation (Anderson et al. 1976). For example, cover types, named by the dominant tree (Eyre 1980), are used as a descriptive land cover classification of forest lands.

Recent land cover systems have increasingly relied on factors that can be characterized through remote sensing imagery (Witmer 1978). Examples include the Gap Analysis (GAP) and the Multi-Resolution Land Characterization (MRLC) Programs. GAP is a national program that uses remote sensing to derive landcover types, which are then used to model the distribution and protection status of animal species. Similarly, the MRLC Program is creating a consistent, remote sensing-based land cover characterization approach across the nation that can be used for both one-time and repeatable resource inventory and environmental assessment objectives.

## 5 USE OF ECOLOGICAL CLASSIFICATION SYSTEMS

Many ecological classification systems use existing biological and environmental information to predict patterns, processes, and occurrences that will support conservation and management objectives. Classifications have been able to integrate more information in recent years, and the resulting products can increasingly be applied to multiple purposes (e.g., strategic planning, environmental monitoring, protected-area identification). Caution must be exercised, however, when using classifications and maps to make resource conservation and management prescriptions when those products were developed through combining preexisting data sets. The products derived from such integrated data sets are, as a rule, much less precise in terms of class and spatial accuracy than their component parts.

Care must be taken to develop classifications tailored to their intended applications and to choose the most appropriate classifications for addressing specific management needs.

### 5.1 Data Issues

Four questions relating to data need to be addressed when developing an ecological classification system: (1) what kind of data are needed, (2) how many data are needed, (3) can existing data be used, and (4) what are the costs associated with preparing data for the classification system? All four questions must be answered in relation to the specific management objective(s) to be addressed.

To respond to the first two questions, it must be recognized that the different applications of ecological classification may not necessarily be achievable through the use of one data set or analysis. Characterization data are used grouping similar patterns and delineating areas of interest. Pattern recovery data are used to extrapolate or interpolate the classification units to areas that have not been sampled. The amount of data needed to satisfy either of these applications is related to the required accuracy and precision of the classification system. Additional data may be needed to characterize the properties of each classification unit. Therefore, the total amount of data needed depends on the objective of the classification; the required accuracy and precision for pattern characterization and recovery; and the type of description required for each classification unit.

To answer the third question, it is important to determine what types of data are already available and whether they will be useful and cost effective before undertaking original research. Existing data sources may include remote sensing imagery, herbarium and museum species records, and plot/transect data that have been collected during past ecological surveys. Data, maps, and charts for vegetation, soils, lakes, geology, fresh waters, estuaries, salinity, temperature, coastal geomorphology, and climate also exist in many geographical areas. Because using existing data often involves compiling disparate and interdisciplinary data, problems related to differences in sampling dates and design, data quality and storage, spatial representation and resolution, and standardization must be solved (see Davis et al. 1991, Davis 1995). Nevertheless, existing data that are well organized and appropriate to the objective can, in certain instances, be an effective way to help develop ecological classifications and maps (e.g, Reid et al. 1995).

The fourth question, regarding costs, can only be answered following the first three. Cost is determined by the amount of equipment, techniques, skills, and time necessary to acquire the type and amount of data needed. It should be noted that, although it might be more expensive, it is often easier to build ecological classifications and prepare maps using new data, rather than attempting to derive them from existing products and data. An example of this concerns the high costs and questionable utility of digitizing old maps of undocumented spatial accuracy into geographic information systems (GIS).

## 5.2 Ecological Classification and Mapping

The relationship between classification and mapping is often confusing for scientists and managers alike. In its purest form, ecological classification is independent of mapping. It is a scientific process of methodically arranging units of quantitative information into classes or groups that possess common properties. Mapping is a representation of these units which is constrained by the scale of the map and by the spatial information that must be related to the ecological units. Additional complexity is introduced with the interpretation of "spectral signatures" or an aggregation of available georeferenced "data layers."

Maps will never exactly represent ecological units unless those units were developed through the aggregation of spatial thematic layers. A map of earth cover that is created through the analysis of remotely sensed data or interpretation of aerial photography will not portray an ecological classification system. If cover types are determined to be an important factor of an ecological classification system, additional research must be conducted to define the relationship between the cover classes and the ecological classification units.

As reviewed earlier in this chapter, ecological classifications may be developed by using a variety of variables. Using spatial data (maps) that include variables needed for a classification system can greatly simplify the process of creating an ecological classification map. Spatial data sets have variable thematic and locational accuracy, so that when a map is compiled from multiple spatial products, its overall accuracy is unpredictable and interpretation must be tentative. Furthermore, the boundaries of separate map themes rarely coincide, largely because of real differences among attributes. This is as true for land units (Bailey et al.1994), as it is for coastal (e.g., physical vs. biogeographic units), and marine units (e.g., marine phytoplankton vs. physical oceanographic regimes (McGowan 1972, Hayden et al. 1984).

In addition to these technical challenges, care must be taken not to substitute a composite map of ecological components for a map of ecological units. Overlay maps do not represent the ecological relationships between the individual layers. An ecological unit in the natural environment is not merely a compilation of independent components, but is functionally integrated.

## 5.3 Ecological Boundaries: What Do the Lines on Maps Mean?

Ecological variables are generally semi-continuous, as opposed to being categorical, and are expressed as gradients. The concept of the "ecotone" reflected this (di Castri et al. 1988, di Castri and Hansen 1992): "...the ecotone is a 'zone of transition' between adjacent ecological systems, having a set of characteristics uniquely defined by space and time scales, and by the strength of the interactions between adjacent ecologic systems." Ecotones may be expressed at any scale. As a result, distinct lines between ecological units are the exception rather than the norm.

The problem of where to draw a line on a map is complex and requires an explicit set of rules and disclaimers. Two untested assumptions are typically made: (1) the mappable attribute is a clearly defined and a predictable property of the ecological unit, and (2) the mappable attribute does not vary over the known distribution range of the ecological unit (Bourgeron et al. 1993). This is difficult enough for simple variables, but when multiple variables are brought into the classification framework, the "fuzziness" of the lines increases dramatically.

Consequently, several important factors must be considered when developing a map of ecological units. First, interpreting ecological units depends in part on knowledge of the distributions of environmental attributes along gradients. Second, the ecological relations among biotic and abiotic components need to be stated and tested explicitly. A clear link between scales, patterns, and processes must be established, and temporal variability must be considered (Bourgeron et al. 1993). The question becomes how much fuzziness in the transition zone is acceptable to meet the classification and mapping objectives. Strategic planning may tolerate more fuzziness, whereas operational planning typically tolerates less. Regardless of the management needs, any ecological map must produce boundaries that are ecologically significant. One way to become explicit is to map the gradients themselves.

The concept of ecoregions is intuitively attractive for all systems — terrestrial, freshwater, and marine. The reconciliation of boundaries among different environmental attributes, as has been done for terrestrial systems (e.g., Bailey 1995), remains problematic. Ecological classifications, especially when presented on maps, give the appearance of ecosystem stasis and offers few opportunities for further analysis.

Furthermore, when classification approaches are dominated by few factors, they tend to address some systems better than others. Each environmental factor and/or system exhibits its own behavior, which in mapped terms may be expressed as gradients across the land- or seascape. Combining these may be possible only through statistical modeling approaches of "dynamic biogeography" (e.g., Hengeveld 1990). The science of ecosystem classification and geography still has a long way to go, especially in the "critical areas of treatment of heterogeneity and scaling" (Hornberger and Boyer 1995).

## 5.4 Key Attributes of Representative Classification Systems

Forty-four classification systems are presented in Table 5, and key attributes in six topic areas are listed for each system. The table can be used to evaluate what each effort classifies (e.g., terrestrial vs. freshwater systems), how the classification was developed (e.g., use of abiotic vs. biotic factors), and at what scale the system is classified (e.g., global vs. local). One classification system often addresses more than one attribute in a topic area; for example, that of Grossman et al. (1994b), Grossman et al. (1998) considers both terrestrial and wetland systems. Each classification effort is described by:

- The ecological system that the effort classifies — terrestrial, freshwater, coastal/marine, estuarine, and/or wetland;
- The geographical coverage that the effort encompasses —global, continental, national, regional, and/or local;
- The overall objectives of the classification effort — research, management, and/or conservation;
- The environmental factors used to develop the classification effort — climate, soils, elevation, geology, landform/land position, and/or hydrology/hydrography;
- The biological factors used to develop the classification effort — existing vegetation structure, existing vegetation composition, potential natural vegetation, zoological guilds, and/or zoological composition; and,
- Other factors that are important in describing the classification effort, such as: are the classification units geographically referenced? Is the classification hierarchical (i.e., do multiple levels exist)? Were a variety of factors used to develop the classification effort? Were extensive data needed to develop the classification?

## 5.5 Management Applications

Regulatory and management agencies at all levels of government, have struggled to apply ecological classifications to resource planning and management challenges. The diversity of ecological classifications has been reviewed in earlier sections. Each system has been developed for a specific set of purposes and many have proven extremely valuable for management applications (see Management Chapter). However, having value for one set of applications does not make a classification system useful for other applications. For example, classification systems that focus on site potential (e.g., ECOMAP Land Type Associations) may be appropriate for gross, strategic analyses of resource management. For operational planning or statistically defensible estimates, analyses that are conducted at the stand level and that integrate existing and potential vegetation are recommended (Roloff 1994).

Table 6 provides an evaluation of the applicability of the major ecological classification systems for different management applications. The management applications are divided into four general areas: land management and use, biodiversity conservation, resource inventory and management, and research and assessments. Additional discussion concerning the applications of ecological classification systems to meet management objective can be found in Carpenter et al. (this volume).

"Land management and use" activities focus on the land unit and its development by humans for a variety of purposes, including

- Infrastructure siting for roads and other "conduits" and buildings;
- Resource planning, site prescription, and management for forestry, wildlife, fisheries, recreation, agriculture, erosion control, minerals and water, and multiple use designations;
- Desired future conditions analysis; and
- Sustainable development.

"Biodiversity conservation" refers to the identification and protection of natural biota as it occurs in ecological systems, including:

- Biological diversity inventory;
- Identification of conservation sites;
- Sustainable design for conservation; and
- Restoration planning.

"Resource inventory and management" includes activities intended to determine the existing number, abundance, and condition of natural resources to

Table 5. Key attributes of major ecological classification efforts (under development).

| Classification Effort | Ecological System | | | | | Geographical Coverage | | | | | Objectives | | | Environmental Factors | | | | | | Biological Factors | | | | | Other Factors | | | |
|---|---|---|---|---|---|---|---|---|---|---|---|---|---|---|---|---|---|---|---|---|---|---|---|---|---|---|---|---|
| | Terrestrial | Freshwater | Marine/Coastal | Estuarine | Wetland | Global | Continental | National | Regional | Local | Research | Management | Conservation | Climate | Soils | Elevation | Geology | Landform/Position | Hydrology/Hydrography | Existing Vegetation Structure | Existing Vegetation Composition | Potential Natural Vegetation | Zoological Guilds | Zoological Composition | Classification Units Are Geographically Referenced | Hierarchical (multi-level) | Multi-factored | Extensive Data Requirements |
| Albert et al. (1995) | √ | | | | √ | | | | √ | | √ | | | √ | √ | √ | √ | √ | √ | | | √ | | | √ | √ | √ | |
| Allee and Schmidt (1937) | √ | | | | | | | | √ | | √ | | | √ | | | | | | | | | √ | √ | √ | | | |
| Anderson et al. (1976) | √ | | | | | | | √ | | | | √ | | | | | | | | | √ | | | | | √ | | √ |
| Bailey (1995) | √ | √ | √ | | √ | √ | √ | √ | √ | | √ | √ | | √ | | √ | | √ | | | | √ | | | √ | √ | √ | |
| Braun-Blanquet (1928, 1932) | √ | | | | √ | √ | | | | | √ | | | | | | | | | | √ | | | | | √ | | |
| Busch and Sly (1992) | | √ | | | | √ | | | √ | | √ | √ | | | | | | | √ | | | | √ | √ | | | √ | |
| Cleland et al. (1994) | √ | | | | √ | | | | √ | √ | √ | √ | | √ | √ | √ | √ | √ | √ | | | √ | | | √ | √ | √ | |
| Cowardin et al. (1979) | | √ | | √ | √ | | | √ | | | √ | √ | | | √ | | | | √ | | √ | | | | √ | √ | √ | |
| Daubenmire (1952) | √ | | | | √ | | | | | | √ | | | | √ | | | √ | | | √ | | | | | | √ | |
| Dinerstein et al. (1995) | √ | | | | √ | | √ | | √ | | | √ | √ | | | | | | | √ | √ | √ | | | √ | √ | | |
| Driscoll (1984) | √ | | | | √ | | | √ | √ | | | √ | | √ | | | | | | √ | √ | | | | | | | |
| ECOMAP (1993) | √ | | | | √ | | | √ | √ | √ | √ | √ | | √ | √ | √ | √ | √ | √ | | | √ | | | √ | √ | √ | |
| Ecoregions Working Group (1989) | √ | | | | | | | √ | | | √ | √ | | √ | √ | √ | | √ | | √ | | √ | | | √ | √ | | √ |
| Frissell et al. (1986) | | √ | | | | | | | | | ? | | | | | | | | √ | | | | | | | √ | | |
| GAP/Scott et al. (1993) | | | | | | | | | | | | | | | | | | | | | | | | | | | | |
| Grossman et al. (TNC) (1994) | √ | | | | | | | √ | | | | √ | √ | √ | √ | | √ | √ | | √ | √ | | | | | √ | | |
| Hayden et al. (1984) | | | √ | | | | | | | | ? | | √ | | | | | | √ | | | | | | | √ | | |
| Higgins et al. (TNC) (in prep) | | √ | | | | | | √ | | | | √ | √ | | | | √ | √ | √ | √ | √ | | √ | √ | | √ | | |
| Holdridge (1947) | √ | | | | | √ | | | | | √ | | | √ | | | | | | | | | | | | | | |
| Illies (1961) | | √ | | | | √ | | | | | √ | | | | | | | | √ | | | | | | √ | | | |
| Ketchum (1972) | | | √ | | | √ | | | | | √ | ? | | | | | | √ | √ | | | | | | √ | | √ | |
| Kotar et al. (1988) | √ | | | | | | | | | √ | | √ | | | | | | | | | | √ | | | | | ? | |

Table 5 (continued)

| | Ecological System | | | | | Geographical Coverage | | | | | Objectives | | | Environmental Factors | | | | | | Biological Factors | | | | | Other Factors | | | |
|---|---|---|---|---|---|---|---|---|---|---|---|---|---|---|---|---|---|---|---|---|---|---|---|---|---|---|---|---|
| Classification Effort | Terrestrial | Freshwater | Marine/Coastal | Estuarine | Wetland | Global | Continental | National | Regional | Local | Research | Management | Conservation | Climate | Soils | Elevation | Geology | Landform/Position | Hydrology/Hydrography | Existing Vegetation Structure | Existing Vegetation Composition | Potential Natural Vegetation | Zoological Guilds | Zoological Composition | Classification Units Are Geographically Referenced | Hierarchical (multi-level) | Multi-factored | Extensive Data Requirements |
| Krajina (1965) | √ | | | | | ? | | | | | √ | | | √ | √ | √ | √ | | | | | √ | | | | √ | √ | |
| Kuchler (1964) | √ | | | | | √ | √ | √ | | | √ | √ | | √ | √ | | | √ | | | | √ | | | √ | √ | √ | |
| Maxwell et al. (1995) | | √ | | | | | √ | | | | | √ | | | | | | | √ | | | | | | √ | √ | | |
| Mitsch and Gosselink (1986) | | | | | √ | | | ? | | | | √ | | | √ | | | | √ | √ | √ | | | | | | √ | |
| MLRC (n.d.) | | | | | | | | | | | | | | | | | | | | | | | | | | | | |
| Moyle and Ellison (1991) | | √ | | | √ | | | | | √ | | √ | | √ | | | | √ | √ | √ | √ | | | | | √ | √ | |
| NAWQA (1990) | | √ | | | | | | √ | | | | √ | | | | | | | √ | | | | | | √ | | | |
| Omernik (1995) | √ | | | | | | √ | √ | | | | √ | | √ | √ | | √ | | √ | √ | | | | | √ | | √ | |
| Pfister and Arno (1980) | √ | | | | | | | | ? | | ? | ? | | √ | √ | | | | | | | √ | | | √ | √ | | |
| Pregitzer and Barnes (1982) | √ | | | | | | | | | √ | √ | √ | | √ | √ | | √ | √ | √ | | | √ | | | √ | √ | √ | |
| Pritchard (1967) and Barnes (1967) | | | | √ | | | | ? | | | | | | | | | √ | √ | √ | | | | | | | | | |
| Ray and Hayden (1992) | | | √ | √ | | | | | ? | | | | √ | | | | | | √ | | | | | | | | | |
| Robins (1992) | | | √ | √ | | | | | | | | | | | | | | | √ | | | | | | | | | |
| Rodwell (1991) | √ | | | | | | | √ | | | √ | √ | | | | | | | | | √ | | | | | | | |
| Rowe (1984) | √ | | | | | | | √ | | | | | | √ | | | | | | | √ | | | | ? | ? | | |
| Schupp (1992) | | √ | | | | | | | | √ | | | | | | | | | √ | | | | | √ | | | √ | |
| Specht et al. (1974) | √ | | | | | | | √ | | | | √ | √ | √ | √ | | √ | √ | | √ | √ | | | | | √ | | |
| Udvarty (1974) | √ | | | | | √ | | | | | √ | √ | | √ | √ | | √ | | | √ | | | | | √ | √ | | |
| UNESCO (1973) | √ | | | | | √ | | | | | √ | √ | | √ | √ | | √ | √ | | √ | √ | √ | | | √ | √ | | |
| USGS (1982) | | √ | | | | | | √ | | | √ | √ | | | | | | √ | √ | | | | | | √ | √ | | |
| Wiken (1986) | √ | | | | | | | √ | | | | √ | | √ | √ | √ | √ | | √ | √ | √ | √ | | | √ | √ | | |
| Zonneveld (1989) | √ | | | | | | | ? | | | √ | √ | | √ | √ | | | ? | | | | √ | | | √ | | | |

Table 6. Management applications of major ecological classification efforts (under development).

| Classification Effort | Land Management and Use | | | | Biodiversity Conservation | | | | Resource Inventory and Management | | | | Research and Assessments | | | | |
|---|---|---|---|---|---|---|---|---|---|---|---|---|---|---|---|---|---|
| | Infrastructure siting | Resource planning, site prescription and management | Desired future conditions analysis | Sustainable development | Biodiversity inventory | Identification of conservation sites | Sustainable design for conservation | Restoration planning | Resource inventory | Resource management | Habitat suitability analysis | Resource reporting and mapping | Environmental impact assessments | Ecosystem and landscape monitoring | Study of pattern and process in relation to scale | Representativeness assessment | Predictive modeling |
| Albert et al. (1995) | √ | | | | | | | | √ | √ | √ | √ | | √ | √ | | √ |
| Allee and Schmidt (1937) | √ | | | | √ | | | | | | | | | | | | √ |
| Anderson et al. (1976) | √ | | | | | | | | √ | √ | | √ | √ | | | | |
| Bailey (1995) | √ | √ | √ | √ | | | | | √ | √ | √ | √ | | | √ | √ | |
| Braun-Blanquet (1928, 1932) | | | | | | | | | | | | | | | | | |
| Busch and Sly (1992) | | √ | | | | | | | √ | | | | | | | | |
| Cleland et al. (1994) | √ | | | | | | | | √ | √ | √ | | √ | √ | | | |
| Cowardin et al. (1979) | √ | √ | | √ | | | | | √ | √ | √ | | | | | | |
| Daubenmire (1952) | | | | | | | | | | | | | | | | | |
| Dinerstein et al. (1995) | | | | | √ | √ | | | | | | | | | | | |
| Driscoll (1984) | √ | | | √ | | | | | | | | | | | | | |
| ECOMAP (1993) | √ | √ | | √ | | | | | √ | √ | √ | | | | | | |
| Ecoregions Working Group (1989) | √ | | | | | | | | √ | √ | | | | | | | |
| Frissell et al. (1986) | | | | | | | | | | | | | | | | | |
| GAP/Scott et al. (1993) | | | | | | | | | | | | | | | | | |
| Grossman et al. (TNC) (1994) | | | | | | √ | √ | | | | | | | | | | |
| Hayden et al. (1984) | | | | | | √ | | | | | | | | | | | |
| Higgins et al. (TNC) (in prep.) | | | | | | √ | √ | | | | | | | | | | |
| Holdridge (1947) | √ | | | | | | | | | | | | | | | √ | |
| Illies (1961) | | | | | | | | | | | | | | | | | |
| Ketchum (1972) | | | | | | | | | | | | | | | | | |
| Kotar et al. (1988) | | | | | | | | | √ | √ | √ | | | | | | |

Table 6 (continued)

| Classification Effort | Land Management and Use | | | | Biodiversity Conservation | | | | Resource Inventory and Management | | | | Research and Assessments | | | | |
|---|---|---|---|---|---|---|---|---|---|---|---|---|---|---|---|---|---|
| | Infrastructure siting | Resource planning, site prescription and management | Desired future conditions analysis | Sustainable development | Biodiversity inventory | Identification of conservation sites | Sustainable design for conservation | Restoration planning | Resource inventory | Resource management | Habitat suitability analysis | Resource reporting and mapping | Environmental impact assessments | Ecosystem and landscape monitoring | Study of pattern and process in relation to scale | Representativeness assessment | Predictive modeling |
| Krajina (1965) | | | | | | | | | | | | | | | | | |
| Kuchler (1964) | | | | | | | | | | | | | | | | | |
| Maxwell et al. (1995) | √ | | | | | | | | √ | √ | | | | | | | |
| Mitsch and Gosselink (1986) | | | | | | | | | | | | | | | | | |
| MLRC (n.d.) | | | | | | | | | | | | | | | | | |
| Moyle and Ellison (1991) | | | | | √ | √ | | | | | | | | | | | |
| NAWQA (1990) | √ | | | | | | | | | | | | √ | √ | | | |
| Omernik (1995) | √ | | | | | | | | √ | √ | | | √ | | | | |
| Pfister and Arno (1980) | | | | | | | | | √ | √ | √ | | | | | | |
| Pregitzer and Barnes (1982) | | | | | | | | | √ | √ | √ | | | | | | |
| Pritchard (1967) and Barnes (1967) | | | | | | | | | | | | | | | | | |
| Ray and Hayden (1992) | | | | | √ | | | | | | | | | | | | |
| Robins (1992) | | | | | | | | | √ | √ | | | | | | | |
| Rodwell (1991) | | | | | | | | | | | | | | | | | |
| Rowe (1984) | | | | | | | | | √ | √ | | | | | | | |
| Schupp (1992) | | | | | | | | | √ | √ | √ | | | | | | |
| Specht et al. (1974) | | | | | √ | √ | | | | | | | | | | | |
| Udvarty (1974) | | | | | √ | √ | | | | | | | | | | | |
| UNESCO (1973) | | | | | | | | | | | | | | | | | |
| USGS (1982) | | | | | | | | | | | | | | | | | |
| Wiken (1986) | √ | √ | | √ | | | | | √ | √ | √ | √ | | | | √ | |
| Zonneveld (1989) | | | | | | | | | | | | | | | | | |

inform management decisions. These activities include:

- Resource sampling;
- Resource inventory;
- Resource management — forestry, wildlife, fisheries, agriculture, recreation, soil, and minerals and water;
- Habitat suitability analysis; and
- Resource reporting and mapping.

"Research and assessments" include the evaluation and prediction of the effects of natural and human changes on natural resources and systems, including:

- Environmental impact assessments;
- Ecosystem and landscape monitoring;
- Study of pattern and process in relation to scale;
- Representativeness assessment; and
- Predictive modeling.

Perhaps the most common application of terrestrial vegetation and site classifications is the determination of appropriate uses for agricultural and forest lands. Integrated units derived from climatic, soil, and vegetation provide a useful tool for the determination of site potential. In addition, the combination of existing and potential vegetation classifications have been used to describe current conditions, predict successional processes, and characterize disturbance regimes (Haufler 1994, Roloff 1994). The limited degree of success using exclusively potential vegetation in accomplishing these goals is based on limited knowledge concerning vegetation–site relationships, and the ability of the dependence on the observers to infer site characteristics from the vegetation.

A standard application for ecological classification systems, terrestrial and aquatic alike, is conservation planning. Ecological classifications have been used extensively to inventory and protect terrestrial and aquatic areas. A direct biological classification will always provide the greatest confidence of capturing the conservation target, but coarser biological and site classifications can be used to develop a high degree of predictive accuracy. Vegetation classifications have been the primary approach for the identification and delineation of terrestrial conservation sites by resource management agencies (Scott et al. 1993, Grossman et al. 1994a). Aquatic researchers and managers have attempted to classify standardized descriptors for habitat sub-units for conservation planning (Busch and Sly 1992). Boundary delineations alone are often used in terrestrial and aquatic systems to protect delineated habitat units that are in short supply, unique, or support known species of special interest.

Especially for terrestrial systems, classifications have been developed to identify boundary conditions which can be employed to impose restrictions on land use. For wetlands connected to larger aquatic systems, the boundary between vegetated and non-vegetated areas is of considerable management interest. The federal government has the mandate to monitor wetland condition and abundance in relation to defined levels for healthy ecosystems and abundant fish and wildlife resources. Wetland classification is the first step in identifying these areas for inventory and management; subsequent steps include protection, regulation, evaluation, restoration, and rehabilitation.

Ecological classification has assisted aquatic ecosystem management in several areas, including water quality assessment (Meador et al. 1993, Paulsen and Linthurst 1994), fish productivity modeling (Fausch et al. 1988), fish habitat requirement modeling (Nelson et al. 1992, Hill and Platts 1995), and adjacent land management (Platts 1980, Maxwell et al. 1995). Because of the complexity of lakes and their typically large size, as well as the historical focus on fisheries management, classification for productivity has received considerable attention.

One example of aquatic ecosystem management is the National Water Quality Assessment Program (NAWQA). In 1990, the U.S. Geological Survey (USGS) initiated this Program as a comprehensive survey of the status and trends of ground and surface water quality in the United States. Physical, chemical, and biological data are being collected from study areas that correspond to hydrologic units based on the drainages of major rivers and aquifers (USGS 1982), which will be further stratified according to the classification framework of Frissell et al. (1986) (Meador et al. 1993). NAWQA researchers are currently assessing the use of ecoregions as a stratification tool for national sampling (J. Higgins, pers. comm.).

For wetlands, management initially focused on waterfowl and furbearer production, which had profound implications for wetland classification. In the past two decades, however, other services provided by wetlands, such as biological functions, habitat, sanctuaries, hydrological functions, and cultural values, have been identified (Greeson et al. 1979, Reppert and Sigleo 1979, Tiner 1984). In addition, society often assigns shoreline marshes a higher ecological value than other types of wetlands because of their direct contribution to fish and wildlife populations, rather than their role in larger ecosystems (Maltby et al. 1983). Estuarine wetlands and wetlands located at river mouths are also highly valued by society because wildlife (bird) and fish usage is usually high, rare plants or unusual wildlife colonies may occur, and these areas are often near urban centers (Crispin 1990).

Descriptive attributes have been used to develop a "taxonomy" of coastal wetlands, including vegetation, hydrology, geography, climate, soils, stratigraphy, and landscape position (topography, aspect, slope). Functional attributes are also used as insights into system dynamics and management, including species life histories, multi-species interactions, landscape interactions, hydrological processes (flood storage and storm-flow modification), nutrient retention and transformation, sediment and toxicant trapping, and sediment stabilization. Both sets are then modeled to provide decisions on wildlife habitat and public use (James E. Perry, Virginia Institute of Marine Sciences, pers. comm.).

Estuaries represent high levels of complexity and conservation urgency. They house major fish populations, and provide vital habitat for early life-history stages of many marine fishes and invertebrates. All anadromous and catadromous fishes must cross estuaries. Estuaries are the sites of most of the world's large cities and ports, and also provide important industrial and commercial sites. As a result, they are centers for coastal pollution and are, in general, heavily disturbed, which has implications for coastal living resource and fishery sustainability. Classification of estuaries requires information on where various types of habitats occur, which are the best examples, and what activities most threaten them.

## 6 PRESENT SITUATION AND FUTURE VISION

The understanding of ecological dynamics at the community, landscape, and ecosystem levels is undergoing rapid growth. Spatial and temporal organization and dynamics are better understood through the application of hierarchy theory and remotely-sensed information. Biogeographic and environmental information are being developed at an unprecedented rate and with a high level of quality control. Technological advances in the management and analysis of spatial data are increasingly available to managers and scientists. This remarkable progress provides a suite of new opportunities for the science and application of ecological classification.

### 6.1 Existing Ecological Classifications

Many ecological classification systems have already been developed. Accounts of the development and application of biogeographic and ecological characterizations and classifications are abundant throughout the scientific literature. However, the existing systems do not fully meet the current expectations or needs o f resource managers.

Most ecological classifications have been descriptive and have focussed on either terrestrial, freshwater, coastal, or marine systems. Up to now, terrestrial systems have dominated ecological classification efforts; these generally do not incorporate adequate information on aquatic systems (e.g., Avers et al. 1994, Bailey 1995). This is partially an artifact of traditional training in ecology and resource management, but also reflects the variable status of data and knowledge associated with the different systems.

Aquatic classifications have primarily been based on biophysical factors, whereas most terrestrial classifications emphasize vegetation (potential and existing), climate, and physiography. For example, Cowardin et al. (1979) focus on wetland topography, while Bailey (1995) emphasizes phytogeography and climate. Coastal classifications are often approached with a more comprehensive "ecosystem" perspective as they include components of both terrestrial and aquatic systems.

The increased need for sound management of natural resources and prioritization of conservation action has resulted in an increased dependence upon existing ecological classification systems. The existing ecological classification systems can address some, but not all, management and conservation concerns. Each system was developed to address a specific set of objectives. It is important to understand the intended and appropriate uses for each classification system and its associated products. No individual system will ever meet the full spectrum of potential applications.

### 6.2 The Development of New Ecological Classifications

Existing classifications provide the framework to go beyond mere description of the distribution of biological species and communities to focus on their relationships to one other and to environmental gradients. There are many challenges associated with the development of the next generation of classification systems.

The dynamics of environmental change are central to the concept of the ecological unit. However, it has been difficult to apply ecosystem concepts and practices to most ecological classification systems, as few of them emphasize ecological process and the dynamics of change. In addition, ecological classifications must reflect our increasing knowledge about the biological and ecological processes that functionally integrate terrestrial, freshwater, and coastal/ marine systems. Marine classification must integrate the coastal systems which, in turn, must integrate terrestrial systems. Until more classifications integrate aquatic and terrestrial features and processes, they cannot be seen as truly "ecosystemic" and they will be restricted in applications and future value.

### 6.2.1 A Set of National Ecological Units

Many managers would derive great benefit from access to one common ecological classification system with explicit standards and application guidelines. In many cases, access to one standard framework of classification units would represent the most efficient and cost-effective solution to many shared resource management challenges.

The successful development of a common set of ecological units will require sufficient consensus on common objectives, appropriate scales of analysis, and the critical ecological processes and biophysical variables that function at those scales. This system ideally would integrate across multiple scales, multiple factors, and be relevant for terrestrial, freshwater and coastal/ marine systems. The development of this common ecological framework would not restrict the development or use of other systems that better address specific needs and applications.

This concept of a common set of ecological units has already gathered momentum. In December 1995, a Memorandum of Understanding (MOU) was signed by the U.S. Department of Agriculture (Natural Resources Conservation Service, Forest Service, and Agricultural Research Service), the U.S. Department of the Interior (Bureau of Land Management, Geological Survey, Fish and Wildlife Service, National Biological Service, and National Park Service), and the U.S. Environmental Protection Agency. This MOU was designed to develop a spatial framework of ecological units for the United States. A National Interagency Steering Committee and a National Interagency Technical Team have been established to implement this MOU. As part of the initial and ongoing effort, maps of common ecological units will be developed and published at standard scales. Digital data sets in formats meeting available Federal Geographic Data Committee standards will also be published.

### 6.2.2 User-Defined Classification Systems

The quantity and quality of ecological and biological data has dramatically increased over the past few decades. More and more of these data are geographically referenced, allowing the information to be represented and interpreted spatially. Simultaneously, there have been remarkable advances in the technological capabilities for managing, aggregating, analyzing, and portraying these data. These advances have stimulated rapid testing, refinement and implementation of numerous classification and assessment approaches that are used to address various conservation and management objectives, as well as basic research questions.

This increased capacity to integrate information for targeted application will greatly augment the development of a common classification approach. Specific classifications can be developed to address specific resource management and conservation objectives. Users can determine the appropriate scales for analyses and identify the specific biophysical variables and ecological processes that apply to their specific questions and objectives. They can draw from multiple data sources to compile and analyze the appropriate spatial data and to develop very specific solutions to their questions.

Where multiple approaches to ecological classifications are appropriate, the ability to share standardized data layers becomes increasingly important. For this to occur, standards for ecological inventory, data management, and analytical approaches must be developed and documented. Ecological classifications and associated products (e.g., keys, assessments, maps, etc.) should be accompanied with appropriate metadata that fully disclose the methods, data sets, scales, variables, analyses, classes, and other information required to fully interpret the utility of the product. All data and data products must meet minimal standards so that they can be broadly interpreted and applied to multiple objectives. Data standards and inclusion of metadata files will allow partners to confidently assess the appropriateness of the data products for addressing their individual objectives.

It is critical that resource managers, conservationists, and researchers will be able to use the information. Potential users will need to have the ability to access all data so they may identify the appropriate information and ecological classification systems that would help them meet their specific objectives. This will also require a high level of terminological consistency; working definitions must be widely agreed upon and implemented to guard against inappropriate applications and faulty interpretation of ecological data, classifications, and associated products.

## 6.3 Human Factors

Many appropriate applications of ecological classification systems have been identified in this report, but perhaps the most important objective for these ecological classification systems has not been specifically mentioned: the ability to use these systems for the purposes of education and communication. Ecological classification systems help managers of ecological systems understand the functions and processes of these ecosystems. This is an important step in breaking the cycle of managing for one component of an ecosystem at a time. Ecological classification systems and associated products can also provide invaluable communications tools conveying critical land management and conservation information to the general public.

Finally, ecological classification efforts have historically avoided the integration of the human dimension into the system, even as reported system attributes. Management and conservation is a human endeavor and people have an increasingly profound impact on the structure and function of ecological systems. Therefore, it is critical that we learn how to integrate key social aspects of politics, economics, anthropology, and sociology into future ecological classification efforts.

## ACKNOWLEDGMENTS

This chapter represents a synthesis of materials that was developed, reviewed and edited by an extensive team of contributors. Some of these contributors provided scientific input for the text, others developed material for figures and diagrams, while others provided invaluable review and editing.

The primary contributors were members of the author team, who provided materials and review as requested. Special recognition is due to Carleton Ray, who, through great effort succeeded in balancing the presentation of terrestrial, freshwater and marine materials throughout the chapter. The authors on the science chapter team benefited from the discourse with and constructive review from the authors in the management chapter team, particularly their team leader Constance Carpenter.

Critical scientific contribution and review were received from Don Faber-Langendoen for the history of vegetation classification, and on freshwater classification systems from Jonathan Higgins and Mary Lammert. Mark Bryer and Xiaojun Li provided constructive scientific review of various sections and helped to develop some of the figures. Cynthia Swinehart provided early editorial assistance. Kat Maybury provided editorial review and coordinated the complex process of getting the final manuscript 'out the door.'

## REFERENCES

Albert, D.A. 1994. Regional landscape ecosystems of Michigan, Minnesota, and Wisconsin: a working map and classification. Fourth revision. USDA Forest Service General Technical Report NC-178, North Central Forest Experiment Station, St. Paul, MN.

Albert, D.A., S.R. Denton, and B.V. Barnes. 1986. *Regional Landscape Ecosystems of Michigan*. University of Michigan, School of Natural Resources, Ann Arbor, MI.

Allee, W.C., and K.P. Schmidt. 1937. *Ecological Animal Geography*. Wiley, New York, NY.

Allen, T.F.H., T.W. Hoekstra, and R.V. O'Neill. 1984. Interlevel relations in ecological research and management: some working principles from hierarchy theory. USDA Forest Service General Technical Report RM-110, Rocky Mountain Research Station, Fort Collins, CO.

Allen, T.F.H., and T.B. Starr. 1982. *Hierarchy: Perspectives for Ecological Complexity*. University of Chicago Press, Chicago, IL.

Anderson, J.R., E.E. Hardy, and J.T. Roach. 1976. Land use and land cover classification system for use with remote sensing data. Geological Survey Professional Paper 964 (a revision of the land use classification system as presented in U.S. Geological Circular 671), U.S. Government Printing Office, Washington, D.C.

Angermeir, P.L., and I.J. Schlosser. 1995. Conserving aquatic biodiversity: beyond species and populations. *American Fisheries Society Symposium* 17402–17414.

Anonymous. 1959. Symposium on the classification of brackish waters. Venice, Italy, April 8–14, 1958. *Archivio di Oceanografia e Limnologia*, Volume 11, Supplemento.

Arno, S.F., and G.E. Gruell. 1985. Douglas-fir encroachment into mountain grasslands in southwestern Montana. *Journal of Range Management* 39: 272–276.

Austin, M.E. 1972. Land resource regions and major land resource areas of the United States. *Agricultural Handbook 296*. Revised edition. USDA Soil Conservation Service, Washington, D.C.

Austin, M.P. 1985. Continuum concept, ordination methods and niche theory. *Annual Review of Ecology and Systematics* 16: 39–61.

Austin, M.P. 1987. Models for the analysis of species response to environmental gradients. *Vegetatio* 69: 35–45.

Austin, M.P. 1991. Vegetation theory in relation to cost-efficient surveys. pp. 17–22. In: C.R. Margules, and M.P. Austin (eds.), *Nature Conservation: Cost Effective Biological Surveys and Data Analysis.* CSIRO Publishing, Collingwood, Vic. Australia.

Austin, M.P., R.B. Cunningham, and P.M Fleming. 1984. New approaches to direct gradient analysis using environmental scalars and statistical curve fitting procedures. *Vegetatio* 55: 11–27.

Austin, M.P., and C.R. Margules. 1986. Assessing representativeness. pp. 45–67. In: M.B. Usher (ed.), *Wildlife Conservation Evaluation.* Chapman and Hall, London, UK.

Austin, M.P., and T.M. Smith. 1989. A new model for the continuum concept. *Vegetatio* 83: 35–47.

Austin, M.P., and G.A. Yapp. 1978. Definition of rainfall regions of southeastern Australia by numerical classification methods. *Archiv fur meterologie, Geophysik und Bioklimatologie, Series B*, 26: 121–142.

Avers, P.E., D.T. Cleland, W.H. McNab, M.E. Jensen, R.G. Bailey, T. King, C.B. Goudney, and W.E. Russell. 1994. National hierarchical framework of ecological units. USDA Forest Service, Washington, D.C.

Bailey, R.G. 1976. Ecoregions of the United States (map, scale 1: 7,500.000). USDA Forest Service, Intermountain Region, Ogden, UT.

Bailey, R.G. 1980. *Descriptions of the Ecoregions of the United States.* USDA Forest Service Miscellaneous Publication 1391, Washington, D.C.

Bailey, R.G. 1983. Delineation of ecosystem regions. *Journal of Environmental Management* 7: 365–373.

Bailey, R.G. 1989a. Ecoregions of the continents (map, scale 1: 30,000,000). USDA Forest Service, Washington, D.C.

Bailey, R.G. 1989b. Explanatory supplement to the ecoregions map of the continents. *Environmental Conservation* 15: 307–309.

Bailey, R.G. 1995. Description of the ecoregions of the United States (with separate map at a scale of 1: 7,500,000). USDA Forest Service Miscellaneous Publication 1391 (revised), Washington, D.C.

Bailey, R.G. 1996. *Ecosystem Geography.* Springer-Verlag, New York, NY.

Bailey, R.G, P.E. Avers, T. King, and W.H. McNab (eds.). 1994. Ecoregions and subregions of the United States (map, scale 1: 7,500,000). USDA Forest Service, Washington, D.C.

Bailey, R.G., and H.C. Hogg. 1986. A world ecoregions map for resource reporting. *Environmental Conservation* 13: 195–202.

Bailey, R.G., M.E. Jensen, D.T. Cleland, and P.S. Bourgeron. 1993. Design and use of ecological mapping units. pp. 105–116. In: M.E. Jensen, and P.S. Bourgeron, editors. Eastside forest ecosystem health assessment. Volume II. Ecosystem management: principles and applications. USDA Forest Service General Technical Report PNW-318, Pacific Northwest Research Station, Portland, OR.

Barnes, B.V. 1984. Forest ecosystem classification and mapping in Baden-Wurttemberg, West Germany. pp. 49–65. In: Forest land classification: Experience, problems, perspectives. *Proceedings of the symposium: March 18–20, 1984, Madison, WI, USA.*

Barnes, B.V., K.S. Pregitzer, T.A. Spies, and V.H. Spooner. 1982. Ecological forest site classification. *Journal of Forestry* 80: 493–498.

Bibby, C.J., N.J. Collar, M.P. Crosby, M.F. Heath, Ch. Imboden, T.H. Johnson, A.J. Long, A.J. Stattersfield, and S.J. Thirgood. 1992. *Putting Biodiversity on the Map: Priority Areas for Global Conservation.* International Council for Bird Preservation (ICBP), Cambridge, UK.

Borhidi, A. 1991. *Phytogeography and Vegetation Ecology of Cuba.* Akademiai Kiado, Budapest. 858 pp.

Bourgeron, P.S., H.C. Humphries, R.L. DeVelice, and M.E. Jensen. 1993. Ecological theory in relation to landscape evaluation and ecosystem characterization. pp. 65–79. In: M.E. Jensen, and P.S. Bourgeron (eds.), Eastside Forest Ecosystem Health Assessment. Volume II. Ecosystem management: principles and applications. USDA Forest Service, Pacific Northwest Research Station, Portland, OR.

Bourgeron, P.S., and M.E. Jensen. 1993. An overview of the ecological principles for ecosystem management. pp. 51–63. In: M.E. Jensen, and P.S. Bourgeron (eds.), Eastside Forest Ecosystem Health Assessment. Volume II. Ecosystem management: principles and applications. USDA Forest Service, Pacific Northwest Research Station, Portland, OR.

Braun, E.L. 1950. *Deciduous Forests of Eastern North America.* Blakiston Press, Philadelphia, PA.

Braun-Blanquet, J. 1928. *Pflanzensoziologie. Grundzuge der vegetationskunde.* Springer, Berlin, Germany. [English translation published in 1932 by McGraw-Hill, New York, NY.]

Briggs, J.C. 1974. *Marine Zoogeography.* McGraw-Hill, New York, NY.

Bright, R.C. 1968. Surface-water chemistry of some Minnesota lakes, with preliminary notes on diatoms. Interim Report 3, Limnological Research Center and Bell Museum of Natural History, University of Minnesota, Minneapolis, MN.

Brown, D.E., C.H. Lowe, and C.P. Pase. 1979. A digitized systematic classification system for the biotic communities of North America, with community (series) and association examples for the Southwest. *Journal of the Arizona-Nevada Academy of Science 14* (suppl. 1): 1–16.

Brown, D.E., C.H. Lowe, and C.P. Pase. 1980. A digitized systematic classification for ecosystems with an illustrated summary of the natural vegetation of North America. USDA Forest Service General Technical Report RM-73, Rocky Mountain Forest and Range Experiment Station, Fort Collins, CO.

Brown, J.H. 1984. On the relation between the abundance and distribution of species. *American Naturalist* 124: 255–279.

Brown, J.H., and A.C. Gibson. 1983. *Biogeography.* C.V. Mosby, St. Louis, MO.

Bulger, A.J., B.P. Hayden, M.E. Monaco, D.M. Nelson, and M.G. McCormick-Ray. 1993. Biologically-based estuarine salinity zones derived from a multivariate analysis. *Estuaries* 16: 311–322.

Burger, D. 1976. The concept of ecosystem regions in forest site classification. pp. 213–218. In: *Proceedings XVI IUFRO World Congress, Division I, June 20–July 2, 1976, Oslo, Norway.*

Busch, W.-D.N., and P.G. Sly, editors. 1992. *The Development of an Aquatic Habitat Classification System for Lakes.* CRC Press, Boca Raton, FL.

Cajender, A.K. 1926. The theory of forest types. *Acta Forestalia Fennica* 29: 1–108.

Carpenter, K.E. 1928. *Life in Inland Waters.* Macmillan, New York, NY.

Christensen, N.L., A.M. Bartuska, J.H. Brown, S. Carpenter, C. D'Antonio, R. Francis, J.F. Franklin, J.A. MacMahon, R.F. Noss, D.J. Parsons, C. H. Peterson, M.G. Turner, and R.G. Woodmansee. 1996. The report of the Ecological Society of America Committee on the Scientific Basis for Ecosystem Management. *Ecological Applications* 6: 665–691.

Cleland, D.T. 1996. Multifactor classification of ecological land units in northeastern Lower Michigan. Unpublished Ph.D. dissertation. Michigan State University, East Lansing, MI.

Cleland, D.T., J.B. Hart, G.E. Host, K.S. Pregitzer, and C.W. Ramm. 1993. *Ecological classification system for the Huron-Manistee National Forest: field guide.* Huron-Manistee National Forest, Cadillac, MI.

Cleland, D.T., J.B. Hart, G.E. Host, K.S. Pregitzer, and C.W. Ramm. 1994. *Ecological classification and inventory system of the Huron-Manistee National Forest.* USDA Forest Service, Region 9, Milwaukee, WI.

Clements, F.E. 1916. *Plant succession: an analysis of the development of vegetation.* Carnegie Institute of Washington Publication, Washington, D.C.

Connell, J.H. 1978. Diversity in tropical rain forests and coral reefs. *Science* 199: 1302–1310.

Constanza, R., L. Weinger, C. Faulk and K. Maler. 1993. Modelling complex ecological economic systems. *Bioscience* 43: 545–555.

Cook, J.E. 1995. Implications of modern successional theory for habitat typing: a review. *Forest Science* 42: 67–75.

Cowardin, L.M., V. Carter, F.C. Golet, and E.T. LaRoe. 1979. *Classification of the wetlands and deepwater habitats of the United States.* U.S. Fish and Wildlife Service, Washington, D.C.

Crispin, S. 1990. A biological database for monitoring the Great Lakes ecosystem. pp. 43–48. In: J. Kusler, and R. Smardon (eds.), *Proceedings of an International Symposium on Wetlands of the Great Lakes: Protection and Restoration Policies.* May 16–18, 1990, Niagra Falls, New York. Association of State Wetland Managers, Berne, NY.

Curtis, J.T. 1959. *The vegetation of Wisconsin: an ordination of plant communities.* University of Wisconsin Press, Madison, WI.

Daubenmire, R.F. 1952. Forest vegetation of northern Idaho and adjacent Washington, and its bearing on concepts of vegetation classification. *Ecological Monographs* 22: 301–330.

Daubenmire, R.F. 1968. *Plant Communities: A Textbook of Synecology.* Harper Rowe, New York, NY.

Davis, F.W. 1995. Information systems for conservation research, policy and planning: better access to data for scientists, policy makers, and managers. *Bioscience* (suppl.) S36–S42.

Davis, F.W., D.A. Quattrochi, and M.K. Ridd. 1991. Environmental analysis using integrated GIS and remote sensed data: some research needs and priorities. *Photogrammetric Engineering and Remote Sensing* 57: 689–697.

Davis, W.M. 1890. The rivers of northern New Jersey, with note on the classification of rivers in general. *National Geographic Magazine* 2: 82–110.

Denton, D.R. and B.V. Barnes. 1988. An ecological climatic classification of Michigan: a quantitative approach. *Forest Science* 34: 119–138.

DeVelice, R.L., G.J. Daumiller, P.S. Bourgeron, and J.O. Jarvie. 1993. Bioenvironmental representativeness of nature preserves: assessment using a combination of a GIS and a rule-based model. pp. 51–58. In: *Proceedings, First Biennial Scientific Conference on the Greater Yellowstone Ecosystem,* September 1992, National Park Service, Mammoth, WY.

di Castri, F., and A.J. Hansen. 1992. The environment and development crises as determinants of landscape dynamics. pp. 3–18. In: A.J. Hansen, and F. di Castri (eds.), Landscape boundaries: consequences for biotic diversity and ecological flows. Springer-Verlag, New York, NY.

di Castri, F., A.J. Hansen, and M.N. Holland, editors. 1988. A new look at ecotones: emerging international projects on landscape boundaries. *Biology International* Special Issue 17, IUBS, Paris, France.

Dinerstein, E., D.M. Olson, D.J. Graham, A.L. Webster, S.A. Pimm, M.P. Bookbinder, and G. Ledec. 1995. *A conservation assessment of the terrestrial ecoregions of Latin America and the Caribbean.* The World Wildlife Fund and The World Bank, Washington, D.C.

Driscoll, R.S., D.L. Merkel, D.L. Radloff, D.E. Snyder, and J.S. Hagihara. 1984. *An ecological land classification framework for the United States.* USDA Forest Service, Miscellaneous Publication 1439, Washington, D.C.

Dunbar, M.J. 1979. The relation between oceans. pp. 112–125. In: S. Van der Spoel, and A.C. Pierrot-Bults (eds.), *Zoogeography and Diversity in Plankton.* Halstead (Wiley), New York, NY.

Ebert, D.J., T.A. Nelson, and J.L. Kershner. 1991. A soil-based assessment of stream fish habitats in Coastal Plain streams. *Proceedings of Warmwater Fisheries Symposium,* June 4–8, 1991, Phoenix, Arizona, USA.

Environment Canada. 1989. *Ecoclimatic regions of Canada, first approximation (with map at a scale of 1: 7,500,000).* Ecological Land Classification Series No. 23, Environment Canada, Ottawa, Canada.

Ekman, S. 1953. Zoogeography of the sea. Sidgwick and Jackson Ltd., London, UK.

Eyre, F.H. (ed.). 1980. *Forest Cover Types of the United States and Canada.* Society of American Foresters, Washington, D.C.

Fausch, K.D., C.L. Hawkes, and M.G. Parsons. 1988. Models that predict standing crop of stream fish from habitat variables: 1950–1985. USDA Forest Service General Technical Report PNW-GTR-213, Pacific Northwest Research Station, Corvallis, OR.

Fenneman, N.M. 1928. Physiographic divisions of the United States. *Annals Association American Geographers* 18: 261–353.

Fenneman, N.M. 1946. *Physical divisions of the United States (map, scale 1: 7,000,000).* U.S. Geological Survey, Reston, VA.

Flahault, C., and C. Schroter. 1910. Rapport sur la nomenclature phytogeographique. *Proceedings of the 3rd International Botanical Congress, Brussels* 1: 131–164.

Forman, R.T.T., and M. Godron. 1986. *Landscape Ecology.* Wiley and Sons, New York, NY.

Franklin, J.F. 1993. Preserving biodiversity: species, ecosystems, or landscapes? *Ecological Applications* 3: 202–205.

Frayer, W.E., L.S. Davis, and P.G. Risser. 1978. Uses of land classification. *Journal of Forestry* 76: 647–649.

Frissell, C.A., W.J. Liss, C.E. Warren, and M.D. Hurley. 1986. A hierarchical framework for stream classification: viewing streams in a watershed context. *Environmental Management* 10: 199–214.

Gallant, A.L., T.R. Whittier, D.P. Larsen, J.M. Omernik, and R.M. Hughes. 1989. Regionalization as a tool for managing environmental resources. EPA/600/3-89/060, U.S. Environmental Protection Agency, Corvallis, OR.

Gauch, H.G., Jr. 1982. *Multivariate Analysis in Community Ecology.* Cambridge University Press, New York, NY.

Gleason, H.A. 1917. The structure and development of the plant association. *Bulletin of the Torrey Botanical Club* 44: 463–481.

Gleason, H.A. 1926. The individualistic concept of the plant association. *Bulletin of the Torrey Botanical Club* 53: 7–26.

Greenacre, M.J. 1984. *Theory and Applications of Correspondence Analysis.* Academic Press, London.

Greeson, P.E., J.R. Clark, and J.E. Clark, editors. 1979. Wetland functions and values: the state of understanding. Proceedings National Wetlands Symposium. American Water Resource Association, Technical Publication No. TPS79-2, Minneapolis, MN.

Grossman, D.H., D. Faber-Langendoen, A.S. Weakley, M. Anderson, P. Bourgeron, R. Crawford, K. Goodin, S. Landaal, K. Metzler, K.D. Patterson, M. Pyne, M. Reid, and L. Sneddon. 1998. International classification of ecological communities: Terrestrial vegetation of the United States. Volume I. The national vegetation classification standard. The Nature Conservancy, Arlington, VA.

Grossman, D.H., K.L. Goodin, X. Li, D. Faber-Langendoen, and M. Anderson. 1994a. Standardized national vegetation classification system. Prepared for the USDI National Biological Survey/National Park Service Vegetation Mapping Program, Denver, CO.

Grossman,, D.H., K.L. Goodin, and C.L. Reuss (eds.). 1994b. *Rare plant communities of the conterminous United States: an initial survey*. The Nature Conservancy, Arlington, VA.

Grumbine, R.E. 1994. What is ecosystem management? *Conservation Biology* 8: 27–28.

Hammond, E.H. 1964. Classes of land-surface form in the forty-eight states, U.S.A. (map, scale 1: 5,000,000). *Annals Association American Geographers* 54: Map Supp. No. 4.

Hansen, A.J., and F. di Castri (eds.), 1992. *Landscape Boundaries: Consequences for Biotic Diversity and Ecologic Flows.* Springer-Verlag, New York, NY.

Haufler, J.B. 1994. An ecological framework for planning for forest health. *Journal of Sustainable Forestry* 2: 307–316.

Hawkes, C.L., D.L. Miller, and W.G. Layher. 1986. Fish ecoregions of Kansas: stream fish assemblage patterns and associated environmental correlates. *Environmental Biology of Fishes* 17: 267–279.

Hayden, B.P., and R. Dolan. 1976. Coastal marine fauna and marine climates of the Americas. *Journal of Biogeography* 3: 71–81.

Hayden, B.P., R.D. Dueser, J.T. Callahan, and H.H. Shugart. 1991. Long-term research at the Virginia Coast Reserve. *BioScience* 41: 310–318.

Hayden, B.P., G.C. Ray, and R. Dolan. 1984. Classification of coastal and marine environments. *Environmental Conservation* 11: 199–207.

Hedgpeth, J.W. 1957. Marine biogeography. *Geological Society America Memoir* 67(1): 359–382.

Hengeveld, R. 1990. *Dynamic Biogeography.* Cambridge University Press, Cambridge, UK.

Herbertson, A.J. 1905. The major natural regions: an essay in systematic geography. *Geography Journal* 25: 300–312.

Herdendorf, C.E., C.N. Raphael, and E. Jaworski. 1986. The ecology of the Lake St. Clair wetlands: a community profile. U.S. Fish and Wildlife Service Biological Report 85(7.7).

Higgins, J., M. Lammert, D. Grossman, and M. Bryer. 1996. A classification framework for freshwater communities: proceedings of The Nature Conservancy's Aquatic Community Classification Workshop, April 9–11, 1996, New Haven, Missouri. Available from The Nature Conservancy, Arlington, VA.

Hill, M., and W.S. Platts. 1995. *Lower Owens River watershed ecosystem management plan.* Ecosystem Sciences, Boise, ID.

Hill, M.O. 1974. Correspondence analysis: a neglected multivariate method. *Journal Royal Statistical Society, Series C*, 23: 340–354.

Hill, M.O. 1979. Twinspan. A FORTRAN program for arranging multivariate data in an ordered two-way table by classification of the individuals and the attributes. Cornell University, Ithaca, NY.

Hills, G.A. 1952. The classification and evaluation of sites for forestry. Ontario Department of Lands and Forests, Resource Division Report 24, Toronto, Ontario, Canada.

Hix, D.M. 1988. Multifactor classification of upland hardwood forest ecosystems of the Kickapoo River watershed, southwestern Wisconsin. *Canadian Journal Forest Research* 18: 1405–1415.

Holdridge, L.R. 1947. Determination of world plant formations from simple climatic data. *Science* 105: 367–368.

Holdridge, L.R. 1967. *Life Zone Ecology.* Tropical Science Center, San Jose, CA.

Holling, C.S., L. Gunderson, and G. Peterson. 1993. Comparing ecological and social systems. Beijer Discussion Paper, Series #36. Beijer International Institute of Ecological Economics, Royal Swedish Academy of Science, Stockholm, Sweden.

Hooper, F.F. 1969. Eutrophication indices and their relation to other indices of ecosystem change. In: *Eutrophication: Causes, Consequences, Correctives: Proceedings of a Symposium.* National Academy of Sciences, Washington, D.C.

Hornberger, G.M., and E.W. Boyer. 1995. Recent advances in watershed modelling. *Reviews Geophysics* (Suppl. July): 949–957.

Horton, R.E. 1945. Erosional development of streams and their drainage basins: hydrophysical approach to quantitative morphology. *Geological Society of America Bulletin* 56: 275–370.

Host, G.E. 1993. Field sampling and data analysis methods for development of ecological land classifications: an application on the Manistee National Forest. USDA Forest Service General Technical Report NC-162, St. Paul, MN.

Host, G.E., and K.S. Pregitzer. 1991. Ecological species groups for upland forest ecosystems of northern Lower Michigan. *Forest Ecology and Management* 43: 87–102.

Host, G.E., K.S. Pregitzer, C.W. Ramm, J.B. Hart, and D.T. Cleland. 1987. Landform mediated differences in successional pathways among upland forest ecosystems in northwestern Lower Michigan. *Forest Science* 33: 445–447.

Host, G.E., K.S. Pregitzer, C.W. Ramm, D.T. Lusch, and D.T. Cleland. 1988. Variations in overstory biomass among glacial landforms and ecological land units in northwestern Lower Michigan. *Canadian Journal of Forest Research* 18: 659–668.

Howard, J.A., and C.W. Mitchell. 1985. *Phytogeomorphology.* John Wiley and Sons, New York, NY.

Hudson, P.L., R.W. Griffiths, and T.J. Wheaton. 1992. Review of habitat classification schemes appropriate to streams, rivers, and connective channels in the Great Lakes drainage basin. pp. 73–107. In: W.-D.N. Busch, and P.L. Sly (eds.), *The Development of an Aquatic Habitat Classification System for Lakes.* CRC Press, Boca Raton, FL.

Hughes, R.M., S.A. Heiskary, W.J. Matthews, and C.O. Yoder. 1994. Use of ecoregions in biological monitoring. pp. 125–151. In: S.L. Loeb, and A. Spacie (eds.), *Biological Monitoring of Aquatic Systems.* Lewis Publishers, Boca Raton, FL.

Hughes, R.M., E. Rexstad, and C.E. Bond. 1987. The relationship of aquatic ecoregions, river basins, and physiographic provinces to the icthyiographic regions of Oregon. *Copeia* 2: 423–432.

Hutchinson, G.E. 1957. *A Treatise on Limnology.* Vol. 1. Wiley, York, NY.

Hutchinson, G.E. 1959. Homage to Santa Rosalia, or why are there so many kinds of animals. *American Naturalist* 93: 45–159.

IEMTF [Interagency Ecosystem Management Task Force]. 1995. *The ecosystem approach: healthy ecosystems and sustainable economies.* Volumes I–III. U.S. Dept. Commerce, National Technical Information Service, Springfield, VA.

Illies, J. 1961. Versuch einer allgemein biozonotischen Gleiderung der Fliessgewasser. Verhandlungen der Internationalen Verinigung fur theroetische und angewandte. *Limnologie* 13: 834–844.

Jensen, M.E., P.S. Bourgeron, R. Everett, and I. Goodman. 1996. Ecosystem management: a landscape ecology perspective. *Water Resources Bulletin* 32: 1–14.

Johnson, P.S. 1992. Oak overstory/reproduction relations in two xeric ecosystems in Michigan. *Forest Ecology and Management* 48: 233–240.

Jones, R.K., G. Pierpoint, G.M. Wickware, J.K. Jeglum, R.W. Arnup, J.M. Bowles. 1983. *Field guide to forest ecosystem classification for the clay belt, site region 3e.* Ministry of Natural Resources, Toronto, Ontario, Canada.

Kay, J.J. 1991. A nonequilibrium thermodynamics framework for discussing ecosystem integrity. *Environmental Management* 15: 483–495.

Kent, M., and P. Coker. 1992. *Vegetation Description and Analysis.* Belhaven Press, London, UK.

Ketchum, B.K. 1972. *The Waters' Edge: Critical Problems of the Coastal Zone.* Massachusetts Institute of Technology Press, Cambridge, MA.

Keys, J. Jr., C. Carpenter, S. Hooks, F. Koening, W.H. McNab, W. Russell, and M.L. Smith. 1996. Ecological units of the Eastern United States — first approximation. Part I: ARC INFO; Part II: Supporting Imagery; Part III: Text. CD-ROM. USDA Forest Service, Atlanta, GA.

Koestler, A. 1967. *The Ghost in the Machine.* Macmillan, New York, NY.

Komarkova, V. 1983. Comparison of habitat type classification to some other classification methods. pp. 21–31. In: W.H. Moir, and L. Hendzel (eds.), *Proceedings of the Workshop on Southwestern Habitat Types, April 8–10, 1983, Albuquerque, New Mexico.* USDA Forest Service, Southwestern Region, Albuquerque, NM.

Köppen, W. 1931. *Grundriss der Klimakunde.* Walter de Gruyter, Berlin, Germany.

Kotar, J., J.A. Kovach, and C.T. Locey. 1988. *Field guide to forest habitat types of northern Wisconsin.* University of Wisconsin, Department of Forestry and Wisconsin Department of Natural Resources, Madison, WI.

Krajina, V.J. 1965. Biogeoclimatic zones and classification of British Columbia. pp. 1–17. In: Krajina, V.J. (ed.), *Ecology of Western North America.* University of British Columbia, Vancouver, British Columbia, Canada.

Küchler, A.W. 1964. *Potential natural vegetation of the conterminous United States (with separate map at a scale of 1: 3,168,000).* American Geographical Society Special Publication 36, New York, NY.

Küchler, A.W. 1967. *Vegetation Mapping.* Ronald Press, New York, NY.

Küchler, A.W. 1988. Ecological vegetation maps and their interpretation. pp. 469–480. In: A.W. Küchler and I.S. Zonneveld (eds.), *Vegetation Mapping.* Kluwer, Boston, MA.

Küchler, A.W. and I.S. Zonneveld, editors. 1988. *Vegetation Mapping.* Kluwer, Boston, MA.

Larsen, D.P., J.M. Omernik, R.M. Hughes, C.M. Rohm, T.R. Whittier, A.J. Kinney, A.L. Gallant, and D.R. Dudley. 1986. Correspondence between spatial patterns in fish assemblages in Ohio streams and aquatic ecoregions. *Environmental Management* 10: 815–828.

Lawton, J.H., S. Nee, A.J. Letcher, and P.H. Harvey. 1993. Animal distributions: patterns and processes. pp. 41–58. In: P.J. Edwards, R.M. May, and N.R. Webb (eds.), *Large-scale Ecology and Conservation Biology.* Blackwell Scientific, Oxford, UK.

Leach, J.H., and R.C. Herron. 1992. A review of lake habitat classification. pp. 27–57. In: W.-D.N Busch, and P.G. Sly (eds.), *The Development of an Aquatic Habitat Classification System for Lakes.* CRC Press, Boca Raton, FL.

Leopold, L.B., and M.G. Wolman. 1957. River channel patterns: braided, meandering and straight. U.S. Geological Survey Professional Paper 422H, Washington, D.C.

Levin, S.A. 1992. The problem of pattern and scale in ecology. *Ecology* 73: 1942–1968.

LOICZ. 1996. Coastal facts. *Land Ocean Interactions Coastal Zone Newsletter* 1: 2.

Lotspeich, F.B. 1980. Watershed as the basic ecosystem: this conceptual framework provides a basis for a natural classification system. *Natural Resources Bulletin* 16: 581–586.

Lotspeich, F.B., and W.S. Platts. 1982. An integrated land-aquatic classification system. *North American Journal of Fisheries Management* 2: 138–149.

Lubchenco, J., A.M. Olson, L.B. Brubaker, S.R. Carpenter, M.M. Holland, S.P. Hubbell, S.A. Levin, J.A. MacMahon, P.A. Matson, J.M. Melillo, H.A. Mooney, C.H. Peterson, H.R. Pulliam, L., S.A. Real, P.J. Regal, and P.G. Risser. 1991. The sustainable biosphere initiative: an ecological research agenda. A Report from the Ecological Society of America. *Ecology* 72: 371–412.

Lyons, J. 1989. Correspondence between the distribution of fish assemblages in Wisconsin streams and Omernik's ecoregions. *American Midland Naturalist* 122: 163–182.

MacKey, B.G., H.A. Nix, M.F. Hutchinson, and J.P. MacMahon. 1988. Assessing representativeness of places for conservation reservation and heritage listing. *Environmental Management* 12: 501–514.

MacKey, B.G., H.A. Nix, J.A. Stein and S.E. Cork. 1989. Assessing the representativeness of the wet tropics of Queensland World Heritage property. *Biological Conservation* 50: 279–303.

Maltby, L.S., G.B. McCullough, and E.Z. Bottomley. 1983. *Baseline studies of 35 selected southern Ontario wetlands.* Canadian Wildlife Service, Ottawa, Canada.

Maxwell, J.R., C.J. Edwards, M.E. Jensen, S.J. Paustian, H. Parrott, and D.M. Hill. 1995. A hierarchical framework of aquatic ecological units in North America. USDA Forest Service General Technical Report NC–176, North Central Forest Experiment Station, St. Paul, MN.

McGowan, J.A. 1971. Oceanographic biogeography of the Pacific. pp. 3–74. In: B.M. Funnell, and W.R. Riedel (eds.), *The Micropaleontology of Oceans.* Cambridge University Press, Cambridge, UK.

McGowan, J.A. 1972. The nature of oceanic ecosystems. pp. 9–28. In: C.B. Miller (ed.), *The Biology of the Oceanic Pacific.* Oregon State University Press, Corvallis, OR.

McNab, W.H., and P.E. Avers, editors. 1994. *Ecological subregions of the United States: section descriptions.* USDA Forest

Service Administrative Publication WO-WSA-5, Washington, D.C.
Meador, M.R., C.R. Hupp, T.F. Cuffney, and M.E. Gurtz. 1993. Methods for characterizing stream habitat as part of the National Water-Quality Assessment Program. U.S. Geological Survey Open File Report 93–408, Reston, VA.
Merriam, C.H. 1898. Life zones and crop zones of the United States. U.S. Department of Agriculture Division Biological Survey, Bulletin 10, Washington, D.C.
Mitsch, W.J., and J.G. Gosselink. 1986. *Wetlands.* Van Nostrand Reinhold, New York, NY.
Monsi, M. 1960. Dry matter reproduction in plants. I. Schemata of dry matter reproduction. *Botanical Magazine* 73: 81–90.
Morrison, D.F. 1976. *Multivariate Statistical Methods.* McGraw-Hill, New York, NY.
Moyle, J.B. 1956. Relationships between chemistry of Minnesota surface waters and wildlife management. *Journal of Wildlife Management* 20: 303–320.
Mueller-Dombois, D. and H. Ellenberg. 1974. *Aims and Methods of Vegetation Ecology.* Wiley, New York, NY.
Muller, P. 1974. *Aspects of Zoogeography.* W. Junk, The Hague, Netherlands.
Naeem, S., L.J. Thompson, S.P. Lawler, J.H. Lawton, and R.M. Woodfin. 1994. Declining biodiversity can alter the performance of ecosystems. *Nature* 368: 734–737.
Naiman, R.J., D.G. Lonzarich, T.J. Beechie, and S.C. Ralph. 1992. General principles of classification and the assessment of conservation potential in rivers. In: P.J. Boon, P. Calow, and G.E. Petts (eds.), *River Conservation and Management.* Wiley, Chichester, UK.
Nelson, R.L., W.S. Platts, D.P. Larsen, and S.E. Jensen. 1992. Trout distribution and habitat in relation to geology and geomorphology in the North Fork Humboldt River drainage, northeastern Nevada. *Transactions of the American Fisheries Society* 121: 405–426.
Nix, H.A. 1982. Environmental determinants and evolution in Terra Australia. pp. 47–66. In: W.R. Barker, and P.J.M. Greenslade (eds.), editors. *Evolution of the Flora and Fauna of Arid Australia.* Peacock Publications, Frewville, Australia.
NOAA [National Oceanographic and Atmospheric Administration]. 1985. *National estuarine inventory: data atlas. Volume 1: physical and hydrologic characteristics.* NOAA-NOS Strategic Assessment Branch, Washington, D.C.
Noy-Meir, I. 1973. Data transformations in ecological ordination. I. Some advantages of non-centering. *Ecology* 61: 329–341.
Odum, F.P. 1971. *Fundamentals of Ecology.* Third edition. W.B. Saunders, Philadelphia, PA.
Ogden, J.C., J.W. Porter, N.P. Smith, A.M. Szmant, W.C. Jaap, and D. Forcucci. 1994. A long-term interdisciplinry study of the Florida Keys seascape. *Bulletin of Marine Science* 54: 1059–1071.
Omernik, J.M. 1986. Ecoregions of the conterminous United States (map, scale 1: 7,500,000). U.S. Environmental Protection Agency, Environmental Research Laboratory, Corvallis, OR.
Omernik, J.M. 1987. Ecoregions of the conterminous United States. *Annals of the Association of American Geographers* 77: 118–125.
Omernik, J.M. 1995. Ecoregions: a framework for environmental management. pp. 49–62. In: W. Davis, and T. Simon (eds.), *Biological Assessment and Criteria: Tools for Water Resource Planning and Decision Making.* Lewis Publishers, Boca Raton, FL.
Omernik, J.M., and A.L. Gallant. 1987. Ecoregions of the southwest states (map, scale 1: 2,500,000). EPA/600/D-87/316, U.S. Environmental Protection Agency, Environmental Research Laboratory, Corvallis, OR.
Omernik, J.M., and A.L. Gallant. 1989. Defining regions for evaluating environmental resources. Unpublished manuscript prepared for Global Natural Resource Monitoring and Assessment, Preparing for the 21st Century, September 24–30, 1989, Venice, Italy. Available from U.S. Environmental Protection Agency, Environmental Research Laboratory, Corvallis, OR.
O'Neill, R.V., D.L. DeAngelis, J.B. Waide, and T.F.H. Allen. 1986. *A Hierarchical Concept of Ecosystems.* Princeton University Press, Princeton, NJ.
Orians, G.H. 1975. Diversity, stability and maturity in natural systems. pp. 139–150. In: W.H. van Dobben and R.H. Lowe-McConnell (eds.), *Unifying Concepts in Ecology.* W. Junk, The Hague, Netherlands.
Paulsen, S.G. and R.A. Linthurst. 1994. Biological monitoring in the Environmental Monitoring and Assessment Program. In: A.L. Loeb, and A. Spacie (eds.), *Biological Monitoring of Aquatic Systems.* Lewis Publishers, Boca Raton, FL.
Peet, R.K. 1980. Ordination as a tool for analyzing complex data sets. *Vegetatio* 42: 171–174.
Peters, R.H. 1976. Tautology in evolution and ecology. *American Naturalist* 110: 1–12.
Pfister, R.D., and S.F. Arno. 1980. Classifying forest habitat types based on potential climax vegetation. *Forest Science* 26: 52–70.
Pfister, R.D., B.L. Kovalchik, S.F. Arno, and R.C. Presby. 1977. Forest habitat types of Montana. USDA Forest Service General Technical Report INT-34, Intermountain Forest and Range Experiment Station, Ogden, UT.
Pielou, E.C. 1977. *Mathematical Ecology.* Wiley, New York, NY.
Pielou, E.C. 1979. *Biogeography.* Wiley-Interscience, New York, NY.
Pinay, G., A. Fabre, P.H. Vervier and F. Gazelle. 1992. Control of C, N, P distribution in soils of riparian forest. *Landscape Ecology* 6: 121–132.
Platts, W.S. 1979. Including the fishery system in land planning. USDA Forest Service. General Technical Report INT-60, Intermountain Forest and Range Experiment Station, Ogden, UT.
Platts, W.S. 1980. A plea for fishery habitat classification. *Fisheries* 5(1): 2–6.
Pojar, J., K. Klink, and J. Meidinger. 1987. Biogeoclimatic ecosystem classification in British Columbia. *Forest Ecology and Management* 22: 119–154.
Pregitzer, K.S., and B.V. Barnes. 1982. The use of ground flora to indicate edaphic factors in upland ecosystems of the McCormick Experimental Forest, Upper Michigan. *Canadian Journal of Forest Research* 12: 661–672.
Prentice, I.C. 1977. Non-metric ordination methods in ecology. *Ecology* 65: 85–94.
Pritchard, D.W. 1967. What is an estuary? pp. 3–5. In: G.H. Lauff (ed.), *Estuaries.* AAAS Press, Washington, D.C.
Raunkier, C. 1937. *Plant Life Forms.* Clarendon Press, Oxford, UK.

Ray, G.C. 1991. Coastal-zone biodiversity patterns. *BioScience* 41: 490-498.
Ray, G.C., and B.P. Hayden. 1992. Coastal zone ecotones. pp. 403–420. In: F. di Castri, and A.J. Hansen (eds.), *Landscape Boundaries: Consequences for Biodiversity and Ecological Flows.* Springer-Verlag, New York, NY.
Ray, G.C., B.P. Hayden, M.G. McCormick-Ray, and T.M. Smith. 1997. Land-seascape diversity of the U.S. East Coast coastal zone with particular reference to estuaries. pp. 357–371. In: R.F.G. Ormond, J. Gage, and M.V. Angel (eds.), *Marine Biodiversity: Causes and Consequences.* Cambridge University Press, Cambridge, UK.
Reid, M.S., P.S. Bourgeron, H.C. Humphries, and M.E. Jensen. 1995. Documentation of the modeling of potential vegetation at three spatial scales using biophysical settings in the Columbia River Basin Assessment Area. Unpublished report prepared for the USDA Forest Service, contract # 53-04H1-6890, by the Western Heritage Task Force, The Nature Conservancy, Boulder, CO.
Reppert, R.T., and W. Sigleo. 1979. *Wetland values: concepts and methods for wetland evaluation.* U.S. Army Institute for Water Resources, Fort Belvoir, VA.
Rhom, C.M., J.W. Giese, and C.C. Bennett. 1987. Evaluation of an aquatic ecoregion classification of streams in Arkansas. *Journal Freshwater Ecology* 4: 127–140.
Ricker, W.E. 1934. An ecological classification of certain Ontario streams. *Publication of the Academy of Natural Sciences of Philadelphia* 101: 277–341.
Robins, C.R. 1971. Distributional patterns of fishes from coastal and shelf waters of the tropical western Atlantic. pp. 249–255. In: *Symposium on Investigations and Resources of the Caribbean Sea and Adjacent Regions.* FAO Papers on Fisheries Resources, FAO, Rome, Italy.
Robins, C.R. 1992. *Saltwater Fish.* Smithmark, New York, NY.
Robins, C.R., and G.C. Ray. 1986. *A Field Guide to Atlantic Coast Fishes of North America.* Houghton Mifflin, Boston, MA.
Rolof, G.J. 1994. Using an ecological classification system and wildlife habitat models in forest planning. Ph.D. dissertation, Michigan State University, East Lansing, MI.
Rowe, J.S. 1961. The level-of-integration concept and ecology. *Ecology* 42: 420–427.
Rowe, J.S. 1972. *Forest regions of Canada (with separate map at a scale of 1: 6,336,000).* Publication No. 1300, Canadian Forestry Service, Ottawa, Ontario, Canada.
Rowe, J.S. 1980. The common denominator in land classification in Canada: An ecological approach to mapping. *Forestry Chronicle* 6: 19–20.
Rowe, J.S. 1984. Forestland classification: limitations of the use of vegetation. pp. 132–147. In: J.G. Bockheim (ed.), *Forest Land Classification: Experiences, Problems, Perspectives.* Proceedings of the symposium, March 18–20, 1984. University of Wisconsin, Madison, WI.
Rowe, J.S., and J.W. Sheard. 1981. Ecological land classification: a survey approach. *Environmental Management* 5: 451–464.
Roy, P.S. 1984. New South Wales estuaries: their origin and evolution. pp. 99–121. In: B.G. Thom (ed.), Coastal geomorphology in Australia. Academic Press, Sydney, Australia.
Rübel, E. 1930. *Die pflanzengesellschaften der Erde.* Huber Verlag, Bern, Switzerland.
Schupp, D.H. 1992. An ecological classification of Minnesota lakes with associated fish communities. Minnesota Department of Natural Resources Investigation Report 417, Minneapolis, MI.
Scott, J.M., F. Davis, B. Csuti, R. Noss, B. Buttterfield, C. Groves, H. Anderson, S. Caicco, F. D'Erchia, T.C. Edwards, Jr., J. Ullman, and R.G. Wright. 1993. Gap analysis: protecting biodiversity using geographic information systems. *Wildlife Monographs,* 123.
Seaber, P.R., F.P. Kapinos, and G.L. Knapp. 1984. State hydrologic unit maps. U.S. Geological Survey Open-file Report 84-708, U.S. Geological Survey, Reston, VA.
Shannon, E.E., and P.L. Brezonik. 1972. Eutrophication analysis: a multivariate approach. *Journal of the Sanitary Engineering Division of the American Society of Civil Engineers* 98 (SA1): 37–57.
Sheldon, A.L. 1972. A quantitative approach to the classification of inland waters. pp. 205–261. In: J.V. Krutilla (ed.), *Natural environments: studies in theoretical applied analysis.* Johns Hopkins Press, Baltimore, MD.
Shelford, V.E. 1911. Ecological succession. I. Stream fishes and method of physiographic analysis. *Biological Bulletin* 21: 9–35.
Shimwell, D.W. 1971. *The Description and Classification of Vegetation.* University of Washington Press. Seattle, WA.
Sims, R.A., W.D. Towill, K.A. Baldwin, and G.M. Wickware. 1989. *Field guide to the forest ecosystem classification for northwestern Ontario.* Forestry Canada, Ontario Ministry of Natural Resources, Thunder Bay, Ontario, Canada.
Sly, P.G., and W.-D.N. Busch. 1992. A system for aquatic habitat classification of lakes. pp. 15–26. In: W.-D.N. Busch, and P.G. Sly (eds.), *The Development of an Aquatic Habitat Classification System for Lakes.* CRC Press, Boca Raton, FL.
Spies, T.A., and B.V. Barnes. 1985. A multifactor ecological classification of the northern hardwood and conifer ecosystems of Sylvania Recreation Area, Upper Peninsula, Michigan. *Canadian Journal of Forest Research* 15: 949–960.
Stearns, S.C. 1976. Life history tactics: a review of the ideas. *Quarterly Review of Biology* 51: 3–47.
Steele, J.H. 1989. The ocean "landscape." *Landscape Ecology* 3: 186–192.
Strahler, A.N. 1957. Quantitative analysis of watershed geomorphology. *American Geophysical Union Transactions* 38: 913–920.
Strong, W.L., E.T. Oswald, and D.J. Downing. 1990. The Canadian vegetation classification system, first approximation. National Vegetation Working Group, Canadian Committee on Ecological Land Classification. Ecological Land Classification Series, No. 25, Sustainable Development, Corporate Policy Group, Environment Canada, Ottawa, Ontario, Canada.
Sukachev, V. 1945. Biogeocoenology and phytocoenology. *C.R. Academy of Science, USSR.* 47: 429–431.
Swanson, F.J., T.K. Kratz, N. Caine, and R.G. Woodmansee. 1988. Landform effects on ecosystem patterns and processes. *Bioscience* 38: 92–98.
Tansley, A.G. 1935. The use and abuse of vegetational concepts and terms. *Ecology* 16: 284–307.
Tiner, R.W., Jr. 1984. *Wetlands of the United States: current status and recent trends.* U.S. Fish and Wildlife Service, U.S. Government Printing Office, Washington, D.C.
Tonn, W.M. 1990. Climate change and fish communities: a conceptual framework. *Transactions of the American Fisheries Society* 119: 337–352.

Tonn, W.M., and J.J. Magnuson. 1982. Patterns in species composition and richness in fish assemblages in northern Wisconsin lakes. *Ecology* 63: 1149–1166.

Trewartha, G.T. 1968. *Introduction to Climate.* 4th edition. McGraw-Hill, New York, NY.

Troll, C. 1941. Studien zur vergleichenden Geographie der Hochgebirge. Bonn: Friedrich Wilhelm Universitat, Bericht 23: 94–96.

Troll, C. 1943. Die Frostwechselhaufigkeit in den Luft - und Bodenklimaten der Erde. *Metеororologische Zeitschrift* 60: 161–171.

Troll, C. 1955. Der jahreszeitliche Ablauf des Naturgeschehens in den verschiedenen Klimagurteln der Erde. *Studium Generale* 8: 713–733.

Troll, C. 1956. *Der Klima- und Vegetationsaufbau der Erde im Lichte neuer Forschungen.* Akademie der Wissencschaften und Literatur, Mainz, pp. 216–229.

Tüxen, R. 1956. Die heutige potentielle natürliche vegetation als Gegenstand der Vegetationskartierung. *Angew. Pflanzensoziol. (Stolzenau, Weser)* 13: 5–42.

Udvardy, M.D.F. 1975. A classification of the biogeographical provinces of the world. International Union for Conservation of Nature and Natural Resources, Occasional Paper No. 18. Morges, Switzerland

UNESCO [United Nations Educational, Scientific and Cultural Organization]. 1973. International Classification and Mapping of Vegetation, Series 6, Ecology and Conservation. UNESCO, Paris, France.

Urban, D.L., R.V. O'Neill, and H.H. Shugart, Jr. 1987. Landscape ecology: a hierarchical perspective can help scientists understand spatial patterns. *Bioscience* 37: 119–127.

USDA Soil Conservation Service. 1981. *Land resource regions and major land resource areas of the United States.* Agriculture Handbook No. 296, USDA Soil Conservation Service, Washington, D.C.

USGS [United States Geological Survey]. 1982. *Hydrologic unit map of the United States.* U.S. Government Printing Office, Washington, D.C.

Valiela, I. 1984. *Marine Ecological Processes.* Springer-Verlag, New York, NY.

Walter, H. 1973. *Die vegetation der Erde I. Tropische and subtropische Zonen.* Third edition. Fisher, Jena, Stuttgart, Germany.

Walter, H. 1985. *Vegetation of the earth, and ecological systems of the geobiosphere.* Third edition. Springer-Verlag, New York, NY.

Walter, H., and E. Box. 1976. Global classification of natural terrestrial ecosystems. *Vegetatio* 32: 75–81.

Warren, C.E. 1979. Toward classification and rationale for watershed management and stream protection. Report No. EPA-600/3-79-059, United States Environmental Protection Agency, Corvalis, OR.

Webb, L.J., J.G. Tracey, W.T. Williams, and G.N. Lance. 1970. Studies in the numerical analysis of complex rainforest communities. V. A comparison of the properties of floristic and physiognomic-structural data. *Journal of Ecology* 58: 203–232.

Wergner, M.J.A. and J.T.C. Sprangers. 1982. Comparison of floristic and structural classification of vegetation. *Vegetatio* 50: 175–183.

Wertz, W.A., and J.A. Arnold. 1972. *Land Systems Inventory.* USDA Forest Service, Intermountain Region, Ogden, UT.

## THE AUTHORS

**Dennis H. Grossman**
*Conservation Science Division,*
*The Nature Conservancy,*
*4245 N. Fairfax Drive, Suite 100,*
*Arlington, VA 22203, USA*

**Patrick Bourgeron**
*Institute for Arctic and Alpine Reserach,*
*University of Colorado,*
*Boulder, CO 80309*

**Wolf-Dieter N. Busch**
*U.S. Fish and Wildlife Service,*
*405 N. French Road, Suite 120A,*
*Buffalo, NY 14228, USA*

**David Cleland**
*Forestry Services Lab,*
*USDA Forest Service,*
*P.O. Box 898*
*Rhinelander, WI 54501, USA*

**William Platts**
*Platts Consulting,*
*1603 Sunrise Rim,*
*Boise, ID 83705, USA*

**G. Carleton Ray**
*Department of Environmental Sciences,*
*University of Virginia,*
*Clark Hall,*
*Charlottesville, VA 22903, USA*

**C. Richard Robins**
*Kansas Museum of Natural History and*
*Biodiversity Research Center,*
*University of Kansas,*
*448 N. 1500 Road,*
*Lawrence, KS 66049-9090, USA*

**Gary Roloff**
*Boise Cascade Corporation,*
*123 Wolf Court,*
*East Lansing, MI 48223, USA*

esthoff, V., and A.J. den Held. 1969. *Plantengemenschappen in Nederland.* Thieme, Zutphen.

Westhoff, V., and E. van der Maarel. 1978. The Braun–Blanquet approach. Chapter 20. In: R.H. Whittaker (ed.), *Classification of Plant Communities.* W. Junk, The Hague, Netherlands.

Whittaker, R.H. 1956. Vegetation of the Great Smoky Mountains. *Ecological Monographs* 26: 1–80.

Whittaker, R.H. 1962. Classification of natural communities. *Botanical Review* 28: 1–239.

Whittaker, R.H. 1972. Evolution and measurement of species diversity. *Taxon* 21: 213–351.

Whittaker, R.H. 1975. *Communities and Ecosystems.* Second edition. MacMillan, New York, NY.

Whittaker, R.H. (ed.). 1978. *Classification of Plant Communities.* W. Junk, The Hague, Netherlands.

Whittaker, R.H. (ed.), 1980. *Classification of Plant Communities.* Volume II. W. Junk, The Hague, Netherlands.

Whittier, T.R., R.M. Hughes, and D.P. Larsen. 1988. Correspondence between ecoregions and spatial patterns in stream ecosystems in Oregon. *Canadian Journal of Fisheries and Aquatic Sciences* 45: 1264–1278.

Wiken, E.B., compiler. 1986. Terrestrial ecozones of Canada. Ecological Land Classification Series No. 19, Environment Canada, Lands Directorate, Ottawa, Ontario, Canada.

Winter, T.C. 1977. Classification of the hydrologic settings of lakes in the north central United States. *Water Resources Research* 13: 753–767

Witmer, R.E. 1978. U.S. Geological Survey land-use and land-cover classification system. *Journal of Forestry* 76: 661–668.

Zak, D.R., G. E. Host, and K.S. Pregitzer. 1989. Regional variability in nitrogen mineralization, nitrification, and overstory biomass in northern Lower Michigan. *Canadian Journal Forest Research* 19: 1521–1526.

Zak, D.R., K.S. Pregitzer, and G.E. Host. 1986. Landscape variation in nitrogen mineralization and nitrification. *Canadian Journal Forest Research* 16: 1258–1263.

Zonneveld, I.S. 1988. Landscape (ecosystem) and vegetation maps, their relation and purpose. pp. 481–486. In: A.W. Küchler, and I.S. Zonneveld (eds.). *Vegetation Mapping.* Kluwer, Boston, Massachusetts.

Zonneveld, I.S. 1989. The land unit — a fundamental concept in landscape ecology, and its application. *Landscape Ecology* 3: 67–86.

# The Use of Ecological Classification in Management

Constance A. Carpenter, Wolf-Dieter N. Busch,
David T. Cleland, Juan Gallegos, Rick Harris, Ray Holm,
Chris Topik, and Al Williamson

**Key questions addressed in this chapter**

- *How do we use biophysical classifications and ecological assessments in decision-making?*
- *How do we use hierarchical classification systems?*
- *How do we develop a common, ecologically based classification that works for all the partners?*

**Keywords: Ecological units, mapping, land cover, landscape, habitat type, desired future conditions**

# 1 INTRODUCTION

Ecological classification systems range over a variety of scales and reflect a variety of scientific viewpoints. They incorporate or emphasize varied arrays of environmental factors. Ecological classifications have been developed for marine, wetland, lake, stream, and terrestrial ecosystems. What are the benefits of ecological classification for natural resource management planning and implementation?

This chapter is written primarily from the viewpoint of the U.S. Forest Service in its efforts to implement multiple public mandates for federal land and draws heavily from USFS experience in ecological classification and mapping. It is also the viewpoint of an agency "standing at the borders looking outward." National Forests recognize they are linked to others by the common goals of a healthy sustainable environment and stable prosperous communities.

This chapter presents examples of ecological classification systems and their use. It includes examples of ecological and other classification systems being used together, generally for landscape and regional planning and examples of public organizations working together to identify common ecological unit boundaries for inventory, monitoring, and management purposes. Grossman et al. (this volume) discuss the development of a variety of classification systems and the technical merits and difficulties of combining them.

## 1.1 Key Questions

Central to the development of this chapter are the following questions posed to the Ecological Classification science and management teams:

1. "How do we use a variety of biophysical classifications and ecological assessments in decision-making?" and
2. "How do partners develop a common, ecologically based classification that works for all the partners?"

## 1.2 Scope of the Management Chapter

This chapter discusses the use of ecological classification in integrated resource inventory, planning and assessment from site specific to national scales. It focuses on management considerations in ecological unit surveys and the use of mapped ecological units. Generally, a few basic concepts underlie the use of ecological units. These are constant and transferable among terrestrial and aquatic ecosystems and from region to region.

The resource manager, newly interested in ecological classification, may be overwhelmed by the wide array of classification systems in use and perhaps by the controversies surrounding their scientific underpinnings. As the state of scientific understanding regarding ecosystems has evolved, so has the science and art of ecological classification and mapping. Indeed each field application has a unique developmental history often based upon the school of thought prevalent at the time and place it was developed.

## 1.3 Approaches to Ecological Classification and Mapping

To understand the use of ecological classification, it is important to understand ecosystem attributes. Ecosystems are complete interacting systems of organisms and their environment. Ecosystems are described at many scales ranging from microsites to the biosphere and vary in composition, structure, and function. All ecosystems grade into others; all are nested within a matrix of larger ecosystems. Ecosystems are continually changing and the changes are not always predictable.

Ecosystem complexity is important for sustaining life but this complexity makes it difficult to determine the function and significance of individual ecosystem components, structures and processes. Ecosystems are more than the aggregate of their parts. The conditions and processes occurring across larger ecosystems affect and often override those of smaller ecosystems, and the properties of smaller systems affect and emerge in the context of larger systems.

What is ecological classification? Ecological classification systems are based on associations among physical and biological factors identified in the classification process. First, interrelationships among biota and the environment are studied at several relevant scales or levels of generality. "Important" variables and scales are identified through data analysis and synthesis. Taxonomic classes are formulated by integrating this information and defining categories based on mutual relationships. These classes allow identification of criteria to meaningfully map geographic areas (Driscoll et al. 1984).

Map units are based on the criteria or relationships identified by the taxonomic system used. Taxonomic systems help us organize our knowledge while mapping units transfer this knowledge to the specific areas where it can be applied. However, the concept of a map unit is not the same as that of a taxonomic class. Taxonomic classes are actually models based on a sample of a larger population. "The advantage of mapping" according to Rowe (1996), "is that every part of the terrain has to be confronted; there is no avoiding those in-between and oddball units that an a priori classification is apt to ignore". In addition, a map unit may be composed of more than one taxonomic class. In this case, the most common case, mapping rules are developed by evaluating the mode, range, relative abundance and distribution of taxonomic classes within the area of interest. Units can then be mapped and used with full knowledge of the important attributes inherent in the taxonomic classification and the mapping rules.

An ecosystem supports vegetation of varied age and community structure over time. Ecological classifications can be separated into two categories based on the way they deal with changes in time. Maps that are used to describe land and aquatic units that behave in a similar manner over time are referred to as biophysical maps. These map boundaries change only when new information indicates they do not reflect long term potential. Existing or historic status maps are used to describe ecosystems or ecosystem components at a point in time. These map boundaries are expected to change every time an area is surveyed.

Ecological types and map unit concepts are three dimensional. They are based on the integration of biotic and abiotic characteristics above and below ground which distinguishes them from classifications of individual ecosystem components such as cover types, soils or remote sensing classifications which are based on spectral or thermal signatures.

Ecological classification and mapping are conducted at a variety of scales or levels of generality (see Haufler et al., this volume). A hierarchical classification can systematically divide the country into progressively smaller areas of land and water having similar physical and biological characteristics and ecological processes. Linkages among units of different scales in a hierarchical system are based upon the dynamics of various energy, water, nutrient and disturbance cycles. Recognizing environmental conditions at a higher level of organization sets a framework for understanding patterns and interactions at lower levels.

Ecological classification and mapping provide a framework for integrating information on the composition, structure and function of ecosystems. It is the explicit integration of information gained through ecological classification, mapping and additional environmental inventory that allows identification of ecosystems and the development of models of ecosystem behavior. It is within this context that ecological units are used to characterize ecosystems over time and space and to test cause-and-effect relationships.

## 1.4 Benefits of Using an Ecological Unit Framework in Management

Ecological classification systems are versatile tools that can be used to resolve issues, determine management direction, and implement ecologically based management approaches. They can be used when managers characterize the environment, inventory resource conditions, conduct environmental analyses, establish desired future conditions, monitor trends in natural resources, and establish priorities for conservation and restoration activities. They provide a means to link models of ecological processes to specific areas. The uses of ecological classification have expanded as understanding of ecosystem needs has progressed.

Ecological classification systems are useful in addressing fundamental management questions such as what constitutes conservation, preservation, restoration and proper management. Ecological unit maps are used to provide an ecological context for planning and management. They contribute to our ability to demonstrate the potential for a variety of alternatives at local, landscape, and regional scales and help establish the logical scope of planning and analysis activities. They add geographic specificity to documents and efforts to communicate the logic underlying management decisions (Avers et al. 1994, Carpenter et al. 1995).

Ecological units provide a framework for describing and understanding ecosystems and an expedient and cost-effective means of ordering and managing information about them. They improve our ability to:

- integrate knowledge from multiple disciplines, traditionally separated,
- develop and share resource data and information across administrative and jurisdictional boundaries, and
- communicate technical information to specialists and lay people through the use of common terminology, common maps, and standardized data.

Hierarchical frameworks, which integrate units of multiple scales by nesting small units within larger ones, have additional benefits (Avers et al. 1994) They:

- help clarify the relationships between ecological patterns and the processes which influence them,
- maximize the use of resource inventory information among multiple geographic scales, and
- foster broad application and appropriate extrapolation of research results.

## 1.5 Overview of Ecological Classification Activity in the Forest Service

The evolutionary histories, the pace and approaches to the development and use of ecological classification in the USFS have varied among National Forests and within Regions. Aquatic and terrestrial ecosystem classification and inventory efforts have evolved independently of each other. Forest-by-Forest efforts have proceeded from the ground level up, generally investing in local and site classification first, then landscape, regional and hierarchical systems. Today, Forests are working toward an integrated global to local hierarchical system that addresses terrestrial and aquatic ecosystem management needs and that supports multipurpose, multiscale inventory, monitoring, and research.

For terrestrial systems, the evolution from single resource classification and mapping toward an integrated hierarchical approach began in the late 1970s. By 1984, the need for some standardization within the agency was recognized (Bockheim 1984). Systems used at that time display differences in terminology and delineation criteria, including differences in how land use information was included. The quality of environmental predictions, the ability to integrate wildlife needs and effects, the ability to assess the relationships among contiguous land units at one or more scales, and the integration of aquatic and terrestrial components varied from system to system as well (Bailey 1984).

An agency-wide review of the status of water unit classification was conducted in 1987 (USDA Forest Service WO-INS 1987). Aquatic ecosystem classification was not in widespread use at that time. The agency task force recommended a four-level information hierarchy be adopted for aquatic classification and information management that is roughly commensurate with the lower four levels displayed in the aquatic hierarchy today. Level one was designated the most general and level four the most site specific. Most of the systems in use were "first cut" systems and based on a single data element. Although systems were found at various levels of resolution, most of the effort was directed at the site level and there was little integration across scales.

Most lake classification conducted at Level 1 (the broadest level) was related to general fish habitat or origin of the lake and underlying geology. Level 2 lake classifications were related to some aspect of water condition or lake bottom structure.

Most wetland classification efforts were based on the U.S. Fish and Wildlife Service method (Cowardin et al. 1979).

Table 1. The Forest Service National Hierarchical Framework for using ecological units (adapted from Avers et al. 1994).

| Planning and Analysis Scale | Purpose, Objectives, General Use | Ecological Units |
|---|---|---|
| Ecoregion<br>Global<br>Continental<br>Regional | Broad applicability for modeling and sampling, strategic planning and assessment, international planning | Domain<br>Division<br>Province |
| Subregion | Strategic and multi-agency scale analysis and assessment, data aggregation. Generating and testing research hypotheses. Technology transfer and data extrapolation. | Section<br>Subsection |
| Landscape | Multiple resource assessment and analysis. Tactical and long term operational planning, data aggregation, research, and monitoring design. | Land Type Association |
| Land Unit | Project planning and implementation, environmental effects analysis, project monitoring and evaluation. | Landtype<br>Landtype phase |

Streams were the primary focus of agency aquatic classification efforts of the time. Parrott et al. (1989) concluded that even though stream classification systems had been developed by several workers, classifications had only been implemented and documented in a few geographic areas. They also pointed out that no hierarchical stream classification system was in widespread use at that time, although Platts (1980) had proposed a hierarchical classification as early as 1980.

The New Perspectives initiative of 1992 highlighted the need to demonstrate the scientific basis for ecosystem management, to conduct more holistic management, and to incorporate biodiversity conservation into planning and management activities. For these reasons, the Forest Service officially adopted the National Hierarchical Framework of Ecological Units (Table 1) in 1993. In 1995, the agency released a companion generic hierarchical framework for characterizing aquatic ecosystems that described the linkages between terrestrial units and aquatic biophysical environment maps (Avers et al. 1994, Maxwell et al. 1995) (Fig. 1).

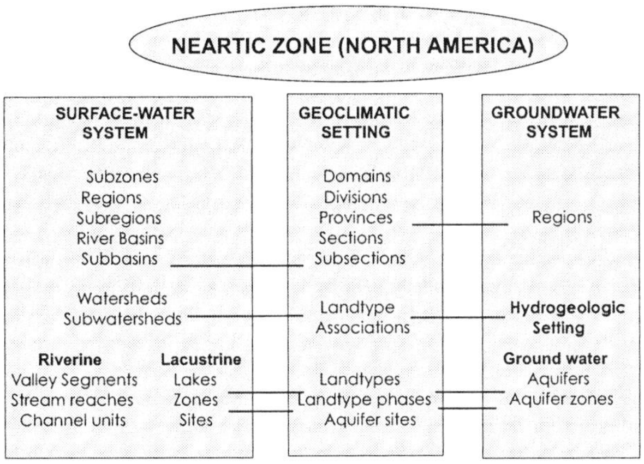


Fig. 1. General framework of aquatic ecological unit hierarchy. Primary linkages between aquatic systems and terrestrial (geoclimatic) systems are shown (Maxwell et al. 1995).

The task of changing from existing conditions of terrestrial and aquatic classification and inventory within the U.S. Forest Service, to that recommended through The National Hierarchy is a challenging one. The Forest Service must be innovative in using existing classifications to the best advantage, while strategically providing additional information and expertise to meet the criteria at multiple levels in the National Hierarchy. The proceedings of the national workshop: *Taking an Ecological Approach to Management* (USFS 1992) provide the most recent descriptions of ecological classification approaches among the Forest Service Regions. The intent of the regional presentations was to provide information that would lead to a strategy to integrate physical, biological, and socio-political information from the National Forest System lands and adjacent lands into an ecological approach to management. Therefore, it is not surprising that classification was a central piece in each presentation.

Since the adoption of the National Hierarchical Framework, the Forest Service has pursued the revision of existing maps and information in a simultaneous top-down and bottom-up fashion. Broadscale efforts are most easily documented. At the national level the USFS publication *Description of the Ecoregions of the United States* (Bailey 1980) was revised (Bailey 1995). All Forest Service regions contributed to a map *Ecoregions and Subregions of the United States* (Bailey et al. 1994) and its companion document *Ecological Subregions of the United States: Section Descriptions* (McNab and Avers 1994). The USFS in cooperation with NRCS published the map *Ecological Units of California: Subsections* (Goudey and Smith, 1994) by subdividing the Sections on the 1994 map (Bailey et al.). Albert (1995) produced *Regional Landscape Ecosystems of Michigan, Minnesota, and Wisconsin: A Working Map and Classification (Fourth Revision: July 1994)* through a cooperative agreement with the North Central Forest Experiment Station by request of the Upper Great Lakes Biodiversity Committee. The most recent publication, which involved two National Forest System regions, three research stations (now organized to two), the Northeastern Area of State and Private Forestry, and numerous collaborators was a map to the subsection level called *Ecological Units of the Eastern United States — First Approximation* (Keys et al. 1995).

Today, technology such as GIS, computer spreadsheets and database programs are important tools to classify, locate, and interpret ecological units at all scales. Regions are experimenting with map overlay and multivariate statistical techniques and the use of modeling to investigate landscape variability within units in response to natural disturbance and management.

Since the last review in 1984, progress has been made toward consistency in classification criteria, although the procedural steps for classification and mapping are not uniform nationally. This is partly because of the need to maximize the use of existing information, partly because the field of geostatistics is going through a period of rapid evolution, and partly because of the wide variation in access to modern technologies. Initial priorities for achieving consistency currently lie among forests at the land type association level, and among federal agencies at the land type association, section, subsection, and province levels.

Clearly, ecological classification systems have become a more integral part of Forest Service planning and management efforts (USDA Forest Service 1992, WO-INS 1993). Ecological surveys have been on-going for several decades; therefore, more Forests now have access to survey information and more Forests have access to multiple levels of classification. The major emphasis is still on terrestrial classification; however, the need for aquatic classification is now widely recognized. The earlier observation of little integration of aquatic and terrestrial components in classification still holds true however.

Cross-boundary, cross-agency cooperation in ecological classification development and mapping is increasing. The Forest Service is increasingly involved with state agencies and organizations to develop units at landscape and regional scales for statewide or watershed-wide planning purposes. In 1996, the Forest Service signed a memorandum of understanding (MOU) relative to "Developing a spatial framework of ecological units of the United States" with eight other federal agencies among the Department of Agriculture, the Department of Interior, and the U.S. Environmental Protection Agency (Case Study 13).

## 2 SITE LEVEL APPLICATIONS

Ecological units are depicted on maps but ecological unit maps alone do not fully characterize ecosystems. Ecological units are used in combination with inventories and maps of existing vegetation, wildlife, aquatic systems, air quality, and human development to characterize the complexes of life and environment we call ecosystems at any point in time. The type of past human use, the intensity of management, and degree of current human development in or around the area you are managing all affect the degree an ecosystem may deviate in composition, structure and/or function from its potential natural state. Ecological units provide information on ecosystem potential and capability but do not substitute for a well thought out prescription based on all available evidence.

### 2.1 Integrated resource inventory

The transition from multiple-use management to ecosystem management is marked by a desire to integrate knowledge from multiple disciplines that have traditionally been separated (see also McCleery et al. this volume). Multidisciplinary cooperation in data gathering and analysis is a prerequisite to implementation of this holistic approach. Ecological units allow managers to provide a consistent context for other spatially referenced information. Ecological units provide a spatial framework for this integration and a conceptual basis for analyzing cause and effect relationships. This ultimately results in an increased ability to analyze resource interactions and management tradeoffs.

Ecological classification and mapping can aid in the integration of resource information in several ways. Ecological classifications by definition integrate multiple environmental factors. Maps of ecological units provide a spatial framework for structuring a variety of resource inventory and monitoring work, thus contributing to the progressive accumulation of knowledge about area ecosystems. Where ecological types have been developed but mapping is not available, single resource inventories can be modified to collect sufficient data to allow identification of ecological types as part of the inventory database.

The use of ecological units, as a standard base for conducting and recording inventory information, is more efficient and cost effective than after-the-fact integration of single resource inventories using a map overlay process. Ecological classification and mapping organize information in a spatial context. Data development and expansion occur in several stages. Data are first collected and analyzed to develop ecological types and to provide interpretations by ecological type for management use. Then, ecological units are mapped and sampling is undertaken to ensure the reliability of the map unit descriptions. Finally, as additional resource information needs are identified, the classification units can be used to stratify additional sampling and then used to look for natural associations or correlation with environmental factors already inventoried.

An ecological unit framework provides an ecological context for assimilating new information which is not provided by merely overlaying maps from multiple independent surveys. The generation of map polygons through a mechanical process of map overlay does not guarantee you can establish the ecological relationships with sufficient accuracy for management at the site level. Many of the boundaries from an overlay approach will not be coincidental, resulting in the presence of map "slivers," which need to be allocated individually to some management or analysis area.

Historically, scientists and managers have inventoried resources by identifying and mapping those characteristics that are important to human use through the development of various classification systems or "taxonomies." Physical components of the environment have been described and classified based upon their morphological characteristics and the associated properties. Similarly living things have been classified according to those morphological and physiological characteristics that affect their adaptability to the environment. Site conditions determine a location's capability or suitability for various uses. Capability and suitability determinations are often referred to as "interpretations" of a classification and are what make classifications valuable to managers.

Monitoring is an important part of ecosystem management. Scientists and managers are encouraged to monitor changes in ecosystem composition, structure and processes to detect changes in ecosystem health. The schematic presented in Fig. 2 provides examples of compositional, structural, and functional characteristic that are often measured in inventory, monitoring, and research programs.

Ecosystem structure is defined here as the vertical and/or horizontal arrangement of ecosystem components viewed in a particular geographic setting at a particular time. Information on composition and structure are needed to describe the functional characteristics of ecosystems. Physical, chemical, and biological processes link multiple resource components to each other and influence their distribution, arrangement, and abundance. Limits to the rate of change include: limited amounts of nutrients, moisture, energy, and space; limitations due to the physiology,

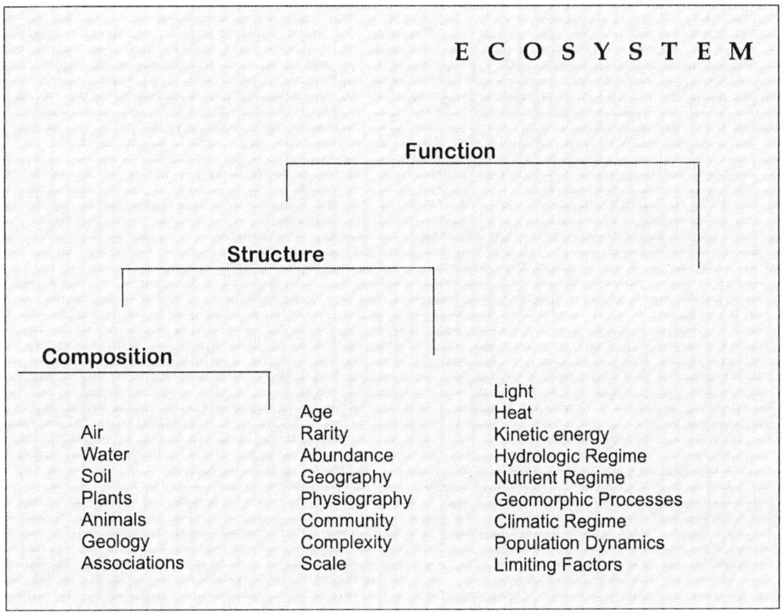


Fig. 2. Ecological characteristics that can be measured or described and the processes which must be examined in determining the response of a given ecosystem to natural disturbance and management.

age, and trophic status of biological organisms; and limitations imposed by disturbance regimes.

Ecological unit maps help identify the spatial interdependence of ecosystems. The inherent capabilities of an ecosystem depend upon the interactions and associations of environmental factors in a given area. This includes the functional linkages such as transfers of heat, moisture, nutrients, sediments, seeds, etc. among diverse but contiguous systems. It includes environmental values that accrue by the mere juxtaposition of diverse areas. For example, wetland ecosystem capability includes the ability to produce biomass in the form of various species assemblages; the productivity of the wetland is influenced by nutrient inputs from the adjacent uplands; and the wildlife habitat value of the wetland is influenced by the juxtaposition of the wetland and upland.

Field inventory represents a significant investment; therefore, steps that enhance the quality and completeness of data collection and mapping should be a high priority for managers. The quality of an ecological unit inventory rests on the ability of the survey crew to allocate land correctly to an ecological type and to recognize the landscape patterns that provide a degree of homogeneity to ecological units. Classification and mapping require personnel with good judgment, training and field experience.

Ecological units are most reliably mapped using relatively permanent features of the environment such as soils, rock, waterbodies, and landforms in combination with vegetation indicators. Using soil, vegetation, and landform indicators together, mappers can compensate for the limitations of any one alone. The use of plant indicators increases the ease and efficiency of mapping if used carefully. Limitations are that plants and plant associations may change over time, especially in areas of repeated or severe site disturbance. Past events can cause certain vegetation patterns that will not persist. Long-lived species may be out of step in the contemporary environment because of a lag in adjusting to environmental change. Herbaceous species do not always reflect soil conditions of importance to deeper-rooted trees or they may reflect features of no importance to trees. After certain kinds of severe disturbance, ground flora may be difficult to interpret accurately. On severely disturbed sites, soil erosion may have eliminated typical soil characteristics and changed site potentials.

Ecological classification and inventory produce information on several ecosystem components at once, an advantage over single resource inventories. Surveys provide information on the size and location of units that often have implications for ecology and management.

The following case studies provide an overview of inventory and mapping activities among a variety of ecosystems and which meet the requirements of a variety of disciplines. Several show the use of completed maps to stratify sampling for additional information that contributes to integrated resource management. Even though the data collected and methods of inventory may vary between terrestrial and aquatic systems, it is apparent that the integration of biotic and abiotic information is a common feature.

*Case Study 1* summarizes a contemporary example of data collection for ecological classification and mapping of terrestrial uplands and shallow water wetlands on the Hiawatha National Forest, Michigan and identifies several uses of this baseline inventory in management. *Case Study 2* provides an example of an integrated stream inventory system. *Case Study 3* provides an evaluation of the use of ecological classification in inventorying wildlife habitat attributes in Michigan.

## Case Study 1 in Integrated Resource Inventory: Hiawatha National Forest Ecological Classification (ECS), and Inventory of Uplands and Shallow Water Wetlands — Contemporary example of data collection and mapping

*Ecological Classification:*

This effort is based upon the National Hierarchical Framework of Ecological Units (Avers et al. 1994). Work on the upland component of the classification was begun in the 1980s. Work on the wetland portion was begun in the 1990s. The Land Type Associations (LTA) range in the 1,000s of acres in size, Ecological Land Types (ELT) range in the 100s of acres, and Ecological Landtype phases (ELTP) range in the 10s to 100s of acres.

*Description:*

1. Initial classification development included the collection of comprehensive plot information on landform, soil, vegetation, and hydrologic factors to develop ELT and ELTP units through the integration of biotic and abiotic data.
2. ECS mapping and inventory has been completed on over 95 percent of the 0.5 million acres in the Hiawatha National Forest western half. The process consists of the delineation of ELTP polygons (minimum size 5 acres) on low level color infrared aerial photography combined with verification plots in each ELTP unit. The plot information is entered into a relational database and consists of a listing of vegetation with relative dominance, a soil characterization to 5 m. or water table, and hydrological information including depth to ground water and water chemistry (pH and electrical conductivity) in wetlands.
3. Polygons identified with ELTP codes are in the process of being transferred from aerial photography to an automated database.

The inventory database with over 4,000 plot entries enables queries to isolate information about specific geographical locations or virtually any combination of species, soil, or hydrologic factors. Soil monitoring for management impacts will be stratified by ELT and ELTP designation. The small scale classification, inventory, and mapping of ELT and ELTP units was used to refine regionalized LTA boundaries and descriptions at a higher scale in the National Hierarchical Framework of Ecological Units.

The ECS mapping and inventory data serve as baseline data and as a basic framework for a variety of management uses and proposed activities. Some specific examples include project area analysis to identify management options in a planning unit of several thousand acres. ECS mapping and ELTP and ELT descriptions are used to compare the present natural community to the potential community and the desired future condition. Existing and potential old-growth communities are stratified by ECS units to identify preferred options in the implementation of an old-growth strategy.

*Contacts*

Greg Kudray, Michigan Technological University, Department of Forestry and Wood Products, Houghton, MI 49931, 906-523-4817, gmkudray@mtu.edu and Kirsten Saleen, Hiawatha National Forest Supervisors Office, Escanaba, MI 906-786-4062

## Case Study 2 in Integrated Resource Inventory: Inventory and Mapping of Freshwater Streams on the Chequamegon National Forest, Wisconsin — Integrated inventory for use in protection, restoration, and assessment of stream environments

*Ecological Classification*

The USDA Forest Service has recently adopted *A Hierarchical Framework of Aquatic Ecological Units in North America (Neartic Zone)* (Maxwell et al. 1995) to group environmental situations in a hierarchical fashion. Within this framework there is category named stream valley segments that is defined by a general set of attributes. Using this framework as a guide, the Chequamegon National Forest collected, over a four-year period, a suite of physical, chemical, and biological data through the range of environmental situations within the national forests of Wisconsin.

*Description*

Effective management of aquatic resources is premised on the notion of optimizing productivity within a particular environmental situation. This optimum or goal is frequently called a reference site. What is lacking in this process is a consistent and logical method of grouping these environmental situations into similar classes. We were able to classify the National Forest streams into 13 discrete valley segment types or classes by statistically analyzing data, such as bank full width, maximum water temperature, alkalinity, and fish and mussel distribution and abundance. With this information we subsequently typed and mapped all Forest streams as one of these 13 classes. We then stratified the stream network by class and sampled those valley segments that had not been sampled previously to verify the efficacy of the classification process. This new data set demonstrated that we had correctly typed and mapped the forest streams with 68 percent accuracy.

In the future, we will use this process to describe the range of variation within each stream type so we can apply established techniques, like the index of biotic integrity, to tailor actions by the Forest Service in a more appropriate and efficient manner. Moreover, we now have the ability to identify the abundance or scarcity of stream types that will enhance our efforts to select special management areas and to stratify monitoring efforts as described in the forest plan. Valley segment types will also be used to identify and prioritize stream segments for restoration or enhancement and as a basis for conducting threatened and endangered species surveys.

This effort will be expanded to determine the relationship of stream valley segments to the next higher (subwatershed) and lower (stream reach) tiers within the framework, so we can begin to understand the form and function of streams within the context of landscape ecology. Armed with such knowledge we should be able to provide some answers to the elusive issue of cumulative effects of forest management on stream environments.

### *Contact*

Dale Higgins, USDA Forest Service, Chequamegon National Forest, 1170 4th Ave. South, Park Falls, WI 54552.

## Case Study 3 in Integrated Resource Inventory: Establishing Wildlife Habitat Capability for Planning — Value in existing and potential natural vegetation

### *Ecological Classification*

Ecological land types (ELT) and ecological land type phases (ELTP) developed for the Huron-Manistee National Forest, MI (Cleland et al. 1993) were evaluated for this study. Groups of ELTPs were chosen as a spatial and logistical compromise between ELTP's and ELTs. The statistical procedures used by Cleland et al. (1994) in development supported the contention that some ELTPs were similar and could be combined into groups without significantly compromising their usefulness.

*Description:* A case study was conducted to evaluate the effectiveness of using an ecological classification system to reduce sample variance in descriptions and predictions of wildlife habitat attributes. This would occur by supplementing inventories of information on existing vegetation (USFS vegetation information system maps) with information on potential site conditions as expressed by groups of USFS ecological land type phases. Existing vegetation conditions were classified according to the U.S. Forest Service's Corporate Database System. The system classified vegetation according to dominant commercial tree species, size, and stocking density. Permanent openings and wetlands were not differentiated. Overlays of the ELTP groups and the existing vegetation classification were used to further stratify the landscape, providing a time referenced template to guide habitat inventory. This template subsequently allowed inferences regarding successional trajectories, historical disturbance regimes, and the effects of management on understory species compositions.

Generally, vegetation attributes associated with the overstory (e.g., canopy cover, tree size, tree stocking) were sufficiently described using only the existing vegetation classification. However, precision was generally enhanced when using the ecosystem template for understory (e.g., shrub cover, shrub species composition) and ground level (e.g., herbaceous cover, downed woody debris) attributes. "Relative efficiency tests" measure the tradeoffs associated with increased costs because of more plots versus increased precision because of a better classification scheme. The results of these tests suggested that this is a cost effective approach for collecting understory and ground level information. Although trends from the relative efficiency tests supported the use of this approach, some tests suggested that microsite variation in physiography and soils further influences forest composition and understory recruitment. Also, the time since the last disturbance and the type of disturbance event have been demonstrated to have tremendous effects on understory and ground level vegetation. These inconsistencies relate to scale and should be addressed relative to the management or planning objectives.

### *Contact*

Gary J. Roloff, Timberland Resources, Boise Cascade Corporation, P.O. Box 50, Boise, ID 83728. 208-384-7761.

## 2.2 Desired Future Conditions

Ecological units are important in devising desired future conditions (DFCs) that can be attained and perpetuated. The varying responses of each ecological unit to an array of management activities means multiple outcomes are possible at any given site. Ecological unit descriptions are used to compare the present natural community to the potential community and the desired future condition. For example, at the end of a 50-year planning horizon, on the same site, a land manager has the option of establishing a 90-year-old longleaf pine forest with an open canopy, dense grass understory, and many snags; or, with a different management regime, a 40-year-old loblolly pine stand with a closed canopy, sparse understory and few snags. The choice made is the desired future condition. In the process of choosing, ecological classification provides information on site potential and response to management which allows each management scenario to be analyzed in terms of its economic efficiency, social and cultural acceptability, and ability to sustain healthy and productive ecosystems.

Ecological classifications evolved partially because of the recognition that several disciplines were collecting similar data for separate purposes such as rating productivity, identifying capability, susceptibility to various hazards, or suitability for specific activities such as road construction, log landings, cold water fisheries, farming, and range among others. Single resource inventories do provide high quality information and interpretations for a limited number of uses. The advantage of integrated ecological and resource inventories is that in the long run they reduce the overall data collection needs associated with multiple resource management, lend themselves easily to extrapolation of information from one unit to similar units, and they facilitate understanding of cause and effect relationships.

Suitability ratings combine information on potential productivity with information on the limitations imposed on management such as the cost of mitigation or decreases in productivity due to soil compaction, pollution, or erosion. Productivity, capability, and suitability ratings group areas that share a quality in common but which are not necessarily similar in other important ways. For example, two areas may be level and suitable for road building but have very different biological capabilities.

In local applications, ecological approaches to classification are desired to overcome limitations in the use of single purpose or limited purpose classifications or artificial rating systems. For example, vegetation based approaches do not give precise estimates of productivity and use constraints. They do not provide models of functional ecosystems necessary for understanding, describing, and predicting environmental effects such as the effect of acid rain impacts on long term site productivity and aquatic resources, etc. Soil and pure landtype or geomorphologic approaches are seldom informative enough to predict habitat and potential forage for range or wildlife management or to predict the presence of rare, threatened and endangered species. Monitoring of non-traditional uses of the forest such as the collection of medicinal herbs, may be helped if the ecological classification is used to establish correlative relations with species to be monitored.

Natural systems provide "reference conditions" which are used as a base of comparison with existing conditions to assess ecosystem health or with conditions predicted as a result of proposed management or natural events. The stability and integrity of ecological systems depend upon many intricate feedback mechanisms among life forms and the environment. Current ecosystem management applications are based on the assumption that the stability and functional integrity of an ecosystem is reflected in its composition and structure. Resilience is defined in light of natural change agents. Hence, ecological classification systems based on "natural conditions" are used to establish standards and criteria for measuring environmental quality and ecosystem integrity.

Ecological classification systems can be used to develop guidance on when management actions or land use allocations will irreversibly or irretrievably affect ecosystem potential. They can provide information useful in determining the costs and benefits associated with (1) managing within existing ecosystem capability, (2) enhancing natural capability through amendments, or (3) managing for species against the natural tendencies of the site. They also provide a yardstick for gauging the success of restoration activities.

Ecological unit inventories provide a framework for organizing, storing, and conveying information on various ecosystem parameters. Databases can be constructed from information associated with ecological types. Each type developed will have a unique list of attributes (e.g., spodic soil, marine deposits, white-cedar swamp), descriptive statistics (e.g., average annual precipitation, fire frequency, flood frequency, infiltration rate, etc.), interpretations (e.g., high erosion hazard, 45–55 site index, suitable for pond development, potential lynx habitat, etc.), and process models (e.g., succession, nutrient transformation pathways) associated with it. Inventoried ecological units provide information about the geographic location, distribution (percent of landscape), and spatial diversity of types within and among units. Ecological units may also be viewed as cartographic entities that can be linked to tabular data in a relational database.

The case studies in this section are drawn from different parts of the country. They provide examples of using ecological classification as a basis for predicting changes over time and predicting what conditions are prevalent where. *Case Study 4* uses empirical data to test FIBER 3.0, a growth and yield model specified and constructed using habitat types (ELTPs) in New England. *Case Study 5* describes studies to test the soil–landform–vegetation relations within landtypes and to verify productivity values associated with landtypes developed for the Interior Uplands in the southeastern United States.

Ecological classification and mapping can be combined with other information to set site specific or area specific standards and expectations. *Case Study 6* demonstrates adaptive management at the site level. Desired future conditions were modified on a specific land type to take advantage of natural ecosystem processes resulting in cost saving and greater environmental protection. *Case Study 7* presents the Boise Cascade Ecosystem Diversity Matrix which is used to establish regional baseline conditions and monitor change over time. *Case Study 8* discusses the use of reference sites to assess functional impairments in open water habitat in Lake Ontario.

## Case Study 4 in Desired Future Conditions: Fiber 3.0: An Ecological Growth Model for Northeastern Forestry Types — A use of ecological type concepts in modeling

### *Ecological Classification*

Habitat types (Leak 1982) in the northeastern US have been defined by landform, soils, and typical climax tree species following the multifactor approach of Hills (Hills and Pierpont 1960). The relationships between tree species and soil/landform conditions vary with climate and bedrock mineralogy and each habitat type exhibits a characteristic successional pattern, indicative of the tree species that will most likely regenerate and compete. Heavy cutting changes the successional stage, but not the characteristic successional sequence or climax forest type. Heavy disturbance, such as agricultural use and fire, may change the relationships of tree cover to soils and landform during the recovery period.

### *Description*

FIBER 3.0 is a revision of FIBER (Solomon et al. 1986b), a stand projection growth model developed to simulate the growth and structural development of forest stands across New England. The acronym stands for "Forest Increment Based on an Ecological Rationale." Predictions such as those from the FIBER 3.0 model are critical in efforts to maintain diversity and habitat conditions. The internal structure of the most recent version was constructed using six habitat types (Leak 1982) which expands the applicability of the model and improves its reliability over a wide range of sites. The habitat types specified are: sugar maple-ash, beech-red maple, oak-white pine, hemlock-red spruce, spruce-fir, cedar-black spruce.

To test the ability of FIBER 3.0 to accurately follow changes in forest structure, species composition, and wildlife habitat, over 700 non-disturbed USFS Forest Inventory and Analysis plots (FIA) across the state of Maine were classified into one of the six habitat types and modeled for 30 years. Comparisons between the actual remeasured and predicted values were made in 1959, 1972, and 1982 show good correspondence, validating the underlying assumptions of the model and, hence, the interpretations associated with each of Leak's habitat types. The comparison of the predicted growth rates and successional changes in species composition are demonstrated in graphic and tabular form (Fig. 3) (Table 2).

### *Contact*

Dr. Dale S. Solomon, USDA Forest Service, NE-4104, P.O. Box 640, Durham, NH 03824, 603-868-7666

## Case Study 5 in Desired Future Conditions: Terrestrial Classification and Inventory in the Highland Rim and Cumberland Plateau Physiographic Regions — Interpreting vegetation for management

### *Ecological Classification*

The system described here (Smalley 1986) was adapted from the Land System Inventory of Wertz and Arnold (1975). The five levels of the system are equivalent to the lower five levels of the National Hierarchical Framework of Ecological Units (Avers et al. 1994). The system is applicable to the Highland Rim/Pennyroyal and Cumberland Plateau physiographic provinces encompassing 29 million acres in parts of Tennessee, Alabama, Georgia, Kentucky, and Virginia. The development of the system can best be described as a process of successive stratification of the landscape based on the interactions and controlling influences of environmental factors — physiography, climate, geology, topography, and soils. Because the current species composition and structure of Rim and Plateau forests is more a function of repeated disturbances than an indi-

Table 2. Changes in species percentages of total stand basal area for spruce fir habitat type on USFS inventory and analysis plots in Maine using actual and FIBER 3.0 predicted values.

| | Year | Species | | | | | | | | | | | | | | |
|---|---|---|---|---|---|---|---|---|---|---|---|---|---|---|---|---|
| | | bf | rs | bs | ws | he | ce | wp | sm | rm | yb | pb | be | wa | as | total |
| Actual | 1959 | 25.1 | 25.7 | 0.4 | 3.4 | 0.6 | 11.0 | 1.3 | 2.6 | 9.1 | 5.6 | 3.9 | 3.0 | 1.1 | 3.4 | 100.0 |
| Predicted | 1969 | 28.0 | 25.6 | 1.4 | 3.6 | 0.8 | 8.1 | 1.2 | 3.1 | 8.1 | 4.9 | 3.5 | 2.5 | 1.4 | 4.0 | 100.0 |
| Actual | 1971 | 30.9 | 28.3 | 0.4 | 4.0 | 0.6 | 8.3 | 1.2 | 2.4 | 7.6 | 3.7 | 3.5 | 1.7 | 1.1 | 3.4 | 100.0 |
| Actual | 1982 | 29.8 | 28.7 | 0.4 | 4.4 | 0.6 | 7.2 | 1.5 | 2.4 | 8.4 | 3.4 | 3.9 | 1.4 | 0.7 | 4.2 | 100.0 |
| Predicted | 1984 | 31.8 | 26.2 | 2.5 | 4.0 | 1.0 | 6.0 | 1.1 | 3.4 | 7.3 | 4.0 | 2.7 | 1.7 | 1.8 | 3.3 | 100.0 |
| Predicted | 2004 | 35.2 | 27.7 | 3.4 | 4.2 | 0.9 | 4.8 | 0.9 | 3.2 | 6.3 | 3.3 | 2.2 | 1.1 | 1.7 | 2.6 | 100.0 |

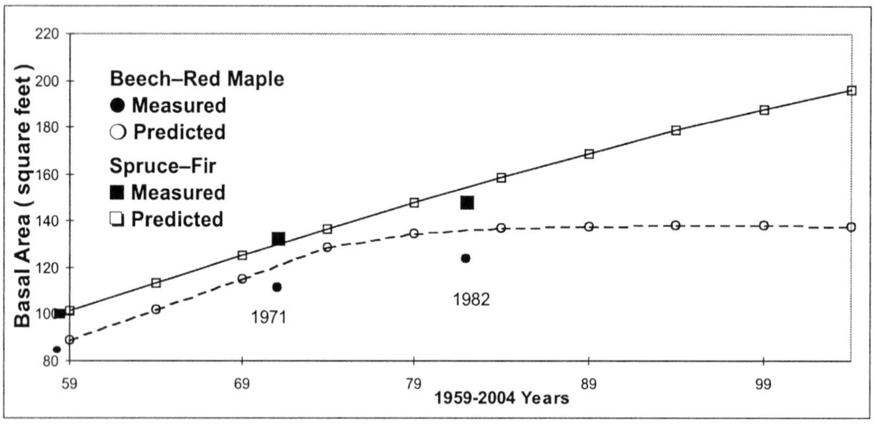


Fig. 3. Comparison of actual and FIBER 3.0 predicted average basal areas on USFS Inventory and Analysis plots in Maine for 2 different forest habitat types.

cation of succession and site potential, vegetation was relegated to a minor role in the development of the land classification system.

The most detailed level (landtype) is mapped at a scale of 1:24,000; individual units may vary from 5 to 100 plus acres depending on topography. To date (Sept. 1996) about 150,000 acres of State Forest and Wildlife management Areas (Smalley et al. 1996) and 300,000 acres of forest industry lands have been mapped at the landtype level. Average cost is about $0.25 per acre. The mapping process (mylar sheets over 1:24,000 quadrangle maps) has reinforced Rowe's maxim "...that every part of the terrain has to be confronted; there is no avoiding those in-between and oddball units..." Thus, as the survey progressed, landtype descriptions were refined and new landtypes identified and described. The system has been extended to the Upper Coastal Plain of west Tennessee and the southern Allegheny Plateau in mid- and northern Kentucky. After the map units are entered in a GIS, the next step in developing management plans will be to merge the landtypes with existing plant community information.

### *Description*

Recent efforts have been directed toward testing the soil–vegetation–landform relationships of the landtypes. An intensive study of the soils and vegetation on three major landtypes on the Mid-Plateau near Crossville, TN revealed that landtypes significantly affected magnitudes of temporal and spatial soil variability (Hammer et al. 1987). The morphological features of soils, when precisely described and interpreted with respect to landtypes, are indicators of patterns of movement and relative amounts of available soil moisture and can be a valuable aid in predicting potential forest site productivity. The land classification system for the Mid-Plateau groups forest soils into landform units having relatively homogeneous chemical and physical properties.

Plant community–landform relationships have been studied on the 26,000 acre Prentice Cooper State

Forest and Wildlife Management Area on the south end of Walden Ridge (Mid-Plateau) west of Chattanooga, TN (Arnold et al. 1996) Although the techniques used did not permit the development of predictive models, relatively discreet plant communities were found to occur on four major landtypes. Apparently, the land classification system for the Mid-Plateau divides the landscape into logical, ecologically distinct units.

Wheat and Dimmick (1987) studied plant community–landform relations on two Western Highland Rim sites. Three ridge landtypes supported similar communities; distinct communities were found on north slopes with limestone chert, south slopes with limestone chert, and in stream bottoms having good drainage.

Clatterbuck (1996) attempted to classify the vegetation on the 19,901-acre Cheatham Wildlife Management Area as a basis for multiple resource planning, including wildlife habitat management. The land classification system for the Western Rim provided a useful initial stratification of the landscape for plant community analysis. However, to gain a better understanding of the diverse upland deciduous forest, it was necessary to further aggregate and segregate the vegetation and landform variables. Probably, the lack of a strong relationship between plant communities and landforms was due to past disturbances by fire and timber harvesting for charcoal production.

### Contact

Glendon W. Smalley, Consultant; Retired Research Soil Scientist, USFS So. Forest Experiment Station; Adjunct Professor, Dept. of Forestry, Wildlife and Fisheries, University of Tennessee and Department of Forestry and Geology, University of the South, Sewanee, TN. 102 Rabbit Run Lane, Sewanee, TN 37375-2753, 615-598-5714.

## Case Study 6 in Desired Future Conditions: Classification, Inventory and Monitoring of Desired Future Conditions on Range Land in the Ashley National Forest, Utah — Managers use units in adaptive management

### Ecological Classification

In the 1970s a Land Systems Inventory was developed on the Ashley National Forest. Elements of this approach are discussed by Godfrey and Cleaves (1991) and Godfrey (1977). The terminology and scale are now consistent with the National Hierarchy of Ecological Units (Avers et al. 1994). Landtype associations (LTAs) included information on geology, geomorphology and geomorphic processes that were useful in this application. Landtypes, smaller units, provided more detail on other features such as slope, soils, and vegetation.

### Description

Components of an ecological approach to management might include the following: classification, inventory, capabilities of land units, values of land units, a published decision stating desired condition and actions to achieve desired condition. Monitoring is included to see if these actions were taken, to see if the desired condition was achieved by those actions, and to see if the desired condition and associated actions are appropriate over time.

Ashley National Forest managers in the 1960s decided that some canyon bottoms should be managed to include graminoid–forb communities with high values for livestock forage and watershed protection. This decision was based upon the dominant traditional use of that portion of the Forest. This same desired condition was specified in the Forest Plan adopted in 1986. Though there was no ecological classification, canyon bottom lands with obviously deeper soils than adjacent slopes were identified because the potential production was higher and these areas were also suitable for cattle grazing.

Later, two landtype phases were identified for these canyon bottoms through systematic land classification and inventory One phase occurs as fans at the base of drainages formed from the sediment washed from steep, erosive side slopes of an adjacent landtype. Another landtype phase is on the wider, flat bottoms where alluvial (water laid) deposition parallels the drainages. Both landtype phases were plowed and seeded in the early 1960s. By the 1990s the seeding on the fans had been 70–90% covered with eroded sediment and the seeded species were replaced by native species well adapted to disturbance including Salina wild rye. On the bottom land type, seeded species persisted as dominants or at least as understory dominants with sagebrush and rubber rabbit brush.

In the 1960s when the seeding was planned and completed, the Ashley National Forest had no information on the rate of sediment deposition on the fans. By the 1990's classification and inventory was available, and monitoring studies documented the contrasting status of the seeding on the two landtype phases. As a result the desired future condition was changed on the fan landtype. Instead of seeding the fans, it was decided that the presence of Salina Wild rye which was naturally abundant there, should be the basic desired condition for watershed and ungulate forage. The change was based on inherent features of the land and economic values. Livestock grazing would likely have been considered the primary factor of vegetation change if comparative information on geomorphic processes and associated plant succession among identified landtypes had not become available. While grazing had *some* influence, this was extremely minor compared to geomorphic processes.

To provide for future decisions, monitoring efforts must remain active. Additional observations and studies indicate a need for greater refinement of the classification and inventory of this canyon bottom landtype. In addition, new values have emerged. Analysis based on a landscape approach in the 1990's validates the value of these bottoms for ungulate forage with adjacent side slopes providing cover for wildlife. Elk, which were absent or rare in the 1960s, have become relatively abundant.

### Contact

Sherel Goodrich, Ashley National Forest, 355 N. Vernal Ave., Vernal, Utah 84078, 801-789-1181, Fax 801-759-1181.

## Case Study 7 in Desired Future Conditions: Boise Cascade Ecosystem Diversity Matrix — Establish regional baseline conditions and monitor change over time

### Ecological Classification

The strategy demonstrated in this application requires the use of a land classification system that identifies inherent variability in the physical environment and then influences the plant and animal associations for any given site. Either ecological land types or habitat types (Daubenmire 1968) may be used as long as the units of land are described in a hierarchical fashion with each succeeding level becoming more homogeneous to satisfy the need for increasing specificity. This example uses habitat types nested within the Southern Batholith of Idaho (Haufler et al. 1996). Boise Cascade is using the Section, and in some areas, groups of Subsections from ECOMAP (Avers et al. 1994) to bound the development and application of individual matrices.

### Description

The ecosystem diversity matrix (Fig. 4) classifies landscapes based on existing vegetation structure (the y-axis), potential vegetation structure (the x-axis), relative moisture and elevation gradients (generally, dry, low elevation to more mesic, high elevation as one proceeds from left to right on the x-axis), and primary historical disturbance regime (note the two successional trajectories depending on disturbance history). Existing vegetation conditions are described in sufficient detail to allow differentiation of biological communities at a scale compatible with land planning objectives (e.g., a forest stand as delineated by homogenous overstory vegetation).

Ecosystem management means blending an understanding of natural disturbance regimes with appropriate management tools to provide for both biodiversity and resource use. Historical ranges of variability provide essential information for understanding natural disturbance regimes and for evaluating the status and health of existing stands of vegetation (with this understanding being a guide rather than a goal for desired future conditions). Information on historic range of variability can be used to help identify successional stages that were in significant abundance or areas that typically supported substantial acreages of old growth. One way of describing a desired future condition for ecosystem diversity is to assign an areal percentage to each type/growth stage combination in the matrix. The matrix forces planners to recognize the dynamic processes at work in the landscape, and to incorporate a temporal component into the planning process. Another use of the matrix may be to track acres meeting certain compositional and structural requirements within each unit (i.e., condition).

### Contact

Gary J. Roloff, Timberland Resources, Boise Cascade Corporation, P.O. Box 50, Boise, ID 83728

## Case Study 8 in Desired Future Conditions: Assessing Open Water Habitat Conditions in Lake Ontario —Using reference sites to assess functional impairments

### Ecological Classification

The Aquatic Habitat Classification System (AHCS) was developed to supplement the Cowardin et al. (1970) approach to classification (Busch and Sly 1992). The AHCS provides information on ecological processes that help us assess the functions performed by habitat units in support of fish or wildlife in the Lake Ontario Basin. Here we discuss the open water and near-shore subsystems (Fig. 5). The near shore subsystem reaches to the 25 m contour as (1) that is the maximum depth to which wave activity exerts its influence (Sly 1991) and (2) thermocline development in Lake Ontario is restricted to the top 25 m (Sly 1991).

### Description

Functional impairments reflected in biological, chemical, or physical stresses were evaluated for 88

# ECOSYSTEM DIVERSITY MATRIX-
# IDAHO SOUTHERN BATHOLITH LANDSCAPE
## FORESTED SYSTEMS

Habitat Type Class

| VEGETATION GROWTH STAGES | Dry Ponderosa Pine Xeric Douglas-Fir: Potential Cover Types | Acres | Warm, Dry Douglas-Fir Moist Ponderosa Pine: Potential Cover Types | Acres | Cool, Dry Douglas-Fir: Potential Cover Types | Acres | Dry Grand Fir: Potential Cover Types | Acres | Warm, Dry Subalpine Fir: Potential Cover Types | Acres | High Elevation Subalpine Fir: Potential Cover Types | Acres |
|---|---|---|---|---|---|---|---|---|---|---|---|---|
| Grass/Forb/Seedling | | | | | | | | | | | | |
| Shrub/Seedling (L, M, H) | *Pinus ponderosa* | | *Pinus ponderosa* (*Pinus contorta*) (*Populus tremuloides*) | | (*Pinus contorta*) (*Populus tremuloides*) | | *Pinus ponderosa* *Pseudotsuga menziesii* (*Populus tremuloides*) | | *Pseudotsuga menziesii* (*Pinus contorta*) (*Populus tremuloides*) | | *Pinus albicaulis* | |
| Sapling; shrub/seedling (L, M, H) | *Pinus ponderosa* | | *Pinus ponderosa* (*Pinus contorta*) (*Populus tremuloides*) | | (*Pinus contorta*) (*Populus tremuloides*) | | *Pinus ponderosa* *Pseudotsuga menziesii* (*Populus tremuloides*) | | *Pseudotsuga menziesii* (*Pinus contorta*) (*Populus tremuloides*) | | *Pinus albicaulis* | |
| *WITH HISTORICAL UNDERSTORY FIRE REGIME* | ***Understory Burn 5 -25 Years*** | | ***Understory Burn 10-22 Years*** | | ***Understory Burn 25-100 Years*** | | ***Understory Burn 10-30 Years*** | | ***Fire Mosaic 50-90 Years*** | | ***Understory Burn 25-70 Years*** | |
| Small Trees Multi-story (L, M, H) | *Pinus ponderosa* | | *Pinus ponderosa* (*Pinus contorta*) (*Populus tremuloides*) | | (*Pinus contorta*) (*Populus tremuloides*) | | *Pinus ponderosa* | | *Pseudotsuga menziesii* (*Pinus contorta*) (*Populus tremuloides*) | | *Pinus albicaulis* | |
| Medium Trees Multi-story (L, M, H) | *Pinus ponderosa* | | *Pinus ponderosa* (*Pinus contorta*) | | *Pseudotsuga menziesii* (*Pinus contorta*) | | *Pinus ponderosa* | | *Pseudotsuga menziesii* (*Pinus contorta*) | | *Pinus albicaulis* | |
| Large Trees Multi-story (L, M, H) | *Pinus ponderosa* | | *Pinus ponderosa* | | *Pseudotsuga menziesii* | | *Pinus ponderosa* | | (*Pinus contorta*) | | *Pinus albicaulis* | |
| *WITH HISTORICAL STAND DESTROYING FIRE REGIME* | ***Stand Destroying Wildfire Unlikely*** | | ***Stand Destroying Wildfire Unlikely*** | | ***Some Stand Destroying Wildfire*** | | ***Stand Destroying Wildfire Unlikely*** | | ***Fire Mosaic 50-90 Years*** | | ***Stand Destroying Wildfire*** | |
| Small Trees Multi-story (L, M, H) | *Pinus ponderosa* (*Pseudotsuga menziesii*) | | *Pinus ponderosa* (*Pinus contorta*) (*Populus tremuloides*) (*Pseudotsuga menziesii*) | | *Pseudotsuga menziesii* (*Pinus contorta*) (*Populus tremuloides*) | | *Pinus ponderosa* *Pseudotsuga menziesii* *Abies grandis* (*Populus tremuloides*) | | *Pseudotsuga menziesii* (*Pinus contorta*) (*Populus tremuloides*) | | *Pinus albicaulis* | |
| Medium Trees Multi-story (L, M, H) | *Pinus ponderosa* (*Pseudotsuga menziesii*) | | *Pinus ponderosa* *Pseudotsuga menziesii* | | *Pseudotsuga menziesii* (*Pinus contorta*) | | *Pinus ponderosa* *Pseudotsuga menziesii* *Abies grandis* | | *Pseudotsuga menziesii* *Picea engelmannii* (*Pinus contorta*) | | *Abies lasiocarpa* *Pinus albicaulis* *Picea engelmannii* | |
| Large Trees Multi-story (L, M, H) | *Pinus ponderosa* (*Pseudotsuga menziesii*) | | *Pinus ponderosa* *Pseudotsuga menziesii* | | *Pseudotsuga menziesii* | | *Pinus ponderosa* *Pseudotsuga menziesii* *Abies grandis* | | *Picea engelmannii* *Abies lasiocarpa* | | *Abies lasiocarpa* *Pinus albicaulis* *Picea engelmannii* | |
| OLD GROWTH (L, M, H) | *Pinus ponderosa* (*Pseudotsuga menziesii*) ≥ 5 trees ≥ 24" dbh/ac. Snags - infrequent | | *Pinus ponderosa* *Pseudotsuga menziesii* ≥ 10 trees ≥ 24" dbh/ac ≤ 1 Snag ≥ 20" dbh/ac | | *Pseudotsuga menziesii* ≥10 trees ≥ 18" dbh/ac ≤ 3 Snags ≥ 16" dbh/ac | | *Abies grandis* ≥ 15 trees ≥ 24" dbh/ac ≥ 2 Snags ≥ 20" dbh/ac ≥ 2 Pieces ≥ 12" dbh/ac | | *Abies lasiocarpa* *Picea engelmannii* ≥ 25 trees ≥ 24" dbh/ac ≥ 2 Snags ≥ 12" dbh/ac | | *Abies lasiocarpa* *Pinus albicaulis* *Picea engelmannii* ≥ 10 trees ≥ 12" dbh/ac | |

Fig. 4. Ecosystem Diversity Matrix for the Idaho Southern Batholith (Haufler, et al. 1996).

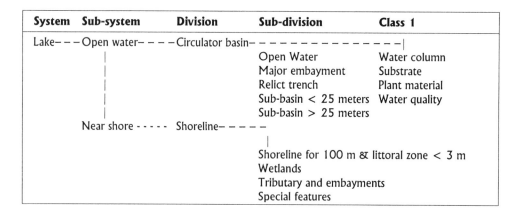

| System | Sub-system | Division | Sub-division | Class 1 |
|---|---|---|---|---|
| Lake | Open water | Circulator basin | | |
| | | | Open Water | Water column |
| | | | Major embayment | Substrate |
| | | | Relict trench | Plant material |
| | | | Sub-basin < 25 meters | Water quality |
| | | | Sub-basin > 25 meters | |
| | Near shore | Shoreline | | |
| | | | Shoreline for 100 m & littoral zone < 3 m | |
| | | | Wetlands | |
| | | | Tributary and embayments | |
| | | | Special features | |

Fig. 5. Simplified diagram of Aquatic Habitat Classification System (AHC).

habitat categories (Class level — Sly and Busch 1992). Stress factors were developed from literature sources and from consultations with natural resource managers (Busch et al. 1993). The criteria for rating each type of stress were (1) the severity of the ecological impact, defined as a significant change or shift in the efficiency or direction of the energy flow between trophic levels, and (2) the expected permanence of the stress defined by time (week, season, year, decades, or permanent). The Lake Ontario habitat information inventory available from Busch et al. (1993) provided the information base and a list of functionally distinct habitat units.

After a specific habitat unit was delineated, the degree of impairment for each category (physical, chemical, biological) was determined using the Delphi technique (Zuboy 1981; Crance 1987). An effort was made to separate the natural from the anthropogenic restrictions. The functional concerns were addressed by comparing impacted areas to reference sites within the basin that have maintained their structure and are able to support ecosystem functions needed for a healthy state (Martin 1994).

The habitats making up the Lake Ontario ecosystem between roughly 1960–1990 (focus 1970–90), were impaired, functioning at 50% of the level of unimpaired habitats (Busch and Lake 1996). The impairments were caused almost equally by biological, chemical and physical stressors. Biological stresses were most severe in the "open water" habitats which comprised 67% of the basin's surface. The cause was a dramatic increase in distribution and abundance of exotic species, most notably the sea lamprey and zebra mussel. (Environment Canada and USEPS 1995). Other contributing factors included artificial changes in primary production and instability within the native fish community caused by loss of native species such as lake trout, Atlantic salmon, blue pike (*Stizostedion vitreum glaucum*), and deepwater sculpin.

Chemical stresses were highest in the "tributary and embayments" habitat category. The impacts include fish tumors, wildlife deformities, and degradation of aquatic biota caused by chemical accumulation from the sediments or watershed (Hartig and Law 1994, Koonce et al.). Physical stresses were primarily from physical and water-flow changes caused by hydropower development, construction of harbor facilities, and maintenance dredging for harbors in the tributaries (Smith 1995). Remaining shoreline littoral and wetland habitats were not identified as being heavily stressed.

*Contact*

W. Dieter N. Busch, Lower Great Lakes Fishery Resources Office, USFWS, 405 N. French Rd., Amherst, NY 14228; Phone: 716-691-6154.

## 3 LANDSCAPE, WATERSHED, AND REGIONAL PLANNING

Ecological units are used to characterize landscapes, watersheds, and regions for planning and to provide a context for their analysis. Ecological units at each scale, internalize vertical and horizontal structure and the functional relationships among ecosystem components over time. Thus, they provide a basis for predicting the response of the ecosystem to various natural events and management over time. Hierarchical frameworks of ecological units integrate units of multiple scales by nesting small units within larger ones. The various patterns recognized in the composition, distribution, and successive arrangement of small units are used in ecosystem management applications to characterize the structure of the larger system they nest within (O'Neil et al. 1986). The theories of landscape ecology are used to interpret the natural variability displayed by this hierarchical ecosystem organization. Ecoregions, landscapes, and watersheds provide a context for understanding dynamic ecological processes such as disturbance regimes and nutrient cycling that the aggregation of smaller units into larger ones by ownership or political boundaries cannot provide.

## 3.1 Ecosystem-Based Planning and Assessment

Although ecological units cannot provide all the information needed for planning and decision-making, they provide a logical basis for examining the complexity, interdependencies, and interactions among societal needs, myriad ecological processes, existing conditions, and the range of possibilities given ecological potentials. Often the stakeholders of public land do not agree on the management objectives of a project or even fully understand the benefit of or "need" for management. Ecological classification, mapping, and integrated inventories provide baselines to model the potential effects of multiple management scenarios at local, landscape, and regional scales. They also provide the spatial and temporal contexts within which "ecologically informed" decisions can be made. Figure 6 identifies ecological units as the basic template for integrating information in ecosystem-based planning.

Both spatial and temporal sources of variability are important when evaluating the environmental effects of various management alternatives; therefore, planning and assessment activities generally require consideration of several scales simultaneously. For example, to discern the cumulative effects of timber harvest over time, we need to examine conditions and processes occurring above and below the level where activity is being considered. Changes in plant community composition and age-class structure are immediately evident within the harvested area. Changes in vernal pool habitat can be detected at micro-sites, the fragmentation of breeding bird habitat would be noticeable at landscape or regional scales, and changes in runoff would be detected within a watershed context.

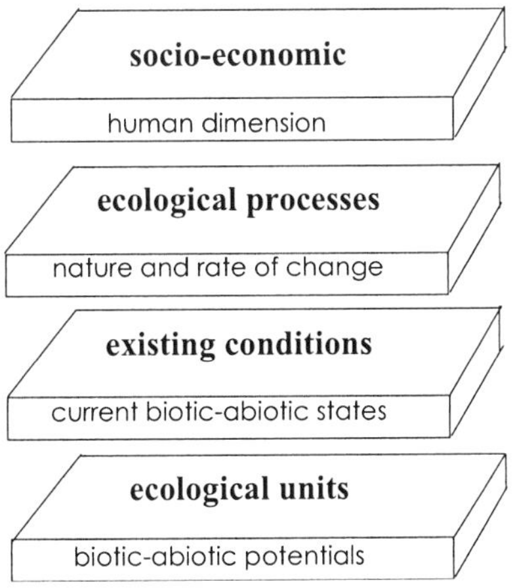


Fig. 6. Ecological units developed at appropriate scales provide an ecological context for examining the complexity, interdependencies, and interactions among societal needs, myriad ecological processes, existing conditions, and the range of possibilities given ecological potentials (Cleland et al. 1995).

Ecological units provide a basis for establishing and testing assumptions related to ecosystem form and function. Ecosystem conditions and thresholds to change defined in this process are used to model and predict natural responses. Then, managers can identify and evaluate cumulative effects related to the timing and distribution of management activities. The accuracy of this effects analysis will depend upon the degree to which (1) the dynamic and functional relationships among biotic and abiotic ecosystem components have been established and (2) they are conveyed through mapping and interpretation of the units. The amount of information conveyed by any map relates to both the level of detail incorporated in the classification and the level of resolution possible at a given mapping scale. As the scale of mapping increases, the amount of variability contained within each unit increases and details are generalized. The ecological unit map compilation scale and the natural scale of the phenomenon to be analyzed should be similar.

A recurring task in implementing ecosystem management is determining the scope and intensity of planning and analysis activities. The areas of analysis must be bounded geographically in order to identify the amount of information and resources necessary for conducting tasks. Commonly this involves considerations of land ownership, legislative or administrative policy, public issues and concerns, and ecological processes or need. Ecological units can be used to determine the geographic extent of planning and analysis activities by linking public issues, management questions, and environmental needs to appropriate units at appropriate scales. Ecological units can be aggregated hierarchically as classified or non-hierarchically according to the characteristics and features significant to resolve an issue.

Comprehensive, intensive ground surveys of environmental conditions and ecosystem potential are not currently available across all ownerships for broad scale, strategic planning. Frankly, support for such an approach is limited due to a number of concerns. One is that important decisions cannot or should not be postponed until comprehensive surveys are completed. There is concern over the high up-front costs of ground survey, worries about waste due to the collection of unnecessary information or information with limited analytical value, and fear of government intrusion.

Probst and Thompson (1996) recommend the use of top-down, holistic, iterative approaches that use concepts developed from detailed studies and analyses as a more economical, pragmatic means to meet the need for information for policy-making and strategic planning. A multi-scaled, spatial framework that establishes an ecological context for data synthesis and analysis is a critical component of this approach. Multi-scaled, because comprehensive assessments of economic, social, and environmental conditions are achieved through a process of successive approximations. Ultimately the underlying assumptions are tested through monitoring and evaluation, including field sampling. The virtue of this approach is that assumptions and data collection needs are clearly stated. Hierarchical frameworks of ecological units provide a vehicle for integrating information spatially and across multiple scales.

Beyond site level planning, in landscape and regional planning, the use of ecological units multiplies. Ecological classification contributes to multiscale and multidisciplinary planning and assessment activities. Time and cost efficiencies are realized and coordination is improved due to common terminology and maps. *Case Study 9* presents an approach to multiscale planning on the Ottawa National Forest, MI. *Case Study 10* from the Lowman Ranger District, ID, demonstrates the use of terrestrial and aquatic classification in watershed analysis. *Case Study 11* provides an example of a multi-scale assessment of neotropical migratory birds distribution using the concept of successive top-down approximation.

## Case Study 9 in Ecosystem Based Planning and Assessment: Ottawa National Forest Planning — An approach to multiscale planning

### *Ecological Classification*

Work on the Ottawa National Forest Ecological Classification and Inventory began in the early 1970s. Initial development followed a multifactor site classification approach. The Eastern Region then adopted the nested hierarchical concept of Wertz and Arnold (1975) which has now evolved to the National Hierarchical Framework (Avers et al. 1994). The principle of simultaneously integrating multiple factors rather than using predetermined classifications (e.g., soil) followed and is continuing. Nearly 1.5 million acres of public and private lands within the forest have been classified and mapped to date.

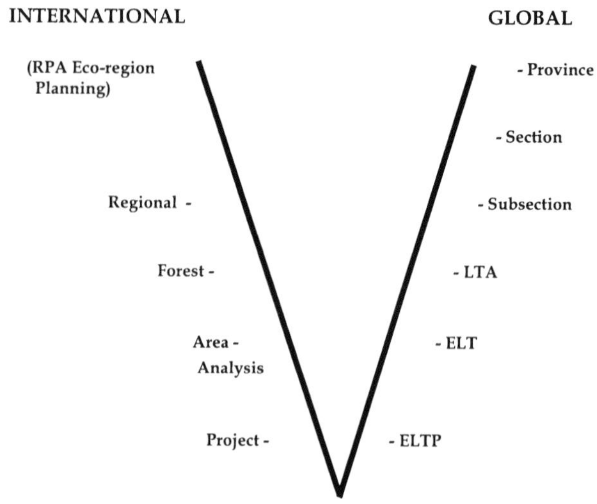


Fig. 7. Decision levels and associated ecological information levels used in the Ottawa National Forest Plan.

### *Description*

Forest Level Planning resulted in the long-term allocation of future forest conditions to large units of land with activities and outputs scheduled by decades. The forest planning model (Fig. 7) was developed to aid forest level, management area, and project level analysis and decision-making (Jordan et al. 1984). The model incorporated the use of ecological units at three scales: the landtype association (LTA), landtype (ELT), and landtype phase (ELTP). The data, maps and interpretations associated with each unit enhanced the ability of the planning team to model existing and potential resource conditions, management practices, management standards, costs, resource yields and environmental effects at multiple scales. The use of a nested geographic system assured that decision choices made at each level were guided by a common set of assumptions and relationships.

Management Areas (MAs) are areas dedicated to a specific set of land uses compatible with a long-term desired future condition (DFC) described in terms of vegetative type composition objectives, planned recreation opportunity spectrum class, desired road density, commodity production and wildlife emphases among others. Prescriptions developed for each area identify the standards, guidelines, and activities to be carried out toward this condition. LTAs were used to identify areas suitable for desired uses and capable of meeting desired future conditions in the manner prescribed. The criteria used in this analysis included: existing vegetative composition, tree species potential, potential productivity, percentage composition by site unit (ELTs), existing recreation opportunity class, existing road density, unique wildlife habitat potential, land ownership pattern, road construction cost, exist-

Table 3. Examples of management area prescriptions and associated LTAs.

| LTA | Acres | Management Prescription | | | | | |
|---|---|---|---|---|---|---|---|
| | | 1.1 | 2.1 | 3.1 | 3.2 | 4.1 | 6.1 |
| 1 | 28,788 | X | | X | X | X | |
| 2 | 178,478 | X | X | X | X | X | X |
| 3,4,7 | 134,910 | | X | X | X | X | X |
| 5 | 50,691 | | X | | | | X |
| 6 | 56,705 | X | X | X | | | X |
| 9,10 | 79.420 | | X | X | X | | X |
| 11 | 47,016 | | X | X | X | | X |
| 12,13 | 66,012 | X | X | X | X | | X |
| 14,17 | 92,243 | X | | X | | X | |
| 14A | 11,100 | | | | | X | X |
| 16,19 | 70,567 | X | | X | | | X |
| 18 | 12,156 | X | | X | | X | X |

ing and potential wildlife habitat, specific public issues, existing sensitivity levels, and existing visual quality objectives.

The Ottawa National Forest made a decision to coincide LTA and management area boundaries for analysis and land allocation purposes. The results of the suitability analysis displayed in *Table 3* illustrates how this worked. Production of high quality hardwoods was a part of the prescription 2.1 and the table illustrated that this objective could be achieved on any of eleven LTAs. High quality saw timber is not likely to occur in LTA 1 where the soils are dry and sandy nor on the rest of the LTAs excluded from Rx 2.1 for a variety of environmental reasons.

From another viewpoint, we know LTA 2 could produce high-quality hardwoods as specified in management prescription 2.1, but it can also meet the conditions specified in any of the other prescriptions. This type of analysis allowed for flexibility in developing alternatives to meet a range of societal and economic needs while eliminating from consideration those areas without a natural capacity to meet a certain need. Another benefit of bounding management areas by LTA is that the standards and guidelines governing management activities could be tailored for a good fit. Key variations within and among management area prescriptions and LTAs were represented in FORPLAN, the optimization model most commonly used during the first round of forest planning. Forest plan analysis was conducted at the LTA scale. Data about resource limits, management costs and product yields were drawn from site, compartment, land type, and land type phase data and aggregated to the LTA scale.

Implementation of the forest plan required additional analysis at the opportunity area level and at the project level. The land types provided capability information at the opportunity area, which helped determine the location of long-term local road corridors, identify operating periods and appropriate road standards, locate areas suited for hardwood saw timber, softwood saw timber, aspen, softwood pulpwood, hardwood pulpwood, and hemlock based on ecological potential; determine areas of even-aged and uneven aged management of northern hardwoods relative to vegetation management objectives for the opportunity area; compare possible wildlife habitat component opportunities and their spatial arrangement.

The land type phase provided detailed information for project layout and design and was used in conjunction with information on existing conditions. It aided in choosing site specific practices, species regeneration options and methods, harvest layout and methods, local road standards, potential productivity by tree species, to identify opportunities for wildlife habitat improvement, etc.

Ecological classification has provided a valuable framework for integrating information and evaluating management alternatives at forest wide, area wide, and project levels on the Ottawa National Forest. Lack of surveys in some areas and database limitations constrained the use of ecological units during the first round of forest planning. The Ottawa NF has continued field survey activities at the land type level in the intervening years. The expanded information and computer databases with enable more integration in the upcoming forest plan revision process.

*Contact*

James K. Jordan, Ottawa National Forest, E6248 U.S. Highway 2, Ironwood, MI 49938; (906) 932-1330.

## Case Study 10 in Ecosystem Based Planning and Assessment: Deadwood Landscape Approach with the National Hierarchical Framework — A ground-up analysis of structure, composition, and function

*Ecological Classification*

Ecological units provide critical information to maintain ecosystems within limits compatible with both present human needs and the capacity of the ecosystem to provide these and future needs. Ecological units representing all scales of the USFS terrestrial hierarchy (Avers et al. 1994) and the riverine portion of the aquatic hierarchy (Maxwell et al. 1995) were used to frame this analysis. The Deadwood Landscape is located within Section M332A - the Idaho Batholith (Bailey 1994). Habitat type classes (Teck and Steele 1995), landtype phases, and channel reach types were nested within larger terrestrial and aquatic units during characterization and analysis.

*Description:* Analyses are seldom carried out at only one scale because one size does not fit all needs. The Deadwood Assessment can be classified as a mid-scale, landscape or watershed assessment that draws upon information at multiple levels of both the terrestrial and aquatic hierarchies for information and context. This analysis complies with NFMA requirements and determines opportunities to be carried into the NEPA process. The analysis process itself combines recommendations from the Federal Guide for Watershed Analysis (USFS 1995) and Forest Landscape Analysis and Design (USFS 1992).

Aquatic and terrestrial ecological units were important in characterizing the watershed and identifying reference conditions to be used in the synthesis and interpretation phase of the project. During synthesis and interpretation, desired future conditions, resource capabilities, sensitive species, sensitive areas, and local constraints and concerns are weighed and balanced against each other. Ecological units will also be critical as the project moves into the design phase and the determination of site specific and cumulative effects.

An experimental approach was developed to link aquatic and terrestrial ecosystem components, from the ground up, into correlated mapping units to be consistently described, mapped, and extrapolated across the 153,000 acre Deadwood Watershed. The various aspects of this analysis are described below.

- Ten habitat types provided the framework to organize vegetation attributes available from timber stand examination and a "Most-Similar-Neighbor" sample inference procedure (Moeur et al. 1995). Twenty-six vegetation growth stages (RMSTAND) were defined using DBH (diameter at breast height), size class, canopy closure class, and vertical structure. Seral stages identified were sorted into early seral, mid-seral, late seral, and climax classes for trees and understory.
- Fire history and fire scar analysis demonstrated the role of fire with a diversity of seral stages. This information, along with knowledge of seral stage by habitat type, was applied to classify areas by historic fire regime. Insect and disease hazard rates were determined. Cumulative effects were calculated using the Prognosis model (FVS) and GIS (Teck and Steele 1995). The GIS query results were incorporated in the Ecosystem Diversity Matrix of the Idaho Southern Batholith Section (See Case Study 6). A diversity matrix is under development for riparian ecosystems.
- Analysis of fishery potential involved a comparison of streams in natural conditions (Overton et al. 1995) to streams with similar geology, landtypes, and channel reach types found in the Deadwood watershed. A hierarchical approach was used to evaluate information by watershed, subwatershed, and channel reach type and by fish habitat attributes.
- Landtypes provided the silviculturist and other resource specialists information on soils, vegetation, hydrology and management qualities such as roads, wood, water, forage, recreation. This provided the silviculturist, hydrologist, and fisheries biologist with a common ecological language.
- Land type associations, nested within the watershed, were attributed with stream and existing vegetation coverages in addition to the performance characteristics identified and associated through the land systems inventory and the channel reach typing. Performance characteristics are measurable attributes such as soil productivity, in stream fine sediment, hill slope erosion, stream width-depth ratios, in stream large woody debris, and structure, composition, and function of terrestrial and riparian vegetative habitat type classes. Dominant habitat types were identified for each of twelve landtype associations. This attribution related vegetation characteristics to the landtype and also soil erosion hazards.
- Four subsections in the Deadwood landscape were evaluated for potential vegetation in a terrain model used in the Columbia River Basin Assessment (1994). This provided a means to evaluate site potential at a larger geoclimatic setting than the LTA scale.
- The Deadwood landscape analysis used GIS to link models and evaluate various geographical orientations and functions of terrestrial components. Through analysis the forest was able to identify reference ranges of variability (RRV) for selected ecological units based on the inherent land capability. These are also useful for mapping ecological units according to the National Hierarchy (Avers et al. 1994).

### Contact

Melody Steele, Lowman Ranger District, Boise National Forest, HC77 Box 3020, Lowman, ID 83637; (208) 259-3361; Fax 364-3366.

## Case Study 11 in Ecosystem based planning and assessment: Assessment of the Ecological Distribution of Midwestern Neotropical Migratory Birds (NTMB) — A multiscale assessment using a top-down successive approximation approach

*Ecological Classification*

Ecological provinces defined as part of the National Hierarchical Framework of Ecological Units adopted by the USDA Forest Service (Bailey et al. 1994, McNab and Avers 1994) were used in this example.

*Description:* Conservation of neotropical migratory birds (NTMBs) is a concern throughout North America. In midwestern North America (defined as 16 states and three Canadian provinces), biologists and conservationists recognize that species viability cannot be insured by evaluating and improving local habitats if conditions and influences outside the region do not support critical life functions. Thus local efforts should fall within a general conservation plan that is applicable throughout much or all of a species range (Thomas et al 1990, Probst and Wienrich 1993). A multi-scale assessment of the geographic and ecological distribution of midwestern NTMBs was conducted to elucidate the relationships among local, regional, and continental conditions and populations at those corresponding scales.

The U.S. Fish and Wildlife Partners in Flight (PIF) database (1980) was modified to identify 187 NTMBs that breed in the Midwest and 47 regional high-priority species for assessment. These priority species represent diverse taxonomic groups using a wide range of habitats. The Breeding Bird Survey (BBS) identified 57 Midwestern species that are declining nationally; trends in the midwest may be important to 47 of these. Ecological provinces provided a meaningful context for aggregating and summarizing data from the PIF physiographic database and provide a basis for relating that to the area and distribution of ecosystems and the trends in vegetation, succession, land use, and landscape structure.

Table 4 summarizes the NTMB numbers by habitat within each province. Eleven habitat classes were developed by pooling vegetation classes from 1-km resolution AVHRR imagery for analysis and these were collapsed into six for this table. The habitat map is a general survey suited only for assessing large-scale patterns as the habitat types contained mixed vegetation types.

Within a regional context cover types, forest types, and their area and distribution are important determinants of animal distributions and populations. At subregional and human landscape scales, major considerations include the distribution of forest types, forest age classes, and non-forest habitats within the context of ecosystem capabilities, disturbance frequency and pattern, and successional pathways (Thompson et al. 1993).

In addition, habitat age and age-distribution are critical determinants to avian habitat associations. Midwestern NTMB showed patterns among upland (dry) versus lowland (wet) ecosystems, conifer versus deciduous forests, and shrub/sapling versus mature forests. Analysis along single and multiple gradients can help explain species distribution and abundance at scales from continental to local if the range of sample variability is reduced by framing the analyses within

Table 4. Number of midwestern neotropical migratory birds and priority species (in parentheses) that breed in land covers and ecological provinces. Species can be associated with more than one land cover class, so rows and columns do not sum to species totals. (From Probst and Thompson, 1996).

| Habitat | Province* | | | | | | | |
|---|---|---|---|---|---|---|---|---|
| | 212 | 222 | 251 | 331 | 332 | M222 | M334 | NTMB |
| Shrub/sapling | 65(14) | 54(16) | 54(11) | 51(8) | 48(8) | 35(8) | 29(3) | 95(22) |
| Forest | 71(19) | 34(10) | 35(6) | 31(3) | 27(4) | 20(5) | 20(2) | 94(24) |
| Agri./Dev. | 38(5) | 38(5) | 40(6) | 39(6) | 39(6) | 31(4) | 29(4) | 47(6) |
| Grassland | 26(6) | 25(6) | 39(15) | 35(9) | 33(8) | 20(5) | 21(4) | 45(16) |
| Savannah | 27(6) | 30(6) | 33(6) | 31(3) | 34(5) | 27(5) | 20(2) | 39(7) |
| Aquatic | 8(1) | 6(1) | 6(1) | 6(1) | 6(1) | 4(1) | 0(0) | 8(1) |
| Totals | 129 (30) | 124 (30) | 136 (33) | 126 (23) | 125 (27) | 93 (21) | 81 (10) | 187(47) |

*Based on Bailey et al. (1994) and McNab and Avers (1994): 212 = Laurentian Mixed Forest Province; 222 = Eastern Broadleaf Forest (continental); 251 = Prairie (Temperate); 331 = Great Plains — Palouse Dry Steppe; 332 = Great Plains Steppe; M222 = Ozark Broadleaf Forest — Meadow; M334 = Black Hills Coniferous Forest.

ecological units or broad vegetation zones. For example, prairie-wetland complexes contain extreme moisture gradients over relatively short distances; bird species' distributions which overlap each other along this gradient may be more effectively assessed within this general context.

Geographical and ecological distribution information derived from multi-scaled assessment are the types of information needed in continental conservation efforts.

*Contact*

John Probst, Research Ecologist, North Central Forest Experiment Station, 5985 Hwy. K, Rhinelander, WI 54501-0898.

## 3.2 Monitoring and Evaluation

Monitoring information is compared with baseline information on the condition, distribution, capability and potential productivity of ecosystems which ecological classification and mapping can help provide. Monitoring is conducted on National Forests to ensure that activities planned are being implemented and to ensure that management is conducted according to the standards and guidelines prescribed. Monitoring is also conducted to determine if the overall plan had the intended results, and to understand and analyze changes in resource conditions and availability over time.

Hierarchical frameworks of ecological units provide information about the geographic patterns in ecosystems. These patterns can be used to identify representative ecological units for sampling. Knowledge gained from such monitoring can then be extended to analogous unsampled ecological units. (Avers et. al. 1994). The stratification provided by a nested geographic system accommodates extensive monitoring needed to track the status of populations and to understand the forces effecting change, as well as, intensive monitoring which is often used to test hypotheses and fine tune measurement techniques.

Ecological classification aids in the interpretation of inventory and monitoring data in a number of ways. Reference conditions, represented by ecological units, are used to add value to conditions and trends. Ecological units internalize vertical structural and functional relationships which reflect the influence of various ecological processes over time and space (Rowe and Sheard 1981). Probabilities associated with various natural disturbances can contribute to risk assessment. Understanding ecosystem functions can assist in interpreting the non-monetary costs so difficult to account for in cost-benefit analyses. Effective inventory and monitoring must focus on critical diagnostic attributes that are comparable over time and space. Stratified sampling schemes test the hypothesis that units with similar attributes behave similarly (Maxwell et al. 1995).

Case studies in this section demonstrate the use of ecological classification as a tool for monitoring and evaluation of environmental conditions and trends. *Case Study 12* describes the value of ecoregionalization for water quality monitoring. In *Case Study 13* bioregions are developed to provide an ecological context for interpreting wildlife habitat data. *Case Study 14* discusses the reporting of the Canadian National Forest Inventory data by Ecozone.

### Case Study 12 in Monitoring and Evaluation: A Protocol to Identify Stream Reference Sites — Using EPA ecoregions in water quality monitoring

*Ecological Classification*

Environmental Protection Agency Ecoregions of the Conterminous United States (Omernik 1986, 1987).

*Description*

The need for an ecoregional/reference site framework to facilitate the development of biological criteria was recognized in the late 1970s. This need was part of a larger concern for a framework to structure the management of aquatic resources in general and increasing awareness that there was more to water quality than addressing water chemistry, which had been the primary focus. Biota must be considered as must physical habitat and toxicity.

Reference sites are selected for each region and subregion to get a sense of the regionally attainable conditions regarding aquatic ecosystems. Attainable quality refers to those conditions that are realistic, rather than "pristine." Therefore candidate streams must be "relatively undisturbed" yet representative of the ecological region they occupy. An initial selection of reference sites is usually accomplished by interpreting 1:1,000,000 and 1:250,000 scale maps with guidance from state resource managers as to minimum stream and watershed sizes for each region or subregion and locations of known problem areas and point sources to avoid. The minimum number of sites necessary for each region or subregion is a function of the size and complexity of the subregion. Small or homogeneous regions may require five or six, complex regions or areas where reference streams represent different stream sizes generally require more.

Once sets of candidate reference sites have been identified for each region, they should be reviewed by state biologists and regional experts. Then field verification of the ecoregion delineations is coupled with visits to representative sets of reference sites. The regions must make sense to those who know and manage the area and are developing the biological criteria for evaluating water resource quality. It is also useful to include experts from adjacent states. Visits to a number of reference sites in each region provide a visual subjective analysis of within-and between-region similarities and differences as well as landscape characteristics within the ecoregion and watershed the streams occupy.

Reference sites representing least-disturbed ecosystem conditions are a moving target of which humans and natural processes are a part. The objective of the reference site network described here is to identify water quality conditions that are attainable within the established pattern of human land use within a region. This differs then from reference sites selected with the objective to study pristine conditions for research and historical purposes. Although the quality of the set of streams reflects the range of best attainable conditions given the current land use patterns in the regions, this does not imply that the quality cannot be improved. A comparison of the difference in the areal patterns of water quality among the reference sites with patterns in natural landscape characteristics should provide a sense for the factors that are responsible for within-region differences in quality.

For the most part, only very small streams have watersheds completely within any one subregion. Larger streams that more closely meet size criteria for reference sites tend to drain areas in two or more subregions. Sets of references sites for these types of subregions must consist of watersheds that have similar proportions in different subregions. In selecting reference sites, care must be taken to avoid including anomalous stream sites and watersheds.

An evaluation of the framework intended to depict patterns in the aggregate of ecosystem components is not an easy task. An appropriate test is not how well patterns of a single ecosystem component, such as fish species richness or total phosphorus in streams, match ecoregions. Alternative approaches appear more effective. Work by Larsen et al. (1988) in Ohio uses principal component analysis to link chemistry with nutrient richness and ionic strength and work by Karr et al. (1986) groups biotic characteristics to express biotic integrity.

*Contact*

Jim Omernick U.S. Environmental Protection Agency, Environmental Research Laboratory, Corvallis, OR.

## Case Study 13 in Monitoring and Evaluation: Regionalization of the California Wildlife Habitat Relationships (CWHR) Database — An ecological context for data interpretation

*Ecological classification*

Sixteen bioregions (Welsh 1994) were developed as an organizational framework for the California WHR information system (Airola 1988) in the state of California. The bioregions are of finer scale than the biotic provinces of Bailey (1976) and Udvardy (1975) and a coarser scale than the 24 developed by Barry (1991) though conceptually consistent with them. The approach used to define these bioregions was grounded in the literature of biogeography with an emphasis on the dynamic nature of "natural" communities along the continuum of ecological to evolutionary processes interacting with regional climates and physiography which determine natural biotic patterns through time.

*Description*

This system of bioregions was developed in response to a lack of regional focus in the statewide WHR database. California contains the most diverse array of habitats in the continental United States. Consequently, the use of a statewide database is often too coarse in resolution when dealing with animal species that may occupy different habitat (i.e. vegetation types) at different times of the year and in different geographic settings within the state. This system of bioregions permits database users to query the database with a focus on regional relationships within California when examining natural resources issues and their potential management impacts on wildlife species. It is important to recognize that the bioregionalized database is not limited to animals endemic to a particular bioregion. Two secondary objectives were to emphasize the value of the bioregional concept in resource planning, and the importance of thinking in terms of dynamic processes in order to reflect accurately how natural systems function through space and time.

*Contact*

Hartwell H. Welsh, Jr. Redwood Science Lab (PSW), 1700 Bayview Dr., Arcata, CA 95521; 707-825-2956

## Case Study 14 in Monitoring and Evaluation: Integration Of The Canadian National Forest Inventory (CanFI) at the Ecoregion Level — National and regional reporting

*Ecological Classification*

The National Ecological Framework for Canada (Ecological Stratification Working Group 1995) covers all of

Canada except for the Great Lakes and marine systems. The scale of application is 1:2,000,000 to 1:10,000,000.

*Description:* Every 5 years the federal Canadian Forest Service compiles the latest forest inventory information available from the ten provinces and two territories into a national compendium. Canada's Forest Inventory 1991 (Lowe et. al. 1994) replaces the 1986 version as the authoritative national statement on the distribution and structure of forest resources. The inventory is a spatially referenced database containing the best information available in 1991. The national inventory is produced with the cooperation of both provincial and territorial forest inventory agencies through the Canadian Forest Inventory Committee (CFIC). Until recently the national inventory data could only be made available spatially by administrative boundaries, e.g., provincial boundaries for analytical purposes. In 1993, Environment Canada began to work with CFS to integrate the National Ecological Framework and its associated databases with CanFI, a grid cell based database. This was completed in 1995 (Hirvonen and Lowe 1996).

The national ecological framework, using CanFI as well as other information, is now used to develop and present indicators of sustainable forest management by ecozone (the most general level of the national ecological hierarchy), for inclusion both in Canada's national set of comprehensive environmental indicators and for tracking certain criteria and indicators of sustainable forest management established by the Canadian Council of Forest Ministers (1995). This integration of databases is also used for general health of the forest reporting.

Data historically have been compiled by administrative units (forest districts, province) and not by ecological units, therefore much effort is required to compile forest data by the ecological units. Incompatibility of data among provinces is a concern (different age classes, methods of compiling attributes measured, etc.). The nature of the forest inventories currently is such that tracking specific attributes over time is difficult.

*Contact*

Harry Hirvonen, Indicator, Monitoring and Assessment, Environment Canada, Place Vincent Massey, 351 St. Joseph Blvd., Hull, Quebec, K1A 0H3, 819-994-1440 hirvonenh@cpits1.am.doe.ca or Steen Magnussen, Forest Inventory and analysis, Canadian Forest Service, 506 West Burnside Rd., Victoria, B.C. V8Z 1M5 604-363-0712, smagnussen@a1.pfc.forestry.ca

## 4 SEEKING A CLASSIFICATION THAT WORKS FOR ALL PARTNERS

How can a common ecological classification system be developed for use by multiple agencies, organizations and landowners? Development of a common classification involves agreement on common objectives. It also involves identification of the appropriate classification concepts, data standards, naming conventions, mapping protocols, and the appropriate multi-scaled hierarchical structure or structures if a multi-scaled system is desired (Grossman et al. this volume).

A number of technical avenues are being explored by scientists and managers to determine the feasibility of a common classification to meet the needs of all partners. Efforts are underway to examine the benefits of merging and linking existing systems taxonomically and spatially. Still others are exploring the proposition that a single ecological classification system may not be necessary if common data standards and map themes are developed.

Grossman et. al. (this volume) evaluated the conceptual similarities and differences among the most prominent abiotic, biotic, and integrated (a.k.a., biotic–abiotic or multifactor) classification systems at a variety of scales and distilled the following basic concepts. The attributes of an ideal system include:

- to be integrated fully with the desired application(s) and products,
- to have a systems orientation; be based on the structure and function of the system,
- to consider spatial and temporal scale properties of the system,
- to be dynamic and allow for environmental and biological change,
- to take hierarchy theory into account.

This challenging agenda calls for integrating new ideas from scientific fields such as conservation biology and landscape ecology. Probably the most complex discussion is whether to integrate aquatic and terrestrial classification objectives, protocols, and hierarchical structures. To date there is no clear statement of what exactly this aquatic–terrestrial integration would mean, the scales where integration should occur, and the anticipated benefits to management of addressing this integration through a common classification rather than through existing separate hierarchical structures, or through modeling, or assessment.

A means to evaluate the potential to merge existing classifications is to classify and map a common area using multiple systems and then work to reconcile the differences. Differences which usually appear are related to the characteristics and attributes defined by the classification, differences in where boundaries are placed on the ground, differences because of variable scales or levels of generality in definitions, and differences in vegetation characteristics because of the time period of reference.

Maps partition environmental gradients and encompass a certain degree of landscape complexity which varies by the intensity of mapping. Key discussions in the development of a common classification will revolve around the logic underlying the delineation of map polygons. Both qualitative and quantitative methods have been promoted although most people agree that the process is a combination of science and art. A measure of successful standardization is inherent in the idea of replicability in the generation of map products (Omernik 1995, Host 1996). Once boundaries are agreed upon, map units can be tested to determine how accurately ecological classification descriptions and interpretations meet their design objectives. Boundary lines should be revised where descriptions or interpretations of ecological units and actual ecosystem response are at odds.

## 4.1 Incentives and Barriers to Cooperation

Should partners collaborate in the development of a common classification system? Each organization will have its own particular reason for participating and its own measure of success. However, some general needs and anticipated benefits can be articulated as a baseline for measuring progress. General incentives are:

- to draw on the combined expertise from many organizations to better understand the ecosystem and its components,
- to help recognize and share solutions to mutual problems,
- to provide a system which consistently holds up to intense scientific and public scrutiny,
- to help communicate to lay persons the basis for differences in management prescriptions among sites and among agencies and organizations,
- to achieve economies of scale when investing time and money in database development,
- to achieve economies of scale when developing ecological models, planning models and decision support systems which incorporate ecological classification,
- to coordinate inventory and monitoring strategies,
- to increase our ability to compare management experiences,
- to establish research in representative areas and extrapolate scientific findings,
- to empower local and regional stewardship and initiative,
- to improve early detection of ecosystem stresses,
- to address better broad-scale issues such as biodiversity conservation, wetland preservation, water quality protection, and ecosystem health.

Barriers exist to cooperation in the development of a common system across agency and organization boundaries. The use of sophisticated statistical processes and the language of ecological classification is often complex and with many nuances so that managers are not sure of the benefits and technical tradeoffs. Organizations with a substantial investment in current resource inventories may be unwilling to provide resources to integrate new concepts and information. Multipurpose classifications do not optimize utility for all purposes; therefore, managers or specialists may be unwilling to allocate resources to a common purpose, or, individuals may have allegiance to a given classification.

Public land cannot provide all the goods and services required by a growing population. The public has different expectations for public land than for private land and for private industrial versus private non-industrial land management. Therefore, increased attention needs to focus on appropriate, cost-effective means of transferring the benefits of ecological classification to private lands. Ecological classification is an efficient tool for education.

## 4.2 Integrating Existing Classifications and Maps

Valuable information is gained by identifying commonalities, strengths and weaknesses among existing classification systems. For example, Table 5 provides a comparison of the ability of soil, vegetation, and integrated approaches to classification to provide capability and forest productivity information at the local scale. The process of comparison identifies how compatible or complementary existing systems are. Advances toward comprehensive data collection and analysis approaches, originally designed to overcome the limitations of existing classifications, have done much to

Table 5. Comparison of the utility and restrictions of several common environmental classification approaches (source Cleland et al. 1994).

| Classification and Objective | Use Considerations |
|---|---|
| **Soil surveys** provide knowledge of soil properties critical in planning for any on-the-ground project; for example, equipment limitations and standards and guidelines for road construction are based on soil properties | They often inadequately predict timber production potential (Carmean 1975,1979; Esu and Grigal 1979) and do not provide enough information about potential natural vegetation to guide management decisions involving the manipulation or preservation of vegetation. |
| **Habitat types** identify areas with similar climax communities and can provide information about plant community composition and succession. Often, there are predictable relationships between ecologically important soil factors and the distribution of ground flora. | Potential forest productivity is not a criterion used in development of the habitat types so that wide ranges in productivity can occur within a type. Areas similarly classified may have different functional attributes. Habitat types do not provide enough information about ecosystem components other than vegetation to develop many capability and suitability ratings. It may be difficult to accurately classify disturbed sites. |
| **Ecological types** provide knowledge of plant community composition, structure, succession, and soil and hydrologic properties.Productivity is an inherent consideration in development. Can better classify disturbed sites. Areas similarly classified have similar functional attributes which allows extrapolation of cause and effect information. | Like all classification and rating systems, the quality assurance and quality control methods used in development will determine the accuracy and ultimate utility to meet the desired use. |

further the emergence of more highly-integrated land classification products (Sims et al. 1996).

Compatibility of objectives, criteria, and resolution must be addressed for systems to be integrated at a given scale. To develop a hierarchical system, the interaction among multiple scale must also be addressed. Ecological classifications can be organized as spatial or taxonomic hierarchies. Both can be used to create maps. Spatial frameworks are map frameworks explicitly designed to partition the landscape based on analysis of environmental gradients and landscape patterns. In spatial hierarchies lower level units are aggregated geographically to form higher level units. The highest units are described by the range of conditions they encompass geographically.

In taxonomic systems, lower level classification types are nested conceptually within the higher levels, but the fact that those classification types are combined at the next higher level does not guarantee that they are geographically associated in a particular landscape. Landscape relationships become evident when mapping taxonomic types at an individual scale but, the geographic patterns are not carried upward to the description of the next level of the hierarchy.

Several major hierarchical classification systems in use and pertinent to the development of a common classification in the United States are presented for comparison below. These systems are organized into spatial and taxonomic hierarchies.

Tables 6 and 7 present the principal map unit criteria and map scale for the spatial hierarchies used for terrestrial and surface-water ecosystems by the USFS. The USFS National Hierarchical Framework of Ecological Units is a regionalization, classification and mapping system for stratifying the earth into progressively smaller areas of increasingly uniform ecological potential. Among the units presented in these tables three types of biophysical environments are recognized: geoclimatic, zoogeographic, and aquatic (see Grossman et al. this volume). The terrestrial, aquatic, and groundwater hierarchies presented in Fig. 1 converge into a Nearctic zone at the global scale. Intended uses were presented earlier in this document.

A major stimulus for the Environmental Protection Agency (EPA) to develop an ecoregional framework has come from a need to assess existing and attainable surface water quality. The most immediate needs being to develop regional biological criteria and water quality standards and goals for non-point source pollution. The EPA has invested in the development of a four-level hierarchy of ecoregions beginning with a first approximation map entitled "Ecoregions of the Conterminous United States" (Omernik 1986, 1987), which shows 76 ecoregions at a scale of 1:7,500,000. Within this hierarchy, Level I is the most general, Level IV is the most detailed.

The premise behind the EPA approach is that ecological regions can be identified by analyzing the patterns and composition of biotic and abiotic phenomena, reflecting differences in ecosystem quality and integrity (1989; Omernik 1987, 1995). These phenomena include geology, physiography, vegetation, climate,

Table 6. USFS National Hierarchical Framework of Ecological Units, scale and principal map unit design criteria (adapted from Avers et al. 1994).

| Map Unit | Criteria | Scale |
|---|---|---|
| **Domain** | Subcontinental area of broad climatic similarity | 1:15,000,000 |
| **Division** | Differentiated by continental climate reflected in common vegetative life forms. | 1:30,000,000 to 1:7,500,000 |
| **Province** | Differentiated primarily by the effects of continental weather patterns interacting with broad landforms and that correspond to broad vegetation regions. Provinces display similarities in geologic age, stratigraphy, lithology, and soil forming processes. Also differentiated are highlands or mountains where changes in elevation correspond with differences in climate, vegetation and soil. | 1:15,000,000 to 1:3,500,000 |
| **Section** | Broad regions of similar geomorphic process, stratigraphy, geologic origin, topography, regional climate and dominant associations of potential natural vegetation. | 1:7,500,000 to 1:1,000,000 |
| **Subsection** | contain common landforms due to common lithology, surficial geology, and/or geomorphic history. also differentiated are mesoscale climatic zones which influence plant community compositions or species dominance. | 1:3,500,000 to 1:250,000 |
| **Land Type Association** | Based upon the effective interaction among landform, geomorphic process, elevation, vegetation and local climate. Display repeatable patterns of soils, plant communities, stream types, lakes, wetlands, and rock types. | 1:250,000 to 1:40,000 |
| **Ecological Land Type** | Unique combinations of soil morphology, soil depth, landscape position, geomorphic process and hydrology are expressed by commonalities in the structure and composition of potential natural communities and basic land capability. | 1:60,000 to 1:24,000 |
| **Ecological Land Type Phase** | Similar to land types but smaller and more narrowly defined. Microclimate, internal drainage, and soil texture, structure and morphology influence the productivity and successional tendencies of the site. | 1:<24,000 |

Table 7. Abbreviated criteria for designing ecological units for aquatic ecosystems (Maxwell et al. 1995).

| Map Unit | Criteria | Map Scale |
|---|---|---|
| **Domain** | Fish family patterns. | 1:7,500,000 |
| **Region** | Fish dispersal and vicariance. | 1:7,500,000 |
| **Subregion** | Fish vicariance and endemism. | 1:7,500,000 |
| **Basin** | Fish endemism and genetics. | 1:2,000,000 |
| **Subbasin** | Physiography. | |
| **Watershed** | Fish genetics watershed and stream network morphology. | |
| **Valley Segment s and Lake Types** | Geomorphology, climatic regime, and hydrologic regime. | 1:63,000 to 1:24,000 |
| **Stream Reach and Lake Zone** | Channel and lake morphology. | 1:24,000 to 1;12,000 |
| **Channel Units and Lake Sites.** | Site specific habitat features, hydraulics, substrate etc. | 1: <12,000 |

soils, land use, wildlife, and hydrology. The relative importance of each characteristic varies from one ecological region to another, regardless of hierarchical level. This approach can be used at each hierarchical level by considering factors to a greater level of detail.

Land Resource Regions (LRR) and Major Land Resource Areas (MLRA) are regional scale classifications developed by The Natural Resource Conservation Service as a basis for making decisions about national and regional agricultural concerns, identifying needs for research and resource inventories, providing a broad base for extrapolating the results of research and as a framework for organizing and operating resource conservation programs (USDA Ag. Handbook 296

1984). LRRs and MLRAs are based on soils, climate, water resources and land use. The delineations draw particularly heavily on concepts underlying Soil Taxonomy and information collected through the National Cooperative Soil Survey.

The U.S. Fish and Wildlife Service's Classification of Wetlands and Deepwater Habitats is intended to "describe taxa, arrange them in a system useful to managers, furnish units for mapping and provide uniformity of concepts and terms" (Cowardin et al. 1979). Five major systems form the highest levels of this classification scheme: Marine, Estuarine, Riverine, Lacustrine, and Palustrine. The first four include both wetland and deep water habitats but the Palustrine includes only wetland habitats.

The Nature Conservancy's mission is to protect biological diversity. They are working to provide a complete listing of all communities that represent variation in biological diversity and to identify communities that require protection. The classification is intended to address protection of all natural systems, rare or not. The terrestrial classification hierarchy is based on existing rather than potential vegetation types which range from early successional through climax associations and include types that are maintained by both natural disturbance regimes and human activity. The terrestrial hierarchy is a modification of UNESCO (1973) and Driscoll et al. (1984).

The Nature Conservancy is also developing an aquatic classification system (Grossman et al., this volume). The Conservancy's classification represents a continental scale approach to setting priorities for freshwater biodiversity protection. The Conservancy's classification system is hierarchical and allows for the characterization of aquatic communities on both abiotic and biotic levels. The abiotic component of the classification framework defines the context and describes the physical structure of aquatic ecosystems at five spatially-nested scales: aquatic province, aquatic section, watershed type, macrohabitat, and microhabitat. The biotic component of the classification framework provides guidelines for identifying, naming and characterizing aquatic communities at two hierarchical levels: alliance and association.

Proper comparisons of the accuracy and compatibility of objectives among systems are easiest when data types and measures are the same. The earlier stated proposition, that a single ecological classification system may not be necessary if common data standards and map themes are developed, is based on the premise that accessibility to extensive standardized databases and accurate spatial information will provide an environment where it will be easier to generate specific ecological classification and interpretive products than to develop a common system to meet all needs (see Grossman et al., this volume). Not everyone agrees that this approach will give the degree of integration necessary to manage ecosystems as a whole rather than as the sum of its parts.

Part of the evolution of existing classifications toward a common classification system is the merging of spatial and non-spatial databases and incorporating temporal information within a spatial framework. Taxonomic and place dependent systems organize spatial and temporal information differently. For example, the U.S. Forest Service National Hierarchical Framework of Ecological Units is a spatial hierarchy. The Forest Service surveys and maps ecological units which are then used to structure information in a database. Potential natural vegetation information is described and bounded by the unit. The varied species composition, plant associations, and vegetative structures which occur over time are carried as attributes of the site mapped. Rarity and abundance values can be generated from information on unit composition throughout all levels of the hierarchical system. In the absence of complete inventories, percent composition can be estimated by applying information on distribution patterns.

The Nature Conservancy conservation database provides an example of a different database structure. TNC describes existing natural communities and associated environmental conditions using their taxonomic system. Differences because of species composition, plant association, or seral stage are each considered significant for distinguishing new classes. In a database, site variables are carried as attributes of the community which has incorporated species composition, plant association, and vegetative structure into its definition. Rarity and abundance are communicated through a state, national, and global ranking system rather than from direct measurement or estimation of areal composition within a defined area.

## 4.3 Promising Partnerships

Activities to link classifications among agencies and organizations are moving forward. For the U.S. Forest Service, the National Hierarchy serves as the central reference point for all efforts to link existing systems. At the national level, nine federal agencies have entered into a Memorandum of Understanding (MOU) to develop a common spatial framework of ecological units for the United States *Case Study 15.* Many federal agencies and national organizations are already working with state partners to achieve consistency in ecological classification and to standardize its use.

Other important partnership efforts revolve around linking federal and state efforts and obtaining consistency in the application of regional and national systems across state boundaries. Collaborative efforts in subsection mapping between the USFS and state natural resource agencies, and others have already been mentioned (Section 1.5). In addition, the Northeast Area Association of State Foresters endorsed the implementation of ecological classification following the USFS Hierarchy as a key component in the implementation of more ecological approaches to management in that twenty state area (Ecosystem Management Strategy Team 1994). Wisconsin and the USFS have signed a formal agreement called the Wisconsin Accord which clarifies the relationship between previous state work in Habitat Type Classification (Kotar et al. 1988, Kotar and Burger 1996) and the USFS National Hierarchy.

Land type associations are being developed or planned with state leadership in New Jersey, Minnesota, Michigan, Wisconsin, New Hampshire, Missouri, Massachusetts, Pennsylvania, New York and Vermont.

In 1992, the Minnesota Department of Natural Resources and the Chippewa National Forest began a cooperative project called the Chippewa Demonstration Area to develop ecological units, descriptions, identification keys and interpretations at all levels on two shared Land Type Associations to demonstrate their use (Hanson and Hargrave 1996). In Indiana, ecological land types and land type phases, developed for the Hoosier National Forest, were presented in field guide format to support application of the system within appropriate natural divisions on adjacent public and private lands at the request of the State Forester (Van Kley et al. 1994).

The EPA Environmental Research Laboratory in Corvallis, Oregon is involved in several collaborative projects with states and EPA regional offices to refine ecoregions, define subregions, and locate sets of reference sites within each region and subregion. This work is being conducted at a 1:250,000 scale. These projects cover Iowa, Florida, Massachusetts and parts of Alabama, Mississippi, Virginia, West Virginia, Maryland, Pennsylvania, Oregon, and Washington (Omernik 1995).

The Nature Conservancy (TNC) and its Natural Heritage Program (NHP) cooperators have been involved in many collaborative efforts throughout the United States including the standardization of vegetation classification protocols and nomenclature for ground survey and remote sensing applications, and the collaborative development of ecological units at a variety of scales. A recent decision to incorporate a bioregional framework into their regional and national conservation planning places them at the center of federal, state, and international efforts to develop a common spatial framework of ecological units. The Nature Conservancy and the US Forest Service have established a cross reference of systems in the Northeastern United States. *Case Study 14* shows the attributions of subsections with TNC regional alliances.

Several contemporary partnership efforts cross the Canadian-U.S. border. The publication *Ecoregions of Alaska* (1994) involved state, federal, and Canadian cooperation (Omernik 1995). Uhlig and Jordan (1996) examined Canadian and American national hierarchical frameworks and proposed a joint project involving the Ontario Ministry of Natural Resources, the Canadian Forest Service, and the U.S. Forest Service in the Upper Great Lakes region. A North American Framework is being developed for Canada, the United States, and Mexico (Omernik, personal communication, 1996).

Broad-scale assessments can contribute considerable information toward the development and refinement of a common classification system. For example, work in the Columbia River Basin, covering parts of Oregon, Washington, Montana, Idaho, Northern California, Nevada, and Utah, involved the collection and synthesis of survey data from the states, the USFS, BLM, NPS, TNC and other sources. Classifications for a variety of purposes have been developed. Management alternatives were constructed and evaluated using models incorporating the biophysical and potential vegetation components of ecological units (Reid et al. 1995).

In 1990, the United States Geological Survey (USGS) initiated NAWQA as a comprehensive survey of the status and trends of ground and surface water quality in the United States. Physical, chemical and biological data will be collected from study areas that correspond to hydrologic units based on the drainages of major rivers and aquifers. They will be further stratified according to Frissell et al.'s (1986) classification framework. NAWQA researchers are currently assessing the use of ecoregions to stratify their national sampling (Higgins, pers. comm. 1996, McMahon, pers. comm. 1996).

Multi-organizational efforts such as the Federal Geographic Data Committee (FGDC) and the USFS Common Survey Data Structure (CSDS) projects are working toward common data and survey standards among multiple agencies and organizations. National standardized databases available today include the Natural Resource Conservation Service STATSGO and SSURGO databases, The Nature Conservancy Conservation Data Base System, the USFWS wetland inventory database and the USGS STORET water quality database.

A key area for future collaboration involves the integration of ecological classification and remote sensing products. A key linkage is being sought in the standardization of vegetation classification nomenclature. The USFWS National GAP Analysis is using remote sensing technology to identify native plant and animal species and natural communities represented on conservation lands (Scott et al. 1993). The land and water classification system for Gap Analysis seeks to link to the United Nations Educational, Scientific and Cultural Organization (UNESCO 1973) system as modified (Driscoll 1984), the USFWS classification (Cowardin et al. 1979), and the remote sensing land cover classification (Anderson et al. 1976).

### Case Study 15 in Promising Partnerships: Memorandum of Understanding on Developing a Spatial Framework of Ecological Units of the United States — Partnerships among agencies of the U.S. Department of Agriculture, the U.S. Department of the Interior, and the Environmental Protection Agency

#### *Ecological Classification*

A common spatial framework for defining ecological units of the United States based on naturally occurring and recognizable features such as soil, geology, geomorphology, climate, water, and vegetation will be developed. Guides for this work will include the National Hierarchical Framework of Ecological Units (ECOMAP, 1993) developed primarily by the Forest Service; the Land Resource Regions and the Major Land Resource Area (MLRA) framework (USDA Agriculture Handbook 296, 1981, revised 1984) developed primarily by NRCS; the EPA Ecoregion Framework (Omernik 1995); and other references, as appropriate, depicting biological and physical components of the environment.

#### *Description*

A Memorandum of Understanding was entered into by the U.S. Department of Agriculture, Natural Resources Conservation Service (NRCS), Forest Service (FS), and Agricultural Research Service (ARS); the U.S. Department of the Interior, Bureau of Land Management (BLM), U.S. Geological Survey (USGS), Fish and Wildlife Service (FWS), National Biological Service (NBS), and National Park Service (NPS); and the U.S. Environmental Protection Agency (EPA).

The MOU documents and defines the responsibilities of the cooperating agencies to develop a common spatial framework for defining ecological units of the United States. It also provides a vehicle for other Federal agencies with natural resource management responsibilities to become part of the cooperative effort nationwide.

The growing interest by federal and state agencies in adopting a more integrated ecological approach to resource management has clarified the need for a common spatial framework for defining ecological units. This common framework will provide a basis for interagency coordination and will permit individual agencies to structure their strategies by the regions within which natural biotic and abiotic capacities and potentials are similar. These ecological units transcend local, state, and national boundaries.

Considering the broad responsibilities and interests of all agencies, it is desirable and mutually beneficial to cooperate and integrate interdisciplinary technical information on environmental factors such as soils, vegetation, geology, geomorphology, water, climate, and others into a common ecological framework, with associate descriptions and digital databases. Development of a common ecological framework will be consistent with standards developed by the Federal Geographic Data Committee (FGDC) according to the Office of Management and Budget (OMB) Circular A-16 and Executive Order 12906 (Coordinating Geographic Data Acquisition and Access: The National Spatial Data Infrastructure) signed April 11, 1994.

Cooperating agencies will use the framework for defining ecological units, with associated narrative descriptions and digital databases to (a) reduce duplication of effort and promote effective, efficient, and scientifically sound management of natural resources; (b) geographically organize and share research, inventory, and monitoring information; (c) facilitate coordinated approaches to characterization and assessment of the Nation's land and water; and (d) enhance program management and technical coordination among parties representing private, tribal, state, and federal interests.

Development of a common spatial framework for defining ecological units will necessitate recognition of the differences and functions of the three existing guides listed above. Commonality and refinement of these guides will be the basis for evolution of the common spatial framework and related databases. Signatory agencies will collaborate on a State-by-State and/or regional project basis using interagency standards and procedures until a set of common and joined ecological units is developed for the entire Nation.

#### *Contact*

James Keys, USDA Forest Service, Auditors Building, 201 14th Street, S.W. at Independence Ave. S.W., Washington, DC 20250. 202-205-1580.

### Case Study 16 in Promising Partnerships: Attribution of USFS Subsections in the Northeastern United States with TNC Regional Alliances — Linking existing classifications for mutual benefit

*Ecological Classification*

This example shows work and outcomes from cross-referencing alliances of the TNC Eastern Regional Community Classification (Sneddon et al 1994) to 88 subsections in the northeastern United States. The subsections were developed according to criteria associated with the USFS National Hierarchy (Avers et al. 1994) and are depicted on the map, *Ecological Units of the Eastern United States: a First Approximation* (Keys et al. 1995). The Nature Conservancy's Eastern Regional Alliance classification describes 128 alliances which share a similar species composition, vegetation structure and environmental setting (Sneddon et al. 1994). The regional classification provides a correlation of alliances identified among states participating in the Natural Heritage Programs in the TNC Eastern Region.

*Description*

The USFS New England/New York Subregional ECOMAP (NE/NY ECOMAP) team was interested in obtaining information on vegetation types and distribution to help in the delineation and characterization of Subsections. Information on the abundance and distribution of potential natural communities (PNC) is important for mapping ecological units, but is not generally available at the subsection scale. The NE/NY ECOMAP team entered into a cooperative agreement with the Eastern Region Office of The Nature Conservancy to attribute existing regional alliances to subsections and identify late successional communities as a first approximation of PNC.

TNC Eastern Region ecologists worked with state natural heritage ecologists, panels of experts and their state biological conservation databases, to determine which regional alliances occurred within each subsection and which approximated PNC. A 3-point scale was used to document the certainty vested in each occurrence. The scale is 1: probably occurs, 2: definitely occurs, 0: definitely not present. This resulted in a matrix of subsections vs. regional alliances. Expert judgment was used to classify each alliance as restricted, limited, widespread or occasional in occurrence within the subsection thus providing qualitative information on distribution and abundance for future reference.

Now both organizations can query the database to determine which and how many alliances occur in each subsection; and which and how many subsections are associated with each alliance. With the attribution completed, it is possible to aggregate and ascribe information to higher levels of the USFS National Hierarchy and to group subsections by any characteristic used as an attribute within the conservation database. For example, coarse distribution maps can be developed for each alliance by querying for presence within the database.

The subsection map provides a geoclimatic context for TNC ecologists to evaluate their classification and correlation efforts among states. Communities which were dissimilar were expected to separate along some ecological unit boundary. Communities classified similarly are expected to either cluster geographically within or among a subsection, or section, or province unless they are associated with some environmental characteristic(s) which explains a discontinuity in distribution. In the latter case, the distribution of calcareous fens is disjunct but logical due to the strong confining influence of mineral and hydrologic factors at a local scale and the broad climatic zone where conditions are suitable for the species.

Work in this area is ongoing and being evaluated by TNC to develop "ecoregional planning units" for Conservancy conservation action across the nation (Anderson et al. 1996). In addition to the benefits of this general characterization, the USFS Eastern Region Research Natural Areas program is also planning to use this information to provide a cross-check of natural area representation beyond National Forest boundaries and to clarify the USFS role in state, regional, and national biodiversity conservation efforts.

*Contact*

Connie Carpenter, EM coordinator, USDA Forest Service, 271 Mast Road, Durham, NH 03824; Mark Anderson, Regional Ecologist, TNC Eastern Region Office 201 Devonshire, 5th floor, Boston, MA.

## 5 SUMMARY

The objective of this chapter was to describe the uses of a variety of biophysical classifications and ecological assessments in decision-making, and to identify ways that partners can work toward use of a common ecologically based classification system. The benefits of using ecological classification in natural resource planning, implementation, and monitoring were presented first for the local level, then for regional and landscape level planning. Case studies were used to highlight the varied uses of ecological classification and to highlight ways that classification supports partnerships and decision-making. Examples were predominantly for management of terrestrial systems on National Forests.

But aquatic amd wetland examples and non-Forest Service examples were included that demonstrate that the objectives and methods of employing ecological units in planning, management, and monitoring are similar.

Ultimately, the performance of any system depends on the degree to which it is compatible with its objectives and the scale of analysis in which it is used. The case studies presented in this chapter provide examples of appropriate uses and demonstrate how to test the validity of the assumptions embedded within the classifications. It is advantageous to use ecological classification systems to solve multiple resource management problems.

## 5.1 Key Science and Management Concepts

Ecological classification and mapping provide a bridge between science and management. It is important for managers to understand the assumptions underlying the definition of ecological types and ecological units if they are to participate in the testing of those assumptions during regular management activities. Table 8 identifies key concepts managers should recognize.

Ecological classification and mapping provide information on ecological potential which is used to establish desired future conditions at single scales, and, in nested geographic systems, at multiple scales. Ecological units can be used to structure integrated resource inventories or integrate existing multidisciplinary information. Ecological classification and mapping provide information useful in program planning and can be used to enhance coordination and cooperation among multiple disciplines and multiple agencies. Table 9 identifies key concepts to help managers use ecological classification and mapping appropriately.

Table 8. Key science concepts related to the use of ecological classification and mapping systems in management.

| Key topic | Concepts |
|---|---|
| Ecological classification systems | exist for terrestrial, freshwater, marine, and wetland ecosystems,<br>integrate multiple biotic and abiotic characteristics in three dimensions, and are used to identify and map areas of different biological and physical potentials. |
| Ecological units | partition environmental gradients,<br>provide a framework for integrating multiple types of resource information,<br>display spatial relationships among ecosystems and<br>follow taxonomic and mapping rules. |
| Hierarchical systems | provide a context for relating landscape patterns to processes, and<br>can be spatial or taxonomic in nature. |
| Criteria for combining classifications | common objectives.<br>common classification concepts,<br>common data standards,<br>common naming conventions,<br>common mapping rules, and<br>common taxonomic and/or spatial hierarchical structure |

Table 9. Key concepts related to the use of ecological units in natural resource management.

| Key topic | Concept |
|---|---|
| Ecological units | must be combined with other environmental, social, and economic information for sound decision-making.<br>provide an expedient and cost-effective means of ordering and managing information about ecosystems.<br>provide a spatial structure for information management.<br>provide the hypothesis that the area within each ecological unit is consistent with the description provided for it,<br>provide the hypothesis that the area within each ecological unit will respond as predicted by the interpretation of its environmental characteristics. |
| A hierarchical framework of ecological units | provides a context for evaluating cumulative effects,<br>allows aggregation of fine scale data into regional databases while preserving ecological meaning.<br>provides a framework for describing the composition, structure and function of ecosystems. and.<br>contributes to our ability to demonstrate the potential for a variety of alternatives at local, landscape, and regional scales. |

## 5.2 Conclusions

It is clear that good classification systems are, without exception, based on sound science. Brief developmental histories are included in each of the case studies presented. Taken as a whole, they illustrate the evolution of ecological classification principles and concepts. Regardless of classification approach, similar environmental factors and variables have emerged as useful discriminators of ecological condition and potential.

This leads to the conclusion that a higher degree of compatibility among existing systems can be achieved as more information is collected and analyzed. Grossman et al. (this volume) discuss the major classification approaches in widespread use today and present a matrix of key attributes that highlight similarities and differences among them.

Where single-resource or single function classifications and maps already exist, managers can evaluate the potential to merge them by mapping a common area. The degree of difference among the maps and the varied effects of using them to make management decisions can then be evaluated. Managers can work with scientists to reconcile the differences and provide assessments of the trade-offs relative to a range of uses.

## ACKNOWLEDGMENTS

The case studies presented in this chapter represent years of work in the field by those dedicated to the stewardship of America's natural resources. Although all those contributing to the projects themselves cannot be individually recognized, the case study authors need to be acknowledged: Greg Kudray, Kirsten Saleen, Dale Higgins, Gary Roloff, Dr. Dale Solomon, Glendon Smalley, Sherel Goodrich, W.-Dieter Busch, Dave Weixelman, Desi Zamudio, Karen Zamudio, Eunice Padley, James Jordan, Melody Steele, John Probst, Jim Omernick, Hartwell H. Welsch, Jr., Connie Carpenter, Mark Anderson, Harry Hirvonen, and Jim Keys.

Thanks are extended to Bill Leak, Marie-Louise Smith, Mark Jensen, and Glendon Smalley for review comments. Helen Thompson provided suggestions for reducing jargon for a jargon rich subject. Helen, I tried.

The science and management teams had significant interaction due in large part to Denny Grossman, science team lead author. W.-Dieter Busch and Dave Cleland worked as authors on both the science and management team papers.

W.-Dieter Busch provided considerable aquatic classification expertise to the team and Harry Parrott provided information on the history of aquatic classification in the Forest Service.

I have personal thanks to extend as lead author to each member of the management team and to Denny Grossman for personal encouragement throughout the process. Greg McClarren, our facilitator from the stewardship project, stayed with us throughout the entire project, working above and beyond the call of duty. It has been a pleasure to work with Nels Johnson and Andrew Malk of the World Resources Institute.

## REFERENCES

Airola, D.A. 1988. *Guide to California Wildlife Habitat Relationships System*. California Resources Agency, Dept. of Fish and Game. Rancho Cordova, CA.

Albert, D.A. 1995. Regional Landscape Ecosystems of Michigan, Minnesota, and Wisconsin, A Working Map and Classification (Fourth Revision: July 1994). Gen. Tech. Rpt. NC-178. USDA Forest Service, North Central Forest Experiment Station. 1992 Folwell Ave., St. Paul, MN 55108.

Albert, D.A., S.R. Denton, and B.V. Barnes. 1986. *Regional Landscape Ecosystems of Michigan*. School of Natural Resources, University of Michigan, Ann Arbor, MI.

Allen, T.F.H., and T.B. Starr. 1982. *Hierarchy: Perspectives for Ecological Complexity*. The University of Chicago Press, Chicago.

Anderson, J.R., E.E. Hardy, and J.T. Roach.1976. *Land use and land cover classification system for use with remote sensing data*. Geological Survey Professional Paper 964. A revision of the land use classification system as presented in U.S. Geological Circular 671. U.S. Government Printing Office, Washington DC.

Anderson, M.G., L. A. Sneddon, M.-L. Smith, and C. Carpenter. 1996. Relationships between geophysical based and biological community based land classification systems. Ecological Society of America Annual Meeting: Oral session on Gap Analysis and Biodiversity. August 11–15, 1995, Providence Rhode Island. *Bulletin of the Ecological Society of America* 77(3): 12.

Arnold, D.H., G.W. Smalley, and E.R. Buckner. 1996. Landtype-forest community relationships: a case study on the Mid-Cumberland Plateau. *Environmental Monitoring and Assessment* 39: 339–352.

Avers, P.E., D. Cleland, W.H. McNab, M. Jensen, R. Bailey, T. King, C. Goudney, and W.E. Russell. 1994. *National Hierarchical Framework of Ecological Units*. USDA Forest Service, Washington DC.

Bailey, R.G. 1984. Integrating Ecosystem Components. In: *Proceedings of the Symposium Forest Land Classification: Experience, Problems, Perspectives*. March 18–20, 1984. University of Wisconsin, Madison, WI. pp. 181–187.

Bailey, R.G. 1976. *Ecoregions of the United States*. USDA Forest Service, Ogden, UT.

Bailey, R.G. 1995. *Description of the Ecoregions of the United States*. (revised). USDA Forest Service Misc. Pub. 1391.

Bailey, R.G., P.E. Avers, T. King, and W.H. McNab (eds.). 1994. Ecoregions and Subregions of the United States (map). Washington, DC: U.S. Geological Survey. Scale 1:7,500,000; colored. Accompanied by a supplementary table of map unit descriptions compiled and edited by W. Henry McNab and R.G. Bailey. Prepared for the U.S. Department of Agriculture, Forest Service.

Barry, W.J. 1991. Ecological Regions of California. Unpublished report, prepared for California Interagency Natural Areas Coordinating Committee, California Department of Parks and Recreation, 15 July, 33 pp.

Bockheim, J.G. (ed.). 1984. *Forest land classification: experience, problems, perspectives. Proceedings of a symposium*. March 18–20, 1984. University of Wisconsin, Madison, WI, pp. 132–147

Busch, W.-D. N., and P.G. Sly (eds.). 1992. *The Development of an Aquatic Habitat Classification System for Lakes*. CRC Press, Boca Raton, FL.

Canadian Council of Forest Ministers (CCFM). 1995. Defining Sustainable Forest Management — A Canadian Approach to Criteria and Indicators. Ottawa. 22 pp.

Carpenter, C.A., M-L. Smith, and S. Fay. 1995. What do ecological unit boundaries mean? The dual role of ecological units in ecosystem analysis: examples from New England and New York states. In: J.E. Thompson (compiler), *Analysis in Support of Ecosystem Management; Workshop III; April 10-13, 1995*, Fort Collins, CO. USDA Forest Service, Ecosystem Management Analysis Center, Washington, DC.

Clatterbuck, W.A. 1996. A community classification system for forest evaluation: development, validation, and extrapolation. *Environmental Monitoring and Assessment* 39: 299–321.

Cleland, D.T., J.B. Hart, G.E. Host, K.S. Pregitzer, and C.W. Ramm, 1994. Field Guide Ecological Classification and Inventory System of the Huron-Manistee National Forests. USDA Forest Service, Huron-Manistee National Forest, Cadillac, MI. 227 pp.

Cowardin, L.M., V. Carter, F.C. Golet, and E.T. LaRoe. 1979. Classification of Wetlands and Deep Water Habitats of the United States. U.S. Fish and Wildlife Service, Department of the Interior, FWS/OBS-79/31. 131 pp.

Driscoll, R.S., D.L. Merkel, D.L. Radloff, D.E. Snyder, and J.S. Hagihara, 1984. *An Ecological Land Classification Framework for the United States*. USDA Forest Service, Miscellaneous Publication 1439, Washington, DC. 56 pp.

Ecological Stratification Working Group 1995. A National Ecological Framework for Canada. Agriculture and Agri-Food Canada, Research Branch, Centre for Land and Biological Resources Research and Environment Canada, State of the Environment Directorate, Ecozone Analysis Branch. Ottawa/Hull. Report and map at 1:7,500,00 scale.

ECOMAP. 1993. National Hierarchical Framework of Ecological Units. USDA Forest Service, Washington, DC, 18 pp.

Ecosystem Management Strategy Team. 1994. The Role of the Northeast Area in Ecosystem Management. USDA Forest Service, Northeast Area State and Private Forestry, Radnor, PA. 15 pp.

Frissel, C.A., W.J. Liss, C.E. Warren, and M.D. Hurley. 1986. A hierarchical framework for stream classification: viewing streams in a watershed context. *Environmental Management* 10: 199–214.

Goudey, C.B., and D.W. Smith (comp. and ed.) 1994. Ecological Units of California: Subsections (map). USDA Forest Service Pacific Southwest Region.

Grossman, D.H., K.L. Goodin, S. Li, D. Faber-Langendoen, and M. Anderson. 1994. Standardized National Vegetation Classification System. Paper prepared for the NBS/NPS Vegetation Mapping Program, USDI Biological Survey and National Park Service. The Nature Conservancy, Arlington, VA.

Hanson, D.S., and B. Hargrave. 1996. Development of a multilevel ecological classification system for the state of Minnesota. In: R.A. Sims, I.G.W. Corns, and K. Klinka (eds.), Global to Local: Ecological Land Classification. Thunderbay, Ontario, Canada, August 14–17, 1994. Reprinted from *Environmental Monitoring and Assessment* 39(1–3), 10 p.

Haufler, J. 1994. An ecological framework for planning for forest health. *Journal of Sustainable Forestry* 2(3/4): 307–316.

Haufler, J.B., C.A. Mehl, and G.J. Roloff. 1996. Using a coarse filter approach with species assessment for ecosystem management. *Wildlife Society Bulletin* 24: 200–208.

Hills, G.A., and G. Pierpoint 1960. Forest Site Evaluation in Ontario. Res. Rep. 42. Ontario, Canada: Ontario Department of Lands and Forestry. 64 p.

Hirvonen, H., and J.J. Lowe. 1996. Integration of Canada's Forest Inventory with the National Ecological Framework for State of the Environment reporting. In: *New Thrusts in Forest Inventory*, Proceedings of the Subject Group S4.02-00, Forest Resource Inventory and monitoring and subject Group S4.12-00, Remote Sensing Technology, vol 1. IUFRO World Congress, August 6–12, 1995, Tampere, Finland. pp. 11–26.

Host, G.E., P.L. Polzer, D.J. Mladenoff, M.A. White, and T.R. Crow. 1996. A quantitative approach to developing regional ecosystem classifications. *Ecological Applications* 6: 608–618.

Jordan, J.K., J.R. Somerville, J.L. Meunier, R.N. Brenner, and R. Stockton. 1984. Ecological classification system, its application in development and implementation of forest planning in the Ottawa National Forest. In: *Proceedings Forest Land Classification: Experience, Problems, Perspectives*, 18–20 March 1984. University of Wisconsin-Madison, Madison, WI.

Karr, J.R., K.D. Fausch, P.L. Angermeier, P.R. Yant, and I.J. Schlosser. 1986. Assessing biological integrity in running waters: a method and its rationale. *Illinois Natural History Survey Special Publication*, 5. Champaign, IL. 28 pp.

Kaufmann, M.R., T. Graham, D.A. Boyce Jr., W.H. Moir, L. Perry, R.T. Reynolds, R.L. Bassett, P. Mehlhop, C.B. Edminster, W.M. Block, and P.S. Corn. 1994. An Ecological Basis for Ecosystem Management. USDA Forest Service, Rocky Mountain Forest and Range Experiment Station, Fort Collins CO 80526. GTR RM-246. 22 pp.

Keys, J., Jr., C. Carpenter, S. Hooks, F. Koenig, W.H. McNab, W. Russell, and M-L. Smith. 1995. Ecological Units of the Eastern United States - First Approximation (map and booklet of map units). Atlanta, GA: U.S. Department of Agriculture, Forest Service, presentation scale 1:3,500,000, colored.

Kotar, J., J.A. Kovach, and C.T. Locey. 1988. *Field Guide to Forest Habitat Types of Northern Wisconsin*. 217 pp. Department of Forestry University of Wisconsin-Madison, 120 Russell Laboratory, 1630 Linden Drive, Madison, WI.

Kotar, J., and T. Burger. 1996. *A Guide to Forest Communities and Habitat Types of Central and Southern Wisconsin*. Department of Forestry UW-Madison, 120 Russell Laboratory, 1630 Linden Drive, Madison, WI.

Larsen, D.P., D.R. Dudley, and R.M. Hughes. 1988. A regional approach to assess attainable water quality: an Ohio case study. *Journal of Soil and Water Conservation* 43: 171–176.

Leak, W.B. 1982. Habitat Mapping and Interpretation in New England. Res. Pap. Broomall, PA: U.S. Department of Agriculture, Forest Service, Northeastern Forest Experiment Station. 28p.

Lotspeich, F.B., and W.S. Platts. 1982. An integrated land-aquatic classification system. *North American Journal of Fisheries Management* 2: 138–149.

Lowe, J.J., K. Power, and S.L. Gray. 1994. Canada's Forest Inventory 1991. Petawawa National Forestry Institute Information Report PI-X-115. 67 pp.

Maxwell, J.R., C.J. Edwards, M.E. Jensen, S.J. Paustian, H. Parrott, and D.M. Hill. 1995. A Hierarchical Framework of Aquatic Ecological Units in North America (Neartic Zone). U.S. Department of Agriculture, Forest Service, N. Central

Forest Experiment Station General Technical Report NC-176: pp. 72.

McNab, W.H., and P.E. Avers (comps.). 1994. *Ecological Subregions of the United States: Section Descriptions*. U.S. For. Serv. Pub. WO-WSA-5. Washington DC.

Moeur, M., N. Crookston, and A. Stage. 1995. Most similar neighbor analysis: a tool to support ecosystem management. In: *Proceedings of Analysis Workshop III: Analysis in Support of Ecosystem Management*. April 10–13, 1995. Forest Service, Fort Collins, CO.

Omernik, J.M. 1986. Ecoregions of the Conterminous United States, map (scale 1:7,500,000). U.S. Environmental Protection Agency, Environmental Research Laboratory, Corvallis, OR.

Omernik, J.M. 1987. Ecoregions of the conterminous United States, a supplement to the map. *The Annals of the Association of American Geographers* 77: 118–125.

Omernik, J.M. 1995. Ecoregions: A spatial framework for environmental management. In: W.S. Davis and T.P. Simon (eds.), *Biological Assessment and Criteria: Tools for Water Resource Planning and Decision Making*. CRC Press, Inc., 2000 Corporate Blvd., N.W., Boca Raton, FL 33431. 49–62

O' Neil, R.V., D.L. Deangelis, J.B. Waide, and T.F.H. Allen. 1986. A hierarchical concept of ecosystems. *Monographs in Population Biology* 23: 1–272.

Overton, K.C., J.D. McIntyre, R. Armstrong, Whitwell, and K.A. Duncan. 1995. *User's Guide to Fish Habitat: Descriptions that Represent Natural Conditions in the Salmon River Basin, Idaho.* USDA Forest Service, Intermountain Research Station.

Parrott, H., D.A. Marion, and R.D. Perkinson. 1989. A four-level hierarchy for organizing wildland stream resource information. In: W.W. Woessner and D.F. Potts (eds.), *Proceedings of the Symposium on Headwaters Hydrology*. AWRA 5410 Grosvenor Lane, Suite 220, Bethesda, MD 20814-2192.

Platts, W.S. 1980. A plea for fishery habitat classification. *Fisheries* 5(1): 2–6.

Probst, J.R., and F.R. Thompson III. 1996. Geographical and ecological distribution of birds in the Midwest. In: F.R. Thompson III, Management of Midwestern Landscapes for the Conservation of Migrant Land Birds. U.S. For. Serv., Gen. Tech. Rep. NC-187. North Central For. Exp. Sta., St. Paul, MN. in press.

Reid, M.S., P.S. Bourgeron, H.C. Humphries, and M.E. Hensen (eds.). 1995. Documentation of the modeling of potential vegetation at three spatial scales using biophysical settings in the Columbia River Basin assessment area. Unpublished report prepared for the USDA Forest Service, contract #53-04H1-6890. Western heritage Task Force, The Nature Conservancy, Boulder, CO. 21p.

Rosgen, D.L. 1985. A stream classification system. In: Riparian Ecosystems and Their Management: Reconciling Conflicting Uses. United States Forest Service General Technical Report M-120. Fort Collins, CO: Rocky Mountain Forest and Range Experiment Station.

Rowe, J.S. 1996. Land classification and ecosystem classification. In: R.A. Sims, I.G.W. Corns, and K. Klinka (eds.), Global to Local: Ecological Land Classification. Thunderbay, Ontario, Canada, August 14–17, 1994. Reprinted from *Environmental Monitoring and Assessment* 39(1–3): 11–20.

Rowe, J.S., and J.W. Sheard. 1981. Ecological land classification: a survey approach. *Environmental Management* 5(5): 451–464.

Scott, J.M., F. Davis, B. Csuti, R. Noss, B. Butterfield, C. Groves, H. Anderson, S. Caicco, F. D' Erchia, T.C. Edwards Jr., J. Ulliman, and R.G. Wright. 1993. Gap Analysis: protecting biodiversity using geographic information systems. *Wildlife Monograph* 123. 141 pp.

Sims, R.A., I.G.W. Corns, and K. Klinka. 1996. Introduction —global to local: ecological land classification. In: R.A. Sims, I.G.W. Corns, and K. Klinka (eds.), Global to Local: Ecological Land Classification, Thunderbay, Ontario, Canada, August 14–17, 1994. Reprinted from *Environmental Monitoring and Assessment*, 39 (1–3) 1–10.

Smalley, G.W. 1986. Site Classification and Evaluation for the Interior Uplands: Forest Sites of the Cumberland Plateau and Highland Rim/Pennyroyal. Tech. Pub. R8-TP9. Atlanta, GA: U.S. Department of Agriculture, Forest Service, Southern Forest Experiment Station and Southern Region. 518 pp.

Sneddon, L., M. Anderson, and K. Metzler. 1994. *A Classification and Description of Terrestrial Community Alliances in the Nature Conservancy's Eastern Region*. The Nature Conservancy - Eastern Region, Boston, MA.

Solomon, D.S., R.A. Hosmer, H.R. Hayslett, Jr. 1986. FIBER Handbook. A growth model for spruce-fir and northern hardwood forest types. Res. Pap. NE-602. Broomall, PA: U.S. Department of Agriculture, Forest Service, Northeastern Forest Experiment Station. 19 pp.

Steele, M. 1995. Forest vegetation simulator as a landscape ecology computer model: a case study of deadwood landscape analysis. In: *Proceedings of Analysis Workshop III: Analysis in Support of Ecosystem Management*. April 10–13, 1995. USDA Forest Service, Fort Collins, CO.

Teck, R., and M. Steele. 1995. Forest Vegetation simulation tools and forest health assessment. *Forest Health Through Silviculture: Proceedings of the 1995 National Silvicultural Workshop*. RM-GTR-267.

Thompson, F.R. III, J.R. Probst, and M. Raphail. 1993. Silvicultural options for Neotropical Migrant Birds. In: D.M. Finch and P.W. Stangel (eds.), Status and Management of Neotropical Migratory Birds. U.S. For. Serv., Gen. Tech. Rep. RM-229. Rocky Mountain For. and Range Exp. Sta., Fort Collins. CO. 353-362.

Udvardy, M. D. F. 1975. A classification of the biogegraphical provinces of the World. IUCN Occasional Paper No. 18, International Union for the Conservation of Nature. Moreges, Switzerland.

Uhlig, P.W.C., and J.K. Jordan. 1996. A spatial hierarchical framework for the co-management of ecosystems in Canada and the United States for the Upper Great Lakes Region. *Environmental Monitoring and Assessment* 39: 59–73.

United Nations Educational, Scientific, and Cultural Organization (UNESCO). 1973. *International Classification and Mapping of Vegetation, Series 6, Ecology and Conservation.* United Nations Educational, Scientific and Cultural Organization, Paris, France, 93 pp.

Urban, D.L., R.V. O'Neil, and H.H. Shugart, Jr. 1987. Landscape ecology: a hierarchical perspective can help scientists understand spatial patterns. *Bioscience* 37: 119–127.

USDA Forest Service. 1992. *Proceedings National Workshop Taking an Ecological Approach to Management*. Salt Lake City, Utah, April 27–30, 1992. U.S. Department of Agriculture, Forest Service, Watershed and Air Management, Washington, DC, WO-WSA-3.

USDA Forest Service. 1992. Forest Landscape Analysis and Design, A Process for Developing and Implementing Land Management Objective for Landscape Patterns, R6, ECO-P-043-92.

USDA Forest Service. 1993. *RMSTAND and RMRIS Users Guide*. USDA Forest Service Intermountain Region, Ogden, UT.

USDA Forest Service WO-INS 1993. Draft Report of the Resource Information Project: Water Units. Region 9 File 2530/1390 Oct 2, 1987.

USDA Forest Service. 1995. *Ecosystem Analysis at the Watershed Scale, Federal Guide for Watershed Analysis Version 2.2*. Regional Ecosystem Office P.O. Box 3623, Portland, OR 97208-3623

Van Kley, J.E., G.R. Parker, D.P. Franzmeier, and J.C. Randolph. 1994. *Field Guide: Ecological Classification of the Hoosier National Forest and Surrounding Areas of Indiana*. USDA Forest Service, Hoosier National Forest. Bedford, IN.

Welsh, H.H., Jr. 1994. Bioregions: an ecological and evolutionary perspective and a proposal for California. *California Fish and Game* 80 (3) 97–124.

Wertz, W.A., and Arnold, J.F. 1975. Land stratification for land-use planning. In: G. Bernier and C.H. Winget (eds.), *Forest soils and forest land management. Proceedings, Fourth North American forest soils conference*, 1973, August; Les Presses de l'Universite Laval, Quebec, Canada, pp. 617–629.

Wheat, R.M., Jr., and R.W. Dimmick. 1987. Forest communities and their relationships with landtypes on the western Highland Rim of Tennessee. In: R.L. Hay, F.W. Woods, H. DeSelm (eds.), *Proceedings Sixth Central Hardwood Forest Conference*; February 24–26, 1987, Knoxville, TN. University of Tennessee, Knoxville, TN, pp. 377–382.

## THE AUTHORS

**Constance A. Carpenter**
*USDA Forest Service,*
*Louis C. Wyman Forestry Sciences Lab,*
*P.O. Box 640,*
*271 Mast Road,*
*Durham, NH 03824, USA*

**Wolf-Dieter N. Busch**
*Atlantic States Marine Fisheries Commission,*
*1444 Eye Street N.W., 6th Floor,*
*Washington, DC 20005, USA*

**David T. Cleland**
*USDA Forest Service,*
*North Central Research Station,*
*5985 Highway K,*
*Rhinelander, WI 54501, USA*

**Juan Gallegos**
*USDA Forest Service,*
*Rocky Mountain Research Station,*
*2205 Columbia, SE,*
*Albuquerque, NM 87106, USA*

**Rick Harris**
*Curecanti National Recreation Area,*
*102 Elk Creek,*
*Gunnison, CO 81230, USA*

**Ray Holm**
*USDI Bureau of Land Management,*
*Jarbridge Resource Area,*
*2620 Kimberly Road,*
*Twin Falls, ID 83302, USA*

**Chris Topik**
*House of Representatives Staff, B-308,*
*Rayburn House Office Building,*
*Washington, DC 20515, USA*

**Al Williamson**
*USDA Forest Service,*
*Chippewa National Forest,*
*Rte 3, Box 244,*
*Cass Lake, MN 56633, USA*

# ♦ Humans as Agents of Ecological Change

# Humans as Agents of Ecological Change — Overview

Nels C. Johnson

No species affects ecosystems as strongly, or as pervasively, as humans. Understanding how humans act as agents of ecological change is fundamental to any concept of ecological stewardship. Natural resource managers need this knowledge for several reasons. First, many management choices today — managing fire, for example — are shaped by ecological legacies created during 12,000 years of human occupancy in North America. Failure to recognize human ecological history can result in costly mistakes and missed opportunities in the sustainable management of ecosystems. Second, the current scale of human populations and the rate of technological change are without precedent. If humanity is to sustain the productivity, diversity, and resilience of ecosystems, it must understand better the ecological consequences of human actions. Third, future prospects for diverse and productive ecosystems are likely to depend on such emerging disciplines as restoration ecology, which may redefine human ecological roles in the twenty-first century. These newer disciplines assert that humans should take their ecological roles more seriously during the next century.

The chapters that follow address three important themes in human interactions with the environment. First, how have human interactions influenced current ecosystems, how do these ecological legacies shape management choices, and how do we apply this knowledge? Second, given that sustainability is a human construct, how have changing notions of sustainability contributed to today's growing focus on ecosystem approaches, what are the key elements of sustainability in an ecosystem management context, and how does one measure sustainability? Third, what are the prospects for humans to more deliberately choose their ecological roles? Ecological restoration promises to help redefine roles, but how do growing populations and consumption affect our ability to choose?

## ECOLOGICAL LEGACIES

To understand human roles, one must start with the past. Agricultural fields, pine plantations, channelized rivers, cities, and suburban subdivisions are obvious examples of how human actions have shaped today's landscapes. But even contemporary wildlands have often been influenced by human activities, many by Native American cultures that disappeared more than a century ago. A long history of human interaction has affected the structure and composition of most ecosystems in the United States today to varying degrees.

Images of European explorers stumbling upon an empty continent filled with untouched wilderness have long fueled the popular imagination. True, Native Americans sparsely inhabited this land, but their nomadic ways and primitive technologies were assumed to have had little impact on their environment. Natural historians and early ecologists often shared these views. Even today, the state of pre-European ecosystems are often referred to as "presettlement" conditions, as if people had not been part of the landscape until after Columbus arrived.

Today, a diverse group of disciplines, including anthropology, archeology, ecology, environmental history and demography, have shattered earlier images of primeval landscapes at European contact. Instead, as Bonnicksen et al. show, twelve millennia of Native American occupation had extensive and cumulative impacts across the continent.

To begin, Bonnicksen et al. show that human populations were almost certainly greater than estimates based on the records and observations of early European explorers and settlers. Smallpox, measles, and other introduced diseases spread rapidly and may have decimated up to 90 percent of the Native American population before Europeans reached the interior of the continent. Estimates vary, but anthropologists and demographers now believe that some 8 to 12 million Native Americans, and possibly more, lived in what is now the United States and Canada before the epidemics spread.

Bonnicksen marshals evidence indicating Native Americans quickly developed the knowledge and the tools to influence the environments in which they lived. At first, they affected ecosystems principally through the selective predation of larger vertebrates and the harvesting of plants. In the millennia that followed, prehistoric peoples refined the use of fire, developed stone and ceramic technologies, and practised agriculture, beginning 5,000 to 6,000 years ago.

Native Americans significantly influenced ecosystems in many ways. They used fire widely to increase the abundance of desirable plant species, stimulate plant growth, attract prey species, reduce undergrowth, facilitate hunting, and minimize the potential for severe wildfires. Fires set by humans became a regular disturbance regime in many areas. As a result, forest, grassland, and other ecosystems adapted and changed. Agricultural practices modified or converted natural habitats in limited areas along river bottoms, diverted water from streams for irrigation, and expanded genetic diversity in a handful of wild plants adapted for cultivation. Intensive hunting and gathering altered wild populations — dramatically in some places — where human populations settled or moved through. By 1492, many of North America's landscapes were systematically harvested for favored wildlife and plants, regularly set on fire by their human inhabitants, and laced with villages, trails, and agricultural fields.

European explorers and settlers ushered in a 500 year succession of rapid environmental changes. They viewed the continent as a storehouse of natural wealth to finance wars, and expand trade networks, and as a source of raw materials to fuel economic development at home and in their new colonies. For Periman et al., European roles in the evolution of North American ecosystems must be understood in this historical context.

European trapping, mining, logging, and agricultural practices soon eclipsed the ecological impacts of the Native Americans they displaced. Growing populations, expanding trade, changing technologies, and local natural resource scarcity meant that many landscapes went through a rapid succession of human uses and practices. Landscapes were transformed not just once, but often two, three, and four times within the span of a century or two. Each transformation — for example, a New England hillside converted from forest to patchy agriculture to an area stripped of trees for fuel ironmaking — carried with it different implications for the ecological composition, structure, and function of the landscape. This history, even more than that of the Native Americans, continues to influence the condition of many ecosystems today.

As Bonnicksen notes, one of the most important ecological consequences of European settlement was the cessation of Native American management practices, e.g., suppression of regular burning. Successional pathways changed, wildfires became less common but more severe, and species composition varied, sometimes dramatically, from the ecosystems used for centuries by Native Americans. The lesson here is that sudden departures from long-established human–ecological interactions can lead to yet another ecological transformation. This new ecological condition may bear little resemblance to a primeval state that disappeared when prehistoric people arrived.

The meaning and significance of human ecological history will vary from place to place. One reason for this is that the understanding of human influences, especially those of Native Americans, is substantially incomplete. The archeological record tends to be concentrated where people lived for extended periods or in large numbers. For much of the rest of the landscape, researchers cannot say with certainty whether and how humans significantly influenced ecosystems. Another reason is that contemporary ecosystem management goals may share little in common with those of our predecessors. Earlier tools and strategies may not always have a place in today's management. But, researchers have learned much about the history of human ecological influences. The tools and strategies to apply this knowledge, documented so well by Periman, should be used more often in the context of ecosystem management.

Which leads to the challenge posed by Sullivan et al.: how can heritage management — that is, the protection and analysis of cultural artifacts and practices — assume a more central role in ecosystem management? If it were, Sullivan speculates that both would be transformed to mutual benefit. Unfortunately, heritage management today is practiced largely alone outside the mainstream of natural resources management. As a result, it is frequently viewed as an obstacle rather than a bridge to management goals. Sullivan boldly asserts that ecosystem managers who ignore historical processes cannot hope to acquire the knowledge needed for informed and effective management.

Sullivan suggests a number of steps to more fully integrate heritage management into ecosystem approaches to natural resources management. To begin, natural resource management agencies should explicitly recognize that ecosystems are products of a specific history, usually with some degree of human activity. This knowledge is needed to understand contemporary ecosystem conditions and trends toward future conditions. Next, understanding cultural landscapes should be emphasized, rather than focusing principally on isolated large, rich archeological sites. The nominating process of the National Register of Historic Places, for example, discounts the value of small archeological sites scattered across the landscape that provide essential clues to understanding past land-use systems. This understanding will make heritage management more relevant to ecosystem management needs, such as developing reasonably accurate reference conditions.

## EXPANDING NOTIONS OF SUSTAINABILITY

For over a century, expanding concepts of sustainability — starting with principles of sustained-yield timber management — have driven changes in natural resource management policy and practice in the United States. Today, ecosystem approaches seek to integrate broad concerns about ecological, social, and economic sustainability into natural resource management. Human choices about what to sustain, where, why, and how have taken center stage in modern natural resource management.

MacCleery and Lemaster's compelling history of natural resource policy and practice makes it clear that sustainability concerns are not a recent phenomena. Rather, evolving concepts of sustainability have been at the center of most important policy changes during the past century. At first, the focus was on sustaining trees and water as a foundation for American livelihoods. Other concerns, such as wildlife and recreation, were important but secondary to Gifford Pinchot and other promoters of "scientific sustained-yield" forestry at the turn of the century. Half a century later, growing public demands for wildlife and recreation on public lands led to broadening natural resource management responsibilities, as embodied in the 1960 Multiple Use Sustained Yield Act. The past decade has seen another transition. Today, natural resource manages must address growing scientific and social concerns about such issues as biodiversity, ecosystem health, and public participation that did not exist 20 years before. Ecosystem approaches represent an integrated, but still evolving, response to the latest expansion of sustainability concerns.

Is ecosystem management a small evolutionary step from earlier multiple-use approaches or is it a revolutionary departure? MacCleery and LeMaster believe the jury is still out. Certainly ecosystem management differs in two major respects: (1) it expands management objectives from a few tangible resource outputs to a broader spectrum of values, uses, and environmental services; and (2) it requires consideration of social, biological, and economic interactions at a variety of spatial scales (local, regional, national, and international) over time. But, society is ambivalent about whether it is willing to trade high volume, low cost natural resource commodities for better protection of biodiversity, ecosystem services, and amenity values. Whether ecosystem management is an evolutionary step or a major departure will depend on whether society internalizes broader notions of sustainability in consumer and political behavior.

According to MacCleery and LeMaster, the current state of ecosystem management deserves a constructive critique. In their view, the fundamental test is whether ecosystem approaches can assist in forging a working consensus on public land management. To do that, ecosystem management must value the history of human experience, define a more explicit role for resource production, develop and monitor better criteria and indicators to evaluate performance, and more fully integrate public participation in the earliest stages of assessment, planning, and policy development.

As Johnson et al. make clear, translating concepts of ecosystem management into daily practice is a complex challenge. Managers must integrate across disciplines, political and ownership boundaries, and the interests of an increasingly fractionated public. Resource management careers have long attracted individuals more interested in managing trees, wildlife, and fish than people. But today, resource managers can no longer be the "lone rangers" their predecessors may have been in simpler times. Reaching out to specialists and the public to form partnerships has become a fundamental part of the job — an inescapable conclusion that flows from Johnson's case studies.

How, where, who, and why are common issues faced by any resource professional adapting an ecosystem management approach to their situation. The case studies indicate that innovation and persistence are essential ingredients in most recipes for effective ecosystem management. But they also reveal the unique circumstances of every project — formulas rigidly defined have limited value. Still, Johnson identifies key issues managers must address to sustain biologically healthy ecosystems and the social and economic benefits that flow from them. Among others, one particularly essential to ecosystem management is understanding the short- and long-term consequences of management actions. Failure to do so is at the root of many conflicts over how to manage natural resources.

Carroll et al. address one of the toughest questions faced in natural resources management: What are the key elements of ecosystem sustainability and how are they affected by management practices? Carroll focuses on ecosystem structure and function as key variables in ecosystem sustainability, not because social and economic dimensions of sustainability are unimportant, rather because ecosystem conditions frame the choices humans make about what to sustain and why. Still, as Carroll notes, sustainability is about the interplay of human values, goals, and behavior with biophysical ecosystem properties.

Carroll singles out five biophysical properties of ecosystems that define the decision space for resource managers. Each has clear management implications, but perhaps the most important is ecosystem resiliency. All ecosystems depend on some level of disturbance and are capable of recovering from perturbations, such as fires and floods, but ecosystems also have limits beyond which important ecosystem elements or processes, such as biodiversity or nutrient cycling, cannot recover. Knowing patterns and limits of ecosystem resiliency provides resource managers with a better sense of what can and can't be done if ecosystem sustainability is a goal. Small patch clearing and understory thinning in forest operations may be well within the resiliency of a forest ecosystem, while large clearcuts on south facing slopes may not. Carroll et al. close by describing desired characteristics of indicators for assessing ecosystem sustainability and summarizing a process used to choose such indicators in Arches National Park. These ideas are a useful contribution to resource managers and researchers developing practical monitoring systems that are essential to an adaptive approach to ecosystem management.

## FREEDOM TO CHOOSE?

Although researchers have only begun to understand how humans influence North American ecosystems, no generation is better equipped to learn from past successes and failures. What's been learned thus far provides an array of knowledge, concepts, tools, success stories, and cautionary tales for ecosystem approaches to natural resources management. Nevertheless, galloping global and domestic population growth and rising consumption threaten to leave society with fewer choices about which ecological roles to play and which to discontinue. This juxtaposition makes it all the more important to understand human ecological roles and apply that knowledge wisely.

Restoration ecology shows that people have enough knowledge to begin undoing ecological damage. As Covington et al. note, the principles and concepts of ecological restoration are not entirely new. Aldo Leopold in the 1930s pioneered the use of ecological knowledge to restore native tallgrass communities and reverse the tide of landscape degradation in his native Wisconsin. But it wasn't until half a century later that a recognizable discipline emerged, drawing broadly from growing knowledge in other scientific fields. Ecological restoration now receives widespread attention from natural resource managers interested in improving environmental conditions and from scientists seeking to develop and test ecological theory.

Like other emerging disciplines, such as conservation biology, restoration ecology is characterized by unsettled definitions and vigorous debate over its scientific and management roles. Much of this debate focuses on how prominent human roles should be in ecological processes. Nevertheless, Covington confidently asserts that restoration ecology will assume an increasingly prominent role in ecosystem approaches. Already, ecological restoration is at the heart of some prominent ecosystem management efforts, from the restoration of hydrological functions in the Everglades to the expansion of late successional forest habitats in the Pacific Northwest.

Kenna et al. explore the iterative processes and adaptive learning that characterize successful ecological restoration efforts. Understanding the range of natural or historic variability within the target ecosystem (including the history of human roles) is a fundamental first step that must be followed by defining restoration project's goals and objectives. Given the limited knowledge about the workings of most ecosystems, Kenna warns that subsequent steps to implement ecological restoration should not be construed as a fixed set of procedures. Rather, ecosystem restoration is an exercise in complex problem solving that employs an array of practices to emulate, restore, or protect key natural ecological processes and functions. Such practices to restore ponderosa pine forests in the Southwest, for example, might include removing young trees, hauling away heavy fuel loads under old-growth trees, lighting prescribed fires, removing introduced weeds, and planting native herbaceous species.

For ecological restoration to be successful, Kenna points out that collaborative partnerships among practitioners, researchers, landowners, and the public are essential. Many restoration tools — for example, prescribed fire or the removal of feral animals — can be controversial if there is no consensus on restoration goals or if public education does not precede planning and implementing the projects. Still, the rapid growth of ecological restoration is an encouraging trend. It signals greater interest by scientists, resource managers, and the public in redefining human roles in ecosystems substantially degraded by past short-sighted or ignorant actions.

But how much freedom to choose ecological roles will there be in an increasingly crowded world? Cohen explores how human use of land, water, and other natural resources is influenced by four factors: population, economics, environment, and culture. Carefully pointing out that past expectations about the future have usually been wrong, Cohen nevertheless concludes that demographic trends will pose American citizens and resource managers with future trade-offs in land use. One reason is that domestic trends — growing populations, increased economic demand for timber, and other resource commodities, and rising social demand for conservation and recreation — will likely sharpen rather than diminish conflicts among competing interests. Another reason is that the United States is becoming progressively less isolated from the rest of the world. As populations and economies grow in resource-poor countries, people around the world are likely to become larger contenders with Americans for the products and services of ecosystems in the United States (just as Americans are major consumers of natural resources elsewhere).

As Cohen suggests, changes in cultural values and innovations in policy and technology can alter the equation linking demographic and economic growth to natural resource consumption and land use. The freedom to widely practice ecosystem approaches to natural resources management will depend as much on culture and politics as it will on science, population, and economics. From Cohen's perspective, an important part of this cultural and political change will have to include greater willingness by the public and policy makers to pay higher prices for products that cost more to produce, but generate fewer unwanted ecological impacts. Willingness to push this economic frontier will help determine the future degrees of freedom humans have to choose their ecological roles.

## THE AUTHOR

**Nels C. Johnson**

*Biological Resources Program,*
*World Resources Institute,*
*10 G St. N.E.,*
*Washington, DC 20002, USA*

# Native American Influences on the Development of Forest Ecosystems

Thomas M. Bonnicksen, M. Kat Anderson, Henry T. Lewis, Charles E. Kay, and Ruthann Knudson

**Key questions addressed in this chapter**

- *Did American Indians have the incentive, technology, and numbers needed to modify and maintain forests and other ecosystems?*
- *How did American Indians exploit and modify ecosystems to obtain the resources they needed to survive, and what can we learn from them that might be useful today?*
- *Did the American Indian's exploitation of resources play a pivotal role in the development of North America's native ecosystems?*
- *How has the removal of American Indian influences contributed to the increase in endangered species and the decline of native ecosystems?*
- *Why are people who use local habitats through day-to-day activities, such as hunting, trapping, fishing, farming, and herding important to science and ecosystem management?*
- *What actions should we take to use the lessons of the past in modern management?*
- *What more do we need to know about native ecosystems and indigenous management practices, and how do we get the answers?*

**Keywords: Fire regimes, hunter-gatherers, agriculture, traditional knowledge, ethnobotany, human populations, ethnography**

# 1 INTRODUCTION

Ecosystem management cannot succeed in promoting stewardship if it fails to recognize that humans are an integral and natural part of the North American landscape. Ecosystem management has the potential for widening the gap between people and nature. Subdividing landscapes into ecosystems could create the false impression that ecosystems are real things. This illusion becomes more dangerous when people think that they live on the outside and nature exists on the inside of ecosystems.

Biologists developed the ecosystem model to describe physical, chemical, and biological interactions at a particular time within an arbitrarily defined volume of space (Lindeman 1942). They usually exclude people because the boundaries are sometimes drawn around small parts of the landscape, such as watersheds. Because management decisions come from outside, ecosystems appear as separate entities. Therefore, ecosystem management may reinforce the myth that nature exists apart from people if it does not explicitly state otherwise.

A corollary myth assumes that climate dictated the structure and function of ecosystems. On the contrary, climate provides either a favorable or unfavorable physical environment for certain plants to grow. It does not dictate which plants grow in that environment. Similarly, climate does not dictate human behavior. It only sets temporary limits. Human innovations in technique and technology can and do push back those limits. Therefore, climate is not the sole determinant nor even in many cases the dominate force in guiding the development of particular ecosystems. American Indians selectively hunted, gathered plants, and fired habitats in North America for at least 12,000 years. Unquestionably, humans played an important role in shaping North America's forest ecosystems.

Interpretations of the impacts made by indigenous people in North America are largely limited to what can be postulated in terms of paleontological, anthropological, and archaeological evidence. None of these approaches have been completely persuasive to skeptics who require more substantial and corroborative evidence before accepting the significance of the environmental changes induced over 12,000 or more years by hunting-gathering societies and, for the last 2,000 years, by indigenous farmers as well. Taken together, however, the evidence shows a clear and convincing pattern of indigenous human influences on prehistoric, historic, and contemporary ecosystems.

In this chapter, we argue that the success of ecosystem management depends on understanding reciprocal relationships between native forests and indigenous peoples. Consequently, we concentrate on the development of forests prior to European settlement. Particular emphasis is placed on American Indians consciously and actively managing landscapes through the selective killing of animals, the cultivation of preferred plants, and the widespread manipulation of habitats with fire.

We also concentrate on what indigenous people did to survive and prosper. We believe it would be inappropriate to use today's ideas and values as standards for judging their actions. Therefore, our chapter focuses on the management practices of indigenous people that succeeded for them and may be useful to us (Rides at the Door and Montagne 1996).

Finally, we argue that local knowledge and practices that followed European settlement provide analogues for reconstructing pre-European settlement conditions as well as for suggesting answers to contemporary management problems. Equally important, we believe that ecosystem management cannot succeed unless current human residents of forests become intimately involved in decisions that affect their lives and surroundings.

# 2 AMERICAN INDIANS SHAPE A CONTINENT

The romantic 20th century idea of a natural area or wilderness as a place without human influence became meaningless in North America when Paleoindians pushed southward between the continental ice sheets and perhaps along the Northwest coast 12,000 or more years ago. They found two unexploited continents with bountiful game. Their populations grew and by 11,200 years ago there may have been millions of Paleoindians living from coast to coast in both North and South America (Fiedel 1987, Roosevelt et al. 1996).

## 2.1 Populations at Contact

The earliest European explorers found the southern United States heavily inhabited by American Indians. At the time of European contact, at least 600 tribes or bands consisted of as many as 12 million American Indians, maybe more, who were actively managing or influencing every corner of the continent (Dobyns 1983).

In some parts of North America, Indian populations were so large, that is before they were decimated by European-introduced diseases (Ramenofsky 1987), that they competed with some wildlife species for food (Neumann 1989). For example, Neumann (1985) found that passenger pigeons (*Ectopistes migratorius*) were relatively rare in pre-European settlement times because human populations in the eastern United States were so extensive that native people consumed most of the seeds and berries, collectively called mast, that the birds needed. It was not until after European diseases decimated native populations ca. 1550 that most of the mast crop became available for wildlife. This led to increased numbers of passenger pigeons, until by the 17th century, flocks darkened the sky. Thus, the tremendous populations of passenger pigeons described in historical journals may have been an artifact of the dramatic decline in American Indian populations (Stannard 1992) and were not representative of earlier times.

Judgments of the significance of the impact of pre-1492 human communities on North American natural resources are always based in part on the estimates of those human populations at any given time. In 1996 North America north of Mexico had a population estimated to be approximately 400 million people, roughly two percent of whom had significant genetic ties to pre-1492 Americans.

Estimates of pre-1492 North American human populations range from approximately 2 million (Ubelaker 1988) to 18 million (Dobyns 1983). Ubelaker (1988) relied on a variety of sources of American Indian population estimates to arrive at his conclusion that only 1,894,350 people were spread across the continent at the time of contact. On the other hand, Dobyns (1983) spent over 30 years studying American Indian demography from pre-European to present conditions, building on previous work by Cook (1937) and others. He calculated human population densities based on the availability of natural resources on which archeological evidence shows groups of native people relied (e.g., edible wild plants, cultivated plants, and sources of animal protein). As a result, Dobyns' (1983) estimate of the 1492 population of North America is 18 million people, with an additional 80 million in Mexico, and Central and South America. Fiedel (1987) used still a different method to derive his intermediate estimate of a 1492 population of 12 million American Indians in North America.

Wide variations in population estimates reflect not only different methods for deriving estimates but also different assumptions about the impact of European diseases. That is, low population estimates usually discount the importance or timing of pandemics on American Indians. From the 1930s on, scholars have been reevaluating estimates of pre-Columbian human populations in the Americas to account for the effect of disease. Old World contagious diseases, such as smallpox, measles, influenza, bubonic plague, diphtheria, typhus, and cholera devastated American Indian populations (Ramenofsky 1987). Perhaps the first documented impact of European disease on American Indians can be found in the Narrative of Alvar Nunez Cabeza de Vaca written in 1528 (Hodge 1990). European diseases may have spread rapidly enough to decimate Indian populations in the Pacific Northwest between 1550 and 1600 (Campbell 1990). However, Henry Dobyns (personal communication with Knudson, October 30, 1995) recently found documentation that swine flu may have been transmitted to America as early as 1493 by crew members of Columbus' second expedition. These diseases spread rapidly because American Indians had regular and extensive trade and exchange systems that covered the continent, and relatively dense populations in some regions. Therefore, low population estimates may be too conservative.

## 2.2 Modifying Landscapes

American Indians were not passive occupants of the land. Regardless of their population at the time of European contact, the landscapes of North and South America described by the first explorers and settlers had already been shaped by millions of people over thousands of years of use and management (Deneven 1992, Fiedel 1987, Gomez-Pompa and Kaus 1992). American Indians had the incentive, technology, and numbers to modify and maintain forests and other ecosystems. By 11,200 years ago the Monte Alege Paleoindians had already begun cutting and burning tropical rainforests in the Amazon to produce fruits, nuts and other foods (Roosevelt et al. 1996). Paleoindians in North America were equally capable of managing their environment. Therefore, the conclusions drawn by Roosevelt et al. (1996) for South America also hold true for North America:

> "Archaeological and ethnobotanical evidence shows that Amazonian forests once thought to be virgin were settled, cut, burned, and cultivated repeatedly during prehistoric and historic times, and that human activities widely altered topography, soil, and water quality. Substantial biodiversity patterning appears to be associated with such human activities."

Unequivocal evidence now exists from numerous sources that American Indians were deliberately managing the land using a variety of techniques for millennia. This is substantiated by contemporary tribal cultures, studies of museum artifacts, archeological remains, paleoecological findings, fire scar studies, and ecological field studies. Numerous early ethnohistoric and ethnographic studies refer to management practices of Indians in different parts of North and South America. Although much is known about Indian management practices in many areas of North America, further research is needed to cover more areas and to enhance our understanding. Nevertheless, enough evidence exists from written documents alone to build an incontestable case that Indians managed the North American landscape.

During the 12 or more millennia in which people have lived in North America, they fished; herded and hunted game; dug roots; collected berries and nuts; cultivated wild plant foods; built structures of earth, rock, wood (including leaves and bark) and bone; quarried stone for making flaked and ground stone tools, or to be used as building material; mined and smelted minerals; strip-mined for coal; burned forests, prairies and wetlands; cleared fields and planted and harvested domesticated crops; built irrigation canals and roads; and dug wells for water, gas and oil. Some managed the North American continent as scattered bands that moved from place to place. Others concentrated by the thousands in sedentary urban masses. Their management practices and day-to-day activities affected nearly all of America's ecosystems in some way, especially forests and prairies.

Many people believe that North America's native forests were all-aged, dense, and self-perpetuating, with a multi-layered canopy that shaded a forest floor strewn with piles of dead trees. Such a view is untenable given the pervasive effects of American Indian use and management. Therefore, only a small proportion of the native forests in the United States fit this image — notably western hemlock (*Tsuga heterophylla*) forests in the Pacific Northwest, and beech (*Fagus grandifolia*) and sugar maple (*Acer saccharum*) forests in the Upper Midwest and East (Daniel et al. 1979, Davis and Mutch 1994). Even all-aged forests burn occasionally, and tornadoes and hurricanes often create large openings in which shade intolerant trees like aspen (*Populus tremuloides*) and birch (*Betula* spp.) can flourish, thus temporarily converting these forests into mosaics that included groups of young trees.

Some of North America's native forests were open and park-like, not dark and dense. For example, Thomas Ashe, an 18th century English traveler, made a statement that was echoed by many other early observers dating back hundreds of years when he said that "The American forests have generally one very interesting quality, that of being entirely free from under or brushwood" (Bakeless 1961). Native forests stayed open because Indian and lightning ignited fires were relatively frequent and extensive. Because these fires were frequent enough to clean up most of the debris on the forest floor in many forest types, fires generally stayed on the ground and burned as surface fires. Some surface fires did reach the canopy and burn small areas of forest, but large crown fires were rare.

Some also believe that native forests consisted of mostly large old trees that covered the landscape like a blanket. Most native forests were dramatically different from this popular image. Instead of a blanket of old trees, native forests consisted of a dynamic mosaic containing patches of young and old trees, interspersed with patches of grass and shrubs. American Indians played a key role in creating and maintaining this mosaic by favoring early successional stages that produced the resources they required. They cleared old forests to make room for crops and new forests, removed fallen trees and debris, thinned overstory and understory trees, and created a diverse landscape of forests, brushlands, and prairies. Indian cultivation and timber harvesting, Indian and lightning fires, and other natural forces such as tornadoes, hurricanes, insect infestations and disease interacted to sustain the mosaic structure of most North American forests. Thus, old trees could only occupy a small part of the landscape in most forest mosaics, but the proportion varied with the frequency and severity of disturbances (Bonnicksen 1994a; Bonnicksen and Stone 1985; Cooper 1960, Teensma et al. 1991).

In short, historical evidence paints a picture of native forests and indigenous peoples that differs markedly from the sentimental views we often nurture today. The management practices of American Indians helped to shape the development of North America's native forests. Understanding native forests and the management practices of American Indians that helped to create and sustain them provides essential lessons for ecosystem management.

## 3 FIRE: CREATING AND MAINTAINING HABITATS

American Indians modified many North American plant communities to meet their needs. They not only engaged in widespread burning but they also cut trees for agriculture, fuel, and building materials. This habitat modification benefited many species, mostly those associated with disturbance, but adversely impacted others, such as those dependent on late successional vegetation or old-growth forests. Most extinctions on Pacific Islands, for example, are thought to have been more the result of habitat alteration than hunting (Olsen 1989). On the other hand, some species that are now in danger of extinction were common in pre-European settlement times because they prospered under various aboriginal land management systems (Blackburn and Anderson 1993).

Fire was the most powerful tool for managing habitats available to American Indians. American Indians used fire to modify forests, brushlands, and grasslands. Missionaries' diaries, settlers' journals, and explorers' and surveyors' field notes document the intentional burning by American Indians as a universal practice. For example, many early explorers commented about Indian burning in southern forests. In 1528, Alvar Nunez Cabeza de Vaca recorded in his narrative that in Texas "The Indians of the interior ... go with brands in the hand firing the plains and forests within their reach, that the mosquitoes may fly away, and at the same time to drive out lizards and other like things from the earth for them to eat. In this way do they appease their hunger, two or three times in the year..." (Hodge 1990).

American Indians also burned forests in California, the Northeast, northern Rocky Mountains, the Southwest, the Northwest, and the Midwest. They burned to stimulate the production of edible wild plants for seeds, leaves, fruits, bulbs and stems; to promote young adventitious and epicormic shoots of shrubs for basketry; to discourage insects, diseases and weeds; to increase palatable grasses and browse for animals; to increase visibility; to facilitate travel; to decrease dense brush and promote more open country; to enhance waterfowl habitat; to create firewood; to use as a weapon of war; to eliminate rubbish and pests in villages; to open and maintain trails; to signal one another; to create fuel breaks; to open meadows along forest trails that attract animals and provide living food reserves during travel; to clear the ground for gathering acorns; to reduce competition for mast trees; to cook insects or seeds before harvesting; to clear fields for planting; to flush and surround game; to clear underbrush that could hide their enemies; and for many other reasons. Some Indian-set fires also went out of control, and signal fires and campfires often were left to burn (Barrett 1980a, b; Barrett and Arno 1982; Blackburn and Anderson 1993; Day 1953; Lewis 1978, 1980; Pyne 1982).

## 3.1 Corrective Fires

The most dramatic changes suggested by paleobotanical evidence are seen in the relative frequencies of types of pollen and correlated increases in charcoal which are associated with the entry of humans into wilderness regions. These include the original discoveries of North America, Australia, New Zealand, Madagascar, and the larger islands of the Pacific such as Hawaii. However, even if it is accepted that the arrival of humans and the correlated changes in pollen frequencies and in charcoal cannot have been merely fortuitous or unique episodic events, the central question remains: How could populations of hunter-gatherers, using relatively simple tools, have brought about changes and subsequently maintained conditions that were different from what would occur under natural circumstances?

Although the environmental impacts made by aboriginal peoples occurred earlier in Australia (50,000–60,000 BP) than in the Americas (12,000 or more BP), paleobotanical studies and their interpretations more clearly indicate a pattern of human influences, both because of the greater amount of work there and the more pyrophytic characteristics of the flora (Kershaw 1986, Singh et al 1981). At the same time, Australia also provides striking examples of artificially induced changes which have been used for constructing analogues of both how and why such changes associated with the arrival of humans would have occurred, as well as the patterns of burning that followed the initial perturbations (Lewis 1994).

The contemporary analogues for Australia are based on what Aborigines in more remote parts of the continent do when, on returning to areas previously occupied and subsequently abandoned, set late dry-season, high-intensity fires to "clean" the environment. These "corrective fires," (Lewis 1994) are set in old-growth, fuel-laden stands of scrub, wetlands, and trees, habitats that have not been burned for 10 or more years. This type of burning, which sometimes results in fairly large conflagrations, is carried out to re-establish conditions within which Aborigines can once again employ early dry-season, low-intensity, and less disruptive "management fires." Part of the overall goal is to re-establish a greater mix of habitats, stands of vegetation at varying stages of ecological succession, induce a greater productivity of selected resources, and reduce the unpredictable disturbances from fires ignited by lightning or accidentally set at the wrong times of year. Conditions for safely and effectively managing resources are virtually impossible to establish and maintain in areas previously not cleared of old-growth stands and large, potentially dangerous accumulations of fuel.

"Corrective fires" are, therefore, considered the necessary, albeit draconian, means for correcting what Aborigines perceived to be the basic environmental problem: fuel. For aboriginal inhabitants of Australia, it was the excessive amounts of old growth and detritus that accumulated because of human neglect, not the requisite need of using fire, that was so highly problematic. These examples, it has been argued (Lewis 1994), represent the archetypes of what hunting–

gathering peoples did in the past when introducing culturally based fire regimes into what were true wilderness areas, beginning the process that significantly changed a range of environments periodically hit by natural fires, such as those originally found in Australia, North America, and elsewhere.

## 3.2 Management Fires

If human-set fires had a goal, it was (and in parts of Australia still is) to establish and then maintain significantly different conditions than those found to exist naturally. The maintenance of such conditions was not the simple replication or even only the intensification of natural fire regimes; to a very large measure it involved their replacement.

Given the composition and characteristics of particular habitats, and especially the culturally defined resources therein, the distinguishing features of cultural fire regimes include: (1) the alternate seasons for burning different kinds of settings, (2) the frequencies with which fires are set and reset over varying periods of time, (3) the corresponding intensities with which fuels can be burned, (4) the specific selection of sites fired and, alternately, those that are not, and (5) a range of natural and artificial controls that humans employ in limiting the spread of human-set fires, such as times of day, winds, fuels, slope, relative humidities, and natural fire breaks. (Lewis 1982).

The considerable literature on hunter-gatherer uses of fire shows that the seasonality of setting fires varied from habitat to habitat. Natural fires primarily derive from late summer or late dry-season lightning storms. They occur at the peak or end of growing seasons, and at times when fires can be highly disruptive and invariably destructive in areas dominated by old growth and detritus. In contrast to natural fire regimes, hunting–gathering peoples set fires that were relatively easy to control and less disruptive, and that initiated plant growth weeks or even months before new growth occurs naturally. A goal of human-set fires was to induce the early emergence of plant growth. Thus, fires were set during the spring in northern and more temperate regions, or in late summer and early autumn in areas where growth took place during much of the winter (e.g., the semi-arid parts of California and in other similar Mediterranean-type regions).

Along with seasonality of burning, hunter-gatherers set fires in terms of different, invariably much greater frequencies than those that occur as a result of lightning storms. In addition, because lightning strikes are partly a function of altitude and are less likely to ignite grasslands, open prairies and brushlands, hunter–gatherers regularly, in some cases annually, set fire to areas as small as forest glades to those constituting large sections of open prairies (Boyd 1986; Lewis 1973, 1982; Morgan 1978; Stewart 1955). In a work summarizing historical accounts of Indian and natural fires on the Northern Great Plains, Higgins (1986:7) stated that,

> "The pattern of seasonality of lightning-set fires ... showed that 73% of the lightning-set fires occurred in July and August. This pattern was different from the two seasonal periods of Indian-set fires reported here-March through May with a peak in April, and July through November with a peak in October. Therefore, it seems that the northern plains Indians did not pattern their use of fire according to lightning-fire pattern but for some other reason."

Other reasons concerned the influence of Indians on local and seasonal movements of bison to induce them to graze in tall-grass areas near their winter encampments along the ecotones separating forests and prairies (aspen parklands and riverine valleys). Then, in burning the tall-grass ecotone areas in early spring, bison were alternately drawn onto the short-grass prairies which had been burned the previous autumn. In these areas new growth emerged earlier and more lush in the spring than it did in unburned areas. The scale of fires set by Indians on the short-grass plains was much greater than in nearby parklands and far greater still than what has been estimated for adjacent boreal and montane forest regions (Lewis 1982).

Other habitat types, such as hardwood and coniferous forest understories, were burned every three or four years, resulting in more intensive mosaics of burned and unburned forest stands (Hammett 1992, Patterson and Sassaman 1988, Day 1953). Stands of sclerophylous brush, such as the chaparral belts in California's coastal and interior foothills, were fired every 12 to 15 years with a resulting patchwork from new to old- growth stands, thus providing much greater diversity, overall productivity, and an essential degree of protection against the disruptions and dangers of wildfires (Shipek 1993; Bean and Lawton 1973; Lewis 1973).

Whereas grassland openings within the boreal forests of western Canada were annually burned whenever possible, surrounding stands of trees were only fired following forest die-off or windfalls, approximately every 80–100 years (Lewis 1982, 1973). As with meadows and small prairies, they were burned as early in the spring as possible, at times and under conditions when surrounding forests were too moisture laden to ignite or cause serious fires. Similarly, within temperate rainforest regions, such as those found from northern California to southern Alaska, forest openings and grassy ridges were annually maintained while surrounding forests were left unburned (Gottesfeld 1994, Turner 1991, Lewis and Ferguson 1989; Lewis 1973). In all boreal and rainforest regions, the maintenance of grasslands and forest openings were critical for concentrating scarce animal resources, for what is described as "yarding," in areas that are rapidly taken over by forest growth in the absence of regular burning.

The frequency of burning depended upon a variety of considerations relating to the resources involved. These include the buildup of fuels, protection of local settlements, the overall pyrophytic characteristics of plant community types, economic strategies (e.g., foraging vs. farming), and, particularly at the time of European contact, the combined effects of introduced diseases, depopulation, and technological changes (e.g., the introduction of the horse and gun on the North American Plains). In turn, these considerations directly affected the intensity of fires because a mix, or mosaic, of habitats at various stages of growth and regrowth, with smaller accumulations of fuels fired at times of the year when burning conditions were low, reduced the impacts of both natural and accidental fires that occurred during drier weather. At the same time, by selecting the times of day, and making use of winds and natural barriers, human-set fires were far less disruptive and much less dangerous than fires ignited accidentally or as the result of lightning at what were considered by hunter-gatherers to be the wrong times of the year.

Unlike natural fires, human-set fires were ignited in selected areas, at particular times, while being excluded or temporarily withheld from other areas for shorter or longer periods. In this respect, lower lying regions and forest openings — areas much less subject to lightning fires than forests and brush at higher elevations — were purposely selected for hunting and gathering advantages, burned with greater frequency, at lower intensities, and at times considered most desirable for one and often more resources involved. Whereas we have yet to understand fully the ecological differences resulting from the introduction of culturally based fire regimes, undoubtedly changes in seasonality, frequency, intensity, and selectivity of burning resulted in changes for natural environments. Human population densities and the associated adaptations of American Indians could not have depended on the erratic occurrence, random distribution, and unpredictable effects characteristic of lightning fire regimes.

A further dimension of foraging and the correlated uses of fire is evidenced from historical and ethnographic descriptions, which note a greater mix of habitat types at varying stages of ecological succession. This is especially noteworthy in the eastern United States where lightning ignitions are relatively few (Patterson and Sassaman 1988). The overall combined subsistence patterns of hunting-gathering and agricultural tribes, further influenced by population densities and later by economic contacts with Europeans, resulted in landscapes that were more cultural in their conformation than they were natural. As Patterson and Sassaman (1988:118) pointed out,

> "Once established, agriculture promoted higher population density and growth, placing a greater demand on limited resource space. Fire became important for two reasons. First, it afforded greater returns on investments in hunting and gathering by increasing the quality and quantity of resources, by reducing the risk of uncertainty, and by promoting cooperation in resource procurement. Second, fire became important to agricultural activity itself as a means of creating arable land and maintaining its fertility."

As with hunters and gatherers elsewhere in the world, the broad spectrum of plant and animal resources utilized by American Indians was enhanced by maintaining a more diversified and complex mosaic of ecosystems at various seral stages than was found under natural conditions. At the same time, the potentially adverse effects of natural fires were much reduced with the presence of culturally based and technologically maintained fire mosaics. Thus, the paradox of humans introducing and subsequently maintaining cultural fire regimes was found in the initial, highly disruptive impacts necessary for establishing conditions under which resources could then be managed. Thus, the worst affects of natural conflagrations could be avoided or at least reduced in severity.

## 4 HUNTING: EXPLOITING AND SUSTAINING GAME

The millions of American Indians (before European-introduced diseases decimated their numbers) had a major impact on wildlife populations, often determining the abundance and distribution of entire species. Certainly American Indians were the ultimate keystone species (Kay 1995). However, researchers have only begun to document the extent and magnitude of aboriginal use and management of wildlife (Alvard 1995, 1994, 1993a,b; Heinen and Low 1992).

### 4.1 Pleistocene Adjustments to Paleoindian Hunting

When the first aboriginal people entered North America 12,000 or more years ago, they found a land populated with four kinds of mammoths (*Mammuthus* spp.), mastodons (*Mammut americanum*), ground sloths (at least four genera), giant beaver (*Castoroides ohioensis*), wild horses (*Equus* spp.), camels (*Camelops hesternus*), big-horned bison (*Bison latifrons*), and other megaherbivores that were prey for an equally impressive collection of megapredators—sabre-tooth cats (*Homotherium* spp. and *Smilodon* spp.), lions (*Panthera leo*), dire wolves (*Canis dirus*), and short-face bears (*Arctodus simus*), among others. Within a few centuries, humans had colonized North and South America, and most of the Pleistocene megafauna were extinct.

In 1860, Sir Richard Owen suggested that these extinctions were the result of the "appearance of mankind on a ... tract of land not before inhabited." In 1911 Alfred Russel Wallace, codeveloper of the theory of evolution by natural selection, wrote, "I am convinced that the rapidity of the extinction of so many large Mammalia is actually due to man's agency" (Leakey and Lewin 1995:172). Their conclusions, however, were largely ignored until revived by Paul Martin who postulated that Paleoindians had unintentionally driven the megafauna to extinction, a contention that has been widely debated ever since (Flannery 1990; Stuart 1991; Martin 1990, 1986; Diamond 1989; Martin and Klein 1984).

Most critics of Martin's overkill hypothesis claim that climatic change, not humans, was primarily responsible for the megafauna extinctions (Grayson 1991; Stuart 1991; Leakey and Lewin 1995; Zimov et al. 1995). However, no small mammals, birds, reptiles, amphibians or plants became extinct—the exact opposite of what would be expected if climatic change was responsible. Furthermore, the climatic changes that occurred when the megafauna became extinct were not unique but had recurred repeatedly throughout the Pleistocene, yet megafauna extinctions are only associated with the last deglaciation.

In addition, megafauna extinctions in other parts of the world are associated with the arrival of aboriginal hunters or with the advent of new hunting technologies. This pattern is especially clear on islands in the Pacific Ocean where the arrival of aboriginal peoples did not coincide with climatic events. Recent studies have shown that prehistoric peoples caused massive extinctions and habitat change throughout the Pacific, including New Zealand and Australia (Cassels 1984; Olson 1989; Flannery 1990; Steadman 1995).

Today the climate paradigm has been expanded to include the effects of human predation. Some scientists now believe that a combination of intense hunting and a loss of habitat from climate change could have caused the American megafauna extinctions (Stuart 1991). Regardless of the cause of extinctions at the end of the Pleistocene, aboriginal hunting and management practices, especially by using fire, tended to maintain a relatively stable mix of wildlife species in North America for thousands of years after the Pleistocene extinctions.

## 4.2 Limiting and Protecting Game Populations

American Indians protected habitats and promoted biodiversity in plant and animal communities by keeping ungulate numbers low (Wagner and Kay 1993, Zimov et al. 1995). Prior to the early 1800s, for example, millions of beaver (*Castor canadensis*) occupied lush riparian zones throughout the West. Beaver were so abundant that in 1825, Peter Skene Ogden's party trapped 511 beavers in only five days on Utah's Ogden River. In 1829, Ogden also reported that his fur brigade took 1,800 beaver in a month on Nevada's Humboldt River (Kay 1994). Yellowstone too once contained large numbers of beaver, but that species is now extinct on the park's northern range. Without American Indian hunters, the park's burgeoning elk population has nearly destroyed the willow and aspen communities that beaver need for food and dam-building materials (Kay and Platts 1997). So American Indian hunting benefited all species by preventing habitat destruction by large populations of ungulates.

Many people believe that before the arrival of Europeans, North America teemed with wildlife, especially ungulates like elk (*Cervus elaphus*), moose (*Alces alces*), and bison (*Bison bison*). Although wildlife were abundant in many places, historical records show that game populations fluctuated in response to local weather, and the hunting and habitat modifying activities of American Indians. Thus American Indians limited game populations in some areas by hunting, and increased populations in other areas by using fire to improve habitats. Changes in weather, predators, and other factors in a particular area aided or interfered with their success. The overall effect of American Indian hunting, however, was to maintain game populations within the carrying capacity of their habitats.

Hunting pressure necessarily depressed game populations in certain places. Birkedal (1993) reported that American Indians armed with no more than spears and hunting dogs once kept Alaskan grizzly bear (*Ursus arctos*) populations at very low levels. In the Intermountain West, early fur trappers seldom reported seeing or killing a moose. When Peter Skene Ogden's fur brigade killed three moose near present-day Philipsburg, Montana, in 1825, he noted that it was the first time any of his men had seen a moose despite having spent a total of nearly 300 man-years in the West during the early 1800s. Kay (1997) concluded that native hunting controlled the numbers and distribution of moose throughout western North America — often keeping moose populations at exceedingly low levels because the species was so easy to hunt.

Although not as rare as moose, elk were also historically uncommon in parts of the Rocky Mountains. Between 1835 and 1872, for example, 20 parties spent a total of 765 days traveling through Yellowstone on foot or horseback, yet reported seeing elk only once every 18 days — today nearly 100,000 elk live in that ecosystem (Kay 1990). The same was true in the Canadian Rockies where early explorers reported seeing elk only once every 31 days (Kay and White 1995). During the 1800s, elk also were rare in parts of Utah, Arizona, New Mexico, as well as other regions of the Intermountain West (Kay 1995, Truett 1996).

Moreover, deer (*Odocoileus hemionus* and *O. virginianus*), antelope (*Antilocapra americana*), and bighorn sheep (*Ovis canadensis*) were also rare in many places when the Rocky Mountains were first visited by Europeans. Accounts of starvation and killing horses for food are common in early journals (Kay 1990). Except for Idaho's Snake River Plain and surrounding areas, few bison were ever seen west of the Continental Divide. Today in Yellowstone National Park there are approximately 4,200 bison, but between 1835 and 1872, early explorers encountered bison only three times despite spending 765 days there (Kay 1990). Game populations also were low on many parts of the Columbia Plateau and in the Great Basin at historical contact (Daubenmire 1985).

Historical photographs also indicate that ungulate populations were low in many places. In Yellowstone, for example, 1870 photographs show that aspen, willows (*Salix* spp.), and other preferred forage species were not as heavily browsed in the early years of the park's existence as those plants are today. Because many plants were 70 to 100 years old when they were first photographed, this means that the enormous numbers of elk that inhabit the northern range today did not exist from the late 1700s through to the 1870s. The same is true in other parts of Intermountain North America.

Archaeological evidence reinforces the idea that ungulates were rare in heavily populated parts of North America before Europeans arrived. Elk presently dominate ungulate communities in most western national parks, yet elk are seldom recovered from archaeological sites. Of over 52,000 ungulate bones excavated from more than 200 archaeological sites in the western United States only 3% were elk and only one moose bone was identified (Kay 1990, 1995).

Both carnivore predation and native hunting limited ungulate numbers (Walters et al. 1981). The age of their respective kills, however, shows that American Indians were more efficient predators than carnivores such as wolves (Temple 1987). The more difficult it is for a predator to capture a particular prey, the more that predator will take weakened individuals and young. So, if two or more predators are preying upon the same species, the least efficient predator will tend to kill fewer prime-age animals. Although wolves (*Canis lupus*) and other carnivores kill primarily young-of-the-year and old animals, American Indians killed mostly prime-age ungulates. For example, ungulates recovered from Intermountain archaeological sites are dominated by prime-age animals, thus suggesting that American Indians were more efficient predators than wolves or other carnivores.

Killing mostly prime-age animals is contrary to a maximum sustained yield strategy (Hastings 1983, 1984). The impact of American Indian hunting was further increased because they killed primarily females (Kay 1994a). Furthermore, because American Indians could prey-switch to small animals, plant foods, and fish, they could take their preferred ungulate prey to low levels without having an adverse effect on human populations. Thus, American Indian hunting probably sustained wildlife habitats by limiting game numbers and preventing large fluctuations in populations.

According to predator-prey theory, prey populations will increase if they have a refugium where they are safe from predation (Taylor 1984). Ungulates that could escape aboriginal hunters in time or space were more abundant. Moreover, refugia do not have to be complete to be effective. Partial refugia will also enable prey populations to survive. This explains why there were larger numbers of ungulates on the Great Plains and in Arctic regions. By undertaking long-distance migrations, bison and caribou (*Rangifer tarandus*) could outdistance most of their human and carnivorous predators (Kay 1994a).

Ungulates also were able to survive in buffer zones between warring tribes (Hickerson 1965). Lewis and Clark (1893:1197), for example noted that, "With regard to game in general, we observe that the greatest quantities of wild animals are usually found in the country lying between nations at war." Thus, American Indians sustained a diverse post-Pleistocene fauna by (1) limiting game populations to what could be supported within the available habitat, (2) using fire to create new habitat, and (3) providing refugia.

## 4.1 Hunting Technology and Ecology

Except from what can be inferred about the impacts that Proto-Indian populations made on North American ecosystems —be it in terms of introducing culturally based fire regimes, the role of humans in the extinction of fauna, or the cultivation of plants — little information is available to reconstruct the strategies whereby prehistoric people influenced, one way or another, the distribution and relative abundance of plants and animals, regionally or locally. Although pollen diagrams and other information provide evidence of changes in the structure and composition of native ecosystems, the magnitude of the contributions that human populations made to those changes continue to be debated.

Historical reports about hunting–gathering technologies have primarily concerned the setting of habitat fires which, for fairly obvious reasons, Europeans and Euro-Americans noted far more frequently than any other ways whereby Indians influenced environments. In contrast, the numerous, much less evident means by which hunting and gathering technological practices could have affected the relative abundance and distribution of plants and animals are seldom found in historical or archival records. For instance the varied tribes of the Basin Plateau were denigrated as "Digger Indians," referring to the fact they were observed using digging sticks to extract edible roots, with no interest or understanding of the indigenous knowledge involved in what would now be recognized as a form of plant cultivation.

At the same time, anthropological descriptions of technology have usually concerned the tools and weapons used, or in some cases the deployment of people in such activities as driving animals into enclosures or over cliffs. That there might have been whole systems of knowledge which informed and guided the production and use of tools, as well as cultural understandings of the environmental consequences of using them, were never considered prior to the final third of this century. Interests in technology were essentially the concerns of collectors involving the cataloging and evolutionary arrangement of "primitive tools," with Native American "material cultures" condensed to fit into museum dioramas. Portraying technologies as little more than inventories of artifacts makes it effortless to assume that the displacement of "primitive tools" by "more advanced" ones fundamentally and irrevocably changes an entire technology (e.g., guns and ammunition for bows and arrows; steel axes for those made of stone; matches for fire drills; or metal pots and pans for woven baskets). However, all technologies, modern or archaic, are infinitely more complex than the sum of their physical parts with knowledge always constituting the infrastructure of what constitutes a society's technology.

As with industrial nations, the technologies of non-industrial peoples involve systems of knowledge — knowledge that people employ for practical purposes in their relationships with particular environments. Technology defined as knowledge for hunting–gathering societies has been discussed by Riddington (1982:471) who wrote that,

> "Perhaps because our own culture is obsessed with the production, exchange, and possession of artifacts, we inadvertently overlook the artifice behind technology in favour of the artifacts that it produces.... I suggest that *technology should be seen as a system of knowledge rather than an inventory of objects*.... The essence of hunting and gathering adaptive strategy is to retain and be able to act upon *information about the*...relationships between people and the natural environment. When realized, *these life-giving relationships are as much the artifacts of hunting and gathering technology as are the material objects that are instrumental in bringing them about*." [emphasis added]

A preoccupation with implements and artifacts at the expense of insight and ingenuity inevitably leads to viewing technological change as merely the perfunctory replacement of one tool or set of tools for another. The result is that the intelligence and skills required for successfully hunting and gathering are considered, if in fact thought about at all, as but epiphenomena of the tools involved. This perspective, which is very much a Euro-American cultural outlook, sees introduced, manufactured artifacts as evidence of what is "modern" and inevitably superior to what is "traditional." It is an ethnocentrism which assumes that the technologies of indigenous peoples were undeveloped rather than differently developed, whether it be the view of sportsmen hunters or environmental scientists. Related to this has been the assumption that the only thing that checked the onslaughts of hunters was the primitiveness of their tools and that, once armed with guns instead of bows and arrows, hunters would decimate animal populations.

The most impressive studies on the technology and ecology of hunting have appeared over the past 20 to 30 years and are, as expected, limited to the remaining more-or-less isolated regions of the world where hunting remains a major or at least important part of human culture. Reports have come from Africa, India, Southeast Asia, Australia, and Central and South America. In North America, almost all research on the knowledge required for hunting and trapping is limited to studies carried out in Canada and Alaska. (A large number of the best but by no means all of these studies can be found in published symposia from the "International Conferences on Hunting and Gathering Societies", with numerous other studies referred to therein (Leacock and Lee 1982; Schrire 1984; Ingold, et al. 1988;

Altman 1989; Meehan and White 1990; Burch and Ellanna 1994).)

Although today virtually all people are influenced by agencies and industries of nation states, cultural understandings exist of how to use and sustain the availability of animal populations through hunting. The degrees to which the hunting strategies of Indians and Inuit in Alaska are consistent with what they were in the past —"traditional" and "unacculturated" by outside forces — are academic matters for the consideration of anthropologists and historians, not ecologists. The important questions for interests in an "Ecologically Based Stewardship" concern how a people's knowledge about hunting technology and animal ecology can today inform us in ways that we either do not understand or that we understand differently.

Even the first anthropological descriptions of North American Indians, the earliest being undertaken in the latter part of the 19th century, followed long after the devastating effects of European contact. And, later still, studies of hunting-gathering ecologies involving the remaining groups of people in isolated regions have only been undertaken well after the most isolated communities had been forced to adjust to the pressures and intrusions of industrial societies and government agencies. Despite this, studies made in recent years have shown both the promise and need for carrying out further research on the processes that constitute the techno-ecologies of hunting and trapping. The anthropologist Harvey Feit (1994), who has focused his research on the James Bay Cree of northern Quebec, has stressed the point that it is inconceivable that hunting harvests are unplanned and unpredictable, or that there are no observable consequences that hunters understand and act upon. As Feit (1994: 429–431) pointed out,

> "One does not have to demonstrate that hunters are conservationists to make the point that their future harvesting activities, and that those connections in many cases can be easily observed in the changing rates of encounters with animals over time.... it would be surprising if hunter–gatherers were completely unaware of all such connections, and if they really gave no thought for the resources or for the morrow. Indeed, much of what they do shows just such considerations, changing hunting areas periodically or on rotation, and burning areas to initiate vegetation and animal population changes.... If hunter–gatherers, in particular circumstances, give no thought to the morrow, then this fact needs as much explanation as does their frequent planning.... the simple fact of immediate response needs to be interpreted carefully, as it may conceal longer-term strategizing."

Generalizing about hunter–gatherer ecologies without having actually studied them is no less uncertain than equivalent, non-empirical generalizations about farmers or industrial workers. As with any studies in ecology, it is the empirical study of particular cases that must precede generalization. Those in anthropology and related disciplines which have, over the past two-to-three decades, focused on particular examples of what hunter-gatherers do — and not merely evolutionary–topological questions of what hunting–gathering societies are assumed to have been and supposed to have done — have provided examples of how foraging societies influenced animal resources. They also illustrate that, in terms of understanding the consequences of their activities, the North American Indian experience was not unique. Such comparisons have shown, and continue to show, important parallels in the ways that hunter-gatherers relate to local environments.

In referring to how Cree hunters have maintained a long-term outlook with respect to resources, Fiet (1994:433) noted,

> "Waswanipi Cree Indian hunters in subarctic Quebec have a profound awareness that they do not control the morrow's harvest, but this is closely linked to the view that what they do today will nevertheless significantly affect that harvest. These two views may be common among circumpolar hunters, although Waswanipi constructs of the issues are certainly distinctive."

The gap between the knowledge systems of indigenous peoples and what scientists know has been characterized by the Aleut historian and environmental consultant Ilarion Merculieff (1994: 405) as follows:

> "Western scientific regimes, which are built upon deeply ingrained linear structures and ways of understanding, currently dominate efforts to address global environmental and economic distress. Difficulties in cross-cultural communication cause the exclusion of knowledge possessed by indigenous peoples because such knowledge draws upon a cyclical understanding of the world. To avoid progressing completely into a global mono-culture and to avoid single-minded, increasingly destructive, and unsustainable approaches to the problems faced by humans today, serious and concerted efforts must be made to include the depth of experience and knowledge indigenous peoples have gained from centuries of intensive interaction with their

environments. We must recognize that cultural bioeconomic diversity are as vital to human survival as genetic diversity is to the survival of the biosphere."

Most anthropological studies on hunting and gathering over the past 30 years have concerned questions regarding the general characteristics of foraging societies — how do they differ from other kinds of societies; how can we extrapolate from contemporary and historical societies to those in the prehistoric past; are foragers true societies; is it a useful category for theoretical purposes? Fortunately, these issues have not concerned topical studies of how people employ technological and ecological systems of knowledge in relating to environments. Whether, for example, Native Alaskans are "traditional hunter-gatherers," living in relatively closed social settings, using a variety of non-traditional tools, or even engaged in the commercial exploitation of some resources (e.g., trapping), need not determine whether or not to undertake studies regarding the technological and ecological knowledge of hunters.

The knowledge that increasingly few individuals in relatively isolated parts of North America still possess is extremely important. The possibilities of learning from people who still understand and, in some cases, still employ such knowledge should be a major concern for ecological restoration. In learning from people, and not merely about them, every effort should be made to include them as not just subjects but as participants in the research.

## 5 CULTIVATION: HARVESTING AND ENCOURAGING WILD PLANTS

North American Indians practiced plant husbandry for prolonged periods, causing positive, negative, and benign effects on the biota in a multitude of vegetation types (White 1984). Detailed observation and experimentation with the natural environment over many generations led to in-depth, native knowledge of how natural systems work. It also improved their understanding of the basic reproductive biology of plant and animal populations crucial to subsistence economies (Uhser 1987, Anderson and Nabhan 1991).

Anthropologists and ecologists have not sufficiently explored aboriginal harvesting strategies and manipulation of wildlands, which contain plant resources for fuelwood, weapons, clothing, basketry, cordage, foods, tools, dyes, and medicines. Yet vegetation manipulation to augment wild plant populations and sustainable resource-gathering strategies for these needs is seldom considered. Nevertheless, the use of fire as a vegetation management tool, plant harvesting strategies, and other horticultural techniques enabled humans to alter systematically the natural environment on a long-term basis, cumulatively affecting vast areas (Gomez-Pompa and Kaus 1992, Blackburn and Anderson 1993).

### 5.1 Plant Gathering

American Indians skilfully gathered plants over long periods of time in different habitat types without depleting their populations. This required intimate knowledge of each species life history. Plants were gathered according to at least six variables: season, frequency, appropriate tool, pattern, scale, and intensity (Anderson 1994a; 1993). If a plant were harvested in the wrong season, too frequently, or without leaving behind plant fragments, the plant population could easily be extirpated, even with low levels of technology. Many examples can be found for compatible relationships between native plant populations and utilization by American Indians. Yet few studies have attempted to define and quantify these harvest variables from careful analysis of existing wild plant systems of indigenous groups (Nabhan, et al. 1983, Chapin 1991, Anderson 1993).

#### 5.1.1 Season of harvest

The time of year a plant is harvested affects both quantity and quality of production, and on a longer term, the productivity of a population. The Nez Perce harvested large quantities of camas (*Camassia quamash*) after it had gone to seed (Harbinger 1964). Because camas does not reproduce vegetatively, digging after seeding ensured future populations of the plant by preparing the seedbed and ensuring that seed fell at the source. The prairie turnip (*Psoralea esculenta*) was harvested by the Indians of the Rocky Mountain region when the tops were browning in July or August after seeding. Because the plant does not produce vegetatively, this timing would be crucial for perpetuation of plant populations through reseeding the site. The harvesting of the tender, immature flower stalks of *Agave* spp. and *Yucca* spp. before flowering, may have stimulated vegetative reproduction in the form of pups. California basketweavers harvested native shrubs, such as sourberry (*Rhus trilobata*), *Ceanothus* spp., and redbud (*Cercis occidentalis*) in the winter after the leaves dropped, or early spring before new leaf formation, a time that was least detrimental to the plant's vital processes (Anderson 1993).

### 5.1.2 Frequency of harvest

Populations of different plant species were harvested by tribes at various intervals, sometimes allowing for rest periods to rejuvenate plant populations. For example, a three-year rest period was specified for the gathering of western red cedar (*Thuja plicata*) roots among the Klikitat basketmakers of southern Washington (Schlick 1994). Pomo Indian basketweavers of California collected sedge (*Carex* spp.) rhizomes along California rivers for lacing in basketry. They allowed a two-year rest period before harvesting again (Peri and Patterson 1979). Navajo medicine men still refrain from harvesting from the same stand two years in succession, granting periods of rest and regrowth between those of tillage and extraction (Anderson and Nabhan 1991).

### 5.1.3 Appropriate tools

Major tools for harvesting plants included digging and knocking sticks, deer antlers, stone knives, and seedbeaters. Frequently the human hand was sufficient to harvest fruits, seeds and leaves. Tools were effective because they were designed to avoid destroying the plant. For example, rather than uprooting plants or breaking off seedheads, many tribes beat seeds from annual and perennial plants into a wide-mouth container with some type of seedbeater (Driver and Massey 1957). Seedbeating protected perennial plants *in situ*, while ensuring that the seed stock was not completely removed from the site. The disturbance caused by the tool often is similar in scale and type to natural disturbances, such as gopher activity or herbivory, which can be necessary for the rejuvenation of certain plant populations.

### 5.1.4 Pattern of harvest

Many people believe that North American Indians and other hunter-gatherers did not stay anywhere long enough to leave lasting impacts. On the contrary, in many areas of the United States tribal groups gathered vegetation at designated sites. These special areas were shaped by continual long-term use, such as for basketry materials, gathering bulbs, collecting seeds, harvesting cordage, and picking greens (Latta 1977; Driver 1936; Turner 1995). The sites contained the best plants for the intended purposes, and they were continually and intensely managed and harvested. Certain restrictions on use also applied so that plant resources could be gathered on a dependable basis.

The most important plant foods were often gathered at traditional collection sites inherited by each new generation. Each band of the Nez Perce had its own camas digging area, and each family had its own spot within this area in which to work (Harbinger 1964). Agave, oak, and mesquite gathering areas were owned by lineages of the Cahuilla in southern California (Bean and Saubel 1972).

### 5.1.5 Scale of harvest

Little information exists on the size of plant harvesting sites in the ethnohistoric or ethnographic literature. According to Indian elders, collection sites were generally small, and the plant populations existed naturally on the site (Anderson 1993b). This did not result in type conversions but rather the maintenance of the general character of the natural habitat. Thus, these areas appeared "untouched" to outside observers. More thorough investigations of size of traditional collection sites for different plant resources need to be conducted.

### 5.1.6 Intensity of harvest

The number of plants or plant parts taken from a given location is a measure of the intensity of collection. It can indicate whether the quantity taken exceeds the biological capacity of the plant population to regenerate or recover. Perennial plants among North American tribes were often harvested *in situ*. Renewable plant parts in the form of fruits, branches, leaves, flowers, corms, and stems were gathered year after year, leaving individual plants and plant parts behind to ensure their replacement. For example, the Hualapai (Walapai) of Arizona harvested the young pads of *Opuntia* spp., leaving the cacti in place. Miwok men climbed tall sugar pine trees in the Sierra Nevada to retrieve pine cones, leaving the trees intact. The Indians of Montana harvested the tubers of arrowhead (*Sagittaria* spp.), which are attached to the fibrous roots, by loosening them with their feet, causing them to float to the top of the water where the Indians could collect them, so the plants were left behind. The Kwakiutl Indians of Vancouver Island, in peeling bark from cedar trees for various uses, never stripped off all the bark. Cultural rules that required the gatherer to spare or pass up certain plants also served a conservation function. The Delaware, for example, were instructed never to take the first plant specimen they saw. Similarly, the Navajo herbalists in gathering deer-plant medicine (*Conioselinum scopulorum*) never took the plant to whom they prayed (Weslager 1973).

## 5.2 Horticulture

Fire was the most powerful, effective and widely employed horticultural tool for manipulating the environment. However, other horticultural techniques were used to change plant abundance, diversity, growth, longevity, yield, and quality to meet cultural needs. These include irrigating, pruning, sowing, tilling, transplanting, and weeding (Anderson 1993).

### 5.2.1 Pruning

Many American Indian cultures pruned shrubs and trees repeatedly, in many cases prolonging their life-spans. For example, the Shoshone and Paiute pruned the older growth of the snowberry (*Symphoricarpus racemosus*) to make small bird arrows. The Pomo pruned elderberry (*Sambucus mexicana*) to induce elongation of young stems for musical instruments. The Timbisha Shoshone pruned honey mesquite (*Prosopis glandulosa*), an important food resource, keeping areas around the trees clear of undergrowth, and of dead limbs and lower branches). Many cultures, such as the Cheyenne, climbed into trees, breaking off and throwing down dead branches for firewood. Special hooked sticks were used by the Miwok to retrieve dead limbs from the canopies of trees for firewood.

Tribes in different parts of the United States purposefully pruned plants to induce the rapid elongation of young shoots for basketry. The Havasupai in northern Arizona pruned sumac (*Rhus trilobata*) and catclaw (*Acacia greggii*) to encourage the new, usable growth. Anthropologist Catherine Fowler reported that the Northern Paiute women at Walker River pruned willows and lemonade berry (*Rhus trilobata*) for basketry fibers (Fowler 1986). The discovery of a 3,500-year-old split-twig animal figurine made from a long, straight shoot of sumac (*Rhus trilobata*) 1.8 meters in height suggests that prehistoric cultures from the Grand Canyon of Arizona and southern Utah knew that pruning or fire stimulated sucker formation in sumac (Bohrer 1983). The branches of redbud (*Cercis occidentalis*) were cut down in the winter or early spring by the Yuki and Pomo to insure suitable material for baskets the following autumn.

A common practice among different California Indian tribes was to use sticks to knock on the branches of oak trees so that the acorns, a highly desirable food, fell to the ground. The common knocking implement was a pole of pine or some other wood. According to Peri and Patterson (1979) knocking acted as a pruning process because

> "...some branch tips and leaves and brittle and dead twigs were also removed in the process, while dead or diseased limbs were intentionally broken off. This autumnal 'pruning' increased the surface area of the canopy and fruit production by stimulating the growth of new branchlets and foliage the following year."

Acorns were also harvested in California by climbing the trees and cutting the limbs.

### 5.2.2 Sowing

Probably long before the invention of domesticated agriculture, many tribes understood how to maintain or augment wild plant populations through collecting and sowing seeds. It was possible to extinguish populations of annual plants by breaking off seedheads or harvesting as many seeds as possible without reseeding. Yet harvesting techniques were designed by some tribes whereby some of the seed was broadcast to replenish gathering sites.

In the area from Lake Winnipeg to Lake Superior, the Ojibway sowed the seed of wild rice (*Zizania aquatica*) in the lakes. Densmore (1974: 314) also noted that "It was not the intention, however, to harvest [by the Ojibway] all the rice, a portion being allowed to fall into the water, or being sowed on the water as seed." The Assiniboine sowed wild rice in the marshes surrounding the Great Lakes to guarantee a harvest the following year. The Paiute sowed the seed of *Mentzelia* and *Chenopodium* plants on burned over ground (Steward 1938). The Yuma broadcast the seeds of native plants in southern California. The Cahuilla and probably other tribes of the Southwest planted the seeds of the desert fan palm (*Washingtonia filifera*) at oases (Nabhan 1985). The Seneca after gathering herbaceous plants would break off the seed stalks and drop the pods into dug holes with leaf mold (Parker 1913 cited in Bierhorst 1994). The seeds of panic grass (*Panicum sonoran*) were broadcast along the lower Colorado River (Nabhan 1985); the Navajo sowed the seeds of *Festuca octoflora*, and the Southern Paiute sowed the seeds of Indian rice grass (*Oryzopsis hymenoides*). When the Potawatomi gathered the roots of plants for medicines, they completed the task by placing the seed heads in the holes from which they removed the roots and covered them with soil so that many plants would grow (Smith 1925). The Lummi Straits, Nooksack, and Nuuwhaha of the Pacific Northwest harvested the bulbs of camas (*Camassia* spp.) and placed the broken stalks bearing the ripe seed capsules into the holes before they were recovered.

### 5.2.3 Tilling

Tilling involves removing earth in the harvest of underground perennial plant organs (e.g., roots, rhizomes, corms, bulbs), frequently followed by dividing these organs and leaving individual fragments in the soil. Underground swollen stems were gathered for foods, medicines, dyes, basketry, and ceremonies. Huge quantities of these plant parts were gathered. This caused considerable soil disturbance, which affected seedbeds, and ultimately the abundance, density, and composition of the vegetation. Digging these bulbs may increase the size of the tract, aerating the soil, lowering weed competition, and severing the tiny bulblets and cormlets to give them a better opportunity to grow.

Different tribes purposefully left bulbs and corms and harvested after the plants had gone to seed. This prepared the seedbed for new plants. For example, the Dena'ina of Alaska left fragments of underground swollen stems behind when digging to insure the growth of new plants (Kari 1987). The Kwakiutl of northern Vancouver Island harvested the tuberous roots of various clovers (*Trifolium* spp.), leaving main root stocks in place to perpetuate populations. The Objibwa took tuberous roots and left some plants undisturbed (Jenness 1935). The Catawba in harvesting roots for medicine, always left reproductive plant parts behind to ensure future abundance (Wilson 1949). The Cahuilla harvested only the large corms of blue dicks (*Dichelostemma capitatum*), replanting the corms to ensure a crop the following year (Bean and Saubel 1972).

### 5.2.4 Transplanting

American Indians widened the ecological amplitude of native species by introducing them to new areas. They transplanted or sowed seed in traditional collection sites away from village sites and adjacent to villages. For example, in California large patches of *Astragalus bolanderi* (a formerly important food that was burned by the Wukchumni Yokuts) may have been transplanted or sown because they occur as discrete areas within red fir forests in the Sierra Nevada. Thus, the ranges and distributions of favored plant species were expanded through human intervention. Cottonwood (*Populus* spp.) and willow (*Salix* spp.) were sometimes transplanted by the Hopi into neighboring washes, and cattail (*Typha angustifolia*) was introduced to other washes (Whiting 1939). Indians of the two Carolinas, Georgia, and Florida transplanted the shrub yaupon (*Ilex vomitoria*) (Hammett 1992).

There are many accounts in the United States of plants established in association with historical human occupation sites. For example, *Agave parryi* exhibits an interesting association with archaeological sites in the western portion of the Apache-Sitgreaves National Forest in Arizona (Minnis and Plog 1976). Three species (*Cleome serrulata*, *Lithospermum caroliniense*, and *Salvia subincisa*) were useful to the Pueblo Indians and all are largely confined to Pueblo ruins in New Mexico (Yarnell 1965). Nettles (*Urtica dioica*) also established in enriched soils around camps of the Salish in Puget Sound.

### 5.2.5 Weeding

Certain wild plants in different plant communities were favored by weeding. The Assiniboine weeded wild rice by pulling out the water grasses that grew among the rice stalks. Beds of sedge (*Carex* spp.) along lowland streams and rivers in California and bracken fern (*Pteridium aquilinum*) at 5,000 to 6,000 feet in the Sierra Nevada were intensively weeded to encourage the production of long, creeping rhizomes. Sedge rhizomes were utilized extensively by many tribes as the weft or lacing in basketry. Dyed bracken fern rhizomes were used extensively as a design material in coiled and twined baskets.

## 6 CONSEQUENCES OF REMOVING INDIGENOUS MANAGEMENT

Humans modify their environments as much as the environment affects human behavior. The relationship between them is reciprocal (Bonnicksen 1991). Thus, the landscapes of North America changed dramatically after the removal of American Indians from traditional economic and land management roles.

Forests cover one-third of the United States, but today's forests are a tattered and unhealthy remnant of historical forests. We have mistakenly assumed that nature operates independently of humans and only requires protection. So, more than a century ago we began to protect forests from fires, and we set some forests aside in parks and reserves to protect them from humans. We did not realize that our forests had evolved with fire and 12,000 years or more of human use. As a result, forests throughout North America grew thicker and the mosaic structure that characterized most of our native forests began to collapse. Wildfires also grew larger and more destructive, some plant and animal species started disappearing and streams began to dry. The stresses caused by increasing tree densities also weakened the forests and helped to spread destruction by insects and diseases. Exotic plants, animals and diseases continue to change North America's forests, and urbanization, forestry and agriculture have also taken their toll. However, the elimination of Indian burning and the suppression of lightning fires produced the greatest changes, even in second-growth forests and prairies (Leach and Givnish 1996).

## 6.1 Disappearing Ecosystems

Succession is replacing one plant community with another in many parts of the United States. It is a natural process of growth that helps to maintain the dynamics and diversity of forests. A forest of trees that started growing in a sunny opening is gradually replaced by other trees that can grow in the thick litter and low light below. Thus, one kind of forest disappears and is replaced by another. As long as new openings form, trees that grow in openings will coexist on different parts of the landscape with trees that grow in shade. This is a natural forest.

If openings stop being formed, a forest of shade-intolerant trees will gradually change into a different kind of forest — one composed of only shade-tolerant species. Uninterrupted succession is changing native forests throughout North America. Pre-European settlement ponderosa pine (*Pinus ponderosa*) savannas throughout the West are becoming denser and shading out the grass underneath the trees (e.g., Cooper 1960). Ponderosa pine forest density on the G.A. Pearson Natural Area near Flagstaff, Arizona, increased from 25 trees per acre in pre-European settlement times to 1,182 trees per acre today. Similar changes are occurring elsewhere in southwestern ponderosa pine forests (Covington and Moore 1994). Mixed-conifer forests in the Sierra Nevada and the Southwest are being replaced by shade-tolerant white fir (*Abies concolor*) trees (Bonnicksen and Stone 1982a,b). Pine trees formerly made up 70 percent of the Boise National Forest in Idaho, but today they only represent 30 percent of the forest. Fir trees replaced pine on the forest. Sugar maple is taking over northern and eastern hardwood forests and oak (*Quercus* spp.) forests are disappearing. Old-growth Douglas-fir (*Pseudotsuga menziesii*) forests set aside in reserves in the Pacific Northwest are gradually being replaced by shade-tolerant western hemlock (Bonnicksen 1994a). Lodgepole pine (*Pinus contorta*) forests also are being replaced by shade-tolerant spruce (*Picea* spp.) and fir in the Rocky Mountains.

Elsewhere in the West, juniper (*Juniperus* spp.) is spreading within pinyon-juniper woodlands and replacing grasslands in the Colorado Plateau and southern Rocky Mountain regions of northern Arizona and northern New Mexico. Because of increases in the density of pine and other conifers, aspen forests in Arizona and New Mexico have decreased by 46 percent since 1962, and they are rapidly disappearing as a distinctive forest type throughout their range (Johnson 1994). This is particularly serious because aspen provides habitat for many species of birds and mammals. Unless something is done now to reverse the deterioration, aspen and many other forest types may cease to exist.

In no other area are the changes in native forests more dramatically displayed than within the mixed-conifer forests of Sierra Nevada national parks. Bonnicksen and Stone (1982a,b) compared a reconstruction of a pre-European settlement giant sequoia-mixed conifer forest with the same forest as it exists today. The results showed that the area of the forest mosaic covered by patches of sapling-size trees and shrubs declined dramatically. On the other hand, a dramatic increase occurred in patches of pole-size trees and mature trees. Thus, today's forest is thicker and older than the pre-European settlement forest. Shrubs, black oak (*Quercus kelloggii*) trees and wildflowers are less abundant, white fir is gradually becoming the dominant tree species, and biodiversity is declining. Even more ominous is an increase in the fire cycle from 129 years to 348 years (Bonnicksen 1994a). This means that the forest will shift from a surface fire regime to a massive fire regime. Heavy cutting will be needed to reduce the density of trees that are choking the understory and to create openings to reproduce trees that require full sunlight. Continued cutting and prescribed burning will be needed to maintain the restored forest or it will simply return to its current deteriorated state.

The decline in forest health is also reducing wildlife and fish habitats, water yields, and biodiversity. The loss of aspen in the Southwest and throughout the Rocky Mountains, especially in Yellowstone and Banff National Parks (Kay 1990, Kay and White 1995), is particularly serious because such forests provide unique wildlife habitat. Today's dense forests also are infected with disease and infested with insects. In 1988-1989 pine bark beetles killed 15,000 acres of pine trees in five East Texas wilderness areas. Tree mortality in the United States increased 24 percent between 1986 and 1991, and forest growth declined by 2 percent during the same period. Competition for water, nutrients, and sunlight among densely packed trees explains some of the decline.

## 6.2 Dangerous Fires

Unlike most of the original native forests, many of today's forests have become dangerous fire hazards. Heavy logs and dead trees clutter these forests and provide massive amounts of fuel. They also are choked with several layers of trees that allow fires to climb easily into the canopy. Fires also can move freely across vast areas because the mosaic structure is disappearing as patches of trees grow to similar sizes. In 1988, a lightning fire that was declared a "prescribed fire" in Montana's Scapegoat Wilderness escaped and burned 240,600 acres, including 40,000 acres outside the wilderness area. Millions of board feet of timber, ranch buildings, fence and hay were lost, and 100 cattle were killed (USDA Forest Service 1989). Wildfires blackened 1.5 million acres in eastern Oregon, eastern Washington, Idaho, Montana and portions of Utah and Nevada during the Summer of 1994. Two million more acres burned in 1996. Many of these fires burned hotter than would have been the case under natural conditions.

The mammoth wildfires that scorched nearly one-half of Yellowstone National Park during the summer of 1988 were significantly larger than any fire that occurred in Yellowstone in the past 350 years. Huge blocks of young trees regenerating on the burned areas, intermixed with dead trees, will grow older and thicker as a unit, becoming a vast unbroken mass of highly flammable fuel. This could create a new cycle of massive wildfires.

Housing developments and other structures scattered through forests and brushlands further enhance the chance of disaster. As a result, wildfires in pine forests destroyed over 1,300 homes between 1985 and 1992 in Florida, Colorado, Idaho, Arizona, California, Oregon, and Washington. Fires in California destroyed 3,000 homes in 1990 and 1991, and the southern California fires of 1993 destroyed an additional 800 homes and cost $1 billion in property losses. There have been 30 deaths from wildfires in California alone since 1990. These losses will grow as many of North America's forests and brushlands become thicker and more homes are built in them..

## 6.3 Endangered Species

The changes occurring in North America's landscapes because of the removal of Indian management also affects endangered plants. The loss of biodiversity in the United States in some cases may be directly tied to past fire suppression policies of federal and state agencies (National Research Council 1992, Leach and Givnish 1996). Researchers who have recently conducted studies on rare and endangered animals and plants in different geographic regions are concluding that frequent fire is necessary to the health and maintenance of habitat for certain endangered biota (Boyd 1987, Soulé and Kohm 1989). Fire suppression policies on public lands were based on a perception of fire as a destructive force without an understanding of the dynamics of fire and its ecological role in ecosystems, and therefore, constituted a real threat to the very resources they were intended to protect. Fire is now a universally accepted management tool in conservation biology (National Research Council 1992). Yet when prescribed burning programs are implemented, they are done with an inadequate understanding of the earlier role of American Indians in setting these fires and creating other kinds of human disturbances. For example, wildfires in the West often occur on a larger scale and are more destructive than the fires in aboriginal times.

In many cases natural disturbance alone, such as lightning, is not sufficient for reestablishing and perpetuating pre-European settlement vegetation patterns (Bonnicksen and Stone 1982a,b, 1985). The removal of American Indians from traditional economic and land management roles in the United States has created its own set of ecological consequences. To assess accurately the health of different ecosystem types requires a thorough understanding of the ecological role of aboriginal people in the dynamics of wild plant populations, communities and ecosystems. This may provide invaluable information for management of now rare and endangered plant and animal species. For example, rare and endangered animals such as the red-cockaded woodpecker (*Picoides borealis*), the northern spotted owl (*Strix occidentalis caurina*) and the great gray owl (*Strix nebulosa*) live in habitats that are no longer being maintained with indigenous burning regimes.

Indigenous burning regimes created a highly suitable habitat (forest structure and valuable food sources from insect and mammal populations) for the California spotted owl (*Strix occidentalis*), which is listed as a sensitive species (Verner et al. 1992). Its primary prey in the Sierra Nevada mountains is the northern flying squirrel (*Glaucomys sabrinus*), which lives in older stands within a forest mosaic, but the owl preys almost exclusively on dusky-footed woodrats (*Neotoma fuscipes*) in Southern California and lowe elevations, which live in younger forests. In the Sierra Nevada, 82 percent of the spotted owl sites were found in mixed-conifer forests (Verner et al. 1992). This forest type cannot be maintained without frequent fires, many of which were set by American Indians during pre-European settlement times (Bonnicksen and Stone 1982a,b, 1985). Thus, the historical forests in which the owl lived was maintained, in part, because of the sophisticated ecological knowledge and management practices of American Indians.

Dry montane meadows in the Sierra Nevada harbored a rich diversity of plant species useful to various tribes. These meadows also are important foraging habitat for the great gray owl. Native grasses for basketry (e.g., *Muhlenbergia rigens*), edible plants and medicinal plants are found there, sometimes in great abundance. "Dry" or woodland meadows are perhaps a later successional stage of the "wet" meadow type and are now being encroached by forests. American Indians prolonged the life of these dry meadows through periodic burning. Setting fires in the ecotone areas surrounding the meadows decreased the more wet-tolerant conifers from encroaching into meadow areas, thus maintaining and perhaps in some cases enlarging meadow areas. By maintaining a string of meadow habitats, American Indians were also maintaining foraging habitat for the owl.

Some of the plants that are now rare and endangered were gathered in great quantities, without contributing to their depletion. Former harvesting strategies and management practices for these species may have maintained and expanded populations, while removal of American Indians from their homeland, causing the discontinuation of these practices, may have contributed to species decline. For example, Different species of clovers were burned periodically to enhance leaf and seed production by the Wukchumni Yokuts, Pomo, the North Fork Mono, and other tribes (Peri et al. 1982, Anderson 1993b).

The Kumeyaay of southern California gathered the nuts of the Torrey pine (*Pinus torreyana*), and ate them raw or roasted. The pine is endangered in a portion of its range because of its shrinking habitat from urban housing encroachment. A significant ecological factor in perpetuation of the species is fire and past indigenous fires perhaps played an essential role in perpetuation of pine stands. In restoring fire cycles to these and other rare conifers, knowledge of how Indians changed the frequency and intensity of fires may be integral to successful management of these species (Anderson 1993).

Indigenous use and management of certain now rare and endangered plants may have led to their sustained yield. American Indians may have positively influenced the persistence of now rare and endangered plants by: (1) maintaining diversity of successional stages and habitat types within the tribal territory with human-set fires, (2) aiding in plant dispersal to other areas through planting of seeds, vegetative cuttings, or transplanting entire plants, (3) introducing repeated disturbance to enhance seed and/or vegetative reproduction, and (4) introducing repeated disturbance to decrease potential plant competitors and destructive agents such as insects and pathogens.

## 7 REVIVING AND USING LOCAL KNOWLEDGE

Environmental knowledge used by people who relate to local habitats through daily activities such as hunting, trapping, fishing, farming and herding are important to scientific and conservation interests. Most people acknowledge that farming and pastoralism were relatively successful and brought about significant changes to the environment before the emergence of agronomy in the early part of the 19th century. Others, however, have difficulty accepting the idea that the actions of hunting and gathering peoples — societies which existed in different environments for hundreds of thousands of years before the emergence of environmental sciences — were relatively successful ecological stewards, or brought about significant environmental changes.

Whether we accept or reject such ideas, the only conclusive answer to such questions is whether or not the more detailed studies of indigenous knowledge and practice — in contrast to uncritical claims that denigrate or extol what aboriginal peoples did — can be verified. Two ways of verifying (or falsifying) are available: (1) showing that the conclusions from indigenous knowledge are replicated in like and, in some cases, even unlike regions, and (2) showing that such knowledge compares well with the conclusions of science. The combination of these two approaches are even more definitive than using either separately.

On the other hand, it is reasonable to claim that the knowledge necessary for long-term adaptations to an environment are actually parallel to scientific knowledge. Arguments pro and con over this, frequently pitting rural vs. urban, native vs. non-native, conservationist vs. technocrat, and even scientist vs. scientist, are being increasingly raised and in too many instances naively asserted, with respect to environmental issues. Supporting the view that traditional knowledge is a form of science, George Hobson (1992: 2), former Director of Canada's Polar Continental Shelf Project, has maintained that the ecological knowledge of the Inuit in northern Canada fits well within the accepted definition of science. He said that

> "Western science has been defined as a systematic approach, a methodological approach to answering questions. Science is equated with knowledge, and it is the development of knowledge that promotes the solution of problems. The 'western' scientist knows that science is based upon the principles of repeatability and predictability. In terms of the northern experience, science also equates to *traditional* knowledge, and southern scientists must never forget that traditional knowledge *is* science."

At the extreme of this position are people who uncritically accept local, indigenous, traditional, or customary knowledge as meaningful and intuitive substitutes for what they see as an impossibly objective and impersonalized science.

Most scientists and government officials are undoubtedly skeptical that there is much scientific significance to be found in non-professional, or layman, explanations of environmental relationships. Even the knowledge behind the more empirical and observable activities of making a living as a hunter-gatherer would be viewed as simply based on workable customs and traditional practices, not the scientific "principles of repeatability and predictability" claimed by Hobson (1992), nor least of all the procedures that characterize scientific inquiry. Even the most open-minded of scientists would probably say that an artisan's knowledge of technology and natural phenomena is at best locally relevant: useful and practical, but inherently idiosyncratic and concerned with much more narrow issues, involving perspectives and interests far more limited and parochial than those that engage the science of ecology. Most scientists would consider Hobson's claim that folk knowledge is actually science as being extravagant; for those adhering to a more iconoclastic scientism, his assertion would be considered absurd.

Local, folk, traditional, or indigenous understandings of natural processes, scientists would argue, simply cannot be considered as counterparts to scientifically based conclusions since they do not involve comparable methods, are not testable, and do not lend themselves to broader generalizations. At the same time, folk beliefs frequently involve outlooks and related activities which include ritual or "superstitious" practices, as well as the use of simple tools and unsophisticated procedures. Using historical examples, it would be noted that simpler tools have always been replaced by more efficient forms of technology (e.g., guns for bows and arrows), resulting in changes which drastically alter the relationships of people to natural systems.

In the continuous debates over folk science vs. modern science, mutually exclusive stereotypes are presented in contrasting knowledge based on custom and practice vs. knowledge based on theory and research; noble vs. ignoble savages; conservationists vs. exploiters; Native traditions vs. Western technology; holistic thinking vs. scientific reductionism; environmental wisdom vs. environmental manipulation; subsistence based life styles vs. those driven by profit; spirituality vs. objectivity. These are simply polemical designations used to suit particular social, philosophical, cultural, personal, or political agendas in the process of either uncritically lauding or automatically disparaging local knowledge, neither of which merits serious consideration.

Largely ignored in these debates is the fact that for almost half a century studies, generally included under the rubric of "ethnoscience," have documented rather than merely guessed about specific cultural understandings of how plants and animals are categorized, how environments function, how natural systems are locally perceived, and how surrounding environments are, or once were, influenced by indigenous technologies. Studies of technological practices and ecological perceptions, like other aspects of a people's culture, require intensive study, which eschew preconceptions about traditional societies as being either innately wise or inherently backward.

To qualify as even rudimentary science there has to be more to justifying local knowledge than providing theoretically unrelated, esoteric case studies. The most general definitions describe science as a form of study which deals with knowledge that is replicable, predictable, and which helps to explain the operation of more general, or nomothetic, principles. In this respect, even the most thoroughly detailed ethnographic descriptions of indigenous beliefs and practices are only culturally relative, specifically local examples of "ethnobotany" or "ethnozoology," anthropologically interesting but of little scientific consequence since they describe, however minutely, singular instances and not general principles.

On the other hand, throughout the history of anthropology, ethnographic data have been tested in order to derive general propositions, specifically through the use of cross-cultural comparisons, with more-or-less general propositions covering a wide diversity of topics such as kinship; religious beliefs; concepts of causality, space, and time; economic exchange and marketing; marriage and exogamy; plus (increasingly) a wide range of parallels in the ways that technologically similar societies have related to similar environments involving parallel strategies of adaptation.

The advantages in studying ecological understandings and related practices, in contrast to less corporeal aspects of culture, is that, in addition to comparisons between societies with like adaptations, such knowledge can also be compared with conclusions and general principles from scientific ecology. In this respect, local ecological knowledge can show varying degrees of correlation with explanations derived from scientific research. If aspects of local knowledge can be shown to parallel scientific ecological knowledge (e.g., understandings of plant and animal successions, animal social life, reproductive cycles, predator–prey relationships, plant and soil associations, the effects of both man-made and natural fires) the scientific significance of folk knowledge can be substantiated. Thus, in addition to cross-cultural replicability, folk knowledge can be said to "connect" with scientific knowledge.

The individual experiments of people coming to terms with environmental systems, while unique in their particular cultural conformations, nonetheless represent test cases that involve replicability, predictability, and which can be linked to what is known, or can be demonstrated, by science. Unlike religious knowledge and ritual practices, in terms of adaptation, the knowledge of hunters and farmers, like that of scientists, ultimately has to come to terms with the ways that the real world operates; environmental users have to adjust how they affect and are affected by local environments. No matter how much a forager may ritually infuse his or her activities with spiritual beliefs and ritual practices — and anthropology is replete with descriptions of such beliefs and practices — they must at numerous points relate to real things and observable events in the world around them. Thus, whereas hunter-gatherers may annotate their skills with mystery and ritual magic, they must in the long run have a broad knowledge of environmental events and relationships. In this way a forager understands a network of things and relationships which enables him or her to act effectively across a range of habitats over time and be able to make reasonable and probable estimates as to day-to-day, season-to-season and year-to-year outcomes. It is in this respect that their behaviors are replicable and, in order to survive, a reasonable number of consequences must be predictable.

Thus, the non-ecological knowledge that a forager employs in the way of myth and ritual in relating to nature — an important part of their cultural cosmology that helps explain, animate, and provide meaning and value to what they do — depends upon how effectively people understand and act upon environmental systems. The fact that these adaptations are mediated through and even shrouded in cultural meanings does not abrogate or nullify the importance of the underlying and interrelated ecological information. Once recognized and analytically separated from the cultural contexts with which they are embedded, such knowledge is directly comparable to similar beliefs and practices in other societies. And, as we maintain, because it will normally have concerned long-term, sustained patterns of adaptation, it can be further compared and substantiated against scientific interpretations.

Despite considerable changes in aboriginal American beliefs and practices, studies of indigenous ecological knowledge continue to show that there still exists — in by no means all but certainly in a variety of settings — important dimensions of understanding about the management of local resources and habitats. The collection from and cooperation with American Indians in sharing this kind of knowledge is particularly important, for American society as a whole but particularly for those people, American Indians, who are the custodians of this knowledge. For environmental agencies it offers a unique opportunity for involving people in a process of acquiring and, in many cases, using knowledge that over time might otherwise be lost.

## 8.0 DOCUMENTING HUMAN HISTORIES AND PAST ENVIRONMENTS

Documenting the history of nonliterate peoples and the environments in which they lived involves using a variety of information sources. These include ethnohistoric and ethnographic literature; museum study of cultural plant products; ethnobotany; archaeology; paleoecology; geology and pedology; pyrodendrochronology, historical photograph interpretation; experimentation and observation.

### 8.1 Contemporary Ethnographic Research

It is urgent to identify and contact elders who are still practicing or remember about native plant harvesting strategies and management practices, and the cultural rules that dictate human–plant interactions. Entré into American Indian communities may be difficult because of the past history of Indian–non-Indian relations, but once rapport is established, valuable information about land-use history and former land management practices can be obtained. A series of ethnographic interviews of selected individuals should be conducted. Questions should be designed in a nontechnical manner, yet carefully constructed to elicit detailed responses useful to ecosystem management.

Ethnographic information is substantiated and supplemented by bringing slides, photographs, or actual specimens of each plant species for their observation and identification. Additionally, site visits are made to key consultants' traditional plant gathering/production sites. Participant observation of selected families should be conducted (Jorgensen 1989). This process will provide more detail by witnessing management practices first-hand. When possible, close-up photographs should be taken of each step. Through visiting collection sites, respondents are viewed in different situations, not just during a brief interview.

Criteria can be established for judging the validity of the findings of ethnographic research. First, the accuracy of oral accounts can be verified by cross-referencing the information with testimony from other American Indian families within and between tribes of the region. Second, ethnographic research can be evaluated for the extent to which it identifies patterns that apply to a wide variety of social phenomena

(Lofland 1974). Third, the research can be judged according to the extent to which it generates formal theories (Athens 1984). Fourth, ethnographic research can be judged by cross-checking its conclusions independently through varied sources of data, such as archeobotanical remains, ecological field experiments, pollen, charcoal deposits, fire scarred tree rings, analysis of museum artifacts and written accounts.

## 8.2 Ethnohistoric and Ethnographic Literature Research

One of these sources, the ethnohistoric and ethnographic literature, is essential in the attempted reconstruction of precontact landscapes and early historical indigenous plant uses and fire management. Ethnographers, explorers, and early settlers wrote detailed accounts of indigenous life and plant material culture in different geographic regions. Additionally, drawings and maps provide valuable information in the depiction of landscapes at the point of Euro-American contact (Hammett 1992). Much of this information is housed in government archives, libraries, museums, and historical societies, in forms perhaps not readily accessible to ecologists and land managers (e.g., obscure journals; unpublished manuscripts; and on notecards). This information, for the most part, has not been evaluated for understanding former horticultural practices, the use of wild plants, or the structure and composition of vegetation from early landscape descriptions (Anderson 1994b). Information from the numerous unpublished manuscripts, journal articles, diaries, and rare books could be collected and systematized in a computer database that is organized in a manner useful to archaeologists, anthropologists, ecologists, and resource managers.

## 8.3 Museum Study of the Cultural Products Made From Plants

As restoration ecology becomes more prominent, cultural museums will become an important source of historical knowledge to practitioners and researchers. Accurate scientific identification of the materials in the collection is the first step. Then the artifacts and the bundles of raw or processed plant materials, invertebrate and vertebrate animal parts, minerals, and rocks, and the finished cultural items, when linked to field notes, will tell us much about American Indian land use history. Researchers also should work closely with the staffs of the museums and the Indian elders or non-Indians who understand the intricacies of product manufacture, and in some cases still make replicates of many of these traditional items. This valuable information will facilitate environmental reconstructions.

Bundles and coils of raw plant materials, "in process" are particularly useful for analysis. These are the plant materials that are the proper grade for manufacturing some item. Working with a plant morphologist and museum curator, the researcher can compare the anatomical and physiological features and plant architecture of the materials (e.g., anthocyanins, lateral branching, length, straightness, diameter) with the wild growth forms observed in the natural environment. If the wild growth forms differ substantially from the plant forms in the museum, this can elucidate former management.

A further, more detailed study would require selecting a subset of cultural items which require large amounts of plant material. The number of plant parts and individual plants needed to complete each of the items selected should be tallied and then compared with the availability of that plant or plant part in an unmanaged plant community. From discussions with elders, estimates of the numbers of plants needed to make an item from both wild and managed populations could be made and compared. That information could be used to calculate the number of managed plant parts and plants needed annually for an average-sized prehistoric village. This figure would give ecologists and anthropologists a much better grasp of the quantities of plant material needed from wildlands, the management practices necessary to achieve sustained-yields, and the strict cultural parameters necessary for the raw plant material to be suitable for use (Anderson 1994b).

## 8.4 Ethnobotanical Research

The ethnobotanies of different tribes have been underutilized in reconstructing past environments. Assemblage and analysis of the lists of useful plants for tribes of a specific geographic region could be used to discern patterns of plant use that are common between tribes. If the same species are used over large geographic areas that cross linguistic boundaries, this pattern may in turn form a land-use complex. The concept of a complex implies a group of species with an apparent common geographic origin and a mutual environmental and cultural association in the area where the complex develops (Ford 1985). These species can then be analyzed in a number of ways to reveal past human-environmental interactions. Specifically, to understand their place in the successional sequence of the ecosystem, their role in the ecology of the ecosystem, and their response to different disturbance regimes. This information may be useful for reconstructing vegetation and the ecological or cultural processes necessary to maintain the biological diversity of a particular ecosystem.

Most ethnobotanical studies tend to focus on plants once they have been removed from their biological context, emphasizing the cultural product rather than the source and dynamics of production. Ethnobotanists must report the connections made between the qualities of a plant used for a particular purpose and how these features are selected for or manipulated in the plants growing in wildland environments. Strong links are usually found between (1) the quality and quantity of the plant material growing in the natural environment, (2) the ease of manufacturing, and (3) the level of functionality of the finished product. For example, the weaving of a water-tight basket is only as good as the skill of its maker and the quality of the shrub branches. This quality can be manipulated through fire management of traditional gathering sites. In those cases where some management information has been recorded by ethnobotanists, it is important to continually relate use to management.

## 8.5 Archaeological and Paleoenvironmental Research

Archaeological studies can provide a long-term, regional perspective on the interactions and ecological impacts of indigenous cultures with their natural environments (Knudson 1997). Length of occupation, permanency of residence, shifts in tools used to process plant and animal parts, and shifts in patterns of resource exploitation can be discerned through careful collection and interpretation of geological, faunal, and floral remains and evidence in archaeological contexts.

Archaeological sites are found in deposits and geomorphic settings whose evidence can provide significant information about the landscape in which the human community lived. Geoarcheology is a interdisciplinary specialty in itself (Waters 1992), including analysis of hydrology, tectonics, soils, and special features such as shorelines.

Regional studies of paleolimnology (lake sediments), palynology (pollen), and plant macrofossils describe the climate of the landscape in which the past human community lived.

Palynology is the use of fossil pollen and spores to reconstruct past environments and cultural activities. The sequence of pollen buildup over time in any sedimentary deposit of (loess, alluvium, colluvium, bogs, lake bottoms or ocean floors) reflects past vegetation. The pollen is identified, counted, and its position in relation to the sediment stratigraphy is dated, analyzed, and interpreted. This information is used to determine changes in vegetation from indigenous land use activities and horticultural practices (Pearsall 1989).

Another relevant paleoecological research tool in understanding the human impact on a landscape is phytolith analysis. Opal phytoliths are microscopic mineral remains of plant cells found in soils, which can be identified taxonomically. This method can help researchers discern vegetation change. For example, soils from areas currently occupied by scrub woodland, or forest, that contain high concentrations of grass opal indicate previous grassland vegetation.

Paleoecological research in archeological deposits or their regional context may include analysis of invertebrate and vertebrate faunal remains. The study of Pleistocene and more recent insect remains (Elias 1994) provides information about the natural landscape in general, and when the remains are from archeological contexts, information about the human landscape and sometimes people's diets and health. In complement, regional studies of the distribution of vertebrate remains (Graham and Mead 1987) provide information about the geographic ranges of animals over time and shifts in direction and rate. Site-specific archaeozoological studies can then use the regional database in interpreting past human ecologies.

More regional archaeological studies are needed, such as those done by the Department of Defense for the Southern Plains and for the Central and Northern Plains (Green and Limp 1995) or by other land-managing agencies. These supplement state historic preservation plans with their cultural resource overviews, which are completed in compliance with the National Historic Preservation Act (National Park Service 1983). As a representative number of archaeological sites are surveyed and excavated, a temporal and spatial picture of former human settlement patterns in a specific region may begin to emerge. Combined with data from other sources, it may be possible to (1) uncover the networks of trade trails, migration and resource exploitation patterns, (2) discern which technologies are indicators of specific, prehistoric economic activities, and (3) discover to what extent different vegetation types were major determinants or coincidental factors of settlement patterns (see Sullivan et al., this volume).

## 8.6 Pyrodendrochronology

Knowledge of the fire history of an area is fundamental to understanding how the mosaic of vegetation types has developed. Fire scar and tree ring analysis is used to reconstruct the fire history of an area. Analysis of the growth rings in trees enables the researcher to date past fire events and variations in climate. New techniques are being developed to distinguish better between lightning and human-set fires. For example, the position of a fire scar in relation to earlywood–latewood formation can assist in discerning the season of burn. If the season is markedly different from the lightning fire season, this might be interpreted as a human-caused fire. Fire scar data gathered from paired forest stands, one of each pair close to former Indian occupation site and the other in an area remote from concentrated use, can also give valuable information regarding past indigenous fire regimes (Barrett and Arno 1982).

### 8.7 Historical Photograph Interpretation

Juxtaposing historical and contemporary photographs has proven to be an effective tool for documenting and interpreting the change in vegetation structure, composition, and pattern or mosaic of a specific area over time. A series of photographs of contemporary landscape scenes from different vantage points are then matched and compared with a sequence of historical photographs to show the extent of vegetation changes (Kay 1990, Kay and White 1995). Another approach involves starting with the historical photographs and then taking contemporary photographs from the same point used by the original photographer (Rogers et al. 1984). In this case, the historical photograph provides the benchmark for comparison.

### 8.8 Experimentation

Ecological field experiments can provide data to test hypotheses regarding the effects of past and present indigenous fire-based management, and other horticultural techniques, on plant species and plant communities. Specifically designed experiments may also substantiate historical and ethnographic reports regarding the claimed effects of certain human disturbance regimes on specific vegetation types (Anderson 1994b). Enough information on indigenous wild plant management is available to examine its role in helping to maintain biodiversity and the conservation of rare and endangered plants. For example, cultural objectives are known for burning in different habitats, the time of year of burning, and in some cases the fire-free interval. The effort to conceive and design such experiments would provide greater understanding and insight than is presently exhibited (Anderson 1993).

## 9 MANAGEMENT RECOMMENDATIONS

### 9.1 Decentralize Decisions

Ecosystem management relies on science and the value judgments that describe the goals or preferred conditions of biological communities at the local level. The courts, Congress, managers and scientists often lack detailed knowledge of local conditions and, unlike local residents, they seldom have to live with the consequences of their decisions. That means that goals should be set with the aid of people who live and work in those local environments.

The coadaptation of people with their environment takes place at the local level (Bonnicksen 1991). Cooperative decision-making involves managers, interest groups, and local residents working together as partners to formulate and carry out decisions. It is based on the idea that it is wiser to include affected groups in making decisions than to try to guess how they may react or what consequences they may suffer. It is also wasteful to ignore the knowledge possessed by people who spend their lives dealing with local conditions. Cooperative decision making at the local level also discourages conflict and fosters teamwork. It also has a proven record of success in resolving complex environmental issues (Bonnicksen 1993b, 1996). It is the best method for carrying out ecosystem management because it uses local knowledge, and it allows local residents to participate in decisions that directly affect their lives.

### 9.2 Use Customary Knowledge and Practice

The consideration of local knowledge systems by environmental agencies remains both possible and important. Although there have been enormous losses of such knowledge and restrictions on the activities of the custodians of such knowledge, there are still people who understand, and in some cases regularly use such knowledge. We argue that existing local knowledge systems can and should be considered as complementary and even as alternative models for relating to the environment.

One of the more difficult problems faced in the collection and (particularly) the utilization of local knowledge is getting scientists and technocrats to recognize and accept the idea that environmental users — fisherman, farmers, pastoralists, and foragers — have frequently developed important understandings about the worlds on which they depend and on which they have variable degrees of influence. The fact that they have done so by different means than those used in science is beside the point. The only valid questions are twofold. Can the knowledge employed by environmental artisans be scientifically verified? Will it work in the setting for which it was designed and in other similar settings? We argue that local knowledge systems can and should be considered as alternative models for relating to the environment.

We thus urge that a high priority be given for establishing program to identify experts in local knowledge (fishermen, farmers, trappers, hunters, pastoralists, loggers). Such a program should be considered and initially verified by ecologists in both the biological and social sciences. And, as local users are identified, the very best of them should be involved in management decisions.

## 9.3 Reintroduce Traditional Fires

The use of fire as a tool is the most dramatic way by which American Indians and other indigenous peoples influenced local environments. However, it was only one *albeit* critically important dimension of the overall strategies by which hunter-gatherers managed environmental resources — and in the process maintained what Europeans naively assumed was an "untouched wilderness." As Pyne (1982:71) pointed out,

> "It is often assumed that the American Indian was incapable of greatly modifying his environment and that he would not have been much interested in doing so if he did have the capabilities. In fact, he possessed both the tool and the will to use it. That tool was fire ... without which most Indian economies would have collapsed."

Traditional management practices are important today in a number of ways. They represent alternative examples of how people managed and, in a few cases and in more limited ways, still manage local resources — examples which cannot be automatically assumed as either appropriate or inappropriate to today's concerns. In almost all instances they show how such strategies changed over time, particularly throughout recent historical times when placed under great external pressures. In a number of instances they provide comparative examples of how people in like regions acted on the environment (e.g., as hunting-gathering societies in temperate rainforests), as well as how unlike traditions acted (e.g., prairie bison hunters and, later, cattlemen). In addition, there are still aspects of traditional knowledge to be found (e.g., hunting strategies, varying uses of fire, fishing practices, plant cultivation) which remain locally important today.

It must be stressed that traditional technology, as with modern technology, is preeminently knowledge and not merely the sum total of tools and techniques that people now or in the past employed. Technology is thus knowledge used for both practical and scientific purposes. Complimenting this, traditional ecological knowledge concerns peoples' understandings of environmental relationships, both those related to natural phenomena (e.g., weather, soils, animal behavior, and plant growth) and those influenced by human activities (e.g., hunting, trapping, burning, collecting plants, and other forms of cultivation). The only question in relating indigenous and scientific knowledge is whether such understandings can be shown to parallel or, in different but testable ways, challenge those of environmental science and practice.

Had American Indians not employed fires they would probably not have adapted at all. Academic as that question may be today, it is clearly the case that they would not have existed in such great numbers, and adaptations would have been far more precarious.

California provides one of the best records on both the extent and variability of burning practices in terms of the types of habitats and preferred resources involved (Blackburn and Anderson 1993; Lewis 1973). As on the Great Plains, California's Central Valley was regularly, in some cases annually, burned during late summer and fall to influence the diversity and productivity of plants and animals. Similar frequencies and seasonal patterns of burning were carried out in adjacent and higher-elevation oak-grasslands. On the other hand, coastal and interior stands of chamise and mixed chaparral were burned to create and maintain a mosaic of young to mature stands of brush, most of which were burned in the late fall and others in early spring, depending upon the preferred resources and kind of plant-animal communities therein.

Sierra-Cascade forests were subjected to regular understory burning, which provided new growth, highly attractive to game, and which at the same time prevented the build-up of dense undergrowth, such as have threatened sequoia and ponderosa forests in this century. In contrast, coastal redwood forests were managed in quite different ways involving the much more limited burning of forest corridors, ridge tops and forest openings, or prairies. In similar ways, these patterns of burning were much more characteristic of what was carried out in the temperate rainforest regions of Washington and British Columbia — as well as functionally similar practices found in the western Canadian boreal forest.

Whereas today fire is generally viewed by the public as only destructive, it is recognized by environmental scientists — as it was by American Indians — as a multi-dimensional tool with one of its most important benefits being that it can be used to limit major conflagrations.

For American Indians it was not in the belated actions of "fighting fire with fire" but, rather, in the preemptive acts of using fire to preclude its most dangerous effects, whether natural or man-made. This is a new perspective — we need more fire rather than less — that the general public, conservation groups, and elected officials are finding extremely difficult to accept.

Again, Indigenous fire regimes were particularly significant in that they were not mere replications of natural fire regimes but involved important differences that we have already discussed in terms of seasonality, frequency, intensity, control, and the purposeful rather than the random selection of sites. Natural fires (i.e., primarily lightning ignitions) helped to create and maintain coarse grained mosaics, being variably important in different biomes. The additive effects of indigenous burning practices, however, created and maintained fine grain mosaics in many forest types. They also reduced the most disruptive effects of natural fires, as is especially evident in semi-arid environments, such as California which today is regularly subjected to fire storms in the absence of either indigenous or management fire regimes.

Because we have accumulated a body of scientific knowledge about the ecology of fire, we can now interpret the environmental significance of what has been learned from historical and ethnographic descriptions of indigenous uses of fire. At the same time, there are still knowledgeable people in parts of Alaska, western and northern Canada, northern Mexico, and isolated areas of the American West who still have understandings about the uses and consequences of managing resources by hunting, gathering, and burning. And, without trying to replicate overall systems within which fire was an important tool for American Indians at the end of the last century, we can consider and perhaps use those features which best suit conditions in the coming centuries.

Indigenous management practices, like other aspects of American Indian cultures, have changed enormously from what they were in the past. No people are more aware of this than American Indians themselves and how, since their first encounters with Europeans, indigenous cultures were changed and had to adapt to new situations. This is as true of hunting, gathering and other habitat management practices as any other aspects of their cultures. In writing about indigenous burning practices in northern Alberta's boreal forests, Lewis (1982) described the perceptions of northern Athabaskan speakers, the Dene Tha, and how changes in burning strategies would have to be considered if they were introduced today on a much larger scale given the many changes in land use practices:

> "Because of pronounced increases in amounts of brush and trees, the environment has been significantly changed — and in (the Dene Tha) view, changed negatively — with the result that all burning, spring, summer or fall is today potentially more dangerous. The elders were also well aware that activities such as logging, drilling, recreation and farming would necessarily involve changes in burning practices. They have grown up with these developments while remembering the traditions of spring burning. Whatever their feelings about these intrusions and enforced changes, they well understand that burning practices would require adjustment, as have other aspects of their lives, to these pronounced alterations in their natural and social environments."

## 9.4 Reintroduce Indigenous Land Use Practices

The utilization of traditional knowledge systems in environmental management is problematic for a number of reasons. In the case of indigenous, or aboriginal, systems much of this knowledge has disappeared, though not entirely. And, where various dimensions of traditional knowledge are still employed in a number of regions, it appears, at least on the surface, to have been drastically altered as a result of exogenous inputs, particularly with respect to industrially produced tools, such as snowmobiles, rifles, etc. However, changes in the elements that make up contemporary techno-ecological practices, plus the losses of and restrictions on the users of such knowledge, do not mean that local ecological perspectives and management practices are either not there or are no longer relevant.

Despite their turbulent history and subsequent acculturation, many North American Indians are still a substantial, yet virtually untapped source of knowledge about present and former traditional plant uses, management practices, and vegetation change (Anderson 1994). In some cases, their knowledge spans the ecological histories of areas that extend before National Park Service and Forest Service charters. In addition, some elders still practice native plant management adjacent to their homes and traditional management is still conducted on some reservations and rancherias. Thus, American Indians could still play a vital role in the conservation of ecosystems (National Research Council 1992).

Through the assemblage, integration, and interpretation of information from tribal elders, and from various disciplines, researchers can identify specific biotic resources, ecosystem types, and whole biomes that were likely to be influenced significantly by indigenous management practices during pre-European settlement times. In some cases, it is possible to reconstruct details of the cultural objectives for management and important variables (e.g., frequency, season, and aerial extent of burning). This information can then be utilized in developing ecological field experiments to investigate alternative management practices. Those methods that meet contemporary needs should then be incorporated into management programs.

### 9.5 Use Timber Harvesting to Simulate Historical Disturbances

The wildfire hazard already is extreme in parts of North America and it is likely to become worse. We must break up blocks of dense trees and heavy fuels by restoring forest mosaics like those that existed during pre-European settlement times. Thus, restoring the historical forest mosaic will reduce the size and destructiveness of wildfires while creating a variety of wildlife habitats that enhance biological diversity.

Without American Indian burning, lightning fires alone cannot be relied upon to restore and maintain native forest mosaics because they occur too infrequently to prevent fuels from building up and causing catastrophic fires (Bonnicksen 1990). The human agents of disturbance are no longer operating, so the effects of American Indian burning must be simulated using either prescribed fire or mechanical methods. As Dr. Leopold said in his 1983 letter to the U.S. Park Service, "A chain-saw would do wonders." Science-based timber harvesting is the safest and most effective way to mimic Indian and lightning fires within an occupied landscape where prescribed fire is too dangerous. It can also be used to meet contemporary resource needs while still restoring an approximation of the historical forest ecosystem. That is, timber harvesting has the added advantage of creating jobs, producing wood, and generating revenue to pay for ecosystem management.

The deterioration of native forests caused by eliminating Indian fires and suppressing lightning fires during the past century demonstrates that diverse and healthy forests can only be sustained by active management. Prescribed fires can be used in many areas and lightning fires in a few areas. However, safety and liability concerns, especially in massive fire regimes, may prevent the widespread and frequent use of prescribed burning to thin forests and restore forest mosaics. A recent interest by Congress and the Environmental Protection Agency in regulating PM 2.5 (particulate matter less than 2.5 microns in diameter) under the Clean Air Act may further reduce opportunities to use prescribed burning. In addition, the density of understory trees is too high and the fuel loading too great in many forests to risk the use of prescribed fire. Restoration-oriented timber harvesting is the safest and most effective way to restore the effects of fire in these forests.

The need to retain living and dead parts of the current forest is an important part of timber harvesting and prescribed burning to mimic historical fires. It is especially important where native forests developed within a massive fire regime. Although massive fires in native forests killed most of the trees, they rarely destroyed all of the trees. Some dead trees were left standing after the fire and others lay in heaps on the ground. During the fire, winds often drove the flames along narrow paths, leaving behind stringers of scorched living trees. Occasionally the flames would leap over, or skip around, protected areas and leave groups of trees untouched. As a result, snags, fallen logs, and patches of live trees usually remained on the site after the fire passed. These remnants of the former forest provided habitat for wildlife and the foundation for a new forest.

Even-aged timber harvesting can mimic natural fires by leaving snags and fallen logs behind, and by creating openings shaped to look as if they were formed by fire. Patch cuts and group selection timber harvesting can mimic surface fires by producing small scale mosaics. Single tree selection can mimic single tree falls in uneven-aged forests composed of shade-tolerant trees. Snags and fallen logs should also be left behind when using these techniques. Thinning and prescribed burning also can keep the forest clear of debris and small trees. Low intensity prescribed fire may also be required following timber harvesting to more closely approximate the ecological effects of natural fires.

## 10 RESEARCH PRIORITIES

### 10.1 Modify Silvicultural Systems to Mimic Natural Disturbances

Modify existing silvicultural systems to more closely mimic the effects of Indian and lightning fires and other natural disturbances. Tailor these modified systems to the fire and other natural disturbance regimes that influenced the development of particular forests types and settings. Conduct experiments to evaluate these silvicultural systems in order to create systems that come as close as possible to the effects of natural processes on plant and animal populations and community structure. The evaluation of these silvicultural systems should include their effects on the quality and quantity of products that society needs from the forests, and the economics of producing them. Special consideration should be given to the use of local residents in management in order to also sustain resource-dependent communities.

### 10.2 Evaluate the Reintroduction of Indigenous Management Methods

Determine whether fire and other vegetation management methods used by American Indians should be reintroduced. Set up a series of long-term field experiments in different geographic regions to simulate indigenous horticultural practices and harvesting strategies and assess the interrelations and impacts of such cultural practices on individual plants, populations, communities, and ecosystem characteristics and dynamics.

### 10.3 Document and Test Local Knowledge Systems

Conduct in-depth studies of local knowledge systems of ranchers, loggers, and other resource users. Determine if extant local knowledge systems can be effective as complimentary and even as alternative models for relating to the environment. Test this knowledge in at least two ways: in terms of existing scientific paradigms and with the knowledge systems of peoples living in separate but similar environments. For biological ecologists, the testing of what knowledge has and can still be gained about American Indian management practices would include evaluations of such knowledge and practice in terms of what is already known in science and, where feasible, field observations and local tests. For ecologists from the social sciences, comparative evaluations would be made in terms of what is known about people in similar circumstances, and further tested with what can be derived from the historical record on North American land use practices.

### 10.4 Document Aboriginal Wildlife Use Systems

Conduct in-depth studies to document the methods American Indian's used to maintain wildlife, and the size and distributions of the resulting historical wildlife populations. Determine the effects of removing American Indian use and management on wildlife, as well as the secondary effects of these changes in wildlife on vegetation. Also assess the potential and desirability of restoring historical wildlife population levels to enhance biodiversity.

### 10.5 Document Knowledge Systems of Tribal Elders

Conduct in-depth ethnographic studies with Indian elders to uncover the details of former harvesting strategies and wildland management practices in different ecosystems. Highest priority should be given to use patterns and knowledge systems that are rapidly disappearing among the elder populations.

### 10.6 Reconstruct Historical Vegetation

Provide an accurate estimate of the plant species composition and community structure of late prehistoric forests and woodlands using phytolith analysis, palynology, ethnographic interviews, early landscape descriptions, comparison photographs, early herbarium specimen collections, the projection of existing vegetation back in time, and other techniques. Such estimates should include the relative proportion of patches of different kinds of vegetation in different stages of successional development that form the forest mosaic; the relative sizes and shapes of patches in the mosaic; the species composition of patches; the ages, sizes, and density of plants within patches; the vertical arrangement of plants within patches; and associated wildlife.

### 10.7 Reconstruct Indigenous Populations

Develop a prehistoric human population estimate for specific regions based upon archaeological site records and models, migration and settlement models, baptismal records, census data, and disease spread models to derive more realistic population numbers than currently exist.

### 10.8 Identify Cultural Use Complexes

Inventory the most important plant species utilized in early historic times by tribes in a specific geographic region for the most significant cultural use categories (requiring the most plant material). Establish complexes (species around which use is clustered). Conduct cross-cultural comparisons between tribes. Tally which plant species have widespread use. These plant species should then be studied in depth to discern their ecological roles in the ecosystem, their reproductive biology, and suites of traits that make them desirable to human cultures.

### 10.9 Assess the Historical Diets of Tribes

Determine the importance of different plant species in the historical diets of tribes. Reconstruct the diets of different tribes in a geographic region through analysis of museum ethnobotanical collections, survey of existing literature, ethnographic interviews, and archeological findings. Assess the importance of each species to the diet, and develop a food complex based upon information from No. 10.8.

### 10.10 Document Habitat Loss of Culturally Significant Plants

Inventory the native plant species that are useful to contemporary American Indian cultures in a specific geographic region and assess the habitat loss of culturally significant plants.

### 10.11 Evaluate the Historical Condition of Rare and Endangered Plants

Compare life history traits and habitat requirements of rare and endangered plants. Concentrate on those species and related common species (same genus) that were managed by a specific tribe. Determine if aboriginal management regimes should be reintroduced or mimicked to maintain rare and endangered plants.

### 10.12 Document and Evaluate Floristic Anomalies in Landscapes

Floristic anomalies in landscapes are islands of vegetation that harbor distinct plant species and assemblages that may be caused by former indigenous land management. These areas are often on the same slope, aspect, and of the same soil type as surrounding vegetation, but they harbor species unique to the area. These islands of unusual vegetation will frequently occur near archeological sites, and are often rapidly disappearing without indigenous management.

### 10.13 Use Interdisciplinary Teams for Regional Studies

Combine archaeological, ethnographic, paleoecological, fire history, and museum research to yield a better understanding of the resource and management needs of a prehistoric tribal village in a designated geographic region. Involve various disciplines, including plant ecology, paleoecology, archaeology, pyrodendrochronology, ethnography, and ethnobotany in the reconstruction of former Indian-nature relationships and cross-check conclusions. With this detailed information it would be possible to estimate the amount of "managed" acreage that would be needed to meet resource requirements of several villages occupying an entire region over long periods of time.

To date, comprehensive regional studies are almost nonexistent in the United States. One such exemplary study is the use of archaeological data, charcoal concentrations and pollen cores to gain a long-term understanding of environmental change in Yosemite Valley (Anderson and Carpenter 1991). An emerging project of the National Biological Service, LUHNA (Land Use History of North America) aims to combine coarse resolution data from paleoecological studies with information derived from original land surveys of the country and U.S. Forest Service data on the fire history of North America to characterize the historic landscape (Sisk and Noon 1995). Even these studies would benefit from the knowledge of other disciplines.

## REFERENCES


Alvard, M.S. 1994. Conservation by native peoples: prey choice in a depleted habitat. *Human Nature* 5: 127–154.

Anderson, M.K. 1993. The experimental approach to assessment of the potential ecological effects of horticultural practices by indigenous peoples on California wildlands. Unpublished Ph.D. dissertation, Department of Environmental Science, Policy, and Management, University of California, Berkeley.

Anderson, M.K. 1994a. The sustainable harvesting and horticultural practices of California indian tribes: linking plant homelands and human homelands. Raise the Stakes. *The Planet Drum Review*. San Francisco.

Anderson, M.K. 1994b. Prehistoric anthropogenic wildland burning by hunter-gatherer societies in the temperate regions: a net source, sink, or neutral to the global carbon budget? *Chemosphere* 29(5): 913–934.

Anderson, M.K., and G.P. Nabhan. 1991. Gardeners in Eden. *Wilderness* 55(194): 27–30.

Anderson, R.S., and S.L. Carpenter. 1991. Vegetation change in Yosemite Valley, Yosemite National Park, California, during the protohistoric period. *Madrono* 38: 1(1–13).

Bakeless, J. 1961. *The Eyes of Discovery: America as Seen by the First Explorers*. Dover Publications.

Barrett, S.W. and S.F. Arno. 1982. Indian fires as an ecological influence in the Northern Rockies. *Journal of Forestry* 80: 647–651.

Bean, L.J. and H.W. Lawton 1973. Some explanations for the rise of cultural complexity in native California with comments on proto-agriculture and agriculture. In: H.T. Lewis (ed.), *Patterns of Indian Burning in California: Ecology and Ethnohistory*, pp. v–xlvii. Ramona, California.

Bean, L.J. and K.S. Saubel. 1972. *Temalpakh Cahuilla Indian Knowledge and Usage of Plants*. Malki Museum Press, Morongo Indian Reservation.

Birkedal, T. 1993. Ancient hunters in the Alaskan wilderness: Human predators and their role and effect on wildlife populations and the implications for resource management. In: W.E. Brown and S.D. Veirs (eds.), *Partners in Stewardship: Proceedings of the 7th Conference on Research and Resource Management in Parks and on Public Lands*, pp. 228–234. The George Wright Society, Hancock, MI.

Blackburn, T.C. and M.K. Anderson (eds). 1993. *Before Wilderness: Environmental Management by Native Californians*. Ballena Press, Menlo Park, CA.

Bohrer, Vorsila L. 1983. New life from ashes: the tale of the burnt bush (*Rhus trilobata*). *Desert Plants* 5(3): 122–124.

Bonnicksen, T.M. 1990. Restoring biodiversity in park and wilderness areas: an assessment of the Yellowstone wildfires. *Proceedings of a symposium, Wilderness Areas: Their Impact*, pp. 25–32. Utah State University, Logan. April 19–20, 1990.

Bonnicksen, T.M. 1991. Managing biosocial systems. *Journal of Forestry* 89(10): 10–15.

Bonnicksen, T.M. 1993a. Social and political issues in ecological restoration. Conference on Sustainable Ecological Systems. Northern Arizona University, Flagstaff, Arizona. July 13–15.

Bonnicksen, T.M. 1993b. The Impact Process: a computer-aided group decision-making procedure for resolving complex issues. *The Environmental Professional* 15(2): 186–197.

Bonnicksen, T.M. 1994a. An analysis of a plan to maintain old-growth forest ecosystems. Forest Resources Technical Bulletin TB 94-3. American Forest & Paper Association, Washington, D.C.

Bonnicksen, T.M. 1994b. Ancient Forests: models for management. Forest Symposium. The World Forestry Institute, Portland, Oregon. March 29.

Bonnicksen, T.M. 1996. Reaching consensus on environmental issues: the use of throw away models. *Politics and the Life Sciences* 15(1): 23–34.

Bonnicksen, T.M. and E.C. Stone. 1982a. Managing vegetation within U. S. National Parks: a policy analysis." *Environmental Management* 6(2): 101–102, 109–122.

Bonnicksen, T.M. and E.C. Stone. 1982b. Reconstruction of a presettlement giant sequoia–mixed conifer forest community using the aggregation approach. *Ecology* 63(4): 1134–1148.

Bonnicksen, T.M. and E.C. Stone. 1985. Restoring naturalness to National Parks. *Environmental Management* 9(6): 479–486.

Boyd, R. 1986. Strategies of Indian burning in the Willamette Valley. *Canadian Journal of Anthropology* 5: 65–86.

Boyd, R. 1987. The effects of controlled burnings on three rare plants. In: T.S. Elias (ed.), *Conservation and Management of Rare and Endangered Plants*, pp. 513–519. Proceedings from a Conference of the California Native Plant Society.

Burch, E.S., Jr. and L. Ellanna (eds.). 1994. *Key Issues in Hunter-Gatherer Research*. Berg Publishers, Oxford.

Campbell, S.K. 1990. *Post-Columbian Culture History in the Northern Columbia Plateau A.D. 1500–1900*. Garland Publishing, New York.

Cassels, R. 1984. The role of prehistoric man in the faunal extinctions of New Zealand and other Pacific islands. In: P.S. Martin and R.G. Klein (eds.), *Quaternary Extinctions*, pp. 741–767. University of Arizona Press, Tucson, AZ.

Chapin, M. 1991. The practical value of ecodevelopment research. In: M. Oldfield and J. Alcorn (eds.), *Biodiversity Culture, Conservation, and Ecodevelopment*, pp. 230–247. Westview Press, Boulder, CO.

Cook, S.F. 1937. The extent and significance of disease among the Indians of Baja California. *Ibero-Americana* 12. Berkeley.

Cooper, C.F. 1960. Changes in vegetation, structure and growth of southwestern pine forests since white settlement. *Ecological Monographs* 30: 129–164.

Covington, W.W. and M.M. Moore. 1994. Southwestern ponderosa forest structure: changes since Euro-American settlement. *Journal of Forestry* 92(1): 39–47.

Daubenmire, R. 1985. The western limits of the range of the American bison. *Ecology* 66: 622–624.

Day, G.M. 1953. The Indian as ecological factor in the northeastern forest. *Ecology* 34: 329–46.

Deneven, W.M. (ed.). 1992. *The Native Population of the Americas in 1492*. 2nd ed. The University of Wisconsin Press, Madison, WI.

Densmore, F. 1974. *How Indians Use Wild Plants for Food, Medine and Crafts*. Dover Publications, New York.

Diamond, J.M. 1989. Quaternary megafaunal extinctions: Variations on a theme by Paganini. *Journal of Archaeol. Sci.* 16: 167–175.

Dobyns, H.F. 1983. *Their Numbers become Thinned: American Indian Population Dynamics in Eastern North America*. University of Tennessee Press, Knoxville, TN.

Dyson-Hudson, R., and E.A. Smith. 1978. Human territoriality: An ecological reassessment. *American Anthropol.* 80: 21–41.

Elias, S.A. 1994. *Quaternary Insects and their Environments*. Smithsonian Institution Press, Washington, DC.

Feit, H.A. 1994. The enduring pursuit: land, time and social relationships in anthropological models of hunter-gatherers in subarctic hunters; images. In: E.S. Burch and L. Ellana (eds.), *Key Issues in Hunter-Gatherer Research*, pp. 421–429. Berg Publishers, Oxford.

Fiedel, S.J. 1987. *Prehistory of the Americas*. Cambridge University Press, New York.

Flannery, T.F. 1990. Pleistocene faunal loss: Implications of the aftershock for Australia's past and future. *Archaeology in Oceania* 25: 45–67.

Ford, R.I. 1985. The patterns of prehistoric food production in North America. In: R.I. Ford (ed.), *Prehistoric Food Production in North America*, pp. 341–367. Anthropological Papers No. 75. University of Michigan Museum of Anthropology.

Fowler, C.S. 1986. Subsistence. In: W.L. D'Azevedo (ed.), *Handbook of North American Indians Volume 11 Great Basin*, pp. 64–97. Smithsonian Institution, Washington, DC.

Gomez-Pompa, A. and A. Kaus. 1992. Taming the wilderness myth. *Bioscience* 42(4): 271–279.

Gottesfeld, L.M. 1994. Aboriginal burning for vegetation management in northwest British Columbia. *Human Ecology* 22: 171–188.

Graham, R.W., and J.I. Mead. 1987. Environmental fluctuations and evolution of mammalian faunas during the last deglaciation in North America. In: W.F. Fuddiman and H.E. Wright (eds.), *North America and Adjacent Oceans During the Last Deglaciation*, pp. 371–402.

Grayson, D.K. 1991. Late Pleistocene mammalian extinctions in North America: Taxonomy, chronology, and explanations. *Journal of World Prehistory* 5: 193–231.

Green, T.J. and W.F. Limp. 1995. Archeological overviews: the Southwest Division overview and the central and northern plains overview and their use in historic preservation. *The George Wright Forum* 12(1): 52–57.

Grinnell, G.B. 1972. *The Cheyenne Indians: Their History and Ways of Life*. University of Nebraska Press, Lincoln, NE.

Hammett, J.E. 1992. The shapes of adaptation: historical ecology of anthropogenic landscapes in the Southeastern United States. *Landscape Ecology* 7(2): 121–135.

Harbinger, L.J. 1964. The Importance of Food Plants in the Maintenance of Nez Perce Cultural Identity. Unpublished Master's Thesis. Anthropology Department. Washington State University.

Hastings, A. 1984. Age-dependent predation is not a simple

process. II. Wolves, ungulates, and a discrete time model for predation on juveniles with a stabilizing tail. *Theor. Population Biol.* 26: 271–282.

Heinen, J., and B. Low. 1992. Human behavioral ecology and environmental conservation. *Environmental Conservation* 19: 105–116.

Higgins, K.F. 1986. Interpretation and Compendium of Historical Fire Accounts in the Northern Great Plains, Resource Publication #161. U.S. Department of the Interior, Washington, D.C.

Hobson, G. 1992. Traditional knowledge is science. *Northern Perspectives* 20: 2.

Hodge, F W. (ed.). 1990. The Narrative of Alvar Nunez Cabeza de Vaca (1527–1537). In: *Spanish Explorers in the Southern United States.* The Texas State Historical Association, Austin, TX.

Ingold, R., D. Riches and J. Woodburn (eds.). 1988. *Hunters and Gatherers 1: History, Evolution and Social Change. 2: Property, Power and Ideology.* Berg Publishers, Oxford.

Kari, P.R. 1987. *Tanaina plantlore Dena'ina K'et'una.* National Park Service Alaska region. Adult Literacy Laboratory, University of Alaska, Anchorage, Alaska.

Kay, C.E. 1990. Yellowstone's northern elk herd: a critical evaluation of the "natural regulation" paradigm. Ph.D. Dissertation, Utah State University, Logan, UT.

Kay, C.E. 1994b. Aboriginal overkill: The role of American Indians in structuring western ecosystems. *Human Nature* 5: 359–398.

Kay, C.E. 1995. Aboriginal overkill and native burning: Implications for modern ecosystem management. *West. J. App. For.* 10: 121–126.

Kay, C.E. 1997. Aboriginal overkill and the biogeography of moose in western North America. *Alces* 32: In press.

Kay, C.E., and C.W. White. 1995. Long-term ecosystem states and processes in the central Canadian Rockies: A new perspective on ecological integrity and ecosystem management. In: R.M. Linn (ed.), *Sustainable Society and Protected Areas*, pp. 119–132. The George Wright Society, Hancock, MI.

Kay, C.E., and W.S. Platts. 1997. Viewpoint: ungulate herbivory, willows, and political ecology in Yellowstone. *Journal of Range Management* 50: 139.

Kershaw, A.P. 1986. Climatic change and Aboriginal burning in north-east Australia during the last two glacial/interglacial cycles. *Nature* 322: 47–49.

Knudson, R. 1997. Using cultural resources to enhance ecosystem management. In: H.K. Cordell and J.C. Bergstrom (eds.), *Integrating Social Sciences and Ecosystem Management*, Southern Research Station, Georgia (in Press).

Latta, F.F. 1977. *Handbook of the Yokut Indians.* Kern County Museum.

Leach, M.K. and T.J. Givnish. 1996. Ecological determinants of species loss in remnant prairies. *Science* 273: 1555–1558.

Leacock, E. and R.B. Lee (eds.). 1982. *Politics and History in Band Societies.* Cambridge University Press, New York.

Leakey, R., and R. Lewin. 1995. *The Sixth Extinction: Patterns of Life and the Future of Humankind.* Doubleday, New York.

Lewis, H.T. 1973. *Patterns of Indian Burning in California: Ecology and Ethnohistory.* Ballena Press Anthropological Papers #1, Ramona, CA.

Lewis, H.T. 1978. Traditional uses of fire by Indians in northern Alberta. *Current Anthropology* 19: 401–402, 1978.

Lewis, H.T. 1982. Fire technology and resource management in aboriginal North America and Australia. In: E. Hunn and N. Williams (eds.), *The Regulation of Environmental Resources in Food Collecting Societies*, pp. 45–67. American Association for the Advancement of Science Selected Symposium Series No. 67. Westview Press. Boulder, CO.

Lewis, H.T. 1994. Management fires vs. corrective fires in northern Australia: an analogue for environmental change. *Chemosphere* 29: 940–963.

Lewis, H.T. 1980. Indian fires of spring. *Natural History* 89: 76–83.

Lewis, H.T. and T.A. Ferguson. 1988. Yards, corridors and mosaics: how to burn a boreal forest. *Human Ecology* 16: 57–77.

Lewis, M., and W. Clark. 1893. *The History of the Lewis and Clark Expedition*, E. Coues (ed.), originally published by Francis P. Harper, NY. Republished by Dover Publications, NY. Vol. I: 1–352, Vol. II: 353–820, Vol. III: 821–1364.

Lofland, J. 1974. Styles of reporting qualitative field research. *American Sociologist* 9: 101–111.

Martin, P.S. 1986. Refuting late Pleistocene extinction models. In: Elliott, D.K. (ed.), *Dynamics of Extinction*, pp. 107–130. Wiley, New York.

Martin, P.S. 1990. 40,000 years of extinctions on the "planet of doom." *Palaeogeography, Palaeoclimatology and Palaeoecology* 82: 187–201.

Martin, P.S., and R.G. Klein. 1984. *Quaternary Extinctions: A Prehistoric Revolution.* University of Arizona Press, Tucson, AZ.

Meehan, B. and N. White (eds.). 1990. *Hunter-Gatherer Demography: Past and Present.* Oceania Monograph No. 39. University of Sydney, Sydney.

Merculieff, I. 1994. Western society's linear systems and aboriginal cultures: the need for two-way exchanges for the sake of survival. In: Burch, E.S., Jr. and L. Ellanna (eds.), *Key Issues in Hunter-Gatherer Research.* Berg Publishers, Oxford, pp. 405–415.

Minnich, R.A., Barbour, M. G., Burk, J.H. and R.F. Fernau. 1993. Sixty years of change in conifer forests of the San Bernardino Mountains: reconstruction of Californian mixed conifer forests prior to fire suppression. Draft Manuscript. Dept. of Earth Sciences, Univ. of California-Riverside.

Minnis, P.E. and S.E. Plog. 1976. A study of the site specific distribution of Agave parryi in east central Arizona. *The Kiva* 41(3–4): 299–308.

Morgan, R.G. 1978. An ecological study of the northern plains as seen through the Garratt Site. *Occasional Papers in Anthropology, #1.* University of Regina Press, Regina, Saskatchewan.

Nabhan, G.P. 1985. *Gathering the Desert.* The University of Arizona Press, Tucson, AZ.

Nabhan, G.P., A.M. Rea, K.L. Reichardt, E. Mellink and C.F. Hutchinson. 1983. Papago influences on habitat and biotic diversity: Quitovac oasis ethnoecology. *Journal of Ethnobiology* 2(2): 124–143.

National Park Service. 1983. Archeology and historic preservation; Secretary of the Interior's standards and guidelines. *Federal Register* 48(190): 44715–44742.

National Research Council. 1992. *Science and the National Parks.* National Academy Press, Washington, DC.

Neumann, T.W. 1985. Human–wildlife competition and the passenger pigeon: Population growth from system destabilization. *Human Ecology* 4: 389–410.

Neumann, T.W. 1989. Human–wildlife competition and prehistoric subsistence: the case of the eastern United States. *J. Middle Atlantic Archaeol.* 5: 29–57.

Olson, S.L. 1989. Extinction on islands: man as a catastrophe. In: Western, D. and M. Pearl (eds.), *Conservation for the Twenty-First Century*, pp. 50–53. Oxford University Press, New York, NY.

Patterson, W.A. III and K.E. Sassaman 1988. Indian fires in the prehistory of New England. In: Nicholas, G.P. (ed.), *Holocene Human Ecology in Northeastern North America*, pp. 107–135. Plenum Press, New York.

Pearsall, D.M. 1989. *Paleoethnobotany: A Handbook of Procedures.* Academic Press, San Diego.

Peri, D.W. and S.M. Patterson. 1979. Ethnobotanical Resources of the Warm Springs Dam-Lake Sonoma Project Area Sonoma County, California. Final Report for U.S. Army Corps of Engineers. San Francisco District. Contract No. DACW07-78-C-0040. San Francisco: U.S. Army Corps of Engineers.

Pyne, S.J. 1982. *Fire in America: A Cultural History of Wildland and Rural Fire.* Princeton University Press.

Ramenofsky, A.F. 1987. *Vectors of Death: The Archaeology of European Contact.* University of New Mexico Press, Albuquerque, NM.

Riddington, R. 1982. Technology, world view, and adaptive strategy in a northern hunting society. *Canadian Review of Sociology and Anthropology* 19: 469–481.

Rides at the Door, R., and C. Montagne. 1996. Holistic resource management meets native culture. *Winds of Change* 11(4): 136–141.

Rogers, G.F., H.E. Malde and R.M. Turner. 1984. *Bibliography of Repeat Photography for Evaluating Landscape Change.* University of Utah Press, Salt Lake City, UT.

Roosevelt, A.C., et al. 1996. Paleoindian cave dwellers in the Amazon: the peopling of the Americas. *Science* 272: 373–384.

Schlick, M.D. 1994. *Columbia River Basketry Gift of the Ancestors, Gift of the Earth.* University of Washington Press, Seattle, WA.

Schrire, C. (ed.). 1984. *Past and Present in Hunter Gatherer Studies.* Academic Press, New York.

Schule, W. 1990. Human evolution, animal behavior, and Quaternary extinctions. *Homo* 41: 229–250.

Shipek, F.C. 1993. Kumeyaay plant husbandry: Fire, water and erosion Control Systems. In: T.C. Blackburn and M.K. Anderson (eds.), *Before the Wilderness: Environmental Management by Native Californians*, pp. 379–388. Ballena Press, Menlo Park, CA.

Singh, G., A.P. Kershaw and R. Clark 1981. Quaternary vegetation and fire history in Australia. In: A.M. Gill et al. (eds.), *Fire and the Australian Biota*, pp. 23–54. Australian Academy of Science, Canberra.

Sisk, T.D. and B.R. Noon. 1995. *Land Use History of North America: An Emerging Project of the National Biological Service.* Park Science.

Soulé, M.E. and K.A. Kohm. 1989. *Research Priorities for Conservation Biology.* Island Press Critical Issues Series. Island Press, Washington DC.

Steadman, D.W. 1995. Prehistoric extinctions of Pacific island birds: Biodiversity meets zooarchaeology. *Science* 267: 1123–1131.

Stewart, O.C. 1955. Fire as the first great force employed by man. In: W.L. Thomas (ed.), *Man's Role in Changing the Face of the Earth*, pp. 115–133. University Press, Chicago, IL.

Stuart, A.J. 1991. Mammalian extinctions in the late Pleistocene of northern Eurasia and North America. *Biological Review* 66: 453–562.

Teensma, D.A., J.T. Rienstra and M.A. Yeiter. 1991. Preliminary reconstruction and analysis of change in forest stand age classes of the Oregon Coast Range from 1850 to 1940. USDI Bureau of Land Management Technical Note T/N OR-9. 9 p. plus maps.

Temple, S.A. 1987. Do predators always capture substandard individuals disproportionately from prey populations? *Ecology* 68: 6699–674.

Truett, J. 1996. Bison and elk in the American Southwest: in search of the pristine. *Environmental Management* 20: 195–206.

Turner, N.J. 1991. Burning mountain sides for better crops: Aboriginal landscape burning in British Columbia. *Archaeology in Montana* 32: 57–73.

Turner, N.J. 1995. Ethnobotany today in northwestern North America. In: R.E. Schulter and S. VonReis (eds.), *Ethnobotany: Evolution of a Discipline*, pp. 264–283. Dioscordies Press, Portland, OR.

Ubelaker, D.H. 1988. North American Indian population size, A.D. 1500–1985. *American Journal of Physical Anthropology* 77: 289–294.

USDA Forest Service. 1989. The 1988 Canyon Creek fire. R1-89-3.

Usher, P.J. 1987. Indigenous management systems and the conservation of wildlife in the Canadian north. *Alternatives* 14(1): 3–9.

Verner, J., R.J. Gutierrez, and G.I. Gould, Jr. 1992. The California spotted owl: general biology and ecological relations. In: Verner, J., K.S. McKelvey, B.R. Noon, R.J. Gutierrez, G.I. Gould, Jr. and T.W. Beck (Technical Coordinators), The California Spotted Owl: A Technical Assessment of Its Current Status, pp. 55–77. USDA Forest Service Gen. Tech. Rep. PSW-GTR-133.

## THE AUTHORS

**Thomas M. Bonnicksen**
*Texas A&M University*
*Department of Forest Science*
*Room 305, Horticulture/Forest Science Building*
*College Station, TX 77843-2135, USA*

**M. Kat Anderson**

**Henry T. Lewis**

**Charles E. Kay**

**Ruthann Knudson**

Wagner, F.H. and C.E. Kay. 1993. "Natural" or "healthy" ecosystems: are U.S. national parks providing them? In: M.J. McDonnell and S.T. Pickett (eds.), *Humans as Components of Ecosystems*, pp. 257–270. Springer-Verlag, New York.

Waters, M.R. 1992. *Principles of Geoarchaeology. A North American Perspective*. The University of Arizona Press, Tucson, AZ.

Weslager, C.A. 1973. *Magic Medicines of the Indians*. Signet/ New American Library, New York.

White, R. 1984. American Indians and the environment. In: W.R. Swagerty (ed.), *Scholars and the Indian Experience: Critical Reviews of Recent Writing in the Social Sciences*, pp. 179–199. Indiana University Press, Bloomington, IN.

Whiting, A.F. 1939. Ethnobotany of the Hopi. *Museum of Northern Arizona Bulletin* 15.

Wilson, E.W. 1949. The American Indian as a conservationist. *Virginia Wildlife* pp. 22–24.

Yarnell, R.A. 1965. Implications of distinctive flora on Pueblo ruins. *American Anthropologist* 67(3): 662–674.

Zimov, S.A., et al. 1995. Steppe–tundra transition: a herbivore-driven biome shift at the end of the Pleistocene. *American Naturalist* 146: 765–794.

# Human Influences on the Development of North American Landscapes: Applications for Ecosystem Management

Richard D. Periman, Connie Reid, Matthew K. Zweifel, Gary McVicker, and Dan Huff

**Key questions addressed in this chapter**

- ♦ *How knowledge of past human influences can aid ecosystem management*
- ♦ *Approaches and techniques for analyzing past human influences*
- ♦ *The need for and utility of a well conceived research design*
- ♦ *Benefit of interdisciplinary analyses*
- ♦ *The need to use paleoscience, history, and ethnography*

**Keywords: Archaeology, reference conditions, prehistoric peoples, historic landscapes, fire, anthropogenic disturbance**

## 1 INTRODUCTION

This chapter provides land managers a guide to acquiring the necessary tools for addressing the complex role of humans in the development of ecosystems in North America. The approaches, methods, and examples are derived from a variety of specialties.Our goal here is to provide the reader with a range of analytical techniques for practical use in a variety of settings and management situations. Example case studies of such alterations are utilized as illustrations.

By approaching both the natural and cultural history of a landscape, with data derived from a wide variety of sciences and sources, we can synthesize an understanding of the human and natural processes that interacted in creating an ecosystem. The more varied the sources of data — the archaeological record, pollen analysis, or written accounts — the more effectively they can be used to cross-check conclusions independently. Integrative qualities of historical ecology, archaeology, and other disciplines can trace the ongoing inter- and intra-relationships between human actions and natural events and processes (Crumley 1994).

The general guidelines in this chapter have been written intentionally without an agency-specific agenda. Ideally, any local, state, or federal agency should be able to utilize the information and procedures presented. Agency archaeologists, historians, paleontologists, and other specialists may be consulted to provide or locate the needed expertise. The material discussed here contains discussions of past human influences on ecosystem dynamics, and relates to Bonnicksen et al. (this volume) — *Native American Influences on the Evolution of Forest Ecosystems*, and Sullivan et al. (this volume) — *Historical Science, Heritage Resources, and Ecosystem Management*.

Among the key issues and themes addressed in this chapter are the following:

1. Human prehistoric and historic activities have been crucial in helping to create North American ecosystems. These human ecological influences need to be taken into account when making management decisions.

2. The archaeological and paleobotanical record has the potential to provide managers with invaluable information concerning the dynamics of these human environmental influences. This valuable information is available from a number of existing documented sources and by collecting new data from archaeological excavation and paleoenvironmental data collection and analysis.

3. Land managers and government agencies can make informed ecosystem management decisions based on knowledge of the human role in shaping landscapes and the environment. Such knowledge may provide more complete knowledge of ecosystem capabilities, giving land managers increased flexibility in the decision-making process.

4. Pertinent, informed research questions concerning the anthropogenic landscape are necessary to save time and money.

## 2 THE IMPORTANCE OF UNDERSTANDING THE HUMAN ROLE IN ECOSYSTEM DEVELOPMENT

When faced with the challenge of ecological stewardship, an overworked manager may ask, "Why is this information important?" or, "Why do I need to know about the ways people affected the environment during the last 12,000 years?" With diminishing budgets and increasing public expectations, such questions are pertinent. When designing land management plans for specific areas, planners could benefit substantially from an awareness of the environmental forces, both natural and cultural, that have shaped ecosystems. They need to consider, for example, that Native Americans manipulated vegetation through burning, thus preventing specific plant species from progressing through later stages of succession, essentially selecting for early seral stages (Lewis 1993). The removal of Native American groups, and the subsequent suppression of fire, influenced population levels of animal species that had previously foraged within burned areas. What appears "natural" now, is actually an environment that has been manipulated and influenced by human activity. The cessation of human-induced processes potentially influences vegetation and ecosystems as much as the introduction of new activities (see Sullivan et al. this volume).

Ecosystem management literature contains an abundance of references to the concepts of "range of historic variation," "reference condition," "restoration," and "pre-settlement conditions." However, many of these documents do not define such "reference conditions" as being influenced by human activity through time (Quigley et al. 1996, Tainter and Tainter 1996). As early as 1864, George P. Marsh recognized the biological impact of human influence on the environment (Marsh 1874). For the latter part of this century the literature of environmental history, environmental archaeology, and historical ecology has definitively placed people in the equation of landscape dynamics (Chambers 1993, Winterhalder 1994, Reitz et al. 1996), yet human influence is often neglected in ecosystem analyses. Because of a lack of understanding concerning environmental history and anthropogenic effects and influences on the landscape, ecosystem reference conditions are often erroneously compared to "pre-settlement" (pre-European) conditions. As a result, the extensive and cumulative effects of 12 millennia of Native American occupation in North America are not considered, or are wrongly believed to have had little long-term effect on a landscape scale. There may be reference to an area's history, listing events such as the building of railroads or the introduction of livestock and homesteading, with little understanding of the effects historic processes may have had, or continue to have, on today's ecosystems.

We need to understand better all the dynamics that created ecosystems over time to identify accurately a "reference condition" or "restore" such systems. The outcome of a restoration effort, for example, could be adversely affected through a lack of knowledge about the ways prehistoric peoples augmented the distribution and health of a plant species, i.e., a field of bitterroot (see Example Case Study 4). A limited understanding of prehistoric use of fire may inadvertently lead agencies to reintroduce fire on a landscape with an intensity and scale that produces unanticipated and unwanted results. In some cases, grazing may have altered the plant community to the point that renewed use of fire may actually accelerate a decline in ecological health. Additionally, the act of defining and separating public lands, as National Parks, Wilderness, and National Forests, alters the course of ecosystem development through the implementation of differing agency priorities regarding recreation and preservation initiatives in contrast to resource development.

Ecosystem management is increasingly practiced as a social process through which people and government agencies work together to manage landscapes of local interest as more responsibility is shifted to participants and away from formal government decision-making. This process is highly dependent upon openly sharing values and information. A wide variety of information is potentially called for, all of which might be thought of as serving three principal purposes: (1) helping participants to find common ground; (2) facilitating their understanding of the ecosystem and landscape health; and (3) fostering a sustainable relationship between people and their environments.

The development of many contemporary values and perceptions concerning natural resource management are rooted in historical context, as are differing opinions concerning public land management. Local ties to cultural landscapes are shaped by personal histories, sociocultural perspectives, and folklore. Landscape values and impressions persist over generations, and are created and transformed by tradition. These sociocultural aspects of ecosystem analysis are discussed further in DeBuys et al. (seeVol. III), Raish et al. (see Vol. III), and Sullivan et al. (this volume).

There is a recognized utility in incorporating a landscape's human and natural history into the planning and decision-making process. However, information is limited concerning the intensity, scale, frequency, and long-term effects of past human influences on ecosystems. Contemporary social and economic conditions can further complicate a manager's decisions. For example, from an ecological and historical perspective the benefits of reintroducing fire in some areas may be acceptable to the public. In other areas, the landscape may be so dramatically altered by historic and modern Euro-American influences that reestablishing fire into an ecosystem, as indicated by archaeological or historical knowledge, may no longer be ecologically feasible. Even when fire can be used as a vegetational management tool with potentially positive ecological results, social concerns may not allow its implementation because of aesthetic and/or air quality concerns. Meanwhile, natural and human processes do not remain static, and environmental change continues.

For history and other information to have a meaningful role in finding solutions, it must be delivered in a more constructive setting than is typical of today's debates over public land management. Argument concerning ecosystem management can be transformed into meaningful dialogue, opening the way for a variety of information to be considered objectively. Historic information can help resolve legitimate questions related to the human role in the environment. To serve that outcome, it must be discussed in the context of the values, attitudes, and policies of its time.

## 3 PREHISTORIC HUMAN INFLUENCES ON ECOSYSTEMS

The entrance of modern humans into the New World at least 12,000 years ago signaled the beginning of anthropogenic changes in an environment that had previously experienced changes induced only by the natural agents of climate, geology, and biological processes. These new inhabitants introduced accidental and intentional use of fire, distributed seeds, exerted selective pressures on various plant species, and affected animal populations and distribution across a landscape (Blackburn and Anderson 1993). The use of fire allowed relatively small human populations to affect the environment on a large landscape scale. Prehistoric groups used fire for a variety of purposes: controlling plant diseases and insect infestations; increasing the frequency and range of plant species; eliciting desirable plant growth characteristics; minimizing the severity and number of uncontrollable wildfires; and facilitating hunting by the reduction of undergrowth (Lewis 1993). Prehistoric peoples in the Americas also affected ecosystems through such activities as sowing and broadcasting of seeds; transplantation of shrubs and small trees; water diversion; pruning and coppicing of plant communities; and construction of water diversion structures for erosion control. The selective harvesting of large bulbs, corms, and tubers (e.g., camas, bitterroot, and other plants) for food may have had the practical effect of thinning naturally crowded plants, aerating the soil, separating and dispersing smaller bulbs, and possibly caused increases in plant sizes. This likely led to the gradual expansion of specific species into new areas (Thoms 1989, Blackburn and Anderson 1993).

In areas where hunting and gathering subsistence strategies prevailed at the time of European contact, such as in the Northwest, elaborate systems of resource procurement were developed, including harvesting resources in large volumes and storing them for the winter months (Shalk 1986). In the Southwest, where agriculture was an important subsistence strategy, people affected their environments on a number of levels, including: the construction of fields; substantial wood harvesting for fuel and construction; construction of earthworks; and complex systems of hydrological manipulation, ranging from the construction of canals to the development of cobble-mulch gardening (Periman 1996; Anschuetz, in press). In eastern forests, large population centers were supported by agriculture and a continent-wide trade network. On the West Coast, people benefitted from a natural abundance of resources, and domesticated and managed entire ecosystems for their use (Blackburn and Anderson 1993, Lawton et al. 1993).

With a pre-contact population estimated to have ranged between 1.04 million (Kroeber 1939) and 18 million (Dobyns 1983), prehistoric peoples had a significant, cumulative influence on North American ecosystems. As the technological means of energy extraction became more efficient and complex, the interaction between people and their environment intensified, and their ability to modify or transform natural ecosystems increased. These adaptive human systems occurred along a complex gradient and/or continuum, involving more intensive interaction between people and their environments. This caused a progressively greater investment of human energy per unit of land, with an expanded capacity to modify or transform natural ecosystems (Blackburn and Anderson 1993).

## 4 POST-1492 HUMAN INFLUENCES ON ECOSYSTEMS

For management purposes, the historic period for North American ecosystems can be defined as having begun at the time of contact in 1492 with the advent of European expansion, settlement, and colonization in the New World. With the landing of Spanish, Portuguese, English, and French explorers and settlers on the shores of the Americas came a 500-year succession of environmental changes, on a scale not encountered since the end of the Ice Age. This new era of human influence, chronicled in the journals and official documents of these early explorers, settlers, military and church officials, is an important part of the 12,000-year continuum of human land use on this continent.

This system evolved from the expansion of European enterprises overseas, between 1500 and 1800, and led to the growth of world trade networks. The efforts to promote a favorable balance of trade and obtain colonies, which produced the raw materials needed by an expanding industrial revolution, developed into the system of multinational trade, banking, and resource development that circles the globe today (Wallerstein 1974). The western hemisphere was regarded as a storehouse of wealth to be exploited for financing wars and expanding European power and influence. The search for precious minerals and agricultural land stimulated this expansion throughout North America.

The demands of the world market directly and indirectly influenced the import of materials and technology, as well as the export of resources from the continent. We can begin to understand the broader scale of human influence on ecosystems through analyzing the layers of historic landscapes, which consist of cumulative influences exerted by world economic systems (Periman 1996)

## 5 THE IMPORTANCE OF DEFINING RESEARCH QUESTIONS

The methods, approaches, and techniques provided here, when used in combination with the appropriate research questions, have the potential to produce a tangible record of human influence on the environment, and a wealth of knowledge about climate change and ecological processes. Gathering data without a well defined research focus is costly, time consuming, and ineffective for management purposes. A simple listing of historical events, with few questions guiding an analysis of their effects on landscapes, may be of little real assistance in the decision-making process. Specific and general questions concerning past human influences need to be developed prior to data collection. The methods chosen will depend on the area to be researched, and the questions asked. For example, research questions focused on the vegetational changes within a landscape that occurred subsequent to the policy of active wildfire suppression could use historic photographs dating prior to fire control for depicting previous vegetational landscapes. To cross-reference and substantiate the landscape conditions depicted in the photos and achieve a higher level of accuracy, additional sources of information, both documentary and paleobotanical, should be used.

## 6 SCALE OF ANALYSIS

The scale and intensity of human action and environmental change has varied with time, space, and culture. The degree to which humans have influenced a specific ecosystem through time is related to how people in an area made their living (e.g., through hunting and gathering, subsistence agriculture, or industry), population density, areal extent of cultural and economic influences, and how long a culture may have persisted. Over hundreds or thousands of years, a number of cultures, each with a potentially different economic system, may have cumulatively altered the dynamics of an ecosystem. It is, therefore, necessary to examine factors of duration, area of influence, intensity, and periodicity at specific temporal and spatial scales. The relationship between human economic systems and particular environments is relative: a certain type of economic activity (e.g., horticulture) may have dramatically different environmental effects depending on ecosystems. A minute change in one environment could be perceived as a major change in another (Crumley 1994). When conducting ecosystem analysis these aspects of scale are extremely important. One size does not fit all, and neither does any single methodology.

### 6.1 Spatial Scale

From a historical perspective, spatial scale in the analysis of human environmental influences is related directly to the population size, economic system, and areal extent of the cultures that may have occupied an area, or a portion of an area, over a given period. Identifying spatial scale for studying anthropogenic influences can be more complex than drawing a boundary on a map. Areas of influence will be different for a small horticultural Anasazi pueblo than for a large scale historic ranching operation. Prehistoric influence, for example, may have effected the entire analysis unit or only portions of the study area depending upon the intensity of use and population size during occupation. As part of a larger economic system and associated technologies, historic ranching and mining may have had ecological effects on a scale that surpasses the geographic extent of the land unit under consideration. During ecosystem analysis the environmental influences of such distinct cultures need to be taken into consideration.

Whether compiling data for a regional study or focusing on a single watershed, having a well-defined scale of inquiry is crucial. When designing a study of a land unit, landscape, or a larger area, it is vital to begin the inquiry at a large, general level. Although not every study of humans and the environment need begin at the regional level, cultural practices, even if localized, usually have a connection to at least a larger sub-regional scale of activity (Ebert 1992). Beginning with a general focus, then narrowing the analysis to a specific landscape or land unit being studied, provides a perspective that is likely to include a human system responsible for environmental change.

Methods for collecting paleoecological data are often limited by spatial scale. Pollen studies, for example, have a finite range, and macrofossil information may be landscape- or site-specific. An appropriate sampling strategy should be designed to include a cross-section of samples derived from a variety of data sources in an ecosystem, as well as from a number of scales.

The USDA Forest Service (1993) National Hierarchical Framework for Ecological Units (NHFEU), is often used to place a study within an appropriate spatial context. However, it should be noted that these units have been developed primarily as biological or ecological units, and do not necessarily match areas occupied by specific cultures or tribal groups. We recommend that the NHFEU units be used only as a means of integrating the "biotic" with the "abiotic" components of an ecosystem. Using the NHFEU to define the units of any study involving human activities could lead to a set of culture area classifications that have little to do with actual human and environmental interaction, social organization, or prehistoric and historic range of effects.

### 6.2 Temporal Scale

Temporal scale pertains to the time span of various processes within an ecosystem, including those influenced by humans. An inherent risk in ecosystem management planning is to regard post-European processes as being the most valid for establishing "reference conditions." When choosing a suitable temporal scale for an analysis, it is important to begin with a general research design before approaching a specific span of human occupation. This moderates the risk of missing an important event, process, or practice related to human land use through time. All ecosystem analyses need not include a study of human effects on the entire globe for the last one million years. But, it is important that the collective effects of the human alterations of a landscape are taken into consideration when establishing the temporal parameters of an investigation. A selection of methods used for age determination is provided later in this chapter.

## 7 ARCHAEOLOGICAL AND PALEOBOTANICAL METHODS

The following methodologies and approaches are derived from a variety of disciplines, including archaeology, geology, climatology, botany, biology, and history. An interdisciplinary approach not only proves useful, but is necessary to understand the data collected. In the context of human exploitation of ecosystems through time, much information has been provided from archaeological excavations and research. Placing the study of humans within an ecological context has been well established in the field of American archaeology since the 1950s (Willey and Sabloff 1993). Ecological analysis is now a common element in archaeological and paleontological investigations. Human behavior and the nature of cultural change is complexly intertwined with environmental factors. Ecosystem research designs and analysis should reflect this dynamic process, and include methods and procedures that extract as much information as possible from archaeological studies and other ecological data sources. Through application of archaeological methods, and those of the other paleosciences, an accurate range of variability, including human and climatic influences, can be understood, and a more complete scope of management options considered.

### 7.1 Archaeological Excavation

Archaeological excavation is one of the most useful methods for acquiring information concerning past human land use. In addition to data relating to human occupation and settlement, archaeological sites contain an abundance of paleoenvironmental data. These time-capsules of human interaction with the environment have the potential to answer many questions, including how people adapted to various climates and ecologies through cultural and technological innovation (Brooks and Johannes 1990).

Excavations are usually conducted with one of the primary goals being the collection of paleoenvironmental information as part of the overall research design. Often data sources, such as microbotanical (pollen) and faunal (animal bone) materials and the various artifacts themselves, yield details concerning the availability and distribution of resources through time. Careful analysis of the artifacts found in archaeological sites can contribute an abundance of information and insight concerning patterns of resource use and distribution. The archaeological record is extensive, considering that the excavations of a number of deeply stratified sites in North America provide evidence of human seasonal use for over 12,000 years (Frison 1991, Stanford and Day 1992).

A single archaeological site usually provides data on a local scale. However, when the findings from a number of sites are analyzed and compared within an ecosystem or region, larger systems of interaction and land-use become discernable. Small scale surface sites, such as lithic scatters, when recorded and plotted over large areas as systems on a landscape or regional scale, have the potential to provide data on resource procurement, availability, and foraging ranges (see Sullivan et al. this volume). The other methods and techniques discussed in this chapter may be applied to analysis of archaeological sites for the extraction of data concerning human environmental interaction and land use. The information collected and the procedures used will depend on the research questions.

#### 7.1.1 Management Considerations

Federal agencies are required by the National Historic Preservation Act (NHPA) to take into consideration the impacts of their actions on archaeological and historic properties that are eligible for listing in the National Register of Historic Places. Compliance with this law not only protects these sources of valuable environmental data, it also presents an opportunity to collect information concerning past human land use and environmental effects to ecosystems. One cost-efficient approach may be to collect landscape level data through reconnaissance and background research during the early planning stages of a project. Additional survey and/or excavation could then be implemented at the project level according to the information needed.

Land management agencies usually have archaeologists on staff, or they have access to contract and academic archaeologists who can provide information and expertise concerning past human land use. Cultural resource managers can access a number of state, federal, and academic archaeological records that contain a range of valuable information at all spatial scales. Geographic Information System data bases of site locations are often maintained by State Historic Preservation Offices.

Archaeological surveys and excavations vary in cost according to the complexity of the environment and sites studied, the size of the area being taken into consideration, and age and depth of the deposits. For landscape level analysis, archaeological data should be compiled from a number of sites within or adjacent to a landscape (Ebert 1992). A single site may not provide the landscape level resolution needed for management and planning.

## 7.2 Paleobotanical Data

Paleobotanical research consists of the analysis of plant remains collected during archaeological excavations, or through sampling conducted by paleontologists. The microscopic materials generally include pollen grains, spores, and plant phytoliths (tiny opaline silica bodies that form in the epidermal cells of growing plants). Pollen grains and spores are resistant to decay under certain anaerobic conditions (e.g., in lake bottom sediments or bogs), or dry conditions (e.g., in protected desert rock shelters). Interestingly, pollen and plant fragments inadvertently incorporated into pottery or adobe bricks have been used in site and artifact analysis (O'Rourke 1983).

Macroscopic remains include large segments of plants such as leaves, stems, seeds, wooden timbers, and other building materials. This type of botanical information may be derived from corn cobs, bow staves, sandals, baskets, arrow shafts, and dwelling construction materials. Charcoal may be found in microscopic and macroscopic sizes. Under certain conditions, some burned botanical remains can be preserved for thousands of years, thus allowing species identification (Pearsall 1989).

Climate and the resulting floras (leading to specific faunas) play a large role in determining how people used and altered their environment. When used in conjunction with and compared to archaeological, ethnographic, and historical data, past human influences on ecosystems may then be inferred within an acceptable range of accuracy (Mehringer et al. 1977).

### 7.2.1 Management Considerations

Paleobotanical information used to reconstruct the vegetational patterns of past environments can help give managers an understanding of: (1) how an ecosystem has evolved through time; (2) the range, frequencies, and distributions of various plant communities through time; and (3) a landscape's range of ecological capabilities.

Paleobotanists and experts in pollen analysis can be found in geography, geology, botany, and related departments at many universities. A small number of contract businesses specialize in paleobotanical analysis, and their costs per sample vary according to the methods used. Agency archaeologists and geologists often maintain a list of experts and companies conducting this type of research.

Pollen provides a generalized reconstruction of past vegetation, whereas macrobotanical remains are more site specific. For large-scale, i.e., subregional and regional, studies, pollen analysis would be the method of choice for environmental reconstruction because of its regional distribution. Both, along with phytolith analysis, should be used for landscape-scale investigations.

## 7.3 Dendrochronology and Dendroclimatology

Dendrochronology (tree-ring dating) is generally considered one of the best known methods for directly determining absolute age of plant materials. This method is based on counting annual growth rings, observable in the cross-section of cut trees. The modern methods of dendrochronology and dendroclimatology (the study of climatic changes through time using tree-ring data) involve a refinement of such tree-ring counts. Cross-linking of ring-growth among trees is used to extend a sequence of growth cycles into the past, beyond the lifetime of a single tree. This computation of a long-term succession of tree-ring growth patterns was first established in the southwestern United States early in this century by the astronomer A. E. Douglass. His original research involved relating past climatic cycles, as reflected in cycles of wider and narrower tree-ring growth, to sunspot cycles. By counting back from a known starting point, a tree-ring segment can be dated by matching it to a part of the known growth sequence. Bristlecone pine, with a chronological record spanning over 8,000 years, has been used to refine the calibration of radiocarbon dating (Scharer and Ashmore 1993, Baillie 1995).

Dendrochronology and dendroclimatology are relatively precise sources of information concerning paleoclimatic change. Some tree species produce observable annual, cortical rings, resulting from a sudden increase in growth beginning in early spring, and ending in late summer or fall. The cell structure produced at the beginning of the growth season is different from that at the end, so a sharp boundary generally separates yearly rings. Starting with living trees, or newly cut trees, from which a small core may be taken, it is possible to count the rings and observe the variation in width of rings produced each year. The environmental limiting factors during the growing season and the location of the trees may affect ring patterns. Accurate use of tree-rings for climatological study requires the examination and cross-dating of valid samples of specimens, grown under climatically stressed conditions. Trees thriving in locations with abundant water, i.e., along riparian areas, will not manifest ring width variations attributable to climate change. Trees used in prehistoric construction, including Piñon pine and Douglas fir, are appropriate for extrapolating information about past climates (Cordell 1984).

In addition to providing information about temperature and precipitation, models of tree growth can address a variety of other environmental conditions. Surface runoff, atmospheric pressure anomalies (Cordell 1984), weather and climate variations, and fire frequencies also may be estimated when the necessary data are available (Touchan and Swetnam 1995).

Dendrochronology and dendroclimatology methods have several limitations that should be considered. Tree-rings are better indicators of low rainfall than of higher precipitation. Ring width varies with environmental conditions other than precipitation, including solar exposure and soil conditions. Also, a ring will not be produced under severe conditions, occurring at the beginning of the growing season, making it difficult to extrapolate the actual extent of climate deprivation (Cordell 1984).

Tree-ring reconstruction emphasizes short-term fluctuations and suppresses low frequency, long-term variations as a result of the standardization procedures used to compile indices from several individual trees. Other sources of archaeological paleoenvironmental information are sensitive to long-term rather than short-term fluctuations, making it difficult to correlate dendroclimatological data with other studies. Generally at the regional scale, accurate dendroclimatological data covers only the past 2,000 years, making it inappropriate and limited for the analyzing the entire range of time people have inhabited North American ecosystems (Cordell 1984, Baillie 1995).

### 7.3.1 Management Considerations

From a management and planning perspective, dendrochronology and dendroclimatology provide another way of determining past landscape conditions and development through time. Also, these methods can give land managers specific information about the spatial and temporal variability of fire regimes, including those regimes instigated by human land use. Tree-ring data can illuminate our understanding of the range of variability in the fire process, providing important guidance for the reintroduction of fire in sustaining forest ecosystems (Touchan and Swetnam 1995).

A wide range of specialization, skill, and familiarity with regional variability exists throughout land management agencies and academic institutions. The Laboratory of Tree-Ring Research at the University of Arizona, Tucson, is one of the world's premier dendrochornology facilities and is a good place to begin a search for this type of information.

## 7.4 Packrat Middens

Packrat middens are unique sources of fossils and have been used to reconstruct 40,000 years of paleoenvironmental progression in western North America. This method of paleoclimatic reconstruction involves the analysis of well-preserved fragments of plants and animals that have been accumulated by packrats (woodrats) in their nests, and are often encased in fragments of crystallized packrat urine. The deposits represent portions of the animal's nest, made of the refuse accumulated along trails or on perch areas (Betancourt et al. 1990). The middens are ubiquitous in caves and crevices throughout the arid West. Packrat midden studies have the potential to provide a record of late Pleistocene and early Holocene (following the end of the Ice Age) environments in western North America, making them one of the best sources of paleoenviromental data (Wells and Jorgensen 1964, Wells 1976, VanDevender and Spaulding 1979).

In protected perch areas, the crystallization of packrat urine through dehydration cements waste remains containing plant fragments, animal bones, fecal pellets, dust, and pollen. After preparation in a laboratory, the organic residue is compared with reference collections. A radiocarbon date also may be obtained from this processed material and applied to the entire assemblage. Middens from different exposures and elevations allow the use of traditional ecological methods for sampling across a landscape. At Chaco Canyon, and a variety of other similar archaeological settings, packrat middens have been used to measure resource availability and the extent of human impact on the prehistoric landscape (Betancourt and Devender 1981).

The association of plant macrofossils to vegetation within a packrat's foraging range, described as roughly one hectare centered at the midden, is complicated by animal behavior and other factors. These complicating variables include dietary preferences, the effect of distance to nearest plants, variation within the midden, duration of depositional episode, postdepositional history, and the ratio of macrofossils to estimates of species located in the surrounding plant community (Spaulding et al. 1990). In addition to macroremains, examination of pollen from middens can provide landscape and regional level paleoenvironmental information (Hall 1988).

### 7.4.1 Management Considerations

For landscape-level planning and management, analysis of packrat midden material, cross-referenced with other paleoenvironmental and archaeological methods, can provide valuable data for reconstructing paleoclimatic models and levels of human effects on ecosystems through time. These natural time capsules of past landscapes help bridge gaps in data collected from other sources, regardless of the complicating factors discussed above.

This type of analysis is highly specialized with only a limited number of scientists conducting packrat midden studies. Some of the institutions doing this type of research include: researchers in various departments at the University of Arizona in Tucson; The Desert Research Institute in Reno, Nevada; the Arizona-Sonora Desert Museum in Tucson; the Department of Geography at the University of Texas at Austin; and the USDA Forest Service, Rocky Mountain Research Station's Pinyon-Juniper Ecology Work Unit located at the University of Nevada in Reno.

Costs will vary according to the level of analysis, the number of middens sampled for a given landscape and whether or not pollen analysis is conducted. In the event that packrat midden derived data are to be used for management and planning, we recommend that this be conducted in the preplanning or early planning stage (for example, at the National Forest Management Act level). One study could provide information relevant to the management of one or more landscapes, thereby giving decision-makers a wider range of project planning alternatives.

## 7.5 Faunal Analysis

Faunal analysis systematically researches the skeletal remains of animals encountered in archaeological and paleontological sites. Like floral analysis, these data can provide a wealth of both cultural and paleoenvironmental information. Faunal material can generate data concerning various animal species exploited by humans, their quantity and therefore availability, and the age classes of the animals utilized. It can also provide implications for predator/prey relationships over time (Stahl 1982).

Faunal analysis begins with the identification, then quantification, of skeletal remains (Grayson 1984). Although this skeletal material may consist primarily of the bones of larger mammals used for food, small mammals, commonly overlooked as a cultural component, were also utilized by prehistoric peoples (Stahl 1982). Analysis of faunal remains associated with cultural materials can provide paleoenvironmental information about past habitat conditions and may indicate the seasons of the year that the site was habitable.

Along with the skeletal material from larger mammals, the remains of numerous smaller organisms have the potential to provide important paleoecological information. The bones of small animals and exoskeletons of insects can be utilized to develop accurate ecosystem, landscape, or land-type models of past environments. Additionally, these tiny sources of information can be used to cross-check other paleoenvironmental methodologies. Careful analysis of insect and rodent remains in archaeological sites should be considered valuable data sources and included in ecosystem level inquiries (Zweifel 1994).

### 7.5.1 Management Considerations

Faunal analysis is generally performed in conjunction with archaeological or paleontological research. In the event that agency or academic excavations are being conducted on public lands, managers should ensure that this type of study be included in research designs, mitigation plans, grant proposals, and/or contracts. It should be noted that unless these data are compiled from a selected number of sites within or adjacent to a landscape, the information gained is most applicable at a site-specific or sub-landscape level.

Specialists who conduct faunal analysis studies are located in biology, paleontology, archaeology, and other related departments at universities and research institutions. Also, several consulting companies employ individuals who specialize in this field. Some agency archaeologists are skilled in faunal analysis, but, if not, can provide managers the information needed to contact people with a background in faunal identification and research.

## 7.6 Geomorphology and Geoarchaeology

Geomorphology and geoarchaeology are closely related fields that are used extensively by archaeologists. Geomorphology is defined as the systematic description and analysis of landscapes and the processes that change them (Bloom 1978). Analyzing these processes can provide an understanding of how and why archaeological sites and other components of an ecosystem developed into the landform patterns observed today. Using this geologically derived information, an archaeologist can understand the formation of sites, features, and landscapes, through natural and human induced alterations. It was with this understanding that the specialty of geoarchaeology developed. Landforms are rarely stable: they constantly undergo processes of erosion, deposition, pedogenesis (soil formation), and pedoturbation (sediment mixing); geoarchaeologists study these processes in relationship to archaeological sites.

The processes by which the sediments have been formed and deposited are indicators of paleoclimatic conditions. The morphology and location of individual sediment particles (boulders, gravels, sand, or silt-sized particles) can be used to reveal how far, and by what agents, they were transported, and under what conditions they were produced (Meirendorf 1984). Analysis of the depositional history in an archaeological site can contribute to our understanding of paleoclimate and the changes induced by flood, drought, and human environmental alterations (Lasca and Donahue 1990).

Understanding the processes that created land forms is important for addressing past land-use practices. Agriculture and grazing can produce both chemical and physical alterations to soils, and specific and identifiable geomorphological features are created by repeated plowing of the same fields over time. These physical and chemical changes can be detected by careful archaeological investigation, and provide valuable insight into past human land-use patterns.

### 7.6.1 Management Considerations

Managers, planners, and anyone conducting landscape level analyses need to be aware that the land surfaces seen today represent only current conditions. Also, the patterns of erosion and sedimentation observable in the geological record may contain information about past human land use. Often deforestation, overgrazing, and industrialization will result in increased erosion and sediment deposition, thereby creating new landforms, river channels, and vegetational patterns (Nir 1981, Butzer 1982).

Soil scientists, sedimentary geologists, and other specialists who can conduct geomorphological analyses generally are located in geology and geography departments at universities, research institutions, and as private consultants. Land management agencies also employ specialists with this background. Archaeologists often have a range of training in geomorphology, but if not, can identify a reliable expert in the field. Like all such specialties, the costs and expertise vary according to the information needed and the temporal and spatial scales of the analysis.

## 8 ABSOLUTE AND RELATIVE DATING METHODS

Dating methods are critical in the establishment of an accurate temporal framework for analyzing human influences on environmental changes through time. This brief discussion of dating techniques, used by archaeologists and other paleoscientists, covers some of the more commonly used methods. These techniques can be separated into two classes: absolute and relative. Absolute dating methods ideally provide an accurate, specific date for an artifact, an occupation, or an event, determining its age on a specific time-scale, as in years before present (B.P.), or according to a fixed calendrical system. Relative dating determinations involve methods for evaluating the age of one piece of data and comparing it with another (for example, artifact A is older than artifact B). Absolute methods are preferred, but are not always available at any given site. A generalized discussion of many commonly used dating techniques can be found in Joukowsky (1986), Michels (1973), or Sharer and Ashmore (1993).

### 8.1 Radiocarbon Dating

Radiocarbon dating has proven to be one of the most useful absolute dating techniques used in archaeology and a range of other disciplines. This method relies on the decay of the radioactive carbon isotope $^{14}C$ into nitrogen. Decay occurs at a predictable rate, when the intake by living organisms of this naturally occurring radiocarbon in the form of $CO_2$ ceases with death, and the proportions of nitrogen and $^{14}C$ begin to change. The ratio of radioactive decay can be calculated, and an accurate date established relative to the death of the organism (Bradley 1985). Radiocarbon dating has been used for a variety of materials including peat, wood, bone, shell, sediments, and atmospheric $CO_2$ trapped in glacial ice. Carbon sample sizes can range from 1.0 to 4.0 grams of carbon for conventional dating techniques, or 0.001 to 0.3 grams for accelerator mass spectrometry (AMS) dating. Soils and other sediments can be dated using either method.

### 8.2 Thermoluminescence

Thermoluminescence is a form of absolute dating used principally on pottery and stone exposed to fire. Most natural minerals are luminescent to some extent, and when heated, release light, which is proportional to the dose of radiation absorbed by the object since it was last heated. The older the material, the greater the amount of thermoluminescence. Accuracy, however, can be problematic, and a large number of parameters must be measured, making the technique more difficult to apply than other common dating techniques (Joukowsky 1986).

### 8.3 Stratigraphic Position

Stratigraphic position is a widely used relative dating method, initially employed in the 1770s by Thomas Jefferson in his excavations of aboriginal burial mounds in Virginia (Willey and Sabloff 1993). This method of dating is based on older materials being found (in undisturbed context) below more recent materials. Although this technique is of undisputed value, its potential is largely limited to relative dating within a single site.

### 8.4 Management Considerations

In the event that archaeological, geographical, or paleoenvironmental studies are being conducted on public lands, managers should ensure that at least radiocarbon dates are established for archaeological sites or soil sample profiles. This type of study should be included in research designs, mitigation plans, grant proposals, and/or contracts. Understanding ecological change through time is meaningless without knowing if events or patterns of events and processes occurred, and obtaining dates is well worth the investment.

Several universities maintain radiocarbon laboratories in the United States. Thermoluminescence is a relatively new technique, compared to radiocarbon dating, and locating a specialist can be challenging. Generally, geomorphologists and archaeologists can determine a relative age through stratigraphic positioning of materials in a soil profile; however, at least a few radiocarbon dates should be established first as reference dates.

Radiocarbon dating being the most common method used is relatively inexpensive when considering the potential information. Costs per sample range from $200 to $400 for the conventional dating techniques to $500 to $600 for AMS dates. Costs, methods, and the time required for processing vary. An agency archaeologist is the best person to contact for advice concerning these and other dating techniques.

## 9 DOCUMENTARY, ARCHIVAL, AND HISTORICAL SOURCES

The methods used for studying historical human effects on ecosystems traditionally have included the use of documentary and archival materials and oral history interviews. The human influence upon ecosystems in the United States since European settlement is often well documented through such records. Additionally, paleobotanical, paleontological, geomorphological, and archaeological techniques have been used with great advantage in studying historic sites and landscapes (Kelso et al. 1995).

History has a relationship to ecological contexts. For example, in 1805, Lewis and Clark's difficulties in finding game during their crossing of the Bitterroot Mountains has been used to support the argument that the West had comparatively less game at that time than today (Lewis and Clark 1893). But, from an ecological perspective, their situation may have occurred because of the fact that historically, wildlife roamed more freely in response to climatic extremes and shortages of forage, unencumbered by highways, fences, and cities.

Although historic sources can provide valuable information about the forces that shaped particular landscapes, there are many gaps in the record, just as there are in archaeological and paleontological records. It is important when using historic data to recognize these limitations. Using a wide variety of sources for cross-checking information and accounts of events helps to reduce this problem.

### 9.1 Documentary and Archival Sources

The distinction between documentary and archival sources of information often causes confusion among specialists unfamiliar with this type of material. Documentary research involves inquiry into published or unpublished books, theses, articles, and other similar materials. Such documents generally are syntheses of primary sources, and are referred to as secondary sources. Archival research is performed using primary source material, including: manuscript collections; government, corporate, and personal correspondence; journals and personal diaries; original photos, survey notes and sketch maps; notes compiled by census-takers; and other uninterpreted information.

Documentary materials can be located in local, state, and federal archives and libraries. These repositories contain an abundance of books and documents that can provide broad, general histories for areas, ranging from single homesteads to cities and states. Historical documents are the best place to begin any type of background research concerning human effects on an ecosystem; although primarily focused upon recorded events, they often mention observations of Native American culture as well.

Historical documents are likely the most pertinent source of written information for regional scale analysis and may contain descriptions of past environmental conditions encountered during specific periods. By using primary materials referenced in these general sources, a researcher may further focus the investigation, answering specific questions concerning past environmental conditions for a more specific area or ecosystem. Local and state historical societies often can provide access to historical documents that are no longer available in city and county libraries, and may maintain archives as well. Many state and federal agencies maintain their own libraries, containing a wealth of historical books and other published material. Every government land management agency has a history and a wealth of archival records. Voluminous in number, all sources of government archival information cannot be listed here. The National Archives, in Washington D.C., is a storehouse for the Nation's records. It is best to contact these agencies and/or societies for a list of pertinent materials. The availability of historical documents varies among agency offices.

In determining the historical human effects on landscapes and locating historical observations of ecosystems, useful archival sources may include, but are not limited to:

- Historical maps and survey notes.
- Land surveys, sketch maps, and notes.
- War Department archives including Native American affairs.
- Military campaign records including journals, correspondence, and maps.
- Hydrological and navigation maps, and survey notes.
- Notes, journals, and sketch maps from original topographic and natural sciences, surveys, and expeditions.
- Early United States Geological Survey notes and correspondence, including soil maps.
- Census maps and Bureau of Agricultural Economics records.
- General Land Office records, correspondence, and maps.
- Homestead entry survey maps and notes.
- Mineral Plat maps containing mineral claim and development information.
- Historic photographs including landscapes and early aerial photos.
- Various government agencies: compiled land inventory records and maps.
- Corporate records (timber, mining, etc.).

## 9.2 Historical Archaeology

Historical archaeology, used in conjunction with existing documentation, can help to fill in the gaps left in the historical record. Frequently, an error committed with historical reconstruction and research of past environmental conditions involves the perception that all important events and occurrences are recorded and stored in a convenient archive. Regardless of meticulous record keeping and documentation, many of the historical activities that helped shape contemporary landscapes were never written about. Even without documentary sources, historical archaeological research has the potential to clarify our understanding of past human processes.

Archaeological investigation of Euro-American sites, originating with contact in the late 15th century, has the potential to provide managers and researchers with a wealth of information concerning historical human alterations to the environment. Historical sites, ranging from early Spanish settlements and British forts, to industrial mining towns and logging camps, contain answers to many questions concerning the evolution of today's landscapes (see Sullivan et al., this volume).

## 9.3 Historical Newspapers

Historical newspapers are informative primary sources and generally may be found in archival collections. Daily and weekly, local and territorial newspapers can provide information and observations concerning political, social, demographic, economic and resource-use trends and events thought pertinent or entertaining at the time of their occurrence. Researching newspapers, which are often preserved on microfiche and microfilm, can be very labor intensive, but the information discovered can be well worth the effort and utilized for ecosystem management. In rural areas especially, newspapers are valuable for understanding local land-use activities, events, and observations, and may be the only documentary source in existence containing such information. Information contained in historical newspapers can illustrate that patterns associated with resource dependent economies appear to be cyclical, even within the same community. The records provided through historical journalism can be used to interface with the local community, serving as a common, neutral source of information for contemporary issues.

## 9.3 Oral History

Oral history interviews can be complementary to documentary and archival research. By interviewing older community members, a researcher can obtain information and descriptions that may not have been previously recorded in written form. Such individuals are often sources of valuable research concerning people, events, and places in the local community and surrounding landscapes. Many may have actually been involved with the historical processes that altered ecosystems. People who grew up on local homesteads, lived in mining camps, and built many of the historical structures in an area can often provide details and insights concerning historical human influences on the land. Community members may own historical photographs dating into the last century, which when compared to photos depicting modern conditions can illustrate substantial alterations of the landscapes depicted.

### 9.4 Historical Landscape Paintings and Drawings

Historical landscape paintings and drawings are important sources of information, but are often overlooked. Although some of these depictions of ecosystems may be idealized, or have little resemblance with actual historical conditions, many artists painstakingly rendered their art to record the landscapes they observed. Artwork, even if it depicts idealized landscapes rather than realistic motifs, may provide the researcher with an understanding about how past cultural values influenced the *representation* of environment and landscape.

### 9.5 Management Considerations

Obtaining the documented, historical, background for a landscape analysis is often one of the more affordable steps in examining the effects of past human. It can be cost-effective to have first a historical overview written for a particular land management unit (National Forest, Ranger District, BLM Management Area), while more focused research can be done at the planning, landscape, and project level. This type of research should be incorporated into cultural resource inventories conducted for compliance with the National Historic Preservation Act.

Several historical archaeologists and a few historians are employed by federal land management agencies. Others are located at universities or work as private consultants. Art historians and scholars specializing in landscape studies are generally located at universities although there are several private consultants in this field as well. Contact a local university, historical society, historical museum, or agency archaeologist for information pertaining to the location of such expertise.

## 10 ETHNOGRAPHY AND ETHNOLOGY

Ethnography and ethnology are two fields of study that use cultural information. Ethnography refers to studies of individual cultural systems consisting of a single society, or a segment of a complex society, as in a particular community. Ethnology assumes a generalizing perspective, using comparisons among ethnographic data in an attempt to understand the processes of culture. By comparing data from many societies, ethnology studies how and why cultural systems operate and change (Sharer and Ashmore 1993). Both approaches can provide an abundance of information regarding past and current uses, manipulations, alterations, and perspectives of human interaction with the environment. Each utilize a culture's knowledge, as preserved in written records (interviews and field notes collected by anthropologists that make up the ethnographical record), or as preserved in the memories of living individuals. These reservoirs of information regarding the past may include Native American tribal elders whose vital cultural and environmental knowledge has been inherited from preceding generations. Also, ethnographical methods use records and oral histories provided by the older Euro-American members of a community who shared their memories of previous land uses and the past conditions of local landscapes.

Ethnological comparisons can greatly enhance the understanding of the processes through which people changed their environments on several temporal and spatial scales. This type of comparative approach is necessary for comprehending cumulative alterations across landscapes, ecosystems, and regions. Understanding the cultural differences and similarities of land use among various groups can be helpful for land managers in the decision-making process. For example, comparing Native American landscape burning practices, such as the Athabaskan's in northern Canada to those of Australian Aborigines, can provide managers with vital information concerning how populations of Native Americans manipulated vegetation through the use of selective fires (see Bonnicksen et al., this volume; Lewis 1993).

Although rich and illustrative, the ethnographic record has limitations, as do other historical, archaeological, and paleoecological methods. When using ethnographical accounts of past or contemporary practices, it is important to remember that our view of the environments we live in is focused through the lens of cultural perspectives and individual perceptions. As with all methods discussed in this chapter, the use of a cross-reference methodology can help moderate this inherent bias. Ethnological comparison should be used judiciously and needs to be well defined when applied to, and between, appropriate cultures and practices (Boas 1940).

## 10.1 Management Considerations

Ethnography and ethnology have the potential to provide decision-makers with crucial information concerning the cultural, economic, and ideological ties various groups of people have had, and continue to have, with landscapes. Ethnographical studies should be done at the landscape and subregional scale during the early planning process. There are occasions when broad-scale ethnographical studies are applicable, for example, at the regional and ecoregional level.

Ethnography and ethnology generally are practiced by anthropologists. Expertise in this field can be found in the anthropology departments of universities, and an increasing number of ethnographers work as private consultants. In general, agency archaeologists are trained under the disciplinary umbrella of anthropology. Some cultural resource professionals, now working as archaeologists, actually specialized in ethnography while obtaining their degrees. Again, one of the best sources for this type of information is an agency archaeologist or cultural resource manager.

# 11 CASE STUDIES

The following cases show how researchers and resource managers have applied various archaeological, ethnographic, paleoenvironmental, and historical methods to questions concerning the human role in the development of a particular North American ecosystem. The projects and methods used in each example are briefly described, followed with a discussion of how the collected information concerning prehistorical and historical human influences on the landscape was applied to resource management.

## Beyond Fire Scar Analysis: Yaak River Valley Archaeology and Paleoenvironmental Study

*Location*

Yaak River Valley, Kootenai National Forest in northwestern Montana.

*Participants*

Heritage personnel from the Kootenai National Forest.

*Reason for Conducting the Project*

This project was initially done to mitigate the effects of a land exchange on a National Register of Historic Places eligible archaeological site on the Kootenai National Forest.

*Project Description*

On the Three Rivers Ranger District of the Kootenai National Forest in northwest Montana, the Yaak River Valley is a north–south tributary to the Kootenai River drainage system. The valley forest is a mixture of true fir, Douglas-fir, spruce, hemlock, cedar, western larch, lodgepole pine, white pine, whitebark pine, ponderosa pine, aspen, and birch.

In the summer of 1993, an archaeological site was discovered along the West Fork of the Yaak River, about one mile above its confluence with the main branch of the Yaak River. Initial testing indicated that the site contained valuable scientific information about past human use of the area and paleoenvironments.

The site record revealed that the general area was apparently affected repeatedly by a long series of periodic fires, combined with subsequent flooding shortly afterwards, thus sealing the burned forest floor under preserving layers of sand. Radiocarbon dates from the charcoal samples showed that a fire history of almost 2,000 years was represented in the stratified sediments. Twelve fires were identified, radiocarbon dates were established for eight samples, and the remaining four were dated according to their locations in the stratigraphic sequence. Fires burned across the site area in A.D.: 20, 420, 690, 970, 1090, 1200, 1510, 1540, 1570, 1610, 1660, and 1770. The radiocarbon date of 1770 A.D., from a charcoal sample collected immediately below a layer of volcanic ash, was identified as originating from the Mt. St. Helens eruption of 1800 A.D., providing an independent confirmation of the radiocarbon date. The oldest identified charcoal sample was western larch, *Larix occidentalis*, whereas the others were identified as ponderosa pine, *Pinus ponderosa*. Although the sample size was small, preliminary indications are that an environmental shift occurred prior to 1,000 years ago that favored ponderosa pine over larch. Correspondingly, about 600 years ago the climate became cooler, in a period known as the "Little Ice Age" (Lamb 1977, Hughs and Diaz 1994).

These dates revealed an increase in the frequency of small, localized fires in the valley floors during the past 2,000 years. The use of maintenance fires by the Kootenai Indians, who occupied the area during that period (Barret 1980), provided foliage for large herbivores, and the valley served as winter range (Thoms and Burtchard 1987). The Yaak Valley was utilized by the Kootenai as a caribou and deer hunting ground (Schaeffer 1940, Turney-High 1941). The use of maintenance fires led to resource intensification in the Kootenai River region. Low intensity burning also has been associated with aboriginal land use practices elsewhere in the inter-mountain West (Mehringer et al. 1977).

*Lessons Learned and Information Gained*
The information yielded by this excavation is being used in conjunction with other data to build a larger, more accurate paleoenvironmental picture of the Yaak Valley. Analyses of pollen collected in this area have provided a record of the large fires that burned in the valley, and micro-fossil identification displays a vegetative history of the general area. This information will enhance an ongoing study of fire history in the valley, utilizing the fire scars present in tree rings. The combination of the fire scar data, the West Fork archaeological information, and the pollen analyses provide a more detailed fire history of the area, building a more complete picture of past ecosystems. The information generated by this project has direct management utility and will be used in conjunction with other data to develop a burning program that better mimics pre-European fire regimes. The historical fire data provide management with a sound rationale for developing a fire maintenance plan based on scientific data. Also, the information collected during this project illustrates the value of archaeological excavation techniques in understanding past environments and human land uses through time, answering questions pertinent to a wide array of disciplines. Wildfires play a vital role in many ecosystems. In forested areas of western North America, fire suppression has dramatically altered vegetation, promoted the spread of tree diseases, and increased the likelihood of catastrophic fire. Archaeological research can provide land managers with data on human instigated fires in landscapes, and provide fire sequence data useful in developing fire management plans.

*Contacts*
For more information regarding this project, contact Connie Reid, USDA Forest Service, North Kaibab Ranger District, 430 S. Main, Fredonia, AZ; or Rebecca Timmons, Forest Supervisor's Office, Kootenai National Forest, 506 Highway 2 West, Libby, Montana.

## The Sheldon Flats Urban Interface Project

*Location*
Sheldon Flats on the Kootenai National Forest in northwestern Montana.

*Participants*
Heritage resource personnel from the Kootenai National Forest and local citizens.

*Reasons for Conducting Project*
During the historical period (after European contact), landscapes in North America were modified by humans at an unprecedented level. Even though prehistoric peoples may have effected the long-term development of landscapes, modern peoples have dramatically influenced the landscapes we observe and manage. Knowledge of this human influence is crucial when attempting to analyze the role of humans in the development of North American landscapes within a historical context. The history of Sheldon Flats area was examined by Kootenai National Forest Heritage Resource personnel in order to determine the effects of historical human activity on the vegetational landscape of the area as part of an ecosystem area assessment. In an effort to gain support for the contemporary management goals of reducing fire hazards and improving timber productivity through controlled burning, timber removal, and reforestation, this historical information was compiled and presented to area residents.

*Project Description*
The Sheldon Flats Urban Interface project on the Kootenai National Forest in Montana is an example of how historical information can be used by land managers. Sheldon Flats is densely forested and presently contains many residential homes. Area residents have become increasingly concerned about potential wildfire hazards. In an effort to address these issues, forest managers conducted a series of meetings, beginning in 1996, to work with residents on the development of desired future conditions for the Sheldon Flats residential area.

To understand better the historical context, data concerning historic vegetation of Sheldon Flats were extracted from the following resources: Government Land Office (GLO) survey notes, historical photographs, Bureau of Forestry timber survey notes, early mining company survey notes, timber company notes, Kootenai National Forest timber harvest and plantation records, and newspaper articles containing descriptions of the area's vegetation from the 1870s to 1904). These data provided managers with pertinent information concerning the historical forces that shaped Sheldon Flats ecosystems.

*Lessons Learned and Information Gained*
The historical research indicated that the dense forests characterizing the area today are a recent phenomenon. Past forests were characterized by open ponderosa pine stands that had likely been burned frequently by Native Americans. Subsequent fire suppression and timber harvest altered the forest composition. Also,

timber-stand inventories indicated that certain areas were characterized by poor regeneration, while adjacent areas were productive. Cross-referencing early aerial photographs and corporate timber records revealed that 60 years ago, the area was logged with techniques that resulted in soil compaction, creating long-term regeneration problems for trees and other vegetation. Field inventory indicated that 1930s era ponderosa pine stands were regenerating poorly, and in many instances had less than a 50% survival rate. The historical record indicated that these stands were not native to the area, as the seedlings had originated in South Dakota under different ecological conditions, and appeared to be unsuitable for the area.

By providing the historical context, the Forest Service was able to gain public consensus for improving ecosystem health for the area. Initial goals include thinning dense stands of trees, prescribed burning, removal of non-native ponderosa pine, and reforestation with suitable seedlings. Historical information and context is, therefore, a valuable aid when addressing public values and resource management issues.

### *Contacts*

For more information regarding this project, contact Connie Reid, USDA Forest Service, North Kaibab Ranger District, 430 S. Main, Fredonia, AZ; or Rebecca Timmons or Mark White, Forest Supervisor's Office, Kootenai National Forest, 506 Highway 2 West, Libby, Montana.

## Tularosa Basin Ecosystems Archaeological Project

### *Location*

The Tularosa Basin in south-central New Mexico.

### *Participants*

The project is a cooperative effort between the U.S. Army White Sands Missile Range and the Bureau of Land Management. The contractor conducting the work is Human Systems Research Inc.

### *Reasons for Conducting Project*

This project is being conducted to provide the U.S. Army with environmental and historical knowledge essential for managing the vast expanses of the White Sands Missile Range. The need for solid data about natural long-term cycles, landscape changes, and the ways humans shaped ecosystems becomes evident as land managers increasingly apply the principles of ecosystem management. Archaeological and historical data about vegetative cycles, fire histories, animal population distributions over time, climatic regimes, and riparian system histories can often provide the information scientists need for ecosystem management projects. Humans have been manipulating ecosystems for many millennia, therefore identifying human-caused changes is essential to understanding how present landscapes are configured.

### *Project Description*

The Tularosa Basin in south-central New Mexico is the focus of a pilot study providing natural resource specialists with data from archaeological, historic, and paleoenvironmental sources relevant to current management issues. The project will produce a database containing annotations about each data source, including a brief abstract, type of data available in the report/record (e.g., pollen, faunal), geographic origin of data, the location where the data or report is available, and a list of species by common and scientific names. A query search will direct researchers to the data sources that could be of assistance in designing, implementing, or choosing a management option. This component of providing a database with application to ecosystem management is the heart of the project.

### *Expected Benefits*

Ancillary summaries of current management issues, data types, and a listing of available natural resource data will facilitate using the database. Because the purpose of the project is to enable managers to make more informed decisions, the current management issues summary is a critical link in the database. Here, a researcher will find which kinds of archaeological, historical, and paleoenvironmental data are relevant to a particular management question. For example, the topic "Animal Species Reintroduction" will include a brief description of how rock art figures and bones from archaeological sites and packrat middens can reveal which species were present in an area at a particular time. The user could then go to the database and construct a query, such as for *Antilocapra Americana* (pronghorn antelope) bone within certain geographic coordinates.

The second summary will be an overview of data types written for noncultural specialists. Each type of datum included in the database will contain a description of its originating context, how it was collected and analyzed, and the primary uses and limitation of such data. For example, there will be explanations of how fossil pollen becomes incorporated into sediments through natural pollen deposition and human activities; conditions required for its place in preservation; how it is gathered, processed, and identified in the laboratory; and conventions of data display. Limitations, such as the species differential for distance traveled from source to deposition, would be included. This would, for example, inform researchers that grass pollen in a sample usually indicates an immediately local occurrence of grass, but pine pollen can be transported hundreds of miles before deposition.

The final summary will be a listing of major collections of natural resource data, such as soil and vegetation inventories, and the media in which the data are available (e.g., paper copy. maps, GIS, database) from land managing agencies within the Tularosa Basin. Eventually all such data will be available on a Geographic Information System, so that a user could construct a query that includes natural resource information. For management issues concerning riparian areas, it could be useful to learn which woody species were present in a particular drainage under various climatic regimes, and compare these data with other areas in the same soil series.

This project was developed with the participation of natural resource specialists and managers, and the design of the database and ancillary summaries reflects their needs and concerns. A follow-up phase, one year after product completion, will focus on the project's practical uses and suggestions for revision of the prototype format. A summary of the Basin's climatic and vegetative history may also be produced. The anticipated outcome of this project is improved ecosystem management, as: (1) decision-makers will have more information concerning the factors that shaped present ecosystems; (2) ecosystem changes from human activity or natural long-term cycles can be discerned, and land management practices adjusted accordingly; and (3) future studies and inventories will determine whether any further data are needed and where such information is available. This ecological stewardship enables informed decisions on how to manage natural resources productively and efficiently.

### *Contacts*

For information on this project contact: Mike Mallouf, White Sands Missile Range, 505-678-8651; Shelley Smith, Army Environmental Center, 410-671-1577 (general information); Meade Kemrer or David Kirkpatrick, Human Systems Research, Inc., 505-524-9456 (database operation and content).

## Traditional Plant Restoration in Western Montana

### *Location*

The Bitterroot National Forest in western Montana.

### *Participants*

Confederated Salish and Kootenai Tribes, Dow Elanco Chemical Company, and the Bitterroot National Forest.

### *Reasons for Conducting the Project*

The goal of this experimental project is to eradicate non-native plant species and to restore a plant that was, and continues to be, culturally important and a food source for the Salish and Kootenai Tribes of western Montana.

### *Project Description*

The edible roots of camas (*Camassia quamash*) and bitterroot (*Lewisia rediviva*) were once heavily utilized by Native American tribes as important food sources, and for traditional, cultural purposes. The core range of camas and bitterroot is the Pacific Northwest and the north central Rocky Mountains. Both early seral species, camas and bitterroot are enhanced by specific disturbance regimes. Native Americans influenced these disturbances on traditional camas and bitterroot gathering sites, and were aware of natural and cultural processes that favored the production of these plants. Plots were seasonally cleared of stones, weeds, and brush (often by controlled burning). During harvest, the soil was systematically lifted out in small sections, the larger bulbs removed, and the soil replaced (Turner 1975).

The decrease of plants such as camas and bitterroot is of great concern to many tribes because their traditional gathering areas have experienced a decline in productivity because of anthropogenic influences in the past century and the proliferation of invasive plants. The Bitterroot National Forest in Montana has begun working with the Confederated Salish and Kootenai Tribes and scientists at the University of Montana in an effort to restore culturally important plants in traditional gathering sites. Competition from exotic weeds, especially spotted knapweed and sulfur cinquefoil, has become the main threat to these native plant populations in many areas of the Forest.

Exotic plant species will be suppressed by using selective herbicides, thereby allowing the native plants to recover from a reduction in competition. DowElanco Chemical Company has offered to provide the herbicides for this study. The site could then be prepared for the reintroduction of designed disturbances (controlled fire, traditional gathering practices) that favor early seral stages, thereby promoting the growth of native plants.

The project's management plan includes chemical testing to measure herbicide residue levels in bitterroot and camas bulbs. This is a critical component of the restoration effort, especially because the plants are intended for human consumption. Restoration plans include integrating traditional, Native American designed disturbance regimes with modern weed management tools (such as herbicides, biological control agents), and evaluating the effect of treatments on target weed species and desired native plants.

### *Expected Benefits*

The expected benefits of this project include: (1) the restoration of native, and culturally significant plant populations at traditional gathering sites to their condition prior to intrusion of exotic species; (2) the production of valuable data about the response of camas and bitterroot to herbicide treatments; and (3) sharing test results and monitoring plans with other agencies and tribes in the Northwest to promote the restoration of traditionally utilized plants.

This project illustrates the value of utilizing the ethnographic record and working with Native Americans to understand how past vegetational communities were maintained in an effort to restore representative native plant communities. Traditional gathering areas on the Bitterroot National Forest will be restored for future gathering use by the Confederated Salish and Kootenai Tribes. The Bitterroot National Forest Camas and Bitterroot Restoration Project demonstrates a successful native plant restoration effort with a partnership between a federal agency, tribal government, academia, and industry.

### *Contacts*

For more information, contact Mary Horstman, Supervisors Office, Bitterroot National Forest, Hamiton, Montana.

## Re-photographing Riparian Zones Along the White River in Colorado

### *Location*

Eighteen selected photo-points along the White River of Colorado.

### *Participants*

Colorado office of the Bureau of Land Management.

### *Reason for Conducting the Project*

Through the use of historical photographs taken in 1907, the purpose of this project was to document human-caused environmental change.

### *Project Description*

Documentation of anthropogenic environmental change during the historical period was conducted by the Bureau of Land Management White River Project (Athearn 1988). Although information concerning Colorado's White River has been available since the Dominguez-Escalante expedition in 1776, journals, survey notes, maps, and newspaper articles, and other sources contain very little environmental data. However, the U.S. Geological Survey (USGA) photographed many areas along the White River at the beginning of the 20th century during a large, general survey which concentrated on mineral and geological information. Using 18 USGS survey photographs and the field notes accompanying them, the BLM relocated earlier photo points and took new photographs of the landscapes in June 1988. Of the 18 selected from the USGS files, 16 original photo points were relocated, and 12 of the points were chosen for comparative study.

### *Information Gained*

The resulting new photographs depict striking changes between 1907 and 1988. For example, a 1907 photo of East Douglas Creek depicts a homestead cabin located next to the stream with considerable riparian vegetation. The 1988 photo displays the same cabin hanging over the creek channel, having been undercut by at least 40 vertical feet of erosion, now with very sparse vegetational cover.

The project report concluded that the White River riparian habitat changed drastically since the beginning of the 20th century as a result of human activities. Based on historical records and photographs and compared to current conditions, the BLM concluded that the White River drainage had a rich riparian area until the 1920s. At that time, erosion increased dramatically and the river channel deepened, becoming a steep narrow trench. The 1988 photos reveal that riparian vegetation is no longer abundant, and massive downcutting is prevalent within stream channels. Much of this degradation was the result of systematic clearing for hay pastures and alfalfa production. Soil erosion increased as vegetation was removed, and the water table dropped as irrigation drained the rivers and streams.

The White River report discusses problems associated with the environmental comparative method of re-photographing. Use of historical photographs is limited. Photographs were unavailable before the 1870s. Additionally, even when historical photos are available, they may not cover the area in question. The report makes a number of suggestions concerning the use of historical photographs for environmental reconstruction of riparian habitats, including: (1) when choosing a stream or river to research, ensure that enough written material provides adequate data for the project; (2) use local source materials whenever possible. (Local historical societies, county records, and residents can often provide much useful information; (3) secondary source materials (books, articles) are as valuable as primary sources (survey records); (4) the use of comparative photos is highly desirable. When choosing areas for re-photography, find images that can be easily relocated (USGS Survey Notes contain geographic descriptions); and (5) local people can be helpful in relocating photo points.

*Contacts*
For further information concerning this project, contact Frederic Athearn, USDI Bureau of Land Management, Colorado State Office, 2800 Youngfield, Lakewood, CO.

## Temporal And Spatial Modeling of Anthropogenic Landscapes in the Rio del Oso, New Mexico

*Location*
The specific study area is located in the Rio del Oso, a tributary of the Rio Chama, on the Santa Fe National Forest in northern New Mexico.

*Participants*
USDA Forest Service, Rocky Mountain Research Station, Cultural Heritage Research Work Unit, and the Rio Grande Ecosystem Project; Española Ranger District and Heritage Resource Program of the Santa Fe National Forest; Linda Scott Cummings, Paleo Research Laboratories, Golden, Colorado; Stephen A. Hall, Department of Geography, University of Texas at Austin; Richard I. Ford and Kurt F. Anschuetz, University of Michigan Museum of Anthropology.

*Reason for Conducting Research*
For more than 12,000 years humans have exerted a lasting influence on the evolution of North American landscapes. Understanding the spatial and temporal dimensions of the interaction between humans and the environment is crucial to the restoration and sustainability of fragile riparian, forest, and grassland ecosystems. An interdisciplinary approach, incorporating Geographic Information Systems, archaeology, paleoecology, plant ecology, and geomorphology, is necessary to identify the continuing effects of past human activities on contemporary ecosystem dynamics. The primary goals of this ongoing research are to develop methods for identifying, measuring, and simulating cumulative anthropogenic effects on North American landscapes.

*Project Description*
This research is examining how prehistoric Archaic (5500 B.C. to A.D. 600), Anasazi (A.D. 1200 to 1600), and historical Hispanic occupations (beginning around 1600) may have influenced landscape development in the lower Rio del Oso drainage for more than 7,000 years. The study focuses on the processes through which these distinct cultures may have affected the vegetation, hydrology, and overall physiography of the Rio del Oso Valley. Three-dimensional Geographic Information System (GIS) models of Rio del Oso vegetational patterns are being developed for specified periods using archaeological, environmental, and paleoenvironmental data. Rather than analyzing individual sites, the history of the drainage area will be viewed as a series of temporally grouped landscape layers. This project will directly address not only the nature and extent of human land use in the drainage, but will also identify the ways in which past human activities have a continuing influence on the ecosystems seen today.

*Information Gained So Far*
During the first year of this project (1996), watershed, timber stand, and vegetation maps were formatted as GIS data layers. Additionally, 290 archaeological site maps were digitized, producing GIS data layers for each estimated period of human occupation. Soil samples for paleobotanical (pollen, phytolith, and microscopic charcoal) and radiocarbon date extraction have been collected from a cutbank in the Rio del Oso drainage where roughly 5 meters of stratified sediments are directly exposed. Four fire hearths were uncovered while preparing the vertical surface of the cutbank for sample collection. Seven distinct paleosols (buried cumulic A-horizons) were identified in the profile as well. The data collected are now being analyzed and interpreted for use in producing the three-dimensional visual simulations of past landscapes.

The methods, techniques, and information developed through this project will provide an interdisciplinary framework for incorporating an ecosystem approach in cultural and natural resource management. This framework is intended to supply a useful planning and decision-making strategy to federal, state, and local land management agencies for the restoration of rangeland and riparian areas, timber sale planning, and other activities.

*Contacts*
For information regarding this project contact Richard Periman, Rocky Mountain Research Station, 2205 Columbia SE, Albuquerque, New Mexico.

## 12 CONCLUSIONS

1. A knowledge of the human role in ecosystem development through time is vitally important for land managers to carry out ecosystem management. Understanding human processes and their extensive and cumulative effects on ecosystems will give land managers better knowledge of the range of an ecosystem's capabilities.

2. Approaching an analysis of the human role in ecosystem development requires a sound framework of pertinent questions, combined with identifying the appropriate temporal and spatial scales, expertise required, and research methods that have the best potential to provide answers. Initiating research with relevant questions and a variety of techniques used to cross-check findings make the entire process more feasible and the gathered data more applicable.

3. Experts from the fields of anthropology, archaeology, history, paleoecology and paleontology, geology, and a variety of other disciplines can be consulted to provide the necessary skills, methods, and analyses for addressing this topic. Government agencies often have staff trained in these fields. Additionally, this expertise is available in colleges, universities, and the private sector.

4. Interdisciplinary and interagency project designs are needed to incorporate effectively our knowledge of human processes into land management planning. Paleoenvironmental, archaeological, and historical research and analyses need to be integrated with the overall, routine planning process. This need not be an added complication in the analysis, evaluation, and public consultation process, but rather a potentially different way of approaching a variety of historical preservation and ecosystem management issues.

5. The incorporation of this type of human prehisorical and historical information into natural resource management is a new endeavor, and developing innovative methods and approaches is important. We are only beginning to understand the complexity of human effects on environmental systems through time. More research needs to be conducted before we can approximate an understanding of the vast extent to which North American ecosystems were altered over the centuries.

6. Despite ongoing advances in various fields specializing in reconstructing the past, we continue to be dependent on sources of data that have managed to survive time's destructive processes. These fragile deposits of environmental and human history have become increasingly valuable for the understanding and stewardship of the ecosystems we depend upon for goods and services, recreation, and personal enjoyment. In this light, the surviving fragments of the natural and human past need to be protected, not only to preserve our human heritage, but also as containers of vital knowledge needed for natural resources management.

## ACKNOWLEDGMENTS

The authors wish to thank the organizers of the Ecological Stewardship Project for inviting them to participate. We also thank the many reviewers, editors, and proofreaders who contributed their time and effort to producing this paper.

## REFERENCES

Anschuetz, K.F., in press. Earning a living in the cool high desert: transformations of the northern Rio Grande landscape by Anasazi farmers to harvest and conserve water. In: R.P. Fish, C.L. Redman, and D. Roberts (eds.), *Human Impact on The Environment: An Archaeological Perspective.* Arizona State University, Tempe, AZ.

Athearn, F.J. 1988. *Habitat in the Past: Historical Perspectives on Riparian Zones of the White River.* U.S. Department of the Interior, Bureau of Land Management, Denver, CO.

Baillie, M.G.L. 1995. *A Slice Through Time: Dendrochronology and Precision Dating.* B.T. Batsford, London.

Barret, S.W. 1980. Indian fires in the pre-settlement forests of western Montana. *Fire History Workshop, October 20–24.* University of Arizona, Tucson, AZ.

Betancourt, J.L., and T.R.V. Devender. 1981. Holocene vegetation in Chaco Canyon, New Mexico. *Science* 214: 656–658.

Betancourt, J.L., T.R.V. Devender, and P.S. Martin (eds.). 1990. *Packrat Middens: The Last 40,000 Years of Biotic Change.* University of Arizona Press, Tucson, AZ.

Blackburn, T., and K. Anderson. 1993. Introduction: Managing the domesticated environment. pp. 1–17. In: T. Blackburn and K. Anderson (eds.), *Before the Wilderness: Environmental Management by Native Californians.* Ballena Press, Menlo Park, CA.

Blackburn, T.C., and K. Anderson (eds.). 1993. *Before the Wilderness: Environmental Management by Native Californians.* Ballena Press Anthropology Papers. Ballena Press, Menlo Park, CA.

Bloom, A.L. 1978. *Geomorphology: A Systematic Study of Late Cenozoic Landforms.* Prentice Hall, Englewood Cliffs, NJ.

Boas, F. 1940. *Race, Language, and Culture.* Macmillan, New York.

Bradley, R.S. 1985. *Quaternary Paleoclimatology.* Allen and Unwin, Boston.

Brooks, R.R., and D. Johannes. 1990. *Phytoarchaeology.* Dioscorides Press, Portland, OR.

Butzer, K.W. 1982. *Archaeology as Human Ecology: Methods and Theory for a Contextual Approach.* Cambridge University Press, Cambridge, MA.

Chambers, F.M. (ed.). 1993. *Climate Change and Human Impact on the Landscape: Studies in Paleoecology and Environmental Archaeology.* Chapman & Hall, London.

Chatters, J.C., and D.M. Leavell. 1995. *Smeads Bench Bog: a 1,500 year history of fire and Succession in the Hemlock Forest of the Lower Clark Fork Valley, Northwest Montana.* U.S. Department of Agriculture Forest Service, Kootenai National Forest, Supervisor's Office, Libby, MT.

Cordell, L.S. 1984. *Prehistory of the Southwest.* Academic Press, San Diego, CA.

Crumley, C.L. 1994. Historical ecology: a multidimensional ecological orientation. pp. 1–16. In: C.L. Crumley (ed.), *Historical Ecology: Culture, Knowledge, and Changing Landscapes.* School of American Research Press, Santa Fe, NM.

Ebert, J.I. 1992. *Distributional Archaeology.* University of New Mexico Press, Albuquerque, NM.

Frison, G.C. 1991. *Prehistoric Hunters of the High Plains.* Academic Press, San Diego, CA.

Grayson, D.K. 1984. *Quantitative zooarchaeology.* Academic Press, New York.

Hall, S.A. 1988. Prehistoric Vegetation and Environment at Chaco Canyon. *American Antiquity* 53(3): 582–592.

Hughs, M.K., and H. F. Diaz. 1994. *The Medieval Warm Period.* Kluwer, Boston, MA.

Joukowsky, M. 1986. *A Complete Manual of Field Archaeology: Tools and Techniques of Field Work for Archaeologists.* Prentice-Hall, Englewood Cliffs, NJ.

Kelso, G.K., S.A. Mrozowski, D. Currie, A.C. Edwards, M.R. Brown III, A.J. Horning, G.J. Brown, and J.R. Dandoy. 1995. Differential pollen preservation in a seventeenth-century refuse pit, Jamestown Island, Virginia. *Historical Archaeology* 29(2): 43–54.

Kroeber, A.L. 1939. *Cultural and Natural Areas of Native North America.* University of California Press, Berkeley, CA.

Lamb, H.H. 1977. *Climate Present, Past, and Future, Volume III: Climatic History and the Future.* Methuen, London.

Lasca, N.P., and J. Donahue (eds.). 1990. *Archaeological Geology of North America.* Centennial special volume: the decade of North American geology project series. The Geological Society of America, Boulder, CO.

Lawton, H.W., P.J. Wilke, M. DeDecker, and W.M. Mason. 1993. Agriculture among the Paiutes of Owens Valley. pp. 329–377. In: T. Blackburn and K. Anderson (eds.), *Before the Wilderness: Environmental Management by Native Californians.* Ballena Press, Menlo Park, CA.

Lewis, H.T. 1993. Patterns of Indian Burning in California: Ecology and Ethnohistory. pp. 55–116. In: T. Blackburn and K. Anderson (eds.), *Before the Wilderness: Environmental Management by Native Californians.* Ballena Press, Menlo Park, CA.

Lewis, M., and W. Clark. 1893. *The History of the Lewis and Clark Expedition.* Republished by Dover Publications, New York, 1970.

Marsh, G.P. 1874. *The Earth as Modified by Human Action.* Scribner, Armstrong & Co, New York.

Mehringer, P.J., S.F. Arno, and K.L. Petersen 1977. Postglacial history of Lost Trail Pass Bog, Bitterroot Mountains, Montana. *Arctic and Alpine Research* 9: 345–368.

Meirendorf, R.R. 1984. Late-pleistocene and early-holocene fluvial processes and their affects on the human use of landscape. *Tebiwa* 21: 15–25.

Nir, D. 1981. *Man, A Geomorphological Agent: An Introduction to Anthropic Geomorphology.* D. Reidel Publishing, Boston, MA.

O'Rourke, M.K. 1983. Pollen from adobe brick. *Journal of Ethnobiology* 3: 39–48.

Pearsall, D.M. 1989. *Paleoethnobotany: A Handbook of Procedures.* Academic Press, New York.

Periman, R.D. 1996. The influence of prehistoric Anasazi cobble-mulch agricultural features on northern Rio Grande landscapes. In: D.W. Shaw and D.F. Finch (eds.), *Desired Future Conditions for Southwestern Riparian Ecosystems: Bringing Interests and Concerns Together.* U.S. Department of Agriculture, Forest Service, Rocky Mountain Forest and Range Experiment Station, Fort Collins, CO.

Quigley, T.M., R.W. Haynes, R.T. Graham eds. 1996. *Integrated Scientific Assessment for Ecosystem Management in the Columbia Basin and Portions of the Klamath and Great Basins.* USDA Forest Service, Pacific Northwest Research Station, Portland, OR.

Reitz, E.J., L.A. Newsom, and S.J. Scudder (eds.). 1996. *Case Studies in Environmental Archaeology.* Interdisciplinary Contributions to Archaeology. Plenum Press, New York.

Schaeffer, C.E. 1940. *The Subsistence Quest of the Kootenai.* University of Pennsylvania, Department of Anthropology, Philadelphia, PA.

Shalk, R.F. 1986. Estimating salmon and steelhead usage in the Columbia Basin before 1850: The anthropological perspective. *The Northwest Environmental Journal* 2(2): 1–29.

Sharer, R.J., and W. Ashmore. 1993. *Archaeology: Discovering Our Past.* Mayfield Publishing, New York.

Simmons, I.J. 1989. *Changing the Face of the Earth: Culture Environment, History.* Basil Blackwell, Cambridge, MA.

Spaulding, W.G., J.L. Betancourt, L.K. Croft, and K.L. Cole. 1990. Packrat middens: their composition and methods of analysis. pp. 59–84. In: J.L. Betancourt, T.R.V. Devender and P.S. Martin (eds.), *Packrat Middens: The Last 40,000 Years of Biotic Change.* University of Arizona Press, Tucson, AZ.

Stahl, P.W. 1982. On small mammal remains in archaeological context. *American Antiquity* 47(4): 822–829.

Stanford, D.J. and J.S. Day (eds.). 1992. *Ice Age Hunters of the Rockies.* University Press of Colorado, Niwot, CO.

Tainter, J.A., and B.B. Tainter. 1996. Riverine settlement in the evolution of prehistoric land-use systems in the Middle Rio Grande Valley, New Mexico. pp. 22–32. In: D.W. Shaw and D.F Finch (eds.), *Desired Future Conditions for Southwestern Riparian Ecosystems: Bringing Interests and Concerns Together.* U.S. Department of Agriculture, Forest Service, Rocky Mountain Forest and Range Experiment Station, Fort Collins, CO.

Thoms, A.V. 1989. *The Northern Roots of Hunter-Gatherer Intensification: Camas and the Pacific Northwest.* Washington State University, Department of Anthropology, Pullman, WA.

Thoms, A.V., and G.C. Burtchard. 1987. *Prehistoric Land Use in the Northern Rocky Mountains: A Perspective From the Middle Kootenai River Valley.* Center for Northwest Anthropology, Pullman, WA.

Touchan, R., and T.W. Swetnam. 1995. Fire History in Ponderosa Pine and Mixed-Conifer Forests of the Jamez Mountains, Northern New Mexico: Final Report to Santa

Fe National Forest and Bandelier National Monument. University of Arizona, Laboratory of Tree-Ring Research, Tucson, AZ.
Turner, N.J. 1975. *Food Plants of British Columbia Indians, Part I: Coastal People.* British Columbia Ministry of Culture, Victoria, BC.
Turney-High, H.H. 1941. *Ethnography of the Kutenai.* American Anthropological Association, Washington, DC.
USDA Forest Service. 1993. *National Hierarchical Framework for Ecological Units.* U.S. Government Printing Office, Washington, DC.
Van Devender, T.R., and W.G. Spaulding. 1979. Development of vegetation and climate in the Southwestern United States. *Science* 204: 701–710.
Wallerstein, I. 1974. *The Modern World-System I: Capitalist Agriculture and the Origins of the European World-Economy in the Sixteenth Century.* Academic Press, New York.
Wells, P.V. 1976. Macrofossil analysis of wood rat (neotoma) middens as a key to the Quaternary vegetational history of arid America. *Quaternary Research* 6: 223–248.
Wells, P.V., and C.D. Jorgensen. 1964. Pleistocene wood rat middens and climate change in the Mohave Desert: a record of juniper Woodlands. *Science* 143: 1171–1174.
Willey, G.R., and J.A. Sabloff. 1993. *A History of American Archaeology.* W.H. Freeman and Co., New York.
Winterhalder, B.P. 1994. Concepts in historical ecology: The view from evolutionary theory. pp. 17–41. In: C.L. Crumley (ed.), *Historical Ecology: Culture, Knowledge, and Changing Landscapes.* School of American Research, Santa Fe, NM.
Zweifel, M.K. 1994. *A Guide to the Identification of the Molariform Teeth of Rodents and Lagomorphs of the Columbia Basin.* Washington State University, Department of Anthropology, Pullman, WA.

## THE AUTHORS

**Richard D. Periman**
*USDA Forest Service*
*Rocky Mountain Research Station*
*Research Social Scientist*
*Forestry Sciences Laboratory*
*2205 Columbia, SE*
*Albuquerque, New Mexico 87106*

**Connie Reid**
*USDA Forest Service*
*North Karbab Ranger District*
*4305 Main*
*Fredonia, AZ, USA*

**Matthew K. Zweifel**
*USDA Forest Service*
*North Karbab Ranger District*
*4305 Main*
*Fredonia, AZ, USA*

**Gary McVicker**
*USDI, BLM*
*Special Assistant to the State Director*
*Colorado State Office*
*2800 Youngfield Street*
*Lakefield, CO 80215-7076*

**Dan Huff**
*Assistant Field Director*
*National Park Service*
*P.O. Box 25287*
*Denver, CO 80225, USA*

# Historical Science, Heritage Resources, and Ecosystem Management

Alan P. Sullivan, III, Joseph A. Tainter, and
Donald L. Hardesty

**Key questions addressed in this chapter**

- ♦ *Because ecosystems are historically constituted, generally as a consequence of human influence, the study of the relationship between culture and environment is essential for ecosystem management.*
- ♦ *Heritage resource research is part of the intellectual core of ecosystem management.*
- ♦ *Anthropogenic sources of ecosystem variation create problems for objectives that focus on ecosystem "restoration" and reestablishment of "natural conditions."*
- ♦ *Ecosystem management is possible only if the human role in ecosystem development is understood.*
- ♦ *Archaeological and historical resources provide unmatched data and interpretive possibilities for understanding ecosystem changes.*
- ♦ *Current heritage management programs need to be reformulated if ecological stewardship of cultural and natural resources is to achieve its full potential.*

**Keywords: Cultural artifacts, archaeology, pre-historic, historic, heritage management, mining, logging**

## 1 INTRODUCTION

The principal purposes of this chapter are to demonstrate that heritage research and management belong at the core of federal approaches to ecosystem management, and to show how heritage management ought to be practiced in an ecological framework (cf. Tainter and Hamre 1988, Jensen and Everett 1993, IEMTF 1995, Anonymous a,b). Regardless of how ecosystem management is defined (Cortner et al. this volume), objectives such as economic sustainability and species diversity maintenance must consider how past human actions have affected the properties of contemporary ecosystems (Lipe 1995). Historical science is fundamental to ecosystem management because ecosystems develop in response to decades, centuries, or millennia of human influence. Understanding the co-evolution of culture and environment, therefore, is essential to achieving the goals of ecosystem management (Barker 1996).

Our contribution outlines a role for heritage management within ecosystem management that is based on the premise that humans have been key factors in shaping ecological structures and processes for millennia (Butzer 1990). Using examples and case studies drawn worldwide, we explain why heritage resources research, whether conducted in prehistoric or historic contexts, is an intellectual and operational prerequisite for achieving the goals of ecosystem management (Manley et al. 1995). We conclude with some recommendations for how the protocols and procedures of heritage resource management should be retooled to ensure that they are compatible with, and contribute to, the long-term missions of ecosystem management.

## 2 THE STUDY OF CULTURE AND ENVIRONMENT

Adoption of an ecosystem approach necessitates that we understand (i) how social scientists have conducted ecologically-based studies and (ii) where the research frontiers are situated today and for the near-term. Since the mid-nineteenth century (Moran 1990), a substantial body of literature has materialized that focuses on the relations between culture and environment (Harris 1968, Preucel and Hodder 1996). Arguments in this literature have taken two forms: either they investigate the hypothesis that "items of cultural behavior function as part of systems that also include environmental phenomena" or the hypothesis that "environmental phenomena are responsible in some manner for the origin or development of the cultural behavior under investigation" (Vayda 1969:xi). Each hypothesis has consequences for modes of inquiry that adopt an ecological perspective because the interaction between culture and environment is not as deterministic as once postulated (see Feldman 1975 for examples). Attempts to make environment the "independent variable" that causes variation in cultural behavior produce generalizations that have too many exceptions to be useful. By themselves, factors such as subsistence or population (Harner 1970), for example, rarely account for much inter-cultural variation.

### 2.1 Human Ecosystems

The human ecosystem may be considered a conceptual framework to investigate the culture-environment dynamic holistically (Heider 1972). Ellen (1978) has argued, moreover, that the anthropological analysis of human ecosystems unifies the study of culture and environment and, when properly executed, provides insights into co-evolutionary processes. In fact, such analyses permit investigation of the distinctive capacity of humans to (i) plan and organize their activities far in advance of their implementation, (ii) modify their environments purposively for a variety of objectives, and (iii) manage elements of their environment selectively (often influenced by non-ecological factors [Brumfiel 1992]).

Human ecosystem analysis is predicated on the premise that interaction between human behavior and environmental variability can be explored in terms of adaptational and mutualistic processes (Boucher et al. 1982). The usefulness of this approach is illustrated in many geographic, ethnographic, and ethnohistoric studies (Lewis 1973, Cooke and Reeves 1976, Ellen 1982, Blackburn and Anderson 1993). Such studies have provided an appreciation for the extent to which humans have altered their surroundings — sometimes catastrophically (Dobyns 1981), sometimes permanently (Sullivan and Downum 1991), invariably profoundly (Cooper 1960). With rare exception, these analyses have indicated how crucial it is to reconstruct the sources of ecosystem change in order to appreciate how contemporary ecosystems have arisen (McDowell et al. 1990). Interestingly, ecosystem variation has been attributed to a surprisingly small set of agents (Dickson 1993). Fire, for instance, is a cultural force that, for hundreds of millennia (Sauer 1975), has affected the distribution of animals (Dobyns 1981, also Mellars 1976) and plants (Martinez 1993:27, see also Wright and Bailey 1982).

### 2.2 Human Ecosystems in Historical Perspective

Given that (i) many characteristics of contemporary ecosystems are historically constituted and that (ii) humans are an integral aspect of the mutualistic co-evolutionary processes that produced contemporary ecosystems, a large proportion of the necessary data on such long-term ecological processes can be supplied by archaeologists, ethnohistorians, and cultural anthropologists (Hardesty 1977). There is, in fact, an intrinsic alliance between the kinds of historical problems that have been investigated with heritage resources and the objectives of ecosystem management (Duke 1995). For example, the Paleoenvironmental Project (Gumerman 1988:1) has focused its efforts on paleoclimatic reconstruction (Dean et al. 1985) and on modeling the effects that climatic change may have had on prehistoric Southwestern cultures (Euler et al. 1979, see additional examples below). With the adoption of an ecosystem perspective (Ellen 1977:37-38), moreover, diachronically-oriented social scientists have explored the effects that people had on the habitats that they created and managed (Savage 1991, Kohler 1992, Keter 1995).

Ecosystem research in archaeology, in particular, has become increasingly more sophisticated as it has moved from "matching" or correlational analyses, which attempted to attribute intersite variation to differences in resource catchments (SARG 1974, Roper 1979, Dennell 1980), to analyses that have focused on the complex interrelations between resource production and resource consumption. Ethnographic (Ellen 1982) and archaeological investigations (Fish and Donaldson 1991) have illustrated that the material record of where and how resources are produced and processed may be highly uncharacteristic of the material record of where and how resources were stored and consumed (Sullivan 1996). Further complicating the interpretive picture are the highly unpredictable relations between settlement occupation modes (e.g., continuous or discontinuous) and abandonment modes (e.g., planned or catastrophic [Cameron and Tomka 1993]), which have important consequences for investigating ecosystem structure and change with archaeological data (Butzer 1982, Jochim 1990, Anonymous b: 5).

Incorporation of an ecosystem approach within archaeology has facilitated the investigation of the origins of cultural landscapes that arose under different conditions (e.g., centuries [Fish and Fish 1990] vs. decades [Sullivan and Downum 1991]). By accentuating the investigation of mutualistic relations between culture and environment (O'Brien 1989), rather than viewing environment merely as a reservoir of calories (Isaac 1990), the ecosystem approach avoids the theoretical pitfalls of "synchronic equilibrium-oriented functionalism" (Moran 1990:12) that hobbled earlier ecological approaches in archaeology (Schoenwetter and Dittert 1968). As we argue in detail, however, the full implementation of an ecosystem approach in heritage management necessitates a consideration of fundamental issues of archaeological theory and method (see Section 4.1 below).

## 3 HERITAGE RESOURCES AND ECOSYSTEM MANAGEMENT

In contrast to the dynamic, integrative nature of ecosystem management, heritage management has been practiced largely as a stand-alone, single-purpose function. Its principal mandate, deriving from such laws as the National Historic Preservation Act (1966) and the Archaeological Resources Protection Act (1979), has been *prioritized conservation*. Archaeological and historical resources, and properties of contemporary cultural value, were to be conserved for the future and managed to minimize conflicts with other land uses (Lipe 1974, Tainter 1987). While this function was and is important, rarely did land managers consider that heritage management could do more, that it could contribute knowledge that is fundamental to the management of ecosystems. A program of heritage management that produces some of the knowledge underlying ecosystem management would have a much more central role in the administration of public lands than is now the case. Rather than being viewed as an obstacle to natural resource and energy production, as (unfortunately and unfairly) it often is, heritage management, if fully incorporated into ecosystem management, would significantly transform both, to their mutual benefit.

In fact, one of the conceptual principles of ecosystem management is that scientific knowledge of ecological structures and processes should guide stewardship. Emphasis should be placed on the term *knowledge*. If ecosystems are complex, in the sense of having great differentiation of structures and integration of processes, then their management must be based on an awareness of that complexity that is as rich and detailed as possible. A major part of understanding complexity of such scope, in particular, must include knowledge of how North American ecosystems have developed, both historically (Hadley and Sheridan 1995) and in the prehistoric period (Hall 1990).

## 3.1 Heritage Resources in the Historical Analysis of Ecosystems

As discussed in greater detail by Bonnicksen and his colleagues (this volume), ecosystem management must appreciate the human role in the development of North American ecosystems (Lipe 1995, also Periman et al., this volume). The once-common notion that small numbers of Native Americans inhabited a primordial garden is giving way to a more realistic view. Many prehistoric occupants of this continent lived at times in dense concentrations (Ramenofsky 1987) and, in nearly all places, they modified environments to increase production of wild, edible resources (Boyd 1986). Native Americans manipulated vegetation, for example, to select for early seral stages and to increase the abundance of particular prey species (Lewis 1973, Cronon 1983, Kay 1995). It now seems clear that Europeans encountered ecosystems that had been evolving for millennia along with, and in response to, Native American populations that grew many times in size and density, and whose societies came to be increasingly complex and to require greater quantities of resources of various kinds (Tainter 1988: 178–187; 1995).

Viewing the prehistory of North American ecosystems in this way leads to significant questions about environmental conditions today. At a fairly obvious level it questions what is meant by terms such as *reference conditions, restoration*, or *range of natural (or historic) variation* (Wright et al. 1995), and whether these terms can apply meaningfully to anthropogenic ecosystems. At a more profound level, if ecosystem management does require knowledge of the structures and processes of ecosystems, that knowledge will remain significantly impoverished without a historical understanding of the human role in shaping ecosystems and their evolutionary processes (McGlade 1995). Lamentably, we lack much fundamental knowledge of how North American ecosystems functioned before A.D. 1500 (Cartledge and Propper 1993). If we are to understand reference conditions, we need to know how ecosystems have developed, during 12,000 years or more, with human societies that ranged, in terms of population and organization, from low-density foragers to high-density village agriculturalists (Fagan 1991).

## 3.2 Heritage Research Contributions to Ecosystem Management

A historical perspective on North American ecosystems suggests a new vision for heritage management, one that eclipses its function solely as the custodian of cultural and historical values. As a central component of ecosystem management, heritage management (i) identifies and conserves sources of data (non-archaeological, archaeological, and historic) that can be studied to understand the trends and influences by which North American ecosystems have developed, and (ii) actively produces knowledge of prehistoric and historic anthropogenic influences on ecological structures and processes, and on related matters, such as past climates (Betancourt et al. 1993).

This new conception of heritage management moves it to the center of federal land management. Indeed, we suggest that ecosystem managers who approach their work without understanding historical processes cannot hope to acquire the knowledge necessary for management. Broadly considered, heritage resources are key sources of information about the evolutionary histories of ecosystems and the roles that humans had in influencing those developmental trajectories (Dean 1996). In the process of investigating ecosystem history, archaeologists and ethnohistorians, among others, have begun to explore the interpretive potential of various data sources (Butzer 1982, Keter 1995). Manley et al. (1995: 83–86), for example, provide some general recommendations regarding types of heritage-resource data sources, such as archival records, dendrochronological analysis, and sedimentological analysis, and their appropriateness for investigating such aspects of human ecosystem history as vegetation successional patterns, variation in past precipitation patterns, and processes of landform modification (see also Glassow 1996, Baisan and Swetnam 1997).

Several specific examples illustrate how heritage resources can contribute to our understandings of ecosystem history and process. In adopting a landscape perspective, abundant ethnographic (Steward 1938), ethnohistoric (Lewis 1973, Dobyns 1981), and archaeological (Chambers 1993) evidence attests to the degree to which humans manipulated and, in many cases, managed (Blackburn and Anderson 1993) their surroundings, thereby sowing the seeds (literally, in some cases) of long-term ecosystem change (Kohler and Matthews 1988). For example, the construction of rock-piles, terraces, and canals on the slopes of the Tortolita Mountains in southern Arizona (Fish and Fish 1990), which were part of an extensive prehistoric agricultural landscape, altered post-occupation drainage patterns and vegetation communities there (see also Waters 1991). Similar landscape modifications have been documented for the Grand Canyon area, where piles of fire-cracked rock and associated processing areas, some replete with *in situ* artifact arrays (Sullivan 1992a), occur in abundance.

In a provocative study that illustrates the interpretive potential of the microscopic study of charcoal recovered from stratified lacustrine deposits in southern India, Morrison (1994) was able to demonstrate variation in the periodicity and the intensity of local and regional fire-history regimes. This study shows, as well, that the identification and preservation of phenomena whose origins are not directly attributable to human activity (e.g., bogs, packrat middens [Betancourt et al. 1990]) may yield crucial data for understanding the tempo and mode of landscape use.

A persistent problem in North American prehistory has been assessing the extent to which extirpation or extinction of animal species was a consequence of human predation (Martin and Klein 1984). In a wide-ranging study that entailed (i) the analysis of wildlife biological data and (ii) age and sex distributional profiles of archaeofaunal assemblages, Kay (1994: 364) concluded that "prior to European influence, predation by Native Americans limited the numbers and distribution of ungulates in the Yellowstone ecosystem and throughout the Intermountain West." The implications of this study tend to support two notions regarding human impacts on landscapes: (i) Native peoples, far from being conservation-oriented stewards and husbanders of animal resources, sometimes outcompeted them for plant resources, and (ii) conservation seldom is, as Kay (1994: 385) notes, an "evolutionarily stable strategy." Kay's study exemplifies how different kinds of data may be employed in unraveling the complex histories of ecosystems and the roles that humans have played, by the application of fire or systematic predation patterns, in influencing the content and structure of plant and animal communities (cf. Mellars 1976).

At least since Braidwood's pioneering Iraq Jarmo Project (conducted during the late 1940s to mid 1950s [Braidwood 1974]), archaeologists who have adopted an ecological perspective have realized the necessity of interdisciplinary collaboration to achieve project research goals (see also Clark 1972). In the American Southwest, for example, archaeologists, palynologists, and geomorphologists have been examining how variation in ecosystem parameters, such as floodplain aggradation, effective moisture, and dendroclimatic spatial and temporal variability (Plog et al. 1988), may have influenced group decision-making regarding short-term or permanent abandonment of local areas (Dean 1988). The methodology employed by this long-term project (the Paleoenvironmental Project, noted above) is especially noteworthy because the investigators used data from non-archaeological sites to evaluate variation among paleoenvironmental samples recovered from archaeological sites, which represent, in many cases, highly selective samples of ecosystem properties (additional studies have shown that humans may transport lithic raw materials [Shelley 1995] and large timbers [Betancourt et al. 1986] considerable distances from their sources). Such a system of checks and balances is essential in the study of ecosystem history because of the extraordinary capacity of humans to remove plants from their evolutionary settings and expose them to new selective pressures (Fish et al. 1985).

Unquestionably, therefore, heritage resources are key data sources about ecosystem history. However, in the absence of controlled studies that involve ecological data from cultural and natural phenomena, scientists and managers alike would be forced to assume that contemporary ecosystems differ little from those of the past and that human uses of these ecosystems have not altered them substantially over time. Such conclusions clearly would be unwarranted in view of compelling evidence to the contrary (Spoerl and Ravesloot 1994).

## 4 ARCHAEOLOGICAL RESOURCES AND ECOSYSTEM MANAGEMENT

Lester Embree, a philosopher of science, portrayed the importance of archeology as follows: "[I]f it has access in principle to far more societies, to all social strata in them, and to greater stretches of time than other historical sciences, then it can be said to be the most basic historical science" (1987:78). Broadly defined, archaeology is the problem-driven (or problem-dependent) study of variation in the archaeological record. As a contemporary (Binford 1983), three-dimensional (Fowler and Givens 1992), material phenomenon (Sabloff et al. 1987), the archaeological record varies substantially in terms of content (e.g., artifact number, artifact density, artifact-type diversity, non-artifact content), extent (aggregate space defined by its content), and visibility (i.e., the degree to which it can be observed using a variety of techniques; Schiffer et al. 1978, Sullivan, in press). Such extensive variation is time-dependent, dynamic, and often ambiguous with respect to causal processes (Patrik 1985, Sullivan 1992b).

### 4.1 Archaeological Theory

To a large extent, contemporary archaeology is emerging from a phase in its history when the discipline's foundation principles conventionally used for interpreting variability have been reexamined and, in many cases, reformulated drastically (Dunnell 1986).

Experimental (Nash and Petraglia 1987), ethnoarchaeological (Yellen 1977), and archaeological studies (Goldberg et al. 1993) have illustrated that the origins of archaeological phenomena are related to complex mixes of functional, occupational, organizational, symbolic, and other factors (Sullivan 1987a). It is part of the archaeologist's task to understand the formation histories of archaeological phenomena with respect to these and other sources of variation (Schiffer 1987).

Because archaeological research is predicated upon ascertaining which sources of variability contributed to the phenomena that are used in developing inferences about the cultural past, archaeologists necessarily have developed a web of theoretical constructs, and appropriate methods for untangling the histories of the remains they investigate (Sullivan 1978, Schiffer 1988). Theory is used here in its most general, and perhaps useful, sense — a conceptual framework for explaining variation of phenomena (Kluckhohn 1940: 43–44, Kelley and Hanen 1988: 370). Archaeological theory, then, consists of "largely unaxiomatized, lawlike generalizations relevant to the subject matter of archaeology: sites (site-formation processes and stratigraphy, for example) and the ancient cultural debris found in, on, and about them" (Watson 1973: 119). Figure 1 illustrates the relations among four principal classes of theory that invariably are entailed in archaeological inference. In addition, as indicated in the left column of the figure, a variety of supplementary modes of inquiry, often referred to as middle-range or controlling studies, test and confirm hypotheses that focus on understanding the factors that contribute to the content and spatial variability of the archaeological record (Daniels 1972, Clarke 1973). Reliable methods for interpreting archaeological phenomena can be developed because such "middle-range" studies (i) establish controls (for identifying those phenomena whose origins are attributable to humans) and (ii) increase the potential range of plausible "cultural pasts."

Research undertaken during the past two decades, in particular, has focused on understanding how different situations contribute to the extensive variability of the archaeological record itself (Schiffer 1987, Rossignol and Wandsnider 1992, Goldberg et al. 1993). For example, ethnoarchaeological and experimental studies revealed that many entrenched notions of such things as how activity areas are transformed into artifact clusters (Binford 1983), how butchered animals are represented by bone fragments (Gifford-Gonzalez 1991), how processed plant foods are incorporated in archaeobotanical assemblages (Miksicek 1987), and how the manufacture of flakes and tools produces lithic scatters were simplistic, at best. These studies matter because they provide a framework for appreciating how easily archaeologists can be misled by the deceptively simple variation of common archaeological phenomena. For example, lithic scatters, which are probably the most common "site type" globally, may originate as consequences of raw material acquisition, core reduction, tool manufacture, tool repair, manufacturing byproduct discard, or various combination of these activities (Sullivan 1987b). To lump all lithic scatters under the category of "limited activity site" obscures potentially significant information about the range of prehistoric technological activities, thereby attenuating our interpretations of the cultural past and limiting our understanding of anthropogenic ecosystem influences. Fortunately, many ethnoarchaeological

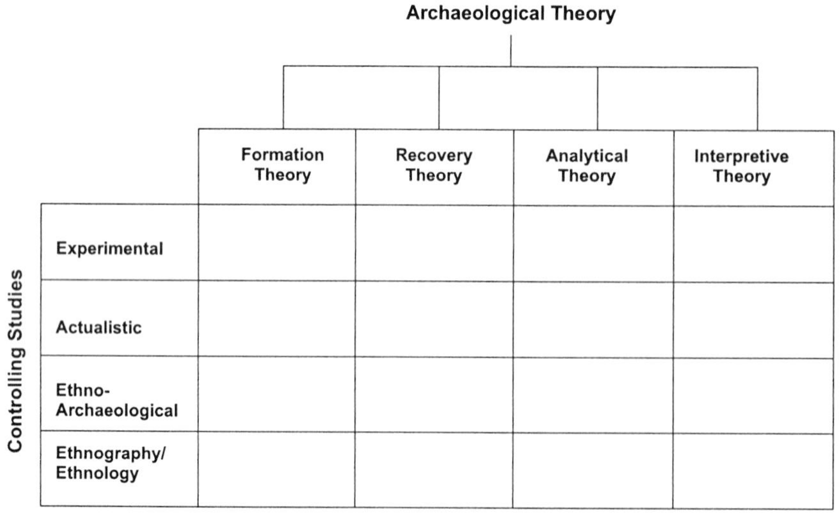


Fig. 1. Conceptual map of theoretical domains that constitute archaeological theory. The boxes under each of the sub-categories of Archaeological Theory represent major areas of research that focus on understanding how various sources of variability (e.g. cultural and non-cultural factors) affect the origins of content, frequency, and spatial differences within the archaeological record itself.

studies were conducted in an ecological framework which helped archaeologists appreciate how variation in resource acquisition, processing, and storage activities comes to be reflected in archaeological phenomena (i.e. artifact scatters, assemblages, etc.).

## 4.2 The Archaeological Study of Cultural Landscapes

Until recently, archaeologists did not concern themselves with the full range of variability expressed by the archaeological record, which often consists of low-density remains that are not clearly patterned, have few if any features, have perhaps been disturbed by later activities, and are located at or near the ground's surface (Dunnell and Dancey 1983). In fact, these kinds of remains often are not recorded sufficiently even to appear in regional data bases (Plog et al. 1978, Tainter 1983).

Archaeology and cultural resource management have tended to focus on remains that are large, obvious, or highly-structured, an approach that has been characterized as the archaeology of salient characteristics (Tainter 1998). Salient characteristics are those that stand out clearly from background noise. Obvious archaeological examples include a Bronze Age tell rising above a level plain, a large Southwestern pueblo, or a site promising stratified deposits or other patterning. Because people respond to such clear signals amidst an otherwise disordered world it is not surprising that archaeologists have concentrated their efforts on the most salient parts of the record. Regrettably, public presentations of archaeological variability, such as the clusters of spectacular ruins at Chaco Canyon and at Mesa Verde, teaching, and popular writing have contributed to the notion that large, prominent sites are highly valuable whereas unobtrusive ones are not (e.g., Fagan 1991).

Human behavior, however, infrequently produces a salient archaeological record. The fact that artifacts can be found nearly everywhere belies such a notion (Ebert 1992). Nor is it likely that accurate descriptions of prehistoric land use can be written from an exclusive focus on salient remains. In the southwestern United States, for example, the most prominent sites are large pueblos, which contain up to 2,000 to 3,000 rooms arranged in blocks around community areas called plazas. Southwestern prehistory conventionally has been written as the rise and fall of such communities, both as individual sites and as a type of regional settlement (Tainter and Plog 1994). Yet they represent but a fraction of the Puebloan-era archaeological record. In some parts of the Southwest, as much as 95 percent of the record consists of small, inconspicuous scatters of lithic or ceramic debris on or near the ground's surface (Plog et al. 1978). For an ecological approach, no reconstruction of prehistoric land and resource use that ignores so much of the archaeological record can claim to be representative or accurate. Inferences about Anasazi diet, for example, are based largely on what has been found at places where food was consumed, at places where the archaeological record is salient and strongly patterned (Sullivan 1996). Because we infrequently incorporate evidence afforded by places where food was produced or processed (e.g., activities that produced piles of fire-cracked rock), our notions of Anasazi economy and land use may be correspondingly incorrect (Sullivan 1995a). The following examples illustrate how the study of non-salient archaeological remains allows heritage management to produce data useful to ecosystem management.

### 4.2.1 The Ethnoarchaeology of Hunting

One of the earliest studies to explore how the archaeological record forms was conducted among Nunamiut Eskimo hunters by Lewis R. Binford (1977). In April 1971, Binford accompanied Nunamiut hunters on 47 trips from their village. The trips were made to recover meat from caches and/or to gather firewood, or to search for game and check traps. The distance of the trips ranged up to 170 miles from the village, and some excursions involved overnight stays. Binford inventoried gear that the hunters took, and the gear that was returned. A total of 647 items was taken on these trips, including guns and metal tools, as well as both traditional and processed foods. Of these items, only 53 were not returned to the village. Of the 53 items left in the field, 36 were disposables such as food containers, shell cases, and bones. Fourteen items were cached in the field for further use. Three items were lost on the trail, including one item that was broken and discarded where it was used. The difference between the two inventories is the potential archaeological record of the hunters' activities. Thus, the archaeological record generated during these excursions amounts to little more than an average of one item per trip.

An archaeologist studying land and resource use would want knowledge of the hunters' behavior, and of the locations where activities took place. Such information could not be fully extracted from the village's archaeological record. Hence, the archaeological record left behind on these trips is crucial for understanding land use, and the factors that influence tool design, caching strategies, and the formation histories of "places" on the landscape (Binford 1982). Yet the archaeological record generated by Nunamiut hunters would be dismissed by many archaeologists as "isolated finds." These items normally would not be considered worthy of management and would not be entered into most data bases. The latter oversight may be the most important because the spatial distribution of low-density remains is crucial to interpreting them (as discussed below). The Nunamiut study shows clearly that significant human behavior does not necessarily produce a salient archaeological record. Conversely, it implies that evaluating archaeological remains on the basis of how salient they are will prevent us from recognizing important aspects of land-use patterns (Camilli 1988).

### 4.2.2 The Archaeology of Prehistoric Foragers

The importance of low-density archaeological phenomena for ecosystem management emerges most clearly in the aggregate (Tainter 1979). Such remains can delineate past cultural landscapes when large numbers of them are analyzed (Allen 1991). Yet isolated artifacts and low-density remains are normally encountered one or a few at a time, in limited geographic areas associated with individual heritage management projects. Encountered in this way, it can be difficult to foresee how such sparse remains can contribute to delineating past cultural landscapes. Yet archaeologists who have worked with such data have accomplished precisely that. Two case studies illustrate this potential.

Luchterhand (1970) analyzed the distribution of Early Archaic period (ca. 8000–6000 B.C.) projectile points (spear points) in the lower Illinois Valley region. Of 289 points, only 26 (9%) were found in the Illinois and Mississippi valley bottomlands, which is where the major Archaic-period sites are located. Fully 91% (263) of the points were found in the upper reaches of secondary stream valleys, or in the uplands between the Illinois and Mississippi river valleys. Thus, the distribution of isolated Early Archaic spear points varies inversely with the distribution of the major Archaic-period sites. To account for this distribution, Luchterhand pointed out that during the winter months, white-tailed deer tend to shelter in secondary valleys. They particularly favor sunny, sheltered slopes of these valleys. Deer wintering in the area could have been spotted from higher ground adjacent to the secondary valleys, and from the uplands. Luchterhand's analysis suggests, therefore, that the distribution of spear points reflects Early Archaic-period hunting of wintering white-tailed deer.

The significance of this study for ecosystem management is that it delineates a part of the Early Archaic cultural landscape, and resource use, in this area. In turn, the study provides a basis on which to assess the early human influence on the distribution and abundance of prey species. One lesson is that this important information came from isolated artifacts, a source of data that heritage management routinely undervalues. Another lesson is that the distribution of these artifacts conveys ecological information that could not have been obtained from large, salient sites.

A more intensive study by Thomas (1971) was designed to evaluate the ethnographic information collected by Julian Steward (1938) on Shoshonean subsistence and land use in the Great Basin. The Shoshone maintained winter villages, which were often located at the ecotone between sagebrush flats and the pinyon--juniper belt. These villages, along with dry caves, form the most salient part of the Great Basin archaeological record. Yet they represent only a fragment of Great Basin archaeology, and of prehistoric cultural landscapes. The Shoshone used a variety of plant and animal resources distributed from lower elevation riverine zones to the highest mountain peaks. In all of these areas, artifacts would have been deposited in the course of resource harvesting. Conversely, these artifacts today comprise the archaeological record from which we can understand Shoshonean land use. Thomas makes the important point that these dispersed subsistence activities would have produced, at best, a faint archaeological record of low-density artifact scatters and isolated items.

Thomas's (1972) approach to evaluating Steward's ethnographic subsistence pattern was to construct a computer simulation that made quantitative predictions about the nature and distribution of archaeological remains. Data to test the model were collected from all pertinent vegetation types. The data consisted largely of the distribution and density of individual examples of artifact classes — isolated artifacts. Statistical evaluation of the data led Thomas (1973) to conclude that the archaeological record of his study area was what one would expect if Steward's ethnographic data depicted the prehistoric land-use system. Thomas's study illustrates how low-density artifact scatters and isolated artifacts may contribute to the investigation of past cultural landscapes and resource procurement systems (cf. Wait 1983). Because such remains often are not recorded, much fieldwork in heritage management has not yet produced data that are relevant for ecosystem management.

## 4.3 Cultural Landscapes and Ecosystem Disturbances

One of the first steps in understanding the human role in influencing past ecosystem processes is to delineate such matters as the landforms and vegetation zones

Fig. 2. Archaic period (ca. 5500 B.C.–A.D. 500) site distribution in the middle Rio Grande Basin. Note the site concentration in the upper Rio Puerco Basin (upper left). Albuquerque, New Mexico is in the lower right of the image.

used at various times. At one time, the term *subsistence-settlement system* was coined to indicate the set of locations from which subsistence resources were derived (Struever 1968, Winters 1969). *Cultural landscape* is a more encompassing notion because it incorporates the physical elements of subsistence-settlement systems, and adds other concepts as well (Wagstaff 1987). These include phenomena that are hard to delineate archaeologically, such as cosmological and metaphorical conceptions of landscapes, as well as matters that are amenable to research, such as anthropogenic environmental processes. Few archaeological studies have yet delineated the last matter in full, but it is central to the role of heritage management in ecosystem management. This is illustrated by considering the distribution of archaeological sites in the middle Rio Grande and Rio Puerco valleys of New Mexico.

Two periods in particular are worth discussing here, for they represent very different types of economy and land use. In the long span of time that archaeologists call the Archaic period (ca. 5500 B.C.–A.D. 500, including the Basketmaker II period), the mode of subsistence was hunting and gathering (Vierra 1994). Some time in the last millennium B.C., small amounts of maize were added to the diet, but it did not become a major part of subsistence until after about A.D. 500 (Matson 1991).

The hunter-gatherer land-use pattern in this area is distinctive. To Euro-Americans, the most desirable landforms within the area are the river valleys. These are where Hispano and Anglo-American farmers settled first. Yet in perhaps 12,000 years of occupation, the Rio Grande and Rio Puerco valleys are where Native Americans settled last. Throughout much of the Archaic period, populations concentrated along the upper reaches of the Arroyo Cuervo, a tributary of the Rio Puerco (Fig. 2) (Irwin-Williams 1973). In the later Archaic period, the broad tableland between the Rio Grande and Rio Puerco valleys, which is called the West Mesa (Reinhart 1967, Tainter and Gillio 1980: 46–48), sustained increasing use. Both areas today are considered marginal even for cattle ranching.

The Archaic-period preference for the Arroyo Cuervo stemmed from two factors. The first is that the heads of side canyons along the Arroyo Cuervo contain seeps that are highly reliable sources of water (Irwin-Williams 1973). The second is that the Arroyo Cuervo has some of the highest topographic diversity in the region. Topographic diversity allows hunter-gatherers to practice a foraging economy with less mobility than is otherwise necessary. By substituting altitudinal for spatial resource variation, the costs of locating, gathering, and transporting resources can be reduced (Tainter and Gillio 1980, Tainter and Tainter 1996).

The cultural landscape of A.D. 1300–1600 (known locally as the Pueblo IV period) was strikingly different (Fig. 3). The Rio Puerco and its tributaries were abandoned and essentially unused. People practiced intensive agriculture in the Rio Grande Valley. This area experienced its most intensive prehistoric occupation and new, aggregated settlements were established on an unprecedented scale (Wendorf 1954, Wendorf and Reed 1955, Cordell 1979). Parts of the valley bottom

Fig. 3. Pueblo IV period (ca. A.D. 1300–1600) site distribution in the middle Rio Grande Basin. Note the abandonment of the Rio Puerco Basin (compare Fig. 2) and the site concentration along the Rio Grande.

were farmed at a level of intensity not experienced again until the 19th century. In 12,000 years of human occupation, it was the most profound transformation in land and resource use.

These changes in land use provide a basis from which to investigate the human role in ecosystem change, particularly the effects of anthropogenic influences. Although definitions vary as to what constitutes an ecological "disturbance" (Lundquist et al. 1995, Wright et al. 1995), one given recently by Russell Graham seems most useful: a disturbance is anything that alters a trajectory or a trend (personal communication to J. A. Tainter, August 1995). The massive convergence of human populations on the northern and middle Rio Grande Valley in the 14th century A.D. would certainly qualify. Within a few generations this area went from supporting a small, dispersed agricultural population, to supporting settlements that ranged up to 2,000–3,000 rooms in size. Lands were cleared for agriculture, and every nearby piece of wood useful for construction, cooking, or heating would have been consumed quickly. The distribution and abundance of native plant and animal species would have been altered in a short time, as would nutrient cycling and the composition of soils. Rio Grande Basin ecosystems would clearly have evolved very differently if this massive disturbance had not occurred (Tainter and Tainter 1996).

The earlier role of hunter-gatherers in ecosystem processes of the Arroyo Cuervo is a subtler problem. For more than 6,000 years, humans were a major, perhaps dominant, ecosystem component. As regional population grew, there would have been increasing pressure to manipulate vegetation (cf. Sullivan 1996). Certainly the hunter-gatherers would have had a controlling influence on such things as the distribution and abundance of seed-bearing plants that they gathered, ungulates that were hunted, rodents that were attracted to their food stores, carnivore populations, tree growth in the pinyon–juniper zone, accessibility of water for wildlife, nutrient cycling, soil formation, and erosion. If disturbance is considered to be the disruption of a trend, then the greatest disturbance to the Arroyo Cuervo ecosystem may have been the *withdrawal* of the human population in the 7th to 8th centuries A.D. Those who equate human use of ecosystems with disturbance may find this inference startling. Certainly, however, a great variety of ecological structures and processes, which had been regulated for 6,000 years by gradually intensifying human use, would suddenly have had to establish new ranges of tolerance and adjust to new conditions (Tainter and Tainter 1996). Although this proposition would benefit from further research, it does suggest that models of pre-European ecosystem structure and process in this area that do not incorporate the prehistoric human influence will be, at best, incomplete and, at worst, misleading.

## 5 HISTORIC HERITAGE RESOURCES AND ECOSYSTEM MANAGEMENT

Abundant heritage resources date to the last 500 years, a period of time that historians refer to as the "modern world." Orser (1996) considers those resources, and the archaeology of the modern world in general, to be associated most strongly with the rise of capitalism, colonialism, Eurocentrism, and modernity. Landscapes of this time period are material expressions of the modern world that link ecosystems to heritage resources (Simmons 1989). Interpreting modern world landscapes, first and foremost, involves consideration of dramatically changed relationships, scales, physical and sociohistorical structures (Marquardt 1992), and boundaries among peoples, places, things, biota, and climates (Kates et al. 1990). Such changes have had dramatic impacts upon the evolution of ecosystems. For example, archaeological studies of the Anahulu Valley, on the Hawaiian island of O'ahu, document enormous change in the valley ecosystem beginning about the third century A.D. with the discovery of the Hawaiian Islands by seafaring Polynesians. The Polynesians introduced exotic domestic plants and animals (e.g., taro, yams, pigs, dogs, fowl) and unintentional species (e.g., skink lizards, snails, weeds) that, among other things, rapidly transformed the pristine mesic forest into a "mosaic of gardens and second-growth vegetation" (Kirch 1992:169). They developed a mode of production based upon irrigation taro farming and husbandry of pigs and dogs. In 1778, Captain James Cook's voyage brought the modern world to Hawaii. The Europeans brought with them an exotic "portmanteau biota" of weeds, domestic plants and animals (e.g., rats, cattle, sheep and goats, maize, beans, sugarcane), and microorganisms (e.g., measles, smallpox, syphilis) (Crosby 1986). Of these, cattle caused the greatest ecological change. Beginning in the 1830s and 1840s, cattle herds, under the protection of chiefs collaborating with ranchers, deforested large areas of the valley, dramatically changing hydrologic patterns, thereby forcing traditional Polynesian farmers to build stone enclosures around their fields. Finally, as Kirch notes (1992:170),

> the sugar plantation developments of the late nineteenth and early twentieth century were to even further disrupt the valley environment, siphoning off the streamflow far inland and diverting it to the tablelands, so that the Anahulu River which once watered the maka'ainana pondfields now flows only intermittently, and flash floods have downcut the stream level by several meters.

## 5.1 Modern World-Landscapes and the World-Economy

Modern world-landscapes, together with their implications for the evolution of ecosystems, are largely a consequence of what Wallerstein (1974) calls the "world-economy," a large scale economic system based upon capitalism that links multiple polities such as nation-states. Wallerstein focuses upon the social relations of exchange as the key to understanding the structure of the world-economy and the operation of large-scale social systems (e.g., the 18th century British empire) within that arena. In Wallerstein's view, world-systems provide the mechanisms for producing and redistributing surplus within the world-economy. The growth of global markets and the resulting global division of labor led to the emergence of specialized regions of surplus concentration (core) and surplus production (periphery) with unequal or asymmetrical exchange relationships.

In contrast to Wallerstein's focus upon the relations of exchange as the key to how the world-economy is structured, Wolf (1982) and others argue that the relations of production are more important to understanding the modern world. Indeed, the relations of exchange probably are much less important to understanding the ecological consequences of the modern world than the relations of production. The "mode of production," which Wolf (1982:75) defines as "a specific historically occurring set of social relations through which labor is deployed to wrest energy from Nature by means of tools, skills, organization, and knowledge," is the key to comprehending the interplay between people and their environment in the modern world. Wolf sees the world-economy as structured by multiple modes of production linked together by capitalistic relations of exchange and dominated by a capitalistic global market. The multiple modes of production are some combination of capitalistic, tributary, or kin-ordered forms, which vary according to the social and cultural principles underlying the relations of production.

## 5.2 Linking Concepts

Modern-world relations of production and exchange bind local-scale ecosystems into large-scale regional and global networks (Weins et al. 1986). The evolution of local ecosystems takes place within the historical context of these networks, up to and including highly correlated changes among the local ecosystems making up the network (Crumley 1994).

### 5.2.1 Sociotechnical Systems

One linking concept associated with modes of production that underlies ecosystem networks is the sociotechnical system, which Pfaffenberger (1992:497) defines as "the distinctive technological activity that stems from the linkage of techniques and material culture to the social coordination of labor." Technique, in turn, is defined as a "system of material resources, tools, operational sequences and skills, verbal and non-verbal knowledge, and specific modes of work co-ordination that come into play in the fabrication of material artifacts" (Pfaffenberger 1992:497). The beliefs, attitudes, and values making up the work culture also play an important part in the system.

A good example of how the sociotechnical system links together local-scale ecosystems into large-scale regional ecosystem networks is the industrial mining system developed in the Comstock Mining District of western Nevada. The Comstock mines in Virginia City and its immediate environs define a mining landscape on a local scale; however, the Comstock Lode served as a control center for a much larger region that encompasses the pinyon–juniper woodlands of the Virginia Range, the Carson River, woodlands in and around the Lake Tahoe Basin, and the farmlands of the Carson Valley, all localities several miles away from the Comstock mines. Each of these areas is a local ecosystem with a distinctive history and ecology linked together into a regional ecosystem by Comstock industrial mining. The Lake Tahoe Basin, for example, is a natural entity with visibly distinct physiographic and other boundaries that define a patch within the regional ecosystem. Although Comstock Era loggers deforested large areas of the basin in the late nineteenth century, their tree cutting cannot be understood outside the context of how industrial mining affected the Comstock Lode many miles away from the lake. Within this geographical framework, then, the cultural resources of mining landscapes in the American West not only provide a material image but also a repository of information about a regional variety of global capitalism.

### 5.2.2 Relations of Exchange and Ecosystem Networks

Modern world-relations of exchange also create large-scale networks of local ecosystems. Jack Williams (1992), for example, illustrates the linkages in his discussion of how archaeological data can be used to test two competing hypotheses about the relations of exchange between Spain and its colonies in America, specifically New Spain. One hypothesis, first proposed by Immanuel Wallerstein, states that New Spain was a fullblown periphery of Spain since the 16th century. The other hypothesis, proposed by French historian Fernand Braudel, contends that New Spain and Spain enjoyed more or less equal relationships. According to this interpretation, surplus accumulated in New Spain and transformed the Colonial economy. Bullion was extracted by private enterprise in New Spain, merchants in New Spain controlled the markets, and both accumulated surplus in the periphery. All that changed with the early 19th century wars of liberation. The new republics established trade with industrial Britain, leading to neo-colonialism in Latin America that created a core-periphery relationship of the type described by Wallerstein.

How can these two models be tested with archaeological data? Wallerstein argues that "essential goods" reflect the unequal relationship between core and periphery. Essential goods are the things, such as tableware, food, and clothing, that are used in everyday life. Peripheries have high percentages of essential goods coming from core regions. According to Williams (1992), Wallerstein's hypothesis entails high percentages of essential goods in New Spain after the 16th century. In contrast, Braudel's hypothesis entails high percentages of essential goods, coming from Britain, only after the Republic Period (1822–1860). Williams uses archaeological data from three southern Arizona presidios (military forts) dating between 1752 and 1856 — Tubac, Tucson, and Santa Cruz — to test the two models. Presidios housed the elite, who accumulated surplus goods in the peripheries and, therefore, should best reflect trade and economic relations. Williams found that the percentage of essential goods coming from outside the region is low in the three presidios, suggesting that they were self-sufficient. Braudel's model, therefore, is supported. After 1860, however, increasing development of transportation, especially railroads, brought more essential goods from the core of the American world system, creating a true periphery.

### 5.2.3 Ecological Niche

Another way of viewing the evolution of modern world-landscapes and ecosystems comes from the concept of the ecological niche (Hardesty 1975). Such components of the "physical" environment as space, minerals, water, topography, and biota are incorporated as dimensions of the human ecological niche. Specific human populations create and occupy niches with a distinctive volume, shape, and size that reflect partitioning of the dimensions not only through biological adaptation but also through social and cultural constructions. For example, the heritage resources of mining-related landscapes in the modern world document the incorporation and transformation of mineral deposits, water, and other geological dimensions into mining niches (Hardesty 1995). Water engineering systems, such as ditches, dams, reservoirs, wells, and holding tanks, which collect and divert water for use in hydraulic and placer mining (Smith 1986), are the material expressions of mineral engineering systems used to explore, extract, and mill or reduce mineral deposits. Such resources reflect not only cultural constructions of "meaningful" minerals but also technical knowledge. In central Nevada's Cortez Mining District, for example, the archaeological remains of limestone quarries and lime kilns document an appropriate technology developed to reduce milling costs by using locally available mineral deposits (Hardesty 1988). The Russell lixiviation technology installed in 1886 at the Tenabo Mill used lime and sulfur to make calcium sulfide as a precipitator rather than the more expensive, if somewhat more effective, sodium sulfide. In this case, the mining niche includes not only the "Nevada Giant Ledge," a large mineral deposit containing silver, among other things, but also limestone, salt, and other minerals used in the technology. Furthermore, the mineral dimension of the Cortez mining niche changed over time, reflecting changing cultural meaning and technical knowledge of mining.

## 5.3 The Archaeology of Modern World-Landscapes

Modern world-landscapes, then, are material expressions of local modes of production and global exchange networks that often have dramatic impacts upon the evolution of ecosystems and that often link together local-scale ecosystems into large-scale regional and global ecosystems. Heritage resources associated with landscapes document those impacts, as the following examples illustrate.

### 5.3.1 Industrial Landscapes

The technology, social relations, and ideology of the Industrial Revolution, which began in England in the 18th century, transformed ecosystems on a geographical scale and with an intensity never before experienced. In the United States, industrialization reflects the outcome of a debate in the early American Republic over whether the future of the nation should be agrarian or industrial (Marx 1964). By the end of the eighteenth century and within the next few decades thereafter, several industrial towns had emerged in the Northeast. The new towns sought to impose upon the labor force the work rules, practices, and mindset of the factory system, both in and out of the workplace. Industrial landscapes and architecture not only reflect but also helped bring about and reinforce the change. In the textile-mill town of Lowell, Massachusetts, for example,

> [T]he original eight blocks of the Boott Mills boardinghouses were erected between 1835 and 1839, contemporaneous with the construction of the mills. The boardinghouses were arranged at right angles to each other and had facades similar to that of the mill. Industrialists thus created a similar architectural setting at home to that which was found in the factory. (Shackel 1994:5)

Studies of yardscapes in the 19th century federal armory town of Harpers Ferry, West Virginia, show the value of archaeological data in interpreting the impact of industrialization upon ecosystem evolution. M.R. Smith (1977) demonstrated that, following its establishment shortly after the American Revolution, the armory first used the traditional craft system to manufacture firearms but after 1829 slowly changed over to an industrial factory system. The armory was abandoned with the outbreak of the Civil War, and the town changed into a regional commercial and tourist center. Recent archaeological studies clearly show that the evolution of the Harpers Ferry landscape closely tracks the town's history (Shackel 1994). Pollen and phytolith data gathered from archaeological excavations in the town, for example, suggest that the pre-armory period landscape was forested with only a few interspersed small grass patches. The earliest armory period landscape showed decreased forest and increased open grasslands, typical of European-style land clearing practices. Ethnobotanical data from the 1820s yard of the master armorer's house strongly suggest a well-groomed, garden-like landscape, a practice consistent with the "agrarian" model of the American Republic and that seems to argue that industry can coexist harmoniously with nature (Shackel 1994:9). By the 1840s, however, the armory no longer maintained the garden landscape. The master armorer's yard appeared to reflect the "unkept" pattern of weeds and other grasses typical of northern industrial towns, such as Lowell, Massachusetts (cf. Kelso 1993).

### 5.3.2 Landscapes of Global Migration

An important characteristic of the world-economy is a global division of labor that involves large-scale migration of people from one part of the world to another. Chinese global migration in the 19th century is a good example. The landscape expression of the migration includes the Chinese terraced garden, which has been recorded in the American West as well as in Canada and New Zealand, often in association with mining activities. Archaeological studies have provided the best evidence of the landform, which often substantially changed the patterning of vegetation as well as landforms of local-scale ecosystems. J.M. Fee (1993), for example, documents 26 acres of Chinese terraced gardens, probably dating to the period between 1875 and 1910, in the steep and rugged terrain of the South Fork of the Salmon River in Idaho. The garden terraces typically appear on mountain slopes as a series of steps engineered to make the most efficient use of space. They vary in width from the size of a hand to 12 feet or more, in length from 60 to 75 feet, and in shape (e.g., rectangles, half-moons, squares, and kidneys). The terraced gardens occur on slopes ranging from 12 to 45 percent and at elevations from 3200 to 4700 feet. Archaeological remains on the terraces, such as pollen, macrobotanicals, ditches, and artifacts, potentially provide a wealth of information about the cultigens grown in the gardens, Chinese gardening practices, engineering of the terraces, and the lifestyles of the gardeners. The interplay between the gardens and local ecosystems also is evident, as Fee (1993:92) observes:

> Erosion is more extensive on steeper slopes with southern exposures where the soil is relatively dry, sandy, and unstable. On north-facing slopes, the rapid succession of natural vegetation has almost totally prevented erosion.

### 5.3.3 Modern World-Vegetation Patterns

Constructing the vegetation history of modern world-landscapes is another use of heritage-resource data. Hattori and Thompson's (1987) study of pinyon–juniper woodlands in the Cortez Mining District of central Nevada provides a good example. Their study focused on the limits and chronology of woodland deforestation by industrial precious-metals mining between 1863 and the early 20th century. Mining technology used pinyon wood for a great variety of purposes, especially as fuel for the steam engines that ran the hoists, air compressors, pumps, furnaces, stamp mills and other equipment and, after its conversion to charcoal, for roasting ovens. In addition, the miners used pinyon logs as mine supports, railroad ties, retaining walls, and for building cabins. Many authorities believe that, as a consequence of high demand during this period, the woodlands were completely clear-cut in a radius of about 60 miles around the district (Young and Budy 1979:117–119, Lanner 1981:124–130). Hattori and Thompson noted, however, that contemporary photographs generally covered only the immediate vicinity of the mines, mills, and towns, leaving invisible what happened to the pinyon–juniper woodlands in the hinterland. Furthermore, some written accounts, such as government reports, suggested that the extent of deforestation was much less dramatic than what most authorities believed. To shed light on these two competing interpretations, Hattori and Thompson conducted a dendrochronological study of the hinterland outside the main mines and settlements. They used the tree-ring record of cut stumps, mine supports, rail ties, stacked cordwood, charcoal, and cabin timbers to construct an absolute chronology of woodland use in the district and concluded that deforestation appeared to be less severe than claimed. The first mining period between 1863 and 1883 was "marked by localized, nonintensive logging to meet the demands of the numerous small mines in the area" (Hattori and Thompson 1987:70). Mining between 1884 and 1891, the first boom period, greatly thinned the pinyon woodland but many trees survived and make up about ten percent of the trees living today in the district. The woodland reestablished itself after 1897, when fuels and materials for mining technology were imported into the district, and continued its growth, and possibly expansion, to the present. Current expansion of the woodland into the valley floor may represent the impact of overgrazing by cattle in the late 19th century and the subsequent successional replacement of grasses by sagebrush and then by pinyon–juniper forest.

### 5.3.4 Modern World-Landform Patterns

Constructing the landform history of cultural landscapes associated with industrial mining is yet another use of heritage-resource data. Examples include the dredging landscapes in Alaska, open-pit mining landscapes, such as at Bingham Canyon in Utah and the Mesabi Range in Minnesota, hydraulic mining landscapes, such as the Malakoff diggings near Nevada City in California, and the strip-mining landscapes of West Virginia. Mechanization of the mining industry brought about the landforms. Power shovels for dredging and open-pit mining were introduced by the end of the 19th century. The first use of power shovels was for dredging placer deposits. Dredging involves building a large pond upon which the dredge floats and excavates mud, sand, and gravel from the bottom of the pond with a large shovel or bucket line. The technology leaves behind a distinctive landscape pattern organized around large "serpentine" tailings piles and remnants of dredging ponds. In some cases, however, the dredge tailings have been removed by later reclamation activities. The "serpentine" piles left behind were "removed, crushed, and used in concrete and as railroad ballast and top dressing for roads" (Smith 1987:93). The use of power shovels for open-pit mining, however, probably had a much more widespread and pervasive impact upon mining landscapes of the 20th century, especially in the American West (Schwantes 1992).

### 5.3.5 The Spatial Organization of Modern World-Landscapes

In providing a framework for understanding how space is used in the modern world, Orser (1996:137), drawing on the ideas of French sociologist Henri Lefebvre (1979) concerning the relationship between space and capitalism, posits that "it is difficult to separate capitalism from its global usage of space and place." From this perspective, the heritage resources of the modern world provide not only a material expression but also a data repository of the interplay between capitalism and space — an interplay that strongly influences the evolution of ecosystems.

The early 20th century Robinson Mining District in eastern Nevada provides a good example of how space is organized by the means of production (Hardesty et al. 1994). In this case, the geographical boundaries of the district are more or less synonymous with the regional community that links together several towns and other settlements, along with copper mines and mills. The mines and mills make up the economic center of the community. Radiating out from the mines are outlying neighborhood settlements, such as company towns (Ruth and Kimberly), satellite settlements (e.g., Riepetown), dairy farms, hay farms, and isolated households. Road networks link together the center and the outliers. As a social interaction network, the regional community is a web of social nodes where people living in the outlying neighborhood settlements congregate not only at the mines and mills but also at the medical facility, the schools, the churches, the company stores, and the union halls. Because the archaeological record of this economic system is clearly dispersed across a landscape, its origins and variations must be analyzed holistically from an ecosystemic perspective.

## 6 ISSUES AND PROSPECTS

Heritage management as practiced today is intended to produce data that (i) are suitable to evaluate whether properties are eligible for the National Register of Historic Places and (ii) can be gathered quickly and economically in a context of constrained budgets, time, and personnel. In some ways this system works well: the conservation mandate of historic preservation is being fulfilled for some classes of properties. It is not clear, however, that the current approach is producing the data needed to ensure that heritage management contributes to ecosystem management.

For instance, consider the following points that have been established by the research discussed in this chapter:

- The bulk of the archaeological record, perhaps worldwide, consists of isolated artifacts or low-density scatters that often have little or no depth or stratigraphic integrity.
- The density of archaeological debris does not reflect in a simple, direct way the intensity or significance of the activities that produced the debris.
- The distribution of surface scatters and isolated finds, in conjunction with that of "salient" remains, is information that is significant to understanding past land and resource use, and thus to ecosystem management.

Heritage management that is conducted within the framework of, and contributes to, ecosystem management, clearly must concern itself with the full range of archaeological phenomena. This concern extends from the highly structured, salient sites that all heritage professionals recognize the importance of, to the faint remains that conventionally have been overlooked (literally, in some cases). A program of heritage management that contributes to ecosystem management must therefore confront the following issues.

### 6.1 Discovery and Definition

Discovery strategies refer to such matters as survey intensity (i.e., inter-surveyor spacing), artificial exposure intensity (e.g., shovel-test spacing), and site definition (Sullivan in press). The standard for archaeological survey crew spacing in the USDA Forest Service Southwestern Region is 15–25 meters. Similar standards are followed in many parts of the United States, and they are actually an improvement over older methods of site discovery (McManamon 1984). Where ground visibility conditions require shovel testing, the interval by which this is done directly affects the number of sites that are found (Kintigh 1988). For example, on the Hoosier National Forest in Indiana, one contractor, expending 4.5 times more effort in surveying one-half the area as another contractor had in an adjacent area, found 35 times more archaeological sites per acre. The obvious lesson here is that discovery techniques that do not enhance the probabilities of discovering low-density artifact scatters and isolated artifacts contribute neither to the goals of heritage management nor those of ecosystem management.

Definitions of what constitutes an archaeological site critically affect data quality. Many site definitions involve artifact-density thresholds, such as five artifacts per square meter (e.g., Plog and Hill 1971). If the surface properties of archaeological remains do not meet or exceed the thresholds, then there is a strong likelihood that those archaeological manifestations will not be incorporated into computer-based distributional studies (Tainter 1983). When several hundred or several thousand such remains are excluded from regional data bases, models of prehistoric land use will be hopelessly inexact. Furthermore, if low-density archaeological remains are not labeled "sites," they typically are not evaluated for the National Register of Historic Places or are considered to be ineligible for it (as discussed below). The site definition itself, therefore, becomes a *de facto* significance evaluation. The problem for heritage management is that the cutoff for calling an archaeological manifestation a "site" may be unjustifiably too high, thereby excluding low-density surface scatters and isolated artifacts from evaluation and further analysis. Thus, site definitions themselves may categorically eliminate many of the manifestations that make heritage management pertinent to ecosystem management (Tainter 1983). Although we have discussed this problem primarily in reference to prehistoric remains, it is an issue of equal importance for historic archaeology (Periman 1995:56).

## 6.2 Significance and Scale

In heritage management, sites typically are evaluated individually on an implicit criterion of how structured or salient they are (Tainter and Plog 1994). Not infrequently, site size and depth have been *de facto* indicators of significance in what may be termed the *National Geographic* approach to heritage management (Tainter 1998). Such a view assumes that significance varies along the same dimensions as the qualities of sites selected for presentation in *National Geographic* magazine, i.e., those with evidence of large size, great depth, spectacular or unambiguous stratigraphy, great age, or other superlative attributes. This bias, it should be clear, selects against just those kinds of manifestations that are important to ecosystem management — low-density surface scatters and isolated artifacts that, in combination with "salient" remains, reflect how people used landscapes. As shown in this chapter, this kind of information may be difficult or impossible to obtain from highly-structured "salient" sites alone (Sullivan 1995b). In an ecosystem framework, hence, site size and debris density are not the only indications of "significance" (Tainter 1979).

Ethnoarchaeological and conventional archaeological studies that have focused on the spatial consequences of regional land-use have discovered that low-density artifact scatters arise under a great range of circumstances (Ebert 1992). However, "site-oriented" approaches have difficulty capturing these kinds of remains (Wandsnider 1998). In "artifact-rich" landscapes, in contrast, the site-oriented approach creates problems for archaeologists and resource managers alike because they often feel obligated to carve sites out of nearly continuous, artifact distributions (cf. Bintliff and Snodgrass 1988).

Further dilemmas arise in considering matters of scale. Although the eligibility criteria of the National Register of Historic Places provide for consideration of entities above the individual site, many heritage studies nevertheless evaluate sites individually and in isolation. Yet for heritage management to contribute to ecosystem management, the appropriate unit of analysis may not be sites but larger entities, such as cultural landscapes (Thomas 1971, Roberts 1987). In the case of historic mining, entire districts have lost both their National Register eligibility and their research potential when parts of these districts were evaluated in a fragmented manner (Periman 1995:iii). Heritage managers, with varying degrees of understanding of the interpretive potential of landscape research, may or may not recommend that low density, ambiguously patterned, surface sites are worthy of preservation. In addition to the issues already discussed, this inconsistency contributes to comparability problems between and among projects.

## 6.3 Methodological Issues

In order to maximize the potential contributions of heritage resources to ecosystem management, we offer the following methodological recommendations.

First, it would be useful to conduct experiments to ascertain optimal inter-surveyor intervals for various problems and conditions to ensure that potentially significant non-salient phenomena (e.g., isolated ground-stone artifacts and projectile points) become members of the resource data base. Along these lines, other experiments could be devised to test which recording protocols enhance efficiency and comparability under various field conditions (Kvamme 1998).

Second, Global Positioning System (GPS) technology has revolutionized the amount, accuracy, and scope of observations that can be obtained about archaeological phenomena. In addition, because GPS can accommodate a great variety of recording protocols, it has the flexibility to accommodate the needs of resource managers, who may still want to enumerate phenomena in terms of site-level data. It can allow for data sharing among resource managers, GIS personnel, and plant ecologists to provide a comprehensive picture of ecosystems for management and interpretation. The widespread, systematic use of GPS will allow archaeologists and resource managers alike to deal with the salient site/non-salient site issue (i.e., low-density surface remains and isolated artifacts) and, thereby, reduce the possibility that potentially significant phenomena are overlooked or excluded from the study of human ecosystems.

Finally, despite a commitment to conservation, some resources may need to be dismantled scientifically in order to provide confirming or independent data regarding the human use of ecosystems. Archaeological and non-archaeological resources (e.g., old trees, bogs, packrat middens, alluvial exposures, etc.) alike are prime sources of paleo-environmental samples, such as macrobotanical remains, pollen, and faunal remains. The content of these resources is crucial for understanding an ecosystem's history and, hence, the decisions that managers will make regarding resource protection, allocation, and investigation. With ecosystem management, such an approach would expand what it means to manage resources—it would entail a shift in orientation from find and protect, to find, protect, and, when appropriate, consume heritage resources.

## 6.4 Rethinking Heritage Management

A complaint sometimes voiced about heritage management is that it preserves many sites and produces vast quantities of information with few clear, immediate benefits other than conservation. One purpose of a heritage management program, however, should be to provide part of the knowledge that underlies ecosystem management. Once this point is understood, several changes to heritage management logically follow.

The salient-site-oriented nature of most heritage projects does not facilitate the compilation of data that are pertinent to studying land use, especially the ecological consequences of how human societies actually worked the cultural landscapes that they created and inhabited. Thus, to incorporate heritage management into ecosystem management requires decisions that challenge how heritage management typically has been conducted. The least costly heritage programs will involve unintensive survey (i.e., large distances between surveyors), little or no shovel testing, site definitions with high thresholds, and piecemeal evaluation of sites based mainly on salient characteristics. Yet a program of heritage management conducted in this way may actually *suppress* considerable data that are relevant to ecosystem management. A heritage program that contributes to ecosystem management, in contrast, should be based upon the following interrelated considerations:

- Expand conceptions of the types of heritage resources that merit management to include low-density surface remains and isolated artifacts;
- Expand discovery protocols to increase the likelihood of finding low-density surface remains and isolated artifacts (e.g., smaller surveyor intervals, more intensive shovel testing where the ground surface is obscured);
- Ensure that low-density surface remains and isolated artifacts are entered into data bases from which distributional analyses can be conducted; and
- Depending on what types of problems are being investigated with heritage resources, consider changing units of analysis from individual sites to landscapes.

Clearly, a program of heritage management that meets these standards will be very different from current practices. To support ecosystem management, for example, compliance with the National Historic Preservation Act and with 36 CFR 800 should be conducted so that salient sites, low-density remains, and isolated artifacts are evaluated jointly as aspects of past land-use systems, rather than as individual entities. Future revisions to the National Historic Preservation Act or to 36 CFR 800, or special arrangements (such as programmatic memoranda) to implement them, should be designed with such principles in mind.

A further change is to recognize that heritage management programs should consist of more than complying with the National Historic Preservation Act and providing public interpretation. Among other topics, these programs should also include a component of scientific research to understand the human role in the development of North American ecosystems. For this reason, the process of selecting heritage professionals for positions in federal agencies should include consideration of an individual's capacity to produce scientific knowledge that is suitable to ecosystem management. For the existing workforce, training programs might be expanded to upgrade skills that are needed to yield ecological knowledge (Tainter 1998).

# 7 CONCLUSIONS

To summarize, several distinct challenges affect the degree to which heritage resources research can contribute to ecosystem management. Archaeologists have tended to study features and sites that one could trip over, the assumption being that the only uncontaminated, reliable data about the cultural past were to be recovered from large, deeply-stratified sites. Current preferences for nominating sites to the National Register of Historic Places favor those properties whose significance depends upon atypical age, size, complexity, or some combination thereof. While such sites will always have high value, these preferences, we argue, may be excluding many of the kinds of heritage resources that could contribute to the implementation of an ecosystem approach.

Ecosystems, it is often noted, are complex, and they have even more complex histories. Programs of ecosystem research and management, to be effective, will have to embrace such complexity, which is always costly (Tainter 1988, 1995). The programmatic changes we have proposed here will appear to be more complex and challenging than traditional heritage management. Yet this will be true of all the disciplines that comprise ecosystem management. If it is done right, ecosystem management is not likely to be less complex or difficult than traditional land management. Heritage management in support of ecosystem management will require standards of work that are more complex and challenging than existing programs. Yet ecosystem management will be fundamentally incomplete unless we accept these challenges and place heritage management at its core.

## ACKNOWLEDGMENTS

The ecological perspective of Dr. William D. Lipe was instrumental in refining the scope and content of this chapter throughout its evolution during the last two years. Critical comments from Dr. John Hanson (Archaeologist, Kaibab National Forest) and Ms. Rebecca A. Hawkins (President, Algonquin Archaeological Consultants) helped us broaden the chapter's usefulness for heritage managers and cultural resource consultants. We thank, as well, Dr. Suzanne Fish, Dr. Clay Mathers, and Dr. Thomas Riley for their reviews of an earlier version of this chapter. Finally, Dr. Marilyn Nickels provided important insights regarding the interpretive value of historical documents in ecosystem management during the formative stages of this chapter.

## REFERENCES

Allen, M.J. 1991. Analysing the landscape: a geographical approach to archaeological problems. In: Schofield, A.J. (ed.), *Interpreting Artefact Scatters: Contributions to Ploughzone Archaeology*, pp. 39–57. Oxbow Books, UK.

Anonymous (a). (no date). The Role of the Northeastern Area in Ecosystem Management. USDA Forest Service, Northeastern Area.

Anonymous (b) (no date). Ecosystem Management: The Heritage Resource Program Interface. USDA Forest Service, Northern Region.

Baisan, C.H., and T.W. Swetnam. 1997. Interactions of Fire Regimes and Land Use in the Central Rio Grande Valley. Research Paper RM-RP-330. USDA Forest Service, Rocky Mountain Range and Forest Experiment Station, Fort Collins, CO.

Barker, J.P. 1996. Archaeological contributions to ecosystem management. *SAA Bulletin* 14: 18–21.

Betancourt, J.L., J.S. Dean, and H.M. Hull. 1986. Prehistoric long-distance transport of construction beams, Chaco Canyon, New Mexico. *American Antiquity* 51: 370–375.

Betancourt, J.L., E.A. Pierson, K.A. Rylander, J.A. Fairchild-Parks, and J.S. Dean. 1993. Influence of history and climate on New Mexico piñon–juniper woodlands. In: Managing Piñon–Juniper Ecosystems for Sustainability and Social Needs edited by Aldon, E.F. and D.W. Shaw. pp. 42–62. Gen. Tech. Rep. RM-236. USDA Forest Service, Rocky Mountain Forest and Range Experiment Station, Fort Collins, CO.

Betancourt, J.L., T.R. Van Devender, and P.S. Martin. 1990. *Packrat Middens: The Last 40,000 Years of Biotic Change*. University of Arizona Press, Tucson, AZ.

Binford, L.R. 1977. Forty-seven trips: a case study in the character of archaeological formation processes. In: R.V.S. Wright (ed.), *Stone Tools as Cultural Markers: Change, Evolution, and Complexity*, pp. 24–36. Australian Institute of Aboriginal Studies, Canberra.

Binford, L.R. 1982. The archaeology of place. *Journal of Anthropological Archaeology* 1: 1–31.

Binford, L.R. 1983. *In Pursuit of the Past: Decoding the Archaeological Record*. Thames and Hudson, New York.

Bintliff, J., and A. Snodgrass. 1988. Off-site pottery distributions: a regional and interregional perspective. *Current Anthropology* 19: 506–513.

Blackburn, T., and K. Anderson. 1993. Introduction: managing the domesticated environment. In: T.C. Blackburn and K. Anderson (eds.), *Before the Wilderness: Environmental Management by Native Californians*, pp. 15–25. Ballena Press, Menlo Park, CA.

Boucher, D.H., S. James, and K.H. Keeler. 1982. The ecology of mutualism. *Annual Review of Ecology and Systematics* 13: 315–347.

Boyd, R. 1986. Strategies of Indian burning in the Willamette Valley. *Canadian Journal of Anthropology* 5: 65–86.

Braidwood, R.J. 1974. The Iraq Jarmo Project. In: G.R. Willey (ed.), *Archaeological Researches in Retrospect*, pp. 61–83. Winthrop Publishers, Cambridge.

Brumfiel, E.M. 1992. Breaking and entering the ecosystem—gender, class, and faction steal the show. *American Anthropologist* 94: 551–567.

Butzer, K.W. 1982. *Archaeology as Human Ecology*. Cambridge University Press, Cambridge, UK.

Butzer, K.W. 1990. A human ecosystem framework for archaeology. In: Moran, E. M. (ed.), *The Ecosystem Approach in Anthropology: From Concept to Practice*, pp. 91–130. University of Michigan Press, Ann Arbor, MI.

Cameron, C.M., and S.A. Tomka (eds.). 1993. *Abandonment of Settlements and Regions: Ethnoarchaeological and Archaeological Approaches*. Cambridge University Press, Cambridge, UK.

Camilli, E. 1988. Interpreting long-term land-use patterns from archeological landscapes. *American Archaeology* 7: 57–66.

Cartledge, T.R., and J.G. Propper. 1993. Piñon–juniper ecosystems through time: information and insights from the past. In: Aldon, E.F. and D.W. Shaw (tech. eds.), Managing Piñon–Juniper Ecosystems for Sustainability and Social

Needs, pp. 63–71. Gen. Tech. Rep. RM-236. USDA Forest Service, Rocky Mountain Forest and Range Experiment Station, Fort Collins, CO.

Chambers, F.M. (ed.). 1993. *Climate Change and Human Impact on the Landscape.* Chapman and Hall, London, UK.

Clark, J.G.D. 1972. *Star Carr: A Case Study in Bioarchaeology.* Addison-Wesley, Reading, MA.

Clarke, D.L. 1973. Archaeology: the loss of innocence. *Antiquity* XLVII: 6–18.

Cooke, R.U., and R.W. Reeves. 1976. *Arroyos and Environmental Change in the American Southwest.* Oxford University Press, London, UK.

Cooper, C.F. 1960. Changes in vegetation, structure, and growth of Southwestern pine forests since white settlement. *Ecological Monographs* 30: 129–164.

Cordell, L.S. 1979. *Cultural Resources Overview, Middle Rio Grande Valley, New Mexico.* USDA Forest Service, Southwestern Region and USDI Bureau of Land Management, New Mexico State Office, Albuquerque and Santa Fe, NM.

Cronon, W. 1983. *Changes in the Land: Indians, Colonists, and the Ecology of New England.* Hill and Wang, New York, NY.

Crosby, A.W. 1986. *Ecological Imperialism: The Biological Expansion of Europe, 900–1900.* Cambridge University Press, Cambridge, UK.

Crumley, C. (ed.). 1994. *Historical Ecology.* School of American Research Press, Santa Fe, NM.

Daniels, S.G.H. 1972. Research design models. In: Clarke, D. L. (ed.), *Models in Archaeology,* pp. 201–229. Methuen, London, UK.

Dean, J.S. 1988. A model of Anasazi behavioral adaptation. In: Gumerman, G.J. (ed.), *The Anasazi in a Changing Environment,* pp. 25–44. Cambridge University Press, Cambridge, UK.

Dean, J.S. 1996. Demography, environment, and subsistence stress. In: Tainter, J.A. and B.B. Tainter (eds.), *Evolving Complexity and Environmental Risk in the Prehistoric Southwest,* pp. 25–78. Addison-Wesley, Reading, MA.

Dean, J.S., R.C. Euler, G.J. Gumerman, F. Plog, R.H. Hevly, and T.N.V. Karlstrom. 1985. Human behavior, demography, and paleoenvironment on the Colorado Plateaus. *American Antiquity* 50: 537–554.

Dennell, R. 1980. The use, abuse, and potential of site catchment analysis. *Anthropology UCLA* 10: 1–20.

Dickson, D.B. 1993. Anthropogenic environmental destruction in prehistory: hard evidence and harsh conclusions. In: Jamieson, R.W., S. Abonyi, and N.A. Mirau (eds.), *Culture and Environment: A Fragile Coexistence,* pp. 289–295. University of Calgary Archaeological Association, Calgary, Canada.

Dobyns, H.F. 1981. *From Fire to Flood: Historic Human Destruction of Sonoran Desert Riverine Oases.* Anthropological Papers No. 20. Ballena Press, Socorro, NM.

Duke, P. 1995. Working through theoretical tensions in contemporary archaeology: a practical attempt from southwestern Colorado. *Journal of Archaeological Method and Theory* 2: 201–229.

Dunnell, R.C. 1986. Fifty years of American archaeology. In: Meltzer, D.J., D.D. Fowler, and J.A. Sabloff (eds.), *American Archaeology Past and Future,* pp. 35–49. Smithsonian Institution Press, Washington D.C.

Dunnell, R.C., and W.S. Dancey. 1983. The siteless survey: a regional scale data collection strategy. In: Schiffer, M.B. (ed.), *Advances in Archaeological Method and Theory,* Vol. 6, pp. 267–287. Academic Press, New York, NY.

Ebert, J.I. 1992. *Distributional Archaeology.* University of New Mexico Press, Albuquerque, NM.

Ellen, R.F. 1977. Ecological models in ethnography and the archaeological analysis of settlement. In: Spriggs, M. (ed.), *Archaeology and Anthropology: Areas of Mutual Interest,* pp. 35–48. Supplementary Series 19. British Archaeological Reports, Oxford, UK.

Ellen, R.F. 1978. Problems and progress in the ethnographic analysis of small-scale human ecosystems. *Man* 13: 290–303.

Ellen, R.F. 1982. *Environment, Subsistence, and System: The Ecology of Small-Scale Social Formations.* Cambridge University Press, Cambridge, UK.

Embree, L. 1987. Archaeology: the most basic science of all. *Antiquity* 61: 75–78.

Euler, R.C., G.J. Gumerman, T.N.V. Karlstrom, J.S. Dean, and R.H. Hevly. 1979. The Colorado Plateaus: cultural dynamics and paleoenvironment. *Science* 205: 1089–1101.

Fagan, B.M. 1991. *Ancient North America: The Archaeology of a Continent.* Thames and Hudson, New York, NY.

Fee, J.M. 1993. Idaho's Chinese mountain gardens. In: Wegars, P. (ed.), *Hidden Heritage: Historical Archaeology of the Overseas Chinese,* pp. 65–96. Baywood Publishing Company, Amityville, NY.

Feldman, D.A. 1975. The history of the relationship between environment and culture in ethnological thought: an overview. *Journal of the History of the Behavioral Sciences* 11: 67–81.

Fish, S.K., and M. Donaldson. 1991. Production and consumption in the archaeological record: a Hohokam example. *Kiva: The Journal of Southwestern Anthropology and History* 56: 255–275.

Fish, S.K., and P.R. Fish. 1990. An archaeological assessment of ecosystems in the Tucson Basin of southern Arizona. In: Moran, E. M. (ed.), *The Ecosystem Approach in Anthropology: From Concept to Practice,* pp. 159–187. University of Michigan Press, Ann Arbor, MI.

Fish, S.K., P.R. Fish, C. Miksicek, and J. Madsen. 1985. Prehistoric agave cultivation in southern Arizona. *Desert Plants* 7: 107–112.

Fowler, D.D., and D.G. Givens. 1992. Preserving the archaeological record. In: Silverman, S. and N.J. Parezo (eds.), *Preserving the Anthropological Record,* pp. 43–52. Wenner-Gren Foundation for Anthropological Research, New York, NY.

Gifford-Gonzalez, D. 1991. Bones are not enough: analogies, knowledge, and interpretive strategies in zooarchaeology. *Journal of Anthropological Archaeology* 10: 215–254.

Glassow, M.A. 1996. *Purisimeño Chumash Prehistory.* Harcourt Brace College Publishers, Fort Worth, TX.

Goldberg, P., D.T. Nash, and M.D. Petraglia (eds.). 1993. *Formation Processes in Archaeological Context.* Prehistory Press, Madison, WI.

Gumerman, G.J. 1988. A historical perspective on environment and culture in Anasazi country. In: Gumerman, G.J. (ed.), *The Anasazi in a Changing Environment,* pp. 1–24. Cambridge University Press, Cambridge, UK.

Hadley, D., and T.E. Sheridan. 1995. Land Use History of the San Rafael Valley, Arizona (1540–1960). Gen. Tech. Rep. RM-GTR-269. USDA Forest Service, Rocky Mountain Forest and Range Experiment Station, Fort Collins, CO.

Hall, S.A. 1990. Holocene landscapes of the San Juan Basin, New Mexico: geomorphic, climatic, and cultural dynamics. In: Lasca, N.P. and J. Donahue (eds.), *Archaeological Geology of North America*, pp. 323–334. Geological Society of America, Boulder, CO.

Hardesty, D.L. 1975. The niche concept: suggestions for its use in human ecology. *Human Ecology* 3: 71–85.

Hardesty, D.L. 1977. *Ecological Anthropology*. Wiley, New York, NY.

Hardesty, D.L. 1988. *The Archaeology of Mining and Miners*. Special Publication No. 6 of the Society for Historical Archaeology, Ann Arbor, MI.

Hardesty, D.L. 1995. The Cultural Resources Legacy of Mining. Paper Presented at the 1995 Annual Conference of the American Society for Environmental History, Las Vegas, NV.

Hardesty, D.L., S. Mehls, and E. Stoner. 1994. Data Recovery at the Riepetown Townsite, White Pine County, Nevada. Report Prepared for Magma Copper Company, Tucson, Arizona, by Western Cultural Resources, Inc., Boulder, CO.

Harner, M.J. 1970. Population pressure and the social evolution of agriculturalists. *Southwestern Journal of Anthropology* 26: 67–86.

Harris, M. 1968. *The Rise of Anthropological Theory*. Thomas Y. Crowell, New York, NY.

Hattori, E., and M.A. Thompson. 1987. Using dendrochronology for historical reconstruction in the Cortez Mining District, North Central Nevada. *Historical Archaeology* 21: 60–73.

Heider, K.G. 1972. Environment, subsistence, and society. In: Siegel, B.J., A.R. Beals, and S.A. Tyler (eds.), *Annual Review of Anthropology*, Vol. 1, pp. 207–226. Annual Reviews, Inc., Palo Alto, CA.

IEMTF (Interagency Ecosystem Management Task Force). 1995. *The Ecosystem Approach: Healthy Ecosystems and Sustainable Economies*. National Technical Information Service, Springfield, VA.

Irwin-Williams, C. 1973. The Oshara Tradition: Origins of Anasazi Culture. *Contributions in Anthropology* 5(1). Eastern New Mexico University, Portales, NM.

Isaac, B.L. 1990. Economy, ecology, and analogy: the Kung San and the generalized foraging model. In: Tankersley, K.B. and B.L. Isaac (eds.), *Early Paleoindian Economies of Eastern North America*, pp. 323–335. JAI Press, Greenwich, CT.

Jensen, M.E. and R. Everett. 1993. An overview of ecosystem management principles. In: Jensen, M.E. and P.S. Bourgeron (eds.), Eastside Forest Ecosystem Health Assessment, Volume II: Ecosystem Management: Principles and Applications, pp. 9–18. Gen. Tech. Rep. PNW-GTR-318. USDA Forest Service, Pacific Northwest Research Station, Portland, OR.

Jochim, M. 1990. The ecosystem concept in archaeology. In: Moran, E. M. (ed.), *The Ecosystem Approach in Anthropology: From Concept to Practice*, pp. 75–90. University of Michigan Press, Ann Arbor, MI.

Kates, R.W., B.L. Turner II, and W.C. Clark. 1990. The great transformation. In: Turner, B.L., II, W.C. Clark, R.W. Kates, J.F. Richards, J.T. Mathews, and W.B. Meyer (eds.), *The Earth as Transformed by Human Action*, pp. 1–17. Cambridge University Press, Cambridge, UK.

Kay, C.E. 1994. Aboriginal overkill: the role of Native Americans in structuring western ecosystems. *Human Nature* 8: 359–398.

Kay, C.E. 1995. Aboriginal overkill and native burning: implications for modern ecosystem management. Paper presented at the 8th George Wright Society Conference on Research and Resource Management on Public Lands. Portland, OR.

Kelley, J.H., and M.P. Hanen. 1988. *Archaeology and the Methodology of Science*. University of New Mexico Press, Albuquerque, NM.

Kelso, G.K. 1993. Pollen-record formation processes, interdisciplinary archaeology, and land use by mill workers and managers: the Boott Mills Corporation, Lowell, Massachusetts, 1836–1942. *Historical Archaeology* 27: 70–94.

Keter, T.S. 1995. Environmental History and Cultural Ecology of the North Fork of the Eel River Basin, California. R5-EM-TP-002. USDA Forest Service, Pacific Southwest Region.

Kintigh, K.W. 1988. The effectiveness of subsurface testing: a simulation approach. *American Antiquity* 53: 686–707.

Kirch, P.V. 1992. *The Archaeology of History, Vol. 2, Anahulu: the Anthropology of History in the Kingdom of Hawaii*. University of Chicago Press, Chicago, IL.

Kluckhohn, C. 1940. Conceptual structure in Middle American studies. In: Hay, C.L., R.L. Hinton, S.K. Lothrop, A.L. Shapiro, and G.C. Vaillant (eds.), *The Maya and Their Neighbors*, pp. 41–51. Appleton, New York, NY.

Kohler, T.A. 1992. Prehistoric human impact on the environment in the upland North American Southwest. *Population and Environment* 13: 255–268.

Kohler, T.A., and M.H. Matthews. 1988. Long-tern Anasazi land use and forest reduction: a case study from southwestern Colorado. *American Antiquity* 53: 537–564.

Kvamme, K.L. 1998. Spatial Structure in Mass Debitage Scatters. In: Sullivan, A.P., III (ed.), *Surface Archaeology*, pp. 127–141. University of New Mexico Press, Albuquerque, NM.

Lanner, R.M. 1981. *The Piñon Pine: A Natural and Cultural History*. University of Nevada Press, Reno, NV.

Lefebvre, H. 1979. Space: social product and use value. In: Freiberg, J.W. (ed.), *Critical Sociology: European Perspectives*, pp. 285–295. Irvington, NY.

Lewis, H.T. 1973. Patterns of Indian Burning in California: Ecology and Ethnohistory. *Anthropological Papers* No. 1. Ballena Press, Ramona, CA.

Lipe, W.D. 1974. A conservation model for American archaeology. *The Kiva* 39: 213–245.

Lipe, W.D. 1995. The archeology of ecology: taking the long view. *Federal Archeology* (Spring): 8–13.

Luchterhand, K. 1970. Early Archaic Projectile Points and Hunting Patterns in the Lower Illinois River Valley. *Illinois Valley Archaeological Program Research Papers* 3. Illinois State Museum, Springfield, IL.

Lundquist, J.E., B.W. Geils, and J.F. Negron. 1995. Integrating applications for understanding the effects of small-scale disturbances in forest ecosystems. In: Thompson, J.E. (comp.), *Analysis in Support of Ecosystem Management*, pp. 77–86. USDA Forest Service, Ecosystem Management Analysis Center, Washington D.C.

Manley, P.N., G.E. Brogan, C. Cook, M.E. Flores, D.G. Fullmer, S. Husari, T.M. Jimerson, L.M. Lux, M.E. McCain, J.A. Rose, G. Schmitt, J.C. Schuyler, and M.J. Skinner. 1995.

*Sustaining Ecosystems: A Conceptual Framework*. USDA Forest Service, Pacific Southwest Region, San Francisco, CA.
Marquardt, W.H. 1992. Dialectical archaeology. In: Schiffer, M.B. (ed.), *Archaeological Method and Theory*, Vol. 4, pp. 101–140. University of Arizona Press, Tucson, AZ.
Martin, P.S., and R.G. Klein (eds.). 1984. *Quaternary Extinctions: A Prehistoric Revolution*. University of Arizona Press, Tucson, AZ.
Martinez, D. 1993. Managing a precarious balance: wilderness versus sustainable forestry. *Winds of Change* (Summer): 23–28.
Marx, L. 1964. *The Machine in the Garden*. Oxford University Press, New York, NY.
Matson, R.G. 1991. *The Origins of Southwestern Agriculture*. University of Arizona Press, Tucson, AZ.
McDowell, P.F., T. Webb III, and P.J. Bartlein. 1990. Long-term environmental change. In: Turner, B.L., II, W.C. Clark, R.W. Kates, J.F. Richards, J.T. Mathews, W.B. Meyer (eds.), *The Earth as Transformed by Human Action*, pp. 143–162. Cambridge University Press, Cambridge, UK.
McGlade, J. 1995. Archaeology and the ecodynamics of human-modified landscapes. *Antiquity* 69: 113–132.
McManamon, F. P. 1984. Discovering sites unseen. In: Schiffer, M. B. (ed.), *Advances in Archaeological Method and Theory*, Vol. 7, pp. 223–292. Academic Press, New York, NY.
Mellars, P. 1976. Fire ecology, animal populations, and man: a study of some ecological relationships in prehistory. *Proceedings of the Prehistoric Society* 42: 15–45.
Miksicek, C.H. 1987. Formation processes of the archaeobotanical record. In: Schiffer, M.B. (ed.), *Advances in Archaeological Method and Theory*, Vol. 10, pp. 211–247. Academic Press, New York, NY.
Moran, E.F. 1990. Ecosystem ecology in biology and anthropology: a critical assessment. In: Moran, E.M. (ed.), *The Ecosystem Approach in Anthropology: From Concept to Practice*, pp. 3–40. University of Michigan Press, Ann Arbor, MI.
Morrison, K.D. 1994. Monitoring regional fire history through size-specific analysis of microscopic charcoal: the last 600 years in south India. *Journal of Archaeological Science* 21: 675–685.
Nash, D.T., and M.D. Petraglia (eds.). 1987. Natural Formation Processes and the Archaeological Record. *BAR International Series* 352. Oxford, UK.
O'Brien, M.J. 1989. Sedentism, population growth, and resource selection in the Woodland Midwest: a review of coevolutionary developments. *Current Anthropology* 28: 177–198.
Orser, C.E., Jr. 1996. *A Historical Archaeology of the Modern World*. Plenum Press, New York, NY.
Patrik, L.E. 1985. Is there an archaeological record? In: Schiffer, M.B. (ed.), *Advances in Archaeological Method and Theory*, pp. 27–62. Academic Press, New York, NY.
Periman, R.D. 1995. Historic Preservation and Management Plan for Historic Mining and Associated Properties, Deerlodge National Forest. *Studies in Heritage Management 14*. USDA Forest Service, Northern Region, Missoula, MT.
Pfaffenberger, B. 1992. The social anthropology of technology. *Annual Review of Anthropology* 21: 491–516. Annual Reviews, Inc., Palo Alto, CA.
Plog, F., G.J. Gumerman, R.C. Euler, J.S. Dean, R.H. Hevly, and T.V.N. Karlstrom. 1988. Anasazi adaptive strategies: the model, predictions, and results. In: Gumerman, G.J. (ed.), *The Anasazi in a Changing Environment*, pp. 230–276. Cambridge University Press, Cambridge, UK.
Plog, F., and J.A. Hill. 1971. Explaining variability in the distribution of sites. In: Gumerman, G.J. (ed.), *The Distribution of Prehistoric Population Aggregates*, pp. 7–36. Prescott College Press, Prescott, AZ.
Plog, S., F. Plog, and W. Wait. 1978. Decision Making in modern surveys. In: Schiffer, M.B. (ed.), *Advances in Archaeological Method and Theory*, Vol. 1, pp. 383–421. Academic Press, New York, NY.
Preucel, R.W., and I. Hodder (eds.). 1996. *Contemporary Archaeology in Theory*. Blackwell Publishers, Oxford, UK.
Ramenofsky, A.F. 1987. *Vectors of Death: The Archaeology of European Contact*. University of New Mexico Press, Albuquerque, NM.
Reinhart, T.R. 1967. Late Archaic Cultures of the Middle Rio Grande Valley, New Mexico: a Study of the Process of Culture Change. Ph.D. dissertation, University of New Mexico. University Microfilms, Ann Arbor, MI.
Roberts, R.W. 1987. Landscape archaeology. In: Wagstaff, S.M. (ed.), *Landscape and Culture: Geographical and Archaeological Perspectives*, pp. 77–95. Basil Blackwell, Oxford, UK.
Roper, D.C. 1979. The method and theory of site catchment analysis: a review. In: Schiffer, M.B. (ed.), *Advances in Archaeological Method and Theory*, Vol. 2, pp. 119–140. Academic Press, New York, NY.
Rossignol, J. and L. Wandsnider (eds.). 1992. *Space, Time, and Archaeological Landscapes*. Plenum Press, New York, NY.
Sabloff, J.A., L.R. Binford, and P.A. McAnany. 1987. Understanding the archaeological record. *Antiquity* 61: 203–209.
SARG (The Members of). 1974. SARG: A cooperative approach towards understanding the locations of human settlement. *World Archaeology* 6: 107–116.
Sauer, C.O. 1975. Man's dominance by use of fire. *Geosciences and Man* 10: 1–13.
Savage, M. 1991. Structural dynamics of a Southwestern pine forest under chronic human influence. *Annals of the Association of American Geographers* 81: 271–289.
Schiffer, M.B. 1987. *Formation Processes of the Archaeological Record*. University of New Mexico Press, Albuquerque, NM.
Schiffer, M.B. 1988. The structure of archaeological theory. *American Antiquity* 53: 461–485.
Schiffer, M.B., A.P. Sullivan, and T.C. Klinger. 1978. The design of archaeological surveys. *World Archaeology* 10: 1–28.
Schoenwetter, J., and A.E. Dittert. 1968. An ecological interpretation of Anasazi settlement patterns. In: Meggers, B.J. (ed.), *Anthropological Archaeology in the Americas*, pp. 41–66. Anthropological Society of Washington, Washington D.C.
Schwantes, C.A. (ed.). 1992. *Bisbee: Urban Outpost on the Frontier*. University of Arizona Press, Tucson, AZ.
Shackel, P.A. 1994. Interdisciplinary approaches to the meanings and uses of material goods in Lower Tow Harpers Ferry. *Historical Archaeology* 28: 3–15.
Shelley, M.S. 1995. Sources of archaeological obsidian in the Greater American Southwest: an update and quantitative analysis. *American Antiquity* 60: 531–551.
Simmons, I.G. 1989. *Changing the Face of the Earth*. Basil Blackwell, Oxford, UK.
Smith, D. 1987. *Mining America*. University of Kansas Press, Lawrence, KS.
Smith, K.L. 1986. *The Magnificent Experiment: Building the Salt*

*River Reclamation Project, 1890–1917*. University of Arizona Press, Tucson, AZ.

Smith, M.R. 1977. *Harpers Ferry Armory and the New Technology: The Challenge of Change*. Cornell University Press, Ithaca, NY.

Spoerl, P.M., and J.C. Ravesloot. 1994. From Casas Grandes to Casa Grande: prehistoric human impacts in the sky islands of southern Arizona and northwestern Mexico. In: DeBano, L.F., G.J. Gottfried, R.H. Hamre, C.B. Edminster, P.F. Ffolliott, and A. Ortega-Rubio (tech. coords.), Biodiversity and Management of the Madrean Archipelago: The Sky Islands of Southwestern United States and Northwestern Mexico, pp. 492–501. Gen. Tech. Rep. RM-GTR-264. USDA Forest Service, Rocky Mountain Forest and Range Experiment Station, Fort Collins, CO.

Steward, J.H. 1938. Basin-Plateau Aboriginal Sociopolitical Groups. *Bulletin 120*. Bureau of American Ethnology, Washington D.C.

Struever, S. 1968. Woodland subsistence-settlement systems in the lower Illinois Valley. In: Binford, S.R. and L.R. Binford (eds.), *New Perspectives in Archaeology*, pp. 285–312. Aldine, Chicago, IL.

Sullivan, A.P., III. 1978. Inference and evidence in archaeology: a discussion of the conceptual problems. In: Schiffer, M. B. (ed.), *Advances in Archaeological Method and Theory*, Vol. 1, pp. 183–222. Academic Press, New York, NY.

Sullivan, A.P., III. 1987a. Probing the sources of lithic assemblage variability: a regional case study near the Homolovi Ruins, Arizona. *North American Archaeologist* 8: 41–71.

Sullivan, A.P., III. 1987b. Artifact scatters, adaptive diversity, and Southwestern abandonment: the Upham hypothesis reconsidered. *Journal of Anthropological Research* 43: 345–360.

Sullivan, A.P., III. 1992a. Pinyon nuts and other wild resources in Western Anasazi subsistence economies. In: Croes, D.R., R.A. Hawkins, and B.L. Isaac (eds.), *Long-Term Subsistence Change in Prehistoric North America*, pp. 195–240. JAI Press, Greenwich, CT.

Sullivan, A.P. 1992b. The role of theory in solving enduring archaeological problems. In: Wandsnider, L. (ed.), *Quandaries and Quests: Visions of Archaeology's Future*, pp. 239–253. Southern Illinois University Press, Carbondale, IL.

Sullivan, A.P., III. 1995a. Artifact scatters and subsistence organization. *Journal of Field Archaeology* 22: 49–64.

Sullivan, A.P., III. 1995b. Behavioral archaeology and the interpretation of archaeological variability. In: Skibo, J.M., A.E. Nielsen, and W.H. Walker (eds.), *Expanding Archaeology*, pp. 178–186. University of Utah Press, Salt Lake City, UT.

Sullivan, A.P., III. 1996. Risk, anthropogenic environments, and Western Anasazi subsistence. In: Tainter, J.A. and B.B. Tainter (eds.), *Evolving Complexity and Environmental Risk in the Prehistoric Southwest*, pp. 145–167. Addison-Wesley, Reading, MA.

Sullivan, A.P., III. in press. Theory of archaeological survey design. In: Ellis, L. (ed.), *Archaeological Method and Theory: An Encyclopedia*. Garland Publishing, New York, NY.

Sullivan, A.P., III, and C.E. Downum. 1991. Aridity, activity, and volcanic ash agriculture: a study of short-term prehistoric cultural-ecological dynamics. *World Archaeology* 22: 271–287.

Tainter, J.A. 1979. The Mountainair lithic scatters: settlement patterns and significance evaluation of low density surface sites. *Journal of Field Archaeology* 6: 463–469.

Tainter, J.A. 1983. Settlement behavior and the archaeological record: concepts for the definition of archaeological site. *Contract Abstracts and CRM Archeology* 3: 130–133.

Tainter, J.A. 1987. Cultural resources management in the United States Forest Service. In: Johnson, R.W. and M.G. Schene (eds.), *Cultural Resources Management*, pp. 49–71. Krieger Publishing Co., Malabar, CA.

Tainter, J.A. 1988. *The Collapse of Complex Societies*. Cambridge University Press, Cambridge, UK.

Tainter, J.A. 1995. Sustainability of complex societies. *Futures* 27: 397–407.

Tainter, J.A. 1998. Surface archaeology: perceptions, values, and potential. In: Sullivan, A.P., III (ed.), *Surface Archaeology*. pp. 169–179 University of New Mexico Press, Albuquerque, NM.

Tainter, J.A., and D.A. Gillio. 1980. Cultural Resources Overview, Mt. Taylor Area, New Mexico. USDA Forest Service, Southwestern Regional Office and USDI Bureau of Land Management, New Mexico State Office, Albuquerque and Santa Fe, NM.

Tainter, J.A., and R.H. Hamre (eds.). 1988. Tools to Manage the Past: Research Priorities for Cultural Resources Management in the Southwest. Gen. Tech. Rep. RM-164. USDA Forest Service, Rocky Mountain Range and Experiment Station, Fort Collins, CO.

Tainter, J.A., and F. Plog. 1994. Strong and weak patterning in Southwestern prehistory: the formation of puebloan archaeology. In: Gumerman, G. J. (ed.), *Themes in Southwest Prehistory*, pp. 165–181. School of American Research Press, Santa Fe, NM.

Tainter, J.A., and B.B. Tainter. 1996. Riverine settlement in the evolution of prehistoric land-use systems in the middle Rio Grande Valley, New Mexico. In: Finch, D.M. and D.W. Shaw (comps.), Desired Future Conditions for Southwestern Riparian Ecosystems: Bringing Interests and Concerns Together, pp. 22–32. Gen. Tech. Rep. RM-272. USDA Forest Service, Rocky Mountain Forest and Range Experiment Station, Fort Collins, CO.

Thomas, D.H. 1971. Prehistoric Subsistence-Settlement Patterns of the Reese River Valley, Central Nevada. Ph.D. dissertation, University of California, Davis. University Microfilms, Ann Arbor, Michigan, MI.

Thomas, D.H. 1972. A computer simulation model of Great Basin Shoshonean subsistence and settlement patterns. In: Clarke, D. L. (ed.), *Models in Archaeology*, pp. 671–704. Methuen, London, UK.

Thomas, D.H. 1973. An empirical test for Steward's model of Great Basin settlement patterns. *American Antiquity* 38: 155–176.

Vayda, A.P. 1969. Introduction. In: Vayda, A. P. (ed.), *Environment and Cultural Behavior*, pp. xi–xvii. Natural History Press, Garden City, NY.

Vierra, B.J. (ed.). 1994. Archaic Hunter-Gatherer Archaeology in the American Southwest. *Contributions in Anthropology* 13. Eastern New University, Portales, NM.

Wagstaff, S.M. (ed.). 1987. *Landscape and Culture: Geographical and Archaeological Perspectives*. Basil Blackwood, Oxford, UK.

Wait, W.K. 1983. Alternate approaches to the analysis of low-density artifact scatters. In: Wait, W.K. and B.A. Nelson (eds.), *The Star Lake Archaeological Project*, pp. 59–94. Southern Illinois University Press, Carbondale, IL.

Wallerstein, I. 1974. *The Modern World-System*. Academic Press, New York, NY.

Waters, M.R. 1991. The geoarchaeology of gullies and arroyos in southern Arizona. *Journal of Field Archaeology* 18: 141–159.

Wandsnider, L. 1998. Landscape Element Configuration, Lifespace, and Occupation History: Ethnoarchaeological Observations and Archaeological Applications. In: Sullivan, A.P., III (ed.), *Surface Archaeology*, pp. 21–39. University of New Mexico Press, Albuquerque, NM.

Watson, P.J. 1973. The future of archaeology in anthropology: culture history and social science. In: Redman, C.L. (ed.), *Research and Theory in Current Archaeology*, pp. 113–124. Wiley, New York, NY.

Weins, J.A., J.F. Addicott, T.J. Case, and J. Diamond. 1986. Overview: the importance of spatial and temporal scales in ecological investigations. In: Diamond, J. and T. J. Case (eds.), *Community Ecology*, pp. 145–153. Harper and Row, New York, NY.

Wendorf, F. 1954. A reconstruction of northern Rio Grande prehistory. *American Anthropologist* 56: 200–227.

Wendorf, F., and E. Reed. 1955. An alternative reconstruction of northern Rio Grande Prehistory. *El Palacio* 62 (5–6): 131–173.

Williams, J. 1992. The archaeology of underdevelopment and the military frontier of New Spain. *Historical Archaeology* 26: 7–21.

Winters, H.D. 1969. The Riverton Culture: A Second Millennium Occupation in the Central Wabash Valley. *Reports of Investigations* 13. Illinois State Museum, Springfield, IL.

Wolf, E. 1982. *Europe and the People without History*. University of California Press, Berkeley, CA.

Wright, H.A., and A.W. Bailey. 1982. *Fire Ecology*. Wiley, New York, NY.

Wright, K.E., L.M. Chapman, and T.M. Jimerson. 1995. Using historic range of vegetation variability to develop desired conditions and model forest plan alternatives. In: Thompson, J.E. (comp.), *Analysis in Support of Ecosystem Management*, pp. 258–266. USDA Forest Service Ecosystem Management Analysis Center, Washington D.C.

Yellen, J.E. 1977. *Archaeological Approaches to the Present: Models for Reconstructing the Past*. Academic Press, New York, NY.

Young, J.A., and J.D. Budy. 1979. Nevada's pinyon-juniper woodlands. *Journal of Forest History* 23: 113–121.

## THE AUTHORS

**Alan P. Sullivan, III**
*Department of Anthropology*
*University of Cincinnati*
*PO Box 210380*
*Cincinnati, OH 45221-0380, USA*

**Joseph A. Tainter**
*U.S. Forest Service*
*Rocky Mountain Research Station*
*2205 Columbia SE*
*Albuquerque, NM 87106, USA*

**Donald L. Hardesty**
*Department of Anthropology (096)*
*University of Nevada*
*Reno, NV 89557, USA*

# The Historical Foundation and Evolving Context for Natural Resource Management on Federal Lands

Douglas W. MacCleery and Dennis C. Le Master

**Key questions addressed in this chapter**

- *How has the national conservation policy framework evolved in the U.S. since 1900; how well has this policy framework performed over time; and how have federal lands and their historic uses and management factored into the conservation policy framework?*
- *How did ecosystem management evolve on federal lands over the last two decades and how does it compare to previous approaches to federal land management?*
- *What are some of the limitations and shortcomings of ecosystem management as it is currently being practiced on federal lands and what specific actions and approaches might be taken that would contribute to strengthening its application?*
- *What are some of the new and emerging ideas for integrating ecological and social objectives on federal lands that could contribute positively to ecological protection while at the same time meeting human economic and material needs in local communities and nationally?*

**Keywords: Multiple-use management, resource production, commodities, consumption, ecosystem services, sustainability**

## 1 INTRODUCTION

Increasing human demand for natural resources and the consequent effects on the ecology of Planet Earth are matters of continuing concern and controversy. The significant question is, can the physical demands, psychological needs, and preferences of a growing human population be provided, while still maintaining the structure and functioning of forested ecosystems?

Ecosystem management, which has also been referred to as an "ecological approach" to federal land and resource management, has emerged as the dominant management strategy for federal lands. Like multiple use–sustained yield management before it, ecosystem management reflects an official recognition of what was already occurring at the field level. Like multiple use–sustained yield before it, ecosystem management has caused a major dialogue on what it means.

For the purposes of this chapter, ecosystem management is a strategy intended for federal lands. Even so, non-federal lands both affect and are affected by what happens on federal lands. On the one hand, the level of resource outputs produced on the one-third of the U.S. land base that is federally administered affects the nature and extent of resources produced on other lands. On the other, the management flexibility available to federal managers can be substantially circumscribed by how adjacent non-federal lands are managed. The spatial scales implied by ecosystem management may have direct implications for non-federal lands as well.

Management of federal lands, particularly in the West, has been locked in public controversy over the past two decades. A key question is whether a working social consensus can be developed on how these lands will be managed. The real test is not whether ecosystem management is a scientifically sound concept, although it must be so to be successful. The real test is whether it can be effective in reducing polarization and assist in developing a working consensus on the objectives of federal land management. To the extent that ecosystem management can do so, it will be successful.

The assigned topic of this chapter is broad. Successfully addressing it within the allotted space requires fitting ecosystem management into the existing overarching framework, a challenging task indeed. Key topics in this chapter are:

- Ecosystem management as it is currently perceived, and whether ecosystem management is evolutionary or revolutionary.
- Some general principles or axioms fundamental to ecosystem management.
- The evolving use and management of U.S. natural resources, the rise of the conservation movement of the 1880s, the environmental movement of the 1960s and 1970s, and the evolution of federal land management in response to recent U.S. economic and social conditions and demands.
- The lessons and implications of the historical record and the performance of conservation policies that have been put in place since 1900.
- Relevant factors and issues relating to producing and consuming goods and services in a broad context, and their relevance to implementing ecosystem management.
- Some of the limitations and short-comings that must be addressed if ecosystem management is to be successful.
- The need for performance measures under ecosystem management.
- The forest health issue in the western United States as a possible litmus test for whether ecosystem management will survive as an operational concept.
- A case study of a landscape approach to producing goods and services on a mixed ownership landscape in western Washington.
- Whether "new ecological thinking" will finally replace the "old ecological thinking" and pose questions on the implications to natural resources policy associated with the change.
- The need for more holistic connections between resource production, consumption, and sustainable natural and human communities.

## 2 ECOSYSTEM MANAGEMENT: EVOLUTION OR REVOLUTION?

Ecosystem management emerged from the management of National Forest System and Bureau of Land Management (BLM) lands under the multiple use–sustained yield strategy. Ecosystem management adds to the number of uses and values for which federal land is purposely managed. To manage for this broader range of uses and values (such as wildlife species having large home ranges), it also gives more emphasis to spatial scales than was the norm in the past.

Like multiple use–sustained yield management, ecosystem management is a means to an end, rather than an end in itself, under which human needs are met while maintaining the health and productivity of ecosystems. But ecosystem management differs in two major respects from multiple use–sustained yield management: (1) it expands the objectives for which federal lands are purposely managed from a relatively few to a broader spectrum of values, uses, and environmental services, and (2) it requires consideration of social, biological, and economic interactions at a variety of spatial scales (local, regional, national, international) and over time.

## 2.1 Evolution of Ecosystem Management

Ecosystem management evolved from multiple use–sustained yield management. Even before the official adoption of ecosystem management by most federal land management agencies in the early 1990s, many of the basic concepts had been developed and implemented to varying degrees by many agencies. Beginning in the early 1970s, research by the Forest Service and a variety of cooperators on the seasonal movements and habitat requirements of Rocky Mountain elk in relation to logging roads led to better understanding of such interrelationships and to the development of management guidelines to reduce conflicts that encompassed multi-ownerships and political jurisdictions (Lyon et al. 1985). In 1979, a groundbreaking publication edited by Jack Ward Thomas provided forest managers an insightful and systematic approach for integrating wildlife management objectives and timber management activities on large ownerships (Thomas 1979). The early 1980s saw the establishment of the Interagency Grizzly Bear Committee, whose objective was to coordinate efforts of the Park Service, Forest Service, Fish and Wildlife Service, and relevant state fish and game agencies to protect the threatened grizzly bear. A similar effort aimed at coordination of federal, state, and local activities was brought together under the Greater Yellowstone Coordinating Committee, which was activated in the mid-1980s.

Because ecosystem management has greatly expanded the objectives for which federal lands are purposely managed, the effect has been a substantial de-emphasis on the role of federal lands in producing commodity outputs. The focus of multiple use–sustained yield has tended to be on resource outputs or "flows," whereas ecosystem management places relatively more emphasis on ecosystem "states" and "conditions" (Grumbine 1994). A corollary to this view is that resource outputs under ecosystem management are often a consequence of achieving biodiversity or other ecosystem-centered objectives, rather than explicit objectives in their own right.

This shift in focus constitutes a substantial change from the past and reflects both a change in the U.S. economy and a greatly expanded interest of the public in how federal lands are managed. It reflects, as well, a substantial reduction in the political influence in the United States of those sectors of the economy dependent upon flows of natural resources.

If a society believes economic growth and vitality are based upon the production and use of natural resources, then maintaining a high-level and sustainable output of such resources is an appropriate and viable strategy. But if a society believes ideas, information services, and other non-natural resource based sectors are the basis for economic growth, then the natural environment becomes a backdrop for such activities. In the latter situation, maintaining the status quo and avoiding changes in the environmental "platform" or backdrop becomes a major focus of attention.

This is the transition that seems to be in progress in the United States. The reality implicit in this transition is that commodities once produced on the federal lands will now be produced elsewhere to a larger extent. Until now, there has been scant consideration given to the environmental consequences on non-federal land associated with this transition.

## 2.2 Comparison of Ecosystem Management and Multiple Use–Sustained Yield Management Strategies

There are similarities as well as differences in ecosystem management as compared to multiple use–sustained yield management. What follows is a chart that attempts to summarize both similarities and differences (Table 1). It is divided between goals or ends and processes or means and, in turn, among three categories of science: biological, socioeconomic, and management. Related features of both management strategies are paired to the extent practicable. Differences in features are in italics, similarities are in boldface.

Ecosystem management is a strategy whose goals for federal lands explicitly include biological diversity, ecosystem and resource sustainability, ecosystem health or integrity, social responsiveness and acceptability, and risk aversion. The processes or means by which these goals are to be accomplished include: (1) a systems approach and a landscape perspective, (2) multiple scales of management, both spatially and temporally, (3) extensive use of public and private partnerships and public involvement in natural resource decision making, (4) multidisciplinary, multi-agency organizational and managerial approaches, and (5) systematic monitoring and adaptive management, that is, adjusting management techniques and technologies on the basis of systematic monitoring and observation to improve management and future planning.

Table 1. Comparison of ecosystem management with multiple use–sustained yield management ( differences in features are in italics, similarities are in boldface.)

| Ecosystem Management | Multiple Use–Sustained Yield Management |
|---|---|
| GOALS OR ENDS | |
| 1. Biological | 1. Biological |
| *Biological diversity* | *Maintain habitats for featured species* |
| **Ecosystem and resource sustainability** | **Sustainability of renewable resources, i.e. achievement and maintenance in perpetuity of high-level outputs of renewable resources without impairment of the productivity of the land** |
| *Ecosystem health or integrity* | *Resource productivity* |
| **Production of resource outputs that "best meet the needs of the American People"** | **Production of resource outputs that "best meet the needs of the American People"** |
| *Integrated management* | *Individual or multi-resource management* |
| 2. Socioeconomic | 2. Socioeconomic |
| *Extension of internal agency resources through public and private partnerships (collaboration) and interagency cooperation* | *No directly comparable goal, but partnerships used as means to achieve programmatic objectives.* |
| *Social responsiveness* | *Commuity stability* |
| **Political acceptability** | **Political acceptability** |
| 3. Management | 3. Management |
| *Risk minimization or aversion* | *Nothing comparable* |
| PROCESSES OR MEANS | |
| 1. Biological | 1. Biological |
| *System, integrated, or holistic approach* | *Multiple-use management or site-by-site consideration of both commodity and non-commodity resources* |
| *Nothing comparable* | *Management strategies organized in terms of individual resources* |
| **Long-term temporal perspective** | **Sustained-yield management** |
| *Landscape perspective* | *Site-specific perspective* |
| *Multiple scales of management* | *Stand-level management focus* |
| *Management within the range of natural variation* | *Management toward efficient production in terms of management objectives and the capabilities of the individual site* |
| *Species presence or populations as indicators* | *Resource outputs and inventories used as indicators* |
| 2. Socioeconomic | 2. Socioeconomic |
| *Ample use of public and private partnerships to accomplish programmatic objectives* | *Sparse use of partnerships to achieve specific programmatic objectives, e.g. fire suppression, insect and disease control, etc.* |
| **Systematic public involvement** | **Public involvement necessary to resolve site-specific issues** |
| **Systematic interagency coordination** | **Interagency coordination sufficient to resolve site-specific issues** |
| *Multidisciplinary management teams* | *Functional and line-staff organizational arrangements with coordination among functional staffs* |
| 3. Management | 3. Management |
| *Development and use of comprehensive, integrated, long-term data sets* | *Development and use of local resource data sets* |
| *Systematic adaptive management* | *Management responsive to research findings; management plans subject to periodic revision* |
| *Risks of management projects to all resources are recognised and considered within the context of the entire ecosystem* | *Nothing comparable* |

Multiple use–sustained yield management is a strategy whose focus is sustained production of resource outputs, uses, and values "that best meet the needs of the American people." The means by which this would be accomplished include: (1) equal consideration of all resources on a site-by-site basis, (2) management organized in terms of individual resources to be yielded on a sustainable basis, (3) public and private partnerships when necessary to achieve programmatic objectives, (4) public involvement when necessary to address site-specific issues, (5) functional line/staff organizational arrangements, and (6) responsive adjustment in management to research findings.

Clearly, ecosystem management is distinctly different from multiple use–sustained yield management even though, as noted before, the former evolved from and is a logical extension of the latter. The move to ecosystem management by federal agencies has occurred in the absence of explicit statutory authority to do so. Rather, it has been an administrative response resulting from a variety of factors, the most important being the requirements of the Endangered Species Act and court cases that have been brought to enforce it. Keitler et al. (Vol. III of this book) discuss the evolving legal basis for ecosystem management.

Some of the most lively debates over ecosystem management center around the issue of whether it raises protection of ecological integrity to a position of primacy as the overriding objective of federal land management. In other words, under ecosystem management, are resource outputs to meet human needs relegated to a subservient position, i.e. merely a secondary consequence of management to maintain ecological integrity? If the answer to this question is "yes," there probably is not an explicit statutory basis for ecosystem management on lands managed under a multiple use–sustained yield mandate. If the answer is that ecosystem management involves managing federal lands to produce a variety of outputs, uses and values, while at the same time maintaining ecosystem processes and functions so as to protect the long-term sustainability of these systems, then current statutory direction would seem to allow such an interpretation.

## 3 AXIOMATIC FOUNDATION FOR ECOSYSTEM MANAGEMENT

The following axioms provide a foundation for ecosystem management:

1. **We have no choice but to engage the resource.** Like all living creatures, humans must act if they are to survive. In order to feed, cloth and house themselves, humans must take life from other organisms. These actions have environmental, economic, social, and ethical consequences that need to be understood to manage intelligently.
2. **Everything humans have done or will do has a natural context and environmental consequences.** Whether we perceive it or not, humans are embedded in nature. We are a part of nature and subject to its processes. We affect the land and the land affects us. This was true in pre-industrial society and remains so today. People have never been ecologically invisible upon the landscape. While there is no such thing as people outside of nature, some cultural traditions act to reinforce that perception.
3. **All concepts of sustainability are social/cultural constructs.** While humankind is embedded in nature, we view both ourselves and what we call nature through a cultural lens. Human views of nature and what is sustainable vary from culture to culture. They also commonly evolve significantly within a given culture in response to improved knowledge and as economic and social changes result in shifts in social preferences about the specific uses, values, and environmental services that are desired.
4. **Seeking to understand linkages is critical.** The move to an ecological approach to management has increased awareness of the linkages between plant and animal communities interacting with each other and with the physical environment at various spatial and temporal scales. This has been an important contribution of the evolving field of conservation biology. Somewhat paradoxically, however, some applications of ecosystem management have ignored human dimension linkages — social and economic connections which profoundly affect ecosystems. If humans act or don't act in one place, it can have substantial effects on the way they act in another. All relevant linkages must be considered for a management strategy to be effective.

   An example is the reduction of timber harvests on federal lands in the Pacific Northwest. This action did not eliminate the ecological effects of timber harvesting. It transferred them in varying degrees to forested ecosystems somewhere else — to private forests in the Pacific Northwest, to public forests in Canada, or to private forests in the southeastern United States. It also resulted in higher consumer prices for wood products and increased use of wood substitutes, such as steel framing, virtually all of which require substantially more energy to produce than does wood.

These human dimension linkages, and their associated environmental and economic effects, have been largely ignored in the ecosystem management literature.

5. **Absolute predictability is impossible.** Nature, society, and human economies are extremely complex systems. Cause and effect relationships follow many pathways in each system. Former Forest Service Chief Jack Ward Thomas has made famous Frank Egler's quote that: "nature is not more complex than we think, but more complex than we can think" (Egler 1977). While predictability is difficult for natural systems, the same can be said about economic and social systems. When economic, social and natural systems interact, uncertainty dramatically increases, making reliable prediction of outcomes most difficult.

   Understanding this difficulty should introduce a strong dose of humility in what we do and reduce any arrogance that might exist that we can predict precisely what is likely to happen.

   The difficulty of predictability is often used as a reason to postpone federal management actions — what might be called the "before we can do anything, we must know everything" syndrome. It is clear that demands for further study are often used as a surrogate for lack of agreement on the proposed management action.

   But failing to act can exact its price as well. Ecosystems change. In some areas, forest fuels continue to build up beyond levels experienced in those ecosystems for over a century, increasing susceptibility to drought and catastrophic wildfire. Failing to act also exacts its toll on local human communities and on ecosystems elsewhere that must take up the slack to meet human resource needs while we study federal ecosystems.

   A systematic approach to adaptive management can both reduce the risk of acting in the face of imperfect knowledge and provide a feedback loop to learn from management experience.

6. **The past has much to tell us, if only we will listen.** There is much to learn from the study of our own past, as well as that of other cultures, that is useful in shaping viable public policies. Much writing of human history has tended to ignore the critical role that nature has played in shaping history and human values. Human history has been written almost like a Shakespearean play: the focus is on the actors, mostly those who are rich and powerful. Often little attention is paid to how the stage affected the play or to how the actors affected the stage.

   In contrast, some of the natural sciences have largely ignored the role of people in shaping natural systems. Much of the literature of the natural sciences might be compared to a coffee table nature picture book: lots of beautiful photographs of natural landscapes, but with no people in them. This ignores the fact that, for tens of thousands of years, virtually every natural system in the world has been substantially affected by people. Even aboriginal hunter/gatherer societies are known to have profoundly affected natural systems (see Bonnicksen et al., this volume; Periman et al., this volume; and Sullivan et al., this volume; for descriptions of pre-European settlement human ecological impacts in North America).

   Today's debate over natural resources is often driven by the polar extremes — by the so-called "anthropocentric" fringe, on the one hand, versus the "biocentric" fringe on the other. While these two extremes, the "human first" versus the "nature first" poles, may seem worlds apart, they are both based upon a common theme and foundation that is fundamentally flawed: that nature and humans are separate and separable. The false dichotomy between views of the natural and human worlds must somehow be bridged. Both views are excessively narrow and create a polarization, based on ideology, that is very difficult to bridge in the arena of federal land management. The emerging fields of environmental history, human and cultural ecology, and others are helping to begin to bridge this gap. If ecosystem management is to be successful, it must seek to do so as well.

7. **Neither nature nor human culture are static.** Indeed, they are inherently dynamic, continually changing over time. The idea of a static, steady-state nature has been largely debunked over the last several decades (Botkin 1990). Similarly, there is overwhelming evidence in the historical and archaeological record that successful human societies have had to adjust their land use and management practices continuously in response to changing conditions of nature, human population growth and decline, and the emergence of new ideas and technologies (both developed internally and imported from outside). The popular mythology that most pre-industrial societies lived over long periods in a steady-state "harmony" with the environment has little objective evidence to support it (Kohler 1992a,b). The practice of "adaptive management," a concept which has reemerged under ecosystem management, is at least two million years old.

8. **The way people think about nature affects what they do about it.** Different cultures can use a given technology in strikingly different ways. The environmental impact of the technology will, of course, vary as well. One example would be the failure of the centrally planned economies of the European communist block to adopt pollution control technologies that were readily available and operational.

9. **Sustainability is a journey we will never complete.** The struggle to live in harmony with the earth is unending — a problem that will never be fully solved. Neither is it one single problem. There is no single strategy for sustainable use and management of natural resources. Pursuing a complex of strategies offers the greatest promise.

The foregoing nine axioms are adapted from a variety of sources. An excellent essay entitled "The Uses of Environmental History" by William Cronon in the Fall 1993 issue of *Environmental History Review* provided the ideas for several of them (Cronon 1993).

## 4 EVOLVING USE AND MANAGEMENT OF U.S. LAND AND RESOURCES

Studying how the use and management of U.S. land and resources have evolved over the years can provide important insights on the complex interrelationships between people and nature that are crucial to developing sustainable policies and practices. This important dimension is discussed in this section (for further discussions of shifting human values and expectations, see Bliss; Cordell et al.; Cohen; and Cinnamon et al., in Vol. III of this book).

Forests were by far the predominant land cover in the eastern United States at the time of European settlement. Forests probably covered at least four-fifths of the land area east of the Mississippi River (Williams 1989). The remaining area was occupied by grasslands and prairies maintained by natural and human-set fire and by Indian agricultural fields.

### 4.1 Rise of a National Conservation Movement

During the settlement period, the United States was, in effect, converting its natural capital, in the form of forests, into economic, social, and transportation capital and infrastructure. Although that was considered to be a socially desirable outcome at the time, the effect was a substantial reduction in forest area, a degradation of many of the forests that remained, depletion of many wildlife populations and species, and other environmental effects, such as destructive wildfires and accelerated soil erosion from farms and forests (Trefethen 1975, Williams 1989).

The rapidly deteriorating forest, wildlife, and other environmental conditions of the late 19th century provided the impetus for the first national conservation movement. One of the first priorities of this fledgling movement was a war against market hunting for meat, hides, and feathers. This effort included a campaign to put in place strong state and federal wildlife conservation laws and the professional agencies needed to enforce them. Concurrent with these efforts were actions to reserve public lands for protection and management, e.g., national forests, national parks, and national wildlife refuges (Trefethen 1975). Also included were efforts: (1) to promote and encourage the protection of forests and grasslands, regardless of their ownership, from wildfire; (2) to acquire scientific knowledge on the management of forests and wildlife and on the more efficient utilization of raw materials; and (3) to improve the management and productivity of agricultural lands and forests (of which 70 percent of the latter were privately owned) through research and technical and financial assistance (MacCleery 1992).

In addition to the conservation policies described above, several fortuitous events combined in the early 1900s to reduce substantially human pressures on forests and wildlife. One was the spectacular increase in agricultural productivity, which after the 1930s rose at a rate much greater than population growth. Others were the conversion from wood to fossil fuels and the shift from draft animals to internal combustion engines. These trends greatly reduced the human pressures on forests and other wildlife habitats and populations for food and energy. Figure 1 shows that the area of U.S. cropland stabilized in the 1920s after rising at the rate of population growth from 1800 to 1920. As the area of cropland stabilized, so did the area of forests. Today the United States has about the same area of forest it had in 1920.

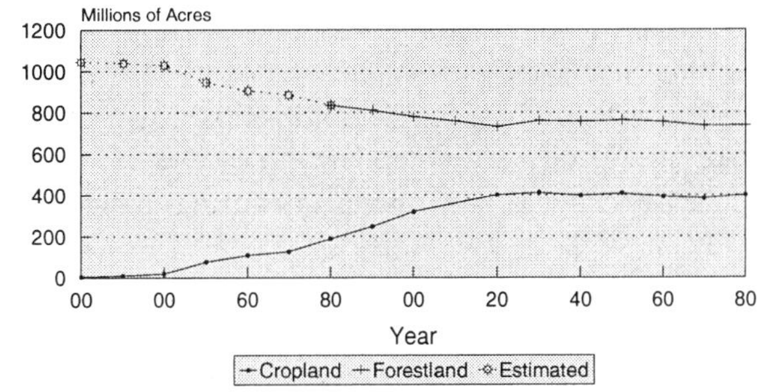


Sources: Fedkiw 1989
USDA/Forest Service 1993a (GTR RM-234)

Fig. 1. Crop and forest land area, 1600–1980.

The policy changes that were put in place, along with conversion to fossil fuels and other fortuitous events, changed profoundly the nature of human impacts on land resources. The United States also had the benefit of a relatively stable and resilient political and social infrastructure, as well as the economic prosperity to make needed changes.

## 4.2 Evolution of Federal Land Management

During the 19th century, more than one billion acres, or over one-half the land area of the United States was transferred from federal to non-federal ownership (Williams 1989). An important element of the conservation policy framework that emerged after 1900 was to retain, rather than dispose of, most of the remaining federal estate. This occurred even though a significant proportion of many western states remained in federal ownership.

### 4.2.1 Consolidation and Custodial Management

The period from 1900 to World War II was in large measure one of consolidation and custodial management for most federal land management agencies. From 1896 to 1910 the area of forest reserves (national forests) rose from 18 to 168 million acres (USDA/Forest Service 1996a). The area of national parks, which was less than 5 million acres in 1900, began to expand rapidly with the establishment of the National Park Service in 1916. By 1920, the National Park System had doubled, but was still mostly in the West, having largely been carved out of former national forest and other public lands. But by 1940, the National Park System included 144 units covering almost 20 million acres, many of which were in the East (DPC 1988).

The primary commercial use of National Forest System and Bureau of Land Management lands during the period before World War II was for livestock grazing. Other than grazing, management of these lands was generally of a custodial nature or was focused on meeting demands for resources primarily in the local area. Efforts to bring livestock numbers down to the carrying capacity of the land was a primary focus of Forest Service and BLM managers during this period (Fedkiw 1989).

Another main focus was seeking to reduce the large area of uncontrolled wildfire that was common prior to the 1930s. Curtailing the 20 to 50 million acres that consistently burned annually across the United States, mostly on private lands, was considered to be a prerequisite to the long-term management of forests and grasslands, public and private. Wildfire prevention and suppression thus became the focus of highly successful cooperative efforts among federal agencies and state and private landowners (Steen 1976).

The work programs of the Great Depression were a stimulus to the planting of trees and the construction of campgrounds, buildings and other facilities on national forests, national parks, and BLM lands, as well as installation of erosion control projects and fire suppression. The current Natural Resource Conservation Service (NRCS) also grew out of "Dust Bowl" years of the Great Depression and focused on cooperative efforts with farmers and ranchers to manage their lands in ways that reduced erosion and stream sedimentation.

While authorized in 1911 by the Weeks Act, acquisition of national forest land in the eastern United States expanded greatly during the Depression years. By 1945, when land acquisition substantially slowed, over 20 million acres of depleted farmsteads and cutover and burned-over woodlands had been incorporated into the eastern national forests.

### 4.2.2 Increased Demands on the Federal Lands After World War II

The period after World War II ushered in a substantial expansion in the demands placed on federal lands for a variety of products and uses. After the war, as millions of service men and women returned home and started families, demand for lumber and other materials for housing rose dramatically. The nation increasingly looked to the national forests and BLM lands in the West to meet that demand. Road access to national forests and BLM lands had improved by the late 1940s and many of the more accessible private lands had been logged to provide wood products for the war effort (Steen 1976).

National forest and BLM timber sale levels increased from 2 to 4 billion board feet in the late 1940s to 11 to 14 billion board feet in the 1960s and beyond. By the 1960s, federal forests were meeting almost 20 percent of the nation's total consumption of wood volume, and over 28 percent of its consumption of softwood sawtimber, the primary source of lumber and plywood for housing (USDA/Forest Service 1993a).

This substantial increase in federal timber harvest not only served to meet a critical national need for timber, it also had the effect of taking pressure off of private forest lands, many of which had been heavily impacted to meet the war effort (Fedkiw 1989).

The 1950s also witnessed a substantial increase in demand for other uses, outputs, and values from the federal lands. An increasingly mobile and affluent population began to look to these lands for outdoor

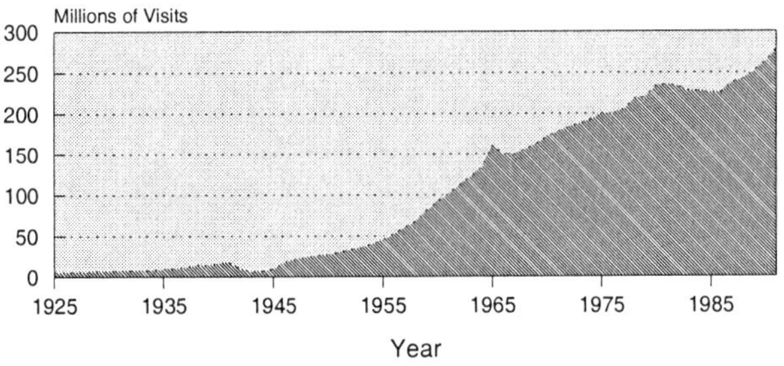


Fig. 2. National forest recreational use 1925–1990. (Source: USDA Forest Service.)

recreation. National forest recreation visitation increased from about 5 million in the early 1920s to 18 million in 1946, 93 million in 1960, and 233 million in 1975 (Census 1975, 1994). National Park visitation increased from 50 million in 1950 to 72 million by 1960, and BLM visitation went from a few million just after WW II to 50 million by 1980. Visitation at state, county, and municipal parks rose even more rapidly than that on federal lands (DPC 1988). Figure 2 illustrates the rise in recreational use of national forest lands.

The increased demands on the federal lands began to be reflected legislatively in the 1960s. The Multiple Use–Sustained Yield Act of 1960 provided that national forests be managed for a variety of uses and values, including outdoor recreation, wildlife, timber, rangeland grazing, and watershed protection. This law largely reflected the uses and management already occurring on these lands.

In 1968, both the Wild and Scenic Rivers Act and the National Trails System Act were passed. These acts created separate systems within which rivers and trails with outstanding scenic, recreational, geologic, cultural, historic, or other values could be designated into national systems (DPC 1988). A Land and Water Conservation Fund was established, financed by oil revenues, to help finance the purchase of land in nationally designated areas.

The 1960s were also a time of growing public controversy over timber harvesting practices on the federal lands. Clearcutting became a particularly controversial practice. The Bitterroot National Forest in Montana, the high elevation national forests in Wyoming, and the Monongahela National Forest in West Virginia, received substantial national attention over the issue.

A growing segment of the public began seeking statutory protection for maintaining federal lands in their "natural" condition. The Wilderness Act, which passed in 1964 after much debate, provided for the designation of significant areas of federal land in their natural and "untrammeled" condition. Most commodity uses were prohibited from these areas. In 1975, legislation was passed to allow designating wilderness in the East (DPC 1988).

In 1974, the Forest and Rangelands Renewable Resources Planning Act (RPA) required the Forest Service to carry out periodic assessments of the national long-term demand and supply situation for all renewable resources, and to lay out a policy and programmatic framework for how agency programs would be structured to address projected resource demands and needs. In 1976, the National Forest Management Act provided detailed guidelines for the management of national forest lands and for the participation of the public in national forest decision-making.

In 1976, the Federal Land Policy and Management Act (FLPMA) gave BLM statutory direction for management of the public lands under multiple use principles. In 1977, the Soil and Water Resources Conservation Act was passed, requiring the Natural Resource Conservation Service (NRCS) to carry out an Appraisal of the condition of the nation's agricultural lands and rangelands and to prepare a Resource Conservation Act Program to guide NRCS activities in response to the Appraisal.

## 4.3 The Environmental Movement of the 1960s and 1970s — A New Agenda

The growing environmental awareness of the 1960s, which had found focus initially on the federal lands, expanded greatly into a general concern over the deterioration of air and water quality and a perceived lack of attention to the environmental and health effects of industrialization. Rachael Carson's *Silent Spring* galvanized public concern over pesticide use. Earth Day 1970 was successful in raising public awareness of environmental issues generally.

Congress responded to these concerns by passing a variety of laws, including the Clean Air Act of 1970; the Clean Water Act of 1972; the National Environmental Policy Act of 1970 (NEPA); amendments to the Federal Insecticide, Fungicide, and Rodenticide Act (FIFRA); the Toxic Substances Control Act of 1976 (TSCA); laws aimed at protecting wetlands; and the Endangered Species Act of 1973 (ESA), to name a few.

One primary focus of the 1970s environmental legislation was to reduce the human health effects of air and water pollution and the use of agricultural pesticides (Clean Water and Clean Air Acts, FIFRA, TSCA). Another focus was on reforming the way decisions affecting the environment were being made by federal agencies. NEPA required federal agencies proposing actions that could have a significant effect on the environment to evaluate a range of alternatives and come to a reasoned choice after providing for public input. Although only a procedural law, no single environmental statute has had so profound an impact on federal decision-making as NEPA.

The Endangered Species Act of 1973 was one of the only laws passed in the 1970s that included a statutory goal for protecting species in jeopardy. It became a powerful tool, which mandated that primacy in federal decision-making be given to endangered species protection, and, by extension, to biodiversity. More than any other law, the ESA is the genesis of the move to ecosystem management. Protection of species with large home ranges virtually mandates an ecosystem approach involving assessments at the scale of multi-ownerships and jurisdictions.

The ESA and the other environmental legislation of the 1970s had a profound influence on the use and management of both federal and non-federal lands. Keiter et al. (see Volume III) describe in detail the legal foundation for ecosystem management.

## 4.4 National Forest System Response to the 1970s Environmental Agenda

The national forests provide an example of the response of one federal agency to the 1970s environmental agenda. Although the focus of much of the environmental legislation in the 1970s was on cleaning up air and water pollution, reducing the risk of pesticides, and similar human health concerns, the response and focus on the national forests was decidedly different. That response can be divided into two separate, but interrelated categories: (1) concerns over land management practices, especially clearcutting, and (2) statutory and administrative allocation of land to special designations that emphasize protection of natural values, recreation, and other uses and that limit or prohibit commodity production.

### 4.4.1 Concerns Over Land Management Practices

The use of clearcutting as a forest management tool increased dramatically on the national forests after World War II. Much of the public concern over national forest land management practices was focused on the visual and ecological effects of clearcutting, and to a lesser degree, on the conversion of hardwood forest types to pine plantations in the southeastern United States Concerns over clearcutting found particular focus on the Bitterroot National Forest in Montana, the high elevation national forests in Wyoming, and the Monongahela National Forest in West Virginia. These controversies led to Congress recommending guidelines for the application of clearcutting on federal lands, and eventually to the passage of the National Forest Management Act of 1976, which provided detailed direction and guidelines for the conduct of national forest planning and timber management practices.

### 4.4.2 Statutory and Administrative Allocation of Land to Designations That Emphasize Protection of Natural Values

A second major thrust of public action was the designation of significant areas of national forest land as statutory wilderness and similar statutory categories that emphasize protection of natural values, recreation, and other uses, and which limit or prohibit commodity production. Beginning with the passage of the Wilderness Act of 1964, this effort gained momentum in the 1970s with the first and second Roadless Area Review and Evaluation programs (RARE I and RARE II), as well as passage of legislation providing for designation of National Recreation Areas, Wild and Scenic Rivers, and similar special areas. By the mid-1980s, Congress had passed omnibus state-wide Wilderness Acts for most states containing national forest lands.

### 4.4.3 The 1980s: Focus on National Forest Land Management Practices and Wilderness Designation Merge

The 1980s saw a merging of focus and linkage between national forest land management practices and wilderness designation brought about by so-called "statutory release" language in omnibus state-wide wilderness acts. This language prevented the Forest Service from considering any more additions to the National Wilderness Preservation System in the first round of land management planning under the National Forest Management Act, but required such consideration when forest plans are revised in 10–15 years. This dramatically shifted the focus of many environmental groups from wilderness designation *per se* to seeking to protect as much undeveloped and unroaded land as possible for future wilderness designation.

Issues emerging strongly in the 1980s that reflected this changed focus included concerns that the Forest Service was selling timber in some areas below its cost of production and the old-growth/northern spotted owl issue in the Pacific Northwest. Although both of these issues reflected important public policy issues in their own right, both also acted as wilderness "proxies" that acted to protect the inventory of undeveloped and roadless areas.

The late 1980s and early 1990s were also the focus of increasing administrative appeals and lawsuits charging that the Forest Service was violating the National Forest Management Act, the Endangered Species Act, and other environmental laws. Such legal challenges became common and were successful often enough to reduce national forest timber and other commodity program outputs substantially.

### 4.4.4 Current National Forest Land Status and Commodity Outputs

Today, 42.8 million acres, or 23 percent of the 191 million-acre National Forest System is statutorily set-aside in various Congressional designations. These include the following: National Wilderness Preservation System, 34.6 million acres; National Monuments, 3.4 million acres; National Recreation Areas, 2.7 million acres; National Game Refuges and Wildlife Preserves, 1.2 million acres; Wild and Scenic Rivers, Scenic, and Primitive areas, 0.9 million acres (USDA/Forest Service 1996b).

In addition to statutory set-asides, large areas of national forest and other federal lands have been made administratively off-limits to commodity production. Since 1985, the National forest System (NFS) land area which is available for planned timber harvest has dropped from about 72 million acres to 49 million acres. Current national forest land management plans provide for timber harvesting as one of the resource objectives on about 26 percent of NFS land area, 35 percent of NFS forestland and 48 percent of NFS productive forestland.

Reduction in national forest lands available for timber production has been particularly significant on the Pacific Coast. The Pacific Northwest Forest Plan provides that only 15 percent, or 3.7 of 24.5 million acres, of federal lands within the range of the northern spotted owl is available for timber production as one of the possible uses. National forest lands comprise 19.4 million acres, or 79 percent of all federal lands in the plan (ROD 1994, USDA/Forest Service 1994b).

The level of national forest timber sales has declined even more dramatically than the available land base, dropping by 70 percent from an average of 10–12 billion board feet during the 1960s, 1970s and 1980s, to 3 to 4 billion board feet today. National forest timber sales in the Pacific Coast states of Oregon, Washington, and California which have been affected by the spotted owl issue have seen the most dramatic declines, dropping by 89 percent between 1987 and 1995, from 6.86 to 0.78 billion board feet. National forest timber sales in other states declined by 53 percent, from 4.46 to 2.10 billion board feed, during the same period (USDA/ Forest Service 1996b).

Between 1989 and 1995, the area of National Forest lands on which timber harvesting occurs annually declined by 44 percent, from 838,000 to 473,000 acres. Timber harvesting now occurs annually on less than one percent of the area identified as suitable for timber production in existing national forest plans. Between 1988 and 1996, the area harvested by clearcutting dropped by 80 percent, from 283,000 to 57,000 acres, and clearcutting as a percentage of all National Forest System area harvested annually declined by over two thirds, from 39 to 12 percent (annual National Forest System Reforestation and Timber Stand Improvement Reports and the TRACS data base).

In addition to the reduced use of clearcutting, the proportion of small sized and salvage timber being offered for sale has increased substantially. Between 1990 and 1996, sawlog-sized material dropped from 77 to 56 percent of total national forest harvest volume and salvage increased from 26 to 47 percent (USDA/ Forest Service Cut and Sold Reports 1990–96).

### 4.4.5 Is Biodiversity the Current Overriding Objective of National Forest Management?

Former Forest Service Chief Jack Ward Thomas has written that the net effect of the administrative and legal challenges to national forest plans has been to make protection of biodiversity the *de facto* overriding goal of the national forests (Thomas 1996):

> "It has become increasingly obvious that the overriding *de facto* policy for the management of Federal lands is the protection of biodiversity. That *de facto* policy has simply evolved through the interaction of laws, regulations, court cases, and expedient administrative direction. This *de facto* policy, I believe, is the very crux of the raging debate over the levels of commodity production that can be expected from Federal lands. Such a dramatically important policy should be examined closely by the American people and the Congress. If that is the policy, it should be clearly stated and the consequences accepted. If such is not the national policy, that should be stated. In recognition of this crux of the issue of Federal land management, and in a clear declaration of policy regarding preservation of biodiversity, lies one key to the (community) 'stability' debate."

While the increased public demands for recreation, scenic and amenity values from the federal lands has reduced the relative role of these lands in producing commodities, as compared to non-federal lands, fed-

Table 2. Role of Federal Lands in U.S. Natural Resource Production.

| Natural Resource | Percent of U.S. Production (1995) |
|---|---|
| Timber | 13 |
| Petroleum | 23 |
| Lead | 52 |
| Natural Gas | 34 |
| Coal | 37 |
| Phosphate | 11 |
| Molybdenum | 70 |
| Silver | 20 |
| Gold | 15 |
| Sodium | 49 |

Note: Timber data are from USDA/Forest Service (1993a); others are from *Mineral Commodity Summaries*, U.S. Department of Interior, and *Monthly Energy Review*, U.S. Department of Energy, all Washington, DC.

eral lands nevertheless continue to produce nationally significant amounts of several important natural resources (Table 2).

## 5 HOW HAVE CONSERVATION POLICIES PERFORMED?: LESSONS AND IMPLICATIONS OF THE HISTORICAL RECORD

With the foregoing brief background on the evolution of use and management of U.S. land and resources, it seems fair to ask how well the conservation policies that have been put in place over the last century have performed. The best way to assess that performance is to look at the land, air and water, and the biological communities associated with them: what are they like today, as compared to what they were like in 1500, 1900, and 1970? First a look at 1500:

### 5.1 The America of 1500 Versus Today

There is no question that when one compares the current biological diversity in the United States with that which is thought to have existed in 1500, there are profound differences. Cropland and pastureland alone account for 553 million acres (25 percent of the U.S. land area) that at one time was forests, grasslands, savannas, and wetlands (Langner and Flather 1994). Most native grasslands and prairies in the humid areas of the United States have been converted to agriculture and other uses. More than one-half of American wetlands have been converted to other uses since 1500. At least 300 million acres of forestland (about one-third the estimated original area) have been converted to non-forest uses since 1500, mostly to agricultural uses.

Most ecosystem transformation in the United States occurred during the settlement period, before 1900. The primary cause of this transformation was agriculture — the clearing of forests and savannas, the plowing of prairies, and the draining of wetlands. Others included: (1) fuelwood harvesting for domestic use and industry, (2) logging for lumber to build the growing cities after 1850, (3) destructive wildfires associated with land clearing and logging, and (4) virtually unrestricted hunting of wildlife for meat, feathers, furs, and hides. Even though these were agents of ecosystem loss, the primary cause of the transformation was a rapidly growing human population, increasing from 5 to 76 million in the 19th century, combined with rather primitive agricultural and resource extraction and utilization technologies. Often unclear and fluctuating land tenure arrangements during the settlement period were also a factor.

Ecosystem management, and especially conservation biology, with its emphasis on natural systems, have focused renewed attention on the nature of pre-Columbian America. In *Endangered Ecosystems of the United States*, Noss et al. (1995) sought to document the loss of ecological diversity that has occurred since 1500, such as the tallgrass prairie and oak savanna ecosystems of the Midwest, the longleaf pine ecosystem of the Southeast, and many others. This is an important effort in seeking to reconstruct the nature of pre-Columbian America. Additional information has been compiled by Langner and Flather (1994).

One of the objectives of ecosystem management is to seek to restore ecosystems at risk. To do so successfully, it is necessary to understand the primary causes of the transformation of American ecological conditions, their historic timing, and the human advantages that accrued from these changes. It is not enough merely to know that the transformation occurred. A prerequisite to the development of effective strategies to reduce or arrest future biodiversity losses is an understanding of the when, where, how, and why of the transformation.

Yet a good information base as would be desirable to characterize pre-Columbian conditions is incomplete. Nor is there a full understanding of the relative roles that natural disturbances and the management practices of aboriginal peoples over at least 15,000 years of human occupation played in shaping the biological diversity of these ecosystems. What evidence there is suggests that many popular public perceptions about the nature of pre-Columbian America are largely incorrect. One of these is the myth of the "forest primeval," the idea that pre-European forests were dominated by a "blanket of ancient forest." This image is one of continuous, closed-canopy, structurally complex, all-aged "climax" forests which nature maintained for long periods in a steady-state, equilibrium balance with the environment. Closely related to the myth of the forest primeval is the "pristine myth" or the popular perception that American Indians lived in the forests and on the plains but never really changed either.

## 5.2 The America of 1900 Versus Today

When one compares U.S. environmental conditions today with those that existed in 1900, unquestionably forest and wildlife conservation has had many successes. The conservation policies and practices put into place in the early decades of the 20th century, as well as some fortuitous events, have been major factors in reducing the rate of biodiversity loss since 1900 (Trefethen 1975). This has occurred in spite of an increase in U.S. population from 76 million to 270 million.

Several wildlife species became extinct because of human actions since 1500. Examples include the great auk, passenger pigeon, heath hen, Carolina parakeet, ivory-billed woodpecker, Bachman's warbler, Labrador duck, and others. Many subspecies were also extirpated, including the eastern elk, Florida red wolf, eastern bison, and Wisconsin cougar. The genetic diversity of many other species and populations has been diminished, as well. Some of the earlier losses were due primarily to overhunting (great auk); many to a combination of overhunting and habitat loss (passenger pigeon, heath hen, Carolina parakeet); and some to habitat loss alone (Bachman's warbler).

Even though several species did become extinct, given the massive assault on wildlife by market hunters and the transformation of ecological conditions that had occurred by 1900, it is perhaps remarkable that more extinctions did not occur. Many species that were severely depleted, or even on the brink of extinction in 1900, have staged remarkable comebacks. While data on wildlife numbers in 1900 leaves much to be desired, there is little doubt that many species that would likely have been on an endangered species list, had one existed in 1900, are today quite abundant (Trefethen 1975, Thomas 1990, Frederick and Sedjo 1991). Examples include: wild turkey; egrets, herons, and many other wading birds; many species of shorebirds; wood ducks, and other species of ducks; tundra swans; Rocky Mountain elk, black bears, beaver, fishers, and most other fur-bearers, pronghorn antelope, bighorn sheep, and white-tailed deer throughout most of its range. Many other species, although not actually threatened with extinction in 1900, are today both more abundant and more widespread than they were in 1900.

Trends in U.S. forest conditions in the eastern U.S., where two-thirds of U.S. forests are, have been largely positive since 1900. U.S. forest area has been generally stable since 1920, when forest clearing for agriculture largely halted. In the northeastern United States, forestland has actually increased substantially since 1900. As millions of acres of agricultural lands were abandoned, forests have increased in this, the most populous region of the nation, from less than one-half the land area to more than two-thirds, an increase of more than 40 percent, or 26 million acres (unpublished USDA/FS stats). The majority of these forests are comprised of native tree species reforesting naturally after agricultural abandonment. Figures 3 and 4 illustrate these trends.

The two primary reasons for the agricultural abandonment and consequent increase in forestland in the Northeast were the conversion of the tallgrass

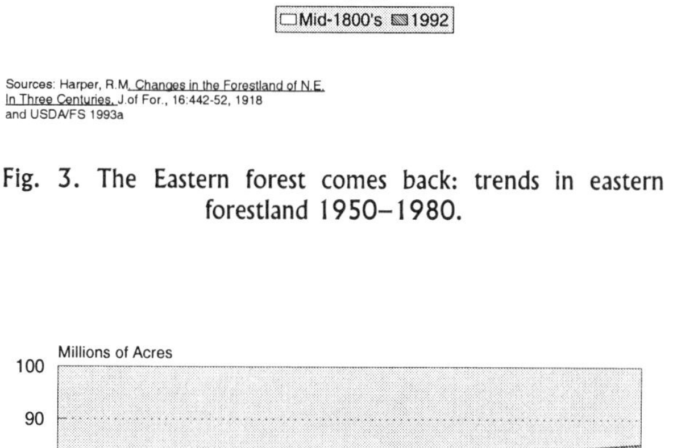

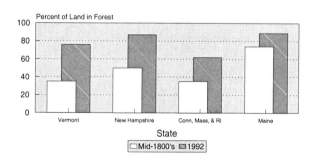

Fig. 3. The Eastern forest comes back: trends in eastern forestland 1950–1980.

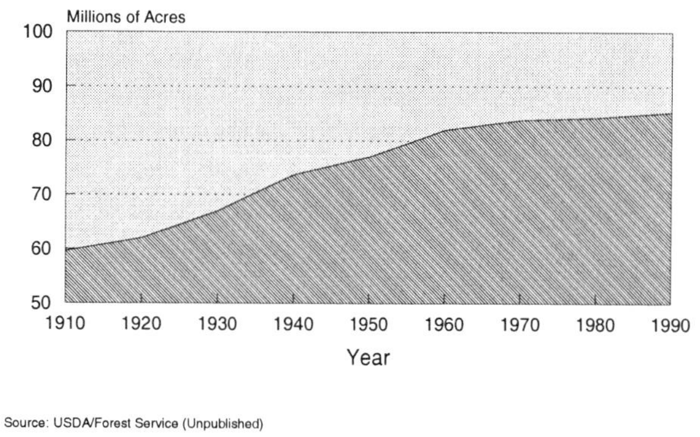


Fig. 4. Trends in forest land in the Northeast 1910–1990.

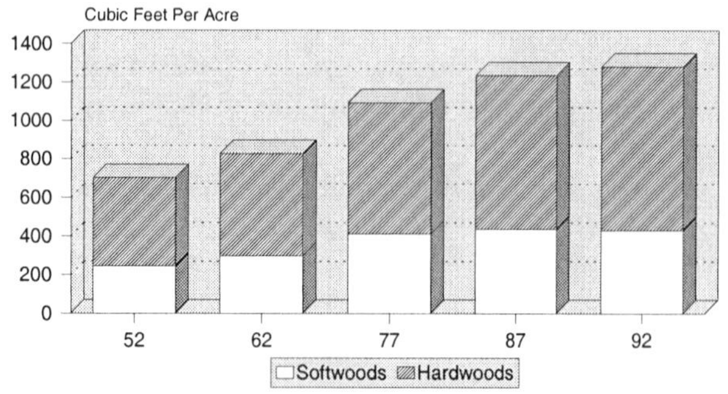


Fig. 5. Trends in standing timber volume per acre in the eastern U.S. 1952–1992.

prairie and oak savanna ecosystems of the Midwest to farmland, and, after 1935, to the improving productivity of American agriculture. Like natural ecosystems, human social and economic systems are linked at various scales. Such linkages can produce both positive and negative effects on natural systems.

Today forest cover encompasses about one-third of the U.S. land area and one-half the land area east of the Mississippi River. This is about two-thirds of the forest cover that is estimated to have existed in 1500 (MacCleery 1992).

American forests are, on balance, more mature than they were a half century ago. Because forest growth nationally has exceeded harvest and forest mortality for at least a half century, average forest biomass per acre has increased by at least one-third since 1950. In the East and South, biomass per acre has almost doubled since 1950 (Fig. 5). A significant exception to these general trends occurs in Pacific Coast forests, where the average biomass per acre has dropped by 4 percent since 1952 and the volume of trees greater than 17 inches in diameter has dropped by one-third (USDA/Forest Service 1993a).

The suite of tree and other woody species that today comprise U.S. forests are largely the same as those present in 1500, although the ages and relative proportions are often substantially changed. However, introduced diseases have relegated some former widespread tree species to minor components, e.g., American chestnut and American elm today are ecologically extinct in the Eastern hardwood forests. Exotic insects and diseases have be a major factor affecting the nature of today's forests. Examples include chestnut blight, white pine blister rust, Dutch elm disease, gypsy moth, beech bark disease, and hemlock wooley adelgid.

As of 1990, plantation forests constituted a relative small proportion (about 5 percent) of U.S. forest area (Brooks 1993). In contrast to many other countries, virtually all U.S. forest plantations are comprised of native species.

## 5.3 The America of 1970 Versus Today

The results of the environmental legislation of the 1970s added considerably to, and enhanced, those gains flowing from previous conservation legislation and policies.

The Clean Air and Clean Water Acts of the 1970s led to actions that resulted in substantial environmental gains: air quality has been steadily improving in U.S. cities. Sulfur dioxide emissions are down over 30 percent; lead emissions are down over 95 percent since 1970; and particulates are down over 80 percent since 1950 (CEQ 1989).

Many, if not most, U.S. rivers and lakes are measurably cleaner than they were two decades ago. Although perhaps justified for human health reasons, improved air and water quality have benefited both the human and non-human inhabitants of the planet, as evidenced by the improving populations of fish and aquatic wildlife in U.S. rivers and lakes. Fish and wildlife have staged comebacks in many rivers and lakes that were severely degraded or even biologically dead two decades ago. There have been increases in the populations of egrets, herons, osprey, geese, largemouth bass, and other fish and wildlife associated with the improved water quality of many rivers and lakes across the country.

Improved water quality, in combination with elimination of the widespread use of persistent chlorinated hydrocarbon pesticides, have led to increasing populations of many raptors, such as bald eagles, peregrine falcons, and osprey, which had seen major declines during the 1950s and 1960s. Trends in 348 U.S. breeding bird species between 1966 and 1991 indicate that 19 percent are increasing (Fig. 6), 20 percent decreasing, and there is no trend for the remaining 61 percent (Langner and Flather 1994).

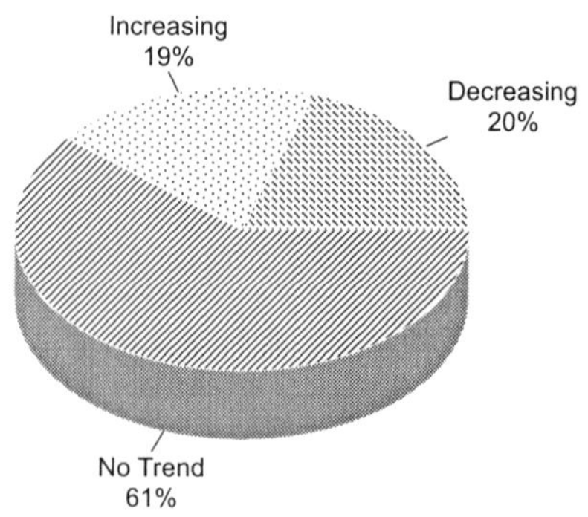


Source: U.S. Fish and Wildlife Service Breeding Bird Survey
348 Species with Adequate Sample

Fig. 6. Trends in U.S. breeding birds 1966-1991.

Many of the bird species that are declining belong to groups or "guilds" dependent upon specific habitat conditions and types, e.g., grassland habitats. Langner and Flather (1994) reported that "(n)ative, endemic grassland birds declined in the past 25 years more consistently, and across a broader geographic range than any other group of birds. The substantial decline in grassland birds was confirmed by Robinson (1997). Meadowlarks, bobolinks, Henslow's sparrow, and upland sandpipers and other grassland birds have shown large and consistent declines. Other species which have shown declines are those that have specialized or niche habitats. For example, freshwater mussels, crayfish, amphibians, and some freshwater fishes lead the list of U.S. species at risk (Stein and Flack 1997). In the United States, habitat loss and alteration and displacement by invasive exotics are the leading causes of species endangerment (Stein and Flack 1997).

Neotropical migratory birds have also been the focus of concern. Of 62 neotropical migrants, 20 have shown statistically significant population declines (Langner and Flather 1994). Habitat fragmentation is suspected as a cause for these declines. The modification of lower canopy layers by excessive populations of white-tailed deer may also be contributing to the decline of those neotropical migrants that nest in these habitats.

More than one-half of the wetlands in the coterminous United States have been converted to other uses since colonial times. Wetlands have been reduced from an estimated 11 percent to about 5 percent of the land area of the coterminous United States (Dahl and Johnson 1991). Most of this loss occurred during the settlement period (Figure 7), but relatively large losses also occurred from the mid-1950s to the mid-1970s, when about 460,000 acres were being lost annually. This amounted to a 0.40 percent average annual loss. From the mid-1970s to today, the rate of wetland loss has been progressively reduced, averaging about 290,000 acres annually from the mid-1970s to the mid-1980s, and 180,000 acres annually between 1982 and 1987, or a 0.17 percent annual loss (Langner and Flather 1994). It is estimated that the rate of loss since 1987 has been reduced even further. Reductions in the rate of wetland loss can be attributed to increased understanding of the environmental value of wetlands (as expressed in individual and regulatory action to protect wetlands), as well as to the fact that many of the most economically attractive wetlands have already been converted to other uses.

The first Forest and Rangelands Conservation Act Appraisal provided important information for conservation policy-makers: that most soil erosion from U.S. agricultural lands was coming from a relatively small percentage of the agricultural land area. The Food

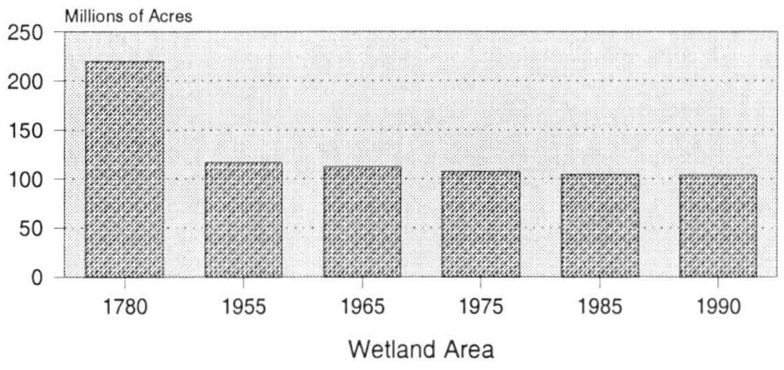


Fig. 7. Trends in U.S. wetland area 1780–1990. (Source: RPA General Technical Report RM-224, April 1994.)

Security Act of 1985 established a number of programs targeted at reducing soil loss from highly erodible lands. The Conservation Reserve Program paid farmers and ranchers to put these highly erodible lands under soil conserving vegetative cover and conservation compliance provisions required farmers having highly erodible lands to adopt a conservation plan or risk losing federal farm benefits. These provisions have been successful in substantially reducing erosion and sedimentation from farmland. The National Resource Conservation Service estimates that by 1992 soil erosion on U.S. cropland was down almost one billion tons annually compared to 1982 (PCSD 1996). Today, the rate of soil erosion from U.S. agricultural land is perhaps lower than it has been in over a century (Fedkiw 1989).

## 5.4 A Brief Snapshot of the United States and World Forest Situation

The pattern of stabilization of forest area and increasing forest biomass per acre that has existed in the United States over the last several decades has also taken place in other developed countries. For example, in Europe the area of forest and other wooded land increased by 11 percent between 1960 and 1990, and the average forest biomass per hectare rose by 43 percent between 1950 and 1990 (Kuusela 1994). Similar patterns have occurred in other developed countries, such as Australia and New Zealand.

These developed-country land-use trends are largely the result of: (1) agricultural productivity increasing much more rapidly than population growth resulting in the abandonment of crop and pasture land and its reversion back to forest, and (2) the volume of forest growth exceeding depletions from natural mortality, fire, and timber removal. It is unlikely, however, that these trends will continue. As agricultural abandonment runs its course, net forest area is likely to stabilize and perhaps even to drop in some developed countries

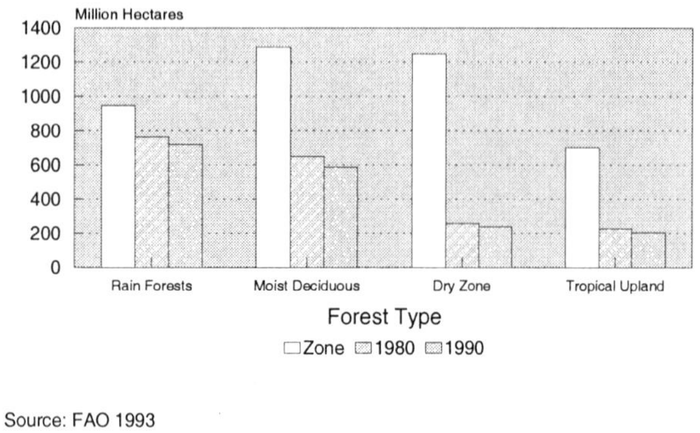


Fig. 8. Trends in tropical forest area 1980–1990. ("Zone" is estimated original extent.)

because of urbanization and other land-use changes. Average biomass per acre is certain to stabilize as forests mature and may even decline in some areas because of forest health problems caused by overstocked stands, the effects of pollution, introduced exotics, and/or disease.

In contrast to the general situation in most developed countries, substantial forest loss and degradation are continuing in many less developed nations. Between 1981–1990, the global area of tropical forests declined by 8 percent, or about 154 million hectares in 10 years. This is an average of over 42,000 hectares (about 165 square miles) per day for the 10-year period (FAO 1993). In addition to deforestation, it has been estimated that an equal area of forest has been degraded by human use, e.g., closed forest has been transformed to open forest or to fragmented forest or to long-fallow shifting agriculture (WB 1995). The largest single cause of forest loss globally is conversion to agriculture, either subsistence or commercial.

Whereas much of the public interest in the loss of topical forests has focused on tropical rainforests, it has been the other tropical forest types that have been impacted most by human use. Figure 8 illustrates trends in loss of tropical forest by type, in comparison to the estimated original extent or "zone" of such forest types.

## 5.5 The Unfinished Agenda

An objective evaluation of the performance of conservation policies since 1900, and of the environmental policies since 1970, suggest some impressive gains. Nevertheless, some environmental trends are not positive and much work remains to be done. Problems include: habitat fragmentation because of residential subdivision and urban development; loss and deterioration of the forest and grassland habitats that once were created by frequent, low intensity fire; reduction and fragmentation of late successional forest habitats because of timber harvesting; loss and degradation of riparian and wetland habitats; effects of air pollution on forests in some areas; and displacement of native species by introduced exotics, to name a few. Of particular concern are rare and unique ecosystem types and species with specialized habitat requirements that are associated with them. Also of increasing concern during the last decade are deforestation and loss of biodiversity in many developing nations.

### 5.5.1 Emerging proposals for conservation networks

Because many of the emerging environmental challenges can effectively be addressed only at the larger landscape scales, public agencies and the general public are beginning to recognize the value of planning at the ecosystem level. Most federal land management agencies, and many at the state level, have now adopted "ecosystem management" as their operating management strategy. While much needs to be worked out as to what ecosystem management means operationally, this is a positive development.

But while thinking and planning at the ecosystem level is a constructive change, it is not going to reduce the tough choices and tradeoffs that lie ahead. Indeed, it only makes them more complex technically and difficult politically, due to the increased number of players involved.

In 1992, a "Global Biodiversity Strategy" was proposed by several international non-governmental organizations (WRI/IUCN/UNEP 1992). This strategy involved several elements, including:

- Protection of a representative range of ecosystem types
- Management of surrounding areas to complement the conservation of biodiversity in adjacent preserves and networks
- Integration of biodiversity conservation into local use of natural resources
- Monitoring trends in environmental performance
- Monitoring national policies affecting natural resources, such as subsidies that act to encourage habitat loss and country institutional capacity to address biodiversity issues
- Establishment of off-site gene banks, such as botanical gardens and zoos
- Development of objective measures of performance for measuring progress towards objectives.

This strategy has much in common with some of those being proposed under what might be called the new environmental agenda (Grumbine 1994, Noss and Cooperrider 1994). Gosz et al. (this volume) and Concannon et al. (this volume) provide a detailed discussion on the evolving theory, concepts, and barriers to maintaining biodiversity at the landscape level.

One of the strategic approaches that is emerging for the conservation of biological diversity on large, multi-ownership landscapes involves a system of preserves or protected areas, often linked by connecting habitat corridors, and surrounded by limited use buffers, with more intensively managed areas beyond (Noss and Cooperrider 1994).

The "Wildlands Project" represents an attempt by a loose coalition of regionally based environmental groups to put such a conservation network in place in North America (Noss 1992). The Wildlands Project would initially seek to identify existing reserves in National Parks, National Wilderness Preservation System lands, existing federal roadless areas, and similar areas. These "core" reserves would be linked by habitat corridors and surrounded by limited access buffer areas on adjacent federal and non-federal lands. The idea is that core reserves, corridors and buffer areas would be expanded over time into a North American biodiversity network (Foreman 1995/96).

A similar approach, not directly related to the Wildlands Project, is being advocated for the conservation of forest biodiversity by the World Wildlife Fund. It would involve a national biodiversity network of forest reserves, supplemented by certification of forests not in reserves as being "sustainably" managed by independent, third-party certifiers (DellaSalla et al. 1997). Section 8 discusses further the issue of green certification.

### 5.5.2 Scientific and Political Barriers to the New Agenda

While the idea of conservation networks as a strategy to protect biodiversity appears sound conceptually, it leaves many important issues to be resolved, including: (1) the size and location of each of the components (reserves, buffers, corridors, etc.), (2) the kinds of management practices to be applied within each, and (3) the mechanisms that will be put in place to encourage cooperation (or require compliance) among all of the landowners and other interests involved.

Some conservation biologists supporting the Wildlands Project have suggested that meeting desirable conservation goals would require between 25 and 75 percent of the total U.S. landscape to ultimately be set-aside in reserves and limited access buffer zones from which no, or very limited, commodity outputs would be produced (Noss and Cooperrider 1994).

Implementing a conservation agenda having such a significant social and economic impact faces a variety of difficulties, to say the least. These include: (1) a basis in the natural sciences which is rapidly evolving, (2) an institutional framework that has not yet demonstrated its capability to address natural resource issues at the larger multi-ownership and jurisdictional landscape scales, and (3) a lack of demonstrated social/political will to make significant adjustments in resource allocation and consumer consumption habits.

A number of these difficulties are readily acknowledged by Wildlands Project proponents. On the subject of rapidly evolving science, Vance-Borland et al. (1995/96) note that "(r)eserve selection and design, like most areas of conservation biology, are rapidly evolving. Methods for designing reserve networks that were considered innovative in the mid-1980s or even early 1990s (for example Noss's [1985, 1987, 1993] work in Florida and the Oregon Coast Range) would generally not be considered scientifically defensible today. And what is not scientifically defensible is unlikely to be taken seriously."

One of the emerging issues associated with conservation networks which extends beyond the selection and design of reserves, is how such reserves will be managed to restore and maintain biodiversity. The current concept of reserves within which no active management will occur is not likely to result in maintaining pre-European settlement conditions in many fire adapted ecosystems. Purposeful management of fire in reserves will be essential in many areas to achieve desirable biodiversity objectives. Acceptance of the need for management of reserves will be difficult for some conservationists (see Section 11, The "new" vs. the "old" ecological thinking, for a discussion of this issue).

Without a substantial change in United States per capita consumption patterns, the likely effects of implementing the recommendations of the Wildlands Project would be: (1) further intensification of management for commodity uses outside of designated areas, and (2) further transfer of the environmental effects of U.S. consumption to ecosystems outside the United States. Another difficulty associated with the Wildlands Project is an apparent inability or unwillingness of at least some of the participating groups to separate science-based biological protection objectives from their traditional political agendas of expanding Wilderness and protecting federal roadless areas from commodity production (see Forman 1995/96).

The reality is that society does not appear to possess either the will nor the resources to save it all. In fact, federal spending on natural resources conservation has been relatively flat in real terms since World War II, while federal spending in other areas, such as social security, health, and related social programs, has increased dramatically (Fedkiw 1993).

An inability or unwillingness to separate the science of biological diversity protection from political agendas on both sides of the political spectrum risks confusing the public and undermining credibility and support for the idea. Rational and practical approaches are needed to decide what it is society will seek to save, and at what cost. Success in the future will increasingly require holistic, science-based, and collaborative approaches.

But even if protecting biodiversity could somehow be placed at the top of the U.S. social agenda, what will result is not a pre-Columbian era version of biological diversity. Rather it will be nature with modern society firmly embedded within it.

### 5.5.3 Potential Role of Conservation Networks in Protecting Biodiversity While Reducing the Uncertainty Associated With Managing Public and Private Lands for Other Objectives

In spite of the substantial scientific and social problems associated with conservation networks, the establishment and maintenance of such networks provides a promising basis for the long-term protection of biodiversity in an American landscape subject to increasing human modification. If protection and maintenance of biodiversity has indeed become the emerging objective of federal land management, as former Forest Service Chief Jack Ward Thomas has suggested (see Subsection 4.4.5), the role of biodiversity networks in advancing that goal should be thoroughly and systematically explored.

Since two-thirds of U.S. land is privately held, even if conservation networks are established on federal and other public land, success in managing for biodiversity will increasingly depend on cooperative, non-regulatory approaches, and creation of tax and other institutional incentives for private landowners to manage their lands to achieve desired objectives (see Section 6.10).

## 6 FEDERAL RESOURCE MANAGEMENT — A CRITIQUE: PROMISE VERSUS PERFORMANCE

As management of federal lands has evolved over the past two decades, there have been a number of short-comings that have rather consistently emerged. The following are some of the issues which must be addressed.

### 6.1 Telling the Story of the Land and People

Both natural and human-induced influences and disturbances have profoundly shaped federal forest and grassland landscapes. Unfortunately, the specific and interconnected roles of natural and human-induced processes and events are often unknown and unappreciated. In the past, telling the story of how local and regional ecosystems have come to be what they are today has been largely ignored.

Every piece of land has both a human history and a natural history, which are inextricably linked. Although traditional history tends to focus on events, leaders and elites, environmental history focuses on the land and the human and ecological factors that influenced it. Because of this perspective, the focus is on the interconnections between the ecological characteristics of the land; the culture and practices of the people who live on, used and influenced it; human population growth and decline; economic and trade linkages; and changing technologies. For further discussion of the human role in shaping the North American landscape, see Bonnicksen et al., this volume; Periman et al., this volume; and Sullivan et al., this volume.

There are many important insights that can be gained from studying the environmental and land use history of the area that are relevant to today's debates over federal land management. These insights include a better understanding of: (1) the resiliency of the land in the face of human and natural disturbances, (2) those ecological conditions and animal and plant communities that existed in the past (the knowledge of which can help inform us about what might be needed to maintain them in the future), and (3) the effects of policies and management practices. Certainly, the natural and human histories of the federal lands are directly relevant to their future management.

Each federal land-planning effort should include a reasonable effort to discuss relevant aspects of both history and prehistory, including the role indigenous peoples had in shaping these ecosystems, which is often not well understood. It includes a description of how European settlement subsequently reshaped the landscape. Like other aspects of ecosystem management, various scales are important. At the regional and national scales, the story should include a discussion of the public policies and other factors, such as trade and evolving technologies, which were influential in shaping our natural landscapes.

At the field level, this effort should include, among other things: (1) maps and descriptions of the historic vegetation that existed in the region, (2) a discussion of the role that indigenous peoples may have had in shaping pre-European contact landscapes, (3) evolution of the landscape since Euro-American contact, and (4) discussion of the major social, cultural, and natural forces and events that shaped the land and people.

These stories should be told in an as objective and value-free manner as possible. Several disciplines can be used in this effort, including the emerging field of environmental history, cultural anthropology, archaeology, ecology, and cultural geography, to name a few. Most of these fields have been largely untapped for this purpose by today's resource managers.

A knowledge of the natural and human influences that shaped today's ecosystems not only is a key element in enlightened federal decision-making, it also can serve as a powerful unifying force in bringing people together on common ground. Seeking to understand and describe the natural and human legacy in the landscape is essential to a better public understanding about how the land and people came to be what they are today and is crucial to informed choices about the future.

## 6.2 Use of Public Policy Tools on Federal Lands

Natural resources are produced from federal lands in the United States within the larger context of producing natural resources from other lands, which are conducted in an economic context that features a comparatively modest degree of government intervention.

Two premises underlie production of natural resources from federal lands: (1) receiving fair market value for federally owned resources, and (2) achieving desired social objectives and values. Sometimes these two objectives may conflict, particularly during periods of rapidly changing social values.

### 6.2.1 Receiving Fair Market Value for Federally Owned Resources

Coggins and Wilkinson stated that: "The policy of the United States toward lands it acquired has changed drastically over the course of two centuries. Until 1976, the official federal policy was to sell or give away the public lands and resources to private owners and to states in order that the Nation would be tamed, farmed, and developed." The Federal Land Policy and Management Act (90 Stat. 2743) was enacted in 1976 and among its policy declarations were: "(1) the public lands be retained in federal ownership, unless...it is determined disposal of a particular parcel will serve the national interest;...(9) the United States receive fair market value of the use of the public lands and their resources unless otherwise provided for by statute;..." (Coggins and Wilkinson 1987).

The trend during the 20th century has been toward retention rather than disposal of federal land. With the exception of the Alaskan Native Claims Settlement Act and Alaskan National Interest Lands Conservation Act of 1980, the policy of retaining public lands in federal ownership has been implemented with only three challenges — specifically, the Sagebrush Rebellion of the late 1970s, the asset management program of the early years of the Reagan administration, and most recently, various grassroots efforts challenging federal sovereignty over public lands in the West (e.g., the Nye County, Nevada, legal challenge) and proposing "devolution" of the administration of federal lands to state agencies. None of these challenges have resulted in a notable change in the federal estate.

Fair market value in the sale of natural resources from federal lands is a basic policy which has been articulated many times by Congress, yet never fully implemented even when dealing with resources that are regularly exchanged in markets. One of the reasons for this failure is because Congress, while articulating the validity for the general concept of fair market value, is ambivalent on the subject, especially when local constituents feel threatened (or inconvenienced) by its application. Theory and experience suggest that exceptions should be allowed only under extraordinary circumstances.

Determining when a subsidy exists is not always as straightforward as it may seem. A distinction should be made between providing goods and services at below market prices and providing them below the cost of production to the federal government. For example, a federal timber sale sold at a competitive price can still generate revenues less than the cost of preparing and administering it. Such sales are referred to as "below cost" timber sales and are often considered subsidies even though the timber is sold at competitive prices. Whether a subsidy exists depends on what timber and non-timber costs and benefits accrue from the sale. If the timber sale constitutes the most cost effective way to achieve desirable resource management objectives, e.g., fuels reduction, forest health, wildlife openings, or the like, then it is not a subsidy, but effective federal land management.

Among the leading resources for which subsidies are a concern are range forage, timber, hardrock minerals, and water, and at issue are grazing fees for private livestock on federal lands, the low fee structure for producing minerals from claims on federal lands, below-cost timber sales from the national forests, and the low prices irrigators pay for water delivered from federal reclamation projects (Humphries et al. 1994). Western ranchers, the hardrock mining industry, the timber industry, and agriculture are all alleged to be subsidized by the federal government at a loss to taxpayers. Other users of federally owned resources are subsidized as well. Recreationists are leading examples. They are allowed access to federal lands for their recreational pursuits at a very small price, if indeed they pay anything at all. Most federal recreation is both below market price and below cost.

Subsidization of natural resources consumption from federal lands is difficult to support, given the existing general level of economic development across the United States, the impacts of natural resources consumption on non-market uses of federal lands, and the current political climate which, ostensibly at least, emphasizes a reduction in the role of government in the economy, fiscal austerity, and efficiency in government operations. Public policy now and in the future should be directed toward achieving fuller implementation of fair market value policy with regard to the private use and consumption of federally owned resources: forage, minerals, timber, recreational opportunities, and water. More precisely, prices for the foregoing natural resources should: (1) reflect fair market value, and (2) generally cover their marginal cost of production, including any external costs of production. Exceptions to covering the marginal cost of production should be permitted only when net public benefits can clearly be demonstrated. This is not a unique prescription. Quite the opposite, it is standard prescription or rule in public finance, well founded in economic theory and experience.

It is recognized that charging fees for use of recreation opportunities on federal lands covering their marginal cost is complicated by problems of accessibility. For example, attempting to charge a user fee in an urban park with unlimited access makes little sense. It may cost more to collect the fee than is provided in revenue. The same is true of national forests with highways running through them with numerous access points. Wellman (1987) lists some other principles, of which three are listed below:

1. Fees should be relatively consistent among federal agencies, but they should be determined by the administration in order to allow flexibility in adjusting them to area characteristics and evolving uses, rather than being congressionally set on a nationwide basis....
2. Fee rates should take into account the prices charged by the private sector and by state and local units of government for comparable services.
3. Fees collected by federal agencies should largely be returned to those agencies for use in paying the costs of operating and maintaining the areas....

To summarize, market resources produced from federal forest lands should be exchanged at fair market value, with exceptions allowed only under extraordinary circumstances where the social value of doing so has been clearly articulated.

### 6.2.2 Role of Non-Market Resources on Federal Forest Lands

Many natural resources produced on forest lands, both public and private, are not exchanged in markets. A forest ecosystem is incredibly, maybe incomprehensibly, complex and yields a multitude of things having human benefit as it moves through various stages of succession. A few of these "things" are exchanged in markets, e.g., timber from some species of trees, forage for domestic livestock, and occasionally, outdoor recreation opportunities, and hence, are called market resources. Most are not, and they are often referred to as non-market resources, and include critical ecological benefits or natural services such as: maintenance of the hydrological cycle, modification of climate, adsorption of pollutants, transformation of toxic chemicals, pest management and control, maintenance of the oxygen and nitrogen cycles, soil fertility and erosion control, water recycling and humidity control, wildlife habitat, and carbon storage and maintenance of the carbon cycle. These essential services are seldom reflected in market prices or national income accounts.

Forests also contain flora and fauna having potential value for human use, for example, as sources of medicine, which are not recognized in markets. Capoten, a drug used to control high blood pressure, developed from the venom of the Brazilian pit viper, and taxol, an extract from the bark of the Pacific yew used in treating cancer, are two such cases.

Private landowners tend to emphasize production of market resources from their land to have income to pay for their consumption of goods and services. They also have a time preference — today as opposed to tomorrow, the short-term as opposed to the long-term — so they do not have to delay consumption or incur debt to consume. Accordingly, the mix of natural resources yielded by purposeful management of private lands tends to be skewed toward market resources. Even so, the aggregate non-market environmental services provided by the more than 70 percent of the U.S. land base that is privately held are considerable.

In contrast, because public land managers, including federal land managers, do not have comparable institutional incentives to generate income as their private counterparts, the mix of market and non-market resources for which public forest lands are managed tends to be oriented more toward non-market resources. Usually their legislative mandate requires them to do so, as well. The time horizon for management also tends to be longer. These differences provide an economic rationale for public forest land ownership, the logic being the combined production from public and private forest lands tends to approximate a more socially desirable mix of market and non-market resources and temporal perspective than would be provided by either category of land ownership by itself (Le Master 1988).

Thus the role of non-market resources and ecological benefits on federal forest lands is a comparatively greater one than it is on private forest lands, and for very good reasons because it provides the rational for the existence of public forest lands.

## 6.3 Promoting Community Stability

Seeking to maintain the economic stability of local communities near federal lands is an enduring, if illusive and enigmatic, objective often associated with federal land management. The concept is used in the sense of programs being conducted and activities being done by federal agencies to promote, achieve, or maintain community stability. The meaning of the term is obscure. No accepted definition exists, and no agreement exists on how to measure it (Le Master and Beuter 1989).

The statutory basis for the policy is explicit and strong for one federal land management agency, namely the Bureau of Land Management in its administration of the revested Oregon and California lands in Oregon. For others, such as the Forest Service, there is no explicit legislative charge.

Implementation of such a policy has severe limitations because it assumes — certainly in the case of timber — a significant measure of economic control over both factor and product markets. While control over factor markets may be possible in areas with abundant federal timber, control over product markets, e.g. lumber, is virtually impossible because it is a commodity exchanged in regional, national, and international markets subject to a variety of market forces. A comparable situation exists with range forage and beef cattle. An agency might exert control over the availability of the range forage resource, but it has no control over the beef cattle markets. Hence, a federal agency might promote community stability in forest- or range-based economies, but it cannot assure such stability because it is beyond the economic power of the agency to do so.

Thus, the fundamental flaw of a community stability policy by federal land management agencies is that it is beyond the grasp of any single federal agency, raising hopes with its promises, and dashing them in the reality of implementation.

On the other hand, actions by federal land management agencies can certainly disrupt local economies. Such actions should be taken only when warranted and implemented in a manner that seeks to minimize economic and social impacts. In other words, federal agency actions that are disruptive to local economies can be necessary for a variety of good reasons, including efficiency of operations and responding to changing values of the larger society. But such transitions can be mitigated or, at least, "phased in" to lessen their local adverse impacts, and done so in the context of a well-informed public. But recent experience suggests federal agencies have a limited capacity even to mitigate adverse impacts on local communities. In recent years, some of the most disruptive agency actions to local communities have been the result of court injunctions that abruptly halted commodity production from federal lands.

Another major flaw in promoting community stability by federal land management agencies is that it can ultimately be inconsistent with the policies of receiving fair market value for federally owned resources and government agencies being responsive to changing social values. This requires a delicate balancing act by federal agencies in carrying out potentially conflicting mandates.

## 6.4 Economic Linkages at National and International Scales

Clearly, economic and social systems are linked at various scales, many of which have significant environmental implications. Federal land managers need to better assess how their actions are likely to act at larger scales to affect economic and ecological conditions on non-federal lands (see Sedjo, Vol. III of this book).

As discussed in Section 3, Item 4, above, resources that are produced (or not produced) in one ecosystem indirectly affect what occurs in other ecosystems, perhaps thousands of miles away. One illustrative example is the linkage between the agricultural conversion of tallgrass prairie and oak savanna ecosystems in the Midwest and recovery of Northeastern forests. To focus only on the adverse environmental effects of the loss of the tallgrass prairie misses the positive effects of the northeastern forest recovery, as well as the more efficient use of land for human objectives. A more recent example is the off-site ecological effects caused by the strategy to protect late successional forest habitats on federal lands in the West. The sharp reductions in federal timber sales, from over 5 billion board feet in 1987 to less than 500 million board feet in 1994, did not eliminate the ecological effects of timber harvesting. It transferred those effects in varying degrees to forested ecosystems somewhere else.

Some analysts predict that declines in federal timber sales will displace the environmental effects from the United States to other countries less able or less willing to maintain environmental safeguards (Sedjo 1995, Perez-Garcia 1995). Since 1991, softwood lumber imports from Canada have increased from 10.5 to 17.8 billion board feet, increasing from 27 to 36 percent of U.S. softwood lumber consumption (Howard 1997, USTR 1996).

Harvesting on private lands in the southern United States also increased after the reduction of federal timber in the West. Today, the harvest of softwood timber in the South exceeds the rate of growth for the first time in at least 50 years (USDA/Forest Service 1993a). Pressure on private forests in the United States will be further exacerbated by a recently concluded U.S.–Canada trade agreement that will reduce U.S. imports of Canadian softwood lumber to near historic levels (USTR 1996).

Reduction in federal timber also results in higher consumer prices for wood products and increases the use of wood substitutes, such as steel framing, virtually all of which require substantially more energy to produce than does wood.

These human dimension linkages, and their associated environmental and economic effects, should be incorporated more fully into decisions made about use and management of federal lands. The appropriate scales for such analyses is at the regional and national levels. Regional assessments (such as the Columbia River Basin Assessment) and national assessments (such as RPA), should, as a normal matter of course, carry out such evaluations (see Lessard et al., Vol. III of this book).

## 6.5 Resource Output Objectives

The statement is sometimes made that one of the most significant differences between the multiple use–sustained yield management of the past and the ecosystem management of today is a shift in focus from the products that will be removed from the land to the condition in which the land is left, e.g., a shift in focus from "flows" to "states." It is often asserted that because of this shift, the establishment of planned resource outputs is inimical to ecosystem management. This is a polarizing and counterproductive position.

The move to ecosystem management has unquestionably focused more attention to the condition of ecosystems, their health and sustainability for multiple values and objectives. But it is incorrect that planning to achieve resource outputs is no longer relevant. Both resource states and flows are important, and are not mutually exclusive.

Land management objectives that give no consideration to product output are certainly mandatory for certain classes of land, such as designated Wilderness. But for many other federal land classifications, consideration can and should be given to both predicted resource outputs and the condition of the land.

Planned resource flows are important in ecosystem management for at least two reasons. For one, planned outputs are important for the sustenance of the human communities that occupy those ecosystems. Human communities, if they are to be socially viable, need reasonable assurance that the resource outputs necessary for their economic health will be produced with a appropriate level of reliability over time. Such assurance is essential for making investments in infrastructure and value-added manufacturing which are necessary to maintaining a viable economic base. Outputs are important, as well, due to the fact that what is done, or not done, in one ecosystem, impacts directly and indirectly ecosystems elsewhere.

No inherent conflict exists between the goal of ecosystem sustainability and that of regular planned outputs of goods and services from those ecosystems. One important tool in achieving sustainability is to focus on the desired future condition (DFC) to be achieved on various components of the ecosystem at some point in the future. Planned outputs can be developed by assuming that forest landscapes will move from their present condition toward a desired condition over a time frame established in the federal agency land management plan. If a well-defined strategy of sequential land treatments is specified, then a rate of product output can be determined and notice can be provided to local communities who depend on these lands for their livelihood.

Federal agencies cannot guarantee the stability of communities or insulate them from the larger social and economic forces that may affect their future. But a relative assurance of a reliable level of resource outputs (set in recognition of all appropriate values and uses) can certainly contribute to a more stable economic future for communities that are surrounded by, or adjacent to, the federal lands. The lack of such assurance can certainly destabilize these communities, as recent history has clearly demonstrated.

Thus, there are serious practical, as well as ethical, questions raised by an ecosystem management paradigm which ignores the importance of planned resource flows and which, by so doing, ignores: (1) the economic and social needs of human communities that occupy those ecosystems, and (2) the effects of exporting the environmental impacts of U.S. consumption to other ecosystems.

## 6.6 Need for Judicious Intensification of Resource Management for Commodity Production

Many in the conservation community have been critical of the intensification of agriculture and forest management due to its resultant site level impacts and ecosystem simplification effects. There has been less recognition of the role of intensification in taking the pressure off other, often more environmentally sensitive lands elsewhere. Experience suggests that lands which are more economically suited for resource intensification are also often less environmentally sensitive than other lands that might otherwise be called upon to produce commodities in the absence of intensification (Huston 1993).

Sedjo (1989) estimated that if an area amounting to about five percent of the current area of the world's forests were placed in high yield plantations, those lands could produce the current volume of world industrial wood products consumption. Just as increasing agricultural productivity has released large areas of land to revert back to forest, so could intensive cropping of forests on appropriate sites potentially release vast areas of forests for other uses.

Some federal lands will likely have a comparative advantage, economically and environmentally, for the production of commodities. Ecosystem management should be holistic enough to accommodate the need for intensive management of those federal lands for commodity outputs in such circumstances. As noted previously, the federal lands currently produce nationally significant quantities of several minerals for that reason. There are also relatively large areas of high productivity forest lands that could have a comparable advantage for producing wood products. About 9 million acres of national forest land are of high site quality (capable of growing 120 cubic feet of wood per acre per year). This is 14 percent of such high site lands (USDA/Forest Service 1993a) in the United States.

## 6.7 Objective Measures of Ecosystem Health and Sustainability

The natural resources community should be more active in development of criteria and indicators for sustainable management. Objective measures of ecosystem trends are needed to assess management performance and to reduce the polarization that the absence of such measures tends to create. This issue is discussed further in Section 7 of this chapter.

## 6.8 Need for a National Biodiversity Strategy

The potential role of conservation networks was discussed in Subsection 5.5.1 Federal and other public lands are an essential part of the conservation networks being proposed by a number of organizations (DellaSalla et al. 1996, Noss and Cooperrider 1994). If a system of conservation networks could be established that is scientifically credible, it could go a long way to reducing both the rate of national biodiversity loss and the uncertainty and polarization that currently exists as to what is necessary to manage for biodiversity at the landscape level. Subsection 5.5.1 also describes some of the barriers that exist to implementing a system of conservation networks. Nevertheless, it would be both desirable and prudent for federal agencies to invest in the research and evaluation necessary to explore the potential design of conservation networks at a regional scale.

Conservation networks could potentially be used, among other things: (1) to help guide the location of and prioritize federal ecosystem restoration efforts (see Covington et al., this volume), (2) identify where it is appropriate for commodity production to occur on federal lands, and (3) provide the foundation for offering positive incentives for the cooperative management of adjacent private lands to achieve biodiversity goals (see Subsection 6.10). Regional scale assessments (see Lessard et al., Vol. III of this book) would seem a logical place to evaluate the implications of alternative conservation network designs.

## 6.9 Research on the Elements of Ecosystem Structure and Functioning

Much of what is new under ecosystem management is based on the emerging field of conservation biology. The theory of conservation biology, which owes its origins to the study of island ecosystems, has evolved rapidly in the last two decades (Noss and Cooperrider 1994).

Unfortunately, the development of the theory has substantially outstripped the data to back it up in specific situations. Today, major resource management decisions are sometimes being made based on data collected in small studies or in ecosystems substantially different than the ones to which they are being applied. This situation must improve if these decisions are to have greater credibility.

Predicting the ecological effects of alternative management approaches at the larger landscape levels is exceedingly difficult. Research to improve such predictions will be both complex and expensive. Some of the most significant research needs include providing a solid scientific basis for: (1) criteria and indicators for sustainable ecosystem management as discussed in Subsection 7.2, (2) the design of conservation networks, as discussed in the previous subsection and in Subsection 5.2 (3) the efficient management of matrix and buffer areas for commodity production compatibly with the non-commodity values of federal lands, and (4) improving the development of land use histories as discussed in Subsection 6.1.

## 6.10 Role of Private Lands in Ecosystem Management

The current pattern of public and private land ownership in the United States is largely a relic of history, rather than the result of a conscious strategy to achieve conservation and economic objectives. Because private lands encompass more than 70 percent of the U.S. land area, the conservation gains described in this chapter have necessarily been achieved through a variety of tax and other public policies designed to encourage the productive management of private lands.

Because ecosystem management is a perspective that encompasses interactions at the landscape level and above, effective strategies to encourage constructive public and private partnerships will be even more important in the future than they have been in the past. Unfortunately, multi-jurisdictional, multi-ownership coordinating institutions do not currently exist at some of the spatial scales needed in ecosystem management.

On the negative side, there are a number of examples of instances in which ecosystem management is considered by private landowners, rightly or wrongly, to be a threat — as a thinly veiled excuse for federal regulation of private lands. There are enough examples of heavy-handed federal approaches to provide fuel to these concerns.

Although the Endangered Species Act (ESA) has proven to be an extremely powerful tool to encourage federal agencies to protect endangered and threatened species, it appears in some cases to be acting as a disincentive to endangered species habitat conservation by private landowners. As noted earlier, the regulatory provisions of the ESA appear to be encouraging some private forest landowners to harvest their forests prematurely lest they become habitat for an endangered species. To the extent that this is occurring, the effect of the ESA is to encourage economically, socially, and environmentally perverse behavior.

Such perverse incentives must be systematically identified and eliminated. In their place, positive incentives to encourage cooperation must be developed. Much work in this area remains. A number approaches are available for developing cooperative strategies and partnerships with private landowners. These are explored elsewhere in this volume by White et al. and M. Johnson et al., and in Vol. III by Bliss; Hummel et al.; and Yaffee.

A variety of instruments exist that could provide positive incentives for private landowner cooperation in ecosystem management (Williams and Lathbury 1996). These include:

1. Property tax preferences for maintenance of open space.
2. Transferable development rights.
3. Purchase and donation of conservation easements.

Considerable additional opportunity exists to develop private sector incentives for habitat conservation. One of the most significant is the reform of federal estate tax codes in ways that reduce the necessity to fragment private properties in order to raise money to pay such taxes.

## 6.11 Federal Agency Cooperation and Coordination

The Endangered Species Act, Clean Air and Clean Water Acts, and other federal environmental laws, by their nature, tend to encourage a piecemeal and ad hoc approach to federal land management. Unfortunately, agency regulations and the behavior of the bureaucracies established to administer them act to further compound this tendency.

Federal land management agencies and the regulatory agencies often have separate and sometimes conflicting objectives. In addition, their perceptions of acceptable risk often differ, which can act to create a sort of federal agency "gridlock." The federal analysis procedures established pursuant to land management planning have become increasingly costly, time consuming, and provide considerable opportunity for interest groups to intervene to block proposed actions.

A compounding problem is a federal agency mindset that focuses on the short-term impacts of a proposed action, rather than the long-term environmental implications of acting or failing to act. An example is the tendency for federal agencies to focus on the short-term environmental effects of forest health projects, such as thinning or removing dead and dying trees, while largely discounting the increased risk of catastrophic, stand-replacing wildfires associated with failure to treat overstocked forests.

In spite of the lip service being given to "adaptive management" under ecosystem management, the reality is that it has become a much more cautious approach than existed under multiple use-sustained yield management. The "burden of proof" has shifted under ecosystem management. Under multiple use-sustained yield management the burden was often upon those who alleged harm would result from a proposed activity to provide reasonable evidence to suggest that it would occur. Under ecosystem management, the burden appears to be on the proponents of an activity to demonstrate, or even prove, that no harm will occur. This is a significant change. It seems to be based on an assumption that not acting is always safer than taking an action. Such is often not the case in fire-prone ecosystems (see Section 8).

Considerable opportunity exists to improve interagency cooperation and coordination. Strong direction and leadership from the Administration will be necessary to accomplish it. On the positive side, the interagency memorandum of understanding on timber salvage has been performing quite well and could provide an example for similar interagency coordination efforts elsewhere. Approaches for encouraging cooperative approaches at a variety of scales are discussed in this volume by White et al. and M. Johnson et al., and in Vol. III by Hummel et al. and Yaffee.

## 6.12 The Elements of Successful Public Participation in Federal Land Management Planning and the Role of the Interdisciplinary Team

The role of public involvement and participation in federal resource management has changed substantially over the last two decades and continues to evolve. The following subsection contains recommendations for improving the effectiveness of this process.

### 6.12.1 Public Participation in Federal Land Management Planning

The role of public participation in federal land management planning has evolved considerably over the last two decades. Many observers view the lack of Forest Service sensitivity to public concerns over the aesthetics of clearcutting in the late 1960s and 1970s to be at least partly responsible for the extended and emotional nature of the issues which eventually lead to passage of the National Forest Management Act. In the 1970s, a variety of laws were passed that caused federal land managing agencies to substantially increase public participation and open consideration and evaluation of management alternatives. This "rational planning model" was based on a rather optimistic premise that federal agency decisions arising out of it would lead to a "working consensus" among diverse interests under which such interests would consent to "share the land" in what constitutes a politically acceptable "social contract." In retrospect, that premise was optimistic, to say the least.

Ecosystem management builds on the experience of past efforts at public participation, and seeks to expand it. The multiple spatial scales encompassed by ecosystem management require even more reliance on cooperation and coordination among federal, state, local, and private entities than ever before. Thus, public participation under ecosystem management is not just a means to arrive at socially acceptable decisions, but in some ways becomes an end in itself, essential to successful implementation of planning decisions.

Experience suggests that, in some cases, federal agencies have been quite successful in encouraging active public participation and local community involvement in the planning process. In other cases, the record is mixed. Experience also suggests that success in the future may require a modified agency role and relationship with the public. In the past, the agency role has been to receive input from a variety of diverse publics on issues to be addressed and on management options that were under consideration. The agency then decided how best to weigh this input in arriving at a decision. The agency role was one of "mediating" competing public and private interests, and it left many in the public less than satisfied.

A modified agency role is now emerging in which a federal agency encourages competing interests to sit down and "reason together" to find ways to accommodate their diverse objectives. The agency role in this case is similar to a "facilitator," rather than mediator. While the decision still rests with the agency, the theory behind this approach is that it will lead to more informed decisions that have broader public ownership and support than in the past. Experience suggests that this approach works best at the local level where the effect of alternative management approaches on specific areas of land can be visualized.

Barriers and questions still remain regarding this emerging approach. One is how federal agencies are to address and balance national interests versus local interests. Another is the Federal Advisory Committee Act, which erects substantial legal barriers to federal agencies accepting the results of a facilitated outcome by any group which has not gone through the rather onerous process of being constituted as a formal federal advisory committee.

In spite of these problems, a considerable body of knowledge exists on the essential elements of effective public participation. Many of these elements have not been effectively institutionalized within federal agencies. Agencies should seek to identify these essential elements and take the necessary steps to encourage all field units to utilize them.

#### 6.12.2 Role of the Interdisciplinary Team in Federal Land Management Planning

There is a particular need to reexamine the traditional role of the interdisciplinary (ID) team in land management planning. Today the role of the federal agency ID team is, among other things: (1) to identify public issues and management concerns, (2) to prepare the analysis of the management situation, (3) to formulate alternatives, (4) to evaluate alternatives, and (5) to recommend a preferred alternative. This role should be reexamined in the light of current experience and decision theory.

A wealth of experience suggests that vesting in one team all of the responsibilities listed above can, at times, encourage a climate of exclusion. Team members sometimes negotiate proposed outcomes amongst themselves in ways that deal the public and even the line officers out of the equation.

Many ID teams have worked well. But others have not. It is likely that those which work well do so in spite of their institutional charge, not because of it. It is important that any institutional change act to encourage, rather than discourage desired behavior. Current decision theory, as described by Oliver and Twery (Vol. III of this book), would suggest that the current role of ID teams must be substantially revised.

Experience suggests that the most effective role for an ID team is to evaluate the environmental, economic, and social implications of alternative courses of action proposed by those outside of the team. This suggests that the task of identifying public issues and management concerns and developing the alternatives that would most effectively address them should be left to others. In addition, experience suggests that identifying the preferred alternative is a responsibility most appropriately given to the management team, not the ID team.

The ID team should continue to play a key role in federal land management planning. That role is likely to be most effective when it serves as an objective evaluator, not as a party with ownership or vested interest in a particular outcome. The current approach often tends to discourage such a role.

## 7 NEED FOR PERFORMANCE CRITERIA FOR MEASURING ACCOMPLISHMENTS UNDER ECOSYSTEM MANAGEMENT

Lack of identified performance criteria is a major barrier to effective implementation of ecosystem management.

### 7.1 Ecosystem Management Lacks Well-Defined Performance Measures

Performance criteria are needed to assess trends in environmental conditions and to better understand the biological and social implications of alternative approaches to sustaining ecosystems at various scales. Robin L. Lawton, a management consultant, writes: "Measurement is management's way of saying 'we care'" (Lawton 1993). Stated another way, what an organization measures in terms of accomplishments shows what is important to it.

In this light, the Forest Service publishes an annual report as required by law. The 1993 *Annual Report of the Forest Service*, titled "Ecosystem Management," contains 37 tables on the National Forest System, 11 on state and private forestry, 4 on research, and 10 on administration (USDA/Forest Service 1994b). Following the title page, the agency lists "Selected FY 1993 Statistics": 17 for the National Forest System, such as recreation use, lands burned by wildfire, wilderness, wildlife and fish habitat improvements, reforestation, livestock grazing, and timber harvested; the number of woodland owners assisted (190,256); research accomplishments (greater than 2,500 books, papers, articles, and reports); and human resource programs (135,556 people served). As indicated by these tables and statistics, what the agency "cares about" is resource outputs and activities in the national forests, funding, and number of employees, including their racial and gender composition. In contrast, the report states on page 2:

> "Ecosystem management has been embraced as the operating philosophy of the Forest Service....
>
> "Using an ecological approach to guide management of the nation's forests and grasslands is an important step in integrating the mission, vision, and guiding principles into agency operations....
>
> "The Forest Service has made some important progress during FY 1993 in defining and implementing the concepts of ecosystem management across the full spectrum of agency programs. Accomplishments include the development of an hierarchical ecosystem mapping and classification system, the completion of some major regional ecosystem assessments, and numerous pilot and demonstration projects."

No performance criteria and data sets are provided in terms of accomplishments in ecosystem management, allowing the question whether the agency really "cares" about it. Of course, a response might be: "Implementation of ecosystem management in its current form is new; forest ecosystems are complex, and much remains to be known about them." Such a response, however, is only acceptable in the near term. Unfortunately, no federal agency has defined performance criteria for ecosystem management.

Ecosystem management is a management strategy, a means to achieve multiple goals, such as biological diversity, ecosystem and resource sustainability, ecosystem integrity, social responsiveness, and risk minimization. These goals must be made explicit and quantifiable objectives established for each of them to measure progress in their achievement. Organizational accomplishments should be organized and presented in terms of these goals and objectives.

Although the lack of knowledge about forest ecosystems is recognized, it is not a justification for not identifying measures of success, performance criteria, for ecosystem management as a management strategy. If ecosystem management is to persist, it must not be an abstract experiment with little or no practical bearing. It is a way of organizing and using resources to address problems and seize opportunities in achieving organizational goals as well as implementing federal statutes. Hence, progress in implementation must be ascertained. Accomplishments in terms of performance criteria associated with identified goals and related objectives must be systematically evaluated.

Failures will occur, inevitable given the considerable voids in knowledge about forest and range ecosystems. On the other hand, this is the very reason adaptive management is a key feature of ecosystem management.

Performance criteria are critical when evaluating accomplishments. Fedkiw (1989), MacCleery (1992), and Frederick and Sedjo (1991) documented in separate studies the many accomplishments in forest management made in the United States during the 20th century, and each acknowledged the prominent role of the Forest Service in these accomplishments. Le Master et al. (1995) considered Forest Service accomplishments while distinguishing between performance criteria of the progressive conservation movement and those of the environmental movement. A different conclusion was reached. If Forest Service accomplishments are evaluated in terms of the performance criteria of the progressive conservation era, the agency has many. If its accomplishments are evaluated in terms of the environmental movement, the agency, arguably at least, has significantly fewer.

But it is difficult to measure progress against any standard without identified performance criteria. This lack causes debates on environmental issues to be more ideological than it ought to be. Information can help narrow the debate and focus it more constructively.

## 7.2 Criteria and Indicators for Sustainable Forest Management Could Form the Beginnings of Performance Measures for Ecosystem Management

Emerging issues related to sustainable forest management have focused on the development of criteria and indicators for determining whether forests are being managed sustainably. These efforts may have application to federal land management.

The ability to analyze and predict the sustainability of forests has been of particular interest since the Earth Summit in 1992 adopted the "Forest Principles." Since then, international and regional efforts to assess sustainable forest management at a national level have resulted in the development of several sets of criteria and indicators. These include the International Tropical Timber Organization Criteria and Guidelines for the Measurement of Sustainable Tropical Forest Management, the Helsinki process on Pan-European Criteria and Indicators for Sustainable Forest Management, the "Montreal Process" on Criteria and Indicators for the Conservation and Sustainable Management of Temperate and Boreal Forests, and the Regional Workshop on the Definition of Criteria and Indicators for Sustainability of Amazonian Forests.

The 12 countries of the "Montreal Process," which include over 90% of the world's temperate and boreal forests (Argentina, Australia, Canada, Chile, China, Japan, Korea, Mexico, New Zealand, Russia, United States, and Uruguay), have agreed to seven criteria for sustainable forest management. These are: (1) conservation of biological diversity, (2) maintenance of the productive capacity of ecosystems, (3) maintenance of forest ecosystem health and vitality, (4) conservation and maintenance of soil and water resources, (5) maintenance of forest contribution to global carbon cycles, (6) maintenance and enhancement of long-term multiple socio-economic benefits to meet the needs of societies, (7) legal, institutional and economic framework for forest conservation and sustainable management (Coulombe 1997).

In addition, 67 indicators have been developed to measure performance under the criteria. The Montreal Process countries just completed a First Approximation Report on the ability of the various countries to report on the Montreal Process criteria (USDA/Forest Service 1997a).

Forest Service Chief Dombeck has recently decided to begin to evaluate the use of these criteria and indicators to measure Forest Service performance under ecosystem management (Dombeck 1997). This is a solid first beginning to the identification and use of performance criteria for federal land management.

### 7.3 Natural and Human Capital Accounts

Successful integration of economic, social, and environmental considerations, as anticipated by ecosystem management, necessitates new approaches in accounting for such interactions. Over the past two decades, considerable attention has been given to developing alternatives to traditional measures of economic performance, such as gross domestic product (GDP). One significant problem with GDP is that it does not account for the depletion of natural capital upon which societies depend for their material sustenance (Daly 1996). Neither does it properly account for the gain or loss of many forms of human capital (e.g., education and a sense of social cohesiveness). Farber and Bradley, and Sedjo et al. (both in Vol. III of this book) discuss the considerable body of work that has focused on the development of natural capital accounts (see also Costanza et al. 1991). Although perhaps less developed in the field of natural resources than the development of natural capital accounts, work is also progressing in the development of social indicators. Parker et al. (Vol. III of this book) discuss the development of social indicators for ecosystem management (see also Machlis et al. 1994). Many countries are depleting their natural capital, at least in part, based on the expectation of increasing their human capital. That is what the United States did on a massive scale in the 19th century. Forests were turned into farms, charcoal for ironmaking, crossties to build the transcontinental railroad, etc. Today's Land-Grant colleges and universities and the State of Washington's school lands are two examples of the legacy of the 19th century U.S. policy of trading natural for human capital. A proper accounting theoretically should be able to track shifts of natural capital into social capital. It would be useful, for example, to be able to assess whether the depletion of tropical forests in a country is actually contributing to an increase in human capital in the area where it is occurring, or ending up in a Swiss bank account. Unfortunately, there remain substantial gaps between the theory and the practical reality of such an accounting. The World Bank has begun to explore approaches for quantifying natural and human capital shifts (Lutz 1993).

A World Bank report (1995) described the preliminary results of a study of the relative value of natural capital, human resources, and produced assets to the wealth of both developed and developing nations (WB 1995). This study found that produced assets, often the primary focus of traditional economic accounts, consistently contributed less to the wealth of nations than either natural capital or human resources. The study found that for high income countries, human resources account for 67 percent of national wealth while natural capital account for 17 percent, with produced assets at 16 percent. In developing nations that were not major raw material exporters, human resources account for 56 percent of national wealth, natural capital for 28 percent, with produced assets at 16 percent. While considerable work remains in developing such accounting methodologies, their use could contribute to more enlightened dialogue on natural resource/social tradeoffs than is possible with the accounting tools currently in use.

## 8 GREEN PRODUCT LABELING AND OTHER MARKET-ORIENTED APPROACHES TO IMPROVED FOREST MANAGEMENT

Over the last several years, there has been a significant increased interest in a variety of product labeling and other market-based approaches as a way to improve the management of forests (Coddington 1993). In a rapidly changing institutional landscape, two general approaches have emerged as mechanisms to assure forest products consumers that the products they are purchasing have come from responsibly managed forests. These are third party certification and self-certification.

Advocates for third party certification feel that conservation goals can be advanced through independent verification of forest ownerships as "well managed" or "sustainably managed." There is also hope that consumers will express a preference for forest products from certified ownerships or will pay a premium for such products.

The largest third-party certification program is led by the Mexico-based Forest Stewardship Council (FSC). FSC was established to seek voluntary, market-based approaches for improving forest management (Upton and Bass 1995). As discussed in Subsection 6.2.1, the World Wildlife Fund is promoting networks of protected areas, combined with third-party certification, as key elements of a global strategy to protect biodiversity and reduce the rate of species extinctions (WWF 1997).

FSC establishes certification standards and accredits those organizations that certify forest properties (FSC 1995). It does not itself certify forest properties. Two organizations are currently accredited by FSC to do business in the United States: the nonprofit Rainforest Alliance or "Smartwood" program, and the for-profit Scientific Certification Systems (Ervin and Pierce 1996). About 1.6 million acres are currently certified in the United States In order for forest products coming from FSC certified forests to wear an FSC label, they must also have a "chain of custody" certification, which is designed to assure consumers that the products actually came from a certified forest property (Ervin and Pierce 1996).

The largest self-certification program in the United States is the American Forest and Paper Association's "Sustainable Forestry Initiative," or SFI (AF&PA 1997). Under SFI, member companies must agree to meet minimum standards of responsible forest management to remain AF&PA members. The SFI is really not a certification program, but a quality assurance program. Among other things, SFI promotes programs to educate forest landowners in proper forest management and to improve the quality of logging by sponsoring logger education and logger certification. A significant percentage of U.S. industrial forest lands are encompassed by the SFI programs.

The quality assurance criteria used by both third-party certifiers and self-certifiers are similar, although the standards and details vary considerably. The criteria generally encompass three general areas: (1) timber sustainability (e.g., growth vs. harvest, protection of site productivity and soil and water resources, wood utilization efficiency) (2) protection of non-timber values (e.g., protecting riparian, wildlife and ecological values), and (3) socioeconomic responsibility (e.g., financial and business responsibility, employee and community relations).

Since certification and green labeling are relatively new, a number of unresolved issues need to be worked out. One of the most significant obstacles to expansion of third-party certification is that U.S. consumers have not yet shown a willingness to seek out and demand or pay a premium for certified forest products in any substantial volume (Vlosky and Ozanne 1997, Hansen 1997). Another issue is that "chain-of custody" requirements are seen by many non-industrial forest landowners as an obstacle to their participation since they commonly market their timber to mills that receive timber from a large number of other landowners (Hansen 1997). A third problem is that the focus of certification at the forest unit level makes it difficult to incorporate landscape scale considerations into local management decisions. FSC is currently developing regional level standards for certification that are intended, in part, to address the landscape scale issue.

Certification of forest products addresses only the management of forests, not the environmental impact associated with the use of wood vs. other materials. "Life-cycle analysis" has been advocated by some as a way to inform consumers of the relative environmental impacts of alternative materials. Lack of both apparent consumer demand, as well as lack of agreement on standards and procedures have been barriers to life-cycle analysis.

## 9 WESTERN FOREST HEALTH: A LITMUS TEST FOR FEDERAL FOREST MANAGEMENT?

The 1994 and 1996 wildfires in the West have highlighted a problem of forest health and fuel buildups that has been growing for decades. In many forest ecosystems in the West, forest biomass per acre has risen substantially since the 1940s. Today many forests have substantially more dense and fire-prone understories than those which existed in the past. Since 1952, forest biomass on national forest lands in the Interior West has increased by more than 44 percent (USDA/Forest Service 1993a).

This buildup is the legacy of fire exclusion and occurred even in the face of timber harvesting that averaged more than 2 billion board feet annually between 1960 and 1990. Today timber harvest levels are substantially reduced, and forest biomass continues to increase. Figure 9 illustrates the trends in forest growth and removals on national forest lands in the Interior West from 1952-1991. The difference between forest growth and removals represents the annual buildup in live forest volume or biomass on national forest lands.

Although past timber harvesting tended to focus largely on larger sized trees, biomass buildups have occurred largely in smaller diameter trees. The volume in trees less than 17 inches in diameter increased by 52 percent between 1952-92; and today such trees comprise two-thirds of total stand volume on all lands in

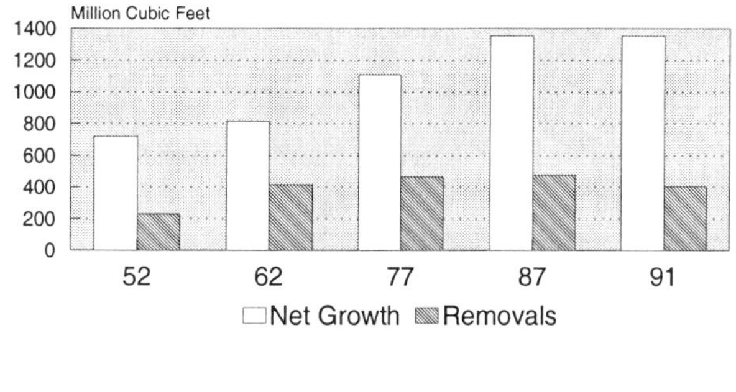


Fig. 9. National forest net forest growth and removals, Interior West 1952–1991.

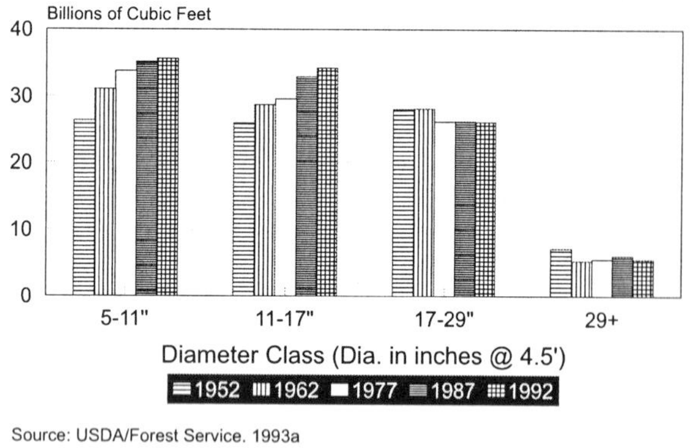


**Fig. 10. Trends in softwood timber volume by size class in all ownerships in the Interior West 1952–1992.**

the Interior West. The volume in trees greater than 17 inches in diameter has been stable since 1952 (USDA/ Forest Service 1993a) (see Fig. 10).

The above figures include live biomass only. There has also been a significant, but undetermined, increase in dead biomass in the Interior West. In many western forest ecosystems, biomass recycles primarily through fire, rather than decay. The large forest biomass increases in the Interior West cannot continue indefinitely. If humans do not make purposeful adjustments, nature will, as the 1994 and 1996 wildfires demonstrated.

## 9.1 Ecological Effects of Changed Forest Conditions

A number of studies have been made which use repeat photography comparing late 19th century landscape photographs to those recently taken from the same photo points (Gruell 1980, Gruell 1983, MacCleery 1994, Progulske 1974, Wyoming State Historical Society 1976). These studies demonstrate the profound ecological changes that have occurred over the last century in many western forest landscapes. These studies indicate not only an increase in the understory density and overstory biomass volume of forest vegetation over the last century but a decrease in both the aspen component and in the herbaceous understory in conifer stands. One study found that between 1962 and 1986, the area of aspen forests in the Southwest had dropped by 46 percent and predicted that if these trends continue, in less than three decades aspen will cease to exist as a distinct forest type in the Southwestern region (USDA/Forest Service 1993b). In addition, grasslands have become woodlands and open woodlands have become dense forests. Other studies strongly corroborate the existence of such changes (Covington and Moore 1994, Sampson and Adams 1993).

The ecological effects of reduced ecosystem fire depend on the nature of the forest. In warm, dry forest ecosystems that historically were characterized by frequent, low-intensity fires, elimination of fire leads rapidly to the development of a dense, multi-storied forest structure — subject to increasing mortality from drought, other forest health problems, and stand-replacing conflagrations, which seldom occurred in the past (USDA/FS 1993b). These historic low intensity fire adapted forest types, which are estimated to comprise more than one-third of the forest area in the Interior West, typically occupy lower elevations (USDA/FS 1993b).

In higher elevation forests, which tend to be cooler and moister, the ecological effects of fire exclusion are typically less profound than in lower elevation forests, at least over the short term. Many cool, moist forest ecosystems were historically subject to infrequent, stand-replacing fires. But even in these forests, reduction in historic fire has had substantial ecological effects in some areas. The aspen component has been substantially reduced, many meadows and openings have diminished in size or disappeared altogether, and existing forest stands have overstory trees which are older on-the-average than historically. The ecological diversity and "patchiness" of the forest landscape has been reduced. Such forests will be subject to increased insect epidemics and to larger and more intense stand-replacing conflagrations than typically would have occurred in the past (Barrett et al. 1991)

There are many gradations between warm/dry and cool/moist forest ecosystems. Many forests were subject both to relatively frequent low intensity fires, as well as occasional stand-replacing conflagrations when weather and forest conditions were right. When viewed from a spatial and temporal scale, the picture can be quite complex. Within the same general vegetation community, there are significant variations in fire frequency and effects over time, depending on periodic drought cycles, topography, and human influences. And within the same general landscape, there can be major variations in vegetation communities and fire behavior depending on soil characteristics, aspect, slope and related factors, e.g., north vs. south slopes. Brown (1994) has proposed a classification scheme for fire regimes that should assist in improving understanding and awareness of the complex processes involved.

These complexities make it difficult to simply characterize the vegetation changes that are occurring as the result of fire exclusion. Much of the debate on the western forest health issue is focused on these differences and opposing interpretations as to their relevance. While more research into the ecological effects of these changes is certainly needed, a general pattern does emerge. Due to changes in forest density, understory composition, and the kinds of tree species that are emerging to replace the existing forest canopy, the forests that are now developing will be decidedly more susceptible to insects, disease, drought, and catastrophic fire than those of the past. When fires do occur in such forests (as they inevitably will), they will tend to be intense, stand replacing, soil damaging fires, beyond that which would have been typical in the past.

Even high elevation forests, where stand replacing fires were the norm in the past, will likely be subject to larger and more intense fires due to increased structural homogeneity and reduced patchiness (Barrett et al. 1991). This will likely lead to reduced biological diversity over time.

There are also immediate watershed effects of the denser forests as well. It has long been known that the density of forest vegetation affects hydrologic functioning, particularly streamflows during the dry season. Today, many streams in western forest areas which were perennial streams a century ago now do not flow year around, and low flows in many others are lower than they used to be. Low flow rates affect water temperature and are a critical factor for fish and other aquatic life.

Acute forest health problems caused by biomass buildups are not universally a problem in Western forests. Such problems tend to be focused in forests in which frequent, low intensity fires were common in the past. The ecological effects of reduced ecosystem fire tend to be less in other ecological zones. But even in other ecological zones, fire was a major ecological factor and the modification of historic fire regimes can be expected to lead to substantial ecological changes over time.

## 9.2 Economic and Social Effects of Changed Ecological Conditions

If current trends continue, there will be: (1) increasing risks to federal and non-federal ecosystems from insects, disease, and conflagration events, (2) a rising toll of loss and degradation of watershed values and wildlife habitats from abnormally intense wildfires, as well as continued losses in the biological diversity of vegetation types that were historically characterized by frequent, low intensity fires, (3) increased risks to the human communities located in forested areas and to the fire fighters sent in to protect them, and (4) significant and increasing losses to taxpayers in fire suppression costs and resource values.

This situation provides a test for the operational application of ecosystem management principles. Although much of the focus of public attention and controversy in the West has been over human activities, such as livestock grazing, dam building, and timber harvesting, it is the reduction in ecosystem fire which has been the one of most significant human interventions of any that has occurred over the last century.

Addressing forest health will require substantially increased use of prescribed fire. But some Forest Service managers have estimated that many existing forests are now so dense in areas once characterized by frequent, low-intensity fire regimes that in less than 10 percent of such forests can prescribed fire be safely introduced without mechanical pre-treatment (Williams 1997). Returning fire to ecosystems alone, even on a carefully controlled basis, simply is not feasible in many situations. Ecosystem restoration will necessitate the use of mechanical treatment, in conjunction with increased use of prescribed fire, in restoring and maintaining ecosystem health (see Covington et al. this volume). Unfortunately, there remains no social consensus on the need for such active management to address forest health.

Forest management practices designed to restore historic ecological conditions will be different than those used to produce timber as a primary objective. Smaller material will be removed. There will be more use of thinnings, salvage, and other treatments that involve removal of only a portion of the trees on a site. Wood product values will be much lower and logging costs higher. There will also be a need for a substantial increase in the use of prescribed fire.

The potential exists for the use of silvicultural operations in support of forest health objectives to become win/win situations. The equipment and technology already exists to "tread lightly" and to carry out the management activities needed to maintain forest health. Some of these operations (but not all) will yield positive economic returns. Even when they do not, capturing timber values can reduce the total cost of the projects. These projects can reduce the substantial economic costs for wildfire suppression and site rehabilitation after severe wildfires. In 1994, the Forest Service spent almost one billion dollars on wildfire suppression and presuppression.

One of the greatest challenges in dealing with the western forest health issue will be to address the issue strategically, while identifying what can be done practically over the short term within existing funding constraints and parameters of social and political acceptability. Within that context, setting clear priorities for what can be done is essential.

It is likely that treating fuels next to at risk residential areas in the wildland/urban interface should be a high priority. Reasons for this include: (1) protecting such areas has accounted for an increasing share of federal and state fire suppression resources in recent years, (2) such areas are the sources of many fire starts during periods of high fire danger, and (3) reducing the risk to these areas will be essential before increased use of prescribed fire on adjacent federal lands is socially and politically feasible. Strong federal agency leadership will be a key factor in successfully forging a strategic view and the necessary social consensus on the use of mechanical treatment and prescribed fire for forest health. Substantial federal, state, and private financial assets will also need to be devoted to forest health issues. The existing and on-going regional ecosystem assessments that lay out the environmental and economic implications of the forest fuel buildups in specific areas could provide a foundation for the priority setting and consensus building for the actions needed to address the problem.

## 10 A LANDSCAPE APPROACH TO PRODUCING AND USING GOODS AND SERVICES UNDER ECOSYSTEM MANAGEMENT: A CASE STUDY FROM WESTERN WASHINGTON

Johnson et al. (this volume) describe several case studies designed to demonstrate the production of goods and services in an ecosystem management context. One of these, the Washington Landscape Management Project, is an example of an important effort to model the implications of ecosystem management at the landscape level (Carey et al. 1996).

The objective of the project was to explore ways in which landscape management can be implemented across an area of mixed ownerships in western Washington to meet the needs of wildlife associated with late seral forests while, at the same time, minimizing the economic and social impacts on the production of commodities. It illustrates how the biological requirements of all known wildlife and fish species might be met while still producing significant levels of timber products and economic values. The key is to protect and manage riparian and landslide-prone areas and to manage forest vegetation actively to minimize the area of forest in conditions that are biologically sterile.

The project was developed and administered jointly by the Washington Department of Natural Resources and the U.S. Forest Service Pacific Northwest Research Station. An interdisciplinary science team was selected to encompass a diversity of disciplines and views. It included Andrew B. Carey, Catherine Elliott, Bruce R. Lippke, John Sessions, Charles J. Chambers, Chadwick D. Oliver, Jerry F. Franklin, and Martin J. Raphael, as members.

The study suggested that managing under the maximum biodiversity option produces a clear net public benefit solution for multiple use public lands and trust lands. Biodiversity objectives can be achieved more rapidly through active forest management than if management is excluded and also produce significant, though not maximized, financial returns. Potential regional benefits from the maximum biodiversity option include: diversification of the wood products industry, increased secondary manufacturing, increased direct employment, increased indirect employment, reduced costs of unemployment compensation, and increased tax revenues. The details of gains vary markedly with assumptions.

## 11 THE "NEW" VERSUS THE "OLD" ECOLOGICAL THINKING: WHAT ARE THE IMPLICATIONS FOR ECOSYSTEM MANAGEMENT?

As already discussed, the last two decades have seen a substantial change in scientific understandings of the nature of the forest and grassland landscapes of pre-European contact North America and of the natural and human-induced influences that shaped them. This has included an increased recognition of the profound role of natural disturbance in those landscapes, as well as the significant role that indigenous peoples played in shaping many of them. This new scientific understanding calls into question many of the views of nature rooted in 19th century Romantic writings which have so strongly shaped the ideas of the modern environmental movement. These new understandings have initiated what John Fedkiw, former USDA natural resources analyst, refers to as a rethinking of the "old ecological thought" and a transition to a "new ecological thinking."

### 11.1 The "Old Ecological Thought"

The "old ecological thought" is based on an assumption of "balance" and "equilibrium" and of a clear and unequivocal separation of the natural from the human. Allan McQuillan wrote that the popular environmental position "sees nature as a virginal otherness, an object of deification, an Eden from which humans must be excluded" (McQuillan 1993). In *Second Nature*, Michael Pollan (1991) wrote of the tendency of the modern environmental movement to separate landscapes in to two categories: (1) those not substantially modified by human action, the pristine or "sacred" and (2) those that have been modified by humans, the secular, profane, or "fallen" landscapes:

> "'All or nothing,'" says the wilderness ethic, and in fact we've ended up with a landscape in America that conforms to that injunction remarkably

well. Thanks to exactly this kind of either/or thinking, Americans have done an admirable job of drawing lines around certain sacred areas....The reason is not hard to find: the only environmental ethic we have has nothing useful to say about those areas outside the line. Once a landscape is no longer 'virgin' it is typically written off as fallen, lost to nature, irredeemable."

A similar view has been expressed by Roderick Nash in *Wilderness and the American Mind* (1967). As previously discussed, since 1970 the environmental movement has vigorously and quite successfully pursued its goal of setting aside large chunks of federal lands in categories that prohibit or severely limit their modification by human activities.

## 11.2 The "New Ecological Thinking"

The "new ecological thinking" challenges the old on several fundamental fronts. It challenges the assumption of long-term natural equilibrium or stasis, for example that the natural or "normal condition" of all (or at least most) forests is a climax or old-growth condition. It challenges the unambiguous separation of landscapes into natural and humanized. For example, it recognizes that huge areas of North America, were, at the time of European contact, very much humanized landscapes. It challenges the implicit assumption that if landscapes are merely left alone and undisturbed by humans long enough, they will, through "natural regulation," return to what they were before European settlement.

But most importantly, the new ecological thinking forces a more holistic view of what is needed to sustain natural and human systems. It expands the previous focus on "protected areas" to encompass an understanding of how various components of the landscape are interconnected at various scales. As the focus expands to these other lands, the previous approach that viewed all human action as "disruptive" to "nature's balance" gives way to a more holistic perspective. This new holistic approach recognizes that setting aside land in reserves, while desirable, is not sufficient. It is also essential to manage lands outside of reserves in ways that maintains environmental values, while at the same time producing goods and services for human use effectively and efficiently (WWF 1997; DellaSalla 1997; M. Johnson et al., this volume; Carroll et al., this volume).

Environmental historian William Cronon in *Uncommon Ground: Toward Reinventing Nature* (1995), describes the growing divergence between the old and new ecological thinking and how it challenges many widely held assumptions about nature in the popular culture:

"(R)ecent scholarship has clearly demonstrated that the natural world is far more dynamic, far more changeable, and far more entangled with human history than popular beliefs about "the balance of nature" have typically acknowledged. Many popular ideas about the environment are premised on the conviction that nature is a stable, holistic, homeostatic community capable of preserving its natural balance more or less indefinitely if only humans can avoid "disturbing" it. This is in fact a deeply problematic assumption. The first generation of American ecologists, led at the start of the 20th century by the Nebraska scientist Frederick Clements, believed that every ecosystem tended to develop toward a natural climax community much as an infant matures into an adult. This climax, according to Clements and his followers, was capable of perpetuating itself forever unless something interfered with its natural balance.

Popular ideas of the natural world still reflect a fairly naive version of this belief, even though professional ecologists began to abandon Clementsian ideas almost half a century ago. By the 1950s, as Michael Barbour explains in his essay for this volume (in M. Johnson et al.), scientists were realizing that natural systems are not nearly so balanced or predictable as the Clementsian climax would have us believe and that Clements's habit of talking about ecosystems as if they were organisms — holistic, organically integrated, with a life cycle much like that of a living animal or plant — was far more metaphorical than real. Furthermore, the work of environmental historians has demonstrated that human beings have been manipulating ecosystems for as long as we have records of their passage. All of this calls into question the familiar modern habit of appealing to nonhuman nature as the objective measure against which human uses of nature should be judged. Recognizing the dynamism of the natural world, in short, challenges one of the most important foundations of popular environmental thought...."

## 11.3 Popular Resistance to the New Ecological Thinking

Many natural resource scientists and managers are now adapting their understanding of ecosystems and operating methods to respond to the new ecological thinking. For many citizens and environmental activists this transition will be a struggle because, as Cronon suggested, it challenges many dearly held views:

> Popular concern about the environment often implicitly appeals to a kind of naive realism for its intellectual foundation, more or less assuming that we can pretty easily recognize nature when we see it and thereby make uncomplicated choices between natural things, which are good, and unnatural things, which are bad. Much of the moral authority that has made environmentalism so compelling as a popular movement flows from its appeal to nature as a stable external source of nonhuman values against which human actions can be judged without much ambiguity. If it now turns out that the nature to which we appeal as the source of our own values has in fact been contaminated or even invented by those values, this would seem to have serious implications for the moral and political authority people ascribe to their own environmental concerns.

One of the factors in the resistance to the new ecological thinking is that once one accepts it, the absolutist standards and values of the past must fall away, and the world immediately becomes more complicated and ambiguous. If one accepts the necessity of some human intervention to maintain desirable ecosystem structure and function, one is faced with the messy question of how much, of what type, and where such intervention should occur. This is a difficult leap to make. Even some recent proposals to establish national biodiversity networks appear to suffer from an inability or unwillingness on the part of at least some of their proponent groups to separate biological science from those groups' traditional political agendas centered on expansion of "protected" areas within which no direct human intervention would be permitted (see Subsection 5.5.1). Considerable indirect human intervention, e.g., influence over fire regimes, would, of course, continue within "protected" areas.

## 11.4 Will Application of the Old Ecological Thought Be Sufficient To Maintain Natural Ecosystems in Their "Natural" or Pre-European Condition?

Huge areas of federal land are now allocated to statutory, administrative, or *de facto* set-asides which permit little or no active human intervention. The assumption that the areas so set-aside will somehow, through "natural regulation," be able to maintain themselves in a "natural balance" situation which approximates their pre-European contact biological conditions is today being tested in what amounts to a massive experiment. But it is an experiment that lacks some of the essential elements needed for success, e.g., both the willingness and the ability to let natural and aboriginal fire assume their primordial role in the landscape.

Even though the new ecology suggests that this experiment will likely fail, the old ecology continues to hold great sway.

A number of important questions are raised by this issue, including:

1. Are the romantic images of nature associated with the set-aside of large areas of federal land a sufficient basis to assume that these lands can be maintained as envisioned? Will some human intervention be needed to maintain the health and biodiversity of these ecosystem? If so, how much and what types of intervention?

2. What are the political, social, and environmental implications of not incorporating the emerging ecological thinking into resource management policy?

3. If biodiversity protection is now a *de facto* goal of the national forests, as suggested by former Forest Service Chief Jack Ward Thomas, can a fundamentally non-interventionist natural resource policy achieve it? What does our current knowledge and understanding tell us about the likely implications of non-intervention for biodiversity and other environmental values?

4. Is there a need to rethink the rationale for, and management of, reserves under ecosystem management? For example, does the new ecological thinking imply that some human intervention (e.g., the use of prescribed fire) is desirable, even essential, to achieve the objective that reserves reflect "natural" or pre-Columbian conditions? Are current federal policies and legal authorities adequate to respond to emerging management requirements for reserves?

5. What are the basic principles of the new ecological thought that can help give better insights into more holistically integrating the human and the natural? Will these new ideas reduce existing barriers to better integration? What would be required to make these new approaches work effectively under ecosystem management? Are even newer ideas and approaches needed before such integration can be successful?

It is not the intent of this chapter to answer these questions. Only to pose them in the hope that they may be the basis for further discussion and debate. The old and emerging ecological thinking must somehow be reconciled to achieve real progress in resolving the current polarization that exists in natural resources policy-making.

## 12 TOWARD MORE HOLISTIC CONNECTIONS BETWEEN RESOURCE PRODUCTION, CONSUMPTION AND SUSTAINABLE NATURAL AND HUMAN COMMUNITIES

For most of human history, people had direct and personal connections with the land as the source of their sustenance. Until the last half of the 19th century, no country on earth had less than half of its people engaged in food and agriculture (Hobhouse 1985). This relationship created well understood connections between what people consumed and the land that produced it. As resources became scarce through use, there existed built-in incentives for conservation. History suggests that, even though these incentives did not always work to eliminate resource degradation, they did tend to reduce it.

Things have changed dramatically in the United States and other developed nations during the last century. The personal connections between resource consumption and production have been severely weakened or even lost for many people. At the turn of the 20th century, the United States had more than half of its population directly engaged in farming, ranching, mining, and other forms of resource production. Today, that figure is less than 5 percent. In a world of farms, forests, and small towns, the linkages between food and fields and between forests and home and hearth were clear and sustained by personal experience. In a world of cities and suburbs, of offices and air conditioning, those linkages have become more obscure, and for many people, virtually non-existent.

Even though there are far fewer resource producers than ever before, we all remain resource consumers. It has sometimes been asserted that the cause of the substantial change in the management of the federal lands over the last two decades is a major "value shift" in the American public vis-a-vis attitudes toward the environment. If this "value shift" is a reality, there exists a major disconnect between those values and the personal behavior of most people as expressed in their consumption choices.

Today the U.S. population consumes more resources than at anytime in its history and also consumes more per capita than any other nation. Yet the resource consumption side of the conservation equation gets scant attention in contemporary debates on natural resources.

The stories of America's past are stories of the struggle to "subdue" the land, and to produce useful products from it (Lee 1994). The stories of America today are the stories of resource consumption — of shopping malls, of mail order catalogues, of fancy automobiles and recreation vehicles. The mottos of today in the United States appear to be "shop until you drop" or "whoever dies with the most toys wins." Since the first Earth Day in 1970, the average family size in the United States has dropped by 16 percent (from 3.14 persons in 1970 to 2.63 persons in 1990), while the size of the average single family house being built increased by 40 percent (from 1,482 to 2,080 square feet during the same period) (Census 1994, Census 1970–1990).

To the extent that land enters into contemporary American stories, it is usually in the nature of a scenic backdrop, a recreation experience, habitat for wildlife, or similar amenity use of land. When land which is used to produce products for human consumption enters into these stories, it is often in the context of resource degradation or damage.

This perceived disjunct between people as consumers and the land from which such products are derived is growing and is reflected in rising discord and alienation between producers and consumers. Loggers, ranchers, fishermen, miners, and other resource producers have all at times felt themselves subject to scorn and ridicule by the very society which benefits from the products they produce (Lee 1994).

What is fundamentally absent from much environmental discourse today is a recognition of the fact that today's urbanized society is no less dependent upon the products of the nation's forests and fields than were the subsistence farmers of America's past. This is clearly reflected in the language used in such discourse. Rural communities traditionally engaged in producing timber and other natural resources for urban consumers are commonly referred to as timber or natural resource "dependent" communities. Seldom are timber and resource dependent communities like Boulder, Denver, Detroit, or Boston ever referred to as such (Lee 1994).

The growing disconnect between resource consumers and producers is contrary to the interests of both and to the objective of responsible land management, as well. The reality is that we are all tied to the land by a variety of connections for both physical and spiritual sustenance. Most resource producers, as well as most environmentalists, are united by an intense love of the land and a desire to protect it and the web of life it supports (Lee 1994). The reality is that our cities and rural areas, often considered as separate and even hostile entities, are not two places but one (Cronon 1991). The human communities of city and country are connected by a myriad of economic and social interdependencies which link them for mutual benefit.

Seeking to help remake the psychological connections between responsible resource consumption and production must be seen as an important task if ecosystem management is to be successful. The solution is not just to seek to diversify the economic base of rural communities, regardless of how desirable that may be. Diversification does nothing for the responsible consumption side of the equation. It may even be environmentally perverse if the effect is to further alienate resource production or merely to shift resource production to ecosystems which have less responsible management practices.

Remaking the connections between consumption and production will be a formidable task. The causes of the disconnects are complicated and the solutions will be as well. Part of the solution will be to adopt the approaches to conservation already discussed in Sections 6 and 7 of this chapter. Public education is also needed on what is encompassed by responsible consumption and responsible production.

But what is needed more than anything else is the emergence of a new conservation perspective and ethic. Botkin has written about how human views of nature have evolved over time, i.e., nature as "living organism," nature as "divine order," nature as "great machine," nature as "biosphere" (Botkin 1990). What is needed today is a new, more holistic, conservation perspective. Both the strict preservation and strict utilitarian sides of the conservation spectrum that today get the bulk of popular attention are flawed at their roots. Both are based upon an assumed dualism between people and nature — that nature and humans are separate and separable. Neither the worship of nature undisturbed nor the worship of unfettered consumption provide useful solutions for the future.

This new perspective would seek to forge a partnership between humans and nature, based on respect and reciprocity — of wisdom and science, rather than ideology. It would merge responsible production with responsible consumption. It would recognize that massive set-asides of land will neither recreate nature as it once was nor serve most human needs. It would acknowledge, as well, that unregulated economic growth will not result in the protection of essential ecological linkages across the landscape or be sufficient to serve human needs in the long run.

This new perspective would seek to explore actions needed to achieve a sustainable future for humans and other living things. It would recognize that the reasons for protecting nature and natural processes are moral and aesthetic, as well as utilitarian. The legacy represented by the natural world is our own legacy as well. It would acknowledge that humans are a product of natural processes and in turn influence those processes. Human actions have influenced nature for so long that artificial separations are difficult or impossible. Human actions have shaped Idaho wilderness as they have Iowa cornfields. The new perspective would understand that it is not just those landscapes that are relatively "undisturbed" by human action that are worthy of respect. It sees an ethical obligation to protect the productivity and soil of Iowa cornfields, as it does the biodiversity of Idaho wilderness. But it also recognizes that human intervention in the form of managed disturbance regimes is sometimes needed to maintain the ecological function of natural areas, to preserve existing characteristics old-growth forests or to restore those characteristics to second-growth forests.

The new understanding would respect those humans who toil to produce the products we all use. It would also establish a personal obligation and commitment on the part of all of us to be responsible consumers. It would thus seek to add a new and critically important dimension to Aldo Leopold's land ethic. The "personal consumption ethic" would recognize the critical linkage between consumption choices and the land. It would seek to build a new partnership between resource producers and consumers — one which combines responsible production and responsible consumption. As with the new partnership with nature, the new partnership between producers and consumers would be based on the concept of respect and reciprocity. It would reject the enemy-making that is too common today, the scapegoating of "environmental radicals" on one hand, or of "corporate greed" on the other.

The history of human interrelationships with nature suggests one of increasing sophistication and understanding. Meeting the challenges of the future on a small planet with ever more people to feed, cloth and house will require an increasingly steep learning curve. Ecosystem management is only the most recent footstep on this path to better understanding. The key to meeting the challenges of the future will be a healthy dose of humility and the creation of new partnerships between humans and nature and between producers and consumers that are based on reciprocity, respect, and good science, rather than doctrine, ideology, and divisiveness.

## 13 SUMMARY AND CONCLUSIONS

Nine principles were listed at the outset to provide a foundation for construction of a basic framework for ecosystem management. They were followed by a discussion of the evolution of the use and management of land and natural resources in the United States, as well as the evolution of federal land and environmental policy. An essential reason for an ecological approach to federal land management is to conserve and restore ecosystems and species at risk. The reasons for ecosystems and species decline, the timing, and the human benefits accruing from the changes need to be understood if conservation and restoration efforts are to be successful.

Ecosystem management was compared with multiple use–sustained-yield management. While there are similarities between the strategies, important differences also exist. Specifically, the goals of ecosystem management for federal land are ecosystem and resource sustainability, maintenance of high levels of biological diversity, ecosystem health or integrity, social responsiveness, and risk aversion. The means by which these goals are to be accomplished, again for federal lands, include a systems or holistic approach and a landscape perspective, multiple scales of management, extensive use of public and private partnerships, public involvement in decision making, multidisciplinary management teams, and adaptive management.

Even though being responsive to changing social values is a frequently heard principle in federal land management, its meaning is ambiguous in an operational sense and a rule-based system is suggested for integrating social values in natural resource decision making.

An actual or *de facto* policy of promoting, achieving, or maintaining community stability by federal land management agencies should be developed to respond to the actual capabilities of such agencies to carry them out. A more realistic policy would be for agencies to avoid actions that disrupt local communities in carrying out their programs and activities. In the event of necessary changes in programs that would potentially disrupt a local community, measures should be taken to phase in such changes to mitigate social impacts.

Performance criteria for evaluation of accomplishments in ecosystem management need to be established soon because it is an important change in the management of federal lands and has significant ecological, economic, and social implications. Nevertheless, it should also be recognized that performance criteria for ecosystem management will change with expanding knowledge about forest ecosystems.

The fundamental test of ecosystem management is whether it can assist in forging a working consensus on federal land management. To do that, ecosystem management must:

- Recognize better the link between federal lands and regional and national economic and social systems;
- Include resource output objectives;
- Recognize the value of knowledge of history and varied human experience;
- Recognize that judicious intensification of resource management for commodity production is both inevitable and desirable on some lands;
- Develop objective measures of ecosystem health and sustainability;
- Encourage further research especially as it relates for ecosystem structure and functioning;
- Address the role of private lands.
- Recognize that federal agency cooperation and coordination must be improved and appropriate consideration must be given to the long term risk of failing to manage landscapes.
- Identify the elements of successful public participation and the proper role of the interdisciplinary team in federal land management planning.
- Properly address the role of timber management in an ecosystem context.
- Provide for more effective incorporation of science into federal resource decision-making.

The current forest health issue in the Interior West associated with buildup of both live and dead forest biomass will be a key test of the effectiveness of ecosystem management. There is promise, however, which is illustrated by a case study of the Washington Landscape Management Project.

But some psychological baggage still remains in the way of development of rational natural resources policies. Not the least of these is a clash between current scientific knowledge of how natural systems operate and the romantic images of the past that still motivate and drive public discourse on the environment. The old and new ecologies must be reconciled before real progress can be made in resolving the current polarization that exists in natural resource policy-making.

Ecosystem management has great promise as a management strategy for federal lands. To achieve that promise, it must be integrated better within the biological, economic, management, and social sciences. Misunderstandings about ecosystem management must also be reduced so it can contribute to the development of a consensus on the goals of federal land management.

There is also an important need to make more holistic connections between resource production, consumption, and sustainable natural and human communities. A new perspective is needed which seeks to establish new partnerships based on respect, humility, solid science, and reciprocity between resource consumers and producers, and between humans and nature.

## REFERENCES


AF&PA. 1994. *1994 Statistics: Paper, Paperboard, and Wood Pulp*. American Forest and Paper Association, Washington, DC.

AF&PA. 1997. Sustainable forestry for tomorrow's world: second annual progress report on the American Forest and Paper Association's sustainable forestry initiative. American Forest and Paper Association, Washington, DC.

Ball-Rokeach, S.J., and W.E. Loges. 1994. Value Theory and Research. In: Edgar F. Borgatta and Marie L. Borgatta (eds.), *Encyclopedia of Sociology, Vol. 4*. Macmillan, New York.

Barrett, S.W., S. Arno, and C.H. Key. 1991. Fire Regimes of western larch–lodgepole pine forests in Glacier National Park, Montana. *Canadian Journal of Forestry Research* 21: 1711–1720.

Botkin, D. 1990. *Discordant Harmonies: A New Ecology for the Twenty-first Century*. Oxford University Press, New York.

Brown, J.K. 1994. Fire Regimes and their relevance to ecosystem management. Presented at the Society of American Foresters/Canadian Institute of Forestry Convention, Sept. 18–22, 1994, Anchorage, Alaska.

Carey, A.B., C. Elliott, B.R. Lippke, J. Sessions, C.J. Chambers, C.D. Oliver, J.F. Franklin, and M.J. Raphael. 1996. Washington forest landscape project: a pragmatic, ecological approach to small-landscape management. Washington Landscape Management Project Report No. 2. Washington Department of Natural Resources, Olympia, WA.

Census 1970–1990. Characteristics of new housing: current construction reports. Series C25. Compilation of 1970-90 C25 reports. U.S. Department of Commerce, Bureau of the Census, Washington, DC.

Census 1975. Historical statistics of the United States from colonial times to 1970. Bicentennial edition, Part 1. U.S. Department of Commerce, Bureau of the Census, Washington, DC.

Census 1994. Statistical Abstract of the United States, 1994. U.S. Department of Commerce, Bureau of the Census, Washington, DC.

Census 1997. Statistical Abstract of the United States, 1997. U.S. Department of Commerce, Bureau of the Census, Washington, DC.

CEQ 1989. Environmental Trends. Council of Environmental Quality, Executive Office of the President, Washington, DC.

Coddington, W. 1993. *Environmental Marketing*. McGraw-Hill, New York.

Costanza, R., H.E. Daly, and J.A. Bartholomew. 1991. Goals, agenda, and policy recommendations for ecological economics. In: R. Constanza (ed.), *Ecological Economics, The Science and Management of Sustainability*. Columbia University Press, New York.

Coulombe, M. 1997. Toward sustainable forest management! Address by Mary J. Coulombe to the Pulp & Paperworkers Resource Council, August 4, 1997. USDA/Forest Service, Washington, DC.

Covington, W.W., and M.M. Moore 1994. Southwestern ponderosa pine forest structure: Changes since Euro-American settlement. *Journal of Forestry* 2: 73–76.

Cronon, W. 1991. *Nature's Metropolis: Chicago and the Great West*. W.W. Norton, New York.

Cronon, W. 1993. The uses of environmental history. *Environmental History Review*, Fall 1993. Environmental History Society.

Cronon, W. (ed.). 1995. *Uncommon Ground: Toward Reinventing Nature*. Norton, New York, 561 pp.

Dahl, T.E., and C.E. Johnson 1991. *Wetland status and trends in the cotermimous U.S.: mid-1970s to mid-1980s*. Fish and Wildlife Service, U.S. Department of the Interior, Washington, DC.

Daly, H.E. 1996. *Beyond Growth*. Beacon Press, Boston.

DellaSalla, D.A., J.R. Strittholt, R.F. Noss, and D.M. Olson. 1996. A critical role for core reserves in managing Inland Northwest landscapes for natural resources and biodiversity. *Wildlife Society Bulletin* 24(2): 209–221.

DellaSalla, D.A., A. Hackman, W. Wettengel, D. Olson, K. Kavanagh, and E. Dinerstein. 1997. Protection and certification: a shared vision of North America's diverse forests. Delivered at the Forests for Life Conference, San Francisco, CA: World Wildlife Fund, May 9, 1997.

Dionne, Jr., E.J. 1996. *They Only Look Dead*. Simon & Schuster, New York.

Dombeck, 1997. Letter from Forest Service Chief Mike Dombeck to Paul D. Frey, President, National Association of State Foresters, dated July 23, 1997, explaining anticipated Forest Service efforts to incorporate Montreal Process Criteria and Indicators into agency programs and activities.

DPC 1988. Outdoor recreation in a nation of communities: Action plan for Americas outdoors. A Report of the Task Force on Outdoor Recreation Resources and Opportunities to the Domestic Policy Council, Washington, DC. 6/88.

Egler, F.E. 1977. *The Nature of Vegetation, Its Management and Mis-management: An Introduction to Vegetation Science*. Aton Forest, Norfolk, CT.

Ervin, J. and A. Pierce. 1996. Status report on forest certification in the United States: a report for the forest stewardship council, U.S. initiative. June 1996.

FAO 1993. Forest resources assessment, 1990: tropical countries. *FAO Forestry Paper 112*. Food and Agricultural Organization of the United Nations, Rome.

FSC. 1995. *Principles and Criteria for Natural Forest Management*. Forest Stewardship Council, Oaxaca, Mexico.

Fedkiw, J. 1993. Natural resources: federal spending and resource performance — 1940–1989. USDA-Office of Budget and Program Analysis, Washington, DC.

Fedkiw, J. 1989. The evolving use and management of the nation's forests, grasslands, croplands, and related resources. General Technical Report RM-175, September 1989, USDA-Forest Servicem Washington, DC.

Fedkiw, J. 1993. Natural resources: federal spending and resource performance 1940–1989, U.S. Department of Agriculture, Office of Budget and Program Analysis, Washington, D.C. 1/93.

Foreman, D. 1995/96. Wilderness: from scenery to nature". *Wild Earth* 5(4), Winter 1995/96, Cenozoic Society, Richmond, Vermont.

Frederick, K.D. and R.A. Sedjo, ed. 1991. *America's renewable resources: historic trends and current challenges*. Resources For the Future, Washington, DC

Gerlak, A.K., H.J. Cortner, and Dennis C. Le Master. 1996. Responding to changing social values in natural resource decision making. Unpublished paper.

Gruell, G.L. 1980. *Fire's influence on wildlife habitat on the Bridger-Teton National Forest, Wyoming*, Vol. I: Photographic record and analysis, INT-252, 207 p. Vol. II: Changes and causes, management implications. INT-252. Intermountain Forest & Range Experiment Station, Ogden, UT.

Gruell, G.E. 1983. Fire and vegetative trends in the northern Rockies: interpretations from 1871–1982 photographs. Intermountain Forest and Range Experiment Station, General Technical Report INT-158, USDA/Forest Service, Ogden, UT. December 1983.

Grumbine, R.E. 1994. What is ecosystem management? *Conservation Biology* 8: 27–38.

Hansen, E. 1997. Forest certification and marketing. *Journal of Forest Products* 47 (3) 3–97.

Hobhouse, H. 1985. *Seeds of Change: Five Plants that Transformed Mankind*. Harper & Row, New York.

Howard, J.L. 1997. U.S. timber production, trade, consumption, and price statistics 1965–1994. Gen. Tech. Rep. FPL-GTR-98. U.S. Department of Agriculture, Forest Service, Forest Products Laboratory, 6/97, Madison, WI.

Huston, M. 1993. Biological diversity, soils, and economics. *Science* 262 (December 10).

Kohler, T.A. 1992a. Prehistoric human impact on the environment in the upland North American Southwest. *Population and Environment: A Journal of Interdisciplinary Studies* 13 (4).

Kohler, T.A. 1992b. Guest editorial: the prehistory of sustainability. *Population and Environment: A Journal of Interdisciplinary Studies* 13 (4).

Kuusela, K. 1994. Forest resources in Europe 1950–1990. European Forest Institute, Research Report 1. Cambridge University Press, Cambridge, UK.

Langner L.L. and C.H. Flather 1994. Biological diversity: status and trends in the U.S. USDA/Forest Service, General Technical Report RM-244, Rocky Mountain Forest and Range Experiment Station, Fort Collins, CO. 4/94.

Lawton, R.L. 1993. *Creating a Customer-Centered Culture*. ASQC Quality Press, Milwaukee, WI.

Lee, R.G. 1994. *Broken Trust, Broken Land: Freeing Ourselves from the War Over the Environment*. Book Partners, Wilsonville, OR.

Le Master, D.C. On expanding the supply of forest resources from Federal forest lands, pp. 44–67. In: C.S. Binkley, G.D Brewer and V.A. Sample (eds.), Redirecting the RPA. Yale School of Forestry and Environmental Studies, New Haven, CT.

Le Master, D.C., J.T. O'Leary, and V.A. Sample. 1995. Forest Service Response to Changing Public Values, Policies and Legislation during the Twentieth Century in the United States. Paper presented during the session of the Forest Law and Environmental Legislation Subject Group of the XX World Congress of the International Union of Forestry Research Organization, Tampere, Finland, August 6–12, 1995.

Le Master, D.C. and J.H. Beuter. 1989. *Community Stability in Forest-based Economies*. Timber Press, Portland, OR.

Lipset, S.M. 1996. *American Exceptionalism*. W.W. Norton and Company, New York.

Lyon, J., T.N. Lonner, J.P. Weigand, C.L. Marcum, W.D. Edge, J.D. Jones, D.W. McCleerey, and L.L. Hicks. 1985. Coordinating elk and timber management: final report of the Montana Cooperative Elk-Logging Study. Montana Cooperative Elk-Logging Study Research Committee. Intermountain Forest and Range Experiment Station, Ogden, UT.

Lutz, E. (ed.). 1993. *Toward Improved Environmental Accounting*. World Bank/UNSTAT, Washington, DC.

MacCleery, D.W. 1992. *American forests: a history of resiliency and recovery*. USDA/Forest Service, FS-540. In cooperation with the Forest History Society, Durham, NC. Also published in 1993 by the Forest History Society.

MacCleery, D. 1994. Repeat photography for assessing ecosystem change: A partial listing of references. USDA/Forest Service. Washington, DC (unpublished).

Machlis, G.E., J.E. Force, S.E. Dalton. 1994. Monitoring social indicators for ecosystem management. Technical paper submitted to the Interior Columbia Basin Project. USDA / Forest Service, Portland, OR.

Magill, F.N. (ed.). 1994. *Survey of Social Science: Sociology Series*, Vol. 5. Salem Press, Englewood Cliffs, NJ.

McQuillan, A.G. 1993. Cabbages and kings: The ethics and aesthetics of new forestry. *Environmental Values* 2(3): 191–221.

Nash, R. 1967. *Wilderness and the American Mind*. Yale University Press, New Haven, CT.

National Journal. 1996. Opinion Outlook, March 2, 1996/No. 9: 502.

Noss, R.F. 1992. The wildlands project: land conservation strategy. *Wild Earth* (Special Issue): 10–25.

Noss, R.F., and A.Y. Cooperrider. 1994. *Saving Nature's Legacy: Protecting and Restoring Biodiversity*. Island Press, Washington, DC.

Noss, R.F., E.T. LaRoe III, and M.J. Scott 1995. Endangered Ecosystems of the U.S., Biological Report No. 28, Washington, DC: U.S. Department of the Interior, 2/95.

Perez-Garcia, J.M. 1995. Global economic and land use consequences. *Journal of Forestry* 93 (7): 34–38.

PCSD (President's Council on Sustainable Development) 1996. *Sustainable America*, Washington, DC.

Pollan, M. 1991. *Second Nature: A Gardener's Education*. The Atlantic Monthly Press, New York.

Progulske, D.R. 1974. *Yellow Ore, Yellow Hair, Yellow Pine: A Photographic Study of a Century of Forest Ecology*. South Dakota Agricultural Experiment Station Bulletin 616.

Robinson, S. 1997. The case of the missing songbirds. *Consequences Magazine*, 6.

ROD 1994. Record of decision for amendments to Forest Service and Bureau of Land Management planning documents within the range of the northern spotted owl, April 1994, p.20, and regional planning documents related to suitable acres. USDA/Forest Service and USDI/Bureau of Land Management.

Rokeach, Milton. 1973. *The Nature of Human Values*. Free Press, New York.

Ross, M., E.D. Larson, R.H. Williams. 1987. Energy demand and materials flows in the economy. *Energy* 12 (10/11): 953–967.

Sampson, R.N. and D.L. Adams (eds.). 1993. Assessing ecosystem health in the Inland West. Overview papers from an American Forests scientific workshop. November 14–19, 1993, Sun Valley, ID. American Forests, Washington, DC.

Sedjo, R.A. 1989. Forests to offset the greenhouse effect. *Journal of Forestry* 87(7): 12–15.

Sedjo, Roger A. 1995. Local Logging — Global Effects. *Journal of Forestry* 93 (7): 25–27.

Steen, H.K. 1976. *The U.S. Forest Service: A History*. University of Washington Press, Seattle, WA.

Stein, B.A., and S.R. Flack. 1997. *Species Report Card: The State of U.S. Plants and Animals*. The Nature Conservancy, Arlington, VA.

Thomas, J.W. 1979. *Wildlife habitats in managed forests: the Blue Mountains of Oregon and Washington*. Agriculture Handbook No. 553, USDA/Forest Service, Washington, DC.

Thomas, J.W. 1990. Wildlife. In: *Natural Resources for the 21st Century*, American Forestry Association. Island Press, Washington, DC.

Thomas, J.W. 1996. The instability of stability. 11 p. (The same argument is expressed in JWT's transcript of the 1993 Forest Conference.)

Trefethen, J.B. 1975. *An American Crusade for Wildlife*. Winchester Press and the Boone and Crockett Club, New York.

Upton, C. and S. Bass. 1995. *The Forest Certification Handbook*. Earthscan Publications, London.

USDA/Forest Service. 1993a. Forest Resources of the United States, 1992. General Technical Report RM-234, Ft. Collins, Colo: Rocky Mountain Forest and Range Experiment Station.

USDA/Forest Service. 1993b. *Changing conditions in Southwestern forests and implications on land stewardship*. Southwestern Region. Albuquerque, NM.

USDA/Forest Service. 1994a. *President's Forest Plan*. Government Printing Office, Washington, DC.

USDA/Forest Service. 1994b. Ecosystem Management: 1993 Annual Report of the Forest Service. Government Printing Office, Washington, DC.

USDA/Forest Service. 1995. The Forest Service Program for Forest and Rangeland Resources: A Long-Term Strategic Plan. Government Printing Office, Washington, DC.

USDA/Forest Service. 1996a. Land Areas of the National Forest System – 1995, FS-383, Washington, DC, 1/96.

USDA/Forest Service. 1996b. National forest cut and sold reports, FY 1950 to date, USDA/Forest Service, Washington, DC.

USDA/Forest Service. 1997a. First Approximation report for sustainable forest management. Report of the United States on Criteria and Indicators for the Sustainable Management of Temperate and Boreal Forests, 6/97, USDA/Forest Service, Washington, DC.

USDA/Forest Service. 1997b. *TRACS-Silva*, Table 20. Forest Management. USDA/Forest Service, Washington, DC.

USTR (Office of the U.S. Trade Representative) 1996. U.S. and Canada Reach Agreement on Softwood Lumber. Press Release, February 16, 1966. Executive Office of the President, Washington, DC.

Vance-Borland, K., R. Noss, J. Strittholt, P. Frost, C. Carroll, and R. Nawa. 1995/96. A biodiversity conservation plan for the Klamath/Siskiyou Region. *Wild Earth* 5(4), Cenozoic Society, Richmond, Vermont.

Vlosky, R.P., L.K. Ozanne. 1997. Forest products certification: the business customer perspective". *Wood and Fiber Science* 29(2), 195–208, Society of Wood Science and Technology.

Williams, C.E. and M.E. Lathbury. 1996. Economic incentives for habitat conservation on private land: applications to the Inland Pacific Northwest. *Wildlife Society Bulletin*, 24(2): 187–191.

Williams, M. 1989. *Americans and Their Forests: An Historical Geography*. Cambridge University Press. New York.

WRI/IUCN/UNEP. 1992. *The Global Biodiversity Strategy*. World Resources Institute, International Union for the Conservation of Nature, and United Nations Environment Programme, Washington, DC.

WWF. 1997. WWF guide to forest certification: 97 (English). WWF-UK, Panda House, Weyside Park, Godalming Surrey GU7 IXR.

Wyoming State Historical Society. 1976. *Re-discovering the Big Horns; a pictorial study of 75 years of ecological change*. Cheyenne, Wyoming.

## THE AUTHORS

**Douglas W. MacCleery**
*USDA Forest Service*
*P.O. Box 96090*
*Washington, DC 20090-6090, USA*

**Dennis C. Le Master**
*Department of Forestry and Natural Resources*
*1159 Forestry Building*
*Purdue University*
*West Lafayette, IN 47907-1159, USA*

# Ecosystem Management and the Use of Natural Resources

Marlin Johnson, James Barbour, David W. Green, Susan Willits, Michael Znerold, James D. Bliss, Sie Ling Chiang, and Dale Toweill

**Key questions addressed in this chapter**

- ♦ *Ecosystem use and sustainability*
- ♦ *Compatability between short-term use and maintenance of long-term sustainability*
- ♦ *Healthy ecosystems and healthy economies*
- ♦ *Resource extraction enhances environmental values*

**Keywords: Forestry, grazing, mining, resource planning, ecosystem health, resource demand**

## 1 INTRODUCTION

Ecosystems have been supporting human life for more than 10,000 years in North America and for even longer in other parts of the world. People have used their environment to extract the goods needed for survival — obtaining food from wildlife, wild plants, cultivated crops, and livestock, and finding shelter from natural areas or making it from plants. Public lands in the United States still supply many of these needs.

In the past, some societies used resources in ways and quantities that were not sustainable. Archeological records from North America indicate that the Anazasi of Mesa Verde and other regions had seriously depleted their forest resources before the sites were abandoned (Cartledge and Propper 1993). Archeological records from other parts of the world tell a similar story — many societies declined after deforestation or improper forest management. For example, the Sumerian empire in Mesopotamia collapsed by 2000 BC, after deforestation of mountains and improper methods of irrigation led to salinity of irrigation water and greatly reduced crop yields. Crete became a commercial power a few centuries later, primarily as a result of its abundant wood supply. Wood was hauled great distances to supply other nations that had lost their forests. Venice, Rome, Cyprus, Egypt, England, and other countries all declined as commercial powers when their forests became depleted (Perlin 1989). Conversely, forest lands that were properly managed enhanced the society's standard of living. In one case, 16th and 17th century Sweden supported its military efforts and empire through use of timber resources (Sundberg et al. 1994).

As outlined in MacCleery and Le Master (this volume), human population growth and resource demand are inextricably linked. Both are increasing at a substantial rate. It is important on both the local and the global scale that ecosystems and resources be maintained and managed to continue to provide a broad array of resources to meet the physical, recreational, and spiritual needs of people.

Management of resources on public lands in the United States has been evolving since the late 1800s, when the conservation movement helped start a system of federal landholdings designed to protect the land while providing for various resource uses. After World War II, an increased level of prosperity created new and greater demands for beef, wood products, and minerals, and a more mobile population created demands for additional recreation opportunities and other services that federal lands could provide. Federal land managers responded to these demands and provided an increasing level and variety of goods and services to the public. During the 1960s and 1970s, a series of legislative mandates provided direction to federal agencies to manage federal lands for a broad array of products and services within a sustainable framework. At the same time, the rise of the environmental movement forced the country to become aware of the effects of increased production of goods and services on the environment. Clean air, clean water, threatened species, biological diversity, and healthy forests and rangeland became increasingly important goods and services provided by federal lands (MacCleery and Le Master, this volume). A new management philosophy, ecosystem management, evolved for federal lands.

Ecosystem management, as currently practiced on public lands in the United States, means managing the lands in ways that ensure, within reasonable limits, that the functionality of damaged ecosystems is restored and healthy ecosystems are sustained. It does not mean that resources are set aside and not used; rather, these lands must continue to satisfy human needs. Its goal is to provide productive biologically diverse ecosystems and ensure quality of life by strengthening the essential connection between economic prosperity and environmental well-being (Interagency Ecosystem Management Task Force 1995).

Under the ecosystem approach, goals are developed based on predictions of sustainability and activities are designed to achieve the desired goals at a landscape (regional) level. This regional or ecosystem scale is one of the major differences between the traditional multiple-use management of the past and the ecosystem management of today (MacCleery and Le Master, this volume). Although scale issues are a critical component of the success of ecosystem management, planning on regional scales also vastly increases the complexity of management and may create problems in development of criteria to evaluate the accomplishments of land managers (see MacCleery and Le Master, this volume).

Resource managers, who operated efficiently and effectively when agency mandates were clearly understood, public participation was limited, and resources were abundant, often find themselves taxed to the limit given the complexities associated with landscape-level decisions and increased demand from growing populations. When establishing standards for healthy functional ecosystems, resource managers must consider a range of biological, geological, climatic, and political factors. At the landscape level, where landownership patterns are often complex, federal, state, and private management mandates vie for consideration. At the same time, interest groups have become more effective at using the courts and politics to bring pressure to bear on the decision-making process. First-line managers from different agencies also operate at widely varying scales, from Ranger Districts to mineral basins, and from state boundaries to vast river basins. With few exceptions, none of these established boundaries interrelates between agencies, leading to frustration as managers attempt to integrate plans and policies.

The goal of this chapter is to provide guidance to first-line natural resource managers and resource specialists charged with "on the ground" implementation of ecosystem management. These are the managers who must simultaneously manage resources under their jurisdiction to produce marketable commodities and provide a wide range of non-market amenities within a framework that ensures sustainability of both on a long-term scale, while maintaining functionality in the ecosystem. The complexities of ecosystem management require managers to deal with new issues and consult with a wide range of specialists. Many agencies have hired people in new fields such as landscape ecology, sociology, and decision support modeling. When the appropriate skills are not readily available on-staff, managers who want to make the best decisions must reach out and find these skills.

Although this chapter concentrates on opportunities to find compatibility among potentially competing resource uses, tradeoffs also must be made. Resource managers charged with making these sometimes difficult decisions can benefit from all the chapters in this publication. From *Data Management* to *Ecosystem Diversity* to *Public Expectations* and everywhere in-between, this book is designed to assist a manager in these decisions.

This chapter also considers some of the various values that people derive from the ecosystems we manage. The products of ecosystem management, as defined in this paper, are all the resources sought from public lands. They include the materials removed from the land as well as desirable recreational and aesthetic values. Some examples of resources sought from public lands are wildlife and fish, recreation, minerals, wood fiber, grazing for livestock, clean water, and many special products such as Christmas trees, mushrooms, and berries, and the healthy sustainable plant communities that produce them.

Examples are presented of approaches that resource managers and scientists have taken to implement the broad policy directions provided by landscape-level assessments. The scale and scope of these examples range from treatments applied to a few acres, to integrated management plans incorporating millions of acres of land managed under multiple ownerships and jurisdictions. Although the examples vary greatly in focus, they share the common theme of attempting to find ways to understand how best to manage ecosystems to provide multiple benefits at the landscape level rather than produce single-commodity outputs with little or no regard for how each treatment unit relates temporally and spatially to the landscape. Additionally, this chapter will demonstrate ways to calculate just what goods and services can be expected now and in the future while managing for these goals. The authors believe it is critical that managers find a way to integrate the planned production of goods and services into ecosystem management, for without that focus, ecosystem management will be guilty of the same narrowness as the systems it has replaced (see MacCleery and Le Master, this volume).

The approaches are grouped into six emphasis areas: (1) fish and wildlife, (2) recreation, (3) riparian wetland areas, (4) rangelands, (5) nonrenewable resources, and (6) forest management. Case studies are summarized in Table 1.

## 2 BASIC CONCEPTS OF ECOSYSTEM MANAGEMENT

Any time public land managers make a decision about actively managing the resources, they must deal with a range of issues and concerns about why the action is being taken and what impact it will have on the biological, economic, and social well-being of the land and associated communities. To deal with these issues effectively, land managers must address the following themes related to implementing ecosystem management:

1. *Identify clear objectives* — The objectives need to be based on biological capacity, long-term disturbance history, and economic and social considerations. These objectives must include meeting society's needs for natural resources, either consumptive or nonconsumptive.
2. *Deal with political boundaries* — Managers must recognize and work with different governments (federal, tribal, state, and local) and agencies (Fish and Wildlife Service, National Marine Fisheries Service, U.S. Geological Service (USGS), USDA Forest Service, and USDA Bureau of Land Management (BLM)), all of which have different jurisdictions, regulations, laws, constituents, and functions. Some agencies are set up to map, monitor, and provide development opportunities (USGS), while others try to manage to provide the broadest spectrum of use and protection (Forest Service); even within agencies, rules and regulations vary (Forest Service vs. BLM).

Table 1. Ecosystem Management Case Studies.

| Resource | Example | Locat ion | Scale/scope | Participants | Approach |
|---|---|---|---|---|---|
| Fish and Wildlife | Boise National Forest | ID | Landscape | Forest Service | Conduct risk assessment of multiple management alternatives. |
| Fish and Wildlife | Feral pig management | HI | Landscape | U.S. Park Service | Restore natural ecosystems while maintaining cultural and recreational values. |
| Recreation | Idaho SCORTP | ID | Statewi de | State and federal agencies, private groups | Develop partnerships by defining common goals. |
| Riparian wetlands | Muddy Creek project | WY | Watershed | Federal, state, and private land managers | Develop consensus to restore, enhance, and maintain proper functioning condition. |
| Rangeland | Arroyo Colorado Allotment | NM | Landscape | Federal, state, and private land managers | Improve and maintain health of rangelands without reducing grazing level. |
| Nonrenewable resources | Columbia River Basin Assessment | ID, MT, OR, WA | Landscape | Federal agencies | Conduct quantitative assessment of mineral deposits within an ecosystem framework. |
| Forest management | Augusta Creek project | OR | Landscape | Forest Service | Evaluate simulations of future landscape and watershed conditions for habitat, timber, and disturbance risk. |
| Forest management | Washington Landscape Study | WA | Landscape | Forest Service, university | Evaluate management alternatives that meet wildlife and timber objectives. |
| Forest management | Ponderosa Pine Forest Partnership | CO | Landscape | County, National Forest, and timber industry | Develop partnership to unite forest health with community sustainability. |
| Forest management | Crowley Project | AZ | Landscape | Forest Service | Enhance aspen, diversity, recreation, and timber products through management. |
| Forest management | Westside examples | OR | Stand | State, federal, and local agencies; private industry; individuals | Use adaptive management scenarios to integrate habitat, timber, structural diversity, and risk. |
| Forest management | Eastside examples | OR, CA | Stand | Federal and state agencies, private companies | Evaluate alternative management regimes in fire origin stands of inland NW for effects on stand structure and health. |

3. *Tie scales together* — Activity planning starts at the landscape and watershed levels but actually gets implemented at the stand, pasture, or campground level. Evaluations at various scales can give completely different results. Decisions need to be made and evaluations done at the correct geographic and temporal scales.

4. *Work closely with public* — The public should be seen as something more than a group of people who must be educated. The public needs to be involved throughout the activity planning process to help establish common goals and action plans and be treated as partners. The combined support and energy of a larger group is a key component to the success of implementing ecosystem management. Lack of understanding of short-term and long-term results and consequences between alternatives is one of the major factors leading to disagreements on alternative management scenarios.

5. *Assess alternatives for multiple resources* — Integration of multiple disciplines, legislative authorities, scales, and time frames is the art of ecosystem management. Managers are responsible for assessing alternative ways to meet varying needs, including some that are in conflict with each other. Conducting risk assessments of multiple management alternatives for a variety of resources is a key element of this responsibility. The ability to develop possible actions that can be combined into alternatives to maximize achievement of multiple goals simultaneously is important.
6. *Deal with economics and social needs* — Cost effectiveness of treatments, economic well-being of communities, and effective use of scarce financial resources need to be included in the overall evaluation of projects.
7. *Implement, adapt, and monitor* — The most effective tool that a manager has is to actually conduct business on the ground as promised and then monitor the results to determine if the activity produced acceptable results. Adapting management and activities to new information learned from research and monitoring is important to producing healthy, diverse, and productive natural resources.

## 3 CASE STUDIES

### 3.1 Fish and Wildlife

Among the most challenging problems facing managers of public lands in the United States is the management of fish and wildlife. Frustrations occur partially because federal land managers do not have direct authority over fish or wildlife populations, yet management practices are supposed to ensure viable populations of these species. As a practical matter, this means that each land manager not only faces constraints imposed by the most sensitive of (typically) 300 to 500 recognized fish, amphibian, reptile, bird, and mammal species, but also must deal with the interests and legal requirements of those agencies that do have management authority. That authority is split among several agencies. State fish and wildlife management agencies have the mandate for managing resident species (i.e., those fish and wildlife species that typically reside within a drainage basin year around, such as most fish species, amphibians, reptiles, nonmigratory birds, and most mammals) and particularly game species (e.g., trout, bass, deer, elk). Most migratory species (and all species classified as threatened or endangered under provisions of the Endangered Species Act of 1969 as amended) are managed by federal agencies. The National Marine Fisheries Service manages threatened and endangered anadromous fish and marine mammals, and the U.S. Fish and Wildlife Service manages migratory birds and threatened and endangered nonmigratory fish, amphibians, reptiles, birds, and mammals. These agencies often have limited flexibility. In the case of threatened and endangered species, constraints are imposed not only by federal law (National Forest Management Act, Endangered Species Act), but also by international law (Convention on International Trade in Endangered Species) and treaties. State agencies typically face economic constraints as well. In most instances, funding for agency operations is based on the sale of licenses and tags for fishing and hunting, which often occur on public lands.

There are other, less obvious, reasons associated with the frustration public land managers often feel when faced with the management of fish and wildlife. The first of these is a recreational (and therefore, social) concern. Fishing and hunting is a major, if not the major, recreational activity on public lands, attracting many people and generating tremendous economic impacts (U.S. Fish and Wildlife Service 1993). Second, public concern is often driven by the perception of scarcity (and therefore, value) of public land, particularly areas that are deemed as special places because of some perceived unique quality, such as a roadless area. Because these services are perceived as both unique and scarce, perceived public value for protection of fish and wildlife resources (usually through protection of particular areas) is often very high. These concerns are provoked and aggravated by commodity production values for timber or livestock forage. As private goods and services, timber and grazing are relatively abundant and market-driven, providing direct economic benefits to a small private sector and indirect benefits to end-users, but at some cost to non-market, public resources. Thus, decisions that may affect fish and wildlife populations are almost always controversial.

In addition to such social concerns, some subtle biological concerns exist. Management implies manipulation, and manipulation of habitats often results in long-term changes to both the terrain and the development of vegetation (i.e., fish and wildlife habitat) through time. Both have important implications. Road development increases sedimentation in area waters and directly affects the degree of silt deposited in stream gravels. Silt, in turn, affects the viability of aquatic insect populations and both the food base and spawning area available to fish populations. Roads, even if closed to human travel, provide attractive travel corridors for most medium and large wildlife species, changing patterns of habitat use. Because roads are also attractive to hunters, they increase the vulnerability of game animals to harvest (an effect vastly multiplied when roads are left open to vehicular access). Changes in vegetation often have a bewildering number of both immediate and long-term impacts on fish and wildlife populations (Maser 1988). Habitats are fragmented (Harris 1984), and because the turnover rates of fish and wildlife populations differ dramatically from stand recruitment rates (and from each other), changes through time will not be in synchrony with current situations.

As a simple example, the harvest of a single tree in a riparian area can have immediate effects: a reduction in stream shading (which contributes to warming of the water temperature); reduction in local food supply (seeds, buds, insects) for a variety of amphibians, reptiles, birds, and small mammals; elimination of critical nesting habitat for songbirds and small mammals; and effects on travel corridors (e.g., screening and resting places) for a wide range of species, from amphibians to birds and large mammals. Over the long term, harvest of this same tree might reduce the contribution of woody debris (essential for fish hiding places and stream thermoregulation) to the stream; eliminate a potential snag necessary for woodpeckers, owls, or other birds or mammals, and subsequently the fallen wet, rotting wood necessary to support a local amphibian population; and (by opening the canopy) change the site to a younger seral stage of willows, supporting an entirely different group of local species.

So what does a manager do? Because these concerns are unavoidable, are there any guidelines? In fact, there are. The following examples demonstrate, at least in part, key factors that affect public land decision-making as identified by Yaffee (1994) in his analysis of the spotted owl controversy in the Pacific Northwest. These factors include the heightened complexity of management issues associated with expanding and conflicting public values, the ambiguous and conflicting norms of collective choice, and the inherently complicated future environmental issues. Yaffee also identified a reduced capacity to meet demands, resulting from declining slack in the natural resource base and in the ability of government to act proactively because of discouraging fiscal realities, unstable coalitions because of fragmented power and interests, and limited vision and guidance from elected and appointed officials and the management institutions they control. Based on his analysis, Yaffee identified four essential components for building more effective agencies and the decision-making process:

1. New mechanisms to bridge the agency–nonagency boundary to build understanding and political concurrence.
2. Altered approaches to organizational management, including updated notions of leadership.
3. Improved means of gathering and analyzing information about resource problems, organizational possibilities, and political and social context.
4. Ways to promote a culture of creativity and risk-taking to generate more effective options for the future.

### 3.1.1 Evaluation of Timber Sales Effects on Forest Birds on the Boise National Forest in Idaho

This case study focuses on approaches used to identify the effects of a proposed timber sale on the long-term viability of two forest bird species, the pileated woodpecker (*Dryocopus pileatus*) and flammulated owl (*Otus flammeolus*), documented by Erickson and Toweill (1994).

*Case study attributes*
Scale: Secondary
Scope: Environmental analysis
Instrument: Formal
Participants: Forest Service and state fish and game agency staff
Duration: 30-year planning horizon

*Background*
The species of interest in this example have specific habitat needs that feature mature stands of timber. Past harvest practices changed the composition of native stands, reducing suitable patch size, altering dominant timber stands, reducing and fragmenting suitable habitat, and increasing susceptibility to further alteration through effects of insect pests, fire, and normal patterns of plant succession. The forest had experienced catastrophic increases in insect and disease infestations, associated with an increase in the number of stems per acre following decades of fire prevention and livestock grazing in a fire-dominated ecosystem. Wide areas were facing high risks of extensive stand-replacing wildfires, and treatment was clearly demanded, both to reduce risks and to increase the potential for future productivity. An analysis of hazards clearly demonstrated that the risks pertained to not only the vegetation in the area but also the continued viability of fish and wildlife populations. In other words, the no-action alternative itself contained a threat of massive ecological change because of clearly identified risks.

*Geographic area*
The area included in this analysis was the 15,000-acre Logging Gulch Timber Sale area, plus adjoining habitats within the potential dispersal range of the bird species of concern.

*Project description*
This analysis examined the amount of suitable habitat and its present distribution, and it identified potential risks and plant succession patterns to evaluate the distribution of suitable habitats to the 30-year planning horizon.

The analyses focused on key species and critical habitats currently available for each (spatial analysis). Direct impacts to critical wildlife habitats were modeled for a variety of alternative management proposals (including no action). Then, a species-by-species population risk model featuring population demographic characteristics [minimum size of required habitat], habitat quality and distribution, and vulnerability to catastrophic events (wildfire, flood, landslides) was applied to the remaining critical habitats. Five- and 30-year projections of vegetation response to potential current hazards and each proposed management alternative were then examined to estimate effects of changes in habitat quality and distribution through time associated with each alternative. Each projection was analyzed to ensure that critical habitats for each species were well distributed and in sufficient proximity to each other throughout the analysis period to maintain viable populations. These models were developed and reviewed with other agencies having management responsibilities, providing each agency a basis for full evaluation of potential risks and benefits of each alternative, including the no-action alternative. Although not reported here, similar analyses were completed for impacts to several populations of game animals.

*Outcome*
The outcome of this approach was to explore graphically the full range of predictable risks of all alternatives, including the no-action alternative, given what was known about the functional ecosystem and specific habitat requirements of the species of concern. This approach significantly reduced confusion about the potential effect of proposed actions and resulted in much interagency consensus and support.

*Lessons learned*
The first and most important lesson is that vegetation regimes are constantly changing naturally, and that lack of direct intervention by human activities does not equate to protection in perpetuity. Any long-range planning to ensure ecosystem sustainability must explicitly identify and strive to predict natural changes in succession as well as changes resulting from human activity. The second lesson is related to the temporal scale of changes — to sustain ecosystems, all critical components for any species must be maintained and be accessible to the organisms of concern throughout the entire planning cycle. Many plants and most animal populations turn over completely (and some many times) within the typical tree life cycle, and loss or inaccessibility of critical habitat components during any portion of this period can result in loss of sustainability of natural resources. The lesson is based, explicitly or intuitively, on risk assessment. Landscapes are exposed to many risks, from fire to invasion by undesirable species. Ensuring ecosystem sustainability demands a conservative approach to landscape alteration on a temporal scale, and it should be accompanied by redundant safety mechanisms such as several discrete areas of habitat for a given species in the event of unforeseen losses.

*Evaluation*
The approach used can be generalized to any fish and wildlife species (or group). However, it is information-intensive, which limits its application to a small number of wildlife species of particular interest if time or funding is limited.

*Contact persons*
John Erickson, Forest Wildlife Biologist, Boise National Forest, Boise, Idaho 83702 (tel. 208-364-4100); Dale Toweill, Wildlife Program Coordinator, Idaho Department of Fish and Game, Boise, Idaho 83707 (tel. 208-334-3180).

### 3.1.2 Removal of Pigs From Hawaiian National Parks

Although it is easy to see that the protection and management of native species is vitally important from an ecosystem management viewpoint, management of non-native species is less clear. This example identifies some approaches used to manage a non-native species.

*Case study attributes*
Scale: Secondary
Scope: Management plan
Instrument: Formal
Participants: Interagency
Duration: Indefinite

*Background*
Pigs were probably initially brought to the Hawaiian Islands by Polynesian settlers between 1,200 and 1,500

years ago. Populations of feral pigs were consequently supplemented by releases of domesticated European pigs (Baker 1979, Vtorov 1993), and feral populations reflect strong influence of European stocks. Feral pigs have been implicated in alteration of native flora and fauna, both through direct actions such as foraging (Singer 1981) and indirect actions such as development of wallows (Baker 1979) and dispersal of non-native plant species. Although removal of pigs has been advocated by restoration biologists, removing pigs from an area will not by itself eliminate problems associated with non-native plant species (Huenneke and Vitousek 1990).

Pigs are important in Polynesian culture and provide recreational hunting opportunities for residents (Anderson and Stone 1993). Support for elimination of feral pigs from the ecosystem is not universal, and removal efforts are strongly opposed by some groups. State forests are managed to provide both recreational hunting and sustained yield of feral pigs, among other things (Katahira et al. 1993), and most control efforts have been limited to National Parks. Past pig control efforts on National Parks included public hunts (Stone and Loope 1987), but hunts proved to be largely ineffective when populations were low, partly as a result of ingress of pigs from other areas. Although poisoning is used elsewhere (Hone and Stone 1989), it is not acceptable in Hawaii because of the potential for adverse secondary poisoning effects and social concerns.

### *Geographic area*

Hawaiian Islands, specifically Hawaii Volcanoes National Park on the island of Hawaii and Haleakala National Park on the island of Maui.

### *Project description*

Two projects are described here. The first, the Hawaiian Volcanoes National Park or HAVO Project (Katahira et al. 1993), was conducted on the island of Hawaii. The second, the Kipahulu Valley Project, was conducted in the Haleakala National Park on the island of Maui (Anderson and Stone 1993).

The HAVO Project focused on three of nine fenced enclosures totaling 19 acres in the Hawaiian Volcanoes National Park. Pig control methods included hunting with dogs, trapping, baiting, and snaring. Eradication was achieved in all nine fenced units over the course of 3 years. Professional hunting with dogs was the most effective control method, although public hunting proved ineffective as a method of eradication. In the Kipahulu Valley Project, snaring was the only method used to eradicate pigs in two fenced units. This project was apparently successful in eradicating pigs from one unit and greatly reducing their numbers in the other. The success of this project led to the adoption of snaring techniques by groups such as the U.S. Fish and Wildlife Service, the Hawaii Department of Forestry and Wildlife, The Nature Conservancy Council of Hawaii, the Maui Land and Pineapple Company, and other landowners who manage natural areas in remote locations.

### *Outcome*

Where pig control is an objective, secure barriers are necessary to confine pigs or restrict their access. Sturdy fences with barbed wire at ground level are effective but require constant maintenance. The methods of fence construction and therefore the costs depend on the types of animals being confined or restricted (Katahira et al. 1993). Maintenance of fences in areas with cattle requires taller and sturdier fences with a second strand of barbed wire along the top (Hone and Atkinson 1983). Snaring pigs in unfenced areas can considerably reduce populations, but may not eliminate the pigs. Without continued control efforts, populations will quickly reach precontrol levels (Anderson and Stone 1993).

However, it is still unclear how, or if, the removal of pigs by itself will contribute to the restoration of natural Hawaiian ecosystems. Available evidence suggests that some characteristics of the natural ecosystems, such as soil microarthropods, may return to normal without further active management. Even when pigs are removed, other non-native species such as rats and snails remain. The ability of these animals to prevent the restoration of natural ecosystems will be an important factor in determining the success of restoration efforts.

### *Lessons learned*

Changes in ecosystem function may be associated with a single invasive species, but they are often associated with impacts of multiple species. Although gains can be made by significantly reducing or eradicating some undesirable species, recovery may be impossible without a multifaceted approach. Eradication of undesirable species, where possible, is time-consuming and difficult. It often demands developing and maintaining impervious barriers to recolonization. Control methods outside of impervious barriers can reduce populations over the short term but only as long as efforts are maintained; populations may recover quickly when control efforts are reduced or terminated.

### *Contact person*

Tim Tunison, Hawaii Volcanoes National Park, Hawaii (tel. 808–967–8226).

### 3.1.3 Idaho Comprehensive Outdoor Recreation and Tourism Plan

The Idaho comprehensive outdoor recreation and tourism plan (SCORTP) provides an example of the many opportunities provided by an interagency, federal–state cooperative framework.

*Case study attributes*
Scale: Primary
Scope: State management plan
Instrument: Formal
Participants: Interagency
Duration: Definite

*Background*
Each state is required to prepare a comprehensive outdoor recreation plan to be eligible for certain matching funds under the Land and Water Conservation Fund Act of 1965.

*Geographic area*
Statewide

*Project description*
Under the leadership of the Idaho Department of Parks and Recreation, development of the Idaho SCORTP was broadened to include participation by other state agencies with a role in recreation (including the Departments of Recreation, Commerce, Fish and Game, and Water Resources), federal agencies (including six Idaho National Forests, BLM, Bureau of Reclamation, and National Park Service), and private groups (including the Idaho Association of Counties, Idaho Association of Cities, Idaho Recreation Initiative, and Idaho Foundation for Parks and Lands). Participants identified 15 recreation goals, such as to improve maintenance and provide recreation and tourism infrastructure and services, to promote a unified communication and marketing program, and to promote and maintain high quality fish and wildlife recreation opportunities. Each goal was reviewed by all participants, who identified whether the goal was central to their organization's mandate (allowing them to assume a leadership role), included within the mandate (allowing formation of partnerships), not excluded by the mandate (allowing a supporting role in some situations), or excluded by the mandate.

*Outcome*
The primary outcome of this exercise was the creation of a partnership framework that identifies opportunities for forming partnerships. Moreover, the synergy created by this process resulted in explicit identification of objectives and resources, which has resulted in pooling of efforts, expertise, timing, and (in some instances) funding. Thus, more is being accomplished under this unified framework than would have occurred had the partners separately pursued their own interests. In addition, this process has enhanced the opportunity to mesh projects; for example, while one agency planned to develop a public boat ramp, another developed a riverside park in a nearby area. This process has also reduced the potential for conflicts; open communication about plans prevent situations such as promoting recreational fishing in an area where fish populations are depressed.

*Evaluation*
This open process resulted in an interagency network of people with differing perspectives but common goals of providing recreational opportunities in Idaho. Once objectives were identified, many potential avenues for future development of partnerships and pooling of limited resources were identified, creating synergy between the partners and smoother delivery of recreational opportunities to a wide range of customers.

*Contact person*
Jake Howard, SCORTP Project Leader, Idaho Department of Parks and Recreation, Boise, Idaho (tel. 208–334–4180).

## 3.2 Riparian Wetland Areas and Effects of Livestock Management

Although riparian wetland areas constitute less than 9 percent of the 270 million acres of public lands being managed by the BLM, these areas are the most economically and environmentally valuable. In 1991, the BLM launched a nationwide program called the Riparian Wetland Initiative for the 1990s (Platts et al. 1987, Debano and Schmidt 1989, Myers 1989, Welsch 1991, Elmore and Kauffman 1994). One of the chief goals of this initiative is to restore and maintain riparian wetland areas so that 75 percent or more of riparian areas are properly functioning by 1997.

The overall objective is to achieve the widest variety of vegetation and habitat diversity for wildlife, fish, and watershed protection. This objective is important to remember because riparian wetland areas will function properly long before they achieve an advanced successional stage. It is also well to remember that the management goals for the area and the corresponding desired plant community may not correspond with the potential plant or natural community.

The functioning condition of a riparian wetland area is a result of interaction among hydrology, land form/soils, and biology. Riparian wetland areas are functioning properly when adequate vegetation is present (1) to dissipate stream energy associated with high water flow, thereby reducing erosion and improving water quality, (2) to develop the filter sediment and flood plain, (3) to improve floodwater retention and groundwater recharge, (4) to develop root masses that stabilize the streambank against erosion, (5) to develop diverse ponding and change characteristics to provide proper habitat and water depth, and (6) to provide shade for extended duration with cool temperatures necessary for fish production, breeding, and other uses and support greater biodiversity.

The definition of proper functioning condition (PFC) is then translated into a set of minimum national standards consisting of checklist criteria for determining the PFC. The checklist criteria are developed by a national-level interdisciplinary team for three components: hydrologic, biological, and erosion deposition. The process of assessing whether a riparian wetland area is functioning properly requires a team of specialists in vegetation, soils, and hydrology. A biologist also needs to be involved because of the high fish and wildlife values associated with riparian wetland areas. After each riparian wetland area is assessed, the area is classified into one of four categories: PFC, functional at risk, nonfunctional, and unknown. For areas that are functional at risk, an assessment should be made of the trend (upward, downward, or not apparent).

Management actions are then developed to consider such factors as critical water quality problems, potential for improvement, risk of further degradation, threatened and endangered species habitat, and fisheries and recreational values. Areas identified as functional-at-risk with a downward trend are often the highest management priority because a decline in resource value is apparent but can usually be restored in a cost-effective manner. As most riparian values have already been lost, restoration of a nonfunctional area is often not cost-effective and usually receives a low priority.

The effectiveness of each action must be assessed as management actions are being implemented through various prescriptions, such as regulating livestock grazing practices while accommodating uses; developing water for dispersed grazing; planting trees, shrubs, and grasses; constructing fences; and conducting prescribed burns. Progress toward meeting PFC must be documented through monitoring. Sites should be revisited periodically as part of the overall monitoring program, which reflects long-term trends. With a change in management, most riparian wetland areas can achieve PFC in a few years, although some will take years to achieve the identified desired plant community or advanced ecological status such as late-seral and natural plant diversity conditions.

When determining whether a riparian wetland area is functioning properly, the condition of the entire watershed is important, including the upland and tributary watershed system. The entire watershed can influence the quality, abundance, and stability of downstream resources by controlling production of sediment and nutrients, influencing streamflow, and modifying the distribution of chemicals throughout the area. Although a healthy riparian wetland area does not necessarily indicate a healthy watershed, an unhealthy watershed will eventually cause damage to downstream riparian areas.

## Muddy Creek Project

The Muddy Creek drainage is located in south–central Wyoming in the upper Colorado River. This watershed encompasses nearly 300,000 acres of mixed federal, state, and private lands in Carbon County and had become rather degraded. Although there was no formal assessment and classification of the riparian area, it was certainly in the functional-at-risk condition, at best, before restoration work began in the early 1990s.

Management plans had been developed for the entire watershed. However, in the 1990s, a coordinated resources management (CRM) group was initiated to focus on management in the upper half of the perennial headwaters of the drainage. The CRM project, one of the original National Seeking Common Ground demonstration projects, was initiated by the local conservation district to promote consensus among all affected interests as opposed to confrontational management of the natural resources in the project areas. To date, more than 25 members are working together to restore, enhance, and maintain the abundant resources in the area while maintaining the economic stability and cultural heritage of the people on the land.

Throughout the watershed, improvements in the health of rangelands (including riparian) have been the result of shorter duration of use and improved management rather than reduction in livestock numbers. The following techniques are being used:

- Water is piped from the creek to a tire trough; the overflow returns to the creek. These sites have reduced the effects of cattle trailing and trampling along stream banks.

- Upland water development is resulting in better distribution of livestock on land and is reducing impacts on riparian areas by both livestock and wildlife.
- Cross fencing is being used to divide large pastures to shorten the duration of livestock use in a given area. The fencing is built to address wildlife concerns; for example, barbed wire fences have a smooth bottom wire to allow small game and antelope to pass under the fence.
- Prescribed burning is used to restore the ecological balance that was lost in the last 100 years of fire suppression. Such burning increases grass cover, which reduces soil erosion, and also increases plant diversity, thus improving habitat and forage for wildlife and livestock.
- Several types of in-stream structures are utilized to repair and improve the riparian zone and fisheries habitat. Although the natural system will eventually repair itself, these structures speed up the process.
- Nearly 10,000 seedlings of many plant species have been planted in the past 2 years to accelerate woody plant revegetation. Woody plants are important for bank stability, stream shading, and wildlife habitat. Revegetation of native woody plants is often a slow process, but it is important to the healing of the riparian habitat.
- Roads are sources of sediment. In addition to educating the public who drive through the area, several measures are being used to address this problem: placing water bars across roads, eliminating or replacing stream crossings with culverts, rerouting roads, signing roads for voluntary non-use, and closing roads.

*Lessons learned*

Wildlife, livestock, and all the associated natural resources, including the proper functioning riparian areas in the watershed, have improved since the initiation of the project. The greatest indication of success is the people story: many people with diverse backgrounds and interests who are working together to develop trust, respect, and commitment to an overall vision and conservation ethic on land management.

*Contact persons*

Eric Luse, Washington Office, Bureau of Land Management, 1849 C Street, N.W., Washington, DC 20240, (202) 452-7743; Wayne Elmore, Prineville District, Bureau of Land Management, 185 East 4th Street, P.O. Box 550, Prineville, OR 97754 (tel. 503–447–4115).

## 3.3 Rangelands

The BLM is steward for 177 million acres of western public rangelands. During the past 30 years, public land-users and managers have learned much about how nature works. The BLM recognizes the progress that has been made in improving public lands, but at the same time it believes that greater success requires a broader approach, one that considers more fully how living things interconnect and affect each other. Success also requires enabling all people who share an interest in public lands to collaborate in finding lasting solutions. The current grazing regulations reflect these ideas (Herbel 1985, Laycock 1991, Cool 1992, Sharpe et al. 1992, National Research Council 1994).

### 3.3.2 BLM Goals and Practices

The goals of BLM rangeland management are:

1. to improve rangeland health to provide lasting benefits for users of public rangelands and future generations,
2. to assist rural western communities in building stable economies on a foundation of sustainable resources, and
3. to ensure that public lands users have a meaningful say in managing public lands.

The grazing regulations require establishment of resource advisory councils (RACs) to provide meaningful participation in BLM resource management programs. Councils represent diverse interests, employ consensus decision-making, and can provide advice to the BLM on land management issues. The RACs play an important role in helping to design state or regional standards and guidelines. Regulations also require the establishment of standards and guidelines for grazing administration, which should be developed at the state or regional level to reflect geographic differences and to involve stakeholders. Standards and guidelines must be based on the fundamentals of rangeland health, which emphasize improving watersheds, restoring areas near streambeds, protecting water quality, and supporting healthy plant and animal communities.

Many rangelands on public lands in the western United States are not healthy by current standards, mainly as a result of improper grazing practices (e.g., overgrazing), lack of adequate facilities (e.g., sources and distribution of water, fencing, and cattle guards), and out-of-date management plans for improving rangeland condition. To restore these lands to a healthy condition could mean a permanent or temporary reduction in grazing levels for the allotment. In the following success story, the allotment has been managed to restore and maintain the health of a rangeland without reduction in animal unit months (AUMs) (an AUM is the amount of forage needed to sustain one cow, five sheep, or five goats for a month) allowing year-long grazing use in the allotment. There have been situations where a reduction in AUMs was necessary to maintain the health of the rangeland. In most of these cases, operators were persuaded to adopt the new practices without appeals. These are success stories as well.

A full AUM fee is charged for each month of grazing by adult animals if the grazing animal (1) is weaned, (2) is at least 6 months old when entering public land, or (3) will become 12 months old during the period of use. For fee purposes, an AUM is the amount of forage used by five weaned or adult sheep or goats or one cow, bull, steer, heifer, horse, or mule. The term AUM is commonly used in three ways: (1) stocking rate, as in *x* acres/AUM, (b) forage allocation, as in allotment A, and (3) utilization, as in *x* AUM consumed from unit B.

### 3.3.3 Arroyo Colorado Allotment

*Geographic area*

The Arroyo Colorado allotment is located 35 miles west of Los Lunas, New Mexico, in Cibola County. The valley of the Arroyo Colorado, which is surrounded by rough, broken topography and mesas, is approximately 16 miles long and 79 miles wide. Elevations range from 5,600 to 7,200 ft. The allotment is 72,165 acres or approximately 113 square miles, including 46,910 acres of public land, 14,135 acres of private land, and 11,120 acres of state land. The allottee has a stock of 670,680 animals, which include cattle and horses. A total of 48 of 8,156 AUMs are reserved for big game (e.g., mule deer, pronghorn antelope); 68 percent of the forage capacity for livestock is on public lands.

*Project description*

The Acoma Pueblo purchased the allotment from Wilson Cattle Company in October of 1978. In 1984, because of their interest in improving the condition of the rangelands and producing a more efficient and economical cattle operation, the Pueblo entered into a cooperative management plan with the Soil Conservation Service and BLM. To implement the plan, which would initially be an eight- to nine-pasture deferred rotation grazing scheme, the Pueblo have constructed pasture fences, a water pipeline, storage tanks, retention dams, and cattle guards. They also developed springs, and maintained existing roads and dams. The Pueblo wants to improve rangeland condition and its cattle breeding program, calf crop, and calf size, which will ultimately increase their profit. The average shipping weight of calves in recent years has been 500 lb.

The 1977 range survey showed that 96 percent of the allotment was in either poor or fair condition. Since 1984, many signs have indicated that the condition has improved, primarily as a result of better facilities and the diligent effort of the range manager and staff in inspecting and monitoring the range condition and frequent herding through pastures to prevent overgrazing.

The new management approach has resulted in the following changes:

- bare areas have been filled in with perennial cover and fewer annuals are growing
- many seedlings of alkali sacaton and fourwing saltbush are growing in alluvial grassland areas
- more vegetation is growing along the eroded banks of the Arroyo Colorado
- more plant and animal litter are accumulating
- vegetation is growing near watering places
- vegetation is holding the soil in place
- increased cover is decreasing the rate of evaporation from the soil surface

During the 1989 drought, the allotment had more forage than did the surrounding allotments, even though it received no more rainfall. Moisture was not a limiting factor because the vigor of the individual plants was at a higher level. The forage withstood drought because soil moisture was held available for a longer time.

*Lessons learned*

The manager has observed plant growth in each pasture and has moved livestock when necessary; livestock have been moved frequently to take advantage of forage quantity and quality, which has improved rangeland health. The trend of range condition has apparently been improving, as indicated by heavier calves and by the fact that the BLM has been able to transplant pronghorn antelope into their historical range.

This project shows that it is possible and practicable to improve and maintain the health of rangelands without a reduction in grazing level. It shows that the level of interest of the local grazing manager is the key to successful implementation of improvements. Lastly, it shows that prompt response to changes in local forage condition are critical to achieving long-term improvement in the condition of the ecosystem.

*Contact person*
Dwain W. Vincent or Hector Villalobos (Area Manager), Rio Puerco Resource Area, 435 Montano Road, NE, Albuquerque, NM 87107 (tel. 505–761–8704).

## 3.4 Nonrenewable Resources

Mineral resources are one of the products that the public demands from public lands. Ecosystem management for nonrenewable resources is different than that for renewable resources. For renewable resources, we assure long-term sustainability by changing or using the resource at a replaceable rate only. This is not possible with nonrenewable resources by definition. Thus, the emphasis in ecosystem management for nonrenewable resources is to take steps to assure compatibility with reclamation of sites and minimum impacts on other values.

Knowing what, where, and how many mineral resources exist, or are likely to exist, in an area can help meet management objectives. Mineral assessments are one way to supply this information. Two mineral assessment approaches are used: qualitative and quantitative. The U.S. Geological Survey and the former Bureau of Mines have conducted qualitative mineral resource evaluations for approximately 44,000,000 acres of federal lands since the Wilderness Act of 1964 was implemented. This work has identified or ranked areas for mineral potential and has helped in recognizing the need to exclude many mineralized areas from wilderness designation. Results of 20 years of qualitative assessments for approximately 80 areas are summarized in Marsh et al. (1984).

Quantitative assessment allows a quantitative comparison of the value of mineral resource development to development of other resources (Singer 1993). Undiscovered resources have been the focus of this approach. Quantitative assessment requires forecasting, an activity most geologists do not relish. Federal geologists rarely have the option of selecting the regions that they will assess. These areas commonly lack obvious signs of undiscovered mineral deposits. The Government uses assessments in multiple ways, and the results are subject to public scrutiny. To facilitate the assessment process, a three-part quantitative assessment approach was developed to allow economic comparison of undiscovered mineral deposits to other competing land uses (Singer 1975) and to satisfy information needs of land management (Fig. 1). This quantitative assessment includes (1) delineation of areas permissive for specific mineral deposit types, (2) estimation of undiscovered mineral deposits using subjective methods or spatial models, and (3) development or use of models of grade, tonnage, and other characteristics of each mineral deposit type (Singer 1993).

Two types of ecosystem management decisions involve mineral resource development, those required for proposed mineral development in the management region and those involved in making land allocations during resource management planning. To make these decisions, tracts of various existing or proposed land-use designations and ecosystems in the management region are superimposed at appropriate scales. Once this has been done, some areas may be found to have ecosystems that are sensitive to mining and are therefore excluded. Other areas may allow mineral development, but mining may be conducted only with constraints. Limitations on mineral development need to be clearly defined to private sector mining corporations during the leasing phase or prior to their entry into an area for locatable minerals. Examples of possible constraints include no surface occupancy or surface occupancy prohibited during certain seasons related to wildlife migration or breeding activity.

If a site is proposed for future mining, the reclamation plan needs to be an integral part of the mine plan. When the mine is decommissioned, will it meet ecosystem management goals such as returning the site to its approximate pre-mining state? If pre-mining conditions are not possible, will surface modification and other modifications be compatible with long-term ecosystem requirements? Mining, as compared to other land uses (e.g., timber production, grazing), affects small areas. Often, the size of the area disturbed by mining is similar in size to areas changed by natural disturbances (Salwerowicz 1994). In addition, active mining is a short-term event when compared to the long life of ecosystems. However, the type of changes to the sites can be very different. It is the state of the post-mining site and its long-term impact in the ecosystem that need careful evaluation (Ripley et al. 1996). Resource extraction may occur, given that ecosystem management sees that the needs of the ecosystem as a whole are met (Salwerowicz 1994).

Mining may be used to help reclaim lands originally disturbed by mining activity by reworking metalliferous mine tailings left by previous operations. Additional processing of the tailings removes more metal and reduces the amount of potentially toxic materials available to the ecosystem. Reducing the volume of tailings remaining from past placer mining activity or of coarse waste rock left by other types of mining is possible if the material is suitable for use as aggregate. Disposal of reworked material can be done in ways that meet ecosystem management goals. Assessments would need to provide appropriate data on tailing and waste material characteristics. No example of this type

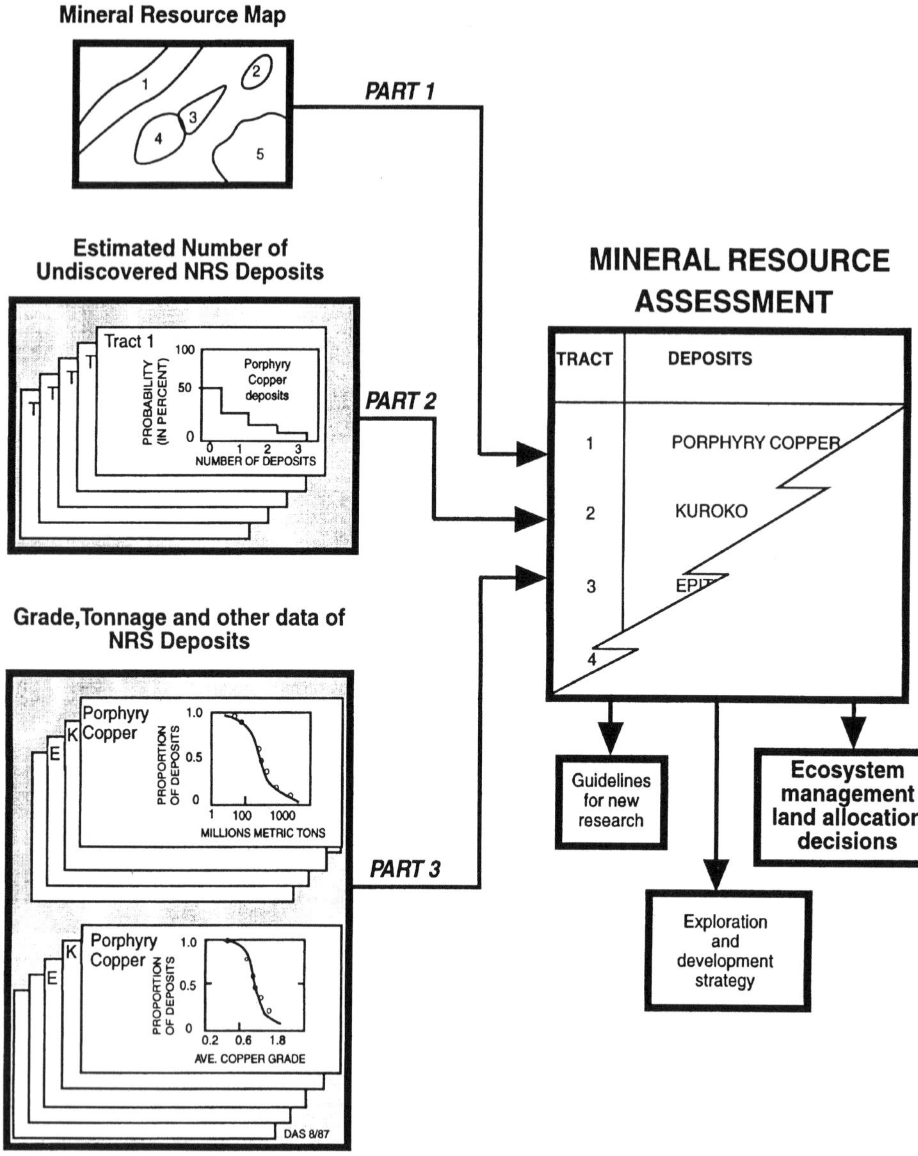


Fig. 1. Quantitative nonrenewable resource assessments comprise three parts, which can be applied to land allocation decision in ecosystem management land allocation (modified after Singer, 1993).

of approach in mineral assessment could be found, but it may represent a promising new direction.

## Example Projects

Spanski (1992) demonstrated the general use of quantitative assessment, and Gunther (1992) described its use in economic analysis. More than 27 quantitative mineral assessments covering more than 1.2 billion acres have been completed (Singer 1993). Most assessment areas were in the United States, but assessments were also conducted in selected areas of Central and South America.

The result of most quantitative mineral resource assessments appears to be the modification of boundaries between lands of different designations. For example, land boundaries affecting State of Alaska and Native Corporation lands were changed after the completion of the mineral resource assessment of Alaska in the late 1970s (D.A. Singer, personal communication, 1996). Assessment reports are frequently used in the evaluation of land for property exchanges or land acquisitions.

### *Tongass National Forest assessment*

The Tongass National Forest assessment in Southeast Alaska (Brew et al. 1992) estimated the gross-in-place value (disregarding costs associated with exploration, development, and extraction) of undiscovered metals at $23.5 billion. The resulting action was the creation of a land-use designation for mineral management prescription. The number of areas so designated increased from 6 to 12 in the draft EIS land management plan.

*Interior Columbia Basin Ecosystem Management Project*
In January 1994, the Chief of the Forest Service and the Director of BLM, under the direction of President Clinton, initiated a study that eventually became known as the Interior Columbia Basin Ecosystem Management Project. Its initial goal was to develop a strategy for dealing with anadromous fish habitat and watershed conservation; the project was eventually expanded to include all of the Columbia River Basin (parts of Idaho, Montana, Oregon, and Washington), plus southeastern Oregon. (Note: Information on the Interior Columbia Basin Ecosystem Management Project is from a written communication from T.P Frost, 1996.)

The overall goals were to provide management tools that can be used to sustain or restore ecosystem integrity, to promote products and services desired by society over the long term, and to provide ways to balance ecosystem conditions, resource uses, and competing needs of stakeholders. Pursuant to these goals, assessments were also made of current and historic landscape conditions, aquatic and terrestrial habitats, species distributions and populations, and economic and social conditions. The project produced scientific assessments of the potential future conditions and possible tradeoffs likely under a number of different disturbance scenarios and management practices.

The Geological Survey was asked to provide estimates of the value of undiscovered mineral resources for the Interior Columbia Basin Ecosystem Management Project using quantitative mineral resource assessment. The results are summarized in Box et al. (1996), Bookstrom et al. (1996), Zientek et al. (1996), and Bookstrom et al. (1995). Knowledge about the presence of existing mineral deposits was used in economic and social assessments and helped to identify sites possibly disturbed by past mining. Information on existing metallic mines and potentially undiscovered deposits (and possible infrastructure related to extraction, benefaction, and processing) was considered in the landscape ecology assessment. A map derived from the lithology map, which showed where sand and gravel was likely, was used to identify areas or tracts currently or likely to be disturbed by mining. The tract boundaries are part of the assessment of aquatic and riparian ecosystems. This map was also found to be useful in the economic assessment. The phosphate mineral resource map was used in a similar fashion for terrestrial ecosystem and economic assessments.

Earth science information was relevant to assessments of past, current, and potential ecological, economic, and social conditions in the area. Bedrock lithology was used to assess aquatic integrity and to identify areas likely to contain some possible roosting sites for cave-dwelling bats (Johnson and Raines 1995). Areas with limestone caves and lava tubes are likely, as are adits and other underground structures of past mining, to contain this habitat (Frost et al. 1996). A number of derivative maps were prepared using bedrock lithology, together with rock chemistry (Raines et al. 1996) and regional geochemistry (Raines and Smith 1996), to help evaluate terrestrial and aquatic ecosystems. Information on hazards associated with earthquake (Algermissen et al. 1990) and volcanic activity (Hoblitt et al. 1987) was explicitly included in evaluation of the landscape ecosystem.

*Lessons learned*
Good mineral assessments can lead to good ecosystem planning and management by providing information on possible future mineral development impacts, so that this information can be integrated in plans for other resource uses. Mineral assessments assist especially in transportation planning, but also in trade-off analysis and land allocation to various uses. They can help with socioeconomic analysis, ecosystem restoration plans, and aquatic integrity. In a few instances they have also led to expanding other ecosystems such as aquatic habitats.

## 3.5 Forest Management

As forest managers make the transition from managing for single product or species outputs and values (e.g., timber, endangered species) to managing for multiple outputs or values on a broader scale (e.g., provenance, landscape, watershed, century-long time-frame), they seek ways to maintain ecosystem diversity and health. This transition concentrates attention during the planning phase on how actions at one location affect ecosystem attributes in other areas and the structure of the landscape in total.

### 3.5.1 Landscape-Level Background

Most first-line forest managers typically encounter questions regarding management on areas ranging from a few to about 100,000 acres (Forest Service Ranger District) and occasionally as large as a National Forest or BLM District (+500,000 acres). The analytical unit is typically a watershed or drainage, and questions revolve around what types of treatments are needed within the watershed to restore or maintain its ecosystem function or how groups of watersheds might be summed to achieve sustainability on larger landscapes. Managers must look at time frames that are decades or centuries long and try to understand how conditions and activities today will play out over time.

The landscape-level case studies are intended to provide ideas about how to address these two types of problems. We outline examples of analyses at different spatial scales, which include a watershed level analysis (Augusta Creek project), implementation of a restoration plan for a Ranger District (Ponderosa Pine Forest Partnership, Crowley Project), and landscape analysis and planning tools for large areas (Washington Landscape Study).

Within these examples and throughout the forest management community, there are recurrent themes, or challenges, that must be addressed to implement ecosystem management successfully in our public forests. The challenges tend to involve restoration of riparian health and function or changes in species composition and stand structure, and they are manifested at the stand or project level rather than the landscape level. These smaller scales are the levels where on-the-ground activities occur and where the landscape-level concepts of ecosystem management become reality. In recognition of this fact, we include information on the state-of-the-art in addressing stand- or project-level issues. These studies add to the tool kit available to resource specialists who advise first-line managers in developing the stand-level prescriptions that ultimately sum to landscape-level decisions. We chose examples that include research-scale implementation of treatments and extend the range of activities well beyond those envisioned for production forestry.

## 3.5.2 Stand-Level Background

In the northwest United States, project-level examples typically apply generically to either coastal areas west of the Cascade Mountains or the interior east of the Cascades. Stand-level approaches to management will be briefly summarized for each area. For the most part, these approaches are being tested as research projects and can be considered as "promising possibilities" at present.

### *Westside examples*

Much of the federal land west of the Cascades is covered by the Northwest Forest Plan for the Recovery of the Northern Spotted Owl (USDA Forest Service and BLM 1994). A recurring theme on these lands is to hasten the development of late-successional structure in areas that were previously managed as single-species plantations intended to maximize timber production.

The scale of this task can be daunting. The Siuslaw National Forest covers about 660,000 acres, with approximately 200,000 acres of plantations less than 30 years old. Only about 60,000 acres are designated as Matrix or areas with primary emphasis on timber production, while the remainder falls into various other land allocations where timber management is restricted or excluded. Although the Siuslaw may be an extreme case, other larger National Forests also have considerable areas of young plantations, e.g., 350,000 acres on the Willamette National Forest (Mayo 1995); many of these National Forests fall within areas with restricted management options. Although no detailed survey exists, it is safe to say that there are millions of acres of young plantations that were originally established to maximize timber production but now will be managed for other objectives. Managers are faced with the problem of changing stand trajectories in an attempt to create a mosaic of species composition and stand structures in a relatively short time-frame. The new objective is to perpetuate a healthy, productive, biologically diverse forest that will continue to have social and economic outputs.

### *Eastside examples*

On the eastside of the Cascade Mountains, problems associated with small-diameter, densely stocked stands are common. These stand types tend to create large, structurally uniform areas; successful implementation of ecosystem management in the West will require workable management strategies for these areas. Some of these stands have a component of larger, older trees but all share a dense small-diameter component. The stands often arose as a result of successful fire suppression efforts, and all contain timber that is of marginal value.

Situations where all the trees in a given area are small diameter might occur when stands arose after stand replacement fires in the early part of the century, followed by successful fire suppression over the past 70 or so years. Situations where there is a component of large-diameter trees in the stand are common in ponderosa pine stands, where periodic low-intensity fires have been excluded. In both cases, late-successional structure might be created by active management, and commercial thinning will sometimes be the appropriate tool.

## 3.5.3 Augusta Creek Project

### *Geographic area*

The Augusta Creek project is located on the Willamette National Forest in western Oregon. It includes areas designated as wilderness, unroaded areas, areas where timber harvest is prescribed, and an aquatic reserve system.

### Background

The goal of ecosystem management on public lands means maintaining native species, ecosystem processes and structures, and long-term ecosystem productivity. However, we currently lack the knowledge necessary to state accurately and completely how native species, ecosystem processes, and productivity can be sustained. Recognition of this condition led to using a relatively conservative approach to human use of ecosystems, which relied on past conditions and natural patterns as guides for future management designs.

### Project description

The Augusta Creek project (Cissel and others, in press) was initiated to establish and integrate landscape and watershed objectives to guide management activities within a 19,000-acre planning area. The preliminary objective was to maintain native species, ecosystem processes and structures, and long-term ecosystem productivity in a federally managed landscape where substantial acreage has been allocated to timber harvest.

A landscape management strategy was developed that uses past landscape conditions and disturbance regimes to provide key reference points and design elements for future landscape objectives. One premise of this approach is that native species have adapted to the range of habitat patterns resulting from disturbance events over thousands of years. The probability of survival of these species is reduced if their environment is maintained outside the range of these historical conditions. Similarly, ecological processes, such as nutrient and hydrological cycles, have historically functioned within a range of conditions established by disturbance and successional patterns. Management activities that move structures and processes outside the range of past conditions may adversely affect ecosystems in both predictable and unforeseen ways. A second premise of the strategy recognizes that existing conditions of human use must be integrated with this historic template to meet long-term objectives.

The analytical process involved five sequential phases. Work in each phase was conducted in the context of the larger surrounding watersheds and was designed to efficiently link to implementation of management objectives.

*Fire history* — A fire history study was conducted within the planning area over the last 500 years. Plot level data were used to map 27 fire events. The maps were used to reconstruct and analyze vegetation patterns within the same 500-year period.

*Analysis of conditions, processes, and uses* — Several approaches were used to analyze the aquatic system and hillslope–to–stream connections. Landslide and debris-flow occurrences and potential for future occurrences were mapped from aerial photographs, existing maps, and field surveys. Relative susceptibility of the landscape to rain-on-snow peak flows and contributions to summer baseflows were mapped. A time–series analysis of aerial photographs spanning 40 years was used to assess riparian vegetation dynamics and disturbance history. Both prehistoric and contemporary human uses were described and mapped. Current human uses included hiking, camping, angling, hunting, and harvest of timber and special forest products.

*Landscape objectives and prescriptions* — The planning area was subdivided into three general categories so that specific landscape management objectives could be developed:

1. large reserves from the Willamette National Forest
2. landscape areas for prescribed timber harvests
3. an aquatic reserve system

*Projection of future conditions* — Maps of future landscape and watershed conditions were developed by simulating the growth of existing forest stands using a simple stand-age model in the Geographical Information System (GIS). Following timber cuts, blocks were reset to specific stand conditions, according to a timber harvest schedule determined by the landscape objectives and prescriptions for the area. Growth was again simulated until the next scheduled cutting. A set of maps depicting future landscape conditions was generated at 20-year intervals for the next 200 years.

*Evaluation* — The Augusta Creek landscape design (ACLD) was evaluated by comparing it to the future landscape generated by application of standards, guidelines, and assumptions in the Northwest Forest Plan (NWFP).

Results from the landscape maps show a gradual change in the landscape from the relatively fragmented forest of today to one dominated by larger blocks and containing a wider array of stand types as described in the landscape objectives. By the year 100, the future landscape appeared significantly different from the existing landscape. Gradual change continued before stabilizing in the year 200. The conclusions of the study are as follows:

1. The ACLD appears superior for most taxa evaluated, especially those dependent on large patches of old forest habitat.
2. Compared to the NWFP, the landscape in the ACLD is much less fragmented and is expected to be less susceptible to wildfire, wind, and insect disturbance.
3. The NWFP is superior with respect to providing more early serial habitat.

4. The board-foot yield of timber is about 6 percent higher under the NWFP scenario. However, this increase is within the error terms of estimates and thus would be considered equal for either plan. Timber value may be higher under the ACLD scenario because of larger, higher value trees.

5. Hydrology and debris slides and flows are expected to differ little between the two scenarios.

*Lessons learned*

The project provides an example of how ecosystem management activities on a project level can be linked to wider objectives, standards, and guidelines established on a much larger scale. Specifically, Augusta Creek can be viewed as a post-watershed analysis implementation of the Northwest Forest Plan. Although the general approach to landscape management should be generally applicable to other landscapes, the mix of specific design elements and the resulting consequences will likely vary considerably. The ability of ecologists and land managers to incorporate new perspectives for ecosystem management is limited by several factors, including the lack of analytical and modeling approaches to landscape-scale problems. Although many of the required components are currently available, or are the subject of ongoing research, more effort should be directed to projecting and evaluating the effects of land-use actions on the sustainability of ecosystem properties from both ecological and social perspectives.

*Contact person*

John Cissel, Blue River Ranger District, Willamette National Forest, Blue River, OR 974113 (tel. 541–822–3317)

### 3.5.4 Washington Landscape Study

*Geographic area*

The Washington Landscape Study is located on state land on the Clallarn River at the western end of the Olympic Peninsula.

*Project description*

The Washington Landscape Study (Cary et al., in press) was initiated by the Department of Natural Resources (DNR) of Washington State. The objective was to evaluate management alternatives across land ownerships that meet the needs of wildlife in late–serial forests while minimizing impacts on the production of commodities. A major reason for choosing this area was that the DNR had developed a substantial database on this landscape and had adapted the database to the SNAP-II landscape simulator.

A conceptual model of landscape management, specific to westside western hemlock–Douglas-fir forests, was developed using ecological theory and concepts. The biodiversity alternatives included conservation of biological legacies at harvest (soil food webs, coarse woody debris), both planting and natural regeneration, precommercial thinning, favorable density thinning, long (70–130 year) rotations, and differing degrees of intervention. Four riparian management schemes were used: the Washington Forest Practices Board (WFPB) regulations, two FEMAT-like approaches, and a variable polygon scheme that emphasized protection of stream banks and thinning to promote development of large trees. Four new indices of forest ecosystem health were developed, as well as several economic measures relating to timber harvest. The SNAP-II simulations were conducted for a 300-year period.

Many alternative landscape management scenarios were developed. The key scenarios were as follows:

- No manipulation, with protection of the entire landscape.
- Protection of wide riparian buffers with maximum net present value (MAX NPV) of timber on remaining areas.
- MAX NPV using protection of riparian buffers, using current WFPB regulations.
- MAX NPV using more frequent intervention (thinning at 30, 50, and 70 years with final harvest at >110 years).
- MAX NPV using less frequent intervention (thinning at 30, 60, and 90 years with final harvest at >130 years).
- Maximization of biodiversity with alternating 70- and 110-year rotations.
- Selection of 30 percent late serial forest (LSF) for biodiversity simulations, including 20 percent in niche diversity and 10 percent in fully functional managed forests.

Some results of the management scenario simulations are as follows:

- No manipulation — 180 years required to meet the 30 percent LSF goal; no commodities produced (NPV = 0); ecological crunches occurred before forest maturity (crunches would lead to continued species declines or extinctions).
- Protection using wide, FEMAT-like buffers — More than 200 years required to meet 30 percent LSF goal; LSFs badly fragmented by intervening intensively managed forest; NPV = $48.5 million.

- Maximization of net present value — No LSFs; inadequate riparian protection; >25 species at risk; NPV = $70.3 million.
- Other intermediate results.

*Results*
Results from the economic analysis of costs showed managers the following:

- Transition costs from present to regulated state can be large.
- NPV depends on timing of incentives.
- Estimated present value cost for each 10 percent increase in LSF could be as low as $100/acre.
- This approach is a net benefit solution for managers of multiple-use public lands.

Salient points about regional benefits are as follows:

- Diversification of wood products industry
- Increased secondary manufacturing
- Increased direct employment
- Increased indirect employment
- Increased tax revenues

All of these regional gains are substantial compared to the suggested incentive programs. However, details of the gains vary markedly with the assumptions made.

*Lessons learned*
This project provides an example of state and federal cooperation to explore alternatives for implementing ecosystem management objectives across multiple land-ownerships while minimizing impacts on the production of both plant- and animal-based commodity projects. This approach highlights the ability to identify a wide variety of ecological, social, and economic benefits under various management alternatives. However, it also highlights the sensitivity of the projected results to the input assumptions. Thus, the project shows how analytical projects can be used not only for making land management decisions but also for identifying key assumptions that require better documentation prior to implementing study results.

*Contact person*
Andy Carey, Olympia Forestry Sciences Laboratory, 3625 93rd Ave. SW, Olympia, WA 98512–9193 (tel. 360–959–2345).

### 3.5.5 Ponderosa Pine Forest Partnership

The Ponderosa Pine Forest Partnership (PPFP) practices community–public lands stewardship by building relationships that unite forest health with community sustainability. This multi-member partnership emerged from a recognition of common needs created by a weakened local timber industry and declining forest health. The PPFP has learned how commercial logging can restore badly needed forest health and how the National Forests can support local communities in the spirit of ecosystem management. The Partnership has replaced gridlock and uncertainty with constructive action.

The PPFP was initiated when Montezuma County submitted a proposal to the USDA Rural Community Assistance program and won a grant for $25,000. An agreement was made among the county, the San Juan National Forest, and the Colorado Timber Industry Association to share time and data, seek markets for small-diameter timber, and hire geographical information system (GIS) mapmakers and ecology researchers.

*Geographic area*
The PPFP is a demonstration of adaptive management techniques on 189,000 acres of southwest Colorado's ponderosa pine forests located on the Mancos–Dolores Ranger District within the San Juan and Rio Grande National Forests. Second-growth pine and a thick understory of Gambel oak dominate the terrain found between 7,500 and 8,500 ft throughout the area. A century of heavy logging, cattle-grazing, and fire suppression have created an unnaturally dense and stagnant ponderosa pine forest at risk of mountain pine beetle infestation and catastrophic wildfire.

*Project description*
From 1950 to 1980, the San Juan Forest timber harvest averaged 45 million board feet (mmbf) per year. Since 1980, it has averaged 24 mmbf, with 12 mmbf harvested in 1994 and 1995. Mill closings marked these later years. About 65 years of timber-related activities, combined with federal agency control of about 75 percent of the land, profoundly shaped local culture and social values. Today, nearly one-third of the District's timberland is second-growth ponderosa pine.

Using tree-ring dating and analysis, ecologists from Fort Lewis College and Northern Arizona University assessed pre-1870 ponderosa pine forest fire history, as well as current ecological conditions across the 189,000-acre study area. Long-time local residents were interviewed as other researchers examined historical uses and past management of these local pine forests. The ecologists speculate that before European settlement, periodic fires, whose frequency averaged 5 to 40 years, created a landscape characterized by large and widely spaced ponderosa pines, ground vegetation dominated by native grasses, and scattered thickets of younger pine regeneration. Stumps still surviving from turn-of-the-century logging show that before 1870, trees were as large as 27 inches in diameter and numbered 40 to 50 per acre. This is considerably different from today's situation. Now, the average size is about 8 inches and there are 280 to 390 trees per acre. Most trees are less than 90 years old. Open grassy areas are uncommon. Wildfires in the pine zone have been actively suppressed for the last 100 years.

The PPFP hallmark is the cooperative development of a GIS map database to facilitate understanding of ecosystem relationships and provide the basis for a strong public involvement process. To develop the vegetation maps, stand exam data were put into GIS format and used to classify and map areas of risk for pine beetle. The Forest's Integrated Resource Inventory (IRI) team provided detailed GIS maps and data about on-site conditions and capabilities. All the mapped data were used to recommend the best sites and priorities for treatment.

The predominance of small-diameter trees makes conventional sawtimber sales and pricing infeasible. The demonstration work is designed to find a feasible and fair approach to ecosystem restoration that maintains timber management as part of the rural culture. Forest Service and Colorado state foresters formulated silvicultural prescriptions to conduct forest restoration at the project sites. The key objective has been to restore vegetative diversity that mimics the diversity that had been caused by natural disturbances before 1870. The prescriptions specify removal of many smaller trees, leaving all trees 16 inches and larger. In the past, only the larger trees would have been harvested. Harvests have been designed to help create a more clumped appearance and create openings for natural regeneration, much like presettlement conditions. The few large trees that remain provide important habitat for plant and animal species not found in the second-growth forests.

The reintroduction of fire is a key element of this restoration project. After locally contracted loggers have completed harvesting these stands, Forest Service crews will conduct prescribed burns. Periodic follow-up burns will be scheduled at various intervals for up to 10 years. Areas will be closely monitored to evaluate the effectiveness of treatments in reducing fire and disease risk and promoting pine regeneration.

Simultaneously, local timber industry representatives are testing new timber harvesting techniques. Colorado State University (CSU) developed a plan to help the timber industry monitor the efficiency of new equipment and logging methods. CSU has also taken an active role in researching alternative product opportunities for small pine material. The PPFP goals hinge heavily on identifying marketable products from small-diameter trees.

The PPFP goals also rely heavily on pricing for the raw materials. Historically, raw materials from small pine do not convert into valuable end-products. The primary appraisal system for valuing timber in the Forest Service does not accurately reflect the much-reduced markets for smaller material. Stewardship contracts have been considered that would allow the contractor to perform needed land management activities and, in turn, be given salvage rights to the raw material.

*Lessons learned*

Federal, state, and local governments and many local cooperators, each with different goals, can all work together to achieve their respective goals through ecosystem management. "The challenge," as one partner says, "is to develop a community stewardship model that allows communities to be active players in making ecological and community sustainability work together." The strategies that reduce pine beetle and wildfire risks, increase plant and animal diversity, and establish a sustainable flow of wood to local communities have become a model for managing second-growth ponderosa pine forests on public lands in the West.

*Contact person*

Mike Preston, Montezuma County Public Lands Coordinator, Administrative Office, Court House, Cortez, CO 81321 (tel. 970–565–8317).

### 3.5.6 Crowley Project

*Geographic area*

The Crowley Project was undertaken on the Cocoino National Forest in Arizona. The objectives were to enhance recreation, vegetative diversity, and visual quality while maintaining production of wood products.

*Background*

Throughout the U.S. inland west, aspen forests ecosystems are aging and are often being replaced by conifer forests. This is part of the natural succession process, which has been going on since the last Ice Age. However, since European people became a dominant force in the area, one major change has occurred: far fewer new aspen stands are being created. Under natural conditions, aspen sprouts rapidly after periodic crown fires, creating new stands of younger age classes to replace older stands that are burned or replaced by conifers. Whether the stand that burns is mostly pure aspen or whether conifers take over, the extra sunlight and heat provided to the forest floor by removing the overstory causes prolific sprouting from the aspen roots. The shoots grow rapidly, resulting in a pure or near-pure aspen forest.

Ungulate grazing (which removes fine fuels and also aspen suckers) in the last half of the 19th and early 20th centuries, fire control, and timber harvest by selective methods have all contributed to the lack of new aspen regeneration. Most recent inventories in Arizona (Connor et al. 1990) and New Mexico (Van Hooser et al. 1993) show declining acreages from previous inventories, but increasing volumes. Photographs previous to the 1950s show aspen stands where conifers predominate today.

Even though aspen is not considered a valuable species from a wood standpoint, ecologists, wildlife biologists, silviculturists, landscape architects, and other resource managers recognize its high value for scenic beauty, wildlife habitat, and biological diversity as well as the variety it provides in forests usually dominated by conifers. The solution to aspen regeneration and overall maintenance of acreage of aspen stands seems fairly obvious: harvest, burn, or a combination of these to remove the overstory and allow new aspen forests to develop. However, in real life, many other factors immensely complicate the implementation of these activities. In the Crowley Project, success has been achieved in improving the ecosystem while also producing wood products for society.

The vegetation in this management block is largely dominated by a sea of ponderosa pine, mostly pole and small sawtimber size. However, scattered through the area are several small aspen clones, ranging from only 10 to 15 trees up to 1 to 2 acres in size. These small groups of aspen provide much-needed diversity and visual variety in an area with little topographic or vegetative diversity. The small aspen clones are uniformly very old and are being crowded out by ponderosa pine. When the aspen try to regenerate with new suckers, they are decimated through grazing, primarily by elk, but sometimes by livestock as well. The result is aspen clones that are dying slowly as the old trees die.

Besides aspen loss, four other environmental concerns needed to be addressed in the Crowley block: (1) invasion of meadows by ponderosa pine, (2) lack of age-class diversity, (3) loss of opportunities to view large yellow-pine trees, and (4) overly dense forests, which are not allowing optimal tree growth and creation of large trees and old-growth conditions over time. Older ponderosa pine with yellow bark (often called yellow-bellies) is highly desirable for viewing. However, over the last 100 years, much of this species has been harvested in the Crowley block. The remaining yellow-bellies are often hidden from view by the sea of smaller pines surrounding them. Additionally, as a result of the dense stocking of the small trees, growth of each tree is extremely slow, so large yellow-bellies are not being developed for the future

*Project description*

A project was designed to improve conditions for the aspen clones as well as for visual quality, tree growth, and diversity. No aspen trees were harvested because of their extremely low number. However, ponderosa pine trees less than 16 inches diameter at breast height (dbh) within and in a circle about 75-ft-wide around each clone were harvested to open the area to sunlight. Harvest disturbed the ground, enhancing aspen sucker production. A total of 20 aspen clones are being treated in a 496-acre area. Each clone is being enclosed in a 6.5-ft-high fence to prevent elk and livestock from browsing on the new aspen shoots. Past research indicated that such fences must be maintained for about 7 years until the new trees are large enough so that elk browsing will not cause significant damage. This fencing is a very expensive (approximately $6,000 per mile of fence), but a necessary part of this project.

Pines are also part of the Crowley Project — removing pines that are invading meadows creates new age-classes for diversity, removing small pines from large pine stands enhances viewing opportunities, and thinning enhances tree growth. An associated recreational value enhanced by the Arizona Department of Game and Fish is increased levels of elk hunting.

Besides these recreational and environmental benefits, the Crowley Project is also producing much-needed raw materials for local industry and consumers. A total of 14,160 hundred cubic feet (CCF) of timber is being harvested. Of this, 6,850 CCF is pulp (5 to 8.9 inch dbh) and 7,310 CCF is sawtimber or trees greater than 9 inches dbh. Total value returned to the Government from this sale is $410,153.

A significant part of increasing aspen regeneration is eliminating or reducing the amount of browsing by elk. In Crowley, the existing population of aspen did not provide the opportunity to create enough new aspen stands to provide more new shoots than the elk population could use. However, one promising possibility is that where there is more aspen, it should be feasible to harvest larger areas and thus eliminate the high cost of fencing. Along with the harvest, another possible measure is to (temporarily) reduce the elk herds in the area through hunting. In northern Arizona this is being done by the Arizona Game and Fish department. In Game Management Unit 7 around San Francisco Peaks, permits have been increased as follows: 1991 and 1992, 1,275 permits; 1993, 1,375 permits; 1994, 1,475 permits; 1995 and 1996, 2,147 permits. In the long run, this will diminish elk numbers, or at least minimize increases, and the consequent impact on aspen regeneration.

*Lessons learned*
The Crowley Project is an outstanding example of enhancing aspen, diversity, and other environmental quality factors, enhancing recreational opportunities and quality, and producing wood for meeting consumer demand. Current forest management activities have made significant progress in moving the area toward a sustainable condition, although it will take 100+ years to get there. The timber sale is the most cost-effective way to achieve needed environmental improvements.

*Contact person*
Jim Rolf, Peaks Ranger District, Coconino National Forest (tel. 520–527–8239). There are many other areas where aspen is being enhanced through ecosystem management projects. In some of these areas, aspen is much more dominant (including pure stands) than in the Crowley area. Two other contacts with expertise and knowledge of projects involving aspen are (1) Wayne Sheperd, Rocky Mountain Station, Fort Collins, CO (tel. 303–498–1259) and (2) Dale Bartos, Intermountain Research Station, Logan, UT (tel. 801–755–3567).

### 3.5.7 Colville Study

The Colville Study (Barbour et al. 1995, Ryland 1996) was an integrated study intended to help natural resource managers understand the silvicultural, operational, and economic implications of performing forest operations in small-diameter, densely stocked stands. This study was a cooperative effort involving the Colville National Forest, Idaho Panhandle National Forest, Ochoco National Forest, USDA Forest Service Forest Products Laboratory, Boise Cascade, Riley Creek Lumber, Vaagan Brothers Lumber, Oregon State University, University of Idaho, University of Washington, Washington State University, and Forest Service Pacific Northwest Research Station.

*Geographic area*
The study focused on the Rocky II timber sale on the Colville National Forest in northeastern Washington. The sale consisted of 18 separate cutting units, totaling 764 acres of thinning that were representative of the densely stocked, small-diameter stands in the forest. A recent inventory had found 115,000 acres of small-diameter, densely stocked stands (Colville National Forest 1994).

*Project description*
The objective of the forest managers was to develop a strategy for changing the trajectories of the small-diameter, densely stocked stands in an attempt to (1) create late-successional structure from large areas of uniform stands, (2) decrease forest health risk, (3) improve wildlife habitat, particularly for white-tailed dear and cavity-nesting birds, and (4) improve stand aesthetics.

Various silvicultural regimes and residual densities were modeled using the Inland Empire variant of the Forest Vegetation Simulator (FVS) for four different stand types. Future stand structures were judged according to their success in providing large-diameter trees, large snags, overstory height, crown height, and other factors. The modeling exercise illustrated that changes in the pattern and rate of stand development could be induced through silvicultural treatment to create desired ecological features and generate timber outputs. The most evident change was the development of large-diameter trees, which could provide large snags for cavity-nesting birds and other wildlife as well as sawtimber. These simulations illustrated the effects of varying degrees of disturbance and suggest that meeting stated ecosystem objectives will require some form of intervention to allow stands to develop the necessary structural and habitat characteristics.

A harvester forwarder system was monitored during harvesting of the Rocky II timber sale, and production functions were developed for the system. Additional work is in progress to develop similar functions for small tractor logging systems. Lumber and veneer recovery studies were conducted to develop grade and volume yield equations for small-diameter logs. Suitability of the material for several composite products and mechanical and kraft pulps was also determined. Reports on these studies are forthcoming.

A financial analysis package is also under development. This package is intended to help timber planners understand what types of treatments are feasible and can be accomplished using timber sales, when thinning contracts are the best option, and when costs will be so high that a hands-off approach is the only option.

*Lessons learned*
The ecosystem objectives outlined by the forest managers would not be met in a reasonable timeframe (less than 200 years) if no treatment was done. Any of the other treatments would meet the objectives sooner, but the economics of harvesting and processing the small-diameter material is extremely sensitive to piece size and market conditions. Harvesting and processing small material is expensive, and the quantity and value of the resulting products are fairly low. Designing timber sales so that the purchaser can react quickly to fluctuating markets is one way to increase the likelihood that timber sales will sell and ecosystem management objectives

will be reached. Finally, the economic evaluation of timber stands for possible sale is very complex. A computer program is needed to understand the interactions of the various components.

*Contact person*
Jamie Barbour, Portland Forestry Sciences Laboratory, Portland, OR (tel. 503–326–4274).

### 3.5.8 Blacks Mountain Experimental Forest Project

In the rush to create stands with diverse species and structures, it is not known whether we can manage old-growth stands to perpetuate their values over time. Information about how old-growth stands have responded to thinning treatments is as useful as information on whether younger stands can be manipulated to accelerate the creation of late-seral conditions. In 1938, research was initiated on thinning old-growth stands of eastside pine types, which contained large trees (31.5 inches dbh, about 5 tpa) at least 300 years old. At that time, the stands were influenced by frequent low-intensity fires and sheep grazing, which kept the understory open and fuel levels low. With the end of sheep grazing and the exclusion of fire, a dense understory of ponderosa pine and white fir has developed.

*Geographic area*
The Blacks Mountain Experimental Forest is located in northeastern California. The study area is roughly 10,000 acres of interior ponderosa pine cover type, locally known as eastside pine. This cover type is found on about 2.3 million acres in California, nearly 14 percent of the total available commercial forest in California.

*Project description*
Six levels of thinning, from a no-thin control to 95 percent removal, were tested on the Blacks Mountain Experimental Forest (Dolph et al. 1995). Measurements were taken at 5, 10, 20, and 50 years after treatment for tree growth, volume production, diameter distribution, and species composition. Results show that the diameter growth was greater on the more intensive treatments, which result in smaller stems; volume production was initially decreased in the intensive thinning, but significantly increased in the 20–50-year period, probably because of in-growth; and diameter distribution showed a consistent increase in trees <27.5 inches dbh, regardless of treatment, and a decrease in trees >27.5 inches dbh in intensive thinning. Finally, although no relationship was found in species composition between treatments, an increase in competition from the in-growth in the understory contributed to the mortality of large trees, even in the control. Major changes on the study plots were observed for both the exclusion of fire and the thinning treatments.

*Lessons learned*
The decline of the old, large-tree component demonstrates an important point that other authors have reported: characteristics or functions of old-growth stands cannot be guaranteed in perpetuity by simply preserving existing old-growth tracts (Debell and Franklin 1987). Like young-growth stands, old-growth stands must be managed for desired attributes.

*Contact person*
Kathy Harcksen, Lassen National Forest, 55 S. Sacramento St., Susanville, CA 96130 (tel. 916–257–2151).

## 4 CONCLUSIONS

Ecosystem management provides the opportunity to produce and use natural resources in ways that ensure, within reasonable limits, sustained ecosystem functions. In fact, ecosystem management includes providing for the needs of humans. We face the continuing challenge of finding ways to forecast how ecosystems are likely to respond to changes related to the production and use of our resources. The resources desired from public lands include wildlife and fish, recreation, minerals, wood fiber, forage for livestock, clean water, and many special products, including Christmas trees, mushrooms and berries.

Planning is foremost in importance as we face the challenge of meeting ecosystem management goals. We found that users of public lands must be involved in developing regional standards and guidelines. Planning at the regional scale must become a collaborative effort, including all levels of government as well as industrial and private cooperators. Planning at the site-specific level must be linked to wider objectives as is shown in the Augusta Creek example. Partnerships must be forged and planning integrated on a landscape basis to mesh agency responsibilities and pool personnel and funding resources. We have shown several examples (e.g., Washington Landscape Study, Recreation, Muddy Creek, Ponderosa Pine Forest Partnership) where such effort has been successful. Resource uses must be monitored to prevent over-use and degradation, or to ensure restoration as in the Rangelands example in the Arroyo Colorado allotment.

Alternative management practices for resource production need to be evaluated to determine possible impacts on the health of the ecosystem. State and federal agencies, working together, are beginning to instill a common land ethic in the public. We believe, and have shown several examples, that resources can be managed to produce marketable commodities as well as provide a wide range of non-market amenities within a framework that ensures sustainability.

Scale, scope, and temporal change are all critical factors in producing resources through ecosystem management. We have shown that a century-plus planning horizon must be considered before we can see what management activities are needed on the landscape today. Vegetative regimes constantly change, and the lack of direct intervention by human activities does not equate to protection in perpetuity. Not doing any vegetation management would have dire consequences in forests that have gone through major changes from their presettlement condition. Native species have adapted to a range of habitat patterns, as shown by historical disturbance events and ecological processes. This is one key to evaluating the impact of resource production on ecosystem health. We have shown in the Fish and Wildlife and other examples that long-range planning must consider not only change resulting from human activity but also natural changes resulting from vegetative succession. One way to assure integration of resource production and use into healthy ecosystems is to see that management activities mimic the patterns of natural disturbance (e.g., Crowley Project, Ponderosa Pine Forest Partnership).

We have shown that management approaches are available that manipulate vegetation and wildlife populations to produce healthy ecosystems. In fact, we have found that in some cases ecosystems must be treated to achieve ecosystem goals such as diversity and long term-sustainability (Crowley Project, Ponderosa Pine Forest Partnership, Blacks Mountain Experimental Forest Project, and others). We have found that traditional products and methods of extraction must sometimes be modified to deal with current ecosystem conditions. We have also shown that exploration and development of nonrenewable resources are possible if exploration is regulated to minimize impacts to ecosystems and if proposed restoration is compatible with long-term ecosystem sustainability.

## REFERENCES

Algermissen, S.T., D.M. Perkins, P.C. Thenhaus, S.L. Hanson, and B.L. Bender. 1990. Probabilistic Earthquake Acceleration and Velocity Maps for the United States and Puerto Rico. U.S. Geological Survey Miscellaneous Field Studies Map, MF-2120, scale 1:7,500,000.

Anderson, S.J., and C.P. Stone. 1993. Snaring to control feral pigs *Sus scrofa* in a remote Hawaiian rain forest. *Biological Conservation* 63: 195–201.

Baker, J.K. 1979. The feral pig in Hawaii Volcanoes National Park. In: R.M. Linn (ed.), *Proceedings of Conference on the Scientific Research in National Parks*, Ser 5(1): 365–367.

Barbour, R.J., J.F. McNeel, S. Tesch, and D.B. Ryland. 1995. Management of mixed species, small-diameter, densely stocked stands. In: *Sustainability, Forest Health, and Meeting the Nations Needs for Wood Products, COFE 1995 Council on Forest Engineering Annual Meeting*, June 5–8, 1995, Cashiers, NC.

Bookstrom, A.A., G.L. Raines, and B.R. Johnson. 1995. Digital Mineral Resource Maps of Phosphate and Natural Aggregate in the Pacific Northwest: A Contribution to the Interior Columbia Basin Ecosystem Management Project. U.S. Geological Survey Open-File Report 96-681.

Bookstrom, A.A., M.L. Zientek, S.E. Box, P.D. Derkey, J.E. Elliott, D. Frishman, R.C. Evarts, R.P. Ashley, L.A. Moyer, D.P. Cox, and S.D. Ludington. 1996. Status and Contained Metal Content of Significant Base and Precious Metal Deposits in the Pacific Northwest: A Contribution to the Interior Columbia Basin Ecosystem Management Project. U.S. Geological Survey Open-File Report 95-688.

Brew, D.A., L.J. Drew, and S.D. Ludington. 1992. The study of the undiscovered mineral resources of the Tongass National Forest and adjacent lands, southeastern Alaska. *Nonrenewable Resources* 1(4): 303–322.

Cartledge, T.R., and J.G. Propper. 1993. Pinon–Juniper ecosystems through time: Information and insights from the past. pp. 63–71. In: Managing Pinon–Juniper Ecosystems for Sustainability and Social Needs. USDA Forest Service General Technical Report RM–236. U.S. Forest Service Rocky Mountain Forest and Range Experiement Station, Fort Collins, MO.

Cary, A.B., C. Elliott, B.R. Lippke, J. Sessions, C.J. Chambers, C.D. Oliver, J.F. Franklin, and M.J. Raphael. A Programatic Approach to Small-Landscape Management: Final Report of the Biodiversity Pathways Working Group of the Washington Landscape Management Project. Washington Landscape Management Project Rep. 2. Washington Department of Natural Resources, Olympia, WA (in press).

Cissel, J., F. Swanson, G. Grant, D. Olson, S. Gregory, S. Garman, L. Ashkenas, M. Hunter, J. Kertis, J. Mayo, M. McSwain, K. Swindle, and D. Wallin. A Disturbance-Based Landscape Design in a Managed Forest Ecosystem: The Augusta Creek Study. USDA Forest Service General Technical Report. U.S. Forest Service Pacific Northwest Station, Portland, OR (in press).

Colville National Forest. 1994. *CROP Creating Opportunities. A Study of Small-Diameter Trees of the Colville National Forest.* Colville, WA: USDA Forest Service, Colville National Forest.

Connor, R.C., J.D. Born, A.W. Green, and R.A. O'Brien. 1990. *Forest Resources of Arizona.* USDA Forest Service Resource Bulletin INT–69. U.S. Forest Service.

Cool, K. L. 1992. Seeking common ground on western rangelands. *Rangelands* 14(2): 90–92.

Debano, L.F., and L.J. Schmidt. 1989. Interrelationship between watershed condition and health of riparian areas in southwestern U.S. In *Proceedings, Riparian Resource Management Workshop*, May 8–11, Billings, MT.

DeBell, D., and J. Franklin. 1987. Old growth Douglas-fir and western hemlock: A 36-year record of growth and mortality. *Western Journal of Applied Forestry* 2(4): 111–114.

Dolph, L.K., S.R. Mori, and W. Oliver. 1995. Long-term response of old-growth stands to varying levels of partial cutting in the eastside pine type. *Western Journal of Applied Forestry* 10(3): 101–108.

Elmore, W., and B. Kauffman. 1994. Riparian and watershed systems: Degradation and restoration. *Ecological Implications of Livestock Herbivory in the West.* Society for Range Management.

Erickson, J.R., and D.E. Toweill. 1994. Forest health and wildlife management on the Boise National Forest, Idaho. pp. 389–409. In: R.N. Sampson and D.L. Adams (eds.), *Assessing Forest Ecosystem Health in the Inland West.* The Haworth Press, New York.

Frost, T.P., G.L. Raines, C. Almquist, and B.R. Johnson. 1996. Digital Maps of Possible Bat Habitats for the Pacific Northwest: A Contribution to the Interior Columbia River Basin Ecosystem Management Project, U.S. Geological Survey Open-File Report 95–683.

Gunther, T.M. 1992. Quantitative assessment of future development of copper/silver resources in the Kootenai National Forest Idaho/Montana. Part II — Economic and policy analysis. *Nonrenewable Resources* 1(4): 267–280.

Harris, L.D. 1984. *The Fragmented Forest: Island Biogeography and the Preservation of Biotic Diversity.* The University of Chicago Press, Chicago, IL.

Herbel, C.H. 1985. Vegetation changes on arid rangelands of the Southwest. *Journal of Range Management* 7: 19–21.

Hoblitt, R.P., C.D. Miller, and W.E. Scott. 1987. *Volcanic Hazards With Regard to Siting Nuclear Power Plants in the Pacific Northwest.* U.S. Geological Survey Open-File Report 87–297.

Hone, J., and W. Atkinson. 1983. Evaluation of fencing to control feral pig movement. *Australian Wildlife Research* 10: 499–505.

Hone, J., and C.P. Stone. 1989. A comparison and evaluation of feral pig management in two national parks. *Wildlife Society Bulletin* 17(4): 419–425.

Huenneke, L.F., and P.M. Vitousek. 1990. Seedling and clonal recruitment of the invasive tree *Psidium cattleianum*: Implications for management of native Hawaiian forests. *Biological Conservation* 53: 199–211.

Interagency Ecosystem Management Task Force. 1995. The Ecosystem Approach: Healthy Ecosystems and Sustainable Economies. Vol. I–Overview. PB95–265583. Available from National Technical Information Service, Springfield, VA.

Johnson, B.R., and G.I. Raines. 1995. Digital Map of Major Lithologic Bedrock Units for the Pacific Northwest: A Contribution to the Interior Columbia River Basin Ecosystem Management Project. U.S. Geological Survey Open-File Report 95–680.

Katahira, L.K., P. Finnegan, and C.P. Stone. 1993. Eradicating feral pigs in mountain mesic habitat in Hawaii Volcanoes National Park. *Wildlife Society Bulletin* 21: 269–274.

Laycock, W.W. 1991. Stable states and thresholds of range condition on North American rangelands: A viewpoint. *Journal of Range Management* 44(5): 427–433.

Marsh, S.P., S.J. Kropschot, and R.G. Dickinson (eds.). 1984. *Wilderness Mineral Potential — Assessment of Mineral Resource Potential in U.S. Forest Service Lands Studied, 1964–1984.* U.S. Geological Survey Bulletin 1300, vol. 1–2.

Maser, C. 1988. *The Redesigned Forest.* R. & E. Miles, San Pedro, CA.

Matter, W.J., and R.W. Mannan. 1988. *Sand and Gravel Pits as Fish and Wildlife Habitat in the Southwest.* Fish and Wildlife Service Resource Publication 171.

Mayo, J. 1995. *Young Stand Thinning and Diversity Study, Willamette National Forest.* Central Cascades Adaptive Management Area. Irregular Publication.

Myers, L.H. 1989. Grazing and riparian management in southwestern Montana. In: *Proceedings, Riparian Resource Management Workshop,* 1989 May 8–11, Billings, MT.

National Research Council. 1994. *Rangeland Health: New Methods to Classify, Inventory, and Monitor Rangelands.* National Academy Press, Washington, DC.

Perlin, J. 1989. *A Forest Journey: The Role of Wood in the Development of Civilization.* W.W. Norton, New York.

Platts, W.S., C. Armour, G.D. Booth, M. Bryant, J.L. Bufford, P. Cuplin, S. Jensen, G.W. Lienkaemper, G.W. Minshall, S.B. Monsen, R.L. Nelson, J.R. Sedell, and J.S. Tuhy. 1987. Methods of Evaluating Riparian Habitats With Applications to Management. USDA Forest Service General Technical Report INT–221. Ogden, UT: U.S. Forest Service, Intermountain Research Station.

Raines, G.L., and C.L. Smith. 1996. Digital Maps of National Uranium Resource Evaluation (NURE) Geochemistry for the Pacific Northwest: A Contribution to the Interior Columbia Basin Ecosystem Management Project, U.S. Geological Survey Open-File Report 95–686.

Raines, G.L., B.R. Johnson, T.P. Frost, and M.L. Zientek. 1996. Digital Maps of Compositionally Classified Lithologies Derived From 1: 500,000 Scale Geologic Mapping For the Pacific Northwest: A Contribution to the Interior Columbia Basin Ecosystem Management Project. U.S. Geological Survey Open-File Report.

Ryland, D. 1996. Evaluating the Impact of Four Silvicultural Prescriptions on Stand Growth and Structure in Northwest Washington — A Modeling Approach. Master's of Forestry Report (unpublished). Oregon State University, Forest Resources Department, Corvallis, OR.

Salwerowicz, F. 1994. Mineral development and ecosystem management. In: *Proceedings, Third International Conference on Environmental Issues and Waste Management in Energy and Mineral Production,* 1994 August 30–September 1, Perth, West Australia. Curtin University of Technology 1–14.

Sharpe, M., D.L. True, J.E. Bowns, J.E. Burkhart, S. Tixier, B. McQuivey, D. Vail, H.J. Box, and D. Boe. 1992. Rangeland Program Initiatives and Strategies: Report of the Blue Ribbon Panel to the National Public Lands Advisory Council. Washington, DC: National Public Lands Advisory Council.

Singer, D.A. 1975. Mineral resource models and the Alaskan Mineral Resource Assessment Program. pp. 370–382. In: W.A. Vogely, (ed.), *Mineral Materials Modeling — A State-of-the-Art Review.* Johns Hopkins University Press, Baltimore, MD.

Singer, D.A. 1993. Basic concepts in three-part quantitative assessments of undiscovered mineral resources. *Nonrenewable Resources* 2(2): 69-81.

Singer, F.J. 1981. Wild pig population in national parks. *Environmental Management* 5: 363–370.

Stone, C.P., and L.L. Loope. 1987. Reducing the negative effects of introduced animals on native biotas in Hawaii: What is being done, what needs doing, and the role of National Parks. *Environmental Conservation* 14: 245–258.

Sundberg, U.L., J. Lindegren, H.T. Odum, and S. Doherty. 1994. Forest energy basis for Swedish power in the 17th century. *Scandinavian Journal of Forest Research*, Supplement 1.

U.S. Fish and Wildlife Service. 1993. *1991 National Survey of Fishing, Hunting, and Wildlife — Associated Recreation*. U.S. Government Printing Office, Washington, DC.

USDA Forest Service and U.S. Dept. of Interior, Bureau of Land Management. 1994. Record of Decision for Amendments to Forest Service and Bureau of Land Management Planning Documents Within the Range of the Northern Spotted Owl. [publisher unknown]

Van Hooser, D.D., D.C. Collins, and R.A. O'Brien. 1993. Forest Resources of New Mexico. INT-79. USDA Forest Service, Intermountain Research Station.

Vtorov, I.P. 1993. Feral pig removal: Effects on soil microarthropods in a Hawaiian rain forest. *Journal of Wildlife Management* 57(4): 875–880.

Welsch, D.J. 1991. Riparian Forest Buffers. Report NA–PR–07–91. Radnor, PA: USDA Forest Service, Northeastern Area, State and Private Forestry.

Yaffee, S.L. 1994. *The Wisdom of the Spotted Owl: Policy Lessons for a New Century*. Island Press, Washington, DC.

Zientek, M.L., A.A. Bookstrom, S.E. Box, and B.R. Johnson. 1996. Future Minerals Related Activity, Interior Columbia Basin Ecosystem Management Project. U.S. Geological Survey Open-File Report 95–687.

## THE AUTHORS

**Marlin Johnson**
*USDA Forest Service*
*Southwestern Region*
*Federal Building*
*517 Gold Avenue, SW*
*Albuquerque, NM 87102, USA*

**James Barbour**
*USDA Forest Service*
*Pacific Northwest Research Station*
*P.O. Box 3890*
*Portland, OR 97208-3890, USA*

**David W. Green**
*USDA Forest Service*
*Forest Products Laboratory*
*One Gifford Pinchot Drive*
*Madison, WI 53705-2398, USA*

**Susan Willits**
*USDA Forest Service*
*Pacific Northwest Research Station*
*P.O. Box 3890*
*Portland, OR 97208-3890, USA*

**Michael Znerold**
*USDA Forest Service*
*San Juan–Rio Grande National Forest*
*Mancos–Dolores Ranger District*
*P.O. Box 210*
*Dolores, CO 81323, USA*

**James D. Bliss**
*U.S. Geological Survey*
*Tucson Field Office*
*Corbett Building*
*210 East 7th Street*
*Tucson, AZ 85705-8454, USA*

**Sie Ling Chiang**
*U.S. Department of the Interior*
*Bureau of Land Management*
*1849 C Street, NW*
*WO-300, LS 202B*
*Washington, DC 20240, USA*

**Dale Toweill**
*Idaho Department of Fish and Game*
*Wildlife Program Coordinator*
*P.O. Box 25*
*660 S. Walnut*
*Boise, ID 83712, USA*

# Ecosystem Sustainability and Condition

C. Ronald Carroll, Jayne Belnap, Bob Breckenridge, and Gary Meffe

**Key questions addressed in this chapter**

- ♦ *Five common properties of ecosystems*
- ♦ *Resiliency is a key constraint for ecosystem management*
- ♦ *Humans are both stakeholders and players in ecosystem processes*
- ♦ *How to select useful indicators for ecosystem management*
- ♦ *The meaning of sustainability in ecosystem management*

**Keywords: Sustainability, biophysical properties, people in ecosystem processes, indicators of ecosystem condition, indicator assessments**

## 1 INTRODUCTION

In this chapter we develop the following five themes.

1. Ecosystems have defined biophysical properties. These properties have characteristic ranges of variability and resiliency to disturbances. Managers must work within these ranges.

   In the first theme, we identify and discuss the key biophysical properties of ecosystems that must be addressed in management plans as well as in the general decision-making processes that involve the use of natural resources. For each of these properties we provide an illustration of the consequences that may occur if managers ignore these essential defining characteristics of ecosystems.

2. The ecosystem: what it means to the researcher and to the manager. Confusion over the meaning of ecosystem management often arises from a failure to distinguish between "ecosystem" used to define a research topic and "ecosystem" used to define a typically large-scale and holistic management unit.

   In developing the second theme, we first discuss the ecosystem as a research construct that is defined for specific spatial and temporal scales to meet particular research objectives. Then we discuss the more inclusive meaning of ecosystem as a holistic management unit the meaning that is of most immediate interest to managers.

3. People as players and stakeholders in ecosystem processes. People are integral parts of most ecosystems and the dominant players in some. People whose lives are affected by the condition of an ecosystem are stakeholders in any decision- making process that may affect the condition of the ecosystem.

   In the third theme, we discuss ways in which people, as consumptive and non-consumptive users of natural resources, interact with ecosystem properties. That is, we treat people both as agents that influence ecosystems and as stakeholders interested in the outcome of ecosystem management.

4. The meaning of ecosystem sustainability. Sustainability is not restricted to biophysical considerations but refers more generally to the interplay of human values, goals, and behavior with biophysical and ecological properties and behavior within an ecosystem.

   In the fourth theme we address the meaning of ecosystem sustainability. We provide examples of how the ways in which we use and manage our natural resources and delineate their spatial and temporal boundaries can affect the sustainability of the ecosystem.

5. Indicators of ecosystem condition: measuring success and avoiding failure. It follows from the previous theme that indicators must be reliable surrogates for both ecological and social/economic attributes of an ecosystem.

   In the fifth and final theme, we identify criteria that are important in the development of useful indicators of the ecological conditions of ecosystems. We then discuss particular indicators and conclude the chapter with an example of how indicators have been developed and used in the management of Arches National Park.

## 2 FIVE BIOPHYSICAL PROPERTIES OF ECOSYSTEMS

Particular ecosystems, of course, have unique properties that may be critical to management decisions, but general properties shared by all ecosystems are also important to managers. In this section we have selected five such general properties that can, at least in principle, be characterized for any ecosystem. Ecosystems are (1) open, (2) comprised of linear and non-linear processes, (3) have spatial and temporal variation, (4) exhibit trophic structure, and (5) have varying degrees of resiliency to disturbances. Other general properties could be added to this list, but these five are of particular concern for ecosystem management. For each of these general properties we briefly discuss their significance to management.

### 2.1 Ecosystems Are Open Systems

The functioning of any particular ecosystem is, to varying degrees, dependent on the surrounding environment for exchange of materials such as nutrients and waste products, energy input and release of dissipative energy, and for the colonization of species. Therefore the management of a particular ecosystem must be cognizant of the various dependencies it has with other ecosystems and the surrounding environment. For example, no one would imagine that rivers, lakes and estuaries could be sustained as natural ecosystems without consideration of surrounding terrestrial drainage basins, watersheds, or floodplains.

*Management significance*: Ecosystems cannot be properly managed without considering the larger environment in which the ecosystem is embedded. In other words, each particular ecosystem is, to varying degrees, linked to other ecosystems in a landscape context. Finding the right mix of spatial and temporal scales for enclosing ecosystems and to establish management boundaries is essential.

## 2.2 Ecosystems Are Comprised of Few Linear and Many Non-linear Processes

Non-linearities are inherent in biological systems because populations grow geometrically and species interact with other species and with the physical environment in non-additive ways. As a result indirect effects and threshold responses are common.

*Management significance*: Surprises, unexpected results and threshold responses will occur. Therefore, management decisions that could result in large and rapid changes to ecosystem functions should be avoided, if possible, and replaced by sequences of decisions that will have smaller, incremental effects on the ecosystem. Monitoring the environmental responses to management decisions and maintaining the flexibility to modify decisions are essential.

## 2.3 Ecosystems Have Considerable Internal Spatial and Temporal Variations

Spatial variation, or heterogeneity, affects biodiversity by creating more possibilities for niche partitioning and refugia from predators, competitors and diseases. High levels of spatial heterogeneity in an ecosystem often lead to more species co-existing and stronger buffering against population extinctions that would otherwise result from local food losses or habitat destruction.

*Management significance*: The normal ranges of variation should not be exceeded nor should the ecosystem be unduly constrained and variance too limited. Unduly constraining variation will lead to catastrophic changes, e.g., suppressing natural fires leads to destructive wildfires, controlling normal flood stages leads to catastrophic floods in unusually wet years. Ideally, a managed ecosystem should enclose sufficient spatial heterogeneity, in the form of alternative habitats, seasonal food sources, successional sequences, and various refugia such that critical resources remain available somewhere during extreme years and that the normal cycles and pulses of ecosystem dynamics are maintained.

## 2.4 Ecosystems Have Characteristic Trophic Structures

For example, relatively simple food webs characterize extremely arid deserts whereas complex webs are found in rainforests and coral reefs. In some ecosystems, the species composition in the higher trophic levels of food webs strongly influences net and secondary productivity and biodiversity at lower trophic levels. This is particularly evident in temperate-zone oligotrophic lakes.

*Management significance*: Hunting, fishing, and other direct and indirect removal of species, especially at higher trophic levels should be conservative. Similarly, the introduction of new predators, such as game fish, into ecosystems should be avoided. Critical "bottom-up" processes of decomposition and nutrient transformation and cycling must be protected, especially from the effects of pollutants and inappropriate extractive resource activities.

## 2.5 Ecosystems Are Resilient

Ecosystems are capable of recovering from perturbations, but only within limits. The resiliency properties of an ecosystem are ecosystem -and stress-specific; resiliency is not a generic property. Ecosystems that are degraded lose resiliency, especially where redundancy is reduced and nutrient transformation rates are slowed.

*Management significance*: Ecosystems can recover from many kinds of perturbations but resiliency is limited and cumulative effects, persistent stresses, and catastrophic events must be avoided. Resiliency defines the upper boundary for rates of consumptive use and therefore should be investigated, at least for the dominant extractive uses of the ecosystem. Although general statements can be made about ecosystem resiliency, resiliency for any particular ecosystem needs to be empirically derived. For this reason, resiliency should be investigated through field experimentation at the appropriate scale for the dominant consumptive practices within the ecosystem.

Lakes have provided some of the best studies of the relationship between biodiversity (principally as the composition of food webs) and ecosystem function (Carpenter and Kitchell 1993). Lakes, therefore, provide good examples of how management decisions can influence the structure and function of ecosystems. Manipulations of the top carnivore in lakes, and perhaps in other ecosystems, can create changes that "cascade" throughout the food web, influencing species composition and nutrient dynamics (Paine 1980). In a seminal study of Peter Lake, a small oligotrophic lake in Wisconsin, the replacement of largemouth bass by rainbow trout had the following consequences. The generalized algal grazer *Daphnia pulex* and the zooplankton *Chaoborus* spp. declined. Populations of a new and more restricted algal grazer, *Daphnia dubia*, irrupted and, because *D. dubia* is cannot feed on large algae, algal blooms followed (Carpenter et al. 1993). Such experimental results raise general concerns about possible ecosystem-wide effects from the removal of top carnivores as game in hunting and fishing activities.

Manipulating higher levels of the food web can generate rapid changes throughout the rest of the food web; however, actions that directly affect lower trophic levels can also slow or even qualitatively change important "bottom-up" processes. For example, acid precipitation can lower the pH in poorly buffered forest soils and, by inhibiting bacterial activities, lead to greatly lowered rates of nitrogen fixation and eventually to severe nitrogen limits to productivity. Phosphate loading in aquatic systems can lead to qualitative changes in algal species composition, to large algal population blooms with subsequent anoxia as bacteria decompose the dying algae.

Of all the natural properties of ecosystems, the properties of change and variation are especially difficult to embrace in management plans. The classical "balance of nature" model that has been humanity's perspective of the natural world for the 19th and first half of the 20th centuries has steadily receded from view in the last two decades. The steady-state, equilibrium view dating back to pre-Victorian times has been replaced by an empirically derived model that clearly recognizes that nature is dynamic, changeable, and, therefore, predictable only at rather large spatial and temporal scales. At the scale of hundreds of hectares, forest structure is highly predictable, unless there is some externally-imposed degradative process at work. At the scale of a hundred square meters, however, tree falls and subsequent successional processes are much less predictable. And, at much smaller scales, the turnover of microbial communities around root systems is highly dynamic and unpredictable. However, these many small- scale and locally unpredictable processes are what fundamentally generate the predictable structure of forests. Natural disturbances, such as fires, floods, earthslides, insect outbreaks, droughts, and storms, are now appreciated as having great influences on all levels of biodiversity ranging from genes to landscapes as well as affecting many ecosystem processes.

Historically, much of natural resource management has been based on an equilibrial view of nature, with systems having set stable points to which they returned after a disturbance. Early approaches for achieving maximum sustained yield and single-species game management were premised on such a perspective. In this view, single components of a system can be manipulated with predictable outcomes for the larger system. Our contemporary perspective of ecosystems as dynamic, changing systems tells us that such prediction is difficult at best, impossible at worst. However, despite this knowledge base, much of resource management by public agencies has proceeded according to historical models, with the view that proper management can restore the "balance of nature" to its pre-disturbance condition after extraction of resources or other manipulations of the system. The target is seen as fixed, and the processes leading to that target are deemed to have predictable outcomes. We now understand that, unfortunately, ecosystems are far from being this simple and predictable, and that management efforts must be modified to account for uncertainty, change, and surprise.

The inherent dynamics of change in ecosystems makes rigid forms of management ineffective. Consequently, management must take a different tack: to be flexible, adaptive, and, as much as possible, predictable.

The limits to ranges of variation in ecosystem structure and function must be identified. All ecosystems have a natural range of variation in structure and processes that they experience over a given time period, and this range is ecosystem-dependent. For example, over the period of a month, the composition and structure of woody species in an old growth deciduous forest will not typically change in a significant way; unless a major disturbance occurs, the forest will consist of the same trees in the same positions. However, over a period of a decade, and certainly over a century, the forest may experience great variation in composition and structure: trees will die, others will replace them, diseases or herbivores will attack the forest, and fire may break out. Range of variation can be expected over the long term for this system.

Because ecosystems experience so much natural variation from so many sources, yet are not destroyed or fundamentally altered by any but the most severe natural (and many human) disturbances, they are said to possess *resilience*. Resilience is the magnitude of disturbance that can be absorbed or accommodated by an ecosystem before its structure is fundamentally changed to a different state. Any ecosystem is expected to be resilient to disturbances and variation that are within the normal "repertoire" of what it has experienced over ecological time (millennia). Consequently, rivers are expected to be resilient to flooding, chaparral and longleaf pine-wiregrass systems to frequent fire, and high-latitude aquatic systems to winter freezing. In fact, ecosystem composition and structure may change significantly when those disturbances do *not* occur, such as when riverine flows are stabilized by dams, or fires are suppressed where they normally occur. Then, the normal ranges of variation are not experienced and ecosystems change.

The spatial and temporal range of variation that an ecosystem experiences naturally is a first approximation for managers who are responsible for setting limits on consumptives uses. For example, the size, frequency, and spatial distribution of forest gaps caused by natural events may provide an estimate of acceptable size and frequency of timber cuts when the goal is to balance timber receipts against rates of regeneration. *However, natural patterns and rates of disturbance may not always reflect the resiliency of a particular ecosystem for three very different reasons. First,* a system's resiliency may be able to cope with disturbance regimes that exceed natural disturbance regimes. In this case, natural disturbance regimes would give an underestimate of the system's resiliency limits for similar kinds of artificial disturbances. *Second,* if the system is fragmented or otherwise degraded it may no longer have adequate resiliency even to cope with natural disturbances. *Third,* disturbances due to consumptive uses may not be analogous to any natural disturbances, hence, natural disturbance regimes would not yield information that is relevant to setting limits on anthropogenic disturbances. Understanding the resiliency properties of an ecosystem is essential for sustainable ecosystem management, especially for those areas that are required to provide multiple uses to society. However, *it seems highly unlikely that resiliency boundaries will be revealed unless mesocosm disturbance experiments, conducted at the appropriate spatial and temporal scales, are designed explicitly to reveal those limits.*

The importance of variation and disturbance is further illustrated in a model developed by Holling (1995), who discussed the *constructive* role that variation and disturbance play in maintaining the integrity of ecosystem function in the face of unexpected events (Fig. 1). Holling's model is, of course, a generalized abstraction of the real world. It provides a useful construct for comparing the behavior of any particular ecosystem. In his model, ecosystems go through four functional stages as they develop (Fig. 1). The first stage is 'exploitation' or early succession after a disturbance. Here, rapidly growing pioneer and opportunist species dominate and exploit the open space. The system then moves toward the 'conservation' stage, where more mature communities develop. In this stage, which can be long-term, complex community structure develops and strong interspecific relationships prevail.

As system organization and 'connectance' increase, the system becomes more 'brittle,' meaning it is more susceptible to biotic and abiotic disturbances such as insect infestations, storms, fires, or pathogens. At some point, a major disturbance will occur, quickly shifting the system into the 'release' phase, whereupon the complexities, structure, and stored energies are quickly released and disorganized. Although release destroys structure it also creates opportunity, in the sense that the system can begin to reassemble; this is called, 'creative destruction' and is an important aspect of disturbance. This rapid and chaotic release phase is followed by 'reorganization,' when the system begins to rebuild. The time spent in each stage in this model is uneven; the exploitative stage is moderate in length, the conservation phase can be very long, the release phase is very rapid, and reorganization is also fast.

During the relatively long exploitation and conservation phases, other, faster, cycles are driven by microbial and invertebrate activities and by smaller, localized releases, such as small patchy fires and gap dynamics. The length of these four major phases varies by the system and locality. Even within ecosystems, considerable temporal variation may be found. For example, after the nine years following the great fires in the Yellowstone Greater Ecosystem, some areas show vigorous sapling growth, other areas are developing dense sapling thickets, still other areas are nearly devoid of any vegetation. Any indicator of terrestrial ecosystem sustainability must be matched with the appropriate part of the four major phases, but, the variation that is generated within the major phases by other faster internal cycles and by spatial heterogeneity needs to be recognized.

Some ecosystems, particularly riparian and coastal ecosystems, may be better characterized by shorter pulsing patterns produced by the hydroperiods. Tidal flow and seasonal changes in water level periodically provide energy and material subsidies to the ecosystem (Odum et al. 1995). In these cases, indicators of sustainability should reflect pulsing events and respond to conditions

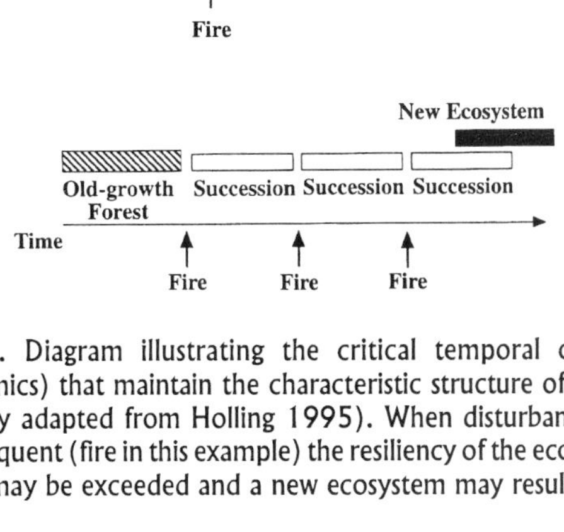

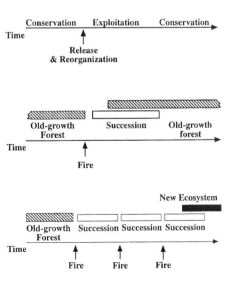

Fig. 1. Diagram illustrating the critical temporal changes (dynamics) that maintain the characteristic structure of forests (loosely adapted from Holling 1995). When disturbances are too frequent (fire in this example) the resiliency of the ecosystem may be exceeded and a new ecosystem may result.

of too much or too little energy and material subsidy. For example, an evaluation of the experimental pulse release of water from Lake Powell into the Grand Canyon would be based on expected amounts of material accumulation in the form of sand bars and other sediment deposits.

The problem in pursuing ecosystem sustainability is in either pushing an ecosystem well beyond its normal range of variation experienced over ecological time, or in suppressing natural variation to the point that the system cannot function in a normal way and in fact loses resilience as a result (Holling and Meffe 1996). Either way, the system may change to a fundamentally different state as a result, and will not continue to function as it normally would. Consequently, identifying the driving disturbances and typical ranges of variation in a given ecosystem is critical to understanding the system and managing it in a sustainable way that does not alter system resilience.

## 3 THE ECOSYSTEM: ITS MEANING TO THE RESEARCHER AND TO THE MANAGER

To a research scientist, an ecosystem is simply a location in which one studies the interaction of biotic and physical processes. An ecosystem may be very large, but for practical reasons, studies generally focus on one or a few compartments within a larger ecosystem. Thus, the forest ecosystem of the coniferous forest biome potentially embraces all biotic activities, from bacteria to higher plants and vertebrates (including humans), their spatial and temporal patterns, and their interactions among themselves and with the physical environment. In practice, of course, any individual ecosystem scientist generally studies only a small sub-set of the larger ecosystem. For research purposes a large ecosystem is usually sub-divided into smaller systems, called compartments, that have strong internal linkages. For example, the large forest ecosystem embraces many smaller systems, such as the soil ecosystem. This, in turn, may be further sub-divided by the researcher into smaller and analytically more tractable ecosystems, such the leaf litter, the microbially active zone around root tips, or even the micro-cavities within soil aggregates; each is chosen to reflect some important set of interactions and processes that contribute to the structure and function of the soil ecosystem and to the larger forest ecosystem.

For natural stewardship agencies, an ecosystem may be used as a hierarchical research construct by the agency's scientists, as discussed above, or as a holistic management unit. Concern for ecosystem sustainability typically takes place at the holistic management unit scale. However, sustainability — as an ecological condition — is determined by processes that take place at various scales; not all of them are easily accessible to the manager. The integrity of an ecosystem is maintained through the aggregate bottom-up processes operating at small scales in ecosystem compartments, through top-down influences from higher trophic levels, through other internal interactions such as competition and mutualism, and through external influences from the surrounding landscape. Human activity may intervene directly or indirectly at any of these levels and affect ecosystem properties.

### 3.1 An Example — The Longleaf Pine/ Wiregrass Ecosystem

A brief introduction to the longleaf pine/wiregrass savanna ecosystem of the southern coastal plain may provide a useful illustration of these general ideas. The sandy coastal plain from Virginia into east Texas was once a nearly continuous savanna with longleaf pine (*Pinus palustris*) and wiregrass (*Aristida beyrichiana* and *A. stricta*) its most conspicuous plant species. Characteristic animals of this savanna include imperiled species, such as, gopher tortoises, indigo snakes, gopher frogs, and Bahkman's Sparrow, and a rich diversity of other less-threatened species. More than 90% of this savanna has been converted to other uses, mainly short-rotation pine plantations, cattle ranches, peanut, sorghum, cotton, and other kinds of agriculture. The few pieces of reasonably intact savanna that remain are isolated from one another in a matrix of agricultural and plantation landscapes. Wiregrass is a key component of this ecosystem because it carries summer fires caused by lightning strikes over large areas. Wiregrass and many other herbaceous species recover quickly from fire and in succeeding years may have enhanced seed production. Longleaf pine is resistant to fire and its seeds germinate and the seedlings grow rapidly where fire has created a good seed bed and removed competing vegetation. Hardwoods, on the other hand, are less tolerant of fire. Thus, fire maintains this ecosystem and prevents its replacement by thickets of hardwoods.

What are the essential processes that lead to and maintain the integrity of this savanna ecosystem? That is, what makes this ecosystem sustainable? First, there must be a minimal ecosystem patch size that will contain sufficient numbers of the characteristic species to ensure population replacement and sufficient genetic variation to continue the process of natural selection and prevent possible inbreeding depression. But, because of the importance of environmental variation, in both space and time, values will have to be given in probabilities not certainties. That is, research might tell us the environmental conditions under which there would be an 80% probability that a particular population of gopher tortoises would persist for at least 50 years, but research could not tell with certainty that the population would persist for at least 50 years.

At the biological level, determining the conditions for sustainability is not absolute and forever, but is hedged by the limits of our research capabilities and by the amplitude and frequency of environmental variation. Our point is that decisions regarding size, distribution, and heterogeneity of management areas should use the best scientific information available but that some level of uncertainty will always characterize management decisions for biological systems. This general approach could be raised to a higher spatial scale in order to assess the quality of a landscape for sets of species. Danielson (1992) suggests using the home ranges of species to define an area in which critical habitat resources must be found in order to provide the minimal needs of individuals. Using habitat classification techniques coupled with remote sensing it should be possible to assess what fraction of a landscape contains home range units with the minimal requisite resources. This approach could be developed for sets of key species and used to define indices of landscape quality.

Ecosystem management certainly includes determining the resource requirements of key species, but ecosystems, as life support systems for all of biodiversity, require an approach that is more holistic than simply summing up the requirements of individual species. Characteristics of ecosystems that are particularly relevant to management include the following:

1. Ecosystems have characteristic ranges of variability. The corollary of this is that ecosystems have characteristic suites of resiliency properties, that may or may not be relevant to anthropogenic disturbances. As we pointed out, the understanding of resiliency limits may generally require specially constructed mesocosm experiments.

2. Regimes of natural disturbances generate much of the variability that characterize ecosystems. As we discussed in the first section, the important effect of these disturbance regimes is to create variable sized patches that are in different stages of development and, depending on the type and intensity of the disturbance, the patches may have different ecological properties.

3. Ecosystems are open; that is, they exchange nutrients, wastes, migrants, etc. with other parts of the landscape. The openness of ecosystems can cause severe management problems because natural ecosystems are generally small relative to the magnitude of areas used primarily for human economic activities and impacts from external sources can be quite large.

In this savanna ecosystem, the occurrence of frequent fires is essential. Small patches of savanna have a low probability of being hit by lightning; large patches have a higher probability of lightning strikes occurring somewhere in the patch. Thus, all other factors being equal, low probabilities of lightning strikes in small patches of pine savanna that are surrounded by hardwoods will result in a higher probability of small patches being displaced by hardwoods and a lower probability of large pine savannas being displaced. From a management perspective, small patches of savanna will have to be maintained by prescribed burning to mimic the frequency of natural fires in larger areas. A further complication arises with the consideration of the effects of invasive species. Over the past few decades the imported red fire ant (*Solenopsis invicta*) has invaded the coastal savanna and established large population densities. Consequently, among the effects we have observed (CRC, pers. observ.), native species of ants have been displaced, successful fledging by ground-nesting birds has been reduced, and the abundant nest sites of the fire ants have reduced soil fertility. Reproductive success in these ants is strongly inhibited by shading; hence, the colonies do poorly when they are shaded by vegetation. This leads to a consideration of a management trade-off. The savanna ecosystem normally requires a fire frequency of approximately every 3–5 years in order to prevent encroachment by hardwoods. The inhibition of colony reproduction by fire ants requires the shading effect of vegetation that is at least five years old. In this case, a manager may decide that a 5–6 year frequency for prescribed burns is an acceptable trade-off between hardwood encroachment that requires frequent burns and fire ant control that requires infrequent burns. A second invasive species, cogongrass (*Imperata cylindrica*), has become established in Florida. This grass is a major weed throughout the Asian and African tropics. It produces copious seeds, reproduces from rhizomes and flourishes in low fertility soils where the vegetation is periodically burned. Thus ecosystem management practices that support the wiregrass-longleaf pine savanna may also favor invasion by fire ants and cogongrass. Sustainability of this savanna ecosystem thus may require a conscious decision to modify a critical natural process, fire frequency. The bottom line message of this example is that management for ecosystem sustainability should be founded on a solid understanding of the key ecological processes that maintain the structure and function of the ecosystem. But, ecological processes are not deterministic, the unexpected will happen, and therefore adaptive management and long-term monitoring will be necessary.

## 4 PEOPLE AS PLAYERS AND STAKEHOLDERS IN ECOSYSTEM PROCESSES

How do we introduce the many dimensions of human activities into our discussion of ecological sustainability? We can estimate the effects that different rates and forms of extraction, exploitation, and degradation of resources have on the biological processes that influence ecosystem sustainability. We can also estimate the effect that extraction, exploitation, and degradation have on the environmental variation of these biological processes. This is the most common approach, reflecting, unfortunately, the generally negative impacts that people have on natural ecosystems. Fortunately, most people have only transient effects on natural ecosystems because they are only temporary visitors to the area. If visitation rates are low, the effects of activities such as seasonal hunting, picking wildflowers, walking on sensitive areas, disturbing wildlife, are minor perturbations without lasting effects. Similarly, pollutants such as sewage and non-persistent toxins can be neutralized by ecosystems as long as they do not occur too frequently or the quantifies are not too great. As visitation rates or pollutant discharges increase, cumulative effects can lead to degradation.

Exploitation of key species can, of course, cause rapid and extensive changes to ecosystems. The removal of top carnivores through hunting, trapping, or poisoning can lead to significant top-down effects that cascade through large parts of the management ecosystem. For example sea otter populations were decimated by the fur trade during the 19th century. Consequently, populations of sea urchins, one of their major prey, greatly increased. The large number of grazing urchins damaged the attachment "hold fasts" of the kelps which resulted in major damage to the kelp beds and their associated organisms (Duggins et al. 1989).

However, through wise stewardship, human activities may also reinforce the ecological processes that maintain sustainability. We can look to a more proactive or prescriptive approach to show how human economic activities might enhance biological and physical processes that underpin the sustainability of ecosystems. The recent experimental pulse release of Lake Powell water was an attempt to mimic the predevelopment spring floods in the Grand Canyon.

Much of the nation's terrestrial biodiversity is in those public lands that are also used for hunting, fishing, lumbering, mining, livestock grazing, military training, and many other consumptive and non-consumptive uses. Most of these lands are managed by the federal agencies: U.S. Forest Service, U.S. Fish and Wildlife Service, the Bureau of Land Management, National Park Service, and the Department of Defense, and state agencies such as departments of natural resources or forestry. On these public lands, the protection of biodiversity and intact ecosystems is only a subset of the total land-use mandates that must be accommodated. Probably the most important goal in managing for sustainability on these exploited lands is to find the appropriate balance for meeting human economic and social needs while protecting the ecological conditions that underpin the sustainability of natural ecosystem function and structure. Consumptive exploitation of ecosystems by humans should not be allowed to increase the normal historical range of variation in ecosystem structure and function. This is a straightforward recommendation for consumptive uses that are analogous to natural processes, e.g., hunting and fishing. Other consumptive use practices, such as clear-cut logging, may have no natural analogue. Hence, defining the appropriate range of variation for ecosystem structure and function for uniquely human consumptive uses will be difficult.

As much as possible, the management decisions that are needed to ensure ecosystem sustainability should reside at the local level. In the context of the USDA Forest Service planning this would normally be at the project level. However, coordination among local decision-makers who work at different sites within a large planning region is necessary to ensure that landscape scale processes influencing stainability are included. The reason for moving ecosystem management as much as possible to the local level is that the greatest knowledge about a particular ecosystem often resides among local researchers, managers, and other stakeholders in the site. Local community participation in the decision-making process is important to ensure that the local stakeholders' role in the ecosystem is included. For example, the development of prescribed burning plans usually involve meetings with the local community in order to find ways to reduce negative effects from drifting smoke, to minimize the chances that private property might be damaged by poorly contained fires, and to ensure that the prescribed burns are compatible with other forest uses such as recreation, logging, hunting and fishing.

By now our discussion should have made abundantly clear the dynamic nature of ecosystems and the need for management decisions to be responsive, ideally anticipatory, to the changing conditions within ecosystems. However historically resource management has been a rigid process with decisions made at the top and carried out at the bottom. This "command and control" decision-making is the antithesis of ecosystem management (Holling and Meffe 1996). To summarize rigid, top-down, management is inappropriate to ecosystem management because:

1. ecosystems are complex and characterized by non-linear processes and indirect effects, threshold responses and unexpected consequences are inherent;
2. local colonization and extinction processes constantly shuffle the biological players in local communities, thereby changing biological interactions; and,
3. because ecosystems are open, economic development around the management area can affect ecological processes through habitat fragmentation, release of toxin, invasion by exotic species, withdrawal of water, and increased frequency of wildfires, among other external threats.

Management therefore needs to be flexible, adaptive, and predictive. This means that managers — and other stakeholders as well — must accept a range of possible, but not equally likely, outcomes from a management action. Management should be seen as a kind of experimental research agenda in which results lead to better decisions. Furthermore, parent institutions should reward innovative managers who learn from mistakes rather than simply acting to protect status quo practices.

Because adaptive management has been applied over several decades in complex systems from industrial optimization to municipal planning for transportation networks, many tools and approaches have been developed. Management of the Greater Everglades Ecosystem provides an instructive example. This holistic approach is being developed through support from the U.S. State Department's Man and the Biosphere Program, Human-Dominated Systems Directorate (Harwell et al. 1996). Land-use and hydrology scenarios were developed for south Florida through historical analyses coupled with simulation and geographic information system models. The scenarios were developed to provide a range of protection to the Everglades National Park and economic development for south Florida. Each scenario was evaluated by teams that represented a range of social, economic, and ecological perspectives in a process known as "scenario-consequence analysis." An interesting and important perspective emerged from this process, modest improvements in agricultural water use efficiency could both maintain water recharge to the Everglades and also contribute to the long-term sustainability of south Florida agriculture. Although this example involves a very complex human-dominated landscape, it contains lessons that are broadly applicable to ecosystem management. First, a combination of information and analytical tools (historical studies, simulation models, geographic information systems) were used to develop predictive and alternative management approaches. Second, the decision-making process was opened up by involving teams of various stakeholder groups to evaluate the management alternatives.

## 4.1 The Social Context for Ecosystem Sustainability

Efforts to manage ecosystems for their long-term sustainability take place in a social as well as an environmental context and the costs and benefits of management decisions are unequally distributed in space and time. Because ecosystem management typically takes place on large landscape scales, many stakeholder groups may be affected by management decisions and the consequences of these decisions have an important spatial dimension. For example, decisions to change land use practices in order to provide increased protection to biodiversity are likely to distribute social benefits broadly while concentrating costs locally. There is a much larger, more global, distribution of people who value biodiversity for any reason, while local people bear most of the costs, usually in the form of opportunity costs when open access to hunting, fishing, extractable resources, or recreation are restricted. This spatial disjunction between costs and benefits makes concensus among stakeholders more difficult to achieve.

Because a long time scale is implicit in sustainability, stakeholders may be asked to forego some immediate benefits in order to achieve sustainability of the ecosystem in the long-term. Because people tend to discount the value of benefits that are realized in the future, it is difficult to explain the rationality of forgoing an immediate benefit, such as the pleasure some find in hunting keystone predators, in order to achieve the more diffuse benefits of a sustainable ecosystem for the future.

In addition to these time and spatial scale problems, we have the additional problem that stakeholders may hold very different values. Stakeholders may have differing value systems; some may value a place or species simply for their existence, others because there are market opportunities, and still others for religious or ethical reasons. How then can people who hold such different value systems ever hope to reach concensus over the long-term management of ecosystems? Fortunately, although the philosophical bases that underpin value systems may differ substantially, the varying levels of importance that people attach to different land use decisions provide some opportunity for negotiation. The approach that economists use when assessing people's attachment to values that are not easily measured in the market place, such as camping in a forest or bird-watching, is known as contingent valuation. One commonly used form of contingent evaluation involves measuring someone's willingness to pay for access to an amenity. Another involves measuring a person's willingness to travel to experience an amenity. For example, the relative "value" of reintroducing wolves into the northern Rocky Mountains could be estimated among different groups of people by determining the distance they would be willing to travel or amount they would pay in order to see or hear wolves in their natural habitat.

These surrogate measures of non-market values may have some utility for assessing the relative importance that stakeholders place on particular land use decisions, e.g., how the "value" of a forest would change if the number of campsites were increased or more hunting licenses issued. However, contingent valuation is most useful when comparing the effects of two or more relatively similar management decisions and may give misleading information when used to compare the relative importance of very different kinds of management decisions. For example, contingent valuation might be appropriate if we wished to assess the relative value that hunters, campers, and nature photographers placed on different practices for wildlife management. Contingent valuation would be highly misleading if it were used to assess the relative values that these same groups attached to a place where they might see a charismatic species such as a grizzly bear versus experiencing a healthy forest ecosystem. Furthermore, measures such as willingness to pay or travel are likely to underestimate values of common, familiar, or local amenities and overestimate amenities that are unique, unfamiliar, or remote. The important point is that there are many different ways to measure the importance that different stakeholders place on management goals and on the social and economic costs to attain those goals.

Another part of the social context in which managers function is within the culture of the agency bureaucracy. Successful ecosystem management for long-term sustainability requires agencies that are innovative, participatory, science-based, adaptive, and capable of long-term commitment of resources. Unfortunately, public land use agencies are bureaucracies with all the attendant decision-making inertia that is inherent in bureaucracies. Bureaucracies do not change easily and in order to develop managers who can successfully pursue the process of ecosystem management, performance review criteria at all levels must be framed to encourage managers to become risk-takers.

## 5 THE MEANING OF ECOSYSTEM SUSTAINABILITY

To this point, we have purposefully avoided a comprehensive definition of ecosystem sustainability in the context of management. Having now discussed the biophysical attributes of ecosystems, the role of scale and landscape heterogeneity, the need to use adaptive management approaches and to include participatory decision-making, we offer the following working definition of ecosystem management with sustainability as its fundamental goal. "Ecosystem management is an approach to maintaining or restoring the composition, structure, and function of natural and modified ecosystems for the goal of long-term sustainability. It is based on a collaboratively developed vision of desired future conditions and integrates ecological, socio-economic, and institutional perspectives, applied within a geographic framework defined primarily by natural ecological boundaries (Meffe and Carroll 1997; adapted from the definition used by the Interagency Task Force 1995)." This may seem like a cumbersome definition, but the concept is necessarily inclusive. Sustainable ecosystem management cannot succeed if ecological perspectives displace human welfare, if good ecological stewardship is sacrificed to meet narrow economic objectives, or if institutions cannot develop the flexibility to adopt better alternative perspectives.

## 6 INDICATORS OF ECOSYSTEM CONDITION: MEASURING SUCCESS AND AVOIDING FAILURE

In this section we explore the development and use of indicators to assess how natural and anthropogenic changes affect the structural and functional properties of ecosystems. From the perspective of resource manager the ecosystem, as a holistic management unit, must include natural structure and functions as well as the influence on them from human activities. The fundamental requirement for sustainable ecosystem management is that the effects from human activities must not exceed the resiliency of the system to maintain essential functions and to retain its characteristic biodiversity. Within these constraints, there remains considerable management flexibility. For example, management requirements for areas designated as "Wilderness Natural Areas" will be strongly weighted towards maintaining natural ecosystem patterns and processes with minimal human influence. Other areas scheduled for timber harvest will, of course, include more economic considerations. In the latter case, ecosystem sustainability might only be achieved at a larger landscape scale in which timber harvests function ecologically as disturbance patches. Similar indicators of ecosystem condition might be used in both cases, but how they are interpreted will be influenced by the management goals for the area.

The first step in identifying appropriate indicators is to classify the ecological resource as it currently exists, e.g., pinyon-juniper rangeland, and define the management boundaries. Once the ecosystem has been classified, its sustainable functions related to physical, biological and social (aesthetics and economics) aspects can be identified based upon environmental and social values. Assessment questions can then be developed that focus on the specific ecosystem in question. Indicators can be selected that provide the connection between ecological processes and the assessment questions. This framework makes it feasible to link values of various stakeholders with associated scientific issues and ecosystem function and structure.

Following is an example of the use of assessment questions to evaluate the condition of public rangelands in the western United States. Stakeholders involved in evaluating the question of whether rangelands are sustainable should first identify what values the ecosystems need to exhibit. For example, using the biological integrity of the ecosystem as a value would result in the development of assessment questions related to species composition and associated ecosystem functions. These would then be evaluated using a set of ecological indicators. The factors contributing to biological integrity can generally be extracted from previous research (Wood 1988, Noss 1990, West 1993) and then refined to meet local needs. Once the assessment questions have been developed, indicators can then be selected that answer or address the assessment questions. For example, the assessment question, "Is the composition of vegetation at a particular site sustainable under current use?" could be addressed by selecting indicators such as plant species composition, ratios of native to exotic species, and amount of bare soil. These indicators would be most useful if they were made relative to reference ungrazed sites, i.e., control sites that were otherwise similar. Other biophysical indicators might include measures of soil fertility, such as cation exchange capacity or amount of soil carbon, if loss of primary productivity was of concern.

When defining ecosystem change in sites that are exploited for economic uses, it is critical that scientists and managers understand what values of the stakeholders are important and relate these to ecological processes through a set of appropriate indicators. These indicators should be selected in the context of the nature of the economic exploitation to evaluate stress to the ecosystem, clearly identify current status, and be useful in trend assessments. Because ecosystems constantly change, important questions are "What is the rate of change? Are economic activities causing the system to change at a greater rate or greater amplitude than under natural conditions? Does the ecosystem have the ability to adjust, that is, has the capacity for resilience been exceeded?"

## 6.1 What Makes a Good Indicator?

Indicators are used to assess the condition of ecosystems because it is not feasible or cost-effective to measure every primary ecological variable. The process of selecting indicators useful for measuring change to major structural and functional aspects of the ecosystem provides a cost-effective and logical approach. The general set of characteristics for selecting indicators discussed here is a culmination of a variety of information in the literature (Hunsaker et al. 1990, Breckenridge et al. 1995). These characteristics have generic utility for most ecosystems, but will, of course, need to be refined to meet the site-specific needs of particular ecosystems.

For most ecosystems, sustainability and overall condition can be analyzed by subsuming assessment questions under the following broad categorical questions.

1. What is the biological integrity of the system?
2. How well does the system capture, store, and use water and energy?
3. How efficient is the system at cycling nutrients?

An ecosystem that is in good condition will have a high degree of biological integrity, will capture water and energy and use it through the growing season, and will cycle nutrients in an efficient manner. We should also add the caveat that these properties should be evaluated in the context of temporal and spatial environmental variation that can be considerable. For example, efficient uptake of water may be a good property in an ecosystem that is generally arid but not necessarily a good property in mesic ecosystems or during unusually wet years.

## 6.2 Minimal Set of Indicator Assessments

The following minimal set of assessment topics is suggested for use in evaluating the indicators for the condition of ecological systems.

- *Characteristic 1*: Ecosystem conceptual approach — the indicator should fit into an ecosystem approach that focuses on interactions of the system and not on a single isolated feature of the environment. To satisfy this characteristic, indicator parameters must relate in a known way to a structure or function of the ecological system to be evaluated so that the information obtained provides a "piece" of the overall puzzle.
- *Characteristic 2*: Usability — the relative completeness and thoroughness of the procedure for measuring the indicator parameter provides the best gauge of indicator usability.
- *Characteristic 3*: Cost-effectiveness — this characteristic can be evaluated by asking the question, "Is the incremental cost associated with the measurement low relative to the information obtained?"
- *Characteristic 4*: Cause and effect — this characteristic focuses on whether there is a clear understanding of the mechanism of the relationship between changes in the degree of stress on the ecosystem and change in the value of the indicator. This characteristic can be evaluated by considering the following questions: (1) Does the indicator respond in a known, quantifiable and unambiguous manner to the stress in question?, (2) Is there dose–response information available for the indicator and the stress?, (3) Are exposure thresholds or trends known for the indicator?, and (4) Will the indicator provide similar information for most potential sampling areas within the same ecological unit?
- *Characteristic 5*: Signal to noise ratio — This refers more specifically to the relative ease with which changes in the indicator caused by the pollutant may be distinguished from changes due to natural variability. In order to apply this criterion, the following questions should be considered: (1) Is the natural spatial and temporal variability associated with the parameter to be measured understood? (2) Are there predictable patterns in the spatial aspect (e.g., slope, soil associations, moisture) or temporal variability (e.g., seasonal) of the indicators identified?, and (3) Does the indicator possess significantly high signal strength in comparison to natural variability to allow detection of statistically significant changes within a reasonable time frame?
- *Characteristic 6*: Alternate approaches — This characteristic would be satisfied if it can be concluded that no other approaches for measuring stress are available that would increase the quantity or quality of information obtained.
- *Characteristic 7*: Quality assurance — This characteristic is satisfied if the quality of the resulting data can be reasonably assessed from a statistical and procedural standpoint. The process must be repeatable between different personnel and only requires a short training time.
- *Characteristic 8*: Anticipatory — Ideally, an indicator for evaluating ecosystem conditions should be selected to provide an early warning signal of widespread changes in ecological condition or processes.
- *Characteristic 9*: Historical records — Ideally, some historical data can be obtained for the parameter of interest from archived databases. Such data can be extremely valuable for establishing natural baseline conditions and the degree of natural variability associated with the parameter for the ecosystem. This information provides an important linkage between work that has been previously conducted and allows for trend assessment.
- *Characteristic 10*: Retrospective — Some parameters allow for retrospective analysis in that new data may be generated that provide information on past conditions. For example, tree rings provide growth indices for each year of life of the tree. Because ecosystems are constantly under stress, it is important to try to evaluate how the stress has changed over time.
- *Characteristic 11*: New information — Can the indicator being selected provide new information rather than simply replicating data that already exist? This is critical for identifying how new data can help meet the goal of better managing ecological resources.
- *Characteristic 12*: Minimal environmental impact — Scientists and managers need to address the question of how much impact will the evaluation of an indicator impart on an environment. In fragile areas, such as desert systems, collecting an extensive amount of data or field sampling can often destroy the resource being protected. In a like manner, cutting down, harvesting, or removing parts of an ecosystem have to be considered in the overall evaluation of indicators for measuring change. Guidance in evaluating this characteristic should consider the question, "Does destructive sampling provide additional useful information that cannot be obtained without removal of the organisms being sampled?"

Indicators can be grouped into three general areas related to physical, biological, and social impacts. Table 1 presents a list of possible indicators for specific issues. Indicators can be combined into indices that relate to overall impacts. These indices are often more useful in presenting information to the public. An index combines data from several measurements into information that is often used in trend analysis. An index can be derived for analysis of a change to a specific indicator like vegetation (Mouat et al. 1993), or can be used to assess change to a habitat type like condition of a riparian zone (Nelson and Andersen 1994). If presented correctly and related in clear terms with examples, indices can provide an effective way to communicate concerns on complex issues. For example, a soil productivity index that combines information from indicators such as erosion, organic matter content, soil crust status, infiltration rate, and soil profile can be more easily understood by the general public.

## 6.3 Where To Find Indicators

There are several approaches to identifying possible indicators for a given ecosystem. These include: (1) searching the literature for indicators used in similar situations and/or finding previous research that highlights possible indicators; (2) conducting research comparing examples of impacted and unimpacted areas for a given ecosystem, and then using identified differences as possible indicators.

Choosing among these options depends on time, money and baseline information available. Finding potential indicators using only a literature search is the cheapest and quickest method, although at least one field season is required for testing identified indicators. Because this approach is based on research from other areas, indicators that are better-suited, or unique to the ecosystem in question, may be missed. It may also result in the selection of indicators that are not appropriate for the ecosystem in question. Employing original research to identify potential indicators requires several field seasons, requires expertise, and is more expensive. Doing both a literature search and original research is the preferred option, given adequate time and funding.

A bias that should be acknowledged in most approaches to indicator selection is that, because of time and money constraints, we generally choose variables for which we have background information, or variables that are readily observable. This may result in excellent indicators being ignored. Also, if particular indicators become widely adopted, it is important that the protocol for measuring and interpreting the indicators be standardized so that cross-comparisons

Table 1. List of Possible Indicators of Ecological and Social Impacts

| | |
|---|---|
| *Physical Impacts* | |
| Soil bulk density | Soil drainage |
| Soil compaction | Soil chemistry |
| Soil pH | Soil productivity |
| Amount of litter and duff | Depth of litter and duff |
| Area of barren core | Area of bare ground |
| Total area of complete campsites | Number of fire rings |
| Size of fire rings | Number of social trails |
| Visible erosion | |
| *Biological Impacts* | |
| Soil fauna and microflora | Ground cover density |
| Percent loss of ground cover | Plant species composition |
| Plant species diversity | Proportion of exotic plant species |
| Plant height | Selected plant species vigor |
| Extent of diseased vegetation | Extent of scarred or mutilated trees |
| Number of tree seedlings | Exposed tree roots |
| Abundance of selected wildlife species | Presence/absence of selected wildlife species |
| Frequency of wildlife sightings | |
| Wildlife reproduction success | Wildlife species diversity |
| *Social Impacts* | |
| Number of encounters with other individuals per day | Number of encounters by activity type |
| Number of encounters by mode of transportation | Number of encounters by size of group |
| Number of encounters with other groups per day | Visitor perception of crowding |
| Number of encounters by location of encounter | Number of visitor complaints |
| Visitor perception of impact on environment | Amount of litter (trash) in area |
| Visitor satisfaction | |
| Visitor reports of undesirable behavior | |

can be made. As an example of the need to standardize protocol, consider the problem of monitoring changes in soil organic matter, which is frequently an important issue in management for erosion control, for land restoration, and for reforestation projects. An easily measured, and therefore commonly used, indicator of soil organic matter dynamics is the "light fraction" or LF (Gregorich and Janzen 1996). This is the lightweight free organic material that is composed largely of partly decomposed plant material and fungal hyphae. LF is a useful fraction to measure because it represents an important stage in the formation of soil organic matter. The LF is easily extracted by flotation of the material in a dense fluid. However, because the amount of LF extracted is influenced by the composition of the flotation fluid and its specific gravity, both need to be specified and an agreed-upon protocol should be developed among the various natural heritage public and private agencies.

### 6.3.1 Literature Search

Literature searches can be conducted by computer, through library work and through phone contacts. In addition, much of the work done in the area of indicator selection has been published in government documents or privately, and are not in most libraries. As a result, calling organizations involved in this type of research, such as the Aldo Leopold Institute (USFS) or the National Parks and Conservation Association is an important part of the literature search as well. Also, contacting individual researchers who have been involved with either indicator selection or the ecosystems in question can also be very valuable, because they are generally well-acquainted with the literature and work done on the topic. A good way to cast the search widely for possible indicators and sources is to post inquiries on appropriate Internet sites and list-servers. The initial responses may be a bit overwhelming and it will take some effort to sort out the ones that have potential utility and even those indicators will require some additional investigative effort to establish their credibility.

### 6.3.2 Original Research

To illustrate a way that original research can be structured to produce a list of potential indicators, an example will be taken from Arches National Park, where indicators of resource condition were recently developed. An underlying assumption of the approach used was that by comparing characteristics of areas of similar vegetation and soils, but different use levels, variables found to be different between the two areas would likely be those sensitive to that type and level of use. Consequently, comparisons of such areas could be used to generate potential indicators for that vegetation and/or soil type. Further refinement of this list would then provide a final suite of indicators.

The process used to choose indicators for Arches National Park consisted of the following steps:

***Step 1****: Major vegetation types were determined, and representative, comparable areas in these vegetation types receiving heavy and light use selected for study. Physical, chemical and biological variables in the two areas were measured.* Major vegetation types in Arches were determined. In each vegetation type, representative adjacent heavily and lightly used areas were selected, and plots established in each area. Within time, money and expertise constraints, vegetation and soil variables were measured. During the project, variables were dropped if it became clear they would not discriminate among sites with different levels of impact. As a result, not all variables were tested at all sites.

***Step 2****: Variables that are different between the compared areas become potential indicators. using criteria established for suitable indicators, a matrix is developed and used to evaluate the potential indicators.* Variables that were significantly different were designated as potential indicators, and evaluated for their suitability, using the criterion previously listed. A matrix was constructed to facilitate this evaluation, with potential indicators and suitability criterion as the axes.

***Step 3****: Potential indicators had to be ecologically relevant in the vegetation type being evaluated.* Potential indicators were then further evaluated for ecological relevancy. An ecologically relevant indicator was defined as an indicator indicative of significant, irreversible or undesirable damage to ecosystem processes or to rare resources. In order to ascertain whether an indicator met this requirement, background research was needed to understand how changes in the identified variable affected the ecosystem being considered, and whether or not the effects were significant enough to be of concern. Based on this background research, further research needs were identified and then additional research was conducted to refine the indicators.

***Step 4****: Indicators were chosen based on suitability ranking and ecological relevancy.* Desirability rankings (Step 2), combined with background information on ecological relevancy (Step 3), led to the chosen indicators. Because of the cost and expertise needed to evaluate some of the most ecologically relevant indicators, a two-tiered system was adopted. Tier 1 included those indicators that could reasonably be expected to be measured annually. Tier 2 were those indicators with high ecological relevancy, but where high costs or limited expertise restricted their use. In addition, tier 2 indicators are expected to act as a check on the more simplistic tier 1 indicators, until it is established that tier 1 is sufficiently sensitive to the resource conditions being monitored.

***Step 5**: Chosen indicators were field-tested, and adjustments made in both indicators and measurement techniques.* Several adjustments were made to Arches indicators after the first field season. Field testing indicated that some measurements were too time-consuming and results too variable. In addition to altering original indicators, a new indicator was added.

An important aspect of this process was maintaining flexibility at all stages. This was especially important at step 1, data collection, and step 5, field testing chosen indicators. During the data collection stage, measured variables were freely added and dropped, based on experience gained in the field. During field testing of chosen indicators, measurement methods were changed several times, until they were adequately modified to meet monitoring needs.

This iterative and flexible approach to indicator selection proved to be both time- and cost-effective. It also provided an opportunity to evaluate the specific habitat in question, rather than having to rely on generic indicators from the literature that were usually developed for very different habitat types and use levels. One of the major problem faced during this process was ways to incorporate variables that were both of high ecological significance and cost. The tiered approach appears to be a good solution to this, but has yet to be tested over time.

## 7 CONCLUSIONS

The biophysical properties of ecosystems have characteristic ranges of variability and resiliency to disturbances. Managers must work within these ranges. The normal ranges of variation should not be exceeded nor should the ecosystem be unduly constrained and variance too limited. Unduly constraining variation will lead to catastrophic changes, e.g., suppressing natural fires leads to destructive wildfires, controlling normal flood stages leads to catastrophic floods in unusually wet years. Ideally, a managed ecosystem should enclose sufficient spatial heterogeneity, in the form of alternative habitats, seasonal food sources, successional sequences, and various refugia such that critical resources remain available somewhere during extreme years and that the normal cycles and pulses of ecosystem dynamics are maintained. Any particular ecosystem is, to varying degrees, linked to other ecosystems in a landscape context. Finding the right mix of spatial and temporal scales for enclosing ecosystems and to establish management boundaries is essential.

Surprises, unexpected results and threshold responses will occur. Monitoring the environmental responses to management decisions and maintaining the flexibility to modify decisions through adaptive management approaches is essential.

Ecosystems can recover from many kinds of perturbations but resiliency is limited and cumulative effects, persistent stresses, and catastrophic events must be avoided. Although general statements can be made about the resiliency of ecosystems, resiliency for any particular ecosystem needs to be empirically derived. For this reason, resiliency should be investigated through field experimentation at the appropriate scale for the dominant consumptive practices within the ecosystem.

Confusion over the meaning of ecosystem management often arises from a failure to distinguish between "ecosystem" used to define a research topic and "ecosystem" used to define a typically large-scale and holistic management unit in which people are generally integral parts of the ecosystem as well as stakeholders affected by management decisions.

The management of large-scale ecosystems will require merging different approaches. First, a combination of information and analytical tools (historical studies, simulation models, geographic information systems,) should be used to develop predictive and alternative management approaches. Second, the decision-making process should be opened up by involving representative stakeholder groups to evaluate management alternatives.

The goal of ecosystem management is to find approaches that meet human needs without degrading the environment; that is, sustainability is a primary goal of ecosystem management. Sustainability is not restricted to biophysical considerations but refers more generally to the interplay of human values, goals, and behavior with biophysical ecological properties and behavior within an ecosystem.

An ecosystem manager must have some metric (indicators) by which to measure the success or failure of decisions that affect the condition of the ecosystem. These indicators must be reliable surrogates for both ecological and social/economic attributes of an ecosystem and they must be sensitive to changes in the condition of an ecosystem that are degenerative or regenerative.

## REFERENCES


Breckenridge, R.P., W.G. Kepner, and D.A. Mouat. 1995. A Process for Selecting Indicators for Monitoring Conditions of Rangeland Health. *Environmental Monitoring and Assessment* 36: 45–60.

Carpenter, S.R. and J.F. Kitchell (eds.) 1993. *The Trophic Cascade in Lakes.* Cambridge University Press, Cambridge, UK.

Carpenter, S.R., T.M. Frost, J.F. Kitchell, T.K. Kratz. 1993. Species Dynamics and Global Environmental Change: A perspective from ecosystem experiments. In: P.M. Kareiva, J.G. Kingsolver, R.B. Huey (eds.), *Biotic Interactions and Global Change*. Sinauer Associates Inc., Sunderland, Massachusetts, pp. 267–279.

Carroll, C.R. 1996. Coarse woody debris in forest ecosystems: an overview of biodiversity issues and consepts. In: Biodiversity and coarse woody debris in southern forests, proceedings of the workshop on coarse woody debris in southern forests: effects on biodiversity; 1993 October 18–20; Athens, GA. Gen. Tech. Rep. SE-94. Asheville, NC: U.S. Department of Agriculture, Forest Service, Southern Research Station. pp. 25–28.

Danielson, B.J. 1992. Habitat selection, inter-specific interactions and landscape composition. *Evol. Ecology* 6: 399–411.

Duggins, D.O., C.A. Simenstad, and J.A. Estes. 1989. Magnification of secondary production by kelp detritus in coastal marine ecosystems. *Science* 245: 170–173.

Franklin, J.F. 1995. Sustainability of managed temperate forest ecosystems. In: M. Munasinghe and W. Shearer (eds.), *Defining and Measuring Sustainability*. The United Nations University and The World Bank, Washington DC. pp. 355–388.

Gregorich, E.G. and H.H. Janzen. 1996. Storage of soil carbon in the light fraction and macroorganic matter. In: M.R. Carter and B.A. Stewart (eds.), *Structure and Organic Matter Storage in Agricultural Soils*. Lewis Publishers, Advances in Soil Science, pp. 167–190.

Hann, W., M.E. Jensen, P.S. Bourgeron, and M. Prather. 1994. Land management assessment using hierarchical principles of landscape ecology. In: M.E. Jensen and P.S. Bourgeron (eds.). Ecosystem Management: Principles and Application, Vol. II. USDA Forest Service, Pacific Northwest Research Station, General Technical Report PNW-GTR-318, February, 1994, pp. 285–297.

Harwell, M.A., J.F. Long, A.M. Bartuska, J.H. Gentile, C.C. Harwell, V. Myers, and J.C. Ogden. Ecosystem management to achieve ecological sustainability: The case of South Florida. *Environmental Management* 20: 498–521.

Holling, C.S. 1995. Sustainability: the cross-scale dimension. In: M. Munasinghe and W. Shearer (eds.), *Defining and Measuring Sustainability*. The United Nations University and The World Bank, Washington DC. pp. 65–76.

Holling, C.S. and G.K. Meffe 1996. Command and control and the pathology of natural resource management. *Conservation Biology* 10: 328–337.

Hunsaker, C.T., D.B. Carpenter, J.J. Messer. 1990. Ecological indicators for regional monitoring. *Ecological Society of America Bulletin* 71: 165–172.

Meffe, G.K. and C.R. Carroll. 1997. *Principles of Conservation Biology*. Sinauer Associates, Inc. 729 pp.

Mouat, D.A., G.G. Mahin, J. Lancaster. 1993. Remote sensing techniques in the analysis of change detection. *Geocarto International* (2) pp. 39–50.

Nelson, S.M. and D.C. Anderson. 1994. An Assessment of Riparian Environmental Quality by Using Butterflies and Disturbance Susceptibility Scores. *The Southwestern Naturalist* 39: 137–142.

Noss, R.F. 1990. Indicators for monitoring biodiversity: a hierarchical approach. *Conservation Biology* 4: 355–364.

Odum, W.E., E.P. Odum, and H.T. Odum. 1995. Nature's pulsing paradigm. *Estuaries* 18: 547–555.

Paine, R.T. 1980. Food webs: Linkage, interaction strength and community infrastructure. *Journal of Animal Ecology* 49: 667–685.

West, N.E. 1993. Biodiversity of rangelands. *Journal of Range Management* 46: 2–13.

Wood, M.K. 1988. Rangeland Vegetation–Hydrological Interactions. In: Tueller, P.T. (ed). *Handbook of Vegetation Science*, Vol. 14, Vegetation Science Applications for Rangeland Analysis and Management. Kluwer, Dordrecht, pp. 469–491.

## THE AUTHORS

**C. Ronald Carroll**
*Institute of Ecology*
*University of Georgia*
*Athens, GA 30602, USA*

**Jayne Belnap**
*Arches National Park*
*Moab, UT 84532, USA*

**Bob Breckenridge**
*INEL*
*Idaho Falls, ID 83401, USA*

**Gary Meffe**
*Department of Wildlife Ecology & Conservation*
*University of Florida*
*Gainesville, FL 32611-0430, USA*

# Ecosystem Restoration and Management: Scientific Principles and Concepts

Wallace Covington, William A. Niering, Ed Starkey, and Joan Walker

**Key questions addressed in this chapter**

- ♦ *Ecosystem restoration in ecosystem management*
- ♦ *Ecosystem restoration defined*
- ♦ *Naturalness and the evolutionary environment*
- ♦ *Examples of ecosystem restoration approaches*

**Keywords: Native species, fire, evolutionary environments, ponderosa pine forest, western hemlock forest, saltmarsh**

*"Once we restore, we are no longer retreating, trying only to slow the wave of destruction. We begin to actually advance, to regain lost ground. Can we really do it, or is the idea only hubris, human arrogance rearing its head one more time? ...The short answer is: yes, we can really do it — to some degree. At worst we can produce something that mimics the real thing and that, given enough time, could become the real thing...."* John P. Wiley, Jr., 1989

## 1 INTRODUCTION

This paper summarizes current thinking regarding ecological restoration from an ecosystem management point of view. The intended audience is natural resource professionals, natural resource interest groups, and interested members of the public. We discuss ecological restoration concepts in the context of three ecological restoration efforts with which we have been involved and which are particularly important to contemporary public land management: ponderosa pine ecosystems, forest ecosystems of the Western Hemlock Zone of the Pacific Northwest, and tidal wetlands of the Northeast. In discussing these examples we emphasize scientific principles and concepts fundamental to ecological restoration. We close our paper with a discussion of ecological restoration and human habitat needs.

Others have presented cogent syntheses of ecological restoration (MacMahon and Jordan 1994), restoration ecology (Jordan et al. 1987), and small group and community-based ecological restoration (Nilsen 1991). Although we draw upon these resources for some of our discussion, our goal is different — to discuss ecological restoration in the light of contemporary ecosystem management concepts.

Concern about the degradation of public lands and associated natural resources has been a driving force in federal land management policy since its inception (Dana and Fairfax 1980). A building consensus suggests that unless something is done to reverse the deterioration of ecosystem health, current and future generations will continue to incur increasing costs while simultaneously enjoying fewer benefits from public lands. Of particular concern is the cumulative effect of ecosystem simplification such that ecosystems are at risk of catastrophic losses of biological diversity and human habitats (Myers 1984). A cornerstone of the federal government's ecosystem management approach to solving these problems is ecosystem restoration. The Report of the Interagency Ecosystem Management Task Force (Anon 1995) stated:

> "The goal of the ecosystem approach is to restore and sustain the health, productivity, and biological diversity of ecosystems and the overall quality of life through a natural resource management approach that is fully integrated with social and economic goals."

In a similar vein, the Ecological Society of America's report, *"The Scientific Basis for Ecosystem Management,"* discussed the importance of ecological restoration in the practice of ecosystem management (Christensen et al. 1995). In a previous report by the Ecological Society of America, Lubchenco et al. (1991) selected ecosystem restoration as one of twelve featured topics for priority research in an ecological research agenda in support of sustaining the biosphere. Restoration and maintenance of ecosystem health is seen as central to ecosystem management by such diverse groups as the Society of American Foresters (1993), the Southwest Forest Alliance (1996), the Sierra Club (e.g., see Berger 1997), and the American Forest and Paper Association Forest Resource Board (1993). Restoration of ecosystem health is, in fact, an international theme. The United Nations (1992) recognized ecosystem restoration as a central concern in the Rio Declaration on Environment and Development in Principle 7 which declares, "States shall cooperate in a spirit of global partnership to conserve, protect and restore the health and integrity of the Earth's ecosystems." But what is ecosystem restoration?

## 2 ECOSYSTEM RESTORATION DEFINED

Modern principles and concepts of ecological restoration began with the thinking of Aldo Leopold (Flader and Callicott 1991). Soon after the beginning of Leopold's professional career as a forester in the Southwestern United States, he recognized the rapid deterioration of forest and range lands because of overgrazing, intensive logging, and predator extirpation (Flader 1974). This spurred Leopold to call for viewing natural resource management as the practice of land health (now termed ecosystem health [Rapport 1995]). Leopold recognized that the practice of ecosystem health required reference points — healthy, intact ecosystems still functioning as they had before disruption by intensive industrialization — and that those reference points were highly limited in the United States. Referring to the need for reference points for the practice of land health, Aldo Leopold said, "The first step is to reconstruct a sample of what we had to begin with."

Upon his re-entry into the academic community as a professor of game management, he joined forces with others at the University of Wisconsin to rebuild approximations of the naturally functioning ecosystems which existed in Wisconsin at the time of Euro-

American settlement (Jordan et al. 1987). Using Civilian Conservation Corps, student, faculty, and community volunteers, Leopold and others began the task of ecological restoration of representative Wisconsin ecosystems at the University of Wisconsin Arboretum in 1935. That same year, Leopold began the restoration of his beloved sand county farmland just an hour's drive north of the campus. Although rehabilitation of degraded land was a widespread goal in the 1930s, the ecological restoration work of Leopold and others was different. The ecological restoration projects in Wisconsin had as their goal the restoration of native ecosystems in contrast to others where the goal was the simple revegetation of derelict lands, stopping accelerated erosion, or improving the productive potential of land.

Over the next 50 years, ecologists working on the arboretum restoration projects learned much about the structure and function of ecological systems through trial and error. In 1981 the University of Wisconsin arboretum began publishing Restoration and Management Notes as a forum for the interchange of ideas and experiences among practicing restorationists. By the 1980s, the synergy between ecological restoration (the practice) and restoration ecology (the biological science) became so apparent that it lead to the recognition of restoration ecology as a focus for developing and testing ecological theory (Jordan et al. 1987). In 1987 a professional society, the Society for Ecological Restoration, was formed; it held its first annual meeting in 1988. In 1993 the Society began publishing its scientific journal, *Restoration Ecology*.

## 2.1 Definitions

One of the first tasks of the Society for Ecological Restoration was defining ecological restoration and its principles and concepts. This task is ongoing and much discussion and debate continue among scientists and practitioners of ecological restoration (Jackson et al. 1995, Aronson and Le Floc'h 1996, Covington and Sampson 1996, Higgs 1997, Dobson et al. 1997). Nonetheless, progress has been made such that basic definitions are generally agreed upon.

The dictionary definition of "restoration" is the act of bringing back to an original or unimpaired condition. Thus, ecological restoration has as its goal the restoration of degraded ecosystems to emulate more closely, although not necessarily duplicate, conditions which prevailed before disruption of natural structures and processes, i.e., environmental conditions which have influenced native communities over recent evolutionary time (see Box 1).

**Box 1**
**Society for Ecological Restoration Mission Statement and Summary of Environmental Policies (Society for Ecological Restoration 1993)**

The mission of the Society is to promote ecological restoration as a means of sustaining the diversity of life on Earth and reestablishing an ecologically healthy relationship between nature and culture. Ecological restoration is the process of reestablishing to the extent possible the structure, function, and integrity of indigenous ecosystems and the sustaining habitats that they provide. To advance its mission, (1) SER serves as a forum for discussion and exchange of ideas; (2) SER raises awareness and promotes the expanded use of ecological restoration; (3) SER works to advance the science and art of ecological restoration. SER welcomes the participation of anyone interested in ecological restoration.

Ecological restoration involves management actions designed to accelerate recovery of degraded ecosystems by complementing or reinforcing natural processes. Ecological restoration has been viewed as ecosystem medicine where the practitioner is helping nature heal (Nilsen 1991), that is, building upon the natural recovery processes inherent in the ecosystem. Restoration ecology, the biological discipline which undergirds ecological restoration, deals with research and management experimentation to determine the mechanisms that control recovery of degraded ecosystems, and with discovering ways for safely restoring degraded ecological systems to more nearly natural conditions.

Ecosystem restoration is founded upon fundamental ecological and conservation principles and involves management actions designed to facilitate the recovery or re-establishment of native ecosystems. A central premise of ecological restoration is that restoration of natural systems to conditions consistent with their recent evolutionary environments will prevent their further degradation while simultaneously conserving their native plants and animals (Society for Ecological Restoration 1993). Practitioners of ecological restoration recognize that a failure to include human interactions with restored systems is not only unrealistic, but also undesirable for their long-term sustainability. In fact, in cases where novel conditions prevent natural system functions, ongoing management may be required to compensate for the unnatural conditions. Examples of such a circumstance are those in which restored sites are too small to support natural predator–prey dynamics or to accommodate natural disturbance regimes.

Ecological restoration is related to other practices of ecological healing such as rehabilitation, reclamation, and bioremediation, although the goals of ecosystem restoration (restoration of natural conditions) are generally more ambitious (MacMahon and Jordan 1994). Although restoration goals are often more ambitious that those of rehabilitation, this does not necessarily imply that restoration is more expensive (see discussion of restoration of Ponderosa Pine Ecosystems, below). The term "reclamation" first came into common usage after the U.S. Surface Mine Control and Reclamation Act of 1977 (Jackson et al. 1995). Reclamation refers to attempts to re-establish elements of the structure and function of ecosystems, but not complete restoration to any specified prior condition. Rehabilitation has as its goal making the land useful again, but, as in reclamation, the goal is not restoration to predisruption ecological conditions (National Academy of Science 1974). Rehabilitation might involve, for example, establishment of agricultural land on a site previously occupied by grassland. Reclamation, rehabilitation, and restoration have as their goals a continuum of outcomes from the least to the most similar to the predisturbance ecosystem (Jackson et al. 1995). Thus, all share to a greater or lesser extent some of the same techniques and can be viewed as closely allied.

In many respects ecological restoration might best be judged by whether the techniques used are setting the ecosystem on a trajectory that will eventually lead to the recovery of original ecosystem structure and function (Bradshaw 1984, MacMahon and Jordan 1994). The underlying assumption of such a view is that facilitating partial recovery of ecosystem structure and function can lead to re-establishment of natural self-regulatory mechanisms which in turn will eventually lead to restoration of the original ecosystem dynamics.

Ecological restoration, therefore, consists of a broad variety of practices designed to restore natural ecosystem structure and function. It is related to reclamation, rehabilitation, and other land recovery practices but has as its goal re-establishment of the original ecosystem structure and function. For example, in the case of southwestern ponderosa forests, ecosystem restoration might consist of removing most of the trees that postdate Euro-American settlement, raking heavy fuels from the base of the old-growth trees, prescribed burning, removing introduced noxious plants, and sowing with native herbaceous seeds. In the case of conifer forests of the Pacific Northwest and tidal wetlands of the northeastern United States, different restoration activities are required (see below).

However, ecosystem restoration should not be construed as a fixed set of procedures, nor as a simple recipe for land management. Rather, it is a broad intellectual and scientific framework for developing mutually beneficial human:wildland interactions compatible with the evolutionary history of native ecological systems. In other words, ecosystem restoration consists not only of restoring ecosystems, but also of developing human uses of wildlands which are in harmony with the natural history of these complex ecological systems.

## 3 WHAT IS "NATURAL"?

"Natural" is one of the most controversial concepts in ecosystem management (Christensen et al. 1995). Although it can be an ambiguous term, it is one that is ecologically, aesthetically, spiritually, and politically important as evidenced by its use in such expressions as natural area, natural range of variability, natural history, and natural processes. When used to connote the evolutionary environment, it is fundamental to conservation biology and restoration ecology. In the context of restoration ecology "natural" implies native species, structures, and processes, in contrast to exotic species, structures, and processes. Indigenous ecological components and processes are natural. Alien ecological components and processes are not. At the heart of these distinctions is the evolutionary ecology principle that species which have interacted over evolutionary time will have developed coevolved regulatory mechanisms and interdependencies that lead them to function as relatively self-regulating ecological systems.

### 3.1 Naturalness

Naturalness is difficult to quantify. Karr (1981) proposed an index of biological integrity for assessing the naturalness of aquatic ecosystems. His proposed index ranged from 12 in areas without fish to 60 in areas with fish composition equivalent to those in undisturbed areas. Karr's index integrates attributes such as fish species richness, indicator taxa (both tolerant and intolerant of pollution), species and trophic guild relative abundances, and the incidence of hybridization, disease and anomalies such as lesions, tumors or fin erosions. Similar approaches could be used in other ecosystems. For example, in forest ecosystems the focus might be on developing an index which integrates plant species richness, indicator animal taxa, and the incidence of hybridization, disease and anomalies in key plant and animal species.

Anderson (1991) wrestled with the problem and suggested that naturalness could be assessed by the proportion of native to non-native species, the amount of human energy needed to maintain current ecosystem conditions, and the relative change in the ecosystem if human inputs to the system cease. The concept of naturalness is sometimes best understood in the context of defining what is unnatural (Hammond and Holland 1995). In this view, a natural ecosystem would be constituted of indigenous (native) species interacting in a self-sustaining manner, i.e., species persistence by natural recruitment as opposed to managed reproduction, population dynamics regulated internally, disturbance regimes functioning within their pre-disruption range of variability, and trophic dynamics that are sustainable over time. An unnatural ecosystem would have a high proportion of non-native species, wide swings in population dynamics requiring management actions to prevent ecosystem simplification, and exotic disturbance regimes far outside those present before ecosystem degradation.

From a restoration ecology point of view the most important definition of "natural" is related to the concept of the evolutionary environment, a key element of defining the reference conditions for ecological restoration projects.

### 3.2 Evolutionary Environment

The concept of the evolutionary environment is central to conservation biology and restoration ecology. The term evolutionary environment refers to the environment in which a species or groups of species evolved — the environment of speciation (sometimes referred to as the habitat of speciation) (Mayr 1942, Smith 1958, Geist 1978).

Over evolutionary time species not only adapt to their evolutionary environment, but they may also come to depend upon those conditions for their continued survival (Mooney 1981, Wilson 1992). Thus, the greatest threat to biological diversity is the loss of evolutionary habitats (Noss 1991), and the greatest hope for reversing the losses is restoration of these habitats (MacMahon and Jordan 1994). But on what time-scale is the evolutionary environment measured?

This question has no simple answer. Evolution is an ongoing process and rates of evolution are a function of generation time, population structure, genetic variability, selection pressure, and other factors. Today's species are the product of millions of years of evolution. However, in the context of contemporary communities the relevant evolutionary environment is generally considered to be that of the past several thousand years (for most forest ecosystems this would approximate 50–100 times the average generation time of the longest lived ecological dominant trees). Based on evolutionary principles, MacArthur (1972) concluded that "...the length of time it normally takes for a species to split and diverge sufficiently to be regarded as two species is a small, uncertain number of thousands of years." A fundamental assumption is that an environmental factor can be considered as part of a species' evolutionary environment when that factor has been of sufficient intensity and duration for the factor to exert selection pressure such that the species has become adapted to it.

For North America, the recent evolutionary environment is typically taken to include Native Americans as participants in evolutionary processes over the past ten thousand years (Parsons et al. 1986, Kay 1995, Bonnicksen et al. this volume). However, the environmental pressures associated with Euro-American settlement, especially the introduction of exotic plants, animals, and land use practices, as well as the disruption of natural disturbance regimes (see White et al., this volume), are unprecedented in the recent evolutionary environment and thus viewed as disrupting evolutionary trajectories (Covington et al. 1994) and leading to pervasive degradation of ecological systems.

Ecological restoration is now seen as an approach for reversing ecosystem degradation and setting ecosystems on a trajectory more consistent with their evolutionary environment. With this approach in mind, we now present an overview of examples of ecological restoration in three major types with which we have detailed experience: ponderosa pine forests of the Southwest, conifer forests of the Western Hemlock Zone of the Pacific Northwest, and tidal wetlands of the Northeast. We use ecological restoration work in southwestern ponderosa pine to illustrate the use of detailed historical and field research based knowledge to design small-scale (1–10,000 acre) ecological restoration experiments. Conifer forests of the Pacific Northwest are used to illustrate the use of ecological restoration research in the design of a regional ecosystem management approach. Our final example, restoration of tidal wetlands in the Northeast, is used to extend our discussion of ecological restoration principles beyond forest ecosystems to aquatic and wetland ecosystems.

## 4 EXAMPLES OF ECOLOGICAL RESTORATION PROBLEMS

### 4.1 Ponderosa Pine Ecosystems of the Southwest

*The Problem*

The evolutionary environment of southwestern ponderosa pine ecosystems is dominated by natural disturbance regimes (e.g., fires, predation, defoliation), which have varied in kind, frequency, intensity, and extent (Covington et al. 1994). These disturbance regimes served as natural ecological checks and balances on populations and insured spatial and temporal habitat diversity (Cooper 1960, Covington and Moore 1994b). Natural fire regimes were particularly important in shaping the communities present at the time of Euro-American settlement.

Previous research has established that southwestern ponderosa pine forests were much more open before Euro-American settlement (ca. 1870) than they are today (Pearson 1950, Cooper 1960, Madany and West 1983, Covington and Sackett 1986, Covington and Moore 1994b). Before settlement, the combination of frequent (every 2–5 years), light surface fires, grass competition, and a climate unfavorable for pine establishment had maintained an open and park-like landscape, dominated by grasses, forbs, and shrubs with scattered groups of ponderosa pine trees. After Euro-American settlement, heavy livestock grazing, fire suppression, logging disturbances, and favorable climatic events favored the invasion of the open park-like vegetation by dense ponderosa pine regeneration.

Various authors (e.g., Cooper 1960, Weaver 1974, Kilgore 1981, Covington and Moore 1994a, 1994b, Kolb et al. 1994) have described symptoms of ecosystem degradation of ponderosa pine ecosystems including increases in tree density, forest floor depth, and fuel loading and consequent problems such as: (1) decreases in soil moisture and nutrient availability; (2) decreases in growth and diversity of both herbaceous and woody plants; (3) increases in mortality in the oldest age class of trees; (4) decreases in stream and spring flows; (5) accumulation of fuels; and (6) increases in fire severity and size. These symptoms are consistent with the general ecosystem health distress syndrome for terrestrial ecosystems as discussed by Rapport (1995) and Rapport and Yazvenko (1996), i.e., reductions in species diversity, leaching of nutrients, reduction in primary productivity, increased amplitude of oscillations of component species, increase in diseases, and reduction in size of dominant organisms.

### *Reference Conditions*

Reference conditions in southwestern ponderosa pine ecosystems come from three lines of evidence: historical records, retrospective ecological analyses, and analogous sites in the Sierra Madre Occidental which continue to burn under a frequent fire regime.

### *Historical Records*

Reports from early travelers illustrate the changes in appearance of ponderosa pine forests since settlement. E.F. Beale who travelled through northern Arizona 1858 is quoted by C.F. Cooper (1960) as follows:

> "We came to a glorious forest of lofty pines, through which we have travelled [sic] ten miles. The country was beautifully undulating, and although we usually associate the idea of barrenness with the pine regions, it was not so in this instance; every foot being covered with the finest grass, and beautiful broad grassy vales extending in every direction. The forest was perfectly open and unencumbered with brush wood, so that the travelling [sic] was excellent."

Cooper (1960) went on to state, "The overwhelming impression one gets from the older Indians and white pioneers of the Arizona pine forest is that the entire forest was once much more open and park-like than it is today."

Before European settlement of northern Arizona in the 1860s and 1870s, periodic natural surface fires occurred in ponderosa pine forests at frequent intervals, perhaps every 2–12 years (Weaver 1951, Cooper 1960, Dieterich 1980, Swetnam and Baisan 1996). Several factors associated with European settlement caused a reduction in fire frequency and size. Roads and trails broke up fuel continuity. Domestic livestock grazing, especially overgrazing and trampling by cattle and sheep in the 1880s and 1890s, greatly reduced herbaceous fuels. Active fire suppression, as early as 1908 in the Flagstaff area, was a principal duty of early foresters in the Southwest. A direct result of interrupting and suppressing these naturally occurring, periodic fires has been the development of overstocked forests.

Cooper (1960) cites the writings of early expedition leaders, Whipple and Beale, both of whom travelled through northern Arizona in the 1850s. They reported that the condition of the southwestern ponderosa pine forest "...was open and park-like with a dense grass cover." These early descriptions of the open nature of presettlement ponderosa pine forests are in agreement with results of recent research which found that canopy coverage by trees of presettlement origin range from 17 percent to 22 percent of the surface area for unharvested sites near Flagstaff, AZ (White 1985, Covington and Sackett 1986, Covington et al. 1997).

### *Retrospective Ecological Analyses*

Cooper (1960) stated that the structure of the southwestern ponderosa pine type in the White Mountains of east-central Arizona is actually that of an all-aged forest composed of even-aged groups. He noted great variation in diameter within a single age class. White (1985), in a study conducted on the Pearson Natural Area near Flagstaff, reported that successful establishment of ponderosa pine in presettlement times was infrequent (as much as four decades between regeneration events). White also determined that stems were strongly aggregated, the aggregation ranged from 3 to 44 stems within a group, with a group occupying an area that ranged from 0.05–0.7 acres. "Ages of stems within a group were also variable with the most homogeneous group having a range of 33 years and the least having a range of 268 years (White 1985)." White's findings of a pattern of uneven-aged groups near Flagstaff are in contrast to the results of Cooper (1960) for the White Mountains.

Madany and West (1983) discuss the effects that many years of heavy grazing and fire suppression have had on ponderosa pine regeneration in southern Zion National Park, Utah. They suggested that ponderosa pine seedling survival was probably greater in the early 1900s than in the presettlement days because of reduced competition of grasses (through grazing) with pine seedlings, and the reduced thinning effect that fires once had on seedlings in presettlement times.

### *The Restoration Process*

A fundamental issue is what treatment or combination of treatments is necessary for rapidly restoring some facsimile of a healthy ponderosa pine ecosystem. The two leading management plans for ecological restoration of ponderosa pine ecosystems are prescribed burning and thinning from below (Williams et al. 1993, Covington and Moore 1994b, Arno et al. 1995, Clark and Sampson 1995).

Previous research has shown that although prescribed burning alone (without thinning or manual fuel removal) can reduce surface fuel loads, stimulate nitrogen availability, and increase herbaceous productivity, it can cause high mortality of the presettlement trees (40 percent mortality over a 20-year period) and lethal soil temperatures under presettlement tree canopies (Covington and Sackett 1984, 1990, 1992, Harrington and Sackett 1990, Sackett et al. 1996). Although some thinning of postsettlement ponderosa pine trees was accomplished by prescribed burning (Harrington and Sackett 1990), results were localized, unpredictable, and difficult to control. Furthermore, reburning, even under very conservative prescriptions (low air temperatures and low windspeed), can produce dangerous fire behavior because the continuing high density of postsettlement trees provides a continuous fuel ladder and thus a high crown fire potential. Clearly, existing research shows that prescribed burning alone (without some mechanical fuel treatments) in today's unnatural ecosystem structure will not restore natural conditions in ponderosa pine/bunchgrass ecosystems. Thus, some combination of thinning, manual fuel removal, and prescribed burning will be necessary for rapidly restoring these systems to natural conditions.

### *Example of Detailed Ecological Restoration Experimentation*

In 1993 a small-scale ecosystem management research project was initiated at the Gus Pearson Natural Area near Flagstaff, Arizona, to test ecological restoration hypotheses (Covington et al. 1997). The research was guided by the general hypothesis that: (1) both restoration of ecosystem structure and reintroduction of fire are necessary for restoring rates of decomposition, nutrient cycling, and net primary production (NPP) to natural (presettlement) levels; and (2) that the rates of these processes will be higher in an ecosystem that is operating within some facsimile of its natural structure and disturbance regime. Specifically, the research hypothesis was that re-establishing presettlement stand structure alone (thinning postsettlement trees) will result in lower rates of decomposition, nutrient cycling, and NPP compared to thinning, forest floor fuel manipulation, and prescribed burning in combination, but that both of these treatments will result in higher rates of these processes compared to controls (see below). They further hypothesized that without periodic burning to hold them in check, pine seedling population irruptions will recur on the thin-only treatment.

Specific questions addressed were:

1. How has ecosystem structure (by biomass component) and nutrient storage changed over the past century of fire exclusion in a ponderosa pine/bunchgrass ecosystem?
2. What are the implications of these changes for NPP, decomposition, nutrient cycling, and other key ecosystem characteristics?
3. Does partial restoration (restoring tree structure alone by thinning postsettlement trees) differ from complete restoration (the same thinning, plus forest floor removal, loading with herbaceous fuels, and prescribed burning) in its effects on ecosystem structure and function?

Using a systematic approach, the authors established

replicated small plot studies to test these hypotheses. Details on the experimental design, variables measured, and analytical techniques are available in Covington et al. (1997). In addition to tree density, herbaceous vegetation cover, and fuel loading, a broad range of ecological attributes related to ecosystem health are being monitored.

Dendrochronological analysis revealed that forest structure had changed substantially in the study area since fire regime disruption. Particularly striking was the population irruption of ponderosa pine from 24.3 trees per acre in 1876 to 1,254 trees per acre in 1992. The irruption of smaller diameter pine trees also created a continuous tree canopy cover at the expense of herbaceous vegetation. In 1876, only 19 percent of the surface area was under pine canopy with the balance (81 percent) representing grassy openings, whereas in 1992 pine canopy covered 93% of the area, with only 7% left as grassy openings.

Thinning resulted in the removal of a total of 5,500 bd.ft./ac (3,700 bd.ft./ac of 9–16 in dbh trees, 1,800 bd.ft./ac of 5–8.9 in dbh trees). Most of the smaller diameter trees (629 trees per acre in the 1–4.9 in dbh class) were utilized as latillas for adobe home construction. A major problem in utilization was what to do with the 37 tons per acre of thinning slash. Because there was no market for this material, it was hauled (70–80, 18-wheel dump truck loads) to a borrow pit and burned.

In the complete restoration treatment, approximately 21.3 tons per acre of duff were removed by raking, some of which was utilized as garden mulch. Additional treatments on the complete restoration treatment (addition of mown grass and prescribed burning) left 4.3 tons per acre. Fire intensities were low with an average flame length of six inches. Overall, soil heating was negligible except under heavy woody fuels and cambial heating was low.

During the 1995 growing season, soil moisture and temperature were consistently higher in treated areas than in the control. These microenvironmental differences between treated and control areas will likely result in higher rates of key soil processes, such as fine root production, litter decomposition, and nitrogen mineralization in treated stands, and these changes in the soil process rates will, in turn, increase herbaceous production and tree growth and resistance to insect attack. In this regard, resin flow of presettlement trees was higher on the thinned area than on the control area, as was foliar toughness, suggesting increased resistance to bark beetles and foliage feeding insects. No changes in populations of turpentine beetles was observed.

Herbaceous production responded markedly to the treatments, with the greatest response to date in the grassy substratum. By 1995, the treated areas were producing almost twice as much herbaceous vegetation as the controls.

Preliminary results from this ecosystem restoration research are encouraging. The combination of thinning and burning changed forest structure such that the restored area has shifted from fire behavior fuel model 9 (Anderson 1982), where crown fires are common, to fuel model 2, where surface fires occur but where crown fires are impossible.

The reduction in tree competition has improved on-site moisture availability and has likely increased insect resistance of presettlement trees. Grasses, forbs, and shrubs are responding favorably as well, indicating a shift away from a net primary productivity dominated by pine toward a more diverse balance across a broader variety of plants.

Ecosystem restoration research requires a long-term, interdisciplinary commitment. Ecosystem attributes being measured are likely to be in transition for the next 10–20 years before they stabilize around some long-term mean. Therefore, Covington et al. (1997) plan to continue this project for the next 24 years (eight 3-year burning intervals), with subsequent burning coinciding with the natural burning season during the spring and summer. Data on other attributes will be used to increase understanding of the restoration treatments on a range of ecosystem characteristics.

The preliminary results from the small plot studies are so encouraging that the authors have joined with the Arizona Strip District of the Bureau of Land Management to test practical ecosystem restoration treatments on an operational scale in the Mount Trumbull Resource Conservation Area, north of the Grand Canyon (Taylor 1996, Covington 1996). At Mount Trumbull, they are working with the BLM using an adaptive ecosystem management approach (Walters and Holling 1990) to restore over 3,000 acres of ponderosa pine to conditions approximating those that existed before Euro-American settlement. They are monitoring a subset of the basic ecosystem health attributes measured in the small plot study, but by virtue of the larger size of the treatment areas, they are able to measure some variables which operate on a larger-scale such as passerine bird populations, community structure of selected insect guilds (e.g., butterflies), and variables indicative of landscape-scale ecosystem health. The hope is that through such a set of integrated, adaptive ecosystem restoration projects many of the symptoms of ecosystem pathology can be alleviated while simultaneously increasing understanding of ecosystem structure and function.

## 4.2 Forest Ecosystems of the Pacific Northwest

### *The Problem*

Timber harvesting has been an important element of the economy of the coastal Pacific Northwest since the arrival of European settlers in the 1800s. Although the extensive forests were initially considered to be obstacles to be cleared to make way for agriculture, their economic value as sources of wood was soon recognized. In the late 1800s and early 1900s, rates of harvest began to increase, and escalated greatly following World War II. By the late 1980s, most forests on private lands had been harvested at least once, and old-growth forests were generally limited to federal lands. More than 80 percent of pre-logging old-growth forests in the region had been removed (Booth 1991).

During the 1970s and 1980s, management of federal forests became increasingly controversial. Much of the concern focused on the habitat requirements of the northern spotted owl (*Strix occidentalis caurina*), a resident of old-growth forests. Largely because of continued loss of suitable habitat and the lack of regulations or policies to protect northern spotted owls, in 1990 the U.S. Fish and Wildlife Service listed the subspecies as threatened under the Endangered Species Act of 1973. Listing of the northern spotted owl greatly reduced the quantity of timber sold from federal lands in western Oregon and Washington, and northern California.

Although preservation of the spotted owl has been a focal point of the debate, the owl also served as a surrogate for other organisms of old-growth forests, and of the forest itself. The real questions were much broader, and included the following: What are appropriate management objectives for federal forests?, How much old-growth forest should be retained?, Is clearcut logging an acceptable harvest method?

In an attempt to resolve the complex issue, on April 2, 1993, President Clinton convened a forest conference in Portland, Oregon, at which he instructed federal agencies to work together and develop a "scientifically sound, ecologically credible and legally responsible" plan to restore, protect, and maintain the long-term health of forests, wildlife, and waterways. This plan would also provide for human and economic concerns and "produce a predictable and sustainable level of timber sales and nontimber resources that will not degrade or destroy the environment."

Following the conference, an interdisciplinary group of scientists was brought together as the Forest Ecosystem Management Assessment Team (FEMAT), to "identify management alternatives that attain the greatest economic and social contribution from the forests of the region and meet the requirements of the applicable laws and regulations..." (USDA et al. 1993). Subsequently the President chose an alternative ("Option 9") which would include approximately 7.4 million acres within Late-Successional Reserves, 2.6 million acres within Riparian Reserves, and 1.5 million acres of Adaptive Management Areas within which application and testing of ecosystem management techniques are encouraged. New, more restrictive, standards and guidelines were also developed for timber harvest within the approximately 4 million acres of federal lands outside of reserves and other areas withdrawn from timber harvest. Following the preparation of a Final Supplemental Environmental Impact Statement (USDA/USDI 1994a), on April 13, 1994, the Secretaries of Agriculture and Interior signed a joint Record of Decision (USDA/USDI 1994b) to implement this plan.

The Record of Decision applies to federal forests within the range of the northern spotted owl. However, in this example we focus on the Western Hemlock Zone, the most extensive vegetation zone in western Oregon and Washington, and the most important in terms of timber production (Franklin and Dyrness 1973). This zone is dominated by western hemlock (*Tsuga heterophylla*), western redcedar (*Thuja plicata*), Douglas-fir (*Pseudotsuga menziesii*), and grand fir (*Abies grandis*).

The Forest Ecosystem Management Assessment Team was specifically asked to develop alternatives for long-term management which met the following objectives (USDA et al. 1993):

- Maintenance and/or restoration of habitat conditions for the northern spotted owl and the marbled murrelet (*Brachyrampus marmoratus*) that will provide for viable populations, well distributed within their current ranges on federal lands.
- Maintenance and/or restoration of habitat conditions to support viable populations, well distributed across their current range, of species known (or reasonably expected) to be associated with old-growth forest.
- Maintenance and/or restoration of spawning and rearing habitat on federal lands to support recovery and maintenance of viable populations of anadromous fish species and stocks and other fish species and stocks considered "sensitive" or "at risk."
- Maintenance and/or creation of a connected or interactive old-growth forest ecosystem on federal lands.

Accomplishing these objectives will require an unprecedented application of principles of ecological restoration across 24.3 million acres of federal lands (USDA et. al 1993), at stand-level, watershed, and regional scales.

### *Reference Conditions*

Pollen records from the Pacific Northwest suggest that forests of modern composition were first established about 6,000 years BP following retreat of lowland glaciers (Brubaker 1991). These forests consist of long-lived conifer species which become massive as they age. With time, these forests develop characteristic Pacific Northwest old-growth attributes: patchy, multi-layered canopies with trees of several age classes, large live trees, and an abundance of snags and fallen logs. Although rate of development for these structural characteristics is variable, old-growth characteristics commonly begin to appear in unmanaged forests at 175–250 years of age (USDA et al. 1993).

These forests provide habitat for an exceedingly rich bird and mammal fauna (Harris 1984), with some species occupying unique ecological niches. One such species, the red tree vole (*Aborimus longicaudus*), is the most specialized vole in the world and the most arboreal mammal in North America (Maser et al. 1981). Northern flying squirrels (*Glaucomys sabrinus*) are the only North American forest mammal that consumes lichens as their primary forage (Harris 1984). Both of these mammals are important prey species for the northern spotted owl, which is closely associated with old-growth forests. Arthropods are also very diverse, with as many as 7,000 species inhabiting these forests (USDA et al. 1993). Although forests of the Pacific Northwest are relatively young, as measured on geological time scales, a rich fauna has co-evolved with these plant communities.

Natural fire-return intervals for these forests are variable, ranging from less than 100 to several hundred years (Agee 1991). Intense fires are important elements of the natural fire regime, but lower severity fires also occur (Agee 1991). Unlike southwestern ponderosa pine forests, fire suppression has only minimally influenced forests of the region. Fire suppression became effective after 1910, and with long fire-return intervals, the effects of 85 years of fire exclusion have been relatively minor (Agee and Edmonds 1992).

The forest was not an unbroken block of old-growth. Prior to European settlement, the regional landscape consisted of a shifting mosaic of forest communities in varying stages of successional development following disturbance. During the early 1800s, prior to extensive fires caused by settlers during the 1840s, nearly 40 percent of forests in the Coast Range of Oregon were less than 200 years old (Ripple 1994).

These younger forests, however, were structurally different than those regenerated after timber harvest. The following discussion focuses on management of young stands to accelerate the development of structural attributes of old-growth forests, and provides an example of restoration challenges faced by forest managers in the Pacific Northwest.

### *Restoration Goals and Treatments*

The heart of the plan is a 10 million-acre federal system (USDA and USDI 1994a) of late-successional and riparian reserves. These reserves were established to provide an inter-connecting network of late-successional and old-growth forests, but at the present time they contain large areas of younger forest. More than 50 percent of the area within late-successional and riparian reserves supports younger forests (USDA and USDI 1994a). Most of these younger stands were usually established following fire or timber harvest. Regeneration of these stands was designed to produce high yields of lumber, not to produce old-growth forest characteristics.

With enough time, some of these forests and associated processes, communities and species will assume old-growth characteristics without intervention. Others may not, or will only do so over greatly lengthened time frames.

Acceleration of the development of old-growth characteristics in these stands is desirable, and would likely improve the long-term potential for success of the Northwest Forest Plan. Northern spotted owls and marbled murrelets may be important beneficiaries of such management. For both of these species, there is concern about population viability during a transition period lasting until habitat conditions improve significantly (USDA et al. 1993).

Most young forests within the reserve system have been managed under an even-aged system. Following clearcutting, regeneration was accomplished with site preparation, planting of nursery-grown seedlings, and control of competing vegetation. Standing snags were commonly removed because of safety concerns. Often these forests were thinned 10-20 years after establishment to control density and ensure uniform spacing. Many such stands subsequently go through a stem exclusion or self-thinning stage during which much of the understory is lost (Oliver and Larson 1990).

The primary objective of these practices is to create a uniform conifer stand which quickly achieves crown closure and dominates nonconifers (Tappeiner et al. 1992). This process of stand development is, however, quite different from that following natural disturbance. In moist Douglas-fir forests, post-fire recruitment may continue for 40–100 years (Agee 1991), rather than as a pulse of regeneration extending for a relatively brief period. Many old-growth trees were apparently established in relatively open conditions with little competition for 100 years or more (Franklin et al. 1981). Such stands may not have gone through the self-thinning stage associated with commercial forests. Understories persisted, and younger trees were established and became intermediate canopies. The dominant old trees retained deeper crowns, thus providing nest sites and thermal cover for wildlife species such as spotted owls and marbled murrelets.

The Northwest Forest Plan encourages the use of silviculture to "accelerate the development of young stands into multilayered stands with large trees and diverse plant species, and structures that may, in turn, maintain or enhance species diversity." Treatment will focus on stands that have been regenerated after tree harvest or on stands that have been thinned and will include, but not be limited to: (1) thinning or managing the overstory to produce large trees, release advanced regeneration of conifers, hardwoods, or other plants, or to reduce risk from fire, insects, disease, or other environmental variables; (2) underplanting and limited understory vegetation control to begin development of multistory stands; (3) killing trees to make snags and logs on the forest floor; (4) reforestation; and (5) use of prescribed fire (USDA et al. 1993, USDA and USDI 1994).

Tappeiner et al. (1992) discussed systems which can be used to accelerate the development of stand structures which are important for northern spotted owls, and to grow habitat in stands where it is unlikely to develop naturally. Simulated outcomes of several silvicultural prescriptions, based on data from actual stands, were included. The following is an example of a prescription and simulated response for a Douglas-fir forest in the Coast Range of Oregon.

The stand contained Douglas-fir, grand fir, and big leaf maple, with an initial density of 881 stems/acre. At age 40 years, 50 percent of the conifers with diameters from 10 to 16 inches were removed and one-hundred conifers/acre were planted. At age 60 years, 50 percent of trees with diameters from 10 to 22 inches were converted to snags, logs on the forest floor, or removed, and 100 conifers/acre were planted. At age 80 years, 60 percent of the conifers 8 to 22 inches in diameter were made into snags, down logs, or removed, with 224 conifers and 70 hardwoods remaining. At age 120 years, the stand was projected to have a multi-story canopy with 53 percent cover, and several trees/acre greater than 42 inches in diameter. Untreated, the stand would had a single layered canopy, no trees with diameters greater than 42 inches, and a sparse understory. Diameter distributions of the treated stand were similar to those measured in stands providing suitable habitat for northern spotted owls (Tappeiner et al. 1992).

Although this example was developed for a Douglas-fir forest in the Coast Range of Oregon, similar approaches could be applied in other areas, including the mixed conifer forests of southern Oregon and northern California. Tappeiner et al. (1992) provide the following suggestions for silvicultural systems designed to emulate natural disturbance and stand development:

- Favor some large trees with numerous limbs for potential nest sites.
- Use hardwoods to help develop a multi-layered stand.
- Encourage the growth of advanced regeneration of shade-tolerant conifer and hardwood species.
- Establish new regeneration by planting or seeding in young stands after making small openings or reducing overstory density in parts of a stand.
- Vary the distribution of overstory trees when thinning. Make openings for new regeneration and to release advanced regeneration.
- When thinning, leave some trees in the smaller crown size classes to help promote a layered stand.
- In stands with irregularly spaced trees, consider a crown thinning to release individual trees while maintaining the irregular spacing.

Although the primary goal of silviculture systems for late-successional reserves is the restoration of old-growth characteristics, important economic benefits can be derived from thinning these stands. Education programs may be required, however, to illustrate the ecological benefits of such activities. Importantly, managers must establish and maintain a high level of public trust when planning silviculture systems for reserves.

### *Adaptive Management and Ecosystem Restoration*

Ecosystems are extremely complex, and often it is difficult or impossible to predict accurately the impact of management actions on future conditions. Failure to act also has unforeseen consequences. The Northwest Forest Plan resolves this dilemma by requiring adaptive management (Holling 1978, Walters 1986).

Adaptive management as envisioned in the forest plan is "a continuing process of action-based planning, monitoring, researching, evaluating and adjusting with the objective of improving the implementation and achieving the goals" of management standards and guidelines. Simply stated, managers and scientists learn from experience and use this knowledge to improve subsequent actions.

Silvicultural activities in late-successional reserves provide a good example of the application of adaptive management. Initially, existing knowledge of silviculture is reviewed and synthesized as a basis for development of a prescription. During and after implementation, a scientifically and statistically credible monitoring and evaluation program determines whether development of old-growth characteristics has been achieved, and, finally, new knowledge and information is incorporated into new prescriptions for similar areas.

Because forest succession occurs slowly from a human perspective, long-term monitoring programs are required. Successional data from these programs will provide important information upon which to refine simulation models of stand development, and more accurately project the outcomes of silviculture systems.

### *Restoration Challenges*

Restoration of old-growth forests and processes will be difficult and exceedingly complex. We have focused on the use of silviculture to restore old-growth characteristics at the stand level. We have not addressed important issues such as landscape level integration, restoration of fire as a natural process, or restoration of anadromous fish habitat. Stand-level restoration will provide key building blocks with which to restore and maintain landscape and regional scale ecosystems. The ultimate ecological test of the Northwest Forest Plan will be its ability to provide an evolutionary environment which maintains native species and natural processes and allows them to continue to evolve.

## 4.3 Salt Marsh Ecosystems of the Northeast

### *The Problem*

Over the last half century the salt marshes along the Northeast coastline have been severely impacted because of human activities (Anon 1961, Tiner 1984). They have been dredged, filled, and impounded, which has either destroyed the marsh vegetation or modified it floristically. In Connecticut, 40 percent of the original salt marshes that fringed Long Island Sound have been destroyed. Thus, the need for restoration is obvious not only to compensate for these losses but also to restore degraded systems. Because of tidal restriction along the valley marshes which characterize many of the New England type salt marshes, vast acreages of *Spartina*-dominated communities have been transformed into monocultures of *Phragmites australis* (common reed) (Haslam 1973, Niering and Warren 1980). Those filled but not developed also exhibit a similar monoculture. There is an urgent need to restore tidal flushing to restricted systems, reclaim filled marshes and encourage *Spartina* plantings in those areas where marshes can be recreated and thus reconnect these productive tidal wetlands with the surrounding estuarine waters.

### *The Reference Marsh*

The reference salt marsh is an undisturbed estuarine, emergent wetland dominated by *Spartina alterniflora* (saltwater cord grass) and *Spartina patens* (salt meadow cordgrass) (Miller and Egler 1950, Niering and Warren 1980, Nixon 1982, Teal 1986, Bertness 1992). The former, growing 1–2 meters in height, characterizes the low marsh and it is flooded by every tidal cycle. It is replaced landward by the high marsh *S. patens* which is flooded periodically by spring high tides. A third belt of *Juncus gerardii* (black grass) forms a border near the upland and is replaced by an upper-most border of *Panicum virgatum* (switch grass) and/or *Iva frutescens* (marsh elder). Where the disturbance occurs along the marsh/upland interface *Phragmites* may form the typical upper border vegetation. This marsh system is flushed by estuarine waters with salinity ranging from 20–30 ppt. It exhibits also a distinctive set of animal populations some of which are restricted to the low marsh whereas others are more typical of the high marsh (Olmstead and Fell 1974).

### *Goals of Restoration*

The major goal is to restore salt marsh systems lost historically and recreate the ecological link between the tidal marsh and contiguous estuary so that the high productivity of the tidal marsh–estuarine system can be restored. These systems carry on a diversity of functional roles or values in terms of finfish and shellfish productivity and shoreline stabilization (Niering 1985, Mitsch and Gosselink 1993).

### *The Restoration Process*

A diversity of restoration strategies can be employed depending upon the nature of site degradation. In the case of tidal restriction, the aim is to restore tidal flushing and attempt to recreate a salinity regime similar to that which previously existed (Rozsa and Orson 1993). With salinities above 20 ppt, *Phragmites* will be killed or sufficiently suppressed to favor the re-entry of the *Spartina* grasses (Capotosto and Spencer 1988, Tiner

1995). In other sites the restoration process may involve removal of spoil or dredged material from a filled salt marsh which may now be *Phragmites*-dominated. Here spoil is removed to expose the original salt marsh peat surface (Capotosto 1993, Waters 1995) and then the creek channels are recreated to restore tidal flushing. Marsh elevations are critical in terms of tidal flooding to favor the establishment of *Spartina alterniflora*. A gentle slope that falls within the range of mean high and mean low tide is required. This is also the prerequisite for any planting of *S. alterniflora* where shoreline stabilization is desired or where new marsh is being created. High-marsh *Spartina patens* can be recreated at slightly higher elevations either under natural conditions or by planting, but high marsh is not as easy to recreate as low marsh because it is controlled by a complex of environmental factors. Another parameter in the restoration process is to avoid flooding of private property during major storms. The use and availability of self-regulating tidal gates invented by Thomas Steinke, Director of the Fairfield Conservation Commission, has greatly aided the marsh restoration process in highly developed areas (Steinke 1986, 1995a,b).

## Case Histories

Several case histories will be briefly described to illustrate the feasibility of salt marsh restoration in various ecological settings (Rosza and Orson 1993). Such efforts also serve as invaluable models where one can actually observe firsthand results of ecological restoration.

The Hammock River (Connecticut) valley marsh on Long Island Sound represents a 250-acre system in which the upper reaches were restricted by tidal gates early in the century, transforming the typical *Spartina*-dominated wetland to a monoculture of Phragmites. To reverse this trend, one of the four tide gates was opened in the summer of 1985. Now more than a decade later *Phragmites* has been dramatically suppressed and the area is now dominated by *Spartina* and other salt marsh vegetation.

Another successful marsh restoration project is at the Barn Island Wildlife Management Area (Connecticut) on the Connecticut/Rhode Island border. Here a valley salt marsh was severely restricted in the late 1940s with only an 0.5-m opening connecting the adjacent estuary (Miller and Egler 1950). The salt marsh was transformed from a *Spartina*-dominated vegetation to a *Typha angustiflolia* (narrow-leaved cattail) marsh with areas of *Phragmites* restricted to the marsh borders. In 1978 a five-foot culvert was installed, followed by another seven foot shortly thereafter in order to increase tidal flushing. Salinity of the adjacent estuary was restored (28–33 ppt) to the impounded area. By the mid 1980s most of the cattail was dead and *Spartina* grasses dominated the area (Sinicrope et al. 1990, Barrett and Niering 1993). With the restoration of the plant community, typical salt marsh animal populations have also become established (Fell et al. 1991). After more than a decade, functional equivalence is being restored, not unlike the productivity and trophic structure in the nearby reference system.

Marsh restoration by planting has also been documented in the Northeast but especially in Southeast (North Carolina) where extensive salt marsh areas have been created on dredged material (Broome et al. 1986, Broome et al. 1988, Broome 1990). In the Northeast, this has been done on a less extensive scale within the intertidal low marsh zone especially to favor shoreline stabilization (Garbisch and Garbisch 1994). Recreating the elevation typical of the low marsh so that the site is flooded by every tidal cycle is, as previously mentioned, the major prerequisite. Planting *S. alterniflora* 20 in apart and protecting it from grazers such as geese until well established is critical. *S. alterniflora* plantings following an oil spill have resulted in successful restoration (Bergen et al. 1995).

## Attaining Functional Equivalence and Monitoring and Establishment of a Buffer Zone

The ultimate goal of salt marsh restoration is to create a self perpetuating ecosystem with its productivity, biogeochemical cycling, and food chain support comparable to the reference or control marsh system. Much literature now exists on this subject from the Southeast, where marsh restoration and creation have been underway since the 1970s (Craft et al. 1988, Sacco et al. 1994, Thompson et al. 1995). Literature is also available from the Northeast (Fell et al. 1991, Allen et al. 1994, Peck et al. 1994, Spelke et al. 1995). It has been shown that various aspects of functional equivalence can be attained within a decade or less depending upon the region. For some functions several decades may be required. Thus, monitoring a minimum of five years upon the completion of the project is basic. This allows time for management correction and assessing the development of functional equivalence.

A further requirement is the establishment of a buffer zone along the upland to accommodate continued sea level rise. Two studies in the Northeast have documented the potential effects of sea-level rise on the marsh vegetation toward a more hydric/less productive phase (Warren and Niering 1993, Nydick et al. 1995). The current rate of sea level rise is one inch per year and this rate is predicted to increase in the future with climate warming (Warrick 1993). Therefore, an undeveloped buffer of 50–100 ft or more is needed to provide for the landward movement of the marsh with sea level rise and overall protection of the restored system.

Finally, it should be noted that the potential for restoration should not substitute for permitting salt marsh destruction because of development. All possible and prudent alternatives should be explored prior to sacrificing a self-perpetuating functional tidal salt marsh ecosystem (Oviatt et al. 1977, Kusler and Kentula 1989, Moy and Levin 1991).

## 5 A SYSTEMATIC APPROACH TO ECOLOGICAL RESTORATION

Far too many ecological restoration projects have been started without clear definition of restoration goals and with little attempt to evaluate the success quantitatively. Bradshaw (1993) argued strongly that ecological restoration should follow a scientific approach: (1) be aware of other relevant work, (2) carry out experiments to test ideas, (3) monitor key indicator parameters, (4) design further experiments and tests based on results of monitoring, and (5) publish peer reviewed results and conclusions. Kaufmann et al. (1994) discussed the importance of a systematic approach in ecosystem management including the determination of reference conditions, determination of current conditions, and using coarse- and fine-filter analyses to determine if goals are met. Oliver et al. (1994) suggested an eight-step systematic approach for achieving forest health which is relevant to ecological restoration. Walters (1987) emphasized the importance of clearly stating assumptions about ecosystem behavior, the building of explicit models that synthesize this knowledge, and testing this knowledge in adaptive learning experiments.

Based on these ideas and other sources, we have developed a stepwise systems analytic approach to the design of ecosystem restoration experiments (Table 1). Most steps are straightforward and broadly discussed in the adaptive ecosystem management literature. Of these steps, step 3, determining reference conditions is most explicit to ecological restoration.

Adaptive ecosystem restoration and management involves a broad variety of practices for designing and testing ecological restoration treatments. A systematic diagnosis of the ecosystem pathology is an essential first step (Rapport 1995). Historical analysis, paleo-ecological techniques, retrospective ecological analysis, and dendrochronology, along with other techniques (see Kaufmann et al. 1994, Morgan et al. 1994), are used to determine the natural structure and function of the ecological system to be restored. Goals and performance measures must be defined in measurable terms. Assumptions about ecosystem dysfunction must be stated. A specific set of scientifically-based alternative treatments for restoring ecosystems to the desired condition must be developed. Finally, monitoring and evaluation procedures are used to determine where the restoration worked and where it did not. A central assessment is whether the ecosystem being restored has been set on a trajectory such that structural and functional equivalency to the reference system will be attained. This information is then fed back into the body of scientific and managerial knowledge for future ecosystem management decisions.

Table 1. A systems analytic approach to adaptive ecosystem restoration.

1. Clearly diagnose the symptoms and causes of the ecosystem health problem. What are the symptoms that suggest the ecological system has been degraded and what are the underlying mechanisms (Rapport 1995, Covington et al. 1994)?
2. Determine reference conditions. What was the condition of the ecosystem before degradation (Kaufmann et al. 1994, Morgan et al. 1994)?
3. Set measurable ecological restoration goals (National Research Council 1992). How close to reference conditions do you intend to get? How will you know if you are moving in the right direction?
4. What factors are most limiting to the restoration process?
5. Develop alternative ecosystem restoration hypotheses (Walters and Holling 1990).
6. Design restoration treatments that will allow you to test the alternative hypotheses (Bradshaw 1993).
7. Monitor ecosystem conditions and evaluate hypotheses.
8. Feed the results back into the design and implementation of ecological restoration treatments — adapting management based on results and changing goals (Kaufmann et al. 1994).

## 6 ECOSYSTEM RESTORATION AND HUMAN HABITAT NEEDS

Philosophically, ecosystem restoration is founded on Leopold's belief that a workable ecosystem conservation paradigm should be defined as "a universal symbiosis with land, economic, public, and private;" as "a protest against destructive land use;" as an effort "to preserve both utility and beauty;" as "a state of harmony between men and land;" and finally, as "a positive exercise of both skill and insight, not merely a negative exercise of abstinence and caution."

Given the wide variety of human needs and goals for federal lands and waters, it seems unlikely that vast areas will be restored to completely natural conditions. In fact, keeping some areas in a somewhat artificial state may be desirable so long as such action does not impair their long-term sustainability. Areas dedicated to wood fiber production, livestock grazing, intensive recreation, and many other human habitat uses might fall into this category.

Setting degraded ecosystems on the path to recovery of natural structure and function seems broadly warranted. Ecological restoration, even though partial, can go a long way toward reducing the undesirable symptoms of dysfunctional ecosystems and benefit not only future generations but also those involved in the restoration process. In fact, some have suggested that the transformation of those involved in ecological restoration is one of the major benefits to be gained from such projects (Dodge 1991, Jordan 1993).

## 7 CONCLUSIONS

Ecological restoration has as its goal the restoration of degraded ecosystems to resemble, or emulate more closely conditions that prevailed before disruption of natural structures and processes. A key concept in restoration ecology is that of the reference conditions defined as the range of ecosystem conditions (including structure and function) which have prevailed over recent evolutionary time. Underlying the idea of reference conditions is the concept of the evolutionary environment—the environment in which species have evolved. Ecological restoration consists of management actions designed to accelerate recovery by complementing or reinforcing natural processes.

Key ecological restoration concepts and principles have been discussed using examples from ponderosa pine ecosystems of the Southwest, forest ecosystems of the Pacific Northwest, and salt marsh ecosystems of the Northeast. A wide variety of methods can be used to determine reference conditions, ranging from historical documentation to paleoecological reconstructions and to locating contemporary ecosystems which can serve as analogs for what the degraded ecosystem might have been like had it not been disturbed.

Once reference conditions have been determined, then sets of management actions must be identified that can accelerate recovery of the ecosystem. Close attention must be given to restoration of both structure and processes, including natural disturbance regimes, if restoration is to be successful. These management actions can be viewed as working hypotheses to be tested in closely monitored management experiments.

For ecological restoration to serve as a viable approach to implementation of ecosystem management, it must not be viewed as a rigid set of procedures nor as a simple recipe. Rather, it must be viewed as a broad intellectual framework for meeting ecological habitat needs for all organisms, including those of humans, by managing in harmony with the natural ecosystem processes and components characteristic of the recent evolutionary environment of the biota.

## REFERENCES


Agee, J.K. 1991. Fire history of Douglas-fir forests in the Pacific Northwest. pp. 25–33. In *Wildlife and vegetation of unmanaged Douglas-fir forests*. USDA Forest Service General Technical Report PNW-GTR-285. Portland, OR: Pacific Northwest Research Station.

Agee, J.K., and R.L. Edmonds. 1992. Forest protection guidelines for the northern spotted owl. In: Final draft recovery plan for the northern spotted owl. Appendix E. U.S. Department of Interior, Washington, DC.

Allen, E.A., P.E. Fell, M.A. Peck, J.A. Gieg, C.R. Guthke, and M.D. Newkirk. 1994. Gut contents of common mummichogs, *Fundulus heteroclitus* L., in a restored impounded marsh and in natural reference marshes. *Estuaries* 17: 462–471.

American Forest and Paper Association. 1993. *Sustainable forestry principles and implementation guidelines*. Forest Resources Board, American Forest and Paper Association, Washington, DC.

Anderson, H.E. 1982. Aids to determining fuel models for estimating fire behavior. USDA Forest Service General Technical Report INT-122. Intermountain Forest and Range Research Station, Ogden, UT.

Anderson, J.E. 1991. A conceptual framework for evaluating and quantifying naturalness. *Conservation Biology* 5: 347–351.

Anon. 1961. Connecticut's coastal marshes: a vanishing resource. *The Connecticut Arboretum Bulletin*, no. 12. Connecticut College, New London, CT.

Anon. 1995. Report of the Interagency Task Force on Ecosystem Management. U.S. Government Printing Office, Washington, DC.

Arno, S.F., J.H. Scott, and M.G. Hartwell. 1995. Age-class structure of old growth ponderosa pine/Douglas-fir stands and its relationship to fire history. USDA Forest Service Research Paper INT-RP-481. Intermountain Forest and Range Experiment Station, Ogden, UT.

Aronson, J., and E. Le Floc'h. 1996. Hierarchies and landscape history: dialoguing with Hobbs and Norton. *Restoration Ecology* 4: 327–333.

Barrett, N.E., and W.A. Niering. 1993. Tidal marsh restoration: trends in vegetation change using a Geographical Information System (GIS). *Restoration Ecology* 1: 18–28.

Bergen, A., M. Levandowsky, T. Gorrell, and C. Alderson. 1995. Restoration of a heavily oiled salt marsh using *Spartina alterniflora* seedlings and transplants: effects on petroleum hydrocarbon levels and soil microflora. Presented to *The 10th Annual Conference on Contaminated Soils; Analysis, Site Assessment, Fate, Environmental and Human Risk Assessment, Remediation, and Regulation.* October 1995. University of Massachusetts.

Berger, J.J. 1997. Nine ways to save our forests. *Sierra* 82(4): 38–39.

Bertness, M.D. 1992. The ecology of a New England salt marsh. *American Scientist* 80: 260–268.

Booth, D.E. 1991. Estimating prelogging old-growth in the Pacific Northwest. *Journal of Forestry* 89: 25–29.

Bradshaw, A.D. 1993. Restoration ecology as science. *Restoration Ecology* 1: 71–73.

Bradshaw, A.D. 1984. Ecological principles and land reclamation practice. *Landscape Planning* 11: 35–48.

Broome, S.W. 1990. Creation and restoration of tidal wetlands of the southeastern United States. pp. 37–72. In: J.E. Kusler and M.E. Kentula (eds.), *Wetland Creation and Restoration: the Status of the Science.* Island Press, Washington, DC.

Broome, S.W., C.B. Craft, and E.D. Seneca. 1988. Creation and development of brackish-water marsh habitat. pp. 197–205. In: J. Zelazny and J.S. Feierabend (eds.), *Increasing Our Wetland Resources.* Proceedings of the Corporate Conservation Council of the National Wildlife Federation, Washington, DC., 4–7 October 1987. National Wildlife Federation, Washington, DC.

Broome, S.W., E.D. Seneca, and W.W. Woodhouse, Jr. 1986. Long-term growth and development of transplants of the salt-marsh grass *Spartina alterniflora. Estuaries* 9: 63–74.

Brubaker, L.B. 1991. Climate change and the origin of old growth Douglas-fir forests in the Puget Sound Lowlands. pp.17–24. In: Wildlife and vegetation of unmanaged Douglas-fir forests. USDA General Technical Report PNW-GTR-285. Portland, OR: Pacific Northwest Research Station.

Capotosto, P. M. 1993. *Restoration of a dredge disposal site in Mumford Cove, Groton, Connecticut.* State of Connecticut. Department of Environmental Protection, Field Services and Boating Division, Hartford.

Capotosto, P.M., and E. Spencer. 1988. *Controlling Phragmites by tidegate management in the State of Connecticut.* State of Connecticut, Department of Health Services, Madison, CT.

Christensen, N.L., et al. 1995. *The Scientific Basis for Ecosystem Management.* Committee on Ecosystem Management, Ecological Society of America.

Clark, L.R., and R.N. Sampson. 1995. *Forest Ecosystem Health in the Inland West: A Science and Policy Reader.* Forest Policy Center, American Forests, Washington, DC.

Cooper, C.F. 1960. Changes in vegetation, structure, and growth of southwestern pine forests since white settlement. *Ecology* 42: 493–499.

Covington, W.W. 1996. Implementing adaptive ecosystem restoration and management in long-needled pine forests. pp. 43–47. In: Conference on adaptive ecosystem restoration and management. USDA Forest Service General Technical Report RM-GTR-278. Rocky Mountain Forest and Range Experiment Station, Fort Collins, CO.

Covington, W.W., R.L. Everett, R.W. Steele, L.I. Irwin, T.A. Daer, and A.N.D. Auclair. 1994. Historical and anticipated changes in forest ecosystems of the Inland West of the United States. *Sustainable Forestry* 2: 13–63.

Covington, W.W., and M.M. Moore. 1994a. Southwestern ponderosa forest structure and resource conditions: changes since Euro-American settlement. *Forestry* 92(1): 39–47.

Covington, W.W., and M.M. Moore. 1994b. Postsettlement changes in natural fire regimes: ecological restoration of old-growth ponderosa pine forests. *Sustainable Forestry* 2: 153–181.

Covington, W.W., and S.S. Sackett. 1984. The effect of a prescribed burn in southwestern ponderosa pine on organic matter and nutrients in woody debris and forest floor. *Forest Science* 30: 183–192.

Covington, W.W., and S.S. Sackett. 1990. Fire effects on ponderosa pine soils and their management implications. pp. 105–111. In: Effects of fire management of southwestern natural resources. USDA Forest Service General Technical Report RM-191. Rocky Mountain Forest and Range Experiment Station, Fort Collins, CO.

Covington, W.W., and S.S. Sackett. 1992. Spatial variation in soil mineral nitrogen following prescribed burning in ponderosa pine. *Forest Ecology and Management* 54: 175–191.

Covington, W.W., and R.N. Sampson. 1996. Restoration of forest ecosystem health: western long-needled pines. *Transactions of the 61st North American Wildlife and Natural Resources Conference.* pp. 511–518.

Covington, W.W., P.Z. Fule, M.M. Moore, S.C. Hart, T.E. Kolb, J.N. Mast, S.S. Sackett, and M.R. Wagner. 1997. Restoring ecosystem health in ponderosa pine forests of the Southwest. *Journal of Forestry* 95(4): 23–29.

Craft, C.B., Broome, S.W., and E.D. Seneca. 1988. Nitrogen, phosphorus and organic carbon pools in natural and transplanted marsh soils. *Estuaries* 11: 272–280.

Dana, S.T., and S.K. Fairfax. 1980. *Forest and Range Policy.* McGraw-Hill, New York.

Dieterich, J.H. 1980. Chimney Spring forest fire history. USDA Forest Service Research Paper RM-220. Rocky Mountain Forest and Range Experiment Station, Fort Collins, CO.

Dobson, A.P., A.D. Bradshaw, and A.J.M. Baker. Hopes for the future: restoration ecology and conservation biology. *Science* 277: 515–522.

Dodge, J. 1991. Life work. pp. 2–4. In: R.L. Nilsen (ed.). *Helping Nature Heal.* Ten Speed Press, Berkeley, CA.

Fell, P.E., K.A. Murphy, Peck, M.A. and M.L. Recchia. 1991. Re-establishment of *Melampus bidentatus* (Say) and other macroinvertebrates on a restored impounded tidal marsh: comparison of populations above and below the impoundment dike. *Journal of Experimental Marine Biology Ecology* 152: 33–48.

Flader, S.L. 1974. *Thinking like a mountain: Aldo Leopold and the evolution of an ecological attitude toward deer, wolves, and forests.* University of Missouri Press, Columbia, MO.

Flader, S.L., and J.B. Callicott (eds.). 1991. *The River of the Mother of God and other essays by Aldo Leopold.* University of Wisconsin Press, Madison, WI.

Franklin, J.F., K. Cromack, Jr., W. Denison, A. McKee, C. Maser, J. Sedell, F. Swanson, and G. Juday. 1981. Ecological characteristics of old-growth Douglas-fir forests. USDA

Forest Service General Technical Report PNW-118. Pacific Northwest Forest and Range Experiment Station, Portland, OR.
Franklin, J.F., and C.T. Dyrness. 1973. Natural vegetation of Oregon and Washington. USDA Forest Service General Technical Report PNW-8. Pacific Northwest Forest and Range Experiment Station, Portland, OR.
Garbisch, E.W., and J.L. Garbisch. 1994. Control of upland bank erosion through tidal marsh construction on restored shores: Application in the Maryland portion of Chesapeake Bay. *Environmental Management* 18: 677–691.
Geist, V. 1978. *Life Strategies, Human Evolution, and Environmental Design: Toward a Biological Theory of Health.* Springer Verlag, New York.
Hammond, M., and A. Holland. 1995. Ecosystem health: Some prognostications. *Environmental Values* 4: 283–286.
Harrington, M.G., and S.S. Sackett. 1990. Using fire as a management tool in southwestern ponderosa pine. pp. 122–133. In: Effects of Fire Management of Southwestern Natural Resources. USDA Forest Service General Technical Report RM-191. Rocky Mountain Forest and Range Experiment Station, Fort Collins, CO.
Harris, L.D. 1984. *The Fragmented Forest: Island Biogeography Theory and the Preservation of Biotic Diversity.* University of Chicago Press, Chicago, IL.
Haslam, S.M. 1973. Some aspects of the life history and autecology of *Phragmites communis* Trin.: A review. *Pol. Arch. Hydrobiol.* 20: 79–100.
Higgs, E. 1997. What is good ecological restoration. *Conservation Biology* 11: 338–348.
Jackson, L.L., N. Lopoukhine, and D. Hillyard. 1995. Ecological restoration: A definition and comments. *Restoration Ecology* 3: 71–75.
Jordan, W.R., III. 1993. Rituals of restoration. *The Humanist*, November/December: 23–26.
Jordan, W.R., III, M.E. Gilpin, and J.D. Aber. 1987. *Restoration Ecology: A Synthetic Approach to Ecological Restoration.* Cambridge University Press, New York.
Karr, J.R. 1981. Assessment of biotic integrity using fish communities. *Fisheries* 6(6): 21–27.
Kaufmann, M.R., et al. 1994. An ecological basis for ecosystem management. USDA Forest Service General Technical Report RM-246. Rocky Mountain Forest and Range Experiment Station, Fort Collins, CO.
Kay, C.E. 1995. Aboriginal overkill and native burning: implications for modern ecosystem management. *Western Journal of Applied Forestry* 10: 121–126.
Kilgore, B.M. 1981. Fire in ecosystem distribution and structure: Western forests and scrublands. pp. 58–89. In: *Fire Regimes and Ecosystem Properties.* USDA Forest Service General Technical Report WO-26. Washington, DC.
Kolb, T.E., M.R. Wagner, and W.W. Covington. 1994. Concepts of forest health. *Forestry* 92: 10–15.
Kusler, J.A., and Kentula, M.E. 1989. Wetland creation and restoration: The status of the science. *Vol. 1: Regional reviews.* EPA/600/3-89/038. U.S. Environmental Protection Agency, Washington, DC.
Lubchenco, J., et al. 1991. The sustainable biosphere initiative: An ecological research agenda. *Ecology* 72: 371–412.
MacArthur, R.H. 1972. *Geographical Ecology: Patterns in the Distribution of Species.* Princeton University Press, Princeton, NJ.
MacMahon, J.A., and W.R. Jordan, III. 1994. Ecological restoration. pp. 409–438. In: G.K. Meffe and C.R. Carroll (eds.), *Principles of Conservation Biology.* Sinauer Associates, Sunderland, MA.
Madany, M.H., and N.E. West. 1983. Livestock grazing — fire regime interactions within montane forests of Zion National Park. *Ecology* 64: 661–667.
Maser, C., B.R. Mate, J.F. Franklin, and C.T. Dyrness. 1981. Natural History of Oregon Coast Mammals. USDA Forest Service General Technical Report PNW-133. Pacific Northwest Forest and Range Experiment Station, Portland, OR.
Mayr, E. 1942. *Systematics and the Origin of Species.* Columbia University Press, New Yorki.
Miller, W.R. and F.E. Egler. 1950. Vegetation of the Wequetequock-Pawcatuck tidal marshes, Connecticut. *Ecological Monographs* 20: 143–172.
Mitsch, W.J., and Gosselink, J.G. 1993. *Wetlands.* 2nd edition. Van Nostrand Reinhold, New York.
Mooney, H.A. 1981. Adaptations of plants to fire regimes: integrating summary. pp. 322–323. In: Fire Regimes and Ecosystem Properties. USDA Forest Service General Technical Report WO-26. Washington, DC.
Morgan, P., G.H. Aplet, J.B. Haufler, H.C. Humphries, M.M. Moore, and W.D. Wilson. 1994. Historical range of variability: A useful tool for evaluating ecological change. *Sustainable Forestry* 2: 87–111.
Moy, L.D., and Levin, L.A. 1991. Are *Spartina* marshes a replaceable resource? A functional approach to evaluation of marsh creation efforts. *Estuaries* 14: 1–16.
Myers, N. 1984. Genetic resources in jeopardy. *Ambio* 13: 171–174.
National Academy of Science. 1974. *Rehabilitation Potential of Western Coal Lands.* Ballinger, Cambridge, MA.
National Research Council, Committee on Restoration of Aquatic Ecosystems. 1992. *Restoration of Aquatic Ecosystems: Science, Technology, and Public Policy.* National Academy Press.
Niering, W.A. 1985. *Wetlands.* The Audubon Series Nature Guides. Alfred A. Knopf, New York.
Niering, W.A., and Warren, R.S. 1980. Vegetation patterns and processes in New England salt marshes. *BioScience* 30: 301–307.
Nilsen, R. (ed.). 1991. *Helping Nature Heal: An Introduction to Environmental Sestoration.* Ten Speed Press, Berkeley, CA.
Nixon, S.W. 1982. The ecology of New England high salt marshes: A community profile. FWS/OBS-81/55. U.S. Fish and Wildlife Service, Office of Biological Services, Washington, DC.
Noss, R.F. 1991. From endangered species to a biodiversity. pp. 227–245. In: K. Kohm (ed.), *Balancing on the Brink of Extinction: The Endangered Species Act and Lessons for the Future.* Island Press, Washington, DC.
Nydick, K.R., A.B. Bidwell, E. Thomas, and J.C. Van Kamp. 1995. A sea-level rise curve from Guilford, Connecticut, USA. *Marine Geology* 124: 13–159.
Oliver, C.D., and B.C. Larson. 1990. *Forest Stand Dynamics.* McGraw-Hill, New York.
Oliver, C.D., D.E. Ferguson, A.E. Harvey, H.S. Malany, J.M. Mandzak, and R.W. Mutch. 1994. Managing ecosystems for forest health: an approach and effects on uses and values. *Sustainable Forestry* 2: 113–133.
Olmstead, N.C., and P.E. Fell. 1974. *Tidal marsh invertebrates of*

*Connecticut.* Connecticut Arboretum Bulletin, no. 20. Connecticut College, New London, CT.

Oviatt, C.A., S.W. Nixon, and J. Garber, 1977. Variation and evaluation of coastal salt marshes. *Environmental Management* 1: 201–211.

Parsons, D.J., D.M. Graber, J.K. Agee, and J.W. van Wagtendonk. 1986. Natural fire management in national parks. *Environmental Management* 19(1): 21–24.

Pearson. G.A. 1950. *Management of Ponderosa Pine in the Southwest.* USDA Monograph 6.

Peck, M.A., P.E. Fell, E.A. Allen, J.A. Gieg, C.R. Guthke, and M.D. Newkirk. 1994. Evaluation of tidal marsh restoration: Comparison of selected macroinvertebrate populations on a restored impounded valley marsh and an unimpounded valley marsh within the same salt marsh system in Connecticut, USA. *Environmental Management* 18: 283–293.

Rapport, D.J. 1995. Ecosystem health: an emerging integrative science. In: D.J. Rapport, C.L. Gaudet, and P. Calow (eds.), *Evaluating and Monitoring the Health of Large-Scale Ecosystems.* NATO ASI Series. Series I: Global environmental change, Vol. 28. Springer-Verlag, Berlin.

Rapport, D.J., and S.B. Yazvenko. 1996. Ecosystem distress syndrome in ponderosa pine forests. pp. 3–9. In: Conference on Adaptive ecosystem restoration and management: Restoration of cordilleran conifer landscapes of North America. USDA Forest Service General Technical Report RM-GTR-278. Rocky Mountain Forest and Range Experiment Station, Fort Collins, CO.

Ripple, W.J. 1994. Historic spatial patterns of old forests in western Oregon. *Forestry* 45–49.

Rosza, R., and R.A. Orson. 1993. Restoration of Degraded Salt Marshes in Connecticut. pp. 196–205. In: Frederick J. Webb, Jr. (ed.), *Proceedings of the Twentieth Annual Conference on Wetlands Restoration and Creation, May 1993.* Institute of Florida Studies, Hillsborough Community College, Plant City, FL.

Sacco, J.N., E.D. Seneca, and T.R. Wentworth. 1994. Infaunal community development of artificially established salt marshes in North Carolina. *Estuaries* 17: 489–500.

Sackett, S.S., S.M. Haase, and M.G. Harrington. 1996. Lessons learned from fire use for restoring southwestern ponderosa pine ecosystems. pp. 54–61. In: Conference on adaptive ecosystem restoration and management. USDA Forest Service General Technical Report RM-GTR-278. Rocky Mountain Forest and Range Experiment Station, Fort Collins, CO.

Sinicrope, T.L., P.G. Hine, R.S. Warren, and W.A. Niering. 1990. Restoration of an impounded salt marsh in New England. *Estuaries* 13: 25–30.

Smith, Maynard. 1958. *Theory of Evolution.* Penguin Books, New York.

Society for Ecological Restoration. 1993. Ecological restoration, conservation and environmentalism: Mission statement and environmental policies of the Society for Ecological Restoration. *Restoration Ecology* 1: 206–207.

Society of American Foresters. 1993. *Sustaining Long-Term Forest Health and Productivity.* Society of American Foresters, Bethesda, MD.

Southwest Forest Alliance. 1996. *Forests forever! A plan to restore ecological and economic integrity to the Southwest's national forests and forest dependent communities.* Southwest Forest Alliance, Flagstaff, AZ.

Spelke, J.A., P.E. Fell, and L.L. Helvenston. 1995. Population structure, growth and fecundity of *Melampus bidentatus* (Say) from two regions of a tidal marsh complex in Connecticut. *The Nautilus* 108: 42–47.

Steinke, T.J. 1986. Hydrologic Manipulation and Restoring Wetland Values: Pine Creek, Fairfield Connecticut. pp. 377–383. In: J.A. Kusler, M. L. Quammen, and G. Brooks (eds.), *National Wetland Symposium; Mitigation of Impacts and Losses.* Association of State Wetland Managers, New Orleans, LA.

Steinke, T.J. 1995a. *The self-regulating tidegate.* Atlantic Waterfowl Council, Delaware Department of Natural Resources and Environmental Control, Division of Fish and Wildlife, Dover, DE.

Steinke, T.J. 1995b. *Restoration of degraded salt marshes in Pine Creek, Fairfield, Connecticut.* Atlantic Waterfowl Council, Delaware Department of Natural Resources and Environmental Control, Division of Fish and Wildlife, Dover, DE.

Swetnam, T.W., and C.H. Baisan. 1996. Historical fire regime patterns in the southwestern United States since AD 1700. In: C.D. Allen (ed.), *Proceedings of the Second La Mesa Fire Symposium, March 29–30, 1994, Los Alamos, NM.* National Park Service, Washington, DC.

Tappeiner, J., et al. 1992. Managing stands for northern spotted owl habitat. In *Final draft recovery plan for the northern spotted owl.* Appendix K. U.S. Department of the Interior, Washington, DC.

Taylor, R. 1996. Mt. Trumbull ecosystem restoration project. pp. 74–78. In: Conference on adaptive ecosystem restoration and management. USDA Forest Service General Technical Report RM-GTR-278. Rocky Mountain Forest and Range Experiment Station, Fort Collins, CO.

Teal, J.M. 1986. The ecology of regularly flooded salt marshes of New England: A community profile. Biological Report 85(7.4). U.S. Fish and Wildlife Service, Washington, DC.

Thompson, S.P., H.W. Paerl, and M.C. Go. 1995. Seasonal patterns of denitrification in a natural and a restored salt marsh. *Estuaries,* in press.

Tiner, R. 1995. Phragmites — Controlling the all-too-common Common Reed. Massachusetts Wetlands Restoration Technical Notes. Wetlands Restoration and Banking Program, Technical Note Number 1: 1–3.

Tiner, R.W. 1984. *Wetlands of the United States: Current Status and Recent Trends.* National Wetlands Inventory, Fish and Wildlife Service, U.S. Department of the Interior, Washington, DC.

United Nations. 1992. Adoption of agreements on environment and development: The Rio Declaration on Environment and Development. UNEP.A/CONF.151/5/Rev.1. 13 June 1992.

U.S. Department of Agriculture and U.S. Department of Interior. 1994a. Final supplemental environmental impact statement on management of habitat for late-successional and old-growth forest related species within the range of the northern spotted owl. Portland, OR. 2 vol.

U.S. Department of Agriculture and U.S. Department of Interior. 1994b. Record of decision for amendments to Forest Service and Bureau of Land Management planning documents within the range of the northern spotted owl. Portland, OR.

U.S. Department of Agriculture, U.S. Department of the Interior, U.S. Department of Commerce, and Environmental

Protection Agency. 1993. *Forest Ecosystem Management: An Ecological, Economic, and Social Assessment.* Portland, OR.

Walters, C. 1987. *Adaptive Management of Renewable Resources.* Macmillan, New York.

Walters, C.J., and C.S. Holling. 1990. Large-scale management experiments and learning by doing. *Ecology* 71: 2060–2068.

Warren, R.S., and Niering, W.A. 1993. Vegetation change on a Northeast tidal marsh; interaction of sea-level rise and marsh accretion. *Ecology* 74: 96–103.

Warrick, R.A. 1993. Climate and sea level change: A synthesis. pp. 2–24. In: R.A. Warrick, E.M. Barrow, and T.M.L. Wigley (eds.), *Climate and Sea Level Changes Observations Projections and Implications.* Cambridge University Press.

Waters, R.F. 1995. Tidal wetland restoration on a dredge disposal site: Patterns of initial plant establishment, Mumford Cove, Groton, Connecticut. M.A. Thesis. Connecticut College, New London, CT.

Weaver, H. 1951. Fire as an ecological force in the southwestern pine forests. *Journal of Forestry* 57: 12–20.

Weaver, H. 1974. Effects of fire on temperate forests: Western United States. pp. 279–319. In: T.T. Kozlowski and C.E. Ahlgren (eds.), *Fire and Ecosystems.* Academic Press, New York.

White, A.S. 1985. Presettlement regeneration patterns in a southwestern ponderosa pine stand. Ecology 66(2): 589–594.

Wiley, J. 1989. Phenomena, comment and notes. A new ecological society will devote itself to the toughest mission of all: the restoration of land left for dead. *Smithsonian* 19(12): 32–33.

Williams, J.T., R.G. Schmidt, R.A. Norum, P.N. Omi, and R.G. Lee. 1993. Fire-related considerations and strategies in support of ecosystem management. USDA Forest Service Staffing Paper, Washington, DC.

Wilson, E.O. 1992. *The Diversity of Life.* W.W. Norton and Company, New York. 424 p.

## THE AUTHORS

**Wallace Covington**
*College of Ecosystem Science and Management*
*Northern Arizona University*
*Flagstaff, AZ 86001*

**William A. Niering**
*Department of Biology*
*Connecticut College*
*New London, CT 06320*

**Ed Starkey**
*USDI National Biological Service*
*3200 Jefferson Way*
*Corvallis, OR 97331*

**Joan Walker**
*USDA Forest Service*
*Southern Research Station*
*Clemson University*
*Clemson SC 29634*

# Ecosystem Restoration: A Manager's Perspective

James G. Kenna, Gilpin R. Robinson, Jr., Bill Pell,
Michael A. Thompson, and Joe McNeel

**Key questions addressed in this chapter**

- *Concepts which underlie design and implementation of restoration action*
- *Relevant considerations when making restoration decisions*
- *Options and techniques that have been applied in restoration efforts*
- *Application of restoration concepts, options and techniques in four different types of settings*
- *Characteristics common to successful restoration efforts*

**Keywords: Rehabilitation, reclamation, wetlands, arid lands, riparian areas, degraded lands**

*Ecosystem management, at root, is an invitation, a call to restorative action.* (Ed Grumbine, 1994)

*The significant problems we face today cannot be solved at the same level of thinking we were at when we created them.* (Albert Einstein)

## 1 INTRODUCTION

Elements of ecological restoration underlie much of what we think of as ecosystem management, and restoration projects on federal lands represent some of the most exciting, challenging, and convincing demonstrations of applied ecosystem management. The Society for Ecological Restoration defined restoration as "the process of reestablishing to the extent possible the structure, function and integrity of indigenous ecosystems and the sustaining habitats that they provide" (Clewell and Covington 1995). Most managers have limited interest in distinctions between restoration and rehabilitation, reclamation, maintenance, or prevention of degradation. Managers focus on problem-solving where current conditions limit ecosystem functions, treating restoration as a broad array of possible actions, which include rehabilitation and reclamation. As a result, a variety of reference conditions is used to define the direction restoration efforts take.

Restoration is a subject of considerable breadth. Restoration projects are occurring across the country, on all types of federal land (in mixed ownerships, as well), and at scales ranging from local habitats to watersheds, landscapes, and bioregions. The level and scope of activities has increased dramatically in the last 10 years, but federal agency involvement in ecological restoration is not new. It is evolving. Restoration has been, and continues to be, a vital part of federal land management.

Integrated actions organized around ecological processes and functions underlie an increasing number of agency efforts to restore, improve or maintain natural resource conditions. Emphasis on ecosystem function requires integrating multiple perspectives and addressing complex physical, biological, social and economic interactions. Additional complexity is added by the scale at which issues like species recovery are addressed and by the accelerating pace of change. Examples of species recovery issues affecting large-scale landscapes include high profile species such as salmon, wolves, northern spotted owls, marbled murrelets, and red-cockaded woodpeckers.

Although much has already been described, understanding the relationships within ecological communities is constantly being improved. As more is learned, new biological and physical linkages are highlighted, thus requiring more study. But land managers must continue to make decisions about the design and direction of activities on public lands in the context of a constantly developing body of knowledge. It is a continuous effort to balance what is known, what could be studied, and what can be done, with the goal of improving ecosystem functions. Baseline data about ecological functions, studies of comparable situations and opportunities for experimentation are all necessary considerations in developing practical solutions. But effective solutions must also consider project-specific factors such as cost, risks, conservation values, potential effects, stakeholder interests and landownership patterns.

The purpose of this chapter is to provide a practical reference for land managers to aid decision-making and spur innovation and experimentation in the field of ecological restoration. Numerous restoration efforts were reviewed to develop an overview of successful projects, barriers to positive results, and promising possibilities. Cross-cutting concepts such as historical context and restoration decision factors are presented first. Then, to illustrate current practices and experience four scenarios are presented:

- Decline in terrestrial systems or conditions
- Impaired stream or riparian function
- Impaired wetland function
- Lands with few biological legacies

Scenarios were used, rather than selected case studies alone, in order to broaden the treatment of applied restoration. These four scenarios were selected based on their broad applicability and importance in public land settings. This chapter represents a search for common approaches and useful findings. It is not a scientific or a perfectly representative sample of all restoration activities or scenarios. Case studies and scenarios are used to illustrate ongoing restoration activity and applied concepts.

### 1.2 Overview of Ecosystem Restoration Approaches

Ecosystem restoration is an exercise in complex problem-solving with an emphasis on the processes and functions of the system. Although restoration goals and objectives are generally based on site conditions, management actions are necessarily designed around land uses or interventions. Potential restoration actions can be grouped into three different types of approaches: structural treatments, management of land uses, and biological intervention. Each approach represents a group of techniques, but the different approaches are frequently combined. For example, grazing use is frequently modified to correspond to prescribed

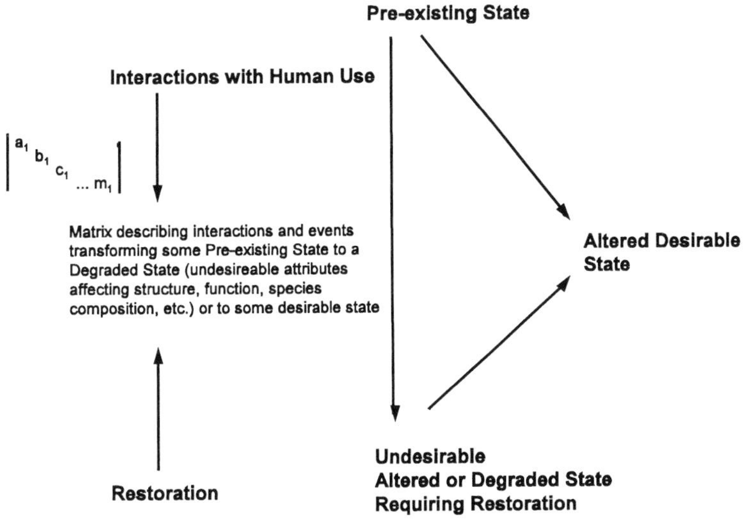


Fig. 1. Conceptual model illustrating the relationships between human use and restoration strategies (modified from Dyer and Ishwaran 1992).

fire activities, seeding, and plant responses. The approaches are based on the type of management action and discussed separately to improve the clarity of the review.

The structural approach includes modification of the physical attributes of the system, or their arrangement and distribution, toward a more desirable state. Examples would be log and rock placements in streams to restore aquatic habitat diversity or introduction of prescribed fire to change plant community structure.

The land use approach modifies the distribution, timing, intensity or duration of uses affecting the landscape. Examples would be changes in domestic livestock grazing or timber practices in sensitive riparian habitats.

The biological intervention approach changes the species composition of an ecosystem, controlling undesirable species or introducing and promoting desirable ones. Examples would be controlling leafy spurge in native grasslands or reintroducing wolves.

The three different approaches are common components of restoration projects, occurring in a wide variety of settings and combinations. Although none of the approaches is applied entirely in isolation, one frequently dominates. Each of the approaches is discussed below in the context of different settings, or scenarios, listed above.

The factors that caused, or are causing, degradation often provide insight into the approach to take for restoration. The factors may include current or past: infrastructure (such as drainage ditches, dikes, dams, levees, roads, culverts, bridges), land uses (grazing, mining, timber harvest, residential or industrial development, hunting, fishing, camping, boating), land-use practices (how and when the land uses are conducted) or management practices (such as fire suppression).

Ecosystems exist at a variety of levels of productivity, integrity, sustainability, and resilience. Although some may be in a relatively undisturbed condition, others require some degree of restoration to improve the functions of ecological processes. If some pre-existing state (a reference condition) is subjected to a complex array of interactions with humans over an extended period of time, some change would be expected. Figure 1 provides a simplified conceptual model highlighting the complex interactions (the matrix) between human uses and ecosystems that have affected all landscapes to some degree. Restoration attempts to affect the interactions and move the system toward a desired set of conditions, often referenced (1) to the range of states that historically occurred on the site or (2) to thresholds between acceptable and unacceptable ecological function. Restoration does not duplicate the pre-existing state (Eddleman 1997).

## 1.2 Historical Overview of Ecological Restoration

Problem-solving with the intent to restore structure and function to ecological systems has a long history. Most of the early efforts were based on trial and error (Odum 1972, Covington et al. this volume). Odum cited the work of George Perkins Marsh in 1864 analyzing the causes of decline in ancient civilizations as one of the first "ecosystematic" views of human interaction with nature. That perspective was developing in a parallel fashion in essays in German, English and Russian through the later 19th century (Odum 1972).

By the 20th century, two general approaches provided most of the foundation for both ecological study and restoration. The first, a holistic approach, focuses on whole systems and works with processes, functions and results. In 1972, Odum illustrated the holistic ecosystem concept as a "black box" with its inputs and outputs. From a manager's perspective in 1997, inputs might include land uses and disturbances acting on an ecosystem. Outputs might include the characteristics and trends in the system, observed through monitoring. The holistic approach acknowledges that the complexity, number, and dynamic nature of interactions and components in an ecosystem prevent complete description of entire systems and all their components.

The holistic approach seeks to understand the dynamics of processes, functions, interactions, and relationships for the purpose of determining how the "black box" will respond. A holistic approach to restoration tends to focus on what might improve characteristics of the "black box" (e.g., ecological site, community, watershed, province), expecting individual species to benefit or react in their own way.

The second approach is more component-based and attempts to decipher and understand specific relationships and interactions among each of the key components of the ecosystem. This commonly leads to more finite studies of individual species or other system components, as they are presented with the array land uses and disturbances represented in an ecosystem at any given time. The component approach seeks to understand the parts and then try "to build up the whole from them" (Forbes from Odum 1972). A component-based approach to restoration tends to focus on improving an individual population, expecting the system to be healthier if the identified population responds favorably.

Although the holistic approach emphasizes physical and community processes, functions, and attributes, as a foundation for species, the component approach (starting with parts of an ecosystem) is more individual-species-oriented and treats ecosystems as a composite of individual parts. Friction between proponents of the different perspectives continues. But managers are likely to argue that neither needs to be universally applied to the exclusion of the other. In fact, a holistic approach provides an excellent framework for applying and interpreting component-based information, particularly on larger scale restoration issues and problems.

Humans have been managing a relationship with land for centuries. In the last century, the recognition of a need to restore and maintain ecosystems was most commonly embodied in preservation strategies such as the establishment of parks, preserves, and reserves (MacCleery 1994). Although this strategy has value and is still applied (e.g., new wilderness areas under the Wilderness Act of 1964), it obviously cannot be applied everywhere. Land uses continue in parks and preserves. Parks and preserves do not occur in isolation from their surroundings and they generally include some lands where ecological functions are impaired. Thus, even parks and preserves need ecological restoration. It was inevitable that concepts would emerge to manage land uses and apply treatments to sustain or restore the health of lands and waters.

Ecological restoration corresponds closely to what Odum called "applied ecology." Applied ecology, as a biological and physical science, probably has no definitive beginning. But if tracked prior to the 1960s, it largely consisted of managing the components of the ecosystem (Odum 1972). Where the objective was more deer, something was changed on the deer winter range, and deer were counted to measure the results. If the concern was flooding or erosion, a structure would be built to control it. Some efforts resulted in a restoration of some improved ecological function or component (even if the objectives were not focused on the ecosystem as a whole). Successful species recoveries in the 20th century include whitetail deer, elk, wild turkey, and pronghorn antelope.

But the results of some problem-solving attempts were not as positive. Some streams were channelized that were meant to meander. Some forests were pruned and manicured, reducing their biological diversity. Fire was suppressed and exotic species were introduced into native rangelands, often with good intentions to produce forage or stabilize streambanks. It should be no surprise that projects designed with narrow objectives or with a limited appreciation for ecosystem complexity could have unintended consequences.

By the 1970s, natural resource management became more integrated and holistic with the advent of interdisciplinary teams to analyze land uses, management actions and impacts, propelled in part by requirements of the National Environmental Policy Act (see also MacCleery and LeMaster, this volume for overview of natural resource management history). By the 1990s, this need for increased integration of perspectives and expertise, using partnerships to develop and test solutions to complex ecological problems, is well recognized. But one agency with considerable biological expertise related to individual species may not have experience in implementing management actions or integrating them, in a sustainable way, with the social and economic systems that are affected.

Many state and federal agencies have missions oriented around individual land uses or selected ecosystem components (fisheries, fish and wildlife, forestry, recreation, agriculture, economic development). Even within agencies with a more complete cross-section of experts and perspectives, individual "programs" are centered around single components of the ecosystem. This same potential for polarity is mirrored in public discussion of natural resource management. The result is sometimes a complex array of competing positions, objectives, and perspectives that retard or block attempts at restoration action (Apfelbaum and Chapman 1994).

Thus, the shift to ecosystem management and restoration is in its infancy. To implement or improve implementation of ecosystem approaches on federally managed public land, the Interagency Ecosystem Management Task Force (1995) listed 31 recommendations for improvements in federal agency coordination, partnership, communication with the public, resource allocation and management, the role of science, and information and data management.

Yet the reasons for restoration can be persuasive: aesthetics, benefits to humans and the health of the systems themselves (Apfelbaum and Chapman 1994). In many respects, progress in restoration hinges on our abilities to learn and understand, to align multiple interests and needs with improving ecosystem health, and to gain agreement on how to begin to act.

## 2 RESTORATION DESIGN

Successful ecosystem restoration efforts take a landscape perspective, gather and assess adequate information, set clear goals and objectives, and monitor to assess results.

### 2.1 Landscape Perspective

A landscape or regional perspective is essential for effective restoration of ecosystem functions (Eddleman 1997, Apfelbaum and Chapman 1994). The reason is obvious. Plant-animal communities in an ecosystem are interlinked and ecosystems are dynamic (Eddleman 1997). Examples include: (1) fire is an important disturbance regime affecting vegetation type, distribution and condition over large areas; (2) soil types, water occurrence, and topography vary, affecting vegetation and wildlife movement; and (3) many fish and wildlife species range over a large area utilizing a variety of habitats and habitat conditions at various times during the year (see also Concannon et al., Gosz et al., this volume).

Ecological classifications and regional data sets provide the framework for understanding disturbances, plant communities, species occurrence, flow events, stream channel types, and susceptibility to erosion (see Grossman et al., and Carpenter et al., this volume). Regional information can then be organized to provide context and integration for restoration goals and objectives across large landscapes. Regional goals and objectives, in turn, create a management direction that can be stepped down to meet both local and regional needs using more site-specific knowledge (Interior Columbia Basin Ecosystem Management Project, Draft EIS, Appendix 3-1). Actions for some species-specific restoration efforts can be very locally focused. For example, actions related to a rare plant species may be applicable only to very specific soil, elevation, aspect, and plant community characteristics. But even then, the larger-scale review contributes to understanding habitat relationships.

The technical tools available to assess ecosystems at the landscape scale include techniques for ecological site inventory and classification. These are generally well-developed for the major federal land management agencies, although tools to better integrate data at larger scales are still developing. Efforts have been initiated to develop national and international classification systems (Eddleman 1997). A number of classification, inventory and monitoring methods have been compiled and summarized by the National Research Council (1994).

Assessing conditions, trends, opportunities, and risks at a regional level is difficult. Quigley et al. (1996) identified some valuable lessons related to regional assessments based on their experience on the Science team for the Interior Columbia Basin Ecosystem Management Project:

- Science needs to provide the types of information that answer questions that can improve decisions, recognizing politics and budget are deciding factors in implementation.
- Involved scientists must be integrative and comfortable with broad policy issues and concerns, rather than functional.
- Better balance is needed among timelines, data quantity and quality, and emerging decision issues. Timely information has more impact than presenting detailed data later.
- Goals must be identified early.
- More cause and effect data are needed, rather than just descriptive material.

Much information and experience exists for restoration. In some cases, generally available sources of information, in combination with existing site data, may be enough for analysis of local needs. In other cases, sources would also need to include local publications and contacts to describe adequately local conditions and experiences, thereby improving the probability of success. Models that simulate landscape changes can be useful analytical tools to relate complex sets of data, but they have limitations related to scale, assumptions used, data quality, and number of simulations (Johnson et al. 1996).

In response to landscape level issues, many recent restoration projects have expanded from localized action to consider strategies, effects, and relationships at a watershed scale. For example, high sediment loads on the Bad River in South Dakota affect ice buildups, dam release capacity, electrical generation, and fishing success for 20 to 30 miles along the Missouri River. Addressing bare ground, accelerated erosion, and poor infiltration along Bad River has the potential to reduce the average sediment production of 1,418 tons per square mile (Platts 1990). The linkage between streams and management of the surrounding uplands is illustrated by the level of sediment transport into the riparian zone (Moore and Flaherty 1996), the role riparian areas play in plant dispersal and wildlife movement (Gregory et al. 1991), and the relationship of activities and conditions in the watershed to hydrologic regimes (Barrett et al. 1991).

Disturbance factors, ecological resiliency, and the suite of ecological processes necessary for ecosystem functions all affect strategies for restoration. Success is generally tied to compatibility of the restoration projects or strategies with the complex interactions of soils, vegetation and water, often at a watershed scale (Eddleman 1997). To understand the physical, chemical, and biological attributes of a site or system practitioners in restoration must also have the ability to analyze diverse sets of data (Hunter 1991).

## 2.2 Setting Goals and Objectives

Setting restoration goals and objectives often leads to a debate over what results should be expected, and how fast they should come. Most successful projects sampled focused on improving trends in the physical, chemical and biological attributes of a site or system. Goals and objectives are often expressed in terms of desired conditions or trends. The conditions selected are intended to gauge improvement in ecological function.

Determining desired conditions and trends is a critical challenge for restoration. The U.S. and Canadian National Park Services routinely target "pre-Columbian settlement" (pre-European) conditions as their restoration objective. Other agencies and organizations target conditions and processes that fall within a range of variability (natural, historic or "reference"), or they aim simply to improve certain conditions or control invasive species and release "natives." The second approach recognizes that variations in ecological conditions are normal and focuses more closely on ecological functions and capacity than on conditions at a point in time. Improvement, a positive direction in the trend, is then generally assessed against reference conditions.

Reference conditions can be estimated in various ways by examining similar areas considered "pristine"; by searching out historical records (Hindley 1996); by examining soil and plant relationships and patterns; by examining the fossil record for the area; by examining adjacent vegetation for signs of flooding and fire; or by talking with local residents who may have some historical knowledge of the area or by relating conditions at multiple sites to a level of ecological function (see also Covington et al. this volume).

Objectives are usually based on practical considerations. What is the distribution of existing conditions across the landscape? How much of a given landscape can be reasonably treated to restore desired conditions over a reasonable length of time? What effects, if any, will such treatments have on existing human uses, jobs, and quality of life? On the Ouachita National Forest, for example, open stands of shortleaf pine and prairie grasses and forbs potentially could cover one million acres of public land. The management plan, however, establishes a pragmatic objective to treat about one-tenth of the area, recognizing both the challenge of sharply increasing prescribed burning and public uncertainty about the proposed changes. As managers and citizens work through these issues, the restoration objective may need to be adjusted (USDA Forest Service 1994, Henderson and Hendrick 1991).

It is possible to set unrealistic or overly rigid restoration objectives. Human ability to predict or control many factors that affect rates of recovery (e.g., precipitation from year to year) is limited. Sometimes the drive to force the rate of ecosystem change by prescribing rigid standards or objectives diverts attention from sustainable solutions that might allow human uses to coexist with properly functioning ecosystems. For example, a preoccupation with the presence, absence or amount of use can undermine creative techniques that focus on timing or methods.

## 2.3 Planning, Assessment, and Evaluation

Planning provides the analytical framework for restoration. It develops alternatives, specifies actions, identifies necessary resources, sets monitoring to provide feedback on effects, and plans for maintaining or adapting in the future. Planning, assessment, and evaluation processes generally draw from the same funding source as implementation for restoration actions across a large area. With limited funds, the cost of these processes affects the level and rate of restoration action. The tradeoff is between (1) the level of knowledge these processes create to abate risk, build support, and add value to restoration designs and (2) their cost, length, and complexity. Planning, assessment, and evaluation are necessary to making informed decisions, but time and effort for these analysis processes must be balanced with restoration action.

Restoration efforts must assess conditions and options with an interdisciplinary team that includes landowners, a variety of specialists, and members from interested organizations. A team facilitates consideration of multiple viewpoints, sources of expertise, and resource values from the start. Participants and proponents then take some responsibility for the outcome. If the team functions well, multiple perspectives and expertise will be integrated in development of the proposed solution. However, care should be taken to evaluate sources of information and expertise the team uses for any inherent biases or other limitations. It is important to distinguish underlying values from facts (Adams and Hairston 1994).

The processes of assessment and evaluation draw on research and monitoring to characterize conditions and trends, so management actions can be adjusted. Although pre- and post-project monitoring is essential to evaluate success (Hunter 1991), it is sometimes desirable to combine a project with research studies to incorporate more complex questions at the watershed, riparian corridor, and community scales (Chambers 1994). Site selection for research should reflect (1) common occurrence (for broad applicability), (2) ability to compare different responses, (3) potential to exhibit varying recovery potentials, (4) community composition and structure, and (5) system or species vulnerability (modified from Chambers 1994, Crow et al. 1994). It may also be relevant to consider research when the information could be broadly useful (Manning et al 1984, rooting characteristics of meadow species), or where restoration of some species has been difficult or inconsistent (Matisse 1994, bitterbrush).

Data collection on each project cannot meet standards for evaluation by research scientists because of cost. Where research is not appropriate or justified, some level of monitoring is necessary. But everything cannot be monitored to the same level without draining valuable resources from the effort to take restoration actions across large landscapes. The translation from ideal to realistic requires tradeoffs concerning when, where, and what types of planning, research, and monitoring are appropriate (see Fig. 2). A sampling

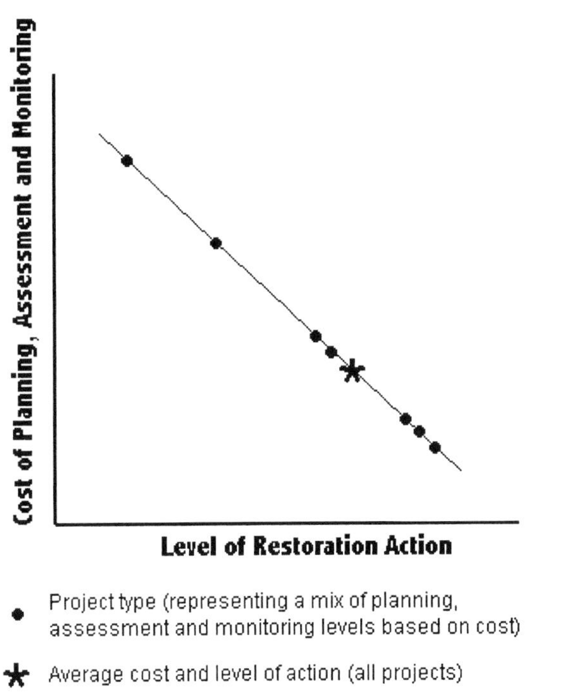


Under the situation illustrated in the graph, different levels of planning, assessment, and monitoring are applied across the variety of projects underway. Research and more complex planning and assessment processes are directed to where they are most appropriate. The simpler, lower cost projects bring the average cost down and allow more restoration action to occur.

Fig. 2. Conceptual relationships between planning, research, monitoring, and implementation under a constant level of available funding for projects across a landscape.

of multiple projects that are similar may be one option to stretch monitoring funds. Monitoring priorities may also be affected by whether the project is well supported by literature and experience, or by the values at risk, at either the project or regional level.

With the practical limitations on data gathering, maintenance, use, and application, it is essential to use data effectively (Callaham 1990, Crow et al. 1994). In many cases, it may be most effective to direct the restoration action and data gathering toward experimentation (adaptive management). An adaptive management approach requires recognizing a project may not be successful. With risk must come commitment to an incremental approach that allows mid-course corrections (Hunter 1991). Restoration through adaptive management must continue because it is the most fertile area for rapid innovation.

Factors that will almost certainly influence a restoration plan are what is technically, physically, and economically possible, as well as what is socially acceptable. Technical solutions or the physical means to carry them out are not usually the primary problem.

But economics are almost always a significant factor in restoration design. Depending on the size of the effort, costs can be astronomical. For example, restoring the 250-acre Ballona wetland (near Los Angeles airport) is expected to cost $50 million, in part because a road through the marsh would have to be elevated to restore tidal flows (National Research Council 1992).

## 3 DECISION FACTORS FOR RESTORATION

While a systematic approach aids in assessing reference conditions, current conditions, and restoration alternatives, it is perhaps even more essential for the decision selecting restoration actions and monitoring strategies. Restoration decisions involve choices about when, where, and how to commit scarce amounts of time, effort, and money. Checklists based on Cairns (1983) and Apfelbaum and Chapman (1994) are included as Appendix A as an aid. This review of restoration projects suggests that, when making restoration decisions, considerations should include level of ecological function, cost-effectiveness, conservation value, potential effects, relationship to other actions and trends, and risk management. However, none of these factors should be considered separately.

### 3.1 Existing Level of Ecological Functions

Although an understanding of conditions is necessary, actively restoring ecosystem health requires focusing primarily on processes and functions, not on conditions (Quigley et al. 1996, Eddleman 1997). Conditions, particularly at broad scales, can be highly variable (Quigley et al. 1996). Many systems have also been altered by human settlement (or other factors) to a degree that options to achieve pre-settlement conditions are not practical.

The level of ecological functions is important because it provides insights into: (1) necessary investments to restore ecological processes, (2) possible responses to restoration measures, and (3) the priority the restoration project should receive. In some cases, systems are altered to a point that a substantial investment of time, effort, and money would be necessary, time effort and money that could be better used in other locations (Pyke and Borman 1993). Conversely, restoration action in systems that are functioning, but at risk of significant decline, could prevent decline for a modest investment.

### 3.2 Potential Effects

A basic consideration is where to work first. In most cases, recovery begins at the top of the watershed (Platts 1990). Watershed scale analysis has been useful to direct restoration projects, provided it is simple and adaptive (Natural Resources Law Center 1996). It is often advantageous to consider the current balance of successional stages represented and give high priority to restoring those which are underrepresented, rather than focusing on some historical stage. Watershed function and vegetative structure (distribution, arrangement) are also frequently key factors. A simple annual review of projects can provide the basis for evaluating potential effects when setting work priorities.

Following Odum's concepts of ecological "outputs," potential effects relate closely to the level of ecological functions. The impact of a restoration investment where a system is at risk of decline can be significantly greater than where a system is already improving or has declined to the point of being non-functional. Natural events or other management actions may be necessary to stabilize a non-functional system before any significant restoration occurs. Increments of restoration can occur in slow, steady increments every year, or may tie closely to climatic cycles or events occurring at longer intervals (see Fig. 3).

Potential effects also relate closely to how directly restoration action addresses the causes of system decline and limiting factors to recovery. Restoration will be ineffective if the causes of degradation are allowed to continue, regardless of the physical or biological investments into the system (i.e., treat the cause, not the symptoms).

The cumulative effects of activities within a watershed are the result of actions on all the ownerships (Reiter and Beschta 1995). Thus, landownership patterns and participation are factors in the expected results. Developing a technically "right" approach to restoration, without the cooperation necessary to implement improvements across ownerships, accomplishes little. Lack of landowner support has prevented or delayed riparian improvement in several cases reviewed by Platts (1990).

Potential effects are not just biological, chemical, or physical. Economic effects are also playing an increasing role in restoration. Domestic livestock grazing is one example where economic outputs and ecological results can be compatible. Improved management practices can yield benefits such as increases in the quantity and quality of forage, increased animal production, control of poisonous plants, increased water yield, control of soil loss, and reduced conflict (Johnson 1992). U.S. Government Accounting Office (USGAO 1988) noted an "...increasing number of ranchers are coming to accept the benefits healthy riparian areas provide their ranching operations." Successful efforts consider both the ranching operation and the bio-

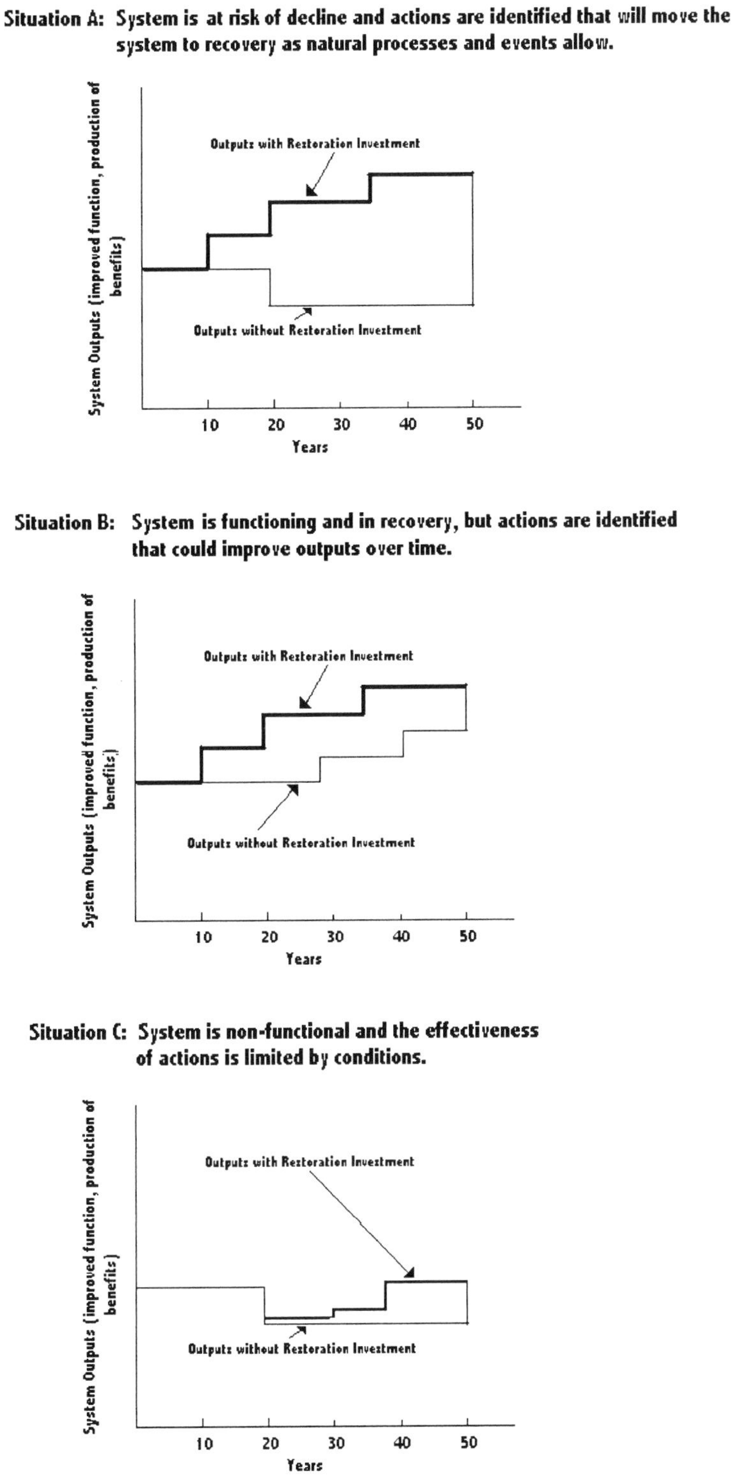


Fig. 3. Comparative ecosystem outputs in three situations, with and without restoration investment over time.

logical/physical factors using an interdisciplinary approach, with wildlife, fisheries, hydrology, range conservation, and soils specialists (USGAO 1988). Riparian objectives clearly can be attained in many cases without loss of livestock forage, diet quality, or weight gain (Moore and Flaherty 1996).

Economic benefits can also be related to other uses, such as recreational fishing or reservoir water quality. Recreation user purchases of licenses, tackle, gas, and food support local communities (Hunter 1991). Even more generally, the public at large benefits when 118,000 cubic feet of sediment are kept out of a reservoir downstream of Sheep Creek in Montana (USGAO 1988).

## 3.3 Risk Assessment

Risk assessment is an a analysis of the likelihood of negative outcomes from various management options, including both action and inaction (see Haynes and Cleaves, Volume III). It can include loss or decline in a species or community. Adams and Hairston (1994) illustrated the tradeoff as shown in Fig. 4.

Risk must be managed at different levels and scales, particularly if flexibility to respond to local conditions and needs is to be retained (Haynes et al. 1996). Assessments at regional scales require "averaging" across a broad array of site conditions (ibid). Some averaging is necessary to assess regional conditions and trends to provide an overall regional direction. For example, assessing and managing risks to a threatened or endangered species that requires broad connectivity (e.g., salmon) must consider conditions and land uses across at least a basin.

Most risk issues have implications at each of the regional, subregional, landscape, and site scales (Haynes et al. 1996). For example, assessing fuel loading in relationship to wildlife habitats at the regional level might lead to a very different conclusion than at a site or landscape level (T.M. Quigley, personal communication). It is not that either assessment is more correct, or more necessary. They simply represent the issue at different scales and both are relevant to decisions.

## 3.4 Conservation Values

Regardless of the degree of degradation a system exhibits, restoration frequently is the best strategy to integrate the multiple ecological, economic, cultural, and social values people place on a particular ecosystem or landscape. But values placed on ecosystem attributes in a diverse society are variable. Some species, communities, or ecological functions may be valued more highly because of their rarity, sensitivity, social role, or economic importance. When the ascribed values are high and shared by a group (e.g., key ecosystem components to ecologists, deer to hunters, salmon to Northwest Indian Tribes, or watershed function in a municipal watershed), the level of action and the priority for action may be greater (MacCleery and LeMaster, this volume).

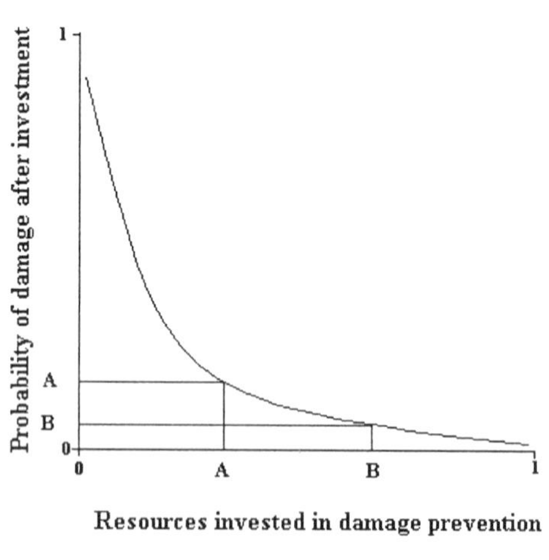


Fig. 4. Examples of Risk Assessment. Note that investment A reduces probable damage about 80 percent but twice the investment, case B, further reduces probable damage only a small amount.

Conservation values assign importance to protection of critical species and ecosystems by evaluating the representation and condition of all vegetation types under some kind of conservation strategy (Burley 1988). But they also have social and economic dimensions that are essential to implementing and sustaining restoration projects. The social and economic values integrate such human needs as wood fiber, food (agriculture, animal forage), road access, and recreation (Interagency Ecosystem Management Task Force 1995, cf. Daily et al. 1997).

Trust in a restoration project develops with sufficiently broad consensus among affected communities of interest, including the scientists, conservationists, and landowners. However, some people may have difficulty visualizing the improvement that is possible, arguing solely on philosophical concepts of what is "natural" or who is "right." For success, all those involved must agree that objective information about the restoration project and its potential effects provide a better basis for dialogue than individual positions. Any party can initiate a vision for restoration, but the intent of the dialogue must be to identify common goals, with acceptable consequences (Interagency Ecosystem Management Task Force 1995).

Some confusion about the complex relationships between proposed restoration actions and a constantly changing landscape is probably inevitable. But neither objective proof of degradation nor assignment of blame for current conditions are necessary prerequisites for restoration action. Success is often promoted by fostering basic understanding of proper functioning condition, disturbance regimes and patch dynamics when explaining or advocating restoration objectives.

Working groups with diverse representation are common, and potentially powerful, tools used to facilitate the necessary communication and understanding (see Yaffee, Volume III). Yet even when diverse interests commit to work together to solve problems, events sometimes occur that break any trust which develops (Natural Resources Law Center 1996). Success requires establishing, and constantly re-establishing, open communication, willingness to change, and willingness to deal with side agendas (Callaham 1990).

Many current working groups are watershed issue forums promoting restoration through communication, education, and conferences (Natural Resources Law Center 1996). At least partially as a result, increasing numbers of restoration projects involve multiple cooperators. The involvement of multiple agencies in Resource Advisory Councils and Provincial Advisory Committees in the West (recently including the Dakotas) in this kind of public dialogue is certainly a promising development. In Oregon, recent federal legislation has recognized public/private partnerships for restoration in two basins.

Odum (1997) suggested that restoration is most successful when four key groups work together in a coordinated manner. These include citizens groups, governmental agencies (local, state, and federal), scientists, and business interests. He also suggested that if any of the groups is not strongly involved, restoration projects may not achieve their goals.

Controversy, even though it can be a barrier to restoration action, sometimes brings restoration issues into public focus. Recent examples include the spotted owl, cutthroat trout, bull trout and anadromous fish issues in the West. Even the perceived risk of a lawsuit or controversy may sometimes affect restoration strategies or promote change (Platts 1990). For example, corridor fencing in a riparian area can be more acceptable to a rancher if it facilitates a "no effect" determination under Section 7 of the Threatened and Endangered Species Act (Munhall 1996). Although progress has been made, much could yet be done to align incentives and management processes to promote consensus on conservation values (Western Governors' Association 1997).

## 3.5 Relationship to Other Actions and Trends

No restoration project can be considered in isolation from the land uses, management projects or resource trends around it (Eddleman 1997). For example, implementing a structural restoration action (check dams, for example) without correcting land uses that are also contributing causes is unlikely to be successful over the long term. Elmore and Kaufman (1994) provide a good conceptual model relating the human caused stresses and natural stresses to the stability and resiliency of a system (as seen in Fig. 5).

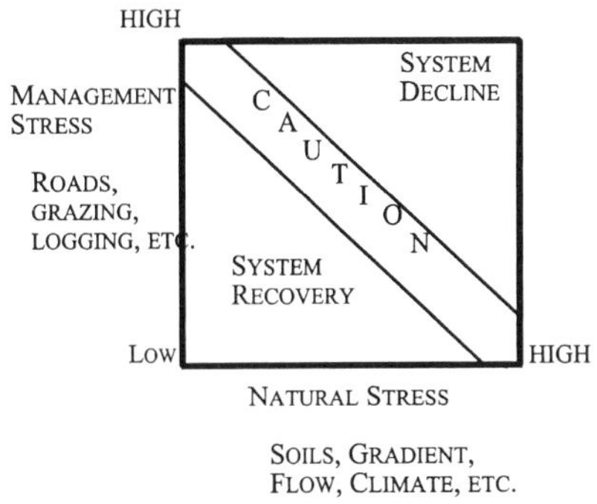


Fig. 5. Stress and stability in ecosystems.

Although the model in Fig. 5 was developed for riparian systems, the concepts apply to restoration in all systems. Systems which are under high levels of natural stress (e.g., a steep stream with highly erodible banks and high flows) can be pushed into decline by moderate levels of human use. Conversely, systems which are stable and resilient can tolerate properly managed human use while continuing to recover. In all cases, restoration action must consider other uses and trends.

Continued monitoring may also identify new limiting factors that require modifications (Hunter 1991). But in natural systems, results are not always immediate, so frequent changes in management strategies may not be appropriate.

## 3.6 Cost-Effectiveness

Assessing cost-effectiveness relates all of the other factors to cost. The intent is to spend where and when the funding will yield the best results or the most benefits. The effectiveness of the same level expenditure varies depending on what it does, where it takes place and the timing of the action (see Fig. 6).

| Potential for Results | | |
|---|---|---|
| | HIGH RESULTS/LOW COST<br><br>Simpler actions<br>Consensus to implement<br>Success in similar applications<br>Frequently at smaller scales<br>Planning costs are low<br>More actions with similar funds<br>Monitoring may be simpler | HIGH RESULTS/HIGH COST<br><br>More complex actions<br>Consensus must be obtained<br>More uncertainty of outcomes<br>Frequently involve larger scales<br>Planning costs are higher<br>Fewer actions with similar funds<br>Monitoring more costly |
| | LOW RESULTS<br><br>System is not ready to respond to the restoration action<br>Causes and limiting factors are incorrectly diagnosed<br>Costs to assess, plan or monitor are out of balance with action<br>Implementation does not meet expectations<br>Lack of consensus prevents or retards necessary action<br>Objectives are set with inadequate or inaccurate information (can be good data but at the wrong scale) | |
| | **Cost to Assess, Plan, Implement and Monitor** | |

Fig. 6. Samples of how cost effectiveness can integrate other decision factors with cost.

Cost and the frequency of intervention or maintenance are also constraints requiring consideration. For example, neutralizing acidic inputs to the Cranberry River in West Virginia requires trips over snow once a week in the winter using 4×4 trucks or snowmobiles at a cost of $80,000 per year (Hunter 1991). These recurring costs must then be weighed against the benefits from the project and the other restoration opportunities foregone to fund the effort. Although not every case involves costs of this magnitude, many do involve the need for constant maintenance to ensure structures function properly or fences are maintained (Hunter 1991).

## 4 APPROACHES TO TREATING TERRESTRIAL SYSTEMS AT RISK

*Scenario: Whole ecosystems have been altered, affecting the representation and sustainability of plant and animal communities within a region or landscape.*

Terrestrial ecosystems are distinct communities of plants, wildlife, insects, and other organisms present in various regions. As presented by Berger (1988), terrestrial systems can include agricultural lands, barrens, coastal areas, prairies, rangelands, and temperate forests. This chapter focuses on terrestrial systems not converted for agricultural use. Terrestrial ecosystems of concern include those most commonly occurring on federally administered public lands such as forests, drylands, prairies and rangelands.

Recognition of the landscape changes that have occurred during the 19th and 20th centuries have heightened concern over the need for ecosystem restoration in the 1990s. Broad interpretations of current conditions fuel concerns over:

- Radical change in fire-maintained ecosystems,
- Invasive plants and animals reducing biodiversity, and
- Widespread habitat change.

As mentioned above, restoration requires clear goals. The Buck Creek Serpentine Woodland Restoration project of the National Forests in North Carolina provides a good example. The restoration goals set for this project include: (1) maintain a physiognomic complex of forest, woodland, and grass-dominated areas while increasing the woodland community coverage to 50 percent of the landscape; (2) reduce frequency of trees greater than 25 cm (diameter at breast height) by 50 percent and trees 2.5–25 cm by 30 percent; (3) reduce shrub coverage by 25–50 percent, resulting in at least 50 percent of the acreage having shrub cover less than 15 percent and increasing the coverage of grasses and forbs; and (4) reduce accumulated leaf/grass litter by 30 percent (Simon, personal communication).

### 4.1 Management Options for Terrestrial Restoration

For this scenario, the options for restoration of terrestrial systems are grouped into the three different approaches discussed above: structural treatments, management of land uses, and biological intervention.

#### 4.1.1 Structural Treatments

One alternative to restoring large scale systems is to alter the structure and composition of the vegetation. Often structural changes in terrestrial systems are achieved by reintroducing processes such as fire. Changes in vegetation structure are important to spatial relationships and physical processes. But structure also directly affects wildlife species abundance and diversity because most wildlife respond more to structure than to species composition (Elmore 1984, Puchy and Marshall 1993).

Some of the most ambitious restoration projects on federal lands are those focusing on re-establishing fire as a key ecosystem process. Examples include efforts to restore longleaf pine-wiregrass, ponderosa pine, quaking aspen, shortleaf pine-bluestem, tallgrass–midgrass–shortgrass prairies, Midwestern oak savannas, and many other once prominent ecosystems (see Case Study 1). In some cases, these efforts dovetail with or complement efforts to increase the future representation of old growth; in others, the efforts are independent.

Upland systems in which fire is to be restored typically represented major elements in pre-suppression landscape on the order of hundreds of thousands to millions of hectares. Restoration work in these systems usually is not a matter of renewing previously existing conditions on a hectare-by-hectare or highly site-specific basis, but more one of re-establishing components of a shifting vegetation mosaic. Partial exceptions to the rule are those cases where restoration objectives are highly site specific, such as for some Midwestern glades. An implicit understanding in either case is that the landscape vegetation mosaic was and remains dynamic, i.e., that the structure and composition of landscape elements shift in space and over time. A good illustration comes from the Midwest Oak Ecosystem Recovery Plan (Leach and Ross 1995: 48):

> "Historically some sites may have cycled between nearly closed oak woodland to very open oak savannah and back again, depending on stochastic fluctuations in fire frequency and intensity and other factors. Recovery efforts should not 'freeze the vegetation in time'..."

## Case Study 1. Simpson Township Barrens, Illinois

This case study summarizes restoration work designed to reestablish Midwestern glades and oak woodlands on the Shawnee National Forest. Cutting and prescribed burning have been used successfully to restore these communities on approximately 235 acres.

*Geographic Area/Location*
The Simpson Township Barrens occupy about 235 acres within the Shawnee National Forest in southern Illinois (Johnson County).

*Participants*
Illinois Department of Natural Resources, Illinois Chapter of The Nature Conservancy, Illinois Nature Preserves Commission, Johnson County Boy Scout Troop, Johnson County Highway Department, and Shawnee National Forest. Some work was accomplished by inmates of the Illinois Department of Corrections.

*Restoration Goals*
Although many restorationists set a goal of reestablishing pre-European settlement conditions, too little information was available to set such a goal in this case. Instead, Shawnee National Forest managers planned to restore the site to something approximating the 1930s landscape, which could then be easily maintained by occasional prescribed fires. Original surveyor section line notes, contemporary literature, and 1938 aerial photos provided a framework upon which to base strategies. Additionally, remnants of plant communities containing plant species characteristic of natural forest openings gave insight concerning species and conditions that may have been present in the 1930s.

*Background*
In the late 1970s, the Illinois Natural Area Inventory recognized the 69-acre "Simpson Township Prairie" as having a significant feature, a "Grade A Limestone Glade" community. The Shawnee National Forest officially designated this site as a Natural Area in its 1986 Land and Resource Management Plan. A review of 1938 aerial photographs revealed that what is today forest was once open oak-hickory woodland and limestone glade; succession, in the absence of fire, has led to gradual loss of the more open conditions. Field visits confirmed that the natural communities of the site were threatened by canopy closure, which suppresses the shade-intolerant plant species. These species were surviving in dense shade as scattered, non-flowering rosettes or etiolated individuals. Occasional "wolf" trees, once open grown, displayed shade-pruning from encroaching woody competition.

Restoration prescriptions were first designed to include a 5-acre management area, then 112 acres and finally 235 acres, a size that allowed for a more comprehensive landscape approach. Managers decided that aggressive work was needed on the portion of the barrens that included the two limestone glade communities. In the winter of 1987–88, six cords of firewood were cut from the glades, restoring more open conditions on approximately five acres. The management area was unequally divided by a roadway, which enabled the site to be treated as two separate units, an east side and a west side, with relative ease. The glades were burned in the spring of 1988 for microsite management.

A landscape burn was implemented the spring of 1989 for both units, and this activity revealed a west-side glade that had been previously overlooked; at this point only 87 acres had been burned on the west side and 25 acres on the east. Since the initial landscape burn, each side has been prescribed burned on a rotational basis to ensure that both sides are never burned during the same season within a given year. Fall burns are conducted when possible and individual units are left unburned for 2 to 3 years depending on the unit's management needs. Since a prescription update in 1994, the Shawnee is managing 106 acres on the east side of the roadway and 129 acres on the west, and the prescription has included additional tree girdling and removal as needed. The west side contains three sizable glades as well as several smaller openings. Prescribed burning is now the main form of management. Fire is allowed to interact with the landscape, resulting in a mosaic of burned and unburned areas within the units providing refugia for insects and other animals. Management prescriptions are updated every 5 years based on results of monitoring and ecological research.

*Results*
Glade conditions have been restored on 10+ acres and open woodland now covers much of the remainder of the site. Several conservative and rare plant species on the Illinois state list of threatened and endangered species have emerged. One sedge appears to be an undescribed species. Several species of butterflies and other insects are attracted to the diversity of flowering plants, which now includes more than 200 native species. The glade on the east-side has become a stop for wildflower and birding enthusiasts.

Exotic plant species are still present in the managed areas, most notably Japanese honeysuckle and white and yellow sweet clovers. While fire seems to temporarily reduce the cover of these species, total eradication has not occurred. Another challenge is that as the dry upland woods have become more open due to management, they have become attractive to users of all terrain vehicles (ATVs). The ATV riders have made some rudimentary trails through the woods as well as using an abandoned wagon trail and burn lines. A third problem encountered was that the ring fire patterns used initially to burn the two glade areas resulted in nearly complete ignition, with little if any unburned areas for plants and animals left. The current landscape burning approach allows for mosaics of burned and unburned portions and provides dispersal corridors or linkages between isolated barrens remnants.

Restoration of Simpson Township Barrens became the model for natural area management across the Shawnee National Forest. Managers have learned not to be afraid to take an aggressive approach in the initial stages of restoration. Despite earlier reluctance to cut and burn areas, the Shawnee has learned to work closely with other agencies, institutions, conservation groups, and individuals in preparing and implementing restoration prescriptions.

*Contact Persons*
Steve Widowski/Dick Johnson (618/658-2111); Beth Shimp (618/253-7114)

Designing effective restoration of fire regimes requires objective information about the likely effects of fire on vegetation, wildlife, and soil, water, and air quality, especially when fire remains a solely destructive force in the minds of many. In keeping with the experimental nature of restoration, fire influence need not, and probably should not, be restored on every acre where it may have played a role historically.

Focusing attention on as little as 10 percent of the total potential restoration area may be appropriate where information about fire effects is tentative and/or logistic considerations limit what can be achieved. Restoring 50 percent or more of the total potential area may be appropriate where supporting data is substantial. One offshoot of landscape-scale planning is that some managers are now experimenting with burning larger areas than they may have previously considered. This option is showing promise, not only for expanding how much prescribed burning can be accomplished but also for significant cost savings. Costs per acre for prescribed burning tend to be inversely proportional to the size of the area burned.

Frequency and seasonality of fires are important factors, and lack of attention to these variables can produce unintended results (Parker 1990). Dormant season fires in the Southeast, for example, often induce coppicing in oaks. Systems in which fire has been excluded for decades will, in all likelihood, have fuel loads, root reserves, stand structures, and regeneration dynamics far different from those in which fire remained a force. Restoration, therefore, is not just a matter of reintroducing fire, but of designing and adapting for a whole suite of variables.

In some cases fires have become more frequent with site conversion. In portions of Idaho, Oregon, and Nevada, large areas have been converted to cheat grass, an exotic annual. Cheat grass is very susceptible to fire when it cures and it thrives with frequent fires. Strategies to break the cycle of frequent fire have included strategically interspersing 30- to 400-foot wide "green strips" of fire resistant species such as crested wheatgrass. This is sometimes combined with planting of native species such as sagebrush to begin to reestablish diversity and structure (USDI NBS 1994). Results have been mixed, perhaps due to drought. But at least two instances have occurred where green strips have stopped fires from advancing (USDI NBS 1994).

Vegetation community structure and composition can be treated directly by techniques such as thinning and brush crushing to affect fire occurrence (frequency, intensity, effects) or to improve watershed health and habitat diversity. The combined use of fire and thinning to reestablish or accelerate the return of more open forest and grassland conditions, or of old-growth characteristics, is on the rise. Covington et al. (this volume) provides an example in the ponderosa pine forest of the Southwest. Other examples include shortleaf pine ecosystems in the Ouachita Mountains (USDA Forest Service 1994) and pinyon–juniper in the Southwest (e.g., Carrizo Demonstration Area, Lincoln National Forest; Alexander, undated publication) (see Case Study 2). Promising attempts at North American grassland restoration are summarized by Allen (1988) and Berger (1990).

Desertification is another restoration problem which is frequently treated using structural techniques. Desertification has been defined as the climatic dryness induced by human disturbances of the topsoil and natural plant communities (United Nations 1977). In many cases, desertification has been caused by improper grazing, introduction of non-native flora and fauna, or destruction of habitat by development. Desertification is a process where the land becomes denuded, with the land smoothing and sealing over. Water infiltration is impeded, as is air transfer from the soil.

Techniques used to restore desert areas by reversing the desertification have been developed and discussed (Dixon 1990; Clary and Johnson 1983). Imprintation is a mechanical process used to roughen and open the desert soil to infiltration by air and water. With imprintation, revegetation is improved during the seed establishment and germination phases (Dixon 1987). In addition, runoff and erosion are reduced through imprinting. Dixon (1990) indicated imprinting with native perennial grasses has been quite successful on several sites. The treatment was also responsible for crowding out tumbleweed and burroweed.

### 4.1.2 Managing Land Uses

In many cases, land uses have caused landscape modifications that affect the structure and function of ecosystems. Many temperate forest ecosystems in North America have been modified by conventional forest management practices. Industrial forest practices, specifically even-aged management for timber production, have created forests comprised of single species with little or no understory present (Horowitz 1990).

Elimination or restriction of land uses is a management option to consider as part of a restoration project, if only to establish baseline areas for comparison with more actively managed portions of restoration areas. Dramatic change, sometimes accompanying an agency designation that restricts or eliminate land uses, may be the only option to achieve restoration in some cases. The most formal Congressional designations — Wilderness, National Wild and Scenic Rivers, National Recreation Areas — are also sometimes paired with

**Case Study 2. Restoration of a Southwestern woodland ecosystem Carrizo Demonstration Area, New Mexico**

This case study summarizes efforts to restore damaged watersheds in the pinyon–juniper woodlands of the Lincoln National Forest. The Carrizo Demonstration Area was established in large part at the urging of local landowners and grazing permittees. Their continued support, together with several active partnerships, has been critical. Over 5,000 acres have been successfully treated since 1989, but thousands more remain in degraded condition.

*Geographic Area/Location*
The Carrizo Demonstration Area is located on the Smokey Bear Ranger District of the Lincoln National Forest in south-central New Mexico. It encompasses 55,000 acres and includes both National Forest and private lands.

*Participants*
USDA Forest Service, New Mexico Department of Game and Fish, New Mexico Division of Forestry and Resource Conservation, New Mexico State University and NMSU Cooperative Extension Service, New Mexico Range Improvement Task Force, thirteen grazing permittees and three private landowners.

*Restoration Goals*
Stop active accelerated soil erosion, stabilize steep gully slopes, and restore permanent riparian vegetation. Goals also include providing for a variety of wildlife habitat, increasing plant and animal diversity, and restoring the natural beauty of the landscape.

*Background*
The Carrizo Demonstration Area was established in 1989 as a pilot effort in watershed restoration. Much of the Carrizo area reflects a history of intensive grazing pressure and fire suppression, which resulted in reduced grass production, increased soil erosion, and proliferation of pinyon pine and juniper trees. Records show that in 1902, 80,000 head of livestock were grazing on what is now the Smokey Bear Ranger District.

The grazing capacity today is about 5,000 head, and pinyon and juniper form a nearly continuous canopy in many places. As tree canopies closed in, grasses and forbs that held soil in place and provided forage continually declined. The extensive gully system that developed helped send silt-laden water into streams and rivers. Many of the area's perennial streams and springs declined in flow.

The impetus for the restoration project came from private landowners and grazing permittees in the area. Both had to contend for years with the deposition of millions of tons of sediment that originated on National Forest land. Grazing permittees were concerned about steady declines in livestock grazing capacity. Wildlife interests also have a stake in the degraded Carrizo area, pointing to the adverse effects on habitat for deer, elk, wild turkey, songbirds, and many other species.

Using the USDA Forest Service's Southwestern Region Terrestrial Ecosystem Survey (a mapping system that combines soil type, potential natural vegetation and climate), the Carrizo area interdisciplinary planning team identified high priority areas for treatment as those with unsatisfactory watershed condition and high soil productivity. Of the 18 map units included in the Terrestrial Ecosystem Survey (ranging in size from 1,345 to 6,284 acres), only half were rated "satisfactory." Prescriptions, with detailed descriptions of desired future condition, were prepared for each map unit. Treatments employed include reseeding, prescribed fire, thinning, and gully reshaping. Many were undertaken as partnerships. (See also USDA Forest Service, Southwest Region (1995).)

*Results/Outcomes*
More than 3,200 acres identified as being in unsatisfactory watershed condition have been treated to increase herbaceous ground cover. Four miles of gullies have received structural improvements or sideslope stabilization treatments, and five miles of roads have been eliminated. More than 1,000 acres have been treated with prescribed fire. Where treatments have been implemented, watershed conditions have improved dramatically. Cool season native grasses and forbs thought to have vanished have now returned in abundance. Springs have begun to flow again in some drainages.

Except under extreme conditions, using prescribed fire to open up areas of pinyon-juniper is very difficult, if not impossible. It is therefore necessary to thin extensively in many areas. Progress toward restoration goals is slower than many stakeholders would prefer.

*Contact Person*
Richard Edwards, Smokey Bear Ranger District, Lincoln National Forest, 901 Melchem, Ruidoso, NM 88345. Phone: 505-257-4095.

restrictions on land use. For example, passive restoration of some forms of old growth is probably taking place on millions of acres of federally designated wilderness, recreation, and special interest areas where commodity production has been eliminated.

Although some would argue that only "nature" can restore old growth, others point out that North American ecosystems are so altered that merely eliminating one activity or another is not likely to lead to restoration (see Bonnicksen et al., this volume). What may in fact develop are ecosystems that never existed before (Botkin 1990). Some federal land managers and conservationists have concluded that several once common community types or system states are unlikely to be restored without active intervention. Management decisions concerning how harvest of forest products occurs affect restoration of forest structure and function. Recent approaches increase emphasis on diversity and the preservation of essential ecological conditions and functions over timber and other commodity production values. "New Forestry" (Franklin 1990) was an early move toward ecosystem management that intertwines commodity production with maintenance of ecological diversity. Timber harvest, however, cannot completely substitute for, or simulate, ecosystem disturbances and processes such as fire. In fact, it can create its own distinct effects through mechanical disturbance (Johnson et al. 1996).

Puleo (1990) discussed techniques used at the H.J. Andrews Experimental Forest in Oregon when actively managing for "New Forestry." These techniques included a strong emphasis on "green tree" retention during any type of harvest. Leaving coarse woody debris on site after any harvest-based treatment was also incorporated. Finally, stand treatment no longer emphasized clearcut-based silvicultural techniques. Instead foresters began prescribing partial harvests, such as shelterwood or seedtree regeneration harvests to mitigate visual concerns and improve structural diversity in the residual stand (see Box 1).

Recently, the Forest Service and the Bureau of Land Management used regional planning and assessment in the Forest Ecosystem Management Assessment Team (FEMAT) report in 1993 to set a regional restoration direction. The report built on the New Forestry movement, evaluating ten options for managing federal forest lands within the range of the northern spotted owl in the Pacific Northwest. Option 9, selected for implementation, focuses on increasing late successional seral stages and maintaining biological diversity through both management activities and natural events such as fire. It promoted broad scale restoration and led to several management process innovations.

Since approval of the Northwest Forest Plan in 1994, the federal agencies in the Pacific Northwest have been striving to restore old-growth conditions over areas ranging from mixed conifer to single-species plantation forests. The predominant methods for managing young forests at the early- or mid-seral stage are partial harvests (such as thinnings) or controlled fire events (prescribed burning). Ecosystem management in forests in the Pacific Northwest is still new, and evidence of success at these early stages is not clearly documented (Covington et al., this volume).

The Interior Columbia Basin Ecosystem Management Project represents the next generation of regional ecosystem planning in the Northwest. It also addresses managing uses on public lands, building on concepts developed for the Northwest Forest Plan. But the Columbia Basin Project covers 72 million acres of public lands in a 144 million acre area spread primarily across eastern Oregon, eastern Washington, Idaho and Montana. In addition to covering an area about the size of France, the project also added innovation for regional assessment and planning (Interior Columbia Basin Ecosystem Management Project 1997) (see Box 2).

**Box 1**
**Options for Restoration Forestry**

1. Managing the land for older forest by preserving existing old growth and other late successional stands, allowing second growth stands to mature, and placing managed stands on longer rotations;
2. Retaining structural diversity, including snags and downed logs, in managed stands;
3. Retaining and restoring large, intact stretches of forest unfragmented by roads, clearcuts and other openings;
4. Retaining and restoring corridors and other linkages between forests;
5. Allowing natural fires to burn, using prescribed fire, or applying silvicultural manipulations that simulate fire and other disturbances in order to maintain a full spectrum of seral stages and structures;
6. Stopping road construction and reconstruction, and obliterating and revegetating most existing roads; and
7. Recovering viable populations of rare species and reintroducing extirpated species.

Noss and Cooperrider (1994)

The scientifically based regional approaches in the Pacific Northwest identified aquatic conditions of streams and watersheds, effects of roads, alterations in composition and structure ecosystems, changes in disturbance regimes (especially fire), exotic species invasion and community resiliency as major considerations for future management action (Quigley and Arbelbide 1997, Quigley et al. 1996). These concerns are related to settlement, population growth and land use practices. They have implications for diverse ecosystem conditions including available plant and animal habitat, water quality, air quality and human quality of life. Our review of case studies found those same concerns, causes and implications to be widespread across the United States. So are the efforts to address them.

One study in the Pacific Northwest, designed to evaluate active management treatments in forested systems, is the Demonstration of Ecosystem Management Options or DEMO. Four treatment options to restore stands to mature-to-late seral stages more rapidly than without treatment (passive management) are being evaluated. Stand treatments are delineated based on the extent of removal and the dispersal pattern used. Harvest removals range from 15 to 40 percent. The study is a collaborative effort designed to evaluate effects on a broad array of variables including forest wildlife, invertebrates, and understory biota. The study is currently in the pre-treatment data collection phase, with data collection and analysis expected to

**Box 2**
**Process Lessons and Innovations for Regional Assessments and Planning in the Pacific Northwest**

*Northwest Forest Plan (FEMAT)*

- amendment of land use allocations and management plans across a regional landscape,
- establishment of strategies to address late successional forest, aquatic conservation and the survey and management of habitat for a wide diversity of species,
- creation of a watershed analysis process,
- creation of advisory committees at the regional and province (generally basins) levels to advise the land management agencies on implementation, and
- creation of a Regional Ecosystems Office to coordinate and facilitate plan interpretation and application among the many affected or involved agencies and offices.

*Interior Columbia Basin Ecosystem Management Project (ICBEMP)*

- a restoration, rather than reserve, strategy can be the basis for the preferred alternative,
- replacement of interim strategies for fish habitat and late successional forest management,
- a very deliberate focus on ecosystems rather than a species-by-species approach,
- inclusion of both forest and range lands,
- improved integration of social and economic issues and information,
- concepts and models that may improve the flexibility, local applicability and cost efficiency of assessment processes and management prescriptions,
- development of separate processes for scientific assessment and land management planning with deliberate ties between the two,
- increased consultation with counties, states, federal agencies, American Indian tribes during development and analysis of alternatives, and
- a very open public participation process during development of the scientific assessment and land management alternatives.

continue for at least the next decade.

Forest product companies with extensive land holdings in western Oregon and Washington are also modifying private land forest management to comply with federal requirements for wildlife habitat protection and maintenance. One company, Plum Creek Timber Company, has established a Habitat Conservation Plan for 170,000 acres of their privately owned timberlands (Jirsa 1995). The plan emphasizes ecosystem restoration and maintenance at the landscape level, rather than single species restoration. It is currently under public review and revisions are expected before implementation occurs.

In forests of the Southwest and the interior of Oregon and Washington, forest health issues drive most restoration efforts. Covington (1994) noted that disruption of natural fire regimes has occurred across the inland West and in the Southwest since European settlers first appeared and began to control wildfires. O'Laughlin (1994) discussed extreme forest health problems on several national forests in Idaho, using net growth to mortality ratios as indicators of the severity of these health problems.

A twofold strategy for forest health improvement was outlined by O'Laughlin. First, restore the dominant species best suited for the sites of concern, such as ponderosa pine, western larch, and western white pine. Second, prevent unhealthy conditions from occurring, primarily through reductions in stand density through thinning operations. This approach emphasizes restoring forest habitat, without concerns over specific forest-based wildlife species. Another alternative is to design actions to promote the presence of a desirable wildlife species. The species is assumed to represent the health of the ecosystem, and successful restoration is then measured by how the species responds.

In Eastern deciduous forests, efforts to restore forest areas to pre-European settlement conditions have been less intensive, primarily because of the fragmented landownership patterns. In addition, past efforts to manipulate conditions in these stands through thinning and prescribed fire have generally failed. However, these methods may still be the most useful techniques for restoring eastern forest ecosystems if used properly.

Forest restoration efforts are increasing in the South as well. Best known are the longleaf-wiregrass, sand pine, and shortleaf pine-bluestem restoration projects. But public land managers and scientists in this region are also experimenting with "new forestry" practices (Baker 1994). The Ouachita and Ozark-St. Francis National Forests have been testing alternative stand level treatments and large scale ecosystem management strategies since 1990. On the Ouachita National Forest, natural reproduction has replaced clearcutting and planting. Uneven aged, modified shelterwood and seed tree systems are used to achieve desired landscape conditions and mixed stands of pines and hardwoods are maintained (USDA Forest Service 1994).

Generally, restoration of forest ecosystems to some desired or needed condition is being either studied or tested on many forests in the western United States. The techniques vary slightly but common methods used include (1) harvest-based treatments to affect stand density, species composition or stand structure, (2) retention of woody debris and/or snags, and (3) use of controlled fire.

The need to increase biodiversity is now commonly integrated into forest health treatments. "Non-managed" areas, where catastrophic disturbances like the 1988 fires in Yellowstone National Park are allowed to occur without intervention, are not common. As a result, forest health treatments in Western forests are increasingly oriented toward active intervention to maintain or restore ecosystem functions and conditions.

At a general level, the options for restoration on rangelands are similar to those for forested systems. However, there can be differences related to their geographic position in most watersheds, their relationship to settlement areas and private lands (agriculture, residential, etc), their land use history and, sometimes, their more open vegetative character or gentler terrain (see Box 3).

As illustrated in Case Studies 1 and 2, restoration of grasslands is also related to fire disturbances and the abundance of woody species. In the Northwest, the encroachment of woody species in rangelands is a primary restoration concern (Quigley et al. 1996).

Vegetation community structure and composition are also a concern in non-forest vegetation types because most wildlife species respond most readily to structure (Maser et al. 1984). The Bureau of Land Management and Forest Service worked with ranchers and interest groups to adjust rangeland management on 250,000 acres in central Oregon to address declines in sage grouse populations, as well as other wildlife and vegetation concerns. Techniques included prescribed burning and changing the timing, duration, and frequency of livestock grazing to improve nesting cover and increase the forb component in the plant community. The intent is that improvement in long-term management of sagebrush communities will allow for potential increases in sage grouse populations, as well as other plant and animal species. There have been some indications of a positive response in sage grouse numbers, although long-term trends cannot yet be established and influences by climate are known to be a factor (USDI BLM 1995).

In northern Arizona, the Prescott National Forest and the Yavapai Ranch have initiated range restoration based on improved grazing management (Whitney 1996). Management objectives included improvement of pronghorn antelope habitat and forage, restoration of critical cover and grassland areas, and improvement of vegetation diversity. Using previously developed best management practices (BMPs) and a coordination structure, the project achieved better vegetative cover and reduced sedimentation. The project used well established management tools like prescribed range fires and controlled grazing to attain these objectives.

Managing recreation use to restore or improve ecosystem function is increasing in importance. With growing populations and recreation demands, more area is affected more frequently by recreation use. Transportation technology has also increased accessibility with the advent of four wheel drive vehicles, specialized motorcycles for trail riding, all terrain vehicles and mountain bikes. Webb and Wilshire (1983) provide options for restoration of vehicle recreation impacts.

While the effects of many vehicle trail uses are different than those on a similar density of roads, leaving large areas open to vehicle use is causing unacceptable levels of impact (USDI BLM 1997; Smith and Pritchard 1992). Millican Valley Off Highway Vehicle Area in Oregon has been open to vehicle use with seasonal restrictions for over 25 years. Almost 500 miles of user created trails have become established in the 65,000 acre area. Drawing from experience and monitoring in a nearby area on the Deschutes National

**Box 3**
**Possible Options for Restoration on Rangelands**

1. Manage land uses and disturbances for a dynamic mosaic of conditions that transitions among different states and retains or restores species richness;
2. Retaining structural diversity, including residual grass cover or regrowth for physical and biological functions, on managed rangeland sites;
3. Retaining and restoring large, intact stretches of rangeland unfragmented by roads;
4. Retaining and restoring corridors and other linkages around settlement areas, along riparian zones, and to forests;
5. Allowing natural fires to burn, using prescribed fire, or applying treatments that simulate fire and other disturbances in order to promote the health of herbaceous and browse plants while maintaining a full spectrum of seral stages and structures;
6. Redesigning road and trail systems, while obliterating and revegetating existing roads and user established vehicle routes; and
7. Recovering viable populations of rare species and reintroducing extirpated species.

Adapted from Noss and Cooperrider (1994);
See also Svejcar and Brown (1992)

Forest, BLM is shifting use to a designated trail system of 262 miles and to seasons better designed to reduce impacts to sage grouse, vegetation, soils and other resources (USDI BLM 1997).

### 4.1.3 Biological Intervention

Efforts oriented around species of concern are common in restoration projects. In some cases, the concern arises because a species appears to be seriously declining or absent, despite apparent availability of suitable habitat. In other cases, the concern arises because an exotic species threatens to displace native species and reduce diversity.

In many regions, invasive species represent one of the greatest threats to ecosystem integrity and one of the greatest challenges for restoration. Federal land managers in many parts of the country can attest to the vigor, reproductive capacity, and tenacity of invasive plants, animals, fungi, and other organisms that were ignored one year and became glaring problems the next. Cheat grass, thistles, Gypsy moth, melaleuca, knapweed, common privet, leafy spurge, hoary cress, Japanese honeysuckle, kudzu, dogwood anthracnose ..., the list is long and growing longer every year (U.S. Congress, Office of Technology Assessment 1993).

In the West, the proliferation of exotic weeds represents one of the greatest changes on arid and semi-arid lands since pre-Columbian times (Allen and Jackson, 1992). Even relict areas are not exempt from invasion by exotic annual plant species (Svejcar and Tausch 1991). Asher et al. (1995) characterized the problem as follows. Exotic plants increased on BLM lands from 2.5 million acres in 1985 to over 8 million acres in 1994. Including exotic populations on Forest Service, National Park Service, and Fish and Wildlife Service lands nearly doubles that infested area. Infestation on these lands is increasing by about 4,600 acres per day.

Exotics as a percentage of total cover have been recognized as an indicator of desertification as aggressive undesirable forbs and shrubs replace natives, spread, or eventually become dominant (Mouat et al. 1993, Dregne 1977). But the impacts are not limited to displacement of native species. Runoff and sediment yield were 56 percent and 192 percent higher for spotted knapweed than for bunchgrass vegetation types (Lacey, 1989). Negative impacts summarized by Huenneke (1995) include influences on native species, soils, nutrient cycling, hydrology, and disturbance regimes (Asher 1996). According to the OTA report (1993: 5):

> "(T)he number and impact of harmful (non-indigenous species, or NIS) are chronically underestimated, especially for species that do not damage agriculture, industry, or human health... From 1906 to 1991, just 79 NIS caused documented losses of $97 billion in harmful effects."

Regional sources of information and assistance in addressing the proliferation of exotic species are generally available. Local county weed supervisors, extension service agents, university weed scientists, and federal/state agency experts can provide information on weed occurrence and effective treatments (Asher 1996). Other groups like The Nature Conservancy and the Illinois Nature Preserves Commission have compiled abstracts on invasive species. The Natural Areas Journal, Restoration and Management Notes, Ecological Restoration, and other periodicals are excellent sources as well. In the West, the "Guidelines for coordinated management of noxious weeds in the Greater Yellowstone Area" and the Lolo National Forest's plan amendment on noxious weed management are good source documents. The Yellowstone document includes detailed guidelines for developing cooperative weed management areas, which can be effective for facilitating cross-boundary action by federal and state land managers and landowners.

The most effective strategies to control exotic plants are generally those designed to prevent establishment and control spread. Cost-effective control techniques are not yet developed for some species once they are well established. An excellent summary of weed control techniques was provided by Harrod et al. (1996). Limiting human activity and surface disturbance reduces opportunities for exotic plants to become established (Asher and Harmon 1995). Potential measures also include using weed-free animal feed, washing vehicles after being in a weed infested area, avoiding seed movement by animals and people, and using weed-free seed (Asher et al. 1995).

Quick detection and control greatly enhances success in treating spot infestations (Asher et al. 1995). Once exotic plants are present, strategies vary depending on species, density, numbers, location, and time of year. Targeted chemical treatment that carefully avoids unintended effects, whether aerial or plant-by-plant, is often the only effective treatment. Mechanical means, most commonly mowing, plowing or discing, can also effectively reduce flowering and seed set (Harrod et al. 1996). Manual removal techniques are most effective when the infestations are small. If native vegetation is not present in adequate densities after treatment of exotics, additional measures may be required to promote native plant establishment and limit re-establishment of exotic species (Asher 1996). In all cases, the timing of control efforts is important to success (see also Case Study 3).

**Case Study 3. Restoration and Maintenance of Native Species and Diversity By Control of Noxious Weeds in the West**

This case study is a compilation of experiences from public lands around the Western United States. The examples cited occur in Idaho, Montana, Oregon, Utah, Washington and Wyoming.

*Geographic Area/Location*

On much of the land in the arid West, infestations of non-native plants are an increasing problem, displacing native species and thereby reducing biodiversity. A sample of locations include: Salmon River Canyon, Idaho; Rock Drainage, Montana; Willowa County, Oregon; Millard County, Utah; private land in Yakima County, Washington; Carbon County Weed Management Area, Wyoming

*Participants*

The participants listed in this case study provide a sample of the people who are working on the problem. They include: private landowners, Idaho County Weed Board, Nez Perce National Forest; Valley County, Montana; Willowa County, Oregon, Oregon Department of Fish and Wildlife; Millard County, Utah's "Scotch Thistle Day" (county, state, private, BLM, USFS, BIA); Yakima County, Washington, Washington State University; Wyoming Department of Transportation, Carbon County, Wyoming.

*Restoration Goals*

Eradication and control of noxious weed infestations. Many of the successes feature prevention, early detection and prompt action. Target species include yellow starthistle, dalmatian toadflax, spotted knapweed, leafy spurge, scotch thistle, featherhead knapweed, and musk thistle.

*Background/Results:*

Idaho: The Forest Service's Slate Creek District is developing a Weed Management Program for the Salmon River Canyon. Over 130 spotted knapweed and ragwort sites were inventoried in the Snake River Canyon in 1993. Several sites were pulled/bagged and one yellow starthistle site is believed to have been eradicated.

Montana: Ranchers contribute about four cents per animal unit month to a county weed control fund creating a steady, reliable and continuous fund for combating weed infestations.

Oregon: Weed prevention efforts include a 1991 Willowa County ordinance that makes it unlawful to transport hay, straw or grass (chopped, baled, etc.) that has not been certified as weed free. Volunteers operate hay exchange stations during hunting season and Oregon Department of Fish and Wildlife informs hunters through their hunting proclamation. Since the ordinance, the number of new tansy ragwort sites detected each year has declined by 85 percent.

Utah: Multiple agencies and levels of government have joined to sponsor "Scotch Thistle Day" mobilizing between 200 and 300 junior high and high school students within a 175 miles radius to work on an infested site and learn more about plant ecology. Beyond the benefits of the weed control itself, the education has caused families to adopt the control effort. Goatstrue was imported to the state in 1891 as an attractive forage plant. It was not utilized but by the 1970s over 40,000 acres was infested. Utah decided to embark on an eradication program in 1980. The plant is relatively easy to control and the project was expected to be successfully completed in 1996 at a cumulative cost of $2 million.

Washington: In 1983, a landowner in Yakima County brought in a couple of plants to the County for identification. The plants were feathered knapweed, the only known infestation in North America. The 20×30 foot patch of rhizomatous plants was eradicated using a tordon spray.

Wyoming: In 1992, Wyoming Department of Transportation explored alternatives to repeated investment in weed control along a highway where adjacent lands were also infested. State and county officials, agency employees (federal, state, county) and landowners joined to create a weed management area and developed a plan with shared goals and priorities. The group works with university expertise and is applying biological controls.

*Outcomes*

The significance of introduced exotic species is being recognized and people are making efforts to address the problem. The weed situation in the Snake River Canyon is better understood and actions to contain and control it have been initiated. Ranchers in Montana have created a stable funding source to address weed infestation problems. Preventing weed infestations is the focus of a local ordinance and volunteer efforts in Oregon. A bridge between eradication efforts and education has been built in Utah. A long term project was sustained in Utah. A weed species unknown in North America was prevented from gaining a foothold in Washington. Biological controls are being tested in Wyoming.

*Contact Person*

Jerry Asher, Oregon State Office, USDI Bureau of Land Management, 1515 S.W. 5th Avenue, Portland Oregon 97201; P.O. Box 2965, Portland, Oregon 97208; Phone: (503)952-6368.

Although not in common use, biological measures to control exotic plants are receiving increasing attention. Biological controls introduce natural enemies of the exotic plant such as fungal pathogens, herbivores, or parasites. The appeal is cost-effectiveness, in some cases, while the risk is that the control agent will attack native species or produce other unintended results. Apparent successes with insects to control weeds such as spotted and diffuse knapweed, St. Johnswort and purple loosestrife have occurred in relatively undisturbed rangelands and aquatic systems (Harrod et al. 1996). Other biological measures being tested include species specific plant-to-plant competition (e.g., squirreltail/medusahead on clay and clay loam sites), competition stimulated by fertilization, and livestock grazing designed and timed to damage the exotic plant (USDI NBS 1994; Hilken and Miller 1994). The Western Society of Weed Science provides a reference on biological control of weeds in the Western United States (Rees et al. 1996).

Barriers to dealing with weed problems include lack of knowledge of the underlying causes for susceptibility to invaders, lack of suitable biological, physical, or chemical control agents, and ignorance or underestimation of the problems. Use of fire or grazing strategies keyed to plant phenology are promising possibilities. To suppress exotic plant species, grazing treatments should be highly controlled and selective, generally high-intensity, short-duration treatments designed to defoliate and/or prevent seed production in the targeted species (Valentine and Stevens 1994).

Prescribed fire apparently reduces seed production in some exotic plant species and can favor competing grasses (Hilken and Miller 1994, Harrod et al. 1996). For some species, fires can be timed to destroy the viability of seeds. Backing fires downhill and against the wind is most effective with medusahead because it more effectively consumes or damages seeds (Hilken and Miller 1994). Experience in the Murderers Creek watershed in Oregon suggests results can be uncertain, even with careful planning. Winds can shift, converting a backing fire to a head fire. Post-fire precipitation affects success but is difficult to predict. The effectiveness of seeding also affects outcomes where seeding is combined with burning. Yet even with the difficulties, reductions in medusahead are occurring in some areas (USDI BLM Central Oregon Resource Area 1996).

Species of concern also include desirable species, but for different reasons. It is common to design restoration in response to the apparent decline of a desired species. Species decline problems are regularly encountered, but examples of apparently successful reversal of the trend toward local extinction are also evident. Breeding for resistance to disease has enabled managers to be more successful at reintroducing certain western conifers to ecosystems where they have been decimated by exotic and/or native pests (Howe and Smith 1994, Mahalovich 1996).

Animal species once nearly eliminated from many parts of their historical ranges and now increasing in large portions of those ranges include the wild turkey, black bear, beaver, American alligator, bald eagle, white-tailed deer, wood ducks, several species of trout, herons, egrets and shorebirds (USDI Fish and Wildlife Service 1987). Habitat restoration on federal lands has played, and will continue to play, a key role in the many states and regions where these success stories unfolded. The desired outcome is usually simple: recovery in a specified area for a self-sustaining population of a species that was nearly eliminated within the past two centuries. In some cases, the desired outcome includes recovery to the point that regulated hunting or fishing is possible.

Success generally comes easier with species that are habitat generalists rather than specialists. State wildlife agencies, in cooperation with federal agencies, have spearheaded efforts involving habitat generalists, efforts which often required the commitment of several agencies operating as partners across state boundaries. Sometimes, federal land managers have been partners, not initiators, a role which is likely to continue to be appropriate. Two obstacles to success are (1) threats or perceived threats to livestock, private land values, and/or management options posed by the reintroduction/recovery effort and (2) insufficient understanding of predator–prey and/or habitat relationships. Lack of local genetic material may also be a problem (see Holthausen, this volume).

As suggested by the listing of recovering species above, much of the habitat restoration focus in previous decades has been on species of social interest, such as game species. For example, few elk were present in eastern Oregon and Washington from 1800 until about 1930 but now they are much more common (Johnson 1994). But as the number of game species has grown, predators have been displaced. This change also has implications for the plant species these large herbivores browse (*ibid;* Blue Mountain Natural Resources Institute 1996).

Species recovery activities in some instances have turned out to be highly compatible with active human use and management, and even produced some surprising side benefits. Peregrine falcons, for example, are becoming more common in urban settings. The Sierra Club issued an award to national forests in Florida for their efforts on behalf of the Florida scrub jay, efforts that included clearcutting and burning. In western Arkansas, the Ouachita National Forest has designated a new management area in which ecosystem restoration and eventual recovery of a red-cockaded woodpecker population are key objectives that can be accomplished without reducing timber sale volume for at least 20 years (USDA Forest Service, Southern Region 1995).

Other species, including perhaps most of the officially listed threatened or endangered species in the United States, have not fared as well. The Nature Conservancy's Annual Report Card for U.S. Plant and Animal Species (The Nature Conservancy 1996) listed 6,246 plants and animals rated either critically imperiled, imperiled, or vulnerable. Of the 3,170 species considered critically imperiled or imperiled, 80 percent are vascular plants. Falk et al. (1996) cover the strategic and legal context for rare plant restoration and provide case studies from across the United States.

Case studies of prominent endangered species recovery programs are profiled by Clark et al. (1994). As of September 30, 1988, six of the ten species delisted were due to extinction and four were due to recovery (Meese 1989). However, these counts may not be an accurate predictor of current and future trends; successes in addressing species rarity and decline are occurring (Clark et al. 1994). More holistic considerations such as species rarity, endemism and richness have not previously received much attention, but they are an increasing focus in the management of federal lands (Quigley and Arbelbide 1997).

Species recovery may be a long-term, costly venture. It can become socially and politically controversial as well, as in the cases of wolf reintroduction, grizzly bear recovery, and red-cockaded woodpecker recovery. Other barriers to restoration include: (1) recovery of habitat specialists is often a slow process, during which political and financial commitments may weaken, and (2) scientific information about the species of concern may be inadequate.

Some of the most promising possibilities for long-term species recovery are the emerging ecosystem-based approaches, e.g., those developed in response to threats to the red-cockaded woodpecker (USDA Forest Service, Southern Region 1995) and anadromous fish (Columbia Basin Salmon Restoration efforts). The likely success of these efforts will be their ability to weave together the social, economic, and ecological elements of a sustainable recovery. The expected solution is likely to include a variety of management techniques across large landscapes, although the emphasis is on structural treatments and managing land uses differently (Interior Columbia Basin Ecosystem Management Project 1997).

### 4.2 Summary for Terrestrial Restoration

The shift to ecosystem management has turned attention to management treatments such as prescribed fire to affect vegetation positively, the importance of managing land uses with ecosystem function in mind, and the potential to carefully intervene to restore or protect biological attributes. These restoration techniques can promote ecosystem health by improving process function and changing the structure and composition of plant and animal communities. Even though social and economic factors continue to affect land management decisions, efforts are underway to integrate them with ecological function to create a sustainable whole. Though some restoration problems, such as controlling aggressive exotic species, are growing in severity, experimentation to develop new and more reliable restoration techniques is also underway.

Habitat changes, particularly in old-growth forests, are receiving considerable emphasis across the nation. Habitat restoration is being addressed at broader scales to consider communities across watersheds, landscapes, and regions. But this change in emphasis is fairly recent. As a result, it is difficult to assess the long-term, large-scale effect of many habitat restoration efforts. Clearly some projects are producing desired results.

## 5 APPROACHES TO TREATING RIPARIAN FUNCTION

*Scenario: Stream systems are impacted by loss of vegetation, bank instability, sedimentation, changes to the channel or elevated water temperatures.*

Impaired or reduced riparian function occurs where historical or present land uses have disrupted physical and biological processes. Nonpoint source pollution is commonly tied to agriculture, forest practices, construction, mining, recreation, urban runoff, and land disposal within the watershed (Johnson 1992). The general categories of problems to be addressed through restoration based on Platts (1989) and Johnson (1992) are:

- Alterations of streamside vegetation and soil conditions
- Changing channel morphology (water velocity, water table)
- Altered water temperatures, nutrient loads, sediment loads, bacterial counts
- Degradation/erosion of stream banks

Lack of instream habitat structure is also important. Riparian restoration to improve water quality, fish and wildlife habitat, and the general health and aesthetics of the stream is becoming widespread and common in every region of the country, although many efforts are in their early stages (Natural Resources Law Center 1996). The National Research Council (1992) has also compiled a very comprehensive text covering restoration of aquatic systems.

The Government Accounting Office, after reviewing 22 stream restoration efforts in the West concluded there were "no major technical impediments" to riparian restoration (USGAO 1988). Systems for classifying riparian community types and channel types are well developed with solid insights into stream function, species composition, fishery potential, and forage production (Gebhardt 1990, Hansen et al. 1995). Practical and useful techniques to assess the physical and biological attributes of the stream are also well documented (Batson 1987, Myers 1989, Hunter 1991, Leonard et al. 1992, Cagney 1993, Pritchard et al. 1993, Pritchard et al. 1996, Hindley 1996). Most current restoration projects are developed in a watershed context.

Successful restoration efforts consider the complex relationships of riparian function and the integral role of vegetation (Elmore and Kaufman 1994). Hunter (1991) suggested that success is dependent on understanding the physical, chemical, and biological attributes of the stream and on the ability to analyze these diverse sets of data. The indicators that recovery is occurring have been described by Elmore and Beschta as follows:

> "As vegetation becomes established along a stream, channels typically begin to 'aggrade', that is the streambed will rise as sediments accumulate... As more sediments are deposited and bank building continues, particularly along low gradient stretches, the water table rises... More water is stored during wet seasons, and its gradual release may allow a stream to flow during the driest of summers... The gradual release of water from increased underground storage can more than offset the water used by streamside vegetation." (Elmore and Beschta 1990, pp. 9–10)

Classifications and rapid assessment techniques for proper functioning condition provide useful tools to evaluate stream channels and systems for setting goals and objectives. Rosgen (1994, 1996) developed a commonly used classification system for understanding stream channel morphology. Assessing proper functioning condition can be used to set restoration priorities by identifying areas which are functioning properly, functioning at risk, or not functioning properly (Pritchard et al. 1993).

## 5.1 Management Options for Riparian Restoration

Options for restorating riparian systems are grouped into the three different approaches discussed above: structural treatments, management of land uses, and biological intervention. In many cases they have been combined and each approach includes techniques which are directed at accelerating positive ecological change in the riparian zone. Compilations of case studies are available to review work done in similar settings (Appendix B).

### 5.1.1 Structural Treatments

Early approaches to restoration often involved engineering the physical attributes of a stream using structures. In 1933–1935, the U.S. Bureau of Sport Fisheries put in 31,084 structures in 406 mountain streams. Many are still in place; some are not. By 1952, the Forest Service Fish Stream Habitat Improvement Handbook recognized that structure work was not a "cure-all," could be overdone, and could cause damage (Hunter 1991).

Experimentation with structural approaches to bank stabilization and instream modification continues today. As expected, experience has taught a few lessons. The most important is that structures are not a substitute for managing land uses (Elmore and Beschta 1989). Structures can be both expensive and ineffective if land uses adversely affecting the stream are not corrected (Elmore and Beschta 1987, Hunter 1991, Beschta et al. 1991) Determining where, and if, to place structures may be best deferred through several years of stream recovery after changes in land use or biological intervention (Elmore and Beschta 1987).

Structures are generally used for bank stabilization, channel aggradation, and instream habitat improvement. Rock rip rap has been commonly used to stabilize banks for many years. It has proven effective when rock is properly sized. However, it generally changes the character of the streambank, is expensive, and may cause problems downstream.

One of the most promising bank stabilization techniques is the use of trees anchored along eroded streambanks to catch sediment and reduce water velocity. Cut juniper trees anchored along eroded streambanks proved beneficial in stabilizing 96 percent of the erosion on eight streams evaluated in Eastern Oregon over a 14-year period, mainly on straight or slightly curved banks. Water velocities were reduced by 65 percent where juniper revetment was evaluated. Failure occurred on outside curves or where poorly anchored (Sheeter and Claire 1988). Tree revetment is a relatively low cost approach ranging from $4 to $40 per meter depending on site specific factors (Sheeter and Claire 1988). Brush deflectors were found to be similarly effective on stream systems in California (Meyer 1988). Even when the revetment fails because the trees were not cabled because of concerns about bank loss, the riparian response may be positive (Munhall 1996).

Success with instream dams to raise stream levels has been mixed. Some have failed because the gradient was too steep, others because the materials used were inadequate, and still others because they were inappropriate to the stream system. Structure failure also can occur where the stream gathers too much velocity upstream of the treated stretch. "Hardening" of some stream channel segments may be inappropriate (Beschta et al. 1991). If a check dam approach is used, it is important 1) to tie the project to channel type and gradient (low) and 2) to develop rock sources a year in advance (Munhall 1996).

Responses to instream structures that enhance fish habitat are well documented. Woody debris is correlated to the quality of fish habitat and lack of woody debris has been suggested as a factor in the decline of wild stocks of coho salmon in the Northwest (House and Boehne 1986). Placement of instream woody debris and other instream fish structures is common and has a history of success in increasing fish numbers in many stream settings (Reeves et al. 1991, Crispin et al. 1993, Solazi 1995). However, the benefits are not always clear and structures may actually cause problems. Cost-effective restoration efforts such as debris placement need not be a separate project and can occur as part of logging operations (Skaugset et al. 1994).

Instream structure projects often involve transporting and placing logs or other flow deflectors in stream channels dominated by riffles and degraded to the substrate. Causes vary from logging and grazing to floods and stream cleaning. Treatment can be a combination of full spanning structures to create pools, partial spanning structures to maintain pool depth, and bank deflectors. Some projects include construction of off-channel areas and cabling of log and rock placements. Structures are effective in catching additional large woody debris (Crispin et al. 1993).

Instream structures are commonly associated with restoration projects, and many have improved fish population numbers. Although instream structures are common and considerable literature is available on "bioengineering" streams to improve fish habitat, structures can be installed that widen the stream, limit gravel bar formation, or create erosion (Beschta et al. 1991). In some cases, woody material has been introduced into meadow systems where it probably was not present before. The best timing for introduction of instream structures is debatable. Some argue for a coordinated approach so all project features can work in concert to reduce the impacts of major runoff events and improve habitat (Terrene Institute 1994). These may be site-specific decisions that need to be based on stream channel characteristics and conditions.

The focus on the relationship between roads and riparian resources has shifted over the last 30 years. Road construction and maintenance used to focus mainly on keeping the road dry, safe, easy to travel, and conveniently located (Terrene Institute 1994). More recently, road construction and maintenance efforts have begun to address environmental problems related to concentrated flow, accelerated runoff, water table effects, sedimentation, and barriers to stream flow (Terrene Institute 1994).

Road design and location are important. A wide gently climbing road in a steep watershed disturbs more soil material than a steeper, narrower one climbing the same slope (Brown 1980). Road drainage water was responsible for about one-quarter of the road related mass soil movements on the H.J. Andrews Experimental Forest, and about 45 percent of the volume of material moved in the winter of 1964–65 (Brown 1980). Poor road design can yield mass soil movement and surface erosion while techniques such as compaction of fill materials, minimizing side cast material and use of excavators for construction can benefit riparian resources (Brown 1980).

Strategies to address road-related problems include closure, realignment, surface changes and crossing changes. The redesign of culverted crossings has shown clear and consistent benefits. But a wide variety techniques can significantly improve bank stability, reduce sedimentation, improve channel stability, and reduce headcutting (see Box 4).

Periodic fluctuations in water quantity and flow are essential to any riparian system. But some systems have been altered by impoundment (dams) or diversion, making water availability an issue. In one of the most interesting riparian restoration cases underway, restoring flows to a previously de-watered stretch of the Owens River Gorge in California represents an attempt to re-establish a functional riparian system. Hill and Platts (1995, p. 16) indicate "The key to ecosystem recovery and continued health in the Gorge is a multiple flow regime that allows nature to build and maintain riparian habitat." The important point here is the system dependence not just on the presence, absence, or amount of water, but also on a flow regime that increased flows to coincide with the reproductive cycle for cottonwoods. Also promising are the releases from Glen Canyon Dam in an effort to restore beaches and fish habitat in the Grand Canyon.

### 5.1.2 Managing Land Uses

In identifying what problems to address, the starting point is to understand the causes of decline (Elmore and Kaufman 1994). That focus may often lead away from engineered solutions to management of land uses. DeBano and Hansen (1989) recommend using a systems approach to evaluate existing conditions, cause–effect relationships, landscape–vegetative potential, land use management, and maintenance. Where the cause is tied to a current land use, the incompatible use may simply need to be "conducted

**Box 4**
**Suggested Actions for Road Management and Riparian Restoration**

1. Disperse drainage rather than concentrating it.
2. Avoid discharge of large amounts of runoff into non-drainage areas.
3. Avoid changing natural drainage patterns when placing culverts or waterbars.
4. Design road surfacing for erosion control during the wettest period of use.
5. Control use on roads deigned only for dry season use.
6. Outslope drainage when possible.
7. Provide frequent drainage relief, avoiding areas prone to gullying, slumping or landslides.
8. Place energy dissipators at drainage outlets and crossings.
9. Provide for vegetative or mechanical stabilization to mitigate for erosion. Consider larger material to support the roadbed or surfacing with gravel, etc.
10. Keep approaches to streams as close to right angles as possible.
11. Minimize streamside disturbance.
12. Provide special design consideration to crossings including stream diversion, disturbance limits, equipment limitations, erosion control and timing.
13. Consider the use of retaining walls where practical.
14. Design culverts for peak flows.
15. Design culverted roads to avoid the potential for flows down the road even if a culvert is plugged.
16. Field check all riparian-related road designs.
17. Consider the benefits of providing for fish passage or isolating a fish population.
18. Close roads near streams or relocate roads away from streams.
19. Reduce road densities within the watershed.

(Modified from Furniss et al. 1991)

differently, relocated or eliminated to allow the riparian vegetation and stream channel to recover" (Hunter 1991, p. 123). On its surface, the process of managing uses for riparian restoration is a simple cycle of steps: highlight problem areas, decide what to work on (objectives), evaluate alternatives, put action into practice, and watch for results (Adams and Fitch 1995).

In identifying a problem's cause, Hunter (1991) and Reeves et al. (1991) suggested trying to find the "bottleneck" by following an orderly review of seasonal land uses, streambank conditions and instream conditions. Often, the clues to a strategy which will be successful are in comparable case studies that address similar land uses in similar settings (see Box 5).

A possible source of confusion is that historical factors affecting current riparian conditions may be difficult to separate from current land uses. For example, fur trappers in the late 19th century eliminated beavers from some streams in the Pacific Northwest. The loss of beaver ponds affected the extent of floodplains, the erosive power of floods, sedimentation, and organic debris (Elmore and Beschta 1987). Another example would be the 1964 flood on the John Day River in Oregon which lowered the water table, caused hay meadows to erode, created the need for irrigation, and led to straightening and "control" measures. The results were increased velocities, cutting of the bed and banks, and loss of fish habitat quality (Hunter 1991). Even though grazing is the current land use, historical practices and modifications are often key restoration considerations (Robbins and Wolf 1994, Todd and Elmore 1997).

Sorting barriers to riparian improvement sometimes takes careful analysis and participation by individuals with different perspectives. The primary focus for stream restoration is restoring streamside vegetation (Elmore and Beschta 1987). It would seem an easy goal to support. Yet even when diverse interests commit to work together to solve riparian problems, events sometimes occur that break any trust which develops (Natural Resources Law Center 1996). See Case Study 4.

The removal of vegetation in the riparian zone or the watershed is generally a primary cause of problems with bank stability, sedimentation, water quality, and changes in instream characteristics. In many cases, livestock grazing and timber harvest practices have contributed to poor riparian conditions on public lands (DeBano and Hansen 1989).

Where grazing has been tied to the problems observed, changes in grazing are an essential part of the solution. Biologically, the simplest change employed has been to exclude cattle from all or a portion of the riparian zone; this action has consistently resulted in recovery (Elmore and Kaufman 1994). Commonly noted exceptions are where trespass has occurred or

**Box 5**
**Guidelines for Identifying Causes that Limit Stream Recovery**

1. Land use is important because it generally is an aspect affecting limiting factors and project planning. Problems related to similar uses are often similar.
2. Landscape setting is valuable because solutions can often be generalized to other forested, agricultural and suburban landscapes.
3. Land use and setting often correlate well to physical stream properties such as gradient and channel type.

(Modified from Hunter 1991)

### Case Study 4. Willow Creek/Whitehorse Creek: Trout Creek Mountains, Oregon

This case study involves evaluation of riparian functions, identification of restoration objectives and management of livestock grazing in the Great Basin of southeast Oregon near the Nevada border. Restoration is being addressed at the larger watershed and landscape scales. But perhaps most importantly, this case involves creating a collaborative solution in the midst of contentious and controversial issues surrounding an endangered fish species and three Wilderness Study Areas.

*Geographic Area/Location*

The Whitehorse Butte grazing allotment includes 127,000 acres of rugged topography, ranges in elevation from 4,000 to 8,000 feet, and receives annual precipitation of 8 to 12 inches. Willow Creek and Whitehorse Creek (and their many tributaries) are the major streams within the allotment. With 64 miles of stream, many canyons are ribboned with riparian zones comprised of willows, alders, sedges, rushes and grasses. The most extensive riparian areas are along Willow, Whitehorse, Little Whitehorse, Doolittle and Fifteenmile Creeks where Lahonton cutthroat trout were documented in 1991.

*Participants*

Vale District of the Bureau of Land Management, Trout Creek Mountain Working Group, Fish and Wildlife Service, Oregon Department of Fish and Wildlife, landowners and ranchers.

*Restoration Goals*

The goal for 64 miles of creek in the Whitehorse Butte Allotment was to improve those with medium or high potential that were in poor or fair condition to good or excellent condition by 2010. Additional objectives included: (1) reversing downward ecological trends in pastures where they were occurring, (2) maintaining or improving ecological trends elsewhere, (3) stabilizing all meadow soils to alleviate head cutting and loss of meadow by 1995, and (4) managing for aspen regeneration and survival to assure perpetuation of the stands by the year 2010.

*Background*

Livestock grazing had occurred on the Trout Creek Mountains during the summer with little change between settlement in the late 1800s and 1970. In 1970, a 2-day horseback tour of the allotment revealed deep gullies with active cutting, little shade from riparian vegetation, and warm water temperatures. BLM initiated projects planting willows and installing 49 trash collector check dams. In 1973, BLM completed a Habitat Management Plan for the Whitehorse and Willow Creek watersheds. Eight miles of stream were then fenced to create a riparian pasture and small watering ponds were built. Substitute livestock forage was developed in three crested wheatgrass seedings. In 1986, several (additional) exclosures were constructed to exclude livestock from three miles of Willow Creek.

In 1987, tours of the Whitehorse Allotment began to occur regularly. In 1988, the Trout Creek Mountain Working Group was formed to include a cross-section of representatives of environmental, ranching and agency interests. The expressed purpose of the working group was to see that management was changed immediately to make a positive difference on the land. The working group sought to create solutions by building trust, respect, credibility and communications among people with very different perspectives on the problems identified.

In July of 1990, BLM issued a grazing decision to balance livestock grazing and other resource values, building on a 1988 "three year rest agreement" reached with Whitehorse Ranch. The ranch had agreed to rest approximately 50,000 acres of the allotment to allow for riparian and watershed improvement. BLM agreed to develop an allotment management plan to maintain positive ecological trends by meeting the physiological needs of plants. The area was also closed to fishing in 1990. BLM and Whitehorse Ranch agreed upon the grazing plan in 1991, reducing livestock numbers as well as changing the timing, duration and frequency of use. To implement the proposed grazing system, three fences (14.75 miles), one creek exclosure, one reservoir, one spring development, two fence removals (6 miles) and four pipelines were proposed to change the timing and distribution of livestock use over the 127,000 acre area. Lahonton cutthroat trout were listed as an endangered species in 1991.

The project area, although very remote, includes several public values of note. Bighorn sheep have been successfully reintroduced into the area. Recreationists visit Willow Creek Hot Springs and use the area for hunting. Three Wilderness Study Areas and important archeological values are also present.

*Results/Outcomes*

The trash collector check dams met with limited success, with 60 percent washing out within three years. Upward trends are apparent on all streams within the allotment. Substantial increases in the size and volume of woody species were noted on all streams. Streambank cover has increased on all streams. Estimates of fish populations in 1994 by the Oregon Department of Fish and Wildlife were 40,000 fish, an increase attributable to riparian improvement and cessation of drought. Recreation use at Willow Creek Hot Springs has doubled, although deer numbers and hunting use have dropped. Monitoring indicates the overall vegetative trend has been upward since 1989.

*Reference:*

USDI Bureau of Land Management, Whitehorse Butte Allotment Management Plan 1991; Whitehorse Butte Allotment Evaluation 1996 and Doc and Connie Hatfield 1994. History of the Trout Creek Mountain Working Group, 1988–1993 and Michael R. Holbert 1991. Whitehorse Butte Allotment — Controversy to Compromise. *Rangelands* 13(3).

*Contact Persons*

Jerry Taylor, Bureau of Land Management, 100 Oregon Street, Vale, Oregon 97918; telephone (541)473-3144 or Doc and Connie Hatfield, Hatfield's High Desert Ranch, Brothers, Oregon 97712, telephone: (541)576-2455.

fences are not maintained. Even one or two head of cattle in the riparian zone at the wrong time can hurt progress on restoration if they are left along the creek too long (A. Munhall, personal communication 1996).

But changes in the management of livestock grazing can also be an important part of the solution, avoiding the large expenditures often required to exclude livestock from riparian areas (Elmore and Beschta 1987, Kinch 1989). The dominant considerations in designing a solution are to meet the physiological needs of the plants and to ensure adequate vegetation growth to stabilize banks during periods of high runoff (USGAO 1988, Elmore and Beschta 1987). This often involves changing the timing, duration, or frequency of grazing use. These strategies and their relationship to riparian plants and bank stability have been evaluated and rated by Elmore and Kaufman (1994). Information on grazing system compatibility with various stream systems has also been compiled by Platts (1990). Johnson (1992) covered the relationship between grazing practices, infiltration, and sediment production. More specific information is also available on the compatibility of grazing systems with willow regeneration (Kovalchik and Elmore 1991).

Evidence indicates that (1) total livestock exclusion is not necessary on many streams, (2) livestock grazing and healthy riparian systems can coexist during recovery, and (3) grazing management changes may be as effective as exclusion on some sites (USGAO 1988, Elmore and Beschta 1987). Improvement in the management of livestock grazing can produce dramatic results. The riparian zones on many creeks in BLM's Prineville District in Oregon have nearly doubled in width and acreage in response to changes in the season, duration, and frequency of grazing over a 15-year period, while the creeks became deeper, clearer, and colder (Rasmussmen 1995). See Case Study 5.

One of the most important technical considerations in managing grazing in riparian areas is the size of the pasture. If it is a riparian pasture with considerable acreage in uplands, the whole pasture is then managed to meet riparian objectives. This could be an important economic and biological constraint depending on the desired season of use for the uplands. A smaller riparian pasture allows more targeted management for both the riparian area and the uplands, but the riparian pasture needs to be large enough to provide for effective livestock management. Disadvantages of riparian corridor fencing include cost of construction, maintenance problems and costs, and loss of forage (Beschta et al. 1991, Munhall 1996). Mitigating for visual impacts and potential for damage to big game animals are also important design considerations (Beschta et al. 1991).

Summer grazing has been continued in several cases, although the tendency is to incur significantly heavier utilization by livestock in the riparian zone during the hotter months. Moving livestock based on utilization levels may necessitate changing forage allocations, additional management measures (e.g., herding), and more intensive monitoring to ensure moves are timely (Munhall 1996). Timing grazing moves to meet the physiological needs of the plants and to allow vegetation regrowth to meet the physical needs of stream function may require use of techniques such as herding, additional fencing, and more frequent moves. Reliance on reducing numbers of livestock or utilization targets may lead to improvements in the upland vegetation, but the riparian zone is not likely to improve (Platts 1990). In cases where changes in the amount of livestock grazing (numbers or utilization) did not produce the desired riparian improvement, a change in timing may have been more successful.

In sampling 22 riparian areas in 10 Western States, the GAO (USGAO 1988) noted that the overriding factor in achieving success was improving the management of livestock to meet the life-cycle needs of native vegetation by herding and/or fencing often in combination with a shorter grazing period, a specific season of use, or limitation to a portion of the area. Moore and Flaherty (1996) also noted that grazing systems need to adapt the frequency and seasonality of use in response to site specific characteristics, year-to-year trends, and seasonal changes.

Logging practices also affect riparian condition, but experience with varied restoration strategies is not as well-developed as with grazing. Logging factors generally include the amount of harvest allowed or the prescribed buffers established to exclude it. Available case studies suggest more complex species- or site-specific strategies are much less common. The logging-related problems most consistently identified have been modified plant composition, loss of woody debris (long term), sedimentation, streambank damage, and erosion.

As with grazing, the streamside zone is a distinct management unit in a forest watershed that deserves special attention based on its sensitivity, variability, and complexity (Brown 1980). Yet little work has been done regarding the effects of forest practices on riparian function, other than the recruitment of large woody debris (Reiter and Beschta 1995). The most widespread forestry approach is to exclude or limit activities in the riparian zone (Moore and Flaherty 1996). This same approach applies to avoidance of disturbances that may damage upslope stability (Swanston 1974). Avoiding use-related impacts allows stream structure and processes to function, promotes higher infiltration rates, and minimizes surface runoff (Brown 1980).

## Case Study 5
## Bear Creek, Crook County, Oregon

This case study involves evaluation of riparian functions, identification of restoration objectives and management of livestock grazing in the Columbia Plateau of Central Oregon. The restoration effort integrated structural treatments such as juniper revetments, but emphasized changes in livestock grazing. Data gathered in the late 1970s and early 1980s allows a longer term assessment of these changes than is available in many locations. Significant improvement in riparian conditions occurred with exclusion of livestock grazing and with grazing management strategies appropriate to the riparian area.

*Geographic Area/Location*

Bear Creek is a 42 mile tributary to the Crooked River that flows into Prineville Reservoir in Central Oregon. The 16 miles of creek managed by the Bureau of Land Management are in a 12 inch precipitation zone at elevations of 3,400 to 4,000 feet. Soils are sandy loams. Grazing management practices are coordinated with adjacent areas managed by the Ochoco National Forest and by private landowners/ ranchers. The stream segments in Bear Creek which were surveyed total approximately 7.5 miles. Historically, the primary riparian communities were willow/birch and sedge/ rush/grass. The creek includes a population of redband trout and beavers are present. Flood events are most commonly due to storms which intensify late winter runoff or summer thunderstorms.

*Participants*

Prineville District of the Bureau of Land Management, Ochoco National Forest, landowners and ranchers.

*Restoration Goals*

The goal for Bear Creek was to restore riparian function while also improving upland vegetation. Problems identified were reduced riparian vegetation, incised channels, and large movements of sediments. Work was to be coordinated with efforts to reduce erosion and sedimentation into Prineville Reservoir. The principal management objective was to protect streambanks against erosion by high flows during spring runoff and high intensity summer thunderstorms.

*Background*

Livestock grazing has occurred on Bear Creek since the 1800s and was typically summer-season long. By the 1970s, most of the stream was in poor condition. In 1973, BLM completed a watershed plan for Bear Creek. The timing and duration of livestock use was changed to better coincide with the needs of the plant species present so they could contribute to hydrologic function. Surveys of riparian conditions, habitat use, water quality and bank stability were conducted on about 400 miles of streams in the Prineville District in the late 1970s and early 1980s. Bear Creek was one of the areas covered by the surveys.

An exclosure was placed along Bear Creek in 1978. Strategies to schedule livestock grazing (e.g: deferred rotation and winter/spring use) were expanded in 1978. Juniper rip rap was placed against selected eroding riparian banks in 1982. BLM cut upland juniper trees from parts of the watershed and divided the riparian area into three pastures in 1985. Juniper treatments for watershed improvement purposes had been occurring in the watershed since mid-1970s.

*Results/Outcomes*

Dramatic changes have occurred in riparian condition between 1978 and 1994. In most cases, grazing has continued but is managed to control when and how long the livestock are present to fit the needs of the riparian zone. Based on the 7.5 miles sampled in four segments, the size of the riparian plant community has increased by 76 percent. The area affected by eroding and damaged banks has declined by over 90 percent. Seventeen to 26 percent increases in the grass-sedge-rush community occurred, with corresponding increases in litter and decreases in forbs and bare ground. In some areas beaver activity has greatly increased.

Considering most of the streams resurveyed in 1994, these results are not unusual. Most other streams also appear to be in better condition in 1994 than they were in 1978. It is important to note that aggrading and expanding riparian systems may experience a loss of shrubs and woody species depending on the substrate. Some occurs because riparian shrubs cannot tolerate year-round saturated soils. But even some riparian species require well drained soils which may not be present under some circumstances. Available data supports the contention that improvement of riparian condition can and will occur with a change to an appropriate grazing management plan if grazing was instrumental in causing the deterioration.

*Reference*

Rasmussen, Christine. 1995. Riparian Community and Bank Response to Management: A Comparison of Old and New Surveys in the Prineville District, Bureau of Land Management; Chaney, Elmore and Platts. 1990. Livestock Grazing on Western Riparian Areas. Environmental Protection Agency.

*Contact Persons*

Wayne Elmore, Bureau of Land Management, National BLM/Forest Service Riparian Team, 3160 East 3rd Street, P.O. Box 550, Prineville, Oregon 97754; telephone (541) 416-6700 or John Swanson, Bureau of Land Management/ Forest Service, Prineville District, same address and phone number.

Historically, forest practices have dramatically altered stream channels with splash dams, stream cleaning, increased frequency of debris torrents, sedimentation, streamside harvesting, and the reduction of large wood in channels and riparian areas (Reiter and Beschta 1995). But with the wide range in forest practices and ecological sites, it is also clear that even though forest practices can significantly alter hydrologic systems in some instances, in others they may have little or no effect (Reiter and Beschta 1995). There is no substitute for site specific analysis. Harvest dispersal can mitigate impacts (Harr 1981). However, modification of harvest levels alone, although attractive in its simplicity, is often not as relevant to mitigating impacts as how and where harvest occurs (Reiter and Beschta 1995).

A "Best Management Practices" (BMP) approach is commonly applied to timber uses. The effects of BMPs on sediment discharge have been documented. Erosion rates from roads can be reduced by 80 percent, whereas rates from harvest areas can be reduced by 35 percent by employing BMPs (Rice 1992). BMPs generally direct special attention to perennial streams and the recognizable area dominated by the riparian vegetation. The intent is to avoid management practices that cause detrimental changes to water, stream channels, erosion, or fish habitat (Harr 1981).

BMPs can reduce sediment delivery to streams by reducing soil compaction, and thus erosion at peak flows (Brown 1980). Examples of alternative harvest systems which can significantly reduce compaction and soil disturbance include small, low-cost cable logging systems on slopes over 30 percent, intermediate supports to extend lift, low ground pressure vehicles, and designated skid trails (Brown 1980). Techniques such as uphill or parallel felling to minimize soil and vegetation impacts also lower the susceptibility to erosion. Yarding systems that lift logs off the ground or uphill yarding to avoid concentrating runoff can also reduce erosion (Moore and Flaherty 1996). Although sedimentation depends on both erosion processes and forest practices, the effects of sediment loading generally decrease as the distance to the activity from the stream increases (Reiter and Beschta 1995).

The effectiveness of buffer strips has been repeatedly demonstrated as a technique to maintain desired stream temperatures. A 50–80-foot buffer strip on Little Rock Creek in Oregon was nearly as effective at maintaining temperature as an uncut forest stand (Brown 1980). Leaving a buffer when applying fertilizer or avoiding slopes where it may readily flow into sensitive areas also helps protect stream water quality (Moore and Flaherty 1996). But buffer strips can also result in increased blowdown, bank damage, and, potentially, increases in temperature (Brown 1980). Buffers also have limited ability to filter sediments resulting from upslope erosion. For buffer strips to be most effective, they must be designed based on the steepness of slopes with gradual transitions between buffer strips and cut areas (Moore and Flaherty 1996).

The degree of shading in a riparian zone, rather than the width of the buffer, is probably the best predictor of effects on water temperature and long-term fish habitat quality (Gregory et al. 1977, Moore and Flaherty 1996). The direction of stream flow is also significant in evaluating shading, with the north side of a stream providing little shading and shading being naturally limited on north–south flowing streams (Brown 1980).

Although forestry practices can potentially alter long-term solar radiation, water temperature, sediment, nutrients, litter inputs, woody debris and channel structure, canopy removal may also cause some short-term increases in fish production (Gregory et al. 1977, House and Boehne 1986). But the ability of the stream to retain algae and litter inputs is closely tied to large woody debris (Gregory et al. 1977). Sedimentation or decreased substrate stability may also decrease abundance of aquatic insects and thus affect the food supply for fish (Gregory et al. 1977).

Site characteristics around the riparian zone are also important. For example, shallow granitic soils after saturation during high intensity rain storms are more prone to slides because of rapid increases in pore water pressure that lubricate the soils and affect sheer strength (Swanston 1974). When timber harvest occurs, sheer strength changes as root strength declines over the following three to five years, particularly on shallow soils and steep slopes (Swanston 1974, Harr 1981). Areas of higher soil instability are often associated with improper placement or design of logging roads (Swanston 1974; see also the discussion of roads above.). With these factors in mind, riparian and stream function restoration begins with forest practices upslope in the watershed.

Intensive recreation use occurs in many riparian zones because they are preferred environments for fishing, camping, picnicking, hiking, and other uses. Problems arise when the location, timing, and intensity yields multiple trails, heavily used campsites, and vehicle routes in the riparian zone. The effects on vegetation and soils include compaction, root exposure, loss or change of vegetation cover, change in the soil profile, reduction in organic matter, increased bulk density, and decreased soil moisture (Clark and Gibbons 1991). Most recreation impacts occur as new uses are introduced into an area with little or no previous use, while continued or increasing recreation use in areas with heavy uses may have little additional effect (*ibid*).

As a result, the most effective approach to managing recreation impacts is to direct use, harden sites and exclude use on other sites. Recreation use density decreases with distance from access points (Clark and Gibbons 1991), and use levels at some sites can be controlled by relocating or closing roads, trails and parking areas. Improving a primary road or trail, while closing other routes, can successfully redirect recreation use (Caffrey and Rivers 1993; USDI BLM Deschutes Resource Area 1996). The response in areas where use is excluded is roughly similar to vegetation and bank stability improvements described for exclusion of other uses.

Successful restoration in managed recreation sites often relies on the ability to direct use to "hardened" sites while closing more sensitive sites. As with roads and trails, facilities such as launch ramps, docks, restrooms, tables, fire grates, shelters, parking areas and signing have tremendous influence over the distribution of use (Clark and Gibbons 1991; Smith and Pritchard 1992). Properly located, they can attract use away from sensitive areas. This allows an outlet for recreation use pressure and localizes impacts to an area that can be surfaced with gravel or some other agent to enhance stability and retard erosion (Moore and Flaherty 1996; USDI BLM Deschutes Resource Area 1996). Reducing the area of bare soil and associated areas where vegetation is dominated by annual species improves streambank stability and riparian area composition as a whole.

Managers tend to be more aware of recreation impacts than visitors (Clark and Gibbons 1991), suggesting that one solution is to do a better job of making information available about riparian potential and trends. Many interest groups have become involved in riparian restoration projects and direct involvement enhances awareness (Clark and Gibbons 1991). Once recreation users understand the need for restoration effort, they are more likely to comply with closures and restrictions.

### 5.1.3 Biological Intervention

In many cases, desired riparian species are absent or declining. Some restoration strategies have centered around reversing declining trends of a specific species or re-establishing a species that is absent. Intervention on behalf of a riparian plant species is similar in concept to wildlife species intervention or reintroduction. Just as steps are sometimes taken to promote one wildlife species by controlling a predator or a competing wildlife species, strategies have also been developed and applied to promote restoration of selected riparian species.

Causes of riparian problems are not limited to human uses. Grazing by ungulates (deer and elk) also contributes to altering plant cover, causing soil compaction or disturbing sensitive soils in some areas (Gifford 1981, Edgerton 1985). Exclusion of deer and elk can increase crown cover and the cover of grasses and sedges (Tiedemann and Berndt 1972; USDA Forest Service, Aspen Working Group 1995; Shepherd 1996).

Although grazing by big game and other species is not frequently addressed in riparian restoration, it can be a factor where predators or other factors are insufficient to limit big game populations (Chadde 1989, Crowe 1996). Foraging by deer and elk can retard establishment of some riparian species (Beschta et al. 1991, Chadde 1989, Emmington and Maas 1994), with subsequent effects on bank stability and riparian function. In some areas, the most common cause of riparian plant mortality or retardation of riparian recovery can be herbivory by beavers or big game (Emmington and Maas 1994).

Given the practical constraints on controlling the timing, duration or frequency of foraging by ungulates, the only likely options are accepting the impacts, fencing to exclude big game or controlling the size of big game populations. Excluding big game has been successfully employed on several streams in Oregon (Beschta et al. 1991, Anderson 1994) and mitigation measures are used in Yellowstone National Park (Chadde 1989). Species-specific exclosures or cutting nurse trees to protect vegetation reproduction from big game are options but fencing to exclude elk can be difficult and expensive (USDA Forest Service, Aspen Working Group 1995).

Underburning is an effective tool to release regeneration in self-perpetuating clones such as aspen. However, there is a risk of losing significant numbers of mature trees to promote resprouting. (USDA Forest Service, Aspen Working Group 1995). Underburning is also a tool to control conifer invasion and restore balance between conifers and aspen (Tewksbury 1996).

Planting desired species to advance bank stabilization or to structure the vegetation community has also been used in many areas, although stabilization with a specific species in mind can sometimes be expensive and difficult (Swanston 1974). In southeast Alaska, a mix of reed canary grass and alder wildlings has been successfully established. In other cases, grass and other species may be seeded simply to establish cover for stability (Brown 1980). Some caution regarding the use of non-native plants is needed. Tamarisk, an aggressive exotic which is also an excellent bank stabilizer, was established in California and now ranges throughout much of the Colorado River system. It is now the target of many eradication efforts.

Technology for re-establishing native woody species like willow and cottonwood from cuttings is well developed. Cuttings can be taken at almost any time during the dormant season, but cuttings from late winter seem to be best and removal of flower buds (or selecting without buds) is desirable (Farmer 1966, Hoag 1995). Cutting, storage, and soaking techniques to enhance success are also reasonably well-documented (Krinard and Randall 1979).

Deep planting with cuttings reaching about 15 cm (6 inches) into the low water table and tall enough above the ground not to be shaded is most effective (Hoag 1995; Randall and Krinard 1977). Various augers and a "stinger" attachment for backhoes can be effective planting tools (USDA SCS 1994). Success rates have increased to 70 to 80 percent after two years for deep planting on some sites (Hoag 1995). Deep planting seems to be particularly useful in arid and semi-arid environments.

Designing planting projects on forested sites requires consideration of both tree growth and vegetation diversity (Chan et al. 1993). Growth may be important to increase the rate at which large wood becomes available, while structure is influenced by diversity. Shade tolerance is sometimes the controlling factor in the growth-diversity tradeoff. Small patch cuts and clearings can reduce competition but may also promote brush species (Emmington and Maas 1994). Partial removal is an alternative to both reduce brush competition and stimulate growth because it can reduce canopy closure and growth is directly correlated to canopy retention (Chan et al. 1993). In some systems, planting is unlikely to succeed without some form of manipulation to the overstory and understory. Additional understory treatments after the initiation of the project are also likely to be necessary to ensure growth.

Success in planting native conifers also requires attention to slope location, site conditions and species selection (Chan et al. 1993; USDI BLM 1993 and 1994). Good survival rates have been attained for western red cedar on Spencer Creek, Johnson Creek and West Fork Smith River in western Oregon, whereas success with western hemlock was poor (USDI BLM 1993; USDI BLM1994; see Box 6).

Vegetation composition in the watershed is one determinant of water availability and riparian function. Watershed studies have shown clearcut timber harvesting can increase water yields for decades and augment low flows for several years (Reiter and Beschta 1995). Entire basins with ongoing timber harvest activities may also show long-term increases in water yield based on stream flow records. Treatment by cutting or burning some non-timber tree species such as western juniper may also increase infiltration, soil retention, and water yield in riparian areas (Bedell et al. 1993). Brush control activities can also extend both the duration and amount of streamflow from the treated watershed. There is potential for improvement in streamflow that will enhance riparian vegetation below treated watersheds (DeBano et al. 1984).

**Box 6**
**Three Year Results for Planting Native Conifers in BLM's Umpqua Resource Area in Oregon**

1. Managing understory vegetation in the first season increases survival and growth of conifer seedlings.
2. Survival and growth of conifer seedlings increases when alders are girdled.
3. Of the treatments applied, managing understory vegetation and girdling 50 percent of the alders provides the best survival and growth for all species of conifer seedlings.
4. Western hemlock had the best overall performance under the experimental conditions of the study.

(Maas and Emmington 1994)

## 5.2 Summary for Riparian Restoration

The key to planning riparian restoration is to avoid blanket prescriptions. The stream setting and ecological site conditions need to be factored into the objectives and the design of the riparian restoration project (Chan et al. 1993). It is also important to ensure that the project addresses the underlying causes that are perpetuating declines in riparian condition.

Many projects have included features designed to accelerate progress to a desired state, in terms of function, plant composition, or habitat quality. Check dams are designed to improve pooling and promote aggradation. Willow and cottonwood plantings are intended to change plant community composition quickly. Instream structures are intended to reintroduce or simulate instream debris. These approaches have been successful under the right circumstances and with the right techniques. But there have also been high failure rates with some "accelerator" techniques such as instream structures. This suggests a thorough investigation is needed before a decision is made on which techniques to use.

Suites of coordinated actions that address riparian restoration are common. Combinations vary but they include grazing management, reforestation, road and trail closure, riparian plantings, instream structures, and facilities to direct recreation use. This may make it difficult to attribute responses in riparian condition to any single action. A number of conclusions may be drawn concerning riparian restoration:

1. Efforts to improve riparian conditions are widespread with varying results and degrees of success.
2. There is a consistent trend toward improvement on most of the riparian projects sampled, although improvement does not always occur at the same rate on different sites.
3. Indicators of improvement are consistent: increased streambank stability, decreased erosion, decreased sedimentation, narrowing and deepening of the stream channel, colder stream temperatures, and improved fish production.
4. Many of the projects address combinations of factors such as species composition, logging effects, recreation effects, impacts related to road design, and grazing effects.
5. Successful projects generally integrate an understanding of the natural and social factors such as climate, topography, soils, land use, history, compliance, and community interests with site-specific factors like stream channel gradient, riparian soils, and flow regimes.
6. Improvement occurs with management of uses, when uses are designed to ensure the system's functional needs are met.

Table 1. Technical References for Riparian and Wetland Areas with Reference Numbers (BLM National Business Center).

| | |
|---|---|
| Riparian Area Management: A Selected Annotated Bibliography of Riparian Area Management (BLM) | TR 1737-1 |
| Livestock Grazing on Western Riparian Areas (EPA) | P-312 |
| Inventory and Monitoring of Riparian Areas (BLM) | TR 1737-3 |
| Grazing Management in Riparian Areas (BLM) | TR 1737-4 |
| Riparian and Wetland Classification Review (BLM) | TR 1737-5 |
| Management Techniques in Riparian Areas (BLM) | TR 1737-6 |
| Procedures for Ecological Site Inventory — With Special Reference to Riparian-Wetland Sites (BLM) | TR 1737-7 |
| Greenline Riparian-Wetland Monitoring (BLM) | TR 1737-8 |
| Process for Assessing Proper Functioning Condition (BLM) | TR 1737-9 |
| The Use of Aerial Photography to Manage Riparian-Wetland Areas (BLM) | TR 1737-10 |
| Process for Assessing Proper Functioning Condition for Lentic Riparian-Wetland Areas (BLM) | TR 1737-11 |
| User Guide to Assessing Proper Functioning Condition and the Supporting Science for Lotic Areas | TR 1737-15 |

Many restoration efforts to improve streamside vegetation, retain soils, raise water tables, improve water quality, stabilize streambanks, and improve fish habitat have clearly been successful where they have been undertaken. But with the number of miles of stream still needing attention, much needs to be done.

A list of references from the U.S. Bureau of Land Management on management techniques applicable to riparian systems is shown in Table 1.

## 6 WETLANDS

*Scenario: Wetland function has been impaired.*

Wetlands areas are inundated by standing, shallow surface or ground water at a frequency and duration sufficient to support, under natural conditions, a prevalence of vegetative and aquatic life dependent on the saturated condition for growth and reproduction (Bridges et al. 1994, Manley et al. 1995). These include, but are not limited to, bogs, muskegs, marshes, swamps, estuaries, riparian areas, wet meadows, and wet forests.

Wetlands perform critical functions as natural sediment and pollution filters, providing clean water to living organisms, including humans. They are very productive areas for both plants and animals because of the abundance of water, nutrients, and minerals that sustain life, often providing critical habitats for many species. Wetlands also serve as buffer zones during storm events, reducing soil erosion and sedimentation.

At one time, wetlands accounted for about 11 percent of the total land area in the United States. This has dwindled to about 5 percent (Dahl 1990, National Research Council 1992). Reasons for reduction in wetlands are many and varied. However, agriculture and development account for a large portion of the total (Frayer 1991; National Research Council 1992). Of the remaining wetlands, many are considered degraded or nonfunctional.

Wetlands are considered degraded when one or more of the following are true:

- water quality has dropped below historic levels
- algal blooms are far in excess of expected levels
- native plant and animal communities have been altered

- hydrogeomorphic processes have been altered
- soil erosion and deposition processes have been altered.

Box 7 lists clues that can be used to quickly identify degraded wetlands.

Returning these attributes and processes to levels that provide for wetland functions, and are more consistent with their natural range of variability, is the essence of wetland restoration. To restore a wetland to proper function, the attributes of, and processes occurring in, a wetland must be well understood. Table 2 presents several wetland attributes and processes from Bridges et al. (1994). Although there are similarities in function between wetland areas, there are also many differences. Therefore, each area must be evaluated relative to its own unique capability (Bridges et al. 1994).

Succession is a natural process in any wetland. The process is generally very slow when allowed to occur at a normal rate. The state of succession represented by the wetland to be restored must be considered throughout the assessment process, and when establishing objectives or desired future conditions. It is often advantageous to consider the current balance of successional stages represented and give high priority to restoring wetlands to underrepresented stages rather than to some historical stage. In the assessment process, attributes and processes (such as those listed in Table 2) should be evaluated against both those considered "natural" for a specific wetland and against criteria for wetland function (Pritchard et al. 1994, Gebhardt et al. 1990; Clemmer 1994).

Table 2. Some Attributes and Processes for Assessing Wetland Function (Bridges et al. 1994).

| | |
|---|---|
| **Hydrogeomorphic** | **Erosion/Deposition** |
| Ground water discharge | Shoreline stability |
| Permafrost | Depositional features |
| Continuous | |
| Discontinuous | **Soils** |
| Flood Modification | Soil type |
| Inundation | Distribution of aerobic and anaerobic soils |
| Depth | Annual pattern of soil and water states |
| Duration | Ponding frequency and duration |
| Frequency | Underlying materials |
| Semipermanently flooded | |
| Shoreline shape | **Water Quality** |
| | Temperature |
| **Vegetation** | pH |
| Community types | Dissolved solids |
| Community type distribution | Dissolved oxygen |
| Surface density | |
| Canopy | **Biotic Community** |
| Community dynamics and succession | Aquatic plant recruitment and reproduction |
| Recruitment and reproduction | Nutrient enrichment |
| Root density | |
| Survival | |

**Box 7**
**Visible clues to decline in wetland function**

- headcutting
- evidence of sheet rill or gully erosion
- evidence of a lower water table such as surface drying
- declining populations of wetland plant species
- encroaching upland plant species
- disappearance of wetland obligate species

(Based on Zeedyk 1996)

The support and partnerships for wetland restoration efforts can often be very difficult, especially in areas where the wetland or the water is highly valued for uses other than habitat for native plants and animals. For example, in the prairie pothole region, wetland areas are very productive for agricultural uses. Resistance to "giving this land back to the birds" is quite natural. However, by showing landowners that profitable agriculture and abundant wildlife can coexist, resistance to restoration of these wetlands is eroding. See also Case Study 6.

## 6.1 Management Options for Wetland Restoration

Options for restoration of wetland systems are grouped into the three different approaches discussed above: structural treatments, management of land uses, and biological intervention. See also Covington et al. (this volume) for additional case histories related to salt marsh restoration in the Northeastern United States.

### 6.1.1 Structural Treatments

A common problem in wetland ecosystems is alteration of hydrologic flows by roads and trails. Until normal

**Case Study 6. Restoration of the Everglades in South Florida**

This case study provides an overview of the long-term, intensive effort in south Florida to restore the Everglades. Channelization, drainage, and filling of wetlands in the south Florida ecosystem have gradually altered the hydrologic regime, resulting in serious changes to natural communities. A large-scale program is now in place to reverse this process and restore balance to the system.

*Geographic Area/Location*
The south Florida ecosystem encompasses about 10,800 square miles, which includes at least eleven major physiographic provinces and a population of six million people.

*Participants*
National Park Service, Fish and Wildlife Service, National Biological Survey, United States Geological Survey, Bureau of Indian Affairs, National Oceanic and Atmospheric Administration, National Marine Fisheries Service, National Ocean Service, Natural Resources Conservation Service, Agricultural Research Service, U.S. Army Corps of Engineers, U.S. Attorney's Office for the Southern District of Florida, and U.S. Environmental Protection Agency.

*Restoration Goals*
To restore and maintain a healthy, balanced and functioning estuarine and marine environment in south Florida. The result should be a system where human activities and actions occur in a manner that supports healthy natural conditions and leads to diversity and abundance of natural biological systems.

*Background*
Prior to the late 1800s, wetlands covered most of central and southern Florida. The landscape was characterized by swamp forest; sawgrass plains; mosaics of sawgrass, tree islands, and ponds; marl-forming prairies; wet prairies; cypress stands; pine flatwoods; pine rocklands; tropical hardwood hammocks; and oak hammocks. The estuarine-coastal system was composed of shallow seagrass beds; riverine and fringe mangrove forests; intertidal flats; coral reefs; hard bottom communities; mud banks; and shallow, open inshore waters. This system was sustained by a hydrologic system that stored and released water on a large scale over a vast area, providing diverse habitat for innumerable plants and animals.

In the late 1800s, efforts began to control and drain water in the south Florida ecosystem. The main objectives were to provide flood protection, produce more arable land, provide land for development, and protect the water supply of a growing population. A multitude of water control structures (dikes, levees, canals, and pumping stations) were built, resulting in major changes to the hydrologic flow regimes and ecosystem processes in the region. Losses in ecosystem function were coupled with overall reduction in the extent of the wetlands as land development progressed. Currently, wetland ecosystems in the region are greatly reduced in extent and highly fragmented, with poor hydrologic connectivity, reduced biodiversity, and reduced biological productivity.

Restoration of this ecosystem to pre-late 1800s conditions is impossible given current land use patterns. This restoration effort is focused on: (1) restoring proper hydrologic flows to critical habitats, (2) recovering healthy populations of plant and animal species, especially those that are threatened or endangered, (3) preventing further wetland loss, (4) recovering as much ecological structure and function as possible, (5) halting and reversing the invasion of exotic plants and animals, (6) preventing pollution, and (7) promoting water conservation and reuse in urban and agricultural areas. A detailed plan aimed at achieving this has been developed and many restoration projects are currently underway.

*Results/Outcomes*
The intended final result of this long-term effort is a sustainable, productive, self-maintaining system that can coexist in harmony with human needs and demands. This will not happen for some time, but progress is being made.

*Reference*
Jacobsen, J.M. 1994. An Analysis of Fifty Years of Success and Failure in Everglades Management. Ochopee FL: The Everglades Institute

*Contact Persons*
Richard G. Ring, Superintendent, Everglades National Park, 40001 State Road 9336, Homestead, FL 33034-6733. Phone: (303)242-7710; Fax: (305)242-7711

flow patterns are returned to the system, attempts to restore native flora and fauna will be ineffective. Restoring hydrologic function to a wetland can be tricky. If the wetland is being drained, a water impoundment structure may be needed. If the area is being inundated with water, some type of drainage structure may be needed. Surface flow problems are usually apparent, but subsurface flows may be more difficult to assess. Subsurface flows are often impeded by roads and trails as a result of compaction of subgrade material. This causes water flow imbalances across the wetland and can result in channelization.

As in the terrestrial and riparian scenarios presented above, changes in road design and maintenance can either be a threat to ecosystem function or an avenue for restoration. Zeedyk (1996) suggests potential effects are tied to road location and alignment, channel crossing location and design, drainage design, road surface materials, sediment abatement and filtration, and wildlife habitat buffers. A variety of technical treatments, including relocation to adjacent sideslopes, are available to fit most situations.

The distribution or effectiveness of water to support wetland functions can also be affected by diking diversion and topographic simplification. For example, in the Warner Wetlands in Oregon, wetlands had been leveled for agricultural use. Dikes had been placed to facilitate flood irrigation. The flow of water onto the area was also affected by a dike to augment water storage in an adjacent lake. Restoring wetland function and sustained habitat conditions required raising and constructing dikes and mounds to increase topographic relief and influence water movement (USDI BLM Lakeview 1990). Technical information on creation of wetlands has been summarized by Scheller-McDonald (1990).

Water availability is a critical, and sometimes controversial, need. In the San Luis Valley in Colorado, much current wetland habitat is supported by wells. Development of water within the valley (some proposed for transport outside the basin) affects the quantity and quality of water available for wetlands and can be complicated by political issues surrounding water rights (Cooley 1991). But the only realistic long-term solution is persistent coordination among all the parties and interests (USDI BLM, Canon City 1992) (see Case Study 7). A similar need for dependable water in the Warner Wetlands of Oregon was resolved by acquiring a water right with a land acquisition and then drilling a well (USDI BLM Lakeview 1990).

Although not intuitively obvious, wetlands do burn. Some wetlands evolved with fire as a significant disturbance regime and, therefore, need fire to maintain conditions. Suppressing fire in these wetlands will cause many changes from the "natural" state. Prescribed burns in these areas at intervals suggested by natural cycles should correct imbalances. If this is not possible for technical or social reasons, some type of vegetative treatment that mimics the effects of fire (such as mowing) might be used to get the same result.

This substitution of one attribute or process for a similar-acting one (i.e., mowing or grazing for fire) can be used whenever the appropriate relationships exist. Care must be taken to validate that the substitute attribute or process is truly interchangeable with the natural one. Examples of attributes or processes that might be substituted for natural processes in wetland restoration projects include water control devices, cultural practices (such as soil working, soil stabilization, mulching, and fertilizing), or manipulating the existing vegetation.

### 6.1.2 Managing Land Uses

Another common problem in wetland ecosystems is sedimentation. Sedimentation is a normal process in wetlands. Agricultural and forestry practices are often implicated in accelerating natural erosion rates. Possible solutions include using appropriate erosion control practices (culverts, water bars, broad-based dips) on lands in the watershed and leaving vegetative filter strips between wetlands and adjacent lands. See also the riparian scenario above.

Chemical loads are a serious concern in wetlands. Wetlands generally act as sinks, causing chemical concentrations to build rather than dissipate. This can affect all organisms in the wetland or only specific sensitive species. Chemicals may come from industrial, agricultural, forestry, or urban sources. Point sources are generally easier to affect than non-point sources. Eliminating the source of the contaminant is the most effective way to deal with the problem. This is often not possible, however. In some cases, adding other substances (e.g., lime to raise pH) or organisms (i.e., bacteria to break down organic chemicals) may help.

Sometimes the main factor degrading a wetland ecosystem is how it is being used. Examples include overharvest of wild rice, clams, or timber resources; recreation use by boats; or food chain disruptions by excessive hunting or trapping. The common method to address wetland uses is to regulate them to a level that can be sustained without degrading the ecosystem. Often regulation includes restricting access or imposing harsh penalties. But in other cases, seasonal limitations will accomplish the desired goals.

Exclusion of uses can sometimes be effectively combined with changing the timing and duration of the remaining use. The Bureau of Land Management found portions of the 51,000 acres of public lands in the Warner Wetlands were best managed without grazing, while grazing was compatible on others. In one segment of the management area, grazing use is designed to maintain habitat for long-billed curlews (USDI BLM Lakeview 1990). See Case Study 8.

### 6.1.3 Biological Intervention

One restoration technique is to intervene through seeding and planting native vegetation or reintroducing wetland species. In concept this is similar to substitution of one attribute or process for another. The appropriate conditions and relationships must exist and care must be taken not to introduce undesirable species or genetic characteristics.

Invasion of exotic plants and animals into wetland ecosystems can be destructive to native species. Depending on the type, exotics can attack native species directly or crowd them out by utilizing scarce resources. Exotics often spread rapidly, with no natural enemies to keep their populations in check. The best approach is to remove them from the system before

**Case Study 7**
**Blanca Wetlands Area, San Luis Valley, Colorado**

This case study involves re-establishment of wetland functions and coordination among restoration programs by adjacent managing agencies with similar restoration objectives. The restoration effort is an attempt to integrate a regional set of objectives for a portion of the Central Flyway. The key is getting management agencies to agree on their roles and then make a commitment to fulfil and maintain those commitments. Management decisions for Blanca Wetlands Area also deliberately incorporated social and economic factors, anticipating future demands on the area.

*Geographic Area/Location*
Blanca Wetlands Area covers 22,363 acres in three tracts, one of which (9,774 acres) is an Area of Critical Environmental Concern (ACEC). It is northeast of Alamosa, Colorado in a broad, high, mountain valley (about 7,500 feet) drained by the Rio Grande River. The core of the Blanca Wetlands is unique to the valley because it is not subject to riverine cycles and influences. Five different ecological sites were identified based on soil vegetation associations. Uplands of greasewood, rubber rabbitbrush, saltgrass, sandhill muhly and sand dropseed are interspersed with wetlands supporting bulrushes, cattail, spike rush, pond weed and watermilfoil. Willows, cottonwoods and hackberry trees add vertical structure.

*Participants*
Bureau of Land Management (BLM), Fish and Wildlife Service, Colorado Division of Wildlife, State of Colorado (Water Resources, Transportation, Parks), Bureau of Reclamation, landowners.

*Restoration Goals*
The goals for Blanca Wetlands are to: (1) provide a place to achieve piece of mind with an opportunity to experience an interaction between mankind and nature, (2) provide restoration of healthy sustainable wetland habitat of sufficient quality to foster viable wildlife and plant populations that will benefit and be accepted by the community, and (3) provide a mosaic of diverse sustainable upland and wetland habitat types which blend into the existing setting. The plan then incorporates specific objectives and land use allocations.

*Background*
Waterfowl populations in the San Luis Valley declined 50 percent in the 1960s and 1970s. The primary cause appears to be loss of habitat, particularly nesting habitat. Publicly-owned wetlands in the valley were being managed primarily by the Colorado Division of Wildlife, Fish and Wildlife Service and the Bureau of Land Management. The Blanca Wetlands (BLM) are at the lower end of the basin in an area known as the "sump" which historically had shallow playa lakes and associated wetlands. As the water table declined, these became dry.

The original BLM Habitat Management Plan was completed in 1968 after a pilot project of three artesian wells and some control dikes was completed in 1966. Since the 1960s, BLM has drilled a total of 51 wells, providing water from an underground artesian aquifer. This water was supplemented by obtaining project water from the Franklin Eddy Canal as mitigation for pumping that occurs in the Closed Basin project area. Legal water issues continue to be a major concern affecting the project.

In 1991, the Colorado Division of Wildlife, Fish and Wildlife Service, Bureau of Reclamation and BLM prepared a joint San Luis Waterbird Plan. The plan agreed to focus management efforts on 25 percent of the wetlands in the San Luis Valley, with the most intensive restoration and management efforts dedicated to 30,000 acres. The plan also presented issues and strategies for waterbird/wetland management, and set shared objectives to guide water and landscape management by the cooperating agencies. In 1992, a review of operations under the original Habitat Management Plan was conducted. In 1995, the BLM plan was updated.

Land uses include livestock grazing and recreation. Grazing by domestic livestock is not a major issue in the Blanca Wetlands. Most of the area, by agreement, is subject to grazing only when requested by BLM to enhance wetland values. The area receives limited amount of recreation use (about 4,000 visits/year), primarily for wildlife observation, sightseeing, fishing and hunting. Limited development of deeper ponds to support sport fishing has been incorporated in the restoration effort, but the objectives have focused more on the shallow emergent wetlands appropriate to the setting. Other forms of recreation are also being managed to conform to appreciation of the wetlands for what they are and to avoid conflict with natural resource objectives.

*Results/Outcomes*
The Blanca Wetlands Area is now recognized by the Colorado Division of Wildlife as a core waterbird production area that is necessary for recovery of the valley's nesting populations of water birds. It is providing habitat for up to 30 wintering bald eagles, as well as white faced ibis, western snowy plovers and the many stemmed spider flower (all sensitive or threatened species).

Since the mid-1980s a severe downward trend in breeding waterbird species in the San Luis Valley has reversed. Nesting pairs of geese in Blanca Wetlands have increased from 10 or fewer in the 1970s to 133 in 1994. In bird counts (by station) 52 to 73 species per wetland area were documented in 1992 and 1993. An estimated 19,000 breeding birds utilized the Blanca Wetlands in 1995.

*Reference*
USDI Bureau of Land Management, Final Blanca Wetlands Integrated Activity Plan/Environmental Assessment 1995; San Luis Valley Waterbird Management Plan 1991.

*Contact Persons*
John Schwarz, Bureau of Land Management, San Luis Resource Area, 1921 State Avenue, Alamosa, Colorado 81101; Phone: (719) 589-4975.

## Case Study 8
## Warner Wetlands, Oregon

This case study included evaluation of wetland functions, identification of restoration objectives and implementation of restoration projects in the Great Basin in southeast Oregon. The restoration effort integrated structural treatments and changes in the management of land uses. Management decisions incorporated social and economic factors to help sustain local communities. Structural treatments (well, dikes, etc) were designed to compensate for historical landscape modification and water availability. While it is early, ecological trends appear to be positive.

*Geographic Area/Location*
Warner Wetlands is a 51,000 acre Area of Critical Environmental Concern (ACEC) designated through the BLM planning process. These remnants of a Pleistocene lake are located just northeast of Plush, Oregon in Lake County within the Warner Valley of the Great Basin physiographic province. The ACEC includes about 19,000 acres of intermingled lakes, ponds, emergent marshes, and meadows. Dunes and sagebrush, greasewood, saltbrush or saltgrass flats are interspersed with the wet areas, completing the mosaic. Annual fluctuations in runoff, as well as longer drought/wet cycles make for a dynamic landscape.

*Participants*
Bureau of Land Management, Ducks Unlimited, North American Wetlands Conservation Council, Oregon Department of State Lands, landowners, local residents and ranchers.

*Restoration Goals*
The goal for the entire ACEC is to emphasize the preservation and protection of unique wildlife, ecological, cultural and geological values. A "Core Wetland Area" of 30,400 acres is managed to improve wildlife resource values, eliminating all conflicting uses, demands and allocations. A 22,618 acre grazed area is managed to increase livestock forage production while improving the composition, vigor and density of plant communities. A 420 acre meadow is managed to provide emphasis on improving wildlife habitat condition or enhancement while providing opportunities for other uses such as recreation (sight seeing, bird watching, etc.).

*Background*
The Wetlands are in a closed basin and are influenced by annual precipitation and runoff. For the most part, water moves through the valley from south to north filling a series of sloughs and channels, depressions and potholes, lakes and ponds. In 1983 and 1984, high runoff brought the wetlands to historical high water levels filling all the lakes in the basin. The high water also flooded many agricultural fields and damaged irrigation dikes. By contrast, all of the valley's lakes have completely dried at least three times in history.

Hart Lake is immediately south of the ACEC. Between Hart Lake and the Warner Wetlands is a large dike which raises the level of the lake by holding back water that would spill to the north. Portions of the area north of the dike had been leveled and converted to agricultural fields. All of the public lands were grazed by livestock, primarily during the growing season before restoration activities began. The area also receives recreation use, primarily for wildlife observation, sightseeing, fishing and hunting. Numbers of nesting waterfowl and water birds were severely depressed due to extended drought and poor nesting habitat.

Completed aspects of the Warner Wetlands project include: (1) 25 miles of engineered dikes and control structures to reintroduce topographic complexity and control water flow, (2) Rebuild, repair, relocate and replace 48 water control structures, and (3) drilling of a 650 foot well (2,000 gpm). The well is located near the Hart Lake dike and is used to supplement the water supply spilling to the north. Land use changes include: (1) elimination of grazing in some areas to enhance nesting cover, (2) changes in grazing in some areas to improve vegetation trends, (3) grazing designed to meet nesting needs for long billed curlews and related species, and (4) facilities to control, direct and facilitate recreation.

The BLM raised concerns over how well the wetlands were functioning in 1987 based on nesting cover studies. A Draft amendment to their land use plan was issued in 1988, generating both interest and controversy. A working group, the District Grazing Advisory Board and the District Multiple Use Advisory Council all participated in efforts to develop a mix of management strategies to integrate wetland restoration and sustainable land uses (especially livestock grazing). One of the outcomes was to provide substitute forage in other areas to mitigate economic impacts.

*Results/Outcomes*
Drought during the first three years of implementation allowed projects to be completed restoring 1,293 acres of wetland and enhancing an additional 2,554 acres. Water conditions in 1993 provided for wetland habitats. The 1993 monitoring studies of 4,673 acres of wetland indicated 7,590 nesting pairs of waterfowl and 3,166 nesting water bird pairs. 1994 monitoring results showed 8,946 nesting pairs of waterfowl and 3,746 nesting water bird pairs.

*Reference*
USDI Bureau of Land Management, Warner Wetlands Area of Critical Environmental Concern (ACEC) Management Plan; Devaurs, Walt. 1993/1994 Completion Reports and Breeding Pair Surveys, Warner Wetlands Habitat Management Plan, Warner Valley, Oregon

*Contact Persons*
Walt Devaurs, Bureau of Land Management, Lakeview District Office, 1000 S. 9th Street, P.O. Box 151, Lakeview, Oregon 97630; Phone: (541)947-2177.

Fig. 7. *Top*: Warner Wetlands in Oregon before restoration; *middle*: re-establishing topographic variation; *bottom*: Warner Wetlands after restoration. (Photos courtesy of James Kenna.)

they spread. This requires local knowledge of exotics, constant vigilance, and a quick, aggressive response. Populations of exotics that get out of control can be difficult and costly to keep in check. Examples affecting wetlands include spruce budworm, purple loosestrife and tamarisk.

### 6.2 Summary for Wetland Restoration

The quantity and quality of some wetland habitats have been improved, at least at the local level, through restoration actions. Restoration efforts commonly center around returning or enhancing hydrologic function and topographic complexity.

Assessments consider all possible sources of change in the system. These may include infrastructure (such as drainage ditches, dikes, dams, levees, roads, culverts, bridges), land-use practices (such as agricultural or forestry practices that are increasing sediment and chemical loads or changing vegetation characteristics), industrial practices (such as wastewater or toxic substance release into air, water, or soil), recreational or subsistence practices (such as excessive hunting, fishing, boating, that have changed ecological balances), and other factors (such as the introduction of exotic plant species and pests, fire suppression).

Tools and techniques to restore proper function and condition to the wetland will be specific to the site and should be well correlated with their expected effects. For example, if lack of water caused by diking is a main cause of degradation, alternatives to return water to the site might include removing the dikes, providing drainage through the dikes, pumping water across the dikes, or redirecting a nearby stream into the area.

## 7 DRASTICALLY DISTURBED LANDS WITH FEW BIOLOGICAL LEGACIES

*Scenario: Lands are degraded with limited species abundance or diversity, frequently with physical or chemical attribute problems.*

Bare lands with few biological legacies are areas which were once biologically productive, but have been disturbed to the point where their biomass production and/or biological diversity are significantly impaired. In such areas, native vegetation, soil microbes and animal life have been killed or removed and most of the topsoil may have been lost, altered, or buried.

Such areas are called "drastically disturbed lands" by Box (1978). Most of these drastically disturbed lands will not completely "heal" themselves within a human lifespan through unassisted ecological processes, and, therefore, are of particular concern in land management (see Box 8). Restoration of drastically disturbed lands almost always requires the incorporation of a blend of structural, human use, and biological approaches referenced in previous sections.

### 7.1 Ecological Problems Associated With Drastically Disturbed Lands

The problems associated with the loss of biological productivity on drastically disturbed lands are intimate-

**Box 8**
**Drastically Disturbed Lands that Frequently Require Restoration**

- Inactive and abandoned mine lands
- Abandoned (exhausted and degraded) agricultural lands
- Abandoned roads, powerlines, transportation corridors
- Derelict industrial or construction sites and areas of urban clearance
- Areas with intense surface disturbance, such as removal of soils, dredging, disposal of solid wastes, and intense erosion
- Areas of natural catastrophes and disturbance, such as severe floods, mudslides, landslides, volcanic eruptions, intense wildfires
- Sites contaminated by industrial or agricultural wastes, chemical spills, polluted waters, or air pollution

ly associated with problems that have arisen from the degradation of physical resources: soils, water, and air. The causes of the loss of biological productivity include:

- disturbance of the land surface resulting in loss of soils, changes in soil structure, or critical changes in geomorphology that alter habitat conditions,
- disturbance of the hydrologic regime resulting in changes in water availability, quality, or flow,
- contamination of lands and waters with toxins that inhibit or limit biological productivity and
- loss of critical nutrients that support biological productivity.

For example, mining and construction activities often create large holes in the ground that are subject to flooding and erosion, altering surface and ground water flow patterns. Earth movement and the disposal of solid wastes may create steep-sided unconsolidated piles that are unstable and subject to erosion. Disturbance of the land by industrial, construction, transportation, and recreational activities may result in compaction and loss of soil structure.

Disturbed surfaces often create conditions of poor drainage or drought. Surface conditions may lead to extreme surface temperatures and erosion by water or wind. In arid areas, sporadic disturbance of soils (e.g., off-road transportation corridors) may not show significant vegetative recovery even after decades of non-disturbance (Webb et al. 1988).

Wastes spread on the surface or buried in the shallow subsurface may lack nutrients or present toxicity problems that inhibit plant growth. On contaminated sites, tolerant plants may concentrate heavy metals, radionucleotides, and other toxins in their plant tissues that ultimately sicken or poison the animal populations that feed upon them (e.g., heavy metals: Rebele et al. 1993, Cook 1981, Gough et al. 1979; selenium: Severson et al. 1991; molybdenum: Erdman et al. 1978). The disturbed surface and surface waste materials may lack soil microorganisms and animals that produce the necessary soil characteristics to increase soil fertility.

The heat of intense wildfires can kill vegetation, nearly sterilize soils, and dramatically reduce the seed reservoir. The resulting conditions significantly slow the rate of biological recovery, leading to increased erosion, sedimentation and decline in water quality (Savage 1974, Lathrop 1994). All of these effects provide an inhospitable environment for plant growth and limit biological productivity.

The ecological effects of drastically disturbed lands are not restricted to the disturbed sites. The presence of drastically disturbed lands alone can have considerable influence over a landscape because they are unsightly and detract from its beauty. The lands may also impair or limit ecosystem processes on a regional scale (see Box 9).

Erodible materials from the disturbed lands may be transported great distances during flood and storm events, changing the sediment character of riparian systems. Windblown materials can produce dust and haze, and can distribute contaminants in the form of fine particulate matter over a large area. Contaminants from acid drainage from abandoned mines can leach into streams and have significant effects on aquatic organisms downstream. See Case Study 9.

Patch effects created by disturbed lands can have undesired ecological consequences by providing suitable habitat for noxious weeds or pests that can later invade surrounding areas. For example, non-native brome grasses and other noxious weeds are common invasive species in disturbed lands throughout the

**Box 9**
**Regional Ecosystem Processes that May Be Affected by Drastically Disturbed Lands**

- Hydrologic flux and storage
- Generation and maintenance of soils
- Biological productivity
- Biogeochemical cycling and storage
- Organic matter decomposition
- Maintenance of biological diversity

### Case Study 9. Restoration of abandoned mine lands in Bonanza Mining District, Colorado

This case study describes a voluntary partnership to restore the water quality of acid mine drainage and contamination from abandoned hard rock mines and mill tailings to improve stream habitat on federal and adjacent private lands. The USFS brokered a voluntary settlement among federal and state agencies with the Potentially Responsible Parties (PRPs) under CERCLA to begin cleanup activities that will reduce legal and taxpayer costs, and achieve faster results than under the standard regulatory procedures.

*Geographic Area/Location*
The Bonanza mining district is located at the northern end of the San Luis Valley of south-central Colorado, partially in the Rio Grande National Forest, centered around upper Kerber Creek and its tributaries and the old mining town of Bonanza.

*Participants*
U.S. Forest Service, U.S. Environmental Protection Agency, Agencies of the State of Colorado and PRPs.

*Restoration Goals*
Restoration of water quality for 10 miles of affected streams. Another USFS goal was to use CERCLA authority to obtain non-USFS funds for restoration.

*Background*
The Bonanza mining district includes a number of abandoned hard-rock mines, tailings piles, ore-processing mills, and drainage tunnels that were developed and worked sporadically from the late 1800s until 1969. Many owners and operators were present in the district over this time, complicating liability and legal issues. Overall production from the district was approximately 600,000 tons containing 158 tons of silver, 14,600 tons of lead, 6,800 tons of copper, 1,300 tons of zinc, and 450 pounds of gold. Many environmental problems on Federal and private land in the Kerber Creek drainage are related to pre-1931 mining activities. These problems include continuous discharge of drainage from mine tunnels, erosion of metal-rich tailings into Kerber Creek at various sites, and acidic seeps and runoff from various waste sites. Due largely to the abandoned mine sites, approximately ten miles of Kerber Creek and other drainages are devoid of fish and other aquatic fauna.

In 1991, the U.S. Forest Service formally notified the Environmental Protection Agency (EPA) of a suspected release of hazardous material on the Rio Grande National Forest-administered lands in accordance with CERCLA. In 1992, the Forest Service completed a CERCLA Preliminary Assessment and the PRP's known at the time were notified of the initiation of a CERCLA action near the town of Bonanza. They were informed of their status as PRP's and were requested to respond to a CERCLA Information Request. The Information Request sought further site information and any additional PRPs. Under CERCLA, PRPs include all current and former owners and operators of a site. Liability under CERCLA is "strict, joint, and several", meaning that PRP's, individually or collectively, are legally liable for all costs (legal, information gathering, and remediation) regardless of (1) whether they performed mining or (2) the size of any "injury" they may have directly caused.

The primary responsibility of the Forest Service is to achieve efficient and cost-effective clean-up of environmental problems on federal lands, but response on private lands was also necessary to effectively achieve restoration. The Forest Service believed the major problems needing to be addressed initially were evident, and that lengthy and costly additional study was unnecessary. The Forest Service convened a PRP meeting in 1993 to review the Bonanza Mining Area CERCLA Site situation, to encourage PRP's to form a group to jointly address issues and to discuss alternatives.

The Forest Service proposed that there were three basic options for CERCLA response actions on the site: (1) The site would be listed as a National Priority List (Superfund) site. EPA would take over both the public and private portions of the site, conduct a thorough evaluation, oversee remediation over a period of several years and would seek cost recovery from the PRPs. (2) The Forest Service would conduct studies, possibly leaving significant environmental problems for the EPA to address in due time. Both the USFS and EPA would seek cost recovery from the PRPs. (3) The PRPs would voluntarily undertake CERCLA response actions on Rio Grande National Forest lands under Forest Service CERCLA authority and oversight, and would initiate clean-up on private lands under State of Colorado permit and oversight.

The Forest Service preferred the third option, believing that it could lead to timely and effective response actions by private parties at minimal taxpayer expense. The third option offered the PRP's the best opportunity to control their costs. Option three also directed funds into actual clean-up rather than study, overhead, and legal fees. The Forest Service encouraged the PRP's to develop voluntary plans for the site areas where they had actually been involved.

As a result of the 1993 PRP meeting, the Forest Service and the State of Colorado received a proposal from the PRP's for the voluntary and public-funded clean-up of several areas within the overall CERCLA site on public and private lands. Remedial action started at the Bonanza CERCLA site in 1994 and continues to the present. More than 100,000 cubic yards of metal-rich mine tailings have been moved into a State-approved waste repository on private land. A small permanent tailings repository was built on National Forest lands to allow treatment and release of more than 150,000 gallons of acidic metal-rich mine-drainage waters that were ponded on the site. The release point of the treated water was shifted downstream to prevent possible degradation to the riparian ecosystem in the immediate vicinity. Other work included construction of a 1.6 million gallon drainage pond for surge protection from tunnel drainage, storm water control, stream channel reconstruction, and revegetation.

*Results/Outcomes*
The Rio Grande National Forest has successfully implemented remedial actions within the Bonanza Mining Area CERCLA site since 1994. As the lead agency, the Forest Service asserted CERCLA authority, sought PRP involvement early, and worked with all parties to remediate both federal and private land. This is the first locale where the Forest Service has enlisted private parties liable under CERCLA to conduct voluntary remediation under Forest Service authority and oversight. The geologic and climatic complexity of the site combined with point source and diffuse contamination (mine workings and waste, tailings, mill sites, drainage tunnels, and downstream transport of contaminants) make review of technical remedial plans difficult. Unexpected field conditions and weather have caused difficulties and minor delays in site actions. The need for continuous and thorough communication with stakeholder agencies and parties, such as the State and EPA, has caused some delays in site work actions.

*Contact Person*
Tim Buxton, Bonanza Mining Area CERCLA site coordinator, San Juan/Rio Grande National Forest, P.O. Box 67, Saguache, Colorado, 81149. Ohone: 719-655-2547.

western United States. (See the terrestrial scenario above.) These species become well-established in both disturbed and undisturbed areas following introduction into a disturbed area (Beatley 1964). By controlling soil moisture levels, opportunistic invasive weeds, which may be at an advantage under the harsh conditions on disturbed lands, may prevent the recovery of the pre-disturbance species and displace adjacent native vegetation (Webb and Wilshire 1988).

## 7.2 Management Objectives, Principles, and Options

Management objectives and guiding principles for drastically disturbed lands are addressed separately below. Although there are similarities to other settings, there are also important differences.

### 7.2.1 Management Objectives

Management objectives for severely degraded lands include improving ecosystem processes and containing or restricting the spread of further ecosystem degradation (see Box 10). Ecological restoration as applied to drastically disturbed lands involves restoring or improving the physical conditions of a site to a point where critical ecosystem processes are operable, allowing re- establishment of plant and animal communities or a critical habitat.

**Box 10**
**Examples of Desired Ecological Conditions to Be Achieved through Restoration**

- The effects of erosion, leaching of heavy metals, toxic materials, and contaminants into and upon surrounding areas (including surface and ground water bodies) should be minimized.
- The system should be physically stable so that heavy rains or rapid snow melt will not destabilize it and cause mud slides, rock falls, and other catastrophic events that block roads or endanger human habitations.
- The restored or rehabilitated ecosystem should achieve and maintain ecological stability as quickly as possible, i.e. the systems should be self sustaining and not dependent on outside manipulation such as periodic irrigation or fertilization.
- The system should be, at best, aesthetically pleasing, and, at least, not displeasing.
- The ecosystem should be compatible with land uses and activities in surrounding areas and contribute to civic and social values as well as biological values.
- Management costs should not be prohibitive or place an excessive burden on the surrounding community so that social and financial pressures to abandon the restored systems are created.

The process of rebuilding the plant community on drastically disturbed lands usually must start from a new soil substrate. The natural ecological processes that reestablish plants and animal communities may need to be enhanced or stimulated artificially through introduction of new plant communities or by altering the conditions under which new plant communities develop.

### 7.2.2 Guiding Principles

The role of the land manager in dealing with drastically disturbed lands is analogous to the role of a nurse in medicine. Human health and ecological conditions are similar in that both are complex and dynamic. Both are capable of self-healing when conditions are not acute. The role of the nurse in medicine is to stabilize the medical conditions of the patient and provide conditions where self-healing can occur. Drastically disturbed lands provide a situation similar to the arrival of a patient in a hospital emergency room. Emergency action may be taken to assess the severity of the problem and to stabilize the patient. Once the patient is stabilized, other options can be evaluated. In both the medical and land management situations, the fundamental ethic guiding intervention is the same: Do no harm.

Restoration of drastically disturbed lands is experimental. The problems are complex and many situations are unique. An effective management approach is to utilize scientific principles in restoration design so that management options can be improved. Scientific principles, such as (1) the use of control areas, (2) utilizing pilot studies first, where possible, and (3) scaling actions so the results can be measured and evaluated are valuable concepts in restoration. Ecological restoration of drastically disturbed lands may involve massive human intervention that has the potential to cause unforseen and unintended consequences. Sufficient background studies are necessary to understand causes of problems well enough to evaluate whether restoration efforts are likely to be successful and to minimize the likelihood that restoration activities may result in significant undesirable physical or biological changes outside of the intended management area. See Case Study 10.

Restoration of drastically disturbed lands is site specific. Before significant intervention occurs, sufficient information should be gathered to define site-specific baseline and background conditions and set achievable restoration goals. Where possible, the use of local soils with their associated microorganisms and

### Case Study 10. Evaluating remediation decisions at the Iron Mountain CERCLA site, California

This case study describes the important role geochemical research played in evaluating remediation options for an abandoned mine site where highly acidic, metal-rich drainage was impacting riparian habitat and water quality in reservoirs that supply water for area communities. Remediation of such sites often requires massive intervention with high costs, as well as complex liability and oversight issues. Additionally, many remedial actions are experimental and unanticipated results may be significant.

*Geographic Area/Location*

The Iron Mountain CERCLA site is located in Shasta County, California, in the Klamath Mountains near the Sacramento River, about 14 km northwest of Redding. It is partly on lands managed by the Bureau of Reclamation. Drainage from the site feeds into the Spring Creek Reservoir, Keswick Reservoir, and the Sacramento River, all of which are drinking water and agricultural irrigation sources for surrounding communities.

*Participants*

U.S. Environmental Protection Agency, U.S. Forest Service, Bureau of Reclamation, U.S. Geological Survey, Agencies of the State of California, Private Parties (potentially responsible parties under CERCLA).

*Restoration Goals*

Mitigation of acid mine drainage into the Sacramento River. Since the 1940s, more than 40 fish kills have occurred in the Sacramento River due to contamination from Iron Mountain and other nearby mines. As many as 100,000 fish (salmon and trout) died from a 3-day rainstorm event that increased metal-rich acidic drainage and remobilized metal-rich sediments in the watershed. Chronic toxicity for salmon and trout from copper, zinc, and cadmium is reported to be the highest in the United States.

*Background*

The Iron Mountain CERCLA site includes three abandoned hard-rock metal sulfide mines that were worked for copper, silver, gold, zinc, and pyrite from 1895 to 1962. Copper production from these mines was the largest in California and the sixth largest in the United States, estimated at about 200 million pounds. The deposits consist of disseminated and massive rock rich in sulfides, mostly pyrite (iron sulfide). When exposed to air and water, these sulfide minerals weather and produce extremely acidic and metal-rich surface and ground waters.

The mines at Iron Mountain, named the Brick Flat, Richmond, and Hornet, worked three different massive sulfide bodies. Brick Flat is the highest in elevation and was mined as an open pit for pyrite to make sulfuric acid. The largest massive sulfide body occurred in the Richmond mine, which was mined underground, and still contains about 8 million tons of pyritic sulfides. The Hornet mine, also mined underground, occurs at the lowest elevation and contains slightly less than a million tons of massive sulfides. These three sulfide ore bodies are offset and cut by faults. Groundwater can reach the sulfide bodies through faults, fractures and caved mine workings. The main portals for both the Richmond and Hornet mines served for both haulage and mine water drainage. The Richmond mine water is extremely acidic (pH of 0 to 1.1) with a discharge rate between 0.5 and 50 liters per second. The Hornet mine drainage has pH values ranging from 0.5 to 2.8 with a discharge between 0.8 and 15 liters per second. At present rates of weathering, it would take 2,000 to 3,000 years to remove the remaining sulfides at Iron Mountain.

The Iron Mountain Mine site was ranked number three on the National Priority List (Superfund) for California. In 1983, a technical advisory committee consisting of technical experts from various state and federal agencies was formed to assist the EPA in their investigations of remediation options. Many remediation alternatives were considered including no action, capping the mountain to prevent water infiltration, mine plugging, air sealing, neutralization of the acidic effluent with lime or limestone, on-site leaching and recovery of metals, surface water diversions, and increasing the size of the debris dam and catchment pond.

A proposal to renovate the subsurface workings of the Richmond mine for human access was adopted by the advisory committee. In 1990, mineral and water samples were collected from inside the underground Richmond mine workings which had not been accessible for about 35 years. The investigators found acid mine waters with pH values from 2 to 3.4 and huge growths of efflorescent soluble iron sulfate salts, formed from weathering of the massive sulfides. These then dissolve in water to form acid metal-rich solutions. Knowledge of the composition, occurrence, and volume of these salts made it possible to estimate the composition of the mine pool in the event the Richmond Mine was plugged. The salts were expected to dissolve completely under two scenarios: (1) infiltration from ground water filling the mine after plugging or (2) injection of clean water to flood the Richmond Mine workings. The efflorescent salts were thought to occupy 1–5 percent of the known volume of the workings. But since this was the most uncertain estimate of the evaluation, it was treated as a variable in the geochemical model developed by the U.S. Geological Survey to calculate the chemistry of the mine pool by simulating the dissolution of the salts. The model showed that regardless of the volume of efflorescent salts, the mine pool would have a pH of 1 or less and would contain several grams per liter of dissolved salts. Thus, there would be 0.6 million cubic meters of highly acidic mine water sitting on top of the ground water table in fracture-flow terrain and in a rock aquifer with almost no neutralizing capacity. Scientists estimated it would take 10 to 50 years for water to seep out because only about a kilometer separates the mined area from the surface along the shortest flow path. The risk from this scenario was much too great. This conclusion became a major turning point in the discussions between the EPA and the PRPs who held the liability. Since the mine workings showed a connection between the Richmond Mine and Hornet mine, the effluent from the Hornet mine was believed to be spillage from the Richmond. If this were true, plugging the Richmond mine workings would alleviate any need to treat effluent from the Hornet. However, if the Hornet mine was producing its own acid drainage, then it would require additional treatment facilities and drive up cost. While it was difficult to answer this question with existing engineering and site information, geochemical modeling proved that most of the water from the Hornet mine was produced on site and could not be leakage from the Richmond mine. The results of the geochemical modeling demonstrated the need for increasing the capacity of the water treatment facility.

*Results/Outcomes*

Geochemical research can be applied to site characterization and evaluation of remediation scenarios with very practical results. Research that focuses on specific remediation issues and objectives can reduce risk, reduce costs, decrease uncertainties, expand possibilities for remediation, improve performance, and avoid mistakes. The Iron Mountain Mine Superfund investigations have demonstrated the critical importance of research in answering some questions.

*Reference*

Nordstrom, D.K., and Alpers, C.N., 1995, Remedial investigations, decisions, and geochemical consequences at Iron Mountain, California, in Hynes, T.P. and Blanchette, M.C., eds., Proceedings of Sudbury '95: Mining and the Environment, May 28–June 1, 1995 Sudbury, Ontario, Canada, v. 2, pp. 633–642, CANMET, Ottawa.

*Contact Persons*

D. Kirk Nordstrom, hydrologist, U.S. Geological Survey, 3215 Marine Street, Boulder, Colorado 80303; Phone: 303-541-3037; and Charles N. Alpers, research chemist, U.S. Geological Survey, Room W-2510, Federal Building, 2800 Cottage Way, Sacramento, California 95825; Phone: 916-979-2615x356.

seed populations, and local vegetation types, greatly aids the restoration effort.

Restoration of drastically disturbed lands can be very expensive. In situations where heavy equipment is needed for major land excavation for site reconstruction or the removal of contaminants, costs quickly escalate into the millions of dollars. It is usually more cost-effective, to treat causes instead of symptoms. Symptom intervention requires chronic maintenance and has a high likelihood of failure.

### 7.2.3 Management Options

Five basic management options exist for ecological restoration on drastically disturbed lands: (a) full restoration, (b) partial restoration of key ecosystem functions, (c) create alternative ecosystems, (d) containment of degradation, and (e) non-intervention with monitoring.

Under option (a), restoration attempts to return the site to its initial ecosystem state. This is a noble goal but seldom attainable in less than decades of time. Partial restoration, under option b, may be considered a pragmatic mix of rehabilitation efforts that mitigate acute problems and allow natural processes to restore the system over some interval of time.

The establishment of an alternative ecosystem under option c attempts to improve the current state of ecosystem functions, without reference to initial ecosystem conditions. Some desirable man-made features of the disturbed site may be retained or enhanced, and some natural features suppressed, because the former provide desirable and possibly unique ecosystem values to the region. Examples include:

- preservation of underground mine and adit openings in the western United States because they are utilized by bat populations as roosting sites (Frost et al. 1995),
- surface mined areas in West Virginia that have been reclaimed as grasslands to represent an ecosystem unique to that state, but similar to the Great Plains in avifaunal composition (Whitmore 1978), and
- a grassland ecosystem in Kansas that has been surface mined and replaced by a lake because the area lacked standing surface waters. The lake environment was more desirable than the initial disturbed condition and might even be regarded by some as more desirable than the initial grassland (Cairns 1983).

Other examples of alternative ecosystem options are artificial wetlands that are established to mitigate acidic and metal-rich discharge from abandoned hard-rock mine sites (Brodie et al. 1988, Dunbabin and Bowmer 1992, Eger 1994). Specific state laws and federal regulations (e.g., Surface Mining Control and Reclamation Act) may limit management options. For example, Florida state law prohibits the construction of wetlands in the phosphate mining district if there were none prior to mining, unless special waivers are obtained.

Containment of degradation under option d is a stopgap intervention intended to stabilize a site and prevent further spread of a problem. This option may be the appropriate management choice in situations where the probability of successful restoration is negligible and/or the costs of intervention greatly exceed the anticipated results and benefits. This option can be used to buy time to understand the causes of the problems on site, to pursue the research, testing, or technical development needed to assure a reasonable likelihood of restoration success, and to develop a better understanding of the probable results of restoration efforts.

The nonintervention with monitoring option (e) may be appropriate when the disturbed site is stabilized and restoration goals are likely to be achieved without intervention in a reasonable period of time or when the costs of intervention greatly exceed the anticipated results and benefits.

## 7.3 Summary for Restoration of Drastically Disturbed Lands

Successful restoration projects generally include some consistent management elements which are adapted and applied to fit the situation (Coats and Williams 1990: Horowitz 1990). These elements include:

1. Clear definition of biological objectives.
2. Translation of biological requirements into hydrologic and landscape conditions that need to be established on the site.
3. Good definition of site conditions.
4. Analysis of physical constraints and opportunities including consideration of surface runoff, water quality, soil characteristics and sediment transport and deposition. The site needs for flood control, erosion control, subsidence mitigation, debris management and treating toxic wastes should be evaluated.
5. Analysis of biological constraints and opportunities on site.
6. Development of restoration design alternatives that can be evaluated using quantitative computer models and empirical geomorphic relationships.

Table 3. Technical references on ecological restoration and rehabilitation of drastically disturbed lands. Listed by citation and topic/title.

**Drastically Disturbed Lands — General**

Bradshaw and Chadwick, 1980, The Restoration of Land

Schaller and Sutton (eds), 1978, Reclamation of Drastically Disturbed Lands

Toy and Hadley, 1987, Geomorphology and Reclamation of Disturbed Lands

**Abandoned Mine Lands**

Ashby and Vogel, 1993, Tree planting on mined lands in the midwest

Brooks, Samuel, and Hill (eds), 1985, Wetlands and Water Management on Mined Lands

Carrier and Bromwell, 1983, Disposal and Reclamation of Mining and Dredging Wastes

Cook, Hyde, Sims, 1974, Revegetation Guidelines for Surface Mined Areas

Hassell, Nordstrom, Keammerer, and Todd, 1992, Mining and High Altitude Revegetation

Hay and Woods, 1980, Minesite Preparation for Reforestation of Strip-mined Lands

Hossner (ed.), 1988, Reclamation of Surface-Mined Lands

Hossner and Hons, 1992, Reclamation of Mine Tailings

Kaemmerer and Hassel, 1993, Mining and High Altitude Revegetation

Karle and Densmore, 1994, Stream and Floodplain Restoration in a Riparian Ecosystem Disturbed by Placer Mining

King (ed), 1995, Environmental Considerations of Active and Abandoned Mine Lands

Law, 1984, Mined Land Rehabilitation

Nawrot, Woolf, and Klimstra, 1982, A Guide for Enhancement of Fish and Wildlife on Abandoned Mine Lands in the Eastern United States

Samuel, Stauffer, Hocutt, and Mason (eds), 1978, Surface Mining and Fish/Wildlife Needs in the Eastern United States

Simpson and Botz, 1985, Concepts and Practices in Replacement of Water Sources on Reclaimed Mined Lands

Soni, Vasistha, and Kumar, 1991, Ecological Rehabilitation of Surface Mined Lands

Veith, Bickel, Hopper, and Norland, 1986, Bibliography on Revegetation of Coal-mined Land

Williams and Schuman (eds), 1987, Reclaiming Mine Soils and Overburden in the Western United States; analytic parameters and procedures

**Degraded Agricultural Lands**

Gough, Hornick, and Parr, 1992, Reclamation of Degraded Agroecosystems

**Degraded Soils**

Hornick and Parr, 1987, Restoring the productivity of marginal soils with organic amendments

Kieft, 1991, Soil Microbiology in Reclamation of Arid and Semiarid Lands

Nawrot, Sandusky, and Klimstra,, 1988, Acid Soils Reclamation

Norland, 1993, Soil Factors Affecting Mycorrhizal use in Surface Mine Reclamation:

Oster, Shainberg, and Abrol, 1996, Reclamation of Salt-affected Soil

Parr, Papendick, Hornick, and Colacicco, 1989, Use of Organic Amendments for Increasing the Productivity of Arid Lands

Shetty, Hetrick, Figge, and Schwab, 1994, Effects of Mycorrhizae and Other Soil Microbes on Revegetation of Heavy Metal Contaminated Mine Spoil

Zak, 1985, Vesicular–Arbuscular Mycorrhizae in the Reclamation of Mined Spoils

Zimovets, 1992, Melioration, Fertility, and Ecology of Soils of the Arid Zone

Zimovets, Zaydelman, Pankova, and Boyko, 1994, Ecological Concept of Soil Reclamation

**Arid Lands**

Call and Roundy, 1991, Revegetation of Arid and Semiarid Rangelands

Fisher and Munshower, 1991, Selenium Issues in Drastically Disturbed Land Reclamation Planning in Arid and Semi-Arid Environments

Goudie (ed), 1990, Techniques for Desert Reclamation

Gough and Severson, 1995, Mine-land Reclamation: the fate of trace elements in arid areas

Kieft,1991, Soil Microbiology in Reclamation of Arid and Semiarid Lands

Severson, Fisher, and Gough (eds), 1991, Selenium in Arid and Semi-Arid Environments

Sigh, Sigh, and Abrol, 1994, Agroforestry Techniques for Salt-Affected Lands

Webb and Wilshire (eds), 1983, Human Impacts and Management in Arid Regions

**Contaminated Land**

Cairney (ed.), 1987, Reclaiming Contaminated Land

Gough, Shacklette, and Case, 1979, Element Concentrations toxic to Plants, Animals, and Man

Logan, 1992, Reclamation of Chemically Degraded Land

Ross (ed.), 1994, Toxic Metals in Soil–Plant Systems

**Natural Catastrophes (volcanic eruptions)**

Collins and Dunne, 1988, Effects of Forest Land Management and Revegetation after the Eruption of Mt. St. Helens

Cook, Barron, Papendickk, and Williams, 1981, Impact on Agriculture of the Mount St. Helens Eruption

Okamura, 1985, (translated title) A Study of Revegetation Methods for Erosion Control in Active Volcanoes

The models need to be calibrated and verified with field data.

7. Review of restoration plans and specifications by a technical review team, including important members of the restoration design team.
8. Site inspection by members of the restoration design team during construction of the restoration project. Design changes proposed during the construction phase should receive approval from the design team.
9. Management of the restoration site to promote successful germination and growth of introduced vegetation. This may include special ground preparation before planting, and use of seedling protection devices, when cost-effective.
10. Post restoration monitoring of both biological and hydrologic parameters.

Because restoration of drastically disturbed lands can be expensive, it is important to follow a fairly rigorous process in the design, implementation, and monitoring of restoration actions. It is possible to spend large amounts of money on actions that produce disappointing results. There may be situations where it is appropriate to short-cut or dismiss some steps to avoid getting caught up in review processes where similar circumstances and conditions have been successfully addressed before. But money is generally well spent when it is dedicated to understanding the site and designing actions with a high probability for success.

Table 3 contains a partial listing of technical references for further reading related to ecological restoration and rehabilitation of drastically disturbed lands organized by category of restoration. Full citations are found in the bibliography. Special attention is given to references pertaining to restoration of abandoned mine lands because they represent the largest number of drastically disturbed sites and the largest acreage of drastically disturbed lands that are managed by federal agencies in the United States (Paone et al. 1978).

## 8 CONCLUSIONS

Restoration is inherently an adaptive process (see Bormann et al., Volume III). Data gathering and analysis alternate with decision and action, guided by goals and objectives. Practitioners revise their thinking based on observation. Sometimes, the goal may need to change if it is unrealistic.

While ecosystem assessments may begin with single resource values and species, the complex interactions and relationships among all the social and biological components eventually surface. To meet this challenge, the most successful approaches actively seek to integrate information from multiple sources. Sometimes direct involvement from research scientists can be secured, so testable hypotheses and experimental designs can be developed and implemented that eventually expand knowledge of restoration (see, for example, Jordan et al. 1987).

Deciding where and when to do restoration can be a complex decision. Most begin with their understanding of levels of ecological function across a landscape and then consider what kinds of beneficial results and risks might be expected over time with various levels of action and investment. Frequently the most difficult aspect of the formulating a restoration decision is establishing sufficient public support to act. This is primarily because conservation values are defined by individuals who perceive the social, economic and ecological aspects of the situation differently. The most common approach is to struggle through the communication necessary to develop shared concepts of conservation value (see Fig. 8).

Fig. 8. Ecosystems can change dramatically due to human activities. This scheme compares the Deschutes River in central Oregon before fire control in approximately 1910 (*top*). The contemporary scene (*bottom*) features a different landscape where fire has been controlled and many homes built. Which landscape do people prefer? (Photos courtesy of James Kenna.)

Any restoration proposal must also be related to events and trends in the area because there may be limiting factors that will affect the success of the project. The final step is to evaluate options based on cost effectiveness. Cost effectiveness relates all the other factors (ecological function, benefits, risks, values, other actions/trends) and to practical considerations like feasibility and affordability.

When you combine the complexity of ecological systems with the complexity of restoration decision making, it is easy to understand why descriptions of "desired future conditions" can be a reference, but not a rigid set of expected outcomes. Rigid expectations presume precise controls and knowledge of the ecosystem that we do not possess. It is better to focus on the degree and direction of change in ecological function.

Examples of restoration on federal lands include work underway in the Strawberry Valley of Utah (Frandsen 1995), the Sierra Nevada of California (Parsons 1995; Parsons 1991; Linquist and Bowie 1988), the South Fork of the Salmon River in Idaho (Megahan et al. 1992), the Blue Mountains in Oregon (USDA Forest Service, Pacific Northwest Region 1992), Sequoia National Park (Parsons 1990, fire in mixed conifers), Redwood National Park (Spreiter 1992), the Rocky Mountains of Colorado (USDI BLM, Colorado State Office 1990), BLM land in Utah (Platts and Nelson 1985, Big Creek; McArthur and Sanderson 1996, Mojave Desert restoration) and jointly with ranches in Montana (Massman 1995). Many others are also referenced in this chapter.

The practice of restoration is really an exercise in complex problem-solving with an emphasis on the processes and functions in the system. Managers attempt to integrate what is possible with what is desired, across multiple scales. While land managers have been able to accomplish restoration on a site-by-site, or even species-specific, basis for years, some of these efforts have not addressed interactions at larger scales.

Restoration requires continuous learning because new and better information will always become available. But it is important not to become too timid, because some of the most important innovations in restoration come from experimentation. Considerable information is also already available covering techniques that have been successfully applied. Where action is better than inaction, the trick is to devise a practical combination of study, monitoring, and action.

This review encountered consistent signs of success and common characteristics of successful restoration projects (see Box 11). Because restoration is a long-term experiment in a complex system, success may not only be difficult to achieve, but also difficult to clearly demonstrate. Given the complex matrix of events and relationships acting in ecosystems, land managers and their partners need the flexibility to learn from observation, adapting actions and expectations based on system responses. For this reason, some level of monitoring from baseline conditions becomes essential to understanding outcomes and trends.

**Box 11**
**Characteristics of Successful Restoration Projects**

- Projects organized around clear objectives designed to work with ecosystem processes and functions. A solid understanding of site conditions and trends is necessary. The depth and detail of necessary monitoring and studies will vary depending on the site and the nature of the problems.
- Integration of physical and biological factors with social and economic considerations during both design and implementation.
- Expectations and monitoring that emphasize the degree and direction of change, rather than a precise set of pre-determined characteristics. Forcing systems to respond faster is not always desirable or possible.
- Consideration of the relationships between ecosystem functions and matrix of actions and conditions affecting species composition, land uses, and physical structures and processes.
- A realistic assessment of physical, fiscal and other constraints. Frequency of expected maintenance efforts to ensure a projects success is an important design factor.
- An interdisciplinary technical team, as well as coordination with constituents in both design and implementation.
- Attention to detail concerning techniques and timing during implementation relative to objectives. Sometimes, not all the variables are controllable (e.g. precipitation after prescribed burning).
- Post restoration monitoring at some level to determine trends and allow for future management adjustments.

Federal land managers initiate restoration projects for many reasons, sometimes even to protect human health, or to avoid forcing legally-ordered actions. More often, the impetus is concern over the decline of a species, water quality, fire hazard, or risks to valuable plants, animals or biotic communities. The drive to restore lands and water is founded in the human desire for environmental, spiritual, aesthetic, and economic health. The translation of this desire into action can generate conflicts and choices that range from relatively simple to extremely complex, so it takes a good deal of knowledge and analysis to make restoration decisions. But even more important, it takes commitment to partnerships with affected interests and integration of multiple perspectives.

Federal land managers and their partners operate across the full range of landscapes and restoration is becoming a major component of public land management. Although the scope and scale of the work to be done is daunting, the number of successes and the quality and intensity of the efforts which are underway are both encouraging.

## REFERENCES CITED

*Note: Suggested reading references are highlighted with a* ***bold phrase*** *indicating the subject area. References listed for "****Riparian and wetlands****" are generally available through the BLM National Business Center in Denver Colorado. See Appendix B for addresses for some information sources.*


Adams, B. and L. Fitch. 1995. *Caring for the Green Zone: Riparian Areas and Grazing Management.* Canada-Alberta Environmentally Sustainable Agriculture Agreement. Graphcomn Printers, Ltd., Lethbridge, Alberta. 36 p. **Grazing in riparian areas.**

Adams, P.W. and A.B. Hairston. 1994. *Using Scientific Input in Policy and Decision Making.* EC 1441. Oregon State University Extension Service, Corvallis, OR.

Allen, E.B. (ed.). 1988. *The Reconstruction of Disturbed Arid Ecosystems.* Westview Press, Boulder, CO. **Arid lands.**

Allen, E. and L. Jackson. 1992. The Arid West. *Restoration and Management News* 10 (2): 56–59

Alexander, S. Undated. Restoring a Pinyon–Juniper Ecosystem: The Carrizo Demonstration Area, Lincoln National Forest. Carizozo NM: South Central Mountain Resource Conservation & Development Council.

Anderson, J.W. 1994. *Wallowa-Whitman National Forest Fisheries Habitat Improvement Annual Report FY 1992-1993* USDA Forest Service and Bonneville Power Administration.

Apfelbaum, S.L. and K.A. Chapman. 1994. Ecological Restoration: A Practical Approach. In: T.R. Crow, A. Haney and D.M. Waller (eds.), *Ecosystem Management,* GTR NC-166. USDA Forest Service, North Central Forest Experiment Station, St. Paul MN. 25 p.

Ashby, W.C. and Vogel, W.G., 1993, *Tree Planting on Mined Lands in the Midwest — A Handbook,* Coal Research Center, Carbondale, IL, 115 p. **Abandoned mine lands.**

Asher, J.E. 1996. Responses to Restoration Questions, Personal Communication, January 14, 1996

Asher, J.E., S. Dewey, C. Johnson, J. Oliveraz and R. Wallen. 1995. Invasive Exotic Plants Destroy Wildland Ecosystem Health Abstract for the Ecological Stewardship Workshop, Tucson, Arizona. December 5–12, 1995

Asher, J.E. and D.W. Harmon, 1995. Invasive exotic plants are destroying the naturalness of U.S. wilderness areas. *International Journal of Wilderness* 1 (1).

Baker, J. 1994. "An overview of stand-level ecosystem management research in the Ouachita/Ozark National Forests." *Ecosystem management research in the Ouachita Mountains: Pretreatment conditions and preliminary findings. Proceedings of a Symposium in Hot Springs, Arkansas, October 26–27, 1993.* USDA Forest Service GTR SO-112. pp. 18–27

Barrett, H., T.E. Bedell, J. Buckhouse, T. DeBoodt, W. Elmore, S. Fitzgerald, C.C. Jacks and P.S. Test. 1991. *Watershed Management Guide for the Interior Northwest* EM 8436. Oregon State University Extension Service, Corvallis OR. **Terrestrial and riparian systems.**

Batson, F.T., P.E. Culpin, and W.A. Crisco. 1987. *Riparian Area Management: The Use of Aerial Photography to Inventory and Monitor Riparian Areas* TR 1737-2. Denver, CO: USDI Bureau of Land Management. **Riparian and wetlands.**

Beatley, J.C. 1966. Ecological status of introduced brome grasses (*Bromus* sp.) in desert vegetation of Southern Nevada. *Ecology* 47 (4): 548–554

Bedell, T.E., T. DeBoodt, L.E. Eddleman, and C. Jacks. February 1993. *Western Juniper: Its Management in Oregon Rangelands.* Oregon State University Extension Service, Corvallis, OR.

Berger, J.J. 1990. Agricultural lands, barrens, coastal ecosystems, prairies, and rangelands. In: J.J. Berger (ed.), *Environmental Restoration,* pp. 3–4. Island Press, Washington, DC. **Terrestrial systems.**

Berger, J.J. (ed.). 1990. *Environmental Restoration —Science and Strategies for Restoring the Earth.* Island Press, Covelo, CA. 398 p. **Terrestrial systems.**

Beschta, R.L., W.S. Platts, B. Kaufman. 1991. Filed Review of Fish Habitat Improvement Projects in the Grande Ronde and John Day Basins of Eastern Oregon, unpublished

Box, T.W. 1978. The Significance and Responsibility of Rehabilitating Drastically Disturbed Land. In: F.W. Schaller and P. Sutton (eds.), *Reclamation of Drastically Disturbed Lands,* pp. 1–10. American Society of Agronomy, Inc., Madison, WI.

Botkin, D.B. 1990. *Discordant Harmonies: A New Ecology for the Twenty-First Century.* Oxford University Press, New York.

Bradshaw, A.D., and M.J. Chadwick. 1980. *The Restoration of Land.* University of California Press, Los Angeles, CA. 317 p. **Terrestrial systems.**

Bridges, C., W. Hagenbuck, R. Krapf, S. Leonard and D. Pritchard. 1994. *Riparian Area Management: Process for Assessing Proper Functioning Condition for Lentic Riparian-Wetland Areas.* TR 1737-11. Bureau of Land Management, Denver CO. 46 p.

Brodie, G.A., D.A. Hammer and D.A. Tomljanovich. 1988. Constructed wetlands for acid drainage control in the Tennessee Valley. *Proceedings: Mine Drainage and Surface Mine Reclamation Conference, Pittsburgh, PA, April 19–21, 1988* Vol. 1: U.S. Bureau of Mines Information Circular 9183 pp. 325–331.

Brooks, R.P. 1992. Wetlands and waterbody restoration and creation associated with mining. In *Wetland Creation and Restoration: the Status of the Science,* pp. 529–548. Island Press, Covelo CA. **Wetlands.**

Brooks, R.P., D.E. Samuel and J.B. Hill (eds.). 1985. *Wetlands and Water Management on Mined Lands; Proceedings of a conference, October 23–24, 1985,* Pennsylvania State University, University Park, PA. 393 p. **Abandoned mine lands.**

Brown, G.W. 1980. *Forestry and Water Quality,* Oregon State University Book Stores, Corvallis OR.

Burley, F.W. 1988. Monitoring Biological Diversity for Setting Priorities in Conservation. *Biodiversity.* National Academy Press, Washington DC. pp. 227–230.

Cagney, J. 1993. *Riparian Area Management: Greenline Riparian-Wetland Monitoring,* TR 1737-8. USDI, Bureau of Land

Management, Denver CO. 45 p. **Riparian and wetland.**

Cairney, T. (ed.), 1987. *Reclaiming Contaminated Land.* Blackie and Son, London. 260 p. **Contaminated land.**

Cairns, J., Jr. 1983. Management options for rehabilitation and enhancement of surface-mined ecosystems. *Minerals and the Environment* 5 (1): 32–38.

Call, C.A. and B.A. Round. 1991. Perspectives and Processes in Revegetation of Arid and Semiarid Rangelands. *Journal of Range Management* 44 (6): 543–549. **Arid lands.**

Callaham, R.Z. (tech. ed.). 1990. *Case Studies and Catalog of Watershed Projects in Western Provinces and States.* Wildland Resources Center, University of California, Berkeley CA.

Callenbach, E. 1995. *Bring Back the Buffalo! A Sustainable Future for America's Great Plains.* Island Press, Covelo CA. **Terrestrial Systems.**

Carrier, W.D., III, and L.G. Bromwell. 1983. Disposal and Reclamation of Mining and Dredging Wastes. *Proceedings of the Panamerican Conference on Soil Mechanics and Foundation Engineering, Seventh Conference, Vancouver, British Columbia, June 1983*, pp. 727–738. **Abandoned mine lands.**

Chadde, S. 1989. Willows and Wildlife of the Northern Range, Yellowstone National Park. *Practical Approaches to Riparian Area Management: An Educational Workshop*, R.E. Gresswell, B.A. Barton, J.L. Kershner and K.A. Hashagen, Jr. USDI Bureau of Land Management, Billings MT.

Chambers, J.C. 1994. Great Basin Interdisciplinary Research and Management Project for Maintaining and Restoring Ecosystem Integrity. *Northwest Regional Riparian Symposium: "Diverse Values: Seeking Common Ground", Boise Idaho, December 8–9, 1994.*

Chan, S., D. Hibbs and P. Giordiano. 1993. COPE Report, Coastal Oregon Productivity Enhancement Program. Vol. 6, no. 3.

Chaney, E., W. Elmore, W.S. Platts. 1990. *Livestock Grazing on Western Riparian Areas*, P-312. U.S. Environmental Protection Agency, Denver, Colorado. 45 p. **Riparian.**

Chaney, E., W. Elmore, W.S. Platts. 1993. *Managing Change: Livestock Grazing on Western Riparian Areas* P-312. U.S. Environmental Protection Agency, Denver, CO. 31 p. **Riparian.**

Clark, R.N., and D.R. Gibbons. 1991. Recreation. *Influences of Forest and Rangeland Management on Salmonid Fishes and their Habitats.* W.R. Meehan (ed.). USDA Forest Service and American Fisheries Society, Bethesda, MD. pp. 459–481.

Clark, T.W., R.P. Reading, and A.L. Clark. 1994. *Endangered Species Recovery: Finding the Lessons, Improving the Process* Island Press, Covelo, CA. **Species recovery.**

Clary, W.P. and R. Johnson. 1983. Land imprinting results in Utah. *Vegetative rehabilitation and equipment workshop, 37th Annual Report.* USDA Forest Service, Missoula, MT. pp. 23–24.

Clemmer, P. 1994. *Riparian Area Management: The Use of Aerial Photography to manage Riparian-Wetland Areas*, TR 1737-10, U.S. Bureau of Land Management, Denver, CO. 55 p. **Riparian and wetlands.**

Clewell, A.F. and W.W. Covington. 1995. Society for Ecological Restoration Mission Statement, Abstract for the Ecological Stewardship Workshop, Tucson, Arizona. December 5–12, 1995.

Coats, R. and P. Williams. 1990. Hydrologic Techniques for Coastal Wetland Restoration. In: J.J. Berger (ed.), *Environmental Restoration: Science and Strategies for Restoring the Earth.* Island Press, Covelo CA. pp. 236–246.

Collins, B.D. and T. Dunne. 1988. Effects of Forest Land Management and Revegetation after the Eruption of Mt. Saint Helens. *Earth Surfaces, Processes, and Landforms* 13 (3): 193–205. **Natural catastrophies.**

Cook, R.J., J.C. Barron, R.I. Papendickk and G.J. Williams III. 1981. Impact on Agriculture of the Mount Saint Helens Eruption. *Science* 211: 16–22. **Natural catastrophies.**

Cook, C.W., R.M. Hyde and P.L. Sims. 1974. *Revegetation Guidelines for Surface Mined Areas.* Colorado State University, Range Science Department, Fort Collins CO. Science Series no. 16. 73 p. **Abandoned mine lands.**

Cook, R.E. 1981. Toxic tailings and tolerant grass. *Natural History Magazine* March: 28–38.

Cooley, C.C. 1991. Water, Waterfowl, and Public Policy in the San Luis Valley, Colorado: Perspectives and Insights on Water Use and Development Affect Waterfowl within a Unique Region of Extensive water Resources and Wetland Habitat, Texas A&M University.

Covington, W. 1994. Changes in forest conditions in Ponderosa pine ecosystems of western North America. *Forest health and fire danger in inland western forests: Proceedings of a conference, Sept. 8–9, 1994.* Spokane, WA. p. 15–17.

Crispin, V., R. House, and D. Roberts. 1993. Changes in Instream Habitat, Large Woody Debris, and Salmon Habitat after the Restructuring of a Coastal Oregon Stream. *North American Journal of Fisheries Management* 13: 96–102.

Crow T.R., A. Haney and D.M. Waller. 1994. *Report on the Scientific Roundtable on Biological Diversity Convened by the Chequamegon and Nicolet National Forests.* North Central Forest Experiment Station, USDA Forest Service, St. Paul MN.

Crowe, E. 1996. Cottonwood and Aspen: Managing for Balance. *Natural Resource News* Special Edition, January: 1–2.

Culpin, P. and R. J. Boyd. 1987. *Riparian Area Management: A Selected, Annotated Bibliography of Riparian Area Management* TR 1737-1. USDI Bureau of Land Management, Denver CO. 67 p. **Riparian and wetlands.**

Dahl, T.E. 1990. Wetland losses in the United States, 1780's to 1980's. US Fish and Wildlife Service, 21 p.

Daily, G.C., S. Alexander, P.R. Ehrlich, L. Goulder, J. Lubchenco, P.A. Matson, H.A. Mooney, S. Postel, S.A. Schneider, D. Tilman and G. Woodwell. 1997. Issues in Ecology. *Ecological Society of America* 2. Spring 1997.

DeBano, L.F., J.J. Bredja and J.H. Brock. 1984. Enhancement of Riparian Vegetation Following Shrub Control in Arizona Chaparral. *Journal of Soil and Water Conservation*, 39 (5): 5.

DeBano, L.F. and W.R. Hansen. 1989. Rehabilitating Depleted Riparian Areas Using Channel Structures. *Practical Approaches to Riparian Resource Management: An Educational Workshop*, R.E. Gresswell, B.A. Barton, J.L. Kershner (eds.). USDI Bureau of Land Management, Billings MT.

Dixon, R.M. 1987. Imprintation: A process for reversing desertification. *Proceedings of the International Erosion Control Conference XVIII*, Reno, NV. pp. 290–306.

Dixon, R.M. 1990. Land imprinting for dryland revegetation and restoration. *Environmental Restoration.* Island Press, Washington DC. pp. 14–22.

Dregne, H. 1977. Desertification of arid lands. *Economic Geography* 53 (4): 322–331.

Dunbabin, J.S. and Bowmer, K.H. 1992. Potential use of constructed wetlands for treatment of industrial wastewaters containing metals. *Science of the Total Environment* 111 (2/3): 155–167.

Dyer, M. and N. Ishwaran. 1992. Redevelopment as a tactic in grazing land rehabilitation. *Ecosystem Rehabilitation.* SPB Academic Publishing, The Hague, the Netherlands. pp. 151–161.

Eddleman, L.E. 1997. Sustainable Restoration. *Range Field Day Annual Report, The Sagebrush Steppe: Sustainable Working Environments* Special Report 979. Eastern Oregon Agricultural Experiment Station, Oregon State University and USDA Agricultural Research Service. Burns OR.

Edgerton, P.J. 1985. Influence of Ungulates on the Development of the Shrub Understory of an Upper Slope Mixed Conifer Forest. *Proceedings: Symposium on Plant-Herbivore Interactions*, USDA Forest Service, Intermountain Research Station, Ogden UT.

Eger. P. 1994. Wetland treatment for trace metal removal from mine drainage — the importance of aerobic and anaerobic Processes. *Water Science and Technology* 29 (4): 249–256.

Elmore, W. 1984. *Juniper Management and the Implications on Wildlife Habitat.* Prineville District Office, U.S. Bureau of Land Management, Prineville, OR.

Elmore, W. and R.L. Beschta. 1987. Riparian areas: perceptions in management. *Rangelands* 9 (6): 260–265

Elmore, W. and R.L. Beschta. 1989. The Fallacy of Structures and the Fortitude of Vegetation. Tech Report PSW-110. USDA Forest Service. pp. 116–119

Elmore, W. and R.L. Beschta. 1990. Restoring riparian areas in eastern Oregon: Grazing management can enhance forage, flows, and fish. *Flow* (Fall): 8–10.

Elmore, W. and B. Kaufman. 1994. Riparian and Watershed Systems: Degradation and Restoration. In: M. Vavra, W.A. Laycock and R.D. Piper (eds.), *Ecological Implications of Livestock Herbivory in the West.* Society for Range Management, Denver CO.

Emmington, W.H. and K. Maas. 1994. COPE Report, Coastal Oregon Productivity Enhancement Program, Vol. 7, nos. 2 and 3.

Erdman, J.A., R.J. Ebens and A.A. Case. 1978. Molybdenosis — a potential problem in ruminants grazing on coal mine spoils. *Journal of Range Management* 32: 159–161.

Falk, D.A., C.I. Miller, and M. Olwell (eds.). 1996. *Restoring Diversity: Strategies for Reintroduction of Endangered Plants* Island Press, Covelo, CA. **Species recovery.**

Farmer, Jr., R.E. 1966. Rooting dormant cuttings of mature cottonwoods. *Journal of Forestry* 64 (3): 196–197.

Fisher, S.E. and F.F. Munshower. 1991. Selenium Issues in Drastically Disturbed Land Reclamation Planning in Arid and Semi-arid Environments. *Proceedings of the 1990 Billings Land Reclamation Symposium on Selenium in Arid and Semi-arid Environments, Western, United States*, R.C. Severson, S.E. Fisher and L.P. Gough (eds.). U.S. Geological Survey Circular 1064. pp. 123–133. **Arid lands.**

Frandsen, J. 1995. *Restoring Strawberry, the Pure Valley: Report on Five Years of Mitigation and Enhancement in the Strawberry Valley.* USDA Forest Service, Uinta and Cache National Forests, Vernal UT.

Franklin, J. 1990. Old growth forests and New Forestry. *Forests — wild and managed: Differences and consequences. Proceedings of a symposium at the University of British Columbia, January 19–20, 1990.* Pearson and Challenger (eds.), pp. 1–21.

Frayer, W.E. 1991. *Status and trends of wetlands and deepwater habitats in the conterminous United States, 1970's to 1980's.* Michigan Technological University, Houghton MI. 31 p.

Furniss, M.J., T.D. Roelofs and C.S. Yee. 1991. Road Construction and Maintenance. In: W.R. Meehan (ed.), *Influences of Forest and Rangeland Management on Salmonid Fishes and their Habitats.* USDA Forest Service and American Fisheries Society, Bethesda, MD.

Gebhardt, K., S. Leonard, G. Staidl and D. Pritchard. 1990. *Riparian Area Management: Riparian and Wetland Classification Review*, TR 1737-5. USDI Bureau of Land Management, Denver CO. 56 p. **Riparian and wetland.**

Goudie, A.S. (ed). 1990. *Techniques for Desert Reclamation.* John Wiley and Sons, New York, NY. 271 p. **Arid lands.**

Gough, L.P., S.B. Hornick and J.F. Parr. 1992. Geosciences, Agroecosystems, and the Reclamation of Degraded Lands. In: M.K. Wali (ed.), *Ecosystem Rehabilitation: Policy Issues, Volume I.* Academic Publishers, The Hague, the Netherlands. pp. 47–61. **Agricultural lands.**

Gough, L.P. and R.C. Severson. 1995. Mine-land Reclamation: the fate of trace elements in arid and semi-arid areas. In: D.J. Swaine and F. Goodarzi (eds), *Environmental Aspects of Trace Elements in Coal.* Kluwer, the Netherlands. pp. 275–307. **Arid lands.**

Gough, L.P., H.T. Shacklette and A.A. Case. 1979. *Element Concentrations Toxic to Plants, Animals, and Man*, U.S. Geological Survey Bulletin 1466. 84 p. **Contaminated land.**

Gregory, S.V., G.A. Lamberti, D.C. Erman, K.V. Koski, M.L. Murphy and J.R. Sedell. 1977. Influence of Forest Practices on Aquatic Production. *Importance, Preservation and Management of Riparian Habitat: A Symposium, July 9, 1977, Tucson, Arizona.* USDA Forest Service

Gregory, S.V., F.J. Swanson, W.A. McKee and K.W. Cummins. 1991. An ecosystem perspective of riparian zones: Focus on links between land and water. *BioScience* 41 (8): 540–551.

Greswell, R.E., B.A. Barton and J.L. Kershner (eds.). 1989. *Practical Approaches to Riparian Resource Management: An Educational Workshop, Billings Montana.* BLM-MT-PT-89-001-4351. USDI Bureau of Land Management. US Government Printing Office, Washington DC. 193 p.

Grumbine, Ed. 1994. *Protecting Biological Diversity through the Greater Ecosystem Concept.* Sierra Institute, University of California Extension, Santa Cruz, CA.

Hansen, P.L., R.D. Pfister, K. Boggs, B.J. Cook, J. Joy and D.K. Hinkley. 1995. *Classification and Management of Montana's Riparian and Wetland Sites.* US Environmental Protection Agency, USDI Bureau of Land Management, Montana Forest and Conservation Experiment Station, University of Montana, Missoula MT.

Harr, R.D. 1981. *Scheduling Timber Harvest to Protect Watershed Values.* Pacific Northwest Range and Forest Experiment Station, USDA Forest Service, Corvallis, OR.

Harrod, R.J., W.L. Gaines, R.J. Taylor, R. Everett, T. Lillybridge and J.D. McIver. 1996. Biodiversity in the Blue Mountains. *A Search for a Solution: Sustaining the Land, People, and Economy of the Blue Mountains.* R.J. Jaindl and T.M. Quigley (eds.), American Forests in cooperation with the Blue Mountains Natural Resource Institute, Washington D C. pp. 107–117.

Hassell, W.G., S.K. Nordstrom, W.R. Keammerer, and J. Todd. 1992. Mining and Revegetation. *Proceedings; High Altitude Revegetation Workshop.* Colorado Water Resources Research Institute Information Series, no. 10. National Park Service, Lakewood, CO. **Abandoned mine lands.**

Hay, R.L. and F.W. Woods. 1980. *Minesite Preparation for Reforestation of Strip-mined Lands*. University of Tennessee, Department of Forestry, Wildlife, and Fish, Knoxville, TN. 24 p. **Abandoned mine lands.**

Haynes, R.W., R.T. Graham and T.M. Quigley (tech. eds.). 1996. *A Framework for Ecosystem Management in the Interior Columbia Basin and Portions of the Klamath and Great Basins*. USDA Forest Service, Pacific Northwest Research Station, Portland OR.

Henderson, D. and L.D. Hedrick. 1991. *Restoration of Old Growth Forests in the Interior Highlands of Arkansas and Oklahoma: Proceeding of a Conference, Winrock International, Morrilton, AR, September 19–20, 1990*. Ouachita National Forest and Winrock International Institute for Agricultural Development.

Hilken, T.O. and R.F. Miller. 1994. *Medusahead (Taeniatherum asperum Nevski): A Review and Annotated Bibliography* Station Bulletin 664. Agricultural Experiment Station, Oregon State University, Corvallis, OR. 18 pp.

Hill, M. and W.S. Platts. 1995. *Restoration and Ecosystem Management of the Owens River Gorge, California*, unpublished, Ecosystem Sciences, PO Box 16444, Boise, ID 83715.

Hindley, E. 1996. *Riparian Area Management: Observing Physical and Biological Change Through Historical Photographs*. TR 1737-13. USDI Bureau of Land Management, Denver CO. 33 p. **Riparian and wetlands.**

Hoag, J.C. 1995. Using Dormant Pole Cuttings to Revegetate Riparian Areas. *Fifth International Rangeland Congress*, Salt Lake City, Utah. July 23–28, 1995.

Hornick, S.B. and Parr, J.F. 1987. Restoring the productivity of marginal soils with organic amendments. *American Journal of Alternative Agriculture* 2 (2): 64–68. **Degraded soils.**

Horowitz, H. 1990. Restoration reforestation. In: J.J. Berger (ed.), *Environmental Restoration — Science and Strategies for Restoring the Earth*. Island Press, Covelo, CA. pp. 84–94. **Terrestrial systems.**

Hossner, L.R. (ed.), 1988. *Reclamation of Surface-Mined Lands*, Vols. I and II. CRC Press, Boca Raton, FL. 219 p.**Abandoned mine lands.**

Hossner, L.R. and F.M. Hons. 1992. Reclamation of Mine Tailings. *Advanced Soil Science* v. 17/Soil Restoration. pp. 311–350. **Abandoned mine lands.**

House, R.A. and P.L. Boehne. 1986. Effects of instream structures on salmonid habitat and populations in Tobe Creek, Oregon, Bureau of Land Management. *North American Journal of Fisheries Management* 6: 38–46.

Howe, G.H. and J. Smith. 1994. The Western White Pine operation Breeding Program: A Progress Report. *Proceedings of the Symposium, Interior Cedar–Hemlock–White Pine Forests: Ecology and Management, March 2–4, 1993, Spokane, WA*. Department of Natural Resource Sciences, Washington State University, Pullman WA.

Huenneke, L.F. 1995. Ecological Impacts of Plant Invasion in Rangeland Ecosystems. *Proceedings of the Alien Plant Invasions: Increasing Deterioration of Rangeland Health Symposium*. Society for Range Management.

Hunter, C.J. 1991. *Better Trout Habitat: A Guide to Stream Restoration and Management*. Montana Land Reliance. Island Press, Washington DC.

Interagency Ecosystem Management Task Force. 1995. *The Ecosystem Approach: Healthy Ecosystems and Sustainable Economies. Volume I — Overview*. K. McGinty, chair. National Technical Information Service, U.S. Department of Commerce, Springfield, VA.

Interior Columbia Basin Ecosystem Management Project. 1997. *Eastside Draft Environmental Impact Statement*. USDA Forest Service, Pacific Northwest Region and USDI Bureau of Land Management, Oregon and Washington, Walla Walla WA.

Jacobsen, J.M. 1994. *An Analysis of Fifty Years of Success and Failure in Everglades Management*. The Everglades Institute, Ochopee, FL.

Jaindl, R.G. and T.M. Quigley (eds.). 1996. *A Search for a Solution: Sustaining the Land, People, and Economy of the Blue Mountains*. American Forests in cooperation with the Blue Mountains Natural Resource Institute, Washington DC. 316 p. **Terrestrial systems.**

Jirsa, R.. 1995. Plum Creek, Fish and Wildlife Service, National Marine Fisheries Service submit habitat planning effort for public comment. Publicity Flyer, Nov. 14, 1995. 5 p.

Johnson, C.G. Jr. 1994. *Forest Health in the Blue Mountains: A Plant Ecologist's Perspective on Ecosystem Processes and Biological Diversity*. T.M. Quigley ed. USDA Forest Service, Pacific Northwest Research Station, Portland, OR. 23 p.

Johnson, K.L. 1992. Management for Water Quality on Rangelands through Best Management Practices: The Idaho Approach. In: R.J. Naiman (ed.), *Watershed Management: Balancing Sustainability and Environmental Change*. NY: Springer-Verlag, New York.

Johnson, K.N., J. Session and J.F. Franklin. 1996. Initial Results from Simulation of Alternative Management Strategies for Two National Forests of the Sierra Nevada. *Sierra Nevada Ecosystem Project: Final Report to Congress Addendum*. Centers for Water and Wildland Resources, University of California, Davis, CA.

Jordan, W.R. III, M.F. Gilpin and J.D. Aber. 1987. Restoration ecology: ecological restoration as a technique for basic research. In: W.R. Jordan and M.E. Gilpin (eds.), *Restoration ecology: a synthetic approach to ecological research* Cambridge University Press, New York, NY. pp. 1–21.

Kinch, G. 1989. *Riparian Area Management: Grazing Management in Riparian Areas* TR 1737-4. USDI Bureau of Land Management, Denver, CO. **Riparian and wetlands.**

Kaemmerer, W.R. and W.G. Hassel. 1992. *Proceedings; High Altitude Revegetation Workshop* Colorado Water Resources Research Institute Information Series, no. 11. National Park Service, Lakewood, CO. **Abandoned mine lands.**

Karle, K.F. and Densmore, R.V. 1994. Stream and floodplain restoration in a riparian ecosystem disturbed by placer mining. *Ecological Engineering* 3 (2): 121–133. **Abandoned mine lands.**

Kieft, T.L. 1991. Soil Microbiology in Reclamation of Arid and Semiarid Lands. *Semiarid Lands and Deserts: Soil Resource and Reclamation*. pp. 209–256. **Degraded soils; Arid lands.**

King, T.V.V. (ed.). 1995. *Environmental Considerations of Active and Abandoned Mine Lands*. U.S. Geological Survey Bulletin 2220. 38p. **Abandoned mine lands.**

Kovalchik, B.L. and W. Elmore. 1991. Effects of Cattle Grazing Systems on Willow Dominated Plant Associations in Central Oregon. *Symposium on Ecology and Management of Riparian Shrub Communities: May 1991*, Sun Valley, ID.

Krinard, R.M. and W. K. Randall. 1979. Soaking Aids Survival of Long Unrooted Cottonwood Cuttings. Reprint from *Tree Planters' Notes* 30 (3): 16–18

Kusler, J.A. and M.E. Kentula (eds.). 1989. *Wetland Creation and Restoration: The Status of the Science Vol. I & II*. Island Press, Covelo, CA. 549 p. **Wetlands.**

Lacey, J. 1989. Influence of spotted knapweed on surface runoff and sediment yield. *Weed Technology* 3: 627–631.

Lathrop, R.G. Jr. 1994. Impacts of the 1988 Wildfires on the water quality of Yellowstone and Lewis Lakes, Wyoming. *International Journal of Wildland Fire* 4 (3): 169–175.

Law, D.L. 1984. *Mined Land Rehabilitation*. Van Nostrand Reinhold, New York, NY. 184 p. **Abandoned mine lands.**

Leach, M. and L. Ross. 1995. *Midwest Oak Ecosystem Recovery Plan: A Call to Action*, no publisher listed, 112 pages

Leonard, S., G. Staidl, J. Fogg, K. Gebhardt, W. Hagenbuck and D. Pritchard. 1992. *Riparian Area Management: Procedures for Ecological Site Inventory — With Special Reference to Riparian-Wetland Sites*. TR 1737-7. USDI Bureau of Land Management, Denver, CO. 137 p. **Riparian and wetlands.**

Lindquist, D.S. and L.Y. Bowie. 1988. Watershed Restoration in the Northern Sierra Nevada: A Biotechnical Approach. *Proceedings of the California Riparian Systems Conference: Protection, Management and Restoration for the 1990s*. GTR PSW-11. USDA Forest Service, Davis, CA.

Logan, T.J. 1992. Reclamation of chemically degraded land. *Advanced Soil Science* 17: 13–35. **Contaminated land.**

Maas, K. and W. H. Emmington. 1994. *COPE Report*. Coastal Oregon Productivity Enhancement Program, Newport, OR. Vol. 7, No. 2 and 3.

MacCleery, D.W. 1994. What on Earth Have We Done to Our Forests: A Brief Overview on the Conditions and Trends of U.S. Forests. USDA Forest Service, Washington DC.

Mahalovich, M.F. 1996. The Role of Genetics in Improving Forest Health. *Proceedings of the National Silvicultural Workshop, May 8–11, 1995, Mecalero, NM*.

Manley, P.N., G.E. Brogan, C. Cook, M.E. Flores, D.G. Fullmer, S. Husari, T.M. Jimerson, L.M. Lux, M.E. McCain, J.A. Rose, G. Schmitt, J.C. Shuyler, and M.J. Skinner. 1995. Sustaining ecosystems: A conceptual framework, version 1. R5-EM-TP-001. USDA Forest Service, Pacific Southwest Region, San Francisco, CA. 216 p.

Manning, M.E., S.R. Swanson, T. Svejcar, and J. Trent. 1989. Rooting characteristics of four intermountain meadow community types. *Journal of Range Management* 2 (4).

Marble, A.D. 1992. *A Guide to Wetland Functional Design*. Lewis Publishers, Chelson, MI. 222 p. **Wetlands.**

Maser, C., J.W. Thomas, R.G. Anderson. 1984. *Wildlife Habitats in Managed Rangelands — The Great Basins of Southeastern Oregon: The Relationship of Terrestrial Vertebrates to Plant Communities and Structural Conditions*. USDA Forest Service, Pacific Northwest Forest and Range Experiment Station, La Grande, OR.

Massman, C. (ed.). 1995. *Riparian Grazing Successes on Montana Ranches*. Conservation Districts Bureau, Montana Department of Natural Resources and Conservation, Helena, MT. 30 p.

Mattise, S.N. 1994. *Bitterbrush Rehabilitation: Squaw Butte Fire Complex*. USDI, Bureau of Land Management, Boise, ID. **Terrestrial systems.**

McArthur, F.D. and S.C. Sanderson. 1996. Twist Hollow: A Mojave Desert restoration project. *Proceeding of the Fifth International Rangeland Conference, July 23–25, 1995, Salt Lake City, Utah, Vol. I*. N.E. West (ed.), Society for Range Management, Denver, CO.

Meese, G.M. 1989. Saving Endangered Species: Implementing the Endangered Species Act. *Preserving Communities and Corridors*. G. Mackintosh (ed.), Defenders of Wildlife, Washington DC. 96 p.

Megahan, W.F., J.P. Potyondy and K.A. Seyedbagheri. 1992. Best Management Practices and Cumulative Effects from Sedimentation in the South Fork Salmon River: An Idaho Case Study. *Watershed Management: Balancing Sustainability and Environmental Change*, J. Robert (ed.). Naiman, Springer-Verlag, New York, NY.

Meyer, M.E. 1988. A Low Cost Brush Deflection System for Bank Stabilization and Revegetation, *Proceedings of the California Riparian Systems Conference: Protection, Management and Restoration for the 1990s* GTR PSW-110. USDA Forest Service, Davis, CA.

Monson, S.B. and S.G. Kitchen (eds.). 1994. *Proceedings: Ecology and Management of Annual Rangelands*. INT-GTR-313. USDA Forest Service, Intermountain Research Station, Ogden, UT. 416 p. **Weed control; Seeding technology.**

Moore, B.C. and L. Flaherty. 1996. Fish, Riparian and Water Quality Issues in the Blue Mountains. *Search for a Solution: Sustaining the Land, People and Economy of the Blue Mountains*, R.G. Jaindl and T.M. Quigley (eds.). American Forests, Washington DC.

Mouat, D., J. Lancaster, T. Minor, K. Mussallem, T. Wade., J. Wickham. 1993. Ecological Risk Assessment Case Study. Report submitted to Environmental Protection Agency. 17 p.

Munhall, A. 1996. Alan's View of Success and Failure of Riparian Projects; Lakeview Resource Area 1978–1995. Personal communication of February 1996.

Myers, L.H. 1989. *Riparian Area Management: Inventory and Monitoring Riparian Areas*, TR 1737-3. USDI Bureau of Land Management, Denver, CO. 81 p. **Riparian and wetlands.**

National Research Council. 1992. *Restoration of Aquatic Ecosystems: Science, Technology, and Public Policy*. National Academy Press, Washington, DC. 552 p.

National Research Council. 1994. *Rangeland Health*. National Academy Press, Washington, DC. 182 p. **Terrestrial systems.**

National Research Council. 1992. *Restoration of Aquatic Ecosystems: Science Technology and Public Policy*. S. Maurizi and F. Poillon (eds.), J. Cairns, Jr., Chairman. Committee on Restoration of Aquatic Ecosystems: Science, Technology, and Public Policy. Water Science and Technology Board. Commission on Geosciences, Environment, and Resources. National Academy Press, Washington, DC. 552 p. **Aquatic ecosystems.**

Natural Resources Law Center. 1996. *The Watershed Source Book: Watershed-Based Solutions to Natural Resource Problems*. University of Colorado School of Law, Boulder, CO. 330 p.

Nawrot, J.R., J. Sandusky, and W.B. Klimstra. 1988. *Acid Soils Reclamation — applying the principles* U.S. Bureau of Mines Information Circular 9184, vol. II. pp. 83–103. **Degraded soils.**

Nawrot, J.R., A. Woolf, and W.D. Klimstra. 1982. *A Guide for Enhancement of Fish and Wildlife on Abandoned Mine Lands in the Eastern United States* FWS/OBS-80/67. USDI Fish and Wildlife Service, Washington DC. 101 p. **Abandoned mine lands.**

Norland, M.R. 1993. *Soil Factors Affecting Mycorrhizal use in Surface Mine Reclamation*: U.S. Bureau of Mines Information Circular 9345. **Degraded soils.**

Noss, R.F. and A.Y. Cooperrider. 1994. *Saving Nature's Legacy: Protecting and Restoring Biodiversity.* Island Press, Covelo, CA. 433 p. **Terrestrial systems.**

Odum, E.P. 1972. Ecosystem Theory in Relation to Man. In: J.A. Weis (ed.), *Ecosystem Structure and Function: Proceedings of the Thirty-First Annual Biology Colloquium.* Oregon State University Press, Corvallis, OR. pp. 11–24.

Odum, E.P. 1997. *Ecology: A Bridge Between Science and Society.* Sinauer Associates, Sunderland, MA. **Terrestrial systems.**

O'Laughlin, J. 1994. Forest Health in Idaho National Forests. *Forest health and fire danger in Inland western forests: Proceedings of a conference, Sept. 8–9, 1994.* Spokane, WA. pp. 33–39.

Oster, J.D., I. Shainberg and I.P. Abrol. 1996. Reclamation of Salt-affected Soil. *Soil Erosion, Conservation, and Rehabilitation* pp. 315–352. **Degraded soils.**

Okamura, T. 1985. (translated title) *A Study of Revegetation Methods for Erosion Control in Active Volcanoes* Report 57. Geological Survey of Hokkaido, Sapporo, Japan. pp. 29–63. **Natural catastrophes.**

Paone, J., P. Struthers and W. Johnson. 1978. Extent of disturbed lands and major reclamation problems in the United States. In: F. W. Schaller and P. Sutton (eds.), *Reclamation of Drastically Disturbed Lands.* American Society of Agronomy, Madison, WI. pp. 11–22.

Parker, V.T. 1990. Problems Encountered While Mimicking Nature in Vegetation Management: An Example from a Fire-prone Vegetation. *Ecosystem Management: Rare Species and Significant Habitats, Proceedings of the 15th Annual Natural Areas Conference.* R.S. Mitchell, C.J. Sheviak, and D.J. Leopold (eds.). Bulletin No. 471. New York State Museum, Albany, NY. pp. 231–233.

Parr, J.F., R.I. Papendick, S.B. Hornick and D. Colacicco. 1989. Use of organic amendments for increasing the productivity of arid lands. *Arid Soils Research and Rehabilitation* 3 (2): 149–170. **Degraded soils.**

Parsons, D.J. 1991. Restoring fire to the Sierra Nevada mixed conifer forest: reconciling science, policy, and practicality. *Proceedings: 1st Annual Meeting of the Society for Ecological Restoration, January 16–20, 1989, Oakland, California* H.G. Hughes and T.M. Bonnicksen (eds.), University of Wisconsin, Madison, WI. pp. 271–279.

Parsons, D.J. 1995. Restoring Fire to Giant Sequoia Groves: What Have We Learned in 25 Years. *Proceedings: Symposium on Wilderness and Park Management, March 30–April 1, 1993, Missoula, Montana.* K. Brown, R.W. Mutch, C.W. Spoon, and R.H. Wakimoto (eds.). USDA Forest Service Intermountain Research Station, Ogden, UT. pp. 256–258.

Platts, W.S. 1989. Compatibility of Livestock Grazing Strategies with Fisheries. *Practical Approaches to Riparian Resource Management: An Educational Workshop, May 8–11, 1989, Billings, Montana.* R.E. Greswell, B.A. Barton and J.L. Kershner (eds.). USDI Bureau of Land Management, Billings MT.

Platts, W.S. 1990. *Managing Fisheries and Wildlife on Rangelands Grazed by Livestock: A Guidance and Reference Document for Biologists.* Nevada Department of Wildlife.

Platts, W.S. and R.L. Nelson. 1985. Stream habitat and fisheries response to livestock grazing and instream improvement structures, Big Creek, Utah. *Journal of Soil and Water Conservation* 40 (4): 374–379.

Pritchard, D. 1998. User Guide to Assessing Proper Functioning Condition and the Supporting Science for Lotic Areas. TR1737-15. USDI Bureau of Land Management, Denver, Co. 126 p. **Riparian and wetlands.**

Pritchard, D., H. Barrett, K. Gebhardt, J. Cagney, P.L. Hansen, R. Clark, B. Mitchell, J. Fogg, D. Tippy. 1993. *Riparian Area Management: Process for Assessing Proper Functioning Condition* TR 1737-9. Denver, CO: USDI Bureau of Land Management. 53 p. **Riparian and wetlands.**

Pritchard, D., C. Bridges, S. Leonard, R. Krapf and W. Hagenbuck. 1994. *Riparian Area Management: Process for Assessing Proper Functioning Condition for Lentic Riparian-Wetland Areas.* TR 1737-11. USDI Bureau of Land Management, Denver, CO. 39 p. **Riparian and wetlands.**

Pritchard, D., P. Clemmer, M. Gorges, G. Meyer, K. Shumac. 1996. *Riparian Area Management: Using Aerial Photographs to Assess Proper Functioning Condition of Riparian-Wetland Areas.* TR 1737-11. USDI Bureau of Land Management, Denver, CO. 43 p. **Riparian and wetlands.**

Puleo, V. 1990. Just do it. *Forests — wild and managed: Differences and consequences, Proceedings of a Symposium at the University of British Columbia, Jan. 19–20, 1990.* Pearson and Challenger (eds.) pp. 135–146.

Putchy, C.A. and D.B. Marshall. 1993. *Oregon Wildlife Diversity Plan.* Oregon Department of Fish and Wildlife, Portland, OR.

Pyke, D.A. and M.M. Borman. 1993. *Problem Analysis for the Vegetation Diversity Project.* Technical Note OR-936-01. USDI Bureau of Land Management, Portland, OR.

Quigley, T.M., R.W. Haynes and R.T. Graham. 1996. *Integrated Scientific Assessment for Ecosystem Management in the Interior Columbia Basin and Portions of the Klamath and Great Basins.* USDA Forest Service, Pacific Northwest Research Station, Portland, OR.

Quigley, T.M and S.J. Arbelbide. 1997. *An Assessment of Ecosystem Components in the Interior Columbia Basin and Portions of the Klamath and Great Basins: Vol. 1.* PNW-GTR-405. USDA Forest Service, Pacific Northwest Research Station, Portland, OR. 335 p.

Randall, W.K. and R.M. Krinard. 1977. *First- Year Growth and Survival of Long Cottonwood Cuttings,* SO-222. USDA Forest Service, Southern Forest Experiment Station, New Orleans, LA.

Rasmussmen, C. 1995. *Riparian Community and Bank Response to Management: A Comparison of old and new surveys in the Prineville District, Bureau of Land Management,* unpublished, September 1995.

Rebele, F., A. Surma, C. Kuznik, R. Bornkamm and T. Brej. 1993. Heavy metal contamination of spontaneous vegetation and soil around the copper smelter Legnica. *Acta Societatis Botanicorum Poloniae* 62 (1–2): 57.

Reeves, G.H., J.D. Hall, T.D. Roelofs, T.L. Hickman and C.O. Baker. 1991. Rehabilitating and Modifying Stream Habitats. *Influences of Forest and Rangeland Management on Salmonid Fishes and their Habitats.* W.R. Meehan (ed.). USDA Forest Service and American Fisheries Society, Bethesda, MD.

Rees, N.E., P.C. Quimby Jr., G.L. Piper, E.M. Coombs, C.E. Turner, N.R. Spencer, L.V. Knutson. 1996. *Biological Control of Weeds in the West.* USDA Agricultural Research Service, Montana Department of Agriculture, Montana State University, Bozeman, MT.

Reiter, M. and R.L. Beschta. 1995. *COPE Report.* Coastal Oregon Productivity Enhancement Program, Newport, OR. Vol. 8, no. 2. May, 1995.

Regional Ecosystem Office. 1995. Ecosystem Analysis at the Watershed Scale: A Federal Guide for Watershed analysis, Analysis Methods and Techniques, Version 2.2. Portland OR. **Terrestrial systems.**

Rice, R.M. 1992. The Science and Politics of BMPs in Forestry: California Experiences. In: R.J. Naiman (ed.), *Watershed Management: Balancing Sustainability and Environmental Change*. Springer-Verlag, New York, NY.

Robbins, W.G. and D.W. Wolf. 1994. Landscape and the Intermontane Northwest: An Environmental History. *Eastside Forest Ecosystem Health Assessment: Volume III (Assessment)* Richard L. Everret, Assessment Team Leader; Paul F. Hessburg, Science Team Leader and tech. ed. USDA Forest Service, Pacific Northwest Research Station, Portland, OR.

Rosgen, D. 1994. A classification of natural rivers. *CATENA: An Interdisciplinary Journal of Soil Science, Hydrology and Geomorphology focusing on Geoecology and Landscape Evolution*. Elsevier, Amsterdam.

Rosgen, D. 1996. *Applied River Morphology* H.L. Silvey, illus. Wildland Hydrology, Pagosa Springs, CO.

Ross, S.M. (ed.). 1994. *Toxic Metals in Soil-Plant Systems*. John Wiley and Sons, New York, NY. 469 p. **Contaminated land.**

Samuel, E.E., J.R. Stauffer, C.H. Hocutt and W.T. Mason, Jr. (eds.), 1978. *Surface Mining and Fish/Wildlife Needs in the Eastern United States* FWS/OBS-78/81. USDI Fish and Wildlife Service, Washington DC. **Abandoned mine lands.**

Savage, S.M. 1974. Mechanism of fire-induced water repellency in soil. *Soil Science Society of America, Proceedings* 38 (July-Aug.): 652–657.

Schaller, F.W. and P. Sutton (eds.). 1978. *Reclamation of Drastically Disturbed Lands*. American Society of Agronomy, Madison, WI. 742 p. **Disturbed lands.**

Scheller-McDonald, K. 1990. *Wetland Creation and Restoration — Description and Summary of the Literature*. USDI Fish and Wildlife Service, Washington DC.

Severson, R.C., S.E. Fisher and L.P. Gough (eds.). 1991. *Selenium in Arid and Semi-arid Environments, Western United States*. U.S. Geological Survey Circular 1064. **Arid lands.**

Sheeter, G.R. and E.W. Claire. 1988. Juniper for Streambank Stabilization in Eastern Oregon. *Proceedings of the California Riparian Systems Conference: Protection, Management and Restoration for the 1990s*. GTR PSW-110. USDA Forest Service, Davis, CA.

Shepperd, W.D. 1996. Silviculture and Management of Aspen in the West. *Natural Resource News*, January 1996, Special Edition. Blue Mountains Natural Resources Institute, La Grande, OR.

Shetty, K.G., B.A.D. Hetrick, D.A.H. Figge and A.P. Schwab. 1994. Effects of mycorrhizae and other soil microbes on revegetation of heavy metal contaminated mine spoil. *Environmental Pollution* 86 (2): 181–188. **Degraded soils.**

Simpson, D.W. and M.K. Botz. 1985. Concepts and Practices in Replacement of Water Sources on Reclaimed Mined Lands. *Proceedings, American Society for Surface Mining and Reclamation*, Denver, CO. pp. 354–364. **Abandoned mine lands.**

Sigh, G., N.T. Sigh, and I.P. Abrol. 1994. Agroforestry techniques for the rehabilitation of degraded salt-affected lands in India. *Land Degradation and Rehabilitation* 5 (3): 223–242. **Arid lands.**

Skaugset, A., L. Kellogg and S. Pilkerton. 1994. *COPE Report*. Coastal Oregon Productivity Enhancement Program, Newport, OR. Vol. 7, no. 1.

Smith, B. and D. Pritchard. 1992. *Riparian Area Management: Management Techniques in Riparian Areas*. TR-1737-6. USDI Bureau of Land Management, Denver, CO. 44 p. **Riparian and wetlands.**

Solazi, M. 1995. *Memorandum: Trapping Summary*. Oregon Department of Fish and Wildlife, June 20.

Soni, P., H. B. Vasistha and O. Kumar. 1991. Ecological rehabilitation of surface mined lands. *The Indian Forester* (June): 485–493. **Abandoned mine lands.**

South Florida Ecosystem Restoration Working Group. 1994. Annual Report. Interagency Working Group. 133 p. **Wetlands bibliography.**

Spreiter, T. 1992. *Redwood National Park Watershed Restoration Manual*. Redwood National Park, Orisk, CA.

Svejcar, T. And J. Brown. 1992. *Is the Range Condition Concept Compatible with Ecosystem Dynamics? Symposium Proceedings: 1992 Annual Meeting, Society for Range Management, Spokane Washington* HC-71 4.51. USDA Agricultural Research Service, Burns OR, and USDA Soil Conservation Service, Davis, CA. 23 p.

Svejcar, T. and R. Tausch. 1991. Anaho Island, Nevada: A Relct Area Dominated by Annual Invader Species. *Rangelands* 13(5).

Swanston, D.N. 1974. *Slope Stability Problems Associated with Timber Harvesting in Mountainous Regions of the Western United States*. TR-PNW-21. Pacific Northwest Forest and Range Experiment Station, USDA Forest Service, Portland, OR.

Terrene Institute, U.S. Environmental Protection Agency, USDA Forest Service. 1994. *Riparian Road Guide: Managing Roads to Enhance Riparian Areas*. Washington DC.

The Nature Conservancy. 1996. *Priorities for Conservation: 1996 Annual Report Card for U.S. Plant and Animal Species*. The Nature Conservancy, Arlington, VA. 16 p.

Tiedemann, A.R. and H.W. Berndt. 1972. Vegetation and soils of a 30-year deer and elk exclosure in central Washington. *Northwest Science* 46 (November 1).

Todd, M. and W. Elmore. 1997. *Historical Changes in Western Riparian Ecosystems*. USDI Bureau of Land Management, Prineville, OR. In publication.

Toy, T.J. and R.F. Hadley. 1987. *Geomorphology and Reclamation of Disturbed Lands*. Academic Press, Orlando, FL. 480 p. **Disturbed lands.**

Tewksbury, J.J. 1996. Cottonwood and Aspen as Wildlife Habitat: A Focus on Birds. *Natural Resource News* Special Edition, January 1996. Blue Mountains Natural Resource Institute, La Grande, OR.

United Nations Secretariat on Desertification. 1977. *Desertification: Causes and Consequences*. Pergamon Press, New York, NY.

U.S. Congress, Office of Technology Assessment. 1993. *Harmful Non-indigenous Species in the United States*. OTA-F-565. U.S. Government Printing Office, Washington DC. 391 pp. **Weeds.**

USDA Forest Service, Pacific Northwest Region. 1992. *Restoring Ecosystems in the Blue Mountains: A Report to the Regional Forester and Forest Supervisors of the Blue Mountain*

*Forests*, 14 pp. Plus appendices
USDA Forest Service. 1994. *Environmental Assessment, Burning and thinning to return old growth characteristics in shortleaf pine ecosystems.* Ouachita National Forest, Hot Springs, AR.
USDA Forest Service, Southwest Region. 1995. *Restoring a Degraded Environment: Understanding and Caring for Pinyon–Juniper Woodlands*, Albuquerque, NM.
USDA Forest Service, Intermountain Research Station. 1995. *Proceedings: Wildland Shrub and Arid Land Restoration Symposium, Las Vegas Nevada, October 19–21, 1993.* Ogden UT. **Terrestrial systems.**
USDA Forest Service, Southern Region. 1995. Renewal of the Shortleaf Pine-Bluestem Grass Ecosystem and Recovery of the Red Cockaded Woodpecker. *Draft Environmental Impact Statement for an Amendment to the Land and Resource Management Plan.* Ouachita National Forest, Hot Springs, AR.
USDA Forest Service NFJD Aspen Working Group. 1995. *A Proposal for the Management, Protection and Enhancement of Quaking Aspen (Populus tremulous Michx.) on the North Fork John Day Ranger District, Umatilla National Forest*, First Draft, February 17, 1995.
USDA Soil Conservation Service. 1994. *Technical Notes: The Stinger*, TN Plant Materials No. 6, Boise, ID.
USDI Bureau of Land Management, Colorado State Office. 1990. *Cows Creeks and Cooperation: Three Colorado Success Stories.* Denver, CO.
USDI Bureau of Land Management, Lakeview District Office. 1990. *Warner Wetlands Area of Critical Environmental Concern (ACEC) Management Plan.* Lakeview, OR.
USDI Bureau of Land Management, Canon City District. 1992. *Blanca Wetlands Area Review.* Canon City, CO.
USDI Bureau of Land Management, Umpqua Resource Area. 1993. *Riparian Enhancement Projects, Seedling Survival Report.* Coos Bay, OR.
USDI Bureau of Land Management, Umpqua Resource Area. 1994. *Riparian Enhancement Projects, Seedling Growth/Survival Report.* Coos Bay, OR.
USDI Bureau of Land Management, Prineville District Office. 1995. *Sage Grouse.* Prineville, OR.
USDI Bureau of Land Management, Deschutes Resource Area. 1996. *Progress Report on Implementation of the Lower Deschutes River Management Plan.* Prineville, OR.
USDI Bureau of Land Management, Central Oregon Resource Area. 1996. Murderer's Creek Treatment Notes. *Murderer's Creek Seeding Information File.* Prineville, OR.
USDI Bureau of Land Management, Prineville District. 1997. *Millican Valley Off Highway Vehicle Management Plan and Decision of Record.* Prineville, OR (unpublished).
USDI Fish and Wildlife Service. 1987. *Restoring America's Wildlife: 1937–1987. The First 50 Years of Aid in Wildlife Restoration (Pittman-Robertson) Act.* U.S. Government Printing Office, Washington DC.
USDI National Biological Survey. 1994. *VDP Update.* Vegetation Diversity Project. National Biological Survey, Corvallis, OR. Vol. 1, no. 2. April–June 1994.
U.S. Environmental Protection Agency. 1993 . *Constructed Wetlands for Watewater Treatment and Wildlife Habitat: 17 Case Studies.* EPA832-R-93-005. EPA, Denver, CO. 174 p. **Wetlands.**
U.S. General Accounting Office. 1988. *Public Rangelands: Some Riparian Areas Restored but Widespread Improvement Will Be Slow.* GAO/RCED-88-105. Washington D.C. June 1988.
Valentine, J.F. and A.R. Stevens. 1994. Use of Livestock to Control Cheatgrass — A Review. *Proceedings: Ecology and Management of Annual Rangelands.* INT-GTR-313. USDA Forest Service, Intermountain Research Station, Ogden, UT.
Veith, D.L., K.L. Bickel, R.W.E. Hopper and M.R Norland. 1986. *Literature on the Revegetation of Coal-mined Lands — an Annotated Bibliography.* U.S. Bureau of Mines Information Circular IC-9048. **Abandoned mine lands.**
Walker, M.M. 1986. Botanical Clubs and Native Plant Societies of the United States. New England Wildflower Society, Farmington MA.
Webb, R.H., J.W. Steiger and E.B. Newman. 1988. *The Response of Vegetation to Disturbance in Death Valley National Monument, California.* U.S. Geological Survey Bulletin 1793. 103 p.
Webb, R.H. and H.G. Wilshire (eds.). 1983. *Environmental Effects of Off Road Vehicles: Impacts and Management in Arid Regions.* Springer Verlag, New York, NY. 534 pp. **Arid lands.**
Western Governors' Association. 1997. Pursuing New Tools to Enhance Future Management of the Bureau of Land Management and Forest Service Lands. Unpublished Draft of 6/1/97.
Whitmore, R.C. 1978. Managing Claimed Surface Mines in West Virginia to Promote Nongame Birds. *Surface Mining and Fish/Wildlife Needs in the Eastern United States, Fish and Wildlife Service*, FWS/OBS-78/81. E.E. Samuel, J.R. Stauffer, C.H. Hocutt and W.T. Mason, Jr. (eds.). U.S. Department of the Interior, Washington DC. pp. 381–386.
Whitney, J. 1996. The Yavapai ecosystem project. *Science and Management Collaboration-Success Stories* no. 7. February, 1996. USDA Forest Service. 2 p.
Williams, R.D. and G.E. Schuman (eds.). 1987. *Reclaiming Mine Soils and Overburden in the Western United States: analytic parameters and procedures.* Soil and Water Conservation Society of America, Ankeny, IA. **Abandoned mine lands.**
Zak, J.C. 1985. The Importance of Vesicular–Arbuscular Mycorrhizae in the Reclamation of Mined Spoils. *Proceedings of the American Society for Surface Mining and Reclamation* Denver, CO. pp. 298–305. **Degraded soils.**
Zeedyk, W.D. 1996. *Managing Roads for Wet Meadow Ecosystem Recovery* FHWA-FLP-96-016. USDA Forest Service, Albuquerque, NM. 73 p. **Roads in wetlands.**
Zimovets, B.A. 1992. Melioration, fertility, and ecology of soils of the arid zone. *Soviet Soil Science* 24 (1): 81–89. **Degraded soils.**
Zimovets, B.A., F.R. Zaydelman, Y.I. Pankova and S.V. Boyko. 1994. Ecological concept of soil reclamation. *Eurasian Soil Science* 26 (3): 73–85. **Degraded soils.**

## APPENDIX A: RESTORATION CHECKLISTS

The purpose of checklists is to bring order to the decision-making and data-gathering processes so that the methods and information presently available can be utilized and deficiencies in the science can be identified. Apfelbaum and Chapman (1994) provide a ten step checklist for successful restoration. Their list is presented below with some modifications.

1. Inventory and map the ecological resources, and describe their condition.

2. Describe the site's history and map it where possible.
3. Develop a hypothesis of how the original system worked. Review technical literature for related ecological studies conducted in the region; visit nearby areas.
4. Develop goals and objectives for each management unit by assessing the potential for restoration with reasonable effort.
5. Develop an implementation plan of actions and costs to accomplish the goals and objectives.
6. Design a monitoring program to evaluate success.
7. Implement the restoration actions.
8. Document results.
9. Evaluate periodically, incorporating new information and ideas, revising goals and objectives as necessary, and modifying actions.
10. Communicate constantly with interested and affected people, providing information as it is available. The Applegate Adaptive Management Area "Learning Summaries" provide an excellent model (USDI BLM, 1997)

The checklist approach of Cairns (1983) is also presented with some modifications to assist the selection of management options for the ecological restoration. These checklists are a series of questions that the manager should be able to answer; the lists should be evaluated and modified as needed to be appropriate for both specific sites and general use by the land management agencies.

### *Checklist for selection of management options (modified from Cairns, 1983)*

The purpose of this checklist is to ensure that adequate data are available before a decision is made:

1. Can any options be excluded because of low probability of success?
2. Is the cost of any option (including cost to society) prohibitive?
3. What time will be required to reach each of the goals?
4. Is further disturbance likely to occur on site?
5. Can the management responsibilities and/or costs be transferred to another group? If so, should this be done?
6. Which option conforms most closely to existing regulatory requirements?
7. What are the comparative costs for monitoring performance?
8. Is cooperation with third parties or stakeholders possible?

### *Option 1: Full Restoration Checklist*

1. Is restoration to a pre-existing or reference condition feasible?
2. Is adequate ecological information available about the pre-existing or reference condition?
3. Is a reference (control) site available?
4. Is the source of original species adequate ?
5. Are management responsibilities (including performance monitoring), cost estimates and duration described?
6. Was the pre-existing or reference ecosystem of local, regional, or national importance? Can societal response to this option be evaluated accordingly?
7. Is there an alternative ecosystem or land use other than the pre-existing or reference condition that is legally permissible and that should be evaluated?

"Full" restoration should be selected: (a) for ecologically unique ecosystems, (b) for aesthetic reasons where the damaged area is part of a larger system of major recreational or ecological value, (c) as a precondition for surface disturbance in unique and irreplaceable ecosystems. This option requires that substantial information about pre-existing ecosystem conditions, as well as how the location fits within the range of variability for the ecological site.

### *Option 2: Partial restoration checklist*

1. Have the desirable parameters or ecological conditions to be restored been explicitly stated?
2. Is there evidence that the site can be rehabilitated without restoring the entire ecosystem to a pre-existing or reference condition?
3. Will the mix of parameters and conditions be ecologically stable? If not, what additional qualities must be added to achieve stability?
4. Is there a significant difference in cost between partial restoration and restoring the ecosystem to its pre-existing or reference condition?

### *Option 3: Alternative Ecosystems checklist*

1. Have the specifications of the alternative ecosystem been explicitly stated?
2. If long-term management is required, who pays for it?

3. What is the likelihood that restoration activities may result in significant undesirable physical or biological changes outside of the intended management area?
4. Is monitoring for potential adverse effects outside of the site boundary required?
5. Is this alternative ecosystem partly or entirely experimental? If so, do the appropriate regulatory and civic authorities know this?
6. If several alternative ecosystems might be established, have the options been discussed with appropriate decision makers?
7. Is there community and stakeholder approval for the proposed actions and associated costs?

*Option 4: Containment option checklist*

1. Can the adverse ecological effects expected from the site be mitigated? How long should they last?
2. Is chronic maintenance of the containment strategy required? At what cost?
3. Are there any adverse health or safety considerations which threaten the general public?
4. Is the likelihood of successfully implementing other restoration options sufficiently low that containment is the most feasible short term management option?
5. How should be appropriate regulatory agencies be informed of the rationale for this decision?
6. Is there community and stakeholder approval for the proposed actions and associated costs?

*Option 5: Non-intervention with monitoring*

1. Is substantive evidence available that natural processes will be more effective than available management practices?
2. Are restoration goals likely to be achieved without intervention in a reasonable period of time?
3. Do the costs of intervention greatly exceed anticipated benefits and results?
4. Can any adverse environmental effects expected from the site be mitigated?
5. Are monitoring objectives and parameters explicitly defined?
6. How should the public and appropriate regulatory agencies be informed of the rationale for this decision?
7. How soon should the management options for the site be evaluated? What additional information is needed to assist this reevaluation of options?

## APPENDIX B: DATA REPOSITORIES

Data Repositories: The following are general repositories of information that may be useful for ecological restoration projects. Local sources, such as district or regional offices of the U.S. Geological Survey, U.S. Forest Service, Bureau of Land Management, U.S. Natural Resource Conservation Service, U.S. Army Corps of Engineers, state geological surveys, or offices of the state climatologists may also possess useful data. The Earth Science Information Centers (ESIC) of the U.S. Geological Survey (listed below) can provide information about air photos, remote sensing data, maps, cartographic, and geographic data from many public and private producers in the United States using automated catalog systems for information retrieval and research services; for further information, call 1-800-USA-MAPS. Some academic departments within the state land grant colleges and universities have research reports, master's theses, doctoral dissertations, and extension service reports that are often very site specific.

*Master Title Plats (Official Status and Historical Indices)*

Official land records for the United States are maintained by all Bureau of Land Management state and district offices for the state(s) or districts they cover. Local county, U.S. Forest Service, Natural Resource Conservation Service, U.S. Fish and Wildlife Service, National Park Service and other agency offices also frequently maintain land status maps.

*Aerial Photography, Remote Sensing Data*

Aerial Photography Field Office, USDA-NRCS, 2222 West 2300 South, P.O. Box 30010, Salt Lake City, UT 84125

Sioux Falls Earth Science Information Center, U.S. Geological Survey, EROS Data Center, Sioux Falls, SD 57198-0001, 605-594-6151; Fax 605-594-6589, TDD 605-594-6933 (Telecommunications for the Deaf)

Reston Earth Science Information Center, U.S. Geological Survey, 507 National Center, Reston, VA 22092, 703-648-6045; Fax 703-648-5548, TDD 703-648-4119 (Telecommunications for the Deaf)

For a detailed description of data available from these and other local services, see: Earth Science Information from the U.S. Geological Survey, U.S. Geological Survey Fact Sheet FS 125-95. Information on these and other USGS products and services is available 24-hours a day from the EARTHFAX fax-on-demand system (Dial 703-648-4888). Product information and select digital cartographic files are available on the Internet through a World Wide Web server. The USGS home page can be accessed using browse tools, such as Mosaic, at URL:http://www.sugs.gov/

*Geologic Maps*

Lakewood Earth Science Information Center, U.S. Geological Survey, Box 25046, Building 810, Denver Federal Center, MS 504, Denver, CO 80225-0046, 303-202-4200; Fax 303-202-4188, or call 1-800-USA-MAPS for further information and other USGS ESIC locations

The USGS Earth Science Information Centers (ESIC) can provide information on sources of geologic and other maps from State offices of geology or conservation, State highway departments, chambers of commerce, and university and public libraries. Address listings of commercial map resources, Earth Science Information Center (ESIC) State Offices, and U.S. Geological Survey map depositories are available from the USGS ESIC's upon request.

*Hydrologic Data*

National Water Data Exchange (NAWDEX): The National Water Data Exchange of the U.S. Geological Survey maintains a computer data system that identifies sources of water data. The NAWDEX Office assists data users in locating sources of water data, identifying sites at which data have been collected, and obtaining specific data.

National Water Data Exchange (NAWDEX), U.S. Geological Survey, 421 National Center, Reston, VA 22092, 703-648-6848; Fax 703-648-5704

National Water Information Clearinghouse (NWIC): The National Water Information Clearinghouse of the U.S. Geological Survey disseminates information on water resources to government agencies, the private sector, and the general public.

National Water Information Clearinghouse, U.S. Geological Survey, 423 National Center, Reston, VA 22092, 1-800-H2O-9000 (1-800-426-9000)

*Land Use, Land Covering, and Associated Maps*

See USGS Earth Science Information Center sources listed under *Aerial Photography, Remote Sensing Data.* Many local county, Bureau of Land Management and Forest Service offices also maintain land use maps and GIS products.

*Meteorologic, Weather, and Climate Data*

National Oceanic and Atmospheric Administration, Environmental Data Service, National Climatic Center, U.S. Department of Commerce, Federal Building, Asheville, NC 28801

*Soil Characteristics Data*

U.S. Department of Agriculture, National Resource Conservation Service, Washington, D.C. 20013

Soil maps are prepared on a county basis and contain a wealth of information but are not available for all counties. The Natural Resource Conservation Service maintains an office in each state and usually several field offices as well. These can be located through telephone directories under the heading for U.S. Department of Agriculture.

*Topographic Maps*

See sources listed under *Geologic Maps.*

*Riparian and Wetland Technical References*

National Business Center, Printed Materials Distribution Section (BC-650B), Bureau of Land Management, Denver Federal Center, Building 50, P.O. Box 25047, Denver, Colorado 80225-0047

*National Wetlands Inventory Maps*

U.S. Fish and Wildlife Service: Local offices of the U.S. Fish and Wildlife Service, Bureau of Land Management, U.S. Forest Service and Natural Resource Conservation Service also frequently maintain copies of these maps.

*Compilations of Riparian Case Studies*

See Natural Resource Law Center 1996; Massman 1995; Callaham 1990; USDI Bureau of Land Management, Colorado State Office 1990; Greswell et al 1989: USGAO 1988. Full citations are in Literature Cited.

## APPENDIX C: NATIVE PLANT SUPPLIERS

The following is a 1997 sampling of commercial and other sources of native and wetland plant supplies that may be useful for ecological restoration projects. Local sources, such as local Natural Resource Conservation Districts, U.S. Fish and Wildlife, Forest Service, Bureau of Land Management, native plant societies, arboretums, etc. may also be able to provide biological supplies for restoration projects. A listing of botanical clubs and native plant societies of the United States is given in Walker (1986). Care should be taken concerning genetic and environmental variability as well as seed quality when selecting a seed source (Eddleman 1997, Pyke and Borman 1993).

Arkansas Valley Seeds Inc., 4625 Colorado Blvd., P.O. Box 16052, Denver, CO 80216

Comstock Seed, 8520 W. 4th Street, Reno, NV 89523

Country Wetlands Nursery, LTD, Box 126, Muskego, WI 53150

Environmental Concern Inc., P.O. Box P, St. Michaels, MD 21663

Fern Hill Farm, Route 3, Box 305, Greenville, AL 36037
Gardens of the Blue Ridge, P.O. Box 10, Pineola, NC 28662
Goble Seed Company, P.O. Box 203, Gunnison, CO 84634
Gooding Seed Company, P.O. Box 57, 103 Main Street, Gooding, ID 83330
Granite Seed, 1697 W. 2100 N., Lehi, UT 84043
Grassland West Company, P.O. Box 489, Clarkston, WA 99403
Horizon Seed Co., 1540 Cornhusker Highway, Lincoln, NB 68500
Horticultural Systems, Inc., P.O. Box 70, Parrish, FL 33564
Idaho Grimm Growers, P.O. Box 276, Blackfoot, ID 83221
Intermountain Seed Company, P.O. Box 62, 445 S. 100 E., Ephraim, UT 84627
Kesters Wild Game Food Nurseries, Inc., P.O. Box V, Omro, WI 54963
Lilypons Water Gardens, Lilypons, MD 21717
The National Wildflower Research Center, Austin Texas (512-929-3600)
Mangelsdorf Seed Co., P.O. Box 327, St. Louis, MO 63166
Mangrove Systems, Inc., 504 S. Brevard Avenue, Tampa, FL 33606
Maple Leaf Industries, Inc. P.O. Box 9-6, 480 S. 50 E., Ephraim, UT 84627
Maughn Seed Company, P.O. Box 72, Manti, UT 84642-0072
North American Revegetation, 1987 Crittenden Loop NW, Albany, OR 97321
The Theodore Payne Foundation, 10459 Tuxford Street, Sun Valley, CA 91352
Plants of the Wild, Box 855, Tekoa, WA 99033
Plummer Seed Company, Inc. P.O. Box 70, Ephraim, UT 84627
Ranier Seeds, Inc. P.O. Box 1549, Port Orchard, WA 98366
San Francisco Bay Marine Research Center, 8 Middle Road, Lafayette, CA 94549
F.W. Schumacher Co., Inc., 36 Spring Hill Road, Sandwich, MA 02563-1023
Sharp Bros. Seed Co., Healy, KS 67850
Siskiyou Rare Plant Nursery, 2825 Cummings Road, Medford, OR 97501
Slocum Water Gardens, 1101 Cypress Gardens Road, Winter Haven, FL 33880
Southern States Coop., 6606 West Broad, P.O. Box 26234, Richmond, VA
Southwest Seed, Inc. 13260 County Road 29, Dolores, CO 81323
Stanford Seed Co., 809 N. Bethlehem Pike, Spring House, PA 19477
Stevenson Intermountain Seed, P.O. Box 2, 488 S. 100 East, Ephriam, UT 84627
Stock Seed Farms, Inc., R.R. Box 112, Murdock, NB 68407
Westland Seed Inc. 1308 Round Butte Road West, Ronan, MT 59864
Wheatland Seed, P.O. Box 513, Brigham City, UT 84302
William Tricker, Inc., 74 Allendale Avenue, Saddle River, NJ 07458
Wind River Seed, Route 1, P.O. Box 97, 3075 Lane 51 2, Manderson, WY 82432
Van Ness Water Gardens, 2460 Euclid Avenue, Upland, CA 91768
Wildlife Nurseries, P.O. Box 2724, Oshkosh, WI 54901

Collecting seed from sites adjacent to the area to be restored has also been employed in restoration. For additional information related to restoration on Mojave and Sonoran desert lands, visit the Desert Lands Restoration Task Force web page at www.serg.sdsu.edu.

## THE AUTHORS

**James G. Kenna**
*Bureau of Land Mangement*
*Palm Springs-South Coast Manager*
*690 West Garnet Avenue*
*P.O. Box 1260*
*North Palm Springs, CA 92258-1260*

**Gilpin R. Robinson Jr.**
*United State Geological Survey*
*Geologist*
*954 National Center*
*Reston, VA 20192, USA*

**Bill Pell**
*Ouachita National Forest*
*Team Leader-Ozark Ouachita Highlands*
*P.O. Box 1270*
*Hot Springs, AR 71902, USA*

**Michael A. Thompson**
*USDA Forest Sciences Laboratory*
*Research Engineer*
*410 MacInnes Drive*
*Haughton, MI 49913-1199, USA*

**Joe McNeel**
*West Virginia University*
*Director, Division of Forestry*
*Morgantown, WV 29506-6125, USA*

# Human Population Growth and Tradeoffs in Land Use

Joel E. Cohen

## Key questions addressed in this chapter

- *Population growth and spatial distribution for the whole world and for the United States*
- *Forest and woodland cover and agricultural extensification and intensification for the whole world and for the United States*
- *Theoretical perspectives on the relations among population change, land-cover change, and human carrying capacity*
- *Tradeoffs in land use and forest use that arise from the combination of population growth, increasing consumption, increasing non-timber values in forest use, and increasing interactions between the U.S. and the rest of the world*

**Keywords: Demographics, technology, culture, economics, consumption, human carrying capacity**

## 1 INTRODUCTION

This chapter focuses on population in relation to land use, and especially on tradeoffs in land use that affect forests. Globally, a growing population is likely to increase the material demands people place on land, including demand for forest products. However, in the past century, changes in prices, technologies, institutions, values and the spatial distribution of the population changed the demands Americans placed on their forests more than the changes in the size of the U.S. population changed these demands.

Human use of land, water and other natural and human-made resources is influenced by four major factors: population, economics, the environment, and culture. A population is described by its size (the numbers of people by categories of age, gender and other characteristics), rate of growth or decline, spatial distribution (for example, urban versus rural, and the distribution of population density) and migration. Economics includes institutions for ownership or common use of land, incentives for land exploitation or conservation, markets or other institutions for dealing in land as well as the products of and inputs to land, labor force availability, and sources and conditions of capital and credit. The environment includes the physical, chemical and biological quality of land, air and water, including climate. Culture includes political institutions; governmental, commercial, and individual policies toward land use; styles of life; expected roles of women, men, children and elderly in paid work and family life; levels of education; and religious and traditional views of relations between humans and their land and water.

American planners, managers, and citizens must consider the global perspective, even if they are concerned only to protect American resources and interests, because the United States is intimately linked to the rest of the world. The United States is linked demographically to populations abroad through migration and competition for jobs (Burtless 1995); economically through international markets and international technologies that affect the demand for commodities and services derived from land; environmentally through atmospheric emissions, introduced forest pests, and global climatic changes; and culturally through the spread of free-market institutions, rising material expectations and consumerism, technologies, political movements, and other values that affect the supply of and demand for products and services derived from land.

In the last thousand years, forests have changed from being superabundant, essentially free and often viewed as an impediment to development (land "improvement" often meant removing the trees) to being priced in many economies. Forests have changed from being valued primarily as a source of land, timber and game to being valued for a host of goods and services. They have changed from being a subject of interest principally to the locality where they are found to a subject often of worldwide interest (for example, as objects of trade, sinks of carbon, reservoirs of biodiversity, or targets of ecotourism). The timber products extracted from forests have changed from primarily logs to lumber to plywood, veneer, composites, chips and pulp (Clawson 1995). If the trend continues toward reconstituting products from smaller components, forests may assume a role as incubators of complex organic molecules. Some of these molecules may replace some of those now derived from fossil fuels.

The innovations in institutions and policies that are developed to deal with problems in land use, particularly forestry, may provide useful models of policies helpful for other environmental concerns, because the state of the world's forests may be a leading indicator of the fate of other human-induced changes in the environment. Kates et al. (1990: 7) estimated human-induced changes in nine aspects of the environment between 10,000 B.C. and the mid-1980s, and tabulated the estimated dates by which half of the total change had been achieved. The environmental variables, and the year by which half the total change was achieved, were: deforested area (1850); number of vertebrate species that became extinct through human action since A.D. 1600 (1880); carbon releases (total mass mobilized by human activity) (1920); lead releases (1950); population size (1950); total annual water withdrawal for human use (1955); carbon tetrachloride production (1960); sulfur releases (1960); phosphorus releases (1975); and nitrogen releases (1975). These estimates suggest that changes in global forests preceded other major environmental changes.

Human population growth and land use interact with economics, the environment and culture. A full treatment of the interactions (Turner et al. 1990, Cohen 1995b) lies far beyond the reach of this chapter. Section 2 of this chapter is devoted to a global perspective, section 3 to a United States perspective, and section 4 to the future tradeoffs that U.S. citizens and managers will have to face in forestry and land use, especially as a result of interaction between domestic U.S. and international factors.

This chapter is intended for front-line land managers who seek a perspective on where daily decisions fit into a larger picture. It is also intended to give upper-level land-use policy-makers in government, business and philanthropy a richer picture of long-term trends and major issues. Finally, this chapter could introduce students of land use and forestry to some of the issues they can expect to confront in their future professional careers.

## 1.1 Historical Context: Theories of Population Change and the Environment

The relationship between human population growth and land use, especially land degradation, is addressed by four major theories or conceptual frameworks: (1) neoclassical economics, (2) classical economics and natural science, (3) dependency theory, and (4) combinations of these approaches that view population as an intermediate variable (Jolly 1994). Each of these approaches has strengths and weaknesses. Here I offer only a cartoon-like summary of each approach, relying largely on Jolly (1994). The purpose is to show the diversity of currently defended views of the relation between population growth and land use.

### 1.1.1 Neoclassical Economics

Neoclassical economics argues that, when markets function well, an economy can provide an increasing population with a steady or rising level of living, given a finite endowment of natural resources. In this view, technology can substitute human-made goods and services for those provided by nature and can make it possible to use the resources provided by nature more efficiently. Rising prices in the market for natural resources, including land and its products, will smoothly elicit technological innovation and shift consumer preferences away from scarce goods and services.

Neoclassical economists often acknowledge that, in many countries and sectors, markets do not function well — especially when governments interfere with markets — and that rapid population growth may make it more difficult for markets to be efficient. In their view, land degradation may be a temporary response to population growth while technology devises a more efficient use of land. Or land degradation may be a response to inefficiencies of markets, as when land resources are commonly owned (Hardin 1968). Or land degradation may be the result of exhausting land resources, a result that is perfectly acceptable to neoclassical economists because the market will call forth equivalent or superior alternatives to use of the land. Slowing population growth would only buy time to find substitutes for land and to correct inefficiencies in markets and institutions. In summary, economic inefficiencies lead to land degradation.

### 1.1.2 Classical Economics and Natural Science

Classical economists, based on an interpretation of the work of Thomas Robert Malthus (1766–1834), and many natural scientists argue that an economy cannot provide a rapidly growing population with a steady or rising level of living, given a finite endowment of natural resources. In this view, additional workers will eventually encounter diminishing returns from a fixed land area and additional consumers will eventually use up enough fixed resources to have a negative impact on the environment. If the population passes a certain level that may be called the carrying capacity of the land, then birth rates must fall or death rates must rise to lower the population to a level that the land can support. In summary, high population growth causes land degradation.

Some proponents of this view acknowledge that the adverse impact on land of rapid population growth is compounded by an unequal distribution of wealth, which may push the numerous poor to the most marginal and fragile lands. Long fallow periods and crop rotation may be abandoned in the face of growing numbers of people to feed. Degradation of land aggravates poverty, which leads to further land degradation. Some proponents also recognize that people's expectations play a crucial role: for a given density of population on land of a given quality, people with higher expectations of what they can extract from the land may degrade it more rapidly than people with lower expectations. Nevertheless, for most proponents of this view, reductions in human fertility are the key to avoiding environmental destruction, including land degradation, and to raising levels of living. Land reform and technological innovations at best buy time until limits are reached, because the human ability to substitute human-made capital for natural resources is limited.

### 1.1.3 Dependency Theory

Dependency theory argues that systems of production and social relations cause poverty, especially the exploitative relations between the now-rich industrialized countries and the now-poor developing countries; and it is poverty that causes both environmental degradation and rapid population growth. In this view, export-oriented production, cash crops to earn foreign currency, inappropriate technologies from industrialized countries, and the influence of multinational corporations all contribute to land degradation in the poor countries. The root problem is seen as the structure of society or political economy within developing countries as well as the international social, political and economic relations between poor and rich countries. Poverty leads to land degradation because the poor countries lack appropriate technology, capital, management skills and educational resources; poor farmers may know their practices are degrading their own land, but lack any resources to rectify the problem. Poverty also leads to rapid population growth because, compared to wealthy families, poor families desire more children as sources of labor (in their youth) and of social security (when parents are old).

For dependency theorists, the key to stopping land degradation is to alleviate poverty. Poverty is to be alleviated, first, by increasing productivity through economic development and, second, by distributing output more equitably through social change (both between rich and poor countries, and within poor countries). If technological innovation only amplifies the relative power of the already wealthy, it may have an adverse impact on both land degradation and population growth; if it is appropriate for the needs and environmental setting of the poor, it can foster greater equity and wealth and contribute to slowing population growth. The Malthusian limits that concern the classical economists and natural scientists lie so far beyond the constraints currently imposed by poverty and inequity as to be of little significance. In summary, poverty and inequity cause both land degradation and high fertility.

### 1.1.4 Population as an Intermediate Variable

"Intermediate variable" theorists argue that a variety of fundamental causes affect land use and land degradation, and that rapid population growth intensifies the environmental effects of these fundamental causes. The fundamental causes may vary from region to region. Examples of fundamental causes include warfare, movements of refugees, polluting technologies, subsidies for inappropriate human settlement, artificial controls on food prices, lack of employment opportunities, absence of rural credit, ineffective extension services, and low agricultural productivity. Higher numbers of people and more rapid population growth aggravate the adverse effects of all of these fundamental causes of land degradation.

For intermediate variable theorists, slowing rapid population growth buys time to address the fundamental causes of land degradation, even though the effects of population policies often take a very long time to appear. Economic, agricultural and silvicultural policies directed at land use are required to address land degradation (Shaw 1989).

### 1.1.5 The Roles of Population Growth and Spatial Distribution

For neoclassical economists, population growth is a neutral factor in land degradation. For classical economists and some natural scientists, population growth is the principal independent cause of land degradation. For dependency theorists, both population growth and land degradation are symptoms of poverty and inequity. For intermediate variable theorists, population growth exacerbates the adverse effects of other ultimate causes of land degradation. Each is a partial, and partially useful, view of the relation between human population growth and land use.

The intermediate variable theory alone allows for spatial variation in the factors that affect land use and land degradation. With that partial exception, none of these theories is situated in a real space of countries and continents with varying economies, environments, cultures, and histories. After reviewing the history that follows, I will describe (in Section 2.6) two other approaches that partially remedy this shortcoming of the theories just summarized.

## 2 GLOBAL HISTORY OF HUMAN POPULATION GROWTH AND LAND USE

Over the centuries, human populations have increased, shifted their distribution, and changed the way they have used natural resources to sustain themselves.

### 2.1 History of Global Population Size

In the last two millennia, global human population growth experienced two major phases (Cohen 1995b). In the first phase, which ended around 1965–70, the rate of increase of the human population steadily increased, not merely in absolute numbers of people added per year but also in the percentage increase per year. (Massive epidemics in the 14th century, and possibly earlier, briefly interrupted the steady increase.) During the interval A.D. 1–1650 the population doubled from roughly 0.25 billion ($10^9$) to 0.5 billion. The next doubling of global population (to 1 billion people) required less than two centuries (roughly 1650–1830), and the next (to 2 billion) about one century (roughly 1830–1930). During the interval 1930–1974, the population doubled from 2 to 4 billion (Fig. 1). Thus the doubling time dropped from roughly 1,650 years to roughly 44 years. These simple facts show that global human population growth cannot be described by an exponential curve or by a logistic curve. An exponentially growing population has a constant growth rate (if measured as percent increase per year) and hence a constant doubling time. In a population that grows according to the logistic curve, the growth rate (again, percent increase per year) always decreases as the population size increases.

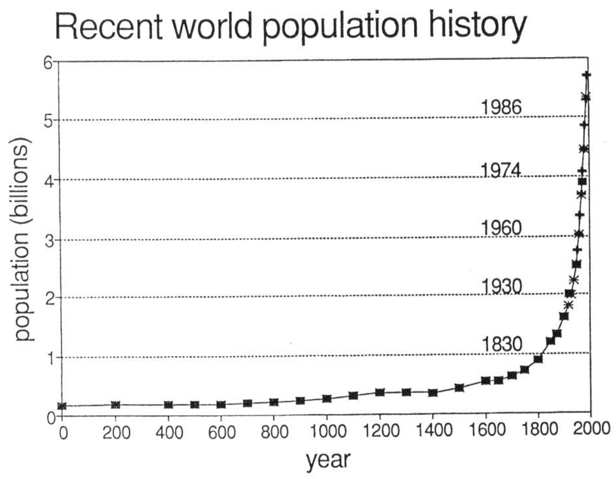


**Fig. 1. Estimated global human population from A.D. 1 to 1997. Source: Cohen (1995b); Population Reference Bureau (1996). Copyright © 1995 by Joel E. Cohen.**

The second phase began around 1965–70 and still continues. In this phase, the population growth rate erratically declined from its peak around 2.1% per year in the interval 1965–70 to an estimated 1.5% per year in 1997. A growth rate of 1.5% per year implies a doubling in 46 years, and is still extremely rapid compared to rates of global population growth experienced before 1945. Temporary dips in the global population growth rate in earlier centuries were due mainly to transient rises in death rates, as consequences of natural or human-induced catastrophes. By contrast, the decline in the global population growth rate since 1965 has been overwhelmingly due to a reduction in the numbers of children born per woman, while death rates mainly continued to decline. Based on the history of fertility in countries where fertility is now low, it seems reasonable to suppose that the global fall in fertility will continue without significant reversals. But the transient, dramatic rise in fertility called "the baby boom" which occurred in the United States and some other countries after 1945 shows that there is nothing inevitable about continuing local and global declines in fertility.

The absolute increase per year in the human population reached its all-time peak of around 90–95 million ($10^6$) additional people per year in the early 1990s. The absolute increase slowly began to decline in the mid-1990s to its present level around 85–90 million additional people per year. These absolute rates of increase are tremendous compared to historical experience. For example, the present absolute increase per year is roughly the same as the century's increase from A.D. 1600 to 1700. More than 90% of all the population increase that has ever occurred has taken place within the last three and a half centuries, and more than 70% within this century alone. While it took from the beginning of time until 1830 to add the first billion people, the most recent billion were added in 12 years.

Global statistics conceal very different stories in different parts of the world. About 1.2 billion people live in the economically more developed and richer regions, where average annual incomes are $18,100: Europe, the United States, Canada, Australia, New Zealand, and Japan. The remaining 4.6 billion live in the economically less developed and poorer regions, where average annual incomes are $1,100 (Population Reference Bureau 1996). In aggregate, the rich one-fifth of the world's population generates and spends about 80% of the world's income.

The population of the rich countries increases perhaps 0.1% per year. This growth, if continued, implies a doubling of population after more than 500 years. The population of the poor countries grows at 1.9% per year, a rate sufficient to double in 37 years if continued. The population of the least developed regions, where the world's poorest half-billion people live, increases by 2.8% per year, with a doubling time of less than 25 years (Population Reference Bureau 1996). Suppose (contrary to what is likely to happen) that the populations of the rich, poor and least developed countries continued to grow at their present rates for a typical lifetime of 74 years, and that no presently poor countries became rich. Then the population of the rich countries would increase roughly 8%, the population of the poor countries would grow 400% (the result of two doublings), and the population of the least developed regions would increase about 800% (about three doublings in 74 years).

## 2.2 Global Population Distribution

In 1994, the world had an average population density on ice-free land of 0.42 people per hectare (ha). In the rich countries, the population density was 0.22 people per ha; in the poor countries, 0.54 people per ha. The poor countries have more than twice the population density of the rich, on average, and their populations are increasing 10 to 20 times faster.

The last two centuries have witnessed a massive movement of people from the countryside to cities. Current statistics on urban populations follow each country's definitions, which vary from country to country. In spite of definitional fuzziness, the overall trend is clear. In A.D. 1800, about 2% of people lived in places with 20,000 or more people. By 1950, about 20% of people lived in places with 20,000 or more people (Cohen 1995b: 100). Today, about 45% of the world's population is urban (Population Reference Bureau 1996). The absolute number of city dwellers rose more than 140-fold from perhaps 18 million in 1800 to 2.5

billion today, while global population increased more than six-fold.

The move from the countryside to cities took place first in the countries that industrialized first; those are today's rich countries. By roughly 1915, more than half the population had left the farm in only one country, Great Britain. Today 75% of the 1.2 billion people in the rich countries (and 75% of people in the United States) live in cities.

Now people in the poor countries are moving to cities, in some cases even when their countries are not industrializing. During 1990–1995, the population of cities in poor countries grew by 3.5% per year, while the urban population of rich countries grew by 0.8% per year (United Nations 1995). In both rich and poor regions, the urban population grew far faster than the total population. But in absolute numbers of people, most of the shift to cities is yet to come: by 1996, in the poor countries, only 35% of people lived in cities (Population Reference Bureau 1996). Rapid urbanization in the poor countries seems likely to continue on a massive scale.

A striking aspect of urbanization has been the rise of megacities, especially in poor countries. A megacity is defined as an urban region with 10 million people or more. In 1950, there was one megacity in the world: New York. In 1994, there were 14 megacities in the world, and 10 of the 14 were in poor countries.

The form of the distribution of population density appears to be remarkably similar at the spatial scales of the whole earth, the United States, and an individual state (New York, in this example). Already, and perhaps increasingly in the future, a relatively small fraction of the land is occupied at a high human population density, while a very large fraction of the land is lightly or very sparsely occupied. As an increasing fraction of people moves into cities, the former direct presence on the land of a dispersed agricultural population is replaced by the remote demands of a city-dwelling populace. This shift may facilitate some aspects of land management and make others more difficult.

### 2.2.1 Self-similarity in the Distribution of Population Density

This subsection is devoted to showing the evidence for the claim in the previous paragraph that the distribution of population density appears to be remarkably similar in form at the spatial scales of the whole earth, the United States, and an individual state (New York, in this example).

I analyzed 1989 estimates of the population and area of 148 countries (World Resources Institute 1992). In 1989, the U.S.S.R. still existed as a political entity. The combined areas and populations of these 148 countries covered more than 13 billion ha, which include almost the entire ice-free land area and human population of the earth. I divided each country's population by its land area to get its population density, then ranked the countries from the least to the most densely populated (Cohen 1995b: 103).

The top left panel of Fig. 2 shows the cumulative area of all countries in which the population density was less than or equal to the population density shown. For example, nearly 13 billion ha had an average population density of 10 or fewer people per ha, and a very small area had a population density greater than 10 people per ha.

To see the cumulative distribution of area at low population densities, I replotted the same data with population density on a logarithmic scale in the second row of the first column of Fig. 2 (Cohen 1995b: 104). More than 11 billion ha (approximately 85% of the land) had one person per ha or fewer, and more than 10 billion ha (over 75% of the land) had on average less than half a person per ha.

To emphasize the distribution of population density among the countries with the lowest population densities, I replotted the same data with both population density and cumulative area on logarithmic scales in the bottom left panel of Fig. 2. Together these three panels give the global pattern of population density in relation to cumulative area on the scale of all nations.

To see the distribution of population density within a nation, I applied the identical treatment to the 1990 populations and areas of the states of the United States plus Washington, DC. The three plots in the middle column of Fig. 2 are remarkably similar to the corresponding plots for the countries of the world.

To see the distribution of population density within a single state of the United States, I applied the identical treatment to the 1990 populations and areas of the 62 counties of New York State. The plots of county population density as a function of cumulative area in the right column of Fig. 2 are similar to the plots for states and countries. The total area of New York State, somewhat less than 13 million ha, is about one-thousandth of the total ice-free land area of the world, roughly 13 billion ha.

For the world's countries, the states of the United States, and the counties of New York State, the distribution of human population density by area is self-similar over a thousandfold range of areas. That is, if the absolute scale is removed from the axis labels, it is not possible to determine the size of units being plotted from the shape of the plotted curves.

These questions remain to be addressed: Does the apparent self-similarity in the distribution of popu-

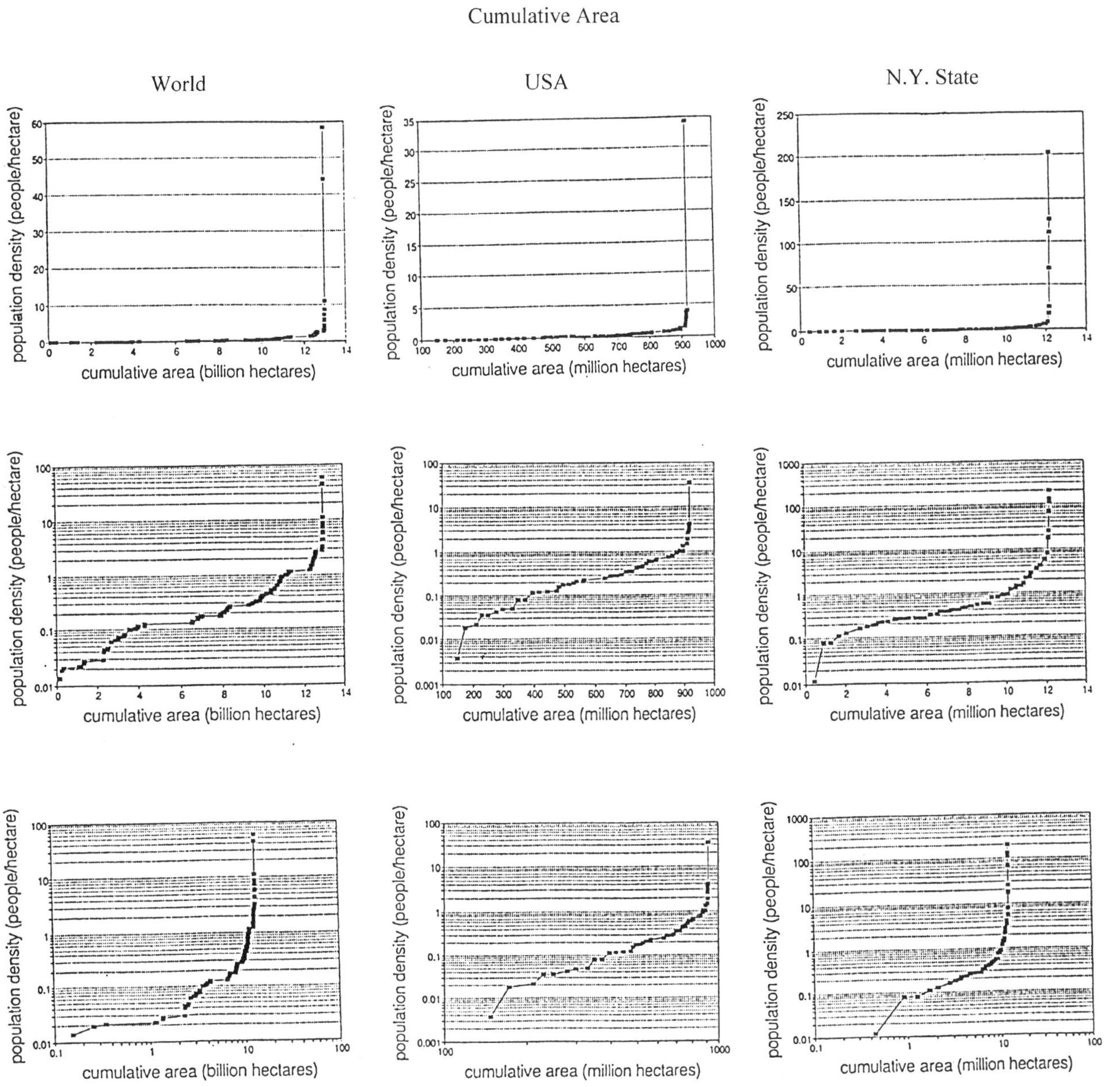


Fig. 2. Cumulative land area (horizontal axis) with population density equal to or less than the value shown on the vertical axis. Left column: countries of the world in 1989 (data: World Resources Institute 1992); middle column: 50 states of the United States plus Washington, D.C. (data: U.S. Census 1990); right column: 62 counties of New York State (data: U.S. Census 1990). First row: both axes are on a linear scale. Second row: vertical axis is logarithmic; horizontal axis is linear. Third row: both axes are on a logarithmic scale. Source: Cohen (1997). Copyright © 1997 by Joel E. Cohen.

lation density hold for other countries and other states? for the human population at earlier times? for non-human species? How can the apparent self-similarity be quantified and evaluated more quantitatively?

In spite of uncertainty about the generality of this finding, visual inspection suggestions that the cumulative distribution of population density is similar on scales of area ranging over three orders of magnitude, from the area of New York State to the land area of the entire earth not covered by ice. Over this range of spatial scales, it seems likely that a growing fraction of the human population will live on a diminishing fraction of the land.

## 2.3 Early History of Land Use

Massive human alteration of grasslands and forests probably began with human mastery of fire hundreds of thousands of years ago. Prior to the evolution or invention of agriculture, closed forest may have covered 4.6 billion ha, and woodland 1.5 billion ha, of the globe, respectively about 35% and 12% of the total ice-free land area. By around 1970 (based on maps dated from 1969 to 1976), about 3.9 billion ha of closed forest and 1.3 billion ha of woodland remained, together covering about 40% of earth's ice-free land (Matthews 1983: 482, Williams 1990: 179). Tropical rainforests had declined about 50 million ha (3.75% of their original extent) while all other forests declined about 650 million ha (19.5% of their original area) (Matthews 1983). These estimates, even for present land cover, are uncertain. For example, Matthews' (1983: 483) estimate of 3.9 billion ha of contemporary forest falls between a 1979 estimate of 3.7 billion ha and a 1975 estimate of 4.9 billion ha; the divergence among published estimates of woodland and shrubland areas are also large. The estimates of pre-agricultural land cover must be still more uncertain.

About half the world's forests and woodlands lie in the tropics, the rest in temperate and boreal regions (Johnson 1996). Six countries — Russia, Canada, Brazil, the United States, Zaire, and Indonesia have just over half the world's forests and woodlands. (For recent surveys of human population growth and land use, see Richards (1990), Turner et al. (1990), Rudel (1991), Grainger (1993), Jolly and Torrey (1993), Pearce and Warford (1993), and Marquette and Bilsborrow (1994).)

Europeans cleared their forests energetically in the centuries up to A.D. 1300 (Cipolla 1994). Clearing the European forests slowed temporarily in the 14th century in the presence of the plague, the onset of the "Little Ice Age" around A.D. 1300 (a drop in global mean temperature of 1.5°C. that lasted to the beginning of the 19th century) (Turekian 1996: 82–83), and economic stagnation. By the 16th century, economic activity recovered, nutrition improved, and population growth rates began to rise. The cutting of European forests was renewed to feed a slowly expanding population, to supply imperial requirements for ships and naval supplies, and to support growing industry, mining, and metal extraction. When England exhausted its own supplies of large timber for shipbuilding, it looked outward to trade with Sweden, Russia, British colonies in North America, India, Burma, and Australia. France and other colonial powers exhausted their forests and looked outward.

Many regions of the world experienced extensive deforestation. While 95% of central and western Europe was originally forested, now about 20% is forested; China, originally 70% forest, is now 5% forested; and the United States lost one-third of its forests between 1790 and 1890 (Ponting 1990: 4).

As human numbers increased, people changed their use of the land by extending their activities to new lands (extensification) and by intensifying their productivity on land already occupied (intensification). These processes are discussed in the following sections.

## 2.4 Extension of Croplands with Population Growth since 1700

Between 1700 and 1980, the area of croplands increased from 0.3 billion ha to 1.5 billion ha, an absolute increase of 1.2 billion ha and a nearly five-fold increase (Fig. 3) (Richards 1990: 164). Over the same period, while grasslands and pasture changed little, the area of forests and woodlands declined from 6.2 billion ha to 5.1 billion ha (nearly 19%). For comparison, the land area of the 50 United States is about 0.93 billion ha; thus the reduction in area of forests and woodlands approximates the entire land area of the United States. Kates et al. (1990: 1) estimated "the net loss of the world's forests due to human activity since pre-agricultural times" at roughly 0.8 billion ha, an area about the size of the conterminous United States (the lower 48 states). Kates' "net loss" may underestimate human impact if it excludes massively disturbed areas that were abandoned and subsequently regrew as well as extensive areas that were disturbed but not destroyed. Meanwhile, the global population rose from around 650 million in 1700 (estimates vary from 610 million to 680 million; Cohen 1995b: 400) to around 4.4 billion in 1980, nearly a seven-fold increase.

This historical association of increasing population with increasing croplands and decreasing forests must be regarded with some caution because the estimated area of agriculture in 1700 was derived from historical population estimates (Richards 1990: 164). However, the changes in croplands versus the changes in forests

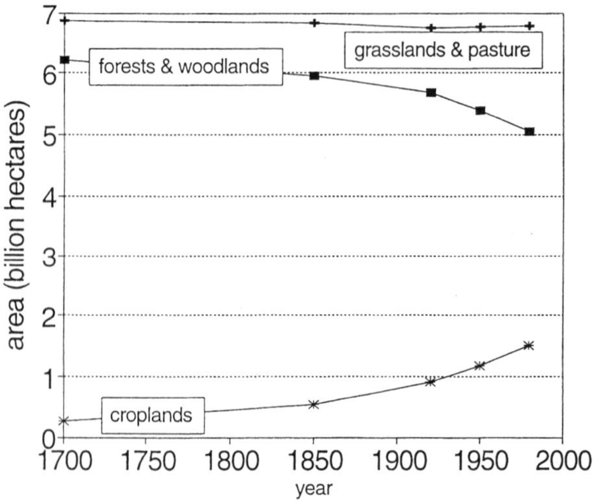


Fig. 3. Land use since 1700. Original figure, based on data of Richards (1990: 164).

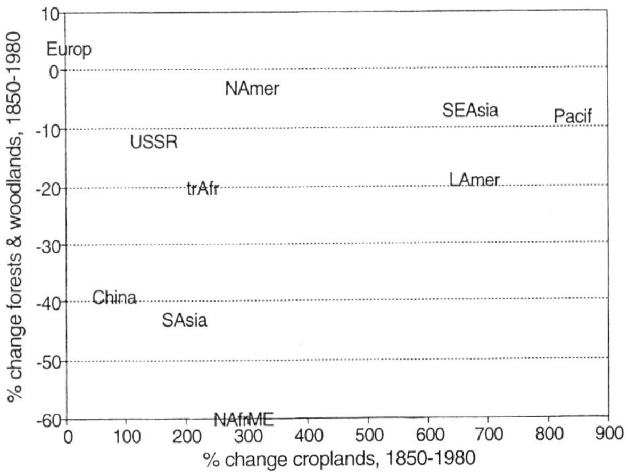


Fig. 4. Changes in areas of croplands versus changes in areas of forests and woodlands, 1850–1980, for 10 regions of the world: trAfr = tropical Africa; NAfrME = North Africa and Middle East; NAmer = North America; LAmer = Latin America; China = China; SAsia = South Asia; SEAsia = Southeast Asia; Europ = Europe; USSR = former Soviet Union; Pacif = Pacific developed countries. Original figure, based on data of Repetto and Gillis (1988: 3–5), which are attributed to World Resources Institute (1987: 272).

and woodlands for 10 regions of the world from 1850 to 1980 confirm, and reveal the diversity concealed by, the global picture (Fig. 4).

In every region except Europe, as croplands increased, forests and woodlands shrank. The ten regions fall into three groups. In the now relatively wealthy regions, Europe, North America, and the USSR, the relative loss of forests and woodlands during this period was small (Europe had a gain), as was the increase in croplands, in comparison with the percentage changes in other regions. In three regions, Southeast Asia, the Pacific developed countries, and Latin America, the relative increase in croplands was enormous compared to the relative decrease in forests. Finally, in three regions that must qualify among the poorest, China, South Asia, North Africa and the Middle East, the percentage loss in forests and woodlands was larger than in the other regions, while the percentage increase in croplands was comparable to that in the wealthy regions. In Fig. 4, tropical Africa is anomalously close to the group of three wealthy regions, perhaps because tropical deforestation has not yet run its full course, or perhaps because the starting base of forest and woodland in 1850 was so large. Even so, Europe alone is the exception to the pattern of increasing croplands and decreasing forests and woodlands between 1850 and 1980.

Kates et al. (1990: 13) stated categorically that "the global transformation of the biosphere is driven first by population growth, followed by technological capacity and sociocultural organization." This assertion has to be evaluated in the light of our earlier observation that the richest one-fifth of today's population commands about 80% of global income and consumption, although the rich countries' population grows very slowly (doubling in five centuries or longer). The rich and poor nations, as well as scholars, disagree whether the levels of consumption of the rich or the numbers of consumers among the poor contribute more to human impacts on the biosphere. Many of the impacts from rich and poor are different.

When population density and forest coverage are measured independently and directly in some subsets of countries today or recently, an inverse relation appears (Preston 1994). For example (Fig. 5), in 60 tropical countries in 1980 (excluding eight arid African countries), the larger the number of people per square

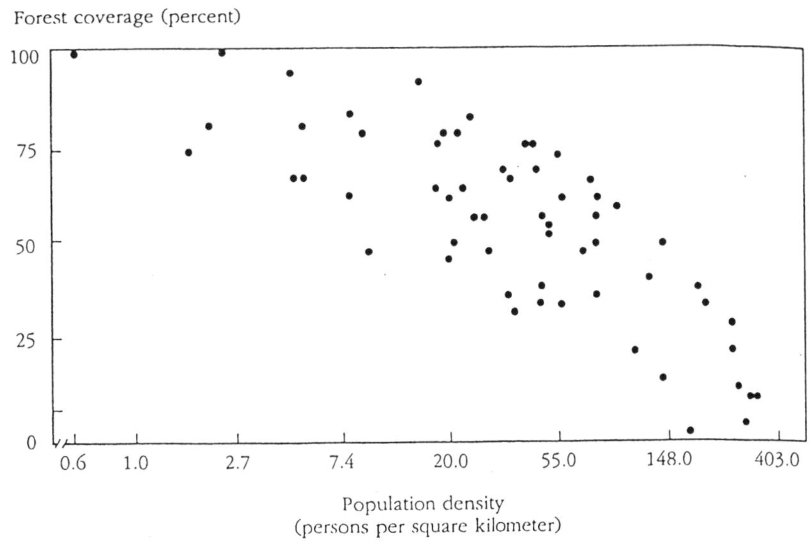


Fig. 5. Relation between forest coverage and population density in 60 tropical countries, 1980. Source: Pearce and Warford (1993: 166).

kilometer, the smaller the percentage of land covered by forest (Pearce and Warford 1993: 166). Although such data suggest that humans are responsible for the smaller forest cover, the data are also compatible with the contrary hypothesis that countries that started off with less forest cover were more easily settled and ended up with higher human population densities. To distinguish between these two hypotheses, data are required on changes over time. When Harrison (1992: 323) ranked 50 countries (of unstated geographic distribution) from high to low percentage of "habitat loss" (not explicitly defined as a measurement of status or of change) in the mid-1980s, the amount of habitat loss decreased with decreasing population density. The top 10 countries had 85% habitat loss and 1.89 people per ha while the bottom 10 countries had 41% habitat loss and 0.29 people per ha.

Statistical associations such as these suggest that rising human numbers increase the demand for agricultural products (including but not limited to food for subsistence), and expanding the area of agricultural production at the expense of forests and grasslands is a frequent response to the rising demand. On the average, when technologies, capital, credit, and farmer skills for intensifying food production are not available, the inferred global historical association of rising population and expanding croplands is largely valid (Preston 1994).

More detailed investigation reveals a more complex interaction between population growth and forest cover. Where relatively small areas of rainforest are surrounded by cleared land, as in Central America, the Philippines, Rwanda, and Burundi, peasants in the cleared areas expand their areas of cultivation, little by little, by nibbling away at the forests. In these cases, variations in rates of deforestation may be explained by variations in local rates of population increase (Rudel 1991: 56).

Where there are large blocks of rainforest, population growth is not enough to explain deforestation. In addition to rapid population growth, substantial capital investment in access roads and an absence of enforced property rights are also necessary for rapid deforestation. For example, rates of deforestation were far higher during the 1970s in Brazil, which was relatively capital-rich, than in capital-poor Bolivia and Zaire. In times of economic hardship, if capital becomes scarce, fewer roads may be built in regions with large extents of rainforest. As these large tracts then remain inaccessible to most migrants from other regions, many potential migrants may stay home and pursue the nibbling form of deforestation. Hence, scarcity of capital may shift the location and nature of deforestation (Rudel 1991).

Mertens (1994: 26) concluded from a review of empirical studies of tropical countries that "if deforestation is to be limited, the most direct policy is to minimize new road construction in the humid tropics, particularly in areas where there are no roads. In some cases, though, as in ... Nepal, improvements in infrastructure can create conditions which make it easier to control deforestation."

When forests are cleared for farmland to feed an increasing population, the rate of cutting depends in part on how much land is required to produce food for one more person. That requirement depends on yields, farmer education, credit for agricultural investments in land and equipment, culturally acceptable crop varieties, soil types, water resources both natural and human-built, and so on through every aspect of culture and economics and the environment. Forests are sometimes cut because governments give land tenure or tax advantages to those who clear trees, and sometimes because domestic and international markets demand wood in quantities determined more by wealth and population density in cities than by human numbers in forested regions. A one-directional causal model like "human population growth causes forest clearing or land conversion" is far too simple in general (Jolly 1994, Marquette and Bilsborrow 1994).

In recent decades as global population has grown and global forests have declined, the price of timber in international commodity markets has increased. According to the World Bank Price Index for the Primary Commodities (revised April 1995) (World Resources Institute 1996: 170), timber was alone among the major categories of commodities to experience an increase in price between 1975 and 1994. In constant prices with the price in 1990 set equal to 100, the price of petroleum fell from 101 in 1975 to 63 in 1994. The prices of metals and minerals fell over the same period from 118 to 77. The price of nonfuel commodities as a whole (including timber) fell from 167 in 1975 to 102 in 1994. Total food commodity prices fell from 224 in 1975 to 98 in 1994; cereal prices in particular fell from 258 in 1975 to 98 in 1994. By contrast, timber prices (separated from other nonfuel commodities) increased from 92 in 1975 to 143 in 1994.

In 1993, a study of population and land use in developing countries offered the following major conclusions (Jolly and Torrey 1993: 9–11): "In the long run, population growth almost certainly affects land use patterns. The effects of population growth occur mainly through the extensification and intensification of agricultural production. ... Most of the changes in land use associated with very rapid population growth are likely to be disadvantageous for human beings. ... Population growth is not the only, or in many cases, the most important influence on land use. Other influences include technological change and changes in production techniques ... inequality itself, however, is in part influenced by rates of population growth ... with clear property rights, robust soils, and efficient markets, population growth is less likely to result in land degradation. ... Rapid population growth is likely to make the survival of other members of the animal and plant kingdom more difficult. Accompanying rapid population growth in the past has been greater species loss and a higher attrition within species than would have occurred in the absence of human expansion."

## 2.5 Increased Farming Intensity with Population Growth

On a long historical time scale, rising population density has been associated with rising farming intensity (Pingali and Binswanger 1987). Hunters and gatherers, who do not cultivate the land, practice a farming intensity of zero. If land is cropped once every year, the farming intensity equals 100%. If multiple crop cycles are completed within a single year and the land is never fallowed, the farming intensity exceeds 100% (Table 1).

A given value of farming intensity between 0 and 100% does not specify the duration of a cycle of cultivation and fallow. For example, a farming intensity of 5% could mean that, on the average, each year of cultivation is followed by 19 years of fallow. It could also mean, in principle, that five consecutive years of cultivation are followed by 95 years of fallow. Quite different amounts of succession and forest recovery can take place under these two regimes. Thus a given value of farming intensity is consistent with very different effects on biological diversity and forest cover.

In the schema of the economist Ester Boserup (1981: 19), forest fallow consists of 1–2 annual crops and 15–25 years of fallow; bush fallow of 2 or more crops and 8–10 years of fallow; short or grass fallow of 1–2 crops and 1–2 years of fallow; annual cropping of 1 crop per year, with fallow for only part of a year; and multicropping of 2 or more crops on the same land each year with no fallow. Bush fallow and more intense farming systems inhibit or prevent forest regeneration.

Boserup (1981: 23) investigated the association between population density and agricultural systems. Her definition of population density is not simply the ratio of people to all land, but rather the ratio of people to potentially arable land (Boserup 1981: 16). Potentially arable land excludes areas under ice, unirrigable deserts and mountains too steep for terracing or pasturing. Potentially arable land includes land that could be developed into agricultural land with suitable

Table 1. Population density, farming intensity, and farming systems in low-technology countries.

| Farming system | Farming intensity* (%) | Population density (people/ha of potentially arable land) | Climate | Tools used |
|---|---|---|---|---|
| Hunter/gatherer | 0 | 0–0.04 | | |
| Pastoralism | 0 | 0–0.04 | | |
| Forest fallow | 0–10 | 0–0.04 | humid | axe, matchet, and digging stick |
| Bush fallow | 10–40 | 0.04–0.64 | humid or semi-humid | above tools plus hoe |
| Short fallow | 40–80 | 0.16–0.64 | semi-humid, semi-arid, high altitude | hoes and animal traction |
| Annual cropping with intensive animal husbandry | 80–100 | 0.64–2.56 | semi-humid, semi-arid, high altitude | animal traction and tractors |
| Multi-cropping with little animal husbandry | 200–300 | 2.56 and up | | |

*Farming intensity (expressed in percent) is defined as 100 times the total number of crops (in one cycle of cultivation and fallow) divided by the total number of years in which the land is cultivated and fallowed (in one cycle of cultivation and fallow).
Sources: Boserup 1981: 9, 19, 23; Pingali and Binswanger 1987: 29.

investments in infrastructure and inputs; land now covered by forests that could be cleared and then farmed; grazing lands that are arable; and long-term fallow lands. This definition of "potentially arable land" is difficult, perhaps impossible, to measure in practice. For example, who knows whether never-cleared tropical forest land will be suitable for agriculture for more than a very few years? Compromising with the available international statistics on land use, Boserup (1981: 16) simply excluded land statistically classified as "other" "only if it is likely to be arctic or desert and accounts for ... a large share of total territory ..." For low-technology countries, she proposed that farming systems are associated with the population densities per area of arable land as shown in Table 1.

The global average population density of 0.42 people per ha would be compatible with bush fallow or short fallow farming if all ice-free land were arable (an unlikely possibility, especially with low levels of technology). Domesticated land (cropland plus permanent pasture) approximated 37% of all land excluding Antarctica during 1986-89 (World Resources Institute 1994: 284). If all domesticated land were potentially arable using low technology, then the global population density per unit of arable land would be 0.42/0.37 = 1.14 people/ha. According to Table 1, annual cropping is required when the population density exceeds 0.64 people per ha of arable land. It follows that nearly all arable land (defined here, for the sake of calculation, as domesticated land) should be cropped at least annually if farmers respond to global population densities rather than to local population densities only, and if farmers use low technology. Because some farmers sell food to remote dense populations, domestic and international trade and transport spread the ecological effects of locally dense populations to less populated regions.

In summary, the effect of global population growth on land use for agriculture, forestry, and other purposes depends in part on domestic and international politics, economics and transport, and in part on the level of technology farmers use. It has been argued that more intensive agriculture would preserve more land for nature (Waggoner 1994, Waggoner et al. 1996). A full accounting of the positive and negative external effects of more intensive agriculture remains to be provided.

## 2.6 Dynamic Theories of Land Use Change

Several theories have been developed to explain how changes in land use take place. These include qualitative models, such as the Richards' center-periphery model, and quantitative models, such as the Malthus–Condorcet–Mill model.

### 2.6.1 Richards' Center-periphery Model: A Qualitative Model

Richards (1990: 165) proposed a dynamic overview of human land exploitation that combines extensification and intensification. This center-periphery model may provide a qualitative basis for the quantitative self-similarity at different spatial scales in the spatial distribution of human population density.

According to Richards (1990: 165), "Intensification of human land use — both conversion and extraction of natural resources — is an essential feature of the spiraling, ever-extending domain of the modern capitalist states and the modern world economy. ...At the heart of this model is the urge to make complementary use of lands at the center and those in the peripheral areas. ...Intensive land use at the center ... relied upon resources extracted ruthlessly from lands in ... dependent regions. ... Urban demands for foodstuffs, energy, water and other commodities dr[o]ve land conversion and resource extraction in their immediate hinterlands. Rising populations and improved access enlarged each city's immediate hinterlands. Highly intensive market-gardening pushed outward extensive grain farming and livestock raising in belts around most early modern European cities. We find therefore a center-periphery model replicated in the regions surrounding each city."

He continued, "Over time, colonial or dependent states moved closer to the European model of intensive land use and control. ... dependent regions were subjected first to heavy resource extraction and commodity production typical of the periphery. In time, indigenous core regions of intensive land use coalesced to form a new land-use hierarchy within each region. At this secondary or intermediary level, core regions directed extraction of resources from their own peripheries as new frontiers of settlement were opened." Richards gave the example of Calcutta, on the periphery of London, becoming a center for extraction from surrounding eastern Bengal, Assam, and Orissa. "At a still deeper level, subregional centers emerged in which the process of urban-dominated land use commenced. Dacca in Eastern Bengal and Assam directed the expansion of settlement, land clearing, timbering, and other exploitative activities in their hinterlands. At this and even lower levels, we can see arrays of smaller frontier regions and subregions merging one into the other. In this fashion, intensified control and productivity on the world's lands gained momentum in each succeeding century."

This hierarchical self-similar pattern of a central city dominating peripheral regions of supply, which spawn their own new cities, provides a qualitative explanation of the observed self-similarity in the distribution of population density. It would be valuable to see how well this model explains the dynamics of population growth and land use in earlier empires of Mesopotamia, China, Meso-America, the Middle East (Ottoman), South Asia (Moghul) and elsewhere. How much of Richards' expanding cycle of intensification and extensification depends on western technologies for transportation and communication and western institutions for administration, accounting and control, and how much is general to the building of empires?

### 2.6.2 The Malthus–Condorcet–Mill Model: A Quantitative Model

Richards' center-periphery model envisions an autocatalytic process in which population growth and growing demands for consumption drive the development or exploitation of additional resources by extensification and intensification, leading to further population growth and consumption. I recently proposed a highly schematic model called the Malthus–Condorcet–Mill model, which presents a possible quantitative version of this process (Cohen 1995a, also see Cohen 1995b, Appendix 6). The model's underlying concepts derive from a debate in the late 18th century between Malthus and Condorcet. The British philosopher John Stuart Mill (1806–1873) contributed to this debate (Mill 1848) by picturing a stationary population as both inevitable and desirable. The views of Condorcet and Malthus are still represented today in the approaches of neoclassical and classical economists (sections 1.1.1 and 1.1.2), respectively.

Malthus described a dynamic relation between human population size and a society's capacity to support itself at a level of living that it defines as satisfactory (Malthus 1798, Chap. VII: 51): "The happiness of a country does not depend, absolutely, upon its poverty or its riches, upon its youth or its age, upon its being thinly or fully inhabited, but upon the rapidity with which it is increasing, upon the degree in which the yearly increase of food approaches to the yearly increase of an unrestricted population." Malthus opposed the optimism of the Marquis de Condorcet (1743–1794), who saw the human mind as capable of removing all obstacles to human progress. Demeny (1988: 232) generalized Malthus's view to incorporate all aspects of economic output, not just food: "Posed in the simplest terms, the economics of population reduces to a race between two rates of growth: that of population and that of economic output."

The Malthus–Condorcet–Mill model is a highly idealized mathematical sketch of the race between the size of the human population and the capacity to provide human well-being, which I shall call human carrying capacity for the moment. Suppose that it is possible to define a current human carrying capacity $K(t)$ as a numerical quantity measured in numbers of individuals. Suppose also that $P(t)$ is the total number of individuals in the population at time $t$ and that

$$dP(t)/dt = r\,P(t)\,(K(t) - P(t)).$$

The constant $r > 0$ is called the Malthusian parameter. I call this the equation of Malthus because it expresses the limitation of population growth by the current carrying capacity, and recognizes, as Malthus did, that the current carrying capacity can change over time. The equation of Malthus is the same as the logistic equation except that the constant $K$ in the logistic equation is replaced by variable carrying capacity $K(t)$ here.

To describe changes in the carrying capacity $K(t)$, let us recognize, in the phrase of former United States President George H.W. Bush (1992) that "every human being represents hands to work, and not just another mouth to feed." Additional people clear rocks from fields, build irrigation canals, discover ore deposits and antibiotics and invent steam engines; they also clear-cut primary forests, contribute to the erosion of topsoil, and manufacture chlorofluorocarbons and plutonium. Additional people may increase savings or dilute and deplete capital; they may increase or decrease the human carrying capacity.

Suppose that the rate of change of human carrying capacity is directly proportional to the product of two factors: (1) the rate of change in population size (the Condorcet factor), and (2) the average resources available per person (the Mill factor). In the language of President Bush, the change in capacity to produce well-being depends on both how many additional hands there are and what those hands have to work with. We assume here, contrary to fact, that each additional pair of hands shares equally in the productive resources available to all existing hands.

Suppose, for example, that there is a constant $L > 0$ such that the productivity of an additional person is $L/P(t)$ (think of land per person); $L$ is the Mill parameter. The assumption that $L/P(t)$ is positive, no matter how big $P(t)$ is, models the dilution of resources, but not their depletion or degradation. The Condorcet–Mill equation supposes that the increment in human carrying capacity equals the product of the productive resources available to an average person, namely, $L/P(t)$, and the increment in population size:

$$dK(t)/dt = (L/P(t))\,(dP(t)/dt).$$

This model assumes no migration and ignores the population's age composition, geographical distribution and distribution of well-being or income. The model ignores stochastic fluctuations in environmental and human factors.

Assume further that $L/P(0) > 1$; this means that, initially, productive resources are so abundant that the average person can provide well-being for more than himself or herself alone. Then the population initially grows faster than exponentially. As $P(t)$ increases past $L$, $c(t)$ passes through 1 and the population experiences a brief instant of exponential growth. Then $c(t)$ falls below 1 and the population size thereafter grows sigmoidally. Population size rises to approach a unique stationary level, which is independent of $r$. The larger the initial carrying capacity $K(0)$ and the larger the supply $L$ of land or other basic natural resource, the larger the stationary level is, other things being equal.

Figure 6 shows a trajectory of human carrying capacity $K(t)$ (dashed upper curve) and population size $P(t)$ (solid lower curve) according to the Malthus–Condorcet–Mill model; $P(t)$ is compared with the estimated human population history (filled boxes) over the past 2,000 years. The theoretical trajectory of population looks sigmoidal on a logarithmic scale. Values of $P(t)$ beyond $t = 1995$ are intended only to illustrate the qualitative behavior of the model, not to predict future human population. Nothing guarantees that the actual human population will reach or remain at the high plateau shown. For example, the model neglects the possibilities that people could increasingly choose to divide the available material resources among fewer offspring, trading numbers for wealth, and that pollution or exogenous climatic changes could diminish human carrying capacity. Further, nothing guarantees that the productive resources available per additional person in the future will be described by a function as simple as $L/P(t)$.

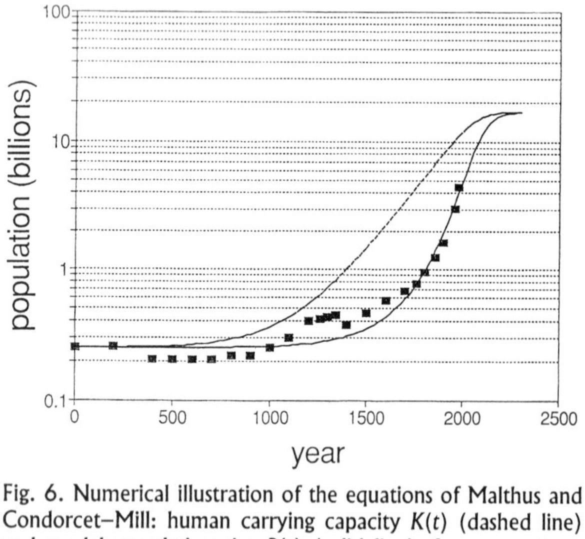


Fig. 6. Numerical illustration of the equations of Malthus and Condorcet–Mill: human carrying capacity $K(t)$ (dashed line) and model population size $P(t)$ (solid line); for comparison, estimated actual human population (solid rectangles). Equations: $P(t + \delta t) - P(t) = rP(t)(K(t)-P(t))\,\delta t$, $K(t + \delta t) - K(t) = Lr(K(t)-P(t))\,\delta t$. Initial conditions and parameters: $\delta t =$ 20 years, $P(0) = 0.252$, $K(0) = 0.252789$, $r = 0.0014829$, $L = 3.7$. $P(0)$, $K(0)$, $L$ are measured in billions. Reprinted with permission from Cohen (1995a). Copyright © 1995 by The American Association for The Advancement of Science.

Up to about $t = 1970$, population sizes (theoretical and actual) are convex on the logarithmic scale; after roughly $t = 1970$, they are concave. The human carrying capacity $K(t)$, initially only slightly above $P(t)$, began to exceed $P(t)$ substantially at times corresponding to the 9th and 10th centuries, and experienced nearly exponential growth (linear increase on the logarithmic scale shown) from the 11th to the mid-20th century. According to the model, the acceleration of population growth in the 17th century was preceded by a long period of increasing human carrying capacity.

Europe grew technologically and economically for a millennium before the Industrial Revolution (Cipolla 1994: 137–159); England developed economically from the 12th century onward (Wilkinson 1973, Hardesty 1977: 209–210). In the 13th century, English forests were cleared, swamps drained, and new lands exploited for cultivation; yields improved as a result of liming, plowing straw ash into the field, and planting new varieties of seeds. At the same time, the mining and smelting of tin, lead and iron; the manufacture of pottery; and the production of salt and wool all increased greatly. Additional surges of economic development occurred in the 16th century.

This allegorical model focuses attention on the factors, natural and human-made, that determine the productivity of each additional person.

## 2.7 Defining Human Carrying Capacity: Natural Constraints and Human Choices

Human transformations of the earth have changed the earth's ability to generate human well-being as well as human definitions of well-being. No concept of carrying capacity in basic and applied ecology is adequate as a concept of human carrying capacity because none of the many variants takes account of the human choices in defining and determining human carrying capacity (Cohen 1995b, Chap. 12).

Estimating how many people the earth or any region of it can support, and defining the variable $K(t)$ in the Malthus–Condorcet–Mill model, involve both natural constraints that humans cannot change and do not fully understand, and human choices that are yet to be made by this and by future generations. Therefore the earth's human carrying capacity is not a single fixed number, now or as long as the earth is habitable (Cohen 1995b, Chap. 13). Because the earth's human carrying capacity is constrained by facts of nature, human choices about the earth's human carrying capacity are not entirely free, and may have consequences that are not entirely predictable. Because of the important roles of uncertainty, natural constraints, and human choices (both individual choices and social decisions), estimates of human carrying capacity cannot aspire to be more than conditional and probable estimates: if future choices are thus-and-so, then the human carrying capacity is likely to be so-and-so.

No sharp line separates human choices and natural constraints. For example, technology obeys the laws of physics, chemistry and biology, but humans choose how, and how much, to invest in creating and applying technology. Hence the technology that people use depends jointly on human choices and natural constraints. The fuzzy zone between choices and constraints shifts as time passes. Changes in knowledge can reveal constraints that had not been recognized previously, and can also make possible new choices. Furthermore, a choice open to rich people may be a constraint for poor people. A rich landowner may choose to leave forest uncut and cropland idle; a subsistence farmer with small holdings may not enjoy the luxury of choosing.

To define and estimate the earth's or a region's human carrying capacity, at least the following questions of human choice need to be answered:

1. What is the desired average level of material well-being?
2. What is the desired distribution of material well-being?
3. What is the desired technology?
4. What are the desired domestic and international political institutions?
5. What are the desired domestic and international economic arrangements?
6. What are the desired domestic and international demographic arrangements?
7. What are the desired physical, chemical and biological environments?
8. What is the desired variability or stability?
9. What is the desired risk or robustness?
10. What is the time horizon?
11. What values, tastes and fashions will people hold?

### 2.7.1 Average Level of Material Well-being

Material well-being includes food (people choose variety and palatability, beyond the constraints imposed by physiological requirements); fiber (people choose cotton, wool or synthetic fibers for clothing, wood pulp or rag for paper); water (tap water or Perrier or the nearest river or mud hole for drinking, washing, cooking and watering your lawn, if you have one); housing (Auschwitz barracks or Thomas Jefferson's Monticello); manufactured goods; waste removal (for human, agricultural and industrial wastes); natural-hazard protection (against floods, storms, volcanoes and earthquakes); health (prevention, cure and care); and the entire range of amenities such as education, travel, social groups, solitude, the arts, religion, and communion with nature. Not all of those features are captured well by standard economic measures.

### 2.7.2 Distribution of Material Well-being

Estimates of human carrying capacity rarely take into account the scatter or distribution of material well-being in a population. Yet people who live in extreme poverty may not know or care that the global average is satisfactory, and the press of present needs may keep them from taking a long-term view. For example, thanks to genetic engineering, any country with a few PhDs in molecular plant biology and a modestly equipped laboratory can insert the genes to create stronger, more disease-resistant, higher-yielding plants. If every region has the scientific and technical resources to improve its own crop plants, the earth can support more people than it can if some regions are too poor to help themselves.

### 2.7.3 Technology

Will people collectively choose to develop technologies for mass transportation or for individual cars? Will people develop new electrical generators based on sunlight, oil, coal, uranium, plant wastes, refuse from cities, agriculture, industry, wind, tidal motion, wave motion, falling water or heat deep in the earth's crust? The human carrying capacity will depend on the choices made by voters, businesses, research organizations, civic groups and governments. Those choices will depend in turn on the economics, environmental effects, cultural acceptability, institutional governability and other features of the technologies.

The complexities of technological choices often disappear in heated exchanges between environmental pessimists and technological optimists:

- *Ecologist*: When a natural resource is being consumed faster than it is being replenished or recycled, an asset is being depleted, to the potential harm of future generations.
- *Technologist*: If new knowledge and technology can produce an equivalent or superior alternative, then future generations may turn out to be better off.
- *Taxpayer*: Which natural resources can be replaced by technology yet to be invented, and which can-

not? Will there be enough time to develop new technology and put it to work on the required scale? Could we avoid future problems, pain and suffering by making other choices now about technology or ways of living?

- (No answer from ecologist or technologist.)

The key to the argument is time. Richard E. Benedick, an officer of the U.S. Department of State who has also served with the World-Wide Fund for Nature, worried (Benedick 1991: 201): "While it is true that technology has generally been able to come up with solutions to human dilemmas, there is no guarantee that ingenuity will always rise to the task. Policymakers must contend with a nagging thought: 'what if it does not, or what if it is too late?'"

### 2.7.4 Domestic and International Political Institutions

Political organization and effectiveness affect human carrying capacity. For example, the United Nations Development Program estimated that developing countries could mobilize for development as much as $50 billion a year (an amount comparable to all official development assistance) if they reduced military expenditures, privatized public enterprises, eliminated corruption, made development priorities economically more rational and improved national governance (Gardner 1992: 30). Conversely, population size, distribution and composition affect political organization and effectiveness.

Political choices that affect human carrying capacity involve a host of questions. How will political institutions and civic participation evolve with increasing numbers of people? As numbers increase, how will people's ability to participate effectively in the political system change? What standards of personal liberty will people choose? How will people bring about political change? By elections and referendums, or by revolution, insurrection and civil war? How will people choose to settle differences between nations, for instance, over disputed borders, shared water resources or common fisheries? War consumes human and physical resources. Negotiation consumes patience and often requires compromise. The two options impose different constraints on human carrying capacity.

### 2.7.5 Domestic and International Economic Arrangements

What levels of physical and human capital are assumed? Tractors, lathes, computers, better health, and better education all make workers in rich countries far more productive than those in poor countries. Wealthier workers make more wealth and can support more people.

What regional and international trade in finished goods and mobility in productive assets are permitted or encouraged? How will work be organized? The invention of the factory organized production to minimize idleness in the use of labor, tools, and machines. What new ways of organizing work should be assumed to estimate the future human carrying capacity?

### 2.7.6 Domestic and International Demographic Arrangements

Almost every aspect of demography (birth, death, age structure, migration, marriage, and family structure) is subject to human choices that influence the earth's human carrying capacity. If global population eventually becomes stationary (unchanging birth rates, death rates and total size), people will have to choose between a long average length of life and a high birth rate. They will also have to choose between a single average birth rate for all regions, on the one hand, and a demographic specialization of labor on the other (in which some areas have fertility above their replacement level, whereas other areas have fertility below their replacement level).

Patterns of marriage and household formation will also influence human carrying capacity. For example, the public resources that have to be devoted to the care of the young and the aged depend on the roles played by families. In China national law requires families to care for and support their elderly members. In the United States each elderly person and the state are largely responsible for supporting that elderly person; his or her relations may choose whether to assume responsibility for care.

### 2.7.7 Physical, Chemical, and Biological Environments

What physical, chemical, and biological environments will people choose for themselves and for their children? Much of the heat in the public argument over current environmental problems arises because the consequences of present and projected choices and changes are uncertain. Will global warming cause great problems, or would a global limitation on fossil-fuel consumption cause greater problems? Will toxic or nuclear wastes or ordinary sewage sludge dumped in the deep ocean come back to haunt future generations when deep currents well up in biologically productive offshore zones, or would the long-term effects of disposing of those wastes on land be worse? The choice of particular alternatives could materially affect human carrying capacity.

### 2.7.8 Variability or Stability

The earth's human carrying capacity depends on how steadily people want the earth to support the human population. If people are willing to let the human population rise and fall, depending on annual crops, decadal weather patterns and long-term shifts in climate, the average population with ups and downs would include the peaks of population size, whereas the guaranteed level would have to be adjusted to the level of the lowest valley. Similar reasoning applies to variability or stability in the level of well-being; in the quality of the physical, chemical and biological environments; and in many other dimensions of choice.

### 2.7.9 Risk or Robustness

The earth's human carrying capacity depends on how controllable people want the well-being of the population to be. One possible strategy would be to maximize numbers at some given level of well-being, ignoring the risk of natural or human disaster. Another would be to accept a smaller population size in return for increased control over random events. For example, if people settle in a previously uninhabited hazardous zone (such as the floodplain of the Mississippi River or the hurricane-prone coast of the southeastern United States), they demand a higher carrying capacity of the hazardous zone, but they must accept a higher risk of catastrophe. When farmers do not give fields a fallow period, they extract a higher carrying capacity along with a higher risk that the soil will lose its fertility (as agronomists at the International Rice Research Institute in the Philippines discovered to their surprise).

### 2.7.10 Time Horizon

Human carrying capacity depends strongly on the time horizon people choose for planning. The population that the earth can support at a given level of well-being for 20 years may differ substantially from the population that can be supported for 100 or 1,000 years. The time horizon is crucial in energy analysis. How fast oil stocks are being consumed matters little if one cares only about the next five years. In the long term, technology can change the definition of resources, as ores that were useless rock 10,000 years ago have been converted to valuable sources of metals today. No one can say whether industrial society is sustainable for 500 years.

Some definitions of human carrying capacity refer to the size of a population that can be supported indefinitely. Such definitions are operationally meaningless. There is no way of knowing what human population size can be supported indefinitely (other than zero population, since the sun is expected to burn out in a few billion years, and the human species almost certainly will be extinct long before then). The concept of indefinite sustainability is a phantasm, a diversion from the difficult problems of today and the coming century.

### 2.7.11 Fashions, Tastes, and Values

The earth's human carrying capacity depends on what people want from life. Many choices that appear to be economic depend heavily on individual and cultural values. Should industrial societies use the available supplies of fossil fuels in households for heating and for personal transportation, or outside of households to produce other goods and services? Do people prefer a high average wage and low employment or a low average wage and high employment (if they must choose)?

Should industrial economies seek now to develop renewable energy sources, or should they keep burning fossil fuels and leave the transition to future generations? Should women (and, by symmetry, should men) work outside their homes? Should economic analyses continue to discount future income and costs, or should they strive to even the balance between the people now living and their unborn descendants?

Humans seem to resolve conflicts of values by personal and social processes that are poorly understood and virtually unpredictable at present. How such conflicts are resolved can materially affect human carrying capacity, and so there is a large element of choice and uncertainty in human carrying capacity.

## 3 HISTORY OF POPULATION GROWTH AND LAND USE IN THE UNITED STATES

Prior to written records, the North American continent was shaped by repeated glaciations and peopled by Amerindians (MacLeish 1994). This long prehistory framed the stage on which immigrant Americans from Europe, Africa, and Asia enacted their settlement.

### 3.1 U.S. Forest History

In outline, the recorded history of American forests has two phases: a decline until 1920, and a gradual, recently variable, recovery since 1920. Likewise, the Federal Government's purpose in dealing with public lands has gone through two phases. At first, the Government aimed to transfer public lands into private hands; later, it aimed to retain and manage public lands. Reviews of land use and forestry in the United States have been written from diverse points of view by Williams (1989, 1990), Sedjo (1991), Alverson et al. (1994), Meyer (1995), Johnson (1996), Diamond and Noonan (1996), Wernick et al. (1997), and MacCleery and LeMaster (this volume).

When Europeans began to settle North America, about half of the conterminous United States was forested. Most of the forests lay in the eastern half of the continent (Meyer 1995: 29). But these were not "forests primeval": many had long been affected by repeated burning, extraction, and other management by an estimated 12 million pre-Columbian Amerindians of North America. Euro-American farmers in the north felled a hectare or so per year, mainly by clearcutting (Williams 1990: 182). As many as 10 million ha of forest were probably cleared to support the population present by 1776 (Williams 1989, 1990).

As might be expected from a dynamic interpretation of the global association between higher population density and less forest cover, the absolute area of American land covered by forests declined steadily during this period; the fraction of forested land declined more rapidly because much of the added territory was not forested. Forested area fell by about 36 million ha between 1780 and 1850 (Williams 1989, 1990). The wooded area of the (then) entire United States declined a further 121 million ha from perhaps 364 million ha around 1850 to about 243 million ha around 1920. (The estimates plotted by Wernick et al. [1997, their Fig. 1], based on other sources, indicate more than 300 million ha of forested land in 1920, and correspondingly higher estimates at other dates. Their numbers may include all of present U.S. territory.)

The wooded area of the conterminous United States slowly increased from 1920 until around 1960. Croplands were abandoned in the east and forests regrew in many cutover areas that were found to be unsuitable for agriculture. From 1930 onward, many timber companies gradually abandoned a frontier style summarized as "cut out and get out" (Williams 1990: 186) in favor of management for sustained yields.

Forested lands in the United States have probably declined slightly since 1960, unlike the forest lands of most industrial democracies, which increased. In 1992, forest lands occupied about 298 million ha (about 32% of the entire United States). By coincidence, the same fraction of the conterminous U.S. is forested. For comparison with the 32% of forest land in the United States, about 10% of the land of the United Kingdom and two-thirds of the area of Japan is forest and woodland. To preserve its mountainous forests, Japan imports far more timber than it exports. Japan's timber imports come mainly from the Americas and from southeast Asia.

Despite the possible decline in U.S. forested area since 1960, between 1952 and 1992 U.S. Forest Service timber inventories reported a 30% increase in timber volume. During this period, the volume of hardwoods rose 80% and that of softwoods 4% (Wernick et al. 1997, citing U.S. Forest Service reports). The carbon stored in U.S. forests increased by one-third (Wernick et al. 1997, citing Birdsey et al. 1993). The increases in timber volume and stored carbon are probably at least partly a result of a very dramatic decline in forest fires: the area burned annually by wildfire dropped an estimated 90% between 1920 and 1990 (Johnson 1996: 13).

In 1850, the public domain covered nearly two-thirds of the conterminous United States. Through the late 19th century, the Federal Government aimed to transfer public land to private holders. This was what Meyer (1995: 28) called "the disposal era." The intent was to raise revenue, to encourage settlement by Euro-Americans, and to "improve" the land, largely by clearing forests and draining wetlands for agriculture. Vestiges of the disposal era remain in present mining laws.

The phase of Federal land management began with the creation of Yellowstone National Park, the world's first national park, in 1872 and gathered force in the 1890s when more than 16 million ha of western lands became Federal forest preserves (Meyer 1995: 28). In recent decades, domestic issues spurred major new laws governing U.S. Forest Service management of National Forest lands, including the Multiple Use–Sustained Yield Act in 1960, the Forest and Rangelands Resource Planning Act in 1974, and the National Forest Management Act in 1976 (Johnson 1996: 1).

According to the 1987 U.S. National Resources Inventory (as reported by Meyer 1995: 27), of the roughly 763 million ha of land (including 40 million ha of wetlands) in the conterminous United States, about 21% is forest, 21% is rangeland, 22% is cropland, and 21% is Federal land. Pasture accounts for about 7%, developed land about 4%, and other land covers and surface water about 3%. Federal land is approximately half forest (thus roughly 10% of the conterminous United States) and half rangeland (another 10% or 11%). Much of this Federal land is leased for private use. Combining Federal and other lands, about 32% of the conterminous United States is forest (as mentioned above), and a roughly equal fraction is rangeland. In each of these categories, about one-third is Federal land. The Federal role in managing forests and rangelands is large, but so also is the role of businesses, individual private owners and other levels of government.

Like American agriculture earlier, many American cities are going through a period of extensification, as transportation has enabled the dispersal of residences and lower densities of settlement (Meyer 1995: 31). Growing urban populations, and higher amounts of land per person in settlements, have converted agricultural and forested land around cities to developed land. Nevertheless, again contrary to what might be expected from the contemporary association in developing tropical countries between population density and forested area, the Northeast region of the United States is both most densely populated and most heavily wooded: three-fifths of its area is forested (Meyer 1995: 28). The intense urbanization of most of the northeastern states makes possible this combination of forest cover and dense population. Cropland dominates the Midwest, rangeland the West. The South has about 40% forest, 20% rangeland, and 20% cropland.

## 3.2 U.S. Population

Between 1790, the year of the first U.S. Census, and 1920, the population of the United States rose from 3.9 million to 106 million, while the land area (excluding inland water) rose from 224 million ha to 769 million ha (U.S. Bureau of the Census 1975: 8). Thus the average population density rose from one person per 57 ha to one person per 7 ha.

In the second phase of American forests, from 1920 to 1990, the population of the conterminous United States more than doubled to about 247 million while the land area of the conterminous United States hardly changed. The population density of the conterminous United States rose from one person per 7 ha to one person per 3 ha.

Current levels of U.S. fertility are below replacement level: the average American woman would bear 2.0 children if she experienced current age-specific birth rates throughout her life. If current levels of fertility were to continue long enough, absent immigration, the U.S. population would peak and gradually decline. However, because the United States currently has many people in their peak years of childbearing, current births exceed deaths by about 1.6 million per year. In 1987, the most recent year for which this information is available, nearly two in five births in the United States resulted from unintended pregnancies, that is, pregnancies that were either mistimed or unwanted at any time (Brown and Eisenberg 1995: 26). In addition, current immigration exceeds emigration by about 1 million per year, although the amount of unauthorized immigration included in this figure can only be guessed. Thus the U.S. population is growing by roughly 1% per year, a rate — if continued — sufficient to double the population in about 70 years. About half the population lives within 80 km of the east or west coast.

Projections of U.S. population in 2020 range from 286 million to 385 million, depending on assumptions about mortality, fertility, and immigration. For 2050, projections range from 280 million to 553 million (Ahlburg and Vaupel 1990: 644). If fertility levels increase and immigration continues at a high level, the U.S. population could more than double by 2050.

Does continued population growth in the United States imply a collision with limited forest resources, especially if public forests will not be managed primarily to produce timber and other commodities? This chapter suggests that the answer depends on economics, technology, politics, and cultural values as much as it depends on rising numbers of humans.

## 3.3 U.S. Forest Economics

Over the long-term, real lumber prices in the United States grew roughly exponentially with a five-fold increase per century since 1800, but with substantial short-term fluctuations. For lumber, the relative producer price index (the actual price index divided by the all-commodities price index) grew from 6 or 7 in the decade 1800–1810 to 35 or 45 in the decade 1900-1910. It fluctuated between 100 and 150 in the years between 1970 and 1990 (U.S. Department of Agriculture 1989: 6).

Rising lumber prices, reflective of relative scarcity, and the availability of less expensive substitutes must explain, at least partially, a remarkable observation: between 1900 and 1993, the aggregate U.S. consumption of all timber products (as fuel, pulp, plywood, and lumber) increased 70%, while gross domestic product rose 16-fold (Wernick et al. 1997). The 16-fold increase was the product of a more than tripled population size and a gross domestic product per person that grew nearly 5-fold. Over the past century, and decade by decade, Americans' impact on their forests, as measured by consumption of timber products, has not varied in proportion to the product of American population size and American affluence, as measured by gross domestic product per person. Future demands on forests cannot reliably be projected by assuming a proportionality between population size and the consumption of forest products.

The consumption per American of all timber products fell by roughly half between 1900 and 1993. Timber ceased to be used to surface roads. Fossil fuels replaced much use of wood for fuel, although fuelwood consumption has had a resurgence since the oil price shocks of 1973. Railroad ties and pier timbers were treated with preservatives to prolong their lives, and sometimes were replaced with other materials such as concrete. Brick, concrete and other materials replaced timber as construction materials. While U.S. demand for wood as a fuel and as a construction material diminished, the demands for plywood and for wood fiber for paper and paperboard rose (Wernick et al. 1997). (Although the annual consumption of timber products per American fell during this century, it does not follow that the annual output of carbon (from timber products plus fossil fuels plus all other sources) to the atmosphere per American fell. Whether environmental impact rises in proportion to, faster than, or slower than population depends on how narrowly or comprehensively the environmental impact is measured.)

The consumption of raw wood per person grew by one-third from 1970 to 1990, to 2.3 cubic meters per person in 1990 (Johnson 1996). Consumption grew for all major categories of wood products, such as paper and paperboard, lumber, and wood-based panels. Currently, average timber productivity is 3.1 cubic meters per ha. If it is harvested to produce a sustained yield at the current average level of timber productivity, the 198 million ha of U.S. forest land that is presently productive enough to harvest and is not legally protected from harvesting can supply domestic wood consumption for roughly 270 million people, at the current level of consumption per person (Johnson 1996: 5). The estimated population of the United States in 1996 was about 265 million and was increasing by roughly 3 million per year (Population Reference Bureau 1996).

In 1993, the United States imported 67 million $m^3$ of forest products (about 10% of total forest product inputs of 646 million $m^3$, including 65 million $m^3$ of recycled material), and exported 65 million $m^3$ (again about 10% of total forest product outputs of 637 million $m^3$ delivered to consumers), according to a synopsis of material flows in the U.S. forest products industry (Wernick et al. 1997). These numbers assume that one metric ton of forest products occupies 2 $m^3$. Thus the volume of forest product imports roughly equals that of exports. It would be of great interest to compare the ecological, economic and social impacts of the growth, extraction, processing and sale of the wood and wood products that the U.S. imports with the impacts of the forest products it exports.

Land use and land cover are most critical to three sectors of the economy: agriculture, livestock, and forest products (Meyer 1995: 32). The share of these sectors in U.S. economic output has steadily declined over the last century and a half. Around 1870, agriculture provided more than one-third of the U.S. gross domestic product (GDP). By 1950, agriculture, forestry and fisheries together accounted for 7% of GDP, and today they represent about 2%. As a fraction of total U.S. energy consumption, wood fell from 90% in 1850 to a few percent today. Meanwhile, Americans began to seek non-timber products, recreation, wildlife habitat, watershed protection, and other values in their forested lands instead of or in addition to timber production. The National Forest system registered 10 times as many recreational visitor days per year in the early 1990s as it did in 1950 (Johnson 1996: 7).

## 4 FUTURE TRADEOFFS FOR AMERICAN CITIZENS AND LAND MANAGERS

American land use and American forestry have never been isolated from demographic, economic, environmental and cultural factors outside of American boundaries. It was the influx of European settlers that launched the large-scale Euro-American onslaught against the forests. With the first settlement of Maine, as British woodlands were depleted and rivalry between European colonial powers mounted, the British Crown claimed Maine's tall pines as masts and spars for its navy.

In future American land use and forestry, it seems likely that purely domestic factors will increasingly have to be balanced against demographic, economic, environmental, and cultural influences that originate outside of American boundaries. Domestic trends — a growing U.S. population, increasing domestic aggregate demand for forest products and non-timber forest services, and rising prices for timber — seem likely to make the tradeoffs more difficult and to sharpen the competing demands that will be made on private and public forest managers.

Without pretending to any completeness in enumerating the external influences and competing demands that will bear on American land use and forestry, I give some current examples from each of the broad categories of demography, economics, environment, and culture.

Observers of the history of human use of land and forests differ greatly about whether existing governmental policies, market institutions, and market forces will permit a smooth adaptation to problems (e.g., Sedjo 1991), or whether fundamental reforms of existing institutions, policies, and practices will be required (e.g., Barber et al. 1994). Historical data can be deployed on both sides, and a resolution seems more likely to emerge from the arena of politics than of science.

One clear lesson from studying past expectations about the future is that those expectations are frequently wrong: unexpected problems arise from unexpected sources. I claim no crystal ball. Instead of relying heavily on forecasts and projections, I will focus on issues that are already of concern.

## 4.1 Population

The U.S. population will probably increase in coming years, but at a rate much less than the rate of population increase in poor countries. How much the U.S. population will increase depends on the future balance of domestic births and deaths and on future authorized and unauthorized immigration. How many people attempt to immigrate to the United States without authorization will depend in part on economic and political conditions in other countries.

Domestic policies can influence the incidence of births resulting from unintended pregnancies, the levels of authorized and unauthorized migration, and settlements in areas vulnerable to natural hazards such as coastal storms, river flooding, and forest fires. If Americans continue to settle preferentially in coastal zones (with the encouragement of federally subsidized disaster aid and insurance), forests inland could experience reduced demand for conversion to urban uses. Forests around urban centers could experience increased demand for conversion at the same time that city dwellers increasingly seek nearby forests for recreation and second homes.

Abroad, population growth is most rapid in the poor countries, but the four-fifths of the world that is poor commands only one-fifth of the world's income. If the income of poor countries rises, their peoples may become contenders with Americans for the products and services of American forests.

## 4.2 Economics

Along with human numbers, wealth drives demand for forest resources, both domestically and internationally, and often by a larger multiple than human numbers alone. For example, in 1993, the average person in the United States had a gross domestic product of $24,000 and consumed about 320 kilograms of paper per year. In Latin America, the average person had a GDP below $2,000 and consumed 30 kilograms of paper per year (Johnson 1996: 8). According to Johnson (1996: 9), "The vast majority of internationally traded forest products are consumed in developed countries. It is the appetite of developed countries that drives the global search for fiber — a search that is now expanding to new parts of the world such as the natural forests of the Amazon Basin and Guyana Shield in South America, Central Africa, the Russian Far East, and Canada, the plantations of Chile, New Zealand, and Brazil, and maturing secondary forests in parts of the United States."

At the same time, wealth enables more productive forestry, more efficient milling, and technological substitutions. Americans burned more wood as fuel in the decade 1936–1945 after the Great Depression and in the decade 1973–1982 after the oil shocks. Wernick et al. (1997) concluded that "in America as in many developing countries, poverty and costly oil cut forests."A full view of the data suggests that both poverty and wealth lead people to cut forests.

The long-term rise in the price of timber domestically and internationally testifies that demand has been and is growing faster than supply. As rich and poor countries seek to increase their wealth, the competition for products derived from forest resources can be expected to intensify. In many tropical countries, the area of this resource base is declining rapidly (Repetto and Gillis 1988: 7). Demands for greater forest productivity, reforestation or other preservation of the global forest stock, and more equitable access to the benefits of forests can be expected to intensify.

Johnson (1996: 18) suggested that it may prove increasingly difficult to satisfy growing expectations of "more wilderness and cheaper 2×4s and toilet paper at the same time." Wernick et al. (1997) suggested optimistically that the present annual timber harvest of roughly 500 million $m^3$ could be grown on only 23% of present timberland if the sites assessed as having the highest "potential" growth achieved that potential (over 5.9 $m^3$ per ha per year, compared with present yields of 3.1 $m^3$ per ha per year). Both the fears and the hopes rest on assumptions that may or may not be realized.

Most markets for land and forestry products and services, and many economic analyses based on market indicators of value, neglect externalities, that is, consequences of market exchanges that fall on parties other than those willingly involved in the exchange. For example, clearcutting a forest may benefit the landowner and the logging operator and all the employees and manufacturers who depend on the timber. At the same time, clearcutting may adversely affect the owners of a downstream hydroelectric project (because accelerated soil erosion increases siltation behind a dam), farmers who depend on the dam for irrigation, future owners of the forest land who acquire a thinner soil cover, future generations who inherit a diminished biodiversity, and people on the other side of the globe who experience an increment in the carbon dioxide concentration of the atmosphere (Sharma 1992). None of the adversely affected parties participated in the exchange that made the clearcut possible. Most economists are well aware of externalities as important market imperfections, but economic ideas about how to counteract the adverse effects of externalities apparently remain insufficiently used in practice. As population growth and rising consumption increase the interdependence among human actions, often across the boundaries of national sovereignty, it seems likely to become increasingly important to account and pay for the externalities of managing lands and forests.

## 4.3 Environment

Domestically, the large-scale suppression of forest fires has invited developers to build homes on the edges of forests with a growing load of flammable dead wood, has interfered with normal patterns of biodiversity that depend on fires started by natural causes, and has led to conflict between some lumbering-based local communities and timber companies, on one side, and some ecologists and conservationists, on the other side. Forest managers sit in the crossfire of these conflicts.

Growing international trade in logs and lumber and growing international travel for purposes unrelated to forests have brought insects and diseases (for example, the Gypsy moth and Dutch elm disease) to forests not prepared by evolution to resist them. Tree plantations of one or a few tree species will become increasingly vulnerable to catastrophic pest outbreaks.

The atmospheric concentrations of carbon dioxide, methane, other greenhouse gases and chlorofluorocarbons, as well as changes in global average temperatures and stratospheric ozone concentrations, affect all countries, although not equally. Acid precipitation and air pollution cross national boundaries in North America and Europe. According to the Intergovernmental Panel on Climate Change (1995: 39–40), "the management of forests, agricultural lands and rangelands can play an important role in reducing current emissions of $CO_2$, $CH_4$ and $N_2O$ and in enhancing carbon sinks. A number of measures could conserve and sequester substantial amounts of carbon (approximately 60–90 GtC (gigatons of carbon) in the forestry sector alone) over the next 50 years. ... Land-use and management measures include: sustaining existing forest cover; slowing deforestation; regenerating natural forests; establishing tree plantations; promoting agroforestry; altering management of agricultural soils and rangelands"; and others. If the recommendations of the IPCC are followed, climate-change mitigation will become another factor for managers of forests and rangelands to consider, along with demands of growing populations and expanding economies for food, fibers, forest products, recreation, biodiversity conservation, and other ecosystem services (Daily 1997).

## 4.4 Culture

Culture includes a society's underlying attitudes toward land and forests. Faced with similar scientific data, Germany and the United States arrived at similar controls on acid precipitation. However, Germany implemented controls a decade before the United States did because Germans saw the threatened forests as central to the origin and myths of German culture, whereas Americans saw their forests more as an exploitable, and for some people an expendable, resource (Tosteson 1994). Today, the U.S. conservationist, the Amazonian aborigine, and the Japanese international timber merchant endow the Amazonian rainforest with three very different cultural overlays. Conflicts of values and attitudes concerning land and forests are likely to intensify as diverse cultures make contact across increasingly permeable national boundaries.

Conflicts over different values arise between different groups within the United States who seek to influence forest management. For example, Donald Waller, a professor of botany at the University of Wisconsin, tried unsuccessfully to make the U.S. Forest Service plan to conserve biological diversity (other than diversity of the ages of tree stands) in its forest management plans for northern Wisconsin in the mid-1980s (Alverson et al. 1994). He wrote of his experience (Waller 1997: 1): "I think a difficulty I had in the mid-1980s was naiveté; ... It was hard for me to wake up to the fact that many trained foresters did not share the same basic values; they were trained in a ... utilitarian mind set, one that placed a lot of emphasis and training on economics, on details of productivity and site conditions, while neglecting in my opinion important ecological characteristics of the rest of the biotic community: the soils, the understory forest herbs, interactions with animals and so on. There was a difference in values, ... grounding, information and training."

Roger A. Sedjo, a forest economist at Resources for the Future in Washington, DC, criticized the U.S. Forest Service from nearly the opposite perspective (Sedjo 1995: 10): "In recent years, ... the leadership of the Forest Service ... has focused on forest ecology — the totality of relationships between forest organisms and their environment. This concern with forest ecology is embodied in the leadership's advocacy of ecosystem management. In accordance with this philosophy, the service has all but abandoned the notion of forests as primarily a vehicle for producing multiple goods (or 'outputs') desired by society. ... More to the point from the perspective of taxpayers, these decisions are being driven almost exclusively by biological considerations, with little attention paid to economic and other concerns. In short, when identifying objectives, ecosystem management ignores the social consensus implicit in the congressionally legislated objective of producing multiple market and nonmarket forest outputs and, instead, attempts to achieve some arbitrary forest condition about which society has little say."

Culture includes political institutions for allocating the benefits of public resources and for resolving conflict. Interest groups attempt to influence these institutions for their own benefit. Interest groups may view their scope as local, national, global, or in between (Sharma 1992). Local people may clear forests for their own subsistence and commercial purposes, or as agents of commercial interests headquartered elsewhere. Nationally, forests may be viewed as a source of employment, foreign exchange, government revenue, and land for other purposes such as industry, mining, agriculture or settlements; economic pressures may favor short-term exploitation. People in one country may look to forests in another country to sequester carbon from the atmosphere and to preserve biodiversity, or as potential lands for profitable ranching and timber extraction. Political change, which shifts the benefits and costs of current practices to different parties, may become increasingly difficult as interests at different local, national, and global scales become increasingly intertwined.

International demand is growing for the products and services that forests provide other than timber (Johnson 1996: 9). Most ferns, mosses and other floral display materials and most mushrooms collected from the forests of the Pacific Northwest are exported to European and Japanese restaurants and other markets. Tourism to natural sites, or ecotourism, has been growing at 7% per year, and may generate as much as $50 billion a year (an amount comparable to all official development assistance). Demand for floral displays and ecotourism is driven by aesthetic values. Forests also have "existence value" to people who value them even when they make no direct use of them (Sharma 1992). Continued conflicts can be expected between those who value forests for timber and those who defend other values, within and outside the United States.

A step that could reduce some of these conflicts, although it is no panacea, would be to make individuals, corporations, and other private interests pay market or closer-to-market prices for the private benefits they receive from forests. As suggested by Repetto (Repetto and Gillis 1988: 380), forest management would apply the same test of economic efficiency to its decisions whether to provide services for timber extraction, for recreation (fishing, hunting, camping, hiking) and for other services (such as fuel-wood extraction): do the marginal benefits, based on recoverable fees, exceed the separable or avoidable costs? If some forests turned out to be simply uneconomic for timber production or recreational use or both, the market test would prevent political conflict over presently free or subsidized services. If some services could be shared between recreational and timber users, economic common interests could be found.

National forests are public institutions in part because they protect public goods (Repetto and Gillis 1988: 380). Some public goods include private beneficiaries. A goal is to set prices that reflect as many as possible of the positive and negative externalities of transactions, and to assure broadly equitable access to the services offered. For example, watersheds serve people and corporations who use the water provided; in this case, water could be priced in a way that covers (at least partially) the costs of protecting the watershed. Some public goods, notably genetic diversity, species diversity, and ecosystem diversity, have uncertain immediate instrumental value but many taxpayers agree there is value in preserving them for the next generation. In such cases, bond issues could distribute the cost of protective services over the present generation and the next. But these measures can be expected to generate opposition. More rational economic approaches to forestry management may change the terms of conflict but will not eliminate it.

## 5 CONCLUSIONS

### Population Growth

The global population growth rate of 1.5% per year in 1996 implies a doubling in 46 years. This growth rate is extremely rapid compared to rates of global population growth experienced before 1945, though it is less than the all-time peak growth rate. While it took until 1830 for human population size to reach 1 billion people, the most recent increment of 1 billion people was added in 12 years. The poor countries have more than twice the population density of the rich, on average, and their populations are increasing 10 to 20 times faster.

In the last two centuries, people moved in large numbers from the countryside to cities. Today's rich countries are 75% urban. Today's poor countries are 35% urban and are urbanizing rapidly. Globally, and at national and regional spatial scales, a relatively small fraction of the land is occupied at a high human population density while most of the land is lightly or very sparsely occupied.

The population density of the conterminous United States rose from one person per 57 ha in 1790 to one person per 3 ha in 1990. The population is growing by about 2.5 million, or roughly 1%, per year.

## Land Use

The fraction of the earth's total ice-free land area covered by closed forest and woodland fell from about 47% prior to the invention of agriculture to about 40% by 1990. As people became more numerous, they extended their activities to new lands (extensification) and increased the productivity of their activities on land already occupied (intensification). In recent decades, the price of timber in international commodity markets has increased, while the price of most other primary commodities has fallen.

In the United States, forested areas declined until 1920, rose gradually from 1920 to 1960 and fell slightly since 1960. Over the last century and a half, real lumber prices in the United States grew roughly five-fold per century, while the share of agriculture, livestock, forest products and fisheries in U.S. economic output declined steadily.

The goal of some timberland owners changed from "cut out and get out" to management for sustainable yield. The Federal Government's purpose in dealing with public lands was originally to transfer public lands into private hands. Its present goal is to retain and manage public lands, which now cover about one-fifth of the conterminous United States.

Between 1900 and 1993, the aggregate U.S. consumption of all timber products (as fuel, pulp, plywood and lumber) increased 70% while gross domestic product rose 16-fold. The 16-fold increase was the product of a more than tripled population size and a gross domestic product per person that grew nearly 5-fold. The consumption per American of all timber products fell by roughly half between 1900 and 1993.

At 1990 levels of U.S. wood consumption (2.3 $m^3$/person) and U.S. timberland productivity (3.1 $m^3$/ha), about three-quarters of a hectare of timberland (2.3/3.1 = 0.74 ha/person) are required to produce the wood consumed by an average American. Americans are increasingly seeking non-timber products, recreation, wildlife habitat, watershed protection and other values in their forested lands instead of or in addition to timber production.

## Relations Between Population Growth and Land Use

On a long historical time scale, rising numbers of people have been associated with an increase in land areas farmed, largely at the expense of a decline in forests, and with rising farming intensity. But a one-directional causal model like "human population growth causes forest clearing or land conversion" is too simple to cover all real cases. What the growing human population puts on the earth and takes from the earth depends not on numbers alone, but also on the human economy, the physical, chemical and biological environment, and the human culture.

For neoclassical economists, population growth is a neutral factor in land degradation. For classical economists and some natural scientists, population growth is the principal independent cause of land degradation. For dependency theorists, both population growth and land degradation are symptoms of poverty and inequity. For intermediate variable theorists, population growth exacerbates the adverse effects of other ultimate causes of land degradation.

Richards' center-periphery model envisions an autocatalytic process in which population growth and growing consumption drive the development or exploitation of additional resources by extensification and intensification, leading to further population growth and consumption. A highly schematic model called the Malthus–Condorcet–Mill model presents a possible quantitative version of this process.

No purely ecological concept of carrying capacity is adequate as a concept of human carrying capacity. Human choices are important in defining human carrying capacity. Choices influence the average and the distribution of the level of material well-being; technology; political, economic, and demographic institutions and arrangements; physical, chemical, and biological environments; variability or stability; risk or robustness; time horizons; and fashions, tastes, and values.

## Tradeoffs

The institutions and policies that deal with problems in land use, particularly forestry, may provide useful models for other environmental concerns, because changes in the forests have led other human-induced changes in the environment.

American planners, managers, and citizens must consider the global perspective, even if they are concerned only to protect American resources and interests, because the United States is and will be intimately linked to the rest of the world. In future American land use and forestry, purely domestic factors will increasingly have to be balanced against demographic, economic, environmental, and cultural influences that originate outside of American boundaries.

Domestic trends — a growing U.S. population, increasing domestic demand for wood products and non-timber forest services, and rising prices for timber — seem likely to sharpen the competing demands that will be made on private and public forest managers. To reduce conflicts, individuals, corporations and other private interests could be required to pay market or closer-to-market prices for the private benefits they receive from forests.

As worldwide and domestic population growth and rising consumption make management decisions increasingly interdependent, often across the boundaries of national sovereignty, it seems likely to become increasingly important to account and pay for the externalities of land use and forest management.

## ACKNOWLEDGMENTS

Jesse H. Ausubel, Nels C. Johnson, Carole L. Jolly, Andrew Malk, Elaine Matthews, Roger A. Sedjo, and anonymous referees provided helpful materials or helpful comments on earlier versions. I am especially indebted to Elaine Matthews for meticulous criticisms of a prior draft. I acknowledge with thanks the support of U.S. National Science Foundation grant BSR92-07293 and the hospitality of Mr. and Mrs. William T. Golden during this work. Section 2.6.2 of this chapter draws on Cohen (1995a), section 2.7 on Cohen (1995b), and section 2.2 on Cohen (1997).

## REFERENCES

Ahlburg, D.A., and J.W. Vaupel. 1990. Alternative projections of the U.S. population. *Demography* 27(4): 639–652.

Alverson, W.S., W. Kuhlmann, and D.M. Waller. 1994. *Wild Forests: Conservation Biology and Public Policy*. Island Press, Washington, DC.

Barber, C.V., N.C. Johnson, and E. Hafild. 1994. *Breaking the Logjam: Obstacles to Forest Policy Reform in Indonesia and the United States.* World Resources Institute, Washington, DC.

Benedick, R.E. 1991. Comment: environmental risk and policy response. In: K. Davis and M.S. Bernstam (eds.), *Resources, Environment, and Population: Present Knowledge, Future Options.* Oxford University Press, New York. Population and Development Review, 16 (Suppl.): 201–204.

Birdsey, R.A., A.J. Plantinga, and L.S. Heath. 1993. Past and perspective carbon storage in United States forests. *Forest Ecology and Management* 58: 33–40.

Boserup, E. 1981. *Population and Technological Change: A Study of Long-Term Trends.* University of Chicago Press, Chicago, IL.

Brown, S.S., and L. Eisenberg (eds.). 1995. *The Best Intentions: Unintended Pregnancy and the Well-Being of Children and Families.* Committee on Unintended Pregnancy, Division of Health Promotion and Disease Prevention, Institute of Medicine. National Academy Press, Washington, DC.

Burtless, G. 1995. International trade and the rise in earnings inequality. *Journal of Economic Literature* (June) 33: 800–816.

Bush, G.H. 1992. Quoted in: Carrying Capacity Network, *Focus: Carrying Capacity Selections* 1(2), 57 (1992) Carrying Capacity Network, Washington, DC.

Cipolla, C.M. 1994. *Before the Industrial Revolution: European Society and Economy, 1000–1700.* 3rd edn. W.W. Norton, New York.

Clawson, M. 1995. Old timber and new growth: an interview with Marion Clawson. *Resources* (Resources for the Future, Washington, DC) 121(Fall): 6–9.

Cohen, J.E. 1995a. Population growth and the Earth's human carrying capacity. *Science* 269(21 July): 341–346.

Cohen, J.E. 1995b. *How Many People Can the Earth Support?* W.W. Norton, New York.

Cohen, J.E. 1997. Conservation and human population growth: what are the linkages? pp. 29–42. In: S.T.A. Pickett, R.S. Ostfeld, G. Likens, and M. Shachak (eds.), *The Ecological Basis for Conservation.* Chapman & Hall, London.

Daily, G.C. (ed.). 1997. *Nature's Services: Societal Dependence on Natural Ecosystems.* Island Press, Washington, DC, and Covelo, CA.

Demeny, P. 1988. Demography and the limits to growth. pp. 213–244. In: Teitelbaum, M.S., and Winter, J.M. (eds). *Population and Resources in Western Intellectual Traditions. Population and Development Review* (Suppl.) 14. Population Council, New York.

Diamond, H.L., and P. Noonan (eds.). 1996. *Land Use in America.* Island Press, Washington, DC.

Gardner, R.N. 1992. *Negotiating Survival: Four Priorities After Rio.* Council on Foreign Relations Press, New York.

Grainger, A. 1993. Population as concept and parameter in the modeling of deforestation. pp. 71–101. In: G.D. Ness, W.D. Drake, and S.R. Brechin (eds.), *Population–Environment Dynamics: Ideas and Observations.* University of Michigan Press, Ann Arbor, MI.

Hardesty, D.L. 1977. *Ecological Anthropology.* John Wiley, New York.

Hardin, G. 1968. The tragedy of the commons. *Science* 162: 1243–1248.

Harrison, P. 1992. *The Third Revolution: Environment, Population and a Sustainable World.* I.B. Tauris, London.

Intergovernmental Panel on Climate Change. 1995. *IPCC Second Assessment Synthesis: Climate Change 1995.* World Meteorological Organization, United Nations Environmental Program.

Johnson, N.C. 1996. Global trends and the future of national forests. Paper presented at: *The National Forest Management Act in a Changing Society, 1976–1996*, September 16–18, 1996, Natural Resources Law Center, University of Colorado, Boulder, CO.

Jolly, C.L. 1994. Four theories of population change and the environment. *Population and Environment* 16(1): 61–90.

Jolly, C.L., and B.B. Torrey (eds.). 1993. *Population and Land Use in Developing Countries.* National Academy Press, Washington, DC.

Kates, R.W., B.L. Turner, and W.C. Clark. 1990. The great transformation. pp. 1–17. In: Turner, B.L., Clark, W.C., Kates, R.W., Richards, J.F., Mathews, J.T., and Meyer, W.B. (eds.), *The Earth as Transformed by Human Action: Global and Regional Changes in the Biosphere Over the Past 300 Years.* Cambridge University Press with Clark University, Cambridge, UK.

MacLeish, W.H. 1994. *The Day before America.* Houghton Mifflin, Boston and New York.

Malthus, T.R. 1798. *An Essay on the Principle of Population.* Complete 1st ed. (1798) and partial 7th ed. (1872) reprinted in *On Population*, Gertrude Himmelfarb (ed.). Modern Library, 1960. New York

Marquette, C.M., and R. Bilsborrow. 1994. Population and the environment in developing countries: literature survey and research bibliography. ESA/P/WP.123. Population Division, United Nations, New York.

Matthews, E. 1983. Global vegetation and land use. *Journal of Climate and Applied Meteorology* 22: 474–487.

Mertens, W. 1994. Population and deforestation in humid tropics. Policy and Research Papers 2. Liege, Belgium: International Union for the Scientific Study of Population.

Meyer, W.B. 1995. Past and present land use and land cover in the USA. *Consequences: The Nature and Implications of Environmental Change* 1(1): 24–33.

Mill, J.S. 1848. *Principles of Political Economy with Some of Their Applications to Social Philosophy*, V.W. Bladen and J.M. Robson, (eds.). University of Toronto Press, Toronto, 1965.

Pearce, D.W., and J.J. Warford. 1993. *World Without End: Economics, Environment, and Sustainable Development*. Published for the World Bank. Oxford University Press, New York.

Pingali, P.L., and H. Binswanger. 1987. Population density and agricultural intensification: a study of the evolution of technologies in tropical agriculture. pp. 27–56. In: Johnson, D.G., and R.D. Lee (eds.), *Population Growth and Economic Development: Issues and Evidence*. University of Wisconsin Press, Madison, WI.

Ponting, C. 1990. Historical perspectives on sustainable development. *Environment* 32(9): 4–9, 31–33.

Population Reference Bureau. 1996. World Population Data Sheet. Population Reference Bureau, Washington, DC.

Preston, S. 1994. Population and the Environment. Distinguished Lecture Series on Population and Development. International Union for the Scientific Study of Population, Liege, Belgium.

Repetto, R., and M. Gillis (eds.). 1988. *Public Policies and the Misuse of Forest Resources*. Cambridge University Press, Cambridge, New York, New Rochelle, Melbourne, Sydney.

Richards, J.F. 1990. Land transformation. pp. 163–178. In: Turner, B.L., W.C. Clark, R.W. Kates, J.F. Richards, J.T. Mathews, and W.B. Meyer (eds). *The Earth as Transformed by Human Action: Global and Regional Changes in the Biosphere Over the Past 300 Years*. Cambridge University Press with Clark University, Cambridge, UK.

Rudel, T.I. 1991. Relationships between population and environment in rural areas of developing countries. *Population Bulletin of the United Nations* 31/32: 52–69.

Sedjo, R.A. 1991. Forest resources: resilient and serviceable. pp. 81–120. In Frederick, K.D., and R.A. Sedjo (eds.), *America's Renewable Resources: Historical Trends and Current Challenges*. Resources for the Future, Washington, DC.

Sedjo, R.A. 1995. Ecosystem management: an uncharted path for public forests. *Resources* (Resources for the Future, Washington, DC) 121(Fall): 10–20.

Sharma, N.P., ed. 1992. *Managing the World's Forests: Looking for Balance between Conservation and Development*. Kendall/Hunt Publishing, Dubuque, Iowa.

Shaw, R.P. 1989. Rapid population growth and environmental degradation: ultimate versus proximate factors. *Environmental Conservation* 16(3): 199–208.

Tosteson, J.L. 1994. Acid rain in Germany and the United States: a comparative study of environmental risk management. B.A. thesis in environmental science and public policy. Cambridge, MA: Harvard University.

Turekian, K.K. 1996. *Global Environmental Change: Past, Present, and Future*. Prentice Hall, Upper Saddle River, NJ.

Turner, B.L., W.C. Clark, R.W. Kates, J.F. Richards, J.T. Mathews, and W.B. Meyer (eds.). 1990. *The Earth as Transformed by Human Action: Global and Regional Changes in the Biosphere Over the Past 300 Years*. Cambridge University Press with Clark University, Cambridge, UK.

United Nations, Department for Economic and Social Information and Policy Analysis. 1995. *World Urbanization Prospects: The 1994 Estimates and Projections of Urban and Rural Populations and of Urban Agglomerations*. ST/ESA/SER.A/150. United Nations, New York.

U.S. Bureau of the Census. 1975. *Historical Statistics of the United States, Colonial Times to 1970*. Bicentennial Edition. U.S. Government Printing Office, Washington, DC.

U.S. Department of Agriculture, Forest Service. 1989. *U.S. Timber Production, Trade, Consumption, and Price Statistics 1960–88*. Miscellaneous Publication No. 1486. U.S. Government Printing Office, Washington, DC.

Waggoner, P.E. 1994. How Much Land Can Ten Billion People Spare For Nature Task Force Report No. 121, February. Council for Agricultural Science and Technology, Ames, IA.

Waggoner, P.E., J.H. Ausubel, I.K. Wernick, and Iddo K. 1996. Lightening the tread of population on the land: American examples. *Population and Development Review* 22(3): 531–545.

Waller, D. 1997. Scientists sue the forest service to protect biodiversity: is the Forest Service taking care of business? *Environmental Review* (Seattle, WA) 4(3): 1–9.

Wernick, I.K., P.E. Waggoner, and J.H. Ausubel. 1997. Searching for leverage to conserve forests: the industrial ecology of wood products in the U.S. *Journal of Industrial Ecology* 1(3): 125–145.

Wilkinson, R.G. 1973. *Poverty and Progress: An Ecological Perspective on Economic Development*. Praeger, New York.

Williams, M. 1989. *Americans and Their Forests: A Historical Geography*. Cambridge University Press, New York.

Williams, M. 1990. Forests. pp. 179–201. In: Turner, B.L., W.C. Clark, R.W. Kates, J.F. Richards, J.T. Mathews, and W.B. Meyer (eds.). *The Earth as Transformed by Human Action: Global and Regional Changes in the Biosphere Over the Past 300 Years*. Cambridge University Press with Clark University, Cambridge, UK.

World Resources Institute. 1987. *World Resources 1987*. Basic Books, New York.

World Resources Institute. 1992. *World Resources 1992–93*. Oxford University Press, New York and Oxford.

World Resources Institute. 1994. *World Resources 1994–95*. Oxford University Press, New York and Oxford.

World Resources Institute. 1996. *World Resources 1996–97*. Oxford University Press, New York and Oxford.

## THE AUTHOR

**Joel E. Cohen**

*Rockefeller University & Columbia University*
*1230 York Avenue*
*Box 20*
*New York, NY 10021-6399*

# Index to Volumes I–III

# List of Abbreviations

| | |
|---|---|
| ACEC | Area of Critical Environmental Concern |
| ACLD | Augusta Creek landscape design |
| AGRICOLA | National Agricultural Library Catalogue |
| AHCS | Aquatic Habitat Classification System |
| AHP | Analytical Hierarchy Process |
| AIHP | American Industrial Heritage Project |
| AIRFA | American Indian Religious Freedom Act |
| ARPA | Archaeological Resource Protection Act |
| AMA | Adaptive Management Area |
| AMS | Accelerator Mass Spectrometry |
| ANILCA | Alaska National Interest Lands Conservation Act |
| ANOVA | Analysis of Variance |
| ANSI | American National Standards Institute |
| APAEA | Asian Pacific American Employee Association |
| ARC | Angeles Restoration Crew |
| ARPA | Advanced Research Projects Agency |
| ARS | Agricultural Research Service |
| ASTM | American Society for Testing and Materials |
| ATLSS | Across Trophic Level System Simulation |
| ATV | All Terrain Vehicle |
| AUM | Animal Unit Month |
| AVHRR | Advanced Very High Resolution Radiometer |
| AVIMS | Annual Vegetation Inventory and Monitoring System |
| BART | Best Available Retrofit Technology |
| BBN | Bayesian Belief Network |
| BBS | Breeding Bird Survey |
| BIA | Bureau of Indian Affairs |
| BLM | Bureau of Land Management |
| BMP | Best Management Practice |
| BOR | Bureau of Reclamation |
| BPR | Business Process Re-engineering |
| BRD | Biological Resource Division |
| CA | Conjoint Analysis |
| CAB | Commonwealth Agricultural Bureau |
| CanFI | Canadian National Forest Inventory |
| CASE | Computer-Aided Systems Engineering |
| CBC | California Biodiversity Council |
| CBSG | Captive Breeding Specialists Group |
| CEN | Committee for European Normalisation |
| CEP | California Environmental Project |
| CEQ | Council on Environmental Quality |
| CERCLA | Comprehensive Environmental Response, Compensation and Liability Act |
| CERES | California Environmental Resources Evaluation System |
| CFIC | Canadian Forest Inventory Committee |
| CODA | Conservation Options and Decisions Analysis |
| COTS | Commercial Off The Shelf |
| CPR | Common Pool Resource |
| CRM | Coordinated Resource Management |
| CSDS | Common Survey Data Structure |
| CUSTOMER | Customer Use and Survey Techniques for Operations, Management, Evaluation and Research |
| CV | Contingent Valuation |
| CWHR | California Wildlife Habitat Relationship |
| CWMA | Cooperative Weed Management Area |
| DBH | Diameter at Breast Height |
| DEM | Digital Elevation Model |
| DEMO | Demonstration of Ecosystem Management Options |
| DER | Department of Environmental Regulation |
| DFC | Desired Future Condition |
| DNR | Department of Natural Resources |
| DOI | Department of the Interior |
| DOQ | Digital Orthophoto Quadrangles |
| DSS | Decision Support System |
| EA | Environmental Assessment |
| ECOMAP | Ecological Classification and Mapping |
| ECS | Ecological Classification System |
| EIS | Environmental Impact Statement |
| ELF | Extremely Low Frequency |
| ELT | Ecological Land Type |
| ELTP | Ecological Land Type Phase |
| EM | Ecosystem Management |
| EMDS | Ecosystem Management Decision Support |
| ENR | Everglades Nutrient Removal |
| EPA | Environmental Protection Agency |
| ERU | Ecological Reporting Unit |

| | |
|---|---|
| ESA | Endangered Species Act |
| ESA | Ecological Society of America |
| ESRI | Environmental Systems Research Institute |
| FACA | Federal Advisory Committee Act |
| FAPRI | Food and Agriculture Policy Research Institute |
| FEMAT | Forest Ecosystem Management and Assessment Team |
| FERC | Federal Energy Regulatory Commission |
| FGDC | Federal Geographic Data Committee |
| FIA | Forest Inventory and Analysis |
| FIFRA | Federal Insecticide, Fungicide, and Rodenticide Act |
| FIP | Forest Incentive Program |
| FIPS | Federal Information Processing Standard |
| FLPMA | Federal Land Policy Management Act |
| FOIA | Freedom of Information Act |
| FS | Forest Service |
| FSC | Forest Stewardship Council |
| FWS | Fish and Wildlife Service |
| GAO | Government Accounting Office |
| GAP | Gap Analysis Program |
| GDP | Gross Domestic Product |
| GDSS | Group Decision Support System |
| GIS | Geographic Information System |
| GLO | Government Land Office |
| GPP | Gross Primary Productivity |
| GPS | Global Positioning System |
| HACU | Hispanic Association of Colleges and Universities |
| HCP | Habitat Conservation Plan |
| HEP | Habitat Evaluation Procedure |
| HSI | Habitat Suitability Index |
| HSI | Hispanic Serving Institution |
| HTM | Habitat Transaction Method |
| ICBEMP | Interior Columbia Basin Ecosystem Management Project |
| ICPSR | Inter-University Consortium for Political and Social Research |
| IDH | Intermediate Disturbance Hypothesis |
| IDT | Interdisciplinary Team |
| IE | Information Engineering |
| IEC | International Electrotechnical Commission |
| IEM | Integrated Environmental Management |
| INA | Information Needs Assessment |
| IPCC | Intergovernmental Panel on Climate Change |
| IRI | Integrated Resource Inventory |
| IRS | Internal Revenue Service |
| ISC | Interagency Scientific Committee |
| ISG | Interagency Steering Group |
| ISO | International Standards Organization |
| IUCN | International Union for Conservation of Nature |
| LACC | Los Angeles Conservation Corps |
| LAN | Local Area Network |
| LCM | Life-Cycle Management |
| LRMP | Land and Resource Management Plan |
| LRR | Land Resource Region |
| LSF | Late Serial Forest |
| LTA | Land-Type Association |
| LTER | Long Term Ecological Research |
| LUHNA | Land Use History of North America |
| LUMP | Land Unsuitable Mining Program |
| MA | Management Area |
| MANOVA | Multivariate Analysis of Variance |
| MLRA | Major Land Resource Area |
| MOA | Memorandum of Agreement |
| MOU | Memorandum of Understanding |
| MRLC | Multi-Resolution Land Characterization |
| MSS | Landsat Multi-Spectral Scanner |
| MUSY | Multiple-Use Sustained Yield Act |
| MVP | Minimum Viable Population |
| NAGPRA | Native American Graves Protection Act |
| NALC | North American Landscape Characterization |
| NAPAP | National Acid Precipitation Assessment Program |
| NAPP | National Aerial Photography Program |
| NASA | National Aeronautics and Space Administration |
| NAWQA | National Water Quality Assessment |
| NBII | National Biological Information Infrastructure |
| NBS | National Biological Service |
| NED | Northeast Decision Model |
| NEP | Net Ecosystem Productivity |
| NEPA | National Environmental Policy Act |
| NFMA | National Forest Management Act |
| NFS | National Forest System |
| NGO | Non-Governmental Organization |
| NHFEU | National Hierarchical Framework for Ecological Units |
| NHPA | National Historic Preservation Act |
| NIPF | Nonindustrial Private Forest |
| NIST | National Institute for Standards and Technology |
| NMFS | National Marine Fisheries Service |
| NOAA | National Oceanographic and Atmospheric Administration |
| NPP | Net Primary Production |
| NPS | National Park Service |
| NPV | Net Present Value |
| NRCS | Natural Resources Conservation Service |
| NSDI | National Spatial Data Infrastructure |
| NSF | National Science Foundation |
| NSFHWAR | National Survey of Fishing, Hunting and Wildlife Associated Recreation |
| NSRE | National Survey on Recreation and the Environment |
| NTMB | Neotropical Migratory Birds |
| NWFP | Northwest Forest Plan |
| OGC | Office of General Council |
| OLA | Opportunity Los Angeles |
| OMB | Office of Management and Budget |
| OPA | Oil Pollution Act |
| PAIS | Public Affairs Information Service |
| PAR | Participatory Action Research |

| | |
|---|---|
| PBRS | Public Benefit Rating System |
| PFC | Paper Functioning Condition |
| PHVA | Population and Habitat Viability Analysis |
| PIF | Partners in Flight |
| PNC | Potential Natural Community |
| PNV | Potential Natural Vegetation |
| PPFP | Ponderosa Pine Forest Partnership |
| PRA | Participatory Rural Appraisal |
| PVA | Population Viability Analysis |
| RAC | Resource Advisory Council |
| RAD | Rapid Application Development |
| RAP | Rapid Assessment Procedure |
| RARE | Roadless Area Review and Evaluation |
| REAP | Rapid Ethnographic Assessment Project |
| RIT | Resource Information Technician |
| RNA | Research Natural Area |
| ROD | Record of Decision |
| RPA | Rangeland and Forest Resources Planning Act |
| RRA | Rapid Rural Appraisal |
| RRV | Reference Range of Variability |
| RSF | Resource Selection Function |
| RU | Random Utility |
| RVA | Range of Variability Approach |
| SAA | Southern Appalachian Assessment |
| SAS | Statistical Analysis System |
| SEPM | Special Emphasis Program Manager |
| SFI | Sustainable Forestry Initiative |
| SIA | Social Impact Assessment |
| SIP | Stewardship Incentive Program |
| SHPO | State Historic Preservation Officer |
| SMNRA | Spring Mountains National Recreation Area |
| SNEP | Sierra Nevada Ecosystem Project |
| SPSS | Statistical Procedure for the Social Sciences |
| SPUDD | Spatial Unified Data Dictionary |
| SQL | Structured Query Language |
| SSI | Social Science Index |
| SUV | sport utility vehicle |
| TC | Travel Cost |
| TCORP | Tahoe Coalition of Recreation Provider |
| TCR | Total Cost of Risk |
| TEK | Traditional Ecological Knowledge |
| TM | Landsat Thematic Mapper |
| TNC | The Nature Conservancy |
| TSCA | Toxic Substances Control Act |
| TSPAS | Timber Sale Program Analysis System |
| TSPIRS | Timber Sale Program Information Reporting System |
| UGI | Urban Greening Initiative |
| UNDP | United Nations Development Programme |
| UNESCO | United Nations Educational, Scientific and Cultural Organization |
| URI | Urban Resource Initiative |
| USDA | U.S. Department of Agriculture |
| USFS | U.S. Forest Service |
| USFWS | U.S. Fish and Wildlife Service |
| USGAO | U.S. Government Accounting Office |
| USGS | U.S. Geological Survey |
| WAN | Wide Area Network |
| WFPB | Washington Forest Practices Board |
| WRI | World Resource Institute |
| WWF | World Wildlife Fund |
| WWW | World Wide Web |
| YCC | Youth Conservation Corps |

# Contents of Volumes I–III

## ♦ VOLUME I: KEY FINDINGS

## INFORMATION AND DATA MANAGEMENT



# ♦ VOLUME II



## INTRODUCTION



## BIOLOGICAL AND ECOLOGICAL DIMENSIONS

## ♦ VOLUME III